Food, Pharmaceutical and Bioengineering Division

Core Programming Topic at the 2011 AIChE Annual Meeting

Minneapolis, Minnesota, USA
16-21 October 2011

Volume 1 of 2

ISBN: 978-1-61839-735-5

Printed from e-media with permission by:

Curran Associates, Inc.
57 Morehouse Lane
Red Hook, NY 12571

CURRAN ASSOCIATES INC. proceedings.com

Some format issues inherent in the e-media version may also appear in this print version.

Printed by Curran Associates, Inc. (2012)

AIChE
3 Park Avenue
New York, NY 10016-5991

Phone: (203) 702-7660
Fax: (203) 775-5177

www.aiche.org

Additional copies of this publication are available from:

Curran Associates, Inc.
57 Morehouse Lane
Red Hook, NY 12571 USA
Phone: 845-758-0400
Fax: 845-758-2634
Email: curran@proceedings.com
Web: www.proceedings.com

TABLE OF CONTENTS

Volume 1

Volume 2

(5a) Engineering Anaerobic Gut Fungi for Lignocellulose Hydrolysis and Biofuel Production

Michelle A. O'Malley[1], Michael K. Theodorou[2] and Chris A. Kaiser[1], (1)Department of Biology, Massachusetts Institute of Technology, Cambridge, MA, (2)Biological and Biomedical Sciences, Durham University, Durham, United Kingdom

Biofuels derived from lignocellulosic materials are an attractive, renewable alternative to petroleum-based fuels, yet the recalcitrance of lignocellulose to enzymatic degradation remains a significant barrier to biofuel development. Anaerobic gut fungi that reside within the rumen and hindgut of large herbivores are among the most efficient digesters of lignocellulose in nature due to their invasive, filamentous growth and broad range of fibrolytic enzymes. Despite the powerful cellulose degrading capacity of gut fungi, remarkably little is known about the cellulolytic enzymes from these organisms due to difficulties associated with their isolation and culture. In order to adapt anaerobic fungi for biotechnology applications, and to explore the attributes of attractive cellulolytic enzymes that they produce, we have implemented methods to isolate and sustain anaerobic gut fungi in batch culture. Following isolation of fungi from horse and cow digestive tracts, we have supported fungal growth in a rumen fluid-rich medium, and coupled cellular proliferation to the generation of fermentation gases that accumulate during growth. Through cultivation of anaerobic fungi and the exploitation of catabolic regulation patterns, we have been able to apply powerful metagenomic strategies to discover new cellulase genes and novel cellulolytic protein complexes in these organisms. En route to adapting anaerobic gut fungi for direct cellulosic bioprocessing, we are also exploring several methods to achieve stable genetic transformation in these microbes.

See more of this Session: Advances In Cell Culture IA: Experimental
See more of this Group/Topical: Food, Pharmaceutical & Bioengineering Division

ENGINEERING CARBON AND ENERGY METABOLISM FOR CHO CELLS FOR IMPROVED PRODUCTIVITY AND EXTENDED LIFESPAN

*Camila Alejandra Wilkens[1,2] and Ziomara Patricia Gerdtzen[1,2]**

[1] Centre for Biochemical Engineering and Biotechnology, Department of Chemical Engineering and Biotechnology, University of Chile, Santiago, Chile
[2] Millenium Institute for Cell Dynamics and Biotechnology: a Centre for Systems Biology, University of Chile, Santiago, Chile.
**correspondence: zgerdtze@ing.uchile.cl*

Background

Several studies have shown that lactate's specific production rate and concentration may inhibit recombinant protein production in mammalian cell cultures. In a previous work we showed that lactate production was associated to a fast consumption of sugar leading to a high pyruvate synthesis rate. Metabolic flux analysis indicated that pyruvate entry into the TCA cycle is limiting, causing pyruvate accumulation inside the cell which leads to lactate production [1].

Improvement in specific protein productivity has been observed after introducing specific changes in key points of the central carbon metabolism through cell engineering. Expression of the yeast pyruvate carboxylase (PYC) gene provided the cell an alternative pathway for pyruvate to enter the TCA cycle [2]. Over expression of the gene that translates into the enzyme malate dehydrogenase II (MDHII) which catalyzes a limiting step in the TCA cycle also improved the cycle's flux [3]. Manipulation of sugar transport by transfecting CHO cells with the fructose transporter gene (Slc2a5) leaded to a reduced carbon source consumption and lower lactate synthesis [4]. Finally, a knock out of the lactate dehydrogenase (LDHa) gene reduced the specific rate of lactate production [5].

In this work we aim at improving cell metabolism through cell engineering by reducing pyruvate accumulation, lactate formation and increasing pyruvate flux towards energy metabolism of a cell line. These changes will be achieved by transfecting CHO DP-12 cells producing Human IgG with genes selected based on the empirical evidence of other studies. The clones produced overexpress one or two genes combining PYC, MDHII and Slc2a5, all involved in the central carbon metabolism. Figure 1 shows a simplified metabolic map of engineered cells. It is possible to observe how fructose enters glycolysis, PYC gene enables carbon input into the TCA cycle and the role of MDH II in the Krebs cycle.

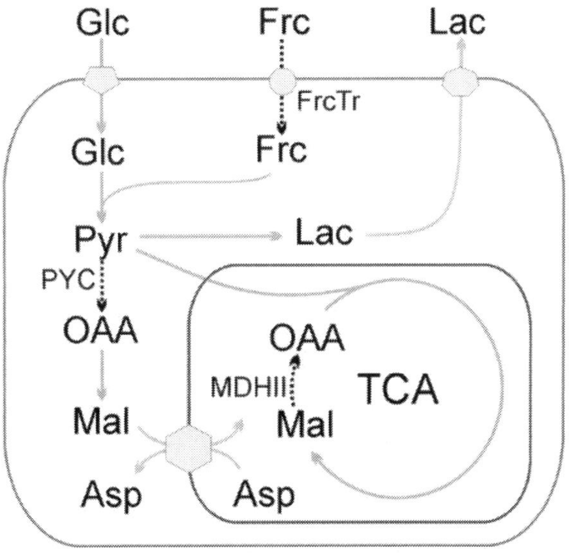

Figure 1: Cell metabolism of cells overexpressing pyruvate carboxylase (PYC), malate dehydrogenase II (MDH II) and Slc2a5 (FrcTr) genes. Enzymes associated to overexpressed genes are shown in dashed lines.

Materials and Methods

IgG producing CHO DP-12 cells were transfected using lipofectamine with vector pcDNA3.1(+) zeo-Slc2a5, pcDNA3.1(-) MDHII and/or PCMVSHE-PYC2+Hygromycine resistance in order to obtain 5 different clones which would overexpress 1 or more genes. Clones were selected with antibiotics and named CHO PYC, CHO MDHII, CHO Slc2a5, CHO PYC-Slc2a5 and CHO MDHII-Slc2a5. Different experiments with 17.5 mM of the main carbon source were designed to study cell proliferation and cells' physiological state. Two control cultures were performed with wild-type CHO cells with glucose and fructose. CHO cells overexpressing PYC2 (CHO PYC) and MDHII (CHO MDHII) were cultured with glucose and CHO cells overexpressing the fructose transporter (CHO Slc2a5), CHO PYC-Slc2a5 and CHO MDHII-Slc2a5 were grown in media containing fructose.

Results

After transfecting the cells with an alternative sugar transporter we observed in the selected clones a reduced flux in the glycolytic pathway which lead to decreased lactate synthesis. Cells transfected with PYC and MDH II also show an improved metabolism with less lactate production and higher cells density. By transfecting the cells with the PYC or MDHII gene additionally to cells overexpressing fructose transporter it is observed that there is further improvement in the metabolism due to a reduction in pyruvate accumulation which in turn decreases lactate production and increases cells' energy metabolism.

Conclusions

Results indicate that cell engineering allows obtaining cells with a more efficient central carbon metabolism, where cells make better use of the main carbon source, cultures reach higher cell densities and have a longer lifespan. Overexpression of PYC and MDHII allows cells to reduce pyruvate accumulation therefore reducing lactate production. Changing the main carbon source to fructose after transfection with Slc2a5 gene allows a slower carbon uptake in a fructose based media than in a glucose based media, but high enough to reach high cell densities but with low lactate synthesis. Finally, cells overexpressing fructose transporter and PYC or MDH II show further improvement in their metabolism with even lower lactate production.

Acknowledgements

We would like to thank Dr. Roland Wagner for the PCMVSHE-PYC2 vector, Dr. Mariella Bollati for the pcDNA3.1(+)zeo vector, Dr. Miranda Yap for the MDHII pcDNA 3.1 Hygro vector and the Genetically Engineered Mouse Facility, M.D. Anderson Cancer Center for the Hygromycine resistance cassette. This work was supported by FONDECYT Initiation Grant 11090268.

References

1. Wilkens, Camila A., Claudia Altamirano and Ziomara P. Gerdtzen (2011), "Comparative metabolic analysis of lactate for CHO cells in glucose and galactose", Biotechnol. Bioprocess Eng, August, (*In press*)
2. Irani, Noushin, Manfred Wirth, Joop van Den Heuvel and Roland Wagner (1999) "Improvement of the primary metabolism of cell cultures by introducing a new cytoplasmic pyruvate carboxylase reaction", Biotechnol Bioeng, pp. 238-46
3. Chong, William P.K., Satty G. Reddy, Faraaz N.K. Yusufi, Dong-Yup Lee, Niki S.C.Wong, Chew Kiat Heng, Miranda G.S. Yap and Ying Swan Ho (2010) "Metabolomics-driven approach for the improvement of Chinese hamster ovary cell growth: overexpression of malate dehydrogenase II" , J Biotechnol, May, pp.116-21
4. Wlaschin, Katie F. and Wei-Shou Hu, (2007) "Engineering cell metabolism for high-density cell culture via manipulation of sugar transport", J Biotechnol, August, pp. 168-76
5. Chen, Keqin, Qian Liu, Liangzhi Xie, Phillip A. Sharp and Daniel I. C. Wang, (2001) "Engineering of a Mammalian Cell Line for Reduction of Lactate Formation and High Monoclonal Antibody Production", Biotechnol Bioeng, January, pp.55-61

(5c) Exploring the Cre/Lox System for Targeted Integration Into the Human Genome

Amanda M. Lanza and Hal Alper, Chemical Engineering, The University of Texas at Austin, Austin, TX

Efficient site-specific genome editing tools would have great utility in biotechnological and genetics applications. To this end, Cre recombinase is commonly employed in mammalian transgene editing however the efficiency for recombination mediated cassette exchange is typically below 1% of surviving cells. One reason for this low efficiency of swapping is the excision function often exhibited by Cre. Regardless, this efficiency is too low for efficient cell line development applications. To address this issue, we sought to increase the propensity of Cre recombinase for swapping by undertaking a simultaneous, unbiased evaluation of several parameters found to influence Cre-mediated swapping and excision in human cells (HT1080). In doing so, we identify parameters that bias total number of recombination events and a separate set of parameters that influence the ratio of swapping to excision events. In particular, a novel mutant *lox* site pairing and delayed introduction of Cre are two critical factors for achieving optimal swapping efficiency. Finally, we create a mathematical model to capture the mechanisms limiting Cre recombinase swapping and highlight two regimes that influence swapping efficiency. By simultaneously optimizing these parameters, we obtain the highest Cre-mediated swapping frequencies without antibiotic enrichment reported in literature, between 8 and 12%. These findings indicate the importance of simultaneously examining variables known to influence Cre activity, rather than evaluating these variables in isolation. Moreover, these results enable more efficient targeted integration into the human genome.

See more of this Session: Advances In Cell Culture IA: Experimental
See more of this Group/Topical: Food, Pharmaceutical & Bioengineering Division

(5d) Enhancing Transient Gene Expression In CHOs Cells Through the Co-Transfection of the Product Gene and Bcl-Xl

Matthew P. Zustiak[1], Yueqing Xie[2], Jianwei Zhu[2] and Michael J. Betenbaugh[1], (1)Chemical and Biomolecular Engineering, Johns Hopkins University, Baltimore, MD, (2)Early Process Sciences, Biopharmaceutical Development Program, SAIC-Frederick, Inc., National Cancer Institute at Frederick, Frederick, MD

Transient gene expression is gaining popularity as a method to rapidly produce recombinant proteins in mammalian cells. Although significant improvements have been made, in terms of expression, more is needed for this method to compete with stable gene expression. Much improvement has come from transfection media optimization, transfection parameter optimization as well as altering culturing conditions to enhance productivity. Recent studies have also included using cell lines engineered for apoptosis resistance through the constitutive expression of an anti-apoptotic protein, Bcl-xL. While significant improvement in yield has been observed there is still a significant time investment needed to generate the stable clones over-expressing Bcl-xL. In this study we examine an alternative method of using the benefits of anti-apoptotic gene expression to enhance the transient expression of bio-therapeutics, namely, through the co-transfection of bcl-xl and the product gene. This method allows for fine tuning the transfection conditions for the optimal Bcl-xL expression level to both protect the cell from the harmful conditions of transfection and enhance specific productivity. For this study we have chosen the commercially relevant suspension CHO cell line together with the in-expensive and effective transfection reagent Polyethylenimine.

We used a YFP vector to monitor transfection efficiencies and relative expression levels through the detection of YFP using flow cytometry and separately utilized a vector containing a gene for an Fc fusion protein to look at overall yield and apoptosis resistance. Several conditions were tested and while there was a negligible effect on transfection efficiencies we found both a reduction in apoptosis and an increase in specific productivity for cultures co-transfected with the bcl-xL gene. Overall, by co-transfecting bcl-xL, minimal cost or complexity is added to a method of transient gene expression and yet significant improvements in product yields are achieved.

See more of this Session: Advances In Cell Culture IA: Experimental
See more of this Group/Topical: Food, Pharmaceutical & Bioengineering Division

(5e) Using In-Situ Raman Spectroscopy to Determine Real-Time Growth and Metabolic Profiles During a Mammalian Cell Culture Process

John Paul Smelko, BioPharm Development, Biogen Idec, Rtp, NC

Abstract Title: Using in-situ Raman spectroscopy todetermine real-time growth and metabolic profiles during a mammalian cellculture process

Abstract Body:

Spectroscopic methods represent an innovative scheme for theanalysis of in vitro processes. Raman spectroscopy offers significant advantagesover other spectroscopies for bioprocess monitoring and was evaluated as ameans of quantifying chemical changes in a dynamic Chinese Hamster Ovary (CHO)cell culture environment. Chemometric analyses of bioprocess data were carriedout in accordance with process analytical technology (PAT) trends utilizedpreviously in other industries. Partial least squares (PLS) calibration modelswere developed from offline reference standards and in-line Raman data for aspecific CHO cell culture process. In the interest of rapidly demonstratingfeasibility of Raman as a PAT tool for biologics manufacturing, the modelingprocess was expedited through the use of an automated sampling system. The autosamplerprovided high quantities of calibration and validation set samples from alimited number of bioreactor batches. High-quality quantitative results wereobtained for glucose, lactate, glutamine, glutamate, ammonium, osmolality andviable cell density. Raman models were developed and validated at both 200L and3L bioreactor scales. Furthermore, a combined model using both 200L and 3L datasets was successfully developed. This talk will cover methods used to generateRaman models for in-line monitoring of mammalian cell culture processes,discuss the challenges of using and implementing Raman spectroscopy, and reviewthe data and results generated to date.

See more of this Session: Advances In Cell Culture IA: Experimental
See more of this Group/Topical: Food, Pharmaceutical & Bioengineering Division

(5f) Effect of Endoplasmic Reticulum Stress On Protein Productivity In Recombinant CHO Cells

Kamal Prashad, Anand Prabhakar and **Sarika Mehra**, Department of Chemical Engineering, Indian Institute of Technology, Bombay, Mumbai, India

Nearly 70% of the recombinant therapeutics used worldwide are produced as secretory proteins in mammalian cells. Productivity of such secretory proteins is a multifaceted phenomenon. Multiple pathways operate in coordination to secrete functionally active proteins out of the cells. The high levels of intracellular protein may disrupt endoplasmic reticulum (ER) homeostasis by increasing the protein folding load on the ER. In response to the high ER stress levels, the cell can respond in different ways: overall translation attenuation, feedback regulation of chaperones or apoptosis. The final choice depends on the interplay between the adaptive and apoptosis pathways.

In this work, we attempt to understand the effect of ER stress on productivity through a combination of experimental and modeling approach. From the available literature on regulation between different genes in the ER, a comprehensive model has been developed for protein secretion. The model accounts for three pathways - protein folding, degradation and unfolded protein response. The model is being simulated to study the effect of genetic alterations on the overall folding efficiency. An optimum chaperone concentration is predicted for enhanced folding efficiency. Depending on the kinetics of folding and degradation, increasing the chaperone concentration above this critical concentration may lead to decrease in folding efficiency. It has also been observed experimentally that over-expression of BiP, a key chaperone, leads to increase in productivity in some cell lines and decrease in others.

In parallel, we have also profiled the dynamics of more than 20 key genes involved in these pathways in different recombinant protein CHO cell lines exhibiting varying productivity levels of IgG. The data shows that the transcript levels of chaperones and UPR transducers vary with productivity levels, with a general increase in overall levels in the high producers. The resulting data is being further analyzed using multivariate analysis tools to identify the key genes influencing productivity. Such a systems level analysis of protein secretion pathway should provide a rational basis to identification of cell engineering targets for enhanced productivity.

See more of this Session: Advances In Cell Culture IA: Experimental
See more of this Group/Topical: Food, Pharmaceutical & Bioengineering Division

(5g) Uncovering the Mechanisms of Methotrexate-Mediated DHFR Amplification In Chinese Hamster Ovary Cells Through Transcriptome Analysis

Nitya M. Jacob[1], **Huong Le**[1], Sze Wai Ng[2], Bernard L. W. Loo[2], Yung-Shyeng Tsao[3], Zhong Liu[3] and Wei-Shou Hu[1], (1)Department of Chemical Engineering and Materials Science, University of Minnesota, Minneapolis, MN, (2)Bioprocessing Technology Institute A-STAR, Singapore, (3) Bioprocess Development, Merck & Co., Union, NJ

In the past decade, the specific productivity of recombinant mammalian cells has increased to a level rivaling that of professional secretors *in vivo*. The development of such high protein-producing cell lines is generally aided by gene amplification, a process that leads to the creation of several copies of the recombinant gene on the chromosome, often resulting in dramatic increases in productivity. However, the physiological mechanisms of the amplification process that can transform a non-secretory cell line to achieve the productivities rivaling professional secretors, remain largely unknown. To this end, we used transcriptome analysis to study the amplification process in CHO cells. A parental CHO cell line deficient in dihydrofolate reductase (DHFR) activity was transfected with a vector expressing a DHFR gene, a hygromycin-resistant gene, as well as antibody product genes. Methotrexate treatment was applied following transfection and initial selection. Clones isolated following transfection showed significant variation in their product expression profiles, as well as varying degrees of gene amplification. As a control, the same cell line was transfected with another vector carrying only the DHFR gene, thus providing a baseline to assess the importance of the expressed product in the transcriptional response. Following amplification, the clones were further subcloned, adapted to suspension and scaled up to fedbatch processes. Transcriptome analysis was performed on clones during the amplification process as well as on the resulting high and low producer subclones in fedbatch cultures. This study provides physiological insights into the mechanisms involved in gene amplification and clonal selection, and allows the design of a screening strategy for the selection of viable production cell line candidates.

See more of this Session: Advances In Cell Culture IA: Experimental
See more of this Group/Topical: Food, Pharmaceutical & Bioengineering Division

(7a) A First Principles Based Structure Prediction Algorithm for Beta and Mixed Alpha/Beta Proteins

Ashwin Subramani, Yanjie Wei and Christodoulos A. Floudas, Chemical and Biological Engineering, Princeton University, Princeton, NJ

We present the latest developments in the prediction of pure beta and mixed alpha/beta proteins using our first principles based structure prediction framework ASTROFOLD 2.0. The framework for the approach has been enhanced to include improvements in secondary structure prediction, beta sheet topology prediction, residue-to-residue contact prediction, tertiary structure prediction and near-native structure identification [1].

For the prediction of secondary structure, two sets of algorithms have been developed which can be employed depending on the sequence similarity of a given protein with the PDB. Firstly, a consensus server prediction algorithm, CONCORD, has been developed, which combines the predictions of 7 well known secondary structure prediction servers using an integer linear optimization model, in order to enhance the prediction accuracy [2]. Alternately, single sequence formulations have been developed, which predict the secondary structure of the protein without the use of profile information. For pure beta and mixed alpha/beta proteins, we use a novel integer optimization prediction based algorithm to derive the most likely beta sheet topologies. These topologies are re-ranked using linear programming formulations to predict the best possible topology for a given predicted secondary structure [3]. Contacts between alpha helices in a mixed alpha/beta protein are then predicted using an integer optimization based algorithm, wherein the contacts between beta strands are fixed to the predictions made at the previous stage [4]. A novel flexible-stem loop structure prediction algorithm is employed to derive tight bounds for the backbone dihedral angles for all loop residues [5]. Separately, ASTRO-FOLD allows for bypassing secondary structure prediction steps by building the secondary structure from 3D-Jury results, and using fixed-stem loop structure prediction algorithms to refine the structure of the loop regions. All of these constraints are used to predict the three dimensional structure of the protein, using a combination of deterministic global optimization, stochastic conformational space annealing and torsion angle dynamics. At each stage of the deterministic and stochastic algorithms, fast side chain optimization steps are introduced to alleviate clashes between side chains and the protein backbone, thus providing better starting points for the optimization algorithms [1]. The final set of predicted structures is clustered using a novel traveling salesman problem based clustering algorithm, ICON [6].

Bibliography

----

[1] Subramani A, Wei Y and Floudas CA (2011) ASTRO-FOLD 2.0: An enhanced framework for protein structure prediction, *submitted.*

[2] Wei Y and Floudas CA (2011) CONCORD: A consensus method for protein secondary structure prediction, *submitted*

[3] Subramani A and Floudas CA (2011), *in preparation.*

[4] Rajgaria R, Wei Y and Floudas CA (2009) Contact prediction for beta and alpha/beta proteins using integer linear optimization and its impact on the first principles 3d structure prediction method ASTRO-FOLD, *Proteins*, 78, 1825-1846.

[5] Subramani A and Floudas CA (2011), *in preparation*

[6] Subramani A, DiMaggio PA and Floudas CA (2009) Selecting high quality protein structures from diverse conformational ensembles, *Biophysical J*, 97, 1728-1736

See more of this Session: Advances In Protein Structure, Function, and Stability I
See more of this Group/Topical: Food, Pharmaceutical & Bioengineering Division

(7b) Hofmeister Effects On Sup35NM Amyloid Formation Kinetics

Jonathan Rubin[1], Sven H. Behrens[1], Yury O. Chernoff[2] and Andreas S. Bommarius[1], (1)Chemical & Biomolecular Engineering, Georgia Institute of Technology, Atlanta, GA, (2)Biology, Georgia Institute of Technology, Atlanta, GA

The addition of salt to a colloidal dispersion reduces the energetic barrier for aggregation; however, in the context of protein, salts do not all have the same effect. The Hofmeister series is a qualitative ranking system for ions describing the extent to which an ion will affect a variety of phenomena in solution including (but not limited to) protein solubility, enzyme activity, and colloidal stability. Since protein stability plays an important role in protein misfolding diseases, we test the influence of a host of Hofmeister salts on the aggregation rate of Sup35NM, a model yeast prion protein. Prion proteins are capable of converting from their soluble forms into highly ordered fibrous cross-β aggregates, known as amyloids. This conversion is associated with certain neurodegenerative conditions in mammals, such as Alzheimer disease, and controls heritable phenotypes in yeast. Previously we have studied Sup35NM aggregation in the presence of select anions using the Congo red binding assay and found an inverse Hofmeister trend, where chaotropes strongly stabilize and kosmotropes destabilize. This study confirmed that result and expands on the topic by examining the effect of a wide range of both anions and cations at varying concentrations on the aggregation rate, lag time prior to aggregation and the extent of aggregation of Sup35NM using the fluorescent amyloid binding dye, thioflavine T. We found significant concentration dependence, long lag time and less overall aggregation for chaotropes; whereas, kosmotropes had little or no concentration dependence, very short lag time and almost complete aggregation. Additionally, we studied the surface charge of Sup35NM to examine the stabilization or destabilization that specific ions offer the protein.

See more of this Session: Advances In Protein Structure, Function, and Stability I
See more of this Group/Topical: Food, Pharmaceutical & Bioengineering Division

(7c) Peptide Probes for Rapid, Specific and Quantitative Detection of Distinct Beta Amyloid Oligomers

Yang Hu, Baihao Su, He Qiu Zheng, Joon Ho Ahn and **Jin Ryoun Kim**, Chemical and Biological Engineering, Polytechnic Institute of New York University, Brooklyn, NY

Aggregation of beta-amyloid (Abeta) is implicated in the pathology of Alzheimer's disease (AD). Development of a robust strategy to detect Abeta oligomeric intermediates, identified as significant toxic agents, would be highly beneficial in the screening of drug candidates as well as enhancing our understanding of Abeta oligomerization. Rapid, specific and quantitative detection, currently unavailable, would be highly preferred for accurate and reliable probing of transient Abeta oligomers. We have previously reported the development of a peptide probe, PG46, for rapid, specific and quantitative detection of Abeta oligomers. PG46 was created by combining Abeta binding sequences and a conformation-sensitive fluorescence dye, FlAsH.

In this presentation, we will report the creation of a second generation of peptide probes by modifying the sequence of PG46. These new peptide probes displayed different conformations compared to PG46, and exhibited novel specificity toward certain Abeta oligomeric forms, which were not recognizable by PG46. These novel peptide probes generated fluorescence signals upon binding to Abeta oligomers through an increase in the quantum yield of bound FlAsH. We anticipate that our peptide probes for Abeta oligomer detection will benefit the understanding of Abeta oligomerization determinants, AD diagnostics and drug development.

See more of this Session: Advances In Protein Structure, Function, and Stability I
See more of this Group/Topical: Food, Pharmaceutical & Bioengineering Division

(7d) Suppressing Protein Aggregation by Reducing Local Concentration Fluctuations

Christopher J. Roberts[1], Erinc Sahin[1], Marco A. Blanco[1] and Tapan K. Das[2], (1)Department of Chemical Engineering, University of Delaware, Newark, DE, (2)Biotherapeutics Research and Development, Pfizer, Inc., Chesterfield, MO

Aggregation of therapeutic proteins is a ubiquitous concern during product purification, formulation, storage and administration. Recent reports in the literature highlight the importance of conformational stability, so as to minimize the concentration of unfolded, aggregation-prone proteins. In contrast, optimizing so-called colloidal stability -- e.g., the osmotic second virial coefficient (B_{22}) -- has been found to be less effective in reducing aggregation rates. In addition, a direct theoretical basis for quantitatively relating B_{22} (or more accurately, the corresponding Kirkwood Buff integral G_{22}) to observed aggregation rates has been lacking. Here we propose a new approach to relate B_{22} and G_{22} to aggregation rates, based on an adaptation of statistical mechanical fluctuation theory. Experimentally, we test the theory with a comparison of conformational stability, laser light scattering, and aggregation kinetics for two model therapeutic monoclonal antibodies as a function of pH and NaCl concentration. The results show that one can achieve dramatic reductions in aggregation rates by maximizing colloidal repulsions, and thereby minimizing local fluctuations in protein concentration that are necessary to nucleate new aggregates. The model also provides a novel means to predict the effects of changes in formulation conditions on aggregation rates quantitatively or semi-quantitatively, and to relate these to molecular-scale fluctuations that can be probed experimentally via techniques such as laser light scattering.

See more of this Session: Advances In Protein Structure, Function, and Stability I
See more of this Group/Topical: Food, Pharmaceutical & Bioengineering Division

(7e) Aromatic Peptides Selectively Target and Remodel Soluble Oligomers of Amyloid

Ali Reza A. Ladiwala, Jonathan S. Dordick and **Peter M. Tessier**, Dept of Chemical & Biological Engineering, Rensselaer Polytechnic Institute, Troy, NY

The misfolding of specific proteins of unrelated sequence into similar toxic aggregates is the hallmark of several conformational disorders, and much effort has focused on identifying molecules that inhibit and reverse protein aggregation. Little is known about how peptides can be used to selectively target and remodel toxic protein aggregates relative to non-toxic conformational isoforms. We find that aromatic-repeat peptides possess striking activity for rapidly remodeling mature toxic soluble oligomers and fibrils of the Aβ42 peptide associated with Alzheimer's disease into insoluble, benign aggregates. Importantly, peptides repetitive in non-aromatic amino acids are inactive for remodeling Aβ oligomers and fibrils. Interestingly, aromatic-repeat peptides do not remodel Aβ non-toxic oligomers (which are indistinguishable in size and morphology relative to toxic Aβ oligomers) nor do they accelerate the aggregation of monomeric Aβ peptide, revealing they are specific for remodeling toxic Aβ aggregates. Moreover, we find that aromatic peptides active for remodeling Aβ soluble oligomers and fibrils do so by converting them into aggregated conformers with strikingly similar structural properties as non-toxic oligomers formed in the absence of aromatic peptides. We expect our findings will guide the design of novel biomolecules that selectively target and remodel toxic aggregates of Aβ and other misfolded proteins.

See more of this Session: Advances In Protein Structure, Function, and Stability I
See more of this Group/Topical: Food, Pharmaceutical & Bioengineering Division

(7f) Mammalian Cell-Based Methods Monitoring Protein Misfolding and Aggregation

Simpson Greogire and **Inchan Kwon**, Department of Chemical Engineering, University of Virginia, Charlottesville, VA

Protein folding is a complex process that converts a linear strand of amino acids to a functional, three-dimensional complex. Point mutations in the primary sequence that encodes for proteins is known to cause aberrant protein conformations. An array of "protein-misfolding" diseases such as Alzheimer's, Amyotrophic Lateral Sclerosis, Parkinson's, are caused by misfolded proteins that form protein deposits in essential organs within the body. With the development of new treatments to help treat these diseases, there is a need to develop a robust method that can accurately monitor the folding of these proteins in the presence of the therapeutics as well as in a physiologically relevant system. Despites the benefits of mammalian cells to study misfolding and aggregation of human proteins associated with neurodegenerative diseases, mammalian cell-based monitoring methods have not been explored much. In order to quantitatively and qualitatively monitor protein misfolding and aggregation in mammalian cells, we utilized three folding reporter proteins, split GFPs, mammalian folding reporter GFP (mfrGFP), and EGFP as a fusion partner. As a test case, we chose human copper/zinc superoxide dismutase (SOD1) associated with Amyotrophic Lateral Sclerosis, also called as Lou Gehrig's disease. There are numerous mutant SOD1s with the different severity of misfolding caused by the point mutations in the amino acid sequence. Fusion of both mfrGFP and EGFP were effective in monitoring mutant SOD1 aggregation within cultured cells, whereas split GFP was not. The cellular fluorescence of transfected HEK293T and NSC-34 cells expressing fusion proteins was determined by flow cytometry. Our findings signify that flow cytometric fluorescence measurement of cells expressing GFP variant fused to mutant SOD1 is a simple, but effective way to monitor mutant SOD1 aggregation in mammalian cells and also has a promise for a high-throughput screening for aggregation modulators.

See more of this Session: Advances In Protein Structure, Function, and Stability I
See more of this Group/Topical: Food, Pharmaceutical & Bioengineering Division

(7g) Identification of Transthyretin Domains Involved with Binding to Beta-Amyloid: In Search of Nature's Anti-Alzheimer's 'Drug'

Jiali Du, Patricia C. Cho, Dennis T. Yang and **Regina M. Murphy**, Chemical and Biological Engineering, University of Wisconsin-Madison, Madison, WI

Alzheimer's disease (AD) is the most common age-associated neurodegenerative disease. Characteristic features include extracellular senile plaques, intraneuronal neurofibrillary tangles, and extensive neuronal cell death. The major proteinaceous component of the plaques is beta-amyloid (Ab). Ab spontaneously self-assembles through a multi-step process into soluble oligomers and fibrillar aggregates. Numerous studies have established that Ab aggregation is causally linked to neurodegeneration; most researchers currently believe that it is the soluble oligomers that are most toxic to neurons.

Transthyretin (TTR) is a homotetrameric transport protein present in both blood and cerebrospinal fluid (CSF). Several studies have demonstrated that TTR binds to Ab, and alters its aggregation. Interestingly, TTR itself is prone to aggregation and is linked to other age-related amyloid disorders. Recent experiments with transgenic AD mice suggest that TTR may protect neurons against Ab-induced damage, raising the intriguing possibility that TTR is a natural anti-AD 'drug'.

The aim of this study is to identify the specific regions of TTR involved with binding to Ab. By doing so, we hope to be able to design mimetics that could replace TTR's natural protective activity that is lost with age or disease. We have used crosslinking and tryptic cleavage coupled with mass spectrometry, along with immobilized peptide arrays, to isolate two regions of TTR, the G strand and the E-strand-E/F loop, as likely domains involved with TTR-Ab association. We have also shown that TTR's natural ligands, thyroxine and retinol-binding protein, which bind to TTR at these identified regions, compete with Ab for binding to TTR. We synthesized peptides corresponding to the putative binding regions of TTR and found that these peptides specifically inhibit Ab binding to TTR, but in a nonlinear fashion. We are using limited alanine-scanning mutagenesis to identify specific residues on TTR involved with binding to Ab. This study is complicated by the fact that the quaternary structure and stability of TTR also strongly influences TTR-Ab association. Finally, we plan to examine the effect of active peptides and TTR mutants on Ab oligomerization and toxicity.

See more of this Session: Advances In Protein Structure, Function, and Stability I
See more of this Group/Topical: Food, Pharmaceutical & Bioengineering Division

(10a) Fermentation Cost Improvement Through Propagation Advances for a Commercially Relevant Biomass-to-Ethanol Process

Kate Brandon Sutton, B. Emme, L. Jones, S. McHatton, C. Kang, C. Corbett, L. Putnam and M. Torry-Smith, Novozymes North America Inc., Franklinton, NC

Compared to their starch-to-ethanol counterparts, pentose-utilizing organisms for biomass-to-ethanol processes are less robust while having to ferment mixed sugars in a significantly more inhibitory substrate. This inhibition can generally be overcome with a high yeast pitch to maximize xylose uptake rates in fermentation. The cost driven trade-off of a high yeast pitch during fermentation is the necessity of dedicated onsite propagation, as opposed to purchase of all yeast. Novozymes' Advance Biofuel cost model suggest the majority of the fermentation cost when using high yeast pitches can be attributed to the propagation variable costs, most significantly from the carbohydrate source. Therefore, maximizing the cost performance of C5 yeast propagation is important to ensure an economically feasible biomass-to-ethanol fermentation step.

Novozymes has identified two approaches to reduce the fermentation cost: first, increase propagation cell yields from sugars to reduce the realized organism cost, preferably on cheap carbohydrate sources, or second, produce a more robust organism from propagation with a higher realized xylose uptake rate in fermentation which would enable fermentation at a lower yeast pitch in a reasonable processing time. This presentation will discuss cost model results in conjunction with reactor data to pinpoint potential savings and opportunities to lower overall fermentation cost for second generation ethanol. Sensitivities on cost from propagation and fermentation process variables targets will be presented. Propagation optimization with respect to higher realized xylose uptake rate in fermentation will be addressed as well. Lastly, the holistic minimum ethanol selling price implications for process optimizations will be given.

See more of this Session: Biological Conversions and Processes for Renewable Feedstocks I
See more of this Group/Topical: Sustainable Engineering Forum

(10b) Combined Computational and Experimental Investigations of Processive Cellulases for Engineering Activity Improvements

Gregg T. Beckham[1], **Larry Taylor**[2], Michael F. Crowley[3], William S. Adney[3] and Michael E. Himmel[3], (1)National Bioenergy Center, National Renewable Energy Laboratory, Golden, CO, (2) Biosciences Center, NREL, Golden, CO, (3)Biosciences Center, National Renewable Energy Laboratory, Golden, CO

The Family 7 cellobiohydrolase (Cel7A) from *Hypocrea jecorina* (*Trichoderma reesei*) is one of the most well-characterized, active cellulases known for deconstruction of cellulose to soluble sugars. With the aim to engineer enhanced cellulase specific activity, we have investigated the Cel7A mechanism on cellulose with molecular dynamics simulations and biochemical experiments. We have elucidated new functions for the carbohydrate-binding module, determined the effect of glycosylation on the stability of the linker and carbohydrate-binding module, examined the role of aromatic and polar residues in the catalytic domain tunnel, and examined the structure of the catalytically-active complex of Cel7A on crystalline cellulose up to the microsecond time scale. These new insights combined with experimental activity data have led to a large-scale experimental screening of new enzymes to examine both natural diversity and engineered enzymes for activity and stability improvements. Our combined computational and experimental approach has demonstrated that unprecedented improvements in specific activities of processive cellulases are possible, which when combined in industrial cocktails, will result in lower enzyme loadings for cell wall deconstruction and lower overall production costs for renewable biofuels.

See more of this Session: Biological Conversions and Processes for Renewable Feedstocks I
See more of this Group/Topical: Sustainable Engineering Forum

(10c) Exploration of Significant Enzyme Activities Required for the Deconstruction of Ionic-Liquid Pretreated Biomass

Christopher J. Barr[1], Jeffery Mertens[2] and Constance Schall[1], (1)Chemical and Environmental Engineering, University of Toledo, Toledo, OH, (2)NCAUR, USDA-ARS, Peoria, IL

Terrestrial biomass is a second generation renewable feedstock, comprised of polysaccharides derived from cellulose and hemicellulose, which can be used to produce carbon-based chemicals and fuels. The recalcitrance to deconstruction of biomass is one of the main impediments in the conversion of biomass to monomeric sugars and the production of sugar platform fuels and chemicals. Ionic Liquid (IL) pretreatment creates unique biomass substrates composed of amorphous cellulose with largely intact hemicellulose and lignin. IL pretreated substrates differ from many other pretreatments which remove hemicellulose and retain cellulose crystallinity. This substrate poses an interesting challenge for enzymatic deconstruction using traditional commercial mixtures.

Mono-component cellulase and hemicellulase enzymes from *Aspergillus nidulans* have been expressed, purified, and characterized. Key enzyme activities and/or interactions that significantly contribute to the conversion of a variety of lignocellulosic biomass substrates to monomeric sugars were examined by enzyme combinations in series of hydrolysis experiments. Auxiliary hemicellulases seem to be essential to the mono-component enzyme mixtures. Investigation of these accessory enzymes also yields insight into an understanding of the differences in hemicellulose structure between feedstocks. Some auxiliary hemicellulases include acetyl xylan esterases (AXE), arabinofuranosidases, and α-glucuronidases. In addition to mono-component enzyme mixtures, next generation commercially-available enzyme mixtures were characterized and formulated using key activities to optimize hydrolysis at low enzyme loadings. These simplified mixtures will be dependent on the substrate. As part of these studies, an analysis of variance (ANOVA) was used to statistically model the main and interaction effects of key enzyme activities for each substrate.

See more of this Session: Biological Conversions and Processes for Renewable Feedstocks I
See more of this Group/Topical: Sustainable Engineering Forum

(10d) Inhibition of ENZYMATIC Hydrolysis by Soluble Sugars ON MODEL CELLULOSE Thin Film USING QUARTZ CRYSTAL Microbalance

Hsin-Fen Li[1], Ravinder Garlapalli[1], Michael D. Flythe[2], Sue E. Nokes[3], Stephen E. Rankin[1] and Barbara L. Knutson[1], (1)Chemical and Materials Engineering, University of Kentucky, Lexington, KY, (2)Department of Animal and Food Sciences, University of Kentucky, Lexington, KY, (3)Department of Biosystems and Agricultural Engineering, University of Kentucky, Lexington, KY

INHIBITION OF ENZYMATIC HYDROLYSIS BY SOLUBLE SUGARS ON MODEL CELLULOSE THIN FILM USING QUARTZ CRYSTAL MICROBALANCE

Hsin-Fen Li [1], Ravinder Garlapalli[1]. Michael D. Flythe [2,3], Sue E. Nokes [4], Stephen E. Rankin[1], Barbara L. Knutson[1]

[1]University of Kentucky, Department of Chemical and Materials Engineering, Lexington, Kentucky 40506

[2]USDA, Agricultural Research Service, Forage-Animal Production Research Unit; Lexington, Kentucky 40546

[3]University of Kentucky, Department of Animal and Food Sciences, Lexington, Kentucky 40546

[4]University of Kentucky, Department of Biosystems and Agricultural Engineering, Lexington, Kentucky 40506

Soluble sugars produced from the hydrolysis of cellulose are one of the most promising feedstocks for production of fuels and commodity chemicals from renewable plant-based resources. The recalcitrance of cellulose to degradation is well documented and attributed to multiple sources, including high cellulose crystallinity and the hydrogen binding framework that is embedded in the matrix substances (lignin and hemicellulose). Cellulase is a cellulytic enzyme that possess a catalytic domain capable of hydrolyzing the beta-1, 4-glycosidic bonds in cellulose. However, during biomass conversion to soluble sugars, its hydrolytic efficiency is limited by the presence of degradation products including cellobiose. To study the inhibitory effect of cellobiose and the interaction of the inhibited enzyme and the cellulose substrate, the change in mass and surface properties of cellulose thin films were measured using Quartz Crystal Microbalance with Dissipation (QCM-D). The cellulose thin film was synthesized from thermochemically dissolved cellulose and was coated on a quartz crystal resonator. The mass of the spin-coated cellulose film was detected by the resonance frequency, and the cellulose surface coverage was observed by atomic force microscopy (AFM). Cellulase adsorption, activity, and inhibition on model cellulose thin films as were measured directly as changes to the piezoelectric properties of the cellulose thin films by QCM-D. At 5 g/L, cellobiose completely inhibits cellulase (0.5 %v/v) activity. The inhibited cellulases bind reversibly to the cellulose and reduce subsequent enzymatic hydrolysis rates by more than 50%. Significantly higher activity is observed when cellulase concentration is decreased from 1 %v/v to 0.1 %v/v, suggesting that cellulase acts progressively on the cellulose chain, and the hydrolysis is restricted by saturated enzyme conditions.

See more of this Session: Biological Conversions and Processes for Renewable Feedstocks I
See more of this Group/Topical: Sustainable Engineering Forum

(10e) Selection and Characterization of Acid-Stable Cellulases Against Pretreated Oil Palm Empty Fruit Bunch From a Colombian Extreme Environment

Luis Miguel Medina[1], Luisa Cabezas[2], Isabella Bahamon[1], Laura Palma[1], Andrs Gonzlez[1] and Silvia Restrepo[2], (1)Chemical Engineering, Centro de Diseo de Productos y Procesos, Universidad de los Andes, Bogot, Colombia, (2)Biological Sciences, Laboratorio de Micologa y Fitopatologa (LAMFU), Universidad de los Andes, Bogot, Colombia

The rapid growth of the oil palm industry around the world generates environmental concerns, with a production of more than 28.0 million tons per year of oil palm around the world, generating a great amount of by-products and wastes, such as liquid effluent, gaseous emission, and solid waste materials. Nowadays, solid waste is used commonly as manure, nevertheless the composting process is reported to be complex[1]. Therefore this agro-industry should take into consideration the use of sustainable technology. A material balance of an oil palm plant yields, in terms of mass, for every Ton of fresh fruit processed 220 kg of oil palm empty fruit bunch (OPEFB), 125-130 kg of fiber, around 50 kg of shell and 195-200 kg of palm oil [2]. Studies on the lignocellulosic content of the OPEFB have shown that display a high cellulose and hemicellulose and low lignin content [2]. This feature allows to use this waste as a substrate to obtain simple sugars, which can be used as carbon source in different fermentation processes to enable the production of alcohols that can be used as fuels. The biodegradation of cellulose is carried out by three classes of enzymes: endoglucanase (carboxymethylcellulase), exogluconase (cellobiohydrolase) and β-glucosidase, which together with a preliminary treatment of cellulose, manage to break the glycosidic bonds to produce simple sugars [3]. However, these enzymes hydrolyze in different parts of the polymer so the relation between the substrate and the enzyme should not be extrapolated due to different lignocellulosic structure [4]. Taking all this into account the objective of this study was to explore the cellulolytic potential of endophytes from a Colombian extreme environment for the saccharification of oil palm empty fruit bunch. It was determined the cellulolytic capacity for the saccharification *in vitro* of the pretreated oil palm empty fruit bunch of one hundred Colombian native endophytes isolated from young leaves attacked by insects (with herbivory) as well with no symptom of attack (without herbivory) from *Espelieta grandiflora* and *E. corymbosa* plants from the Páramo Crúz Verde. The microorganisms were screened to asses which of them have cellulolytic capacity on solid agar with CMC as substrate followed by Congo red staining method [5]. The microorganisms that showed cellulolytic activity were grown on cellulase inducting medium, using pretreated OPEFB as substrate, the activity of the three main components of cellulase (CMCase, β-glucosidase and Cellobiohydrolase) was measured, and total cellulolytic activity (Filter paper activity. FPA), using the procedures described by Zhang *et. al (2009)*. Candidates obtained, (two entophytes), were partially purified through partial precipitation with ammonium sulfate (90%) [6] and chromatography using DAE-shepadex A-50 column previously equilibrated with sodium acetate buffer (pH 5, 0.05 M) and the fractions were withdrawn with a linear NaCl gradient, the fractions with cellulolytic activity were pooled and then concentrated by ultrafiltration with a centriprep (Amicon) having a 10 kDa molecular mass cut-off. The screenings allowed us to identify 28 endophytes with cellulolytic activity, and the best obtained were a *Penicilium sp.* and an *Aspergillus sp.* with maximum enzyme activities, CMCase, FPase, exogluconase and β-glucosidase, of 0.50-0.16, 0.035-0.032, 0.87-0.99 and 0.068-0.01 U/ml respectively. CMCase and the FPA activities of the partially purified cellulase was assessed at pHs ranging from 1 to 10, and temperatures ranging from 30 to 80 °C in order to find the optimum conditions for the cellulases. The cellulases from *Penicillium sp. and Aspergillus sp.* were stable at low pH which could be a valuable tool for hydrolysis of cellulose under acidic conditions.

References

1. Yusoff, S., *Renewable energy from palm oil - innovation on effective utilization of waste.* Journal of Cleaner Production, 2006. **14**(1): p. 87-93.

2. Umikalsom, M.S., et al., *The treatment of oil palm empty fruit bunch fibre for subsequent use as substrate for cellulase production by Chaetomium globosum Kunze.* Bioresource Technology, 1997. **62**(1-2): p. 1-9.

3. Sun, Y. and J. Cheng, *Hydrolysis of lignocellulosic materials for ethanol production, a review.* Bioresource Technology, 2002. **83**(1): p. 1-11.

4. Zhang, Y.H.P., M.E. Himmel, and J.R. Mielenz, *Outlook for cellulase improvement: Screening and selection strategies.* Biotechnology Advances, 2006. **24**: p. 452-481.

5. Kasana, R., et al., *A Rapid and Easy Method for the Detection of Microbial Cellulases on Agar Plates Using Gram's Iodine.* Current Microbiology, 2008. **57**(5): p. 503-507.

6. Scopes, R.K., *Protein purification principles and practice* Second ed. Spinger Advanced texts in chemistry, ed. C.R. Cantor. 1982: Springer-Verlag.

See more of this Session: Biological Conversions and Processes for Renewable Feedstocks I
See more of this Group/Topical: Sustainable Engineering Forum

(10f) Effect of AFEX Pretreatment Degradation Products On Enzymatic Hydrolysis and Microbial Fermentation by Saccharomyces Cerevisiae 424A (LNH-ST)

Xiaoyu Tang[1], Leonardo da Costa Sousa[1], Shishir Chundawat[1], Mingjie Jin[1], Lau Ming[1], James Humpula[1], Nirmal Uppugundla[1], Kevin Chambliss[2], Ramin Vismeh[3], A. Daniel Jones[3], Zeyi Xiao[4], Bruce Dale[1] and **Venkatesh Balan**[1], (1)Chemical Engineering and Material Science, Michigan State Univeristy, Lansing, MI, (2)Chemistry and Biochemistry, Baylor University, Waco, TX, (3) Biochemistry and Molecular Biology, Michigan State University, East Lansing, MI, (4)School of Chemical Engineering, Sichuan University, Chengdu, CA, China

Several degradation products are formed or released during AFEX pretreatment process, which can be divided into five groups: 1) carbohydrates, 2) aliphatic acids, 3) nitrogenous compounds, 4) furans, 5) aromatic compounds. Some of these degradation products are expected to be inhibitory to *Saccharomyces cerevisiae* 424A(LNH-ST), thus affecting the xylose fermentation, as well as various saccharolytic enzymes. In this work, the effect of various AFEX degradation products on co-fermentation by *Saccharomyces cerevisiae* 424A(LNH-ST) were tested individually and in combinations. A synthetic media composed of similar nutrient composition as AFEX corn stover hydrolyzate was designed and successfully used as a control fermentation media in this study. Nitrogenous compounds showed the inhibitory effect on xylose fermentation, followed by carbohydrates, aliphatic acid and aromatic compounds. The inhibitory effect caused by nitrogenous compounds and carbohydrates were attributed to the presence of amides and oligomeric xylan, respectively. To our knowledge, this is the first report of inhibitory effect of oligomeric xylan and amides on xylose fermentation by *Saccharomyces cerevisiae*. When testing the amides (feruloyl amide, coumaroyl amide and acetamide) and their corresponding acids (ferulic acid, coumaric acid and acetic acid) individually, the amides were less inhibitory than their acids with respect to cell growth and xylose fermentation. From this point, the shift of degradation products from organic acids to amides may be responsible for the high fermentability of AFEX pretreated lignocellulosic biomass. We also evaluated the extent of enzymatic inhibition by various molecular fractions of AFEX treated corn stover degradation products. The pretreated biomass extracted products were fractionated using ultracentrifugation, gel filtration, and C18 resin solid phase extraction and further analyzed by mass spectrometry. The results indicate that both large and small molecule compounds play a role in enzyme/ microbial inhibition.

See more of this Session: Biological Conversions and Processes for Renewable Feedstocks I
See more of this Group/Topical: Sustainable Engineering Forum

MODELLING SIMULTANEOUS SACCHARIFICATION AND FERMENTATION OF MICROCRYSTALLINE CELLULOSE USING COMPUTATIONAL FLUID DYNAMICS

Josebus Maree van Zyl, Stellenbosch University, Stellenbosch, RSA
Thomas Michael Harms, Stellenbosch University, Stellenbosch, RSA
Eugéne van Rensburg, Stellenbosch University, Stellenbosch, RSA
Willem Heber van Zyl, Stellenbosch University, Stellenbosch, RSA
Lee Rybeck Lynd, Dartmouth College, Hanover, NH

Presentation ID#: 233146

Biography

J.M. van Zyl
email: jmvz@sun.ac.za
Tel: +27 (0)21 808 4272

Extended Abstract

Cellulose is a polysaccharide and one of the most abundant organic compounds found on the Earth surface [1]. It holds the potential to reduce human reliance on fossil derived fuels by producing ethanol through the hydrolysis and fermentation of cellulose. Ethanol is a highly combustible liquid fuel which has been shown to be a viable alternative to petrol. Producing ethanol from cellulose however has proved most challenging, particularly overcoming the recalcitrance of cellulose to enzymatic hydrolysis.

Cellulose to ethanol conversion has successfully been demonstrated at laboratory scale for a variety of feedstock and production methods and in a few cases at demonstration scale. Commercialization of cellulosic ethanol is however limited by the lack of large capital funds required to construct such facilities. A compounding effect is that investors still consider the technology as immature and high risk and as such are hesitant to commit. Obtaining the necessary capital funding requires de-risking the technology as well as policy frameworks and incentives from government.

Engineers have recently begun the development of numerical models for implementation with computational fluid dynamic (CFD) software that will allow investigation of secondary effects of cellulosic ethanol production scale-up. The advantage of CFD is the modelling of the flow patterns and conditions that could exist in a commercial scale fermenter. This includes shear-rates, power requirements, degree of particle suspension and through user-coded modules the reaction-rates of biological or chemical processes. This methodology could allow a significant reduction in the risk associated for investors at a reasonably low financial expense.

Presented in this paper are results of the reaction kinetic modelling (Figure 1) of microcrystalline cellulose to ethanol conversion using cellulases from the fungus *Trichoderma reesei* and an industrial grade yeast *Saccharomyces cerevisiae*, commonly used in the food industry [2]. Observations from the results indicated that the number of available bonding sites on the particle surface appeared to control the rate at which hydrolysis occurred. Enzymes appear to catalyse a reaction at a constant rate thus more sites available to attach should result in higher reaction rates. However due to the fibril structure of microcrystalline cellulose a large number of bonding sites are available resulting in a large number of exoglucanase and endoglucanase enzymes initially attaching to sites on the cellulose chains. Once the initial available sites diminish, a large percentage of the exoglucanase enzymes detach from the particle surface and enters the fermentation broth, effectively reducing the rate of hydrolysis. It thus suggests that if the work rate of the endoglucanase enzyme could be significantly increased, high hydrolysis rates could be achieved as more reduction ends become available for the exoglucanase enzymes.

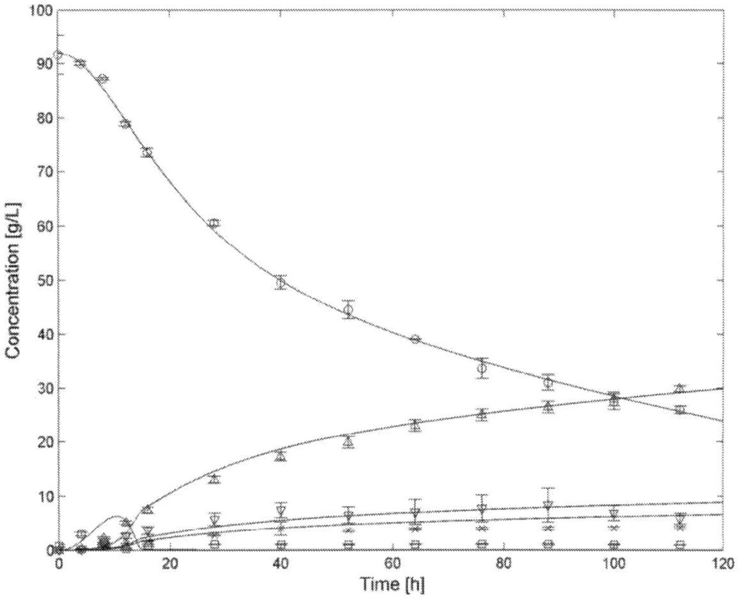

Figure 1: Kinetics model for hydrolysis and fermentation of Avicel PH-101
SSF of Avicel (○) forming glucose (□) fermented to ethanol (Δ) and glycerol (x). Additional yeast cells grown (∇). Simulated results are presented with solid lines.

Microcrystalline cellulose is an insoluble compound which needs to be modelled as particles. It is further required to determine the effects of each substance in the fermentation broth on the viscosity of the system. Results indicated microcrystalline particles occur in various shapes and sizes and exhibit significantly higher drag resistance when the average particle sized is in the order of 50 μm and smaller. This effect is possibly due to the combination of particle topology, structured water [3] and near surface effects [4]. Furthermore the presence of particles has the greatest influence on the apparent dynamic viscosity of the fermentation broth (Figure 2). A numerical model was developed for use in CFD codes which could predict the effects of the particles along with ethanol and glycerol on the apparent dynamic viscosity of the fermentation broth with an average error of 11.1 %.

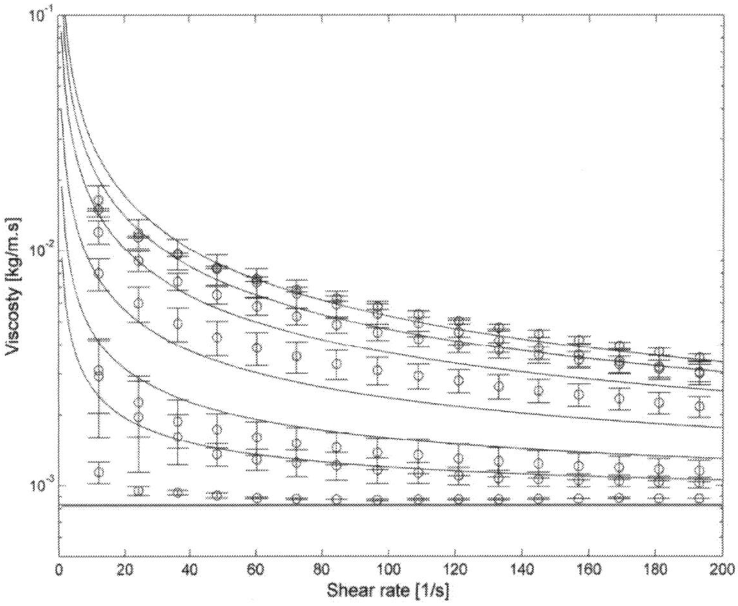

Figure 2: Dynamic viscosity influence of Avicel PH-101 particles

Experimental results (○) with error-bars indicating the standard deviation compared to the numerical predictions (–)

The above mentioned models were incorporated into the commercial CFD package STAR-CCM+ [5] to evaluate the dominant flow patterns and particle distribution within a typical batch mixing vessel. The kinetic models were used to calculate the efficiency of the mixing vessel in-terms of ethanol production rate based on the substrate and reactants concentrations. Results indicated that as cellulose conversion occurs the dynamic viscosity of the fermentation medium increases while the particle sizes decrease. This leads to improved particle suspension and homogeneity which in-turn approaches the optimal conditions for biological hydrolysis and thus ethanol production.

References

1. Lynd LR, Weimer PJ, van Zyl WH, and Pretorius IS, 2002, Microbial cellulose utilization: Fundamentals and Biotechnology, Microbiology and Molecular Biology Reviews, 66 (3): 506-577.
2. van Zyl JM, van Rensburg E, van Zyl WH, Lynd LR, and Harms TM, 2010, A kinetic model for simultaneous saccharification and fermentation of avicel with saccharomyces cerevisiae, Biotechnology and Bioengineering. 108:924-933.
3. J.F. Matthews, C.E. Skopec, P.E. Mason, P. Zuccato, R.W. Torget, J. Sugiyama, M.E. Himmel, J.W. Brady, Computer simulation studies of microcrystalline cellulose Iβ, Carbohyd. Res. 341 (2006) 138-152.
4. J. Zheng, G.H. Pollack, Long-range forces extending from polymer-gel surfaces. Phys. Rev. E, 68 (2003) 031408.
5. STAR-CCM+, version 6.02.007, CD-Adapco, London, UK

(14a) Substrate Elasticity Regulates the Biophysical Properties of Hematopoietic Stem and Progenitor Cells

Ji S. Choi, Tyler Leonard and Brendan A. Harley, Chemical and Biomolecular Engineering, University of Illinois at Urbana-Champaign, Urbana, IL

Substrate elasticity regulates thebiophysical properties of hematopoietic stem and progenitor cells

Ji Sun (Sunny) Choi[*], Tyler GlennLeonard[*], Brendan Harley[*,#]

[*] Dept. ofChemical and Biomolecular Engineering, University of Illinois at Urbana-Champaign,Urbana, IL 61801

[#] Institutefor Genomic Biology, University of Illinois at Urbana-Champaign, Urbana, IL61801

Hematopoietic stem cells(HSCs) are adult stem cells that give rise to the blood and immune cells in thebody. These cells reside in a local microenvironment known as the stem cell *niche,* whichconsists of stromal cells, extracellular matrix (ECM), and other soluble andgrowth factors. Evidence suggests thatthe HSC niche provides critical spatial and temporal extrinsic signals to modulateHSC biology: quiescence, self-renewal,differentiation, and migration (*1*). To elucidatethe underlying mechanisms, which still remain largely unknown, it is importantto decouple the extrinsic variables in a controlled manner. In this project, weaim to assess the cell-matrix interactions on HSC biophysical properties andearly fate decisions (proliferation,motility, differentiation) by using flexible collagen hydrogel systems with varyingmechanical properties. As a first step,we fabricated four different 2D collagen-coated polyacrylamide (PA) gels (moduli:0.71 0.24 kPa, 3.48 0.59 kPa, 8.95 1.40 kPa, 196 54.1 kPa) and two different3D collagen hydrogels (1.45, 2.9 mg/ml;moduli: 14.8 6.1, 44.2 10.7 Pa)according to published protocols (*2, 3*). HSCs were then harvested from 4-10-week-old C57Bl\6mouse femurs and tibias and cultured on top of 2D PA substrates or on top of 3Dcollagen hydrogels for 24 hours. Alternatively, they were encapsulated in collagenhydrogels to create 3D collagen constructs, and cultured for 24 hours. At theend of culture, cells were assayed for their viability or fixed and stained tovisualize F-actin structure and nuclei with a multiphoton laser scanningmicroscope to create projected 2D images of the cells. Cell spread area andcell shape index (CSI), which is a measure of the circularity of a cell, werethen quantified and plotted (Figure 1a). PA gel collagen coating densities werevaried and its effect on HSC behavior was also observed (Figure 1b, c). Inaddition, fluorescence intensity across the cytoplasm revealed differences in Factin organization in cells based on the substrate stiffness. This data wascompared to 32D cells (IL-3 dependent murine myeloid progenitor cell line)grown in identical culture conditions, where 32D cells behaved in a differentmanner due to decreased differentiation potential. In conclusion, differencesin cell morphology, spread area, and viability were observed in HSCs culturedon top of or within collagen hydrogels. Ongoing work is investigating the influence of matrix chemistry,integrin blocking, and myosin inhibitorson HSC behavior. Further experiments willincorporate cell division tracing assays, colony forming and competitiverepopulation assays to assess the frequency of HSC differentiation,self-renewal, quiescence in these defined matrix environments.

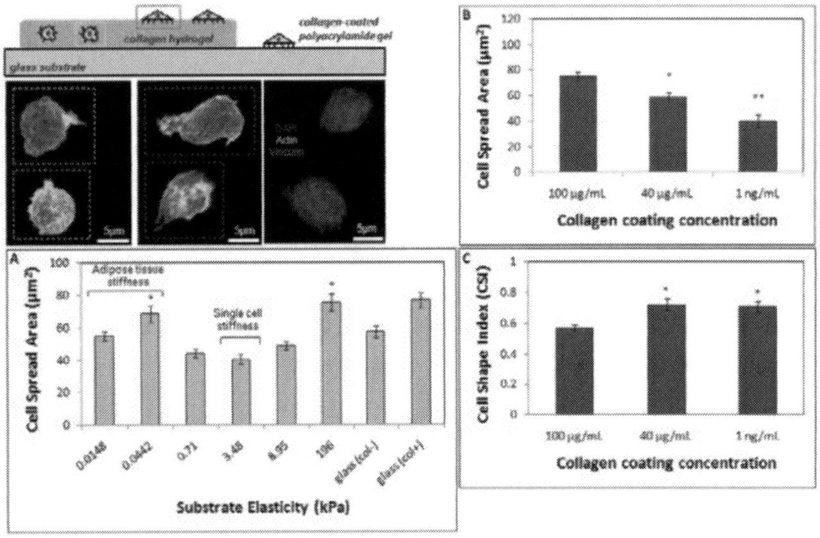

Figure 1. Morphological differences in cells cultured on 2D PAsubstrates with varying stiffness

References:

1. A. Ehninger, A.Trumpp, *The Journal of Experimental Medicine* **208**, 421 (March 14, 2011).
2. J. R. Tse, A. J. Engler, *Curr Protoc Cell Biol* **Chapter10**, Unit 10 16 (Jun, 2010).
3. E. L. Baker, R. T. Bonnecaze, M. H. Zaman, *BiophysicalJournal* **97**, 1013 (2009).

See more of this Session: Cell Biomechanics
See more of this Group/Topical: Food, Pharmaceutical & Bioengineering Division

SOLUBLE GROWTH FACTORS MEDIATE MESENCHYMAL STEM CELL MIGRATION TO TUMORS

Deepraj Ghosh, Michelle Dawson
School of Chemical and Biomolecular Engineering, Georgia Institute of Technology

Mesenchymal stem cells (MSCs) have emerged as a potential candidate for the development of cell-based therapeutics and engineered tissues. In recent years, human and animal MSCs have been derived from different sources, including bone marrow, placenta, and umbilical cord blood. Bone marrow-derived MSCs are an extensively characterized, highly proliferative, multi-potent population, which migrate spontaneously to different organs for tissue regeneration[1]. In vivo, MSCs have been shown to migrate and accumulate in tumor tissues. Several studies have demonstrated their immunosuppressive characteristics, which makes them an attractive source for the development of allogeneic cell-based cancer therapeutics. However, ex vivo culture, expansion and manipulation affects the mobility of MSCs limiting their migration after reinfusion and drastically reducing their accumulation in tumors. In vivo, tumor secreted soluble factors, including transforming growth factor-β1 (TGF-β1), vascular endothelial growth factor (VEGF), and platelet derived growth factor (PDGF), mobilize bone marrow-derived MSCs into the circulation[2]. We hypothesized that TGF- β1, VEGF, and PDGF mobilize MSCs into the circulation by altering their mechanical properties[3]. We systematically characterized the microscopic mechanical properties of MSCs, including their cytoskeletal rheology, cell adhesion molecule expression, cell elongation, and migratory behavior. In vitro migration studies were used to determine the concentrations of soluble growth factors to be used in mechanical characterization. At the concentrations selected all three growth factors increased MSC migration; however, only TGF-β1 altered microscopic mechanical properties, including cytoskeletal rheology and cell adhesion molecule expression.

1. Chapel, A., Bertho, J. M. *et al.* Mesenchymal stem cells home to injured tissues when co-infused with hematopoietic cells to treat a radiation-induced multi-organ failure syndrome. *J Gene Med* **5**, 1028-1038, (2003).
2. Ferrara, N. & Kerbel, R. S. Angiogenesis as a therapeutic target. *Nature* **438**, 967-974, (2005).
3. Ponte, A. L., Marais, E. *et al.* The in vitro migration capacity of human bone marrow mesenchymal stem cells: comparison of chemokine and growth factor chemotactic activities. *Stem Cells* **25**, 1737-1745, (2007).

(14c) Mechanical Characterization of Adult Bone Marrow Stem Cells for Injection or Implantation

Alexandre J.S. Ribeiro, Stanford University, Stanford, CA, Richard W.E. Taylor, Biomedical Engineering, Carnegie Mellon University and **Kris Noel Dahl**, Chemical Engineering, Carnegie Mellon University, Pittsburgh, PA

Therapies using adult stem cells often require mechanical manipulation such as injection or incorporation of cells into scaffolds, but force-induced rupture and mechanosensitivity of cells during manipulation is largely ignored. We perform biophysical characterization of bone marrow CD34+ hematopoietic and bone marrow mesenchymal stromal cells. We use micropipette aspiration to characterize cell mechanics and quantify deformation of subcellular structures under force. Our results suggest that CD34+ cells are mechanically suitable for injection systems since cells transition from solid- to fluid-like at constant aspiration pressure, probably due to a fluid actin cytoskeleton. Conversely, mesenchymal stromal cells are more suitable for seeding into biomaterial scaffolds since they are mechanically robust and have developed cytoskeletal structures. We confirm the relevance of our measurements by injecting cells, and we find a 16 fold increased survival of CD34+ cells. Since different types of adult stem cells may be used for similar applications, we suggest considering optimal mechanical characteristics for therapies.

See more of this Session: Cell Biomechanics
See more of this Group/Topical: Food, Pharmaceutical & Bioengineering Division

(14d) Effects of Substrate Mechanics On Yield and Contractility of Cardiomyocytes Generated From Pluripotent Stem Cells

Laurie Hazeltine[1], Chelsey Simmons[2], Max Salick[3], Mehmet Badur[1], Tetsuro Wakatsuki[4], Wendy Crone[3], Beth Pruitt[2] and Sean Palecek[1], (1)Chemical & Biological Engineering, University of Wisconsin-Madison, Madison, WI, (2)Mechanical Engineering, Stanford University, Stanford, CA, (3) Engineering Physics, University of Wisconsin-Madison, Madison, WI, (4)Physiology, Medical College of Wisconsin, Milwaukee, WI

The mechanical microenvironment plays a role in proper development and function of heart cells, but implementing mechanical cues in an appropriate context during differentiation of human pluripotent stem cells (hPSCs) to cardiomyocytes remains challenging. We used polyacrylamide hydrogels as a model system to investigate the effects of substrate mechanics on differentiation and functionality of cardiomyocytes generated from hPSCs. By changing the concentration of crosslinker in our polyacrylamide hydrogels, we obtained substrates with a physiologically relevant range of stiffnesses. We plated hPSCs undergoing embryoid body-based differentiation onto the hydrogels and observed that cardiogenesis peaks on substrates of intermediate stiffness. To investigate the effects of substrate stiffness on contractility, we added fluorescent beads to the surface of the hydrogels, which act as optical trackers for the movement of contracting cells. We used traction force microscopy to quantify the contraction stress of singularized cardiomyocytes cultured on a range of stiffnesses. In primary rat cardiomyocytes, we observed that contraction stress increases with substrate stiffness. We observed the same trend in cardiomyocytes derived from hPSCs, providing an indication of their similarity to native, fully differentiated cardiomyocytes. These results demonstrate that hPSCs and their cardiomyocyte derivatives can respond to substrate mechanics. Our approaches for investigating and quantifying this response may aid in the development of novel culture systems for effective differentiation of functional cardiomyocytes from hPSCs.

See more of this Session: Cell Biomechanics
See more of this Group/Topical: Food, Pharmaceutical & Bioengineering Division

(14e) Correlating Effect of Gel Microstructural Features with Specific Differentiation Patterning of Embryonic Stem Cells

Keith Task[1], Antonio D'Amore[2], Satish Singh[1], Maria Jaramillo[2], Prashant Kumta[2] and Ipsita Banerjee[3], (1)Chemical Engineering, University of Pittsburgh, Pittsburgh, PA, (2)Bioengineering, University of Pittsburgh, Pittsburgh, PA, (3)Chemical and Petroleum Engineering, University of Pittsburgh, Pittsburgh, PA

Embryonic stem cells (ESCs) have the potential to be used in many therapeutic applications due to their ability to differentiate into cells of any of the three primary germ layers (ectoderm, endoderm, and mesoderm). There are several different routes by which pluripotent ESCs can be guided to differentiate into specific germ layers, including genetic manipulation, chemical cues, and mechanical stimuli. With the latter, it has been observed that substrate stiffness plays a significant role in governing phenotype specific differentiation.

In our previous study we investigated the effect of mechanical stiffness of fibrin, selected as a model natural polymer based substrate, on ESC differentiation. Our results indicated that while ectoderm and mesoderm germ layers responded weakly to the change in fibrin substrate stiffness in the chosen range (2 – 250KPa), endoderm markers were strongly responsive, with the softer substrates up-regulating endoderm specific markers. Although cells respond differently to substrates of varying stiffness, this macroscopic property is perhaps not what the cell truly experiences; modification of stiffness arises from changes in the gel microstructure which also directly interacts with the cells. However, it is not clear what specific microstructural features are the most influential in inducing cellular differentiation, and how they affect ESC behavior.

In the current work, we have developed an integrated experimental and computational approach to investigate the effect of microstructural features of fibrin gels on the differentiation of mouse ESCs. Twelve different fibrin gels were fabricated by varying the fibrinogen concentration (1, 2, 4, and 8 mg/mL) and fibrin to thrombin crosslinking ratio (10, 2.5, and 1.25). The fibrin gels were used to induce ESC differentiation employing both 2D and 3D cultures. For the former, cells were seeded on top of the fibrin gel; for the 3D culture, cells were suspended within a fibrinogen solution before the addition of thrombin to induce subsequent polymerization. After the differentiation protocol the ESCs were analyzed for phenotypic commitment by performing qRT-PCR for the specific germ layer markers. Each of the 12 different fibrin gels was imaged with Scanning Electron Microscopy (SEM). Microstructural features of each of these gels were quantified using an image analysis tool developed at the University of Pittsburgh [1] for the characterization of fibrous scaffolds. Specific features which were characterized include fiber diameter, fiber orientation, pore aspect ratio, porosity, and pore size distribution. Such characterization enabled correlation of gel macroscopic properties with its microstructural features. In the next step the gel microstructural features were correlated with the ESC differentiation pattern using regression analysis. Two variants of least-squared regression, the LASSO and Ridge Regression, were utilized in which penalties are imposed on regression coefficients with high values, a method known as shrinkage. The variability of the system is handled by bootstrapping which creates the feature space for each regression. Performing regression at each of the bootstrap points results in a range of correlation coefficients from which confidence intervals of the parameters were obtained. This analysis reveals the sensitivity of cellular phenotype commitment and differentiation patterning to each of the microstructural features and paves the way for conducting further mechanistic study. Moreover, such information can be used to help guide the design of scaffolds with specific properties for tissue engineering applications.

[1] A. D'Amore *et al.* "Characterization of the complete fiber network topology of planar fibrous tissues and scaffolds." Biomaterials, Vol. 31, 2010, 5345-5354.

See more of this Session: Cell Biomechanics
See more of this Group/Topical: Food, Pharmaceutical & Bioengineering Division

(14f) Monitoring the Effect of Mechanical Load On Fibronectin Conformation Using 3-D Magnetic Tweezers

Jack Chai, Arjun S. Adhikari, Alice C. Chang and Alexander R. Dunn, Chemical Engineering, Stanford University, Stanford, CA

The extracellular matrix protein fibronectin plays critical roles during cancer progression, wound repair, and cell differentiation. A variety of evidence suggests that fibronectin experiences mechanical stretch as part of its physiological function, and that this stretch may lead to partial unfolding of individual fibronectin domains. In this study, we exert mechanical load on single fibronectin molecules using magnetic tweezers, and observe the resulting conformational changes in domain unfolding using a 3D particle tracking algorithm capable of nanometer resolution. Ongoing studies using these techniques examine the complex conformational dynamics of native fibronectin and its binding partners.

See more of this Session: Cell Biomechanics
See more of this Group/Topical: Food, Pharmaceutical & Bioengineering Division

(20a) The Polymorphs of ROY: Application of Crystal Structure Prediction Techniques

Manolis Vasileiadis, Andrei V. Kazantsev, Panagiotis (Panos) G. Karamertzanis, Claire S. Adjiman and Constantinos C. Pantelides, Department of Chemical Engineering, Imperial College London,Center for Process Systems Engineering, London, United Kingdom

Any compound can crystallise in more than one crystal forms, a phenomenon known as *Polymorphism*. Although different polymorphs have the same chemical properties, their physical properties (e.g. solubility) differ significantly as a result of the different spatial arrangement of the molecules[1]. This phenomenon has substantial industrial implications. For example, the production of Ritanovir, a HIV drug, was stopped as a consequence of the unexpected appearance of an unknown polymorph with poor therapeutic effectiveness, affecting patients and leading to large financial losses for its manufacturer. Thus the ability to reliably predict the number and the structure of polymorphs would be a valuable tool for the better design of products and processes.

Over the last two decades, there has been great progress on the development of computational methods for the prediction of the polymorphic forms in which a compound can crystallise. These predictions are mainly based on the lattice energy minimisation of a large number of trial structures. Success is dependent upon the accuracy of the lattice energy calculations, especially in terms of electrostatics and the impact of molecular flexibility, and the ability to optimise a hundreds of thousands to millions of putative structures. Such approaches have been proven to be successful in predicting the polymorphs of compounds with one or two polymorphs in many cases[2].

In this study, predictions are made for the first time for the molecule with the largest number of known polymorphs (ten), 5-methyl-2-[(2-nitrophenyl)amino]-3-thiophenecarbonitrile (also known as ROY because of the Red, Orange and Yellow colours of its polymorphs)..For seven of these polymorphs the structure is known. In addition to the large number of polymorphs, its exceptional properties make ROY a widely used molecule in experimental studies related to polymorphism[3].

The objective of the study is to determine whether known polymorphs can be identified as low-energy structures in the computational energy landscape of ROY crystals. We use a multi-stage methodology. A model is first chosen based on existing experimental information (seven known structures and relative stability of the seven solved polymorphs). Then a large number of structures are generated and locally minimised with a computationally inexpensive model, using the *CrystalPredictor*[4,5] algorithm. Then the most promising structures (lowest in energy) are further minimised with *CrystalOptimizer*[6,7], a program performing local minimisation, adopting a highly accurate though computationally expensive model.

The seven known structures are predicted as lattice energy minima. Moreover five of the known structures are within the fifteen lowest energy structures. Some of the remaining low energy structures are likely candidates for the unresolved polymorphs.

Bibliography

1. Bernstein, J., "Polymorphism in Molecular Crystals". *Clarendon Press*, Oxford, **2002**

2. G.M. Day et. al. , "Significant progress in predicting the crystal structures of small organic molecules-a report on the fourth blind test", *Acta. Cryst.*, B65:107-125 (**2009**)

3. Yu, L., "Polymorphism in Molecular Solids: An Extraordinary system of Red, Orange and Yellow crystals", *Acc. Chem. Res.*, Vol. 43, No. 9 , 1257-1266

4. P.G. Karamertzanis and C.C. Pantelides, "*Ab initio*-Crystal Structure Prediction – I. Rigid molecules", *J. Comput. Chem.*, 26:304-324 (**2005**)

5. P.G. Karamertzanis and C.C. Pantelides, "*Ab initio*-Crystal Structure Prediction – II. Flexible molecules", *Molecular Physics*, Vol. 105 , Nos. 2-3, 273-291 (**2007**)

6. A.V. Kazantsev et.al., *J. Chem. Theory Comput.*, submitted (October 2010)

7. A.V. Kazantsev et.al., "CrystalOptimizer: An efficient algorithm for lattice energy minimisation of organic crystals using quantum mechanical calculations", Molecular Systems Engineering, C.S. Adjiman and A. Galindo, Wiley-VCH

See more of this Session: Crystallization of Pharmaceutical and Biological Molecules I
See more of this Group/Topical: Separations Division

(20b) Spiral Growth Model for Organic Crystals of Real Complexity: How to Account for Unstable Spiral Edges

Zubin B. Kuvadia and Michael F. Doherty, Department of Chemical Engineering, University of California, Santa Barbara, Santa Barbara, CA

Crystal morphology influences the end-use efficacy of solid products as well as the downstream performance of the entire manufacturing and formulation process. The desired crystal size and shape distributions is a function of the final application, hence a general methodology for the prediction and improvement of crystal morphology is one of the major goals in the field of crystallization. The classical Burton-Cabrera-Frank (BCF) spiral growth model fails to work satisfactorily for many non-centrosymmetric organic molecules such as APIs, nonlinear optical compounds, etc., due to the inherent assumption of the Kossel crystal structure in the classical model. These complex molecules warrant a more rigorous treatment for successful prediction of crystal shapes. We develop a general mechanistic spiral growth model that can be used to predict crystal morphology of any organic molecule, centrosymmetric as well as non-centrosymmetric.

This work discusses in detail the stable and unstable periodic bond chain (PBC) edges that arise due to complex asymmetric bonding patterns. Modified step velocity expressions are derived that account for unstable edges. The entire model is successfully applied to industrially relevant systems like acetaminophen and lovastatin and the results are discussed with experimental data from the literature and from our laboratory.

References

Kuvadia, Z and Doherty M. F. 2011. Spiral Growth Model for Faceted Crystals of Non-Centrosymmetric Organic Molecules Grown from solution. Crystal Growth and Design

Burton,W.K., Cabrera, N. & Frank, F.C. 1951. The growth of crystals and the equilibrium structure of their surfaces. Phil. Trans. R. Soc. A 243, 299-358

Snyder, R. C. & Doherty, M. F. 2009. Predicting crystal growth by spiral motion. Proceedings of the Royal Society A: Mathematical, Physical and Engineering Science 465, 1145-1171

See more of this Session: Crystallization of Pharmaceutical and Biological Molecules I
See more of this Group/Topical: Separations Division

(20c) Modeling of Morphology Transformations In Crystalline Materials: *A Generalized Framework*

Meenesh R. Singh[1], Christian Borchert[2], Kai Sundmacher[2] and Dorwaiswami Ramkrishna[1], (1)School of Chemical Engineering, Purdue University, West Lafayette, IN, (2)Max Planck Institute for Dynamics of Complex Technical Systems, Magdeburg, Germany

This talk will present a generalized population balance model (PBM) to predict dynamic behavior of morphology (shape) transformations in populations of crystals. The model utilizes symmetry of crystal to reduce the degree of freedom in shape representations. The main crux of the model lies in its ability to describe the evolution of morphology distributions for any type of crystals in least possible dimensions (or dynamic variables). It will be shown subsequently that the dynamic behavior of morphology transformations for any crystalline material can be described by *f*-dimensional PBM, where *f* is the number of energetically favorable families of crystal planes.

The current problem carries immense opportunities for semiconductor, catalyst and pharmaceutical-based industries to produce crystals of desired shapes and shape-related properties. Many efforts have been made in past to understand the dynamics of crystal morphology at molecular level via various approaches like Phase-Field Modeling, Molecular Dynamics simulations, Monte Carlo simulations etc. However, their applications in determining the dynamics of morphology distributions are non-trivial due to the requirement of tremendous computational efforts. Conversely, the mechanistic models like PBMs are efficient in large scale applications as they require relatively less computational efforts and can give reasonably accurate results in no time. The talk will provide the recipe to formulate PBM for any type of crystalline material, which can be solved to predict the evolution of morphology distributions.

The generalized framework has the following few basic elements, which will be discussed thoroughly in the talk:

1. Identification of Maximum Degenerate State of a Crystal Shape.
2. Identification of Boundaries of Crystal State Space.
3. Developing Morphology Net.
4. Writing Multi-dimensional PBM.
5. Reduction to *f*-dimensional PBM.

This model provides the most efficient description for the morphology transformations in population of crystals, whose usefulness is quite apparent in different applications.

See more of this Session: Crystallization of Pharmaceutical and Biological Molecules I

(20d) Microfluidic Platform for Evaporation-Based Solid Form Screening of Pharmaceutical Compounds

Sachit Goyal[1], Michael R. Thorson[1], Araz G. Zarkhah[1], Yuchuan Gong[2], Geoff G.Z. Zhang[2], Charles F. Zukoski[1] and Paul J.A. Kenis[1], (1)Department of Chemical & Biomolecular Engineering, University of Illinois at Urbana Champaign, Urbana, IL, (2)Solid State Sciences, Abbott Laboratories, North Chicago, IL

We present a microfluidic platform for solid form (salts and co-crystals) screening of pharmaceutical parent compounds (PC) via regulated evaporation of the solvent. Solid form screening of PCs is critical to the drug development process. Reduced sample volumes of PC solutions allow screening of multiple conditions with limited amount of PC (≈ 10 mg) available during early stages of development enabling early identification of PCs with high propensity for success. This helps reduce the time and money spent on developing PCs. Microfluidics has the potential to enable efficient solid form screening using limited PC by reducing sample volumes (≈ 200 nL) and allowing controlled mixing of reactants [1].

The microfluidic platform enables combinatorial mixing of PC and counterion (salt or cocrystal formers) solutions on-chip in arrays of 24/48 200 nL wells, which is a drastic reduction in the amount of compound needed per condition compared to traditional screening approaches (~5 to 100 ul per condition [2]). Controlled mixing and controlled solvent evaporation (10 – 50 hours) provide consistent control over supersaturation and hence high propensity to form crystals on-chip. The platform is compatible with several polar solvents used in the pharmaceutical industry, long-term evaporation experiments, and on-chip Raman and X-ray analysis (enabling *in situ* identification of the solid forms) due to incorporation of X-ray transparent material and reduction in the thickness of elastomeric absorbent material. A solvent resistant layer with evaporation channels of different geometries permits predetermined rates of evaporation of the solvent (10 – 50 hours) [2].

In this presentation, the design, fabrication, and application of the platform will be covered. Model compounds like Tamoxifen, Ephedrine, and Caffeine have been used to validate the platform's ability to screen for salts and cocrystals. On-chip Raman and X-Ray analysis were employed to identify the salts and cocrystals.

References

[1] C. Hansen, E. Skordalakes, J. Berger, S. Quake, Proc. Nat'l Acad. Sci., 2002, 99, 16531-6.

[2] S. Talreja, D.Y. Kim, A.Y. Mirarefi, C.F. Zukoski, P.J.A. Kenis, J. Appl. Crystallography, 2005, 38,

988-995.

See more of this Session: Crystallization of Pharmaceutical and Biological Molecules I
See more of this Group/Topical: Separations Division

(20e) Microfluidic Platforms for Crystallization Screening and In Situ Phasing & Structure Determination of Proteins

Sudipto Guha[1], Sarah L. Perry[1], Ashtamurthy Pawate[1], Satish K. Nair[2] and Paul J. A. Kenis[1], (1) Department of Chemical & Biomolecular Engineering, University of Illinois at Urbana-Champaign, Urbana, IL, (2)Biochemistry, University of Illinois at Urbana-Champaign, Urbana, IL

An ideal crystallization platform should be able to screen hundreds of crystallization conditions, allow for optimization of a successful hit which gives diffraction quality crystals and enable in situ structure determination from data collected on the platform preferably without having to harvest the crystal.

Our X-ray compatible, array-based microfluidic chip functions as a complete structure determination platform. To ensure X-ray compatibility we have eliminated as far as possible the use of polydimethylsiloxane (PDMS) which attenuates X-rays because of its silicon content. Instead our hybrid chip consists of different layers of materials: cyclic olefin copolymer (COC) for the valve layer, a thin, flexible PDMS fluid layer and a COC/Duralar substrate. The PDMS layer is needed for valve actuation to manipulate fluid flow on-chip. The scattering of X-rays by PDMS and the COC/Duralar material is minimal and does not affect the diffraction data.

The use of X-ray compatible materials in the chip eliminates manual harvesting of crystals thus allowing in situ analysis. This prevents damage to fragile protein crystals which might occur during harvesting. We have validated this by crystallizing on-chip lysozyme, thaumatin and ribonuclease A. We take advantage of the small scale and fluid handling capabilities of multilayer microfluidics to screen multiple crystallization conditions using minimal amount of protein (<2μL for 96 conditions). Once a successful condition is identified, the array based design allows growing multiple crystals on a single chip, following which slices of data can be collected from each crystal and later be merged together to give the complete data set.

To solve the structure of novel proteins it is essential to get phase information. Single wavelength anomalous dispersion (SAD) is one of the methods used for phasing to solve a protein structure. Our microfluidic platform allows the accurate measurement of the weak, anomalous diffraction differences between the heavy atoms (Selenium) used for phasing and the protein itself. We have successfully determined the structure of a novel bacterial lyase, PhNA to 2.3Å resolution using SAD phasing, solely using data collected on chip.

See more of this Session: Crystallization of Pharmaceutical and Biological Molecules I
See more of this Group/Topical: Separations Division

(20f) Microfluidic Platforms for In Situ Crystallographic Analysis

Sarah L. Perry[1], Sudipto Guha[1], Ashtamurthy Pawate[1], Robert B. Gennis[2], Zhong Ren[3] and Paul J.A. Kenis[1], (1)Department of Chemical & Biomolecular Engineering, University of Illinois at Urbana-Champaign, Urbana, IL, (2)Biochemistry, University of Illinois at Urbana-Champaign, Urbana, IL, (3) Department of Biochemistry and Molecular Biology and the Consortium for Advanced Radiation Sources, University of Chicago, Chicago, IL

Protein crystallography is an important aspect of structural biology, which provides insight into protein structure as well as further information about their function. There are many steps in the process which finally leads to the structure of the protein and subsequently unravels how it functions. These involve expression, purification, crystallization, harvesting of the crystals for diffraction and finally X-ray data analysis. Growing diffraction quality crystals and manually harvesting them still remains a major challenge. The difficulty in obtaining quality crystals stems from the fact that crystallization conditions for novel proteins are not known *a priori*. Traditionally, sparse matrix screens need to be set up to obtain suitable crystallization conditions and this requires large amounts of protein and is a labor intensive process. Recent developments in robotics have allowed researchers to set up thousands of trials in a short duration of time (1). However, the ability to set up a large number of trials does not overcome challenges associated with the reproducibility of crystallization events or crystal quality. Furthermore, these efforts are limited by the need to manually harvest fragile crystals.

Microfluidics offers exquisite control over transport phenomena, which provides precise control over composition and kinetics of a crystallization experiment than what is possible at the macroscale (2). However, to take advantage of these benefits, it is necessary to couple complex fluid handling capabilities with the ability to perform on-chip crystallographic analysis.

The multilayer microfluidic crystallization platforms described here fabricated out of X-ray transparent materials which allow *in situ* analysis and can be used for high throughput experiments. Our 96-well screening and optimization chips can screen eight different protein-to-precipitant ratios for twelve different precipitant conditions using less than 2µL of protein. A suitable condition once identified, can be scaled out in an array format to reproducibly grow a large number of crystals.

We are currently working on extending the present chip architecture, which has been validated for *in situ* screening, to perform time resolved studies on protein dynamics. Laue crystallography is a technique which is used to perform time resolved studies on protein dynamics and we have validated our chip architecture for Laue diffraction. Here we aim to couple *in situ* analysis of a large number of crystals with the capabilities of a microfluidic flow cell to enable the controlled introduction of various chemical stimuli and thus trigger biologically relevant conformational changes in the structure which can then be analyzed *in situ* in our microfluidic chips.

References:

1. Stevens RC (2000) High-throughput protein crystallization. *Current Opinion in Structural Biology* 10(5):558-563.

2. Zheng B, Tice JD, Roach LS, & Ismagilov RF (2004) A droplet-based, composite PDMS/glass capillary microfluidic system for evaluating protein crystallization conditions by microbatch and vapor-diffusion methods with on-chip X-ray diffraction. *Angewandte Chemie-International Edition* 43 (19):2508-2511.

See more of this Session: Crystallization of Pharmaceutical and Biological Molecules I
See more of this Group/Topical: Separations Division

(30a) Development and Optimization Process Deriving Fermentable Sugars From Lignocellulosic Application In Biofuels and Chemicals

Sarad Parekh, Research and Development, Sweet Water Energy, Rochester, NY

Ethanol as a liquid transportation fuel is currently produced from various renewable lignocellulosic resources . Ethanol and other advanced biofuels production from cellulosic biomass offer an attractive environmentally sustainable solution. Lignocellulosic biomass consists of a complex network of rich hemi-cellulose and cellulose bound by lignin. The process of converting lignocellulosic biomass to fermentable-ready sugars involves a pretreatment step that disrupts the complex polymer of lignin, cellulose, and hemi-cellulose, making it amenable for enzymatic saccharification. Using a special combination of enzymes on the pretreated biomass, it is possible to extract a variety of C5 and C6 sugars that can be subsequently fermented by yeasts and proprietary microbes to ethanol, butanol, advanced fuels like jet fuels, and chemicals and bio-plastics. However, one of the challenges in commercializing the biomass-to-biofuels process is, the logistics of biomass collection, supply and processing the material for fermentation

Sweetwater Energy, in Rochester, NY, is developing, designing and fabricating feedstock-flexible, decentralized, modular bioprocessing machines that can be operated at the biomass-collection site. Various pretreatment options, such as auto-hydrolysis, dilute acid, and de-lignification were studied to test the effectiveness of pretreatment that rendered the biomass easy for enzymatic hydrolysis. Small-scale and preliminary data on processing corn stover and sorghum indicate that one can recover sugars from biomass at approximately 70% of theoretical value. The sugar syrups rich in C5 and C6 sugars were tested for ethanol and advanced biofuels production. This presentation will review Sweetwater Energy's technology and overall approach to process development, and will highlight recent advances made at its R&D site in the cellulosic bioprocessing technology and the conversion to biofuels. The importance and the prospects of developing cost-effective pretreatment and integrated sugar platform technologies for the production of biofuels from biomass will be discussed.

See more of this Session: Industrial Forum: Recent Advances and Development In Biotech Industry
See more of this Group/Topical: Food, Pharmaceutical & Bioengineering Division

(30b) Application of Physiologically Based Pharmacokinetic Model for Drug Development

Haiying Zhou, Viera Lukacova, Michael Bolger and Walt Woltosz, Simulations Plus, Inc., Lancaster, CA

Physiologically-based pharmacokinetics (PBPK) modeling has become well-accepted in pharmaceutical research for *in silico* prediction of *in vivo* pharmacokinetics of new drugs with small to intermediate molecular size. This is a science-based, mechanistic approach that accounts for tissue size, perfusion rates, and partitioning of unbound drug into physically relevant compartments. GastroPlus™ (Simulation Plus, Inc.) is a software tool that simulates the rate and extent of drug absorption coupled with pharmacokinetics and pharmacodynamics. The application of its advanced compartmental absorption and transit (ACAT) model with its PBPK module (PBPKPlus™) can provide great insight into how a drug distributes in the body, which can help to guide decisions in drug development. PBPK modeling is also a useful method for interspecies scaling and early prediction of human pharmacokinetics. This presentation will show a PBPK model that was developed for terbinafine in rats based on literature data using GastroPlus™ v7.5, which successfully predicted human plasma concentration-time profile based only on inter-species and *in vitro/in vivo* extrapolation (IVIVE). Human clearance was estimated by allometric scaling using the familiar ¾-power law.

This presentation will also discuss the feasibility of applying PBPK modeling for therapeutic macromolecules. Macromolecules pose challenges for PBPK modeling due to their size, hydrolytic instability and physico-chemical properties. The lymphatic system plays an important role in the absorption and disposition of macromolecules after extravascular administration due to permeability limitations through blood capillaries. Binding between macromolecules and specific proteins also affects drug transport and clearance. Potential improvements to current PBPK models to simulate the pharmacokinetics of macromolecules will be discussed.

See more of this Session: Industrial Forum: Recent Advances and Development In Biotech Industry
See more of this Group/Topical: Food, Pharmaceutical & Bioengineering Division

(30c) Chromatography Applications In the Purification of Whey Proteins

Ng, P. K., Bio-Rad Laboratories

Alpha lactalbumin, beta lactoglobulin, albumin, immunoglobulin and lactoferrin are major protein components in whey. The objective of this study is to develop separation processes to fractionate these proteins from whey.

In recent years, significant progress has been made in enhancing the binding capacity of ion exchangers. Newly developed ion exchangers, Nuvia S and Nuvia Q, are of high dynamic capacity and are suitable for the fast capture of proteins. Purification of whey proteins using these new resins will be shown.

One challenge with regard to lactoferrin is that it can be difficult to obtain in high purity using conventional process chromatography. This is due to the presence of lactoperoxidase that is substantially similar to lactoferrin in both molecular weight and isoelectric point. We will describe a ceramic hydroxyapatite-based method for direct capture of lactoperoxidase, lactoferrin and other alkaline proteins from whey. Purified lactoferrin was then obtained by isocratic elution in a single step.

Finally, we will discuss the separation of a mixture containing albumin and alpha lactalbumin. They could be obtained as an eluate in anion exchange chromatography of whey proteins. The two proteins with isoelectric points difference of <0.5 units were separated by induced pH gradient chromatography on a cation-exchange column containing weak acid groups

(30d) Controlling Glycation Levels In Antibodies Produced In Fed-Batch Cell Culture Processes

Sharat Varma, Natarajan Vijayasankaran and Steven Meier, Late Stage Cell Culture, Genentech, Inc., South San Francisco, CA

This poster will focus on a case study of cell culture process development for the synthesis of a monoclonal antibody that experienced glycation levels exceeding 25%, when produced in a standard platform cell culture process. Glycation is the non-enzymatic addition of reducing sugars such as glucose to specific lysine resides. Glycation levels observed in recombinant monoclonal antibodies can vary depending on the presence or absence of a "glycation hot spot". In this specific study, such a hot spot was apparently caused by the spacial proximity of lysine 49 and aspartate 31 in the light chain. Cell culture process development efforts were focused on reducing the fraction of glycated antibody without making major modifications to the process platform. To aid this effort, a dynamic mathematical model was developed to simulate the relationship between certain process parameters and observed glycation levels. The model simulated the growth of cells, substrate consumption, protein synthesis and glycation. After it was confirmed that previously collected experimental data were in general agreement with the results of these simulations, the simulation was used to guide the design of subsequent experiments. Using this approach, glycation levels as low as 15% were achieved by lowering the glucose concentration in the nutrient feeds as well as the overall glucose feeding strategy.

See more of this Session: Industrial Forum: Recent Advances and Development In Biotech Industry
See more of this Group/Topical: Food, Pharmaceutical & Bioengineering Division

(30e) Use of Statistical Modeling Tools to Characterize the Design Space for a Late-Stage Mammalian Cell Culture Process

James Lambropoulos, Lia Tescione, Ram Paranandi, Rhiannon Quirk, Barbara Woppmann, Helena Yusuf-Makagiansar, Marty Sinacore and Thomas Ryll, Cell Culture Development, Biogen Idec, Cambridge, MA

We describe here the design space characterization for the production of a recombinant human protein from a mammalian cell culture process. These activities included (1) the establishment of a 3L scale-down model representative of the 2000 L manufacturing process, (2) a fractional factorial parameter screening study, (3) a response-surface modeling study, which examined pH, seed density, harvest day, and temperature and their impact on growth, productivity, and product quality attributes, and (4) model verification. We found that several response parameters were well modeled and showed a high degree of predictability, while others had poor-fitting models. We made a number of observations based on these individual models, including: (1) that temperature set point was optimized for titer; (2) that decreases in pH set point or extension of harvest day led to decreased levels of Glycan Attribute 1; and (3) that decreases in temperature or harvest day led to increased levels of Glycan Attribute 2. The various mathematical models that were generated from this work allowed for the identification of process optima, the setting of action limits for the Process Validation campaign, and the identification of critical, key, and non-key parameters. Lessons learned from these studies will be reviewed and recommendations made to streamline the workflow for cell culture process characterization and optimization using multivariate, factorial design of experiments.

See more of this Session: Industrial Forum: Recent Advances and Development In Biotech Industry
See more of this Group/Topical: Food, Pharmaceutical & Bioengineering Division

(30f) Proof-of-Concept of a Continuous-Flow Heck Coupling and HBr-Catalyzed Oxidation for Pharmaceutical Lab-Hood Manufacturing

Nikolay Zaborenko[1], Jonas Y. Buser[2], Adam D. McFarland[2], Douglas P. Kjell[3], Bret A. Astleford[3] and Rui Lin Gu[3], (1)Chemical Product Research and Development, Eli Lilly and Company, Indianapolis, IN, (2)Analytical Sciences Research and Development, Eli Lilly and Company, Indianapolis, IN, (3) Chemical Product Design and Development, Eli Lilly and Company, Indianapolis, IN

Continuous-flow GMP manufacture of APIs with demand of less than 5 tons is feasible using only equipment and setups of laboratory fume hood scale. The development of a continuous-flow method for the Heck coupling and bromic-acid-catalyzed oxidation steps in an API GMP sequence is presented. This process was designed with the goal of increasing reproducibility, robustness, operational safety, and manufacturing cost. The Heck coupling was demonstrated in flow, with ongoing work towards telescoping it via continuous crystallization to the subsequent step. The oxidation reaction has been modified to provide a homogeneous starting solution while minimizing the number and quantity of solvents/reagents. Performing it in a CSTR followed by a reactive crystallizer maximized reaction yield by increasing the residence time of certain reactive intermediates relative to the starting material. Direct crystallization of the diketone free base was also possible, eliminating one process crystallization step. Overall, significant progress was made towards continuous synthesis of several GMP steps in a laboratory fume hood at manufacturing scales to meet the anticipated 500-kg/year API demand.

See more of this Session: Industrial Forum: Recent Advances and Development In Biotech Industry
See more of this Group/Topical: Food, Pharmaceutical & Bioengineering Division

(32a) Correlating Powder Flowability to Particle Size Distribution Using Chemometric Techniques

Christopher L. Burcham, Chemical Product Research and Development; Particle Design Laboratory, Eli Lilly and Company, Indianapolis, IN and Robert T. Roginski, Eigenvector Research, Inc., Wenatchee, WA

A compound terminated in Phase III clinical trials was being developed as a 140 mg dose. In order to meet the needs of the patient population, the tablet must be no larger than 450 mg, resulting in a high drug substance loading. The drug product was manufactured using a roller compaction process. The high drug load required that the drug substance have good flow properties as a dry powder. An understanding of the impact that the particle size distribution of the drug substance on the powder flowability was needed to allow for the establishment of particle size specifications on the drug substance. Principal Components Analysis (PCA) was applied to the particle size distribution data generated from numerous crystallization conditions and scales to identify clustering. A predictive model for powder flowability was built based on the particle size distribution data using Partial Least Squares (PLS). This model was further refined using local weighting techniques to deliver better prediction metrics. The results of these modeling efforts will be presented.

See more of this Session: Integrated API and Drug Product Development
See more of this Group/Topical: Food, Pharmaceutical & Bioengineering Division

(32b) Optimization of API Attributes and Flow Properties Via Crystallization and Size Reduction Techniques

Plamen Grigorov, Azzeddine Lekhal and Dimitrios Zarkadas, Chemical Process Development & Commercialization, Merck and Co. Inc., Rahway, NJ

The flowabilityof final Active Pharmaceutical Ingredients (APIs) is an important property thatvery often impacts formulation process performance as well as Drug Product (DP)performance. The more poorly the API flows, the more troublesome is itsbehavior during formulation. Therefore, it is beneficial to developprocesses/protocols that result in API with optimized attributes and flowproperties. Bulk density is usually a good indicator of flow properties withlow values indicating poor flow. This paper describes our efforts to improvebulk density and hence flow properties of a needle shape API with a very lowbulk density by a combination of controlled crystallization and differentmilling techniques.

We attempted to increasebulk density by exploring several alternative approaches. These included: 1) attempts to alter crystal morphology by changing solvent composition, 2) controlof crystals aspect ratio (a) through pin-milling and (b) via controlledcrystallization coupled with different milling techniques and 3) change the crystalaspect ratio and improve packing by using high pressure homogenization (HPH).The last three of these methods increased API bulk density from its originalvalue. The most drastic impact was observed with HPH.

The original material,obtained by seeded cooling crystallization, was needle like powder with verypoor flowability and "Styrofoam-like" appearance. Its bulk and tap densitieswere 0.10 g/ml and 0.28 g/ml respectively. The bulk density value is typical ofmicronized materials, which generally tend to have very poor flow properties. Pin-millingincreased the bulk and tap densities to 0.14 g/ml and 0.34 g/ml respectively. Showingsome promising results, this milling approach was taken one step further. Thecrystallization procedure was optimized to include temperature cycling, whichpromoted crystal growth in length as well as in width. Size reduction of thesecrystals resulted in API with reduced aspect ratio and ultimately higher bulk andtap densities (0.25 g/ml and 0.50 g/ml respectively), a two-fold increase fromthe original values. Flowability and formulation tests showed that the newmaterial exhibited acceptable properties for final formulation.

The last methodinvestigated for bulk density increase involved HPH and demonstrated thebiggest improvements. High pressure homogenization is a technique commonly usedin the pharmaceutical industry for cell breakage, emulsion generation orparticle size reduction. It can be also viewed as a wet micronization techniquecapable to achieve micron and submicron particle sizes. When the slurry fromthe API crystallization is run through the homogenizer, the original needlesare broken down to several microns in size. The resulting particles are sosmall that the initial needle-like morphology ceases to exist and the resultingaspect ratio is very close to one. During subsequent filtration and dryingthese small particles agglomerate into well-packed hard solid blocks. Dependingon the degree of de-lumping during a co-milling step, the final bulk densitycan be finely tuned. Using this approach the bulk density of the API underdevelopment was increased to 0.34 g/ml a three-fold increase from theoriginal value and the highest ever achieved for this API.

See more of this Session: Integrated API and Drug Product Development
See more of this Group/Topical: Food, Pharmaceutical & Bioengineering Division

(32c) Crystallization and Milling Development Enabling A High Drug Load Formulation

Philip Hoffman, Chemical Product Research & Development, Eli Lilly & Co., Inc., Indianapolis, IN and Jeffrey Vicenzi, Chemical Product Research and Development, Eli Lilly & Co., Inc., Indianapolis, IN

Dry milling of an Active Pharmaceutical Ingredient (API) provides a reliable means for particle size reduction. However, dry milling can present a challenge for containment of highly potent or toxic drug substances and can also can result in the generation of a large number of fines which may lead to undesirable processing behavior. Slurry milling alleviates many of the containment issues associated with particle size reduction. When combined with controlled crystallization strategies, precise particle size control and favorable processing attributes can be obtained.

See more of this Session: Integrated API and Drug Product Development
See more of this Group/Topical: Food, Pharmaceutical & Bioengineering Division

(32d) Analysis of Bayesian Linearity (BL) to Improve Performance for API to DP Integration

Dr. Albert Ivan Charles Einstiene, Wooster Technical Foundation (WTF), Cambridge, MA

Using of a mixture of Bayesian statistics, data mining, and modeling it was determined how absolute temperature (AT), population density, (PD) and equipment coefficient (EC) affect yield. This fundamental chemical engineering problem was the focus of our group's research, the Wooster Technical Foundations (WTF). We have asked questions that have brought significant insight in developing solutions with Bayesian Linearity (BL).

We discovered that a reaction that occurs at cold average temperatures of 285°K significantly produces a low response of ~20% compared to the yield that occurs at temperatures averaging 295°K. This relatively small difference in temperature is an important, but not the sole contributor to lower yield: we also observed a pollution balance effect in which an inverse correlation exists between the temperature and the relative population density, even at average temperatures below 285°K, so that one can overcome yield by a higher population density. The temperature to yield factor is well known in many fields but also one additional factor was discovered, which was classified as EC. The EC factor can also improve yield even at low temperatures and low populations densities. Using data mining techniques, along with the fundamental mathematics, experimental methods applied in other fields and a computational method, we developed a model demonstrating how performance of DS/DP integration is improved though seemingly simple factors of population balances, temperature differential and equipment.

See more of this Session: Integrated API and Drug Product Development
See more of this Group/Topical: Food, Pharmaceutical & Bioengineering Division

(32e) API Particle Engineering As a Strategy to Address Sticking/Picking During Drug Product Development

Todd McDermott[1], Ahmad Sheikh[2] and Jorge Gandarilla[2], (1)Global Pharmaceutical Sciences, Abbott Laboratories, Abbott Park, IL, (2)Process Research & Development, Abbott Laboratories, North Chicago, IL

APIparticle engineering as a strategy to address sticking/picking during drugproduct development

Todd S. McDermott

Abbott Laboratories, Global Pharmaceutical Sciences

Ahmad Sheikh

Jorge Gandarilla

Abbott Laboratories, Process Research & Development- Solid State Chemistry

Sticking of material to tablet punch surfaces is a commonproblem encountered during tablet development. APIs that have sticking issuesalso tend to have poor flow and density, which add to the formulationchallenges. For particularly difficult APIs, traditional strategies may not besufficient to address the severe sticking and flow issues. An alternativeapproach is to fundamentally change the API properties by particle engineering.Liquid-liquid phase separation crystallization (LLPS) was used to greatlyimprove the solid-state properties of a model API. The process was optimizedand demonstrated on pilot-plant scale. The resulting material was formulatedand characterized in terms of flow, density, compaction properties, and sticking.The resulting material shows significant improvement relative to the originalAPI.

See more of this Session: Integrated API and Drug Product Development
See more of this Group/Topical: Food, Pharmaceutical & Bioengineering Division

(32f) Utilizing Drop-On-Demand Printing In the Manufacture of Transdermal Dosage Forms

Marlena Brown, Chemical and Biochemical Engineering/Food Science, Rutgers University, Ocean, NJ

We propose to utilize Drop-on-Demand technology to manufacture transdermal drug delivery systems (TDDS) in an innovative and systematic approach. The primary goal is to decrease the amount of layers currently used in the traditional system and create new dosage forms by printing the API, skin enhancers, additives, and adhesive components onto the same film with distinct patterns and configurations to achieve the desired bioavailability and controlled release rates. Previous research has detailed that manipulation of substrate properties can be used to control the morphology of the API deposits and obtain API forms ranging from amorphous to crystalline structures. This study establishes the relationships between the API and polymer film properties, kinetics of the evaporation process, final stain morphology, solubility properties of the fabricated dosage, and API dissolution. Ultimately this research will enable the development of scientific fundamentals which allow us to accurately create novel API dosage forms where the final droplet structure exists as defined two and three dimensional architecture with a controllable release profile.

See more of this Session: Integrated API and Drug Product Development
See more of this Group/Topical: Food, Pharmaceutical & Bioengineering Division

(32g) API Excipient Co-Crystal Screen to Optimize Robustness of An Oral Solution Drug Product

Suman Luthra, Brendan Murphy, Brian Samas and Joseph Krzyzaniak, Pfizer, Groton, CT

Purpose: When designing a liquid formulation, it is important to understand the solubility of the lowest energy (least soluble) API form to prevent undesired precipitation. During manufacturing and storage, observation of a new API form composed of formulation excipients, made it necessary to improvise screening approaches to enable definition of the thermodynamically stable form within the drug product.

Methods: Extensive polymorph screening of the ingoing API found no additional crystalline anhydrous forms. Multiple crystalline salts were found through salt screening. Components of this formulation (API, sodium benzoate, phosphoric acid) demonstrate an ionic equilibrium in the target pH range. Slow evaporation and binary diffusion of solutions of different combinations of components of oral formulation were conducted at room temperature and 5°C to find stable API forms. Single crystals isolated from this sample set were solved. Slurries of the stable form as a function of pH and temperature in aqueous and binary solvents were also conducted.

Results: Formulation robustness scr eening identified the stable solid form as a function of formulation composition, pH and temperature. Multiple stable forms were identified within the drug product design space, including salts and excipient co- crystals.

Conclusion: An understanding of the stable API Form as a function of the drug product design space enabled the definition of the low energy form within the drug product. It also enabled drug product design to mitigate the risk of precipitation during clinical supplies manufacture within a short time frame.

See more of this Session: Integrated API and Drug Product Development
See more of this Group/Topical: Food, Pharmaceutical & Bioengineering Division

(48a) Quality by Design (QbD) Opportunities, Challenges and Future Direction

Moheb Nasr, DHHS/FDA/Cder/OPS/Ondqa, FDA, Silver Spring, MD

QbD is a systematic approach to pharmaceutical development, manufacturing and product lifecycle management. QbD provides new opportunities to utilize state of the art science, quality risk management and modern engineering tools. This presentation will discuss current regulatory guidelines and implementation progress. In addition, Dr. Nasr will outline some of the challenges and provide suggestions to move forward. The regulatory role need to be better defined to encourage innovation, facilitate drug development and provide high assurance of quality.

See more of this Session: QbD Plenary
See more of this Group/Topical: Topical I: Comprehensive Quality by Design in Pharmaceutical Development and Manufacture

(48b) QbD - It's Not Just for Breakfast Anymore

Paul C. Collins, Chemical Product Research and Development, Eli Lilly and Co, Indianapolis, IN

We are approaching the ten-year mark for the Quality by Design initiative in the pharmaceutical industry. Much has been accomplished and demonstrated, particularly with regard to enhanced understanding of processes and better definition of processing parameters such as CPPs. Design space has become an often utilized term when process scientists talk about their projects and the concept has led to increased use of modeling approaches in late stage development. Despite the interest generated by the initiative and the progress made by it's participants, the true power of QbD may yet to be unleashed. Human-centered design of pharmaceuticals, with a more intimate tie to patient need and patient performance, should be the next step in how we more forward the QbD initiative, and represents a new way for engineers to influence drug design.

See more of this Session: QbD Plenary
See more of this Group/Topical: Topical I: Comprehensive Quality by Design in Pharmaceutical Development and Manufacture

(48c) Challenges and Opportunities for Process Systems Engineering Approaches to QbD and Design Space

G. V. Reklaitis, Pharmaceutical Technology & Education Center, Purdue University, W Lafayette, IN

The key components of the QbD concept, namely, the Design Space and PAT, do draw extensively for their realization on tools from the domain of process systems engineering methodology. Both the challenges and the opportunities in adapting core PSE methodology, such as process modeling and simulation, process dynamics and control, monitoring and fault diagnosis, quantification of process uncertainty and risk, and product and process optimization, to the development and manufacture of pharmaceutical products are significant. In this presentation we will review the progress made in the application of pse methods to this domain and outline some of the important developments that remain to be realized.

See more of this Session: QbD Plenary
See more of this Group/Topical: Topical I: Comprehensive Quality by Design in Pharmaceutical Development and Manufacture

(48d) Evolution of QbD From a Drug Substance Scientist's Perspective

Timothy J. Watson, Pfizer Inc, Groton, CT

tobe added

See more of this Session: QbD Plenary
See more of this Group/Topical: Topical I: Comprehensive Quality by Design in Pharmaceutical Development and Manufacture

(53a) Regulating Calcitonin-Like Receptor Signaling Through Transmembrane Domain Interactions

Pin-Chuan Su, Chemical Engineering, Lehigh University, Bethlehem, PA, William Si, Bioengineering, Lehigh University, Bethlehem, PA, James Wu, Cardiothoracic Surgery, Lehigh Valley Hospital, Allentown, PA and **Bryan W. Berger**, Chemical Engineering and Program in Bioengineering, Lehigh University, Bethlehem, PA

Introduction: Protein-protein and protein-lipid interactions in membranes play a critical role in regulating transmembrane signal transduction. Receptors, channels and other integral membrane proteins reside in equilibrium between resting and active, ligand-bound state, the latter of which is responsible for activating distinct intracellular signaling pathways. A key mechanism by which this equilibrium is regulated is through transmembrane-mediated receptor oligomerization (1, 3). We are interested in studying the regulated oligomerization of a class of membrane co-receptors, receptor activity modifying proteins (RAMPs), with the G-protein coupled receptor calcitonin-like receptor (CLR) during early heart development, focusing on the role of transmembrane domain interactions in regulating signaling. RAMP1-CLR plays numerous roles in heart development and vascular maintenance in adults, and dysregulation of the RAMP1-CLR signaling has been linked to numerous disease states, including acute migraines and acute coronary syndrome (ACS), and small-molecule therapeutics targeting CLR-RAMP heterooligomers are currently in clinical trials as a treatment for migraines (2). Thus, understanding the structural basis for CLR-RAMP heterooligomerization is central to understanding the basis for CLR signaling in vascular development as well as in developing effective approaches to treating diseases associated with CLR.

Results and Discussion: Using sequence-directed searches of transmembrane structural databases, we identified multiple interfaces common to all co-receptors and CLR and developed a model linking these interfaces to the ability to heterooligomerize with CLR. Biochemically, we confirmed the importance of one interface in transfected cells using oligomerization assays, monitoring cAMP expression levels and *in vivo* FRET measurements, all of which demonstrated that targeted mutations to the predicted interface inhibited oligomerization of RAMP-CLR and attenuated cAMP expression. MO-mediated knockdown and RAMP1 rescue studies using zebrafish also confirm the importance of RAMP1 during heart development as well as the key role of the predicted interface in regulating multiple, distinct signaling events during cardiovascular development. Our qPCR results from our human aortic valve samples and zebrafish studies both point to multiple signaling pathways affected downstream of RAMP1-CLR as well as potential differences in expression between control and disease samples.

Conclusions: Overall, our results point to a central role for transmembrane-mediated interactions in regulating RAMP1-CLR heterooligomerization as well as a specific interface responsible for regulating signaling in both human aortic valve samples and zebrafish embryos. Our current and future work is focused on developing a robust, large-scale expression system that will allow us to investigate in detail the structural basis for RAMP1-CLR heterooligomerization, which will give us greater insight into the molecular details of transmembrane-mediated CLR signaling.

References:

1. Moore D, Berger B, & DeGrado W (2008) Protein-protein interactions in the membrane: sequence, structural, and biological motifs. *Structure* 16(7):991-1001.

2. Brain SD & Grant AD (2004) Vascular actions of calcitonin gene-related peptide and adrenomedullin. *Physiological Reviews* 84(3):903-934.

3. Berger BW, *et al.* (2010) Consensus motif for integrin transmembrane helix association. *Proceedings of the National Academy of Sciences* 107(2):703-708.

See more of this Session: Receptor-Mediated Phenomena
See more of this Group/Topical: Food, Pharmaceutical & Bioengineering Division

(53b) Spatiotemporal Membrane Organization At the Single Molecular Level During T Cell Receptor Triggering

Kyung-Ho Roh, Howard Hughes Medical Institute, Stanford University, Stanford, CA and Mark M. Davis, Microbiology and Immunology, Stanford University, Howard Hughes Medical Institute, Stanford, CA

In the adaptive immune system, critical antigen-specific T cell responses are triggered by the specific recognition of an antigenic peptide-major histocompatibility complex (pMHC) by the T cell receptor (TCR). Despite extensive studies on this TCR engagement and subsequent activation of the T cell, the spatiotemporal dynamics between the molecules involved in the process are still poorly understood, especially at the single molecular level. In this study, simultaneous dual-color imaging of the molecules that have critical roles in T cell activation was performed at unprecedented high resolution. For this purpose, we employed dual-color photoactivated localization microscopy (dc-PALM) and direct stochastic optical reconstruction microscopy (dSTORM). T cell membrane organizations in both latent and activated states were examined in detail.

T cell blasts prepared from the lymph nodes of 5c.c7 TCR transgenic mice were examined under microscope 7 days after initial activation by 10 μM Moth cytochrome C (MCC, 88-103) peptide. As fluorescent labels for dc-PALM, either PSCFP2 or PAmCherry was genetically tagged onto the CD3ζ, the linker for activation of T cells (Lat), and the CD4 molecule, a pair of which was co-expressed in a cell. For dSTORM, Alexa488 and Alexa 568 were conjugated to the specific antibodies and used for the immunofluorescence staining. Lasers of 405 nm, 488 nm and 568 nm were used for the activation of both fluorescent proteins, the excitation of PSCFP2/Alexa488, and the excitation of PAmCherry/Alexa568, respectively. Both excitation lasers were used in the total internal reflection fluorescence microscopy setup. The localization of individual molecules, reconstruction of dual color images, and spatial analyses were performed using in-house software written in MATLAB.

One of the foremost interests of this study is on how the TCR, Lat, and CD4 molecules reside in the T cell membrane before and after T cell activation. On the poly-L-lysine coated glass slides, which were used as non-stimulatory surfaces, molecules reside in their own domains which are within the size of a couple of hundred nanometers, and these domains are well distributed over the cell membrane. On the stimulatory glass surfaces coated with I-E^k/MCC and B7.1, these protein domains concatenate to form bigger microclusters. But surprisingly the degree of co-localization of protein domains seems moderate at molecular level, and mainly occurs in the interface between the two microclusters. Detailed statistical spatial analyses were employed to characterize the behaviors of protein domains quantitatively. In the talk, discussions also will be made to connect these findings to the T cell receptor activation process or other membrane receptor behavior in general.

(53c) Spatial Dynamics In Bacterial Signal Transduction

Kang Wu, Energy Biosciences Institute, University of Illinois, Urbana, IL and Christopher V. Rao, Chemical and Biomolecular Engineering, University of Illinois, Urbana Champaign, Urbana, IL

Chemotaxis is the process by which cells sense chemical gradients in their environment and then move towards more favorable conditions. In bacteria, the chemotaxis pathways all appear utilize a modified two-component system to sense and respond to chemicals, where the receptors form ternary complexes with the CheA histidine kinase and the CheW adaptor protein. The clustering of these ternary complexes into semi-ordered hexagonal lattices has been documented in multiple organisms and is presumably conserved in all species of chemotactic bacteria. These arrays are thought to amplify the response to attractant binding.

Previously, these signally clusters were thought to be predominantly clustered at the cell poles. They were also thought to be static and reorganize only locally in response to attractant binding. However, recent studies have shown that these structures may in fact be much more fluid. In this study, we investigated the localization of the chemotaxis signaling complexes in the bacterium *Bacillus subtilis* using immunofluorescence and live-cell fluorescence. We found that the localization and composition of these signaling clusters change upon exposure to attractant. Moreover, these changes in clustering and localization are reversible with the prestimulus distribution reestablished in adapted cells. Finally, we also found that these changes are phosphorylation dependent.

These results collectively demonstrate that bacteria employ spatiotemporal mechanisms in signal transduction, something previously thought to occur only in eukaryotes. While we still do not fully understand the physiological significance of these changes, these result nonetheless demonstrate that signal transduction in bacteria is far more complex than initially envisioned with many new complexities still to be resolved.

References:

Wu K, Walukiewicz HE, Glekas GD, Ordal GW, Rao CV. Attractant binding induces distinct structural changes to the polar and lateral signaling clusters in *Bacillus subtilis* chemotaxis. Journal of Biological Chemistry, 286(4):2587-95, 2011.

See more of this Session: Receptor-Mediated Phenomena
See more of this Group/Topical: Food, Pharmaceutical & Bioengineering Division

(53d) Investigation of the Residue Specific Interactions of Aβ with Neuron-Like Cell Surface Molecules: Implications for the Mechanism of Pathogenesis of Alzheimer's Disease

Arundhathi K. Venkatasubramaniam and Dr. Theresa A Good, Chemical and Biochemical Engineering, University of Maryland Baltimore County, Baltimore, MD

Alzheimer's disease (AD) is a progressive neuro-degenerative disorder. Memory impairment, as well as problems with language, decision-making ability, judgment, and personality are the various symptoms of this disease. AD primarily affects people over 65 years of age. In view of the United States' ageing population, it is especially important to discover new therapeutic approaches to combat this disease.

Senile plaques and neuro-fibrillary tangles are observed in the brains of AD patients at the histopathological level. Beta-amyloid (Aβ) peptide is a major component of these senile plaques. Aggregated Aβ is believed to be toxic to neurons and is implicated in Alzheimer's disease. However, the mechanism by which Aβ leads to neurodegeneration is still unclear. Elucidating steps in the mechanism of Aβ-neuron interaction that leads to AD pathology may provide insight into new therapeutic avenues to treat AD.

The neurotoxicity of Aβ is thought to be dependent upon its spontaneous adoption of an aggregated structure. A number of residues in the beta amyloid peptide have been found to be linked to peptide neurtoxicity, including Methoionine 35 (M35), residues 16-22, Glycine 33 (G33) and Asparagine 27 (N27). Some of these residues like M35 and G33 has specifically been linked to Aβ toxicity via their oxidative potential or their affect on neuron electrical activity (long term potentiation or LTP), respectively . Other residues such as region 17-21 are thought to be important because of the role of this hydrophobic region in Aβ aggregation.

We hypothesize that the sequence of Aβ has a role to play in its aggregated structure, its association with cells and its resultant toxicity. We test this hypothesis by analyzing if particular residues of the structure of Aβ are important for binding to cell components and for inducing changes in cell signaling linked to Aβ pathology.

We used fluorescence resonance energy transfer (FRET) and confocal microscopy, along with flow cytometry of both mutated and chemically modified Aβ-40 sequences to probe the importance of different residues on Aβ in the interaction of aggregated Aβ with neuron-like cells. Initially we showed that Aβ bound near an integrin receptor (α6) on the cell surface. The first 7 residues of Aβ contain a possible integrin binding sequence, RHD, thus we examined the possibility that these residues were important in Aβ-40 interaction. Results suggest that the 8-40 mutant of Aβ-40 , has significantly higher energy transfer to the integrin than wild type Aβ40 leading to the speculation , that the 1-7 residues and the RHD sequence that occurs in 1-8 residue sequence are not crucial for the binding of Aβ-40 to integrin. Further, we show that methylation of lysines on the Aβ sequence (K16 and K28), and mutation of N27 to proline significantly decreased FRET between Aβ and the α6 integrin receptor. Our results suggest that the sequence of beta amyloid has an important role to play in how it binds to cellular molecules, possibly in turn influencing signaling events in the cell.

Given our results to date, we believe that it is probable that the binding of the aggregated Aβ peptide to the cell is regulated by electrostatic interactions. We also have multiple lines of evidence that residues 27-33 may be important in Aβ cell interaction. We will continue to investigate these hypotheses using both mutated and chemically modified Aβ.

Our FRET studies suggest Aβ binds near the α6 integrin receptor subunit. We also have demonstrated that Aβ activates a tyrosine kinase (possibly from the Src family) upon interaction with the neuron-like cell. It is possible that activation of the integrin receptor, followed by tyrosine kinase activation, could lead to cellular changes that may impact LTP or other molecular components of neuron learning and memory. We will present our data showing how the same residue specific changes in Aβ that affect energy transfer with the integrin receptor alter tyrosine kinase activity and our understanding of the mechanism of this signaling cascade in cells exposed to aggregated Aβ.

See more of this Session: Receptor-Mediated Phenomena
See more of this Group/Topical: Food, Pharmaceutical & Bioengineering Division

(53e) Stochastic Modeling of BMP Signaling Pathways Identifies Mechanisms of Noise Suppression In Cell Signaling

Mohammad Shahriar Karim, Electrical and Computer Engineering, Agricultural and Biological Engineering, Purdue University, West Lafayette, IN, Gregery T. Buzzard, Department of Mathematics, Purdue University, West Lafayette, IN and David M. Umulis, Agricultural and Biological Engineering, Purdue University, West Lafayette, IN

Title: Stochastic modeling of BMP signaling pathways identifies mechanisms of noise suppression in cell signaling

Abstract: Bone Morphogenetic Proteins (BMPs) are growth factors that are often utilized during organismal development to direct cell differentiation. Their induction properties have also led to their widespread use to direct stem cell differentiation *in vitro*. BMP-mediated development is highly reproducible, with low cell-to-cell variability, but BMP-mediated control of stem cell differentiation is highly variable and exquisitely sensitive to BMP dose and duration. During development BMP signaling is regulated by BMP binding proteins that modulate ligand-receptor interactions such as the secreted binding protein Crossveinless-2 (Cv-2). To determine whether are nor extracellular regulation by Cv-2 like molecules could attenuate signaling noise, we developed a large-scale parameter screen of mathematical models for Cv-2 action. In the absence of Cv-2, low molecule numbers and slow reactions rates, consistent with measured *in vitro* kinetic data yields high-amplitude, long-duration departures from the deterministic mean levels of BMP signaling. We find that if BMP signaling is regulated by Cv-2 like molecules, intrinsic noise can be greatly reduced depending on the specific choice of model parameters. To fully characterize the behavior of Cv-2 on signaling, we used a truncated state space approximation to the Chemical Master Equation (CME) that approximates the full probability distribution of each species over time. Efficient solution of the CME via the truncated state space method allowed us to carry-out an exhaustive parameter screen with nested continuation in 4 parameters and fully characterize the biological process. Cv-2 like molecules reduce the amplitude of stochastic noise for a subset of the parameters and significantly increased the frequency of fluctuations for a high percentage of parameter space. High frequency fluctuations are rapidly suppressed by the intracellular network providing a simple, yet reliable mechanism to reduce noise in cellular differentiation.

See more of this Session: Receptor-Mediated Phenomena
See more of this Group/Topical: Food, Pharmaceutical & Bioengineering Division

(53f) Implication of Particle-Cell Dynamics In Physiological Blood Flow On the Efficacy of Vascular-Targeted Drug Carriers

Phapanin Charoenphol and **Omolola Eniola-Adefeso**, Chemical Engineering, University of Michigan, Ann Arbor, MI

p.MsoNormal, li.MsoNormal, div.MsoNormal { margin: 0in 0in 0.0001pt; font-size: 12pt; ""; }div.Section1 { page: Section1; }

Introduction: Nanospheres are commonly used as drug carriers for intravenous delivery due to their ability to pass through microcirculation without occlusion and their ability to avoid eliciting immune response. In addition to these abilities, ideal targeted drug carriers must also effectively adhere at the targeted location to release the therapeutic agents. Recent studies focused on elucidating the vascular wall binding efficiency of targeted spheres with sizes ranging from nano- to micrometer in reconstituted human blood (platelet and leukocyte depleted blood) demonstrated that particle binding increases with increasing particle size, i.e. nanospheres exhibit minimal adhesion to inflamed endothelium from blood flow relative to microspheres. Since many vascular wall target molecules (e.g. selectin and ICAM) are also expressed on other cells of the blood, it becomes important to understand the possible physical/ molecular interactions that exist between blood cells and vascular-targeted drug carriers. Thus, this study aims to elucidate the influence of leukocytes (WBCs) and platelets on the margination and adhesive dynamics of vascular-targeted drug carriers under physiological blood flows as a function of particle size and targeting moiety.

Materials and Methods: Dynamics of 0.2, 0.5, 2 and 5 mm spherical particles in blood were assessed in a parallel plate flow chamber equipped with a straight or step channel. Human Umbilical Vein Endothelial Cells (HUVECs) activated by IL1-b were employed as inflamed endothelial cells. These cells were attached at the bottom of the chamber and used in flow adhesion assays. Spheres were either targeted to endothelial selectins via surface-linked sLe^A or ICAM-1 via anti-ICAM-1 antibody (surface density for each targeting ligand was fixed for all particle sizes). Particle adhesion was observed in human reconstituted blood (plasma and RBCs, at ~ 40% Hct), whole blood, platelet-depleted blood and leukocyte-depleted blood flow with steady, pulsing or recirculating profile.

Results and Discussion: The level of particle binding in whole blood flow is a function of particle size and type of blood flow. Specifically, in laminar and pulsatile whole blood flow, 2 mm spheres showed the highest binding over other sizes whereas in steady recirculating whole blood flow, 5 mm had maximum binding. Overall, adhesion of nanospheres (0.2 and 0.5 mm) was minimal in all flow conditions compared to microspheres. We find that the higher adhesion observed for 2 mm spheres over 5 mm ones in laminar and pulsatile whole blood flow was due to their physical and molecular interactions with blood cells in whole blood that served to improve their recruitment to the wall. However, the presence of leukocytes in blood appears to negatively affected the adhesion of 5 mm spheres in laminar and pulsatile flow via: (1) competition for the available binding space and (2) physical interaction with the bound particles that removes them from the adhesive state. We propose that the higher adhesion of 5 mm spheres in recirculating flow is due to the rotational motion induced by the step channel that cause the depletion of white blood cells in the vortex area and thus eliminate the competition for binding between particles and white blood cells.

Conclusions: The binding efficiency of drug carriers in blood flow depends on the interplay between particle size and other factors such as the blood hematocrit, interaction with non-erythrocytes blood cells and type of flows. In addition to elucidating the physical and molecular dynamics of spherical particles in blood flow, our results can serve as a road map for designing efficient drug carrier systems for targeted delivery applications in many human diseases.

See more of this Session: Receptor-Mediated Phenomena
See more of this Group/Topical: Food, Pharmaceutical & Bioengineering Division

(53g) cAMP Stimulation Regulates Aquaporin-1 Mediated Transendothelial Water Transport In Bovine Aortic Endothelium Monolayers

Chirag B. Raval[1], John M. Tarbell[1], Kung-Ming Jan[2] and David S. Rumschitzki[3], (1)Department of Biomedical Engineering at The City College of New York, Gradute Center at the City University of New York, New York, NY, (2)College of Physician and Surgeons, Columbia University, Bronx, NY, (3) Department of Chemical Engineering, The City College of The City University of New York, New York, NY

Abstract — The pressure inside large arteries is typically ~100 mmHg higher than outside of them. This difference, ΔP, drives a flow of plasma, mainly water and advected/swept-along solutes, that would otherwise only transport by (much slower) diffusion [1], across the porous vessel wall. It is natural to ask whether this flow enters the wall across the endothelium solely through inter-endothelial cell (EC) junctions or whether a portion of the water traverses the ECs via membrane proteins from the ubiquitous aquaporin (AQP) family [2]. Data from our group show that bovine aortic ECs (BAECs) express AQP-1 in cultured monolayers and that rat aortic ECs (RAECs) express AQP-1 in culture and in excised vessels. Chemically blocking AQP-1 or knocking down AQP-1 expression significantly reduces endothelial hydraulic conductivity (L_{pe}). Upregulating AQP-1 expression on L_{pe} in vessels and monolayers might be a route towards washing LDL out of the subendothelial intima (SI) before it can bind significantly, a process that can lead to lesions.

Belkacemi *et al.* [3] showed a 4-fold protein expressional change of AQP1 in trophoblast cells after treatment with arginine vasopressin (AVP). Additionally, expression of vasopressin type-2 receptors (V2R) has been confirmed in human lung microvascular endothelial cells [4] and rat aortic strips [5]. We sought to investigate the effect of V2R stimulation on L_{pe} and AQP-1 expression in BAEC monolayers for both short and long term treatments.

We found that AVP increased L_{pe} 34% and AQP-1 expression 52% in BAEC monolayers as compared to untreated monolayers.

AVP treatment had no effect on albumin transport across BAEC monolayers under either convective or diffusive conditions, indicating no apparent affect on junctional tightness or length. Since albumin transports significantly by the paracellular route, this indicates that AVP treatment does not affect this pathway and the change in L_{pe} is likely due to the transcellular AQP-1 route.

Short term forskolin (F) treatment is known to increase intercellular cAMP, which causes membrane shuttling of internal AQP-1 to the cell membrane. We found that short term F treatment increased L_{pe} 172% but did not significantly change AQP-1 expression, which is consistent with the shuttling hypothesis. Satavaptan (S), an inverse V2R agonist, treatment decreased L_{pe} -46% and AQP-1 expression -44%. F/S treatment did not significantly change L_{pe} and decreased aquaporin-1 expression -29%. A/S treatment decreased L_{pe} -38 and caused no significant change in AQP-1 expression.

[1] Huang, Y., Rumschitzki, D., Chien, S. and Weinbaum, S., "A fiber matrix model for the filtration through fenestral pores in a compressible arterial wall," Amer. J. Physiol., 272 (41), H2023(1997).

[2] Preston, G. M. and Agre P. (1991). "Isolation of the cDNA for erythrocyte integral membrane protein of 28 kilodaltons: member of an ancient channel family." Proc Natl Acad Sci U S A **88**(24): 11110-4.

[3] Belkacemi, L., Beall, M.H., Magee, T.R., Pourtemour, M., Ross, M.G. (2008). "AQP1 gene expression is upregulated by arginine vasopressin and cyclic AMP agonists in trophoblast cells." Life Sci. **82**(25-26): 1272-80.

[4] Kaufmann, J.E. and Vischer, U.M. (2003). "Cellular mechanisms of the hemostatic effects of desmopressin (DDAVP)."J Thrombosis and Haemostasis, **1**: 682–689.

[5] M°Chaly I, Laurent F, Portet K, Serrano J, Cros G. (1999). "Vasopressin V2 (SR121463A) and V1a (SR49059) receptor antagonists both inhibit desmopressin vasorelaxing activity." Eur J Pharmacol. **383** (3):287-90.

See more of this Session: Receptor-Mediated Phenomena
See more of this Group/Topical: Food, Pharmaceutical & Bioengineering Division

(53h) Antibiotic Pumps In Bacteria Export Endocrine Disrupting Chemicals During Wastewater Treatment

Xinhua Li, Civil and environmental engineering, University of Utah, Salt Lake City, UT and Otakuye Conroy-Ben, Civil and Environmental Engineering, University of Utah, Salt Lake City, UT

Along with the presence of pharmaceuticals and endocrine disrupting chemicals (EDCs) in the environment, drug resistance can be found in wastewater effluent, receiving waters, and produced biosolids. The classes of resistance genes found in wastewater and treated effluent vary significantly due to the nature of antibiotics present. Some classes include: b-lactam resistance (ampicillin), polyketide resistance (tetracycline), and aminoglycoside resistance (kanamycin). It was recently discovered that the major antibiotic pumps AcrAB-TolC and AcrAD-TolC in *e. coli* export human hormones in addition to the common substrates antibiotics, bile salts, and surfactants. AcrAB-TolC and AcrAD-TolC belong to a class of antibiotic pumps known as the Resistant Nodulation and Division (RND) exporters. RND proteins span the entire inner and outer membranes of gram-negative bacteria and consist of an inner membrane component, a membrane fusion protein, and an outer membrane channel. During export, an antibiotic (or other chemical) enters the protein complex through periplasm, and is pumped out of the bacterium.

To determine if multidrug resistance efflux pumps export endocrine disrupting chemicals in addition to common substrates, minimum inhibitory concentrations (MIC) were measured. The MIC is the concentration of a chemical at which no growth of a bacteria strain is observed. Two major Resistant Nodulation and Division (RND) efflux pumps, AcrAB-TolC and MexAB-OprM, were expressed in *Escherichia coli* and exposed to EDCs. These include hormones, synthetic hormones, alkylphenols, platsticizers, and phytoestrogens. Hormones such as 17β-estradiol, estrone, and estriol, showed no inhibitory effect on the *E. coli* with the chemical concentration up to 100 mg/L. Meanwhile, the tripartite complexes of AcrAB-TolC and MexAB-OprM efflux pumps were revealed as important functions in resistance of 17α-ethynylestradiol, octylphenol, nonylphenol, bisphenol-A.

The presence of EDC resistant bacteria in local wastewater effluent and river samples was qualified. Water samples were plated on LB agar with appropriate concentrations of EDCs determined from MIC tests above (17α-ethynylestradiol, octylphenol, nonylphenol, bisphenol-A). Results showed that various EDC-resistant bacteria survive UV disinfection, and reach the Jordan River, which feeds into the Great Salt Lake. Such bacteria have also been detected in fresh-water tributaries leading to the Jordan River.

Bacteria are ever-adapting to chemical stressors in the environment, leading to biochemical responses such as drug resistance. The persistence of endocrine disruptors in the environment supports the findings of this research, in that EDCs are substrates of multidrug efflux pumps. Since EDCs are exported from the bacterial cell, they may not reach oxidizing proteins which would eliminate them from wastewater. One reason for the lack of degradation during secondary wastewater treatment could be that EDCs are exported from cells before coming into contact with degrading enzymes.

See more of this Session: Receptor-Mediated Phenomena
See more of this Group/Topical: Food, Pharmaceutical & Bioengineering Division

(62a) Optimization of a Louisiana Native Algal Co-Culture for Biofuel Production

Rong Bai[1], Maria Teresa Gutierrez-Wing[2], **Michael G. Benton**[1] and Kelly A. Rusch[2], (1)Cain Department of Chemical Engineering, Louisiana State University, Baton Rouge, LA, (2)Civil and Environmental Engineering, Louisiana State University, Baton Rouge, LA

Microalgae have been reported to possess significantly higher lipid productivity per unit area than traditional oil crops, a benefit for biofuel production. Microalgal strains such as *Nannochloropsis sp., Botryococcus braunii,* and *Chlorella vulgaris* have been reported to contain high lipid/oil content. However, these results are mostly at the lab scale. In large-scale, commercial culture systems, tolerance to local environmental conditions and invasive algal species significantly affect lipid/oil production. Therefore, native microalgal strains with high lipid productivity are likely better options for large-scale biofuel production.

A native fresh water microalgae strain from Louisiana was investigated. This work measures the impacts of irradiance and nitrate levels on total lipid yield and the fatty acid profile of this strain. The culture was harvested in the late exponential-stationary phase to maximize the lipid content, and Soxhlet extraction with chloroform/ methanol (2:1, v/v) was used for total lipid content evaluation. Microwaves and nanoparticle assisted Folch's extraction methods have been used to optimize lipid extraction. The fatty acid profiles of neutral lipids were analyzed by gas chromatography combined with mass spectrometry.

The results indicated the lipid yield and content at 400 and 800 mmol $m^{-2}s^{-1}$ was the highest and no significant differences (32.13 and 32.76 mg $L^{-1}d^{-1}$, 31% and 30%) between the two treatment levels were observed. Results also indicate that although the lower nitrate level increased lipid content, the increase did not compensate for a decrease in growth rate. The fatty acid profiles indicate the major components are 16- and 18-carbon fatty acids, and ~35% are saturated fatty acids. The fatty acid profiles do not vary significantly with the irradiance and nitrate levels.

See more of this Session: Advances In Algal Biorefineries I
See more of this Group/Topical: Sustainable Engineering Forum

(62b) Controls of Microalgal Biomass and Lipid Production In Municipal Wastewater-Fed Bioreactors

Belinda S.M. Sturm[1], Val H. Smith[2], Marie-Odile Fortier[1], Edward Peltier[1] and Frank Jerry deNoyelles[3], (1)Civil, Environmental & Architectural Engineering, University of Kansas, Lawrence, KS, (2)Ecology and Evolutionary Biology, University of Kansas, Lawrence, (3)Kansas Biological Survey, Lawrence

The use of wastewater as a nutrient source for algal growth as a biodiesel feedstock could increase the viability of this process and reduce wastewater discharge of N and P compounds. In this study, four algal bioreactors were fed continuously with municipal wastewater effluent over a six month period; these were monitored weekly to determine algal biomass and lipid production, nutrient removal, and carbon and nutrient availability. Algal biomass production ranged from 0.78 to 15.9 g dry weight m^{-2} d^{-1} over the experimental period. Algal reactors removed 19% of dissolved nitrogen and 43% of dissolved phosphorus from wastewater effluent. A stoichiometric analysis of particulates revealed that algal biomass had far lower C:P ratios (67:1) and N:P ratios (6:1), but higher C:N ratios (17:1), than Redfield values. These data suggest that the algae produced were rich in phosphorus, and that N was likely to be growth-limiting in these systems. Numerous studies also suggest that algal biomass can be enhanced by providing CO_2 in excess of normal atmospheric concentrations, and that algal biofuel production can contribute to CO_2 sequestration efforts. In order to optimize nutrient removal in wastewater-fed algal reactors, and to determine the potential for CO_2 sequestration, future reactor engineering research should continue to incorporate the expertise of ecological stoichiometry.

See more of this Session: Advances In Algal Biorefineries I
See more of this Group/Topical: Sustainable Engineering Forum

(62c) Effects of Total Inorganic Carbon and Nitrogen Concentrations On Lipid Formation In Chlorella Vulgaris

Jinsoo Kim, Department of Chemical & Materials Engineering, University of Cincinnati, Cincinnati, OH, Joo-Youp Lee, Department of Chemical and Materials Engineering, University of Cincinnati, Cincinnati, OH and Kaniz F. Siddiqui, The Metropolitan Sewer District of Greater Cincinnati, Cincinnati, OH

Effects of totalinorganic carbon and nitrogen concentrations on lipid formation in *Chlorella vulgaris*

Jinsoo Kim[1], Joo-YoupLee[1], Kaniz F. Siddiqui[2]

[1] Chemical EngineeringProgram, University of Cincinnati, OH 45221-0012

[2] The Metropolitan SewerDistrict of Greater Cincinnati, Cincinnati, OH 45204-2002

Microalgae have recently received greatattention as a promising alternative bioenergy source because they can produce asubstantial amount of triacylglycerols (e.g. 20-50 % dry cell weight) and canbe readily converted into biodiesel. *Chlorella vulgaris* has been selected forthis growth kinetic study since it is one of the fastest growing microalgae andcontains a reasonable amount of triacylglycerols (e.g. 14-22 % dry cell weight). Dissolved carbon dioxide (CO_2(aq)) and bicarbonate ion (HCO_3^-) areessential inorganic carbon forms required for autotrophic algal growth, but fewstudies have been systematically conducted on the impact of inorganic carbon onalgal growth. In this respect, thegrowth kinetics of *Chlorella vulgaris*was previously investigated dependent on the total inorganic carbonconcentration (i. e., CO_2(aq) and HCO_3^-) As a result, it was observed that the growthkinetics was increased when the total inorganic carbon concentration wasincreased in the culture medium.

Under many nutrient limiting conditions,lipids are accumulated as a means of storage when energy (i.e. light) andcarbon (i.e. CO_2 for autotrophic growth) sources are available. However, many of these nutrients areessential for photosynthesis responsible for metabolic flux generation intolipids formation. Thesenutrient-deficient conditions decrease biomass growth rate, and offset thebenefit of lipid accumulation. Thegrowth of algal biomass such as *Chlorella vulgaris* is dependent on bicarbonate ion concentrationsavailable in the culture medium. Bicarbonateion also plays a critical role as a catalyst in the first step of fatty acidsbiosynthesis for the synthesis of malonyl-CoA.In this study, a change in total lipids content has been investigatedusing the Nile red stain method when the total inorganic carbon concentrationin culture medium is increased. This isexpected to provide fundamental understanding of the utilization of bicarbonateions for the design and operation of bioreactors.

Keywords: *Chlorella vulgaris*, total inorganic carbon, nitrogen deficiency,lipid content.

See more of this Session: Advances In Algal Biorefineries I
See more of this Group/Topical: Sustainable Engineering Forum

(62d) Medium Optimization for Enhanced Lipid Production by Chlorella Protothecoides UTEX25

Kuan-Chen Cheng[1], Ming Ren[1] and Kimberly Ogden[2], (1)Chemical and Environmental Engineering, The University of Arizona, Tucson, AZ, (2)University of Arizona, Tucson, AZ

Algal biodiesel as the third generation biofuels has received considerable attention in recent years because of its renewable, biodegradable, and nontoxic fuel properties. Algal also exhibit several advantages over other energy crops, such as higher photosynthetic efficiency, higher biomass production and lipid content, and faster growth rate. The composition of algal-produced hydrocarbons vary among the microalgae species resulting in many different lipids, hydrocarbons, and other complex oils.

In the present study, a fresh water microalgae, *Chlorella protothecoides* (UTEX25), was evaluated for its lipid productivity through medium optimization. A one-factor-at-a-time approach was applied to determine the influence of each ingredient of proteose medium on both biomass and lipid production. $MgSO_4$°E7H_2O, $NaNO_3$, $CaCl_2$°E2H_2O, K_2HPO_4 and KH_2PO_4, and NaCl were removed individually from the culture medium, and the results were compared with the one obtained from the original medium. After 11-day cultivation, the results demonstrated that the removal of MgSO4°E7H_2O has positive effect on both biomass growth and lipid production. The biomass reached 0.95 g/L (the biomass production obtained from original medium was 0.54 g/L). The lipid content also increased to from 4.6 to 9.5 % when MgSO4 was removed. On the other hand, the removal of $NaNO_3$ and proteose adversely reduced biomass growth. The biomass production decreased from 0.54 to 0.37 and 0.29 g/L, respectively, when $NaNO_3$ and proteose was removed. Therefore, MgSO4°E7H_2O, $NaNO_3$, and proteose were chosen for further evaluation.

A statistical experimental design, response surface methodology (RSM), was applied to determine the optimal concentration of the three nutrients for both biomass and lipid production. The predicted results of these three nutrients are validated in comparison with experimental data. The change of the lipid profiles obtained from original and optimized medium are evaluated using GC-FID. Results are applied to minimize nutrient requirements in large-scale pond cultivation.

See more of this Session: Advances In Algal Biorefineries I
See more of this Group/Topical: Sustainable Engineering Forum

(62e) Single-Cell Growth Kinetics of Algae, Chlorela Vulgaris

Alim Dewan[1], Jihye Kim[2], Swastika S. Bithi[2], Marci Kerls[2], Siva A. Vanapalli[2] and M. Nazmul Karim[2], (1)Chemical Engineering, Texas Tech University, Lubbock, TX, (2)Chenical Engineering, Texas Tech University, Lubbock, TX

Single-cell growth kinetics of algae, *Chlorela vulgaris*

AlimDewan, JihyeKim, Swastika S. Bithi, Marci Kerls, Siva A.Vanapalli, M. Nazmul Karim*

Departmentof Chemical Engineering, Texas Tech University, Lubbock, Texas 79409, USA.

*E-mail: naz.karim@ttu.edu

Abstract

Traditionally,the growth kinetics of algae is investigated using cell culture equipment thatcontains large number of cells. The kinetic parameters are measured fromrepresentative samples of the cell culture and the measured parametersrepresent the statistical average of all the cells in the samples. In thisstudy, we have investigated the growth kinetics of single-cell of *Chlorellavulgaris* using droplet based PDMS microfluidic chip. The kinetic parameterswere determined by counting the total number of cells under investigationduring the entire period of the experiment. We found that there is significantheterogeneity in the growth and size of single cells and kinetic parameters aredifferent even if the initial cell numbers are the same. We found that thereis significant heterogeneity in the growth and size of single cells and kineticparameters are different even if the initial cell numbers are the same. Figure1exmplifies the heterogeneity of single-cell growth kinetics. We hypothesizethat the heterogeneity occurs due to the difference in initial cell conditionssuch as the age of the cell, and the size.

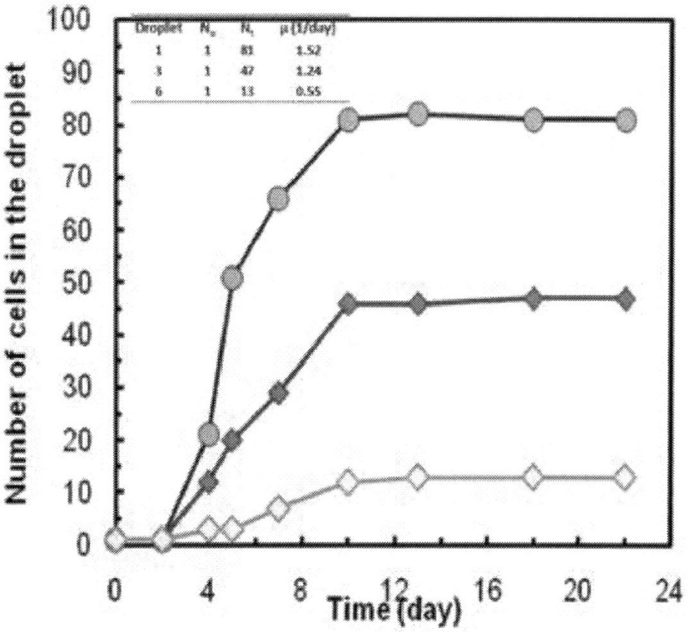

Figure1.Example of single-cell growth curves. Inset shows the initial cell number (N_o),final cell number (N_t) and the specific growth rate ().

See more of this Session: Advances In Algal Biorefineries I
See more of this Group/Topical: Sustainable Engineering Forum

(62f) Modeling and Optimization of Algae Based Biodiesel Production

Soumya Yadala, Chemical Engineering, The University of Tulsa, Tulsa, OK and Selen Cremaschi, Department of Chemical Engineering, The University of Tulsa, Tulsa, OK

Processingof biodiesel from algae biomass has attracted much interest in recent times dueto many reasons. Significant amongst them are global warming and scarcity infossil fuels. It has been suggested that this third generation biofuel, acarbon neutral alternative, is capable of meeting the existing demand fortransport fuels [1]. However, substantial challenges are encountered inderiving fuels from algae biomass. Primary amongst them is the relativelyhigher cost of cultivating and harvesting algae species. Due to this reason,currently, algae based biofuels cost much higher than the biofuels from firstgeneration and second generation biofuel feedstock [2].

Thiswork focuses on modeling and optimization of algae based biodiesel production. Designparameters, for cultivation of algal species, at a geographical location are affected by that location's climatic conditions such as annual solar irradiance.Hence, geographical location, at which algae is cultivated, plays a vital rolein the biodiesel production. Another significant factor to be considered for producingmaximum amount of biodiesel is the selection of algal species. There are avariety of algal species present throughout the world, however, not all algalspecies are suitable for biodiesel production because of varying amounts of oilcontent and specific growth rates. Hence, selection of algal species is also animportant factor in determining the cost of biodiesel production. There are anumber of technologies for each process involved in the production of biodieselfrom algae biomass but the question of which technology, for which species, andat which location, to attain maximum profits is still an open challenge to theresearchers. Figure 1 presents the tasks that should be completed to producealgae-based biodiesel and options/technologies (which can be used to achievethese tasks) considered in this work. Using Figure 1, a disjunctive programmingmodel is developed to determine the algal species, the location for the cultivationof algae, and the optimum route from growth of algae to production of biodieselby selecting one technology for each process with the objective of minimum costof biodiesel production. The resulting Mixed Integer Nonlinear Programming(MINLP) model is solved using GAMS 23.6.5/SBB solver. With the model parameters used, 8000 tons/year of biodiesel can be produced at a cost of about $153 per gallon for a period of ten years using the selected technologies. Accordingly,Tricornutum species was selected to grow at all the given locations by means of artificial growing technique-tubular photo bioreactor. The cultivated algae were harvested through flocculation which was then concentrated using centrifugation.The algae biomass was dried by low pressure shelf drying followed by extraction of algal oil by expeller press and finally, the algal oil was sent to transesterification for biodiesel production.

REFERENCES

1. Y. Chisti, Biodiesel from microalgae, *Biotechnol. Adv.* **25** (2007), pp. 294-306.

2. Sheehan, J., Dunahay, T., Benemann, J., Roessler, P. A Look Backat the U.S. Department of Energy's Aquatic Species Program: Biodiesel fromAlgae; Close-Out Report. U.S. Department of Energy (1998).

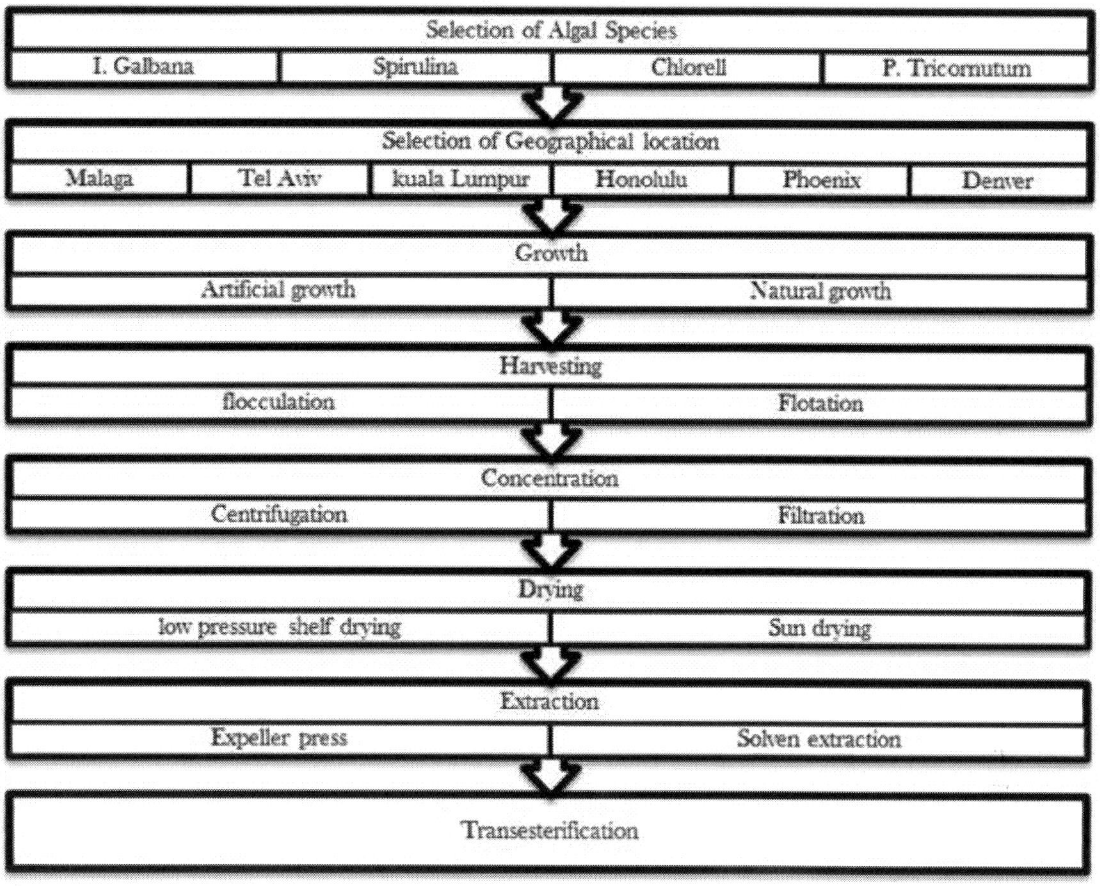

Figure 1. Tasks and corresponding options/technologies for biodiesel production from algae

See more of this Session: Advances In Algal Biorefineries I
See more of this Group/Topical: Sustainable Engineering Forum

(63a) Enzymatic Reduction of Nitro Compounds to Amines with Nitroreductase

Jonathan T. Park, Chemical and Biomolecular Engineering, Georgia Institute of Technology, Atlanta, GA and Andreas S. Bommarius, School of Chemical and Biomolecular Engineering, Georgia Institute of Technology, Atlanta, GA

Nitroreductases are enzymes that reduce nitro compounds to corresponding nitroso, hydroxylamine, and amino compounds. In the past 20 years, they have gathered interest in several applications such as degradation of xenobiotic nitroaromatics, pharmaceutical production of amines, and activation of prodrugs in medical applications.

In previous studies, nitroreductases have shown to produce the hydroxylamine equivalent of the substrate. The instability and carcinogenicity of this product has been a barrier in using this enzyme. More recently, a nitroreductase from *Salmonella typhimurium* has been identified to produce the amine compounds from nitro aromatics.

Through this study we will prove the pathway of product and byproduct formation. With thorough understanding of the reaction cascade, we will show increased amine yields through optimization of the reaction conditions and mutational work.

See more of this Session: Advances In Biocatalysis and Biosynthesis I
See more of this Group/Topical: Food, Pharmaceutical & Bioengineering Division

(63b) How to Tie Peptide Knots

A. James Link, Chemical & Biological Engineering, Princeton University, Princeton, NJ

Lasso peptides are unique ribosomally synthesized natural products that exist in a topologically constrained conformation. Recently we have uncovered some mechanistic details about the biosynthesis of these peptides in their bacterial hosts. This talk will discuss the role of the leader peptide in lasso peptide biosynthesis as well as the promiscuity of the maturation enzymes.

See more of this Session: Advances In Biocatalysis and Biosynthesis I
See more of this Group/Topical: Food, Pharmaceutical & Bioengineering Division

(63c) Implementation of Metabolic Engineering to Construct a Robust Platform for Resveratrol Production and Protein Engineering Study of Stilbene Synthase

Namita Bhan[1], Chin Giaw Lim[2], Zachary Fowler[2], Thomas Hueller[3], Steffen Schaffer[3] and Mattheos A. G. Koffas[2], (1)Chemical and Biological Engineering, Rensselear Polytechnic Institute, Troy, NY, (2) Chemical and Biological Engineering, University at Buffalo, the State University of New York, Buffalo, NY, (3)Evonik Degussa GmbH, Marl, Germany

Resveratrol (3,4',5-Stilbenetriol) is a promising polyphenol, reported to have various anti-cancer, copper chelation, anti-oxidant scavenging of free radicals and anti-aging effects. It is synthesized by a plant type III polyketide synthase: Stilbene synthase, via the phenylpropanoid pathway, from a tetraketide intermediate. Plant extraction or chemical synthesis; both are economically non feasible methods for producing higher titers to meet the current market demands. The bacterial production platform for Resveratrol was optimized, by trying out combinations of heterologous expression methods, like different expression systems, and bacterial production strains. Also enhancement of intracellular Malonyl-CoA led to the final titers of 2.3 g/L. We are also attempting to produce a library of compounds similar to Resveratrol, by systematic mutation of the functionally important residues in Stilbene Synthase.

See more of this Session: Advances In Biocatalysis and Biosynthesis I
See more of this Group/Topical: Food, Pharmaceutical & Bioengineering Division

(63d) Refactored Biosynthetic Pathways: Delineating Pathway Expression From Sophisticated Regulation Cascades Via Synthetic Biology

Zengyi Shao, Chemical & Biomolecular Engineering, University of Illinois at Urbana-Champaign, Urbana, IL and Huimin Zhao, Department of Chemical and Biomolecular Engineering, Department of Chemistry and Biochemistry, Institute for Genomic Biology, University of Illinois at Urbana-Champaign, Urbana, IL

Microorganisms have been evolved to produce a myriad array of complex molecules known as natural products or secondary metabolites of pharmaceutical and biological importance. The production of natural product is subjected to sophisticated regulation cascades in which target gene and multiple regulators interact with each other, resulting in a complex regulatory network, and responding to a variety of physiological and environmental signals. This highly regulated process complicates the discovery and characterization of new compounds with important biological activities and hampers sustainable production of the ones with high-value. Here we developed a general synthetic biology-based strategy to resolve the complexity and delineate the pathway expression from sophisticated regulation cascades.

Three important biosynthetic pathways from Streptomyces including the spectinabilin gene cluster, the fosfomycin gene cluster, and the FR900098 gene cluster were chosen for study. Spectinablin, isolated from *Streptomyces spectabilis* and *Streptomyces orinoci*, is a nitro-phenyl substituted polyketide exhibiting antimalarial and antiviral activities. Although both hosts were reported to produce spectinabilin, only the pathway from *S. spectabilis* can be functionally expressed in *S. lividans*, making the silent spectinabilin biosynthetic pathway a perfect model for studies of cryptic pathways. These two clusters undergo different regulations as the cluster from *S. spectabilis* contains SpnD as an activator and the cluster from *S. orinoci* contains NorD as a repressor. Real-time PCR was used to analyze the expression level of each *nor* gene in both the native host and the heterologous host. It was shown that 7 out of the 10 enzymes involved in spectinabilin biosynthesis were expressed at an extremely low level in the heterologous host. This may have resulted in the silence of the spectinabilin biosynthesis in *S. lividans*.

To activate the spectinabilin pathway in *S. lividans*, we propose to construct a refactored pathway by inserting a constitutive promoter in front of each gene. Approximately 40 constitutive promoter candidates in front of house-keeping genes were cloned from the genomes of Streptomyces and other related actinobacteria and fused with reporter genes. Currently, they are being evaluated in different growth conditions. The strong constitutive promoters will be inserted in front of the *nor* genes. The success in isolating strong and constitutive promoters will be very useful in activating numerous uncharacterized biosynthetic pathways.

In addition, the proposed strategy for refactoring biosynthetic pathways can also be used to improve the production of valuable natural products in Streptomyces. Previous effort mainly focused on testing various growth conditions to optimize pathway expression and precursor availability. The overall process is tedious and labor-intensive as there is no universal solution and pathways have to be examined one by one in an array of media. Here we aim to demonstrate that the optimization process can be greatly shortened if the sophisticated regulation is removed. Two example pathways, the fosfomycin biosynthetic pathway and the FR900098 biosynthetic pathway, both having important biomedical applications are chosen and will be reconstructed with strong and constitutive promoters for higher production levels.

See more of this Session: Advances In Biocatalysis and Biosynthesis I
See more of this Group/Topical: Food, Pharmaceutical & Bioengineering Division

(63e) Constructing Building Blocks for a Cell-Free Synthetic Pathway Biotransformation: a High Yield and Low-Cost Bio-Manufacturing Platform

Suwan Myung and Y-H Percival Zhang, Biological Systems Engineering, Virginia Polytechnic Institute and State University, Blacksburg, VA

The production of hydrogen from low-cost renewable biomass is appealing because biomass is an enriched, collectable renewable energy source. The use of carbohydrates as high-density hydrogen carriers and energy sources for hydrogen production is possible due to cell-free synthetic biology technology. Synthetic pathway biotransformation (SyPaB) is the implementation of complicated biochemical reactions by the *in vitro* assembly of a large number of enzymes and coenzymes. Hydrogen can be produced from carbohydrate with an overall reaction of $C_6H_{10}O_5$ (aq) + $7H_2O$ (l) --> $12H_2$ (g) + $6CO_2$ (g). As a result, nearly 12 moles of dihydrogen are produced per glucose equivalent of polysaccharides and water by this non-natural synthetic pathway, which contains 13-14 enzymes. For facilitating cascade reactions among three enzymes: *Thermus thermophilus* triose phosphate Isomerase (*Ttc*TIM), *Thermotoga maritima* fructose bisphosphate aldolase (*Tm*ALD), and *Thermotoga maritima* fructose 1,6-bisphosphatase (*Tm*FBP), we need find optimal conditions and proper enzyme ratios for co-immobilized enzymes (combi-CLEAs). In this talk, we will present our recent progresses in enzyme purification, immobilization, and substrate channeling among cascade enzymes.

See more of this Session: Advances In Biocatalysis and Biosynthesis I
See more of this Group/Topical: Food, Pharmaceutical & Bioengineering Division

(63f) Engineering E. Coli Whole-Cell Catalysts for Consolidated Production of Lactic Acid From Cellodextrin

Charles Rutter, Nimmy Mathew and Rachel Chen, Chemical and Biomolecular Engineering, Georgia Institute of Technology, Atlanta, GA

Engineering *E. coli* whole-cell catalystsfor consolidated production of lactic acid from cellodextrin

Charles Rutter, NimmyMathew, Rachel Chen

School of Chemicaland Biomolecular Engineering, Georgia Institute of Technology

Reduction of production cost forbio-based fuels and chemicals is an important engineering objective yet to beachieved. Consolidated bioprocessing is one of the approaches being pursued bymany laboratories around the globe and offers an opportunity for significantcost reduction by combining cellulase production, cellulose depolymerization,and product synthesis in a single step. Engineering microbial catalyst capableof performing these three tasks is, however, challenging. As a consequence,consolidated processing remains largely laboratory adventures, despitesignificant progresses in the past few years.

In order to develop a system to be used for consolidatedbioprocessing of cellulose, expression and secretion of β-glucosidase andcellodextrinase must be achieved. This presentation describes the use of aheterologous cellodextrinase to enable *E. coli* cells to grow oncellodextrins and cellobiose. This enzyme has been shown to release glucosefrom cello-oligomers across a range of degrees of polymerization with highestactivity on cellotetraose. Additionally, we have shown that this heterologouslipoprotein is properly transolcated, acylated, and inner membrane associatedin the *E. coli* host. Finally, periplasmic displayed enzyme conferredthe cells the ability to metabolize cellodextrins and ferment them to lacticacid at nearly 75% of theoretical yield.

This study marks the first attempt at using aninner membrane anchored lipoprotein as an enzyme for degradation of substrateslocated in the extracellular medium. The ability of this enzyme to hydrolyzeextracellular without being translocated across the outer membrane helps to reducethe energetic requirements of this strain for use in consolidatedbioprocesses. Additionally, this study demonstrates that heterologouslipoproteins can be properly modified post-translationally by *E. coli*and localized to the periplasmic space.

See more of this Session: Advances In Biocatalysis and Biosynthesis I
See more of this Group/Topical: Food, Pharmaceutical & Bioengineering Division

(63g) Inceasing Organic Acid Tolerance In Cyanobacteria

Matthew B. Begemann, Microbiology Doctoral Training PROGRAM, UNIVERSITY OF WISCONSIN-MADISON, Madison, WI and Brian F. Pfleger, Chemical and Biological Engineering, University of Wisconsin Madison, Madison, WI

Due to escalating petroleum and commodity prices, there is increased interest in engineering photosynthetic microorganisms to directly produce fuel and chemicals from light and CO_2. Of particular interest are the organic acids acrylate and 3-hydroxypropionate. Acrylate is used in the manufacturing of plastics and coatings, while 3-hydroxypropionate can be catalytically converted to a variety of useful compounds, including acrylate. One obstacle to engineering cyanobacteria to produce commodity chemicals, in particular organic acids, is toxicity. The cyanobacterium *Synechococcus* sp. PCC 7002, a well studied organism with a relatively fast growth rate and amendable genome, is completely inhibited by 50μM acrylate. In comparison E. coli is inhibited by 25mM. Attempts to increase the tolerance of PCC 7002 to organic acids resulted in the identification of a mutant strain capable of growth in greater than 100mM acrylate, a 2000-fold increase. In addition, this mutation resulted in 100 fold increases in tolerance to other commercially relevant organic acids including propionate and 3-hydroxypropionate. This mutation was mapped to the *acsA* gene, which encodes an acetyl-CoA ligase. A direct knock out of acsA resulted in a similar phenotype as the spontaneous mutant. The utility of this strain is currently being demonstrated by engineering a 3-hydroxypropionate production pathway.

See more of this Session: Advances In Biocatalysis and Biosynthesis I
See more of this Group/Topical: Food, Pharmaceutical & Bioengineering Division

(64a) Extrinsic Growth Signals Driving Drosophila Imaginal Cell Proliferation In Vitro

Jeremiah J. Zartman, Simon Restrepo and Konrad Basler, Institute of Molecular Life Sciences, University of Zurich, Zurich, Switzerland

Insect cell lines are used both commercially in the production of heterologous recombinant proteins and as biochemical tools that complement research in established genetics systems such as *Drosophila* embryonic and larval development. Recently, a few studies have been successful in identifying novel extrinsic growth factors that stimulate the growth of a cell line, Cl.8, which was derived from wing imaginal disks. The utility of Cl.8 cells for growth control research derives from their strong dependency on growth factors to proliferate. Motivated by the success of previous studies and recognizing that much is still unknown about extrinsic regulators of insect growth, we have undertaken a detailed characterization of the culture medium supporting proliferation of Cl.8 cells. Our approach has been to use design-of-experiment methodology to screen conditions and optimize the current culture medium to gain further insight into extrinsic growth regulation of imaginal cells. We have found that of three supplements included in the original media formulation, a raw extract derived from Drosophila was the only essential factor. As a result we have optimized the extraction process to improve the stimulatory effect and are characterizing the biochemical properties of components from the extracts, which affect both proliferation and cell-cell adhesion. Efforts to formulate a more defined culture medium for Cl.8 cells will further increase the utility of this cell-line in Drosophila genetics studies on growth control. The identification of novel growth factors stimulating Cl.8 cells also has the potential to inform efforts toward increasing the yield of insect cells in commercial applications.

See more of this Session: Advances In Cell Culture IB: Experimental
See more of this Group/Topical: Food, Pharmaceutical & Bioengineering Division

GENE EXPRESSION PROFILING OF CD133^{+} PROSTATE CANCER STEM CELLS INDUCED FROM PC3 CARCINOMA CELL LINE CULTURES

Portillo-Lara R,[1] *Alvarez MM*[1,2]

[1] *Centro de Biotecnología-FEMSA, Tecnológico de Monterrey at Monterrey, Monterrey, NL. México;*
[2] *Escuela de Medicina, Tecnológico de Monterrey at Monterrey, Monterrey, NL. México.*

Abstract

According to the stem cell model of cancer progression, the disease emerges through mutations and epigenetic modifications that disrupt normal function in adult stem cells. These cancer stem cells (CSCs) constitute a very small subset of the tumor and are responsible for establishing and maintaining the disease. The identification of potential therapeutic targets that are differentially expressed in the normal and malignant stem cell populations has been limited by the lack of suitable *in vitro* experimental models. We report the implementation of a custom cell culture protocol for the enrichment of the CD133^{+} CSC population from PC3 carcinoma cells. When analyzed by flow cytometry, using anti-CD133 (293C3) labeled antibodies we observed that this side population becomes significantly enriched (up to 5.6% positive events) as compared to standard protocols (0.2%). We further characterized both CD133^{+} and CD133^{-} fractions using novel gene expression-profiling technology. Functional annotation of differentially expressed gene patterns reveals up-regulation of pathways comprising several genes associated with cell fate as well as DNA repair mechanisms, cell cycle regulation, inhibition of differentiation and metastasis.

Key words: Prostate cancer, CSCs, Gene expression profiling, cell culture.

Introduction

Prostate carcinoma is the most commonly diagnosed male malignancy in the western hemisphere. It accounts for one quarter of all reported cancer cases in men [1]. Given the heterogeneity observed in prostatic malignancies, it has been suggested that this type of cancer originates from the neoplastic transformation of normal prostatic stem cells (PSCs) or transit amplifying (TA) cells [2]. PSCs possess strong self-renewal capacities but proliferate relatively slow, dividing to renew themselves (symmetric division) and at the same time, generate a more differentiated progeny (asymmetric division) [3].

According to the stem cell model of cancer progression, the disease emerges through mutations and epigenetic modifications occurring in adult PSCs. These "cancer stem cells" (CSCs) share the same ability of PSCs to divide, self-renew and differentiate. However, CSCs proliferate through abnormally expressed pathways and so, generate aberrant tumor tissues. Current methodologies for the isolation of CSCs subpopulations from human prostate tumors are based on the expression of molecular markers previously used to isolate normal PSCs (CD44^{+}/integrin $\alpha_2\beta_1^{hi}$/CD133^{+}) [4]. Isolation and *in vitro* expansion of CSCs offers the unique opportunity to analyze cellular processes participating in malignant transformation and tumor propagation separately from those mechanisms present in non-tumorigenic PSCs populations. This specific targeting of oncogenic gene mutations, altered pathways or aberrant proteins present exclusively in CSCs has the potential to spare non-tumorigenic populations, minimizing the adverse effects observed with current indiscriminative therapies.

However, the identification of potential therapeutic targets has been limited by the lack of suitable experimental models.

It has been reported that growing carcinoma cell lines in anchorage-independent culture conditions promotes the formation of spheroid structures that express a CSC-like phenotype [5-7]. These structures display increased tumorigenicity and exhibit gene expression profiles consistent with those of CSCs isolated directly from patients. In addition to this, cell plating at low densities under serum-free media seems to preserve and even increase the expression of stem cell markers like CD133 [8]. This particular protein, encoded by the PROM1 gene, is a member of a pentaspan transmembrane glycoprotein family localized to membrane protrusions that has been associated with early stages of differentiation [9, 10]. More specifically, expression of antigens CD133-1 (AC133) and CD133-2 (293C3) has been used as a selection criterion for the isolation of populations exhibiting a stem-like phenotype throughout a variety of normal and malignant tissues.

Modern transcriptomics technologies are capable of characterizing genetic expression patterns occurring in malignant tissues in a quantitative and accurate fashion. Several groups have reported the use of microarray-based technology as a gene discovery tool to identify genetic markers that discriminate between normal and cancerous prostate tissues [11-13]. Even though microarrays are a remarkable tool for genome-wide expression profiling, they are often limited by the relative insensitivity of this technique to detect transcripts expressed at low levels [14]. The quality and quantity of RNA represent then a restrictive factor since the natural complexity of some samples limits the amount of RNA that can be isolated. In addition to this, QRT-PCR validation must be carried out because of the cross-hybridization issues that are frequently observed with this technique. Alternative technologies that address these issues have become recently available [15]. The nCounter NanoString System allows the simultaneous measurement of up to 800 genes with a remarkable sensitivity, coupling high multiplex capability and digital readout. Samples are processed from 100ng of RNA without the need for linear amplification and with a sensitivity that is equal to that of TaqMan RT-PCR.

Here we present an experimental platform for the study of $CD133^+$ prostate CSCs induced and isolated from a commercially available PC3 carcinoma cell line. Also, we further characterized these cells using the NanoString gene expression profiling system in order to identify key regulatory mechanisms that could potentially lead to the development of new, more specific, therapeutic strategies.

Materials and Methods

Cells and culture protocols. PC3 cells were obtained from the American Type Culture Collection and maintained in three different culture protocols: a) monolayer cultures grown in DMEM/F12 50/50 (Mediatech) supplemented with 5% fetal bovine serum (FBS) and 100U/100 µg/ml Penicillin-Streptomycin; b) non-adherent cultures grown in DMEM/F12 50/50 supplemented with 20 ng/µl hEGF (Gibco), 20 ng/µl bFGF (Gibco), 0.5 µg/ml Hydrocortisone (Sigma), 5 µg/ml Heparin (Sigma), 1X B27 supplement without Vitamine A (Gibco), 1X Insulin-Transferrin-Selenium-A (Gibco), 100U/100 µg/ml Penicillin-Streptomycin; and c) non-adherent cultures grown in EpiGRO Human Prostate Complete Media Kit (HPCMK) without the addition of rhTGF-α and supplemented with 20 ng/µl hEGF and 20 ng/µl bFGF. Monolayer cultures were maintained and passaged at a maximum confluence level of 80%, culture media was changed every 48hrs. Non-adherent cultures were grown in ultra-low cluster plates at a density between $2X10^5$ and $4X10^5$ cells/ml, 1 ml of fresh media was added directly to the culture every 48hrs. Non-adherent cultures were dissociated and passaged every 96 hrs. All of the cultures were maintained at 37°C with 95% relative humidity in a 5% CO2 and 95% air atmosphere for two weeks before processing. The morphology of the cultures was monitored by microscopic examination using an Axiovert 200 inverted microscope.

Immunofluorescent labeling, cell sorting and flow cytometric analysis. Monolayer and non-adherent cultures were enzymatically dissociated using StemPro Accutase (Gibco®). All antibody incubations and flow cytometric analyses were performed in PBS buffer pH 7.2, 0.5% bovine serum albumin and 2 mM EDTA. Cells were count and labeled separately with Phycoerythrin (PE)-labeled anti-CD133/2 (293C3) (Miltenyi Biotech®), PE-labeled anti-CD133/1 (AC133) mAb (Milyenyi Biotech) or PE-Cy7 labeled anti-CD44 mAb (eBioscience®) following the instructions from the manufacturer. For cytometric analysis, labeled cells were processed on a FACSCanto II flow cytometer (BD Biosciences®) and a minimum of 50,000 events was recorded. Magnetic-activated cell sorting (MACS) was performed via magnetic on-column cell immobilization using anti-PE microbeads (Miltenyi Biotech®) and the CD133/2 (293C3) mAb.

RNA isolation and genetic expression analysis. Total RNA was extracted and purified from both CD133+ and CD133- fractions using the RNeasy® mini or RNeasy® micro kit (Qiagen®) depending on the amount of starting material. The concentration and quantity of the purified RNA were evaluated using the NanoDrop® ND-1000 spectrophotometer at 260/280 nm. Detection of mRNA transcripts was then carried out in multiplexed hybridization reactions using the NanoString® system. For this experiment we used the commercially available nCounter Human Cancer Reference kit (NanoString® Technologies). This panel is comprised of sequence-specific probe pairs (capture/reporter) targeting 230 highly curated cancer-related human genes and six internal reference genes.

Results and discussion

Proliferation of $CD44^{+}/CD133^{+}$ PC3 carcinoma cells can be induced manipulating in vitro culture conditions. The development of cell culture protocols that specifically aid to preserve the undifferentiated phenotype in the small percentage of CSC populations in established cell lines is a very critical line of research. *In vitro* expansion of these cells in such conditions would allow putative CSCs to undergo symmetric (self-renewal) and asymmetric (differentiation) divisions. This in turn would provide a remarkably useful experimental platform for their study. In order to elucidate a comprehensive medium for the growth and maintenance of $CD133^{+}$ prostate CSCs, we tested the ability of two serum-free media formulations developed particularly for epithelial stem cell maintenance. Anchorage-independent cultures growing in EGF and bFGF supplemented serum free DMEM/F12 and EpiGRO HPCMK formed proliferating spheres that could be serially passaged and exhibited increased expression of putative stem cell marker CD133 as compared to cells grown in monolayer cultures. The CD44 marker particularly did not exhibit significantly different levels of expression between treatments, ranging from 90 up to 98% of positive events for all tree culture protocols (data not shown). **Figure 1** shows the percentage of $CD133^{+}$ (measured with the 293C3 mAb) events when analyzed by flow cytometry. FBS supplemented adherent cultures exhibited 0.2% positive events (B), which coincides with the expected low expression levels of this marker reported in established cell lines such as PC3 [5]. For the anchorage-independent, serum-free DMEM/F12 formulation and the EpiGro HPCMK this percentage increased to 0.7% (C) and 1.3% (D) respectively. Using a parallel strategy consisting of direct addition of fresh media to the existing culture instead of continuous serial passaging, this percentage increased dramatically to 5.6% of positive events (E). This more successful strategy suggested the presence of soluble secreted factors residing in the culture media that were playing a critical role in the proliferation of the CD133^{+} side population via autocrine or paracrine signaling.

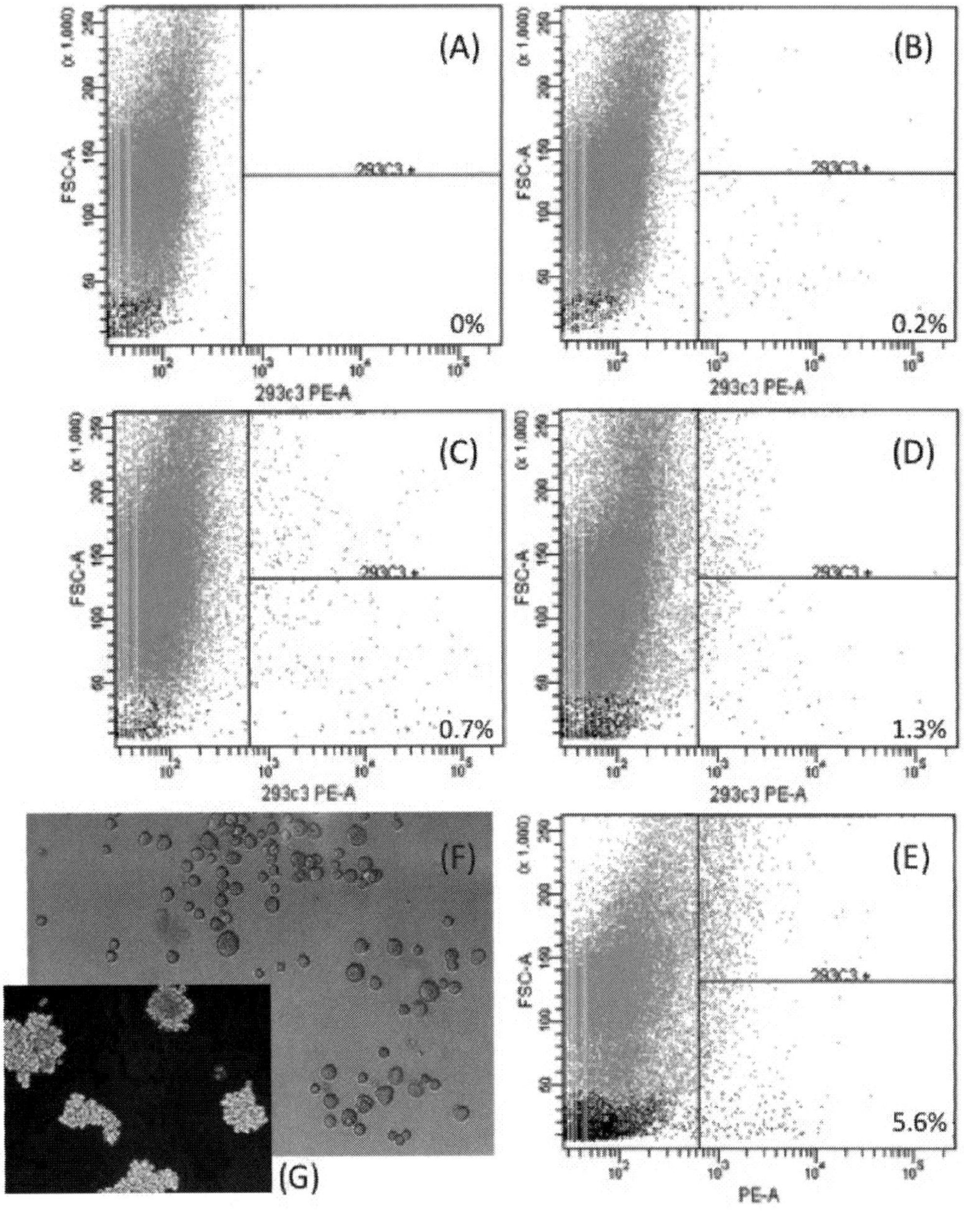

Figure 1. PE-coupled CD133 (293C3) fluorescence detection via flow cytometry. (A) Isotype control; (B) FBS-adherent cultures exhibit the lowest percentage of CD133 positive events (0.2%); (C) DMEM/F12 50/50 serum-free formulation with 0.7%; (D) EpiGRO HPCMK with 1.3%; and (E) EpiGRO HPCMK being added directly to the culture instead of serial passaging with 5.6% (a minimum of 50000 events were recorded). (F) Single cell suspension of CD133^{+} cells obtained via MACS. (G) PC3 cells growing in ultra-low attachment cluster plates.

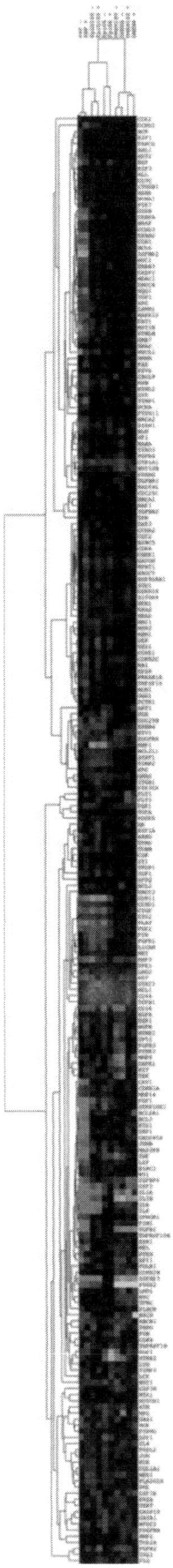

Figure 2. Heat map representing the hierarchical organization and clustering of the analyzed genes. This heat map illustrates the 230 genes from the Cancer Reference Panel grouping together according to their levels of expression throughout the tree protocols tested. CD133$^+$ and CD133$^-$ fractions corresponding to the same culture strategy tend to cluster together with the exception of the DMEM/F12 formulations. Abbreviations: SBF (Fetal Bovine Serum); EpiG (EpiGRO HPCMK); DMEM (DMEM/F12 50/50 mix).

CD133⁺ PC3 carcinoma cells exhibit known prostate cancer-associated gene expression profiles. In order to elucidate the molecular basis of proliferative pathways being used by $CD133^{+}$ PC3 cells exclusively, we isolated total RNA from both $CD133^{+}$ and $CD133^{-}$ fractions once these were dissociated and sorted out via MACS. Targeting the CD133 surface marker with a PE-conjugated anti CD133 (293C3) mAb we are able to obtain a single cell suspension that is around 90 to 95% pure for the desired cell type. Gene expression profiles of $CD133^{+}$ and $CD133^{-}$ PC3 cells growing in ultra-low attachment cluster plates were compared to the reference cell line growing in adherent substrates under FBS-supplemented media. **Figure 2** shows the hierarchical organization of the 230 genes contained in the Cancer Reference Panel. It can be observed that most of the positive and negative fractions from the same protocol cluster together, with the exception of the DMEM/F12 formulation. Changes in gene expression were visualized as ratios and only those exhibiting a fold change equal to or greater than 50% (ratio= 1.5 or 0.66) were considered for further analysis.

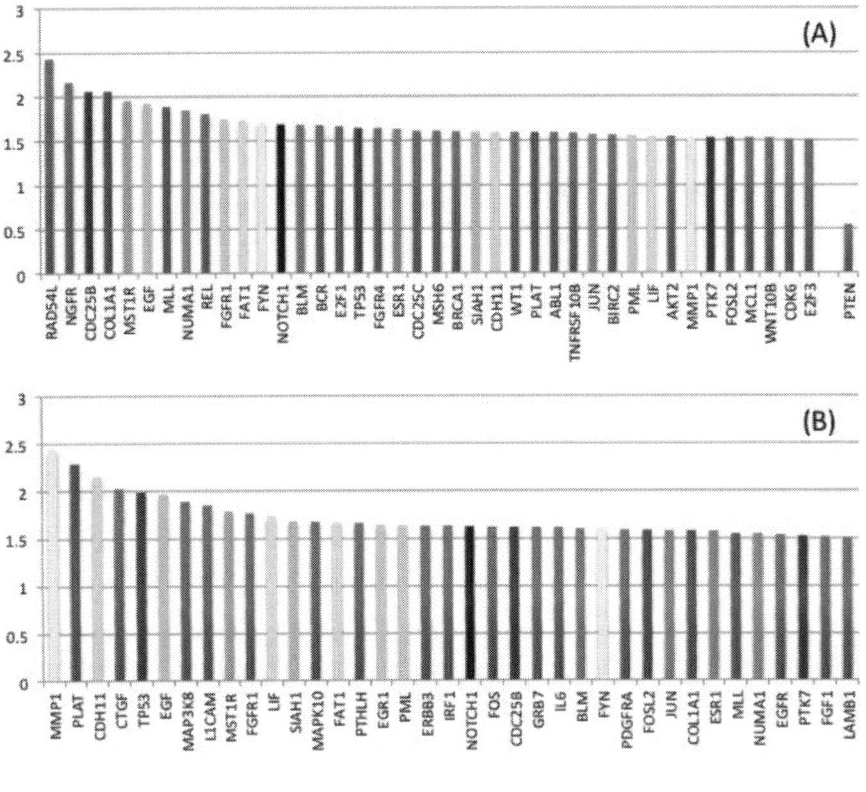

Figure 3. Differentially expressed genes in the $CD133^{+}$ side populations. Up-regulated genes in the $CD133^{+}$ fraction from (A) FBS-supplemented and (B) EpiGRO HPCMK cultures as compared to the $CD133^{-}$ fraction. Only PTEN from the FBS-supplemented media was found to be down-regulated for all tree protocols. This demonstrates that the differences between fractions within the same protocols are mainly because of up-regulated genes.

Shown here in **Figure 3** are genes that were found to be differentially expressed between the $CD133^{+}$ and $CD133^{-}$ fractions for the FBS-supplemented media (A) and the EpiGRO HPCMK (B). PC3 cells growing in the DMEM/F12 formulation exhibited a more homogeneous expression profile and only three genes matched the parameters to be considered as differentially expressed: BCL2 a potent suppressor of cellular apoptosis, NGFR a mediator of cell survival and death, and BLM a DNA-helicase that participates actively in DNA replication and repair. The FBS and EpiGRO formulations exhibited a more heterogeneous nature, with similar profiling between them. Both CD133$^+$ fractions exhibited a common up regulation of 22 genes as compared to their corresponding CD133$^-$ reference fractions. Several genes that were previously reported to be related to CSCs maintenance were found to be up regulated including key modulators of gene transcription and inhibitors of apoptosis (CD25B, CDC25C, EGF, FGF1, MLL, REL, FAT1, TP53, JUN, FOS, FOSL2, PML, CTGF, E2F1, E2F3, WT1, BIRC2, MCL1, IRF1); membrane proteins for signal transduction (NGFR, MST1R, FGFR1, FGFR4, NOTCH1, ESR1, ERBB3, TNFRSF10B, PTK7, PDGFRA, EGFR); proteins involved in protein degradation, tissue remodeling and metastasis (SIAH1, MMP1, PLAT, LAMB1); intracellular protein kinases and secreted factors (FYN, ABL1, AKT2, LIF, WNT10B, CDK6, MAP3K8, MAPK10, BCR, PTHLH, GRB7, IL6); molecules involved in cell-cell adhesion (L1CAM, CDH11); and regulators of DNA repair (RAD54L, BLM, MSH6, BRCA1). Only PTEN, a gene identified as a tumor suppressor that is mutated in a large number of cancers at high frequency, was found to be down regulated according to the criteria employed. In general, the only differences observed between fractions within the same protocol are originated from up-regulated genes. In contrast, when we compare the CD133$^+$ fraction from the EpiGRO HPCMK with the CD133$^-$ fraction from FBS-supplemented formulations, the differences between them comprise not only up-regulated, but also down-regulated genes (Figure 4).

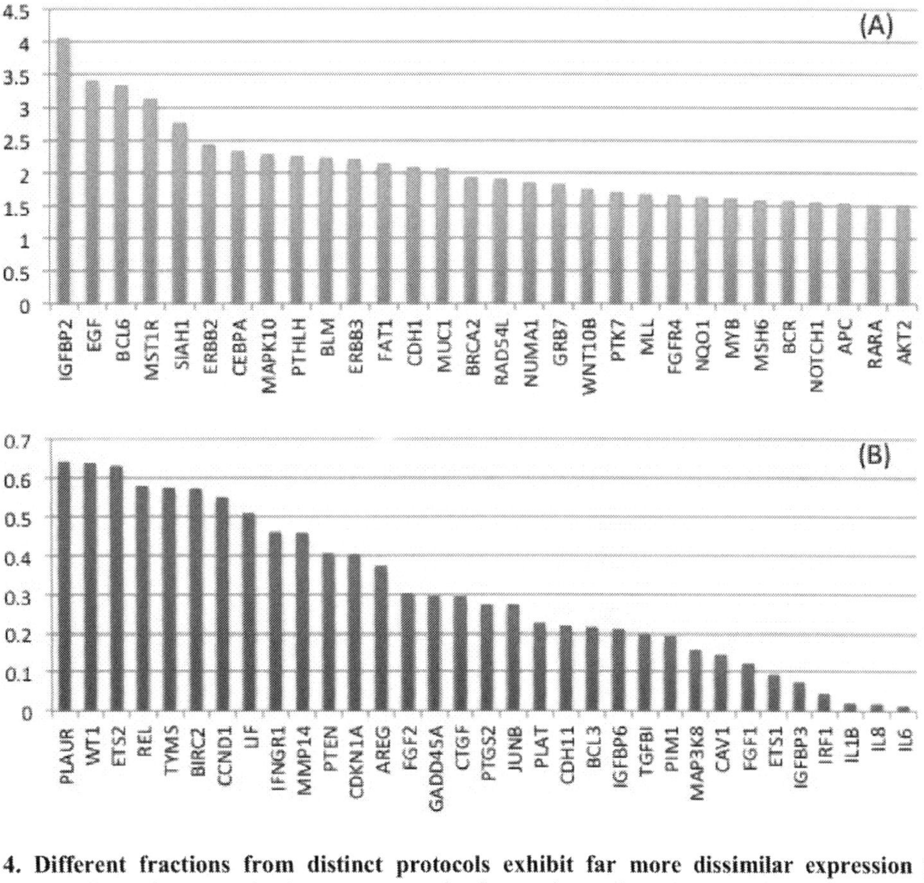

Figure 4. Different fractions from distinct protocols exhibit far more dissimilar expression profiles reflecting the effect of the media formulations. The figure shows the (A) up-regulated and (B) the down-regulated genes in the CD133$^+$ fraction from EpiGRO HPCMK as compared to the CD133$^-$ fraction from the FBS-supplemented formulation. Differences between CD133$^+$ and CD133$^-$ from different media formulations are greater than those observed within the same protocol.

$CD133^+$ cells from EpiGRO cultures exhibit a profile of up-regulated genes similar to the ones described above with the inclusion of the IGFBP2 modulator of the Insulin-like growth factor receptor, the BCL6, CBPA, and MYB transcription factors, the ERBB2 and RARA receptors, the membrane-bound glycosylated phosphoprotein MUC1, the APC tumor suppressor and the putative oncogene AKT2. Particularly, the expression of BCR2 and NQO1 would confer these cells superior survival capabilities in the presence of DNA damaging agents and reactive oxygen species. Panel B from Figure 3 shows that there are also genes whose expression appears to be down-regulated when compared to the parental cell line from which they originated. This set of genes includes a variety of transcription factors, growth factor, regulators of cell cycle progression, cytokines and interleukins whose expression is no longer being favored because of the absence of media components such as phenol red, antibiotics and more importantly FBS.

Conclusions

We have developed an experimental platform for the *in vitro* expansion and analysis of the $CD133^+$ side population from PC3 carcinoma cell line cultures. We provided experimental evidence that demonstrate that these cells exhibit genetic expression profiles that support their CSC identity. Results suggest that serum-free formulations favor experimental cell culture models that are more specific and less subjected to random effects that those obtained using FBS. This contribution demonstrates the use of a novel culture media, designed for the maintenance of prostate epithelial stem cells, and modified for the induction and expansion of prostate CSC from a commercially available cell line. These cells are not only being generated at larger rates, but also exhibit a more invasive phenotype ideal for studies involving metastasis, as well as superior cell survival mechanisms that can be useful in cell viability assays for chemotherapeutic drugs and other compounds.

References

1. ACS. *Cancer Statistics 2011.* Available from: http://www.cancer.org/Cancer/ProstateCancer/DetailedGuide/prostate-cancer-key-statistics.
2. Uzgare, A.R. and J.T. Isaacs, *Prostate cancer: potential targets of anti-proliferative and apoptotic signaling pathways.* Int J Biochem Cell Biol, 2005. **37**(4): p. 707-14.
3. Vander Griend, D.J., et al., *The role of CD133 in normal human prostate stem cells and malignant cancer-initiating cells.* Cancer Res, 2008. **68**(23): p. 9703-11.
4. Collins, A.T., et al., *Prospective identification of tumorigenic prostate cancer stem cells.* Cancer Res, 2005. **65**(23): p. 10946-51.
5. Pfeiffer, M.J. and J.A. Schalken, *Stem cell characteristics in prostate cancer cell lines.* Eur Urol, 2010. **57**(2): p. 246-54.
6. Rappa, G., et al., *Growth of cancer cell lines under stem cell-like conditions has the potential to unveil therapeutic targets.* Exp Cell Res, 2008. **314**(10): p. 2110-22.
7. Wei, C., et al., *Cancer stem-like cells in human prostate carcinoma cells DU145: the seeds of the cell line?* Cancer Biol Ther, 2007. **6**(5): p. 763-8.
8. Shi, X., J. Gipp, and W. Bushman, *Anchorage-independent culture maintains prostate stem cells.* Dev Biol, 2007. **312**(1): p. 396-406.
9. Weigmann, A., et al., *Prominin, a novel microvilli-specific polytopic membrane protein of the apical surface of epithelial cells, is targeted to plasmalemmal protrusions of non-epithelial cells.* Proc Natl Acad Sci U S A, 1997. **94**(23): p. 12425-30.
10. Yu, Y., et al., *AC133-2, a novel isoform of human AC133 stem cell antigen.* J Biol Chem, 2002. **277**(23): p. 20711-6.
11. Birnie, R., et al., *Gene expression profiling of human prostate cancer stem cells reveals a pro-inflammatory phenotype and the importance of extracellular matrix interactions.* Genome Biol, 2008. **9**(5): p. R83.
12. Gerald, W.L., *Genome-wide gene expression analysis of prostate carcinoma.* Semin Oncol, 2003. **30**(5): p. 635-48.
13. Shepherd, C.J., et al., *Expression profiling of CD133+ and CD133- epithelial cells from human prostate.* Prostate, 2008. **68**(9): p. 1007-24.
14. Russo, G., C. Zegar, and A. Giordano, *Advantages and limitations of microarray technology in human cancer.* Oncogene, 2003. **22**(42): p. 6497-507.
15. Geiss, G.K., et al., *Direct multiplexed measurement of gene expression with color-coded probe pairs.* Nat Biotechnol, 2008. **26**(3): p. 317-25.

(64c) Breaking the Habit: Overcoming Acute Sensitivities of Antibody-Producing Mammalian Cell Lines to pH and Raw Materials

Pratik Jaluria, Hunter Malanson, E. Todd Sorensen, Anne Kantardjieff and Derek Adams, Upstream Development, Alexion Pharmaceuticals, Cheshire, CT

As part of a broad effort to understand and refine our existing scale-down bioreactor model for several monoclonal antibody producing CHO and NS0 cell lines, we examined several possible sensitivities including pH set-point and raw materials. Initially, cells were grown at pH set-points ranging from 6.8 to 7.1 using the same bioreactor configuration. As the pH increased, cells produced more lactate and alanine. In addition, at lower pH set-points, ammonia production decreased. Follow-up studies examined feed composition and feeding strategy in several different bioreactor scales. We found that subsurface feed addition resulted in dramatic improvements in final antibody titer (~25%) while also reducing lactate production by more than 50%. Similarly, we examined the impact of an animal-origin derived raw material in the performance of our scale-down model and were successful in re-creating large-scale trends in growth and productivity. However, our scale-down model also exhibited tremendous variability, as replicate conditions diverged in performance. Careful review of historical data suggested several avenues to explore including: cell bank source, age of media supplements, cleaning procedures, feeding strategy, agitation and sparging flow rates. Several experimental blocks were constructed to investigate these factors and our results indicated the cell source, a single raw material and overall feeding strategy were most influential in terms of process variability. Subsequent work has reduced the variability in our small-scale model while preserving the ability to re-create large-scale trends in performance. Collectively, understanding these sensitivities has enabled us to improve our cell culture performance in terms of both productivity and robustness.

See more of this Session: Advances In Cell Culture IB: Experimental
See more of this Group/Topical: Food, Pharmaceutical & Bioengineering Division

(64d) Process Characterization for Commercial Cell Culture Process

Natarajan Vijayasankaran[1], Kara Calhoun[1], Sharat Varma[1], Adam Winchester[2] and Steven Meier[1], (1)Late Stage Cell Culture, Genentech, Inc., South San Francisco, CA, (2)Manufacturing Sciences and Technology, Genentech, Inc., South San Francisco, CA

When commercial manufacturing processes are modified to take advantage of the latest cell culture medium and process technologies to improve productivity, they need to be characterized to ensure that they can robustly produce the product. This presentation will focus on a specific case study in which multivariate experiments were combined with univariate experiments to design an overall process characterization strategy for a new version of the cell culture process for an existing commercial product. The process parameters to test were selected based on a risk-based approach after extensive analysis of the collected process development data. Designing multivariate experiments to characterize cell culture processes can be a challenge since establishing a multivariate design space requires that the acceptable range for one or more parameters be narrowed to mitigate failures due to the additive effects of multiple factors. The test ranges for these experiments needs to be established by balancing the competing priorities that require that the ranges be narrow enough to successfully establish a multivariate design space without excessive experimentation and be broad enough to meaningfully characterize the process. This can be especially challenging for existing commercial products for which it is desired that all product quality attributes meet previously established specifications. To simplify the design of the multivariate experiments and to improve the interpretability of the results, parts of the cell culture process with different growth and protein synthesis characteristics were treated as two different unit operations, which were studied in separate multivariate experiments. The ranges for the these parameters were carefully established by using stochastic simulations that utilized data collected during process development to ensure that the test ranges were sufficiently broad to characterize the process without risking excessive number of failures. The results of these studies will be discussed along with opportunities and pitfalls associated with this approach.

See more of this Session: Advances In Cell Culture IB: Experimental
See more of this Group/Topical: Food, Pharmaceutical & Bioengineering Division

(64e) Improved Process Development Based On Increased Understanding of Metabolism

Alan Gilbert, Kyle McElearney, Rashmi Kshirsagar, Marty Sinacore and Thomas Ryll, Cell Culture Development, Biogen Idec, Cambridge, MA

The drive towards high titer and chemically defined medium based processes has introduced new challenges. We have demonstrated a robust, scalable process platform with titers over 3 g L^{-1}. However, large variations in metabolism were observed at extended cultivation times when applied to an antibody program. These metabolic shifts are undesirable and limit process options. We investigated the causes of the metabolic variability and have identified several triggers which ultimately result in high lactate production rates. By implementing an intracellular fluorescent staining protocol, a likely mechanism inducing the variability was elucidated. Risk mitigation techniques were implemented taking direct advantage of the mechanistic knowledge, and these mitigation steps will also be discussed.

See more of this Session: Advances In Cell Culture IB: Experimental
See more of this Group/Topical: Food, Pharmaceutical & Bioengineering Division

(64f) Effect of Operating Parameters On the Product Quality of Monoclonal Antibody

Marija Sarenac, Benjamin Neunstoecklin, Miroslav Soos and Massimo Morbidelli, Institute of Chemical and Bioengineering, ETH Zurich, Zurich, Switzerland

Among biopharmaceutical drugs, monoclonal antibodies (mAbs) have evolved to the most successful protein-drugs in terms of clinical benefit delivered and revenue generated in the biopharmaceutical industry. To date, nearly 30 recombinant immunoglobulin G (IgG) antibody therapeutics have been approved to the market for the treatment of various diseases, particularly different forms of cancer. The efficacy of mAbs arises from their specificity to the target antigen and their ability to trigger immune responses by activating effector functions towards the targeted antigen. Especially the ability to elicit the immune response is critically dependent on appropriate posttranslational modifications of the antibody. For this reason, mAbs are commonly produced in mammalian cell culture. The use of mammalian cell-based production processes has significantly improved in the last years. Still, these processes remain rather sophisticated and require fundamental studies to be fully understood. In particular, the effect of various external factors on cell performance and the quality of the mAb has to be further investigated. External factors are, on the one hand, process operating parameters that can vary due to heterogeneities in the bioreactor or due to changes in the control strategy. Process operating parameters usually include dissolved oxygen levels, pH, hydrodynamic forces induced by agitation or sparging, osmolarity, dissolved CO_2 levels and temperature. On the other hand, cell performance in the bioreactor can be externally influenced by differences in the composition of the nutrient media. This involves different starting concentrations of the basic metabolites glucose and glutamine and all other proteinogenic amino acids.

Here we investigate, on the one hand, how external factors induce changes in cell metabolism that can for example result in the accumulation of waste metabolites, and how external parameters affect the productivity of cells. On the other hand, we aim to understand the effect of external factors on product quality attributes such as glycosylation, aggregation, C-terminal lysine truncation and deamidation, which can affect the pharmacokinetics of the antibody. Finally, the gained knowledge is used in the development of a multi-scale metabolic model that is able to describe the effect of external factors on cell growth kinetics, metabolism and on cell productivity.

See more of this Session: Advances In Cell Culture IB: Experimental
See more of this Group/Topical: Food, Pharmaceutical & Bioengineering Division

(64g) In Vitro Aggregation Behavior of a Lambda Light Chain Dimer Expressed In Mammalian Cells

Paolo Arosio[1], Marta Owczarz[1], Thomas Mller-Spth[1], Marten Beeg[1], Hua Wu[1], Mario Salmona[2] and Massimo Morbidelli[1], (1)Chemistry and Applied Biosciences, ETH Zrich, Zurich, Switzerland, (2) Department of Molecular Biochemistry and Pharmacology, Mario Negri, Milano, Italy

The aggregation of the immunoglobulin light chain fragment is responsible for light chain amyloidosis (AL) and light chain deposition diseases (LCDD). In this work, we produce a IgE λ light chain dimer from human mammalian cells U266, which originated from a patient suffering from multiple myeloma, and we investigate the effect of several physicochemical parameters on the *in vitro* stability of this protein. The dimer is stable in physiological conditions and aggregation is observed only when strong denaturating conditions are applied (acidic pH with salt at large concentration or heating at melting temperature T_m at pH 7.4). The produced aggregates are spherical, amorphous oligomers. Despite the larger β-sheet content of such oligomers with respect to the native state, they do not bind Congo Red or ThT. The impossibility to obtain fibrils from the light chain dimer suggests that the occurrence of amyloidosis in patients requires the presence of the light chain fragment in the monomer form, while dimer can form only amorphous oligomers or amorphous deposits. No aggregation is observed after denaturant addition at pH 7.4 or at pH 2.0 with low salt concentration, indicating that not a generic unfolding but specific conformational changes are necessary to trigger aggregation. A specific anion effect in increasing the aggregation rate at pH 2.0 is observed according to the following order: SO_4^{--} >> Cl^- > $H_2PO_4^-$, confirming the peculiar role of sulfate in promoting protein aggregation. It is found that, at least for the investigated case, the mechanism of the sulfate effect is related to protein secondary structure changes induced by anion binding.

See more of this Session: Advances In Cell Culture IB: Experimental
See more of this Group/Topical: Food, Pharmaceutical & Bioengineering Division

(65a) Mutagenesis of the Lasso Antibiotic Peptide Microcin J25: Discovery of Variants with Increased Potency and New Insights Into the Structure-Activity Relationship

Jessica Si-Jia Pan, Chemical and Biological Engineering, Princeton University, Princeton, NJ and A. James Link, Chemical & Biological Engineering, Princeton University, Princeton, NJ

Infections caused by drug-resistant Gram-negative bacteria pose a rising threat as some strains become ever more resistant to virtually all existing antibiotics. Microcin J25 (MccJ25) is a ribosomally-synthesized antibacterial peptide isolated from *Escherichia coli* that exhibits potent mechanisms of action against pathogenic strains of *Salmonella*, *E. coli*, and *Shigella*. MccJ25 has an unusual threaded lasso structure in which the C-terminal "tail" of the peptide is fed through a macrocyclic "ring" formed by the N-terminal residues. Production of MccJ25 in *E. coli* is dependent on a four-gene cluster encoding the structural gene *mcjA*, two maturation enzymes *mcjB* and *mcjC*, and an immunity factor, *mcjD*, in the form of an MccJ25 export pump. Here we have developed a system for orthogonal control of the expression of the structural gene and the export pump thus permitting independent control of MccJ25 production and export/immunity in *E. coli*. We used this system to screen saturation mutagenesis libraries targeted to the ring and tail portions of MccJ25. While multiple amino acid substitutions in the tail portion of the peptide were well-tolerated, mutagenesis of the ring portion of the peptide was detrimental to the antimicrobial function of MccJ25. We demonstrated that substitutions in the ring portion of the peptide likely disrupt the transport of these variants into the cytoplasm of susceptible strains. Additionally, we found several MccJ25 variants from the tail library with up to nearly 5-fold improvement in potency toward the strains *E. coli* and *Salmonella enterica* serovar Newport.

See more of this Session: Advances In Protein Structure, Function and Stability II
See more of this Group/Topical: Food, Pharmaceutical & Bioengineering Division

(65b) Deep Mutational Scanning As a New Tool for Protein Structure/Function Relationships

Timothy Whitehead, Chemical Engineering and Materials Science; Biosystems and Agricultural Engineering, Michigan State University, East Lansing, MI

What does the sequence-function fitness landscape of a protein look like? It has been appreciated for some time that most single point mutations are neutral to deleterious for overall protein fitness, but the fine details at the individual mutation level are scant in the literature. In this talk I present a new method called deep mutational scanning, which assesses the binding and thermodynamic folding energies of all possible single point mutations in a protein sequence using the enabling technologies of synthetic DNA libraries, yeast surface display, and deep sequencing. I show the application of the method to two proteins that were designed to bind a neutralizing epitope of the influenza hemagglutinin (HA) from the 1918 H1N1 pandemic virus. From the scanning results, new variants of these proteins were constructed to bind specific HA subtypes as well as to rapidly tighten affinity by over an order of magnitude to the 1918 H1 HA. Such construction and subsequent traversing of fitness landscapes should have wide application in the engineering of useful proteins as well as aid in the understanding of the determinants of protein sequence to function.

See more of this Session: Advances In Protein Structure, Function and Stability II
See more of this Group/Topical: Food, Pharmaceutical & Bioengineering Division

(65c) Broadly Protective Influenza Vaccines Based On the Conserved Hemagglutinin Stalk Region

Yuan Lu, John P. Welsh and James R. Swartz, Department of Chemical Engineering, Stanford University, Stanford, CA

The emergence of pandemic H1N1 subtype influenza in 2009 emphasizes the need for rapid methods to manufacture large quantities of highly potent influenza vaccines, which would broadly protect against multiple influenza virus strains.

All commercial influenza vaccines are produced by propagating the virus in embryonated chicken eggs. This technology is slow and requires one embryonated egg per vaccine dose. Cell-free protein synthesis (CFPS) system developed recently in the Swartz lab now allow us to produce large amounts of vaccine proteins at short notice during an epidemic pandemic.

We developed highly potent vaccines for seasonal and pandemic influenzas, which were based on a conserved region of the hemagglutinin (HA) protein. Current commercial vaccines are effective against only a narrow range of influenza virus strains. It is the reason why new vaccines must be generated and administered each year. The novel vaccines in this study would broadly protect against multiple influenza strains. Different methods have been tried to express and get correctly folded trimeric HA stalk domains.

To elicit strong immune responses, we further covalently attached trimeric HA proteins and other immunostimulatory molecules into immune-stimulating virus-like particles (VLPs), using click chemistry. CFPS was used for the production of VLPs with incorporated non-natural amino acids for covalent attachment of other proteins. These VLP constructs containing trimeric HA offer the potential for highly efficacious and broadly protective influenza vaccines.

See more of this Session: Advances In Protein Structure, Function and Stability II
See more of this Group/Topical: Food, Pharmaceutical & Bioengineering Division

(65d) Designing Viral Capsid Nanoreactors for the Study of Compartmentalized Processes

Danielle Tullman-Ercek, Chemical and Biomolecular Engineering, University of California, Berkeley, Berkeley, CA and Jeff Glasgow, Chemistry, University of California, Berkeley, Berkeley, CA

Bacterial microcompartments (BMCs) are bacterial organelles comprising a large icosahedral protein shell that houses metabolic processes, and have great potential to serve as scaffolds for both *in vitro* and *in vivo* nanoreactors. For such applications, it is critical that we are able to control the diffusion of small molecules into BMCs. In fact, this may be the primary natural function of the BMC; it is speculated that the protein shell of one BMC, the carboxysome of photosynthetic cyanobacteria, serves to enable an increased local CO_2 concentration near the carbon-fixation enzyme, and sequester that enzyme from oxygen. Due to the difficulty of working with carboxysomes, this theory has gone untested. Here, we will discuss our work on developing the well-characterized MS2 viral capsid as a model system to investigate protein membrane permeability. The capsid is spherical with icosahedral symmetry, and has 1.8 nm pores at its 5-fold and quasi-6-fold axes. When expressed recombinantly in *E. coli* 180 monomers of capsid protein assemble into intact capsid that is stable to a wide range of pH's and denaturants, in addition to mutations and chemical modifications. We hypothesize that the pores of the capsid (and BMC shell) are the primary control point for small molecule diffusion. To investigate this idea, we are working towards encapsulating enzymes in the capsid while varying the charge around the pore, so that we can use enzymatic activity as a readout of diffusion. This encapsulation requires disassembly of the capsid followed by reassembly around a specific nucleic acid, which is conjugated to our protein of interest. With this presentation, we will describe 1) the effect of various nucleic acids on MS2 reassembly, 2) how we greatly enhanced reassembly yields with the addition of certain small molecules, 3) the bioconjugation methods used to attach the nucleic acids to our model protein, and 4) effects of pore-residue alterations on capsid reassembly.

See more of this Session: Advances In Protein Structure, Function and Stability II
See more of this Group/Topical: Food, Pharmaceutical & Bioengineering Division

(65e) Endothelial Modulation of Adiponectin Oligomer Transport and Function

Joseph M. Rutkowski and Philipp E. Scherer, Touchstone Diabetes Center - Department of Internal Medicine, UT Southwestern Medical Center, Dallas, TX

Adiponectin has exhibited positive effects in regulating of systemic metabolism and insulin sensitivity through tissues such as liver, skeletal muscle, and centrally in the brain. With the ability to form high molecular weight (HMW) oligomers and circulating molecules of 30 to potentially 1080 kDa, adiponectin must cross the endothelium to have metabolic effects. Our laboratory has reported differential efficacy of these oligomers on insulin sensitivity and clinical data suggest that the HMW form is most important for metabolic health. Recent data has also demonstrated oligomer size-dependence on adiponectin clearance. Here, we have produced all oligomers of adiponectin in mammalian cells and used FPLC to fractionate adiponectin by molecular size. We have measured the Stokes radii of adiponectin to be approximately 5, 10, and 25nm for the trimeric, hexameric, and high molecular weight (HMW) forms, respectively. These molecular transport sizes, particularly that of the HMW, present transport limitations across endothelial barriers with low intercellular permeabilities such as in muscle or in the brain. Model murine MS1, bEnd.3, and EOMA endothelial cell lines were used in *in vitro* transwell permeability studies to calculate the relative transport of oligomers across different endothelium. We demonstrate size-dependent exclusion of adiponectin oligomers in MS1 and bEnd.3 cells, and in EOMA cells, a microvasculararture model, equivalent transport of trimeric and HMW adiponectin for their diffusive sizes. Each cell type had notable differences in the amount of intracellular adiponectin at the experiments' end; expression of adiponectin receptors correlated with internalization. Cells transfected with T-cadherin demonstrated increased surface binding. In total, these findings suggest active adiponectin transport mechanisms. *In vivo*, increasing vessel permeability pharmacologically, or through inducible transgenic overexpression of VEGF, reduced the circulatory half-life of adiponectin. These data demonstrate that the effects of different adiponectin oligomers on tissue metabolism may depend on the distinct endothelial character of a given tissue.

See more of this Session: Advances In Protein Structure, Function and Stability II
See more of this Group/Topical: Food, Pharmaceutical & Bioengineering Division

(65f) Engineered, Peptide BINDING Chaperones for Membrane PROTEIN Crystallization

Jennifer A. Maynard[1], Jennifer C. Pai[1], Kevin Enztminger[2] and Raquel L. Lieberman[3], (1)Chemical Engineering, University of Texas, Austin, Austin, TX, (2)Biochemistry, University of Texas, Austin, Austin, TX, (3)Chemistry & Biochemistry, Georgia Institute of Technology, Atlanta, GA

Hydrophobic membrane proteins perform a variety of important functions in the cell, but their structures are notoriously difficult to solve. Thus, new strategies to obtain crystals of membrane proteins for structure determination are critical. We aim to develop a toolbox of orthogonal single-chain antibody fragments engineered for hyper-crystallizability and specific recognition of short peptides. These peptide sequences can be introduced into flexible internal loops or at the termini of membrane proteins without interfering with protein function. The resulting scFv-membrane protein complex is expected to form a crystal lattice mediated by chaperone interactions, resulting in high-resolution crystal structures.

We have developed candidate scFv chaperone proteins binding hexa-histidine, EYMPME (EE) and FLAG tags. The scFv's were engineered via site-directed and random mutagenesis in conjunction with phage display to improve biophysical features influencing crystallization propensity, including peptide affinity and specificity, stability, solubility and expression level. Selected variants exhibit high solubility (up to 16.6 mg/ml) and nanomolar peptide binding affinities; complexes of one chaperone with the signal peptide peptidase integral membrane proteins harboring an internal EE tag has been isolated by gel filtration. Three of these scFv chaperones have been crystallized. The 3.1 Å resolution structure of this chaperone reveals a binding surface complementary to the EE peptide and a ~52 Å channel in the crystal lattice. These engineered scFvs represent a new class of chaperones that may eliminate the need for de novo identification of candidate chaperones from large antibody libraries.

See more of this Session: Advances In Protein Structure, Function and Stability II
See more of this Group/Topical: Food, Pharmaceutical & Bioengineering Division

(65g) Enhanced Affinity of An Insulin Degrading Enzyme Mutant towards Insulin

Bilal Cakir, Halil Kavakli and Seda Kizilel, Chemical and Biological Engineering, Ko University, Istanbul, Turkey

Insulin-degrading enzyme (IDE), Zn $^{+2}$ metalloprotease, is key enzyme in the catabolism of both amyloid beta (Aβ), and insulin. IDE is known to exist as an equilibrium mixture of monomers, dimers and higher oligomers with dimer being in predominant form. Distal sites of IDE also have key roles in the catabolism other than catalytic site and exosite. Therefore, the identification of distal sites which have roles in the activity of IDE is crucial for the development of drugs to regulate the activity of IDE. We generated mutant IDE by random mutagenesis with the aid of Taq polymerase error rate. The mutation T797A increases the affinity of the enzyme for insulin binding with a Km value of 47.5 nM compared to the wild type IDE Km value of 87.1 nM. This mutant IDE has also been observed to have lower catalysis towards the degradation of substrate V by about 20 %.This site represents new regions that may take part in the activity of IDE, and shows the possibility of modulating the proteolytic activity of IDE at sites other than its catalytic site or exosite.

See more of this Session: Advances In Protein Structure, Function and Stability II
See more of this Group/Topical: Food, Pharmaceutical & Bioengineering Division

Practical Applications of QbD for a Parenteral Drug Product

Nancy J. Harper, Venkat Koganti and Gautam R. Ranade, Pharmaceutical Research and Development, Pfizer, Groton, CT

This presentation addresses the application of Quality by Design principles to two common unit operations encountered in the manufacture of parenteral products. Examples will be described for 1) the application of mixing models to facilitate the scale-up of the API dissolution step, and 2) an approach for establishing a sterilization design space based on chemical kinetics and sterilization theory.

API Dissolution During Solution Compounding

In this example, mixing conditions to achieve complete dissolution of an API are dependent on several process variables, including API particle size, temperature, agitation rate, agitation time, and tank/agitator configuration. The fluid dynamics of the mixing vessel are scale-dependent, but can be modeled to enable predicted mixing conditions on scale-up. The approach taken in this example was to develop a lab-scale mixing vessel that was geometrically similar to the production vessel. The key aspect ratios of the production equipment (impeller diameter/tank diameter, liquid height/tank diameter, distance between impeller/tank diameter) were reproduced in a lab-scale mixer. Based on fluid dynamic principles, maintaining a similar power per unit volume (P/V) during mixing at either scale will result in similar solid-liquid mass transfer coefficient and therefore similar mixing times to achieve dissolution. Determinations of P/V were made using computational fluid dynamic (CFD) modeling of the mixing systems. Dissolution studies were performed in the lab-scale mixer to evaluate the combined influences of agitation rate and batch temperature on dissolution time, using worst-case API (a batch with particle size parameters near the maximum control limits). Using CFD modeling to determine agitation rates that will achieve the same P/V in the commercial tank, along with a constant scaling factor statistically determined during development of the model for this tank geometry, predicted agitation rates were determined that would achieve complete dissolution within the same time interval as observed in lab-scale. The suitability of the predicted agitation rates was determined for one full-scale batch, intentionally manufactured at the lower limit of the temperature range, representing worst-case conditions for dissolution. Successful dissolution within the target time interval served to confirm the model predictions and verify the minimum agitation rate and time for commercial production.

Terminal Sterilization

The molecule for this example is heat sensitive, and when exposed to pharmacopoeial (overkill) sterilization conditions (15 minutes at 121 °C), the extent of degradation exceeds an acceptable impurity level. Therefore a temperature below 121°C was evaluated for the sterilization cycle. Both microbiological lethality and degradant formation are directly dependent on cumulative thermal exposure, and therefore sterilization conditions are well suited for the development of a design space. To establish the design space, sterilization temperature and exposure time were explored in terms of conditions that simultaneously provide 1) a minimum $\log_{10}$ reduction of 8 (assuring sterility), and 2) maximum degradant formation according to the specified limit (assuring appropriate purity).

This approach to the establishment of a design space is unique in that it allows the use of first principles (chemical kinetics and sterilization theory) to define acceptable boundaries for process parameters. While some experimentation is required (i.e., determination of degradation rate constants and sterilization D-value), these studies are scale-independent and can be conducted at the lab bench. The experimentally determined constants are incorporated into theoretical equations that are used to define the boundaries of the design space.

The degradation rate constants were determined in development studies by traditional chemical kinetic experimentation utilizing drug product exposed to elevated temperatures in the range of sterilization conditions (above 100°C). Rate constants were determined at worst-case conditions of pH and headspace oxygen content based on targeted in-process control limits, thus representing worst-case stability conditions within the formulation design space. The Arrhenius equation was then used to define a time/temperature profile that results in a pre-defined maximum limit of degradation. For sterilization requirements, the D-value of a thermally resistant spore-former (*Geobacillus stearothermophilus*) was determined in the drug product solution. Based on sterilization theory, a time/temperature profile was derived that would result in an 8 log reduction of microorganisms. Superposition of these two time/temperature profiles then defines the design space, shown in Figure 1.

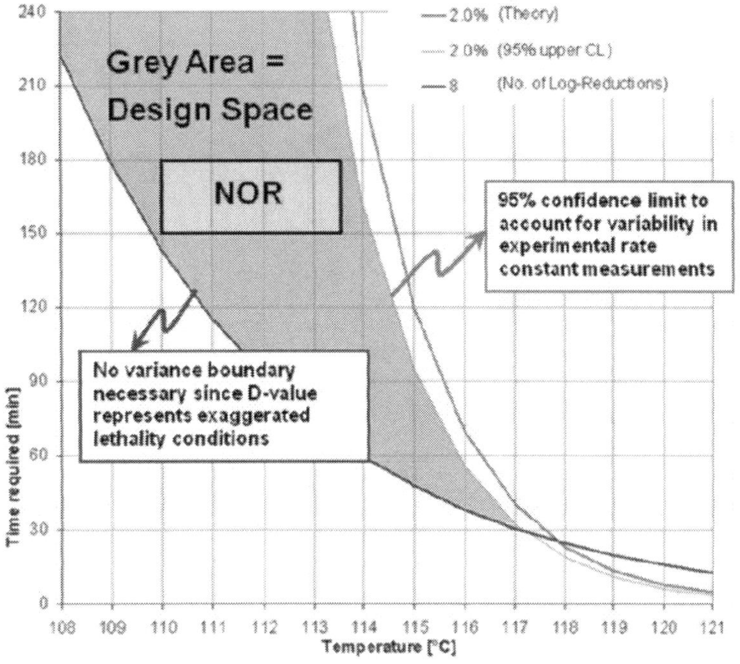

Figure 1. Moist Heat Sterilization Design Space and Normal Operating Ranges

The design space definition includes the 95% confidence limits for degradation. Statistical variance considerations for the lethality input (log-reduction) were not applied since the lethality experiments (D value generation) were performed with highly heat-resistance spores (*Geobacillus stearothermophilus*) to reflect exaggerated conditions.

Acknowledgements

Mixing Model: Fred Carroll, Rich Ferraina, Mark Berry, Yang Liu, Ken Norris, Robert Leasure, Jeff Gaudio (Pfizer); Rick Falk, Yogesh Waghmare (Bend Research)

Sterilization: EFPIA Mock P2 Injectables group: Heinz Wiederkehr (Roche), Thomas Backensfeld (Bayer Schering), Ritva Haikala (Orion Pharma), Christian DeMuynck (Nycomed), Brian Withers (Abbott)

(68b) Rational Design and Scale-up of Batch Lubrication Processes for Immediate Release Formulations

Joseph Kushner IV, Daniel O. Blackwood, Bruce C. MacDonald, Mark Polizzi, Francis M. Moore and Bruno C. Hancock, Pharmaceutical Development, Pfizer Worldwide Research and Development, Groton, CT

Purpose: To provide stepwise guidance for lubrication of pharmaceutical powders through the development of a unifying model which captures lubrication effects during blending and compression.

Methods: Literature models of powder properties following lubrication blending over a range of scales and conditions were reviewed. Additionally, a new model was developed by lubricating pharmaceutical powders in a tablet press feed frame. Lubricated powders were compressed on a development-scale rotary tablet press (Kilian T100) with varying throughput and feed-frame paddle rotation speeds. The resulting tablets were measured for tensile strength and a feed frame lubrication sensitivity model was developed. The two models were then combined to create an integrated lubrication understanding framework.

Results: Lubrication process design and scale-up guidelines were developed that addressed the selection of the appropriate amount of lubricant, the impact of equipment design (scale & principle of operation), and the need to select processing conditions that will result in the consistent manufacture of product possessing the desired extent of lubrication. To provide quantitative guidance to this end, a new unified model was developed which captured the effect of lubrication on pharmaceutical powders during blending and compression.

Conclusions: The proposed stepwise approach outlines the current state-of-the-art for conventional tablet blend lubrication processes, and embodies the principles of Quality-by-Design and risk-based drug product development. As such it is recommended as the starting point for scientists undertaking the development of immediate release tablet formulations.

See more of this Session: Application of Quality by Design In Drug Product Process Development I
See more of this Group/Topical: Topical I: Comprehensive Quality by Design in Pharmaceutical Development and Manufacture

(68c) Automation and Control of Drug-On-Demand Technology

Laura Hirshfield, Arun Giridhar, Gintaras V. Reklaitis, Venkat Venkatasubramanian and Michael T. Harris, Chemical Engineering, Purdue University, West Lafayette, IN

After the implementation of initiatives such as PAT and QbD, there has been a drive towards efficient manufacturing in the pharmaceutical industry. The concept of "mini-manufacturing" of drugs, as opposed to mass production of drugs, has shown promise in this area. We describe one such "mini-manufacturing" process, "drug-on-demand," which uses drop-on-demand printhead technology to deposit active pharmaceutical ingredients (API) onto edible substrates. This method also allows for the layering of different APIs and the creation of personalized dosage forms, in which the amount of API can be varied depending on the patient. In this work, we focus on the automation and control strategy for the drug-on-demand pilot facility done as part of the Engineering Research Center for Structured Organic Particulate Systems. A proper control strategy is necessary so that the system can run at optimal setpoints and adjust in real-time, allowing for an efficient manufacturing process and a precise on-spec drug product. The control strategy for our "drug-on-demand" setup allows us to execute an automated, optimized, and controlled print cycle while closely monitoring drop size, drug morphology, and drop deposition pattern.We summarize our achievements and current research in this area.

See more of this Session: Application of Quality by Design In Drug Product Process Development I
See more of this Group/Topical: Topical I: Comprehensive Quality by Design in Pharmaceutical Development and Manufacture

Dynamic simulation of the dissolution of pharmaceutical tablet integrated with continuous manufacturing process

Presenter: Lukasz Mioduszewski

Research Participants: Fani Boukouvala, Marianthi Ierapetritou, Rohit Ramachandran

ENGINEERING RESEARCH CENTER FOR
STRUCTURED ORGANIC PARTICULATE SYSTEMS
RUTGERS UNIVERSITY
PURDUE UNIVERSITY
NEW JERSEY INSTITUTE OF TECHNOLOGY
UNIVERSITY OF PUERTO RICO AT MAYAGÜEZ

Scientific Goals

- Develop a dynamic model of dissolution of pharmaceutical tablets as part of a dynamic flowsheet simulation of a continuous tablet manufacturing process
- Correlate operating and design parameters of upstream processes to properties of tablet dissolution through the integration of processing units
- Develop a model for tablet dissolution using Fick's Second Law with a non-constant diffusion coefficient

Introduction

Flowsheet simulation:

❑ Flowsheet simulation is used for **process design**, dynamic optimization and flexibility analysis of complex large-scale integrated systems.
❑ Application of simulators for solid processes is still very primitive compared to state of the art for fluid processes.
❑ Through dynamic flowsheet simulation of an integrated tablet manufacturing line, operating conditions and design of the system can be linked to properties of final product
❑ The effects of possible perturbations during operation can be and the necessary control scheme to alleviate them can be identified

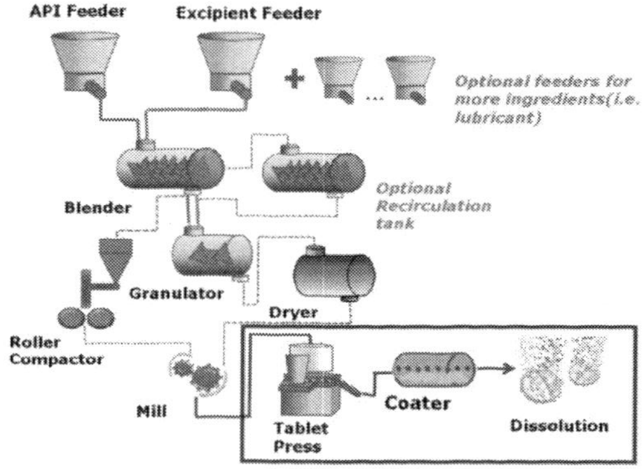

Tablet Dissolution:

A tablet is ingested and passed through the esophagus to the stomach, which is an aqueous environment. This is the first place where the tablet may dissolve. The rate of dissolution is a key target for controlling the duration of a drug's effect.

Methods:

Representation of tablet:

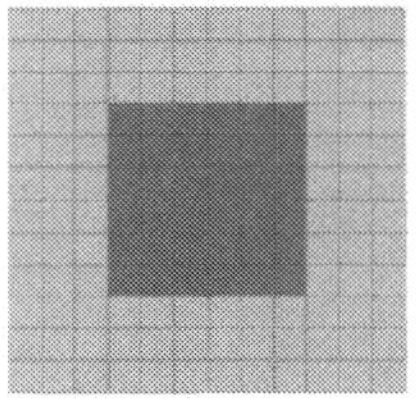

We initialized the void fractions with the equation for porosity. We assigned the center of the tablet a value of void fraction 20% larger than that for the outer tablet volume

Governing equations dissolution:

Fick's Second Law

$$\frac{\partial c_i}{\partial t} = -\nabla \cdot (-D_i \nabla c_i) + s_i \quad i = A,B$$

where

Source term

$$s_i(\mathbf{x}) = \begin{cases} k_i(c_i^{sat} - c_i(\mathbf{x})) & \forall \mathbf{x} : \phi_i(\mathbf{x}) > 0 \\ 0 & \text{otherwise} \end{cases}$$

Diffusion coefficient

$$D_i = D_{i0}\left(1 - \sum_{j=1}^{n} \phi_j\right)^{\alpha_i}$$

Evolution equation

$$\frac{\partial \phi_i}{\partial t} = -\frac{s_i}{\rho_i}$$

Where:
C_i is the concentration of component i
D_i is the diffusion coefficient
s_i is the source term
A,B are components (API and Excipient)
k_i is the dissolution rate constant
φ is the phase volume fraction (porosity)
α is diffusivity exponent
ρ is density

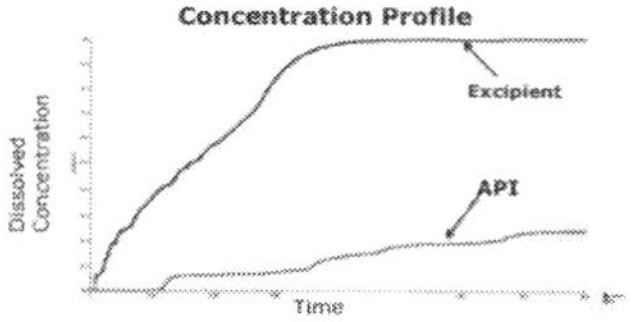

Integration of tablet press and dissolution

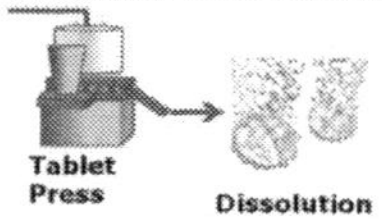

Correlation between compression force (F) of tablet press to porosity of tablet

φ= 0.5921*EXP(-0.0002*F)

Developed from empirical analysis (Nokhodchi, A. Ford, J. Rowe, P. Rubinstein, M.)

Correlation between mean particle size and Diffusion coefficient

$D_{i0} = kT / 6\pi\eta R_H$

Boltzmann constant; Temperature; Viscosity; Particle radius

Brownian Motion

Results:

Graphical Profiles of Results:

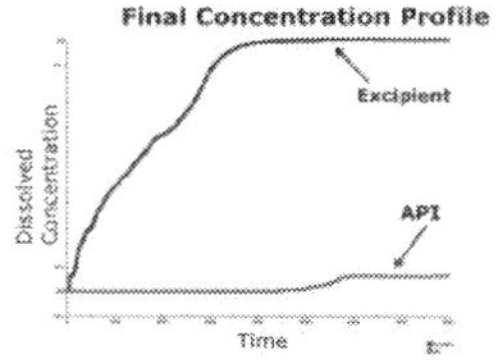

The concentration profiles for both API and Excipient increase until entire quantity is dissolved

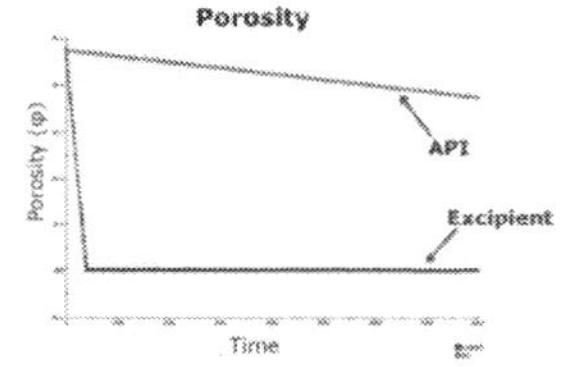

The porosity tends to zero for both API and Excipient. Profile for API tends to zero slowly because API volume fraction is initially higher than that of Excipient. Also, the API is located in the middle of the tablet, therefore its dissolution takes longer

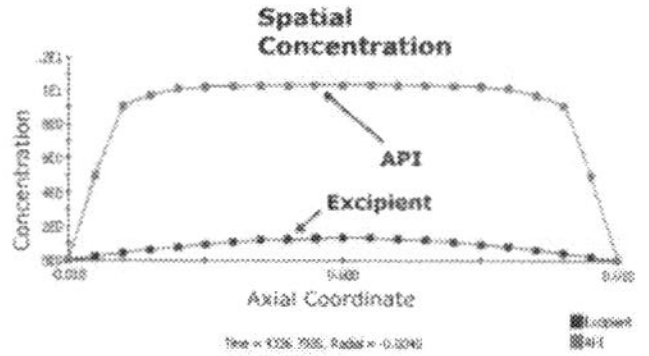

The spatial concentration profile of the tablet is shown. The concentration is maximum in the middle of the tablet and tends to zero when approaching the edges of the tablet

Conclusions

1. Dissolution model is the last step in the flowsheet simulation and determines many important properties of the final product
2. Tablet press and dissolution models were connected using Brownian motion and empirical data available for hydroxypropylmethylcellulose in regard to the compaction properties and void fractions
3. Flowsheet simulation of integrated system can help identify optimum design, operating conditions and sequence of tablet manufacturing process in order to produce high quality product with consistency

Future Work

1. Conducting experiments with the actual tablet press and obtaining real data for a specific pharmaceutical component
2. Modifying model parameters for a specific drug
3. Modifying the empirical equations in the model to customize the flowsheet for particular compounds

(68e) Innovative Techniques for Planning Experiments Across Mulitple Unit Operations

Kelly Canter, Pharmaceutical Sciences, Pfizer Inc., St. Paul, MN and Leonard Perry, University of San Diego, San Diego, CA

Many well-established analytical tools and methodologies exist for developing and characterizing a product for manufacture. However, the application of these tools all too often can fall short of their objective or full potential to gain process understanding – even in the hands of a well trained chemometrician or statistician. Another essential and uncommon taught skill is in the planning of experiments over the lifecycle of a products development and subsequent manufacture. The selection of the proper experimental approach and planning activities that precede the actual experimentation are critical to successfully establish process understanding. The application of proper risk management tools along with established experimental planning principles aide significantly in effectively establishing a product's experimental development plan. There are well documented sources that aid in the planning and selection of designed experiments to achieve a desired outcome, especially when looking at a single unit operation at a time. Yet minimal attention is given to the planning and sequential approach required for experimentation across multiple units operations. This paper highlights a set of tools and guidelines for assessing technical issues and establishing an appropriate experimental plan when looking across a products manufacturing process. A pharmaceutical case study involving the experimental strategy taken to develop the design space for a Quality-by-Design filing is provided, and the trade-offs between resource limitations and statistical consequences is included.

See more of this Session: Application of Quality by Design In Drug Product Process Development I
See more of this Group/Topical: Topical I: Comprehensive Quality by Design in Pharmaceutical Development and Manufacture

(68f) Real-Time Process Management and Knowledge Management for Continuous Pharmaceutical Manufacturing

Girish Joglekar, **Arun Giridhar**, Gintaras V. Reklaitis and Venkat Venkatasubramanian, Chemical Engineering, Purdue University, West Lafayette, IN

The manufacture of pharmaceutical drug products has benefited significantly from a QbD approach, including the enabling of continuous production over batch processing. Developing and improving processes for continuous manufacture requires the combined use of process data with process and formulation knowledge. Process data can be generated across multiple unitops simultaneously, and correcting deviations in one critical process parameter or critical quality attribute may require manipulations of a totally different process unit. Such integrated control must be done in a manner that is both easy to scale to different sizes and easy to transfer to different materials, requiring an integrated approach in gathering and managing process knowledge and using it flexibly for new processes.

In this work, we present the results and case studies from the Engineering Research Center for Structured Organic Particulate Systems (ERC-SOPS), in developing a real-time process management system for continuous pharmaceutical manufacture, that makes use of an ontological informatics framework to manage process knowledge of multiple types, including material properties (such as stress-density relationships), process parameters (such as process setpoints), equipment parameters (such as die and punch sizes in a tablet press), and process control knowledge (mathematical models for PID and model predictive control, process signatures for fault diagnosis, mathematical models to plan future setpoint changes, etc). The models taken together operate on a wide range of time scales, and the knowledge management framework is designed to be flexible for all uses. Further, we show how the system architecture allows for the use of various control systems and modeling programs in a flexible manner. Finally, we present case studies in tablet manufacturing and drop-on-demand systems, and summarize our current results.

See more of this Session: Application of Quality by Design In Drug Product Process Development I
See more of this Group/Topical: Topical I: Comprehensive Quality by Design in Pharmaceutical Development and Manufacture

(75a) Collagen Binding Domain Fusion Proteins for Therapeutics: Expression and Isolation

McKinzie Fruchtl[1], Hannah Moreland[2], Robert R. Beitle[1] and Joshua Sakon[3], (1)Ralph E. Martin Department of Chemical Engineering, University of Arkansas, Fayetteville, AR, (2)Department of Biological and Agricultural Engineering, University of Arkansas, Fayetteville, AR, (3)Department of Chemistry and Biochemistry, University of Arkansas, Fayetteville, AR

Interest in the production of recombinant proteins consisting of collagen binding domain (CBD) fused to a bioactive material has increased due to the targeting / attachment capabilities of CBD. For example, CBD fusions can be applied to reversing of bone density loss and repair of eardrum, specifically, by choosing an appropriate fusion partner (parathyroid hormone or epidermal growth factor). The production of CBD fusions was examined using batch and fed batch culturing of *Escherichia coli* to express the fusion proteins, and affinity chromatography to isolate the final product. Different medium formulations and induction strategies were tested in order to develop a production strategy lacking yeast extract or other difficult-to-validate materials, and various methods of recovery were verified to be effective. Data indicated that high productivity of these novel CBD fusion proteins can be achieved for use during cancer treatment and tympanoplasty, respectively.

See more of this Session: Biomolecular Engineering
See more of this Group/Topical: Food, Pharmaceutical & Bioengineering Division

(75b) Identification of Macromolecular Designs for Inhibiting the Atherogenic Effect of Endothelial Scavenger Receptor, LOX-1

Kubra Kamisoglu[1], Dawanne Poree[2], Sarah M. Sparks[3], Kathryn E. Uhrich[2], William Welsh[4] and Prabhas V. Moghe[1], (1)Chemical and Biochemical Engineering, Rutgers University, Piscataway, NJ, (2) Chemistry and Chemical Biology, Rutgers University, Piscataway, NJ, (3)Chemistry, Stonehill Collage, Easton, MA, (4)Pharmacology, University of Medicine and Dentistry of New Jersey, Piscataway

Scavenger receptors (SRs) are among the key cell surface proteinsrelated to the progression of atherosclerosis. SRs mediate unregulated uptakeof oxidatively modified forms of lipoproteins which initiate the pro-inflammatorysignaling cascade and cause excessive lipid accumulation in vascular cells. Vascular endothelial cells predominantly express lectin like oxidized LDL(oxLDL) receptor (LOX-1), a member of the scavenger receptor family. OxLDLrecognition and subsequent internalization by LOX-1 has been shown tocontribute to all stages of atherosclerosis by 1) promoting reactive oxygenspecies production, 2) activating proinflammatory signaling, 3) impairing theregulation of vascular tone, and 4) inducing apoptosis. Therefore, LOX-1 is a promisingtarget for atherosclerosis therapies currently under development.

Previously our laboratories demonstrated that a class of amphiphilicmacromolecules can competitively inhibit oxLDL uptake by blocking macrophagescavenger receptors. These molecules consist of a sugar (mucic acid) backbone thatis covalently linked to four aliphatic arms and poly(ethylene glycol). Usingmolecular modeling, by docking and scoring approach, a new library ofmacromolecules has been screened for efficiency of binding to LOX-1. Thelibrary includes amphiphilic macromolecules featuring different backbone and/orcharge characteristics in an effort to identify the key molecular properties thataffect binding to LOX-1. Molecules were built *in silico* and docked tooxLDL binding domain of LOX-1. Binding energies were calculated by evaluatingthe highest ranking poses. Modeling results were compared with *in vitro* competitiveoxLDL uptake inhibition experiments utilizing transfected CHO cells that haveinducible expression of human LOX-1.

Binding energy values obtained *in silico* correlated wellwith *in vitro* results. Molecules with the most favorable LOX-1 binding(lowest binding energy) exhibited most efficient inhibition of oxLDL uptake. Hydrophobicinteractions played a more significant role in binding to LOX-1 than to chargepresentation. Due to the multi-faceted importance of LOX-1 in atherogenesis andathero-inflammation, the exhibited inhibition of oxLDL binding to LOX-1 is animportant step towards the development of novel macromolecular based therapies.

See more of this Session: Biomolecular Engineering
See more of this Group/Topical: Food, Pharmaceutical & Bioengineering Division

(75c) The Missing Links: Elucidating Mechanisms of Heterologous GPCR Expression and Trafficking In S. Cerevisiae

Carissa L. Young[1], Zachary T. Britton[1], Emily C. McCusker[2], Jeffrey Caplan[3], Kirk J. Czymmek[3] and Anne S. Robinson[1], (1)Department of Chemical Engineering, University of Delaware, Newark, DE, (2) Department of Crystallography, Institute of Structural and Molecular Biology, London, United Kingdom, (3)DBI BioImaging Center, Delaware Biotechnology Institute, Newark, DE

Within their native host, Gprotein-coupled receptors (GPCRs) are ubiquitously expressed, and assist thecell in querying the extracellular environment, initiating cellular responses todiverse sensory and chemical stimuli. Consequently, this large superfamily ofmembrane proteins that consist of seven transmembrane domains destined for theplasma membrane of eukaryotic cells regulate most physiological processes. Inorder to reach the plasma membrane, these proteins must navigate the secretorypathway, where they are translocated into the ER, properly fold, and undergopost-translation modifications en route to the cell surface. In recent years,it has been shown that the production of GPCRs in yeast is hindered by hostcellular responses, including the unfolded protein response (UPR). Conventionally, the UPR is believed to be triggered when the folding capacityof the endoplasmic reticulum (ER) is exceeded. We hypothesize that the UPR ispredominantly induced by an abundance of heterologously expressed GPCRs in theER, and not due exclusively to improperly folded proteins, which yields animproper balance of protein biosynthesis/maturation or defects in secretorypathway trafficking.

To improve functional production(i.e. ligand-binding yields indicative of active receptors) of GPCRs, we havegenerated a versatile yeast expression cassette designed to optimizetranscription/translation rates; analyze protein structure/function; andincorporate multiple tags for identification and purification. To evaluatesub-cellular localization of GPCRs, we created fluorescent protein variants andcodon-optimized fluorophores of organelle targets analyzed by four-colorimaging using live-cell confocal microscopy, cryo-TEM,and correlative imaging techniques. Time course analysis, quantitative PCR, co-immunoprecipitation of select proteins, andyeast deletion strains in combination with novel high-resolution imagingtechniques have shown differences in GPCR trafficking and quality control initiation, including the UPR, autophagy, and ER associated degradation (ERAD) pathways. Wehave evaluated GPCR expression profiles; optimized conditions to minimize UPRinduction; determined colocalization with organelles and sub-compartments; andconfirmed the activity of human adenosine receptors (i.e. hA_2aR, hA_1R,hA_2bR, hA_3R). Furthermore, we engineered domains andidentified motifs altering localization and functional production by generatingrational chimeric receptors. We continue to analyze the expression,trafficking, and activity of additional GPCR families, including the neurokininreceptors (hNK_1R, hNK_2R, and rNK_2R) in orderto determine the generalizability of our observations.

(75d) Computational and Experimental Study of Palmitate Binding to the IRE1 Protein

Hyun Ju Cho[1], Liang Fang[2], Pratheeba Palasuberniam[2], Chris Bush[1], Michael Feig[2] and Christina Chan[3], (1)Chemical Engineering and Materials Science, Michigan State University, East Lansing, MI, (2)Biochemistry and Molecular Biology, Michigan State University, East Lansing, MI, (3)Chemical Engineering and Materials Science; Biochemistry and Molecular Biology, Michigan State Uiversity, East Lansing, MI

IRE1 (inositol-requiring enzyme 1) is an ER (endoplasmic reticulum) transmembrane Ser/Thr protein kinase/endoribonuclease involved in the unfolded protein response (UPR), and is an adaptive intracellular signal transduction pathway induced upon ER stress. Activation of IRE1 by dimerization of the N-terminal luminal domain induces autophosphorylation of its cytosolic kinase domain, thereby activating its C-terminal nuclease domain. Palmitate, a saturated fatty acid, has been recognized to induce ER stress, and as an IRE1 activator will upregulate UPR target genes, i.e. pro-apoptotic genes. Recently, our lab found that palmitate interacts with PKR (protein kinase R) to regulate the PKR kinase activity. Given the similarity in the back-to-back kinase domains of PKR and IRE1, we hypothesize that palmitate also directly binds to the IRE1 protein. We confirmed, using a fluorescence polarization (FP)-based binding assay, that palmitate indeed interacts with the cytosolic domain containing the kinase and nuclease domains, but not the luminal domain of IRE1. To identify potential palmitate-binding sites, we performed molecular dynamics simulations. The computational results suggest that palmitate locates near several sites (i.e. the activation loop and ATP binding site in the kinase domain and the nuclease active site in the nuclease domain) which control IRE1 enzymatic activities. Site-specific mutation experiments further confirmed that the predicted residues are important for palmitate binding and in turn IRE1 enzymatic function. The computationally assisted biochemical studies help confirm the molecular mechanism by which palmitate directly activates the IRE1 functions as well as possibly the mechanism by which palmitate initiates ER stress through the UPR sensor proteins.

See more of this Session: Biomolecular Engineering
See more of this Group/Topical: Food, Pharmaceutical & Bioengineering Division

(75e) A Library of Inducible Synthetic Devices for Phenotype Regulation In Prokaryotes

Katherine G. Volzing, Konstantinos Biliouris and Yiannis Kaznessis, Chemical Engineering and Material Science, University of Minnesota, Minneapolis, MN

Implementing an original workflow, we have designed, constructed and characterized a library of new synthetic molecular devices to achieve an assortment of inducible gene expression phenotypes in prokaryotes. We initially focused our efforts on two simple synthetic transcription activators, prokaryotic-TetOn and prokaryotic-TetOff. These synthetic devices switch gene expression on in response to the presence or the absence of anhydrotetracycline (respectively). Using these devices, in combination with naturally occurring regulatory components, we subsequently designed, built and tested other synthetic devices to achieve a range of functionalities.

For each desired phenotype, we first selected the appropriate device type and the protein and DNA components to use. Next, we optimized the geometries of these components and the overall device using molecular modeling. Through this, we achieved an unprecedented combination of controllable and transactivating functions with each system. We then built the devices and tested their behavior using flow cytometry. Last, we implemented stochastic simulations to quantify the strength of the key biomolecular interactions arising in each system.

The unique functions of our systems were found to be robustly tight and covering a wide range of dynamic gene expression phenotypes. When induced, they activate gene expression many-fold relative to their uninduced states. The new devices presented here may become useful additions to the molecular toolboxes used by biologists to control bacterial gene expression. Our methodology may also be a foundation for the design, development and characterization of additional synthetic device libraries.

See more of this Session: Biomolecular Engineering
See more of this Group/Topical: Food, Pharmaceutical & Bioengineering Division

(75f) Designing Highly Active siRNAs From Asymmetry-Based Selection Algorithm Predictions

Amanda P. Malefyt[1], Ming Wu[2], Christina Chan[1] and S. Patrick Walton[1], (1)Chemical Engineering and Materials Science, Michigan State University, East Lansing, MI, (2)Computer Science and Engineering, Michigan State University, East Lansing, MI

In the development of RNA interference (RNAi) therapeutics, selecting siRNA sequences that complement the messenger RNA (mRNA) target does not guarantee silencing. Factors such as 5'-end stability are known to be critical for ensuring the correct strand is preferentially incorporated into the RNA induced silencing complex (RISC). Two methods for determining this asymmetry between strands are terminal sequence and relative terminal thermodynamic stability. Through the analysis of large siRNA databases, we have shown that highly active siRNA sequences are more likely to have large asymmetry between the sense and antisense 5'-ends in both end sequence nucleotides as well as thermodynamic stability.

We used this information to create an algorithm for predicting highly active siRNA sequences against desired proteins using only the mRNA sequence of the target. The algorithm uses end sequence and thermodynamic stability parameters, trained from existing siRNA activity databases, to rank the probability that an siRNA sequence has high, medium, and low activity for its target gene. We will discuss the applicability of the algorithm for predicting highly active sequences for enhanced green fluorescent protein, EGFP. Additionally, we will highlight comparisons between our technique and other selection approaches.

See more of this Session: Biomolecular Engineering
See more of this Group/Topical: Food, Pharmaceutical & Bioengineering Division

(75g) Streamlined Protocol for mRNA Display

Pamela A. Barendt, Casey N. McQuade and Casim A. Sarkar, Department of Bioengineering, University of Pennsylvania, Philadelphia, PA

In vitro directed evolution techniques such as ribosome display and mRNA display are powerful tools for protein engineering, capable of handling libraries containing trillions of members. mRNA display uses an mRNA-DNA-puromycin fusion as template for *in vitro* translation, which results in a covalent puromycin linkage between the mRNA and the corresponding nascent peptide, thus creating a highly stable selection particle which is useful for protein/peptide stability selection experiments. Our objective in this study was to establish a streamlined protocol for mRNA display based on a ribosome display protocol currently used in our laboratory. The mRNA display protocol is significantly streamlined and optimized, which should enable faster and easier directed evolution experiments, thus improving the *in vitro* development of therapeutic proteins and peptides. Advancements to the protocol included: ultrafiltration rather than gel-purification to purify the ligation reaction between mRNA and DNA-puromycin; use of PURExpress *in vitro* protein synthesis kit (New England Biolabs) rather than a lysate-based system to maximize the number of functional selection particles; and optimization of the conditions used to generate and purify the selection particles. PURExpress offers better reproducibility, lack of nucleases, and a high concentration of ribosomes. Relative yield of mRNA-peptide was determined by quantitative RT-PCR following affinity selection. We were able to produce consistent results with selection particles displaying a peptide (3xFLAG) or a protein (off7, a designed ankyrin repeat protein). The crucial steps at which efficiency may typically suffer, including ligation of mRNA to the DNA-puromycin linker and covalent binding of puromycin to the nascent polypeptide chain, have been found to be adequately efficient in our system. Ongoing experiments in our laboratory are utilizing this streamlined method to select peptide binders from a naïve library.

See more of this Session: Biomolecular Engineering
See more of this Group/Topical: Food, Pharmaceutical & Bioengineering Division

(77a) Micropatterned Growth Factor-Containing Surfaces to Study Hepatic Response to Injury

Dipali Patel, Caroline Jones and **Alexander Revzin**, Biomedical Engineering, University of California, Davis, Davis, CA

The goal of this project is to investigate the effects of growth factors (GF) during in vitro liver injury. Micropatterned surfaces were created by mixing GF such as HGF, BMP7 and TGF-β with collagen I in solution and printing this ECM/GF mixture onto silane modified glass slides. Primary rat hepatocytes were seeded on top of the GF/ECM arrays forming clusters corresponding in size to the underlying protein spots (300 to 500 mm). Cell arrays were then injured in culture by exposure to 100 mM ethanol for 48h. Hepatocytes residing on GF spots were found to have less apoptosis then cells cultured on collagen-only spots. Least apoptosis (0.3 % as estimated by TUNEL assay) was observed on HGF/BMP7/collagen spots whereas most apoptosis (17.3%) was seen on collagen-only arrays. Interestingly, the extent of alcohol-induced apoptosis in hepatocytes varied based on the concentration of printed GF. In addition to preventing apoptosis, printed GFs contributed to maintenance of epithelial phenotype during alcohol injury as evidenced by higher levels of E-cadherin expression in HGF-protected hepatocytes.

Importantly, GF microarrays could be used to investigate heterotypic interactions in the context of liver injury. To highlight this, stellate cells -nonparenchymal liver cells involved in fibrosis - were added to hepatocytes residing on arrays of either HGF/collagen or collagen-only spots. Exposure of these cocultures to ethanol followed by RT-PCR analysis revealed that stellate cells residing alongside HGF-protected hepatocytes were significantly less activated (less fibrotic) compared to controls. Overall, our results demonstrate that micropatterned surfaces and GF microarray format can be used to screen anti-fibrotic and anti-apoptotic effects of growth factors as well as to investigate how signals delivered to a specific cell type modulate heterotypic cellular interactions.

See more of this Session: Cell Adhesion and Migration I
See more of this Group/Topical: Food, Pharmaceutical & Bioengineering Division

(77b) Microfluidic Assay for Neutrophil Chemotaxis towards Chemoattractant Gradients

Caroline N. Jones and Daniel Irimia, Harvard Medical School, Charlestown, MA

Chemoattractants represent an essential group of molecular guidance signals that regulate the transit of leukocytes out of the mainstream of blood and into tissues at sites of inflammation. Neutrophils are uniquely sensitive to a vast array of chemoattractants including complement fragments (C5a), lipid mediators (Leukotriene B4 (LTB4)), a multitude of chemokines (Interleukin 8 (IL-8)), as well as exogenous mediators produced by pathogens (Formyl-Methionyl-Leucyl-Phenylalanine (fMLP)). Cell adhesion molecules, including fibronectin, E-selectin, and P-selectin, are also known to play an important roll in neutrophil capture from the bloodstream and subsequent transmigration into the target tissue during inflammation. The goal of this project is to validate a robust biomimetic assay for neutrophil chemotaxis towards gradients of chemoattractants and cell adhesion molecules using enclosed microfluidic channels previously developed by our group. These devices consist of non-planar, enclosed microenvironments of increasing complexity, from straight channels, to channels with posts and simple mazes. The cross-section of the channels (3 × 6 μ m) is smaller than the human neutrophils, so that cells are mechanically confined and allowed to move only along the axis of the channels toward a gradient of chemoattractactant and cell adhesion molecules.

For the study, neutrophils were isolated from the whole blood of at least six healthy donors and cells were loaded into an array of twelve microfluidic devices allowing a combinatorial analysis of four chemoattractants and three cell adhesion molecules simultaneously. Cell motility was recorded on a Nikon TiE inverted microscope with a heated tissue incubating chamber set at 37 °C. Individual frames recorded at 2 min intervals and the speed of at least fifty cells per condition was measured using ImageJ. In straight channels, neutrophils undergo sustained, unidirectional motion towards a chemoattractant source. In more complex maze-like geometries, neutrophils are able to select the most direct route over 78% of the time for all chemoattractants tested.

LTB4 [100 nM] was shown to be the most potent chemoattractant and neutrophil migration speed was increased two-fold compared to fMLP [100 nM] on fibronectin [25 nM] (41.1 μm/min vs. 19.2 μm/min). The device was also used to create dose-response curves with LTB4. Interestingly, neutrophil migration speed was shown to increase in a dose-dependent manner to LTB4 concentration. P-selectin [50 ng/mL] was also shown to have a significant negative affect on neutrophil migration decreasing speed to both fMLP [100 nM] (19.2 μm/min vs. 10.7 μm/min) and LTB4 [100 nM] (41.1 μm/min vs. 11.7 μm/min). In this study, we have validated a robust and easy to use microfluidic device for the analysis of neutrophil chemotaxis. In the future, this device will be used to investigate impaired neutrophil migration in burn and transplant patients.

See more of this Session: Cell Adhesion and Migration I

(77c) Neutrophil Migratory Response In Competing Endogenous Chemotactic Gradients

Ashish Kapoor[1], Yuki Kimura[1], Matthew Byrne[1], Yuan He[2], Fei Wang[2], Paul J.A. Kenis[1] and Christopher V. Rao[1], (1)Chemical and Biomolecular Engineering, University of Illinois, Urbana Champaign, Urbana, IL, (2)Cell and Developmental Biology, University of Illinois, Urbana Champaign, Urbana, IL

Neutrophils, the most abundant type of white blood cells, play an essential role in the primary stages of the cellular immune response. These cells efficiently sense and migrate up concentration gradients of chemical attractants toward sites of infection and inflammation in a process termed chemotaxis. Such chemoattractants include end-target chemoattractants produced at or proximal to the source of infection, as well as endogenous chemoattractants produced by inflamed host tissues. While much is known about the molecular interactions and signaling pathways that regulate cellular responses to individual cues, little is known about how cells process multiple cues to effect migration in the appropriate direction *in vivo*.

The aim of this work is to understand how neutrophils integrate and prioritize multiple chemotactic signals to navigate toward pathogens. We investigate neutrophil migration in response to multiple chemoattractant gradients through the generation of precisely controlled microenvironments using microfluidic platforms. By analyzing the migratory response of the cells in response to single and dual gradients of endogenous chemoattractants, we have begun to elucidate how neutrophils respond to multiple chemotactic cues. The peculiar chemotactic behavior of primary neutrophils in competing endogenous gradients suggests the role of directional persistence and cellular memory in guiding cells to their targets. These results provide insight into key mechanisms underlying inflammatory response and should aid in the design of novel immunotherapeutic strategies.

See more of this Session: Cell Adhesion and Migration I
See more of this Group/Topical: Food, Pharmaceutical & Bioengineering Division

(77d) A Combined Experimental and Theoretical Approach to Model Cell-Cell Interactions In Cancer Cell Migration

Saheli Sarkar, Case Western Reserve University, Cleveland, OH, Bethany Bustard, MS&T Analytical Forensics, Eli Lilly & Company, Indianapolis, IN, Jean F. Welter, Department of Biology, Case Western Reserve University, Cleveland, OH and **Harihara Baskaran**, Chemical Engineering, Case Western Reserve University, Cleveland, OH

Migration of cancer cells is a key determinant of metastasis, which is correlated with poor prognosis in patients. Evidence shows that cancer cell motility is regulated by stromal cell interactions. To quantify the role of homotypic and heterotypic cell-cell interaction on migration, a two-dimensional migration assay has been developed by microfabrication techniques. Two breast cancer cell lines, MDA-MB-231 and MDA-MB-453, were used to develop micropatterns of cancer cells (cell islands) that revealed distinct migration profiles in this assay. Although the individual migration rates of these cells showed only a seven-fold difference, MDA-MB-453 islands migrated significantly lower than MDA-MB-231 islands, indicating differential regulation of migration in isolated cells vs. islands. Island size had the greatest impact on migration, primarily for MDA-MB-231 cells. Migration of MDA-MB-231 islands was decreased by interaction with homotypic cells, and significantly more by heterotypic non-cancer associated fibroblasts. In addition, a mathematical model of island migration in multi-cellular population has been developed using Stefan-Maxwell's equation. The model showed qualitative agreement with experimental results and predicted a biphasic relation between cell densities and island sizes. The combined experimental and mathematical model can be used to quantitatively study the impact of cell-cell interactions on migration.

See more of this Session: Cell Adhesion and Migration I
See more of this Group/Topical: Food, Pharmaceutical & Bioengineering Division

(77e) Comparing Monocyte Adhesion and Transendothelial Migration Between a 3D Vascular Tissue Model and a 2D Cell Culture System Under Normal Conditions and Activated with TNF-&alpha

Neda Ghousifam, Krisada Leemasawatdigul and Heather Gappa Fahlenkamp, School of Chemical Engineering, Oklahoma State University, Stillwater, OK

Abstract:

Atherosclerosis is known as an inflammatory disease which is mainly initiated by accumulation of lipid substances in the subendothelial layer of major arteries, followed by adhesion and transmigration of monocytes. This adhesion and transmigration involves several steps such as slow rolling, adhesion strengthening, crawling, paracellular and transcellular migration, and transmigration through the basement membrane, which are mediated by many bioactive molecules. The bioactive molecules include cellular adhesion molecules (CAMs) and chemotactic cytokines (chemokines). CAMs are proteins expressed on the surface of the cells and are involved in the adhesion of monocytes to the endothelial cell layer. Monocyte Chemoattractant Protein-1 (MCP-1) is one chemokine that plays an important role in monocyte trafficking across the endothelial layer. Studies have shown that chemokine concentration gradients drive and possibly control cell migration. Many studies investigate the effect of chemokines, such as MCP-1 on monocyte migration by using a 2D experimental system. Endothelial cells are grown on a thin, microporous membrane, the chemokine is added to the aqueous media below the membrane, monocytes are added to the apical surface of the endothelial cells, and transmigration of the cells across the endothelial cell layer and the microporous membrane is observed. The 2D system may not adequately predict *in vivo* cell behavior due to the lack of the third dimension. A 3D tissue model consisting of a collagen matrix would be a better experimental model to mimic the subendothelial extracellular matrix (ECM). The 3D tissue model provides the added dimension that is important for the creation of a diffusive concentration gradient formed in the ECM, which is responsible for the control of many cellular mechanisms. The concentration gradient formed in the 3D model is distinctly different compared to the one in the 2D cell culture system, where the secreted factors from the endothelial cells dissolve quickly into the surrounding media. Apart from this advantage, when focusing on the transmigration of monocytes, this matrix can also provide an area for monocytes to localize and differentiate into macrophages after the transendothelial migration. We hypothesized that the chemokine concentration gradient within the ECM of the 3D tissue model will drive a different cellular response than that in the 2D system. The goal of this research was to measure the formation of MCP-1 concentration gradients within a 3D vascular tissue model and determine the effect on monocyte migration. The first objective was to characterize the effect of key cellular adhesion molecules on the surface of endothelial cells, known to be important for monocyte adhesion and transmigration. This was necessary in order to determine if the endothelial cell surface characteristics (kinetics, level of expression) are similar between 3D and 2D models. The experimental results for this part show that the CAMs expression of Human Aortic Endothelial Cells (HAECs) after stimulation with tumor necrosis factor-α to mimic inflammatory conditions is significantly different between the two systems. The next objective was to compare the monocyte adhesion and transmigration through the endothelial layer between the two models. Experimental results show that the expression of MCP-1 on HAECs in the 2D system was significantly higher than in the 3D system after stimulation. However, monocyte transmigration was significantly higher in the 3D system compared with the 2D system, suggesting that other mechanisms controlling cell migration may be at play. The final results of this research will provide new information about the relationship between MCP-1 concentration gradients and monocyte transendothelial migration, and will lead to the development of an improved model to study transendothelial monocyte migration associated with inflammation.

See more of this Session: Cell Adhesion and Migration I
See more of this Group/Topical: Food, Pharmaceutical & Bioengineering Division

Optimization of Mesenchymal Stem Cell Migration in the Wound Bed

Kevin Rodriguez, Deepraj Ghosh, Kathleen McAndrews, Daniel McGrail, Michelle Dawson
Georgia Institute of Technology, Atlanta, GA

Upon injury to the skin, the epidermal barrier is disrupted and the complex process of wound healing begins. Growth factors are essential in mediating the stages of wound healing, which proceed continuously from the time of epidermal damage to scar formation (Martin 1997). Elderly patients and patients suffering from chronic illness, including diabetes, ischemia, or hypertension, develop chronic wounds when the wound healing process is arrested in a state of chronic inflammation. Two of the factors that have been associated with the formation of chronic wounds are: (1) impaired growth factor production and (2) reduced angiogenesis (Wu, Chen et al. 2007). Mesenchymal stem cells (MSCs) have been shown to secrete cytokines, which promoted more rapid wound healing in normal and diabetic mice (Wu, Chen et al. 2007). Our studies are focused on optimization of the mechanical properties of MSCs for more rapid migration in wound tissue. We hypothesize that migrating MSCs that disseminate throughout the wound will contribute to the formation of granulation tissue, which will constrict the wound for more rapid wound closure.

Cell-based therapies often suffer from in vivo transport limitations. Previous studies have demonstrated that MSC treatment with soluble growth factors results in increased in vitro cell migration; however, little is known about the effects of these proteins on the mechanical properties and in vivo migration of MSCs. We systematically characterized the effects of tumor conditioned medium (TCM), a cocktail of soluble factors secreted by tumor cells in culture, and TGF-b1, a soluble factor involved in wound healing, on the mechanical properties of bone marrow-derived murine MSCs. Treatment of MSCs with TCM and TGF-b1 resulted in increased cytoskeletal elasticity, measured with multiple particle tracking microrheology, dramatic cell elongation, and increased cell migration. In vitro characterization of the mechanical properties of treated and control MSCs was used to design an in vivo murine wound healing model. Two full-thickness 5-mm skin biopsy punch wounds were created on either side of the dorsal midline on Balb-C mice. Treated and control fluorescently-labeled MSCs were injected at the periphery of the wound. Wound healing was monitored for 7 or 14 days, and the migration of MSCs in the wound bed was quantified with intravital epifluorescent microscopy. We found that TCM treatment significantly increased MSC migration in the wound bed and increased the rate of wound closure. The optimization of MSC migration may be an important step in the development of more effective MSC-based wound healing therapeutics.

Martin, P. (1997). "Wound healing-aiming for perfect skin regeneration." Science **276**(5309): 75-81.

Wu, Y., L. Chen, et al. (2007). "Mesenchymal stem cells enhance wound healing through differentiation and angiogenesis." Stem Cells **25**(10): 2648-59.

(77g) Metastatic Cancer Cells Optimize Space-Searching

Sabil Huda[1], Siowling Soh[1], Konstantin Tretiakov[1], Didzis Pilans[1], Kristiana Kandere-Grzybowska[1] and Bartosz Grzybowski[2], (1)Northwestern University, Evanston, IL, (2)Chemical and Biological Engineering, Northwestern University, Evanston, IL

Cell motility is integral to cancer metastasis. Developing new therapeutics to target the motility process requires the study of the molecular processes involved in cell motility. We have developed a novel class of cell biology tools for the quantification of cytoskeletal organization and dynamics in populations of cancerous cells with low variability. Anisotropic Solid Microetching (ASoMiC) was used to constrain cells onto transparent linear tracks. This methodology creates a micro-environment resembling *in vivo* conditions and is conducive for studying cell motility in molecular detail.

By constraining cells to 1-D tracks we found that metastatic cells were "superdiffusive" vs. "diffusive" non-metastatic cells. This reveals that metastatic cells are moving faster and more persistently than non-metastatic cells. More importantly, the superdiffusive behavior of metastatic cells arises from executing Lvy walks – which is an optimal search strategy.

Our study highlights that metastatic and non-metastatic cells are intrinsically different in their motility strategy. By understanding these differences we hope to identify new therapeutic targets for the treatment of metastatic cancer.

See more of this Session: Cell Adhesion and Migration I
See more of this Group/Topical: Food, Pharmaceutical & Bioengineering Division

(77h) Focal Adhesion Maturation and Dynamics Regulate Cell-to-Cell Variability In EGF-Stimulated Migration

Ian Schneider and Yue Hou, Chemical and Biological Engineering, Iowa State University, Ames, IA

Cell migration plays an essential role in many biological processes, such as cancer metastasis, wound healing and immune response. Cell migration is mediated in part through adhesion via focal adhesions (FAs), which are dynamic, macromolecular structures that serve as mechanical linkages and centers of intracellular signal transduction. FAs exist in different maturation states, which have been traditionally tied to either protein composition or morphology. Different maturation states regulate cell migration by either promoting local protrusion, quiescence or retraction. Epidermal growth factor (EGF) is known to enhance migration rate; however it is not known how FA maturation contributes to this control.

Our first objective is to determine the response of a population of cells to different doses of EGF at the level of migration. Our second objective is to use the cell-to-cell variability in migration rate at each EGF dose to understand which FA maturation and dynamics characteristics drive fast migration under EGF stimulation. We expressed a GFP fusion protein that localizes to FAs of various maturation states, paxillin-GFP, and imaged cells using total internal reflection fluorescence (TIRF) microscopy. Computer vision software was used to segment and track FA characteristics and quantify cell migration rate under different doses of EGF.

We found that there was considerable cell-to-cell variability in the effect of EGF on cell migration rate. The mean speed of cells only increased slightly with increased dose of EGF, while the distribution widened, indicating subpopulations of cells that both migrated much faster and much slower than those at lower EGF concentrations. We examined the FA characteristics, such as size, integrated intensity and lifetime under different doses of EGF and found that the average values and distributions of these characteristics were not significantly different among each concentration of EGF. However, when we grouped cells according to migration rate and examined the FA characteristics, we found that these characteristics differed between slow and fast moving cells. Slow cells had a higher probability of having both small and large FAs, while fast cells had a higher probability of having intermediate sized FAs. Additionally, slow cells had FAs with high intensities while fast cells had FAs with low intensities.

These findings indicate that there is high cell-to-cell variability in the response of cell migration to EGF. However, this variability can be leveraged to understand how cell migration rate is regulated by FA maturation and dynamics. Interestingly, some maturation characteristics such as FA area may not contribute to enhanced cell migration in a monotonic fashion. Rather, intermediate values of these characteristics may be optimized for fast cell migration. By understanding how FA maturation and dynamics correlate with migration rate, we can manipulate the cell migration behavior through the intracellular regulation of FAs, providing a potential way to control migration in artificial matrices or during pathologies such as cancer metastasis.

See more of this Session: Cell Adhesion and Migration I
See more of this Group/Topical: Food, Pharmaceutical & Bioengineering Division

(77i) Mucin-16 and Podocalyxin Are Functional Selectin Ligands Expressed by Pancreatic Cancer Cells

Shih-Hsun Chen[1], Matthew Dallas[1], Susan N. Thomas[2] and Konstantinos Konstantopoulos[1], (1) Chemical and Biomolecular Engineering, Johns Hopkins University, Baltimore, MD, (2)Institute for Bioengineering, cole Polytechnique Fdrale de Lausanne (EPFL), Lausanne, Switzerland

Metastatic dissemination of circulating tumor cells requires a multi-step cascade initiated by the interactions between selectins (E-, P-, and L-selectin) expressed by host cells and the selectin ligands displayed by cancerous cells. Following our recent identification of CD44 variant isoforms, carcinoembryonic antigen (CEA), and podocalyxin (PODXL) as functional selectin ligands in LS174T colon cancer cells we discovered ~200 and >460 kDa sialofucosylated glycoproteins that mediated selectin binding in SW1990 pancreatic carcinoma cells. Utilizing immunoaffinity chromatography and tandem mass spectrometry we identified Mucin-16 (MUC16) and PODXL as putative selectin ligands. Blot rolling and cell-free flow-based assays confirmed that MUC16 and PODXL expressed by SW1990 cells possess E- and L-selectin binding activity. In addition, the selectin binding capacity of SW1990 cells was dramatically decreased after sequential immunodepletion of MUC16 and PODXL. To assess and further investigate the functional role of MUC16 in selectin-mediated binding, we generated stable MUC16-konckdown SW1990 and E3LZ10.7 pancreatic cancer cells using MUC16 short hairpin RNA. Knocking down of MUC16 expression markedly reduced the cell binding to E- and L-selectin relative to control-transfected cells. Moreover, by treatment of glycosidases and carbohydrate metabolic inhibitors, the selectin-binding determinants on MUC16 are found as sialofucosylated epitopes displayed on both O- and N-linked glycans. Taken together, these findings offer new insight on the observed metastatic potential associated with MUC16 overexpression as well as the role of selectins in metastasis. Targeted therapies against MUC16-expressing cells could address multiple aspects of the metastatic cascade, including the invasive phenotype of metastasizing cells as well as blood-borne tumor/host cell interactions.

See more of this Session: Cell Adhesion and Migration I
See more of this Group/Topical: Food, Pharmaceutical & Bioengineering Division

(83a) The Unexpected Nucleation/Polymorphic Transformation-Promoting Effect of Glycine on Crystallisation of L-Glutamic Acid

Michal Rusin, Bruce C.R. Ewan and Radoljub I. Ristic, Department of Chemical and Biological Engineering, The University of Sheffield, Sheffield, United Kingdom

There has been a widely accepted view that molecular modifiers in solution are capable of directing growth morphology, but have only neutral or inhibitory effects on primary nucleation and growth kinetics. While the former belief still remains unchallenged, the latter one was recently questioned by in-situ atomic force microscopy (AFM) studies on biomineralisation of calcite, which have shown that two acidic proteins dramatically accelerate calcite growth while still altering the surface morphology through interactions with specific atomic steps on the (104) face[1,2]. In spite of scarcity of data that exists, one may ask a more general question, 'Should accelerating effects occur for the primary nucleation of polymorphic materials in the presence of selectively chosen molecular modifiers, and how these initial conditions may, at the later stage, influence the kinetics of solution mediated transformation of metastable to stable polymorphic form'?

Remarkably, for the first time, we found in our research on the influence of various amino acids on crystallisation of dimorphic L-glutamic acid (LGA) that glycine used as a doping material significantly enhances both nucleation and solution-mediated polymorphic transformation rates with respect to the pure system.

The effect of the additive level was investigated using laser-light scattering and concentration evolution monitoring techniques. It was observed that the LGA nucleation/polymorphic transition rates increase with increasing concentration of the doping agent. Furthermore, investigation of the initial product crystal samples using scanning electron microscopy (SEM) indicated that glycine promotes primary nucleation of the β form of L-glutamic acid. Several mechanisms to explain why this previously unreported effect may occur were suggested.

We believe that our findings would be interesting and stimulating for both industrial and scientific communities. The insights derived from these studies could facilitate the design of potent new synthetic nucleation/transformation modulators for diverse polymorphic systems. Ultimately, this would allow better control of the transformation process between polymorphs of crystalline materials in a wide range of consumer products, from foods, through cosmetics to pharmaceuticals.

References:

1. Elhadj, S., De Yoreo, J.J., Hoyer, J.R. and Dove, P.M., *Proc. Natl. Acad. Sci. USA*, 2006, **103**, 19237.
2. Fu, G., Qiu, S.R., Orme, C.A., Morse, D.E. and De Yoreo, J.J., *Adv. Mater.*, 2005, **17,** 2678.

(83b) Implementation of a High-Resolution Population Balance Solver to Model Pharmaceutical Crystallizations

Jose E. Tabora[1], Saravanababu Murugesan[1], James Vernille[1], Richard Braatz[2], Mitsuko Fujiwara[3] and Michael L. Rasche[3], (1)Chemical Development, Bristol Myers Squibb, New Brunswick, NJ, (2) Chemical Engineering, Massachusetts Institute of Technology, Cambridge, MA, (3)Department of Chemical and Biomolecular Engineering, University of Illinois at Urbana-Champaign, Urbana, IL

Implementing a successful Quality-by-Design (QbD) strategy involves the construction and evaluation of adequate models describing the quantitative relationships between process parameters and intermediate/API quality. In the crystallization of the API, the quality of the product is dictated not only by the chemical purity of the product but also by its physical properties. Therefore, high fidelity models that accurately describe the API properties such as surface area and particle size distribution as a function of the crystallization parameters in the final isolation are indispensable resources of any QbD strategy. Population balance models (PBM's) have been shown to be powerful tools in the description of crystallization processes. This work implements a high resolution PBM solver to model the crystallization of an API. The data used for parameter regression were obtained from crystallizations incorporating a constant supersaturation control strategy that allows the crystallizations to operate under predominantly growth or secondary nucleation regimes. The model was subsequently used as a guide for a design space that ensured control of the desired particle size distribution of the API.

See more of this Session: Crystallization of Pharmaceutical and Biological Molecules II
See more of this Group/Topical: Separations Division

(83c) Optimal Control of Antisolvent and Cooling Crystallization

Lifang Zhou[1], Kwun-Ning Ma[2], Hao Feng[3] and Richard D. Braatz[1], (1)Chemical Engineering, Massachusetts Institute of Technology, Cambridge, MA, (2)Chemical and Biomolecular Engineering, University of Illinois at Urbana-Champaign, Urbana, IL, (3)University of Illinois at Urbana-Champaign

Quality requirements for crystallization products in the pharmaceutical industry are specified by the demands of the drug administration (e.g., bioavailability, size uniformity, and polymorphic form) and manufacturing considerations for downstream processes such as filtration, drying, and flowability. The large effect of the crystal size distribution (CSD) on crystal properties motivates the design of processes with reproducible and optimal control of the CSD [1-6].

This study investigates a new process for CSD control that combines antisolvent crystallization in a dual impinging jet crystallizer with cooling crystallization in a mixing tank, which is an extension of an idea published in an earlier theoretical investigation [7-9]. Specification of the inlet velocities to the dual impinging jet crystallizer as a function of time enables the controlled variation of the size distribution of seed crystals continually added to the mixing tank (that is, control of the crystal nucleation). Operation of the cooling crystallization under closed-loop feedback control of the solute concentration enables the specification of the supersaturation in the mixing tank as a function of time, which is equivalent to control of the crystal growth due to the direct algebraic relationship between supersatuation and crystal growth. This proposed semi-continuous process provides a higher degree of control of the CSD than has been obtainable with past crystallizer designs. The proposed approach uses the same equipment as described in Ref. [9], but with the extension that the supersaturation is time-varying instead of specified at a constant controlled value.

The ATR-FTIR spectra are first collected to construct a calibration model, which was used along with the infrared spectra to measure the solubility of paracetamol in isopropanol-water solution at various temperatures and solvent ratios, in a similar manner as earlier studies [1, 4, 10-12]. This information was used in subsequent simulations of open and closed-loop controls for the integrated crystallization process. Various time variations for the continual seeding produced by real-time antisolvent crystallization to various time variations in the supersaturation profile in the well-mixed cooling tank were simulated, to demonstrate the rate of product crystal size distributions that can be manufactured using this system. The design provides a higher degree of control of the CSD by decoupling crystal nucleation and crystal growth throughout the crystallization process. Optimization of the temperature profiles to achieve targeted crystal size distributions is presented. The simulation results motivate future experimental implementations.

[1] M. Fujiwara, P. Chow, D. Ma, and R. Braatz, "Paracetamol crystallization using laser backscattering and ATR-FTIR spectroscopy: Metastability, agglomeration, and control," *Crystal Growth*

& Design, vol. 2, pp. 363–370, 2002.

[2] H. Gron, A. Borissova, and K. Roberts, "In-process ATR-FTIR spectroscopy for closed-loop supersaturation control of a batch crystallizer producing monosodium glutamate crystals of defined size," *Industrial & Engineering Chemistry Research*, vol. 42, pp. 198–206, 2003.

[3] V. Liotta and V. Sabesan, "Monitoring and feedback control of supersaturation using ATR-FTIR to produce an active pharmaceutical ingredient of a desired crystal size," *Organic Process Research & Development*, vol. 8, pp. 488–494, 2004.

[4] G. Zhou, M. Fujiwara, X. Woo, E. Rusli, H. Tung, C. Starbuck, O. Davidson, Z. Ge, and R. Braatz, "Direct design of pharmaceutical antisolvent crystallization through concentration control," *Crystal Growth & Design*, vol. 6, pp. 892–898, 2006.

[5] C. Lindenberg, M. Kraettli, J. Cornel, M. Mazzotti, and J. Brozio, "Design and optimization of a combined cooling/antisolvent crystallization process," *Crystal Growth & Design*, vol. 9, pp. 1124–1136, 2009.

[6] N. C. S. Kee, R. B. H. Tan, and R. D. Braatz, "Selective crystallization of the metastable alpha-form of L-glutamic acid using concentration feedback control," *Crystal Growth & Design*, vol. 9, pp. 3044–3051, 2009.

[7] X. Y. Woo, R. B. H. Tan, and R. D. Braatz, "Precise tailoring of the crystal size distribution by controlled growth and continuous seeding from impinging jet crystallizers," *CrystEngComm*, vol. 13, no. 6, pp. 2006–2014, 2011.

[8] X. Y. Woo, Z. K. Nagy, R. B. H. Tan, and R. D. Braatz, "Adaptive concentration control of cooling and antisolvent crystallization with laser backscattering measurement," *Crystal Growth & Design*, vol. 9, pp. 182–191, 2009.

[9] X. Y. Woo, *Modeling and simulation of antisolvent crystallization mixing and control*. PhD thesis, University of Illinois at Urbana - Champaign and National University of Singapore, 2007.

[10] T. Togkalidou, M. Fujiwara, S. Patel, and R. Braatz, "Solute concentration prediction using chemometrics and ATR-FTIR spectroscopy," *Journal of Crystal Growth*, vol. 231, pp. 534–543, 2001.

[11] Z. Q. Yu, P. S. Chow, and R. B. H. Tan, "Application of attenuated total reflectance-Fourier transform infrared (ATR-FTIR) technique in the monitoring and control of anti-solvent crystallization," *Industrial & Engineering Chemistry Research*, vol. 45, no. 1, pp. 438–444, 2006.

[12] Z. K. Nagy, M. Fujiwara, and R. D. Braatz, "Modelling and control of combined cooling and antisolvent crystallization processes," *Journal of Process Control*, vol. 18, pp. 856–864, 2008.

(83d) Super-Saturation Control In API Crystallization for Optimal Impurity Rejection and Particle Size Distribution

Lei Cao, David Li and Yubo Yang, Abbott Laboratories, North Chicago, IL

Understanding and control of crystallization process plays an important role in pharmaceutical API manufacturing. Major API quality attributes such as impurity profile, crystal form, and particle size distribution (PSD) are usually controlled in the crystallization process.

We present an integrated platform for real time super-saturation detection and control to meet the target impurity profile and final crystal PSD under certain solvent systems. High sensitive Mid-IR ATR probe (Mettler Toledo ReacIR^{TM} 45) is used for quantitative measurement of API concentration in liquid phase. Crystal size is characterized using a Focused Beam Reflectance Measurement (FBRM) probe (Mettler Toledo Lasentec D600). The software package developed by Professor Richard Braatz and his group is capable of integrating the signals from both IR and FBRM probes to be used in the platform[1,2].

By controlling a heating/cooling system and a solvent charge pump, this platform is automated to realize the following objectives: 1) determining meta-stable limit of an API compound for a given solvent system; 2) building a robust chemometric IR model for concentration quantification; 3) determining the quasi-equilibrium solubility as a function of temperature and solvent composition; 4) imposing feed-back control to follow a pre-determined super-saturation profile during cooling/anti-solvent crystallization process.

The above-mentioned platform is applied to two case studies. In the first case study, small particle sizes are desired, however, the current crystallization protocol produces large crystals that require a harsh milling. The correlation between desuper-saturation rate and final crystal PSD is achieved using the platform, enable us to obtain the target PSD without the need for milling. The second case study is to relate nucleation and initial crystal growth with impurity rejection. Previously knowledge about the API crystallization from its crude solution indicates that one or two impurities tend to co-crystallize with the API under certain conditions. It is desired to establish the operating range where this phenomenon can be suppressed, and we successfully applied this crystallization platform to achieve the objective.

References:

1. P. Barrett and B. Glennon, Characterizing the metastable zone width and solubility curve using Lasentec FBRM and PVM.
2. T. Togkalidou, H.-H. Tung, Y. Sun, A. Andrews, and R. D. Braatz. Solution concentration prediction for pharmaceutical crystallization processes using robust chemometrics and ATR

FTIR spectroscopy. *Org. Process Res. Dev.*, 6:317-322, 2002.

See more of this Session: Crystallization of Pharmaceutical and Biological Molecules II
See more of this Group/Topical: Separations Division

(83e) Design of Nanostructured 'catalyst' for Nucleation of Molecular Crystals

Ying Diao, Matthew E. Helgeson, Patrick S. Doyle, Allan S. Myerson, T. Alan Hatton and Bernhardt L. Trout, Chemical Engineering, Massachusetts Institute of Technology, Cambridge, MA

Despite its widespread occurrence in nature and broad application in industrial practice, nucleation of crystalline materials remains largely unpredictable and therefore difficult to control. In this work, we demonstrate a new method to control nucleation with polymer microgels by tuning their nanostructure to vary systematically the degree of nanoscopic confinement and its effects on nucleation. We find that the polymer nanostructure has a significant impact on nucleation kinetics. Moreover, there exists an optimum polymer mesh size at which the rate of nucleation is dramatically enhanced, the degree to which depends on the extent of polymer-solute interactions. Besides, we also found that these polymeric particles play a role in determining the crystal polymorphism of pharmaceutical compounds, which is of great importance in the pharmaceutical manufacturing process.

See more of this Session: Crystallization of Pharmaceutical and Biological Molecules II
See more of this Group/Topical: Separations Division

(83f) Turbidity and Endoscopy Assisted Monitoring of Solvent Mediated Polymorphic Transformation Processes

Levente L. Simon, Institute for Chemical and Bioengineering, Safety and Environmental Technology Group, ETH Zurich, Zurich, Switzerland and Konrad Hungerbuehler, Chemical Engineering, ETH Zurich, Zurich, Switzerland

Polymorphism is a feature which characterizes most of the activepharmaceutical ingredients and influences the solubility, shape and color ofcrystals. The process of solvent mediated polymorphic transformation is aseries of events such as crystal dissolution, nucleation and growth and severalon-line PAT sensors have been evaluated to characterize the transformationprocess by monitoring the solid or liquid phase properties.

The focus of this work is to compare low-cost on-line PAT sensors with morecomplex instruments such as the FBRM, FT-Raman, ATR-FTIR and FT-NIRspectroscopy. The target of the investigation is to evaluate whether theseprobes are able to detect changes in solid concentration due todissolution-nucleation-growth events and it is not expected that these giveinformation about the ratio of the polymorphic forms. In the family of low-costsensors are included the turbidity sensor and the bulk video imaging methodwhich can be implemented using an endoscope. Since the turbidity sensor is alreadyfound in many industrial crystallizers and it is used to detect nucleation andgrowth its evaluation for polymorphic transformation monitoring is obvious.Moreover, note that no additional investment and installation costs occur.Recently, GSK presented a case-study where turbidity and FT-Raman spectroscopywas successfully used to follow the polymorphic transformation of an API [1].

The endoscopy based process monitoring is a low-cost PAT tool inevaluation phase [2]. It is based on imaging and recently it was proposed fornucleation detection applications. Depending on the image analysis method usedit can detect the first formed crystals as objects; alternatively, when themean gray intensity of the images is calculated its operation is more similarto that of the turbidity probes. Note that one image can provide the equivalentinformation of several thousands of spatially distributed turbidity probes. Inthis work the endoscopy images are analyzed using texture analysis basedmethods. Simple image texture descriptors have been previously evaluated in theform of mean, minimum and maximum intensities [3,4] for the purpose ofnucleation detection. Since the turbidity probe operates at 880 nm which is inthe NIR range FT-NIR spectroscopy was used to gain further information aboutthe operation of this probe. This work presents a quantitative evaluation ofthe detection sensitivity of the dissolution and nucleation/steps of thesensors mentioned above. Furthermore, parameters and sensor design issues whichinfluence the monitoring sensitivity are discussed in detail.

References:

[1] Barnes, S., Gillian, J., Diederich, A., Burton, D., Ertl, D., In ProcessMonitoring of Polymorphic Form Conversion by Raman Spectroscopy and TurbidityMeasurements, American Pharmaceutical Review, http://americanpharmaceuticalreview.com/ViewArticle.aspx?ContentID=3386.

[2] Simon, L. L., Nagy, Z. K., Hungerbuhler, K., 2009b, Endoscopy-Based inSitu Bulk Video Imaging of Batch Crystallization Processes, Organic ProcessResearch & Development, Special Issue on Polymorphism and Crystallization,13: 1254-1261.

[3] Simon, L. L., Abbou Oucherif, K., Nagy, Z. K., Hungerbuhler, K., 2010,Bulk Video Imaging Based Multivariate Image Analysis, Process Control Chart andAcoustic Signal Assisted Nucleation Detection, Chem Eng Sci, 65: 4983-4995.

[4] Simon, L. L., Nagy, Z. K., Hungerbuhler, K., 2009a, Comparison ofExternal Bulk Video Imaging with Focused Beam Reflectance and UltraViolet-Visible Spectroscopy for Crystallization Nucleation Detection andMetastable Zone Identification, Chem Eng Sci, 64: 3344-3351.

See more of this Session: Crystallization of Pharmaceutical and Biological Molecules II
See more of this Group/Topical: Separations Division

(89a) Transcriptional Regulatory Network In Pluripotent Stem Cells

Shikha Sharma[1], Andrew Yongky[1], Catherine Verfaillie[2] and Wei-Shou Hu[1], (1)Department of Chemical Engineering and Materials Science, University of Minnesota, Minneapolis, MN, (2)Stem Cell Institute and Department of Medicine, Katholieke Universiteit Leuven, Leuven, Belgium

Stem cells maintain their pluripotent state through a robust gene network consisting of complex feedback and feedforward interactions between key transcription factors. In self-renewing stem cells, the gene network maintains expression of pluripotency-associated genes such as Oct4 and Nanog, while suppressing the expression of genes associated with differentiation to somatic lineages. In the context of embryonic development, this network is actively maintained in pluripotent cells of the inner cell mass while the trophoblast and primitive endoderm lineages lose expression of key genes like Oct4 and Nanog, respectively. Primitive endoderm-precursor cells, derived from E4.5 blastocyst, are separated from the pluripotent stem cells of the inner cell mass through a single differentiation event. These cells express many key pluripotency-associated genes including Oct4 and Sall4, but also express genes representative of the primitive endoderm lineage, such as Gata6 and Sox17. This unique gene expression profile, developmental proximity to pluripotent stem cells and, genetic tractability make these cells ideal candidates for probing the dynamics of pluripotency gene network. With these precursor cells as our model system, we have used a combinatorial approach involving experimental gene over-expression/ knockdown studies and, a theoretical boolen model, to explore interactions within this network. Using this approach, we have been able to predict outcomes of experimental perturbations within the network. This has provided us with novel insights into how pluripotency is established and maintained in stem cells.

See more of this Session: Experimental Approaches In Systems Biology
See more of this Group/Topical: Food, Pharmaceutical & Bioengineering Division

(89b) A Systems Approach to Identify Effects of Chronic Alcohol Intake On the Dynamic Gene Regulatory Response Patterns During Liver Regeneration

Rajanikanth Vadigepalli, Biswanath Patra, Lakshmi Kuttippurathu, Egle Juskeviciute and Jan Hoek, Department of Pathology, Anatomy and Cell Biology, Thomas Jefferson University, Philadelphia, PA

The goal of the present study is to characterize the transcriptomic regulatory dynamics operational in the regenerating rat liver affected by chronic alcohol intake. While transcriptomic data on the gene expression changes during normal liver regeneration is increasingly becoming available from multiple studies, the effect of chronic alcohol consumption on these changes is less clear. Such an alcohol-mediated disruption of repair mechanisms presumably contributes to the onset of liver damage in alcoholic liver disease.

We take an integrated experimental and computational approach to investigate the dynamics of the system-wide transcriptional changes, use sensitive bioinformatics methods to infer regulatory modules comprising of transcription factors (TFs) and target genes, and validate these for changes in the activity of TFs and their binding at target gene promoters. In our experimental approach, rats were fed a liquid diet containing 36% of total calories derived from ethanol; corresponding pair-fed calorie-matched controls received liquid diets in which ethanol calories were replaced either by carbohydrate or by fat. After five weeks, rats were subjected to 2/3d PHx by surgical excision of left-lateral and medial (LLM) lobes. Remnant liver tissues were harvested after 1, 6 and 24h to represent the priming phase, early G1, and G1/S transition, respectively. LLM tissues obtained at t=0 served as within-animal controls. We used Affymetrix Rat Gene 1.0 ST arrays with ~25000 probe sets to obtain global gene expression data from each liver sample (n=4 replicate animals, 72 arrays total). Analysis of the gene expression data using a 3-way ANOVA revealed a total of 6893 transcript clusters as responsive to PHx in at least one of the diet groups at either 1, 6 or 24h post PHx. Clustering the sample replicates indicated that the gene expression response to PHx is the primary factor separating the groups.

We developed a novel dynamic response pattern analysis approach in which the average differential gene expression data was discretized to three levels (+1, 0, -1) based on a fold change threshold of 1.5 up or down regulation. Within each diet group, this discretization yielded a dynamic response pattern for each gene, encoded by one of 27 possible ordered sets of the three levels +1, 0, and -1. Pairs of diet groups were compared to count the number of genes that follow each of the possible 27 * 27 (=729) comparative dynamic response patterns. Only a select set of the possible 729 patterns contained more than 25 genes indicating that the ethanol effects are mediated through specific mechanisms. The largest set of genes with different patterns between the diet groups were those with lack of differential regulation in ethanol group. This was followed by another gene set with de novo differential expression in ethanol group but not in that of the control diets. Other patterns involved alterations of dynamic gene expression from a transient response at one of the time points in control groups to a more persistent differential gene expression in ethanol group, and vice versa. Notably, ethanol group showed a marked lack of up regulation of cell cycle related gene expression at 6 and 24h.

In complementary experiments, we detected genome-wide NF-kB binding targets using Roche NimbleGen ChIP-chip microarray platform using Chromatin Immunoprecipitated (ChIP) samples. Our results indicate that PHx induced significant increase in NF-kB genome-wide binding at 1h, more so in chronic ethanol samples (5300 genes) than in controls (4000 genes), with several common targets (~3300 genes). At 6h post PHx, only the control samples showed a further increase in NF-kB promoter binding (~5500 genes). Pathway analysis revealed that NF-kB target genes specific to the chronic ethanol group participate in key processes such as DNA methylation, cell death, proteolysis, histone modification, and regulation of cell cycle. Our dynamic response pattern analysis of NF-kB promoter binding revealed similar alterations seen in gene expression response patterns, notably the conversion of persistent response to a transient response and vice versa in the ethanol group as compared to the control diet groups.

Taken together, our results provide a first system-wide view of the regulatory network dynamics in liver regeneration as it is affected by alcohol, and implicate combinatorial regulatory modules as participating in the adverse liver regeneration response. The above results indicate a phenomenon in which compensatory mechanisms that lead to an adaptive state in alcoholic rats play a role in suppressing the sensitivity of the system to certain perturbations, affecting the ability of the liver to initiate effective repair responses in the face of additional environmental challenges.

Supported by: NIH AA008714, AA014986, AA017261, AA016919, and AA018873.

See more of this Session: Experimental Approaches In Systems Biology
See more of this Group/Topical: Food, Pharmaceutical & Bioengineering Division

(89c) Targeted Metabolomics of Cofactors In Liver Drug Metabolism

Michael Yi, Kimberly L. Stachenfeld, Gautham V. Sridharan, Katherine A. Carson and Kyongbum Lee, Chemical and Biological Engineering, Tufts University, Medford, MA

Metabolomics, the comprehensive quantification of small molecules in biological samples, offers a powerful complement to genomics and proteomics in systems biology. In conjunction with mathematical models, the metabolome of a cell or tissue can be used to quantitatively and comprehensively characterize the effects of disease or other biological and chemical stresses on the engagements of metabolic pathways. Quantifying pathway engagements is particularly useful when a clear phenotype and/or mechanism remain to be established for a complex disease (e.g. cancer) or chemical insult (e.g. adverse drug reaction). In this work, we apply targeted metabolomics to characterize the metabolic burden in liver cells treated with combinations of drug chemicals. Drug metabolizing enzymes in the liver depend on the availability of cofactors such as ATP/ADP, NADH/NAD+, NADPH/NADP+, and GSSG/GSH (oxidized and reduced glutathione) to continuously operate xenobiotic functionalization (phase I) and conjugation (phase II) reactions critical for safe elimination. The goal of our analysis is to investigate the hypothesis that drug chemicals not known to directly interact (e.g. through mutual inhibition of corresponding phase I enzymes) could generate an unexpected metabolic burden through cofactor depletion.

In recent years, coupled liquid chromatography-tandem mass spectrometry (LC-MS/MS) has emerged as a promising analytical platform to selectively detect and quantify a variety of non-volatile small molecule metabolites without requiring costly chemical derivatization of the target analytes. However, MS/MS analysis of cofactors, due to the presence of labile bonds susceptible to oxidation or reduction, is highly sensitive to the buffers and conditions used for sample preparation and chromatographic separation. To date, development of robust protocols for single platform analysis of intracellular cofactors remains insufficiently addressed.

In this paper, we describe a streamlined, single platform methodology for the extraction, detection, and quantification of intracellular cofactors and other small molecule metabolites in mammalian cells. To avoid the use of detergents in sample preparation, a series of methanol/chloroform extractions were performed on freeze-quenched cell pellets. The extracts were analyzed on a hybrid triple quadruple linear ion trap mass spectrometer (3200 QTRAP, AB Sciex, Foster City, CA) coupled to a binary pump liquid chromatography system (1200 Series, Agilent Technologies, Santa Clara, CA). Compound detection was optimized in Multiple Reaction Monitoring (MRM) mode for each target metabolite. Chromatograph separation was achieved on a C-18 reverse phase column (Synergy Fusion, Phenomenex, Torrance, CA). To account for ion suppression and other matrix effects, the method of standard additions was used for establishing correlations, i.e. standard curves, between peak areas and metabolite concentrations.

The extraction and LC-MS/MS analysis protocols were used to determine differences in intracellular cofactor concentrations between untreated control liver cells (HepG2 hepatocarcinoma) and cells treated with combinations of anti-diabetic agents (troglitazone and metformin). Our results suggest that metformin exerts a stronger effect on intracellular cofactor concentrations compared to troglitazone. Interestingly, the effects were differential across the various cofactors. A significant difference in the ATP concentration relative to the control was observed only when both anti-diabetic agents were administered, consistent with our observations on cell growth.

See more of this Session: Experimental Approaches In Systems Biology
See more of this Group/Topical: Food, Pharmaceutical & Bioengineering Division

(89d) Metabolic Flux Analysis Using 13C-Tracers and Tandem Mass Spectrometry

Maciek R. Antoniewicz, Chemical Engineering, University of Delaware, Newark, DE, Jungik Choi, Chemical Engineering, Univerisity of Delaware, Newark, DE and Matt Grossbach, Chemical Engineering, University of Delaware, Newark

The Antoniewicz lab is pioneering the development of a new technique for metabolic flux analysis (MFA) based on stable-isotope labeling and tandem mass spectrometry (MS/MS). Here, we present the computational, analytical and experimental techniques we have developed for ^{13}C-MFA using GC-MS/MS. As a practical example, we demonstrate for the first time the application of tandem MS for metabolic flux analysis in *E. coli*. We show that tandem MS offers considerable advantages for in vivo flux analysis compared to conventional mass spectrometry techniques (GC/MS and LC/MS). The number of independent measurements is greatly increased with tandem MS, as is the information content of the labeling measurements. As a result, metabolic fluxes are determined with much higher precision and accuracy. The application of tandem MS allowed us to quantify accurately metabolic pathway fluxes in *E. coli* that were previously unobservable with conventional MS and NMR techniques. In summary, our new tandem MS technique allows tracing of isotopic labeling in complex metabolic networks with much higher resolution and ultimately improves measurement of metabolic fluxes.

See more of this Session: Experimental Approaches In Systems Biology
See more of this Group/Topical: Food, Pharmaceutical & Bioengineering Division

(89e) Quantitative Assessment of Cell Signaling Pathways Affecting Stem Cell Differentiation Using Lentiviral Arrays

Stella Alimperti[1], Jun Tian[2], Pedro Lei[2] and Stelios T. Andreadis[2], (1)Department of Chemical and Biological Engineering, State University of New York-SUNY at Buffalo, Amherst, NY, (2)Department of Chemical and Biological Engineering, State University of New York -SUNY at Buffalo, Amherst, NY

Loss of gene function is avaluable tool for screening genes in cellular processes such as MesenchymalStem Cell (MSC) differentiation. However,the criteria for evaluating gene knockdown are usually based on end-pointanalysissuch as immunostaining andtherefore, quantitative, real-time measurement of the process is difficult. Toovercome these limitations, we created live cell **L**enti**V**iral **A**rrays(LVA) for assessing the effects of certain pathways on MSC differentiation. Inparticular, we engineered a shRNAencoding **d**ual **p**romoter lentiviral vector (sh-LVDP) that allowsreal-time monitoring of MSC differentiation and simultaneous gene knockdownthereby, providing quantitative assessment of the effect of various pathways inthis process. Inthis vector, the activity of the aSMA promoter was used as anindicator of myogenic differentiation and constitutive expression of DsRed was used to measure transduction efficiency and to normalizethe promoter activity. Finally, different shRNAs were encoded by a doxycyclineregulatable H1 promoter within theviral 3' long terminal repeat (LTR).

We found that normalized promoteractivity (GFI/RFI) measured by the sh-LVDP vector remained unaffected bylentiviral titer allowing quantitative measurements of MSC differentiation asdifferent genes were knocked down.To demonstrate the utility of this approach we chose to knock down known effectors of the TGF-β1 or Rho signaling pathway (SMAD2,SMAD3, SMAD4, RhoA, Rock1, MRTFB/MLK2, PKN1, SRF, myocardin), which decreased αSMA promoter activity significantlyas compared to control (scramble shRNA) (Fig. 1B). On the other hand, knockingdown myogenic differentiation inhibitor such as KLF4 increased aSMApromoter activity significantly. Notably, some proteins e.g. SMAD7 or KLF4 showeddifferential regulation of myogenic differentiation of bone marrow as comparedto hair follicle derived MSC. Finally, doxycyline regulatable shRNA expressionprovided dynamic monitoring of myogenic differentiation as a function of thedegree of gene knockdown. Our data indicates that real time monitoring ofgene expression profiles using LVA with concomitant gene knockdown provides auseful tool for deciphering gene regulatory networks of complex biologicalprocesses such as stem cell

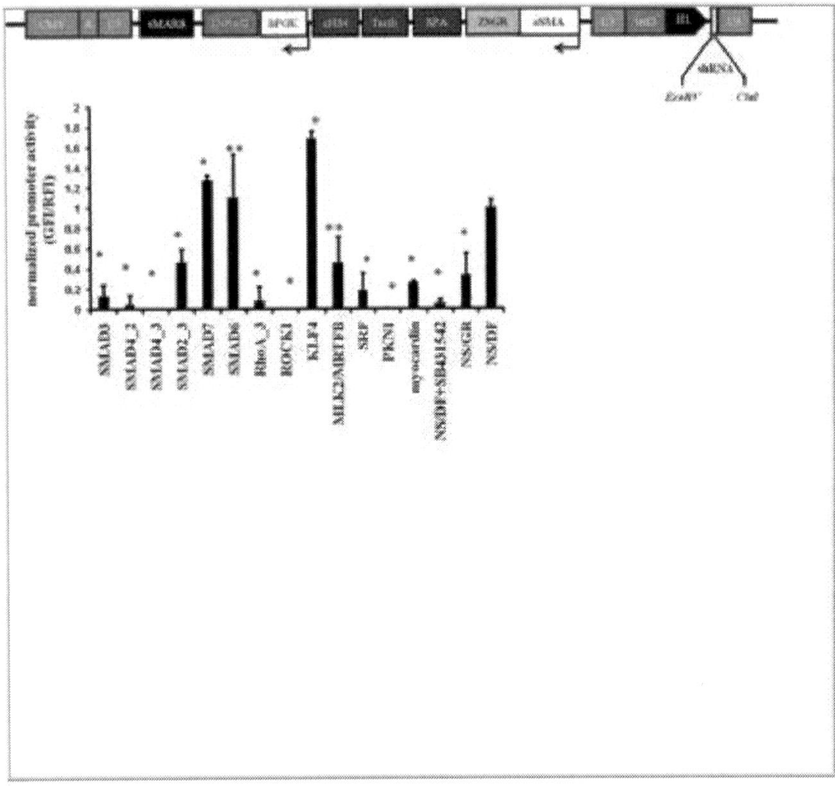

Figure 1: Lentiviral MicroArray (LVA) with sh-LVDP for differentiation toward smooth muscle cells in response to TGFβ. (A) Schematic of sh-LVDP vector carrying the aSMA promoter. (B) Normalization of fluorescence intensity (GFI/RFI) indicates the a-actin promoter activity affected by gene knock-down in Human Aortic Smooth Muscle Cells (HASMCs).

See more of this Session: Experimental Approaches In Systems Biology
See more of this Group/Topical: Food, Pharmaceutical & Bioengineering Division

(89f) A Novel Experimental Approach to Investigate Cytokine Signaling Networks of Human T Cells

Qing Han, Chemical Engineering, The David H. Koch Institute for Integrative Cancer Research, MIT, Cambridge, MA, Neda Bagheri, Biological Engineering, MIT, Cambridge, MA, Douglas Lauffenburger, Department of Biological Engineering, MIT, Cambridge, MA and J. Christopher Love, Department of Chemical Engineering, The David H. Koch Institute for Integrative Cancer Research, Massachusetts Institute of Technology, Cambridge, MA

T cells are important for adaptive immune responses, and understanding the nature of their contributions are critical to identify immunological signatures for diseases, as well as define successful responses to therapeutic interventions. In recent years, it has become evident that the functional responses of differentiated T cells may be more dynamic than previously considered, but existing tools for characterizing activated human T cells (i.e., ELISpot and intracellular cytokine staining) cannot monitor the evolution of responses by individual cells or how their responses modulate each other. In this work, we present an application of microengraving to track the functional evolution and cytokine regulation of T cells. Our experimental platform evaluates over 10,000 individual cells in parallel, and generates up to 30 data points per cell that describe the cytokine secretion from eight distinct subsets of primary CD3+ T cells. Resulting dynamic measurements quantify the evolution of single-cell cytokine secretion following stimulation, offering a unique multi-dimensional approach to investigate the dynamics of specific immunophenotypes. Using this technology, we explore the effect of different autocrine and paracrine signals on the dynamic secretion profiles of individual cells. Such experimental analysis may offer insight on the extracellular cytokine networks as well as intracellular regulation of immune signaling pathways.

See more of this Session: Experimental Approaches In Systems Biology
See more of this Group/Topical: Food, Pharmaceutical & Bioengineering Division

(89g) Pairwise Agonist Screening and Microfluidics Reveal Platelet Function Phenotype

Thomas Colace, Manash Chatterjee and Scott L. Diamond, Chemical Engineering, University of Pennsylvania, Philadelphia, PA

Pairwise agonist screeningand microfluidics reveal platelet function phenotype

T. Colace, M. Chatterjee, and S.L. Diamond

In this study we consider the relationship between ahigh-throughput approach designed to investigate platelet activation viacalcium mobilization and a microfluidic model used to study platelet function. Chatterjeeet al. (2010) have demonstrated that calcium responses to a variety of plateletagonists can serve as a platelet activation phenotype in an assay calledPairwise Agonist Scanning (PAS), but did not illustrate a connection toplatelet function. Using PAS data collected for 3 healthy donors we attemptedto predict changes in platelet function in the presence of the anti-plateletagent Indomethacin, a non-selective COX inhibitor, and to confirm them in amicrofluidic model of platelet aggregation on a collagen surface. Thecombination of these techniques represents a link between the complex andunique activating environment experienced by a platelet under flow conditionsand the traditional methods of measuring platelet activation in a static, wellmixed setting.

Whole blood, anticoagulated with PPACK (100 μM), was treatedwith Indomethacin (140 μM), Iloprost (a stable prostacyclin analog thatglobally inhibits platelet activation, 1 μM), both agents together, or wasuntreated (control). Each condition was loaded in duplicate into a separateinlet of an 8 channel microfluidic device with a single outlet [**Figure 1A**].The samples were perfused over a fibrillar collagen surface at a shear rate of200 s^{-1} for 8 min [**Figure 1B**]. Platelet accumulation at thecollagen strip was measured in each channel simultaneously in 15 sec intervals.To perform the PAS study platelet rich plasma from the same blood draw was loadedwith a calcium sensitive dye and Indomethacin (to prevent autocrine activationvia TXA_2). Samples were treated with all pairwise combinations of 4concentrations (zero, low, medium, high) of the platelet agonists convulxin (acollagen mimetic), ADP, and U46619 (a stable TXA_2 mimetic), in thepresence or absence of Iloprost. Platelet calcium mobilization was measured viafluorescence for 200 sec.

The platelet collagen response was recorded for 4 trials per donoron two separate days, with corresponding PAS data. While Iloprost was effectivefor all donors (mean aggregation inhibition of 70%), Indomethacin showedeffectiveness only for donors 1 and 2 (35% and 20%, respectively), and caused asignificant increase in platelet accumulation for donor 3 (20%) [**Figure 1C**].Platelet calcium mobilization in response to U46619 alone illustrated a reducedsensitivity to the TXA_2 analog for donor 2 as compared to donor 1 (~44%reduction in Ca^{2+} mobilization) and for donor 3 as compared to donor2 (~80% reduction) [**Figure 1D**]. In order to confirm an activating rolefor high doses of Indomethacin, the study was repeated at 14 μMIndomethacin. At this concentration donor 2 showed a 34% reduction in plateletadhesion while donor 3 showed at 13% reduction (donor 1 has yet to be investigated).PAS data was not affected by the change in Indomethacin concentration. Furthermore,we have observed a 2-fold increase in platelet aggregation for donor 1 ascompared to donors 2 and 3. We believe this result is reflective of a universalincrease in Ca^{2+} mobilization as measured under all conditions inthe PAS output.

These data reveal that donor sensitivity to anti-platelet agentscan be predicted using PAS and confirmed in a microfluidic model of plateletadhesion. The combination of these techniques bridges the gap between traditionalmeasures of platelet activation (Ca^{2+} mobilization) and studies ofplatelet function under physiologic flow conditions. We believe these methodshold promise as a novel diagnostic tool for evaluating platelet function andefficacy of anti-platelet therapy.

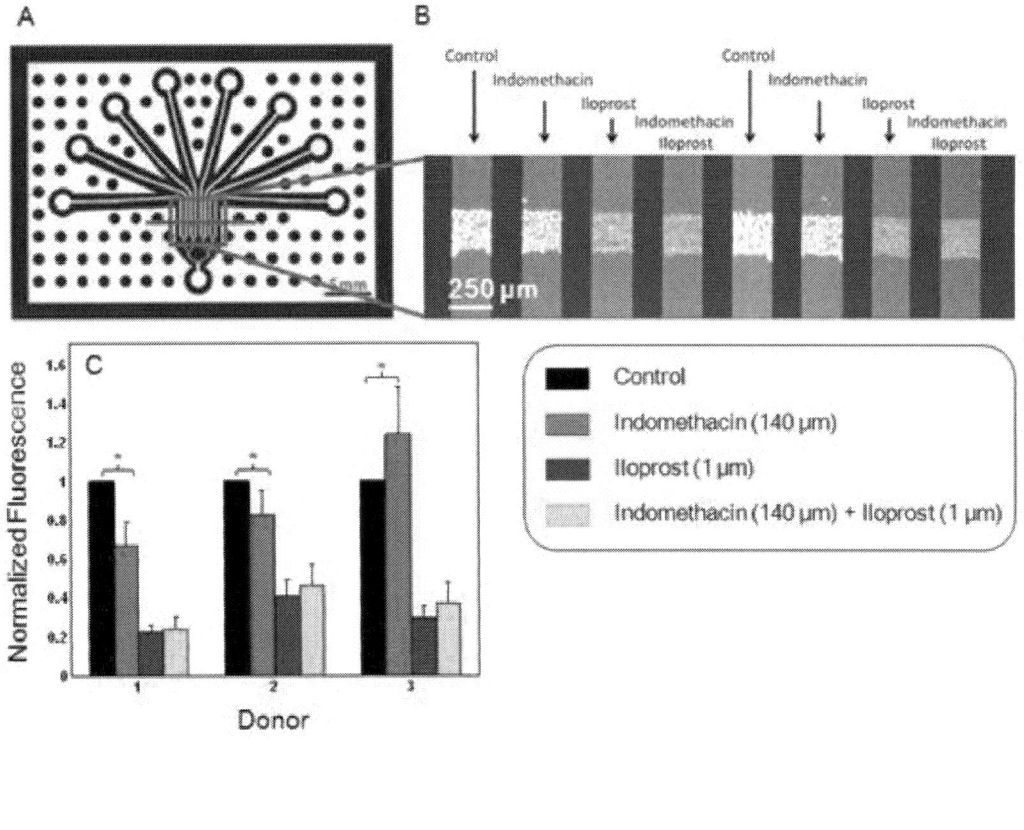

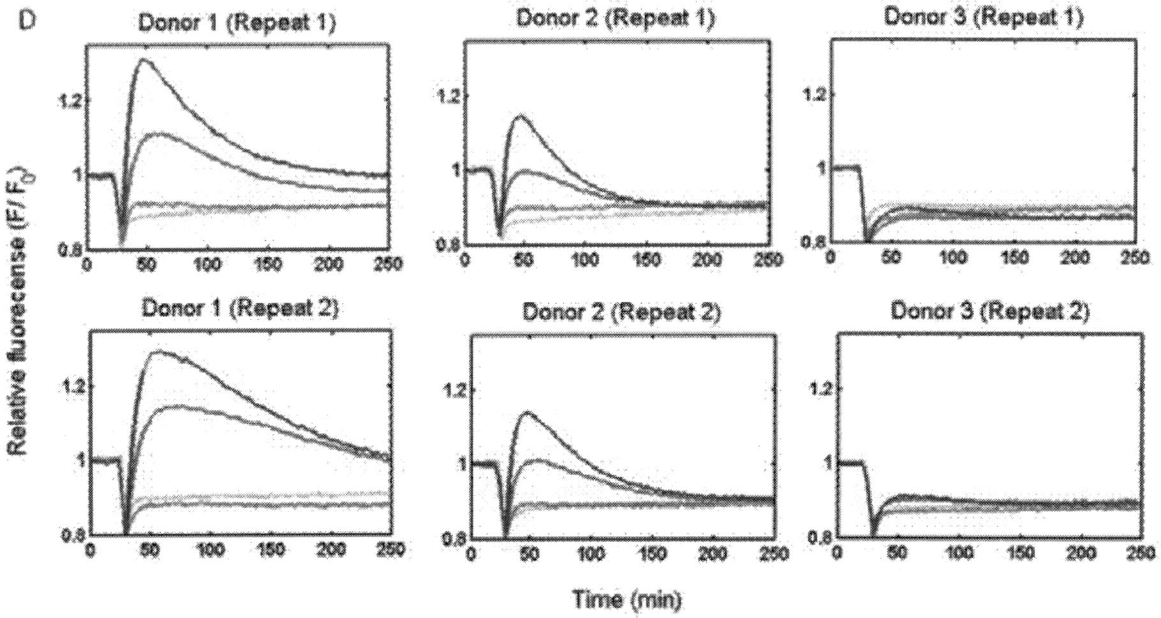

Figure 1 (A) Schematic of the 8 channel microfluidic device. 8 separate channels converge on one patterned collagen strip (blue) that can be imaged in a 2x field of view. **(B)** Whole blood (PPACK 100 μM), treated with the indicated anti-platelet agents, is perfused at 200 s^{-1} and platelet accumulation on the collagen surface is measured in 15 sec intervals. **(C)** Normalized platelet accumulation for three separate donors (n=16). Iloprost shows effectiveness for all 3 donors while Indomethacin shows effectiveness for donors 1 and 2 and an activating effect for donor 3. (* $p < .01$) **(D)** Calcium responses to 4 doses of U46619 (green:zero, pink:low, red:medium, blue:high) are presented for 3 donors in duplicate.

See more of this Session: Experimental Approaches In Systems Biology
See more of this Group/Topical: Food, Pharmaceutical & Bioengineering Division

(102a) Determination of Drop Size Distribution In a Centrifugal Contactor

Nicholas B. Wyatt[1], **Timothy J. O'Hern**[2] and Bion Shelden[1], (1)Sandia National Laboratories, Albuquerque, NM, (2)Engineering Sciences, Sandia National Laboratories, Albuquerque, NM

Centrifugal contactors are widely used in separation and mass transfer applications involving immiscible phases. Quantifying drop size distributions in such processes is critical to understanding and tuning mass transfer characteristics between immiscible phases. Many studies have focused on the complex flows in both the annular and separation regions of the contactor. However, the magnitude and distribution of droplet sizes in the mixing region has not been characterized. For mass transfer applications, drop size and, thus, surface area are key factors. Through the use of high speed imaging, we measure drop size distributions in the annular mixing region for mixtures of silicone oil and water. We measure the evolution of the drop size distribution as the material travels from the mixing region to the vane region. Further, we quantify size distributions for air bubbles that are entrained in the flow. The distributions are examined over a range of rotor speeds from 1100 to 2100 rpm for silicone oil viscosities from 5 to 20 cSt. Drop sizes for both air and oil (dispersed phase) are found to be normally distributed at locations near the surface of the mixing zone. Near the bottom of the mixing zone, the distributions become log-normal and the mean drop diameter shifts to lower values.

Sandia National Laboratories is a multi-program laboratory operated by Sandia Corporation, a wholly owned subsidiary of Lockheed Martin Corporation, for the U.S. Department of Energy's National Nuclear Security Administration under contract DE-AC04-94AL85000.

See more of this Session: Mixing In Multi-Phase Systems II
See more of this Group/Topical: North American Mixing Forum

(102b) Effect of Impeller Submergence On Flow Characteristics and Minimum Agitation Speed for Solids Suspension In Partially Filled Stirred Vessels for Different Impeller off-Bottom Clearances and Impellers Sizes

Shilan Motamedvaziri and Piero Armenante, Otto H. York Department of Chemical, Biological and Pharmaceutical Engineering, New Jersey Institute of Technology, Newark, NJ

In many industrial applications, mixing vessels have liquid height-to-tank diameter ratio H/T equal to, or larger than, 1. However, there are many instances where this ratio is lower than 1, as in all those cases in which the vessel is emptied. Even when H/T<1, sufficient agitation must still be provided in order to attain the desired process objectives. When the impeller submergence is reduced as a result of lowering the liquid level, the fluid dynamics of even a single-phase stirred liquid can become quite complex, with different regimes possibly existing depending on the geometric characteristics of the system (such as impeller clearance, liquid height, or liquid head above the impeller). The objective of this work is to determine the minimum liquid levels and the critical impeller submergence for different impeller off-bottom distances (C_T/T=0.12, 0.18, 0.37, 0.54 and 0.68) and different impeller diameters (D/T=0.26, 0.31 and 0.42) where adequate mixing process can still be achieved, both in a single liquid phase and in solid-liquid suspensions.

Computational Fluid Dynamic (CFD) simulations as well as Particle Image Velocimetry (PIV) were used here to study the system's hydrodynamics in flat-bottom baffled vessels equipped with a single Disk Turbine (DT) for different height-to-tank diameter ratios when the liquid level is smaller than the tank diameter. A commercial mesh generator (Gambit 2.4.6) coupled with a CFD software package (Fluent 6.3.26) was used to simulate the flow field. A standard κ-ε model as well as a realizable κ-ε model coupled with enhanced wall treatment was used to model the turbulence flow. Results were obtained using a multiple references of frames (MRF) approach. For the case where air entrainment occurred, a Volume of Fluid (VOF) model coupled with MRF simulation was used. Experimental work was also carried out with a single liquid phase as well as with solid-liquid suspensions with 0.5 wt% dispersed solids. The minimum agitation speed for complete solid suspension, N_{js}, at different filling ratios was obtained visually and then they compared with CFD predictions.

In general, good agreement between the experimental data and the predicted results for the velocities distribution, pumping number and N_{js} were obtained. Both the experimental and the computational results, for different impeller off bottom clearances and impellers sizes, show that there is a minimum liquid level (or minimum impeller submergence) below which: (1) the macroscopic flow pattern changes substantially; (2) the pumping number drop significantly; (3) air entrainment occurs. When this occurred, it was observed that the average velocity field and turbulence intensity close to the tank bottom decreased substantially, which could be the reason why solid suspension becomes problematic.

See more of this Session: Mixing In Multi-Phase Systems II
See more of this Group/Topical: North American Mixing Forum

(102c) Aggregation In Homogeneous, Isotropic Turbulence

Jos Derksen, Department of Chemical & Materials Engineering, University of Alberta, Edmonton, AB, Canada

Aggregation in homogeneous, isotropic turbulence

Aggregation of solid particles suspended in a carrier fluidis an important particle growth mechanism. In many cases it is an unwantedphenomenon as it interferes negatively with processes designed for generatingparticle size distributions based on nucleation and growth. In other casesaggregation is desired as it for instance improves the filterability of a slurry. Aggregation and flow dynamics are tightlyconnected. For aggregation events to occur, particles need to collide and sticktogether. For the bigger particles we are interested in, collisions are mainlybrought about by velocity gradients in the flow field the particles aresuspended in (orthokinetic aggregation); not so muchby Brownian motion. The same velocity gradients that bring together particlesalso exert mechanical forces on aggregates that could break them again. Westudy this dynamic process of the generation of aggregate size distributions(ASD's) by means of numerical simulations that fully resolve the fluid andparticle dynamics. Turbulence with well-defined properties is created by meansof linear forcing in a fully periodic, three dimensional domain. In the domainwe release uniformly sized spherical primary particles that we make sticky byletting them interact through a square-well potential. We study the ASD and itsdynamics as a function of the strength of the turbulence, and as a function ofthe strength of the square-well potential. The figure shows some impressions ofthe simulations. In the left panel all particles in the domain are shown; theyare colored by the size of the aggregate they are part of (red: small; blue:big). The right panel shows the four biggest aggregate at the same moment intime (each aggregate has a different color; the periodic conditions show in thesense that the red aggregate is connected through the side boundaries).

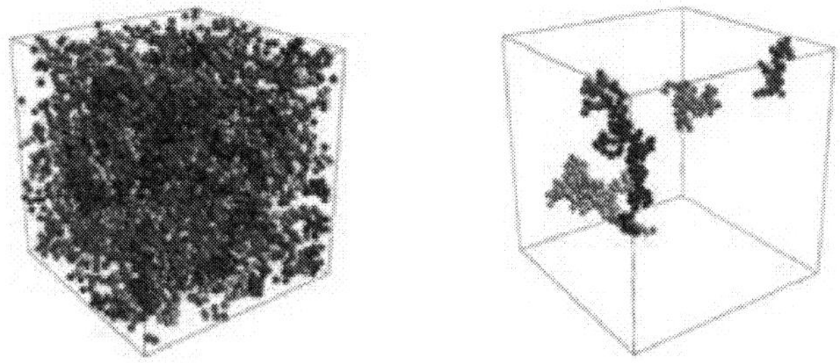

See more of this Session: Mixing In Multi-Phase Systems II
See more of this Group/Topical: North American Mixing Forum

(102d) Towards the Modelling of the Stabilization Process In Pickering Emulsions

Louis Fradette, URPEI, Chemical Engineering Department, Ecole Polytechnique de Montreal, Montreal, QC, Canada and Emir Tsabet, URPEI, Chemical Engineering, Ecole Polytechnique de Montreal, Montreal, QC, Canada

The replacement of traditional molecular surfactants by nano- or micrometer-sized solid particles in emulsification processes offers many advantages and creates new opportunities in the process industry. The emulsion stabilization mechanism with solid particles can be visualized in three steps: 1) the particle first approaches and reaches the fluid/fluid interface; 2) the particle adsorbs and gets trapped at the interface; 3) the adsorbed particles form a network that stabilizes the emulsion. At the moment, the state of knowledge does not allow for process design at the industrial scale mainly because of the lack of information about the overall stabilization mechanism. Being interested in emulsification process design, we are developing a methodology to predict the solid stabilized emulsions properties from the process conditions. The present work aims at developing a model describing the droplets stabilization on the basis of particle adsorption forces measurements using Atomic Force Microscopy. This approach allows us to identify the main parameters affecting the particles adsorption mechanism and therefore the stabilization rate, a condition to predict the droplets distribution size during the emulsification process.

See more of this Session: Mixing In Multi-Phase Systems II
See more of this Group/Topical: North American Mixing Forum

(102e) Effect of Oxygen Bubble Distributions On Rare Earth Precipitates In Molten Salts

Ryan W. Bezzant, Nuclear Engineering, University of Idaho-Idaho Falls, Idaho Falls, ID, Supathorn Phongikaroon, Department of Chemical Engineering and Nuclear Engineering Program , University of Idaho, Idaho Fall, ID and Michael F. Simpson, Pyroprocessing Technology Department, Idaho National Laboratory, Idaho Falls, ID

Effect of Oxygen BubbleDistributions on Rare Earth Precipitates in Molten Salts

Ryan W. Bezzant and Supathorn Phongikaroon

Universityof Idaho-Idaho Falls

Center forAdvanced Energy Studies

1776 Science Center Dr.

IdahoFalls, ID 83402 Michael F. Simpson Pyroprocessing Technology Department Idaho National Laboratory Idaho Falls, ID 83402

Abstract:

Pyroprocessing technologyhas been developed at the Idaho National Laboratory for treating spent nuclearfuel from Experimental Breeder Reactor-II. An electrorefiner (ER) is the keyunit operation of this process, recovering uranium metal via anodic dissolutionand electrodeposition at a cathode. During this process, minor actinides,fission products and rare earths (RE) form chlorides and accumulate in the LiCl-KCleutectic salt. This results in excess heat generation and other effects whichlower the system efficiency. To maintain ER performance, it is necessary toremove contaminated salt and refill with fresh salt. But this approach iscostly and generates a high waste volume.

Researchers at the KoreaAtomic Energy Research Institute have demonstrated oxygen sparging as aneffective way of separating rare earth chlorides from molten salt to partiallyprepare it for reuse. Rare earth chlorides react with oxygen at hightemperatures. Rare earth oxides and oxychlorides form and precipitate to thebottom of the salt container. After the salt cools, the salt can bemechanically separated from the rare earth fission products. Aside from ademonstration of the process, no fundamental aspects of this system are well understood.

Thus, to gain insight intophysiochemical parameters for systematic scale-up, such as mass transfer rateand gas holdup fraction, a transparent furnace with an oxygen sparging systemwas constructed with the capability of operating under an argon environment at500 °C. LiCl-KCl-$RECl_3$ternary salt was loaded in the quartz crucible to studythe effect of oxygen bubble sizes on rare earth precipitate in the system. Asimple two blade, downward pitch turbine was used for mixing the bubbles in themedium with a rotational speed varying between 100 and 250 RPM. The oxygenflowrate was varied between 0.05 L/min and 0.25 L/min. Oxygen concentrationwas detected by an oxygen sensor. Bubble sizes were captured by using a highspeed digital camera. Detailed results will be presented and discussed.

See more of this Session: Mixing In Multi-Phase Systems II
See more of this Group/Topical: North American Mixing Forum

(102f) Power Dissipation and Impeller Power Number In USP Dissolution Testing Apparatus 2

Yimin Wang and **Piero M. Armenante**, Otto H. York Department of Chemical, Biological and Pharmaceutical Engineering, New Jersey Institute of Technology, Newark, NJ

Dissolution testing is routinely used in the pharmaceutical industry to provide critical *in vitro* drug release information for quality control purposes and especially to assess batch-to-batch consistency of solid oral dosage forms such as tablets. The United States Pharmacopoeia (USP) lists several types of dissolution apparatuses. Of these, the most commonly used for solid dosage forms is the UPS Dissolution Testing Apparatus 2 consisting of an unbaffled, hemispherical-bottomed vessel equipped with a 2-blade radial impeller.

Significant information can be found in the literature of the power consumption of impeller-agitated, baffled systems. However, despite extensive industrial use and a large body of work on Apparatus 2, little or no information can be found in the literature on the power dissipated by this apparatus. Power consumption is a key parameter to predict for the rate of the dissolution process in such a system, especially for the prediction of mass transfer rate between the solid oral dosage form and the dissolution medium, since it has been shown previously that the power dissipated by an impeller rotating in a fluid correlates directly to the mass transfer coefficient experience by suspended particles. Thus, the determination of power dissipation is critical to quantify the mass transfer rate and hence be able to predict eventually the drug release rate from an oral dosage form during dissolution tests. Therefore, the main objective of this work was to experimentally determine and computationally predict the power dissipated by the impeller in the USP Dissolution Testing Apparatus 2 and the associated power number Po, and to relate these values to the mass transfer coefficient experienced by suspended drug particles.

A sensitive torque measurement apparatus was specifically built to measure the small torque applied to the liquid in the dissolution vessel when stirred by the paddle. To do so, the dissolution vessel and its content were floated in a shallow tank, and the force required to prevent rotation of this assembly was experimentally measured, from which torque and the power dissipation could be determined knowing the system geometry and the agitation speed.

A commercial CFD software package (FLUENT, 6.3.26; Gambit, 2.4.6) was used to predict the velocity distribution profiles in the USP 2 vessel, as well as the pressure distribution on the surface of the impeller blades from which the torque applied to the impeller and hence the impeller power dissipation were numerically calculated. Furthermore, the power dissipated by the impeller was additionally predicted by calculating, via CFD, the distribution of the local energy dissipation rate per unit mass in the USP 2 vessel, and then by integrating this variable over the entire liquid mass in the vessel.

The power dissipation was computationally and experimentally determined for a number of agitation speeds and liquid volumes. In general, good agreement was found between the experimental measurements and the computational results obtained with both methods. When Po was plotted against the impeller Reynolds number Re, a similar good agreement was typically found. As in other unbaffled system, Po was found to decrease slightly with Re, and was found to be in the range 0.25-0.38 for the case of the 900 mL volume fill level and in the range 0.20-0.29 for the 500-mL volume case. The values of the power dissipation per unit liquid mass were used to predict the mass transfer coefficient to suspended particles of different size and under different agitation conditions.

See more of this Session: Mixing In Multi-Phase Systems II
See more of this Group/Topical: North American Mixing Forum

(120a) A Combined Experimental and Model-Based Approach to Prediction of Particle Size Attrition In Agitated Dryers

Alexander J. Marchut[1], Ann M. Czyzewski[2], Raimundo Ho[3], Carlos Orihuela[3], Ahmad Sheikh[3], Fredrik L. Nordstrom[3], Thomas B. Borchardt[3] and Shailendra V. Bordawekar[3], (1)Process Research and Development, Abbott Laboratories, North Chicago, IL, (2)Process Engineering, Abbott Laboratories, North Chicago, IL, (3)Process Research & Development, Abbott Laboratories, North Chicago, IL

The final crystallization of an active pharmaceutical ingredient often requires meeting a target particle size specification in order to facilitate downstream processing of the material in formulation of the final dosage form. Agitated filter dryers, often used to prevent exposure of operations personnel to hazardous compounds, can alter the particle size produced from the crystallization. During filtration and drying, agitation can result in particle size attrition due to breakage of the particles or agglomeration of the particles into undesired lumps. In order to successfully scale up processes from small to large filter dryers while minimizing particle attrition and agglomeration, the mixing in filter dryers must be understood. Discrete Element Model (DEM) simulations can be used to characterize the mixing in filter dryers at various different scales and determine the forces and velocities that the particles experience, which can be linked to particle breakage via mechanical property assessment of the powder particles. We will report our results on simulations of mixing in filter dryers, the relation of particle size attrition to the mechanical properties of the powder particles, and our attempt to predict particle size attrition using the simulations and understanding of the fundamental mechanical properties of the API.

See more of this Session: Scale-up of Pharmaceutical Isolation and Drying Operations
See more of this Group/Topical: Food, Pharmaceutical & Bioengineering Division

(120b) Investigation of Acoustic Dryer for API Processing

David Lamberto[1], **Peter Lucon**[2], Dwynn R. Nyquist[3] and Josh E. Yarrington[3], (1)Chemical Process Development & Commercialization, Merck & Co., Inc., Rahway, NJ, (2)Resodyn Corporation, Butte, MT, (3)Resodyn Acoustic Mixers, Butte, MT

Abstract for AIChE

4/29/2011

Investigation of Acoustic Dryer for API Processing

David J. Lamberto, Merck & Co., Inc., Rahway, NJ

Peter Lucon, Dwynn Nyquist, Josh Yarrington, Resodyn, Butte, MT

The development of novel drying technologies which improve drying rates and avoid particle breakage and loss of crystallinity would be beneficial to many industries. In particular, isolation of crystalline products in the pharmaceutical industry involves operations that frequently change the physical characteristics of the solids being isolated. Particle size reduction, undesired agglomeration, and the generation of amorphous material as a result of applied mechanical stress have been observed. Thus far, procedures to avoid, or minimize these issues can lead to reduced process efficiencies and increased development times.

The use of the ResonantAcoustic(R) mixer technology has the potential to lead to advances in the drying of crystalline materials through the use of low frequency sonic energy for material fluidization with minimal mechanical stress on the particles. The new technology avoids the negative aspects of mechanical agitation and does not have the complications associated with traditional fluidized bed systems for Geldart Group C materials. Mixing is based on intrinsic material properties and is less dependent on extrinsic equipment parameters, and therefore scale-up development efforts can be minimized and transition to production facilities is simplified.

In this study, the performance of a prototype acoustic dryer is compared with a conventional conical dryer using two pharmaceutically relevant compounds which have widely different drying behaviors.

Data were obtained under similar operating conditions of temperature and pressure for the two technologies. Mixing in the acoustic dryer was conducted within a fixed acceleration range for all cases while both continuous and the more typical intermittent agitation conditions in the conical dryer were explored.

As expected, for the compounds investigated, the use of continuous agitation in the conical dryer enhanced the drying rates as compared with the use of intermittent agitation. In one case the rates achieved were higher than that achieved in the prototype acoustic dryer while for the other the rates were equivalent. The acoustic dryer showed less temperature and moisture variations across the dryer as compared with the conical dryer. In addition, placement of the temperature probe directly into cake inside the acoustic dryer versus at the wall increased the accuracy of the actual cake readings throughout the drying process. The acoustic dryer achieved drying rates 2-3 times higher as compared with the conical dryer operated with the less aggressive intermittent agitation (commonly used to minimize particle attrition). Particle size reduction or aggregation within the acoustic dyer was minimal and similar to that achieved with intermittent agitation in the conical dryer.

The new acoustic dryer technology has the potential to be a platform drying technology and provide a technical competitive advantage over existing technology. The enhanced heat and mass transport and reduced impact on physical properties will result in reduced cycle times, higher product quality and enhanced process portability. The acoustic dryer will find wide application for products where mechanical agitation is currently limited in traditional drying equipment. Continued optimization of the acoustic dryer should lead to improved performance, better process control, and increased applicability across a wider portfolio of compounds.

See more of this Session: Scale-up of Pharmaceutical Isolation and Drying Operations
See more of this Group/Topical: Food, Pharmaceutical & Bioengineering Division

(120c) Near-Infrared Spectroscopy for On-Line Drying End-Point Determination In a Small-Scale Fluidized Bed

Nicolas Heigl, Research Center Pharmaceutical Engineering GmbH, Graz, Austria, Johannes G. Khinast, Institute for Process and Particle Engineering, Graz University of Technology, Graz, Austria, Benjamin Glasser, Dept. of Chemical and Biochemical Engineering, Rutgers University, Piscataway, NJ and Fernando J. Muzzio, Chemical and Biochemical Engineering, Rutgers, The State University of New Jersey, Piscataway, NJ

A method based on near-infrared spectroscopy (NIRS) with partial least squares regression (PLSR) was developed to determine the drying end-point of fluid bed dried dibasic calcium phosphate anhydrous (DCPA). In a first approach, parameters were adjusted according to experimental designs to induce structured variations to (i) the processing, i.e. changing air flow-rates and amounts of DCPA, and (ii) the material properties, i.e. changing DCPA particle size compositions. Response surface modeling was consequently applied to predict the drying end-points at varying process and material characteristics within the confined design regions. Laser diffraction measurements of the DCPA before and after the fluidized bed drying revealed particle size changes due to the processing, i.e. potential particle agglomeration, attrition, breakage or fracture. In a second approach, NIRS was implemented to quantify the residual water content of the DCPA. Therefore the NIR spectra were recorded on-line with a fiber optic probe from outside the processor and off-line, i.e. samples were consecutively thief-probed from the processor, put into glass vials, and analyzed with an integrating sphere device. Furthermore, background spectra of dry DCPA which was fluidized at respective process air velocities were collected in an effort to correct for potential absorption interferences from the processor material (polymethylmethacrylate) and the glass vials. Subsequently, the performances of the thereby developed PLSR models with traditionally developed PLSR models (Spectralon® as a background material) were compared. Loss-on-drying (LOD) was used as a reference method and confirmed the actual moisture content of the DCPA. The experiments showed that (i) the thief-probed sample amount from the processor and thus the sample weight, and (ii) the down-time between taking a sample and its actual analysis with NIR and LOD, can bias the moisture content predictions. Consolidation variations of the DCPA inside the glass vials did not play a significant role. Continuous mixing, and thus the collection of representatively averaged spectra for the PLSR modeling and for further predictions, and a maximum available sample amount during the spectral data acquisition are considered the main advantages of dynamic or on-line NIR monitoring over NIR off-line analysis.

[1] N. Heigl et al; Quantitative On-Line vs. Off-line NIR Analysis of a Fluidized Bed Drying Process with Consideration of Sub-sampling Issues; Manuscript in preparation.

[2] N. Heigl et al; Near-infrared Spectroscopy with Experimental Design for Characterizing the

Fluidized Bed Drying Process of Dibasic Calcium Phosphate Anhydrous; Manuscript in preparation.

See more of this Session: Scale-up of Pharmaceutical Isolation and Drying Operations
See more of this Group/Topical: Food, Pharmaceutical & Bioengineering Division

(120d) Developing a Robust Isolation and Drying Protocol for An API with Unique Physical Properties

Amanda Rogers, Saravanababu Murugesan, Alan Braem, Prashant Deshpande, William Bartels and Jose Tabora, Bristol-Myers Squibb Company, New Brunswick, NJ

The powder properties of APIs have the potential to impact downstream processing and affect the quality of the drug product. For this reason, it is important to deliver material with consistent physical characteristics for drug product manufacture. This requires that material attributes including form, particle size distribution and crystallinity be controlled throughout the isolation and drying processes of the API manufacture. The evaluation and design of isolation and drying unit operations is challenging due to their inherent sensitivity to scale. Performance can vary significantly as a function of equipment selection, geometry, and size. In addition, material behavior changes throughout the drying process, as wetted granular flows are often quite different from those of dried powders.

The present work describes a case study of a compound with unique material properties including a low melting point onset, potential for desolvation under shear, low bulk density and a propensity to bridge during solids conveying operations. This amplified the sensitivity of process performance to equipment configuration and scale. The phenomena observed at the pilot plant and manufacturing scales were not reproducible in the laboratory, so experimentation at manufacturing scale was required to develop an effective and robust isolation and drying protocol. Equipment selection, configuration and operating conditions along with the implementation of PAT were important factors in the development and ultimate success of the designed process. The developed process was demonstrated in numerous pilot plant batches and manufacturing scale trials. The final process was successfully transferred to a manufacturing site.

See more of this Session: Scale-up of Pharmaceutical Isolation and Drying Operations
See more of this Group/Topical: Food, Pharmaceutical & Bioengineering Division

(120e) Developing Dynamic Models of Pharmaceutical Drying Operations Using On-Line Mass Spectroscopy

Aditya Gopi, Department of Chemical Engineering, University of Massachusetts, Amherst, Amherst, MA, Kostas Saranteas, Process Chemistry and Engineering, Sunovion Pharmaceuticals, Marlborough, MA and **Michael A. Henson**, Department of Chemical Engineering, University of Massachusetts Amherst, Amherst, MA

Drying is a critical downstream process in pharmaceutical manufacturing. This type of liquid-solid separation is subject to stringent constraints because the active pharmaceutical ingredient (API) can degrade if the drying temperature is too high while long batch times can result if the temperature is too low. Because pharmaceutical manufacturing is typically accomplished through batch operation, overall process productivity is strongly affected by the batch times of the individual unit operations. Drying often represents a process bottleneck because the moisture content of the wet cake is difficult to measure in real time. As a result, the drying process may be run far longer than necessary to ensure that the moisture content drops to an acceptable level. We believe that the combination of on-line moisture content analysis and dynamic process models offer the potential to reduce and ultimately minimize dryer batch times.

The objective of this study was to use on-line mass spectroscopy to develop dynamic models capable of predicting cake moisture content in a vacuum tray dryer. To demonstrate generality of the approach, we considered both diffusion-type models and combined heat-mass transfer models. Unlike many simpler models available in the literature, these models allowed the prediction of both the constant and falling rate drying regimes without resorting to artificial switching between individual models. As a first step towards pharmaceutical applications, we first developed models for a simple experimental system consisting of glass beads as the solid matrix and different solvents. We studied a wide range of particle sizes (10-200 microns) and explored the effects of particle size, initial moisture content, cake depth, drying temperature and solvent type on drying behavior with moisture content profiles resolved on a second time scale by on-line mass spectroscopy. The data was used to develop and validate diffusion and heat-mass transfer models for predicting cake moisture content from gas phase moisture measurements. We believe that this technology holds considerable promise for monitoring and optimizing drying operations in the pharmaceutical industry.

See more of this Session: Scale-up of Pharmaceutical Isolation and Drying Operations
See more of this Group/Topical: Food, Pharmaceutical & Bioengineering Division

(120f) Development of Laboratory Methods ("Predictive Tools") to Preserve Particle Properties of APIs Upon Scale-up to Agitated Filter Driers

Mark T. Maloney[1], David J. am Ende[1], Steven J. Brenek[1] and Brent Maranzano[2], (1)Chemical R&D, Pfizer, Inc, Groton, CT, (2)Analytical R&D, Pfizer, Inc, Groton, CT

Agitated Filter Driers (AFD's) are commonly used in the pharmaceutical manufacturing for performing both filtration and drying operations. Successful scale-up from lab to manufacturing AFD equipment requires that the physical properties be met. These properties include product quality such as purity and particle size. Depending on the API-solvent system and other operational parameters, undesired attrition or agglomeration may occur impacting the desired particle properties. Of importance is increased understanding and better prediction of scale-up performance. The focus of this presentation will be to discuss recent advances in laboratory methods and predictive tools to help us better assess the risk of scale-up for agglomeration and attrition potential upon scale-up. In addition progress on enhancing cycle times for difficult to dry materials will also be presented.

See more of this Session: Scale-up of Pharmaceutical Isolation and Drying Operations
See more of this Group/Topical: Food, Pharmaceutical & Bioengineering Division

(132a) Rapid Characterization and Engineering of Natural Product Biosynthetic Pathways Via DNA Assembler

Zengyi Shao[1], Yunzi Luo[1] and Huimin Zhao[2], (1)Chemical & Biomolecular Engineering, University of Illinois at Urbana-Champaign, Urbana, IL, (2)Department of Chemical and Biomolecular Engineering, Department of Chemistry and Biochemistry, Institute for Genomic Biology, University of Illinois at Urbana-Champaign, Urbana, IL

Microorganisms and plants have been evolved to produce a myriad array of complex molecules known as natural products or secondary metabolites of pharmaceutical and biological importance. Sequenced genomes and metagenomes provide a tremendously rich source for discovery of new genes and pathways. However, the discovery and sustainable production of valuable natural products are often hampered owing to our limited ability to manipulate the corresponding biosynthetic pathways. Here we developed a synthetic biology strategy based on the DNA assembler method we developed previously for discovery, characterization, and engineering of natural product biosynthetic pathways.

By taking advantage of the highly efficient yeast *in vivo* homologous recombination mechanism, this strategy synthesizes an entire expression vector containing the target biosynthetic pathway and the genetic elements needed for DNA maintenance and replication in *Saccharomyces cerevisiae*, *Escherichia coli*, and a target heterologous expression host in a single-step manner. Since all the fragments are prepared by PCR, various manipulations including site-directed mutagenesis, scar-less gene deletion and artificial gene cluster construction can be easily achieved. Such a strategy with unprecedented flexibility and versatility has many applications in heterologous expression, pathway functional studies, and combinatorial biosynthesis.

As proof of concept, two important biosynthetic pathways from Streptomyces including the aureothin and spectinabilin biosynthetic pathways were selected for characterization and engineering. Aureothin, isolated from *Streptomyces thiolutes*, is a nitro-aryl containing polyketide exhibiting antitumor, antifungal, and insecticidal activities. Using DNA assembler, we reassembled the aureothin biosynthetic pathway in the heterologous host *Streptomyces lividans*, and mutagenized the active site residues of the dehydratase domains and the ketoreductase domains in AurA and AurB. As a result, two aureothin derivatives were generated by the mutated pathways, with their corresponding polyketide chain modified. Spectinablin, isolated from *Streptomyces spectabilis* and *Streptomyces orinoci*, is a nitro-phenyl substituted polyketide exhibiting antimalarial and antiviral activities. Although both hosts were reported to produce spectinabilin, only the pathway from *S. spectabilis* can be functionally expressed in *S. lividans*. Examining the two gene clusters revealed two major differences. First of all, the cluster from *S. spectabilis* contains three extra genes, *spnK*, *spnL* and *spnM*, which are not found in *S. orinoci*. Secondly, these two clusters undergo different regulations as the cluster from *S. spectabilis* contains

SpnD as an activator and the cluster from *S. orinoci* contains NorD as a repressor. Previous bioinformatics study showed that these three genes are only involved in up-regulating substrate concentration and product transport. To investigate the mechanisms for the failure of heterologous spectinabilin production, *spnK* was inactivated and *spnM* and *spnL* were deleted from the cluster *via* DNA assembler. As a result, all three mutants could still synthesize spectinabilin in *S. lividans*, indicating these three genes are not essential for spectinabilin biosynthesis, supporting our bioinformatics study. Therefore, we hypothesized the silencing of the spectinabilin biosynthesis is due to the altered regulation in *S. lividans.*

See more of this Session: Advances In Biocatalysis and Biosynthesis II
See more of this Group/Topical: Food, Pharmaceutical & Bioengineering Division

(132b) Engineering the Microalgae Strain *Chlamydomonas Reinhardtii* for Xylose Utilization

Azadeh Pourmir and Tyler Johannes, Chemical Engineering, The University of Tulsa, Tulsa, OK

Microalgae have been recognized as potentially great sources for the production of biofuels because of their high oil content and rapid biomass production. For achieving the full potential of microalgae, genetic improvements are necessary in order to enhance the capabilities of native strains to make algae-based processes more economically attractive. Microalgae are well known for their ability to be grow photoautotrophically; however higher biomass yields can be achieved when algae is grown heterotrophically. Currently a great amount of research effort is being devoted to breaking down lignocellulosic biomass into fermentable sugars. The fermentable sugars derived from cellulosic feedstocks vary significantly depending on the feedstock and the conversion process. The pentose sugars xylose and arabinose typically constitute a substantial portion of the total sugar content from these different feedstocks. This work seeks to engineer microalgae strain *Chlamydomonas reinhardtii* to metabolize the pentose sugar xylose by introducing a fungal D-xylose utilization pathway into the chloroplast genome of *C. reinhardtii*.

See more of this Session: Advances In Biocatalysis and Biosynthesis II
See more of this Group/Topical: Food, Pharmaceutical & Bioengineering Division

(132c) Kinetic Modeling of Isobutanol Fermentation by Recombinant Escherichia Coli

Yi Xiao, Xueyang Feng, Huifeng Yu, Arul Varman and Yinjie Tang, Washington University in St. Louis, Saint Louis, MO

This study explored the kinetics of isobutanol fermentation from an engineered *E. coli* strain which was constructed by introducing a *kivd* gene encoding 2-ketoisovalerate decarboxylase and an *adhA* gene encoding aldehyde reductase. Using a well-controlled bioreactor, we performed fermentation under aerobic and oxygen-limited conditions with minimal and rich media. From these experiments, four general observations emerged. 1) Isobutanol production was correlated with biomass growth, and it reached 0.95 g/L in a closed oxygen-limited bioreactor. 2) Oxygen-limited conditions generated significant amount of byproducts (lactate and ethanol); 3) Byproduct acetate reduced alcohol fermentation and biomass growth via non-competitive inhibition; 4) Nutrients enhanced biomass growth and offered building blocks for faster isobutanol production (inferred from isotopic experiments), but they also generated more byproducts and reduced the yield of isobutanol from glucose. Based on the experimental data, we then built an empirical Monod-based model that described the time-dependent bioprocess variables, including glucose consumption, biomass growth, isobutanol production, and metabolic byproduct secretion (ethanol, lactate, and acetate). Such a kinetic model can provide important knowledge underlying the design of optimal bioprocess conditions for higher alcohol production.

See more of this Session: Advances In Biocatalysis and Biosynthesis II
See more of this Group/Topical: Food, Pharmaceutical & Bioengineering Division

(132d) Uncovering Cryptic Biosynthetic Pathways Using Synthetic Biology

Yunzi Luo[1], Zengyi Shao[1] and Huimin Zhao[2], (1)Chemical & Biomolecular Engineering, University of Illinois at Urbana-Champaign, Urbana, IL, (2)Department of Chemical and Biomolecular Engineering, Department of Chemistry and Biochemistry, Institute for Genomic Biology, University of Illinois at Urbana-Champaign, Urbana, IL

Computational analysis of sequenced genomes and metagenomes has revealed numerous gene clusters that are potentially involved in the biosynthesis of natural products. However, many of these gene clusters produce unknown products. Hence they are called cryptic biosynthetic pathways. We recently developed a highly efficient one-step method, so-called "DNA assembler", for rapid construction of a multi-gene biochemical pathway via *in vivo* homologous recombination in *Saccharomyces cerevisiae*. Here we extended this synthetic biology approach to study cryptic biosynthetic pathways. As proof of concept, one pathway from *Streptomyces griseus*, SGR810-815, containing a PKS/NRPS hybrid gene, was chosen as a model system. To decipher this target gene cluster, seven overlapping DNA fragments covering the entire gene cluster and three helper DNA fragments carrying the genetic elements needed for DNA maintenance and replication in *S. cerevisiae*, *E. coli*, and the target heterologous expression host- *Streptomyces lividans* were PCR-amplified from the genomic DNA of the native producer and the corresponding vectors, respectively. All of the resulting DNA products were subsequently co-transformed into *S. cerevisiae* for assembly into a single DNA molecule. The isolated plasmids were transformed into *E. coli* for plasmid enrichment and verification. Three verified constructs were selected and transformed into *Streptomyces lividans* for heterologous expression of the target gene cluster. At the same time, two modified constructs, one with an ErmE promoter added in front of the whole gene cluster, the other with a different promoter added in front of each of the six pathway genes, were assembled. First, qPCR was performed to track the transcription of each gene. The results indicated that the original gene cluster showed poor expression of every gene and the native promoter was in-active while the ErmE promoter construct showed enhanced level of gene expression and was compatible to the native producer. LC-MS analysis of the cell extracts from all the constructs indicated that a peak with molecular weight of 510 may be the final product of the gene cluster. Further production and structural characterization is in progress.

See more of this Session: Advances In Biocatalysis and Biosynthesis II
See more of this Group/Topical: Food, Pharmaceutical & Bioengineering Division

(132e) Substrate Specificity Change of a Dehydrogenase

Michael J. Abrahamson and Andreas S. Bommarius, School of Chemical and Biomolecular Engineering, Georgia Institute of Technology, Atlanta, GA

Biocatalysts are increasingly used in industry to create enantiomerically pure compounds. Enantiomeric compounds have a broad range of uses including an utmost importance in pharmaceuticals. Biocatalysis employs enzymes to produce enantiomerically pure compounds circumventing the difficulties of current production methods. Contemporary methods of protein engineering, such as applying rational design guided by mechanistic and structural knowledge, have greatly increased the ability to create novel enzyme functionality. We have developed a novel amino acid dehydrogenase (AADH) with broad substrate specificity.

The novel amino acid dehydrogenase was developed from an existing amino acid dehydrogenase scaffold. Expansion of its substrate specificity was achieved through several rounds of focused mutagenesis. Constraint of mutant library size was guided by mechanistic and structural knowledge, ultimately reducing the screening requirements while maintaining an increased chance of generating alternate substrate specificity.

See more of this Session: Advances In Biocatalysis and Biosynthesis II
See more of this Group/Topical: Food, Pharmaceutical & Bioengineering Division

Asymmetric reduction of α–hydroxy aromatic ketones to chiral aryl vicinal diols using carrot enzymatic system

Xiang Liu [1,*], Yi Wang [1] and Jian He Xu [2]

[1] School of Chemical and Material Engineering, Jiangnan University, Wuxi, China
[2] State Key Laboratory of Bioreactor Engineering, East China University of Science and Technology, Shanghai, China

Abstract

Asymmetric reduction of α–hydroxy aromatic ketones was carried out by using carrot enzymatic system and the corresponding chiral vicinal diols with special functional group were received. The optimal reaction conditions were gained after the investigation of various influencing factors. Chiral aryl vicinal diols were obtained with good yield and excellent enantiomeric excess under the appropriate conditions. Meanwhile, the steric factors and electronic effects of the substituents at the aromatic ring were shown to have an interesting effect on the yield and enantioselectivity.

Keywords: asymmetric reduction; chiral vicinal diol; aromatic ketone; carrot cell

1 Introduction

Exploitation of synthetic procedures for useful chiral building blocks is one of the main subjects in the area of organic chemistry [1]. Due to its high chemo-, regio-, and stereo-selectivity and non-toxicity, enzyme catalysis has become an ideal tool for asymmetric and green syntheses [2-3]. Moreover, using whole cells instead of isolated enzymes as biocatalysts is a relatively simple handling that does not require complex protein separation procedures or the addition and/or recycling of exogenous cofactors. There are extensive works on the asymmetric reduction catalyzed by whole cells with excellent enantioselectivity and yield [4-6]. However, the whole cells catalytic systems that previous reported were mainly generated from microbial (fungal, bacterial, and yeast) sources [7-11]. By comparison, less attention has been given to the use of plants as potential sources of enzymes that could conduct selected organic reactions [12-15].

Chiral aryl vicinal diols with special functional group are potentially valuable synthetic intermediates for the preparation of pharmaceuticals, agrochemicals and pheromones [16-18]. The reduction of biocatalytic carbonyl to enantiopure secondary alcohols is one of the most convenient methods for establishing chirality. Herein we explored the asymmetric reduction of prochiral α–hydroxy aromatic ketones catalyzed by carrot (*Daucus carota* L.) and obtained corresponding chiral aryl vicinal diols (Scheme 1).

E-mail address: liuxiang@jiangnan.edu.cn (X. Liu)

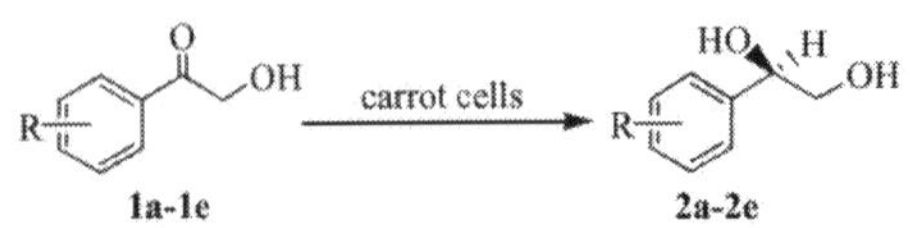

Scheme 1. Asymmetric reduction of **1a–1e** catalyzed by carrot cells

2 Experimental

Fresh carrot roots were purchased from a local market. The external layer of these roots was removed and the rest was carefully cut into small thin pieces (approximately 5 mm ×5 mm × 2 mm). Alpha hydroxy aromatic ketones **1a–1e** were synthesized by transformation of α-bromo aromatic ketones catalyzed by cesium formate in dry methanol [19]. The racemic diol standards were prepared by chemical reduction of the **1a–1e** with sodium borohydride in ethanol [20]. Other chemical reagents were of analytic purity and purchased from local suppliers.

General procedure for the asymmetric reduction of α–hydroxy aromatic ketones to chiral aryl vicinal diols: The small thin pieces of fresh carrot roots (10.0 g) were placed in flasks with 50 mL of 0.1 mol/L phosphate buffer or distilled water. To this suspension, 30-70 mg of **1a** was added and shaken at 28-36℃ for 20-36 h in an orbital shaker (120 r/ min). The suspension was filtered and the filtrate was extracted with ethyl acetate (3 ×20 mL). The organic phases were dried over anhydrous sodium sulfate and evaporated in vacuum. The final products were purified by column chromatography. The enantiomeric composition of product was established by HPLC equipped with Chiralcel OB-H column (4.6 mm × 25cm; Daicel Chemical Ind, Ltd., Japan) or Chiralcel OD-H column (4.6 mm × 250mm; Daicel Chemical Ind, Ltd., Japan). Configuration of product was determined by the sign of the specific rotation.

3 Results and discussion

3.1 Effect of substrate concentration on yield and ee of (*R*)–1–phenyl–1,2–ethanediol

The effect of substrate concentration on the yield and the ee of **2a** were investigated. As shown in Fig.1, the yield of **2a** was seriously influenced by the substrate concentration. The activity of plant cells were readily inhibited by substrate **1a** when the substrate concentration was equal or superior to 1.0 g/L. The yield rapidly decreased with increase of substrate concentration. The probable reason for this is that the intracellular activity of carrot cells gradually decreased until inactivated because of the toxicity of substrate to the cells. On the other hand, the ee of **2a** gave no noticeable differences in the concentration range between 0.6 g/L and 1.4 g/L. The ee of **2a** kept above 96.7% in all tested cases. Thus, the 1.0 g/L substrate concentration was appropriate when considering the factors of both yield and ee in this experiment.

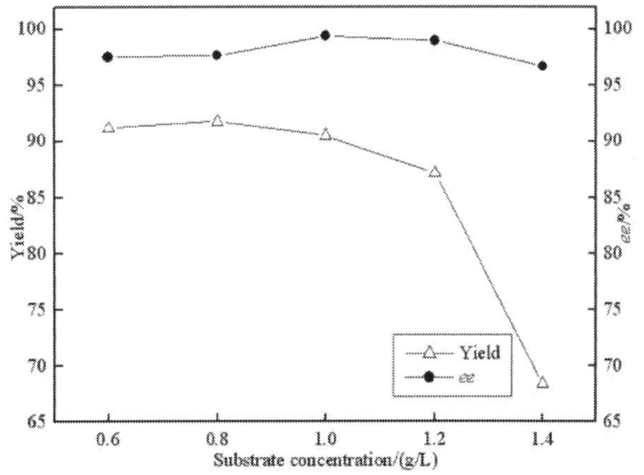

Fig.1 Effect of substrate concentration on the yield and the ee of **2a**.

Reduction conditions: various substrate concentration, 32℃, distilled water, 28h.

3.2 Effect of temperature on yield and ee of (*R*)–1–phenyl– 1,2–ethanediol

Fig.2 demonstrates that both the yield and ee rose as the reaction temperature increased. The yield and ee reached the peak value at 32℃ and 30℃, respectively, but after that, both of them gradually diminished. In a certain reaction time, the yield have a nearly relationship with reaction rate. On the one hand, the reaction rate increased as the temperature raised up, which enhanced the yield; but on the other hand, too high temperature would make enzyme inactivation that diminished the reaction rate, as a result, the yield of **2a** declined. Compare to yield, the enantiopurity of **2a** had less influence by the reaction temperature. However, the ee of product decreased with increasing of temperature because the interaction between enzyme and substrate was gradually destroyed. When the temperature decreased, the enantiopurity of product also was not high for the molecular movement rate slowing down which affected the activity of enzyme. The temperature for asymmetric reduction at 32℃ was chosen because of good yield while maintaining relatively high enantioselectivity.

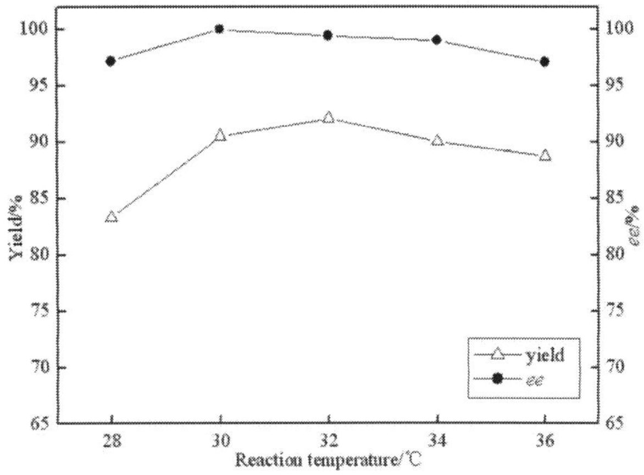

Fig.2 Effect of reaction temperature on the yield and the ee of **2a**.
Reduction conditions: various reaction temperatures, 1.0 g/L **1a**, distilled water, 28h.

3.3 Effect of reaction medium on yield and ee of (*R*)–1–phenyl–1,2–ethanediol

As can be seen in Fig.3, the pH of reaction medium exhibit noticeable effect on yield and enantiopurity of **2a**. The yield markedly rose as the pH of reaction medium increased. The interesting results are that both yield and enantiopurity of **2a** reached the peak value in distilled water not in aqueous buffer system. **2a** was obtained with 99.4% ee and 90.5% yield owning to high activity of the carrot cells. The pH of reaction medium affected reaction mainly as a result of influencing the enzymatic activity, it can damage the spatial structure of enzyme protein if too high or too low, causing changes in enzyme conformation and thus affecting the enantioselectivity and activity. Furthermore, the pH affects the ionization of substrates which influence the interaction between substrates and enzyme, or the products can not be generated. In our work, the distilled water of pH 7.0 was chosen as the reaction medium.

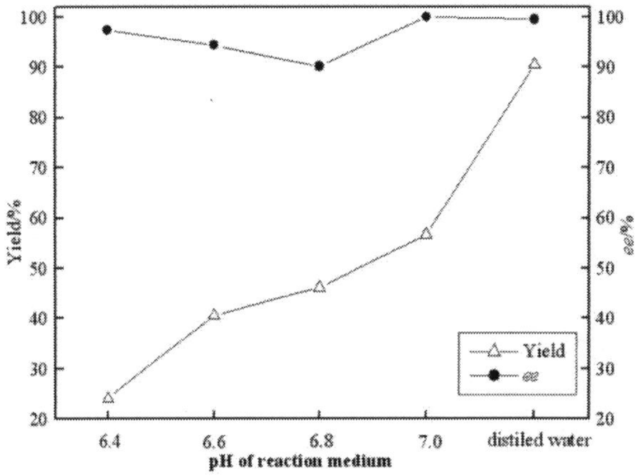

Fig.3 Effect of pH of reaction medium on yield and ee of **2a**.

Reaction conditions: various pH, 1.0 g/L **1a**, 32°C, 28h.

3.4 Effect of reaction time on yield and ee of (*R*)–1–phenyl–1,2 –ethanediol

Fig.4 shows the yield and ee changed depending reaction time. Both the yield and *ee* didn't instantly went-up with the reaction time's lengthening until at 32h, and after that, the yield and enantiopurity of the product gradually started to go down. The reason for this may be assumed that some adverse reactions happened in such conditions and product inhibition happened as the over much reaction time. The phenomenon indicates that the reaction time must be moderate. The 28 h reaction time was selected in this experiment.

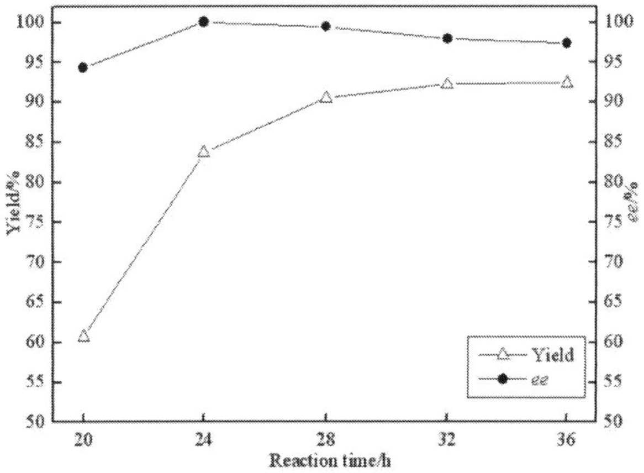

Fig.4 Effect of reaction time on yield and ee of **2a**.

Reaction conditions: various reaction time, 1.0 g/L **1a**, 32℃, distilled water.

3.5 Effect of substituted groups on yield and ee of aryl vicinal diols

The optimal reaction conditions were applied to the asymmetric reduction of a series of α-hydroxy aromatic ketones **1b-1e**, and the results are shown in Fig.5.

As can be seen in Fig.5, a series of substrates was successfully reduced under the appropriate conditions, giving the corresponding diols **2a-2e** in good yield and excellent ee. As the best model substrate, α-hydroxyacetophenone (**1a**) can be effectively reduced to (*R*)–1–phenyl–1,2–ethanediol(**2a**), the yield and ee can reach to 90.5% and 99.4%, respectively, which indicated that steric factors of substituted obstructed interaction between substrates and enzyme, leading yield declined. However, the results of reduction α-hydroxy-3-chloroacetophenone(**1d**) and α-hydroxy-4-chloroacetophenone(**1e**) are more attractive in term of the yield than the rest, which inferred substrates with electron-withdrawing substituted groups are more acceptable to plant tissue. Besides, the substrates with electron-donating groups afforded their products (**2b,2c**) with higher ee than those of others, it seems that the electron-donating groups are benefit to enantiopurity. Therefore, the electronic effects, caused by the substituted group, makes the carbonyl reductases present in plant tissue more efficient.

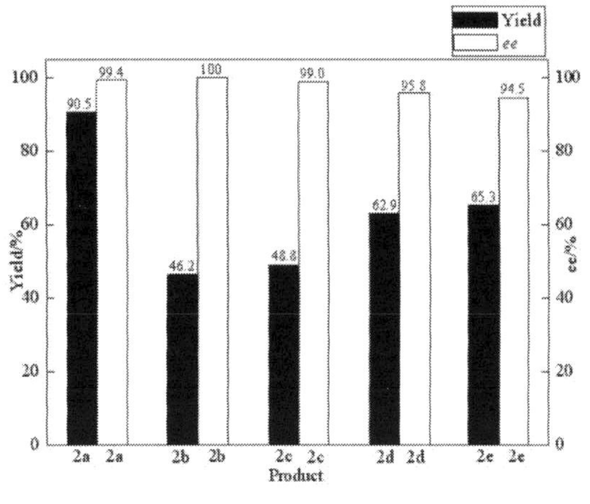

Fig.5 The variational trend of yields and ee of different products.

Reaction conditions:1.0 g/L substrates, 32℃, distilled water, 28h.

4 Conclusions

Bio-reduction catalyzed by dehydrogenases enzymes present in carrot cells provides an attractive approach in reducing various types of α–hydroxy aromatic ketones with good yields and excellent enantioselectivities. The enantioselectivity was the same as Prelog's rule as expected. The steric factors and electronic effects of substituted groups had a clear effect on the yield and enantioselectivity of the reaction. This provides a very clean, simple and convenient method for preparation chiral aryl vicinal diols which are important chiral building blocks often involving in the production of a wide range of fine chemicals and pharmaceutical molecules.

Acknowledgments

This research was financially supported by the National Special Fund for State Key Laboratory of Bioreactor Engineering (No. 2060204) and the National Natural Science Foundation of China (No. 20672037).

References

[1]Ohkuma T, Utsumi N, Watanabe M, Tsutsumi K, Arai N, Murata K. Org. Lett., 2007, 9(13): 2565–2567
[2]Schmid A, Dordick J S, Hauer B, Kiener A, Wubbolts M, Witholt B, Nature., 2001, 409: 258
[3]Woodley J M, Trends Biotechnol., 2008, 26: 321
[4]Rodrigues J A R, Moran P J S, Fardelone L C. Food Technol Biotechnol., 2004, 42: 295–303
[5]Kroutil W, Mang H, Edegger K, Faber K. Curr Opin Chem Biol 2004, 8:120–126
[6]Nakamura K, Yamanaka R, Matsudab T. Tetrahedron Asymmetry., 2003, 14:2659–2681
[7]Ni Y, Xu J-H. J Mol Catal B Enzym., 2002, 18:233–241
[8]Yang Z, Zeng R, Wang G, Yao S., Chem Eng China, 2007, 1: 416–420
[9]Yang Z-H, Yao S-J, Guan Y-X. Ind Eng Chem Res., 2005, 44(15): 5411–5416
[10]Ema T, Yagasaki H, Okita N, Takeda M, Sakai T. Tetrahedron., 2006, 62: 6143–6149
[11]Chin-Joe I, Haberland J, Straathof A J J, Jongejan J A, Liese A,Heijnen J J. Enzyme Microb Technol., 2002 31:665–672
[12]Andrade L H, Utsunomiya R S, Omori A T J, Porto A L M, Comasseto J M. Molec. Catal . B : Enzym., 2006, 38: 84–90
[13]Bruni R , Fantin G, Pedrini P, Maietti S, Medici A, Sscchetti G. Tetrahedron: Asymmetry.,2006, 17: 2287–2291
[14]Yang Z H, Zeng R, Yang G, Wang Y, Li L Z, Lv Z S, Yao M, Lai B. J Ind Microbiol Biotechnol., 2008, 35: 1047–1051
[15]Liu X, Pan Z G, Xu J H, Li H X. Chin. Chem. Lett., 2010, 21: 305–308
[16]Schoemaker H E, Mink D, Wubbolts M G. Science., 2003, 299: 1694–1697
[17]Pye P J, Rossen K, Weissman S A, Maliakal A, Reamer R A, Ball R, Tsou N N, Volante R P, Reider P J. Chem. Eur. J., 2002, 8(6): 1372–1376

[18]Haukaas M H, O'Doherty G A. Org. Lett., 2002, 4(10): 1771– 1774
[19]Wong F F, Chang P W, Lin H C, You B J, Huang J J, Lin S K. Journal of Organometallic Chemistry., 2009, 694: 3452–3455
[20]Andrade L H, Polak R, Porto A L M, Schoenlein–Crusius I H, Comasseto J V. Lett. Org. Chem., 2006, 3(8): 613–618

(132g) Phylogenetic and Experimental Characterization of An Acyl-ACP Thioesterase Family Reveals Significant Diversity In Enzymatic Specificity and Activity

David C. Cantu[1], Fuyuan Jing[2], Jarmila Tvaruzkova[2], Jay P. Chipman[1], Basil J. Nikolau[2], Marna D. Yandeau-Nelson[2] and Peter J. Reilly[1], (1)Chemical and Biological Engineering, Iowa State University, Ames, IA, (2)Biochemistry, Biophysics, and Molecular Biology, Iowa State University, Ames, IA

Acyl-ACP thioesterases (TEs) catalyze the hydrolysis of the thioester bond that links the acyl chain to the sulfhydryl of the phosphopantetheine prosthetic group of ACP. This reaction terminates acyl-chain elongation of fatty acid biosynthesis. It is also the biochemical determinant of the fatty acid compositions of storage lipids in plant seeds. To explore enzymatic diversity and to identify novel TEs, product distributions and activities of 31 TEs in an acyl-ACP TE family were determined. These TEs were chosen by two different approaches: 1) selected from public databases on the basis of a phylogenetic analysis and fatty acid profile knowledge of their source organisms; and 2) molecularly cloned organisms that produce medium-chain and short-chain fatty acids in their seeds. The *in vivo* substrate specificities of all the TEs were determined in *E. coli*. Based on their specificities, these TEs were clustered into classes: 1) Class I TEs that act primarily on 14- and 16-carbon acyl-ACP substrates; 2) Class II TEs that have broad substrate specificities, with major activities toward 8- and 14-carbon acyl-ACP substrates; and 3) Class III TEs that act predominantly on 8-carbon acyl-ACPs. Several novel acyl-ACP TEs produced short-chain and unsaturated acyl-ACP or 3-ketoacyl-ACP substrates, indicating the diversity of enzymatic specificity in this enzyme family. These acyl-ACP TEs can potentially be used to diversify the fatty acid biosynthesis pathway to produce novel fatty acids.

See more of this Session: Advances In Biocatalysis and Biosynthesis II
See more of this Group/Topical: Food, Pharmaceutical & Bioengineering Division

(133a) The Genome Sequence of the Chinese Hamster:Ushering In An Era of CHO Genome Engineering

Nitya M. Jacob[1], **Nandita Vishwanathan**[1], Faraaz Yusufi[2], Ju-Xin Chin[2], Terk Shuen Lee[2], Kathryn C. Johnson[1], Huong Le[1], Thiru Ramaraj[3], Jimmy Woodward[3], Ernest F. Retzel[3], John A. Crow[3], Song Gao[4], Niranjan Nagarajan[4], Xiaoan Ruan[4], Yijun Ruan[4], Bernard L. W. Loo[2], Dong-Yup Lee[2], George Karypis[5], Miranda G. S. Yap[2] and Wei-Shou Hu[1], (1)Department of Chemical Engineering and Materials Science, University of Minnesota, Minneapolis, MN, (2)Bioprocessing Technology Institute A-STAR, Singapore, (3)National Center for Genome Resources, Santa Fe, NM, (4)Genome Institute of Singapore A-STAR, Singapore, (5)Department of Computer Science and Engineering, University of Minnesota, Minneapolis, MN

CHO cells, the workhorses of the biopharmaceutical industry, are derived from the Chinese hamster, arguably making it the most economically important industrial organism. The synergistic application of high-throughput sequencing technologies, along with the existing CHO EST collection as backbone, enabled the efficient assembly of the Chinese hamster genome. The current assembly (~2.5Gb), constituting over two billion sequence reads, includes more than 25,000 annotated genes across a range of functional classes. This has allowed a global comparative analysis with the mouse, rat and human genomes. Furthermore, the investigation of regulatory features including promoters, CpG Islands and microRNAs has opened up new avenues for manipulating individual gene expression as well as genome level interventions.

In addition, this work aims to study the genetic variation underlying economically important productivity traits in CHO cells, by a comparative genomics approach, with diploid hamster DNA as reference. Further, cell line-specific functional polymorphisms have been identified utilizing RNA-Seq data from several different recombinant lines. The availability of a well-annotated Chinese hamster genome will open up many new opportunities for cell engineering and metabolic intervention for process enhancement.

See more of this Session: Advances In Cell Culture II : Metabolic Flux Analysis and Modeling of CHO Cells
See more of this Group/Topical: Food, Pharmaceutical & Bioengineering Division

(133b) Metabolic Flux Analysis of CHO Cells In Semi Fed-Batch Culture

Maciek R. Antoniewicz and Woo Suk Ahn, Chemical Engineering, University of Delaware, Newark, DE

Mammalian cell culture is of key importance for biopharmaceutical companies for the production of recombinant proteins, monoclonal antibodies and vaccines. In recent approvals in US and EU, the majority of biotherapeutics were produced by mammalian cell lines and of those a large fraction by Chinese Hamster Ovary (CHO) cells. Metabolism of CHO cells is characterized by high rates of glycolysis and glutaminolysis. As a result, large amounts of lactate and ammonium accumulate during the culture from the conversion of glucose to lactate and the decomposition and metabolism of glutamine. The cellular by-products inhibit cell growth and protein production and can deteriorate the glycosylation quality of the proteins. Relatively little is known about the intracellular metabolism of CHO cells in cell culture. In this work, the intracellular metabolism of CHO cells was studied in detail in a typical semi fed-batch culture. Intracellular metabolism was characterized by extracellular uptake and excretion rates of metabolites combined with a metabolic network model for CHO cell metabolism. To obtain additional information ^{13}C-tracers were applied and ^{13}C-labeling was measured using mass spectrometry. The flux results revealed significant rewiring of intracellular metabolic fluxes in the transition from growth to non-growth including changes in energy metabolism and redox metabolism. The results presented here provide a solid foundation for future studies of CHO cell metabolism for applications such as cell line development and medium optimization for high-titer production of recombinant proteins.

See more of this Session: Advances In Cell Culture II : Metabolic Flux Analysis and Modeling of CHO Cells
See more of this Group/Topical: Food, Pharmaceutical & Bioengineering Division

(133c) The Metabolic Effects of An Anti-Apoptotic Protein Bcl-2Δ Overexpressing CHO Cell Line

Neil Templeton and Jamey Young, Chemical and Biomolecular Engineering, Vanderbilt University, Nashville, TN

Incorporating apoptotic resistance into production celllines holds remarkable potential for several industrial applications, especiallythat of monoclonal antibody (mAb) production. However, the associated impacts on cellularmetabolism are only beginning to be understood, especially in regards to howanti-apoptotic proteins regulate lactate production and mitochondrialmetabolism. We are thereforeinvestigating the global metabolic reprogramming that takes place in ChineseHamster Ovary (CHO) cells that have been engineered to overexpress theanti-apoptotic Bcl-2Δ protein. Ourprevious work has shown that limiting glutamine availability can have profoundimpacts on growth and lactate accumulation in CHO cell cultures. Specifically,when initial glutamine concentration was halved, the outgoing lactate flux wasreduced by more than 50% while growth rate was increased by roughly 30%. Whenoverexpressing Bcl-2Δ, outgoing lactate flux and incoming glucose flux werereduced by roughly 40%, and the IVCD over the life of the culture was increasedby over 30%. We are currently applying ^{13}C metabolic flux analysis(MFA) to investigate how Bcl-2Dinteracts with glucose and glutamine availability to control metabolic fluxesin CHO cell cultures. Our goal is togenerate comprehensive flux maps that depict the flow of carbon through centralmetabolism under conditions of varying glucose and glutamine availability todemonstrate the impact of Bcl-2Δ overexpression. We will present our current view of hownutrient availability and anti-apoptotic proteins interact to control lactateproduction and central metabolic fluxes in CHO cells.

See more of this Session: Advances In Cell Culture II : Metabolic Flux Analysis and Modeling of CHO Cells
See more of this Group/Topical: Food, Pharmaceutical & Bioengineering Division

(133d) Proteomic and Metabolic Flux Analysis of Recombinant CHO Cell Cultures

John A. Morgan, School of Chemical Engineering, Purdue University, West Lafayette, IN, Neelanjan Sengupta, Chemical Engineering, Purdue University, West Lafayette, IN and Min Zhang, Eli Lilly & Co., Indianapolis, IN

Chinese Hamster Ovary (CHO) cells cultures are the major source for producing therapeutic proteins that require post-translational modification. The cell culture process can dramatically influence the cell culture performance, including cell growth, protein yield and quality. In this presentation, we report fluxes under two process conditions that yield different lactic acid profiles. In one process, the lactic acid production was highest during the growth phase of the cell culture, and decreased as the cell culture progressed. Later, it changed to lactate consumption around mid stationary phase, but then changed to lactate excretion during the death phase. In contrast, the second process had lower lactic acid production during growth phase and complete reutilization during the stationary phase. Fluxes were evaluated for the two processes using ^{13}C-MFA using isotopically labeled glucose. The lower lactic acid producing process was found to have lower flux towards the pentose phosphate pathway as compared to higher lactic acid producing process. Further, oxygen uptake profiles and TCA cycle fluxes for the two processes indicated that for the higher lactic acid producing process, the mitochondrial capacity was limited as compared to the glucose uptake. An untargeted proteomic analysis was also performed to understand the causes of the underlying shifts in these fluxes and lactic acid production profiles. Partial least squares analysis was used to correlate the changes in metabolic proteins to key fluxes and split ratios. Some key proteins were identified that were correlated to the changing flux profiles. For the high lactic acid production process, the changes in the protein profiles related to mitochondrial function (Pyruvate dehydrogenase, alpha- keto dehydrogenase) and ATP production were correlated to the TCA cycle flux and lactic acid production. The lactate dehydrogenase expression pattern was not correlated or had a minor effect on lactic acid production flux, in both processes. This indicated the control of the lactic acid production, was allosteric and dependent on the substrate availability. Further, for both the processes, proteins that play a role in mitochondrial oxidative stress and oxidative stress in general were found to be correlated to the flux profiles. This indicated that the proteins that affect redox potential and energy charge of the cells play a role in changing flux profiles over the cell culture.

See more of this Session: Advances In Cell Culture II : Metabolic Flux Analysis and Modeling of CHO Cells
See more of this Group/Topical: Food, Pharmaceutical & Bioengineering Division

(133e) Bistability In Glycolysis Pathway As a Physiological Switch In Energy Metabolism

Bhanu Chandra Mulukutla, **Andrew Yongky**, Prodromos Daoutidis and Wei-Shou Hu, Chemical Engineering and Materials Science, University of Minnesota, Minneapolis, MN

Mammalian cells in culture consume large amounts of glucose and divert most of it towards lactate production. Such high glucose consumption rates are in general typical of rapidly proliferating cells including embryonic stem cells and tumor cells. In contrast, non-proliferating tissues consume glucose at much slower rates and oxidize most of it to carbon dioxide in the mitochondria. Such differences in glucose metabolism can be attributed to the coordinated expression of different glycolytic isozymes, which can form mutually regulated networks. The regulations governing such networks are mediated through the allosteric control of the expressed isozymes by reaction intermediates upstream and downstream of the glycolysis pathway. These regulatory networks can result in different glycolytic behavior ranging from sustained oscillations to bistability.

Here, using kinetic modeling we focus on the importance of these regulations in the occurrence of bistability in glycolytic activity. Results demonstrate that bistability is mainly limited to conditions when cells express isozymes specific to rapidly proliferating cells including muscle isozyme of phosphofructokinase (PFKM), hexokinase 1 (HK1) or hexokinase 2 (HK2) and pyruvate kinase M2 (PKM2) isozyme. We identified fructose-1,6-bisphosphate (F16BP) as a critical regulator of the bistable behavior through its modulation of the above mentioned isozymes, directly or indirectly. Further, the physiological significance of this bistable switch and its modulation by various signaling pathways and other metabolic intermediates will be discussed. These novel findings will enhance our understanding of metabolic control in mammalian cells and help in designing strategies for process enhancement through manipulation of cellular energy metabolism.

See more of this Session: Advances In Cell Culture II : Metabolic Flux Analysis and Modeling of CHO Cells
See more of this Group/Topical: Food, Pharmaceutical & Bioengineering Division

(133f) Dynamic Model for CHO Cell and Process Engineering

Ryan P. Nolan, Culture Process Development, Pfizer, Andover, MA and **Kyongbum Lee**, Chemical and Biological Engineering, Tufts University, Medford, MA

Industrial CHO cell fed-batch processes have progressed significantly over the past decade, with recombinant protein production titers consistently reaching the gram per liter level. Such improvements have largely resulted from separate advances in process and cell line development. In contrast, metabolite overproduction using microbial cells has spectacularly benefited from metabolic engineering efforts combining model-based design and genetic manipulations. Model-based selection of targets for genetic engineering in CHO cells is confounded by the dynamic nature of the fed-batch process. Due to transitions such as the lactate shift, inferior targets may be selected by relying on steady-state models used to design and optimize microorganisms during log-phase growth.

In this work, we use a dynamic model of CHO cell metabolism to simultaneously identify both process and cell modifications that improve antibody production. Model simulations were used to characterize the dynamic responses of metabolic gene knockdowns in the context of variations in process variables over the full course of a multi-day fed-batch. The simulations explored ca. 9,200 combinations of process variables (shift temperature, shift day, seed density, and harvest day) and knockdowns (8 metabolic enzymes). The simulation results were screened for improvements in product titer as well as other quality control criteria, including peak cell density and final lactate and ammonia concentrations in the medium. Approximately 0.5% of the tested combinations met these criteria and produced an antibody titer that is an increase greater than 1.5-fold compared to the control condition (unmodified cell and base process). Among these combinations, 88% involved knockdown of enzymes involved in lactate metabolism. The improvement in antibody titer with reduced lactate production was corroborated by experimental observations. Interestingly, depending on the process conditions, modulating the lactate enzymes yielded varying productivities, ranging from a reduction in final titer to greater than 2-fold improvement.

Our results underscore the benefit of combining process and cell optimization, and highlight the need to develop dynamic models capable of accurately simulating experimentally documented metabolic shifts in fed-batch cultures of mammalian cells.

See more of this Session: Advances In Cell Culture II : Metabolic Flux Analysis and Modeling of CHO Cells
See more of this Group/Topical: Food, Pharmaceutical & Bioengineering Division

(133g) Towards the Implementation of QbD to the Production of Therapeutic Monoclonal Antibodies with Desired Glycosylation Patterns

Ioscani Jimenez del Val and Cleo Kontoravdi, Chemical Engineering and Chemical Technology, Imperial College London, London, United Kingdom

The US FDA and the European Medicines Agency have recently proposed the implementation of the Quality by Design (QbD) paradigm to the manufacture of biopharmaceuticals. Its implementation requires the use of all available knowledge of a given product, including the parameters that affect its quality, for the design, optimization and control of the manufacturing process. The goal is to ensure that quality is built into the product at every stage of the manufacturing process. Most licensed monoclonal antibodies (mAbs) are based on the immunoglobulin G isotype and contain a consensus N-linked glycosylation site on the Cg2 domains of their heavy chains. Studies have found that the oligosaccharides attached to this site dramatically influence the efficacy of mAbs as therapeutics either by reducing their serum half-life or by directly affecting the mechanisms they trigger *in vivo*, thus defining glycosylation as a critical quality attribute of mAbs under the QbD scope. It has been recently proposed that detailed mathematical models will play a critical role in the design, control and optimization of biopharmaceutical manufacturing processes under the QbD scope. To our knowledge, there are currently no mathematical models that relate mAb glycosylation with cell culture conditions.

Several reports have shown that glycosylation is directly affected by the intracellular availability of nucleotide sugar donors (NSDs), which are the co-substrates for the glycosylation reactions that occur in the Golgi apparatus. During culture, cells synthesize all the relevant NSDs from glucose through the nucleotide sugar metabolic pathway. In an effort to relate process conditions with mAb glycosylation, we have generated a dynamic mathematical model for this metabolic pathway. The kinetic model was reduced based on the methodology described by Nolan and Lee (Methods in Bioengineering: Systems Analysis of Biological Networks, 2009). In order to estimate the unknown parameters of the reduced model, the intracellular concentration of glucose and eight NSDs was monitored through the course of a batch culture of CHO cells. The cells were harvested and their metabolism quenched. The cells were subsequently lysed and their contents lyophilized for analysis. The intracellular glucose concentration was determined using a fluorescence-based assay kit and the intracellular NSD concentrations were determined with an ion-pairing reverse-phase liquid chromatographic technique. Finally, a series of fed-batch cell culture experiments were performed in order to validate the reduced model and assess confidence in the estimated parameters.

Our mathematical model for NSD metabolism was then coupled to a model for mAb N-linked glycosylation in the Golgi apparatus to generate a direct link between extracellular glucose concentration, which is a readily measurable process variable, and protein glycosylation. The model for mAb glycosylation approximates the Golgi apparatus to a plug flow reactor so that cisternal maturation is represented. Recycling of Golgi-resident proteins (glycosylation enzymes and transport proteins) was also considered. The rate expressions for glycosylation reactions were derived based on their reported kinetic mechanisms and transport of nucleotide sugar donors from the cytosol to the Golgi lumen was modelled to serve as a link between glycosylation and cellular metabolism. Optimization-based methodologies were developed for estimating enzyme and transport protein concentration profiles. The resulting model is capable of reproducing glycosylation profiles of commercial mAbs. It can further reproduce the effect gene silencing of the FucT glycosylation enzyme and cytosolic nucleotide sugar donor depletion have on the mAb oligosaccharide profile.

Such a combined model has great potential for the design, control and optimization of manufacturing processes that produce mAbs with *built in* glycosylation-associated quality as proposed under the QbD paradigm.

See more of this Session: Advances In Cell Culture II : Metabolic Flux Analysis and Modeling of CHO Cells
See more of this Group/Topical: Food, Pharmaceutical & Bioengineering Division

(143a) Transport Modulation of Nanocarriers Across the Pulmonary Epithelium: Potential Strategy for the Local and Systemic Delivery of Drugs to and Through the Lungs

Balaji Bharatwaj, Denise S. Conti and Sandro R. P. da Rocha, Chemical Engineering and Materials Science, Wayne State University, Detroit, MI

While it has long been recognized as an attractive non-invasive mode for delivering therapeutics to treat regional disorders, the potential for administering medicaments to the lungs via oral inhalation (OI) for targeting systemic ailments is rapidly gaining momentum due to the proximity of the organ to the bloodstream, low proteolytic activity and a large surface area. Incorporating versatile delivery agents like polymeric nanocarriers (PNCs) into OI regimen can greatly enhance the applicability of such therapies for regional delivery, as PNCs have several advantages that include ease in cellular internalization, specificity in targeting and sustained release of encapsulated therapeutics. The versatility of PNCs can be further enhanced by tailoring their surface properties, which may potentially assist in overcoming cellular barriers and aid in the modulation of transport of drug molecules.

In this work a series of surface-modified dendrimer nanocarriers (DNCs) were synthesized, characterized and their interaction with a model pulmonary epithelium was investigated. Calu-3 cell monolayers were grown to confluence under an air-interface culture (AIC) and were characterized. The transport and uptake of DNCs across these monolayers was studied as a function of their surface properties. Our results indicate that the epithelial translocation of DNCs and their uptake into the cell cytosol can be efficiently modulated by varying their surface properties. These results suggest that incorporating DNCs as drug delivery vehicles in OI strategies can potentially enhance the applicability of OI therapy not only as an effective route for targeting regional ailments but also as an approach to access the systemic portal.

See more of this Session: Bionanotechnology
See more of this Group/Topical: Food, Pharmaceutical & Bioengineering Division

(143b) High Throughput Evaluation of the In Vivo Biodistribution and Dispersion of Polyanhydride Nanoparticle Adjuvants

Latrisha Petersen[1], Lucas Huntimer[2], Katharine Walz[1], Amanda Ramer-Tait[2], Michael J. Wannemuehler[2] and Balaji Narasimhan[3], (1)Chemical and Biological Engineering, Iowa State University, Ames, IA, (2)Veterinary Microbiology and Preventive Medicine, Iowa State University, Ames, IA, (3)Department of Chemical and Biological Engineering, Iowa State University, Ames, IA

While vaccination remains the best strategy for disease prevention, current vaccine strategies often require large doses of poorly immunogenic antigens, repeated immunizations, and undesirable side effects, which can all hinder vaccine efficacy. Furthermore, most current vaccine administration routes differ from that of the natural infection which could limit immune protection. Thus, there is a need for the development of biocompatible vaccine adjuvants capable of multi-route administration and prolonged antigen release.

Polyanhydrides are a class of biodegradable, non-toxic, non-mutagenic materials that are capable of encapsulating and delivering biological molecules *in vivo*. These polymers are capable of being formulated into nanoparticles enabling administration parenterally or intranasally. Nanoparticles based upon copolymers of sebacic acid (SA), 1,6-bis-(*p*-carboxyphenoxy hexane) (CPH), and 1,8-bis-(*p*-carboxyphenoxy)-3,6-dioxaocatane) (CPTEG) have been shown to exhibit tunable properties, including tailored protein release kinetics, protein stabilization, and immunomodulatory adjuvant behavior. In this work, a combinatorial platform was utilized to investigate the *in vivo* biodistribution, depot effect, and mucoadhesion of polyanhydride nanoparticles as a function of nanoparticle chemistry and administration route. This was achieved by employing a live animal imaging system with near infrared fluorescent markers to simultaneously investigate each parameter. All nanoparticle chemistries resulted in rapid dispersal when delivered intranasally and provided a long-term depot when administered parenterally. Chemistry-dependent trends were identified in which intranasally administered amphiphilic and hydrophobic polyanhydride nanoparticles, 50:50 CPTEG:CPH and 50:50 CPH:SA, demonstrated superior mucoadhesive properties enabling prolonged residence in lung tissue. It was observed that parenterally administered amphiphilic 50:50 CPTEG:CPH nanoparticles demonstrated the longest injection site residence time. These results provide new insights into the biodistribution and tissue-specific targeting of these nanoparticle-based vaccine adjuvants. The mucoadhesive polymer nanoparticles identified in this work provide opportunities for their use as adjuvants in intranasal and parenteral vaccination. Furthermore, the insights gained from these studies will enable the rational design of vaccine and therapeutic drug delivery systems for targeted and localized delivery over extended time periods.

See more of this Session: Bionanotechnology

(143c) Polyelectrolyte-Gold Nanorod Assemblies for Transgene Delivery

James Ramos[1], **Huang-Chiao Huang**[2] and Kaushal Rege[2], (1)Biomedical Engineering, Arizona State University, Tempe, AZ, (2)Chemical Engineering, Arizona State University, Tempe, AZ

Gold nanorods (GNRs) have emerged as promising nanomaterials for biosensing, imaging, photothermal therapy and transgene delivery for cancer diseases. We have generated polyelectrolyte-functionalized gold nanorods (PE-GNRs) using a layer-by-layer deposition approach; a cationic polyelectrolyte library recently synthesized in our laboratory was employed to generate the PE-GNR assemblies. PE-GNR assemblies demonstrate long-term colloidal stability as well as the capacity to bind plasmid DNA by means of electrostatic interactions. Sub-toxic concentrations of PE-GNRs were employed to deliver plasmid DNA to prostate cancer cells in vitro. PE-GNRs generated using cationic polymers from our laboratory demonstrated higher transgene expression and exhibited lower cytotoxicities when compared to PE-GNRs generated using polyethylenimine (pEI), a current standard for polymer-mediated gene delivery. The role of polyelectrolyte chemistry, particle size, and zeta-potential on transgene expression efficacies of PE-GNRs was investigated. Plasmonic properties of gold nanorods were exploited for dark-field imaging of the delivered PE-GNRs. In addition, the stable photothermal properties of gold nanorods were exploited to investigate the role of GNR-induced hyperthermia in enhancing polymer-mediated transgene expression. Our results indicate that PE-GNRs are a promising engineered platform that demonstrate high stabilities, low cytotoxicities and enhanced transgene expression efficacies. PE-GNRs also have the potential to be used simultaneously for photothermal ablation, photothermally enhanced drug and gene delivery, and biological imaging, thus making them a powerful theranostic platform.

See more of this Session: Bionanotechnology
See more of this Group/Topical: Food, Pharmaceutical & Bioengineering Division

(143d) Nanoparticle/Heat-Responsive-Polymer Complex for Hyperthermia and Sustained Drug Release

Sarah Wang[1], Michelle C. Liu[1] and **Kyung A. Kang**[2], (1)DuPont Manual High School, Louisville, KY, (2)Chemical Engineering, University of Louisville, Louisville, KY

Iron oxide nanoparticles (IONPs) can get heated under alternating electromagnetic (AEM) field. They are non-toxic and have been used as a heat-delivering medium for cancer hyperthermia. Pluronic F-127 (PF127) is a biocompatible polymer that possesses both temperature and concentration-dependent gelation properties in aqueous solutions. We have combined these two unique materials (IONPs and PF127) to develop both tumor-specific hyperthermia and long-term, sustained drug delivery. IONPs are mixed with the PF127 solution of a particular concentration, whose phase is liquid at room temperature and is semi-solid (gel) at or above the human body temperature. The mixture of IONPs and cancer-drug containing PF127 solution is to be injected to the site of the cancer in human body. Then the mixture becomes a gel at body temperature. AEM field is then applied to the cancer site to perform low temperature (42-45°C) hyperthermia. For the tumor cells surviving hyperthermia, the PF127 releasing cancer-drug will treat them in a longer term.

Our study results showed that 30 nm IONPs appeared to be highly effective for hyperthermic treatment. The gelation temperatures of PF127 at various concentrations were determined to select the PF127 concentration that is liquid at room temperature (during injection) but becomes a gel phase at body temperature. 18% PF127 (w/w) was ideal for our purpose because it was liquid at room temperature and its gelation temperature was ~28°C. It should be injectable at room temperature and should form a gel upon injection into human body (37°C). Then the effect of the IONPs in the PF127 phase transition temperature was also studied. Study results on the IONP-mixed 18% PF127 showed that the IONPs in the PF127 solution had little effect on gelation temperatures at the IONP concentration range of 0.01-0.03%. The effect of PF127 on the IONP heating was also studied and with the addition of PF127, IONPs showed a slight decrease in heating.

This design would be minimally-invasive and can be only to the cancer area. It provides both hyperthermia and zero-order, sustainable drug release at the cancer site, enabling the cancer treatment in dual cancer treatment modalities and also minimizing the negative side effects of traditional, systemic cancer treatments.

Acknowledgements: We acknowledge the BASF Global for generously supplying PF127.

See more of this Session: Bionanotechnology

(143e) Elastic Mapping of Epithelial Cells Treated with Nanomaterials

Chenbo Dong[1], Linda Q. Sargent[2], David H. Lowry[2] and **Cerasela Zoica Dinu**[1], (1)Chemical Engineering, West Virginia University, Morgantown, WV, (2)National Institute for Occupation Safety and Health, Morgantown, WV

Topological analysis of cells and subcellular structures is one of the major trends in biology. Since the spatial and temporal changes of the mechanical properties of living cells reflect complex underlying physiological processes, there have been considerable efforts in following these changes and providing valuable insight into the biological importance of cellular mechanics and its regulation. In here we propose to study the cell mechanics and regulation in response to exposure to nanomaterials. Nanomaterials have been recently used in biolabeling, biodetection, biomolecule delivery, bioseparation or regenerative medicine and tissue engineering. Specifically, we use Atomic Force Microscopy (AFM) for probing cellular topography and studying mechanical properties of immobilized and live cells, and cells exposed to nanomaterials. Our aim is to produce robust, internally quantitative maps of relative elasticity of fixed cells and compare those maps to those of live cells, and cells incubated with nanomaterials. Our experimental results aim to demonstrate that spherical mapping of cells can be correlated with cell dynamics and thus provide the possibility to develop further topological-based approaches to probe the structure-function relationship in live cells exposed to nanomaterials, all with high spatial resolution and increased sensitivity.

See more of this Session: Bionanotechnology
See more of this Group/Topical: Food, Pharmaceutical & Bioengineering Division

(143f) Enhanced Biodesulfurization by Magnetic Immobilized Rhodococcus Erythropolis LSSE8-1-Vgb Assembled with Nano-Al2O3

Jianmin Xing, Ting Zhang, Zelong Wang and Wangliang Li, National Key Lab of Biochemical Engineering, Institute of Process Engineering, Chinese Academy of Sciences, Beijing, China

Biodesulfurization activity can be enhanced by assembling nano -Al2O3 particles on the magnetic immobilized Rhodococcus erythropolis LSSE8-1-vgb. The cells can be collected and reused conveniently by an external magnetic field. Firstly, cells were magnetic immobilized by coating with Fe3O4 nano-particles. The optimal ratio of cells to magnetic Fe3O4 nano-particles was determined to be 50:1 (g/g). Then nano-c-Al2O3 adsorbents were assembled onto the cells to enhance the desulfurization activity. The nano-c-Al2O3 adsorbent had the largest pore volume as well as specific surface area, and the strongest electrostatics interaction with microbial cell, and cells assembled with this nano-adsorbent performed the highest desulfurization activity. The activity of magnetic immobilized cells assembled with adsorbents was tested in desulfurization of model oil. The desulfurization rate was raised by nearly 20% when the amount ratio of magnetic particles to adsorbents was 1:5 (g/g). These cells can be reused. The activity decreased less than 10% through out three desulfurization- activation-reuse recycles.

See more of this Session: Bionanotechnology
See more of this Group/Topical: Food, Pharmaceutical & Bioengineering Division

(143g) A Multiplexed Microfluidic Platform Utilizing Normally Closed Valves for Investigation of Microbial Gene Expression

Ritika Mohan, Arnab Mukherjee, Charles M. Schroeder and Paul J. A. Kenis, University of Illinois, Urbana Champaign, Urbana, IL

In this work, we describe a microfluidic platform that utilizes normally closed valves to enable quantitative biological measurements of gene expression and stimulus driven responses of living cells such as *E. coli* and *P. aeruginosa*. Multiplexing on-chip is achieved by having multiple cell and reagent loading compartments on the same device, which also enables single cell analysis [1]. We built a 4x6 microfluidic chip capable of high-throughput screening and on-chip mixing within each sub-chamber in combinatorial format. The device permits controlled exposure of cells in ~50 pL volume compartments to different chemical stimuli loaded in contiguous compartments. The device design is convenient for biological assays due to increased detection sensitivity since the analysis is done at microscale. In addition, the device facilitates portability and exhibits multiple advantages as a miniaturized biological assay, including low sample and reagent volumes for analysis and integration of assay steps in an "on-chip" format.

Using this device in a proof-of-concept experiment, we study bistability in gene expression associated with positive regulation in a bacterial genetic network, specifically the gene network mediating microbial growth on a lactose carbon source. Stimulation of *E. coli* bacterial cells in the device with varying concentrations of a lactose analog in the different compartments initiates a bistable response [2]. The cells display mixed ON-OFF response, and at high concentrations of inducer, the cells are uniformly either ON or OFF. Overall, this result emphasizes the utility of the microfluidic platform for single cell measurements over a wide range of stimuli on a single integrated chip. In ongoing work, we are using the multiplexed microfluidic device to study the effects of antibiotics on mediating cell death as a function of the growth stage of cell cycle. Overall, this research demonstrates the utility of a microfluidic platform for single cell analysis in variable environmental conditions, including antibiotics, hormones, toxins and growth factors.

References:

[1] Boedicker, J., Vincent, M., and Ismagilov, R. (2009), *Microfluidic Confinement of Single Cells of Bacteria in Small Volumes Initiates High-Density Behavior of Quorum Sensing and Growth and Reveals Its Variability*. Angewandte Chemie International Edition, 48: 5908–5911.

[2]Ertugrul M. Ozbudak, Mukund Thattai, Han N Lim, Boris I. Shraiman, and Alexander van Oudenaarden (2004), *Multistability in the lactose utilization network of Escherichia coli*. Nature, 427

(145a) Influence of Nucelar Stiffness On Cell Migration

Takamasa Harada and Dennis E. Discher, Chemical and Biomolecular Engineering, University of Pennsylvania, Philadelphia, PA

Nucleus is the largest organelle in cell, and therefore considered to be the biggest obstacle for cell migration in constrained tissue. The stiffness of cell nucleus is largely determined by filamentous protein network located at inner nuclear membrane called nuclear lamina, which is mainly composed of lamin proteins. Lamins, which are type V intermediate filament proteins, can be divided into two subtypes namely A-type and B-type lamins. Unlike ubiquitously expressed B-type, A-type lamins are found in most of differentiated cells with involvement in chromatin organization, transcription, translation, cell differentiation and many other functions. The purpose of this study is to investigate a correlation between cancer cell migration in physically constrained environment and the expression of lamin A and C proteins (lamin A/C), motivated by the fact that many types of cancer cell are known to have lower expression of lamin A/C compared to their healthy conterpart.

Lamin A/C expression was manipulated by the use of short interfering RNA (siRNA) in human lung carcinoma cell A549. The effect of lamin A/C knockdown on nuclear stiffness was investigated by micropipette aspiration. The migration capacity of the lamin A/C knockdown A549 cells were evaluated in 2- and 3-dimentional environment by wound scratch healing assay and trans-well migration assay, respectively. Furthermore, lamin A/C down-regulated A549 cells were injected in subcutaneous tissue on one side of flank site of the immuno-deficient mouse, along with scrambled siRNA-treated control cells on the other side, in order to monitor the effect of Lamin A/C knockdown on tumor propagation *in vivo*.

Micropipette aspiration measurement showed more than 2-fold increase in softness of the nucleus after treatment with siRNA against lamin A/C. Lamin A/C knockdown only affected 3-D migration of the cells, not 2-D, where modification in nuclear shape is not involved in the process. These in vitro results strongly supports the in vivo observation, which showed that tumors formed by lamin A/C siRNA-treated A549 cells propagated in higher rate compared with control tumors (scrambled siRNA-treated or cells without any treatment).

In conclusion, the present study showed nuclear softening caused by knockdown of nuclear protein lamin A/C increased the cell migration capacity of A549 cells *in vitro* and also more effectively promoted the tumor proliferation in subcutaneous xenograft model due to more rapid expansion of the tumor. This result may explain the alleged lower lamin A/C expression or softer nuclei of cancer cells and their capability to efficiently migrate in the tissue, which ultimately leads to the metastasis to other part of the body.

(145b) JNK Regulates Cell-Substrate and Cell-Cell Adhesion On Substrates of Varying Compliance

Hui You, Aishwarya Ranganathan and Stelios T. Andreadis, Department of Chemical and Biological Engineering, University at Buffalo, the State University of New York, Amherst, NY

It is well accepted that substrate rigidity can guide the behavior ofcells including spreading, migration, proliferation and differentiation *invitro* (1, 2) and tissue formation and maintenance *in vivo* (3). Substrate rigidity also has been shown to regulatethe cross talk betweencell-cell and cell-substrate adhesion (4). Our group demonstrated JNK phorphorylates b-catenin and regulates adherentjunction formation, a-catenin/b-catenin interactions and actinre-organization in epithelial cells (5, 6). Here we provide strong evidencesupporting the role of JNK as a regulator of cell-cell vs. cell-substrateadhesion on substrates of varying compliance.

Our *in vitro* studies show that the substrate rigidity regulates JNKactivity of human primary keratinocytes. The levels of p-JNK were significantlyreduced in the keratinocytes within 24 hrs after seeded on soft substrates (p-JNK1,127%; p-JNK2, 239% of plastic wells). In the cells seeded on soft substrate,E-cadherin and bcells protrusion were reduced.Keratinocytes with constitutive active JNK could not form adherent junction onthe soft substrate, and the focal adhesion was still detectable. However, keratinocytesknocked down JNK1, JNK2, or JNK1/2 form adherent junction even on the rigid plasticsurface. *In vivo*, the negative correlation (R=-0.89) between adherentjunction formation and JNK activity has been demonstrated in human foreskin andwild type mouse skin tissue. Keratinocytes closer to basal layer express morep-JNK and p-c-Jun, however, the adherent junctions did not formed properly. Thefocal adhesion markers, p-FAK and p-paxillin, were prominently expressed in thebasal layer keratinocytes of skin epidermis, which is consistent with b-/-> mouse skin epidermisand siRNA JNK1 bioengineered skin contained fewer layers of keratinocytes (mouseskin 3 *vs.*7, skin equivalents 6 *vs.*12). And compared to wildtype, the epidermis is much thinner (mouse skin 332%, p<0.05; skin equivalents272%, p<0.05 of WT). However JNK2$^{-/-}$ mouse skin and siRNA JNK2 skinequivalent contain similar layers compared to wild type, and the epidermis isthicker than WT (mouse skin:1165%, p>0.05, skin equivalents:1308%, p<0.05of WT). In either JNK1 and JNK2 deficient mouse skin or bioengineered skin, adherentjunction is properly formed even in the basal layer confirming our results *invitro*.

In conclusion,keratinocytes on rigid substrate (*in vitro*) or close to dermis (*invivo*) express high levels of p-JNK, which inhibits adherent junctionformation. On the contrary, keratinocytes on soft substrates or at suprabasalcell layers showed low p-JNK level, lack of focal adhesions and enhancedadherent junction formation. In summary, JNKplays a role as a switch regulating the cross talk between adherent junctionand focal adhesion both *in vitro* and *in vivo* models.

Acknowledgements: This work was supportedby grants from NIH (R01 EB000876) and NSF (BES-0354626) to S.T.A. We thank Dr.Roger J. Davis (UMass Medical School) for kindly providing the JNK1$^{-/-}$and JNK2$^{-/-}$ mouse skin samples.

References:

1. **Buxboim A, *et al.*** 2010 Nat Methods 7:695-697

2. **JoyA, *et al.*** 2011 Langmuir 27:1891-1899

3. **GuoWH, *et al.*** 2006 Biophys J 90:2213-2220

4. **TsaiJ, *et al.***2009 Biophys J 96:L39-41

5. **LeeMH, *et al.*** 2009 FASEB J 23:3874-3883

6. **LeeMH, *et al.*** 2010 FASEB J 25:613-623

7. **KoriaP, *et al.*** 2006 J Invest Dermatol 126:1834-1841

See more of this Session: Cell Adhesion and Migration II
See more of this Group/Topical: Food, Pharmaceutical & Bioengineering Division

(145c) F-Actin and Myosin IIa-Dependent Microtubule Targeting of Focal Adhesions

Sabil Huda[1], Siowling Soh[2], Marta Byrska-Bishop[1], Didzis Pilans[1], Jiwon Kim[3], Kristiana Kandere-Grzybowska[2] and Bartosz Grzybowski[2], (1)Northwestern University, Evanston, IL, (2)Chemical and Biological Engineering, Northwestern University, Evanston, IL, (3)Chemistry, Northwestern University, Evanston, IL

Spatial-temporal targeting of Focal Adhesions (FA) by Microtubule (MT) is thought to be responsible for mediating cell-substrate de-adhesion and promote cell translocation. However, the mechanism behind the targeting phenomenon remains largely unstudied. Here we use Anisotropic Solid Microetching (ASoMiC) to study MT targeting of FA. Combining micropatterning of living cells with high-resolution in-cell imaging permitted the study of the targeting process in quantitative and molecular detail.

Triangular microislands spatially segregate cytoskeletal components and allow for study of MT growth trajectories in defined geometric confines. We show that growing MTs are targeted towards FA. RNAi studies reveal that myosin IIA dependent f-actin bundles are integral to this process.

This study shows that not only are MT guided towards FA, but it also provides methods of quantifying the guidance and investigates the molecular mechanisms behind this process.

See more of this Session: Cell Adhesion and Migration II
See more of this Group/Topical: Food, Pharmaceutical & Bioengineering Division

(145d) Control On Glioblastoma Multiforme Dynamics Using Aligned Fiber Networks

Kevin Sheets, School of Biomedical Engineering and Sciences, Virginia Tech, Blacksburg, VA and **Amrinder S. Nain**, Mechanical Engineering, Virginia Tech, Blacksburg, VA

Glioblastoma multiforme (GBM) is one of the most infiltrative and aggressive forms of brain cancer with an estimated life expectancy of just over one year after diagnosis. Contributing to grim prognosis is the high migration speed of GBM cells such as DBTRG's (Denver Brain Tumor Research Group). These cells, which attach to and interact with their immediate fibrous microenvironment known as the extracellular matrix (ECM), receive region-specific biochemical and biophysical cues from their underlying substrates. The ability to manipulate cell motion and perhaps increase the migration rates of these cells may contribute to an enhanced survivability of GBM by providing drug therapies a "worst-case scenario" testing platform. Here we present DBTRG cell behavior on the STEP (Spinneret-based Tunable Engineered Parameters) platform, which deposits high aspect ratio nano/micro fibers with tight control on diameter, spacing, porosity, and orientation. The resulting fibrous scaffolds, which mimic the mechanistic environment of ECM by providing biomechanical cues, can be tuned to alter migration speed for therapeutics applications.

In this study, DBTRG cells are seeded onto fibrous STEP scaffolds and their migration speeds are shown to directly result from varying substrate stiffness in forms of fiber lengths of 4, 8 and 16mm. Fibrous scaffolds were spun using the previously-reported STEP technique. Polystyrene fibers 500 nm in diameter were spun onto two substrate types (suspended single-layer (SS) and suspended double-layer (SD)) in addition to flat glass control. Using standard cell culture techniques, DBTRGs were seeded onto the scaffolds and attached within 4-6 hours. After attaching, 2 mL media was added to keep cells viable over a 24 hour period. During this time, cells on fiber networks were time-lapse imaged by an incubating microscope. Cell migration speeds were calculated by tracking the position of the cell's nucleus every hour and noting the maximum distance traveled per hour in a 16-hour window.

Cells were observed to attach to and elongate along the fiber axis. Cells constricted to SS fibers wrapped themselves entirely around the fiber, and traveled at speeds of 35 μ m/hr on the 4mm substrates. As the length of the fiber was increased, the substrate stiffness decreased and focal adhesions become temporally less stable, leading to an increased migration speed. For example, DBTRG's traveled at approximately 60 μ m/hr on the 8mm substrates and over 80 μ m/hr on the 16mm substrates. STEP platform provides a unique tool to fine-tune the mechanical environment to which native cells attach and interact with so that scaffolds and drug delivery therapies may be improved upon in the future.

See more of this Session: Cell Adhesion and Migration II

(145e) Platelet Adhesion to Fibrinogen Patterned Substrates

Jeffery Fontenot[1], Aurore B. Van De Walle[2], Travis Spain[2] and **David W. Schmidtke**[2], (1) Bioengineering, University of Oklahoma, Norman, OK, (2)School of Chemical, Biological, and Materials Engineering, University of Oklahoma, Norman, OK

Platelet adhesion to the vessel wall during vascular injury is mediated by platelet glycoproteins binding to their respective ligands on the vascular wall. In this study we investigated the role that ligand spacing plays in regulating platelet interactions with fibrinogen under hemodynamic flow. To regulate the size and distance between patches of fibrinogen we developed a photolithography based technique to fabricate patterns (e.g. lines, dots) of proteins surrounded by a protein repellant layer of poly(ethylene glycol). We demonstrate that when mepacrine labeled whole blood is perfused over substrates patterned with 4 micron wide lines of fibrinogen, platelets selectively adhere to the areas of patterned fibrinogen. Using fluorescent microscopy we demonstrate that the amount of platelet coverage (5% - 35%) and the ability of thrombi to grow laterally was dependent upon the distance (3 - 30 microns) between parallel lines of fibrinogen. We also report on the effects of fibrinogen patch size on platelet adhesion by varying the size of the protein patch (2 – 8 microns) available for adhesion. The technique described in this paper provides an effective way to fabricate patterned protein surfaces for use in novel platelet adhesion assays under physiologic flow conditions.

See more of this Session: Cell Adhesion and Migration II
See more of this Group/Topical: Food, Pharmaceutical & Bioengineering Division

(145f) Engineering Bioadhesive Polyvinyl Alcohol Hydrogels for Aneurysm Occlusion

Marjan Rafat, Lisa Rotenstein, Jin-Oh You and Debra Auguste, School of Engineering and Applied Sciences, Harvard University, Cambridge, MA

Aneurysmspose a major health risk; 30,000 aneurysms rupture each year in the UnitedStates, causing stroke, permanent nerve damage, or subarachnoid hemorrhage. Thecurrent endovascular embolization method usingplatinum coils cannot treat wide-necked, large, or giant aneurysms. To create anew treatment paradigm, we synthesized photopolymerizablepolyvinyl alcohol (PVA) gels to fabricate a space-filling material withengineered bioadhesion to inflamed endothelial cells.PVA was chosen for its low adhesiveness, non-degradability, and chemical tunability. PVA was modified to include methacrylateand amine side groups for photopolymerization and bioconjugation, respectively. We functionalized PVAsurfaces with varying ratios of antibodies against cell adhesion molecules(CAMs) highly expressed on inflamed human umbilical vein endothelial cells(HUVECs). We validated gene expression profiles of endothelial leukocyteadhesion molecule-1 (ELAM) and vascular cell adhesion molecule-1 (VCAM) onHUVECs treated with interleukin-1α(IL-1α) tosimulate an inflammatory environment. PVA gels were either modified withanti-ELAM only, anti-VCAM only, or a 1:1 ratio of anti-ELAM to anti-VCAM.HUVECs were stimulated with IL-1αbefore seeding onto the gels, and attachment was quantified after 2 and 24 hr.Cell adhesion was assessed by calculating the ratio of cells retained aftercentrifugation over cells retained after inversion. Preliminary results showedstrong and reversible cell attachment to PVA gels conjugated with eitheranti-ELAM or anti-VCAM. Though anti-ELAM and anti-VCAM showed strong adhesionat early times, this adhesion was reversed as a function of time. Synergisticbinding of both anti-ELAM and anti-VCAM showed stable adhesion at 24 hrpost-seeding. Endothelial cells were retained on 1:1 anti-ELAM:anti-VCAM presenting surfaces at a centrifugal forceof 300xg; adherence decreased steadily with increased centrifugation speeds. Anin vitro microfluidic aneurysm was designed toevaluate in situ polymerization and adhesion at physiological shear stresses ofup to 15 dynes/cm^2. Our findings suggest that cytokine-activated CAMexpression may be used to engineer bioadhesive gelsfor the specific and effective adhesion of inflamed endothelial cells. Such materialshave significant potential for use as an embolic agent for cerebral aneurysmsas they may conform to any aneurysm geometry and maintain adhesion to preventrupture and dislodging.

See more of this Session: Cell Adhesion and Migration II
See more of this Group/Topical: Food, Pharmaceutical & Bioengineering Division

(145g) Influence of CD151 Expression and Endothelial Cell Phenotype On Selectivity of Cancer Cell Adhesion to Endothelial Monolayers

Jennifer Fischer[1], Adrianne Shearer[1], Xinyu Deng[2], Xiuwei Yang[2], Richard Eitel[1] and Kimberly W., Anderson[1], (1)Chemical and Materials Engineering, University of Kentucky, Lexington, KY, (2) Molecular and Biomedical Pharmacology, University of Kentucky, Lexington, KY

Though cancer has been widely studied for some time, it continues to be the second leading cause of death in the United States. In the majority of cases, it is not the primary tumor that results in fatality, but a secondary tumor resulting from the dissemination of circulating tumor cells (CTCs) to other tissues and organs throughout the body. The basic steps of the metastatic cascade, from initial tumor formation to growth at a secondary site, are a linear sequence of steps composed of a network of mechanisms that are not well understood. One of these steps, the adhesion of the cancer cells to the endothelial cells during dissemination is the focus of this research. Two main hypotheses have been suggested to explain the mechanism of cancer cell adhesion to endothelial cells. Honn and Tang proposed a "docking and locking" mechanism, which describes the ligand-receptor interactions that take place as the tumor cells roll along the endothelium. Alternatively, other investigators have proposed that the cancer cells must first lodge in small microvessels before firmly attaching to the endothelium. In this study, a parallel plate flow chamber was employed to investigate the role of initial adhesion vs. lodging in the ability of cancer cells to attach to endothelial cells. This system allows us to test the "docking and locking" mechanism by visualizing initial adhesion of cancer cells to the endothelial cell monolayer as the cells roll along the monolayer at controlled shear stresses. Detachment of cells under controlled shear stresses after the cancer cells are allowed to settle on the monolayer can also be observed, hence simulating the lodging mechanism. In addition, the role of endothelial cell phenotype was studied by investigating the interactions of the cancer cells with endothelial cells originating from veins and arteries. Specifically, the "docking and locking" and lodging mechanisms were investigated and compared in both human umbilical vein endothelial cells (HUVECs) and human umbilical arterial endothelial cells (HUAECs). Finally, the influence of the gene CD151 on these interactions was quantified. CD151 is a gene that encodes a cell-surface glycoprotein that is known to complex with integrins and enhance cancer cell motility, invasion and metastasis. This was accomplished by comparing adhesion assays using human breast and prostate cancer lines (sum149 and PC3, respectively) as well as sum149 and PC3 lines in which CD151 expression was lowered using shRNA. Preliminary results have suggested that lodging plays a larger role than initial attachment in the interactions between breast cancer cells and HUVECs. Other results presented will include a comparison of the initial attachment and lodging of these cells on HUVECs vs. HUAECs as well as the effects of CD151 expression in these adhesion assays.

See more of this Session: Cell Adhesion and Migration II

(150a) Engineering the Lymph Node Microenvironment with Controlled Release Polymer Particles for Enhanced Vaccination

Christopher M. Jewell, Departments of Bioengineering, Materials Science and Engineering, and The Ragon Institute of MGH, MIT, and Harvard, Koch Institute for Integrative Cancer Research, Massachusetts Institute of Technology, Cambridge, MA, Sandra C. Bustamante Lopez, Departments of Bioengineering, Materials Science and Engineering, Koch Institute for Integrative Cancer Research, Massachusetts Institute of Technology, Cambridge, MA and Darrell J. Irvine, Departments of Bioengineering, Materials Science and Engineering, Koch Institute for Integrative Cancer Research, Ragon Institute of MGH, MIT, and Harvard, and the Howard Hughes Medical Institute, Massachusetts Institute of Technology, Cambridge, MA

Recent studies have demonstrated a simple, potentially universal strategy to enhance vaccines via intra-lymph node (*i.LN.*) injection. Lymph nodes (LNs) are a key site of immune response generation, where dendritic cells (DCs) process and present antigens to resident T cells and B cells, which then migrate to the periphery to carry out mechanisms of adaptive immune response. Since direct administration of antigens or adjuvants in LNs concentrates vaccines at this key site, *i.LN.* immunization with soluble vaccines has demonstrated great promise in recent human clinical trials. However, these strategies require complex, multiple injection regimens owing to rapid vaccine clearance from LNs. We hypothesized that combining *i.LN.* injection with controlled release biomaterials permitting sustained, concentrated dosing to local LN environments might augment the function of adjuvants or other drugs, providing a route for improved vaccination or immunotherapy. In this talk I will discuss two components of a new materials-based vaccination strategy aimed at 1) increasing vaccine potency and 2) tailoring specific aspects of immune response.

We first encapsulated poly(inosinic:cytidylic acid) (polyIC) – a potent nucleic acid adjuvant undergoing clinical trials – in biodegradable poly(lactide-co-glycolide) (PLGA) microparticles (MPs) or nanoparticles (NPs) designed to release polyIC in LNs over several days. We hypothesized particle size would allow control over intra-LN particle fate, with NPs being internalized by LN-resident cells, and MPs remaining extracellular and releasing encapsulated cargo within the LN. MPs and NPs were synthesized by double emulsion and lipids were used to stabilize the aqueous/organic interface, and to incorporate functionalized lipids (e.g., poly(ethylene glycol) to control aggregation). Properties such as particle size (0.2-15 μm), polyIC encapsulation (0.5-4% wt/wt), and release kinetics (2-25 days) were tunable by controlling emulsion conditions. To identify the target LN injection sites we administered a non-toxic tracer at the tail base of mice. Trafficking of the dye to the draining LN allowed subsequent identification of vaccine injection sites. Following injection of fluorescently-labeled particles, live animal imaging and histological analysis confirmed particles were localized within LNs. We used this approach to inject polyIC *i.LN.* in soluble form or encapsulated in MPs (MP-polyIC) or NPs (NP-polyIC) and discovered that MP-polyIC conferred a 4.5-fold increase in polyIC persistence four days after injection compared to soluble polyIC, with polyIC-NPs mediating an intermediate enhancement less than MP-polyIC, but greater than soluble polyIC. The enhancement observed with MP-polyIC mediated a dramatic increase in polyIC uptake by LN-resident DCs over 4 days, in contrast to soluble polyIC which was flushed away one day after injection. Mice vaccinated with antigen mixed with MP-polyIC exhibited antigen-specific T cell (CD8+) levels as high as 18% (7.6-fold greater than MP-polyIC given intramuscularly, *i.m.*; 4.0-fold greater than soluble polyIC given *i.LN.*). This increase indicates potent "arming" of the immune system, unmatched by *i.LN.* injection of antigen mixed with the same dose of NP-polyIC, or of a 10-fold greater doses of soluble polyIC. When T-cells were restimulated with antigen, MP-polyIC enhanced secretion of key cytokines (i.e., IFN-γ, TNF-α) involved in generating immunity, suggesting a functional or protective capacity that we are now exploring.

We have extended this initial approach to tailor immune response by sustained delivery of small molecule immunomodulators – molecules able to bias the specific nature or duration of immune response. A number of such agents have been described, but the potential of these compounds is limited by poor water solubility and rapid clearance *in vivo*. Thus treatment with these drugs typically involves high doses administered at frequent intervals. Since many immunomodulators act in LNs, polymeric particles designed to slowly release encapsulated small molecules in the LN could allow sustained stimulation, enabling this class of drugs to be effectively employed without continuous re-dosing. Daily, low-dose peripheral injections of Rapamycin –a potent small molecule drug – have recently been found to greatly enhance T cell response. Using modified solvent evaporation, we prepared rapamycin PLGA MPs (MP-rapa) and conducted biodistribution studies using *i.m.* or *i.LN.* injections of soluble rapamycin or MP-rapa. HPLC analysis of drug extracts prepared from blood and excised LNs demonstrates MP-rapa permits continuous dosing of rapamycin in LNs for 1 week, whereas injections of soluble rapamycin are undetectable at times longer than 24 hours after injection. In vaccination studies with model antigens, MP-rapa significantly increased the duration of CD8 T cell response, and induced a 6-fold increase in central memory T cells – cells critical for long-term vaccine protection. Strikingly, this enhancement was mediated by just three *i.LN.* injections of MP-rapa, and these benefits were unmatched by 150 daily peripheral injections of rapamycin in soluble form – the "gold" standard injection regimen for rapamycin. In preliminary studies with clinically-relevant HIV gag antigens, MP-Rapa induced a 2.2-fold increase in CD8 memory populations compared with soluble drug formulations, and these studies are currently being extended to small animal challenge models of HIV.

See more of this Session: Disease Models and Therapies
See more of this Group/Topical: Food, Pharmaceutical & Bioengineering Division

(150b) Design and Characterization of Self-Assembling, Peptide-Based Biomaterials As Immunotherapeutics

Amanda Trent, Biomolecular Science and Engineering, University of California, Santa Barbara, Santa Barbara, CA, Matthew J. Black, Chemical Engineering, University of California, Santa Barbara, Santa Barbara, CA, Colleen Olive, Immunity and Vaccinology Laboratory, The Queensland Institute of Medical Research, Herston, Queensland, Australia and Matthew Tirrell, Institute for Molecular Engineering, University of Chicago, Chicago, IL

With the goal of exploring strategies for the rational design of immunotherapeutics, we propose using peptide amphiphile (PA) molecules as building blocks for novel peptide antigen delivery systems. PAs can form versatile macromolecular constructs that combine the biofunctionality of peptides with a controllable self-assembly character obtained by synthetic chemical conjugation of peptide head groups with hydrophobic hydrocarbon tails (i.e. a fatty acid). In aqueous conditions, the hydrophobic effect drives PAs to self assemble into hierarchical structures, including spherical, cylindrical and worm-like micelles. PAs confer a number of attractive properties for peptide-based immunotherapy, including the multivalent presentation of peptide antigens, the enhanced peptide secondary structure often seen upon aggregation, and the potential of the hydrophobic tails to act as adjuvants. Peptides derived from the model tumor antigen ovalbumin and from an antigenic sequence found in Group A Streptococcus have been incorporated into two different PA vaccines. Their physical characterization, as well as their immunogenicity both *in vitro* and in mouse studies will be presented.

See more of this Session: Disease Models and Therapies
See more of this Group/Topical: Food, Pharmaceutical & Bioengineering Division

(150c) siRNA Based Gene Therapy In Alzheimer's Disease

Aparna Chaudhary and **Sanjeev Garg**, Chemical Engineering, IIT Kanpur, Kanpur, Uttar Pradesh, India

Alzheimer's disease (AD) is one of the most common forms of dementia. Existing therapeutics in AD are devoted to symptomatic relief. However, after the emergence of amyloid hypothesis, disease modifying therapeutics are being explored. According to amyloid hypothesis, a fundamental pathological event in AD is the sequential cleavage of the trans-membrane amyloid precursor protein (*APP*) by β-Site APP-Cleaving Enzyme1 (***BACE1***) and γ-*secretase*, generating toxic C-terminal fragments (*CTF*) and *Amyloid*β (Aβ). Aβ deposition occurs in the extracellular space which leads to plaque formation.

The present work aims to inhibit the expression of gene responsible for production of BACE1 enzyme through RNA interference. RNA interference is an intracellular mechanism that can be used as a powerful tool for gene silencing mediated by small interfering RNAs (siRNA). Suitable siRNA is the critical factor for effective silencing which can be achieved by a design tool. A siRNA design tool was developed and its viability was established with the reported siRNA constructs for genes implicated in cancer. In AD, it is proposed to inhibit the cascade of reaction which is responsible for Aβ aggregation. The siRNA constructs generated were compared with validated datasets for BACE1. The results obtained for BACE1 were better in thermodynamic score, GC content, positional preferences as compared to validated datasets. The off-target silencing effects were also comparable with the published results. It is to be noted that the results generated are *in silico* and need to be verified on mammalian cell lines where *BACE1* is expressed.

See more of this Session: Disease Models and Therapies
See more of this Group/Topical: Food, Pharmaceutical & Bioengineering Division

(150d) Identifying Novel Bacterial Proteases Involved In Enhanced Virulence During Chronic Infection

Logan MacDonald[1], Justin Nice[1], Sean O'Keefe[1] and **Bryan W. Berger**[2], (1)Bioengineering, Lehigh University, Bethlehem, PA, (2)Chemical Engineering and Program in Bioengineering, Lehigh University, Bethlehem, PA

Host protease-antiprotease balance has emerged as a major factor in regulating the extent of inflammatory response by neutrophils, macrophages and endothelial cells to bacterial infections. In response to pro-inflammatory cytokines, activated neutrophils release a series of extracellular proteases, including cathepsins, MMPs and neutrophil elastase (NE), all of which exhibit strong elastolytic activity, the largest percentage of which is attributed to cathepsins. Activated cathepsins are inhibited by cystatins, and dysregulation of the cathepsin-cystatin balance has been strongly linked to chronic bronchial inflammation during *P. aeruginosa* infection, especially in cystic fibrosis (CF) patients. We recently identified a family of conserved bacterial peptidases present in several obligate pathogens, including *P. aeruginosa*, that we believe play a critical role in disturbing the cystatin-cathepsin balance during chronic infection. We plan to investigate peptidase expression levels from clinical isolates to correlate expression with severity of chronic infections, and utilize a novel chemical biology approach to design peptide-based inhibitors of protease function as a potential therapeutic and diagnostic for *P. aeruginosa* infection. Our work will advance the understanding of how chronic inflammation in the endobronchial space, exacerbated by the dominant neutrophil inflammatory response, can lead to host-induced tissue damage caused by pathogens such as *P. aeruginosa.*

See more of this Session: Disease Models and Therapies
See more of this Group/Topical: Food, Pharmaceutical & Bioengineering Division

(150e) Interplay of Flow and Oncotically Active Solute Transport Across the Arterial Endothelium: Hydraulic Conductivity Masking and Relevance to Atherogenesis

Shripad D. Joshi, Chemical Engineering, City College and Graduate Center of City University of New York, New York, NY, Kung-Ming Jan, College of Physician and Surgeons, Columbia University, Bronx, NY and David S. Rumschitzki, Department of Chemical Engineering, The City College of The City University of New York, New York, NY

Macromolecular transport across the vessel wall, apparently the earliest pre-atherosclerotic event, is known to occur due to advection by transmural (i.e., across the vessel wall) pressure-driven plasma (water and solutes) transport through leaky junctions around rare endothelial cells that are temporarily widened and leaky, some because they are either dying or dividing. A wall parameter, the hydraulic conductivity (L_p), defined as the ratio of the transmural water flux to (hydrostatic-oncotic, the latter universally taken as zero in large artery walls) pressure difference, is central to the understanding of this transmural water transport in detail. On one hand, it advects macromolecules like low-density lipoprotein (LDL) into the arterial subendothelial intima (SI) through these leaks and spreads it there, thereby giving it the chance to bind to extracellular matrix and possibly to trigger the start of atherosclerotic lesion formation. On the other hand, the overall water flow across the normal (non-leaky) endothelium dilutes the lo-cal SI LDL concentration, thereby likely slow-ing binding kine-tics, and washing not-yet-bound lipid further into the wall.

Our group's discovery of the presence of Aquaporin-1 (AQP) in rat aortic endothelial cells suggests a new possibility of water transport across the endothelial cell, alongside the generally accepted paracellular route. Interestingly, we found that chemically blocking AQPs changes rat aortic wall Lp in a strongly pressure dependent manner. We then proposed a new theory that agrees well with this otherwise perplexing experimentally-observed pressure-dependence. It suggests that AQPs contribute about 30% to the phenomenological endothelial Lp at low transmural pressures. However, since the lumen is isotonic, an AQP-mediated, transendothelial cell, pure water flow into the arterial SI should set up an oncotic pressure gradient $\Delta\Pi$ that opposes and quickly halts the ΔP-driven flow through the cell. How then could trans-AQP flow persist for many hours, as our chemical blocking of AQP experiments show? We have developed a filtration and advection-diffusion model for water and the transport of the small, oncotically-active solute albumin across the glycocalyx layer, across endothelial interface and through the entire vessel wall. The filtration flow and the mass transfer problem are intimately coupled: The flow advects tracer while albumin differences across the endothelium set up a $\Delta\Pi$ that sucks fluid across the endothelium. And it is nonlinear: the advective term is the product of two dependent variables, velocity and (the gradient of) the concentration. We simultaneously solve the combined steady state problem as the long-time solution of an unsteady problem that we attack with finite difference methods. We use parameters taken from the literature, mostly from experiment.

Our results show that, since the media retardation coefficient ~ 0.3, the media acts to strongly filter the transport of the solute albumin relative to the water solvent, causing a build-up of albumin in the media that backs up into the SI by the time steady state is achieved. In contrast to our naïve suggestion above, this increased albumin concentration results in *higher* osmotic pressures in the SI than in the glycocalyx, which drives the water flow into the SI from the luminal side of the EC, and not the other way around. Thus, even though oncotic gradients are normally assumed negligible across the walls of large arteries, we find that they indeed play a role *within* those vessel walls. We show to what extent these oncotic gradients mask the true wall and layer hydraulic conductivities. We look at the effects of vesicular albumin transport and of varying the junctional boundary conditions on these results. Given AQP's presence in the high-pressure cardiac endocardium, heart, lung and renal epithelia, resolving and understanding how mass-transfer-induced osmotic gradients interact with the flows that advect those solutes may have far broader and illuminating implications.

See more of this Session: Disease Models and Therapies
See more of this Group/Topical: Food, Pharmaceutical & Bioengineering Division

(150f) Molecularly Imprinted Polymeric Nanoparticles for Treating Viral Infections - A Synthetic Antibody Approach

Yen Wah Tong, Department of Chemical and Biomolecular Engineering, National University of Singapore, Singapore, Singapore and Niranjani Sankarakumar, Chemical and Biomolecular Engineering, National University of Singapore, Singapore, Singapore

Molecular imprinting is an effective method that has been used commercially for more than a decade in chemical purification, analysis and sensor technologies. The working principle behind this technique is the creation of recognition sites in polymers through the specific 3-dimensional arrangements of monomers around the template molecules to be imprinted. This makes the imprint highly specific in recognizing the template molecule, and this behaviour is a chemical analog of biological antibodies.

In this research, our goal is to create molecular imprints of viruses on the surfaces of polymeric nanoparticles to "capture" and remove any active, disease-causing viruses. Virus transmitted diseases such as chikungunya, dengue and hepatitis pose significant social and economic problems throughout the world every year. Hence, the removal of viruses could be the ultimate approach for a vast range of virally transmitted diseases. Typical virus removal methods include ultrafine filtration or high speed centrifugation which is usually slow and expensive. Therefore, virus specific molecularly imprinted polymers may be an attractive and cheaper alternative to existing techniques involving the design of the **"viral sponge"** which would specifically and selectively remove or kill pathogenic virus from blood or sera.

We prepared surface imprinted polymers for viruses and tested its capability of specifically adsorbing viruses and preventing viral infections under *in vitro* conditions. For this proof of concept study, a simple bacteriophage was used as the model virus template. Virus imprinted (vMIPs) and Non-imprinted polymeric nanoparticles sized about 40nm were synthesized using miniemulsion polymerization. Methyl methacrylate (MMA) and ethylene glycol dimethacrylate (EGDMA) were employed as the functional and cross-linker monomers, respectively. The imprinted polymers exhibited higher preferential binding to template virus as compared to controls. *In vitro* experiments in a model host-virus system proved that our vMIPs was able to partially inhibit phage infection in *E.coli*. This research is the first of its kind to demonstrate the usage of molecularly imprinted polymers to catch viruses in the presence of its host cells.

See more of this Session: Disease Models and Therapies
See more of this Group/Topical: Food, Pharmaceutical & Bioengineering Division

(150g) Breast Cancer CSC Display Novel In Vivo Imaging Features, Reminiscent of Development, During Early Tumor Progression

Natesh Parashurama[1], Neethan Lobo[2], Michael Clarke[2] and Sanjiv Sam Gambhir[3], (1)Department of Radiology- Molecular Imaging Program at Stanford, Stanford University, Stanford, CA, (2)Institute for Stem Cell Biology and Regenerative Medicine, Stanford University, Stanford, CA, (3)Radiology and Molecular Imaging Program, Stanford University Schools of Engineering and Medicine, Stanford, CA

Virtually nothing is known about how nascent tumors form from cancer stem cells (CSC) and mammary stroma, yet this is essential for advancing breast cancer stem cell biology and developing molecular imaging and therapeutic strategies directed towards CSC. With this in mind, we developed a novel in vivo intravital imaging system designed to both image across the entire mammary gland and stroma. Importantly, this allowed us to identify new, rapidly developing, in vivo, quantitative imaging features at high resolution/sensitivity, not appreciable with conventional techniques. Using the developing mouse as a model for early tumor growth, we imaged and mapped the same fluorescent mammary gland at 3 and 4.5 weeks, and demonstrated rapid, spatially restricted growth; certain branches displayed a 400% increase in area over 1.5 weeks and maintained branching architecture, while neighboring ducts within 500μm displayed a 200% decrease. While examining stromal elements responsible for duct growth, we identified that normal mammary buds in virgin females displayed frequent microvascular interactions, and surprisingly, 77.7 % ± 6.4 (mean ± SD) of buds strongly interacted with the microvasculature (n=3, total terminal end buds =120). Notably, this stroma imaging feature was "conserved" in the 3 week old developing mouse, where we found 33/40 (83%) of mammary buds displayed analogous microvascular interactions. We then applied these imaging features to interpret a CSC tumor model designed to simulate early tumor formation from CSC. Two weeks after direct injection of 2500 CSC (or Tumorigenic) or non-CSC (non-Tumorigenic) cells from FvB MMTV-Wnt1 mice derived tumors into the developing GFP mammary gland, we observed new imaging features from the host mammary glands and host vasculature of the CSC populations only, including a 170% increase ($p < 0.28$) in side branching of endogenous, eGFP+ ducts, with maintained host microvascular-bud interactions. Lastly, two weeks after transplantation of dye-labeled CSC and non-CSC , we observed a 240% increase in overall angiogenesis ($p < 0.027$), and quantifiable imaging features of normal growth ($p < 0.005$), including selective bud and duct outgrowths supplied with vasculature, and invasion by loosely organized neovasculature. Overall, our data suggests that CSC populations display several rapidly developing imaging features involving CSC and the underlying stroma, some of which are seen in developing mammary gland, which can be potentially useful for early radiologic identification of CSC in tissues, identification of new targets for molecular probes, monitoring therapy, or understanding the biology of CSC-stromal interactions.

See more of this Session: Disease Models and Therapies

(157a) What a Difference a Letter Makes: Aggregation of PolyA and PolyQ

Joseph P. Bernacki, Robert H. Walters and **Regina M. Murphy**, Chemical and Biological Engineering, University of Wisconsin-Madison, Madison, WI

Polyglutamine (polyQ) and polyalanine (polyA) are two of the three most prevalent homopeptide repeats in eukaryotes. Abnormally expanded polyQ domains are associated with at least nine neurodegenerative diseases, including Huntington's disease. Abnormal expansion of the polyA repeat is linked to at least 9 human diseases, eight of which are congenital disorders. In both cases, expansion of the glutamine or alanine domain facilitates aggregation of the impacted protein, and the disease mechanism likely involves the aggregation triggered by the increase in length of the repeat unit. As with other aggregation-related disorders, soluble oligomers are suspected to serve as intermediates in the aggregation process, and to be more toxic than mature aggregates.

Expanded polyA diseases share some characteristics with the better-known expanded polyQ-mediated disorders, but there are distinct differences as well. All known polyQ diseases are late-onset disorders, while symptoms in all but one polyA disease appear at birth. The normal length of polyA, and the threshold expansion for disease phenotype, tend to be smaller than polyQ (normally 9-20 for polyA with +1-14 expansion, compared to 4-44 with +1-271 expansion for polyQ). Some evidence suggests that expansion of polyA may be more damaging than expansion of polyQ.

Studies of synthetic peptides have contributed substantially to our understanding of the mechanism of aggregation. We have previously presented work in which we examined the length-dependent aggregation of polyQ peptides. In work presented in this talk, we inserted interrupting residues into a peptide containing 20 glutamines, and examined the impact on conformational and aggregation properties. A peptide with 2 alanine residues formed laterally-aligned fibrillar aggregates which were similar to the uninterrupted Q20 peptide. Insertion of 2 proline residues resulted in soluble, nonfibrillar aggregates, which did not mature into insoluble aggregates. In contrast, insertion of the β-turn template $_{D}$PG rapidly accelerated aggregation and resulted in a fibrillar aggregate morphology that lacked the lateral alignment between fibrils observed in Q20. These results were interpreted to indicate that (a) nonspecific interactions between glutamines lead to the formation soluble oligomers, while insoluble oligomers form following an increase in β-sheet content and dehydration, and (b) that soluble oligomers dynamically interact with each other, while insoluble oligomers are relatively inert. Kinetic analysis revealed that the increase in aggregation caused by the $_{D}$PG insert is inconsistent with the nucleation-elongation mechanism of aggregation featuring a monomeric nucleus. Rather, the data support a mechanism of polyglutamine aggregation by which monomer collapse drives formation of soluble oligomers, which then undergo slow structural rearrangement to form sedimentable aggregates.

We also synthesized polyA peptides containing 6 to 24 alanines, and characterized their length-dependent conformation and aggregation properties. Helical content increased with increasing length. Measurements of end-to-end distance demonstrated that physiological buffers are theta solvents for shorter polyA peptides and poor solvents for longer peptides. At moderate concentrations, some soluble aggregates were observed in all peptides, with a sharp transition in the aggregate physical properties between 18 and 24 alanines. All aggregates were globular, none progressed to fibrillar morphologies, and none were insoluble, in sharp contrast to the behavior of polyglutamine peptides. The data suggest that, under physiologically relevant conditions, polyA peptides assemble into soluble oligomers due to hydrophobic collapse, but do not undergo structural rearrangement to form fibrils. The data were interpreted in light of a simple thermodynamic model, to explain why polyQ sometimes forms fibrils while polyA does not.

See more of this Session: Fundamentals of Protein Folding In Diseases
See more of this Group/Topical: Food, Pharmaceutical & Bioengineering Division

(157b) Synthetic YE8 Peptide Triggers Secondary Heterogeneous Nucleation of Insulin

Mirco Sorci[1], Gaius Takor[2], Seiichiro Higashiya[2], Dmitri Kurouski[2], Timothy Gehan[1], Shuqi Zhang[1], Igor K. Lednev[2], John T. Welch[2] and Georges Belfort[1], (1)Chemical and Biological Engineering and Center for Biotechnology and Interdisciplinary Studies, Rensselaer Polytechnic Institute, Troy, NY, (2) Department of Chemistry, University at Albany, State University of New York, Albany, NY

Amyloid diseases are characterized by misfolding of normally functioning proteins, which then aggregate into ordered β-sheet rich oligomers, protofibrils and fibrils. What triggers this process is still not completely understood. Researchers have suggested that specific amino acid segments are responsible for protein fibrillation, while others propose that amyloidogenic oligomerization is a generic property of the polypeptide backbone. To investigate this, $GH_6[(GA)_3GY(GA)_3GE]_8GAH_6$ (YE8), a synthetic amyloid-like peptide, was used to seed solutions of native insulin, a model amyloid protein. The aggregation was followed by both absorbance at 600 nm (A_{600}) and Thioflavin-T (ThT) measurements. YE8 addition increased the fibrillation magnitude (asymptotic values of A_{600} and ThT) and decreased the pre-fibril incubation period (lag-time) by 2- and 0.9-fold, respectively. Fibril morphology was characterized using atomic force microscopy, deep UV resonance Raman (DUVRR) spectroscopy and fluorescent probes. In particular, DUVRR spectroscopy showed that seeded fibrils have a chemical structural intermediate between the unseeded insulin and YE8 fibrils. These results suggest the existence of a secondary *heterogeneous* nucleation pathway, in which already formed YE8 fibrils are able (i) to catalyze the formation of new insulin fibrils, and (ii) to influence its misfolding.

See more of this Session: Fundamentals of Protein Folding In Diseases
See more of this Group/Topical: Food, Pharmaceutical & Bioengineering Division

(157c) Engineering of Non-Aggregating Alpha Synuclein Variants

Michael Hernandez, Sofia Golbert, Li Guo Zhang and **Jin Ryoun Kim**, Chemical and Biological Engineering, Polytechnic Institute of New York University, Brooklyn, NY

Aggregation of a 140 amino acid protein, alpha synuclein (alphaS), is implicated in the pathology of Parkinson's disease (PD). Monomeric alphaS is mostly irregularly structured, and aggregation of alphaS monomers to oligomers and fibrils is associated with conformational changes from disordered to beta-sheet rich structures. Aggregation propensity of alphaS and its naturally existing mutants has been found to be determined by combination of physicochemical properties of amino acids, such as hydrophobicity, secondary structure propensity and charges.

In this presentation, we will report that 1) one can create nearly non-aggregating alphaS variants using simple mutations without changing sequence physicochemical properties, such as secondary structure propensity, hydrophobicity and charge state, and 2) the resulting alphaS variants can inhibit alphaS aggregation by specific binding to alphaS aggregates. Our results suggest the importance of the sequence context in alphaS aggregation. Moreover, we anticipate that similar mutations may directly benefit development of aggregation modulators or aggregation state-specific binders for other amyloidogenic proteins.

See more of this Session: Fundamentals of Protein Folding In Diseases
See more of this Group/Topical: Food, Pharmaceutical & Bioengineering Division

(157d) Curcumin Attenuates Amyloid Beta Peptide Toxicity by Forming off-Pathway Intermediates

Arjun Thapa, Cell Biology, The University of Oklahoma Health Sciences Center, Oklahoma City, OK and Eva Y. Chi, Chemical and Nuclear Engineering, University of New Mexico, Albuquerque, NM

Amyloid beta peptide (Aβ) aggregation plays a central role in the pathogenesis of Alzheimer's disease (AD). Therefore, inhibition of Aβ aggregation using small molecule inhibitors is considered as an attractive therapeutic strategy to suppress AD. On the other hand, small molecules inhibitors of Aβ aggregation have been useful to categorize the distinct and complex processes of amyloid formation occurring in heterogeneous protein aggregation. A recent study on small molecule inhibitors of Aβ aggregation showed that polyphenolic antioxidant small molecule, curcumin, a previously known potent inhibitor of Aβ oligomerization/fibrilization and fibrils disrupting agent, inhibits Aβ oligomerization but not fibrilization. This finding also generated a doubt over the previously purposed mechanism(s) of some of the other potent anti-fibrillogenic small molecules to reduce Aβ toxicity. To understand how curcumin mediates cytoprotective effect by inhibiting oligomerization but not fibrilization, we analyzed the Aβ aggregates formed in the presence of curcumin by thioflavin-T assay, circular dichroism study and transmission electron microscopy and found that curcumin competes with thioflavin-T binding without influencing the fibril formation of Aβ40. Size-exclusion high performance chromatography (SE-HPLC) revealed that curcumin populates Aβ oligomers; presumably partially folded metastable off-pathway intermediates assembled in parallel with on-pathway fibril formation and stabilizes them to nontoxic conformations. Our data also show that curcumin is also effective against various toxic species of Aβ, including oligomers and fibrils though they may be distinctly different in sizes, structures, conformations and toxicities. Moreover, SE-HPLC fractionated curcumin accumulated Aβ intermediates demonstrate that these species modulate generic Aβ fibril formation to off-pathway by generating unusual protofibrillar-like assemblies. Taken together, our data suggest that curcumin captures the Aβ intermediates, most likely off-pathway assemblies, and transforms them into a nontoxic form without significantly altering on-pathway fibril formation.

See more of this Session: Fundamentals of Protein Folding In Diseases
See more of this Group/Topical: Food, Pharmaceutical & Bioengineering Division

(157e) Structure-Function Analysis of the Novel Small Molecule Modulators of Amyloid-Beta Aggregation and Cytotoxicity

Hann-Chung Wong, Erik E. Fernandez and **Inchan Kwon**, Department of Chemical Engineering, University of Virginia, Charlottesville, VA

Alzheimer's disease (AD) is a debilitating and ultimately fatal disease. Currently, 5.3 million people are affected in the U.S., with the number projected to rise to 13.5 million by 2050. A hallmark of AD is the accumulation of neurotoxic amyloid-beta peptide aggregates. Growing evidence suggests that soluble amyloid-beta oligomers are the primary toxic pathological species whose presence strongly correlates with dementia. Therefore, logically, a promising therapeutic strategy to inhibit disease progression is the reduction of neurotoxic amyloid-beta oligomers. To this end, numerous small molecules have been investigated for their potential to reduce and control the physiological concentrations of neurotoxic amyloid-beta species in the brain. However, the discovery and design of effective drug molecules has been hindered by both practical limitations, such as the low blood-brain-barrier permeability and intrinsic toxicity of small-molecules, and to limited understanding of their biophysical and biochemical functionality. Recently, we discovered that several small molecules with the demonstrated safety profiles significantly modulate amyloid-beta aggregation and reduce amyloid-beta-associated neurotoxicity. The small molecules identified were shown to bind to the specific sequence of amyloid-beta reducing cytotoxicity by generating non-toxic oligomers or promoting fibril formation. We have also identified critical structural features of the small molecules for their modulating capabilities on amyloid-beta aggregation and cytotoxicity through structure-function studies, which is expected to open new avenues into rational drug design for combating AD.

See more of this Session: Fundamentals of Protein Folding In Diseases
See more of this Group/Topical: Food, Pharmaceutical & Bioengineering Division

(157f) Mechanistic Inhibition of Alzheimer's-Associated Aβ Aggregation by Gold Nanoparticles

Kelly A. Wilson[1], Mihyun Lim[2], Kaliah Jackson[3], Rahina Mahtab[3] and Melissa A. Moss[4], (1)Biomedical Engineering Program, University of South Carolina, Columbia, SC, (2)Department of Biological Sciences, University of South Carolina, Columbia, SC, (3)Department of Chemistry, SC State University, Orangeburg, SC, (4)Chemical Engineering, University of South Carolina, Columbia, SC

Alzheimer's disease (AD) is currentlythe most common type of dementia and the 6th leading cause of deathin the United States. One pathological hallmark of AD is amyloid plaques,which deposit around the neurons in the brain. These plaques are composedprimarily of amyloid-beta (Aβ), which is a 40-42 amino acid protein thatforms from the cleavage of the amyloid precursor protein (APP). Aβ is anaturally occurring protein in the body; however it self-assembles to createaggregated structures and ultimately insoluble fibrils in a process that ishypothesized to be closely tied to disease progression. During self-assembly,monomeric Aβ forms nuclei, which progress to form soluble aggregates.These soluble aggregates can then associate or elongate to produce insolublefibrils. In association, soluble aggregates laterally bind to one another toform fibrils of an increased diameter. In elongation, monomeric Aβ bindsto the ends of the soluble aggregates to form fibrils with an increasedlength. Inhibition of Aβ aggregation has emerged as a therapeuticstrategy for AD. We have examined gold nanoparticles for their ability to haltspecific Aβ aggregation mechanisms. Nanoparticles are of particularinterest because they cross the blood-brain barrier and have been used for drugdelivery. By employing TEM images to evaluate aggregate morphology as well asassays that isolate Aβ aggregation mechanisms, we have tested themechanistic-specific inhibitory capabilities of gold nanospheres with citrate,cetrimonium bromide (CTAB), poly acrylic acid (PAA), and poly allylaminehydrochloride (PAH) overcoatings.

The effect of gold nanoparticles on $A\beta_{1-40}$monomer aggregation was first evaluated. Subsequently, mechanistic-specificassays were employed to assess the effect of gold nanoparticles on soluble $A\beta_{1-40}$aggregate elongation and association. These assays utilized thioflavin T tomonitor increases in amyloid material, dynamic light scattering to monitorincrease in aggregate size, and TEM to observe changes in aggregate morphology.

Gold nanoparticles overcoated withcitrate, CTAB, and PAH were observed to have little effect on the quantity ofamyloid material formed during $A\beta_{1-40}$ monomer aggregation. TEMimages (Figure 1), however, demonstrate that aggregations performed in thepresence of CTAB overcoated nanoparticles favored aggregate association overelongation. Mechanistic-specific assays confirm these trends and directly tieaggregation mechanisms to aggregate morphology. PAA nanoparticles abrogatedthe formation of amyloid material in monomer aggregation assays. Furthermore,this inhibition was observed at a substiochiomeric ratio of nanoparticles to Aβof 1:200. Elongation and association assays further define the mechanism ofthis inhibition.

Together these results demonstrate thatgold nanoparticles serve as effective inhibitors of Aβ self-assembly thatmodify aggregation at substiochiometric concentrations. Furthermore, inhibitionof Aβ aggregation by gold nanoparticles is mechanistic-specific in amanner that is dependent upon the nanoparticle coating. Determining themechanisms by which these various nanoparticles inhibit Aβ self-assemblymay be used to develop an inhibitor capable of targeting multiple mechanisms.

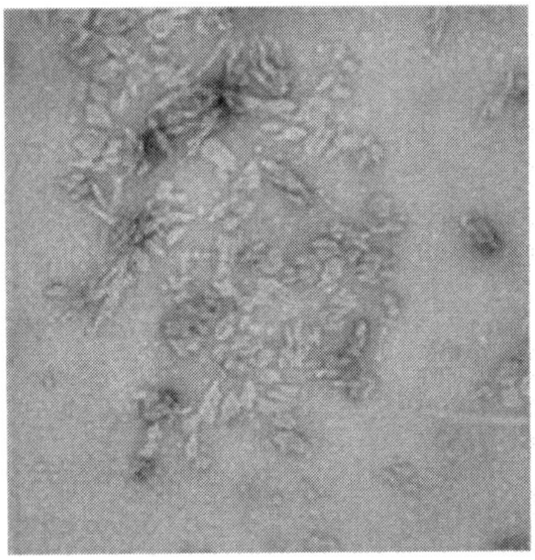

Figure 1: TEMimage (100,000X), 40M Aβ monomer aggregated in the presence of0.2nM CTAB nanoparticles. Presence of the CTAB nanoparticles allows for thedevelopment of associated and non-elongated aggregates.

See more of this Session: Fundamentals of Protein Folding In Diseases
See more of this Group/Topical: Food, Pharmaceutical & Bioengineering Division

(157g) Conformational Differences Between Toxic and Non-Toxic Oligomers of Amyloid

Ali Reza A. Ladiwala and **Peter M. Tessier**, Dept of Chemical & Biological Engineering, Rensselaer Polytechnic Institute, Troy, NY

Many amyloidogenic proteins misfold into soluble oligomers that mature into amyloid fibrils. In addition to this pathway, we have discovered that soluble oligomers of the amyloid β (Aβ) peptide associated with Alzheimer's disease can mature into a second oligomeric isoform of identical size that possesses a combination of biochemical properties distinct from highly toxic Aβ soluble oligomers and mildly toxic fibrils. Importantly, these uniquely folded Aβ oligomers are non-toxic to mammalian cells. In this presentation, we will discuss our characterization of Aβ soluble and non-toxic oligomers using a panel of biochemical, immunological, imaging and cell toxicity assays. Importantly, we find that the lack of toxicity for Aβ non-toxic oligomers relative to soluble oligomers and fibrils is not due to simple differences in size, secondary structure, hydrophobicity and conformational stability. We will discuss our recent site-specific analysis of Aβ soluble and non-toxic oligomers aimed at identifying the size-independent conformational determinants of Aβ-mediated toxicity.

See more of this Session: Fundamentals of Protein Folding In Diseases
See more of this Group/Topical: Food, Pharmaceutical & Bioengineering Division

(160a) QbD Principles for Packaging Design and Stability: A Regulatory Perspective

Sharmista Chatterjee, Ondqa/Cder/FDA, FDA, Silver Spring, MD

Quality by Design (QbD) is a science and risk based approach for pharmaceutical manufacturing and development. Enhanced product understanding gained from following a QbD based approach, is increasingly being used to select packaging design parameters and to determine optimum storage conditions.The aim of this presentation is to illustrate, using relevant examples, various QbD based approaches that may be implemented for packaging design, determination of storage conditions and shelf life estimation and to offer a regulatory perspective regarding considerations for successful implementation of such approaches throughout the product life cycle.

See more of this Session: Implementation of the QbD-Paradigm for-Packaging Design and Estimation of Product Shelf Life
See more of this Group/Topical: Topical I: Comprehensive Quality by Design in Pharmaceutical Development and Manufacture

(160b) Stability In a QbD Framework

Theodora Kourti, GSK, Hamilton, ON, Canada

Multivariate Data Based Empirical models can be used to predict future product performance in terms of Stability, and anticipate product shelf life. So far Quality by Design is being used to assess the quality of a product at release based upon manufacturing data but this concept can be extended to investigate stability patterns based on multivariate interactions of raw material characteristics, manufacturing process parameters, and packaging characteristics. The DoEs for Design Space determination could be planned to include stability studies. Process Understanding derived from such studies would lead to appropriate Control Strategy that takes into account stability requirements

See more of this Session: Implementation of the QbD Paradigm for Packaging Design and Estimation of Product Shelf Life
See more of this Group/Topical: Topical I: Comprehensive Quality by Design in Pharmaceutical Development and Manufacture

(160c) Science-Based Stability and Packaging Development Under the Umbrella of QbD

Rey Chern, Pharmaceutical Commercialization Technology, Merck & Co, Inc., West Point, PA

Abstract

Science-Based Stability and Packaging DevelopmentUnder The Umbrella of QbD

Rey T Chern

Global Pharmaceutical Commercialization

Merck Sharp & Dohme

Asystematic approach for characterization of product stability risks, selectionof primary packages and utilization of first-principle modeling to managestability outcomes is presented. Theoverarching framework of this approach focuses on three areas. First, controlled Product CharacterizationStudies (PCS) are performed to gain quantitative understanding of the impact ofthe factors including environmental conditions (e.g. temperature, relativehumidity, oxygen, and light), formulation ingredients (e.g. API form, critical excipient impurities, etc.), and processing parameters (e.g.mechanical stress, drying, etc.) on the stability of the product.

Second,the primary containers under consideration are characterized quantitatively fortheir protective performance, specifically relevant to the stability drivers(e.g. moisture or oxygen transmission rate). Often, the critical environmental factorscontrolling the stability of oral products are temperature & water moisture. A model for quantitatively describing thetime-course of water activity that the product is exposed to over the expectedshelf life at the ICH testing conditions has been developed. This model, labeled as "**S**imulationof **H**eadspace **M**oisture **A**ctivity in **P**ackages(SHMAP)", is derived from the first principles of thermodynamics, mass balance,and mass transport dynamics. It has beenverified by experimental data collected for >20 development projects in thepast 10 years.

Third,the knowledge gained from PCS and SHMAP is combined to predict stability performanceof the various product/primary-container combinations intended for registration. This approach allows for improved design ofregistration stability studies and enables projection of thestability-performance of new product-packaging configurations with a highdegree of assurance since all needed scientific data input are available andcontrolled independently.

Thisapproach for stability & primary package development is a good example ofthe application of

Quality by Design principles.

See more of this Session: Implementation of the QbD Paradigm for Packaging Design and Estimation of Product Shelf Life
See more of this Group/Topical: Topical I: Comprehensive Quality by Design in Pharmaceutical Development and Manufacture

(160d) Using QbD to Asses the Factors That Affect Stability and Packaging

Stephanie Krogmeier, Regulatory Affairs, Vertex Pharmaceuticals, Cambridge, MA and Kevin J. Bittorf, CMC Process & Alliance Management, Strategic Development Networks, Vertex Pharmaceuticals, Cambridge, MA

Packaging is often the last step in drug product (DP) development; however packaging is extremely important in determining the marketability of a product. The ability to predict whether a product can be stored at room temperature early in development can affect how the drug is developed. In order to determine packaging requirements, and estimate shelf-life of the drug product it is important to understand the factors that impact DP degradation, which include storage temperature, water content, and starting materials.

Using quality by design (QbD) principles, an experimental plan and subsequent model were developed to systematically evaluate these factors. From the developed stability model, a strategy was determined to facilitate packaging selections, storage conditions and justification of specifications. The ability to use this model to determine the product packaging and justify the product specifications is crucial to patient safety and critical for a successful regulatory filing. Finally, discussed are the regulatory implications and how this information can be used in conjunction with ICH stability studies.

See more of this Session: Implementation of the QbD-Paradigm for-Packaging Design and Estimation of Product Shelf Life
See more of this Group/Topical: Topical I: Comprehensive Quality by Design in Pharmaceutical Development and Manufacture

(160e) Designing Pharmaceutical Packaging Using a Product-Packaging Stability Modeling

Seung-yil Yoon, Eli Lilly and Company, Indianapolis, IN

DesigningPharmaceutical Packaging using a Product-Packaging Stability Modeling

ABSTRACT

Pharmaceutical solid oral dosage forms are degradedby environment hazards such as moisture, oxygen, and temperature as a functionof time. One of functions of packagingis to protect products from those environment hazards. When a package is designed, variables such asproduct degradation kinetics, package barrier, sorbentsorption and environment conditions should be understood and integrated. These variables could be understood using a traditionaltrial and error approach where products are put in various combinations ofpackaging on stability screening for 6-12 months. However, it is time-consuming, costly and it can not be usedto explain a variety of unique situations that can arise during development andmanufacturing. Product-packagingmodeling integrates experimentally-determined product and packaging data, and themodeling provides estimated product stability throughout dynamically changingconditions inside the package during shelf life storage.

In this presentation, theories and key data inputsfor the product-packaging modeling will be explained and several successfulmodeling applications in support of QbD paradigm willbe shared.

See more of this Session: Implementation of the QbD-Paradigm for-Packaging Design and Estimation of Product Shelf Life
See more of this Group/Topical: Topical I: Comprehensive Quality by Design in Pharmaceutical Development and Manufacture

(160f) Implementation of QbD Paradigm In Sterile Dosage Form Packaging - Some Practical Considerations

Dr. Hemant N. Joshi, Tara Innovations LLC, Parsippany, NJ

Out of all the pharmaceutical dosage forms, the Quality Target Product Profile (QTPP) for the sterile dosage form is the strictest for obvious reasons. One has to pay attention to the critical quality attributes (CQA) to a great extent. This presentation is focused only on the packaging of sterile dosage forms. There are several types of sterile dosage forms such as small and large volume parenteral liquids, lyophilized products, ophthalmic solutions/suspensions etc. The packaging is different for each type. The Quality by Design (Qbd) approach should be applied during the packaging operation (identification, management and control of packaging operation). Leachables and extractables are critical in other dosage forms, but are most important in sterile dosage forms. QbD should also consider other aspects including – uniformity of distribution, pH, sterility, particulate matter, water content, antimicrobial and antioxidant preservative content, osmolarity, and endotoxins/pyrogens. In this seminars topics such as development of design space for packaging, influence of scaling up, design of clean room, environmental monitoring and aseptic processing will also be discussed. Important, design and interpretation of media fills will also be discussed to build in quality by design.

See more of this Session: Implementation of the QbD-Paradigm for-Packaging Design and Estimation of Product Shelf Life
See more of this Group/Topical: Topical I: Comprehensive Quality by Design in Pharmaceutical Development and Manufacture

(165a) Advances In Microalgal Biotechnology

Feng Chen, Institute for Food & Bioresource Engineering, College of Engineering, Peking University, Beijing, China

Microalgae are "old" organisms on earth but the recognition of their usefulness in terms of industrial application is relatively recent. Microalgae are a wonderful and unique source of high-value biological active compounds and can be exploited as functional foods (or functional food ingredients) as well as pharmaceuticals. In this lecture, I will summarize the applications of microalgae in a number of existing and new areas, and the advancement of technology employed. The two fundamental components of biotechnology, namely, bioprocess engineering and molecular biotechnology that can be used for microalgae research and application will be reviewed and discussed.

See more of this Session: International Forum: Biotechnology In China
See more of this Group/Topical: Food, Pharmaceutical & Bioengineering Division

(165b) Biobutanol R&D In China

Yin Li and Yanping Zhang, Institute of Microbiology, Chinese Academy of Sciences, Beijing, China

Fermentative production of acetone-butanol-ethanol (ABE) was initiated from the beginning of the twentieth century. It was once one of the largest fermentation process, ranking the second in scale next to ethanol fermentation by yeast. In the latter half of twentieth century, fermentative ABE production declined because of its weak economic competitiveness compared with the petroleum-based processes. Recently, the increasing crisis of petroleum and environmental pollution bring scientists and industries back to the ABE production derived from biomass, especially to the microbial butanol production. China is one of the few countries, which maintained the fermentative ABE production for several decades. Starting from 2006, several newly established ABE fermentation plants, with the capacity amount to over 300,000 tons, have resumed production in China. We will discuss recent progress on biobutanol in China, including government policy and funding, technical breakthrough (genetic tools, metabolic engineering, systems biology, process optimization), and commercial opportunities.

See more of this Session: International Forum: Biotechnology In China
See more of this Group/Topical: Food, Pharmaceutical & Bioengineering Division

(165c) Organic Acid In China: Progress and Prospect

He Huang, College of Biotechnology and Pharmaceutical Engineering, Nanjing University of Technology, Nanjing, China

A modern biotechnology industry has been established successfully over the last 30 years in China. Presently, China is a major producer of many products through biotechnology processes. For example, China has been the world's top producer of some organic acid such as glutamic acid, L-lysine and citric acid. This presentation would introduce the development status and trends of organic acid fermentation industry in China.

In this presentation, the organic acid fermentation industry in China firstly was introduced. The content includes the supply and demand status, market size and growth potential, major producing company of some organic acid such as lactic acid, fumaric acid, citric acid, L-lysine, glutamic acid, itaconic acid, etc. Secondly, the technology advantages and disadvantages of organic acid industry have been presented, including up-steam fermentation and down-stream separation technology. Thirdly, the developing technology trend has been introduced in detail. The research and manufacture of bio-based fumaric acid and malic acid has been selected as an example to demonstrate that China would pay attention to the food application of the bio-based organic acid. In order to broaden the range of raw materials so that non-food biomass can be utilized efficiently, some researches have been devoted to the pretreatment of ligno-cellulose and the mutant strain breeding, which would accelerate the efficient conversion of low-quality biomass. Based on the organic acid, the new extended product chain have been researched and established. For example, non-toxic citrate esters plasticizer from citric acid, acrylic acid from catalytic dehydration of fermented lactic acid, 1,5-Pentanediamine from L-lysine through decarboxylation process, poly(butylene succinate) from bio-based succinic acid, novel polylactic acid manufacture and processing technology, and so on. Lastly, a concluding remark about organic acid in China has been presented.

See more of this Session: International Forum: Biotechnology In China
See more of this Group/Topical: Food, Pharmaceutical & Bioengineering Division

(165d) Systematic Analysis and Decoupling of Yeast Tolerance to Combined Inhibitors In Liginocellulose Hydrolytes

Ying-jin Yuan, Key Lab of Systems Bioengineering, MOE; Department of Pharmaceutical Engineering, School of Chemical Engineering and Technology, Tianjin University, Tianjin, China and Ming-zhu Ding, Key Lab of systems bioengineering, MOE, Tianjin University, Tianjin, China

Inhibitors produced during lignocellulose pretreatment including furan derivatives, weak acids and phenolic compounds impose severely synergistic effects on *Saccharomyces cerevisiae* during ethanol fermentation. Tolerant yeast which could exhibit better characteristics in the presence of combined-inhibitors ((1.3 g/L furfural, 5.3 g/L acetic acid and 0.5 g/L phenol)) was obtained. The complex relationship between combined inhibitors and yeast cells was analyzed systematically on mRNA, proein and metabolite levels.

GC-TOF/MS based metabolomics investigation indicated that acetic acid played the most important role in the combined inhibitors. Tolerant yeast exhibited better tolerance to inhibitors in growth, fementation and metabolism, reflecting on less affected amino acid metabolism and carbon central metabolism. 2-DE and MALDI-TOF/MS based proteomic study suggested that oxidative, osmotic and unfolded protein stress responses were induced in yeast by combined inhibitors, but the mechanisms of stress response in parental and tolerant yeasts were different. Higher stress response and detoxification realated proteins were important for tolerant yeast to protect itself. And lower levels of amino acid and nucleotide metabolism related proteins and higher levels of energy related proteins were necessary for tolerant yeast to defend the inhibitors stresses. Glycolysis and gluconeogenesis related proteins were upregulated in parental yeas to produce more energy to resist the inhibitors stresses. Metabolomic and proteomic results indicated that protein degradation was increased by combined inhibitors, inducing oxidative stress and inhibiting growth and production of yeast.

Microarray based transcriptomics study found that RNA and protein synthesis related genes were repressed, while electron transport and membrane-associated energy conservation related genes were upregulated by combined inhibitors. It indicated that biosynthesis in yeast was inhibited, and more energy was needed to defend the inhibitors stresses. In addition, inhibitors caused protein degradation in cells, resulting in oxidative detoxification and autophagy. LC-MS/MS based lipidomic study revealed that the variation of chain length in PC and PI was induced by inhibitors. PE and PS were significantly different for parental and tolerant yeasts in response to combined inhibitors. Transcriptomics and lipidomics results indicated that PC synthesized was inhibited by inhibitors, resulting in increased membrane curvature and pore size, which facilitated the import of inhibitors to cells to affect normal metabolism.

(165e) How Polymer Enhances the Stability and Functionality of the Conjugated Protein

Cheng Yang[1], Binbin Zhu[2], Diannan Lu[2] and **Zheng Liu**[2], (1)Chemical Engnieering, Institute of Biochemcial Engineering, Beijing, China, (2)Chemical Engineering, Institute of Biochemcial Engineering, Beijing, China

This review summarizes our recent efforts on protein stabilization by polymer conjugation. The first part focuses on molecular dynamics simulation of PEGlyated insulin, in which an annealing procedure was proposed that enabled an all-atom level examination of the interaction between PEG polymers of different chain lengths and insulin. It was shown that PEG entangled around protein surface through hydrophobic interaction and concurrently formed hydrogen bonds with the surrounding water molecules. In addition to an enhanced structural stability, conjugation gave an increased size while a decreased the solvent accessible surface area. These favored a prolonged circulation life despite kidney filtration, proteolysis, and immunogenic side effects. The simulation that reproduced the experimental results in literates provided a molecular insight into the PEGlayted protein.

The second part presents a novel procedure to fabricate uniform protein-polymer conjugate with improved stability under adverse conditions. This method was based on *in situ* aqueous activators generated by electron transfer for atom transfer radical polymerization (AGET-ATRP) in air, using enzyme as macroinitiator. The procedure started by grafting 2-bromoisobutyryl bromide (BIB) as initiator onto the protein surface, followed by in situ AGET ATRP polymerization in air using CuBr2/1,1,4,7,7-pentamethyldiethylenetriamine (PMDTA) as a catalyst and ascorbic acid as a reducing agent. The size of conjugate was shown to be a function of monomer loading and reaction time. The effectiveness of this method has been demonstrated using horseradish peroxidase (HRP) as a model protein and acrylamide as the monomer. The yielded HRP conjugates essentially reproduced the catalytic behavior of HRP in free form but exhibited significantly enhanced thermal stability against high temperature and trypsin digestion. One noteworthy advantage of this procedure that it prevented the formation of copolymer and thus facilitated purification of the protein conjugate.

See more of this Session: International Forum: Biotechnology In China
See more of this Group/Topical: Food, Pharmaceutical & Bioengineering Division

(165f) Prodrugs AS Multifunctional Drug Nanocarriers

Youqing Shen, Department of Chemical and Biochemical Engineering, Zhejiang University, Hangzhou, China

Nanomedicines mainly contain two parts, the inert carriers and the bioactive agents. The carriers are typically inert and their only role is to make the vehicles. For instance, the lipid and cholesterol in conventional liposomes have no other functions but to make the vesicles carrying drugs. The amphiphilic block copolymers used to make nanoparticles are just for forming the core-shell structure encapsulating hydrophobic drugs in the core. But generally, the carriers are the major while the bioactive agents are minor components in nanomedicine. For instance, in nanoparticles or liposomes, the drug contents are generally not greater than 10%. Polymer–drug conjugates only contain a few percents of drugs to keep the conjugates water-soluble. Therefore, large amounts of carriers are needed to deliver required doses of drugs, bringing difficulty and problems for parenteral biodelivery. Furthermore, the multistep fabrication processes of most nanomedicines raise batch-to-batch variation and quality concerns. Herein, we demonstrated a new concept employing the drug molecules themselves as components of nanocarriers to substantially increase the drug loading content, minimize the use of inactive materials, and suppress the premature burst drug release.

Taking advantage of the hydrophobicity of CPT, two CPT molecules were conjugated to a very short PEG chain (OEG), forming an amphiphilic phospholipid-like molecule, OEG-DiCPT (Figure 1). Once in water OEG-DiCPT formed stable liposome-like nanocapsules with a fixed CPT loading content as high as 58 wt% without burst release in aqueous solution. OEG-DiCPT itself was a prodrug releasing CPT once inside cells, which showed high *in vitro* and *in vivo* antitumor activity. Meanwhile, the formed nanocapsules could be loaded with a water-soluble drug such as DOX·HCl with high loading efficiency. The DOX·HCl–loaded nanocapsules delivered two anticancer drugs producing a synergetic cytotoxicity to cancer cells. Similarly, curcumin was used to surfactant-like prodrug formed nanoparticles as carriers of curcumin itself and other drugs and polymer-drug conjugates with high drug loading contents.

See more of this Session: International Forum: Biotechnology In China
See more of this Group/Topical: Food, Pharmaceutical & Bioengineering Division

(197a) Genome-Scale Metabolic Network Modeling Results In Minimal Interventions That Cooperatively Force Carbon Flux towards Malonyl-CoA

Peng Xu, Chemical and Biological Engineering, Rensselaer Polytechnic Institute, Troy, NY, Sridhar Ranganathan, Huck Institutes of Life Sciences, The Pennsylvania State University, University Park, PA, Costas D. Maranas, Department of Chemical Engineering, University Park, PA and Mattheos A. G. Koffas, Chemical and Biological Engineering, University at Buffalo, the State University of New York, Buffalo, NY

Malonyl-coenzyme A (malonyl-CoA) is an important precursor metabolite for the biosynthesis of polyketides, flavonoids and biofuels. However, malonyl-CoA naturally synthesized in microorganisms is consumed for the production of amino acids, fatty acids and phospholipids leaving only a small amount available for the production of other metabolic targets in recombinant biosynthesis. During the past decade, metabolic pathway modeling has aided various metabolic engineering efforts to design strains of bacteria and yeast that overproduce high-value and commodity chemicals. Here we present an integrated computational and experimental approach aimed at improving the intracellular availability of malonyl-CoA in *Escherichia coli*. We used a customized version of the recently developed OptForce methodology to predict a minimal set of genetic interventions that guarantee a pre-specified yield of malonyl-CoA in *E. coli* strain BL21 Star^{TM}. In order to validate the model predictions, we have successfully constructed an *E.coli* recombinant strain that exhibits a 4-fold increase in the levels of intracellular malonyl-CoA compared to the wild-type strain. Furthermore, we demonstrate the potential of this *E. coli* strain for the synthesis of plant-specific secondary metabolites (i.e., flavanones) that are the precursors of promising agents in the treatment of chronic diseases such as cardiovascular disorders and diabetes. Specifically, a titer of 474 mg/L of the flavanone naringenin was observed which, so far, is the highest yield achieved in a lab-scale fermentation process. Quantitative analysis of kinetic profiles from these strains highlights the synergistic effect of combining beneficial mutants (Δ*fumC* and Δ*sucC*) and overexpression targets (*acc*, *pgk*, *gapA* and *pdh*) predicted by OptForce. More importantly, the presented strategy can also be readily expanded to the production of other malonyl-CoA-derived compounds like polyketides and fatty acids.

See more of this Session: Advances In Cell Culture III : Metabolic Flux Analysis and Modeling
See more of this Group/Topical: Food, Pharmaceutical & Bioengineering Division

(197b) Validation of Metabolic Network Model for E. Coli and Assumptions of 13C-Metabolic Flux Analysis and Using [U-13C]Glucose Tracer

Maciek R. Antoniewicz, Chemical Engineering, University of Delaware, Newark, DE and Robert Leighty, Department of Chemical Engineering, Univerisity of Delaware, Newark, DE

Currently, the most powerful technique for measuring intracellular metabolic fluxes in living cells is based on the application of ^{13}C-tracers and measurement of isotopic labeling by mass spectrometry and/or NMR. Two key assumptions of ^{13}C-metabolic flux analysis (MFA) are that 1) enzymes doen't discriminate between ^{12}C and ^{13}C-atoms, i.e. no kinetic isotope effect; and that 2) the metabolic network model that is used for fitting of isotopomer data is accurate and complete. In this study, we validated both assumptons experimentally for wild-type *E. coli* using mixtures of natural and [U-^{13}C]glucose. *E. coli* were grown in parallel cultures on media containing 0%, 20%, 40%, 60%, 80% and 100% [U-^{13}C] glucose (with balanced natural glucose). Extracellular measurements didn't reveal any significant differences between the cultures. The growth rate, the biomass-specific glucose uptake rate and the specific product secretion rates were identical for all cultures. Thus, these results provide support for the assumption that presence of ^{13}C-atoms in the substrate doesn't affect growth of *E. coli*. To investigate the potental impact on intracellular metabolism, GC-MS measurements were collected for proteinogenic amino acids and the isotopomer data were fit to a previously established metabolic model of *E. coli*. The initial fits did not provide statisitically acceptable results. Further inspection of the isotopomer data and the metabolic network model revealed that the network model was missing important reactions. After complementing the model with these additional, previously missed, reactions we obtained consistent flux results for all experiments. Thus, these results provide additional support for the hypothesis that the presence of ^{13}C-atoms in intracellular metabolites doesn't affect enzymatic reactions and that the outcome of metabolic flux analyses is independent of the amount of ^{13}C-tracer used.

See more of this Session: Advances In Cell Culture III : Metabolic Flux Analysis and Modeling
See more of this Group/Topical: Food, Pharmaceutical & Bioengineering Division

(197c) Metabolic Flux Analysis of the Central Metabolism of Laboratory and Industrial Saccharomyces Cerevisiae Strains Using 13C-Labeling Experiments

Byoungjin Kim, Energy Biosciences Institute, Institute for Genomic Biology, Urbana, IL and Huimin Zhao, Department of Chemical and Biomolecular Engineering, Department of Chemistry and Biochemistry, Institute for Genomic Biology, University of Illinois at Urbana-Champaign, Urbana, IL

Recombinant *Saccharomyces cerevisiae* capable of utilizing xylose as a carbon source contains fungal or bacterial xylose catabolic pathways. In Saccharomyces cerevisiae is highly regulated in a coo Recombinant yeast strains expressing identical xylose catabolic pathway show differences in cell ^{13}C-labeled substrate and mass isotopomer distributions of central metabolites ar*Saccharomyces cerevisae* would provide valuable metabolic engineering strategies to create recombinant yeast strains efficiently utilizing xylose for bioethanol production.

See more of this Session: Advances In Cell Culture III : Metabolic Flux Analysis and Modeling
See more of this Group/Topical: Food, Pharmaceutical & Bioengineering Division

(197d) Metabolic Flux Analysis Using Designed Isotope Labeling Experiments Reveals Effects of Light On Metabolic Fluxes In Heterotrophic Plant Suspension Cells

Shilpa Nargund[1], Max E. Joffe[1], Vitali Tugarinov[2] and Ganesh Sriram[1], (1)Chemical and Biomolecular Engineering, University of Maryland, College Park, MD, (2)Chemistry and Biochemistry, University of Maryland, College Park, MD

It is well known that light regulates plant metabolism substantially and in several different ways. For example, energy from light is crucial to the generation of ATP and the electron carrier nicotine adenine dinucleotide phosphate (reduced) (NADPH) during photosynthesis. The important role of light in plant primary metabolism is exemplified by its induction of glutamine synthetase (which plays an essential role in nitrogen assimilation; Peterman and Goodman, Mol. Gen. Genet. 230: 145-154, 1991) and repression of glucose-6-phosphate dehydrogenase (which catalyzes the first, NADPH-providing step of the oxidative pentose phosphate pathway; Scheibe et al., Arch. Biochem. Biophys. 274: 290-297, 1989; Hutchings et al., J. Exp. Bot. 56: 577-585, 2005). Furthermore, the absence of light causes accumulation of vegetative storage proteins (Berger et al., Plant Mol. Biol. 27: 933-942, 1995) whereas its presence causes accumulation of starch (Hendriks et al., Plant Physiol. 133: 838-849, 2003). Hence light affects various facets of plant metabolism including carbon and nitrogen fixation, the accumulation of various metabolites as well as growth and development, at the systems level (Bläsing et al., The Plant Cell 17: 3257 -3281, 2005). To gain a systemwide perspective on the metabolic changes occurring in plant cells under differential light treatments we performed isotope-assisted metabolic flux analysis (isotope MFA) on heterotrophically grown plant suspension cells that were subjected to contrasting light conditions.

Isotope MFA uses isotopic tracers to map carbon traffic (fluxes) through metabolic pathways which provide valuable information in understanding cell physiology. We used statistical design methods to judiciously choose the isotopic tracers (in this study, differently ^{13}C-labeled isomers of glucose), and determined the isotopically labeled metabolite measurements that can provide the most information on primary metabolic fluxes. On the basis of these results, three independent tracer experiments with 100% 1-^{13}C, 100% 1,2-^{13}C or 30% U-^{13}C glucose were conducted on the plant suspension cells grown under continuous light or dark. On attainment of isotopic and metabolic steady state, several intracellular metabolites were analyzed for ^{13}C labeling patterns using both gas chromatography-mass spectrometry and 2-D [^{13}C, ^{1}H] NMR. The experimental data thus obtained from the three independent tracer experiments were collectively fitted to a multicompartmental plant metabolic network model consisting of primary carbon metabolic pathways. In this presentation we will compare the metabolic flux maps obtained under the two contrasting light conditions and discuss the implications of the differences. We expect that these flux maps will complement and augment the information obtained by previous MFA studies on plant suspension cells (Williams et al., Plant Physiol. 148: 704-718, 2008; Masakapalli et al., Plant Physiol. 152: 602-619, 2009).

See more of this Session: Advances In Cell Culture III : Metabolic Flux Analysis and Modeling
See more of this Group/Topical: Food, Pharmaceutical & Bioengineering Division

(197e) Metabolic Response of Perfused Livers to Various Oxygenation Conditions

Mehmet A. Orman[1], John Mattick[1], Marianthi G. Ierapetritou[1], Ioannis P. Androulakis[1] and Francois Berthiaume[2], (1)Department of Chemical and Biochemical Engineering, Rutgers, The State University of New Jersey, Piscataway, NJ, (2)Department of Biomedical Engineering, Rutgers, The State University of New Jersey, Piscataway, NJ

Isolated liver perfusion systems have been used to characterize intrinsic metabolic changes in liver as a result of various perturbations, including systemic injury, hepatotoxin exposure, and warm ischemia. Most of these studies were done using hyperoxic conditions (95% O_2) but without the use of oxygen carriers in the perfusate [1-4]. Inadequate oxygenation could by itself alter gene expression levels [5] and/or promote anaerobic pathways, including glycolysis. Prior literature data do not clearly establish the impact of oxygenation, and in particular that of adding oxygen carriers to the perfusate, on the metabolic functions of the liver. Thus, the importance of using oxygen carriers in liver perfusion studies remains controversial, and the full ramifications of the impact of oxygenation have not been well characterized.

In this study, the effects of three modes of oxygen delivery to perfused livers were compared: normoxic (arterial) perfusate (21% O_2), hyperoxic perfusate (95% O_2), and hyperoxic perfusate with oxygen carriers (95% O_2 + 10% hematocrit using bovine red blood cells). In all cases, the rat liver was perfused in-situ via the portal vein at constant flow rate, 3 mL/min/g liver. The hepatic artery and the suprarenal vena cava were ligated, and the liver outflow from the hepatic vein collected through the catheter which was cannulated into the inferior vena cava via the right atrium. The concentration changes of important metabolites in the perfusate were analyzed in order to determine net extracellular fluxes which have been applied to a stoichiometric-based optimization problem incorporating flux balance and metabolic pathway analyses (elementary mode analysis). The flux distribution vectors as well as possible active elementary modes were uniquely identified by maximizing the activity of short pathways [6].

We found that perfused livers consumed oxygen at in vivo rates only when the perfusate contained RBCs. Even when using 95% O_2, in the absence of oxygen carriers, oxygen uptake was only half the in vivo rate, urea and ketone body production were significantly decreased, and metabolic pathway analysis suggests that significant anaerobic glycolysis occurred. Conversely, when RBCs were used, glucose production from lactate and glutamate, as well as pathways related to energy metabolism were upregulated. The improved physiological relevance of a perfusion system using oxygen carriers makes it a more attractive tool to investigate the effect of various perturbations on liver metabolism, including the response to toxicants and drugs, as well as disease conditions known to alter liver metabolism, such as burns and trauma leading to systemic hypermetabolism.

References

1. Banta, S., et al., *Contribution of gene expression to metabolic fluxes in hypermetabolic livers induced through burn injury and cecal ligation and puncture in rats.* Biotechnology and Bioengineering, 2007. **97**(1): p. 118-137.

2. Arai, K., et al., *Intrahepatic amino acid and glucose metabolism in a D-galactosamine-induced rat liver failure model.* Hepatology, 2001. **34**(2): p. 360-371.

3. Lee, K., et al., *Profiling of dynamic changes in hypermetabolic livers.* Biotechnology and Bioengineering, 2003. **83**(4): p. 400-415.

4. Orman, M.A., et al., *Metabolic Flux Determination in Perfused Livers by Mass Balance Analysis: Effect of Fasting.* Biotechnology and Bioengineering, 2010. **107**(5): p. 825-835.

5. Rosmorduc, O. and C. Housset, *Hypoxia: A Link between Fibrogenesis, Angiogenesis, and Carcinogenesis in Liver Disease.* Seminars in Liver Disease, 2010. **30**(3): p. 258-270.

6. Orman, M.A., et al., *Pathway analysis of liver metabolism under stressed condition.* Journal of Theoretical Biology, 2011. **272**(1): p. 131-140.

See more of this Session: Advances In Cell Culture III : Metabolic Flux Analysis and Modeling
See more of this Group/Topical: Food, Pharmaceutical & Bioengineering Division

(197f) Probing Cancer Metabolism Using Isotope Tracers

Christian M. Metallo, Paulo Gameiro and Gregory Stephanopoulos, Chemical Engineering, Massachusetts Institute of Technology, Cambridge, MA

Tumors and proliferating tissues exhibit a unique metabolic profile which fuels cell growth and division. As a result, there is a renewed interest in exploiting metabolic pathways for the therapeutic treatment of cancer. In conjunction with established techniques for analyzing cell signaling and transcriptional regulation, we have applied stable isotope tracers, chromatography coupled to mass spectrometry, and computational tools for flux analysis to identify activated metabolic pathways in transformed cells. Together, these tools offer researchers the ability to quantitatively characterize metabolism in unprecedented detail. To highlight the utility of this approach I will describe results in three systems of cellular transformation: lipogenesis in renal cell carcinoma, metabolic transformation upon K-ras mutation, and overexpression of the ErbB2 oncogene in breast cancer. Our findings demonstrate the unique capabilities of isotopic tracers and metabolic flux analysis (MFA) to uncover the regulation of central carbon metabolism and substrate utilization by oncogenes and tumor suppressors. Moving forward, application of such methods to model systems of cancer is of paramount importance as we attempt to mitigate tumor growth by targeting metabolic pathways.

See more of this Session: Advances In Cell Culture III : Metabolic Flux Analysis and Modeling
See more of this Group/Topical: Food, Pharmaceutical & Bioengineering Division

(197g) Isotopically Nonstationary Proteinogenic MFA As a Tool for Analysis of Tumor Cell Metabolism

Taylor A. Murphy and Jamey Young, Chemical and Biomolecular Engineering, Vanderbilt University, Nashville, TN

Understanding the metabolic consequences of specific genetic and therapeutic perturbations is of critical interest in cancer research. Metabolic flux analysis (MFA) is a tool that allows for the quantification of intracellular fluxes of many organisms and cell types. The application of MFA to understanding the metabolism of cancer has been increasing in recent years, but it still remains a challenging task. Traditional MFA requires isotopic steady state of either protein extracts or intracellular metabolites prior to extraction. This requirement, especially for protein extracts, makes its application difficult in certain situations: mainly with cells that have a low growth rate either due to normal physiology or through genetic or chemical inhibition. Isotopically nonstationary MFA (INST-MFA) can be used to overcome this requirement using measurements of labeling during the transient period prior to steady state. INST-MFA allows for shorter experiments to be conducted, and it returns similar, and sometimes better, precision of the estimated fluxes.

In our recent work, we show that traditional methods of achieving proteinogenic steady-state via multiple re-platings of an *in vitro* cell culture are invalid. Reintroduction of unlabeled material during each re-plating creates a false isotopic steady-state. Our INST-MFA method gives a better approximation of what the true isotopic steady-state labeling would be, and it generates additional time-course data that is used to better estimate the fluxes in our model. Experiments starting at low cell densities show that a higher level of labeling is achieved when cells are grown continuously in a single flask rather than when multiple re-platings are conducted. Our INST-MFA method requires samples to be acquired approximately every half doubling of the cell culture. Isotopic labeling is then measured using mass spectrometry. To simulate the intracellular fluxes, we use an in-house program written in Matlab called Isotopomer Network Compartmental Analysis (INCA). We compared, using a highly proliferative B-cell line, the traditional steady-state method to our transient method. We found that, using a simpler model of metabolism within INCA, the steady-state methodology gave erroneous flux estimates that we disproved using additional tracer studies and protein expression analysis. A more complex model of central metabolism, including amino acid catabolism, gave results that were similar in both instances. This provided a validation of the proteinogenic INST-MFA method. We then applied INST-MFA to compare the flux maps generated for a B-cell model of lymphoma under two different expression levels of the c-Myc oncoprotein.

See more of this Session: Advances In Cell Culture III : Metabolic Flux Analysis and Modeling
See more of this Group/Topical: Food, Pharmaceutical & Bioengineering Division

(202a) Application of Design Space Mapping Accounting for Model Uncertainty and Common Cause Variability In Drug Substance Development

Shankar Vaidyaraman, Isabel Figueroa, Sze Wing Wong and Shekhar Viswanath, Chemical Product Research and Development, Eli Lilly and Company, Indianapolis, IN

This talk will focus on the methodology for mapping probabilistic design space based on first principles models and illustrate it with applications within drug substance development. The uncertainties in a model can be classified into two categories (1) Uncertainty in model parameters fitted from experimental data (2) Common cause variability in input factors of the model. The incorporation of these uncertainties present some statistical and computational challenges in the context of first principles models which will be highlighted in the talk. The application of design space mapping will be illustrated using case studies on reaction kinetics and reactive distillation unit operations.

See more of this Session: Application of Quality by Design In API Process Development
See more of this Group/Topical: Topical I: Comprehensive Quality by Design in Pharmaceutical Development and Manufacture

(202b) Design Space Development for the Final Recrystallization Process for Axitinib, a Complex Solvating API

Steven M. Guinness, Engineering Technologies - Chemical Research & Development, Pfizer Inc., Groton, CT, Brian P. Chekal, Chemical Research & Development, Pfizer Global Research & Development, Groton, CT and Gregory S. Steeno, Non-Clinical Statistics, Pfizer Global Research & Development, Groton, CT

This example for axitinib, an oncology candidate recently submitted to the FDA, will examine the development of the final processing step (a seeded recrystallization) to deliver final drug substance to meet the desired solid form and quality specifications. The process development work for the recrystallization is a textbook case of the use of sequentially designed multivariate experiments coupled with first principles to establish a robust design space.

See more of this Session: Application of Quality by Design In API Process Development
See more of this Group/Topical: Topical I: Comprehensive Quality by Design in Pharmaceutical Development and Manufacture

(202c) Impurity Fate Mapping In the Development of Pazopanib

Michael A. McGuire, Synthetic Chemistry, Glaxo SmithKline, King of Prussia, PA

Pazopanib hydrochloride is a VEGF antagonist developed by GSK which was approved for the treatment of renal cancer by the FDA and EMEA in 2009 -2010. Quality by design principles were used in the development of pazopanib hydrochloride. Impurity fate mapping was used extensively to understand and control impurities. This paper will discuss the impurity fate mapping strategy for pazopanib hydrochloride and how this strategy helps guarantee patient safety.

See more of this Session: Application of Quality by Design In API Process Development
See more of this Group/Topical: Topical I: Comprehensive Quality by Design in Pharmaceutical Development and Manufacture

(202d) A Model-Centric Solution to Link Content Uniformity Targets with API Particle Size Specifications and Process for a QbD Excercise

Salvador Garca-Muoz[1], Weili Yu[2], Mark A. Pinto[3] and Sean K. Bermingham[3], (1)Process Modeling & Engineering Technology, Pfizer Worldwide Research & Development, Groton, CT, (2) Pharmaceutical Development, Pfizer Worldwide Research & Development, Groton, CT, (3)Process Systems Enterprise, London W6 7HA, United Kingdom

One of the contributing factors in determining targets for the particle size of an API can be the content uniformity of the dosage forms. This work presents a framework established with a model-based specification for particle size of the API that can be used in a bi-directional mode to: a) determine center point processing conditions for the API given a target in content-uniformity for the dosage form, and b) to propagate expected sources of variability in the API process onto the final quality of the product.

This work discusses the necessary considerations and advantages when a model-based specification is considered for the particle size distribution of the API and how this compares to common practices involving discrete points of the size distribution (such as D[10], D[50], D[90]).

Our discussion is centered on the need of a structural specification for the API particle size that can be effectively used to guide the design of the last stages of the drug substance manufacturing train. We also emphasize the need to standardize the descriptors of particle size in order to link the effects of the variability in the API process as it affects the drug product train and eventually the quality of the dosage form.

See more of this Session: Application of Quality by Design In API Process Development
See more of this Group/Topical: Topical I: Comprehensive Quality by Design in Pharmaceutical Development and Manufacture

Process QbD Case Study: An Integrated PAT Approach for a Dynamic Pharmaceutical Co-precipitation Process Characterization and Process Design Space Development

Huiquan Wu*, Maury White, Mansoor Khan, Food and Drug Administration, Silver Spring, MD
(*Corresponding author: Huiquan.wu@fda.hhs.gov; Tel: 301-796-0022)

Abstract: Previously we had explored the concepts of developing an integrated PAT approach to monitor and map the state of a dynamic pharmaceutical co-precipitation process in real time, and to determine the nucleation and growth mechanisms occurred during a pharmaceutical co-precipitation process. The aim of current work was to develop a process QbD case study that demonstrates the feasibility of using an integrated PAT approach for a dynamic pharmaceutical co-precipitation process characterization and design space development at laboratory scale. A dynamic co-precipitation process by gradually introducing water to the ternary system of naproxen-Eudragit L100-alcohol was monitored at real-time *in situ* via Lasentec FBRM and PVM. The 3-D map of count-time-chord length revealed three distinguishable process stages: incubation, transition, and steady-state. The effects of high risk process variables (slurry temperature, stirring rate, and water addition rate) on both derived co-precipitation process rates and final chord-length-distribution were evaluated systematically using a 3^3 full factorial design. Critical process variables were identified via ANOVA for both transition and steady state. General Linear Models (GLM) were then used for parameter estimation for each critical variable. Clear trends about effects of each critical variable during transition and steady state were found by GLM and were interpreted using fundamental process principles and Nyvlt's transfer model. Neural network models were able to link process variables with response variables at transition and steady state with R^2 of 0.88~0.98. PVM images evidenced nucleation and crystal growth. Therefore, this work illustrated the possibility of combining first-principle modeling, mechanistic modeling, and empirical modeling approaches for dynamic process characterization and process design space development. Secondly, the technical feasibility of using an integrated PAT approach for in-depth process understanding and process design space development was demonstrated. Finally, the regulatory science perspective around the core concepts was briefly discussed.

1. Introduction

According to the definition of ICH Guideline Q8(R2) (FDA/ICH, 2009), "Quality-by-Design (QbD) is a systematic approach to development that begins with predefined objectives and emphasizes product and process understanding and process control, based on sound science and quality risk management." During the past few years, the pharmaceutical quality (Woodcock, 2004) and QbD implementation strategies (Yu, 2008; Rathore and Winkle, 2009; Smith et al., 2009) for various pharmaceutical product areas have received a lot of attentions. The FDA's PAT Guidance (FDA, 2004) and the ICH Q8, Q8(R2), Q9, and Q10 guidelines (FDA/ICH, 2006a,b, 2007, 2009) have collectively set up a roadmap for the implementation of the science-based and risk-based regulatory pathway. This can be characterized as pharmaceutical quality paradigm shift: from testing into quality alone to building the quality into the process and product.

In the QbD framework, understanding the impact of raw material attributes and process parameters on the critical quality attributes (CQAs) as well as identification and control of sources of variability are essential. However, pharmaceutical products and processes are complicated and multivariate in nature. Understanding of the relevant multi-factorial relationships among formulation parameters, process variables, and product quality attributes usually requires the use of multivariate approaches, such as statistical design of experiments (DOE) and multivariate data analysis (MVDA). Although both DOE and MVDA as standard techniques have been widely used for many years in various areas, the integrated use of both of them are relatively few when applied to pharmaceutical development (Naelapaa et al., 2008; Huang et al., 2009).

Over the past few years, some internal efforts (Wu and Hussain 2005; Wu and Khan 2009; Wu et al., 2009; Xie et al., 2008) have been made in the aspect of integration of DOE and MVDA for pharmaceutical product and process understanding, process monitoring and control. In principle, this integrated approach in conjunction with appropriate real time PAT process analyzer should provide ample opportunities for enhanced process understanding, process monitoring, and process control of other pharmaceutical processes (such as crystallization, co-precipitation, etc.) under the QbD/PAT framework.

Previously (Wu & Khan, 2010, 2011) we had explored the concepts of developing an integrated PAT approach to monitor and map the state of a dynamic pharmaceutical co-precipitation process in real time, and to determine the nucleation and growth mechanisms occurred during a pharmaceutical co-precipitation process. A uni-variate thermodynamic approach had been used for those works. Three thermodynamically important formulation and process variables were studied, including the drug/polymer ratio, water usage, and alcohol usage, etc. It demonstrated that integration real-time PAT process monitoring with first-principles modeling and off-line product characterization could enhance understanding to co-precipitation process dynamics and nucleation/growth mechanisms, which is impossible via off-line techniques alone. The knowledge gained through the above studies is essential for rationale design of both the co-precipitation process and co-precipitation vehicle. Furthermore, they can provide scientific evidence as for setting up practical operation range during the stage of process and product development.

The aim of current work was to develop a process QbD case study that demonstrates the feasibility of using an integrated PAT approach for a dynamic pharmaceutical co-precipitation process characterization and design space development at laboratory scale. In this work, a variety of independent process conditions were studied via a 3^3 full factorial design. For each set of process conditions, the co-precipitation process was monitored in real time using an integrated PAT approach. The effects of high risk process variables (slurry temperature, stirring rate, and water addition rate) on both derived co-precipitation process rates and final chord-length-distribution were evaluated systematically. Various models such as ANOVA Models, General Linear Models (GLM), and Neural Network (NN) models were able to link process variables with response variables at transition and steady state. A co-precipitation process design space was established.

2. Experimental and Data Analysis

2.1. Materials and Experimental Protocol for a Dynamic Co-precipitation Process

Naproxen USP (lot No: NPX 368) was obtained from Albemarle Corporation (Orangeburg, SC). Eudragit L 100 (lot No: 1221203048) was obtained from Röhm America Inc. (Somerset, NJ). Solvent reagent alcohol (HPLC grade, lot No: 053546) was purchased from Fisher Scientific (Fair Lawn, NJ). Non solvent DI water was obtained from a FDA in-house facility (Silver Spring, MD) and was kept in refrigerator at 3^0C prior to use. All of these chemicals, solvent, and non solvent were used without any further processing or purification prior to the use for this experimental work.

In this work, the drug/polymer ratio of 2 (4.0g of naproxen and 2.0g of Eudragit L100) was selected due to the fact that nice co-precipitates with regular morphology could be obtained via the dynamic co-precipitation process designed in this work. The naproxen and Eudragit were dissolved separately in 150 and 100 mL of reagent alcohol, respectively. These solutions were then added to the 1L reaction vessel of Chemglass reaction kit (ChemGlass,Vineland, NJ) via a funnel. The stir speed was set via the OptiChem digital overhead stirrer. The temperature was set via the ThermoScientific Neslab RTE 740 thermoregulator. The solution in the reaction vessel was allowed to reach the pre-defined temperature prior to data acquisition. Once this temperature had been achieved, real-time monitoring began using the Lasentec FBRM system and PVM system, prior to any water addition. The first addition of water (125 ml) was made to the reaction vessel at the pre-defined flow rate via a flexible tubing (Tygon R-3603, Fisher Scientific), a MasterFlex solid state speed control device (Chicago, IL), and a Cole-Parmer peristaltic pump (Chicago, IL). Ten minutes later, the 2nd addition of water (50 ml) was made to the system. Depending on the temperature of the individual experiment (15^0C, 25^0C, or 35^0C), the system was introduced with water for the 3rd, 4th, and 5th addition (each time with 50 ml) respectively. The time interval between each addition of water was kept as 10 minutes. To ensure both the experimental consistency and good dispersion of water to the ternary system, the flexible tubing was kept vertically by two supporting claps attached to a lab frame. Furthermore, the end portion of the tubing was immersed in the liquid phase during the entire course of experiment.

2.2. Risk-Based Approach to Select Variables for the DOE Study

A risk assessment tool was applied to identify potential variables that can have an impact on the desired quality attribute. In this work, a Fishbone diagram or Ishikawa diagram was created to structure the process of identifying possible causes that could have impacts on co-precipitation process rate, as shown in Figure 1.

Although six categories of factors (materials, process, measurement, equipment, people, and environment) could have impact on the co-precipitation process rate, given a research laboratory setting, a large number of factors can actually be eliminated safely. For example, co-precipitation formulation components selection and pre-dissolution in solvent, co-precipitation vessel selection, errors due to operators and analytical, environmental variables, and measurement variables, could be pre-specified in the study protocol. Therefore, this risk mitigation and risk management strategy would allow us to focus on evaluating the impact of main process variables or high risk variables (such as slurry temperature, slurry stirring rate, and non solvent addition rate) on process dynamics. Furthermore, this simplified situation will enable us to exemplify the concept proposed in this work.

2.3. Design of Experiments (DOE), Real-time Process Monitoring, and Data Analysis

Based on risk assessment results above, three main process variables (slurry temperature (^{0}C), slurry stirring rate (RPM), non solvent addition rate (ml/sec) were selected as independent process variables. Each factor was tested with three levels. A 3^3 full factorial design was used for experiments as shown in Table 1.

A FBRM D600L (Mettler-Toledo AutoChem, Columbia, MD) system was used to monitor the dynamic co-precipitation process *in situ* at real-time for a total of 27 batches based on a 3^3 factorial design. The FBRM D600 L system was set to collect Chord-Length-Distributions (CLDs) *in situ* with a laser speed of 2 m/s, applying the standard 'F-electronic' mode and using measurement duration of 2 seconds. A Lasentec Particle Vision Microscopy system PVM819 (Mettler-Toledo AutoChem, Columbia, MD) was used to monitor the co-precipitation process *in situ* at real-time for the selected batches. The PVM image was acquired every 2 seconds.

A 3-D count-time-size map was constructed. It was used to illustrate the progress of a co-precipitation process and to identify various distinguishable process stages. The impacts of main process variables on the process characteristics such as derived co-precipitation process rates and final CLD were assessed. ANOVA was conducted to identify statistical significant variables. General Linear Models were used for parameter estimate. Neural network algorithm was utilized to link the process variables with response variables.

3. Results and Discussion

From a process engineering perspective, various process variables could impact thermodynamics, kinetics, and mass-transfer of a co-precipitation process to various extents, depending on the combination of process conditions. Therefore, the effects of main process variables on the co-precipitation process dynamics and quality attributes of final co-precipitate slurry were examined systematically. The following chord length (CL) range codes were used throughout this paper: CL1: 0-10 μm; CL2: 10-19.953 μm; CL3: 19.953 -50.119 μm; CL4: 50.119-100 μm; CL5: 100-251.189 μm.

3.1. 3-D Count-Time-Size Plot of the Dynamical Co-precipitation Process

A 3-D map of count-time-size was constructed to visualize the co-precipitation process dynamics as a function of time. It can illustrate the process progress and distinguish various process stages (Wu and Khan 2010a), as shown in Figure 2 for process run 132.

It was found for most runs, the 3-D maps consist of three stages: (1) incubation period when there was no significant particle counts detected, its time length depending on the combinations of process conditions; (2) transition period when significant nucleation and growth occurs thus significant counts were detected; (3) steady state when the number of counts per second for small chord length ranges (CL1 and CL2) remains relatively constant throughout the remainder of the process, which indicates no significant nucleation occurs. In the CL3 range, there is a slight increase after the 2nd addition of water and a slight decrease after the 3rd addition of water for run 132; there is a slight decrease after the 2nd and 3rd addition of water for run 123.

3.2. Derived Process Rates for a Co-precipitation Process During the Transition Period

Although CLD is a relative value, it can be used to represent the process characteristics of a particulate system because it depends on both PSD and shape of the particles (both are a function of process time). Based on the profile of the FBRM counts/s vs. time at specific CL range during the transition period, a new procedure was developed in this work to derive the co-precipitation process rates from FBRM counts/s vs. time profiles. The slopes were based on the FBRM counts/s over time in five different CL ranges. They were calculated by (1) first identifying the general time range where the maximum slope occurred for a given size range during a single addition; (3) then calculating the slope of that CL range over the preceding 20 seconds for each individual time point within that CL range; (3) the maximum of these calculated slope values for each addition and each CL range was used as the derived co-precipitation process rate for the corresponding process conditions at specific single addition and CL range. Following this procedure, the slope of BBRM counts/s over time for each run after each addition of water was obtained. The following derived process rate codes were used throughout the paper: Rate1, Rate2, Rate3, Rate4 are corresponding to the derived rates at CL1, CL2, CL3, and CL4 after the 1st water addition; Rate5, Rate6, Rate7, Rate8 are corresponding to the derived rates at CL1, CL2, CL3, and CL4 after the 2nd water addition. These derived process rate data as the designated response variable values were listed in Table 2.

3.3. CLD at the End of Steady State Period

To characterize the effects of main process variables on the steady state, the CLDs at the end of steady state of the 27 runs were plotted. Figure 3 is a representative example for run 111, 112, and 113. All of the CLDs have bell shapes. As characteristic parameters of this bell- shaped CLD, its mode and peak frequency were taken as additional designated response variables and listed in the last section of Table 2.

3.4. Analysis of Variance (ANOVA) and General Linear Modeling for Both Transition Period and Steady State Period

Statistical data analysis and general linear modeling were performed on data from both transition period and steady state period.

3.4.1. ANOVA for the DOE Results of the Co-precipitation Process: Identifying critical process variables and interactions

Let X1, X2, and X3 be the slurry temperature, stirring rate, and water addition rate for the 3^3 full factorial design, respectively. With 27 runs of 3^3 factorial design of experiment, the analysis of variance (ANOVA) indicates that all main effects and two-way interactions among these main effects are estimable and the three-way interaction is used to estimate the underlying error. Therefore, the following statistical model was used:

$$y_{ijk} = \mu + \alpha_i + \beta_j + \gamma_k + (\alpha\beta)_{ij} + (\alpha\gamma)_{ik} + (\beta\gamma)_{jk} + \varepsilon_{ijk}$$
$$i = 1,2,3 \quad j = 1,2,3 \quad k = 1,2,3 \qquad (2)$$

where y_{ijk} is the observation (either derived process rate value, mode, or peak frequency) in the ith level of variable X1 and the jth level of variable X2 and the kth level of variable X3, μ is the overall mean, α_i is the ith X1 effect, β_j is the jth X2 effect, γ_k is the kth X3 effect, $(\alpha\beta)_{ij}$ is the two-way interaction between the ith X1 effect and the jth X2 effect, $(\alpha\gamma)_{ik}$ is the two-way interaction between the ith X1 effect and the kth X3 effect, $(\beta\gamma)_{jk}$ is the two-way interaction between the jth X2 effect and the kth X3 effect, ε_{ijk}, the random error, is normally independently identically distributed with mean 0 and variance σ^2, which is written as $\varepsilon_{ijk} \sim iid\ N(0,\sigma^2)$.

This statistical model was used for ANOVA to screen significant factors and interactions. It reveals that, at significance level α=0.05, X1 and X3 are statistically significant for both Rate1 and Rate7; X1, X2, and X3 are statistically significant for Rate5, Rate6, Rate8, and peak frequency; X1 and X2 are statistically significant for mode. No statistically significant variables or interactions were found for Rate2, Rate3, and Rate4.

For the initial addition during transition period, only X1 and X3 stand out as critical process variables to the derived co-precipitation process rate for the smallest CL range of [0, 10] μm which is most likely related to the nucleation process. Since X1 is directly related to the degree of super-saturation (*S*), X3 is directly related to the creation of *S* in the co-precipitation vessel, the ANOVA results for the 1st water addition suggest that the initial nucleation process is thermodynamically controlled. No statistically significant process variables and interactions were found from ANOVA at other CL ranges after the 1st water addition. This suggests that the competition among various process mechanisms (such as thermodynamics for super-saturation creation, nucleation kinetics, and multi-phase mass transfer, etc.) is at such state that none of them becomes dominant.

For the 2nd addition during the transition period, all of three process variables (X1, X2, and X3) stand out as critical variables to Rate5, Rate6, and Rate8. This may suggest all of three process variables can contribute directly to the co-precipitation process rates at CL5, CL6, and CL8. X1 and X3 stand put as critical process variables to Rate7, which may suggest both the thermodynamic factor (X1) and a mass-transfer factor (X3) are playing equally important role.

At the end of steady state of the co-precipitation process, X1 and X2 are critical process variables to the mode of the final CLD. This suggests both the thermodynamic factor (X1) and mass-transfer factor (X2) are most influential to the particle size of the final co-precipitate slurry. X1, X2, and X3 are critical process variables to the peak frequency of the final CLD. This suggests that all of the main process variables directly contribute to the number of counts generated. Therefore, the ANOVA results for the characteristic parameters of the final CLD demonstrated the critical importance of process control (i.e., controlling process parameters as studied in this work) to achieve the desired product quality attributes at the end of the co-precipitation process.

3.4.2. Development of Statistical Predictive Models Using General Linear Modeling

Based on the screening results of statistically significant process variables, further statistical analyses were carried out using general linear models (GLM). The general linear models and parameters' estimations are listed in Table 3.

To illustrate the GLM model formulation, let us discuss the linear model for Rate1.

$$(Rate\,1)_i = \beta_0 + \beta_1 (X1)_i + \beta_3 (X3)_i + \varepsilon_i, i = 1,....,27, \quad (3)$$

where $(\text{Rate1})_i$ is the ith observation of Rate1, $(X1)_i$ and $(X3)_i$ are the ith values for X1 and X3, respectively, error term ε_i is normally independently identically distributed with mean 0 and variance σ^2, which is written as $\varepsilon_i \sim iid\ N(0, \sigma^2), i = 1,...,27$. The GLM modeling was carried out using SAS9.2. The modeling results were summarized in Table 4. From the parameter estimate listed in the 5th and 6th columns, it is easily appreciated that the effects of main process variables on derived process rates, mode and peak frequency of the final CLD have certain trends as discussed below.

3.4.2.1. Slurry temperature (X1)

It has a significant effect (at significance level α=0.05) on the derived process Rate1, Rate5, Rate6, Rate7, and Rate8. Furthermore, since parameter β_1 estimate is negative for these rates, the lower slurry temperature will result in greater derived process rates. This could be explained from a thermodynamic point of view, based on the fact that both solubility and degree of super-saturation are function of slurry temperature.

On the other hand, X1 was a significant factor (at significance level α=0.05) on mode of final CLD at the end of steady state. However, parameter β_1 estimate is positive for mode. Therefore, the lower slurry temperature will result in smaller mode or finer co-precipitate crystal.

3.4.2.2. Slurry stirring rate (X2)

It was a significant factor (at significance level α=0.05) for Rate5 and Rate6. The parameter β_2 estimate is positive for Rate5, Rate6, and Rate8. This indicates that a higher X2 will result in a higher Rate5, Rate6, and Rate8. This trend is likely attributed to the possible mixing effect caused by stirring, as discussed below.

From a fluid dynamics perspective, local and mean values of the specific power input caused by the rotation speed of the impeller determine the degree of macro-mixing and micro-mixing. The bulk mixing inside the co-precipitation vessel must be ensured that its elements participate equally in the mass transfer process and hence the entire volume is sufficiently utilized. In addition to this bulk mixing, sufficient micro-mixing has to be present in the vicinity of contacting zone where water is introduced and fed to the slurry via the feed inlet, in order to even out quickly any local super-saturation peaks generated A lower X2 may not be strong enough to guarantee an effective mixing at both macro and micro level. On the other hand, a higher X2 could promote micro-mixing in the vicinity of contact zone thus reduce the super-saturation there. Furthermore, a higher X2 can facilitate mass transfer process including both bulk transfer and diffusion across the diffusion layer. Therefore, according to the Nyvlt's model (Nyvlt et al., 1985), the apparent overall derived process rates covering the three steps of multi-phase mass transfer process are faster.

X2 was a significant factor (at significance level α=0.05) for both mode and peak frequency of the final CLD at the end of steady state. However, the parameter β_2 estimate is negative for mode but positive for peak frequency. This indicates that higher stirring rate will lead to smaller particles but more counts in its final CLD.

3.4.2.3. Addition rate of water (X3)

It has a significant effect (at significance level α=0.05) on Rate1, Rate5, Rate6, and Rate7, with a higher X3 resulting in a greater derived process rate, as indicated by the GLM modeling results (the parameter β_3 estimate is positive). Similar trend was observed for the derived process rate 8 although it did not meet the significance criteria. This trend could be explained by the generation of super-saturation and the re-distribution of various species in the vicinity of contacting zone. Firstly, the addition of water into the slurry system will directly change the concentrations of various species, which in turn will change the degree of super-saturation in the vicinity of contacting zone; secondly, it indirectly impacts the re-distribution of various species within the vessel. The generation rate of super-saturation in the vicinity of contacting zone due to a faster X3 could be larger than the consumption rate of the super-saturation due to micro-mixing caused by the stirring. Therefore, a trend of a higher X3 leading to a greater derived process rate was observed.

Similar trend was observed for peak frequency of the final CLD at the end of steady state of the co-precipitation process, although it did not meet the significance criteria.

3.5. Neural Network Modeling for Both Transition Period and Steady State Period

As revealed by ANOVA screening results, there are certain two-way interactions existing, although not necessarily statistically significant at significance level α=0.05. Therefore, there is certain nonlinearity for the co-precipitation process. A neural network (NN) algorithm implemented in JMP 7.0 software was utilized to construct the linkage between the three normalized process variables and response variables that include derived process rates at transition stage and mode and peak frequency of CLD at the end of steady state. Our modeling results show that the NN was able to generate quantitative models with decent correlation coefficients, as shown in Table 4. Most of the R^2 values are pretty high (0.88~0.98), Furthermore, the predicted response variable values agree well with the actual response variable values, as shown in Figure 4 for Rate1.

In summary, our NN modeling results suggest that quantitative correlations between the process variables and the derived process rate based on Lasentec FBRM CLD vs. time profile do exist. Therefore, it is possible to utilize these quantitative correlations for better process design and process control.

3.6. PVM Images Evidenced Nucleation and Crystal Growth during the Dynamic Course of Adding Water

In this work, 6 co-precipitation runs were selected for in-situ real-time monitoring using PVM during the entire course of co-precipitation process. First of all, it demonstrated that for most cases, no appreciable sizes of particles were detected via PVM before and immediate after the 1st addition of water to the ternary system. However, about 10 minutes after the 1st addition of water, right before the 2nd addition of water, co-precipitates with appreciable sizes were detected via PVM. This is qualitatively consistent with what FBRM data revealed that relative fewer counts of fine particles were detected after the 1st addition of water. Secondly, significant increases were observed via PVM for both the number and dimensions of the co-precipitates before and right after the 2nd addition of water. Again, this is qualitatively consistent with what FBRM indicated that significant counts at various CL ranges were detected after the 2nd addition of water. Further analysis demonstrated the trends of the dimension changes of co-precipitates at both length and width dimensions during the course of a co-precipitation process.

3.7. Discussion from a Regulatory Science Perspective

ICH Q8(R2) (FDA/ICH, 2009) defines the design space as "The multidimensional combination and interaction of input variables (e.g., material attributes) and process parameters that have been demonstrated to provide assurance of quality. Working within the design space is not considered as a change. Movement out of the design space is considered to be a change and would normally initiate a regulatory post-approval change process." In addition to identifying the main process variable effects, multi-factorial DOE study as conducted in this work permits the quantification of variable interactions. The information on the presence or absence of interactions might be valuable for process optimization, identifying cause of variability, or informing a failure modes and effects analysis (FMEA) for defining a design space. As a proof-of-concept, this simplified case study was mainly focused on the effects of process variables while keeping other variables fixed based on risk analysis and initial formulation development results. It is important to keep in mind that other variables such as formulation variables can certainly play vital roles during process characterization and design space development. Based on the GLM modeling results, both 2-D and 3-D contour plots were made for certain response variables of the dynamic co-precipitation process, such as Rate 7 for the transition period as shown in Figure 5 (a)-(b). Figure 5(a)-(b) demonstrated that maximum Rate 7 was obtained around the following process window: $-1<X1<0.1$, $-0.25<X3<0.95$.

As demonstrated in this and previous works, process knowledge and in-depth process understanding are essential to establish appropriate process control strategy and operational space, such that co-precipitation product with desired quality attributes and properties (such as PSD, morphology, etc.) could be achieved eventually. Therefore, they are key elements for developing appropriate process design space. Risk analysis, identification of critical formulation/process variables and interaction, understanding the effects of these critical variables and interactions on the key product quality attributes are essential steps to achieve enhanced process understanding. Validating and challenging design space are important steps to ensure a robust design space developed for a pharmaceutical manufacturing process. In this work, various PAT tools and QbD principles had been integrated together to generate a process QbD case study at laboratory scale. Development of other approaches and effects of scale on this dynamic pharmaceutical co-precipitation system are ongoing.

4. Conclusions

In this work, a dynamic pharmaceutical co-precipitation process of gradually introducing water (non solvent) to the ternary system of naproxen (durg) - Eudragit L100 (polymer) - alcohol (solvent) was monitored in real-time and *in situ* by using Lasentec FBRM and PVM technology. Based on the Fishbone analysis result, the impact of high risk process variables such as co-precipitation slurry temperature, slurry stirring rate, and non solvent addition rate on both the FBRM counts/s vs. time profile and chord-length-distribution(CLD) at the end of co-precipitation process were examined systematically, using a 3^3 full factorial design. A 3-D count-time-size plot was constructed to map the entire co-precipitation process and distinguish various process stages: incubation, transition, and steady state. A new method was developed to extract a derived process rate from the Lasentec FBRM counts/s vs. time profile by calculating the maximum slope of the FBRM conuts/s vs. time after each addition of water. Critical process variables were identified via ANOVA for both transition and steady state periods. General Linear Models (GLM) were used for parameter estimate of each critical variable. The GLM results demonstrated clear trends about the effects of each critical variable, which can be interpreted using principles of fundamental process phenomena and Nyvlt's diffusion layer model. Furthermore, contour plots demonstrated the process windows bordered by ranges of critical process variables where maximum Rate 7 and maximum mode can be obtained at the transition period and at the end of steady state, respectively. Neural network models were able to link process variables with response variables both at transition period and steady state with R^2 of 0.90~0.98. PVM images evidenced nucleation and crystal growth during the dynamic co-precipitation process.

Acknowledgements

The following persons and organizations are acknowledged: Dr. Meiyu Shen in CDER of FDA for strong statistical supports on DOE analysis and statistical modeling; Dr. Vincent Vilker at the Office of Testing and Research, CDER, FDA for insightful internal review;Mr. Eric Dycus at the Mettler Toledo AutoChem Inc.(Columbia, MD) for technical support to our Lasentec systems; Dr. San Kiang at the Bristol-Myers Squibb Company (New Brunswick, NJ), Mr. Steve Ware and Mr. Brendan Simon at the Chemglass Inc. (Vineland, NJ) for their technical discussion and assistance to our crystallization system; Dr. Richard D. Braatz at MIT (Boston, MA) for technical discussion on topics of crystallization;The FDA BioSciences Library for literature supports.

Disclaimer: ***The views and opinions expressed in this paper are only those of the authors, and do not necessarily reflect the views or policies of the FDA.***

References

FDA, 2004. "Guidance for Industry. PAT—A Framework for Innovative Pharmaceutical Development, Manufacturing, and Quality Assurance". Available at: http://www.fda.gov/cder/guidance/6419fnl.pdf
FDA/ICH, 2006a. Guidance for Industry. Q8 Pharmaceutical Development. Available at: http://www.fda.gov/cder/guidance/6746fnl.pdf
FDA/ICH, 2006b. "Guidance for Industry. Q9 Quality Risk Management". Available at: http://www.fda.gov/cder/guidance/7153fnl.pdf
FDA/ICH, 2007. Q10. Pharmaceutical Quality System. Draft Consensus Guideline. Available at: http://www.fda.gov/cder/guidance/7891dft.pdf
FDA/ICH, 2009. Guidance for Industry, Q8(R2) Pharmaceutical Development. Draft available at: http://www.fda.gov/downloads/Drugs/GuidanceComplianceRegulatoryInformation/Guidances/ucm073507.pdf
Huang, J., Kaul, G., Cai, C., Chatlapalli, R., Hernandez-Abad, P., Ghosh, K., Nagi, A., 2009. Quality by design case study: An integrated multivariate approach to drug product and process development. Intl J Pharm. 382, 23-32.
Naelapaa, K., Alleso, M., Kristensen, H.G., Bro, R., Rantanen, J., Bertelsen, P., 2008. Increasing process understanding by analyzing complex interactions in experimental data. J. Pharm. Sci. 98, 1852–1861.
Nyvlt, J., Sohnel, O., Matuchova, M., Broul, M., 1985. The Kinetics of Industrial Crystallization. Elsevier, New York.
Rathore, A.S. and Winkle, H., 2009. Quality by design for biopharmaceuticals. Nature Biotechnology 27(1), 26-34.
Smith, J., Moore, C., Nasr, M., 2009. Quality Trio Takes Final Shape. Pharmaceutical Technology 33(6), 66.
Tavare, NS. 1987. Batch Crystallizations: A Review. Chem. Eng. Comm. 61:259-318.
Woodcock, J., 2004. The Concept of Pharmaceutical Quality. American Pharmaceutical Review. 7 (6), pg. 10-15.
Wu, H. and Hussain, A.S., 2005. Integration of multivariate statistics and design of experiments to identify critical process variables for pharmaceutical process analytical technology (PAT) applications. In: American Statistical Association 2005 Proceedings of Joint Statistical Meetings, Minneapolis, Minnesota, August 7-11, 2005. CD-ROM , Mira Digital Publishing.
Wu, H. and Khan, M.A., 2009. Quality-by-Design (QbD): An Integrated Approach for Evaluation of Powder Blending Process Kinetics and Determination of Powder Blending End-Point. J Pharm Sci. 98(8), 2784-2798.
Wu, H., Tawakkul, M., White, M., and Khan, M.A., 2009. Quality-by-Design (QbD): An Integrated Multivariate Approach for the Component Quantification in Powder Blends. Int J Pharm. 372(1-2), 39-48.

Wu, H. and Khan, M.A., 2010. Quality-by-Design (QbD): An Integrated Process Analytical Technology (PAT) Approach for Real-time Monitoring and Mapping the State of a Pharmaceutical Co-precipitation Process. Journal of Pharmaceutical Sciences 99(3): 1516-1534.
Wu, H. and Khan, M.A., 2011. Quality-by-Design (QbD): An Integrated Process Analytical Technology (PAT) Approach to Determine the Nucleation and Growth Mechanisms during a Dynamic Pharmaceutical Co-precipitation Process.
Journal of Pharmaceutical Sciences 100(5): 1969-1986.
Xie, L., Wu, H., Shen, M., Augsburger, L., Lyon, R.C., Khan, M.A., Hussain, A.S., Hoag, S.W., 2008. Quality-by-Design (QbD): Effects of Testing Parameters and Formulation Variables on the Segregation Tendency of Pharmaceutical Powder Measured by the ASTM D 6940-04 Segregation Tester. J. Pharm. Sci. 97(10),4485-97.
Yu, L., Lionberger, R.A., Raw, A.S., D'Costa, R., Wu, H., Hussain, A.S., 2004. Applications of process analytical technology to crystallization processes. Advanced Drug Delivery Reviews. **56**(3): 349-369.
Yu, L.X., 2008. Pharmaceutical Quality by Design: product and Process Development, Understanding, and Control. Pharmaceutical Research 25(4), 781-791.

Table 1 The 3^3 full factorial design for the co-precipitation study

Run No	**Pattern**	**X1(^{0}C)**	**X2(RPM)**	**X3(ml/sec)**
1	133	15	400	6.66
2	213	25	50	6.66
3	312	35	50	3.66
4	132	15	400	3.66
5	331	35	400	0.83
6	123	15	200	6.66
7	323	35	200	6.66
8	221	25	200	0.83
9	313	35	50	6.66
10	222	25	200	3.66
11	321	35	200	0.83
12	212	25	50	3.66
13	311	35	50	0.83
14	131	15	400	0.83
15	211	25	50	0.83
16	332	35	400	3.66
17	223	25	200	6.66
18	112	15	50	3.66
19	122	15	200	3.66
20	231	25	400	0.83
21	232	25	400	3.66
22	121	15	200	0.83
23	233	25	400	6.66
24	322	35	200	3.66
25	113	15	50	6.66
26	111	15	50	0.83
27	333	35	400	6.66

Table 2. Derived process rates during transition and final CLD parameters at the end of steady state for the 27 co-precipitation runs based on real-time FBRM data (CL1: 0-10 μm;CL2: 10-19.953 μm;CL3: 19.953-50.119 μm;CL4: 50.119-100 μm).

Run #	Pattern	During the transition period (Rate: counts/s/s)								At the end of steady state period	
		After the 1st addition of water				After the 2nd addition of water					
		Rate1	Rate2	Rate3	Rate4	Rate5	Rate6	Rate7	Rate8	Mode(μm)	Peak frequency(counts/s)
1	133	10.5736	10.7710	10.5172	1.6980	24.6889	22.4311	21.5892	3.7845	27.123	305.72
2	213	2.2180	3.9841	8.4474	0.8013	1.4502	3.1170	8.1841	2.4616	46.416	91.581
3	312	0.3826	0.4129	0.9374	0.6608	5.2500	2.9591	4.9887	2.3592	54.117	46.169
4	132	2.7907	3.4472	4.2661	1.1092	29.2702	28.5301	26.8955	3.8570	25.119	270.67
5	331	1.6878	1.2320	2.7800	1.3865	2.8786	1.8109	3.3692	1.5301	42.987	60.755
6	123	17.4587	14.5936	12.1450	0.9303	31.2612	38.0034	55.6426	5.0592	31.623	392.19
7	323	1.2708	2.1529	3.9911	2.0179	6.3940	7.0410	11.3126	4.9157	39.811	109.45
8	221	0.8667	1.2557	3.3192	1.2608	0.7273	0.9842	0.4499	1.5104	46.416	92.221
9	313	0.3736	0.6492	0.9462	0.7991	1.8012	2.9355	5.7506	2.0944	54.117	59.715
10	222	4.7627	5.3115	5.7942	0.3398	12.8510	17.5157	35.4949	9.5555	36.869	254.68
11	321	1.0028	1.4100	3.4853	1.6461	1.7586	3.0196	4.7413	2.2973	39.811	87.882
12	212	16.8696	28.8235	59.4750	4.7923	10.3219	10.9985	27.4774	4.0631	36.869	260.91
13	311	0.5023	0.5972	1.4905	0.9732	0.9189	1.2675	2.4570	1.2153	50.119	34.156
14	131	2.4071	3.3990	3.2181	0.5674	9.6399	9.4964	15.6298	3.2423	34.145	205.99
15	211	0.2351	0.3937	0.9423	0.2103	1.0915	1.8768	5.8460	2.7091	50.119	88.607
16	332	4.2665	5.7420	8.6329	2.5536	5.8941	6.4683	11.9191	4.6014	34.145	127.26
17	223	5.4219	6.9866	8.5837	0.4530	10.4084	15.8936	34.5608	7.6965	34.145	252.77
18	112	20.9567	30.1534	62.3363	8.2928	17.1882	18.1251	36.7147	5.7329	31.623	285.74
19	122	3.8570	0.1250	0.1682	0.0324	14.4302	22.9464	35.9995	9.6199	34.145	292.63
20	231	1.0080	0.6780	1.0058	0.7622	9.3001	10.1723	17.7417	5.2822	31.623	155.29
21	232	0.8753	0.5902	0.8450	1.0472	12.5681	14.7266	26.3863	7.0553	31.623	171.35
22	121	4.3244	6.5958	10.1978	1.0433	5.3310	6.1365	7.4282	2.6867	31.523	243.11
23	233	1.5458	1.4770	1.7359	0.4937	11.0354	14.8249	31.6042	7.8285	29.286	203.1
24	322	0.8057	1.4403	2.7305	1.3534	3.8065	4.0625	9.3274	3.4756	39.811	82.84
25	113	17.8998	26.7439	50.1879	5.4916	2.2388	2.7659	6.1132	2.1597	36.869	201.59
26	111	1.2927	2.2658	4.0293	0.5283	3.5820	4.1017	10.2035	3.0273	34.145	131.14
27	333	11.7237	16.2410	26.0520	3.6388	5.0155	4.0808	8.1521	4.4261	36.869	159.24

Table 3 Development of statistical predictive models using general linear modeling

Process Period	Response Variable	Model	Coefficient of Correlation, r	Parameter	Estimate	Standard error	t-Value	P-value
Transition period after the 1st water addition	Rate1	$(Rate1)_i = \beta_0 + \beta_1(X1)_i + \beta_3(X3)_i + \varepsilon_i, \varepsilon_i \sim N(0,\sigma^2)$	0.595	β_0	5.09	1.02	5.01	<0.0001
				β_1	-3.31	1.24	-2.66	0.0138
				β_3	3.06	1.24	2.46	0.0214
Transition period after the 2nd water addition	Rate5	$(Rate5)_i = \beta_0 + \beta_1(X1)_i + \beta_2(X2)_i + \beta_3(X3)_i + \varepsilon_i, \varepsilon_i \sim N(0,\sigma^2)$	0.751	β_0	8.93	1.14	7.85	<0.0001
				β_1	-5.77	1.39	-4.15	0.0004
				β_2	3.69	1.39	2.65	0.0143
				β_3	3.28	1.39	2.36	0.0273
	Rate6	$(Rate6)_i = \beta_0 + \beta_1(X1)_i + \beta_2(X2)_i + \beta_3(X3)_i + \varepsilon_i, \varepsilon_i \sim N(0,\sigma^2)$	0.748	β_0	10.23	1.29	7.96	<0.0001
				β_1	-6.61	1.57	-4.19	0.0003
				β_2	3.58	1.57	2.27	0.0328
				β_3	4.01	1.57	2.55	0.0180
	Rate7	$(Rate7)_i = \beta_0 + \beta_1(X1)_i + \beta_3(X3)_i + \varepsilon_i, \varepsilon_i \sim N(0,\sigma^2)$	0.631	β_0	17.26	2.19	7.87	<0.0001
				β_1	-8.57	2.68	-3.19	0.0039
				β_3	6.39	2.68	2.38	0.0256
	Rate8	$(Rate8)_i = \beta_0 + \beta_1(X1)_i + \beta_2(X2)_i + \beta_3(X3)_i + \varepsilon_i, \varepsilon_i \sim N(0,\sigma^2)$	0.511	β_0	4.23	0.42	10.15	<0.0001
				β_1	-0.68	0.51	-1.33	0.1956
				β_2	0.88	0.51	1.72	0.0994
				β_3	0.94	0.51	1.84	0.0786
At the end of steady state period	mode	$(\text{mode})_i = \beta_0 + \beta_1(X1)_i + \beta_2(X2)_i + \varepsilon_i, \varepsilon_i \sim N(0,\sigma^2)$	0.854	β_0	37.83	0.83	45.76	<0.0001
				β_1	5.86	1.01	5.79	<0.0001
				β_2	-5.64	1.01	-5.57	<0.0001
	Peak frequency	$(peak)_i = \beta_0 + \beta_1(X1)_i + \beta_2(X2)_i + \beta_3(X3)_i + \varepsilon_i, \varepsilon_i \sim N(0,\sigma^2)$	0.856	β_0	172.84	10.07	17.16	<0.0001
				β_1	-86.74	12.33	-7.03	<0.0001
				β_2	25.54	12.33	2.07	0.0498
				β_3	37.57	12.33	3.05	0.0057

Table 4 Neural Network models' performance summary

Process Period	Response variable	SSE	RMSE	SSE Scaled	RMSE Scaled	R^2
After the 1st water addition during the transition period	Rate1(counts/s/s)	19.7528	0.9267	0.4963	0.1469	0.9809
	Rate2(counts/s/s)	54.0231	1.5326	0.6703	0.1707	0.9742
	Rate3(counts/s/s)	151.4385	2.5660	0.4891	0.1458	0.9812
	Rate4(counts/s/s)	6.2611	0.5217	1.7930	0.2792	0.9310
After the 2nd water addition during the transition period	Rate5(counts/s/s)	86.5533	1.9399	1.2218	0.2305	0.9530
	Rate6(counts/s/s)	121.0679	2.2943	1.3498	0.2422	0.9481
	Rate7(counts/s/s)	587.4973	5.0540	2.9544	0.3584	0.8864
	Rate8(counts/s/s)	11.6128	0.7106	2.0667	0.2998	0.9205
At the end of steady state period	Mode(µm)	99.5207	2.0801	1.5845	0.2625	0.9391
	Peak frequency (counts/s)	14200.851	24.8481	1.5675	0.2611	0.9397

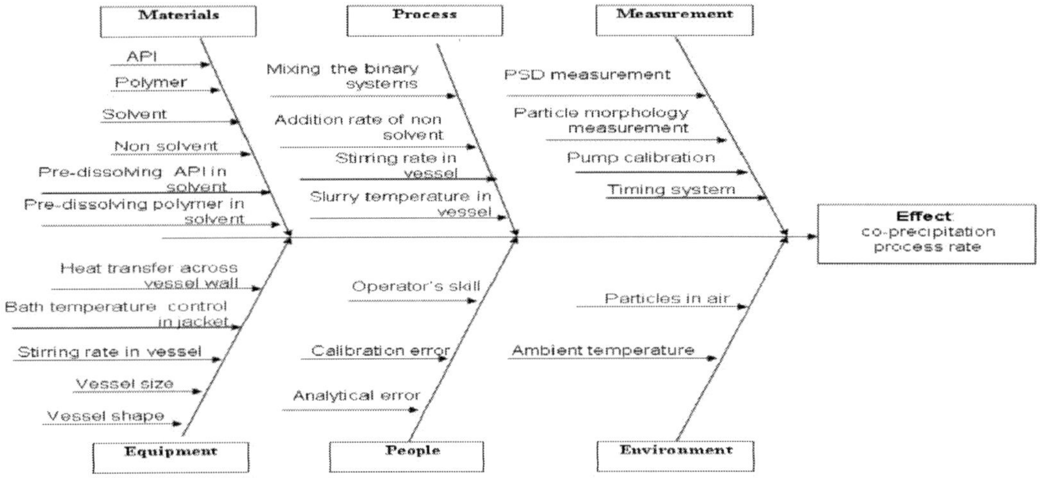

Figure 1.Fishbone diagram for risk analysis

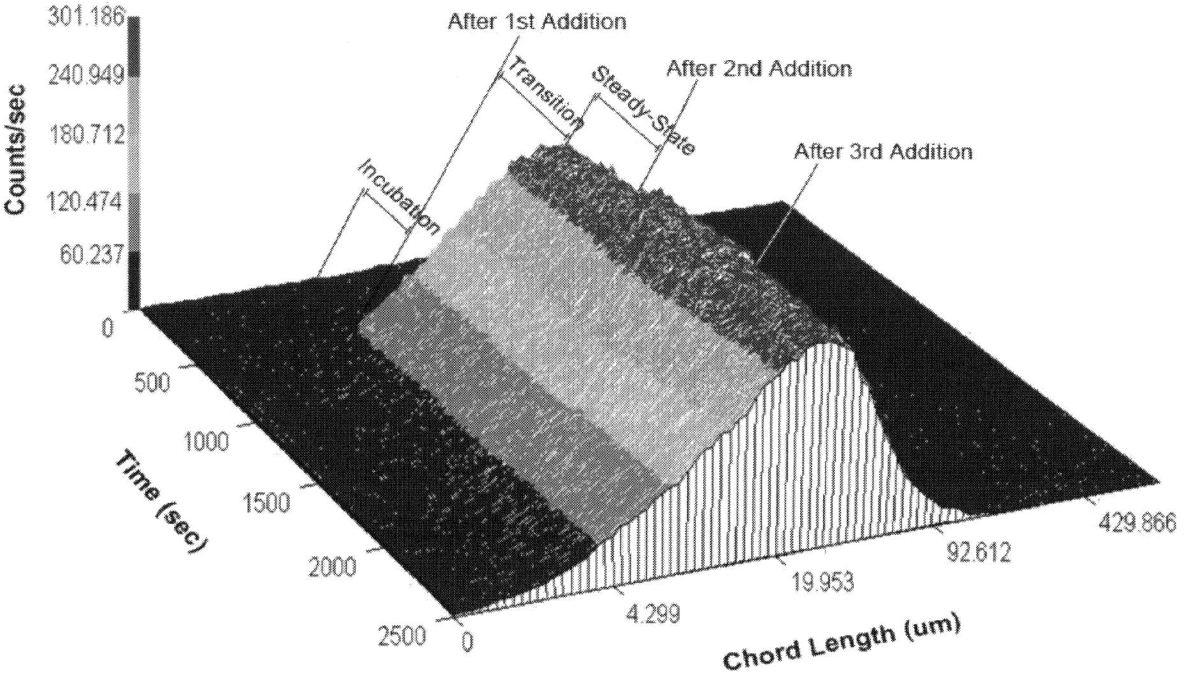

Figure 2. 3-D Count-Time-Size plots of a dynamic co-precipitation process run 132

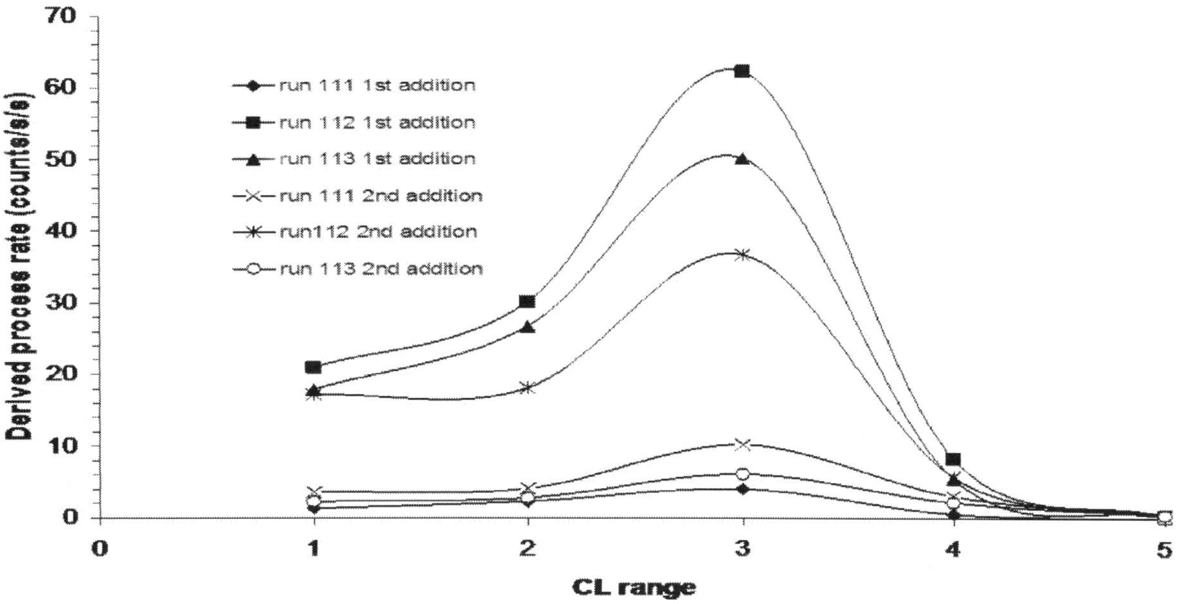

Figure 3. Plots of derived process rate vs. CL range

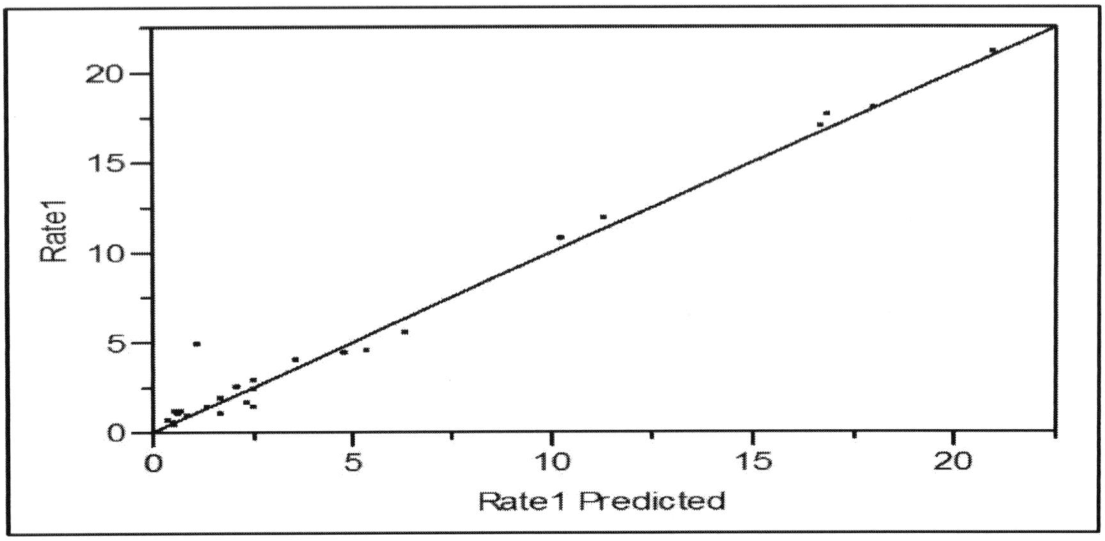

Figure 4. NN predicted response variable value vs. the actual response variable value for Rate1

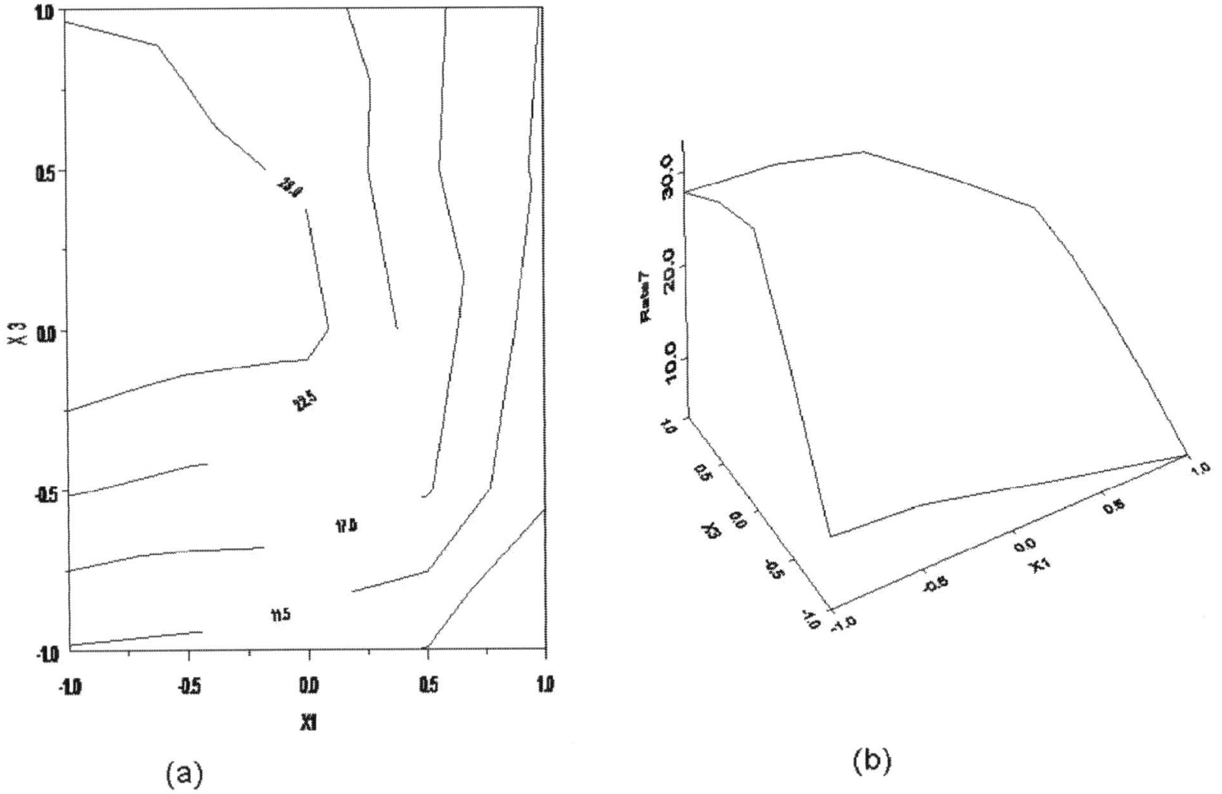

Figure 5. 2-D and 3-D contour plots for Rate7

(202f) A QbD Approach to Improve Tablet Coating Uniformity

Atul Dubey[1], Fani Boukouvala[1], Golshid Keyvan[1], Richard Hsia[2], Kostas Saranteas[2], Dean Brone[2], Tushar Misra[2], Marianthi Ierapetritou[1] and Fernando. J Muzzio[1], (1)Chemical and Biochemical Engineering, Rutgers University, Piscataway, NJ, (2)Chemical Process Research and Development, Sunovion Inc., Marlborough, MA

A combination of analytical and statistical methods is used to improve a tablet coating process guided by quality by design principles. A solid dosage form product was found to intermittently exhibit bad taste. A suspected cause was the variability in coating thickness which could lead to the subject tasting the active ingredient in some tablets. A number of samples were analyzed using a Laser Induced Breakdown Spectroscopy (LIBS) based analytical method and it was found that the main variability component was the tablet-to-tablet variability within a lot. Hence it was inferred that the coating process (performed in a perforated rotating pan) required optimization. A set of designed experiments along with Response Surface Modeling and Kriging method were used to arrive at an optimal set of operating conditions. Effects of the amount of coating imparted, spray rate, pan rotation speed and spray temperature were characterized. Results were quantified in terms of the relative standard deviation of tablet-averaged LIBS score and a coating variability index which was the ratio of the standard deviation of the tablet-averaged LIBS score and the weight gain of the tablets. The data driven models developed based on the designed experiments predicted that the minimum value of this index would be obtained for a 6% weight gain, for a pan operating at highest speed at the maximum fill level while using the lowest spraying rate and temperature from the chosen parametric space. This systematic application of QbD based method resulted in an enhanced process understanding and reducing the coating variability by more than half.

See more of this Session: Application of Quality by Design In API Process Development
See more of this Group/Topical: Topical I: Comprehensive Quality by Design in Pharmaceutical Development and Manufacture

(216a) Evaporation-Driven Fluid Flow Enhances Sensitivity of An *In Vitro* Influenza Virus Drug-Susceptibility Assay

Stephen M. Lindsay and John Yin, Chemical and Biological Engineering, University of Wisconsin - Madison, Madison, WI

InfluenzaA in the United States causes around 36,000 annual deaths and $10 billion in economiclosses. Rapid evolution through multiple mechanisms leads to periodic novel strainswith elevated virulence, transmissibility, and/or drug resistance. Althoughvaccination is the primary public health measure designed to fight influenza,antiviral drugs are very important and effective for individual cases ofelevated risk or compromised immune system. The traditional plaque reductionassay can be used to quantify the ability of drug to inhibit virus spread in infectedcell culture monolayers under agar, which inhibits convection flows. Becausethere are no flows in this assay, the virus spreads only to neighboring cells.These circular plaques are counted, and the effectiveness of the drug isquantified as the ability to inhibit plaque formation. Modifying this assay byculturing infections under liquid media rather than agar gel so that spontaneousnatural convection flows arising within the fluid spread the infections incomet-like patterns radiating outward from the center of the plate. Byleveraging these spontaneous flows and quantifying comet area rather thansimple counts, we have reduced assay time and increased the drug-sensitivity byan order of magnitude relative to the plaque reduction assay.

Wehave also shown that the morphology of the comets depends strongly on the relativehumidity level during incubation, providing evidence that the driving forcebehind the spontaneous flows arising in culture wells arise from evaporativecooling at the air-liquid interface. By confocal imaging of atemperature-sensitive dye solution, we characterized the three dimensionaltemperature profile within the fluid and predicted the resulting flow patternsusing computational fluid dynamics. The improved sensitivity of flow-enhancedassays allows drug sensitivity assays or screens to be performed with lowerdrug concentrations, minimizing the toxicity to cells. In addition, othergroups conducting *in vitro* biological assays should be aware of flowpatterns in wells and the affects of small changes in relative humidity.

See more of this Session: Drug Discovery
See more of this Group/Topical: Food, Pharmaceutical & Bioengineering Division

(216b) Single-Cell Kinetics of Infection by An RNA Virus: a Step Toward Drug-Targeting Strategies That Resist Escape

Andrea Cascio Timm and John Yin, Chemical & Biological Engineering, University of Wisconsin - Madison, Madison, WI

Viruses that carry RNA genomes define some of the most challenging targets for the development of effective anti-viral drugs. Such RNA viruses, which include HIV-1, influenza A, and hepatitis C, often exhibit high mutation rates and large population sizes, enabling their adaptation to new host environments. The presence of anti-viral drugs that target specific virus functions, such as the reverse transcriptase of HIV, the neuraminidase of influenza A, or the protease of hepatitis C, select for virus escape mutants --- variant strains that are less sensitive or more resistant to the inhibitory effects of the drugs. The accumulation of multiple drug-resistance mutations in RNA viruses significantly contributes to the persistence of diverse viral diseases that include AIDS, influenza, and liver cancer. As an alternative to the more common approach of targeting single molecular functions of a virus, we are exploring more holistic approaches that acknowledge the importance of multiple interacting functions in defining the highly coordinated and potentially conserved processes of virus production within cells. Toward this end, we have employed a model RNA virus, vesicular stomatitis virus (VSV) and its infection of baby hamster kidney (BHK21) cells, to quantitatively characterize the kinetics of virus production from single cells. Our approach employs serial dilution of infected cell populations, identification of single cells by microscopy, repeated total sampling and replenishment of the infected-cell supernatant, and quantitative analysis of these samples for infectious virus progeny. Our observations highlight several key parameters for virus production from single cells: the time required for an infected cell to initiate production of virus (delay time), following this delay, the number of virus particles released per unit time (rise rate), and the total number of virus particles released by an infected cell (yield). Here we show that cell-to-cell differences in rise rates and yields for virus production vary over two orders of magnitude, while delay times are relatively constant. Our results suggest that virus-cell interactions during the earliest stages of infection are more constant, predictable and potentially targetable than interactions associated with the more variable and less predictable later stages of infection. This observation has important consequences for rethinking strategies for drug targeting of virus-host interactions. Generalization of our approach to other RNA viruses may enable treatment of diverse infectious diseases with drug strategies that resist escape.

See more of this Session: Drug Discovery
See more of this Group/Topical: Food, Pharmaceutical & Bioengineering Division

(216c) Engineering Hetero-Bivalent Peptidic Ligands to Inhibit Mast Cell Degranulation In Allergies

Michael W. Handlogten[1], Tanyel Kiziltepe[2] and **Basar Bilgicer**[1], (1)Chemical and Biomolecular Engineering, University of Notre Dame, Notre Dame, IN, (2)Advanced Diagnostics and Therapeutics / Chemical and Biomolecular Engineering, University of Notre Dame, Notre Dame, IN

Allergy is a disorder of the immune system which is elicited by the allergen-mediated-clustering of the immunoglobulin E (IgE) molecules on the surface of mast cells and basophils. Allergenic diseases are a major cause of illness and disability in the US, and there is broad consensus that the prevalence of type I allergy is increasing worldwide. The majority of the current therapies used to treat allergy related adverse immunologic responses involve non-specific suppression of parts of the immune system. This results in the increased risk of infectious diseases (i.e. parasitic infections). Development and use of selectively targeted therapeutics would decrease these risks. Multivalent binding interactions provide higher affinity (termed avidity), and result in stronger interactions. Our overall goal in this project was to engineer hetero-bivalent peptidic ligands that would competitively and selectively inhibit allergen from binding to IgEs on mast cells, thereby inhibiting IgE clustering and allergic responses. This was accomplished via designing hetero-bivalent molecules that simultaneously targeted two nearby sites located on the Fab domain of an IgE, the antigen binding site, and the less renowned "unconventional nucleotide binding site". Simultaneous bivalent binding to both these sites provided the hetero-bivalent ligand with enhanced avidity and selectivity for the target IgE, and enabled competitive inhibition of allergen binding to the IgEs. The hetero-bivalent ligands were designed to have two moieties – the antigenic molecule itself, and the ATP analog- covalently linked by a flexible linker molecule. Specifically we have: i) designed and synthesized hetero-bivalent ligands with various linker molecules; ii) characterized the biophysical properties of these hetero-bivalent ligands and their binding properties to IgE antibodies; iii) evaluated the potential of these hetero-bivalent ligands to inhibit allergic responses by using an *in vitro* basophilic cell assay. The long-term goal of this project is to develop a broadly applicable technology that can selectively inhibit the allergic responses induced by an allergen via competitively inhibiting its binding to the IgE antibodies with engineered hetero-bivalent ligands. In addition, this approach has the potential to discover the utilization of a non-conventional binding site on antibody Fab regions by design of hetero-bivalent ligands.

See more of this Session: Drug Discovery
See more of this Group/Topical: Food, Pharmaceutical & Bioengineering Division

(216d) Model-Based Clustering Applied to Combinatorial Libraries

Andrew D. White and Shaoyi Jiang, Chemical Engineering, University of Washington, Seattle, WA

Peptide and organic combinatorial libraries are essential to screening for new drugs as well as other properties. Often, however, it is difficult to effectively treat false positives. This is especially true when searching for specific peptide sequences. Some sequences will non-specifically bind to proteins or specifically bind to classes of proteins. One approach to dealing with this problem is to cluster the data onto a structure-activity relationship (SAR) surface in order to discover the classes of molecules or peptides. One technique that allows the simultaneous discovery of the SAR surface and the clusters is model-based clustering. We've shown that this semi-parametric technique works well on peptide libraries and, when combined with other techniques, can handle over-specified QSAR models. The inherent advantage of this system, however, is that the Bayesian formulation of model-based clustering provides confidence intervals on the model parameters. For example, it can provide marginal likelihood for each SAR dimension. This provides a new way of analyzing imperfect complex combinatorial libraries. We used this technique on previously published results and on peptide library data in our own group. The model identified a number of clusters and further testing of sequences from those clusters revealed some to be non-specific and other clusters to be specific.

See more of this Session: Drug Discovery
See more of this Group/Topical: Food, Pharmaceutical & Bioengineering Division

(216e) Acetylcholinesterase Inhibitors Can Interfere In Amyloid-β Self-Assembly: Potential As Multi-Target Drugs for Alzheimer's Disease

Jui-Heng Tseng[1], Chen Suo[1], Darien Davda[2], Jie Gao[1], Alvin Terry[3], James Chapman[4] and Melissa A. Moss[5], (1)University of South Carolina, Columbia, SC, (2)Biomedical Engineering, University of South Carolina, Columbia, SC, (3)Pharmacology and Toxicology, Georgia Health Science University, Augusta, GA, (4)University of South Carolina School of Pharmacy, Columbia, SC, (5)Chemical Engineering, University of South Carolina, Columbia, SC

Alzheimer's disease (AD) is one of the most widespread neurodegenerative diseases. As the disease advances, symptoms include confusion, language skill breakdown and long-term memory loss. Amyloid-β (Aβ) is a small protein produced at low levels by normal brain metabolism. In AD, the amount of Aβ becomes disorderly because of abnormal clearance or overproduction. Some researchers have proposed the amyloid cascade hypothesis to illustrate the pathogenesis of AD. In this hypothesis, aggregated forms of Aβ are considered to induce AD pathogenesis.

Aggregation of Aβ involves a multi-step pathway with several interconnected steps. From monomer, the protein presents a lag phase indicative of nucleation, which is considered the rate limiting step, and then rapidly polymerizes to form small soluble aggregates, including oligomers and protofibrils. These intermediates grow by elongation via monomer addition and direct lateral association to form insoluble fibrils. Insoluble fibrils eventually deposit as plaques in the brains of AD patients.

The cholinergic hypothesis suggests that a dysfunction of acetylcholine-containing neurons in the brain contributes considerably to AD cognitive impairment and progressive memory loss as a result of decreased acetylcholine levels. To date, there are three FDA-approved drugs that improve AD symptoms by reducing the activity of acetylcholinesterase. This tactic raises the level of acetylcholine in the synaptic cleft to maintain stable neurotransmission. If acetylcholinesterase inhibitors could also stop Aβ aggregation, they could function as dual-action drugs.

Here, we introduced several analogs of naphthalimide, a known acetylcholinesterase inhibitor, to investigate the interaction between these compounds and different steps of Aβ self-assembly. In our experiments, we focus on four distinct steps: monomer aggregation, oligomer formation, protofibril elongation, and protofibril association. Several of these novel compounds were capable of slowing Aβ self-assembly. These novel naphthalimide analogs are promising compounds for a multitarget therapeutic strategy to confront AD.

See more of this Session: Drug Discovery

(216f) Identification of Mitochondria Damaging Agents Using Parallel Screening

David J. Taylor, Chemical Engineering, Arizona State University, Tempe, AZ, Arul Jayaraman, Department of Chemical Engineering, Texas A&M University, College Station, TX and Kaushal Rege, Chemical Engineering Department, Arizona State University, Tempe, AZ

Prostate cancer is the most commonly diagnosed cancer in men in the US and results in the second largest number of fatalities. Similarly, pancreatic cancer ranks as the fourth most fatal malignancy in both, men and women; the five-year survival rate is less than 5%. Unregulated growth and prevalence of cancer cells can be attributed to the shift in the apoptotic mechanism toward cell survival. Mitochondria are key regulators of apoptosis and the release of pro-apoptotic proteins by the mitochondria is considered "the point of no return" of apoptosis. Causing direct damage to the mitochondria can facilitate the release of proapoptotic proteins while overcoming upstream resistance to apoptosis. A small library of antineoplastic drugs was screened in order to identify leads that induced mitochondrial depolarization. Chemotherapeutic leads were further characterized for their dose-dependent activity in several pancreatic and prostate cancer cell lines. In addition, we also performed a detailed mechanistic investigation including the kinetics of mitochondrial depolarization, ability to induce reactive oxygen species (ROS), protein release from the mitochondria, and correlation of depolarization with cancer cell apoptosis. Finally, nanoparticles were investigated for delivery of mitochondria damaging drugs to cells with an eye toward improved delivery and targeting.

See more of this Session: Drug Discovery
See more of this Group/Topical: Food, Pharmaceutical & Bioengineering Division

(216g) Targeted Nitric Oxide Donor Enhances Efficacy of Chemotherapeutics Used to Treat Glioblastoma Multiforme

Shahana Safdar, Chemical and Biomolecular Engineering, Georgia Tech, Atlanta, GA and Lakeshia J. Taite, Georgia Institute of Technology, Atlanta, GA

Nitric oxide (NO) plays aubiquitous role in human physiology. Research has shown that at sufficient concentrations NO can induce apoptosis as well as increase chemosensitization in tumor cells. However, thus far NOrole in cancer therapy has been limited because of its short half-life and thevariety of effects it can have on numerous biological function. Previouslywe have shown that by derivatizing chlorotoxin (CTX), a protein specific to Glioblastoma multiforme (GBM)cells, we can synthesize targeted NO donors, which deliver NO only to thetherapeutic target, thereby significantly decreasing their viability. GBM isthe most common malignant central nervous system tumors and patients diagnosedwith the disease have very poor prognoses. The two most popularchemotherapeutics used in treatment are temozolomide(TMZ) and carmustine (BCNU), which have both shown only modest improvements inpatient survival. In this study we investigatedthe effect of the targeted NO donor, CTX-NO on the chemosensitivityof GBM cells to the chemotherapeutics BCNU and TMZ.

To assess theeffect of NO on cell viability, T98G and U-87MG human glioblastomacells were incubated with varying concentration of CTX-NO, with and withoutBCNU for 48 hours, trypsinized, and counted. At a NOconcentration of 2 M, CTX-NO alone reduced glioma cell viability to 81.4 5.5% of the control cell number. When cells were incubated with only BCNU,glioma cell viability was not significantly reduced. However when cells wereincubated with a combination of CTX-NO and BCNU, the glioma cell viability wasdecreased to 45.3 1.5% (Fig 1A). When these cells were reseeded in freshmedia their proliferation was hindered by the prior exposure to the combinationtherapy whereas the proliferation of cells previously exposed to only CTX-NO oronly BCNU was not significantly affected (Fig 1B). In contract BCNU decreasedthe viability of U-87MG to 33.8 - 2.1% and whereas the combination therapy ofBCNU and the NO donor decreased cell viability to 31.2 2.7%. Thus the NO donor does not affect thesensitivity of U-87MG cells to BCNU.

We have shown that the targeted NOdonor, CTX-NO, can enhancethe effect of carmustine on T98G cells but not in U-87MG cells. Additionallythe combination therapy affected the proliferation ability of glioma cells evenafter the therapeutic agents had been removed. To further elucidate themechanism of this NO induced chemosensitivity, theaffect of NO on O6-methylguanine DNA methyltransferaseactivity will be examined.

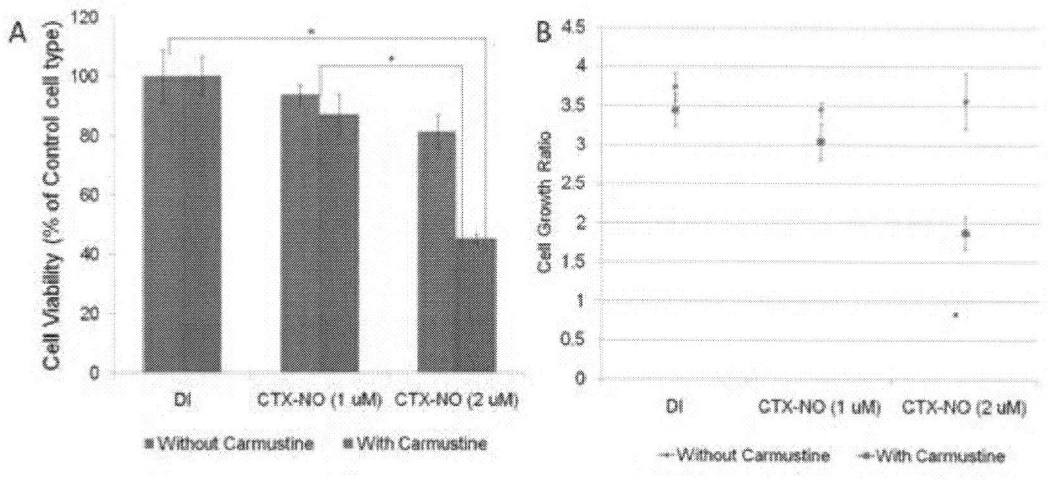

Figure1: (A) Cell viability of T98G cellsincubated with varying concentration of CTX-NO, with and without carmustine.Tumor cell viability is decreased dramatically when cells are exposed to acombination of CTX-NO and carmustine. (B) Cell proliferation of cells that wereexposed to CTX-NO, with and without carmustine, for 48 hours, trysinized, incubated with fresh media for 48 hours andthen counted. The ability to proliferate was severely hindered by thecombination therapy.

Error barsindicate standard deviation, *p <0.05

See more of this Session: Drug Discovery
See more of this Group/Topical: Food, Pharmaceutical & Bioengineering Division

(230a) Living Polymerization/Reaction Control: Innovative Strategies for Enhancing Drug Delivery From Imprinted Thin Films

Vishal D. Salian and Mark E. Byrne, Biomimetic & Biohybrid Materials, Biomedical Devices, & Drug Delivery Laboratories, Department of Chemical Engineering, Auburn University, Auburn, AL

Molecular imprinting allows for greater control of drug transport from hydrogels by the creation macromolecular memory for the drug/template molecule. Using living polymerization to create imprinted polymer networks is a more recent technique to enhance the effect of molecular imprinting. Here we will demonstrate that living polymerization leads to significantly enhanced binding parameters and delayed transport properties for drug molecules in weakly crosslinked aqueous polymers. Poly (HEMA-DEAEM-PEG200DMA) polymer gels were used as the drug delivery system with diclofenac, a non-steroidal anti-inflammatory drug, as the template molecule. Multiple gel compositions were created using both living and conventional free radical polymerization by varying the drug concentration present during polymerization, concentration of crosslinking monomer and solvent content in the pre-polymerization mixture. Template binding affinity and capacity were calculated using the Langmuir isotherm while the analysis of the polymer network was carried out using digital scanning calorimetry and gel permeation chromatography.

We demonstrate that using living polymerization results in a 38% increase in binding affinity and a 67% increase in capacity over the conventionally created hydrogel. By controlling the conversion further increases in capacity or affinity can be achieved adding greater tailorability to the hydrogel properties. Lastly, the drug release profile was extended two fold over conventionally created imprinted hydrogels and four fold over the non imprinted hydrogels. The observed enhancement in binding and transport properties may be explained by an extension of the chemically controlled chain propagation mechanism by the living polymerization technique. Thus, the results confirm that improved structural homogeneity of template binding sites and a global energy minimum of the spatial arrangement of polymer chains lead to enhanced binding parameters and delayed template release. This work will lead to a new generation of tailorable drug delivery biomaterials which can be designed from the molecular scale to yield exquisite control over controlled release of therapeutic drug molecules.

See more of this Session: Innovations In Drug Delivery Technology I
See more of this Group/Topical: Food, Pharmaceutical & Bioengineering Division

(230b) Stripfilms As Oral Dosage Platform for the Delivery of Nano-Particles of Active Pharmaceutical Ingredients (APIs)

Rajesh N. Dave, Chemical Engineering, New Jersey Institute of Technology, Newark, NJ, Lucas Sievens-Figueroa, New Jersey Center for Engineered Particulates, New Jersey Institute of Technology, Newark, NJ and Ecevit Bilgili, Chemical, Biological and Pharmaceutical Engineering, New Jersey Institute of Technology, Newark, NJ

We describe formation of stripfilms loaded with nano and low micron sized API particles to achieve desired delivery properties through maintaining of the high surface area of particles via particle engineering. As a platform to incorporate particle engineering, this approach will allow for a fundamental shift in the design of pharmaceutical products containing nano-particles without the need for first drying or recovering nano-particles in dry form. Our approach consists of forming stable nano-suspensions of BCS Class II drugs via three top-down and bottom-up approaches to achieve size range of 100–2000 nm, with API loading of 0.1 to 20 wt% in the suspensions. The API suspensions are stabilized via judicious use of polymer-surfactant combinations to achieve size stability ranging from 2 hours to 2 days. These suspensions are further processed to increase their polymer content by mixing with higher polymer containing aqueous gels or solutions. These film-precursors containing nano-APIs are formed into films, whose thickness is in the range of 50 to 200 microns, via the solvent-casting approach utilizing convective drying.

We have analyzed the retention of large API surface area in these films via re-dispersion in water, showing that this integrated particle engineering approach allows for maintaining high surface area of the API particles, which has been a non-trivial problem faced by the pharmaceutical industry. Films characterization confirms good API uniformity with API content linearly dependent on film thickness. Dissolution of films also proves to be much faster than other dosage forms mainly due to the high surface area associated with thin films in combination with fast, full-recovery of small API particles during dissolution. This approach may be suitable for BCS Class II drugs requiring enhancements in bioavailability for producing clinical supplies. Stripfilms can also serve as a quick-to-market platform for flexible, uniform dose manufacturing of BCS Class II drugs.

See more of this Session: Innovations In Drug Delivery Technology I
See more of this Group/Topical: Food, Pharmaceutical & Bioengineering Division

PARTICLE SCALE STUDY OF THE DISPERSION OF DRY POWDER IN PHARMACEUTICAL AEROSOL INHALERS

Z.B. Tong, University of NSW, Sydney, AU
R.Y. Yang, University of NSW, Sydney, AU
A.B. Yu, University of NSW, Sydney, AU
S. Adi, University of Sydney, Sydney, AU
H.-K. Chan, University of Sydney, Sydney, AU

Introduction

Inhalation aerosols offer the distinct advantage of rapid and convenient delivery of bioactive substances to the site of action. Significant research effort has been directed towards the development of innovative technologies for the generation and delivery of dry powder aerosols for inhalation. Dry powder aerosols require no propellant (detrimental to the environment), have superior chemical stability and bacterial growth resistance compared with solution, and are easy to use (hence good patient compliance).

Existing DPIs have low efficiency and high variability of dosing, and generally less than 30 % of the dose loaded to the device can deposit in the lungs [1]. This is due to the strong cohesion between pharmaceutical particles which are usually less than 5 μm and hence make them difficult for dispersion. In drug delivery, the agglomerates have to be dispersed to form a fine respirable aerosol cloud in order to minimise deposition in the mouth and upper airways, and to maximise delivery to the more distal parts of the respiratory tract. To improve the aerosol performance, much research effort in the past two decades has been directed at the powder formulation [2]. However, powder dispersion also depends strongly on the prevailing flow conditions in an inhaler device and their effects on powder deagglomeration remain unclear because it is difficult to obtain experimental details and quantitative information about the dispersion process due to the small agglomerate size (~1 mm) and the short duration (5s <) of dispersion.

We have developed a coupled computation fluid dynamics (CFD) and discrete element method (DEM) model to simulate the dispersion process in DPIs. The model incorporated powder properties obtained from experiments and was calibrated carefully. Here we briefly review these studies, aiming to provide a base for future development in this area.

Model development

CFD-DEM model

The model treats the dispersion process as a multiphase flow where the discrete powders and the continuous air flow are modelled by DEM and CFD, respectively. In DEM model, the translational and rotational motions of a particle in a system are governed by its interactions with neighbouring particles, walls and surrounding fluid described by Newton's second law of motion. The governing equations for the continuous fluid in CFD model are the conservations of mass and momentum in terms of the local mean variables over a computational cell. The CFD-DEM coupling is achieved by combining an in-house built DEM with the commercial software ANSYS Fluent®. At each time step, DEM calculates particle-related information based on fluid flow field, such as the positions and velocities of individual particles, to determine porosity and volumetric particle-fluid interactions in the individual computational cells. CFD then uses these data to determine the air flow field which then yields the particle-fluid forces acting on individual particles. Incorporation of the resulting forces into DEM produces information about the motion of individual particles for the next time step. Such procedure continues until the whole process is finished [3-5].

Model validation

The numerical model was validated by comparing the simulated results, such as the trajectory of the agglomerates and the amount of fine particles less than 5μm (FPF) generated, with those from the experiments using the setup as shown in Fig. 1. In the setup, air of different flow rates entered the impaction throat from the right side air inlet and an agglomerate was introduced via the feed apparatus (as indicated by "B" in Fig. 1). The agglomerate moved along with the air flow and impacted with throats containing two angles (Figs. 1a and 1b) or a single angle (Figs. 1c, 1d and 1e).

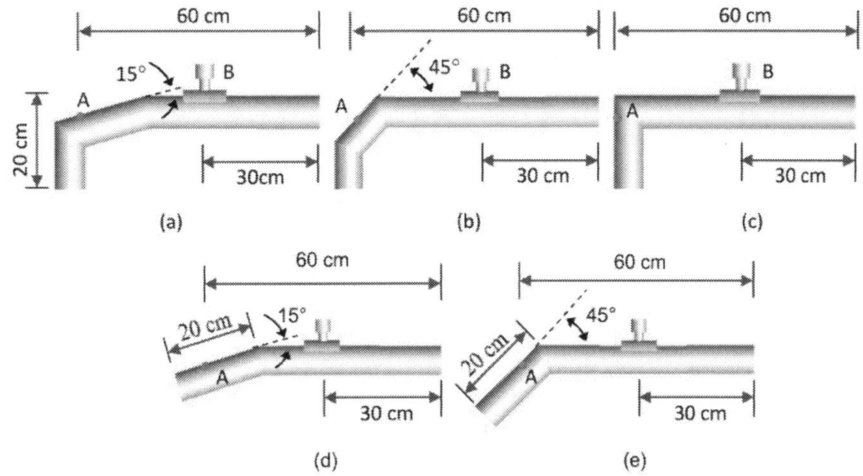

Fig. 1. Schematics of impaction throats A-E.

Fig. 2 shows the simulated and experimentally recorded trajectories of the agglomerate in the 90° impaction throat at the flow rate of 120 l min^{-1}. The trajectories are similar, both indicate that mannitol agglomerate initially drops to the bottom of the pipe after being fed into the impaction throat, then is accelerated and lifted by the airflow and finally impacts on the throat.

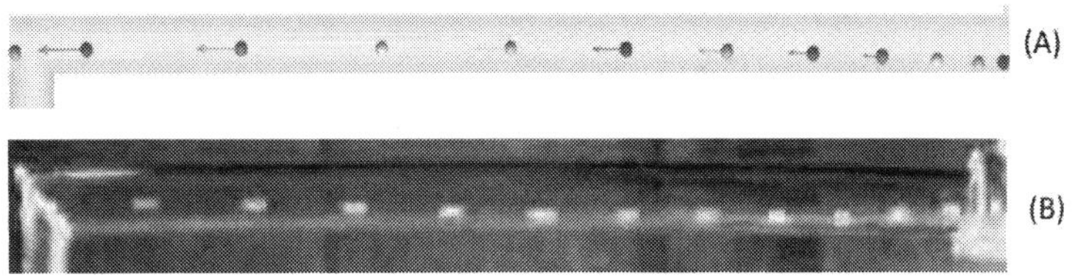

Fig. 2. Comparison of the trajectory of the agglomerate from simulation (A) and experiment (B) in the one angle impaction (90° and 120 l min^{-1}).

Fig. 3 shows the simulated and measured FPF and powder depositions under different impaction conditions. Again the simulations are quite comparable with experimental results, further confirming the validity of the model. For the two-angle throats (A and B) and the single angle throat C with 90-degree impaction angle, FPF increases with impact angles at low flow rates (60 and 120 l min^{-1}) but decreases at the high flow rates of 150 l min^{-1}. For the single angle throats D and E, FPF increases monotonically with flow rate. As shown later on, the decrease in FPF at high flow rate for throats A, B and C is caused by the increased powder deposition. In general, the throats with two angles (A and B) result in better dispersion than those with a single angle (C, D and E), showing that the multiple impactions can improve powder de-agglomeration.

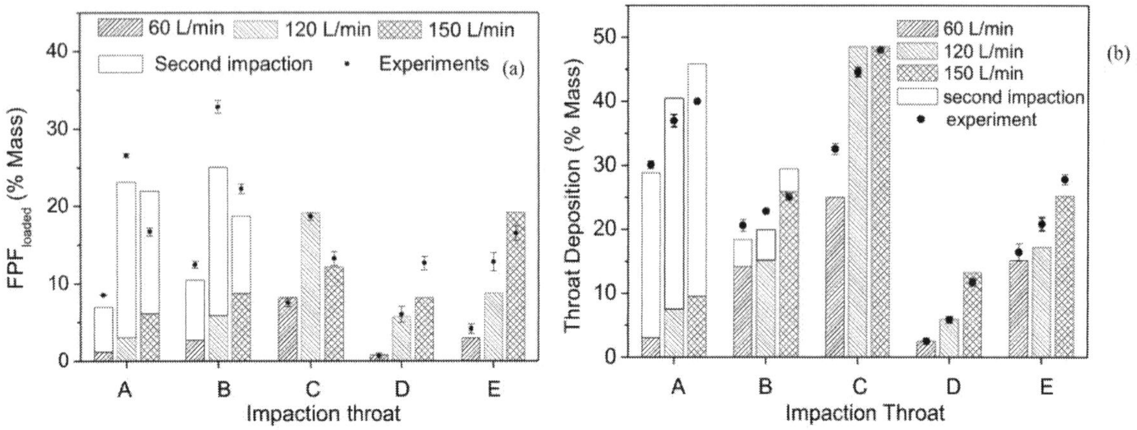

Fig. 3. (a) FPF generated at different impaction conditions and flow rates; (b) Mass fraction of powder deposition at different flow rates. The experimental data [6] are also plotted for comparison.

Dispersion mechanisms

Particle-wall interaction

The powder dispersion due to the mechanical impaction was simulated by impacting an agglomerate with a target wall with different velocities and angles [7]. The agglomerate was formed with 5 μm mannitol powders and had final diameter of about 100 μm with a packing fraction of 0.552 and a relatively weak strength. When impacted with the wall, the agglomerate experiences a large plastic deformation on impact before disintegrating into many small pieces (Fig. 4a). By analyzing the structural and force evolutions (Fig. 4b) of the agglomerates during the impacts and the final breakage patterns, the work showed that the impact can be divided into three periods in which the microscopic structures of the agglomerates (i.e. porosity, velocity profile and force) evolve differently.

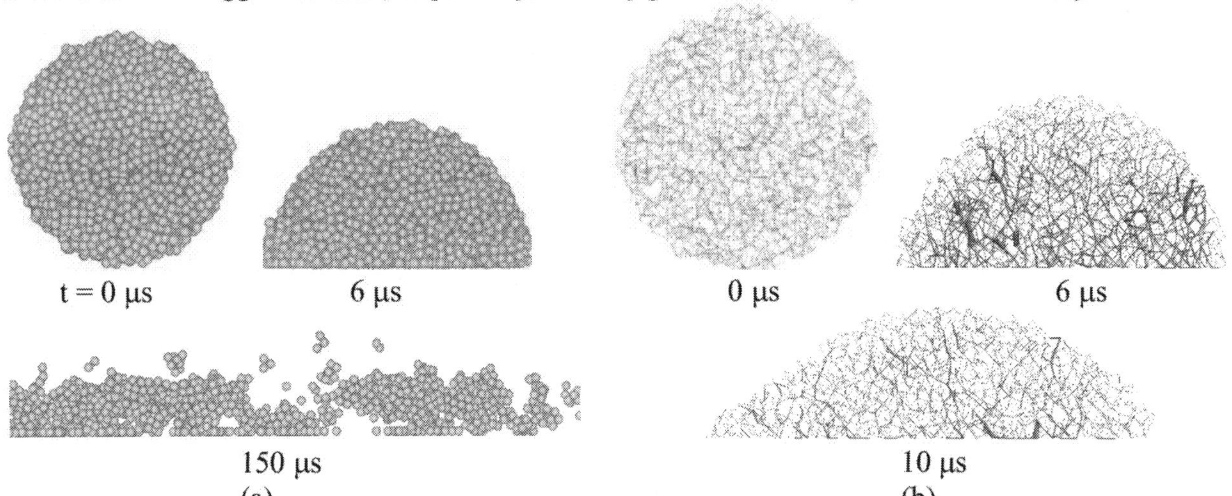

Fig. 4. Snapshots of (a) the agglomerate; and (b) spatial distribution of contact force different times with $\theta = 45^{o}$ and v = 10 m/s (only part of the fragments are shown at t = 150 μs).

Unlike strong agglomerates, a shear band can be observed in the velocity fields of the agglomerates. Larger forces are developed within the agglomerate near the contact region and propagate upwards. The size distribution of fragments shows that while increasing impact velocity generally has better breakage performance, impact angle at 45-degree can give the maximum breakage performance for a given impact velocity. The particle-wall impact energy can be used to interpret the breakage pattern of agglomerates. For breakage of loose agglomerates, the impact energies in normal and tangential directions are comparable and both should be considered in order to describe breakage performance properly. As shown in Fig. 5, the breakage of loose agglomerate is closely related to the total impact energy from the wall, and the breakage efficiency is governed by the combined effect of normal and tangential interactions between the particles and the wall.

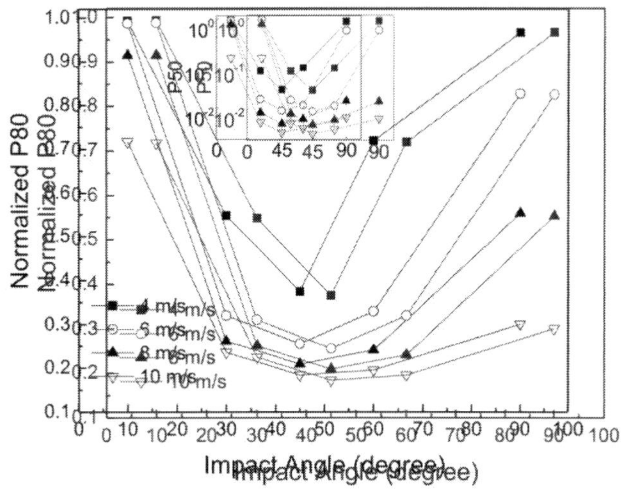

Fig. 5. Normalised P_{80} as a function of impact angle for different impact velocities. The inset shows the variation of the normalised P_{50} values.

Particle-flow interaction

This work [8] investigated the dispersion process of single agglomerate in a cyclonic model which has two air inlets and one outlet (Fig. 6a). In the simulation, an agglomerate is introduced via the left inlet with zero initial velocity while air flow enters from both inlets with a pre-set velocity. The agglomerate starts to accelerate due to the flow drag force and, due to flow shearing and turbulence and impact with the device, is dispersed into small fragments which are then discharged from the outlet. Fig. 6b shows size distributions and morphology of formed agglomerates. Agglomerates A, B and E have similar spans (0.412) with decreasing D_{50}, (4.14 – 2.6μm) while agglomerates B, C and D have similar D_{50} (3.27 μm) with increasing span (0.412 – 1.168) [9].

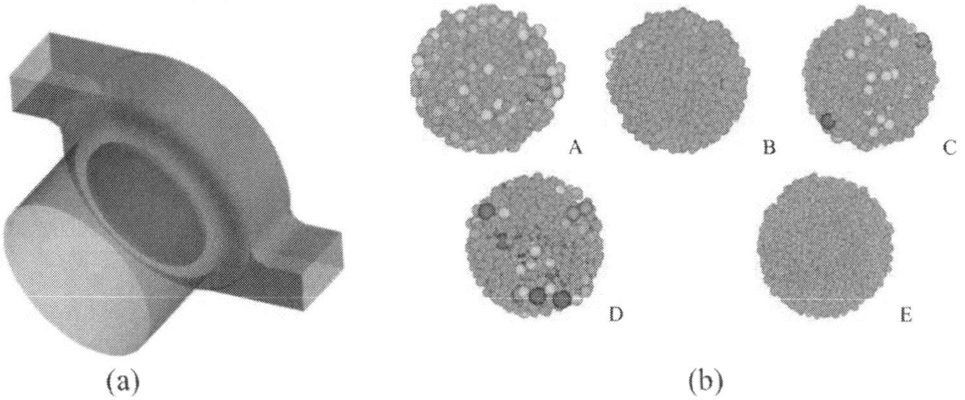

(a) (b)

Fig. 6. (a) 3-dimensional view of the cyclonic model; and (b) Morphology of the formed agglomerates (the colours represent particle diameter).

Fig. 7a indicates a typical cyclonic flow pattern with the velocity having the maximum of 34 m/s near the two inlets, gradually slowing down along the flow direction and reaching the lowest velocity of 2 m/s at the centre. Fig. 7b shows the velocity and breakage pattern of agglomerate B at the flow velocity v = 30 m/s. It shows that increasing flow velocity decreases the agglomerate residence time t_e which is defined as the duration of the agglomerate in the device before being discharged from the outlet.

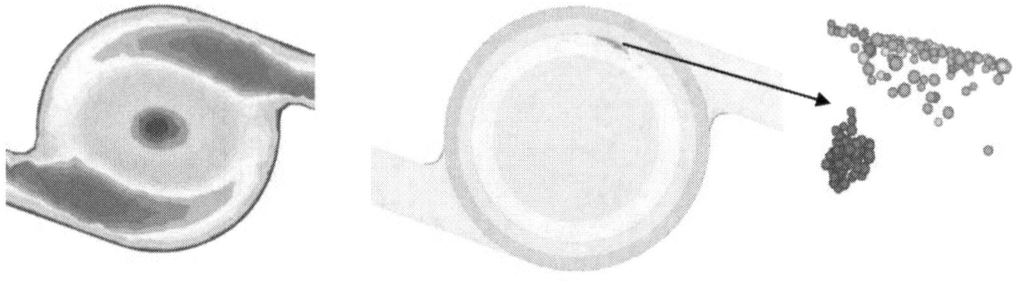

(a) (b)

Fig. 7 Top views of (a) fluid velocity; and (b) particles velocity and breakage pattern of an agglomerate at t = 0.0015 s. Air flow velocity v = 30 m/s

Fig. 8 shows the FPF of the agglomerates at various flow velocities. In general, larger flow velocities generate larger FPFs for an agglomerate. However, different agglomerates behave differently with flow rates. While agglomerate A has the best dispersion efficiency at the low flow velocity of 20 m/s, agglomerates B and E, which have similar spans but large D_{50}, generate more fine particles at the median and high flow rates, respectively (Fig. 8a). Fig. 8b shows the agglomerates with smaller span (narrower size distribution) have larger FPF. However, the differences in FPF for the three agglomerates are very small, indicating that the effect of powder polydispersity on dispersion efficiency is less significant than particle size.

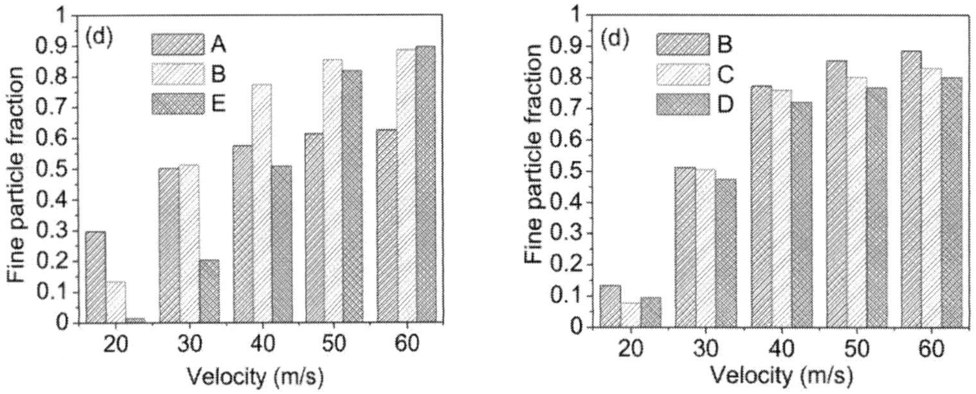

(a) (b)

Fig. 8 Effects of (a) particle size; and (b) span of distribution on the dispersion process at different flow velocities.

Particle-particle interaction

This work simulates powder dispersion in a T-shape pipe with two air inlets and one outlet [10]. In the simulations, the air enters from both inlets and two agglomerates are introduced separately via the left and right inlets. The simulations are considered to study the effects of airflow rate, agglomerate strength and agglomerate-agglomerate impact angle. Fig. 9 shows the dispersion of the agglomerates in the pipe at different times with the air flow velocity 15 m/s. The two agglomerates start to accelerate due to the flow drag force, then due to agglomerate-agglomerate interaction, are broken into small fragments, and finally are discharged from the outlet. The fluid flow and agglomerates are coloured by the magnitudes of the airflow velocity and the radius of primary particle, respectively. The influence of agglomerates on the surrounding fluid flow can be observed from the change of fluid velocity at different times.

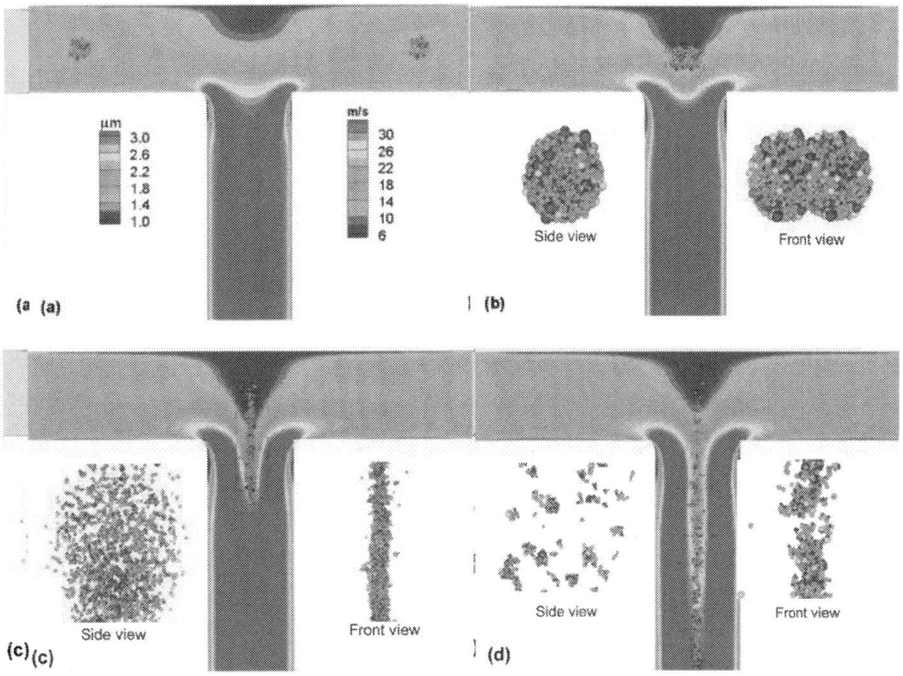

Fig. 9. Contour of airflow velocity (only central slices are shown here) and breakage pattern of agglomerate at airflow velocity of 15 m/s: (a) $\tau = 0$; (b) $\tau = 0.15$; (c) $\tau = 0.26$; and (d) $\tau = 0.33$.

Fig. 10 shows the FPF of the agglomerates at different flow rates (Fig. 10a), agglomerate strength (Fig. 10b) and impaction angles (Fig. 10c). In general, FPF increases with increasing flow rates, decreasing agglomerate strengths and increasing impaction angles.

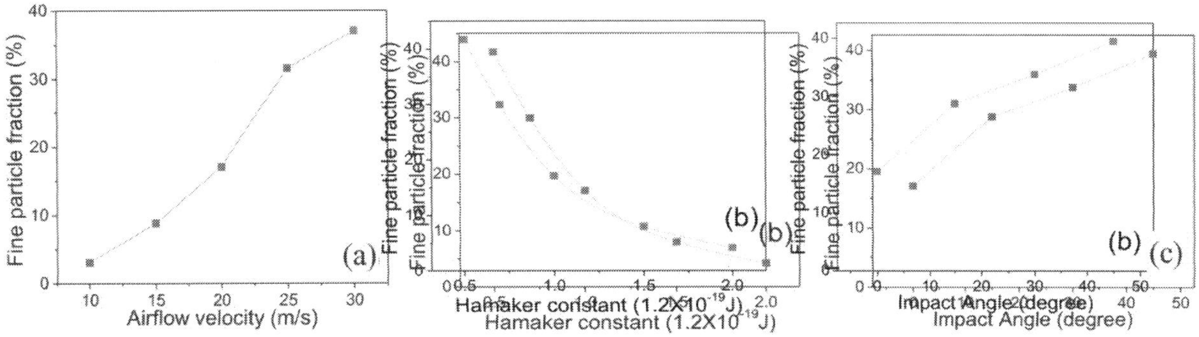

Fig. 10. Fine particle fraction after dispersion with different (a) airflow velocity; (b) agglomerate strength; and (c) impact angle.

Dispersion in dry powder inhaler

This work focuses on the dispersion process of multiple agglomerates in a dry powder inhaler (Fig. 11a) which includes all types of interaction mentioned above. In the simulations, multiple agglomerates were sprayed out from the capsule. Then agglomerates were accelerated due to the fluid drag force, and due to flow shearing and turbulence and impact with the device were dispersed into small fragments which were then discharged from the outlet. Fig. 11b shows the flow field at the flow velocity of 40 m/s. This indicates a typical cyclonic flow pattern with the velocity field inside the chamber having significant change with rotating capsule.

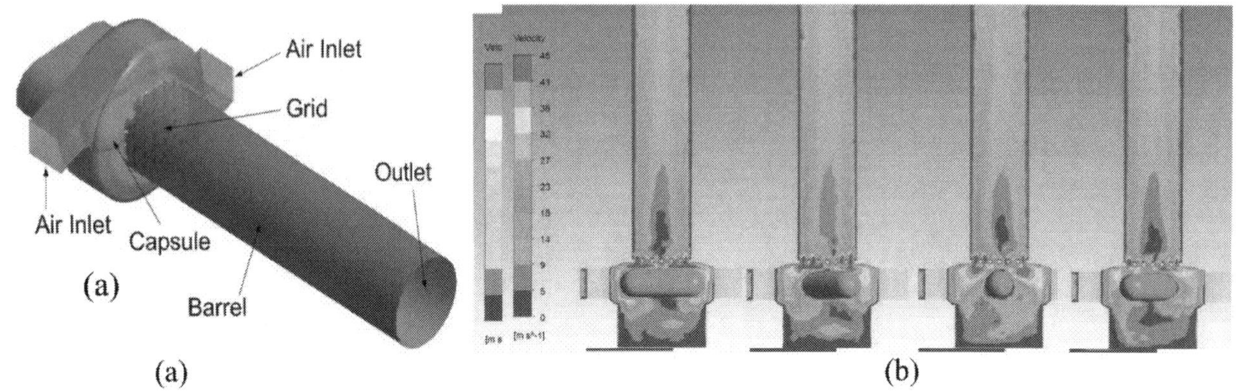

Fig. 11. (a) 3-dimensional view of the device; and (b) Side view of the velocity pattern at different times schematic and numerical grid representation.

Agglomerate dispersion normally is more efficient with increased flow rate as the total impact energy increases (Fig. 12a). However, FPF_{laoded} sharply decreases with further increasing to 50 m/s as the throat deposition increases sharply. As shown on Fig. 12b, FPF_{laoded} also increases with the number of agglomerates from 2 to 8 agglomerates, then has little changes with further increasing the number of agglomerate. So FPF_{laoded} can be explained by a balance between de-agglomeration rate and deposition with increasing number of agglomerates and airflow rate.

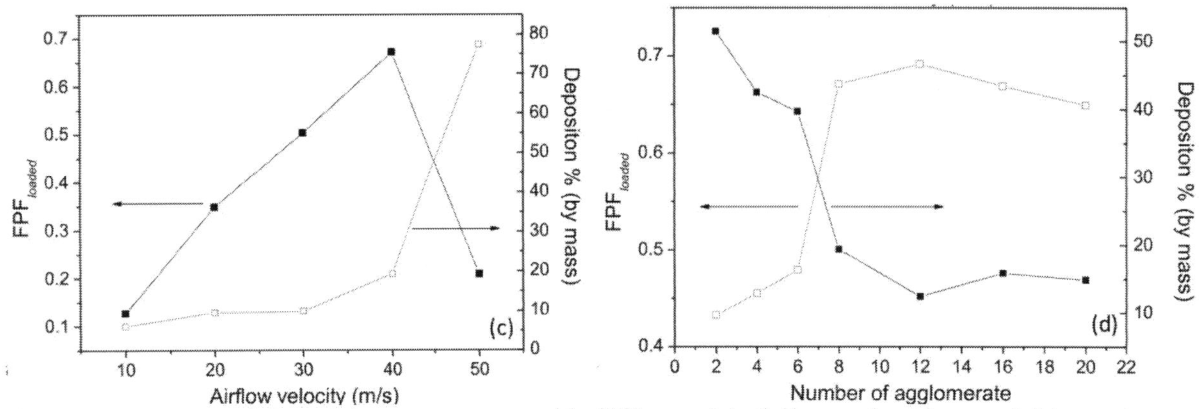

Fig. 12. FPF_{loaded} and deposition percentage with different (a) airflow velocities; and (b) numbers of aggloemrates.

Dispersion quantification

As demonstrated above, agglomerate breakage is mainly determined by particle-particle cohesion and particle-wall impact. Since the particle-wall impact is governed by the particle-fluid interaction related to flow velocity, the effect of flow rate can also be reflected from the impact interactions. Therefore, establishing a quantitative link between the interaction forces and dispersion performance can provide a way to characterise and optimise dispersion processes based on simulation results. Fig. 13 shows FPF_{loaded} obtained under different conditions as a function of the ratio of impact and tensile energies. The previous results under the swirl flow [8], T-pipe and throats [11] are also plotted for comparison. It is observed the results collapse on a single master curve. The results further indicate that agglomerate breakage is governed by the same mechanisms even the breakages occur at different conditions.

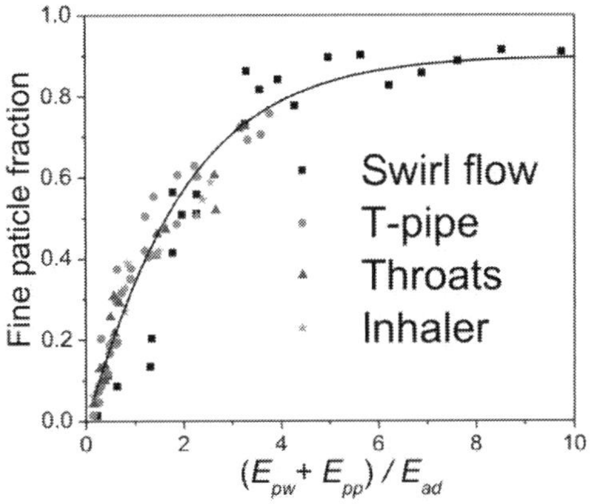

Fig. 13. Correlation between relative fine particle fraction (FPFLoaded) and the ratio between the impact energy and agglomerate cohesion energy (ϕ) in the inhaler and previous models [7, 8, 11].

Conclusions

CFD-DEM model has been developed to simulate the dispersion process of pharmaceutical powders in dry powder inhalers (DPIs). The model can generate the flow and powder related information and reliably describe the dispersion of drug powders. Analysis of microdynamic results such as particle-particle, particle-fluid and particle-inhaler interactions along with trajectory of and transient forces acting on individual particles are useful to understanding mechanisms governing particle agglomeration and de-agglomeration in relation to flow conditions and inhaler geometry. It also has the wide applicability to a variety of inhalers and operational conditions. The results can be used to improve the performance of existing DPI or designing new inhalers with desirable dispersion.

Acknowledgements

Authors are grateful to the Australia Research Council (ARC) for the financial support for this work.

References

1. I. Gonda. Physicochemical principles in aerosol delivery. Topics in pharmaceutical sciences 1991. Stuttgart, Germany: Medpharm Sci Publ 1992:95-115.
2. H.-K. Chan. Dry powder aerosol drug delivery-opportunities for colloid and surface scientists. Colloids and Surfaces, A: Physicochemical and Engineering Aspects 284-285 (2006) 50-5.
3. R.Y. Yang, R.P. Zou, A.B. Yu. Computer simulation of the packing of fine particles. Physical Review E 62 (2000 Sep) 3900-8.
4. K.W. Chu, B. Wang, A.B. Yu, A. Vince. CFD-DEM modelling of multiphase flow in dense medium cyclones. Powder Technol 193 (2009) 235-47.
5. Z.Y. Zhou, S.B. Kuang, K.W. Chu, A.B. Yu. Discrete particle simulation of particle-fluid flow: model formulations and their applicability. Journal of Fluid Mechanics 661 (2010 Oct) 482-510.
6. S. Adi, Z.B. Tong, H.K. Chan, R.Y. Yang, A.B. Yu. Impact angles as an alternative way to improve aerosolisation of powders for inhalation? Eur J Pharm Sci 41 (2010 Oct) 320-7.
7. Z.B. Tong, R.Y. Yang, A.B. Yu, S. Adi, H.K. Chan. Numerical modelling of the breakage of loose agglomerates of fine particles. Powder Technol 196 (2009) 213-21.
8. Z.B. Tong, R.Y. Yang, K.W. Chu, A.B. Yu, S. Adi, H.K. Chan. Numerical study of the effects of particle size and polydispersity on the agglomerate dispersion in a cyclonic flow. Chem Eng J 164 (2010) 432-41.

9. S. Adi, H. Adi, H.-K. Chan, W.H. Finlay, Z. Tong, R. Yang, A. Yu. Agglomerate strength and dispersion of pharmaceutical powders. Journal of Aerosol Science 42 (2011) 285-94.
10. Z.B. Tong, R.Y. Yang, A.B. Yu, S. Adi, H.K. Chan. Numerical Study of Effects of Powder Size and Polydispersity on the Dispersion of Fine Powders in a Cyclonic Flow. In: Nakagawa M., Luding S., eds. Powders and Grains 2009. Melville: Amer Inst Physics 2009:807-10.
11. Z.B. Tong, S. Adi, R.Y. Yang, H.K. Chan, A.B. Yu. Numerical investigation of the de-agglomeration mechanisms of fine powders on mechanical impaction. Journal of Aerosol Science In Press, Accepted Manuscript (2011).

(230d) A Microfluidic Platform for Aerosol and Nanoparticle Drug and Gene Delivery

Aisha Qi, Anushi Rajapaksa, Peggy Chan, James Friend and **Leslie Y. Yeo**, Micro/Nanophysics Research Laboratory, Monash University, Clayton, VIC, Australia

Pulmonary drug administration requires the direct delivery of drug formulations into the deep lung region in the form of inhaled particles or droplets, providing a distinct advantage over other methods for the treatment of respiratory diseases: the drug can be delivered directly to the site of inflammation, thus reducing the need for systemic exposure and the possibility of adverse effects. Nevertheless, it is difficult to produce aerosol droplets of a drug solution within a narrow polydisperse size range between 1 and 5 μm to achieve maximum deposition in the lower pulmonary tract and alveoli. We first demonstrate the use of surface acoustic wave (SAW) microfluidic atomization as an efficient means to generate aerosols containing the short-acting β_2-agonist salbutamol for the treatment of asthma. The 2.84 μ m mean diameter of the aerosol droplets produced lies well within the optimum size range. This is subsequently confirmed using a twin-stage impinger *in vitro* lung model, in which we observe that approximately 70% to 80% of the drug is deposited within the lung; such lung dose efficiencies are considerably higher than those possible through currently available metered dose inhalers and nebulizers. Thus, even with a modest power range of around 1 W, which is a tenth of that required with ultrasonic nebulizers, the SAW atomization technology provides a viable and efficient generic nebulization platform for drug delivery via the pulmonary route. In addition, the potential for the platform to be used for non-viral gene delivery is demonstrated through high levels of gene expression observed in Western blot analysis of the COS-7 cells transfected with post-atomised plasmid DNA encoded with a merozoite surface protein 4/5, which is a potential malaria vaccine candidate. This is subsequently verified through the expression of yellow fluorescent protein in lung sections of Swiss mice following intratracheal instillation of the plasmid DNA containing the protein. Further, we show that the technology is a very rapid, efficient and straightforward means for the synthesis of biodegradable polymeric particles with dimensions down to 100 nm within which therapeutic molecules such as nucleic acids, proteins and peptides can be encapsulated. Finally, the ability to synthesize multiple coatings of polyelectrolyte layers encapsulating these biomolecules is demonstrated using sequential atomization-suspension steps as a fast and efficient alternative to conventional layer-by-layer polyelectrolyte assembly. These multilayer nanocapsules offer the exciting possibility for tuning the drug release profile for controlled delivery over a prolonged period or for targeting the delivery to a specific location within the body. In all of the above, the control offered over the aerosol/particle size, the possibility for flexible and multilayer encapsulation, the low power requirement, the high delivery efficiency, and the miniaturization of the system altogether suggest that the proposed SAW atomization platform represents an attractive alternative to current nebulizers and inhalers, which we envisage could constitute next-generation devices that revolutionize pulmonary drug and gene delivery for needle-free vaccination or for the treatment of various diseases in the near future.

(230e) Development of Anti-Tubercular Nanoscale Drug Cocktails for Targeted Therapy

Suzanne M. D'Addio[1], Venkata Reddy[2], Philip Kwok[3], Leo Einck[2], Hak-Kim Chan[3] and Robert Prud'homme[1], (1)Chemical Engineering, Princeton University, Princeton, NJ, (2)Sequella, Inc, Rockville, MD, (3)Faculty of Pharmacy, The University of Sydney, Sydney, Australia

While tuberculosis (TB) is a curable intracellular disease, it remains a global health issue with over 1 million deaths reported in 2007. Causes of failed treatment include patient non-compliance and the emergence of drug resistance over the 6 month therapeutic course. Advanced technologies are needed to improve the efficiency of drug delivery to reduce dosing frequency and limit drug resistance, while ensuring patient compliance. Block copolymer directed rapid precipitations by Flash NanoPrecipitation enable formation of nanoparticles which have two unique advantages: (1) multiple drugs can be simultaneously assembled in the same nanoparticle and (2) targeting ligands can be introduced efficiently in drug formulations. Using nanoparticles as a drug delivery platform, we have created nanoscale drug cocktails to colocalize a novel, potent anti-tubercular drug with an efflux pump blocker to ensure therapeutic doses are achieved intracellularly. Targeted polyethylene glycol protected nanoparticles with variable mannose density on the surface have been studied *in vitro* to achieve optimized active targeting vehicles for anti-TB drug delivery specifically to macrophages. To promote patient compliance, aerosol carrier particles have been developed for inhalable targeted delivery of nanoparticles to the lung, with predictable fine particle fractions. The combination of these functionalities is a promising route to overcome some of the current treatment limitations in TB therapy.

See more of this Session: Innovations In Drug Delivery Technology I
See more of this Group/Topical: Food, Pharmaceutical & Bioengineering Division

(230f) Engineering Multifunctional Nanoparticles for Overcoming Drug Resistance In Multiple Myeloma

Tanyel Kiziltepe[1], Jonathan D. Ashley[2], Jared Stefanick[2] and **Basar Bilgicer**[2], (1)Advanced Diagnostics and Therapeutics / Chemical and Biomolecular Engineering, University of Notre Dame, Notre Dame, IN, (2)Chemical and Biomolecular Engineering, University of Notre Dame, Notre Dame, IN

Multiple myeloma (MM), a B-cell malignancy characterized by proliferation of monoclonal plasma cells in the bone marrow (BM), is the second most common type of blood cancer in the U.S. Despite the recent advances in treatment strategies and the emergence of novel therapies, it still remains incurable due to the development of drug resistance with a median survival of 4-5 years. A major factor that contributes to development of drug resistance in MM is the interaction of MM cells with the BM microenvironment. It has been demonstrated that the adhesion of MM cells to the BM stroma via $\alpha 4\beta 1$ integrins leads to cell adhesion mediated drug resistance (CAM-DR), which enables MM cells to gain resistance to drugs such as doxorubicin (Dox)–a 1st line chemotherapeutic in the treatment of MM. The overall objective of this proposed project is to engineer "smart" nanoparticles that will deliver and exert the cytotoxic effects of the chemotherapeutic agents on MM cells, and at the same time do it in such a manner to overcome CAM-DR for improved patient outcome. To enable this, we engineered micellar nanoparticles that are (i) functionalized with $\alpha 4\beta 1$-antagonist peptides as well as Dox conjugates, and (ii) designed to show the adhesion inhibitory and the cytotoxic effects in a temporal sequence. When the nanoparticles are delivered to the MM cells, as a first step they interact with the cell surface $\alpha 4\beta 1$ integrins and inhibit MM cell adhesion to the stroma, thereby preventing development of CAM-DR. In the second step, Dox exerts its cytotoxic effects after cellular uptake, as the nanoparticles will be designed to require a low pH environment such as the endocytic vesicles, to release active Dox. This way, the "smart" nanoparticles act on the MM cells in a temporal fashion and prevent development of CAM-DR for improved patient outcome.

See more of this Session: Innovations In Drug Delivery Technology I
See more of this Group/Topical: Food, Pharmaceutical & Bioengineering Division

(230g) Melt Mixing Active Pharmaceutical Ingredient and Polymer to Achieve Different Release Profiles

Adeyinka Adegoke, Ashish Sarode and **Peng Wang**, Department of Chemical Engineering, University of Rhode Island, Kingston, RI

Traditional blending processes in the pharmaceutical industry are operated at room temperature. In recent years, melt mixing has received a great deal of interest, mainly due to the challenge associated with the ever increasing number of poorly soluble drugs. In addition to the dissolution enhancement, melt mixing also possesses several other advantages. It is a solvent free process; it can be realized by using a continuous extruder, which is commonly used in the plastic and processed food industries. Our experimental data show that API and polymeric excipients can form one phase at the elevated temperature. The drug may remain in amorphous form or recrystallize after the mixture is cooled to room temperature, depending mostly on the formulation. In addition, our results suggest that melt mixing can be applied to significantly increase or slow down drug's release rate, which cannot be achieved if the same components are blended at room temperature.

See more of this Session: Innovations In Drug Delivery Technology I
See more of this Group/Topical: Food, Pharmaceutical & Bioengineering Division

(231a) A Novel SHF Process for Conversion of AFEX-CS to Ethanol Featuring High Ethanol Productivity, Enzyme Recycling, and Yeast Cells Reuse

Ming Jie Jin, Christa Gunawan, Nirmal Uppugundla, Shishir Chundawat, Bruce Dale and **Venkatesh Balan**, Chemical Engineering and Material Science, Michigan State University, East Lansing, MI

There are two major problems in current cellulosic ethanol production process: low ethanol productivity and high enzyme loading. Based on the current Separate Hydrolysis and Fermentation (SHF) process, it takes at least 96h to perform enzymatic hydrolysis to convert AFEX-CS to fermentable sugars at 6% glucan loading. It will then take another 96-168h to ferment those sugars to ethanol. By simply integrating enzymatic hydrolysis and fermentation, which is SSCF (simultaneously saccharification and co-fermentation), it still requires at least 168h to reach comparable results as SHF. Here we will present a novel SHF process. In this process, the easily hydrolyzed part of biomass is hydrolyzed first to fermentable sugars in 24h. The hydrolysate was then separated from residue solids and fermented to ethanol in 24h using high yeast cell inoculation. The residual solids (difficultly hydrolyzed part of biomass (DHPB)), which bound enzymes, are then transferred to the next cycle of hydrolysis. This way, enzymes are largely reused and enzyme loadings in the next cycles can thus be reduced. The yeast cells after 24h fermentation are also recovered and reused for the next cycle of hydrolysate fermentation. The time for each cycle hydrolysis/fermentation is 24h/24h. After five cycles, the difficult hydrolyzed part is collected and enzymatic hydrolysis is performed without adding additional enzymes. Since enzymes are accumulated to a relatively high concentration and part of the decomposition products are removed with liquid during the five cycles, the hydrolysis of DHPB would be easier to achieve high sugar conversion. Profiles of enzyme recycled in each step were characterized.

See more of this Session: Integrated Processes for Biochemical Conversion of Renewable Feedstocks to Fuels and Chemicals I
See more of this Group/Topical: Sustainable Engineering Forum

(231b) Concentration of a Solids-Free Lignocellulose Hydrolysate for Very High Gravity Fermentation: Inhibition Effects

Steven J. Schneiderman, Todd J. Menkhaus and Patrick C. Gilcrease, Chemical and Biological Engineering, South Dakota School of Mines and Technology, Rapid City, SD

A major problem in the efficient conversion of lignocellulose hydrolysates to ethanol is low ethanol concentrations in the fermentation broth due to low initial sugar concentrations. This is undesirable because low ethanol concentrations lead to increased separation costs downstream. Distillation is energy intensive, and constitutes more than 50% of the total energy consumption of a typical ethanol plant. An obvious solution is to start with a high initial sugars concentration, which is referred to as a very high gravity (VHG) fermentation. By increasing the glucose concentration in the hydrolysate from 40 to 100 g/l, the distillation energy consumption would be reduced by 60%. In the first step of our integrated process, polyelectrolyte flocculants facilitate the efficient removal of biomass solids from a pine wood hydrolysate. Once the hydrolysate is solids-free, an evaporation or reverse osmosis step can be readily used to remove water and concentrate the fermentable sugars. VHG fermentations present some challenges of their own, including higher energy costs associated with concentrating the sugars, simultaneous concentration of some or all of the fermentation inhibitors, and osmotic stresses on the fermentation organism at higher substrate concentrations. Because some of the inhibition effects are synergistic, this could become more problematic at VHG conditions.

The goal of this study was to investigate the potential problems associated with fermenting a concentrated lignocellulose hydrolysate, and to determine operating conditions that maximize productivity and minimize costs for the combined sugars concentration-fermentation-distillation process. A batch culture design of experiments was used to quantify the inhibition effects of five common inhibitors on the productivity of our *S. cerevisiae* D_5A strain. The inhibitors studied were glucose, ethanol, furfural, HMF, and acetic acid at concentration ranges of 25 – 300 g/l, 0 – 150 g/l, 0 – 5 g/l, 0 – 5 g/l and 0 – 10 g/l, respectively. Our experimental results were used to regress kinetic model parameters, and define an acceptable operating window for VHG fermentation. The hydrolysate concentration step (evaporation/reverse osmosis) was then tailored to minimize total separation costs (sugars concentration and ethanol distillation) while maintaining a high level of ethanol productivity. Finally, the fermentation of a concentrated pine hydrolysate at optimum conditions was demonstrated at the bench scale.

See more of this Session: Integrated Processes for Biochemical Conversion of Renewable Feedstocks to Fuels and Chemicals I
See more of this Group/Topical: Sustainable Engineering Forum

(231c) Separate Processing of Hemicellulose and Cellulose to Optimize Ethanol Production From Corn Stover

Alexandre Chapeaux[1], Nancy Dowe[1] and Daniel J. Schell[2], (1)National Bioenergy Center, National Renewable Energy Laboratory, Golden, CO, (2)National Renewable Energy Laboratory, Golden, CO

Separate processing ofhemicellulose and cellulose to optimize ethanol production from corn stover

Alexandre Chapeaux, NancyDowe, Daniel J. Schell

Using a series of thermochemical (pretreatment)and biochemical steps, lignocellulosic biomass is converted to sugars that subsequentlycan be made into fuels and chemicals. One such process uses heat and dilute sulfuric acid to hydrolyzes the hemicellulosic fractionof the biomass to monomeric and oligomeric sugars.The removal of hemicellulose and to a lesser extent lignin from the biomass significantlyimproves the performance of the next step in the process, enzymatic hydrolysisof cellulose to glucose. However, enzymatic cellulose hydrolysis can beperformed in the presence or absence of the hemicellulosic sugars producedduring pretreatment. It is well known that removing the hemicellulose sugarsand other compounds produced during pretreatment improves enzymatic celluloseconversion yields. In previous work, we assessed the potential of a processthat removes the hemicellulosic sugars prior to enzymatic cellulose hydrolysis byexploring the interaction between biomass solids loading, yields, and processcosts. However, this process is more expensive because of the additional costfor solid-liquid separation equipment required to remove th-richsugar liquor from the cellulosic solids.

In this study, the goal was to deepen ourunderstanding of the advantages and disadvantages of the separatehemicellulose/cellulose process. The feedstock was dilute-acid-pretreated cornstover produced in a 200 kg dry biomass/d continuous horizontal reactor. Thepretreated slurry was separated into a liquor stream (hemicellulosic sugars)and a washed cellulosic solid stream. Enzymatic cellulose hydrolysis was investigatedat four enzyme loadings (10, 20, 30 and 40 mg protein/g cellulose) and atinsoluble solids concentrations from 9% to 21% (w/w). The resulting glucose wasfermented to ethanol using several yeasts and a glucose-xylose co-fermentingbacterium, *Zymomonas mobilis* 8b. The sugars in the liquorstream were fermented to ethanol with three strains of co-fermenting *Z. mobilis* after dilution of the streamwith water to lower sugar concentrations. Finally, the overall performance ofthe process was assessed after partitioning water between the two phases in away that optimizes overall ethanol yields. We then compared these results to aprocess configuration in which the streams are not separated. In the latterprocess, enzymatic cellulose hydrolysis is performed in the presence of thebackground hemicellulosic sugars produced during pretreatment of the biomass.

See more of this Session: Integrated Processes for Biochemical Conversion of Renewable Feedstocks to

(231d) Ethanol and Furfural Production From Corn Stover Using Hybrid Fractionation Process with Zinc Chloride and Simultaneous Saccharification and Fermentation (SSF)

Chang Geun Yoo[1], **Tae Hyun Kim**[1] and Monlin Kuo[2], (1)Agricultural and Biosystems Engineering, Iowa State University, Ames, IA, (2)Department of Natural Resources Ecology and Management, Iowa State University, Ames, IA

Two-stage hybrid fractionation process was investigated to produce cellulosic ethanol and furfural from corn stover. Zinc chloride ($ZnCl_2$) was used to selectively solubilize hemicellulose in the first stage. During the second stage, the remaining treated solids were converted into ethanol using commercial cellulase and *Saccharomyces cerevisiae* or recombinant *Escherichia coli*, KO11.

Hybrid fractionation process recovered >90% of glucan, ~90% of xylan, and ~75% of lignin at optimal reaction conditions: 1st stage: 5% acidified $ZnCl_2$, 7.5 ml/min, 150 °C (10 min) and 170 °C (10 min); 2nd stage: Simultaneous saccharification and fermentation (SSF). Furfural yield from hemicellulose hydrolysates was 58 %. SSF of $ZnCl_2$ treated solid with *Saccharomyces cerevisiae* resulted in 69-98 % of theoretical maximum ethanol yields based on glucan present in the treated solids. After fermentation, solid residues contained mostly lignin. Based on total lignin in untreated corn stover, the lignin recovery yield was ~75%.

See more of this Session: Integrated Processes for Biochemical Conversion of Renewable Feedstocks to Fuels and Chemicals I
See more of this Group/Topical: Sustainable Engineering Forum

(231e) Fischer Tropsch Synthesis Via Biomass Derived Synthesis Gas

Syed Ali Gardezi, Chemical Engineering, University of South Florida, Tampa, FL, Babu Joseph, Chemical and Biomedical Engineering, Clean Energy Research Center, University of South Florida, Tampa, FL and John T. Wolan, University of South Florida, Tampa, FL

Due to the recent surge in energy costs and uncertainties in fuel prices, there is significant interest in seeking alternative sources of energy particularly from renewable such as biomass. Indirect liquefaction of biomass (via the thermo-chemical conversion to syngas followed by liquefaction via Fischer-Tropsch Synthesis or FTS) offers a commercially viable route for meeting the challenge of producing renewable fungible liquid fuels. Apart from biomass, FTS is also the critical step for converting natural gas and coal to liquid fuels.

A FTS technology has been developed for producing middle distillates (diesel and aviation fuel) by using pine chips as the main feedstock. In this "two step" process, pine chips were subjected to an entrained flow gasification process developed by Pearson et al. [1]. The resulting tar free gas is comprised mainly of carbon monoxide, hydrogen, carbon dioxide, methane and some organic contaminants such as benzene, toluene, and naphthalene along with substantial amount of water. For further liquefaction this gas was cleansed using multiple adsorbents. An inline "Shaw Moisture Meter" was used to continuously monitor the moisture content of the feed to the liquefaction process.

For the production of liquid fuels, eggshell cobalt catalysts were used. Cobalt catalysts have several advantages e.g. they produce linear paraffinic hydrocarbons and minimize undesirable water gas shift reaction. This eggshell design is a well known strategy for producing a narrow hydrocarbon distribution. In this regard previous research work conducted by Iglesia et al. was also consulted [2]. A bench scale fixed bed reactor was used for the conversion of syngas. Modeling of the reactor startup and of adjustment of the reactor packing ensured successful operation resulting in a high conversion and narrow distribution of hydrocarbons in the range of diesel and aviation fuel.

References :

[1] Pearson et al., U.S. Patent 4,857,076, issued Aug 15, 1989.

[2] E. Iglesia, S. L. Soled, J. E. Baumgartner, S. C. Reyes, *J. Catal.* **153**, 108 (1995).

See more of this Session: Integrated Processes for Biochemical Conversion of Renewable Feedstocks to Fuels and Chemicals I
See more of this Group/Topical: Sustainable Engineering Forum

(231f) Integrations of Biosurfactant Production Into Advanced Biorefineries

William Colonna[1], **Mustafa E. Marti**[1], Michelle Pynn[2], Gabriel Reznik[2], Kevin Jarrell[2], Buddhi Lamsal[3] and Charles E. Glatz[1], (1)Chemical and Biological Engineering, Iowa State University, Ames, IA, (2)Modular Genetics, Inc., Woburn, MA, (3)Food Science and Human Nutrition, Iowa State University, Ames, IA

Several classes of biosurfactants are produced by microorganisms. One such surfactant is FA-Glu, which is related to the natural biosurfactant, surfactin. FA-Glu consists of a long-chain β-hydroxy fatty acid coupled to glutamic acid. It is produced by a strain of *Bacillus subtilis* genetically-engineered by Modular, Inc. It has several potential applications including use as a dispersant for oil spills, where biodegradability is desirable.

Economic production and eventual application will be favored if substrate costs can be reduced and low-cost recovery methods developed. Several such opportunities arise in advanced biorefineries. Product value would be further enhanced if such production could achieve favored "green" rating by avoiding solvents other than water and simple alcohols and effect emulsification and low toxicity could be demonstrated. This report focus on production considerations.

Production of FA-Glu by fermentation on glucose is compared to results using lower cost substrates available in advanced biorefineries, including fiber components of soy and glycerol. FA-Glu yields from soybean hulls (SBH) ranged from 29-51% of that from glucose on a substrate weight basis.. When SBH polysaccharides were hydrolyzed (~65%) by pre-treatment with carbohydrases, FA-Glu levels were only ~21-37% of those from glucose. The cells consumed the enzymatically-released sugars; however, very little of this was converted into FA-Glu. Addition of glucose to the enzymatic SBH hydrolysate improved FA-Glu yields, but only slightly. The low FA-Glu yields from SBH could be due to inhibitors released from the hulls by the enzymes or inhibitory substances in the enzyme preparations.

Glycerol, a by-product of biodiesel production, proved to be a suitable carbon source for FA-Glu production. Cells grown on 2% w/v glycerol produced ~25% more FA-Glu than cells grown on 2% glucose. Glycerol is abundant and low in cost. Its use as a feedstock for production of microbial biosurfactants would provide a means for efficient utilization of soy by-products and improving the ag economy.

The FA-Glu was purified by collecting the foam from the aerated broth, clarification by centrifugation or filtration, and adsorption and elution from an anion exchange column to a purity of 97.9%. Final product color can be improved by liquid-liquid extraction, but this may compromixe "green" status. A precipitation based alternative process is being examined as a means of eliminating solvent use.

(251a) Structure and Application of Nucleic Acid-Surfactant Films for Transfection In Human Cells

S.L. Perry[1], Surekha Gajria[2], T. Neumann[1], J. Ni[1], J. Weinstein[1], D.V. Schaffer[1] and M.V. Tirrell[1], (1) Department of Bioengineering, University of California at Berkeley, Berkeley, CA, (2)Chemistry and Biochemistry, University of California, Santa Barbara, Santa Barbara, CA

Cationic surfactants have tremendous potential as non-viral transfection reagents. Although studies to date have not been able to achieve the levels of transfection attainable using viral vectors, they have significant advantages including ease of preparation and a lack of immunogenic and oncogenic concerns. Nucleic acid (NA)-surfactant complexes (lipoplexes) form via electrostatic interactions between the negatively charged backbone of NAs and the cationic surfactant headgroup. While the majority of transfection studies have focused on the use of lipoplexes in solution, we present here the use of solid NA-surfactant films for localized transfection. This method has the potential to enable both higher transfection efficiencies compared to solution-based methods and allows for the controlled release of transfection agents via the use of a layered film structure.

Characterization of both lipoplexes and NA-surfactant films has demonstrated that a rich variety of self assembled structures can be achieved; including lamellar bilayers, inverted hexagonal rods, and bicontinuous cubic geometries. In the case of DNA-surfactant films, however, the DNA has been observed to undergo a fast and reversible transition from double-stranded DNA to single-stranded DNA, coupled with a corresponding rearrangement of the surfactant layers upon drying. In films composed of DNA-dimethyldidodecylammonium bromide (DDAB) which forms a lamellar structure, the change from double-stranded to single-stranded DNA is accompanied by a change in the surfactant layers, transitioning from a bilayer structure to a monolayer structure. The particular advantage of the change from double-stranded to single-stranded NA is that it allows for long-term storage of dry films in a biologically inactive state while also imparting an improved resistance against nuclease degradation. [1-3] When needed, these films can then be made biologically active once more by simple rehydration.

In addition to benefits associated with long-term storage, solid films have the potential to increase transfection efficiency by localizing the transfection agents (i.e. surfactant complexed DNA/siRNA) close to the cell surface. However, improvements in transfection efficiency due to increased local concentrations of the transfection agents must be balanced with the potential for commensurate increases in cytotoxicity from the complexing surfactant. Preliminary results have indicated the potential for very high levels of GFP transfection in HeLa cells (>90% cells transfected). Further experiments to validate these results, examine the delivery of siRNA for gene silencing, and extend the utility of these films to enable the controlled and/or sequential release of transfection agents are ongoing.

References:

(1) Neumann, T.; Gajria, S.; Bouxsein, N. F.; Jaeger, L.; Tirrell, M. *J Am Chem Soc* **2010**, *132*, 7025.

(2) Neumann, T.; Gajria, S.; Tirrell, M.; Jaeger, L. *J Am Chem Soc* **2009**, *131*, 3440.

(3) Smitthipong, W.; Neumann, T.; Gajria, S.; Li, Y.; Chworos, A.; Jaeger, L.; Tirrell, M. *Biomacromolecules* **2009**, *10*, 221.

See more of this Session: Nucleic Acid Delivery I
See more of this Group/Topical: Food, Pharmaceutical & Bioengineering Division

(251b) Analysis of siRNA Transport and Delivery within 3-Dimensional Collagen Matrices

Amanda P. Malefyt[1], Elizabeth Hinds[1], Arul Jayaraman[2], Catherine Kuo[3], Kyongbum Lee[3], Christina Chan[1] and S. Patrick Walton[1], (1)Chemical Engineering and Materials Science, Michigan State University, East Lansing, MI, (2)Texas A&M University, (3)Tufts University

Laboratory research studies involving nucleic acid therapies, such as delivery of short, interfering RNAs (siRNAs) for RNA interference (RNAi), are typically performed on two-dimensional (2D) cell cultures. However, this is not an ideal model of the three-dimensional (3D) *in vivo* environment. As such, we are seeking to optimize transfection of siRNAs in 3D, specifically to cells cultured in collagen gels. Commercial transfection reagents with proven efficacy in 2D siRNA transfections have difficulty silencing cells suspended within hydrogel matrices, primarily due to interactions between the transfection complexes and the matrix.

In this study, we used multicolor fluorescence confocal microscopy and flow cytometry to analyze the cellular uptake and efficiencies of siRNAs delivered through 3D collagen matrices. Delivery was compared using Lipofectamine 2000 (LF2K) and 3-D FectIN, a transfection reagent developed specifically for hydrogel delivery. Although our results showed siRNA uptake under multiple conditions, effective silencing has only been achieved through using 3-D FectIN for delivery during gel formation. Additionally, treatments after gel formation demonstrate significant siRNA accumulation at the top of the matrix with minimal penetration throughout the gel, even at longer transfection times (e.g., 72 h). We will discuss our recent work explaining this phenomenon as well as other methods to aid in delivery throughout the collagen scaffolds.

See more of this Session: Nucleic Acid Delivery I
See more of this Group/Topical: Food, Pharmaceutical & Bioengineering Division

(251c) Design, Synthesis and Evaluation of Poly (amino ethers) for Efficient Transgene Delivery

Thrimoorthy Potta[1], Lucas Vu[2] and Kaushal Rege[2], (1)Chemical Engineering, Arizona State University, Tempe, AZ, (2)Chemical Engineering Department, Arizona state university, Tempe, AZ

Design, Synthesis and Evaluation of Poly (amino ethers) for efficient Transgene Delivery

Thrimoorthy Potta, Lucas Vu, Kaushal Rege

Chemical Engineering

Arizona State University

Tempe, AZ 85287-6106

A combinatorial library of novel branched cationic poly(aminoethers) was synthesized based on the ring opening polymerization between diglycidyl ethers and poly(amines) with an eye towards rapidly identifying candidates for transgene delivery. Parallel screening of the resulting polymers led to identification of several leads that demonstrated higher transgene expression efficacies and lower cytotoxicities compared to branched pEI (25 kDa) in different cancer cell lines. Polymers were characterized using ^{1}H NMR, FTIR and GPC analyses. The role of polymer and polyplex physicochemical properties, including pH-buffering capability, DNA binding efficacy, polyplex size and zeta potential, on transgene expression efficacies was investigated. Lead polymers were interfaced with chemotherapeutic drugs (e.g. histone deacetylases inhibitors) in order to further enhance their transgene expression efficacies. Furthermore, polyaminoethers were also employed to enhance virus-mediated transgene expression in several cancer cell lines. These studies indicate that branched cationic poly (aminoethers) are promising candidates for transgene expression using by themselves as non-viral vectors and as polymer-virus hybrids.

See more of this Session: Nucleic Acid Delivery I
See more of this Group/Topical: Food, Pharmaceutical & Bioengineering Division

(251d) Microarray Analysis of Gene Expression Profiles Characteristic of Successfully Transfected Cells

Timothy M. Martin[1], Sarah A. Plautz[1], Jean-Jack M. Riethoven[2] and Angela K. Pannier[1], (1) Biological Systems Engineering, University of Nebraska-Lincoln, Lincoln, NE, (2)Bioinformatics Core Research Facility, University of Nebraska-Lincoln, Lincoln, NE

The design of efficient nonviral gene delivery systems offers potential uses in therapeutic applications such as tissue engineering, biomaterials, and gene therapy. Nonviral gene delivery involves the delivery of exogenous gene(s) to cells, facilitated by the electrostatic complexation of the gene (as plasmid DNA) with a nonviral vector such as a cationic polymer or lipid. Unlike viral based delivery systems, nonviral gene delivery systems can carry a larger genetic payload, overcome several safety concerns, and are easily modified to allow for cellular targeting. However, current nonviral gene delivery systems are inefficient, prohibiting their use in therapeutic applications. Much emphasis has been placed on increasing transfection efficiency through the physiochemical modification of the delivery vector. While this empirical approach has resulted in some improvements, transfection efficiency has not been increased to levels desired. Our research focuses on enhancing transfection through the identification and exploitation of endogenous intracellular mechanisms characteristic of successfully transfected cells. We transfected HEK293T cells with a plasmid encoding EGFP and luciferase complexed with 25 kDa branched polyethyleneimine (PEI) and then separated successfully transfected cells (those expressing the green fluorescent protein, GFP+) from untransfected cells (those not expressing the green fluorescent protein, GFP-) using flow cytometry. Subsequently, cDNA microarrays were used to identify the endogenous gene profile differences among GFP+, GFP-, and control cells using traditional cutoff limits of expression levels (greater than two fold difference in expression between data sets and False Discovery Rate (FDR) adjusted p-value less than 0.05). Comparing gene profiles between control cells and GFP- cells revealed nine coding genes to be differentially expressed, including RAP1A, HIST1H2AM, KRTAP19-1, HIST1H3H, HIST1H2AC, HIST2H4A, HIST1H2BD, SLC16A6, and EGR1 . Comparing gene profiles between GFP+ cells and control cells revealed 405 coding genes to be differentially expressed. Finally, comparing gene profiles between GFP+ cells and GFP- cells revealed 113 coding genes to be differentially expressed. To better understand why one cell transfects (GFP+) while another cell does not (GFP-), as both groups of cells are treated identically in the same population, these 113 genes were further evaluated using Exploratory Gene Association Networks (EGAN) analysis, which revealed the majority (40.7 %) of the genes to be implicated in highly enriched Gene Ontology (GO) Processes of apoptosis, cell cycle signaling, cell response to stress, cytoskeletal signaling, DNA/ RNA maintenance, motility, and transport. Among the 113 genes, eight key genes (RAP1A, ACRC, ATF3, SCG5, ACTA1, MYH3, DDIT3, and PLK2) exhibited three-fold greater expression in GFP+ versus GFP- cells and were implicated in the identified GO pathways and five key genes (CHORDC1, WDR78, IREB2, PGAP1, NEB) were highly expressed in GFP+ versus GFP- cells (greater than fivefold upregulated). Together, these 13 key genes were then further studied for their role in transfection using chemical and molecular activators and inhibitors of each gene pathway. The concomitant elucidation of intracellular signaling pathways implicated in successful nonviral gene delivery combined with alteration of the endogenous gene profile as an adjuvant for delivery offers a novel approach to increase gene transfer for therapeutic applications.

See more of this Session: Nucleic Acid Delivery I
See more of this Group/Topical: Food, Pharmaceutical & Bioengineering Division

(251e) Dynamics of DNA Release From Polyplexes Induced by Inter-Polyelectrolyte Exchange

Yi Zou and Guangzhao Mao, Chemical Engineering and Materials Science, Wayne State University, Detroit, MI

Polymeric gene delivery vectors are attractive alternatives to viral carriers because of their lower safety risk, higher gene carrying capacity, and more targeting cell types. A major problem of polymeric gene carriers is their short circulation half-life. Polymeric gene delivery carriers are unstable in extracellular environment, which results in low transfection efficiency. We carried out a mechanistic study of DNA release by polyelectrolyte exchange using heparin and serum. The study was carried out at the single polyplex level using real-time atomic force microscopy (AFM) imaging. Bioreducible hyperbranched poly(amido amine)s (PAA)s and modified bioreducible PAAs were studied. The AFM results are correlated with fluorescence data as well as cell transfection data. A critical heparin concentration above which significant DNA release occurs was determined. The critical concentration was found to be a function of polycation charge density, chain architecture, and amine to phosphate (N/P) ratio. AFM captured morphological release pathway with distinct polyplex structures. These characteristic morphologies include core-shell, toroid, nanoparticle decorated toroid, and loose chains held by a compact core structure. In addition, PEG and hyaluronic acid modification were shown to be effective against inter-polyelectrolyte exchange and improve gene delivery efficiency. This study contributes to basic understanding of DNA release dynamics from single polyplex particles in simulated physiologic conditions and provides synthetic strategies to improve gene delivery efficiency of polymeric vectors.

See more of this Session: Nucleic Acid Delivery I
See more of this Group/Topical: Food, Pharmaceutical & Bioengineering Division

(251f) Interplay of Matrix Rigidity and Cell Type for Non-Viral Gene Delivery

Cathy Chu, Chemical and Biomolecular Engineering, University of Illinois at Urbana-Champaign, Urbana, IL and Hyunjoon Kong, Department of Chemical and Biomolecular Engineering, University of Illinois Urbana-Champaign, Urbana, IL

Non-viral gene delivery has the potential to treat a wide array of ailments through the transfer of genetic material. Towards this end, extensive efforts have been made to synthesize vectors for plasmid DNA delivery in order to improve the uptake and expression of delivered genes. Unfortunately, non-viral gene delivery has been hampered by limited expression *in vivo* following injection to target tissues, possibly due to a limited understanding of the interactions between target cells and their extracellular microenvironment. Recent studies have reported that the spacing of cell adhesion ligands and stiffness presented by a synthetic extracellular matrix regulate non-viral gene delivery for pre-osteoblasts.[1,2] However, it is unclear whether such influence from the extracellular microenvironment is universally valid to various cell populations that reside within a target tissue.

This study examines the role of matrix rigidity on gene delivery as mediated by cell type. In this study, poly(ethylene glycol) diacrylate (PEGDA) hydrogels of varying stiffness modified with fibronectin were used as synthetic extracellular matrices. The elastic moduli of the hydrogels were varied over a wide range to mimic both soft and hard tissues. Plasmid DNA (pDNA) encoding bone morphogenetic protein (BMP)-2 was delivered to NIH3T3 fibroblasts, D1 bone marrow stromal cells, and C2C12 myoblasts cultured on hydrogels of varying stiffness.

Both the cellular uptake of pDNA and the resulting gene expression were found to increase with increasing matrix stiffness. Gene expression from fibroblasts exhibited the greatest dependency on matrix stiffness as compared with that from bone marrow stromal cells and myoblasts. The effects of matrix stiffness and cell type on non-viral gene delivery were further related to cellular proliferation and nuclear morphology. Overall, the results of this study may be useful for the development of gene delivery strategies and the design of tissue engineering scaffolds for gene-based tissue regeneration therapies.

1. Kong, H.J., et al. *Nat Mater* 2005, 4, 460 – 464.

2. Kong, H.J., et al. *Nano Lett* 2007, 7, 161 – 166.

See more of this Session: Nucleic Acid Delivery I
See more of this Group/Topical: Food, Pharmaceutical & Bioengineering Division

(251g) Gene Activated Matrix Transfection of Human Mesenchymal Stem Cells by Chitosan-DNA Nanoparticles In Chitosan-Based Scaffolds

Kimberly T. Hunter and Teng Ma, Chemical and Biomedical Engineering, Florida State University, Tallahassee, FL

Human mesenchymal stem cells (hMSCs) are a potent osteogenic cell source, but hMSC based bone constructs require a scaffold that provides both mechanical stability and the capability to promote hMSC proliferation and to direct osteogenic differentiation. Gene activated matrices (GAMs) create a singular environment where hMSCs can adhere, proliferate, be continuously transfected with plasmid, and exhibit sustained protein expression. We are investigating a scaffold that incorporates BMP2 plasmid DNA for sustained delivery. Naked BMP2 pDNA or chitosan-BMP2 pDNA nanoparticles inside a bulk material of hydroxyapatite-chitosan-gelatin (HCG) hydrogel will serve as the GAMs. This composite material has many advantages, including similarity to natural bone ECM, biocompatibility and long term protection of plasmid DNA from nucleases.

Our previous studies have demonstrated that HCG hydrogel absorbs a mediating ECM protein layer that supports hMSC attachment and proliferation, and possesses intrinsic osteoinductive properties (1). In the current study, hMSC transfection in the HCG GAM is achieved by directly incorporating BMP2-GFP pDNA and by incorporating chitosan encapsulated BMP2-GFP pDNA. This fusion vector provides both the functional BMP2 gene that promotes osteogenic differentiation and the GFP reporter gene that allows direct evaluation of transfection efficiency. Using bolus administration of chitosan-DNA nanoparticles as controls, hMSCs exhibit a dose dependent transfection to bolus chitosan-DNA nanoparticles. Confocal analysis revealed uniform distribution of chitosan-pDNA nanoparticles in the HCG scaffolds with comparable bulk morphology to the HCG hydrogel controls. hMSC exhibited extensive proliferation in both the HCG GAMs and control during first three weeks but the rate decreased in the GAMs from weeks 3 to 5. Protein expression kinetics revealed that HCG hydrogel matrix is continuously enriched with bone marker proteins throughout the culture period and that this enrichment is more extensive in the GAM hydrogels. PCR results confirm expression of the GFP reporter protein after 35 days in culture, indicating the sustained delivery of DNA from the HCG hydrogel GAM. Ongoing experiments focus on elucidating hMSC's dose-dependent response to bolus transfection kinetics, and evaluating the impact of nanoparticle encapsulation on the scaffold mediated hMSC transfection.

Bibliography

1. Zhao F, Grayson W, Ma T, Bunnell B, Lu W. *Effects of hydroxyapatite in 3-D chitosan gelatin polymer network on human mesenchymal stem cell construct development.* Biomaterials, 2006, Vol. 27,

pp. 1859-1867.

See more of this Session: Nucleic Acid Delivery I
See more of this Group/Topical: Food, Pharmaceutical & Bioengineering Division

(332a) FDA Perspective On QbD for Analytical Methods

Elaine Morefield, DHHS/FDA/Cder/OPS/ONDQA, FDA, Silver Spring, MD

QbD is a systematic approach to pharmaceutical development and product lifecyclemanagement. QbD provides new opportunities and modern tools to utilize in analytical research and development. ICH guidelines and implementation strategies focused mainly on formulation and manufacturing design and development. However, many of the QbD approaches and tools are applicable to analytical methods as well. The linkage between QbD and analytical method development and validation needs further development. Inclusion of enhanced pharmaceutical development information, including analytical method development and validation studies, in the submissions will provide the applicants the opportunity for flexible regulatory approaches, recognizing however, the degree of regulatory flexibility is predicated on the level of demonstrated product, analytical method and manufacturing process understanding. A summary of the progress made to date and the continued challenges facing the industry and FDA will be presented.

See more of this Session: QbD In Analytical Development
See more of this Group/Topical: Topical I: Comprehensive Quality by Design in Pharmaceutical Development and Manufacture

(332b) Co-Operative Implementation of Quality-by-Design Based Experimentation Between Product Development and Analytical R&D Laboratories - A Recipe for Success!

Graham Shelver, S-Matrix Corporation, Concord, CA

Abstract to be added

See more of this Session: QbD In Analytical Development
See more of this Group/Topical: Topical I: Comprehensive Quality by Design in Pharmaceutical Development and Manufacture

(332c) Quality by Design for Analytical Methods In Early Development

Jackson Pellett, Small Molecule Analytical Chemistry & Quality Control, Genentech Inc, South San Francisco, CA

In Phase 1 and 2 a high attrition rate and continuously changing targets can make spending resources of QbD activities seem very unattractive. However, applying many of the QbD principles early in the lifecycle can payoff with decreased method development time, fewer changes required due to poor method performance and a better understanding of method capability and risks. Column screening systems and modeling tools are often used in early development to identify and optimize separation conditions. Other tools such as risk assessments and method performance monitoring are less commonly used. Examples of early analytical development QbD approaches will be presented.

See more of this Session: QbD In Analytical Development
See more of this Group/Topical: Topical I: Comprehensive Quality by Design in Pharmaceutical Development and Manufacture

(332d) Examples of Applying Quality by Design Principles to Analytical Methods

Kimber Barnett, Analytical Research & Development, Pfizer Global Research & Development, Groton, CT, Timothy Graul, Analytical R&D US, Pfizer Global Research & Development, Groton, CT and Melissa Hanna-Brown, Dev Analy & ICH Stability, Pfizer Inc, Sandwich, Kent, United Kingdom

The application of Quality by Design principles to analytical method development practices is a science and risk based approach that begins with the identification of method requirements. Once method requirements have been identified, an analytical technique is selected based on these requirements. A risk assessment is then conducted to identify parameters that could potentially affect method performance. Subsequent experiments are conducted to develop a more complete understanding and control of sources of variability. Ultimately, this leads to the implementation of a robust measurement system that can support process evaluation and improvement throughout the product lifecycle. Examples that illustrate these concepts will be presented.

See more of this Session: QbD In Analytical Development
See more of this Group/Topical: Topical I: Comprehensive Quality by Design in Pharmaceutical Development and Manufacture

(332e) Quality by Design for Particle Size Analysis

Zhigang Sun, Office of Generic Drugs, OPS, CDER, FDA, Rockville, MD

Particle size analysis of pharmaceutical powders or dispersions is of great importance due to the potentially significant impact of particle size on safety, efficacy and quality of the drug product. The development of an appropriate particle sizing method is fundamental to establishing particle size specifications, which may be critical for control of drug product performance as well as ensuring manufacturing consistency. This presentation will describe a quality by design (QbD) approach to development of particle sizing methods. While QbD primarily focuses on product design and process development, it can be extended to development of analytical methods as well. In this approach, several QbD strategies (e.g., analytical target profile, method selection, risk assessment, control strategy, etc.) have been applied during the method development to ensure that the particle sizing method is rugged and robust and provides optimum performance. As a result, this more systematic approach could facilitate continual improvement throughout the method's lifecycle of use.

See more of this Session: QbD In Analytical Development
See more of this Group/Topical: Topical I: Comprehensive Quality by Design in Pharmaceutical Development and Manufacture

(332f) Setting CQA Acceptance Criteria for a Quality by Design Approach to Development

Paul Motchnik, Protein Analytical Chemistry, Genentech, Inc, South San Francisco, CA

To be added

See more of this Session: QbD In Analytical Development
See more of this Group/Topical: Topical I: Comprehensive Quality by Design in Pharmaceutical Development and Manufacture

(336a) High-Throughput Evolutionary Substrate Screening for Embryonic Stem Cell Attachment and Long-Term Maintenance

Michael R. Zonca Jr.[1], Philip S. Yune[2], Georges Belfort[3] and Yubing Xie[1], (1)College of Nanoscale Science and Engineering, University at Albany, State University of New York, Albany, NY, (2)Howard P. Isermann Dept of Chemical & Biological Engineering and The Center for Biotechnology & Interdisciplinary Studies, Rensselaer Polytechnic Institute, Troy, NY, (3)Department of Chemical and Biological Engineering, Rensselaer Polytechnic Institute, Troy, NY

Due to their self-renewal capacity and differentiation potential, embryonic stem (ES) cells have great potential in the fields of drug discovery, tissue regeneration and cell therapy. To achieve the full potential of these cells, strategies must be implemented to maintain them in a pluripotent state. Cell attachment is essential for maintaining ES cells in an undifferentiated state and preventing them from forming unwanted embryoid bodies. Fully defined, synthetic substrates are an ideal candidate for the development of a simple, broadly applicable, uniform culture system for pluripotent stem cells. The use of polymeric materials as scaffolds has been well documented and offers several advantages for supporting ES cell maintenance. However, little is known about how surface chemistry influences stem cell pluripotency.

In this work, we identify a defined substrate from a comprehensive library of synthetic chemically modified surfaces. It supports long-term cell attachment of mouse ES cells and sustains their pluripotency. Using an established high-throughput synthesis and screening platform, we have been able to create a library of 66 synthetic chemical surfaces in three 96-well plates. Mouse ES cells were pre-stained with Cell-Tracker Green and seeded into 96-well plates containing the library of 66 synthetic chemical surfaces. The cell attachment on these chemically modified surfaces was measured by the fluorescence intensity, and a cell attachment index was calculated by normalizing the fluorescence intensity with that of a control group. The surface chemistry with the highest attachment index was our new surface and the lowest attachment was with polyethylene glycol. The cell attachment on the "hit" surface chemistry was further confirmed by scanning electron microscopy (SEM) observation. The pluripotency of ES cells grown on the "hit" surface was confirmed by retaining pluripotency markers of stem cells, Oct4 and stage-specific embryonic antigen (SSEA-1), the ability to form embryoid bodies, and the maintenance of differentiation potential. From our initial results, we selected 11 representative monomers and further investigated cell attachment using two different cell lines. One surface from this set exhibited the highest cell attachment for both cell lines. We also investigated the potential to use this same chemistry for long-term maintenance of ES cells. The mechanism of ES cell attachment to the "hit" surface will be discussed. This monomer-based, chemically defined, scalable, sustainable and controllable, polymeric substrate provides a new avenue for understanding and manipulating surface chemistry for stem cell expansion and differentiation.

(336b) Bioactive Peptides for Mesenchymal Stem Cell Differentiation In Cartilage Tissue Engineering

Julie N. Renner, Yeji Kim and Julie C. Liu, School of Chemical Engineering, Purdue University, West Lafayette, IN

Osteoarthritis is characterized by degradation of articular cartilage, and there is strong demand for long-lasting and minimally invasive cartilage replacements. One promising approach for cartilage regeneration is to use the patient's own bone marrow-derived stem cells (MSCs) to seed an artificial protein matrix embedded with bioactive differentiation cues. This method could allow MSCs embedded in the matrix to be directly implanted, which would reduce ex vivo culturing time. In addition, this technique could lead to the development of materials with spatially patterned differentiation cues, which would aid in the development of complex tissues such as osteochondral grafts.

We are developing an artificial protein matrix that directs human MSC (hMSC) differentiation and maintains the desired chondrogenic phenotype based on embedded biochemical cues. Short peptide sequence candidates were screened and characterized for their effect on hMSC differentiation to cartilage. To evaluate the efficacy of agonist peptides derived from growth factors, cells were cultured utilizing a modified pellet culture in a 96-well plate.

The most promising of the screened peptides is derived from the knuckle epitope of bone morphogenetic protein-2 (BMP-2). To study the effect of concentration, the BMP-2 peptide was added to medium at a range from 0.02 to 200 μg/mL. After two weeks of culture, we found that glycosaminoglycan (GAG) production of hMSCs cultured with >100 ug/mL of the peptide was 74% of the positive control (cells cultured with 200 ng/mL BMP-2 growth factor). Further characterizations were performed with the peptide at 100 μg/mL and included histology, measurements of matrix production over time, and gene expression. From one to four weeks, the peptide promoted GAG production at levels comparable to the BMP-2 positive control. The peptide had little effect on the production of collagen but did not increase the alkaline phosphatase (AP) activity over time, which suggests that it does not promote hypertrophy. In addition, the BMP peptide did not enhance GAG production in TGF-β3 mediated chondrogenesis, whereas BMP-2 does, suggesting a different mechanism for the bioactivity. Finally, the peptide promoted more homogenous distribution of extracellular matrix in the pellets compared to BMP-2, making the peptide an interesting new tool in hMSC differentiation to cartilage.

See more of this Session: Stem Cells In Tissue Engineering II
See more of this Group/Topical: Food, Pharmaceutical & Bioengineering Division

(336c) An Injectable, Self-Healing Protein Hydrogel for Stem Cell Delivery

Kyle J. Lampe[1], Brian Aguado[2], Andreina Parisi-Amon[2], Widya Mulyasasmita[3], James Su[2] and Sarah C. Heilshorn[1], (1)Materials Science and Engineering, Stanford University, Stanford, CA, (2)Stanford University, Stanford, CA, (3)Bioengineering, Stanford University, Stanford, CA

Stem cells are a key focus in tissue engineering strategies for clinical applications. However, the ability to inject and localize them to a target site while maintaining predictably high cell viability is critically lacking. We have shown that the viscoelastic properties of an injection matrix in part determine the relative viability of a variety of different stem cells after injection. When suspended in PBS, the simple act of injection through a 28 gauge syringe needle leads to compromised cell viability. When cells are instead suspended within a shear-thinning alginate hydrogel, ejected cell viability of human umbilical vein endothelials cells (HUVECs), human adipocyte stem cells (hASCs), rat mesenchymal stem cells (rMSC), and murine neural stem cells (mNSC) can be improved by 10-30%.

Common shear-thinning, reversibly gelling matrices such as collagen and Matrigel utilize non-physiological environmental triggers (e.g., pH and temperature shifts) detrimental to cells. We have recently developed a mixture-induced two component hydrogel (MITCH) that undergoes gelation under constant physiological conditions. The two components are both recombinantly-engineered protein polymers that include either the WW domain or its proline-rich ligand. These two domains hetero-assemble upon mixing through simple hydrogen bonding. Microrheology experiments confirm gelation within one minute after mixing to produce moduli ranging from ~10 Pa to ~50 Pa, dependent on the choice of molecular recognition binding partners. These gels undergo rapid shear-thinning and can re-form within 5 minutes, exemplifying their self-healing nature. This physically gelling material allows biocompatible cell encapsulation; for example, hASC viability immediately post-encapsulation is 97 ± 2%. MITCH mechanical properties are also suitable for neural encapsulation, and mNSCs encapsulated within MITCH remain highly viable and proliferative. Encapsulated and differentiated mNSCs stain positive for Map2 and GFAP, indicating neuronal and glial cells, respectively. These cells adopt 3D-branched morphologies with elongated dendrites extending over 100 μ m.

Here we show that the viscosity of an injection material strongly influences the viability of cells ejected through a clinically relevant syringe needle. Additionally, MITCH materials promote the differentiation of neural progenitors into neuronal phenotypes, which adopt a 3D-branched morphology within the hydrogels. The modular hetero-assembly of MITCH materials also permits the tethering of neuroprotective peptides to the gel, facilitating further optimization for the prevention and treatment of central nervous system injuries.

See more of this Session: Stem Cells In Tissue Engineering II

(336d) Decellularization of Tissue Engineered Leaflets with Adult Stem Cell Recellularization

Zeeshan Syedain, Biomedical/Chemical Engineering, University of Minnesota, Minneapolis, MN, Allison Bradee, Biomedical Engineering, University of Minnesota, Minneapolis, MN and Robert T. Tranquillo, Biomedical Engineering and Chemical Engineering & Materials Science, University of Minnesota, Minneapolis, MN

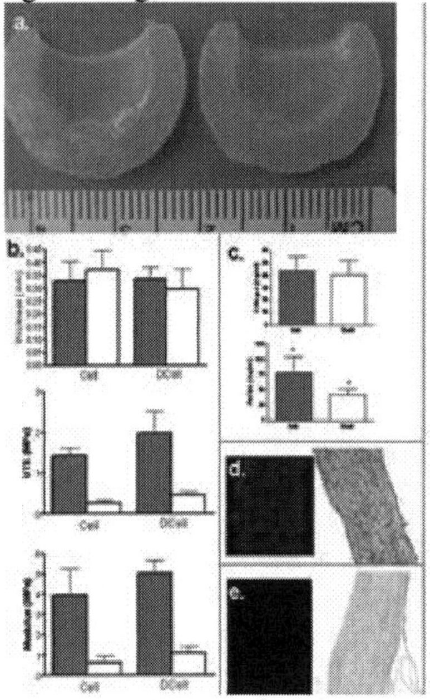

Fig 1: Image of leaflet (a.) and thickness, UTS and modulus of circumferential (black) and radial (white) strip of tissue before and after decellularization (b.). Collagen and protein concentrations (c.), and DAPI and trichrome stains of control (d.) and decell leaflet (e.).

Introduction: Tissue engineered heart valves(TEHV) can revolutionize valve replacements, especially for juvenile patients.While much progress has been made toward implanting TEHV, a major limitation todate has been retraction of the leaflets by the cells transplanted with theTEHV, especially those developed from fibrin (Syedain et al. 2011, Flanagan etal. 2009). Here, we present a novel approach of decellularizingan engineered leaflet made from fibrin that is remodeled by dermal fibroblast,and subsequently recellularized with adult human mesenchymal stem cells (ahMSC).The approach has potential to deliver off-the-shelf engineered leaflet that canbe recellularized with autologous stem cells.

Materialand Methods: Fibrin-basedengineered leaflets were cast with 4mg/ml fibrin gel and 1 million/ml neonatalhuman dermal fibroblast (nhDF). For decellularization, leaflets were incubated in 1% SDSover-night followed by 30 min rinse in 1% triton-X. Leaflets were characterizedfor mechanical and biochemical properties. For recellularization,leaflets were seeded with ahMSC. Reseeded sampleswere evaluated for cell proliferation and invasion up to 3 weeks.

Resultand Discussion: Engineeredleaflets had thickness, mechanical properties (UTS and Modulus), and anisotropysimilar to native sheep pulmonary valve leaflets (the standard large animalmodel). No change in mechanical properties was seen after decellularization(Fig.1). Decellularization removed 99% of DNA with a drop in total protein(indicative of cellular protein loss) but no change in collagen concentration. Recellularized leaflets showed MSC proliferation over timewith gradual invasion of MSC into the leaflet tissue (Fig.2). The mechanical propertiesafter 3 weeks incubation with MSC showed no detrimental effects of MSC seedingand subsequent culture, and there was no evidence of tissue contraction (nochange in thickness). MSC maintained positive staining for CD29, CD44, andCD90, and negative staining for CD45 and CD34 after 3 weeks.

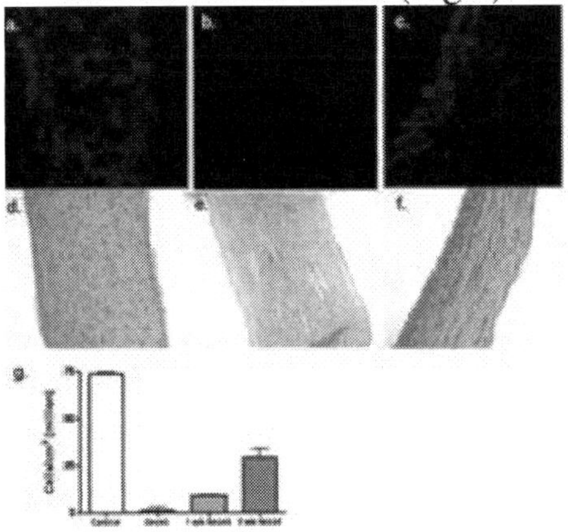

Conclusion: We present herean approach to remove the tissue-forming but also tissue-contractingfibroblasts from TEHV, thereby creating a readily available biological scaffoldfor reseeding with autologous stem cells. Preliminary work with recellularization shows good penetration and increased celldensity over 3 week without loss in tensile properties of the de/recellularized leaflets.

References:

Syedain et al. Cardio Vas EngrTech 2011: *in press* Flanagan et al. Tissue Engr2009:15(10) p2965-76

Fig 2: H&E stain and DAPI (cell nuclei) stain of an engineered leaflet pre-decell (a&d), post- decell (b&e), and 3 weeks after recell with MSC (c&f). Cell concentration in leaflet pre- and post-decell, after 1 and 3 weeks of recell (g).

Acknowledgment:Funding from NIH R01 HL107572 (to R.T.T.)

See more of this Session: Stem Cells In Tissue Engineering II
See more of this Group/Topical: Food, Pharmaceutical & Bioengineering Division

(336e) Guided Cardiogenic Differentiation of Human Pluripotent Stem Cells In Static and Stirred-Suspension Cultures

Donghui Jing, Abhirath Parikh and **Emmanuel (Manolis) S. Tzanakakis**, Chemical and Biological Engineering, State University of New York at Buffalo, Buffalo, NY

Human pluripotent stem cells (hPSCs) can be utilized as a source of cardiomyocytes for heart therapies. However, realization of the clinical potential of hPSCs will require efficient and cost-effective bioprocesses for the large-scale generation of implantable cardiomyocytes. We demonstrated the production of cardiac cells from encapsulated embryonic stem cells cultured in stirred-suspension bioreactors[1]. More recently, we established differentiation protocols for directing the fate of hPSCs along cardiac cell lineages utilizing physiological factors. Importantly, application of these protocols was investigated for the scalable production of heart muscle cells in stirred-suspension microcarrier vessels.

A 4-stage differentiation scheme was established for guiding hPSCs to mesendoderm (ME), mesoderm (MS), early cardiac (EC) cells and cardiomyocytes (CM). Cells were coaxed with physiological factors such as TGF-β ligands and bone morphogenetic proteins (BMPs) in the absence of serum. Design-of-experiment methods were implemented to determine combinations of factors, their concentrations and length of cell treatment after measuring the expression of specific markers at each stage. Differentiating hPSC progeny was analyzed by quantitative reverse transcription-PCR, immunocytochemistry, western blotting analysis and flow cytometry. Besides static cultures, hPSCs were also seeded on microcarriers and expanded in stirred-suspension bioreactors as described[2,3]. The concentration, viability and potential lysis of cells cultured in stirred suspension were determined.

Human PSC monolayers exposed to activin A in serum-free medium gave rise to ME cells as we reported[2] resulting in the expression of brachyury (T) by over 90% of the cells after 1 day. Further exposure of ME cells to activin and BMP resulted in upregulation of MS markers including FLK1. Commitment to EC cells expressing NKX2.5 and GATA4 was achieved by inhibition of the canonical Wnt pathway. After 2 weeks of differentiation, cells emerged exhibiting heart cell markers such as cardiac troponin I (cTnI) and α-actinin and organizing into beating foci responding to drugs in a chronotropic manner.

Translation of this differentiation scheme to a scalable microcarrier bioreactor culture of hPSCs was further explored. Operating conditions such as the agitation rate, seeding cell density and cell/microcarrier ratio were selected for the cardiogenic differentiation of hPSCs. After expansion under non-differentiating conditions, cells were treated with the same protocol described above for static cultures of hPSCs. The differentiating cells expressed stage-specific markers in an orderly fashion and eventually formed clusters with contractile activity. Beating activity was modulated organotypically upon treatment with diltiazem and phosphodiesterase inhibitors. The final fraction of cardiomyocytes was determined by the expression of cardiac markers such as myosin heavy chain and cardiac troponins via flow cytometry. Approximately 7 cardiomyocytes were generated in the bioreactor culture per stem cell seeded compared to ~3.5 cardiomyocytes/hPSC in dishes. Ongoing efforts focus on optimizing the final concentration of hPSC-derived cardiomyocytes in stirred-suspension microcarrier cultures.

Our results demonstrate the stirred-suspension cultivation of hPSCs and their en masse differentiation along heart cell lineages in serum-free media with physiologically relevant factors. These findings support the development of bioprocesses based on this bioreactor system for producing clinically-relevant quantities of cardiomyocytes.

References:

1. Jing, D. et al. Cell Transplant. 2010, 19:1397-412.

2. Lock, L.T. et al. Tissue Eng. Part A 2009, 15:2051-63.

3. Kehoe, D. et al. Tissue Eng. Part A. 2010, 16:405-21.

See more of this Session: Stem Cells In Tissue Engineering II
See more of this Group/Topical: Food, Pharmaceutical & Bioengineering Division

(336f) Controlling the Electrophysiological Maturation of Stem Cell-Derived Cardiomyocytes Via Nitric Oxide Supplementation

Alexander J. Hodge, James H Morris and Elizabeth A Lipke, Chemical Engineering, Auburn University, Auburn, AL

Heart disease is responsible for approximately one in four deaths annually in the United States. Because donor organs remain scarce, the ability to generate physiologically-accurate heart muscle cells (cardiomyocytes) *in vitro* provides an alternative tissue source for the treatment of cardiovascular disease. Pluripotent stem cells have demonstrated the capacity for developing mature cardiomyocytes for use in both clinical and diagnostic research applications. Unfortunately, current methods for generating therapeutically-relevant quantities of cardiomyocytes from stem cells are hindered by the creation of cardiac cells with immature electrophysiological phenotypes. Furthermore, inherent heterogeneities among stem cell-derived cardiomyocyte populations in individual batches may contribute a host of arrhythmogenic effects post-implantation in native myocardium. Nitric oxide (NO), a small, free radical molecule, is involved in a multitude of different cell signaling mechanisms including neurotransmission, vasodilatation, and differentiation. The supplementation of NO-donating materials into differentiation media has been shown to up-regulate processes responsible for cardiogenesis. While changes in gene and protein expression of stem cell-derived cardiomyocytes grown under such culture conditions have been shown, the electrophysiological effects of nitric oxide addition have not yet been examined. In this study, populations of embryoid bodies (EBs) were exposed to either S-nitrosothiol (CysNO, at 10, 50, or 100μM) NO-donors or nitric oxide synthase inhibitor (L-NAME, 1mM) every two days after the onset of differentiation; spontaneous contractile activity was observed and quantified over a period of 18 days. Both the normalized beating percentage and beat frequency of EBs exposed to CysNO demonstrated higher rates of spontaneous beating than either the control or L-NAME populations from day 10 to 14. Additionally, beating frequency among NO-supplemented groups increased between days 8 and 14, whereas control and L-NAME groups retained a constant beating frequency. Further work will investigate the protein and genetic expression profiles of the stem cell-derived cardiomyocytes. Furthermore, calcium transient analysis and whole cell patch clamp will elucidate the ion channel activity of maturing cardiomyocytes. The results of this study provide insight into the mechanisms of stem cell-derived cardiomyocyte maturation and will aid in establishing a platform for improving electrophysiological homogeneity of pluripotent stem cells differentiated in large-scale culture environments.

See more of this Session: Stem Cells In Tissue Engineering II
See more of this Group/Topical: Food, Pharmaceutical & Bioengineering Division

(336g) Functional Smooth Muscle Cells Derived From Induced Pluripotent Stem Cells for Cardiovascular Tissue Engineering Applications

Vivek K. Bajpai and Stelios T. Andreadis, Chemical & Biological Engineering, State University of New York at Buffalo, Amherst, NY

Tissueengineered vessels (TEV) represent attractive alternative to the shortage ofautologous vascular grafts for coronary artery disease. Less proliferative andsenescent autologous vascular cells from the aged patients have motivatedresearchers to seek alternate autologous sources. In this light, inducedpluripotent stem cell (iPSC) technology bears great promise for regenerativemedicine as it may provide a potential inexhaustible source for autologouscells. However, iPSC have not been studied in detail in the context of vasculartissue engineering particularly iPSC derived smooth muscle cells (iPS-SMC)haven't been rigorously characterized for their functionality in-vitro andin-vivo which is imperative for their successful application in cardiovasculartissue engineering. In the present work, we developed novel monolayer cultureprotocol of human (h)iPSC differentiation into smooth muscle using extracellular matrix molecules and diffusible signals and characterized them with abattery of biochemical and functional assays. To assess differentiation inreal-time and monitor its progression, two hiPSC lines (generated from Yamanakafactors (OCT4, SOX2, cMYC, KLF4) and Thomson factors (OCT4, SOX2, NANOG, LIN28))were transduced on matrigel (to limit contamination by MEF) with dual promoterlentiviral construct (LVDP). LVDP encodes for DsRed2 under the constitutivehuman phosphoglycerokinase (hPGK) promoter and ZsGreen under the smooth musclealpha actin (αSMA) promoter. Thus DsRed2 expression provides a measure oftransduction efficiency while ZsGreen expression works as differentiationselection marker. Transduced hiPSC were sequentially cultured on matrigel (in MEFconditioned hESC medium (MEF-CM) plus bFGF) followed by culture on collagen IVor gelatin (in presence of SMC medium (M231 supplemented with 5%FBS, bFGF, EGF,Insulin and heparin) and finally in maturation media (M231 plus TGF- β1 andheparin)). hiPS-SMC were characterized for known biochemical markers at thetranscriptional and translational level. TEV were fabricated from hiPS-SMC andfibrin and were tested for mechanical strength and vasoreactivity to determinefunctional status of hiPS-SMC. Histological analysis of TEV was performed toexamine collagen and elastin synthesis by hiPS-SMC. We found that by day 7 onmatrigel, ~25% of transduced cells (DsRed+) were also green (ZsGreen+)suggesting that αSMA promoter was active and this fraction furtherenriched to >95% by treatment with SMC medium for 8 days as determinedthrough fluorescence microscopy and flow cytometry. RT-PCR confirmedupregulation of SMC specific genes and concomitant downregulation of pluripotent,mesendodermal and ectodermal genes during differentiation. Immunostaining forearly (αSMA), intermediate (calponin) and late (myosin heavy chain, MHC)SMC markers showed that only small fraction of cells exhibited filamentousprotein organization after SMC medium treatment, suggesting that thecontractile apparatus of enriched hiPS-SMC was still immature. Indeed vascularcontractility of TEV made with hiPS-SMC at this stage of differentiation waslow as compared to mature SMC. However, after treatment with maturation media for5 days, immunostaining for the same proteins showed markedly increasedfilamentous organization. TEV fabricated from mature iPS-SMC showed increasedcontractility in response to vasoagonists like KCl, U46619 (Thromboxane A2agonist) and endothelin-1 suggestive of well developed receptor mediated andnon-receptor mediated pathways of contractility. Additionally, hiPS-SMCembedded in the vascular wall aligned circumferentially and synthesized collagenand tropoelastin as evidenced by histological analysis. Figure 1 depicts phasecontrast, fluorescent microscopic and immunocytochemistry images of hiPSC andmature hiPS-SMC.

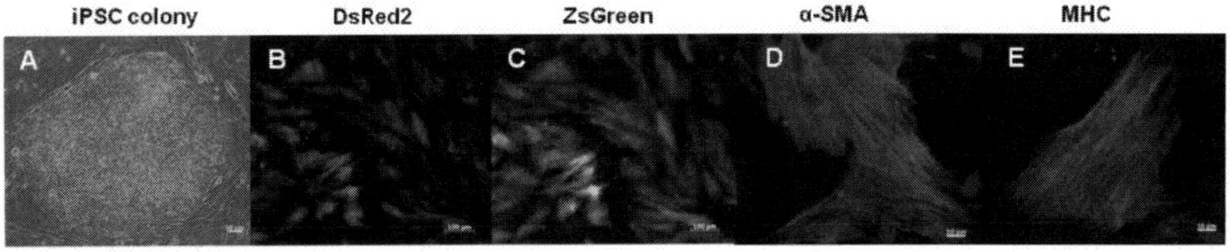

Figure1. hiPSC differentiation into SMC. iPSC colonyon feeder layer (A); Mature hiPS-SMC fluorescence microscopy (B and C); MaturehiPS-SMC Immunocytochemistry for αSMA (D) and MHC (E). Bars, A 50μm; B and C 100μm; D and E 10μm.

Our resultssuggest that hiPSC can be efficiently differentiated into functional SMC usingmonolayer protocol and can serve as an untapped autologous cell source forpotential application in cardiovascular tissue engineering and regenerativemedicine.

See more of this Session: Stem Cells In Tissue Engineering II
See more of this Group/Topical: Food, Pharmaceutical & Bioengineering Division

(371a) High Throughput Screening for Biomolecule Discovery and Detection of Food Borne Pathogens

S.T. Yang, Chemical Engineering, The Ohio State University, Columbus, OH

abstract to follow

See more of this Session: Food, Pharmaceutical and Bioengineering Plenary Session
See more of this Group/Topical: Food, Pharmaceutical & Bioengineering Division

(371b) "Pathogen-Mimicking" Nanoparticles for Prevention and Treatment of Respiratory Infectious Diseases

Balaji Narasimhan, Department of Chemical and Biological Engineering, Iowa State University, Ames, IA

abstract to follow

See more of this Session: Food, Pharmaceutical and Bioengineering Plenary Session
See more of this Group/Topical: Food, Pharmaceutical & Bioengineering Division

(371c) Manipulating Cellular Reducing Equivalents Through Metabolic Engineering

Ka Yiu San, Department of Bioengineering, Department of Chemical and Biomolecular Engineering, Rice University, Houston, TX

abstract to follow

See more of this Session: Food, Pharmaceutical and Bioengineering Plenary Session
See more of this Group/Topical: Food, Pharmaceutical & Bioengineering Division

(371d) Cell Migration in 3D

Denis Wirtz, Chemical and Biomolecular Engineering, Johns Hopkins University, Baltimore, MD

abstract to follow

See more of this Session: Food, Pharmaceutical and Bioengineering Plenary Session
See more of this Group/Topical: Food, Pharmaceutical & Bioengineering Division

(371e) A Fantastic Voyage Through Bioengineering

Martin Yarmush, Biomedical Engineering, Rutgers - The State University of New Jersey, Piscataway, NJ

abstract to follow

See more of this Session: Food, Pharmaceutical and Bioengineering Plenary Session
See more of this Group/Topical: Food, Pharmaceutical & Bioengineering Division

(388a) Metabolic Flux Analysis of *Escherichia Coli* MG1655 Under Octanoic Acid Stress

Yanfen Fu[1], Jong M. Yoon[1], Liam A. Royce[1], Maria Rodriguez-Moya[2], Ramon Gonzalez[2], Laura Jarboe[1] and Jacqueline V. Shanks[1], (1)Department of Chemical and Biological Engineering, Iowa State University, Ames, IA, (2)Departments of Chemical & Biomolecular Engineering and Bioengineering, Rice University, Houston, TX

Metabolic engineering has evolved to the point of fulfilling the dream of having our industrial chemicals produced renewably. Carboxylic acids (e.g., short chain fatty acids (SCFAs) such as octanoic acid (C8)) are one such chemical intermediate that can be produced with *Escherichia coli* engineered with a short chain thioesterase. However, C8 is toxic to *E. coli* growth at concentrations greater than 10 mM. To design a better host strain for the production of large amount of those SCFAs, metabolic flux analysis of *E. coli* grown under C8 stress was performed. When *E. coli* MG1655 was grown in M9 minimal media supplemented with 35 mM C8 under aerobic conditions, cell growth was inhibited by 55.1% ± 10.0% (mean ± standard deviation, n=4) compared to control (without C8). To study fluxes in central carbon metabolism, 10 g/L of a mixture of labeled and unlabeled glucose (25% uniform label glucose, 25% 1-label glucose and 50% natural glucose) was used as the sole carbon source for bacterial growth. Proteinogenic amino acid isotopomers were measured by 2D (^{13}C, ^{1}H) NMR. Notable differences of several amino acid isotopomer enrichments were observed between control condition and C8 stress condition, suggesting the precursor nodes of these amino acids in metabolic pathways are responding to the stress. More specifically, the difference suggested that the flux distribution among tricarboxylic acid (TCA) cycle, pyruvate node and α-ketoglutarate node and oxaloacetate node was changed. Under C8 stress, acetate production flux increased by ~50% and the flux from phosphoenolpyruvate to pyruvate increased by ~50%, while the TCA cycle pathway, the malic enzyme pathway, and the pentose phosphate pathway decreased by 25%, 90%, and 20% respectively. Downregulation of the TCA cycle can occur if an excess of NADH occurs, due to feedback inhibition of isocitrate synthase. Since our flux data shows that less NADH is produced via central carbon metabolism under C8 stress, an excess of NADH could occur if C8 stress inhibits regeneration of NAD^+ from NADH, via membrane disruption. Our transcriptome and proteomic data suggest that carboxylic acids may negatively impact the function and/or integrity of the cell membrane. Flux data combined with transcriptomic and proteomic analysis allow the development of hypotheses in context with the mechanism of C8 toxicity, which in turn helps to engineer strains with increased tolerance of C8.

See more of this Session: Microbial Engineering for the Synthesis of Bulk and High-Value Products
See more of this Group/Topical: Sustainable Engineering Forum

(388b) Metabolic Engineering of Rhizopus Oryzae for Enhanced Production of Fumaric Acid From Sugars

Baohua Zhang, WILLIAM G. LOWRIE Department of Chemical and Biomolecular Engineering, The Ohio State University, Columbus, OH and Shang-Tian Yang, The Ohio State University, Columbus, OH

Fumaric acid is a dicarboxylic acid used extensively in resins, food acidulants, and other applications including oil field fluids, esters etc. Some strains of *Rhizopus oryzae*, which is a filamentous fungus, can produce fumaric acid from sugars under aerobic conditions but the fumarate yield is relatively low due to the co-production of ethanol and other fermentation byproducts. In this work, metabolic engineering technique is applied to optimize the fermentation for enhanced fumaric acid production. Overexpression of native enzymes fumarase and pyruvate carboxylase showed no significant effect on the fumaric acid production, suggesting that the biosynthesis of fumaric acid is not limited by these two enzymes. On the other hand, overexpression of heterologous enzyme phosphoenolpyruvate carboxylase, which converts pyruvate with CO_2 fixation to oxaloacetate, led to increased fumaric acid production compared to the wild type control. The fermentation kinetics and activities of enzymes in the fermentation pathway of these *Rhizopus* mutants were studied and will be reported in this paper.

See more of this Session: Microbial Engineering for the Synthesis of Bulk and High-Value Products
See more of this Group/Topical: Sustainable Engineering Forum

Metabolic Engineering ■ (■■■■) ■■■–■■■

Contents lists available at ScienceDirect

Metabolic Engineering

journal homepage: www.elsevier.com/locate/ymben

Styrene biosynthesis from glucose by engineered *E. coli*

Rebekah McKenna, David R. Nielsen*

Chemical Engineering, Arizona State University, 501 E. Tyler Mall, ECG 301, Tempe, AZ 85287-6106, USA

ARTICLE INFO

Article history:
Received 18 March 2011
Received in revised form
2 June 2011
Accepted 14 June 2011

Keywords:
Styrene
Phenylalanine ammonia lyase
E. coli
Cinnamic acid
L-Phenylalanine
Aromatic

ABSTRACT

Styrene is a large volume, commodity petrochemical with diverse commercial applications, including as a monomer building-block for the synthesis of many useful polymers. Here we demonstrate how, through the *de novo* design and development of a novel metabolic pathway, styrene can alternatively be synthesized from renewable substrates such as glucose. The conversion of endogenously synthesized L-phenylalanine to styrene was achieved by the co-expression of phenylalanine ammonia lyase and *trans*-cinnamate decarboxylase. Candidate isoenzymes for each step were screened from bacterial, yeast, and plant genetic sources. Finally, over-expression of *PAL2* from *Arabidopsis thaliana* and *FDC1* from *Saccharomyces cerevisiae* (originally classified as ferulate decarboxylase) in an L-phenylalanine over-producing *Escherichia coli* host led to the accumulation of up to 260 mg/L in shake flask cultures. Achievable titers already approach the styrene toxicity threshold (determined as ~300 mg/L). To the best of our knowledge, this is the first report of microbial styrene production from sustainable feedstocks.

1. Introduction

Styrene is a versatile, large commodity chemical for which 60% of its global annual consumption supports the production of numerous, industrially important polymers, and co-polymers (SRI, 2010). In 2006, over 6 million metric tons of styrene were produced by U.S. manufacturers, a market that was valued at nearly $28 billion and projected to grow by 4.3% per year through 2010 (SRI, 2010). Today, all commercially available styrene is derived from the world's dwindling petroleum resources. Conventional styrene synthesis is achieved through the chemocatalytic dehydrogenation of petroleum-derived ethylbenzene (Wu et al., 1981) which requires over 3 metric tons of steam per metric ton of styrene produced. This exorbitant requirement renders styrene production as the most energy-intensive among commodity chemical production routes, consuming nearly 200 trillion BTU of steam for its domestic annual production alone (DoE, 2002). With that being said, the goal of this study was to engineer a biocatalyst capable of synthesizing styrene from renewable resources as a more sustainable and greener source of styrene and styrene-derived polymers.

Beyond its petrochemical origins, styrene has been observed as a trace metabolite in foods, in particular cheeses, where it acts as an aroma defect. For instance, the yeast *Penicillium camemberti* has been reported to be capable of synthesizing low levels of styrene from excess L-phenylalanine, but neither a defined pathway nor the requisite genes have thus far been elucidated (Pagot et al., 2007). Styrene is also known to be naturally synthesized by select plant species, including several trees in the *Styracaceae* family (including several *Styrax* sp.). Here, styrene is also synthesized from excess L-phenylalanine where it then subsequently accumulates as a minor constituent (< 0.55% of total dry weight) within benzoin resins (which are predominantly composed of benzoic acid) (Fernandez et al., 2005). Again, however, neither the enzymes nor genes associated with said pathway have been identified to date. Although it is possible to purify styrene from plant resins via distillation or liquid–liquid extraction (Clark, 1990), considering the extremely low productivity, poor net yields, and low inherent value of styrene, its potential, large-scale biological production by such a mechanism is rendered as completely uneconomical and unsustainable. A more sustainable and inexpensive approach, however, would involve the engineering of microorganisms that possess the unique ability to synthesize styrene directly from renewable resources.

In recent years, a variety of additional, novel synthetic routes have been proposed and engineered in microorganisms for the production (from renewable substrates such as glucose) of a number of other useful, functionalized monoaromatic compounds with structural similarity to styrene. For example, a biosynthetic pathway for the production of *p*-hydroxystyrene (a monomer used in synthesis of photo-resist polymers) from renewable sugars has been constructed using both *Escherichia coli* (Qi et al., 2007) and *Pseudomonas putida* (Verhoef et al., 2009) as host platforms. Meanwhile, both phenol (a precursor and monomer

* Corresponding author. Fax: +1 480 727 9321.
E-mail address: David.R.Nielsen@asu.edu (D.R. Nielsen).

1096-7176/$ - see front matter © 2011 Elsevier Inc. All rights reserved.
doi:10.1016/j.ymben.2011.06.005

Please cite this article as: McKenna, R., Nielsen, D.R., Styrene biosynthesis from glucose by engineered *E. coli*. Metab. Eng. (2011), doi:10.1016/j.ymben.2011.06.005

for phenolic resins) (Wierckx et al., 2005) and p-hydroxybenzoate (a precursor to parabens, which are used as preservatives) (Verhoef et al., 2007) have also been synthesized as individual products from glucose by engineered strains of *P. putida*. Interestingly, each of the above non-natural metabolites were derived using L-tyrosine (or its immediate precursor, 4-hydroxyphenylpyruvate) as a pathway precursor, thereby making them each phenolics (Boudet, 2007). To date, there remain few examples of engineered biosynthetic pathways for the production of non-phenolic, monoaromatic compounds using microbial biocatalysts. Moreover, there exist no previous reports regarding the development of a styrene biosynthetic pathway or the engineering of microbes capable of synthesizing styrene from renewable resources. With this in mind, the present study describes the *de novo* design and development of a functional styrene biosynthetic pathway and the engineering of *E. coli* strains capable of styrene biosynthesis from glucose.

The proposed styrene biosynthesis pathway utilizes endogenously synthesized (from glucose) L-phenylalanine as an intermediate precursor which is converted to styrene by a series of two enzymatic steps, as in Fig. 1. First, endogenously occurring L-phenylalanine is converted to *trans*-cinnamic acid (tCA) through its deamination, as catalyzed by phenylalanine ammonia lyase (PAL). Said activity and substrate specificity has been previously reported for a number of PAL isoenzymes that have been identified and characterized in yeast (Gilbert and Tully, 1982; Qi et al., 2007; Vannelli et al., 2007b), plants (Cochrane et al., 2004; Young et al., 1966), and (although less prevalently) bacteria (Moffitt et al., 2007a; Xiang and Moore, 2005; Xiang and Moore, 2006; Young et al., 1966). Among previously characterized PAL isoenzymes, considerable variability with respect to both substrate specificity and activity has been reported. The most commonly studied PAL isoenzyme in recombinant systems, including *E. coli*, remains the bifunctional PAL/TAL (TAL: tyrosine ammonia lyase, which also catalyzes the deamination of L-tyrosine to *p*-coumaric acid) of the yeast *Rhodotorula* sp. (Cui et al., 2008; Gilbert et al., 1985, 1983; Gilbert and Tully, 1982; Santos et al., 2011; Vannelli et al., 2007a). Meanwhile, a number of PAL isoenzymes have been also been studied from plant sources (Jones, 1984) where, for example, *Arabidopsis thaliana* has been characterized as possessing four distinct PALs (encoded by *PAL1*, *PAL2*, *PAL3*, and *PAL4*) (Cochrane et al., 2004). More recently, a number of prokaryotic PALs have been isolated and characterized, beginning with that which is encoded by *encP* from *Streptomyces maritimus* (Xiang and Moore, 2002, 2005). Soon after, two additional PAL isoenzymes were discovered in the cyanobacteria *Nostoc puntiformes* and *Anabaena variabilis* (Moffitt et al., 2007a; Xiang and Moore, 2005, 2006). Interestingly, each of these prokaryotic PALs was also found to be highly specific for L-phenylalanine, and thus do not also display TAL activity that is so common among yeast PALs.

The second step in the proposed styrene biosynthesis pathway involves the subsequent decarboxylation of tCA by a phenylacrylate decarboxylase (PADC) displaying *trans*-cinnamate decarboxylase activity to yield styrene as the final product. Several genes encoding PADC activity have been characterized and reported in the literature, including *pdc* from *Lactobacillus plantarum* and *padC* from *Bacillus subtilis* (Tran et al., 2008). It currently remains unknown, however, if their enzyme products could specifically decarboxylate tCA to produce styrene. Meanwhile, the yeast *Saccharomyces cerevisiae* has demonstrated the ability to synthesize styrene when supplied with exogenous tCA (Clausen et al., 1994); however, the underlying mechanism and genetic basis for said activity presently remains unclear. For instance, previous characterizations of *PAD1* have shown that its over-expression in *S. cerevisiae* results in increased resistance to tCA (as achieved by its conversion to styrene) (Larsson et al., 2001). However, it has also been demonstrated that the expression of *PAD1* alone is insufficient for achieving PADC activity (Clausen et al., 1994), and it was later speculated that the expression of a second enzyme was necessary to convert tCA to styrene (Jiang et al., 2005). Most recently, it was reported that tCA decarboxylase activity in *S. cerevisiae* maintains an essential dependence on the co-expression of both *PAD1* and *FDC1*, the latter a gene previously characterized as encoding ferulic acid decarboxylase (note that ferulic acid is also a phenylacrylic acid) (Mukai et al., 2010).

The present study describes the *de novo* design of a styrene biosynthetic pathway, as supported through the comprehensive screening of composite pathways enzymes from various genetic sources. The synthesis of styrene from glucose was ultimately achieved through the co-expression of PAL and tCA decarboxylase enzymes in an L-phenylalanine over-producing *E. coli* host platform.

2. Materials and methods

2.1. Chemicals

Chemicals used in this study include L-phenylalanine (98.5%, VWR, Westchester, PA), *trans*-cinnamic acid (99%, MP Biomedicals, Solon, OH), styrene (99%, Alfa Aesar, Ward Hill, MA), methanol (99.8%, VWR, Westchester, PA), and trifluoroacetic acid (99.5%, EMD, Darmstadt, Germany). All other chemicals used in this study are from Sigma-Aldrich (St. Louis, MO).

2.2. Strains and media

All strains, plasmids, and oligonucleotide primers used in this study are listed in Table 1. Custom oligonucleotide primers were synthesized by Integrated DNA Technologies (Coralville, IA). cDNA of *N. puntiformes* and *A. variabilis* were gifts from Prof. Bradley Moore (UCSD). Strains of *B. subtilis*, *L. plantarum*, and *S. cerevisiae*, as well as the plasmids pSTV28 and pTrc99A were all gifts from Prof. Kristala Prather (MIT). Genomic DNA was prepared from whole cells using the ZR Fungal/Bacterial DNA MiniPrep (Zymo Research, Irvine, CA) according to vendor protocols. Strains were

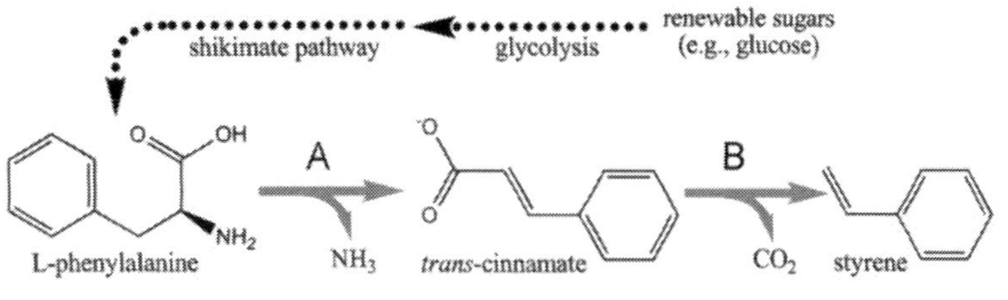

Fig. 1. Enzymatic pathway to convert the precursor L-phenylalanine to the product styrene via the intermediate *trans*-cinnamate. The two-step pathway from L-phenylalanine is achieved by the co-expression of one or more genes which encoded phenylalanine ammonia lyase (PAL) activity (A), and one or more genes which encoded *trans*-cinnamic acid decarboxylase (CADC) activity (B).

Please cite this article as: McKenna, R., Nielsen, D.R., Styrene biosynthesis from glucose by engineered *E. coli*. Metab. Eng. (2011), doi:10.1016/j.ymben.2011.06.005

Table 1
Strains, plasmids, and oligonucleotide primers used in the present study.

	Description	Source
Strains		
E. coli NST74	*aroH367, tyrR366, tna-2, lacY5, aroF394(fbr), malT384, pheA101(fbr), pheO352, aroG397(fbr)*	ATCC 31884
E. coli BL21Star(DE3)	*F-ompT hsdS$_B$ (r$_B$-m$_B$-) gal dcm rne131 (DE3)*	Invitrogen
E. coli NEB-10 beta	*araD139 Δ(ara,leu)7697 fhuA lacX74 galK16 galE15 mcrA f80d(lacZΔM15)recA1 relA1 endA1 nupG rpsL rph spoT1Δ(mrr-hsdRMS-mcrBC)*	New England Biolabs
S. cerevisiae W303	Source of *pad1* and *fdc1*	Prather Lab, MIT
L. plantarum	Source of *pdc*	Prather Lab, MIT
B. subtilis 3610	Source of *padC*	Prather Lab, MIT
Plasmids		
pTrc99A	P_{trc}, pBR322 ori, *lacI^q*, *Amp^r*	Prather Lab, MIT
pSTV28	P_{lac}, pACYC184 ori, *Cm^r*	Prather Lab, MIT
pSpal1At	*PAL1* of *A. thaliana* inserted into the *EcoRI* and *SphI* sites of pSTV28	This study
pSpal2At	*PAL2* of *A. thaliana* inserted into the *EcoRI* and *SphI* sites of pSTV28	This study
pSpalAv	*pal* of *A. variabilis* inserted into the *EcoRI* and *BamHI* sites of pSTV28	This study
pSpalNp	*pal* of *N. punctiforme* inserted into the *EcoRI* and *BamHI* sites of pSTV28	This study
pSencPSm	*encP* of *S. maritimus* inserted into the *EcoRI* and *BamHI* sites of pSTV28	This study
pTpad1Sc	*PAD1* of *S. cerevisiae* inserted into the *NcoI* and *XbaI* sites of pTrc99A	This study
pTpad1Sc-fdc1Sc	*FDC1* of *S. cerevisiae* inserted into pTpad1Sc	This study
pTfdc1Sc	*FDC1* of *S. cerevisiae* inserted into the *SalI* and *HindIII* sites of pTrc99A	This study
pTpadCBs	*padC* of *B. subtilis* inserted into the *BamHI* and *SbfI* sites of pTrc99A	This study
pTpdcLp	*pdc* of *L. plantarum* inserted into the *BamHI* and *SbfI* sites of pTrc99A	This study
pUN15-pal1	Clone U10120 containing AT2G37040 (*PAL1*) from *A. thaliana*	ABRC
pUN15-pal2	Clone U12256 containing AT3G53260 (*PAL2*) from *A. thaliana*	ABRC
Primers (5′→3′)		
F_pal_AV_EcoRI	TAAGAATTCAAGGGGATAAATAATGAAGACACTATCTCAAGC	
R_pal_AV_BamHI	ATTGGATCCTTAATGCAAGCAGGGT	
F_pal_NP_EcoRI	TAAGAATTCAAGGGGATAAATAATGAATATAACATCTCTACAAC	
R_pal_NP_BamHI	ATTGGATCCTTACGTTGACTTTAAGCT	
F_pal1_AT_EcoRI	TAAGAATTCAAGGGGATAAATAATGGAGATTAACGGGGCAC	
R_pal1_AT_SphI	ATTGCATGCTTAACATATTGGAATGGGAGCTC	
F_pal2_AT_EcoRI	TAAGAATTCAAGGGGATAAATAATGGATCAAATCGAAGCAATG	
R_pal2_AT_SphI	ATTGCATGCTTAGCAAATCGGAATCGGAG	
F_encP_SM_EcoRI	TAAGAATTCAAGGGGATAAATAATGACCTTCGTCATAGAGCT	
R_encP_SM_BamHI	ATTGGATCCTTAGTGCGCCGCCACG	
F_padC_BS_BamHI	AAAGGATCCCGACTAAGGGAGGATAAGATGGAAAACTTTATCGGAAG	
R_padC_BS_SbfI	ATACCTGCAGGATGTTTATTATAATCTTCCCGCG	
F_pdc_LP_BamHI	ATAGGATCCCTCTGGAGGCAGTTCTAATGACAAAAACTTTTAAAACACT	
R_pdc_LP_SbfI	ATACCTGCAGGCCAGAATGTTTCACGTGAA	
F_pad1_SC_NcoI	TTACCATGGAGGAACCTAGGCACACAATGGTCCTATTTCCAAGAAGAA	
R_pad1_SC_XbaI	ATTTCTAGATTACTTGCTTTTTATTCCTTCCC	
F_fdc1_SC_SalI	ATAGTCGACAGACATCAAAGGACGGTTCATGAGGAAGCTAAATCCAGCT	
R_fdc1_SC_HindIII	ATTAAGCTTTTATTTATATCCGTACCTTTTCCAAT	

routinely cultured in Luria-Bertani (LB) broth (supplemented with antibiotics as necessary). Cultures were assayed for their ability to synthesize L-phenylalanine, *trans*-cinnamate, and styrene by cultivation in phosphate-limited minimal media (herein referred to as "MM1") with glucose. MM1 was adapted from Qi et al. (2007), and is composed of glucose (nominally 15 g/L), $MgSO_4 \cdot 7H_2O$ (0.5 g/L), $(NH_4)_2SO_4$ (4.0 g/L), MOPS (24.7 g/L), KH_2PO_4 (0.3 g/L), K_2HPO_4 (0.7 g/L), and 5 mL/L ATCC Trace Mineral Supplement (Catalog No. MD-TMS) (EDTA (0.5 g/L), $MgSO_4 \cdot 7H_2O$ (3 g/L), $MnSO_4 \cdot 7H_2O$ (0.5 g/L), NaCl (1 g/L), $FeSO_4 \cdot 7H_2O$ (0.1 g/L), $Co(NO_3)_2 \cdot 6H_2O$ (0.1 g/L), $CaCl_2$ (0.1 g/L), $ZnSO4 \cdot 7H_2O$ (0.1 g/L), $CuSO_4 \cdot 5H_2O$ (0.01 g/L), $AlK(SO_4)_2$ (0.01 g/L), H_3BO_3 (0.01 g/L), $Na_2MoO_4 \cdot 2H_2O$ (0.01 g/L), Na_2SeO_3 (0.001 g/L), $Na_2WO_4 \cdot 2H_2O$ (0.10 g/L), and $NiCl_2 \cdot 6H_2O$ (0.02 g/L)).

2.3. *Toxicity assays*

To determine extent and effects of metabolite toxicity on *E. coli*, the impacts of the exogenous addition of tCA and styrene (at increasing final concentrations) to growing cultures was investigated. Seed cultures of *E. coli* NST74 were prepared in 5 mL of LB broth and grown at 32 °C overnight while shaking at 250 rpm. The seed culture (1 ml) was then used to inoculate 50 ml of LB broth in a 250 mL shake flask. Cultures were grown to an optical density (OD_{600}) of either ~0.6 or ~2.0, at which time either tCA or styrene was added to the flasks at an array of final concentrations between 0 to 1 g/L. Two different initial OD_{600} levels were used to investigate the potential effects of product toxicity as a function of cell density and/or growth stage. In all cases, culturing then resumed at 32 °C for another 6–8 h while cell growth, as determined by OD_{600} measurements, was periodically monitored using a UV/vis spectrophotometer (Beckman Colter DU800, Brea, CA).

2.4. *Cloning of candidate genes encoding PAL activity from A. variabilis, N. punctiforme, and A. thaliana*

All genes used in this study were PCR amplified using a BioRad iCycler system with Phusion DNA Polymerase (Finnzymes, Espoo, Finland) using custom oligonucleotide primers. PCR cycling and reaction conditions were standardized according to manufacturer instructions. Candidate PAL encoding genes were amplified from the cDNA of *A. variabilis* and *N. punctiforme*. Candidate PAL encoding genes from *A. thaliana* were derived from cDNA library plasmids containing the specific loci of interest (Table 1) obtained from the ABRC (Ohio State University, Columbus, OH). Whereas *A. thaliana* posses four distinct PALs, we focused on those encoded by *PAL1* and *PAL2* as these have displayed the greatest activity when expressed in recombinant *E. coli* (noting also that PAL3 was

Please cite this article as: McKenna, R., Nielsen, D.R., Styrene biosynthesis from glucose by engineered *E. coli*. Metab. Eng. (2011), doi:10.1016/j.ymben.2011.06.005

found to be of 'very low activity') (Cochrane et al., 2004). Amplified linear DNA fragments were purified using the Zyppy Clean and Concentrator kit (Zymo Research, Orange, CA). Purified fragments were treated by endonuclease digestion using appropriate restriction enzymes (all from New England Biolabs, Ipswich, MA). Amplified fragments containing the *pal* from both *N. punctiforme* and *A. variabilis* were digested with *BamHI* and *EcoRI* whereas fragments containing *PAL1* and *PAL2* from *A. thaliana* were digested with *EcoRI* and *SphI*. All digestions were performed at 37 °C according to manufacturer's protocols. The expression vector pSTV28 was similarly digested with either *BamHI* and *EcoRI* or *EcoRI* and *SphI*. Digested fragments were gel purified using the Zyppy Gel DNA recovery kit (Zymo Research, Orange, CA). Linearized fragments were ligated using T4 DNA ligase (New England Biolabs, Ipswich, MA) at 4 °C overnight. Chemically competent *E. coli* NEB10-Beta (New England Biolabs, Ipswich, MA) cells were used for all transformations. Transformants were selected by plating on LB solid agar containing 34 mg/L chloramphenicol and culturing at 37 °C overnight. The transformant pool was subsequently screened according to both colony PCR (employing the same primers as used in the initial amplification) and restriction digest mapping of the resultant plasmids to identify those clones harboring the successful construct. This approach resulted in construction of plasmids pSpalAv, pSpalNp, pSpal1At, and pSpal2At, as listed in Table 1.

2.5. *Cloning of candidate genes encoding PADC activity from L. plantarum, B. subtilis, and S. cerevisiae*

Candidate PADC encoding genes, including *pdc*, *padC*, *PAD1*, and *FDC1*, were amplified via PCR using genomic DNA templates derived from *L. plantarum*, *B. subtilis*, and *S. cerevisiae*, respectively. PCR amplified DNA fragments were purified before treatment by endonuclease digestion. Fragments containing *padC* and *pdc* were each digested with *BamHI* and *SbfI*. The *E. coli* expression vector pTrc99A was similarly digested with *BamHI* and *SbfI*. Meanwhile, the amplified fragment containing *PAD1* was digested with *NcoI* and *XbaI* whereas the *FDC1* containing fragment was digested with *SalI* and *HindIII*. The *E. coli* expression vector pTrc99A was similarly digested with either *NcoI* and *XbaI* (for the insertion of *PAD1*) or *SalI* and *HindIII* (for the insertion of *FDC1*). All digested fragments were gel purified then ligated at 4 °C overnight before their transformation into chemically competent *E. coli* NEB10-Beta. Selection was then achieved by plating transformed cells on LB solid agar containing 100 mg/L ampicillin and culturing at 37 °C overnight. After confirmation of the correct transformant, these works resulted in the generation of plasmids pTpadCBs, pTpdcLp, pTpad1Sc, and pTfdc1Sc, as listed in Table 1. The *SalI–HindIII* digested *FDC1* fragment was then also cloned into the same sites in the newly generated plasmid pTpad1Sc by an analogous protocol, resulting in the plasmid pTpad1Sc-fdc1Sc.

2.6. *Assaying PAL activity in crude lysates of recombinant E. coli*

Each of the newly created PAL harboring plasmids (namely pSpalAv, pSpalNp, pSpal1At, and pSpal2At) were individually transformed into *E. coli* BL21(DE3) (Invitrogen, Carlsbad, CA). Seed cultures of each of the resultant strains were prepared (in triplicate) in 5 mL LB broth supplemented with 34 mg/L chloramphenicol and cultured at 32 °C while shaking at 250 rpm overnight. 50 µl of each seed was then used to again inoculate 5 ml LB broth. These cultures were grown until reaching an OD_{600} of ~0.6, at which point each was induced by the addition of isopropyl β-D-1-thiogalactopyranoside (IPTG) at a final concentration of 0.2 mM. Induced cultures were incubated for an additional 6 h after which an equal number of cells (determined by OD_{600} measurement) were collected by centrifugation at 1400 × g for 4 min. The cell pellet was re-suspended in 900 µL distilled water. Cell lysis was achieved using the FastBreak Cell Lysis Reagent kit (Promega, Madison, WI) and the supernatant collected after centrifugation at 11,000 × g for 2 min. PAL activity was analyzed at room temperature in pH 7.5 50 mM Tris–HCl buffer containing 250 mM L-phenylalanine. Activity assays were initiated by the addition of 5 µL of crude cell lysate. The production of tCA was followed at 290 nm on a Beckman Colter UV/vis spectrophotometer for a total of 5 min at 20 s intervals. A molar extinction coefficient of 9000 $M^{-1} cm^{-1}$ and a 1 cm path length were used to establish enzyme activity in terms of U mg^{-1} protein. The PAL protein content in each crude lysate was determined via first separation by SDS-PAGE using Mini-PROTEAN TGX 4–20% precast gels (Bio-Rad, Hercules, CA) and standard protocols. Concentration was then analyzed using the ImageJ software package (NIH, Bethesda, MD) and calibrated vs. Precision Plus Unstained Protein Standards (Bio-Rad, Hercules, CA).

2.7. *Assaying PAL activity in recombinant E. coli whole cells*

Seed cultures of *E. coli* BL21(DE3) harboring one of pSpalAv, pSpalNp, pSpal1At, or pSpal2At, were prepared in 5 mL LB broth and grown overnight. Shake flasks (250 mL) containing 50 mL of LB were inoculated with 1 mL of each seed culture. Cultures were grown until an OD_{600} of ~0.6 was reached, at which point the cultures were induced by IPTG addition at a final concentration of 0.2 mM. Cultures were then incubated for an additional 6 h (resulting in an OD_{600} of ~2) before an equal number of cells were collected and centrifuged at 1400 × g for 5 min. The pellet was washed once with PBS (phosphate buffered saline) at pH 7 before being re-suspended in 12 ml PBS buffer. Finally, the appropriate substrate, L-phenylalanine or L-tyrosine at a final concentration of 1 g/L, was added to the suspension. The suspensions were then shaken at 32 °C for a total of 3 h. Samples (1 mL) were taken every hour, centrifuged, and 750 µL of supernatant was collected for HPLC analysis to monitor the production of either *tCA* or *p*-coumaric acid.

2.8. *Assaying PADC activity in recombinant E. coli whole cells*

The plasmids pTpad1Sc-fdc1Sc, pTpad1Sc, pTfdc1Sc, pTpdcLp, and pTpadcBs were each individually transformed into chemically competent *E. coli* BL21(DE3). A seed culture of each strain was then grown in LB broth overnight. Shake flasks (250 mL) containing 50 mL of LB were inoculated with 1 ml of seed culture. Cultures were grown at 32 °C until an OD_{600} of ~0.6, at which point they were induced by adding 0.2 mM IPTG and then incubated for an additional 6 h. Cells were then collected and re-suspended in 12 ml PBS buffer (as previously described) and the substrate (*tCA* or *p*-coumaric acid) was added at a final concentration of 1 g/L. Samples (1 mL) were removed from the culture at both the time of initiation as well as after 12 h of incubation at 32 °C and analyzed by HPLC using the methods described herein.

2.9. *Co-expression of PAL and CADC isoenzymes in E. coli NST74 to convert L-phenylalanine to styrene*

The L-phenylalanine over-producing strain *E. coli* NST74 (Table 1) was co-transformed with the plasmids pSpal2At and pTfdc1Sc and selected for on LB agar supplemented with 100 mg/L ampicillin and 34 mg/L chloramphenicol. The resultant transformant was then grown overnight at 32 °C in 5 mL LB broth. Shake flasks (250 mL) containing 50 mL LB were inoculated with 1 mL of seed culture. The culture was then grown at 32 °C until

Please cite this article as: McKenna, R., Nielsen, D.R., Styrene biosynthesis from glucose by engineered E. coli. Metab. Eng. (2011), doi:10.1016/j.ymben.2011.06.005

an OD_{600} of ~0.6, at which time it was induced with 0.2 mM IPTG and then incubated for an additional 8 h. Cells were then collected and re-suspended in 12 ml PBS buffer (as previously described) and L-phenylalanine added at a final concentration of either 400 or 900 mg/L. Samples were taken periodically and analyzed by HPLC to determine the content of L-phenylalanine, tCA, and styrene, according to the methods presented below.

2.10. *Co-expression of PAL and CADC isoenzymes in E. coli NST74 to convert glucose to styrene in shake flask cultures*

The L-phenylalanine over-producing strain *E. coli* NST74 was co-transformed with each of the following combinations of plasmids: pSpalAv and pTfdc1Sc, pSpalNp and pTfdc1Sc, pSpal1At and pTfdc1Sc, and pSpal2At and pTfdc1Sc.Single colonies were then selected from the resulting transformants and grown in 5 mL LB broth for 12 h at 32 °C while shaking at 250 rpm to prepare a seed culture. Each seed (1 mL) was then used to inoculate 50 mL MM1. These cultures were performed in 100 mL serum bottles outfitted with septa caps that were tightly sealed upon inoculation. A closed system was used to avoid volatile product (i.e., styrene) losses. A large headspace was used to preclude the depletion of oxygen from the bottle. Cultures were grown for 10 h prior to being induced by the addition of IPTG at a final concentration of 0.2 mM. Culturing continued for 29–48 h post-induction while 1 ml samples were periodically taken from each culture and analyzed for relevant metabolite contents via HPLC, using the methods described herein.

2.11. *Metabolite analysis by HPLC*

Samples were prepared by removing 1 mL of culture from a shake flask culture and pelleting the cells at 11,000 × g for 2 min. The supernatant (0.75 mL) was then transferred to a glass HPLC vial and sealed with a Teflon-lined cap. HPLC analysis was carried out using a Hewlett Packard 1100 series HPLC system equipped with an auto sampler, diode array (UV/vis) detector, and reverse-phase Hypersil Gold SBC18 column (4.6 mm × 150 mm; Thermo Fisher, USA). Samples (5 μL) were injected for analysis at a total constant flow rate of 1.0 ml/min and constant column temperature of 45 °C. The column was eluted with 'solvent A' (consisting of double-distilled water) and 'solvent B' (consisting of methanol (99.8% grade) plus 0.1% trifluoroacetic acid (TFA)). The eluent began as a mixture of 95% solvent A and 5% solvent B before a linear gradient was applied over 8 min to then reach a mixture of 20% solvent A and 80% solvent B. This eluent composition was then held constant for 2 min before a second linear gradient was then applied over the course of 4 min to achieve a final mixture of 95% solvent A and 5% solvent B. The eluent was monitored at each of 215 nm for L-phenylalanine and 258 nm for *tCA*, *p*-coumaric acid, hydroxystyrene, and styrene. Under these conditions L-phenylalanine, *p*-coumaric acid, *tCA*, *p*-hydroxystyrene, and styrene were eluted at 4.5, 6.7, 8.67, 8.78, and 10.4 min, respectively.

2.12. *Confirmation of styrene biosynthesis by GC–MS*

Culture supernatant (1 ml) was added to hexane (1 ml) and vortexed for 20 min at maximum speed. The biphasic mixture was centrifuged for 1 min at 11,000 × g to settle. 750 μl of the hexane layer was removed for analysis by GC–MS. GC–MS analysis was performed on a Hewlett Packard 5890 Series II gas chromatograph with a flame ionizing detector and Supelco MDN-5S (30 m × 0.25 mm id) fused-silica capillary column using helium as the carrier gas. The injector, column, and detector temperatures were initial set at 280, 240, and 220 °C, respectively. The column temperature was then increased from 40 to 320 °C at 14 °C/min.

3. Results and discussion

3.1. *Assaying styrene toxicity*

Under the studied conditions, maximal titers of L-phenylalanine produced by *E. coli* NST74 after 48 h of culture in MM1 (with 1.5% glucose) reached about 700–1000 mg/L (results not shown). Whereas the theoretical yield of L-phenylalanine on glucose is 0.55 g/g (Baez-Viveros et al., 2004), achievable yields by *E. coli* NST74 under the present conditions in shake flask cultures were only 0.052–0.074 g/g (or about 10% of theoretical). Assuming complete conversion of all endogenously produced L-phenylalanine to styrene were possible (for example, if our engineered pathway could achieve a particularly high flux), we would expect to be able to synthesize up to 440–630 mg/L of styrene under the same culture conditions (corresponding to a glucose yield of 0.051–0.072 g/g), barring any physiological limitations such as product toxicity. Through a preliminary screening study, however, we have found that the products *trans*-cinnamate ($\log K_{ow}$=2.13) and styrene ($\log K_{ow}$=3.05) can each inhibit *E. coli* growth when added to cultures at final concentrations above ~800 and ~300 mg/L, respectively, as demonstrated in Fig. 2. For reference, *p*-hydroxystyrene ($\log K_{ow}$=2.28) has previously been shown to inhibit *P. putida* at concentrations of ~540 mg/L (Verhoef et al., 2009). Solvent hydrophobicity is typically a good

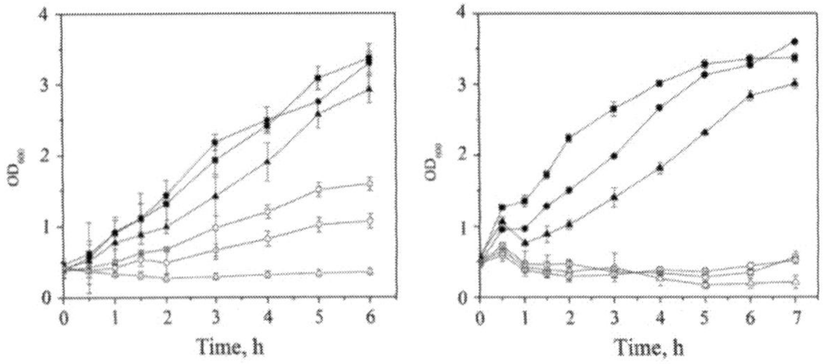

Fig. 2. Left: growth response of *E. coli* NST74 to tCA at concentrations of 0 mg/L (solid square), 200 mg/L (solid circle), 400 mg/L (solid triangle), 600 mg/L (open square), 800 mg/L (open circle), and 1000 mg/L (open triangle). Right: growth response of *E. coli* NST74 to styrene at concentrations of 0 mg/L (solid square), 100 mg/L (solid circle), 200 mg/L (solid triangle), 300 mg/L (open square), 400 mg/L (open circle), and 500 mg/L (open triangle). Error bars reported at one standard deviation from triplicate experiments.

Please cite this article as: McKenna, R., Nielsen, D.R., Styrene biosynthesis from glucose by engineered *E. coli*. Metab. Eng. (2011), doi:10.1016/j.ymben.2011.06.005

indicator of its toxicity towards microorganisms and may be quantified by $\log K_{ow}$, the log of the octanol–water partition coefficient (Ramos et al., 2002). Even at low concentrations, compounds with a $\log K_{ow}$ between 1.0 to 5.0 tend to be very toxic (Heipieper et al., 1994). Accumulation of hydrophobic aromatics within the cytoplasmic membrane has been shown to disrupt its integrity, resulting in the leakage of ions, metabolites, lipids, and proteins, as well as affecting the cells ability to maintain its internal pH and an appropriate transmembrane proton gradient (Ramos et al., 2002; Weber et al., 1993). The measured toxicity thresholds were found to be consistent at both low (OD_{600} 0.6) and high (OD_{600} 2, data not shown) initial cell densities (reflecting early and late exponential growth, respectively), indicating growth inhibition was independent of cell density and growth stage. As a result of the inherent toxicity of styrene we anticipated that its biosynthesis using an NST74 host platform would be limited to below its maximum potential under our culture conditions. While no prior studies have specifically explored either the effects of styrene toxicity on *E. coli* or strategies to improve its tolerance, several *Pseudomonas* sp. have been shown to display enhanced styrene tolerance characteristics (Weber et al., 1993). Whereas the present study is solely focused on the prototyping of a novel pathway for styrene biosynthesis, it is clear from our preliminary assays that styrene toxicity must eventually be overcome or effectively circumvented if renewable styrene production is ever to become viable or sustainable.

Table 2
Specific activity of PAL isoenzymes from *A. variabilis*, *N. punctiforme*, and *A. thaliana* on L-phenylalanine and L-tyrosine when expressed in recombinant *E. coli* BL21(DE3). Errors reported at one standard deviation from triplicate experiments.

Strain	Substrate	Activity (U mg^{-1} protein)
BL21(DE3)	L-Phenylalanine	n.d.
	L-Tyrosine	n.d.
pSencPSm	L-Phenylalanine	n.d.
	L-Tyrosine	n.d.
pSpalAv	L-Phenylalanine	2.38 ± 0.64
	L-Tyrosine	n.d.
pSpalNp	L-Phenylalanine	0.91 ± 0.32
	L-Tyrosine	n.d.
pSpal1At	L-Phenylalanine	2.42 ± 1.07
	L-Tyrosine	n.d.
pSpal2At	L-Phenylalanine	4.08 ± 0.11
	L-Tyrosine	n.d.

n.d.—not detected.

3.2. Screening candidate PAL isoenzymes for activity in recombinant E. coli

The activities of recombinant PALs from various genetic sources were analyzed according to both *in vitro* (crude lysate) and *in vivo* (whole, resting cell) assays. All recombinant PALs, expressed from plasmids pSencPSm, pSpalAv, pSpalNp, pSpal1At, and pSpal2At (Table 1) in an *E. coli* BL21(DE3) background were recovered as crude lysates. According to our *in vitro* assay results, all of the tested PALs showed comparable levels of activity on L-phenylalanine as substrate (Table 2), with the exception of EncP from *S. maritimus* whose activity was non-measureable. It has previously been shown through kinetic studies on prokaryotic PALs that those derived from *N. puntiformes* and *A. variabilis* possessed 500–1000 times greater activity than EncP from *S. maritimus*; furthermore, the k_{cat}/K_m value of PAL from *A. variabilis* was found to be greater than that of *N. puntiformes* (72.2 and 43.8 $mM^{-1}\,s^{-1}$, respectively) (Moffitt et al., 2007b). Thus, it is possible that EncP was in fact functionally expressed in our study; however, its activity was too low to measure according to the protocols employed. The relative activities of candidate PALs were then further explored through the use of whole-cell assays in recombinant *E. coli*. Resting cells suspended in PBS buffer (pH 7) were supplemented with L-phenylalanine or L-tyrosine and product (tCA or *p*-coumaric acid, respectively) formation was monitored over the course of 3 h. Although the PALs from *A. thaliana* were found to possess the greatest specific activity in crude lysates (Table 2), assays of *in vivo* function provided a more stark contrast into the relative activities among all candidate PALs, as shown in Fig. 3. Under the studied conditions, the PALs of *A. thaliana* were found to display the greatest activity (being significantly higher that of either prokaryotic PAL) although being comparable for both isoenzymes. This result is more consistent with previous studies wherein k_{cat}/K_m values of PAL1 and PAL2 were estimated as 25,500 and 51,200 $M^{-1}\,s^{-1}$ (or six orders of magnitude greater than those reported for *N. puntiformes* and *A. variabilis*) when expressed in recombinant *E. coli* (Cochrane et al., 2004). The elevated relative activity of PAL2 is also consistent with the observation of higher initial rates of tCA accumulation when expressed from BL21(DE3) pSpal2At, as seen in Fig. 3.

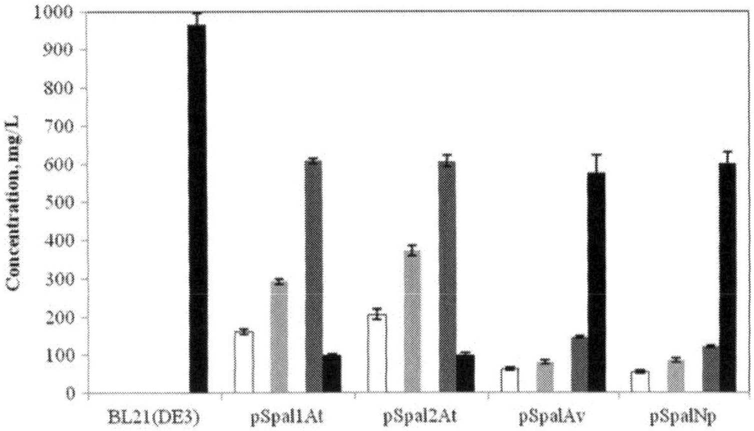

Fig. 3. Phenylalanine ammonia lyase activity from candidate genes cloned from *A. thaliana*, *A. variabilis*, and *N. punctiformes* in recombinant *E. coli* BL21(DE3) whole cells. 50 ml cultures were grown for 8 h (induced with 0.2 mM IPTG after 1.5 h), spun down and re-suspended in 10 ml PBS buffer. The conversion of 1 g/L L-phenylalanine (black) to *trans*-cinnamic acid after 1 h (no color), 2 h (light gray), and 3 h (dark gray). Error bars reported at one standard deviation from triplicate experiments.

Please cite this article as: McKenna, R., Nielsen, D.R., Styrene biosynthesis from glucose by engineered E. coli. Metab. Eng. (2011), doi:10.1016/j.ymben.2011.06.005

Again, however, none of the tested PAL isoenzymes were found to display any activity on L-tyrosine as the substrate (data not shown), which is consistent with the results of our *in vitro* assays (Table 2). Taken together, these analyses demonstrate that all of the studied PALs display rigid substrate specificity for L-phenylalanine. This important result contrasts numerous prior works on aromatic pathway development in recombinant *E. coli*, which have solely relied upon the use of bifunctional yeast PAL/TALs (Cui et al., 2008; Gilbert et al., 1985; Gilbert et al., 1983; Gilbert and Tully, 1982) and which was specifically used to synthesize *p*-hydroxystyrene (Qi et al., 2007). As these results illustrate that the first committed step in our pathway is highly specific for the intended substrate (L-phenylalanine) alone, we anticipate that this advantageous outcome will ultimately help to control product purity (specifically styrene over *p*-hydroxystyrene or a mixture of products) while also improving the activity and flux of our desired pathway. Meanwhile, the same cannot be assured had the first committed step of our pathway been catalyzed by PAL/TAL, as was the case for all engineered *p*-hydroxystyrene pathways reported to date (Qi et al., 2007; Verhoef et al., 2009).

3.3. Screening candidate PADC isoenzymes for tCA decarboxylase activity in recombinant E. coli

Candidate PADC isoenzymes from *L. plantarum*, *B. subtilis*, and *S. cerevisiae* were screened for their ability to decarboxylate tCA to produce styrene when expressed in *E. coli*. Plasmids harboring the candidate PADC-encoding genes (Table 1) were individually transformed into *E. coli* BL21(DE3), as described above. Whole, resting cells were prepared in PBS buffer supplemented with 1 g/L of tCA acid or *p*-coumaric acid. The production of styrene or *p*-hydroxystyrene, respectively, was then followed periodically, and the results after 12 h of culture are compared in Fig. 4. With the exception of the strain expressing *PAD1* from *S. cerevisiae* alone, all other strains displayed decarboxylase activity on *p*-coumaric acid, leading to *p*-hydroxystyrene biosynthesis. These results are consistent with previous reports on the functional expression of *pdc* and *padC* in recombinant *E. coli* (Qi et al., 2007) to support p-hydroxystyrene biosynthesis from glucose. However, our findings further and importantly show that the enzymes from *L. plantarum* and *B. subtilis* cannot catalyze the conversion of tCA to styrene (thus making them more specific for *p*-coumaric acid). In the present study it was instead found that the sole expression of *FDC1* from *S. cerevisiae* (which was previously characterized to encode a ferulic acid decarboxylase) was sufficient for achieving decarboxylase activity on either tCA or *p*-coumaric acid (with perhaps a slight preference towards tCA, based on overall conversion). Though it was previously reported that the co-expression of both *FDC1* and *PAD1* is necessary to achieve tCA decarboxylase activity in the native *S. cerevisiae* (Mukai et al., 2010), we now report that functional tCA decarboxylase activity in *E. coli* depends solely upon *FDC1* over-expression and is not dependent upon the co-expression of *PAD1*. Furthermore, as can be seen in Fig. 4, comparable styrene titers were achieved when either *FDC1* was expressed alone or together with *PAD1*, indicating that the co-expression of *PAD1* does not increase (or otherwise alter) tCA decarboxylase activity in *E. coli*, as has previously been suggested in the native *S. cerevisiae* (Larsson et al., 2001). It is, however, plausible that the expression of *ubix* (which has been shown to be 50% similar to *PAD1* (Mukai et al., 2010)) in our *E. coli* background could have served to compensate for the absence of *PAD1* expression, enabling tCA decarboxylase activity to be achieved in *E. coli* expressing *FDC1* alone. Future studies will explore the deletion of *ubix* from the background of *E. coli* to further investigate the recombinant function of *FDC1*. Most importantly, these results have demonstrated that *FDC1* over-expression in *E. coli* uniquely enables the decarboxylation of tCA to styrene.

Although the proposed styrene biosynthesis pathway is somewhat analogous to that which was previously developed for

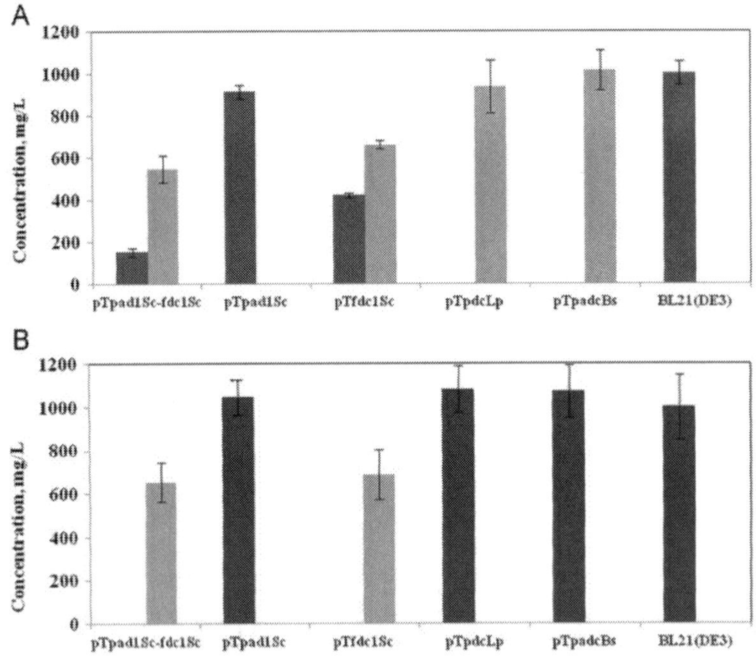

Fig. 4. Phenylacrylic acid decarboxylase activity from candidate genes cloned from *S. cerevisiae*, *L. plantarum*, and *B. subtilis* in recombinant *E. coli* BL21(DE3) whole cells. The conversion of (A) 1 g/L *p*-coumaric acid (dark gray) to *p*-hydroxystyrene (light gray) and (B) 1 g/L *trans*-cinnamic acid (dark gray) to styrene (light gray) after 12 h. Error bars reported at one standard deviation from triplicate experiments.

Please cite this article as: McKenna, R., Nielsen, D.R., Styrene biosynthesis from glucose by engineered *E. coli*. Metab. Eng. (2011), doi:10.1016/j.ymben.2011.06.005

p-hydroxystyrene biosynthesis (Qi et al., 2007; Verhoef et al., 2009), it is essential to realize that the specific enzyme 'parts' used in those two prior studies are wholly inadequate for achieving the present objective of styrene biosynthesis from glucose. In other words, said previously characterized enzymes were not simply transferrable among the two (styrene and *p*-hydroxystyrene) distinct pathways, but rather that styrene biosynthesis could only be supported through the identification and characterization of a specific tCA decarboxylase.

3.4. Biosynthesis of tCA and styrene from glucose in recombinant E. coli

The biosynthesis of tCA from glucose was first investigated using strains of *E. coli* NST74 that individually carried the plasmids pSpalAv, pSpalNp, pSpal1At, and pSpal2At. Cultures were grown in MM1 media with 1.5% glucose. 48 h after induction with 0.2 mM IPTG, the production of tCA was detected in each culture at final titers of 600, 473, 648, and 918 mg/L, respectively. Neither styrene nor *p*-coumaric acid were detected in any samples.

To test for styrene biosynthesis from glucose, *FDC1* from *S. cerevisiae* was then co-expressed with each of the PAL-encoding genes according to the following constructed strains: NST74 pSpalAv pTfdc1Sc, NST74 pSpalNp pTfdc1Sc, NST74 pSpal1At pTfdc1Sc, and NST74 pSpal2At pTfdc1Sc. Conversion of glucose to styrene was followed after IPTG induction and results for the strain NST74 pSpal2At pTfdc1Sc are shown in Table 3 (other results not shown as this dataset was representative of the results and trends observed with all other strains). Substantial L-phenylalanine accumulation was not observed until 17 h post-induction, after which time styrene titers then also rose considerably, reaching a final titer of 260 mg/L in the culture medium 29 h post-induction. The final styrene titers for the strains NST74 pSpalAv pTfdc1Sc, NST74 pSpalNp pTfdc1Sc, and NST74 pSpal1At pTfdc1Sc were found to be 210, 183, and 188 mg/L, respectively. As styrene-producing strains were cultured, a strong 'hydrocarbon' aroma was readily detected upon opening the sealed culture bottles. The high volatility of styrene necessitated the use of sealed jars for culturing. To ensure that sufficient oxygen remained available to the culture, a large headspace volume (200 mL headspace vs. 50 mL media) was used. Thus, cultures were maintained under aerobic conditions throughout the culture (as confirmed by the inclusion of the indicator dye resazurin). However, the use of such a large headspace volume also allows for the significant accumulation of styrene vapor, which should also be accounted for when assessing productivity. Under dilute and near-ambient conditions, the equilibrated headspace vapor composition can be estimated by application of Henry's Law. Using a dimensionless Henry's Law constant of 0.113 (Yang, 1992), it can be predicted that a headspace equilibrated with an aqueous phase containing 260 mg/L styrene will contain an additional 29 mg/L styrene, for a total of 18.9 mg in the 250 mL flask (note this would be equivalent to an aqueous titer of 376 mg/L if no volatilization had occurred). In an analogous manner, the final styrene production by strains NST74 pSpalAv pTfdc1Sc, NST74 pSpalNp pTfdc1Sc, and NST74 pSpal1At pTfdc1Sc would be estimated as 15.2, 13.4, and 13.6 mg, respectively.

Table 3
Biosynthesis of L-phenylalanine, *trans*-cinnamic acid, and styrene by *E. coli* NST74 pSpal2At pTfdc1Sc from glucose in MM1 media. Errors reported at one standard deviation from triplicate experiments.

Time (h)	L-Phenylalanine (mg/L)	*trans*-Cinnamic acid (mg/L)	Styrene (mg/L)
0	13.9 ± 0.03	0	0
13	56.4 ± 0.13	3.7 ± 0.01	7.3 ± 0.12
17	236.5 ± 0.57	6.3 ± 0.01	23 ± 0.38
21	167.4 ± 0.40	9.1 ± 0.02	205 ± 3.40
25	152.2 ± 0.36	13.5 ± 0.02	243 ± 4.03
29	179.5 ± 0.43	14.0 ± 0.03	260 ± 4.31

Following these fermentation studies, culture supernatants were extracted using hexane, and the extracts were analyzed by GC–MS to confirm that it was in fact styrene that was being synthesized by our cultures. Comparing the spectra of the dominant metabolite peak which was recovered from the extract to the NIST08 spectral database (Babushok et al., 2007), styrene was found to be the most probable compound. As can be seen from Fig. 5, the mass spectra of the extracted sample and the library reference provide an excellent match, providing confirmation that styrene was in fact synthesized by our engineered strains of *E. coli*.

Throughout the duration of culture, never was all of the L-phenylalanine observed to be fully assimilated into the styrene pathway (as seen in Table 3). Furthermore, tCA titers were observed to remain low throughout, indicating that almost all of the synthesized tCA could be quickly converted to styrene. Taken together, these observations suggest that low PAL activity presently remains as the flux limiting condition in the engineered styrene biosynthesis pathway. As the first committed step in the styrene pathway, high PAL activity is essential and must be

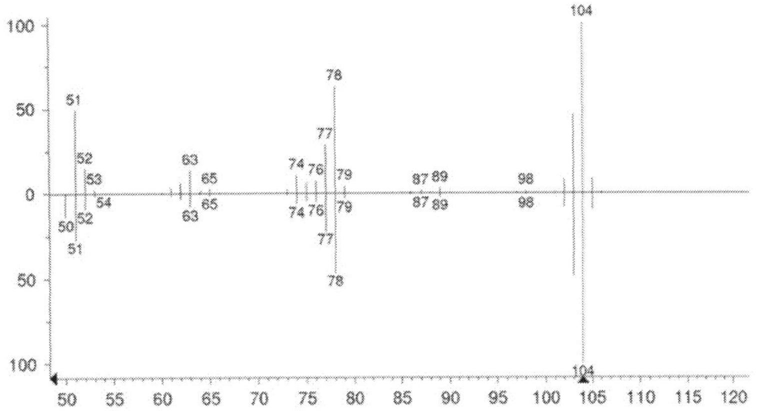

Fig. 5. Head to tail comparison of the standard mass spectra showing the relative abundance of the mass-to-charge ratio of styrene from the NIST08 library (lower) with that of the dominant metabolite peak obtained in hexane extractions of the culture broth (upper).

Please cite this article as: McKenna, R., Nielsen, D.R., Styrene biosynthesis from glucose by engineered *E. coli*. Metab. Eng. (2011), doi:10.1016/j.ymben.2011.06.005

improved in subsequent generations of our strains. To start, the expression of PAL can likely be improved through the use of codon optimized variants. This approach has shown to be useful, particularly in cases for which pathways involve the expression of plant-derived genes in *E. coli*. For example, codon optimization was applied to the amorphadiene oxidase from *Artemisia annua* (a plant species) as part as a synthetic pathway to produce artemisinic acid (Keasling, 2008). Whereas the native protein originally showed neither *in vivo* nor *in vitro* activity when expressed in *E. coli*, a codon optimized variant resulted in its functional expression in *E. coli*, contributing to successful pathway development.

Whereas improvements in PAL expression can lead to enhancements in the specific metabolite flux, the net flux can also be improved by increasing the pathway 'driving force'. That is, by promoting the increased availability of the pathway's immediate precursor, L-phenylalanine. This notion was tested in resting cells assays wherein *E. coli* NST74 pSpal2At pTfdc1Sc cell suspensions were supplemented with exogenous L-phenylalanine at initial concentrations of either 400 or 950 mg/L. As shown in Fig. 6, the conversion of 400 mg/L L-phenylalanine to styrene occurred rapidly and completely (i.e., no residual L-phenylalanine or tCA was detectable), yielding a final styrene titer of 250 mg/L after 30 h. Meanwhile, when 950 mg/L L-phenylalanine was added to the resting cell cultures, nearly 500 mg/L styrene could be produced as all of the L-phenylalanine was consumed. However, in this case, nearly an additional 250 mg/L tCA also remained in the culture. Noting that such a final styrene titer actually surpasses the toxicity threshold of growing cells, it is not altogether surprising that the reaction was unable to proceed to completion. It is speculated that the final styrene titer was only capable of surpassing the previously-determined toxicity threshold in resting cell studies because the biotransformation in this case was no longer growth dependent. That is to say that the stationary phase resting cells had previously assembled the necessary pathway enzymes and were not reliant upon endogenously synthesized L-phenylalanine as a precursor. Thus, the impediment of conversion observed at further elevated styrene titers most likely represents an inhibition imposed at the level of the pathway enzymes rather than an impairment of cell growth, as was observed in our previous characterization experiments. Further experiments will be required to characterize and understand this potential inhibitory mechanism. These results demonstrate that if the L-phenylalanine precursor pool can be enhanced the net production of styrene can also be increased. Clearly, however, the addition of exogenous L-phenylalanine is not a sustainable approach towards enhancing styrene biosynthesis, but rather endogenous L-phenylalanine production must be enhanced in our host platform. Subsequent studies will explore the incorporation of additional modifications into the genome of the *E. coli* NST74 host that have previously been shown by other groups to result in L-phenylalanine yield enhancements. For instance, it has been reported that the over-expression of endogenous transketolase I (encoded by *tktA*) supports elevated L-phenylalanine biosynthesis (Gosset et al., 1996). TktA over-expression enables the enhanced biosynthesis of erythrose 4-phosphate (E4P) which, when condensed with phosphoenolpyruvate (PEP), yields 3-deoxy-D-arabino-heptulosonate 7-phosphate (DAHP), the first committed intermediate in the aromatic amino acid biosynthesis pathway (Flores et al., 1996; Lütke-Eversloh and Stephanopoulos, 2008). Meanwhile, disruption of the global carbon storage regulatory system of *E. coli* through deletion of *csrA*, increases PEP biosynthesis, leading to up to a 2-fold enhancement in L-phenylalanine yield (Tatarko and Romeo, 2001).

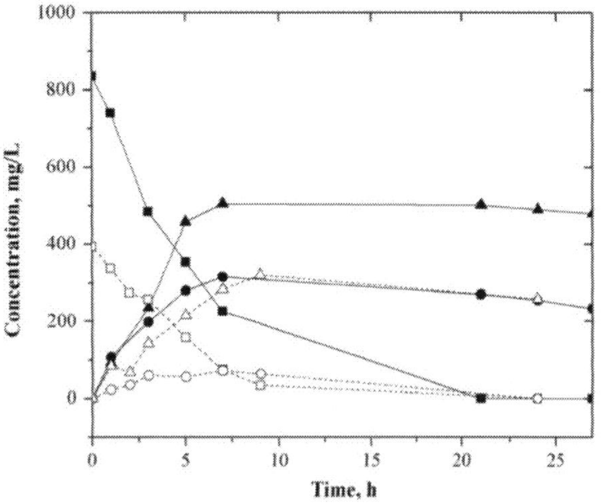

Fig. 6. Whole cell production of styrene (triangle) from *trans*-cinnamic acid (circle) after exogenous L-phenylalanine (square) addition by *E. coli* NST74 pSpal2At pTfdc1Sc. Open and filled shapes correspond to the initial addition of 400 and 950 mg/L L-phenylalanine.

For microbially derived, renewable styrene to become an economically viable and sustainable alternative to petroleum-derived predecessor, titers and productivity must ultimately be improved. Although the styrene titers achieved by our 1st generation strains were modest, in comparison to the toxicity assays performed above it can be seen that they are already approaching the inhibitory threshold. Thus, whereas increasing L-phenylalanine yields could eventually translate into elevated styrene production, product toxicity would soon become the subsequent limiting factor and must be addressed. As complex phenotypes such as solvent tolerance are simply not monogenic in nature (Alper et al., 2006), numerous specific mutations must applied in concert to achieve the specifically desired result. Since little is presently known of styrene toxicity or tolerance, rational approaches to engineering more tolerant strains may be less suitable than higher throughput, combinatorial strategies. Although commonly employed, combinatorial approaches towards enhancing desired phenotypes often rely upon the aggressive use of chemomutagenesis, such techniques are less desirable as they can also result an unforeseen and difficult to understand negative impacts on host fitness and/or productivity (Bonomo et al., 2006), in addition to requiring laborious screening and selection procedures. A more effective approach for the present application might involve the use of an alternative host platform which inherently boasts greater tolerance (Fischer et al., 2008), a strategy which has been applied for the production of biofuels, such as n-butanol (Nielsen et al., 2009). The bacterium *P. putida* S12 has been engineered, for example, as a solvent tolerant platform for the biosynthesis of both *p*-hydroxybenzoate and *p*-hydroxystyrene (Verhoef et al., 2007, 2009), and might also make excellent starting point for styrene production. Since our pathway is derived from the ubiquitous, proteinogenic amino acid L-phenylalanine, its transference to an alternative host platform remains wholly compatible. Alternatively, improved product tolerance can also be achieved through genome evolution, as accomplished through serial adaptations to increasing styrene concentrations. Such an approach has worked well for achieving tolerance to biofuels like n-butanol and could also be coupled with genomic library screens to provide the first comprehensive view of styrene inhibition and tolerance in *E. coli* (Reyes et al., 2011). In an interesting example, prior works have shown that *Pseudomonas* sp. may be adapted to styrene when also grown in the presence of acetate or similar carboxylic acids as the sole

Please cite this article as: McKenna, R., Nielsen, D.R., Styrene biosynthesis from glucose by engineered *E. coli*. Metab. Eng. (2011), doi:10.1016/j.ymben.2011.06.005

carbon source (Weber et al., 1993). It has been suggested that this evolved phenotype may be associated with genetic changes leading to reductions in membrane fluidity, as well as through the activation of genes believed to be specifically associated with enhanced tolerance to aromatic compounds. As an alternative strategy, combinatorial procedures such as the global Transcription Machinery Engineering (gTME) (Alper et al., 2006; Nicolaou et al., 2010) may be implemented to improve styrene tolerance in *E. coli* or alternative host platforms selected due to their elevated tolerance baseline. The principals of gTME have been successfully employed to enhance ethanol tolerance in both yeast (Alper et al., 2006) and *E. coli* (Alper and Stephanopoulos, 2007).

In addition to the development of styrene tolerant phenotypes, product toxicity can also be enhanced through the use of in situ product recovery (ISPR). Common ISPR approaches involve solvent extraction (Gyamerah and Glover, 1996; Malinowski, 2001; Weilnhammer and Blass, 1994), adsorption (Nielsen et al., 2010; Nielsen and Prather, 2009), gas and vacuum stripping (Loser et al., 2005), and membrane pervaporation (Vane, 2005). Each of these approaches has been successfully applied for the continuous recovery of biofuel compounds such as ethanol, for example, and even aromatics like L-phenylacetylcarbinol (Khan and Daugulis, 2010) and benzaldehyde (Jain et al., 2010). The latter three approaches may be particularly well-suited for styrene recovery given its volatile nature, as previously discussed.

4. Conclusion

For the first time, the present study has demonstrated the biosynthesis of styrene from renewable resources using an engineered microbial platform. Whereas low activity of pathway enzymes and product toxicity remain as challenges which limit the productivity of our styrene-producing strains, continued improvements will lead to the development of robust biocatalysts for the sustainable production of this important, large volume, commodity chemical.

Acknowledgments

We thank Prof. Bradley Moore (UCSD) and Prof. Kristala Prather (MIT) for their kind gifts. We also thank Tom Colella (ASU) for his skillful technical assistance in GC–MS analysis.

References

Alper, H., Moxley, J., Nevoigt, E., Fink, G.R., Stephanopoulos, G., 2006. Engineering yeast transcription machinery for improved ethanol tolerance and production. Science 314, 1565–1568.

Alper, H., Stephanopoulos, G., 2007. Global transcription machinery engineering: a new approach for improving cellular phenotype. Metabolic Engineering 9, 258–267.

Babushok, V.I., Linstrom, P.J., Reed, J.J., Zenkevich, I.G., Brown, R.L., Mallard, W.G., Stein, S.E., 2007. Development of a database of gas chromatographic retention properties of organic compounds. Journal of Chromatography A. 1157, 414–421.

Baez-Viveros, J.L., Osuna, J., Hernandez-Chavez, G., Soberon, X., Bolivar, F., Gosset, G., 2004. Metabolic engineering and protein directed evolution increase the yield of L-phenylalanine synthesized from glucose in *Escherichia coli*. Biotechnology and Bioengineering 87, 516–524.

Bonomo, J., Warnecke, T., Hume, P., Marizcurrena, A., Gill, R.T., 2006. A comparative study of metabolic engineering anti-metabolite tolerance in *Escherichia coli*. Metabolic Engineering. 8, 227–239.

Boudet, A.M., 2007. Evolution and current status of research in phenolic compounds. Phytochemistry 68, 2722–2735.

Clark, G., 1990. Phenethyl alcohol. Perfumer Flavor 15, 37–44.

Clausen, M., Lamb, C.J., Megnet, R., Doerner, P.W., 1994. PAD1 encodes phenylacrylic acid decarboxylase which confers resistance to cinnamic acid in *Saccharomyces cerevisiae*. Gene 142, 107–112.

Cochrane, F.C., Davin, L.B., Lewis, N.G., 2004. The Arabidopsis phenylalanine ammonia lyase gene family: kinetic characterization of the four PAL isoforms. Phytochemistry 65, 1557–1564.

Cui, J.D., Jia, S.R., Sun, A.Y., 2008. Influence of amino acids, organic solvents and surfactants for phenylalanine ammonia lyase activity in recombinant *Escherichia coli*. Letters in Applied Microbiology 46, 631–635.

DoE, Steam system opportunity assessment for the pulp and paper, chemical manufacturing, and petroleum refining industries. In: Energy, O. o. E. E. a. R. E. U. S. D. o. (Ed.), 2002.

Fernandez, X., Lizzani-Cuvelier, L., Loiseau, A.M., Perichet, C., Delbecque, C., Arnaudo, J.F., 2005. Chemical composition of the essential oils from Turkish and Honduras Styrax. Flavour and Fragrance Journal 20, 70–73.

Fischer, C.R., Klein-Marcuschamer, D., Stephanopoulos, G., 2008. Selection and optimization of microbial hosts for biofuels production. Metabolic Engineering 10, 295–304.

Flores, N., Xiao, J., Berry, A., Bolivar, F., Valle, F., 1996. Pathway engineering for the production of aromatic compounds in *Escherichia coli*. Nature Biotechnology 14, 620–623.

Gilbert, H.J., Clarke, I.N., Gibson, R.K., Stephenson, J.R., Tully, M., 1985. Molecular-cloning of the phenylalanine ammonia-lyase gene from rhodosporidium-toruloides in *Escherichia-coli* K-12. Journal of Bacteriology 161, 314–320.

Gilbert, H.J., Stephenson, J.R., Tully, M., 1983. Control of synthesis of functional messenger-rna coding for phenylalanine ammonia-lyase from rhodosporidium-toruloides. Journal of Bacteriology 153, 1147–1154.

Gilbert, H.J., Tully, M., 1982. Synthesis and degradation of phenylalanine ammonia-lyase of rhodosporidium-toruloides. Journal of Bacteriology 150, 498–505.

Gosset, G., Yong-Xiao, J., Berry, A., 1996. A direct comparison of approaches for increasing carbon flow to aromatic biosynthesis in *Escherichia coli*. Journal of Industrial Microbiology 17, 47–52.

Gyamerah, M., Glover, J., 1996. Production of ethanol by continuous fermentation and liquid-liquid extraction. Journal of Chemical Technology and Biotechnology 66, 145–152.

Heipieper, H.J., Weber, F.J., Sikkema, J., Keweloh, H., de Bont, J.A.M., 1994. Mechanisms of resistance of whole cells to toxic organic solvents. Trends in Biotechnology 12, 409–415.

Jain, A.N., Khan, T.R., Daugulis, A.J., 2010. Bioproduction of benzaldehyde in a solid–liquid two-phase partitioning bioreactor using *Pichia pastoris*. Biotechnology Letters 32, 1649–1654.

Jiang, H., Wood, K.V., Morgan, J.A., 2005. Metabolic engineering of the phenylpropanoid pathway in *Saccharomyces cerevisiae*. Applied and Environmental Microbiology 71, 2962–2969.

Jones, D.H., 1984. Phenylalanine ammonia-lyase: regulation of its induction, and its role in plant development. Phytochemistry 23, 1349–1359.

Keasling, J.D., 2008. Synthetic biology for synthetic chemistry. ACS Chemical Biology 3, 64–76.

Khan, T.R., Daugulis, A.J., 2010. Application of solid–liquid TPPBs to the production of L-phenylacetylcarbinol from benzaldehyde using *Candida utilis*. Biotechnology and Bioengineering 107, 633–641.

Larsson, S., Nilvebrant, N.O., Jonsson, L.J., 2001. Effect of overexpression of *Saccharomyces cerevisiae* Pad1p on the resistance to phenylacrylic acids and lignocellulose hydrolysates under aerobic and oxygen-limited conditions. Applied Microbiology and Biotechnology 57, 167–174.

Loser, C., Schroder, A., Deponte, S., Bley, T., 2005. Balancing the ethanol formation in continuous bioreactors with ethanol stripping. Engineering in Life Sciences 5, 325–332.

Lütke-Eversloh, T., Stephanopoulos, G., 2008. Combinatorial pathway analysis for improved L-tyrosine production in *Escherichia coli*: identification of enzymatic bottlenecks by systematic gene overexpression. Metabolic Engineering 10, 69–77.

Malinowski, J.J., 2001. Two-phase partitioning bioreactors in fermentation technology. Biotechnology Advances 19, 525–538.

Moffitt, M.C., Louie, G.V., Bowman, M.E., Pence, J., Noel, J.P., Moore, B.S., 2007a. Discovery of two cyanobacterial phenylalanine ammonia lyases: kinetic and structural characterization. Biochemistry 46, 1004–1012.

Moffitt, M.C., Louie, G.V., Bowman, M.E., Pence, J., Noel, J.P., Moore, B.S., 2007b. Discovery of two cyanobacterial phenylalanine ammonia lyases: kinetic and structural characterization. Biochemistry 46, 1004–1012.

Mukai, N., Masaki, K., Fujii, T., Kawamukai, M., Iefuji, H., 2010. PAD1 and FDC1 are essential for the decarboxylation of phenylacrylic acids in *Saccharomyces cerevisiae*. Journal of Bioscience and Bioengineering 109, 564–569.

Nicolaou, S.A., Gaida, S.M., Papoutsakis, E.T., 2010. A comparative view of metabolite and substrate stress and tolerance in microbial bioprocessing: from biofuels and chemicals, to biocatalysis and bioremediation. Metabolic Engineering 12, 307–331.

Nielsen, D.R., Amarasiriwardena, G.S., Prather, K.L.J., 2010. Predicting the adsorption of second generation biofuels by polymeric resins with applications for in situ product recovery (ISPR). Bioresource Technology 101, 2762–2769.

Nielsen, D.R., Leonard, E., Yoon, S.H., Tseng, H.C., Yuan, C., Prather, K.L., 2009. Engineering alternative butanol production platforms in heterologous bacteria. Metabolic Engineering.

Nielsen, D.R., Prather, K.J., 2009. In situ product recovery of n-butanol using polymeric resins. Biotechnology and Bioengineering 102, 811–821.

Pagot, Y., Belin, J.M., Husson, F., Spinnler, H.E., 2007. Metabolism of phenylalanine and biosynthesis of styrene in *Penicillium camemberti*. Journal of Dairy Research 74, 180–185.

Please cite this article as: McKenna, R., Nielsen, D.R., Styrene biosynthesis from glucose by engineered *E. coli*. Metab. Eng. (2011), doi:10.1016/j.ymben.2011.06.005

Qi, W.W., Vannelli, T., Breinig, S., Ben-Bassat, A., Gatenby, A.A., Haynie, S.L., Sariaslani, F.S., 2007. Functional expression of prokaryotic and eukaryotic genes in *Escherichia coli* for conversion of glucose to p-hydroxystyrene. Metabolic Engineering 9, 268–276.

Ramos, J.L., Duque, E., Gallegos, M.-T., Godoy, P., Ramos-González, M.I., Rojas, A., Terán, W., Segura, A., 2002. Mechanisms of solvent tolerance in gram-negative bacteria. Annual Review of Microbiology 56, 743–768.

Reyes, L.H., Almario, M.P., Kao, K.C., 2011. Genomic library screens for genes involved in n-butanol tolerance in *Escherichia coli*. PLoS ONE 6, e17678.

Santos, C.N.S., Koffas, M., Stephanopoulos, G., 2011. Optimization of a heterologous pathway for the production of flavonoids from glucose. Metabolic Engineering 13, 392–400.

SRI, 2010. Styrene. Access Intelligence LLC Inc.

Tatarko, M., Romeo, T., 2001. Disruption of a global regulatory gene to enhance central carbon flux into phenylalanine biosynthesis in *Escherichia coli*. Current Microbiology 43, 26–32.

Tran, N.P., Gury, J., Dartois, V., Nguyen, T.K.C., Seraut, H., Barthelmebs, L., Gervais, P., Cavin, J.-F., 2008. Phenolic acid-mediated regulation of the padC gene, encoding the phenolic acid decarboxylase of *Bacillus subtilis*. Journal of Bacteriology 190, 3213–3224.

Vane, L.M., 2005. A review of pervaporation for product recovery from biomass fermentation processes. Journal of Chemical Technology and Biotechnology 80, 603–629.

Vannelli, T., Wei, Qi, W., Sweigard, J., Gatenby, A.A., Sariaslani, F.S., 2007a. Production of p-hydroxycinnamic acid from glucose in *Saccharomyces cerevisiae* and *Escherichia coli* by expression of heterologous genes from plants and fungi. Metabolic Engineering 9, 142–151.

Vannelli, T., Xue, Z.X., Breinig, S., Qi, W.W., Sariaslani, F.S., 2007b. Functional expression in *Escherichia coli* of the tyrosine-inducible tyrosine ammonia-lyase enzyme from yeast Trichosporon cutaneum for production of p-hydroxy-cinnamic acid. Enzyme and Microbial Technology 41, 413–422.

Verhoef, S., Ruijssenaars, H.J., de Bont, J.A., Wery, J., 2007. Bioproduction of p-hydroxybenzoate from renewable feedstock by solvent-tolerant *Pseudomonas putida* S12. Journal of Biotechnology 132, 49–56.

Verhoef, S., Wierckx, N., Westerhof, R.G., de Winde, J.H., Ruijssenaars, H.J., 2009. Bioproduction of p-hydroxystyrene from glucose by the solvent-tolerant bacterium *Pseudomonas putida* S12 in a two-phase water-decanol fermentation. Applied and Environmental Microbiology 75, 931–936.

Weber, F.J., Ooijkaas, L.P., Schemen, R.M., Hartmans, S., de Bont, J.A., 1993. Adaptation of *Pseudomonas putida* S12 to high concentrations of styrene and other organic solvents. Applied and Environmental Microbiology 59, 3502–3504.

Weilnhammer, C., Blass, E., 1994. Continuous fermentation with product recovery by in-situ extraction. Chemical Engineering & Technology 17, 365–373.

Wierckx, N.J., Ballerstedt, H., de Bont, J.A., Wery, J., 2005. Engineering of solvent-tolerant *Pseudomonas putida* S12 for bioproduction of phenol from glucose. Applied and Environmental Microbiology 71, 8221–8227.

Wu, C., Koylinski, T., Bozik, J., 1981. Preparation of styrene from ethylbenzene US Patent 4,255,599.

Xiang, L.K., Moore, B.S., 2002. Inactivation, complementation, and heterologous expression of encP, a novel bacterial phenylalanine ammonia-lyase gene. Journal of Biological Chemistry 277, 32505–32509.

Xiang, L.K., Moore, B.S., 2005. Biochemical characterization of a prokaryotic phenylalanine ammonia lyase. Journal of Bacteriology 187, 4286–4289.

Xiang, L.K., Moore, B.S., 2006. Biochemical characterization of a prokaryotic phenylalanine ammonia lyase (187, pg4286, 2005). Journal of Bacteriology 188, 5331.

Yang, C.L.Y. a.H.C., 1992. Henry's Law Constant for Compound in Water. Gulf Publishing Company, Houston.

Young, M.R., Towers, G.H.N., Neish, A.C., 1966. Taxonomic distribution of ammonia-lyases for L-phenylalanine and L-tyrosine in relation to lignification. Canadian Journal of Botany 44, 341.

Please cite this article as: McKenna, R., Nielsen, D.R., Styrene biosynthesis from glucose by engineered *E. coli*. Metab. Eng. (2011), doi:10.1016/j.ymben.2011.06.005

(388d) Enhanced Xylitol Production Through Simultaneous Co-Consumption of Cellobiose and Xylose by An Engineereed Saccharomyces Cerevisiae Strain

Eun Joong Oh[1], Suk-Jin Ha[1], Soo Rin Kim[1], Jonathan M. Galazka[2], Jamie H.D. Cate[2] and Yong-Su Jin[1], (1)Department of Food Science and Human Nutrition, University of Illinois at Urbana-Champaign, Urbana, IL, (2)Department of Molecular and Cell Biology,, University of California at Berkeley, Berkeley, CA

Xylitol is a five-carbon sugar alcohol and has been used as a sugar substitute in the food industry because of its low caloric and anti-carcinogenic characteristics. In addition, xylitol is a building block for a variety of commodity chemicals. Biological routes for producing xylitol from cellulosic hydrolysates have been developed through metabolic engineering of *Saccharomyces cerevisiae. S. cerevisiae* cannot utilize xylose as a carbon source, but the expression of a xylose reductase (XR) gene from *Pichia stipitis* facilitates the production of xylitol from xylose with high yields. However, insufficient supply of NAD(P)H for XR, and inhibition of xylose transport by glucose, are major constraints for high productivity production of xylitol.

To overcome these problems, we engineered *S. cerevisiae* to convert xylose into xylitol through the simultaneous consumption of cellobiose and xylose. Specifically, xylose reductase from *P. stipitis* was chromosomally integrated into the *S. cerevisiae* genome, while a cellobiose transporter (*cdt-1*) and an intracellular β-glucosidase (*gh1-1*) from *Neurospora crassa* were introduced on a multi-copy plasmid. The resulting transformant was able to produce xylitol from a mixture of cellobiose and xylose. In contrast to glucose, cellobiose did not repress the transport of xylose, allowing the co-consumption of xylose and cellobiose. The co-consumption and metabolism of cellobiose regenerated NADPH for XR. As a result, an engineered *S. cerevisiae* strain (D-10-BT) exhibited 40% higher xylitol productivity when cellobiose was used as a co-substrate as compared to glucose. This result suggests that co-consumption of cellobiose and xylose allows efficient xylose uptake and cofactor regeneration without catabolic repression for the production of xylitol.

See more of this Session: Microbial Engineering for the Synthesis of Bulk and High-Value Products
See more of this Group/Topical: Sustainable Engineering Forum

(388e) α-Olefin Biosynthesis In the Cyanobacterium *Synechococcus Sp.* PCC7002

Daniel Mendez-Perez[1], Matthew B. Begemann[2] and Brian F. Pfleger[1], (1)Department of Chemical and BIOLOGICAL Engineering, UNIVERSITY OF WISCONSIN-MADISON, Madison, WI, (2) Microbiology Doctoral Training PROGRAM, UNIVERSITY OF WISCONSIN-MADISON, Madison, WI

Biological hydrocarbons are a promising alternative to petroleum-based liquid transportation fuels as they would be compatible with current engines and distribution systems, yet be sourced from renewable substrates. Cyanobacteria have received significant attention because the direct conversion of sunlight and carbon dioxide to lipids will by-pass recalcitrant intermediates found in terrestrial biomass. Although it has been known for decades that some species of cyanobacteria, including *Synechococcus sp.* PCC7002, can synthesize hydrocarbons of different lengths, very little is known about their biosynthesis. In this study, we analyzed the hydrocarbon profile of *Synechococcus sp.* PCC7002 and demonstrate the involvement of a gene in the biosynthesis of two C19 α-olefins. The gene encodes a large multi-domain protein with homology to Type I polyketide synthases, suggesting a new route for hydrocarbon biosynthesis from fatty acids via an elongation-decarboxylation mechanism. The combinatorial nature of polyketide synthesis raises the possibility of using this type of genes to produce a wide range of industrially relevant α-olefins and high value chemical building blocks.

See more of this Session: Microbial Engineering for the Synthesis of Bulk and High-Value Products
See more of this Group/Topical: Sustainable Engineering Forum

(388f) Building a Better Biofuel Production Platform: Engineering Metabolic Control Techniques to Control the Photosynthetic Cyanobacterium Synechocystis Sp. PCC 6803

Stevan Albers, Department of Cell and Molecular Biology, Colorado State University, Fort Collins, CO and Christie A.M. Peebles, Chemical Engineering, Colorodo State University, Fort Collins, CO

Building a Better Biofuel Production Platform:

Engineering metabolic control techniques to control the photosynthetic cyanobacterium *Synechocystis sp.* PCC 6803

Stevan C. Albers[1], Christie A.M. Peebles[1,2]

[1]Cell and Molecular Biology Graduate Program, [2]Department of Chemical and Biological Engineering

Colorado State University, Fort Collins, Colorado

Biofuel production in microorganisms has great potential to fulfill this nation's need for fungible "green" transportation fuels. Because of the simplicity of bacterial cells, photosynthetic cyanobacteria make good candidates as a platform for biofuel production. Many metabolic engineering techniques and procedures have been optimized for various heterotrophic bacteria, but few techniques have been designed with autotrophic cyanobacteria in mind. Designing rational control systems in cyanobacteria is the first step in controlling pathways that utilize carbon dioxide to produce long chain hydrocarbon like molecules. Understanding and controlling the metabolic pathways that produce these molecules is critical to demonstrate the true functionality of these photosynthetic organisms.

Synechocystis sp. PCC 6803 is a very versatile organism for a variety of reasons. It is a photoautotrophic bacterium for which it's haploid genomic DNA has been sequenced. Because of the simplicity of cyanobacterial cells, higher system efficiencies can be reached when compared to other photosynthetic algal cells. Manipulation techniques like natural uptake, electroporation, and homologous recombination are well annotated in this species. The organism is capable of rapid doubling times while harboring such metabolic pathways as flavonoid, clavulanic acid, and fatty acid biosynthesis while also harboring pathways capable of toluene, atrizine, DDT degradation. Our group plans to focus on understanding specific pathways endemic to *Synechocystis* and utilizing these pathways to further the molecular engineering toolset available for this versatile organism. Specifically, we will be discussing the design, construction, and expression of various promoter constructs capable of rational gene control in *Synechocystis sp.* PCC 6803. This talk will focus on our progress to date.

See more of this Session: Microbial Engineering for the Synthesis of Bulk and High-Value Products
See more of this Group/Topical: Sustainable Engineering Forum

(390a) Proteomic, Physiological, and *In Silico* Testing of Economic Tradeoffs In Metabolic Networks

Ross P. Carlson, **Reed Taffs** and James P. Folsom, Chemical and Biological Engineering, Montana State University, Bozeman, MT

Microbes conserve energy and perform carbon conversions using robust and often redundant networks of enzymatic machinery. Competition for resources in nature has likely selected for phenotypes wherein operational metabolic pathways optimize tradeoffs between catabolic efficiency of an enzyme set and the investment of anabolic nutrients required to synthesize the enzymes. An *in silico* economics-based analysis quantified molecular-level, resource allocation tradeoff strategies that permit competitive cellular functioning under a continuum of nutrient scarcities. The analysis decomposed a metabolic network into a complete listing of non-divisible, mathematically-defined biochemical pathways which were then used to identify all strategies for investing limiting resources like iron and nitrogen into the genome encoded metabolic machinery. Nitrogen, for example, is required for the synthesis of every enzyme; as nitrogen availability decreases, tradeoff analysis predicts a testable shift from catabolism optimized phenotypes to nitrogen investment-optimized phenotypes. *In silico* predictions were evaluated experimentally using physiological and proteomic data collected from iron- or nitrogen-limited *Escherichia coli* chemostat cultures. Experimental chemostat data trends were consistent with *in silico* theory and illustrated that under iron- and nitrogen-limited conditions *E. coli* regulates its metabolism to invest scarce resource competitively at the cost of optimal biomass yields on electron donor. The study highlights a fundamental evolutionary and metabolic design paradigm for competitive network structure and control with applications to bioprocess engineering, environmental microbiology and treatment of microbial pathogens.

See more of this Session: Modeling Approaches In the Life Sciences II
See more of this Group/Topical: Topical A: Systems Biology

(390b) Pattern Formation In the Acto-Myosin Cortex: An Active Reaction-Diffusion-Advection Process

Justin S. Bois[1], Frank Jlicher[2] and Stephan W. Grill[2], (1)Chemistry and Biochemistry, UCLA, Los Angeles, CA, (2)Biological Physics, Max-Planck Institute for the Physics of Complex Systems, Dresden, Germany

The acto-myosin cortex, which resides immediately below the eukaryotic cell membrane, is a thin network of crosslinked actin filaments on which myosin motors exert stress though active processes involving ATP hydrolysis. The cortex is continuously remodeled, so that on long time scales it may be treated as a viscous fluid. Regulators of myosin activity and actin dynamics, and therefore active stress, are also present in the cortex. To capture this biochemical and mechanical coupling, we have developed a general hydrodynamic description of the cortex as a thin film of an active viscous fluid in which active stresses are under regulation of biochemical species which undergo diffusion and chemical reactions. The coupling of active mechanical stresses with a reaction-diffusion system results in biochemical pattern formation with steady viscous flow. In many cases, active fluid motion promotes patterns in systems which would otherwise remain homogeneous. Such patterns are prevalent in embryonic development, cell motility, and polarization, and this theory provides a simple, general framework in which they may be understood. We apply the theory to the dynamics of the polarization of the one-cell *C. elegans* embryo, a system for which mechanical and chemical coupling is essential.

See more of this Session: Modeling Approaches In the Life Sciences II
See more of this Group/Topical: Topical A: Systems Biology

(390c) SHREDing a Biochemical Network Into Hierarchical Modules

Gautham V. Sridharan[1], Soha Hassoun[2] and **Kyongbum Lee**[1], (1)Chemical and Biological Engineering, Tufts University, Medford, MA, (2)Computer Science, Tufts University, Medford, MA

Hierarchical modularity has emerged as anorganizational principle of biochemical networks, providing insights into thecoordinated regulation of reactions within and across pathways. In principle, themodularity of a biochemical network should allow the system to be partitioned intominimally interdependent parts, which in turn can facilitate detailed analysisof each part in the context of the overall system. In practice, modularityanalysis has often relied on *ad hoc*decisions to mainly corroborate existing knowledge. While there is generalagreement that a module should consist of a biologically meaningful group ofconnected components in the network, there is little consensus on the metricneeded to quantitatively evaluate the quality of a partition. The goal of thisstudy is to investigate a novel metric that can be used to *systematically* partition a biochemical network into functionallyrelevant groups of reactions. The metric, termed the Shortest RetroactiveDistance (ShReD),characterizes the retroactive connectivity between any two reactions in anetwork arising from potential feedback interactions, thereby grouping togthernetwork components which mutually influence each other. We evaluate the metricusing two test networks: epidermal growth factor receptor (EGFR) signaling and liverdrug transformation.

Each test network was abstracted as areaction-centric directed. An edge was drawn betweentwo reaction nodes existed if one reaction produced either a reactant orallosteric effector of the other reaction. The ShReDbetween two reactions was calculated as the path-length of the shortest cycle thatinvolves both reaction nodes. Using ShReD as the measure of connectivity, weadapted Newman's algorithm[1] to obtain a set of hierarchical partitions.Briefly, Newman's original algorithm dividesan *undirected* network such that theresulting partition maximizes the number of connections *within* each sub-network *relativeto* the number of expected connections between two randomly chosen nodes inthe network. In this study, the partitions were performed based on the ShReD metric, rather than the undirected connectivity usedin Newman's work.

For the EGFR network, thepartitions obtained using Newman's original algorithm and ShReDwere largely similar (not shown). Importantly, the ShReDpartition generated reaction modules with larger numbers of cyclicalinteractions. For the metabolic network, which included cofactors, the ShReD partition again generated hierarchical modules whosecompositions compared favorably with canonical associations based on textbookbiochemistry (Fig. 1). The Newman partition was unable to generate anyhierarchy. Interestingly, the ShReD partitionrevealed a 'redox' module involving reactions of xenobiotic transformation,glucose metabolism, pyruvate metabolism, and lipid metabolism interactingthrough shared production and consumption of NADPH.

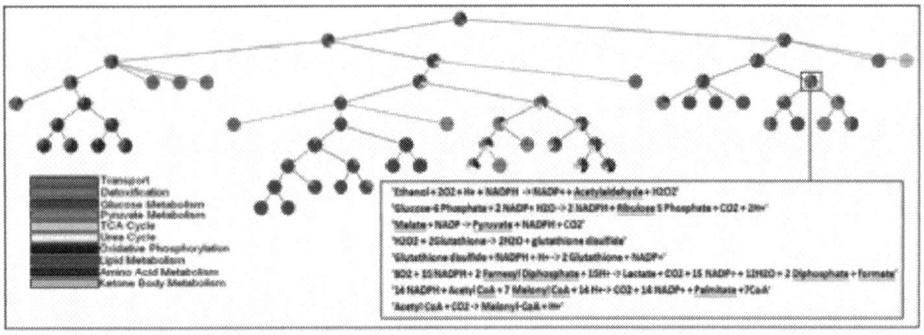

Fig. 1. ShReD-based hierarchical partition of a hepatocyte metabolic network.The pie charts represent modules of reactions. Each pie represents the fractionof

Inconclusion, our novel metric ShReD, combined with Newman's algorithm, is to ourknowledge the first modularity analysis technique that partitions a biochemicalnetwork to preserve cylical interactions between reactions.

1. Newman, ME (2006). Proc Natl Acad SciU S A. 103(23): 8577-82.

See more of this Session: Modeling Approaches In the Life Sciences II
See more of this Group/Topical: Topical A: Systems Biology

(390d) Shewanella Oneidensis MR-1 Population-Level Chemotaxis In a Diffusion Gradient Chamber

Rui Li[1], Jennifer M. Auchtung[2], James M. Tiedje[2] and R. Mark Worden[1], (1)Chemical Engineering and Materials Science, Michigan State Univeristy, East Lansing, MI, (2)Center for Microbial Ecology, Michigan State University, East Lansing, MI

Shewanella species are famous for their broad range of terminal electron acceptors, nearly ubiquitous presence in various environmental niches, and the ability to perform chemotaxis towards both organic and inorganic electron acceptors. Knowledge of *Shewanella's* chemotactic properties is important to understand its competitiveness and roles in elemental cycling and bioremediation (e.g., precipitation of soluble uranium oxides). Population-level microbial chemotaxis involves a complex interplay of cellular process (growth, metabolism, chemotaxis, and random motility) and molecular processes (diffusion of electron donors and acceptors that serve as chemoattractants). This paper describes use of a diffusion gradient chamber (DGC) to study population-level growth and chemotaxis of *Shewanella oneidensis* MR-1 in response to applied and cell-generated gradients of chemotactic electron acceptors. A mathematical model describing the key cellular and molecular processes governing *S. oneidensis'* chemotaxis was developed and used to analyze the experimental results. Population-level chemotactic responses of wild-type *S. oneidensis* MR-1 and electron-acceptor reductase mutants to uniformly distributed electron acceptors, pre-established unsteady or quasi-steady state gradient of electron acceptors were explored in the DGC. The model was able to reproduce key trends in the observed cell growth and migration patterns. The predicted velocities of chemotactic cell bands closely matched those measured in the DGC. These results validate the use of the DGC and its mathematical model to measure and simulate *Shewanella's* chemotaxis in response to electron acceptor gradients.

See more of this Session: Modeling Approaches In the Life Sciences II
See more of this Group/Topical: Topical A: Systems Biology

HIGHER ORDER GENE DELETION MUTANTS GIVE INSIGHTS ON LATENT PATHWAY ACTIVATION

Pratish Gawand, University of Toronto, Toronto, ON

Radhakrishnan Mahadevan, University of Toronto, Toronto, ON

It has been well established that deletion of metabolic genes causes a local redistribution of fluxes which results in a sub-optimal growth rate of the mutants (Segre *et al.*, 2002). However, the mutants can adaptively evolve to achieve the optimal FBA predicted growth rates (Fong and Palsson, 2004). Observations on temporal flux profiles of single gene deletion mutants have shown that latent pathways are transiently activated during the course of adaptive evolution (Fong *et al.*, 2006). The reason for the activation of these pathways, however, still remains speculative. Recently, it has been proposed that the presence of these pathways adversely affects the growth rates during the sub-optimal growth phase (Cornelius *et al.*, 2011). Furthermore, it is not well understood whether the activation of latent pathways is a suboptimal response to the genetic perturbations or whether it is a global non-specific response.

We attempted to understand the phenomenon using higher order gene deletion mutants, which showed different growth phenotypes on different carbon sources. Two *Escherichia coli* mutants, one with three gene deletions (LMSE2) and one with five gene deletions (LMSE5), were used in the study. The strategic gene deletions in the two mutants were expected to give insights on the latent pathways activation under different media conditions. Using the genome-scale model iAF1260, the latent pathways for both the mutants were identified as the reactions that were activated during the sub-optimal growth phase (predicted using MOMA) and but not active during optimal growth phase (predicted using FBA) (Cornelius *et al.*, 2011). Abilities of the mutants to recover on different carbon sources after gene deletions, was used to predict the activation of the latent pathways in these mutants.

Our experimental observations suggest that latent pathways are either not activated or are unable to support growth in an event of major rerouting of fluxes. In addition to this, our results indicate that the ability of the mutants to return to the original growth state plays a major role in the recruitment of latent pathways and hence, determines the mutant's survival. It has been previously suggested that mutants do not recover in the event of major flux re-routing. Our results are in agreement with this observation and show that in *E. coli*, the recovery of the mutant strongly correlates to the ability of the mutants to restore the original flux distribution.

References

1. Segrè, D., Vitkup, D. & Church, G. M. (2002) "Analysis of optimality in natural and perturbed metabolic networks", *Proc. Natl. Acad. Sci. U. S. A.* **99**, 15112-15117.

2. Fong, S. S. & Palsson, B. Ø. (2004) "Metabolic gene-deletion strains of Escherichia coli evolve to computationally predicted growth phenotypes", *Nat. Genet.* **36**, 1056-1058.

3. Fong, S. S., Nanchen, A., Palsson, B. O. & Sauer, U. (2006) "Latent pathway activation and increased pathway capacity enable Escherichia coli adaptation to loss of key metabolic enzymes", *J. Biol. Chem.* **281**, 8024-8033.

4. Cornelius, S., Lee, J. & Motter, A. (2011) "Dispensability of Escherichia coli's latent pathways", *Proc. Natl. Acad. Sci. U. S. A.* **108**, 3124-3129.

(390f) Derivation of a Kinematic Model Coupling Stress Fiber Dynamics with JNK Activation In Response to Matrix Stretching

Roland Kaunas, Department of Biomedical Engineering, Texas A&M University, College Station, TX, **Zuyi (Jacky) Huang**, Department of Chemical Engineering, Villanova University, Villanova, PA and Juergen Hahn, Department of Chemical Engineering, Texas A&M University, College Station, TX

Adherent cells in many tissues, including arteries, bladder, muscle, and tendon, to name a few, are subjected to stretch as the tissue deforms in response to external forces. Mechanical stretch regulates many cellular functions, including proliferation, apoptosis, migration and morphology that occur in response to stretch-induced changes in intracellular signaling and gene expression (Haga et al., 2007). While the mechanisms that regulate mechanotransduction remain unclear, focal adhesions and the actin cytoskeleton clearly play a role (Bershadsky et al., 2003). It is well established that focal adhesions and stress fibers are interdependent. This is supported by the fact that focal adhesions anchor stress fibers to the matrix and the integrity of the actin cytoskeleton is necessary to maintain focal adhesions (Volberg et al., 1994). Integrins have been implicated as mechanosensors responsible for stretch-induced intracellular signaling (Thodeti et al., 2009). Katsumi et al. (2005) reported that c-jun N-terminal kinase (JNK) activation in response to stretch requires the formation of new integrin–matrix bonds. We have shown that cells, along with their stress fibers and focal adhesions, orient perpendicular to the direction of cyclic uniaxial stretch (Kaunas et al., 2005). The time course of cyclic stretch-induced JNK activation corresponds with that of stress fiber realignment (Kaunas et al., 2006). Specifically, cyclic uniaxial stretch causes the transient activation of JNK that subsides as the stress fibers become oriented perpendicular to the direction of stretch. Stretch-induced stress fiber alignment occurs via disassembly and reassembly of stress fibers (Hayakawa et al., 2001), which would result in the formation of new integrin–matrix bonds at the associated focal adhesions. Together, these studies suggest that stretch-induced JNK activation is upregulated during times when stress fibers and their associated integrin–matrix adhesions are undergoing relatively high rates of disassembly and reassembly. It is of interest to investigate intracellular signaling that is associated with JNK activation in response to different patterns of stretch stimulation using a mathematical modeling approach, as it can provide systems-level insight into the complex relationships between matrix stretch pattern, stress fiber dynamics, integrin turnover, and JNK signaling.

In order to link the kinetics of intracellular signaling with different patterns of stretch, a model is needed that describes the temporal and spatial evolution of stress fiber organization in response to stretch. Recently, computational modeling has been used to elucidate the complex interplay between matrix deformation, cytoskeletal and integrin bond dynamics and stress fiber reorganization. For example, Kaunas and Hsu (2009) have formulated a computational model of stress fiber dynamic reorganization based on constrained mixture theory (Humphrey, 2008) that successfully describes published time courses of stress fiber reorientation in response to cyclic uniaxial and equibiaxial stretches. Based on the interdependence of stress fibers, integrins, and JNK activation, we coupled the stress fiber dynamics model presented by Kaunas and Hsu (2009) to a kinetic model of JNK activation in this work. Specifically, we developed a mathematical model coupling the dynamic disassembly and reassembly of actin stress fibers and associated focal adhesions to the activation of JNK in cells attached to deformable matrices. The model is based on the assumptions that stress fibers are pre-extended to a preferred level under static conditions and that perturbations from this preferred level destabilize the stress fibers. Numerical solutions of the developed model equations predict that different patterns of matrix stretch result in distinct temporal patterns in JNK activation that compare well with published experimental results. In the case of cyclic uniaxial stretching, stretch-induced JNK activation slowly subsides as stress fibers gradually reorient perpendicular to the stretch direction. In contrast, JNK activation is chronically elevated in response to cyclic equibiaxial stretch. A step change in either uniaxial or equibiaxial stretch results in a short, transient upregulation in JNK that quickly returns to the basal level as overly stretched stress fibers disassemble and are replaced by fibers assembled at the preferred level of stretch. In summary, the model describes a mechanism by which the dynamic properties of the actin cytoskeleton allow cells to adapt to applied forces through turnover and reorganization to modulate intracellular signaling.

References

Bershadsky, A.D., Balaban, N.Q., Geiger, B., 2003. Adhesion-dependent cell mechanosensitivity. Annu. Rev. Cell Dev. Biol. 19, 677–695.

Haga, J.H., Li, Y.S., Chien, S., 2007. Molecular basis of the effects of mechanical stretch on vascular smooth muscle cells. J. Biomech. 40, 947–960.

Hayakawa, K., Sato, N., Obinata, T., 2001. Dynamic reorientation of cultured cells and stress fibers under mechanical stress from periodic stretching. Exp. Cell Res. 268, 104–114.

Humphrey, J.D., 2008. Vascular adaptation and mechanical homeostasis at tissue, cellular, and sub-cellular levels. Cell Biochem. Biophys. 50, 53–78.

Katsumi, A., Naoe, T., Matsushita, T., Kaibuchi, K., Schwartz, M.A., 2005. Integrin activation and matrix binding mediate cellular responses to mechanical stretch. J. Biol. Chem. 280, 16546–16549.

Kaunas, R., Hsu, H.J., 2009. A kinematic model of stretch-induced stress fiber turnover and reorientation. J. Theor. Biol. 257, 320–330.

Kaunas, R., Usami, S., Chien, S., 2006. Regulation of stretch-induced JNK activation by stress fiber orientation. Cell Signal. 18, 1924–1931.

Kaunas, R., Nguyen, P., Usami, S., Chien, S., 2005. Cooperative effects of Rho and mechanical stretch on stress fiber organization. Proc. Nat. Acad. Sci. USA 102, 15895–15900.

Thodeti, C.K., Matthews, B., Ravi, A., Mammoto, A., Ghosh, K., Bracha, A.L., Ingber, D.E., 2009. TRPV4 channels mediate cyclic strain-induced endothelial cell reorientation through integrin-to-integrin signaling. Circ. Res. 104, 1123–1130.

Volberg, T., Geiger, B., Citi, S., Bershadsky, A.D., 1994. Effect of protein kinase inhibitor H-7 on the contractility, integrity, and membrane anchorage of the microfilament system. Cell Motil. Cytoskeleton 29, 321–338.

See more of this Session: Modeling Approaches In the Life Sciences II
See more of this Group/Topical: Topical A: Systems Biology

(390g) A Systems Biology Approach to Identify Immune Targets That Control Tuberculosis Granuloma Function

Mohammad Fallahi-Sichani[1], Denise E. Kirschner[2] and Jennifer J. Linderman[1], (1)Chemical Engineering, University of Michigan, Ann Arbor, MI, (2)Microbiology and Immunology, University of Michigan Medical School, Ann Arbor, MI

Tuberculosis (TB) is a common and deadly infectious disease caused by a highly successful bacterium, *Mycobacterium tuberculosis* (Mtb). Multiple host immune factors control the formation of a self-organizing aggregate of immune cells termed a granuloma, after inhalation of Mtb. One such factor, tumor necrosis factor-α (TNF), is a protein that regulates inflammatory immune responses. TNF (in conjunction with the cytokine IFN-γ) induces macrophage activation, enhances immune cell recruitment to the site of infection, and augments chemokine and cytokine expression by macrophages through activation of the NF-κB signaling pathway. However, there are differences among dynamics of each of these NF-κB-associated responses as a result of differences in molecular processes (e.g. RNA and protein degradation rates) acting downstream of TNF-induced NF-κB activation. The impact of these dynamical differences on formation and function of granulomas and thus the outcome of Mtb infection is not known.

We have developed a multi-scale computational model that describes the immune response to Mtb in lung over three biological length scales: tissue, cellular and molecular. Using this model, we predict the impact of NF-κB associated response dynamics on the outcome of infection at the level of a granuloma. Our model suggests that the timing of these responses, in addition to the extent of response, plays a critical role in control of infection and inflammation. Manipulations of the dynamics of these responses lead to different outcomes, including clearance of bacteria, containment of bacteria within a stable granuloma, uncontrolled growth of bacteria, or excessive inflammation. The intracellular NF-κB associated signaling molecules and processes involved in TB immunity that we identify as crucial may be new targets for therapy.

See more of this Session: Modeling Approaches In the Life Sciences II
See more of this Group/Topical: Topical A: Systems Biology

(390h) Dissecting Gene Regulation Through DNA Loop Formation Using Statistical Mechanical Models and Experiments

James Boedicker, Applied Physics, Caltech, Pasadena, CA, Hernan Garcia, Physics, Caltech, Pasadena, CA and Rob Phillips, Applied Physics and Mechanical Engineering, Caltech, Pasadena, CA

Testing ourunderstanding of genetic circuits requires models which make quantitativepredictions that can be experimentally validated. A common motif of generegulation involves the interactions between transcription factors, RNApolymerase, and DNA. Although many examples of gene regulation through suchinteractions are well known, our understanding of even the simplest suchsystems is often qualitative, at the level of biological °Cartoons. Thesecartoons only predict whether interactions between the regulatory components increase,decrease, or have no effect on gene expression. Such models fail to makequantitative predictions about specific realizations of such genetic circuits, evenfor well-studied systems such as the lac operon.

Our approach hasbeen to construct models of gene expression based on equilibrium statisticalmechanics. The models require only a few input parameters. The modelparameters include binding energies of proteins, copy numbers of components,and the mechanical properties of DNA. These parameters are measured in simple controlexperiments. Once these inputs have been determined, the models contain nofree parameters and make quantitative predictions about the level of geneexpression over a wide range of parameter space. We have constructed andexperimentally validated such models to test our understanding of the lacoperon.

In the lac operon,a dimeric transcription factor, lac repressor, can bind to three differentbinding sites on DNA, called operators. Gene expression is reduced byrepressor binding to the center operator just adjacent to the promoter. It isthought that binding to the outer, auxiliary sites alone cannot directlyinfluence gene expression. Instead, the auxiliary sites can participate inloop formation, in which one dimeric repressor can simultaneously bind to boththe center and an auxiliary operator, bending the intervening DNA into a loop.The looped state leads to enhancement of repression due to the loop locking inthe repressor on the center operator.

We use thisgenetic circuit to make predictions about how the specifics features of the lacoperon lead to a specific level of gene expression. The model we developed forthe lac operon has six states, with the probability of each state weighted by aBoltzmann factor calculated from the change in free energy associated with eachstate. Some of the states lead to repression while others enable binding ofthe RNA polymerase and transcription of the gene. Using the model, we makequantitative predictions about how the level of gene expression is modulated byeach component of the system. We model and test how changes in the bindingenergies of the repressor to the operators, the number of repressors per cell, thelength of the DNA loop, and the flexibility of the DNA in the loop alter thelevel of gene expression.

The models weretested experimentally by creating genetic circuits containing the generegulatory components of the lac operon. These circuits were incorporated intothe genome of *E. coli*. Fluorescent reporters were used to quantify thelevel of gene expression in each circuit. The readout of the experiments wasthe absolute level of fluorescent protein. Using this approach, thepredictions of the thermodynamic equilibrium models were experimentally testedfor many genetic circuits containing different combinations of the regulatoryelements of the lac operon.

The experiments revealthat the standard model described above does not hold. There is a contributionfrom the downstream auxiliary operator to repression, even in the absence ofloop formation. This result shows that binding events downstream of thepromoter can still interfere with the activity of RNA polymerase, and do so ina manner which is dependent upon the orientation of the promoter and operator.Additionally, the intrinsic flexibility of the looping DNA has no effect on thelevel of repression. DNA sequences that have previously been shown to easilyform loops *in vitro* do not lead to increased repression. The apparentlack of dependence of repression on the flexibility of the looping DNA suggeststhat other *in vivo* factors such as DNA bending proteins are dominant indetermining the probability of the looping in vivo. These results highlightthe ability of quantitative models of gene expression, such as the zero free parameterthermodynamic equilibrium model applied here, to test and expand our understandingof gene regulatory mechanisms. This approach should lead to both mechanisticinsights into gene regulation *in vivo*, and also enable the constructionof synthetic genetic circuits with fine-tuned control of gene expression.

See more of this Session: Modeling Approaches In the Life Sciences II
See more of this Group/Topical: Topical A: Systems Biology

(390i) Regulation of Epidermal Growth Factor Receptor Phosphorylation and Endocytosis Dynamics by Protein Tyrosine Phosphatases

Calixte S. Monast and Matthew J. Lazzara, Chemical and Biomolecular Engineering, University of Pennsylvania, Philadelphia, PA

Cell signaling pathways initiated by the epidermal growth factor receptor (EGFR) are broadly involved in normal physiology and pathology. Phosphorylated EGFR levels, which dictate receptor signaling strength and persistence, are the net result of two opposing processes, auto-phosphorylation of cytoplasmic tyrosines by the receptor kinase and dephosphorylation of those tyrosines by protein tyrosine phosphatases (PTPs). Whereas the processes leading to tyrosine phosphorylation have been studied in great detail, relatively little is known about the kinetics of EGFR tyrosine dephosphorylation or the coupling of this process to other receptor-level phenomena which depend upon receptor phosphorylation, such as endocytosis. Here, we developed a novel mechanistic model of EGFR phosphorylation dynamics to determine the rates of EGFR dephosphorylation at the plasma membrane and in the cell interior and applied the model to generate predictions for the effects of these dephosphorylation kinetics on EGFR endocytosis dynamics. Our model was constructed to allow for the possibility of different rates of receptor dephosphorylation at the plasma membrane and cell interior and accounted for interactions of EGFR with itself (dimerization and phosphorylation), ATP, kinase inhibitors, PTPs, and adapter proteins involved in receptor trafficking processes. To determine rate constants for EGFR tyrosine dephosphorylation at the plasma membrane and cell interior, the model was regressed against a diverse set of experimental data for receptor phosphorylation response to EGFR ligands, PTP inhibitors, and EGFR kinase inhibitors as well as data for steady EGFR phosphorylation levels in the absence of ligands. Our regression results led to the novel conclusion that EGFR dephosphorylation is significantly more rapid at the plasma membrane than in the cell interior, a finding which contradicts the classical view of EGFR regulation by PTPs and several previous computational studies of EGFR phosphorylation dynamics which were developed by considering data for receptor response to ligand alone. In fact, our model results suggest that the process of EGFR dephosphorylation is so rapid that receptor tyrosines may go through a complete phosphate cycle (phosphorylation and dephosphorylation) as many as thirteen times per minute. Thus, our model results suggest that, as with other kinase-phosphatase systems, EGFR and its associated PTPs participate in a "futile cycling" process wherein phosphate cycling occurs on a time scale much smaller than that for other important receptor processes such as endocytosis and degradation. Since the most dominant pathways for EGFR endocytosis require phosphorylation of key EGFR tyrosines, a feature encoded within our model, our results further suggest that regulation of EGFR by PTPs at the plasma membrane may play a significant role in determining rates of ligand-mediated EGFR endocytosis. In fact, for an EGFR expression level characteristic of epithelial cells, our model predicted that PTP activity at the plasma membrane may reduce the rate constant for ligand-mediated EGFR endocytosis by up to six-fold from the value it would otherwise be in the absence of PTP activity at the plasma membrane. This predicted effect of PTPs on EGFR endocytosis could be reduced by the inclusion of the trafficking adapter protein referred to previously in the model calculations. Experimental measurements of the effects of PTPs on EGFR endocytosis were made for cultured fibroblasts and epithelial cells using radiolabeled EGFR ligands or antibodies and inhibitors of PTP activity. Overall, comparison between our measurements and model calculations suggested that PTP activity at the plasma membrane influences EGFR endocytosis kinetics but that the enhanced binding of adapter proteins in the context of PTP inhibition reduces the observed effect from the maximum predicted effect.

See more of this Session: Modeling Approaches In the Life Sciences II
See more of this Group/Topical: Topical A: Systems Biology

(391a) Evaluating Metabolic Network Elements Via a Forced Coupling Algorithm

Christopher J. Tervo, Department of Chemical and Biological Engineering, University of Wisconsin - Madison, Madison, WI and Jennifer L. Reed, Department of Chemical and Biological Engineering, University of Wisconsin-Madison, Madison, WI

The development of genome-scale models (GEMS) has allowed for the interrogation of microbial reaction networks for a variety of purposes such as predicting growth phenotypes, guiding metabolic engineering strategies, elucidating network properties and facilitating hypothesis-driven discovery [1]. However, despite the many applications of GEMS, there remains room for improvement and development, both in terms of the networks themselves, as well as the algorithms used to interrogate such systems. The success of *in silico* cellular models depends on the ability of both the metabolic and regulatory networks to capture the underlying characteristics of the microbe of interest. Unfortunately, the introduction of error into these networks during construction is often unavoidable due to errors in databases or functional annotations, or to reactions added during pathway refinement.

To facilitate model testing and validation, we have developed a forced coupling algorithm (FCA) that can design experiments which test for the existence of network elements, whether these elements are hypothesized reactions or gene to reaction associations. The FCA makes use of the concept of flux coupling [2] to propose mutant strains and media conditions under which a particular network element is conditionally essential for cell growth or coupled to another measurable reaction (e.g. chemical production). By testing such conditions experimentally, hypothesized or questionable network elements can be confirmed or dismissed and network models can be further refined to better represent the true physiology of the microbial organism.

This algorithm has been applied to a variety of genome-scale models, including *Escherichia coli* (iJR904) [3], *Bacillus subtillis* (iYO844) [4] and *Pseudomonae putida* [5], and has been successful in coupling over 80 percent of their accessible metabolic network to biomass production. For *E. coli*, the FCA results were compared to existing experimental datasets [6, 7] to confirm or refute gene to reaction relationships based on measured gene essentiality. For reactions that were coupled to biomass using only media coupling conditions (without requiring any gene deletions) the experimental results supported the FCA predictions (in ~80% of cases), indicating that these reactions are used under these particular conditions.

The FCA coupling conditions can be used to validate existing cellular network annotations such as models produced using the Model SEED architecture.[8] Moreover, these conditions can be used as screens to identify genes responsible for orphan reactions (reactions whose associated genes are unknown), thus filling important knowledge gaps in metabolism. Additionally, we have integrated the FCA algorithm into metabolic strain design algorithms to propose unique *E. coli* mutants that if evolved would co-utilize multiple carbon substrates and produce ethanol at high yields. The FCA algorithm can be used for a variety of potential applications (model testing, experimental design, and metabolic engineering) and these will be presented.

Acknowledgement:

CT is supported by an NHGRI training grant to the Genomic Sciences Training Program (5T32HG002760).

References:

1. Oberhardt, M.A., B.O. Palsson, and J.A. Papin, *Applications of genome-scale metabolic reconstructions.* Mol Syst Biol, 2009. **5**.

2. Burgard, A.P., et al., *Flux Coupling Analysis of Genome-Scale Metabolic Network Reconstructions.* Genome Research, 2004. **14**(2): p. 301-312.

3. Reed, J., et al., *An expanded genome-scale model of Escherichia coli K-12 (iJR904 GSM/GPR).* Genome Biology, 2003. **4**(9): p. R54.

4. Oh, Y.-K., et al., *Genome-scale Reconstruction of Metabolic Network in Bacillus subtilis Based on High-throughput Phenotyping and Gene Essentiality Data.* Journal of Biological Chemistry, 2007. **282**(39): p. 28791-28799.

5. Oberhardt, M.A., et al., *Reconciliation of Genome-Scale Metabolic Reconstructions for Comparative Systems Analysis.* PLoS Comput Biol, 2011. **Accepted**.

6. Baba, T., et al., *Construction of Escherichia coli K-12 in-frame, single-gene knockout mutants: the Keio collection.* Mol Syst Biol, 2006. **2**.

7. Joyce, A.R., et al., *Experimental and Computational Assessment of Conditionally Essential Genes in Escherichia coli.* The Journal of Bacteriology, 2006. **188**(23): p. 8259-8271.

8. Henry, C.S., et al., *High-throughput generation, optimization and analysis of genome-scale metabolic models.* Nat Biotech, 2010. **28**(9): p. 977-982.

See more of this Session: Modeling Approaches to Examine Issues In Life Sciences - II
See more of this Group/Topical: Food, Pharmaceutical & Bioengineering Division

(391b) Robust Selection of Gene Modifications

Mona Yousofshahi[1], Michael Orshansky[2], Kyongbum Lee[3] and **Soha Hassoun**[1], (1)Computer Science, Tufts University, Medford, MA, (2)Department of Electrical and Computer Engineering, The University of Texas at Austin, Austin, TX, (3)Chemical and Biological Engineering, Tufts University, Medford, MA

In recent years, increasingly sophisticated computational methods have been developed to identify optimal genetic modifications to achieve a desired metabolic engineering objective. Notable examples include OptReg [1], OptKnock [2], and GDLS [3]. The problem of identifying optimal genetic modifications can be formulated in terms of either operating state variables such as reaction flux or control variables such as gene expression. The optimal design "tunes" these variables such that the solution meets the engineering objective while satisfying several different constraints reflecting physicochemical considerations, experimental observations and assumptions about the physiology of the cell or organism. Typically, the number of optimization model variables far exceeds the number of constraints, resulting in an underdetermined system with a large number of degrees of freedom. Previously, this type of model uncertainty was addressed by estimating the possible ranges of the flux or gene expression variables. For example, flux balance analysis was used to characterize the maximal and minimal flux through each reaction in a metabolic network.

Our work addresses another type of uncertainty, which arises from the imprecision of engineering interventions, which rely on statistical events such as cellular transformation (or transfection) and stochastic steps in gene expression. Generally, an optimal solution may be sensitive to perturbations in the parameters of the problem, such that small deviations in the parameters render the solution suboptimal or even infeasible. In this paper, we describe a *robust* formulation of a computational metabolic engineering optimization problem. Specifically, we investigate a robust optimization problem with the objective of maximizing a desired metabolite product through gene up-regulation operations. The uncertain parameters in this formulation are the flux capacities, which are set by the expression levels of the corresponding genes. Our formulation introduces a probabilistic constraint on the flux resulting from a corresponding gene up-regulation operation: $\text{Prob}\{v<y.\text{Cap}\}\geq\alpha$. Here, *v* is the flux, *y* is the binary decision variable for the up-regulation operation, and *Cap* is the uncertain flux capacity. The problem of optimizing the decision variables *y* (whether to up-regulate a gene or not) can then be re-formulated as a mixed-integer linear program (MILP) in a chance-constrained framework [4].

To evaluate the robust formulation, we performed a series of comparisons with a deterministic formulation, where the flux capacities were treated as known constants, rather than distributions. As the test system, we used a previously developed metabolic model of the fat cell, for which detailed metabolic flux data were at hand. The test metabolic engineering objective was to maximize the lipid storage of the fat cell subject to physiologically meaningful uptake of glucose, amino acids and fatty acids. The distributions of flux capacities were estimated by computing the maximal fluxes resulting from changes in enzyme activities using enzyme control flux (ECF) analysis [5]. Our results suggest that the gene up-regulation operations identified by the robust formulation yield a higher expected increase in lipid storage compared to the operations identified by the deterministic formulation.

[1] P. Pharkya and C. D. Maranas, "An optimization framework for identifying reaction activation/ inhibition or elimination candidates for overproduction in microbial systems," *Metab. Eng.*, vol. 8, pp. 1-13, 1, 2006.

[2] A. P. Burgard, P. Pharkya and C. D. Maranas, "Optknock: A bilevel programming framework for identifying gene knockout strategies for microbial strain optimization," *Biotechnol. Bioeng.*, vol. 84, pp. 647-657, 2003.

[3] D. S. Lun, G. Rockwell, N. J. Guido, M. Baym, J. A. Kelner, B. Berger, J. E. Galagan and G. M. Church, "Large-scale identification of genetic design strategies using local search," *Mol Syst Biol,* vol. 5, 08/18/print, 2009.

[4] F. S. Hillier, "Chance-Constrained Programming with 0-1 or Bounded Continuous Decision Variables," *Management Science,* vol. 14, pp. pp. 34-57, Sep., 1967.

[5] H. Kurata, Q. Zhao, R. Okuda and K. Shimizu, "Integration of enzyme activities into metabolic flux distributions by elementary mode analysis," *BMC Systems Biology,* vol. 1, pp. 31, 2007.

See more of this Session: Modeling Approaches to Examine Issues In Life Sciences - II
See more of this Group/Topical: Food, Pharmaceutical & Bioengineering Division

(391c) Flux Balance Analysis of Plant Suspension Cell Culture

Ashish Misra[1], Margaret Simons[2], Matthew Conway[2], Gary D. Coleman[3] and Ganesh Sriram[2], (1) Chemical and Biomolecular Engineering, University of Maryland, College Park, College Park, MD, (2) Chemical and Biomolecular Engineering, University of Maryland, College Park, MD, (3)Plant Science and Landscape Architecture, University of Maryland, College Park, College Park, MD

We report the use of flux balance analysis (FBA) for reconstructing flux maps in plant and tree suspension cells. The utility of plants as renewable sources for producing energy and other chemicals merits extensive study of their complex metabolism. FBA uses optimization of constraint-based models of biochemical networks to determine fluxes through metabolic pathways. It has been successfully applied in many cases for estimating fluxes in prokaryotic organisms. Its use in estimating fluxes in eukaryotic systems, especially plants and trees, has been less widespread.

Plant metabolism is much more complex than that of prokaryotes with cellular compartmentation and replication of pathways across compartments. In plants, a couple of pioneering studies have applied FBA for estimating fluxes using plant cell culture (Williams et al., Plant Physiol. 154: 311-323, 2010; Belau et al, Plant Physiol. 149: 585-598, 2009). We have applied FBA for estimating fluxes in cell cultures of *Arabidopsis thaliana* and poplar (*Populus trichocarpa*). As part of the work, we will present investigation of the effect of different objective functions for estimating fluxes using FBA. The growth maximization objective function commonly used for prokaryotic organisms is not suitable for eukaryotic organisms due to their complex growth requirements and pathways. By comparing fluxes obtained from FBA with those from isotopic MFA, we will determine the best objective functions that are critical for accurate flux estimation by FBA. Using the optimized objective function, we will present metabolic landscapes obtained for the networks in both systems, and specifically present changes in carbon-nitrogen interactions/metabolism in poplar. Our studies will represent the first application of FBA to a model tree system, and additionally provide insight on the metabolic network in model plant and tree organisms.

See more of this Session: Modeling Approaches to Examine Issues In Life Sciences - II
See more of this Group/Topical: Food, Pharmaceutical & Bioengineering Division

(391d) Immersed Boundary Model of Biofilm Deformation In Response to Fluid Flow

Jeffrey J. Heys, Chemical and Biological Engineering, Montana State University, Bozeman, MT and Garret D. Vo, Mechanical Engineering, Montana State University, Bozeman, MT

Biofilms are complex mixtures of microorganisms and extracellular matrix that exist on many wetted surfaces. Recently, magnetic resonance microscopy (MRM) has been used to measure fluid velocities near biofilms that are attached to the walls of capillary channels. These velocity measurements showed unexpectedly high secondary velocities (i.e., high velocity magnitudes perpendicular to the direction of bulk flow and perpendicular to the surface that the biofilm is attached), and the presence of high secondary velocities near a biofilm could increase the delivery of substrates to the biofilm. A mathematical model, based on the immersed boundary method, is used here to examine the physical interaction between a biofilm and a moving fluid in a capillary and to analyze possible factors that may contribute to the elevated secondary velocities observed experimentally. The simulation predicts the formation of a recirculation downstream of a biofilm, and this recirculation deforms and lifts the biofilm upward from the surface to which the biofilm is attached. Changing the mechanical properties (i.e., stiffness) of the biofilm impacts both the lifting of the biofilm and the magnitude of the secondary velocities. The maximum lifting of the biofilm occurs when the biofilm properties are similar to previous experimental measurements, which indicates that the mechanical properties of the biofilm may be tuned for the generation of maximum secondary velocity magnitude and transport of substrates to the biofilm.

See more of this Session: Modeling Approaches to Examine Issues In Life Sciences - II
See more of this Group/Topical: Food, Pharmaceutical & Bioengineering Division

(391e) Modeling Population Dynamics of Cancer Stem Cells with Epithelial-Mesenchymal Transition

Haewoo Lee and Seongkyu Yoon, Chemical Engineering, University of Massashusetts, Lowell, MA

Cancer stem cells (CSCs) are tumor cells that possess stem cell-like characteristics of self-renewal and multipotency. In CSC model, CSCs are differentiated into heterogeneous population of tumor cells, and these differentiated non-tumorigenic cells constitute bulk of tumor with lack of self-renewal capacity and limited proliferation potential. In this context, studying cancer cell population during tumor initiation/progression/metastasis with the concept of CSCs might bring further insight into molecular drug target and clinical strategies. However, it has been recently revealed that there exists plasticity in certain tumors in that the phenotypically differentiated cells can de-differentiate and acquire stem cell features. Induction of epithelial-mesenchymal transition (EMT) is thought to being crucial to this regain of CSCs phenotype, and can be regarded as one of the important tools in maintaining CSCs pool in tumors. Despite the importance of EMT in tumor hierarchy and metastasis, so far, there is no clear explanation on the role of EMT in population dynamics of cancers and little is known about therapeutic implication of it due to difficulties as well as uncertainty in identifying CSCs experimentally. Under these circumstances, mathematical model can give great insight to unravel complicated nature of hierarchical tumor heterogeneity. Therefore, in this study, we first tried to construct comprehensive population dynamics model of tumors by introducing EMT and other known phenomena, such as dormancy of CSCs or regulatory feedback loops. To investigate the effects of various biological factors, sensitivity analysis of model parameters was conducted and we could obtain reasonable prediction of tumor hierarchy driven by heterogeneous populations of CSCs and non-CSCs. Then, the tumor population dynamics were explored under different therapeutic scenarios of conventional chemotherapy, chemotherapy targeting CSCs, or combinational chemotherapy employing both of them. The results clearly illustrate the improved efficacy of combinational chemotherapeutic agents on heterogeneous populations and address several issues of central importance to the success treatment of human malignancies.

See more of this Session: Modeling Approaches to Examine Issues In Life Sciences - II
See more of this Group/Topical: Food, Pharmaceutical & Bioengineering Division

(391f) Intra-Thrombus Transport and Local Shear Stress Distributions Modeling Based On In-Vivo Imaging with Single Platelet Resolution

Roman S. Voronov[1], Timothy J. Stalker[2], Lawrence F. Brass[2] and Scott L. Diamond[1], (1)Chemical and Biomolecular Engineering, University of Pennsylvania, Philadelphia, PA, (2)School of Medicine, University of Pennsylvania, Philadelphia, PA

The assembly of platelet deposits and fibrin polymerization results in over one million heart attacks and strokes each year in the US. Conversely, deficiencies in these processes result in bleeding risks that confront surgeons on a regular basis. In order to quantify thrombus formation and thrombus break-up, it is necessary to first measure the thrombus structure and the biological and physical factors controlling structure. Recent findings from in-vivo laser injury studies demonstrate that the structure of the thrombus is non-uniform with the porosity varying as a function of both time and space. Also, the flow field around a growing thrombus exerts forces on the cells that make up the thrombus, thereby straining the thrombus mechanically. The former phenomenon affects intra-thrombus chemo-dynamics, while the latter affects clot mechanics. Therefore, the overall objective of this work is to develop a framework for fundamental understanding of thrombogenesis and the role that both mechanical and chemical cues play in determining the thrombus structure and integrity.

In-vivo laser injury experiments in mice are performed in order to image thrombus formation and platelet activation in 3D using confocal fluorescent microscopy. Optical Doppler velocimetry is used in order to measure the centerline flow velocity in the mouse blood vessels. The rigidity of the platelet aggregates and the strength of their binding to each other are varied by blocking actin and fibrin polymerization, respectively. Porosity of the platelet aggregates is controlled by inhibiting platelet activation via blocking thrombin function or fibrin polymerization. Local stress fields within thrombi structures are calculated via flow dynamics. Inhibitors of platelet activation pathways, fibrin polymerization, and clot retraction are used in order to chemically vary the structure of the thrombus. Once the thrombus structure and the platelet activation states are measured, chemo-transport simulations are performed in order to calculate permeability, fluxes and diffusivities of the key soluble species within the thrombus.

The presented innovative multidisciplinary approach is expected to yield an enhanced fundamental understanding of how thrombus structure is affected by mechano- and chemo-perturbations. Collectively, the outcome of this work is expected to provide insight towards understanding clot generation, structure and stability relevant to thrombosis, thromboembolism and coagulopathy, ideally to reduce risks of occlusion and embolism. Moreover, it is planned to perform a similar analysis for in-vitro platelet aggregation in microfluidic chambers, as such systems provide for greater experimental control and more accessible imaging.

(391g) Kinetics and Modeling of Disproportionation Reactions Catalyzed by Clostridium Acetobutylicum and Ruminococcus Flavefaciens Endoglucanases

Christopher D. Warner[1], Clark Ford[2] and Peter J. Reilly[1], (1)Chemical and Biological Engineering, Iowa State University, Ames, IA, (2)Food Science and Human Nutrition, Iowa State University, Ames, IA

Cellotetraose, cellopentaose, and cellohexaose were separately incubated with a *Clostridium acetobutylicum* glycoside hydrolase family 44 endoglucanase, and the resulting reactant and product concentrations were obtained by capillary electrophoresis chromatography. Cellotetraose was slowly converted to more cellotriose than glucose, with no cellobiose being formed and with most of the cellotetraose remaining at long incubation times. Cellopentaose was hydrolyzed more quickly to cellotetraose, cellobiose, cellotriose, and glucose, in order of near-equilibrium concentrations, with a very small amount of cellopentaose remaining. Cellohexaose was even more quickly converted to cellotetraose, cellobiose, cellopentaose, cellotriose, and glucose, with cellopentaose concentration decreasing after reaching a peak. No cellohexaose remained after a short time. A kinetic model, including both hydrolysis and transglycosylation reactions yielding products larger than the substrates, was formulated based on these results and on previously reported results from a *Ruminococcus flavefaciens* glycoside hydrolase family 44 endoglucanase, each set showing asymmetric attack on substrates as well as product dispropor-tionation. The model fit data obtained for digestion of the three substrates by *C. acetobutylicum* endoglucanase and by *R. flavefaciens* endoglucanase with just a catalytic domain. It fit data obtained for the *R. flavefaciens* endoglucanase with a catalytic domain/ carbohydrate binding module complex somewhat less well. It appears that the absence of observed larger transglycos-ylation products in capillary electrophoresis chromatographs is due to product insolubility or to concentrations near or below the detection limit.

See more of this Session: Modeling Approaches to Examine Issues In Life Sciences - II
See more of this Group/Topical: Food, Pharmaceutical & Bioengineering Division

(405a) Inflamed Leukocyte-Mimetic Nanoparticles for Molecular Imaging of Inflammation: Application to Detection of Tumor and Acute Inflammation

Xiaoyue Chen*, Richard Wong*, Ildar Khalidov†, Y. Andrew Wang‡, Jeerapond Leelawattanachai*, Yi Wang*† and Moonsoo M. Jin*†

*Department of Biomedical Engineering, Cornell University, Ithaca, NY 14853

†Department of Radiology, Weill Cornell Medical College, New York, NY 10022

‡Ocean Nanotech, LLC, Springdale, AR 72764

I. SUMMARY

This study presents physiology-inspired design of superparamagnetic iron oxide (SPIO) nanoparticles for *in vivo* detection by optical imaging and MRI, mimicking activated leukocyte in its ability to recognize inflamed endothelium. Nanoparticles *in vivo* will experience hydrodynamic force induced by the blood flow, requiring simultaneous molecular interactions with sufficient adhesion strength with the cells for nanoparticles to remain on the cell surface. Therefore, the design of nanoparticles with tunable affinity and avidity of physiological interactions would be critical to the selectivity and efficiency of targeting. Here we demonstrate specific accumulation of systemically-delivered nanoparticles mimicking the behavior of inflamed leukocytes into tumor cells overexpressing ICAM-1 and the surrounding vascular microenvironment. Further findings of our leukocyte-mimetic nanoparticles (LMN) localizing to the inflamed vasculature of tumors expressing low level of ICAM-1 highlight a potential use of LMN for detection of tumor growth by their accumulation into the inflammatory tumor microenvironment, irrespective of the type of tumor surface antigen.

II. DESCRIPTION OF DESIGN OBJECTIVE

Current studies investigating targeted nanoparticles for the detection and treatment of cancer mainly employ antibodies or small peptides specific to tumor surface antigens or tumor-associated microenvironmental factors. Among these molecules, ICAM-1 has caught our particular interest not only because of its upregulation in several carcinomas compared to respective normal epithelium, but also its induction observed in tumor vasculature due to the inflammatory network conjoining tumor and tumor microenvironment [1-3]. Upon induction of inflammation, adhesion molecules, including ICAM-1, are upregulated in endothelium, with which immune cells interact using integrins like LFA-1 to adhere to endothelium and initiate diapedesis. Inspired by the physiological interaction between leukocytes and endothelium under inflammation, we envision nanoparticles designed to mimic inflamed leukocytes can accumulate to both tumor with high ICAM-1 expression and its inflamed milieu, providing more efficient and sensitive targeting. Potentially, they can also function as a universal tumor targeting strategy through the inflammatory microenvironment, irrespective of the type of tumor surface antigen, therefore providing a wide-ranging impact on cancer diagnosis and therapeutics.

III. DESCRIPTION

Our nanoparticles are developed for *in vivo* detection by both optical imaging and MRI, designed to mimic activated leukocyte in its ability to recognize inflamed endothelium using LFA-1 integrin (Fig. 1a). Selective binding of LMN to overexpressed ICAM-1 is conferred by surface coating at an optimal density with the I domain of human LFA-1, engineered for high affinity by mutations of F265S/F292G (denoted as F265S/F292G, K_D = 6 nM) [4]. In order to fine-tune the coating density of F265S/F292G and present targeting moiety in a most functional orientation, oleic acid capped SPIO nanocrystals were encapsulated with a layer of phospholipid consisting of 80% n-poly(ethylene glycol) phosphatidylethanolamine (DPPE-PEG) and 20% dioleoyl-glycero-succinyl-nitrilotriacetic acid (DOGS-NTA). DOGS-NTA was used for non-covalent conjugation of targeting moieties with His tag in a robust and reproducible manner via high affinity binding to nickel ions chelated by NTA (Ni-NTA) [5]. DPPE-PEG was included to maintain solubility, stability, and for its low immunogenicity and non-specific binding to cells and tissues *in vivo* [6]. F265S/F292G crossreacts with both human and murine ICAM-1, allowing LMN to detect ICAM-1 induction on both human tumors and murine vascular microenvironment in our animal models. Transmission electron microscope (TEM) images revealed monodispersed SPIO nanocyrstals with an outer layer of micelle-like structure (Fig. 1b; dark ring density corresponds to uniformly distributed nickel ions (the black arrow in top right) and diffuse dark density to His-tagged proteins attached to Ni-NTA (the white arrow in bottom right)). Hydrodynamic size of SPIO with or

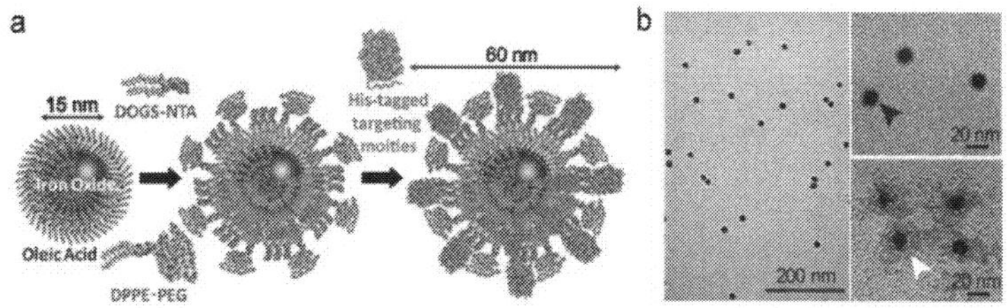

Figure 1. (a) Scheme of nanoparticle assembly. (b) Transmission electron microscopy (TEM) images of SPIO at different synthesis steps, with (right panels) and without (left panel) 2% uranyl acetate staining. Black arrow: Ni^{2+} iron. White arrow: Protein and PEG layer.

without protein conjugation was measured to be 60±10 nm by dynamic light scattering (DLS), an increase from 15 nm diameter SPIO core mainly due to the addition of phospholipid and PEG.

IV. ANALYSIS AND PERFORMANCE EVALUATION

Prior to detecting ICAM-1 with nanoparticles *in vivo*, we first examined the binding of nanoparticles coated with different I domain variants labeled with Alexa Fluor 488 (AF488) to ICAM-1 expressed in monolayer culture of cervical cancer cells (HeLa), by immunofluorescence flow cytometry (Fig. 2a) and fluorescence microscopy (Fig. 2b). Consistent with our prediction, the level of nanoparticle binding varied in order from highest to lowest: SPIO-F265S/F292G, which is abbreviated as Leukocyte-Mimetic Nanoparticle or 'LMN'; SPIO-wt (wildtype I domain, K_D = 1.7 mM); and SPIO-D137A (inactive I domain mutant with abolished ICAM-1 binding), which is abbreviated as Non-Targeted Nanoparticle or 'NTN'. Markedly, specific binding of LMN was not inhibited by competition with soluble I domain, presumably due to multivalent interaction between nanoparticles and HeLa cells (Fig. 2a). ICAM-1-mediated binding of LMN but not NTN led to significant cell surface labeling and internalization into the cells, evidenced by fluorescence microscopy and Perl's Prussian blue staining (Fig. 2b). To further quantify specific nanoparticle labeling, ^{3}H-labeled phospholipid was additionally incorporated into the outer phospholipid layer of SPIO, so that the presence of nanoparticles can be measured by radioactivity. When HeLa cells were incubated with LMN for 4 hr at 37°C, an increase in concentration led to an increase in cellular uptake of nanoparticles, reaching a plateau at 450 ng/10^6 cells (Fig. 2c). LMN delivery was specific to ICAM-1 expression, again evidenced by little accumulation into HeLa with NTN and much lower delivery with LMN into 293T, a cell line with no or little expression of ICAM-1.

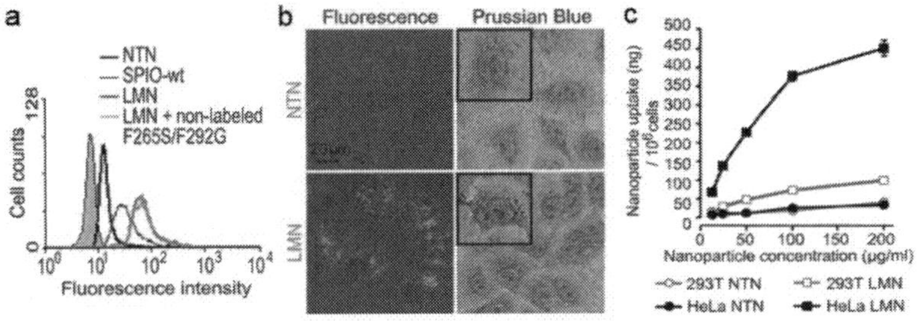

Figure 2. (a) Flow cytometry measurements of HeLa cells stained with nanoparticles (25 μg/ml of SPIO and 10 μg/ml of I domains) conjugated with different I domains. Non-labeled I domains as a competitor were used at 50 μg/ml. (b) ICAM-1 dependent internalization of SPIO into HeLa cells was confirmed with confocal fluorescence microscopy and Perl's Prussian blue staining. (c) The amount of SPIO internalized into HeLa or 293T was measured by radioisotope measurement of 3H-phospholipid incorporated into SPIO nanoparticles.

With the knowledge that ICAM-1 upregulation has been observed in several carcinomas and tumor vasculature, we then examined if systemically delivered nanoparticles would accumulate to the tumor and inflamed stroma by ICAM-1 targeting. To validate that nanoparticle localization is ICAM-1 specific and not due to an increased permeability of the tumor vasculature, NTN and ICAM-1 negative 293T cell xenograft were used as controls. The growth of HeLa and 293T xenograft in mice was confirmed by whole body imaging of GFP, which was stably expressed by both HeLa and 293T cells through a lentiviral transduction system (Fig. 3a). At 50 hr after intravenous injection of nanoparticles, whole body imaging of near-infrared (near-IR) fluorescence (AF750 attached to the I domains) indicated accumulation of LMN into HeLa but much less into 293T xenograft (Fig. 3a,b). Subsequent *ex vivo* imaging of the tumor and the major organs harvested from the mice further confirmed a greater level of delivery into HeLa xenograft with LMN (Fig. 3c). The signal from the kidney was by far greater than those from other organs both for LMN and NTN, indicating ICAM-1 independent clearance through the kidney (Fig. 3c). Interestingly, higher fluorescence was also detected in the liver with LMN, presumably caused by persistent, low-level inflammation in the liver. Whole body imaging of nanoparticles localized to HeLa and 293T xenograft over the time course of 30 min to 50 hr post-injection showed that the peak accumulation occurred at 1-3 hr post-injection, followed by a gradual decrease over 3 days and a complete clearance by 7 days (Fig. 3d). The presence of LMN into HeLa was also confirmed by direct staining of iron with Prussian blue (Fig. 3e,f). To map the distribution of LMN by fluorescence microscopy, SPIO nanoparticles conjugated with AF594-labeled I domains were intravenously injected into the mice with HeLa/293T xenograft. When the xenograft tissue was examined 4 hr after nanoparticle injection, specific accumulation of LMN into HeLa tumor was observed, judging from colocalization between GFP expression in HeLa and AF594 fluorescence (Fig. 3g,h). Importantly, a high percentage of CD31 positive cells in HeLa as well as in 293T xenografts were also targeted by LMN. The localization of ICAM-1 specific nanoparticles within the tumor-associated vasculature, therefore, was likely responsible for higher signals detected within 293T xenograft at earlier time points (Fig. 3d), despite the fact that 293T itself exhibited almost no binding of LMN (Fig. 3e,f). This finding highlights a potential use of LMN for detection of tumor growth by their accumulation into inflamed tumor vasculature, irrespective of the type of tumor surface antigen.

V. DISCUSSIONS

In this study, we designed MRI-compatible SPIO nanoparticles, and demonstrated a successful detection of constitutive expression of ICAM-1 in tumor, as well as ICAM-1 induction in tumor-associated vasculature, where tumor growth and angiogenesis are active. An increasing number of studies have begun to focus on the crosstalk between the immune activation of vascular niche, angiogenesis, and tumor progression [7, 8]. Upregulated levels of ICAM-1 in tumor have been linked to two different contexts, one serving as a marker for the recruitment of effector immune cells and tumor killing, while it was also observed in malignant and metastatic tumors with poor prognosis. Seemingly contradicting roles of ICAM-1 may be due to the complexity of inflammation in various phases of tumor development, which can be better examined by in vivo imaging tools. Our nanoparticles did indeed show the localization into the tumor vasculature, while their localization into the vasculature elsewhere was non-detectable. With human tumor xenograft model, we observed that the majority of tumor mass was comprised of tumor cells with poor vascularity present therein. Higher vascularity was found in the periphery of the tumor, often called the invasive tumor front, where ICAM-1 overexpression associated with higher immune activity has also been reported in many carcinoma cases. Notably, we have demonstrated that intravenously injected nanoparticles targeting ICAM-1 specifically localized into the vasculature associated with the tumor progression. With further improvement of detection sensitivity in addition to more native tumor models containing not only tumor cells themselves but also fully-developed vasculatures and other stroma cells such as macrophages, our nanoparticles may provide a universal tumor imaging strategy not by tumor surface markers limited to specific cancer types but by the inflamed microenvironment which is associated with almost all cancer development.

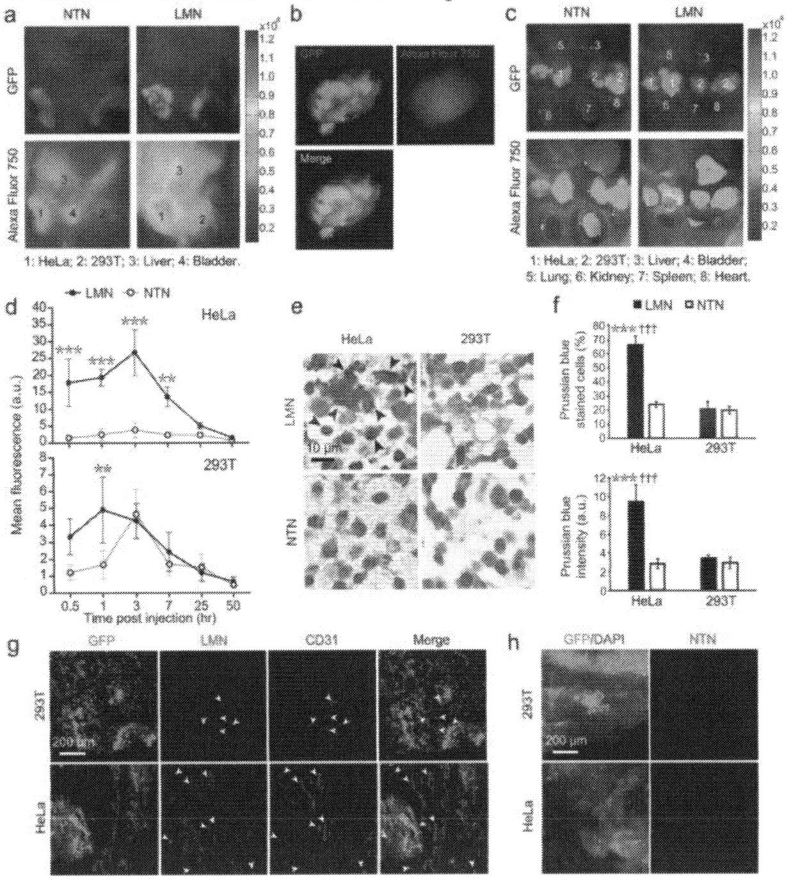

Figure 3. *In vivo* (a, b) and *ex vivo* (c) near-IR imaging of mice at 50 hr after intravenous injection of NTN vs. LMN. GFP indicates the growth of HeLa ('1') and 293T ('2') tumors. The distribution of nanoparticles into the major organs ('3'-'8') were also examined. (d) Near-IR fluorescence intensities of HeLa and 293T tumors at different time points after intravenous injection of LMN vs. NTN (n = 4; $^{**}P < 0.01$, $^{***}p < 0.001$). (e) Perl's Prussian blue staining of tumor sections collected at 50 hr after the injection of LMN vs. NTN. Stained iron is marked with black arrows. (f) Percentage of cells stained in Prussian blue (top) and the intensity of Prussian blue in the field of view (bottom) within tumor sections (n = 3; $^{***}p < 0.001$ between LMN vs. NTN in Hela tumor. $^{\dagger\dagger\dagger}p < 0.001$ between HeLa and 293T tumor using LMN). (g) Immunofluorescence imaging of tumors at 4 hr post-injection of LMN. Tumor sections were also stained with anti-CD31 antibody for delineating vasculature. LMN localization into the tumor vasculatures was indicated with white arrows. (h) Fluorescence imaging of tumors at 4 hr post-injection of NTN.

IV. REFERENCES

[1] Kelly CP, O'Keane JC, Orellana J, Schroy PC, 3rd, Yang S, LaMont JT, et al. Human colon cancer cells express ICAM-1 in vivo and support LFA-1-dependent lymphocyte adhesion in vitro. Am J Physiol 1992;263:G864-70.

[2] Maurer CA, Friess H, Kretschmann B, Wildi S, Muller C, Graber H, et al. Over-expression of ICAM-1, VCAM-1 and ELAM-1 might influence tumor progression in colorectal cancer. Int J Cancer 1998;79:76-81.
[3] Hayes SH, Seigel GM. Immunoreactivity of ICAM-1 in human tumors, metastases and normal tissues. Int J Clin Exp Pathol 2009;2:553-60.
[4] Jin M, Song G, Carman CV, Kim YS, Astrof NS, Shimaoka M, et al. Directed evolution to probe protein allostery and integrin I domains of 200,000-fold higher affinity. Proc Natl Acad Sci U S A 2006;103:5758-63.
[5] Kang S, Park T, Chen X, Dickens G, Lee B, Lu K, et al. Tunable physiologic interactions of adhesion molecules for inflamed cell-selective drug delivery. Biomaterials 2011;32:3487-3498.
[6] Dubertret B, Skourides P, Norris DJ, Noireaux V, Brivanlou AH, Libchaber A. In vivo imaging of quantum dots encapsulated in phospholipid micelles. Science 2002;298:1759-62.
[7] Rajashekhar G, Willuweit A, Patterson CE, Sun P, Hilbig A, Breier G, et al. Continuous endothelial cell activation increases angiogenesis: evidence for the direct role of endothelium linking angiogenesis and inflammation. J Vasc Res 2006;43:193-204.
[8] Franses JW, Baker AB, Chitalia VC, Edelman ER. Stromal endothelial cells directly influence cancer progression. Sci Transl Med 2011;3:66ra5.

(405b) Stable High Affinity Streptavidin Protomers for Cell Biology Application

Kok Hong Lim[1], Cheng Kuo Hsu[1], Heng Huang[2], Arnd Pralle[2] and Sheldon Park[1], (1)Chemical and Biological Engineering, University at Buffalo, Buffalo, NY, (2)Physics, University at Buffalo, Buffalo, NY

While the streptavidin-biotin interaction is the basis ofmany biotechnology and medical applications, the large size of streptavidintetramer and the multivalency of the molecule present obstacles in somesituations. The focus of the current study is thus to engineer smallerstreptavidin variants that have sufficient stability and affinity to make thetechnology accessible to other applications, including in vivo cell imaging. Tothis end, we adopted a rational design approach to engineer a monomer and ahairline dimer mutant with improved stability and affinity. We tested monomermutants containing one or more engineered disulfides to stabilize thebiotin-bound conformation. The best mutant was then engineered further byintroducing surface mutations to create an extended stabilizing salt bridgenetwork. The T_m = 47 C and K_d = 60 nM ofthe resulting monomer, which compare favorably with those of wt monomer (T_m= 32 C and K_d = 130 nM), allow the molecule to be used in acell biology experiment performed at 37 C. It has long been suspected thathairline interaction across the dimer-dimer interface is important for highaffinity biotin binding, but engineering such a dimer has been difficult due tolimited interface between the subunits. To engineer a stable hairline dimer, wemodeled the structure and introduced an interchain disulfide between twodesigned monomer subunits. We show that the purified dimer, which elutes as asingle peak by gel filtration and has high monodispersity by dynamic lightscattering, is more stable than the substituent monomer by ~ 6 C and has an 8fold higher affinity compared to wt monomer, making it an ideal molecule forapplications that require a stable labeling reagent.

In addition to rational design ofstreptavidin protomers, we will present results from two other parallel works forimproving the biophysical properties of streptavidin monomer. First, we haverecently demonstrated that functional streptavidin monomer can be displayed onthe yeast surface using biotin as a molecular chaperone. This allowed us toimprove the stability of rationally designed molecule by directed evolution,which has not been possible until now due to the difficulties of foldingstreptavidin in yeast. Second, we used homology modeling of streptavidin and streptavidin-likemolecules from other organisms to create a stable hybrid molecule that isroughly 20% of the mw of wt streptavidin. Toward improving the thermodynamicand binding properties of streptavidin monomer, we will present a comparativeanalysis of designed chimeras that reveal the challenges of engineering anoptimized a monomer sequence.

See more of this Session: Protein Engineering III - Applications
See more of this Group/Topical: Food, Pharmaceutical & Bioengineering Division

(405c) Engineering High-Affinity, Aggregation-Resistant Domain Antibodies

Joseph M. Perchiacca, Moumita Bhattacharya and **Peter M. Tessier**, Dept of Chemical & Biological Engineering, Rensselaer Polytechnic Institute, Troy, NY

High-affinity antibodies are critical for numerous detection and therapeutic applications, yet their utility is limited by their variable propensities to aggregate either at low concentrations for antibody fragments or high concentrations for full-length antibodies. Therefore, determining the sequence and structural features that differentiate aggregation-resistant antibodies from aggregation-prone ones is critical to improving their activity. We have investigated the molecular origins of antibody aggregation for human domain antibodies that differ only in the sequence the loops containing their complementarity determining regions (CDRs), yet such antibodies possess dramatically different aggregation propensities in a manner not correlated with their conformational stabilities. We find the propensity of these antibodies to aggregate after being transiently unfolded is not a distributed property of the CDR loops, but can be localized to aggregation hotspots within and near individual CDRs. Moreover, we have identified charged mutations within and adjacent to CDRs that endow poorly soluble variants with biophysical properties of aggregation-resistant antibodies. Guided by these findings for germline domain antibodies, we have sought to engineer a panel of anti-amyloid antibodies that possess hydrophobic CDRs with extreme resistance to aggregation without diminishing their affinity and specificity. We find that the CDR sequences of anti-amyloid antibodies strongly influence their solubility. We are currently investigating the impact of charge mutations adjacent to hydrophobic CDRs within these antibodies with the goal of engineering high-affinity, aggregation-resistant antibodies.

See more of this Session: Protein Engineering III - Applications
See more of this Group/Topical: Food, Pharmaceutical & Bioengineering Division

(405d) Isolation and Characterization of Novel Antibodies That Target the Blood-Brain Barrier

Charles C. Stutz[1], Angela R. Jones[2], Sue-Zanne Tan[2] and Eric Shusta[3], (1)Chemical and Biological Engineering, University of Wisconsin - Madison, Madison, WI, (2)University of Wisconsin - Madison, Madison, WI, (3)University of Wisconsin, Madison, WI

Central nervous system (CNS) disorders affect 1.5 billion people worldwide, and these numbers will continue to increase with an aging population. Thus, it is important that CNS therapeutics continue to be developed and optimized. One important challenge in the treatment of brain disease is the limitations on drug delivery imposed by the blood-brain barrier (BBB). The BBB is a layer of tightly connected endothelial cells that prevents almost all molecules from entering the brain by diffusion. In attempts to circumvent the BBB, the endogenous nutrient transporters present in the brain have been suggested as potential drug delivery portals. The most studied of these transporters are the insulin receptor and transferrin receptor, both of which undergo receptor-mediated transcytosis (RMT). Antibodies against these receptors have been conjugated with either drugs or tracers molecules and have been shown to penetrate the brain via the RMT pathway. The drawback of using these two well-studied systems is that these transporters are also widely expressed elsewhere in the body. This allows for significant accumulation of the antibody and payload in peripheral organs, which can both increase the risk of side effects and lowers the blood availability of the payload. Thus, our goal has been to isolate novel RMT transporters and corresponding targeting antibodies that allow for more efficient and targeted drug delivery across the BBB. To this end, a two-step combinatorial screen was performed using a phage-displayed library of human scFv. In the first step, two rounds of *in vitro* endocytosis screening were performed on cultured primary rat brain endothelial cells or freshly isolated rat brain capillaries. Prior to the internalization selection the libraries were pre-subtracted against peripheral endothelia to help promote BBB specificity. The *in vitro* step serves to focus the antibody pool to clones that have the capacity to endocytose via RMT systems. The phage pools resulting from the second round of the *in vitro* selection were then used in *in vivo* transport screens (tail vein injection) and binding screens (transcardial perfusion). While the transport screens led to some promising antibody clones displaying BBB selectivity and brain uptake, the targeted receptors were only sporadically expressed throughout the brain. In contrast, the binding screen substantially enriched the antibody pools for clones that bind the luminal side of the rat brain capillaries (14/27 clones tested). Promising phage-displayed antibody clones have been used in immunohistochemical staining of rat organ tissue sections to predict biodistribution, and BBB binding was confirmed by transcardial perfusion. Several scFv have been also been produced as soluble proteins and are being used to test both *in vivo* biodistribution and to expression clone the cognate RMT system. Taken together, these studies have resulted in a panel of BBB-targeted antibodies that could ultimately expand the repertoire of brain drug delivery options.

See more of this Session: Protein Engineering III - Applications
See more of this Group/Topical: Food, Pharmaceutical & Bioengineering Division

(405e) Using An Orthogonal Ligand-Receptor Pair to Control Endogenous Gene Expression In Mammalian Cell

Jing Liang, Department of Chemical and Biomolecular Engineering, University of Illinois at Urbana-Champaign, Urbana, IL, Michael McLachlan, Cellular Dynamics, Madison, WI and Huimin Zhao, Department of Chemical and Biomolecular Engineering, Department of Chemistry and Biochemistry, Institute for Genomic Biology, University of Illinois at Urbana-Champaign, Urbana, IL

Small molecule regulated gene expression systems (gene switches) are powerful tools for gene therapy, functional genomics, metabolic engineering, and tissue engineering. By combining an engineered estrogen receptor ligand-binding domain (LBD), an activation domain, and an artificial zinc-finger DNA-binding domain (DBD), we have constructed a gene switch that is able to regulate the endogenous VEGF-A (vascular endothelial growth factor A) expression in mammalian cell. The gene switch is specifically and reversibly controlled by 4,4'-dyhydroxybenzil (DHB), a small molecule, non-steroid synthetic ligand, which acts orthogonally in a mammalian system. After optimization of the gene switch architecture, a VEGF-A induction ratio of ~200X can be achieved in HEK293 cells at micromolar concentration of DHB. This gene switch can potentially be combined with other artificial DBDs to control other endogenous genes. Furthermore, we can also substitute the LBD with one from another orthogonal ligand-receptor pair to create gene switches that can simultaneously regulate multiple genes in a mammalian system.

See more of this Session: Protein Engineering III - Applications
See more of this Group/Topical: Food, Pharmaceutical & Bioengineering Division

(405f) De Novo Protein Design of Agonists and Antagonists of the C3a Receptor: Theoretical Predictions and Experimental Validation

Meghan B. Peterson[1], Hoki Fung[1], Christodoulos A. Floudas[1], Dimitrios Morikis[2], Peter N. Monk[3] and Trent M. Woodruff[4], (1)Department of Chemical and Biological Engineering, Princeton University, Princeton, NJ, (2)Department of Bioengineering and Center for Bioengineering Research, University of California, Riverside, Riverside, CA, (3)Department of Infection and Immunity, Sheffield University Medical School, Sheffield, United Kingdom, (4)School of Biomedical Sciences, The University of Queensland, Brisbane, Australia

Protein design, also known as theinverse folding problem, seeks the amino acid sequence that will fold into agiven 3-dimensional template. The protein design problem exhibits degeneracydue to the fact that many amino acid sequences fold into a given template. Itis therefore important to examine all the possible sequences for a giventemplate and rank them based upon specific properties that are being designed(activity, specificity, etc.).

A de novo design framework with a ranking metricbased on fold specificities was applied to the design of C3a receptor (C3aR)agonists and antagonists. Thedesign is based upon the structure of C3a, which activates C3aR. C3a is a 77-residue peptide that mediates the pro-inflammatoryactivities in the human complement system and possibly has opposingimmunological roles in some cellular systems (34). Improper activation of thecomplement system can cause tissue injury in various pathological conditionsand contributes to several immune diseases, including stroke, heart attack,Alzheimer's disease, asthma, rheumatoid arthritis, and rejection ofxenotransplantation. Thecrystal structure resolved by Huber et al. [1], as well as flexible templatestructures generated by MD simulations with explicit solvation via watermolecules were employed as the design template.

The framework consists of twostages: a sequence selection stage and a validation stage. The sequenceselection stage produces a rank-ordered list of amino acid sequences with thelowest energies by solving an integer programming sequence selection model [2].The sequence selection model incorporates backbone flexibility into the designprocess by utilizing the set of structures obtained from the MD simulationswith implicit solvation and with explicit water molecules. In doing so, thepairwise energy between two residues can take on a range of values dependingupon the range of distances obtained from the structures.

The second stage re-ranks thesequences from stage one using either a fold specificity or an approximatebinding affinity [3-4]. Since structural information of the C3a:C3aR complexwas unknown, only fold specificity calculations could be employed. Fold specificitymeasures how likely a given sequence will fold into the design templatestructure [5]. Thus the design was driven by the hypothesis that structureimplies function, and novel sequences of C3a that adopt the C3a fold arepotential candidates for C3aR agonists or antagonists.

Hundreds of sequences weregenerated and ranked using the de novo protein design framework. The top sevensequences were selected for synthesis and experimental validation. Of the seven designed peptides tested,two were prominent agonists while two others were prominent antagonists. Theagonists showed a up to a five-fold improvement in EC50 compared with the previouslydiscovered "superagonist" by Ember et al. [6], with EC50s of 33.4 and 66.16 nM.The antagonists were also very potent, with IC50s of 26.13 and 63.04 nM.

Thiswork highlights the success of the protein design framework. In this case,structural information on the binding site was unknown, so the design wasdriven by fold specificities. Since structure often implies function, this is agood metric to use when structural binding data is unavailable.

[1] Huber,R., H. Scholze, E. Paques, and J. Deisenhofer, 1980. Crystal Structure Analysisand Molecular Model of Human C3a Anaphylatoxin. Hoppe Seylers Zeitschrifft fuerPhysiologische Chemie 361:1389–1399.

[2] H. K. Fung, M. S. Taylor, and C. A. Floudas. Novel Formulations forthe sequence selection problem in de novo protein design with flexibletemplates. *Optim. Methods & Software*,22:51-71, 2007.

[3] Bellows,M. L., H. K. Fung, C. A. Floudas, A. Lopez de Victoria, and D. Morikis, 2010.New Compstatin Variants Through Two De Novo Protein Design Frameworks. BiophysJ 98:2337–2346.

[4] Bellows,M. L., M. S. Taylor, P. A. Cole, L. Shen, R. F. Siliciano, H. K. Fung, and C.A. Floudas, 2010. Discovery of entry inhibitors for HIV-1 via a new de novoprotein design framework. Biophys J 99:3445–3453.

[5] Fung,H. K., C. A. Floudas, M. S. Taylor, L. Zhang, and D. Morikis, 2008. TowardFull-Sequence De Novo Protein Design with Flexible Templates for HumanBeta-Defensin-2. Biophys. J. 94:584–599.

[6] Ember,J., N. Johansen, and T. Hugli, 1991. Designing Synthetic Superagonists of C3aAnaphylatoxin. Biochem. 30:3603–3612.

See more of this Session: Protein Engineering III - Applications
See more of this Group/Topical: Food, Pharmaceutical & Bioengineering Division

(405g) Directed Evolution of Bright Mutants of a Flavin Dependent Anaerobic Fluorescent Protein From Pseudomonas Putida

Arnab Mukherjee and Charles M. Schroeder, Chemical and Biomolecular Engineering, University of Illinois at Urbana-Champaign, Urbana, IL

We report the construction of bright variants of an **anaerobic fluorescent protein** (AFP) reporter from *Pseudomonas putida* [1]. Fluorescent reporter proteins are widely used as non-invasive cellular probes to study gene expression, protein-protein interactions, subcellular localization, and protein trafficking. The green fluorescent protein (GFP) and its related analogs are routinely employed for *in vivo* imaging. However GFPs require molecular oxygen for fluorescence, thereby restricting their use to anaerobic niches. Recently, a novel class of flavin mononucleotide (FMN) dependent fluorescent proteins from *Pseudomonas putida* and *Bacillus subtilis* was reported [2]. The flavin-dependent fluorescent proteins employ a buried FMN chromophore for fluorescence in an oxygen-independent manner. Therefore FMN based fluorescent proteins are attractive candidates for live cell imaging in anaerobic conditions.

In this work, we describe the engineering of two spectrally improved mutants of the flavin-binding anaerobic fluorescent protein from *Pseudomonas putida* that show a two-fold increase in fluorescence emission relative to the parent protein. To isolate these bright mutants, we randomly mutagenized key amino acids by saturation mutagenesis with degenerate nucleotides and screened the mutant library by spectrofluorometric measurements for spectral shifts and improved brightness. The amino acids were selected for mutation based on their proximity (up to 0.3 nm) to the buried FMN chromophore in a molecular model of the three-dimensional protein structure constructed by homology modeling. We found that brightness enhanced mutants were both mutated at the same amino acid – a flavin proximal phenylalanine residue (F37), and we designated these mutants as AFP F37S and F37T. Aromatic amino acids are known to quench aromatic fluorophores through stacking interactions. We therefore hypothesize that the two-fold enhancement in fluorescence that results from the F->S and F->T mutations in the bright variants is a consequence of a decrease in quenching of flavin fluorescence by the aromatic phenylalanine residue in the wild type protein. We corroborate this claim by demonstrating that mutating the wild type F37 residue to the aromatic amino acids tryptophan (F37W) and tyrosine (F37Y) abrogates the native fluorescence significantly.

To the best of our knowledge, our study is the first report of engineering of a flavin-based anaerobic fluorescent protein for prokaryotic expression. Further engineering of anaerobic fluorescent proteins for spectral enhancements would enable studies of real time gene expression, protein localizations, and dynamics in obligate anaerobes as well as in hypoxic niches of the human body (*e.g.* hypoxic tumors), that have thus far been hindered by the strict requirement of O_2 for GFP to fluoresce.

References:

1. A. Mukherjee, C.M. Schroeder, submitted to *Biotechnology and Bioengineering* (2011).
2. T. Drepper, et al. *Nature Biotechnology*, 25: 443:445 (2007).

See more of this Session: Protein Engineering III - Applications
See more of this Group/Topical: Food, Pharmaceutical & Bioengineering Division

(406a) Ampicillin Synthesis Using a Two-Enzyme Cascade System with Both &alpha-Amino Ester Hydrolase and Penicillin G Acylase

Janna Blum[1], Andria L. Deaguero[2], Carolina V. Perez[1] and Andreas S. Bommarius[1], (1)School of Chemical and Biomolecular Engineering, Georgia Institute of Technology, Atlanta, GA, (2)Chemical and Biomolecular Engineering, Georgia Institute of Technology, Atlanta, GA

The current enzymatic production of semi-synthetic penicillin antibiotics requires isolation and purification of the intermediate 6-aminopenicillanic acid (6-APA), which adds cost and complexity to the manufacturing process. In this work we take advantage of the unique substrate specificity of α-amino ester hydrolases (AEH, EC 3.1.1.43), which are specific for α-amino containing β-lactam antibiotics, to perform a purely aqueous one-pot production of ampicillin directly from penicillin G, catalyzed by both AEH and penicillin G acylase (PGA, EC 3.5.1.11). The synthesis was performed in both a one-pot, one-step synthesis with 39% maximum conversion to ampicillin, and a one-pot, two-step process resulting in a maximum of 47% conversion to ampicillin. This two-enzyme cascade reaction to semi-synthetic antibiotics is a promising alternative to the current two-step, two-pot enzymatic process with intermittent isolation of 6-APA.

See more of this Session: Reaction Engineering In API Synthesis
See more of this Group/Topical: Food, Pharmaceutical & Bioengineering Division

(406b) Continuous Processing As a Process Development Tool for Fast Reactions

Mourad Hamedi, Chemical Process Development & Commercialization, Merck and Co. Inc., Rahway, NJ, Vijay Gnanadesikan, Process Chemistry, Merck and Co. Inc., Rahway, NJ, Dimitrios Zarkadas, Chemical Process Development & Commercialization, Merck & Co. Inc., Rahway, NJ, Christopher Pridgen, Pharmaceutical Commercialization Technology, Merck and Co. Inc., Summit, NJ and George Wu, Process Chemistry, Merck & Co. Inc., Rahway, NJ

Continuous processing is rarely seen as an investigational tool in the early development of chemical reactions in the pharmaceutical industry although it is sometimes most fit to provide needed kinetics and impurity profile related information. This is particularly the case for fast reactions that are sensitive to water and/or oxygen for which sampling may be challenging.

Our approach to reaction process development using a continuous process is showcased through a loop formation reaction in the presence of NaHMDS. The aim of the investigation is to optimize the reaction operating conditions by varying temperature and base composition. Process optimization has been achieved through a design of experiment approach encompassing eight operating conditions. The DOE was concluded in half a day and utilized only four grams of substrate. A statistical analysis of the experimental data allowed establishing correlations between the operating conditions and the impurity generation as well as obtaining reaction kinetics related information, thus leading to identifying optimal operating conditions. Conclusions drawn from this work have allowed a speedy and successful scale up to pilot plant scale.

See more of this Session: Reaction Engineering In API Synthesis
See more of this Group/Topical: Food, Pharmaceutical & Bioengineering Division

(406c) Kilos and Beyond: Using Continuous Chemistry to Speed Early Candidate Delivery and the Hurdles Incumbent throughout the Development Process

John G. Van Alsten, Chemical Research & Development, Pfizer Inc., Groton, CT, Bryan Li, Research-API, Pfizer Inc., Groton, CT, John Lucas, Chemical Technolgy-API, Pfizer Inc., Groton, CT and David Place, Engineering Technology-API, Pfizer Inc., Groton, CT

We will present an honest, "warts and all" history of our experience in bringing flow processing to the development of early stage candidate APIs. To become accepted flow technologies must enable the development chemist to undertake transformations deemed "unbatchable" for reasons of safety or extreme processing conditions, and we will present several case histories of these. However, we have found that technical issues are only one component in acceptance of the technology-the formidable skills of the synthetic chemist and the justifiably risk-averse development environment are steep barriers, for example. Considerable impediments exist for late-stage candidates as well. We will discuss our strategies for overcoming these barriers.

See more of this Session: Reaction Engineering In API Synthesis
See more of this Group/Topical: Food, Pharmaceutical & Bioengineering Division

(406d) Synthesis of 7-Methoxy -4'-Substituted Flavones As Potential Pharmaceutically Active Compounds

Jingyun An, **Xiang Liu** and YI Wang, School of Chemical and Material Engineering, Jiangnan University, Wuxi, China

A series of 7-methoxy-4'-substituted flavones were synthesized by the Baker-Venkataraman rearrangement approach, starting from 2-hydroxyl-4-methoxy acetophenone and 4-substituted benzoyl chlorides. High selectivity, short reaction times and good yields were reached. Several new flavones involving amino group were also synthesized with 7-methoxy-4'-nitroflavone as raw material. 7-methoxy-4'- aminoflavones were found to possess significant antioxidation, antitumor and anti-inflammatory effects, as well as the 7-methoxy-4'-nitroflavone.

See more of this Session: Reaction Engineering In API Synthesis
See more of this Group/Topical: Food, Pharmaceutical & Bioengineering Division

(407a) Progress In Quality by Design Implementation In Fda's Office of New Drug Quality Assessment

Christine Moore, DHHS/FDA/Cder/OPS/Ondqa, FDA, Silver Spring, MD

Since the launch of the Office of New Drug Quality Assessment (ONDQA) CMC pilot program in July 2005, over 40 Quality by Design (QbD) containing applications have been submitted, reviewed, and approved by ONDQA, both within and outside of the pilot. Although challenges still remain, many approaches for preparation and review of QbD containing applications have been established and efforts are underway toward international harmonization. This presentation will discuss the progress made to date in QbD and remaining scientific and regulatory challenges. Additionally, an overview will be given of harmonization efforts, including the recently announced pilot between FDA and the European Medicines Agency (EMA) for QbD containing applications.

See more of this Session: Round Table Invited Session On QbD: QbD Filings
See more of this Group/Topical: Topical I: Comprehensive Quality by Design in Pharmaceutical Development and Manufacture

(407b) QbD Approaches for Continuous Manufacturing

Fernando J. Muzzio, Chemical and Biochemical Engineering, Rutgers University, Piscataway, NJ

In this talk, we discuss multiple examples of application of QbD ideas and principles to the design of continuous manufacturing systems. Examples include the use of DoE methods for the characterization of system subcomponents (feeders, blenders, feed frames, tablet press) and the identification of critical process variables, examination of the effect of process parameters (blend composition, shear, strain) on the material properties of intermediate and finished products (blend uniformity and flowability, tablet hardness and dissolution) and the analysis of the integrated system.

See more of this Session: Round Table Invited Session On QbD: QbD Filings
See more of this Group/Topical: Topical I: Comprehensive Quality by Design in Pharmaceutical Development and Manufacture

(407c) Experiences with Quality by Design Filings and Prior Approval Inspections

Chris Balducci, Technical Research & Development, Novartis Pharmaceuticals Corporation, East Hanover, NJ

Experiences with Quality by Design Filings and Prior Approval Inspections

Abstract:

Quality by Design has been embraced by the Pharmaceutical Industry yielding much progress in interpreting ICH Guidelines, implementing risk assessment strategies and constructing Design Space philosophies in an effort to improve scientific understanding during product development. These activities are then assessed at the time of dossier filing, when years of development work are evaluated for alignment with Health Authorities expectations.

This short presentation will introduce the audience to Novartis' experiences regarding dossier module preparation and prior approval inspections conducted in the US and EU. Examples will be given highlighting the importance of source document quality and content and successful cross functional team interactions. The goal of this presentation is to provide enough detail to foster round table discussion and sharing of experiences regarding QbD filings.

See more of this Session: Round Table Invited Session On QbD: QbD Filings
See more of this Group/Topical: Topical I: Comprehensive Quality by Design in Pharmaceutical Development and Manufacture

(407d) Quality by Design for Generic Drugs

Yingxu Peng, Office of Generic Drugs, FDA CDER, Rockville, MD

To facilitate the implementation of Quality by Design (QbD) for abbreviated new drug application (ANDA), Office of Generic Drugs (OGD) has developed mock QbD examples for immediate release (IR) and modified release (MR) dosage forms. This presentation will provide insight into OGD's current thinking on QbD filing requirement and definition of some QbD nomenclatures. The Question-based Review (QbR) system is to be revised in 2011 and 2012 and is expected to be fully implemented by January 2013. This presentation will also discuss some practical tools and types of pharmaceutical development studies ANDA applicants may use as they implement QbD in their pharmaceutical development process.

See more of this Session: Round Table Invited Session On QbD: QbD Filings
See more of this Group/Topical: Topical I: Comprehensive Quality by Design in Pharmaceutical Development and Manufacture

(407e) To Be Updated

Stefanie Pluschkel, Pfizer Inc, Groton, CT

tobeadded

See more of this Session: Round Table Invited Session On QbD: QbD Filings
See more of this Group/Topical: Topical I: Comprehensive Quality by Design in Pharmaceutical Development and Manufacture

QUALITY BY DESIGN (QBD), BIOPHARMACEUTICAL MANUFACTURE.

Kurt Brorson, Ph.D., Office of Biotechnology Products, CDER/FDA, 10903 New Hampshire Ave., Silver Spring MD 20903. Views expressed are those of the author and not necessarily those of the US FDA.

In the biopharmaceutical industry, consistently manufacturing a safe and high quality protein drug substance can be challenging. Recently, a regulatory structure supporting Quality by Design (QbD) for pharmaceuticals and biopharmaceuticals has been outlined in International Conference on Harmonization (ICH) Quality documents (Q8 through Q11; available at www.ich.org). Many of the concepts espoused in these ICH documents [e.g., quality target product profile (QTPP), risk-based approaches, product and process understanding, design spaces, control strategies with process analytical technologies (PAT), product life cycle, continuous process verification] are universal for pharma/biotechnology manufacturing and can readily be applied to bulk protein production.

The application of QbD to biotechnology development and manufacturing traces its origins to the early 2000's. In this timeframe, QbD concepts were being discussed and evolved during the implementation of CDER's GMP initiative and ICH discussions. These discussions identified the need for science-and risk-based approaches to improve quality and consistency of both traditional pharmaceuticals and biotechnology products. In 2005, CDER's Office of New Drug Quality Assessment (ONDQA) initiated a pilot program allowing sponsors to submit quality by design based applications (70 FR 40719, July 14, 2005). In 2008, CDER's Office of Biotechnology Products (OBP) follow suit, initiating a similar pilot program for biotech products (73 FR 37972, July 2, 2008). These applications were submitted to Federal Register Docket No. FDA-2008-N-0355, with a deadline to request participation by September 30, 2010. OBP's goal was to facilitate quality by design, risk-based approaches for manufacturing of complex molecules. As OBP consists of two divisions, the Division of Monoclonal Antibodies or the Division of Therapeutic Proteins, this program covers applications for monoclonal antibody, antibody fusion protein, cytokine, enzyme, hematologic proteins and other recombinant proteins (other than hormone or insulin products which are regulated by ONDQA).

To date, the program has successfully drawn applications from several firms. Applications accepted in QbD Pilot include five Original Applications (4 Monoclonal Antibodies and 1 Fc Fusion Protein) and four Post-approval Supplements (2 Monoclonal Antibodies, 1 Therapeutic Protein, and 1 multi-product). Two of these applications involve production site transfers; thus, OBP is working closely with CDER's Office of Compliance (OC). To enhance this cooperation, a CDER manual of policy and procedure (MAPP 4730.3) was recently developed to define and coordinate interactions between OBP & OC. The pilot program is intended to be highly interactive; on average, there are 3-5 meetings for each application/supplement. This level of interaction between CDER and regulated firms is critical due to the complexity of QbD applications and the need to proscriptively define goals, targets and agreements before initiating the large volume of work required in a QbD application.

Beyond the QbD applications submitted to CDER, hypothetical and real examples have been published. This is very useful from an educational standpoint because the submitted QbD dossiers are trade secret and thus the structure and detail is not publically available. In contrast, the hypothetical case studies are public, present a possible framework for QbD application and contain a high level of detail. The best known of these examples is "A-Mab: a case study in bioprocess development" written in 2009 by the CMC Biotech Working Group, a consortium of firms involved in biotech manufacture, with input from CDER (http://www.casss.org/associations/9165/files/A-Mab_Case_Study_Version_2-1.pdf). This case study is highly comprehensive, covering implementation of QbD from upstream manufacture to final fill/finish. It contains example risk analyses of individual unit operations and follow-up design of experiment studies. It also discusses aspects of QbD germane to molecule design, QTPP identification, linkage analysis, control strategy, life cycle management and regulatory. Two other case studies have been published which are restricted to one aspect of one unit operation, viral clearance by anion exchange chromatography (Shah et al., 2010, Encyc. Ind. Biotechnol. Wiley Press; Strauss et al., Biotechnol. Prog., 2010, 26:750). Both follow a similar format QTPP identification → risk analysis → DoE studies → control strategy.

Manageability and cost-benefits need to be considered when implementing QbD and designing DoE studies. The cost of very large DoE studies needs to be weighed against the benefit of the processing flexibility. As the industry moves forward, lab automation and high-through put methods like Q-PCR can tip the equation towards the QbD approach. It must be remembered that there is no single, mandatory approach for QbD, a point stressed by the above mentioned case studies. In fact, many risk tools are available (see the list in ICH Q9). The existing product class knowledgebase can be leveraged for risk assessment purposes and DoE design, especially for products with shared unit operations and where cross-product experience is available like antibodies, hematologic products or cytokines. Finally, it must be remembered that QbD itself is optional, and may not always be the best approach (e.g. very unique or small volume products).

(425a) Optimal Strain Design of Saccharomyces Cerevisiae for Bioethanol Production

Joshua C. Quarterman[1], Yong-Su Jin[1], Nathan D. Price[2] and Pan-Jun Kim[3], (1)Food Science and Human Nutrition, University of Illinois at Urbana-Champaign, Urbana, IL, (2)Chemical and Biomolecular Engineering, University of Illinois at Urbana-Champaign, Urbana, IL, (3)Institute for Genomic Biology, University of Illinois Urbana Champaign, Urbana, IL

For several thousand years, yeast has been used as a biocatalyst for the fermentation of sugar into ethanol in a wide variety of food applications. More recently, yeast has been employed for large-scale ethanol production to replace petroleum fuels and meet a growing demand for homegrown energy, mainly using sugars from food products such as corn or sugarcane as a raw material. In the present study, we attempt to identify gene targets for improving ethanol production in the yeast *Saccharomyces cerevisiae* by using two mathematical modeling methods: flux balance analysis (FBA) and elementary flux mode (EFM) analysis. The results demonstrate the usefulness of an *in silico* analysis for identifying genetic perturbations in yeast for improved bioethanol production.

Flux Balance Analysis (FBA) is a mathematical tool for calculating the flow of metabolites through a metabolic network with given constraints and an objective function. Using the concept of FBA, an *in silico* gene deletion study was conducted for the iND750 genome-scale yeast model and gene targets were identified for improving ethanol production with sufficient biomass for cell growth. Experimental validation confirmed the effectiveness of the *in silico* analysis and led to identification of novel gene targets for a 20-30% improvement in ethanol production during glucose fermentation.

EFM analysis is another useful mathematical framework for defining and describing all metabolic routes that are both stoichiometrically and thermodynamically feasible for a group of enzymes. The analysis can decompose a complex metabolic network of many highly interconnected reactions into uniquely organized backbone pathways. In this study, a novel strategy was developed to identify knockout targets for improved phenotype using the concept of EFMs. Subsequently, the strategy was applied to a small-scale 58-reaction metabolic network for *S. cerevisiae* with the objective of high ethanol yield and sufficient biomass production for cell growth. METATOOL 5.1 was used to calculate the 1918 EFMs for the yeast metabolic network and the calculated EFMs were evaluated based on their ethanol-producing capabilities. Three sequential gene knockout targets have been identified for eliminating over 99% of the inefficient pathways for ethanol production through the network and thereby improving ethanol production.

See more of this Session: Advances In Metabolic Engineering and Bioinformatics IA - Biofuels
See more of this Group/Topical: Food, Pharmaceutical & Bioengineering Division

(425b) Towards Cellulosic Biodiesel: Engineering Xylose and Lipid Metabolism In Yeast

Mitchell Tai and Gregory Stephanopoulos, Chemical Engineering, Massachusetts Institute of Technology, Cambridge, MA

The future of biofuels faces significant technological challenges to achieve sustainable and widespread use. Microbial production of biofuels seeks to address some of these issues by utilizing biological platforms which can readily be engineered for the specific task of fuel production. Engineering efforts range from the utilization of lignocellulosic materials to the overproduction of naturally-rare fuel products. The oleaginous yeast, *Yarrowia lipolytica*, has been growing in relevance as a platform for fatty acid-based biofuel production, naturally accumulating large amounts of triacylglycerides. As one of the more heavily studied 'unconventional' microorganisms, it is well equipped for implementing production pathways for yeast oil, which can be converted into biodiesel or other fatty acid derivatives. Here we show progress in using metabolic engineering to optimize the lipid accumulation pathway as well as to introduce a xylose-consuming pathway in *Y. lipolytica*. Higher flux towards lipid accumulation is achieved with the overexpression of two key rate-limiting steps which divert and drive flux towards lipid synthesis: acetyl-coa carboxylase (ACC) and diacylglycerol acyltransferase (DGA). Furthermore, the xylose reductase-xylitol dehydrogenase (XYL12) and xylose isomerase (XYLA) pathways are both examined for enabling pentose sugar utilization. Our findings highlight the excellent lipid production capacity of *Y. lipolytica* and demonstrate its promise as a platform for cellulosic biodiesel production.

See more of this Session: Advances In Metabolic Engineering and Bioinformatics IA - Biofuels
See more of this Group/Topical: Food, Pharmaceutical & Bioengineering Division

(425c) Metabolic Engineering of Fatty Acid Pathways In Synechocystis PCC 6803 for Biofuel Production

Yi Ern Cheah, Chemical and Biological Engineering, Colorado State University, Fort Collins, CO and Christie A.M. Peebles, Chemical Engineering, Colorodo State University, Fort Collins, CO

Cyanobacteria are photosynthetic microorganisms that can consume carbon dioxide and convert them into useful products such as alcohols, fatty acids, and biopolymers. The cyanobacterium *Synechocystis PCC6803* is a good candidate for biofuel production because it has a well established molecular biology toolbox for genetic manipulation and its complete genome has been sequenced. There are two variants of *Synechocystis* 6803; one being the Glucose Tolerant (GT) mutant strain and the other being the wild type (WT) strain. The methods for gene knockouts and insertions developed thus far have primarily focused on the GT mutant strain. We want to utilize the WT strain because it has a higher growth rate compared to the GT strain. To engineer the WT strain, we developed a novel method for counter selection using the general protein synthesis inhibitor mazF from *E.coli*. We demonstrate the effectiveness of our method by inserting genes to manipulate the fatty acid biosynthesis pathway in the WT strain of *Synechocystis*.

Prokaryotic fatty acid biosynthesis is a well studied pathway especially in *E.coli*. In *Synechocystis*, the pathway generally functions to synthesize fatty acid chains which are mostly used for the formation and maintenance of the bacterial cell membrane. The fatty acid pathway can be manipulated to produce a variety of biochemicals and biofuels that could replace gasoline and diesel. We are particularly interested in utilizing the techniques of metabolic engineering and synthetic biology to engineer the WT strain of *Synechocystis* for hydrocarbon production. This talk will focus on our efforts to date.

See more of this Session: Advances In Metabolic Engineering and Bioinformatics IA - Biofuels
See more of this Group/Topical: Food, Pharmaceutical & Bioengineering Division

(425d) Predicting Immediate Behaviors of Engineered Microbial Strains for Biofuel Production

Joonhoon Kim, Christos T. Maravelias and Jennifer L. Reed, Department of Chemical and Biological Engineering; DOE Great Lakes Bioenergy Research Center, University of Wisconsin-Madison, Madison, WI

Computational modeling and analysis of metabolic networks has been successful in metabolic engineering of microbial strains for biochemical production including biofuels [1,2]. Among different computational approaches, constraint-based models allow for the use of genome-scale metabolic networks, and they can predict metabolic flux distributions in microbial cells. For example, flux balance analysis (FBA) predicts metabolic flux distributions in optimally growing microbial cells [3]. FBA can also predict the effects of gene additions or deletions on metabolic behaviors by adding or removing the associated reactions, respectively. Previous studies have shown that FBA accurately predicts the behaviors of *Escherichia coli* strains that have undergone adaptive evolution process to optimize their fitness [4,5].

FBA assumes that cells grow optimally; however, un-evolved mutant strains often exhibit suboptimal growth behaviors before adaptive evolution due to regulatory restrictions or other limitations. Therefore, mutant strains designed using FBA would need to be subjected to adaptive evolution to achieve high biochemical production, which may take several weeks to months [6]. An alternative to FBA is minimization of metabolic adjustment (MOMA), which predicts the behaviors of un-evolved mutants by minimizing the changes in flux distributions between the mutant strains and the parental strains [7], and it has been used to design mutant *E. coli* strains for improving lycopene [8] and L-valine [9] production. Another approach, regulatory on/off minimization (ROOM) minimizes the number of significant flux changes in mutant strains relative to the parental strain [10]. Both MOMA and ROOM can better predict qualitative behaviors of un-evolved (e.g., growth phenotypes), but their capabilities are still limited in predicting quantitatively accurate flux distributions.

We developed a new approach (RELATCH) to more accurately predict the quantitative flux distributions in un-evolved mutants. With this approach, we first utilize metabolic flux analysis measurements and gene expression data for the parental strain to determine a reference flux distribution. Then, the flux distribution for a knockout mutant strain is predicted by minimizing relative flux changes and additional enzyme usage from the reference flux distribution. We have used RELATCH to predict flux distributions of un-evolved mutants and compared the model predictions to experimental datasets for *E. coli*, *Saccharomyces cerevisiae*, and *Bacillus subtilis*. The results indicate that RELATCH more accurately predicts flux distributions, as well as growth rates, for un-evolved mutants, as compared to existing approaches (MOMA and ROOM). For example, for un-evolved *E. coli* mutants [11], the sum of squared errors (between model predictions and experimental results) for the fluxes predicted by RELATCH were 13 or 15 times smaller than the sum of squared errors when MOMA or ROOM was used, respectively.

We have also developed a new bi-level mixed-integer programming strain design approach, which considers un-evolved behaviors of mutants. This approach uses a quadratic inner objective function, such as MOMA or RELATCH, to identify mutant strains with improved biochemical production. Therefore, it allows us to design mutant strains which would not require adaptive evolution to achieve high biochemical production. The developed approach extends the scope of computational strain design, and can be used to identify novel metabolic engineering strategies for biofuel production.

1. Park JM, Kim TY, Lee SY (2009) Constraints-based genome-scale metabolic simulation for systems metabolic engineering. Biotechnol Adv 27: 979-988.

2. Blazeck J, Alper H (2010) Systems metabolic engineering: genome-scale models and beyond. Biotechnol J 5: 647-659.

3. Edwards JS, Ibarra RU, Palsson BO (2001) In silico predictions of Escherichia coli metabolic capabilities are consistent with experimental data. Nat Biotechnol 19: 125-130.

4. Ibarra RU, Edwards JS, Palsson BO (2002) Escherichia coli K-12 undergoes adaptive evolution to achieve in silico predicted optimal growth. Nature 420: 186-189.

5. Fong SS, Palsson BO (2004) Metabolic gene-deletion strains of Escherichia coli evolve to computationally predicted growth phenotypes. Nat Genet 36: 1056-1058.

6. Fong SS, Burgard AP, Herring CD, Knight EM, Blattner FR, et al. (2005) In silico design and adaptive evolution of Escherichia coli for production of lactic acid. Biotechnol Bioeng 91: 643-648.

7. Segre D, Vitkup D, Church GM (2002) Analysis of optimality in natural and perturbed metabolic networks. Proc Natl Acad Sci U S A 99: 15112-15117.

8. Alper H, Jin YS, Moxley JF, Stephanopoulos G (2005) Identifying gene targets for the metabolic engineering of lycopene biosynthesis in Escherichia coli. Metab Eng 7: 155-164.

9. Park JH, Lee KH, Kim TY, Lee SY (2007) Metabolic engineering of Escherichia coli for the production of L-valine based on transcriptome analysis and in silico gene knockout simulation. Proc Natl Acad Sci U S A 104: 7797-7802.

10. Shlomi T, Berkman O, Ruppin E (2005) Regulatory on/off minimization of metabolic flux changes after genetic perturbations. Proc Natl Acad Sci U S A 102: 7695-7700.

11. Fong SS, Nanchen A, Palsson BO, Sauer U (2006) Latent pathway activation and increased pathway capacity enable Escherichia coli adaptation to loss of key metabolic enzymes. J Biol Chem 281: 8024-8033.

See more of this Session: Advances In Metabolic Engineering and Bioinformatics IA - Biofuels
See more of this Group/Topical: Food, Pharmaceutical & Bioengineering Division

(425e) Functional Genomics Study of Fatty Acid Overproduction In Escherichia Coli

Rebecca M. Lennen[1], Max A Kruziki[2], Kritika Kumar[2], Robert A. Zinkel[3], Kristin E. Burnum[4], Mary S Lipton[4], Spencer W. Hoover[1], Ruwan Ranatunga[5], Tyler M. Wittkopp[5], Wesley D. Marner II[5] and Brian F. Pfleger[2], (1)Chemical and Biological Engineering, University of Wisconsin-Madison, Madison, WI, (2)Chemical and Biological Engineering, University of Wisconsin Madison, Madison, WI, (3) University of Wisconsin Biotechnology Center, Madison, WI, (4)Fundamental and Computational Sciences Directorate, Pacific Northwest National Laboratory, Richland, WA, (5)U.S. Department of Energy Great Lakes Bioenergy Research Center, Madison, WI

Next generation renewable biofuels that are compatible with current infrastructure and support the heavy transportation sector, where ethanol is non-viable, are strategically important targets for biofuel research. One route to highly reduced fuel molecules is by microbial conversion of hydrolyzed cellulosic sugars to free fatty acids, which are a useful precursor for downstream catalytic or enzymatic conversions to alkanes, olefins, fatty alcohols, or fatty acid esters. In particular, expression of an acyl-acyl carrier protein thioesterase in *E. coli* results in increased flux through fatty acid biosynthesis and the accumulation of moderate titers of free fatty acids. However a next-generation strain has not yet been engineered that can achieve greater than 30% of the maximum theoretical yield on any carbon source. It was found that endogenous fatty acid production specifically results in large reductions in cell viability, compromised inner membrane integrity, and changes in cell morphology. To determine a functional basis for these observations, a differential transcriptomic, proteomic, and lipidomic analysis was undertaken comparing a fatty acid overproducing strain to a non-overproducing strain, with the goal of determining potential stress responses and metabolic perturbations. Some key findings were induction of the genes encoding phage shock proteins, which are induced under conditions that cause dissipation of the proton motive force; induction of the MarA/Rob/SoxS regulons, which are induced by small toxic molecules or oxidative stress; and a greatly increased degree of unsaturation in long chain fatty acids accompanied by down-regulation of genes involved in unsaturated fatty acid biosynthesis. These findings and successive iterations of strain engineering based on these results will be discussed.

See more of this Session: Advances In Metabolic Engineering and Bioinformatics IA - Biofuels
See more of this Group/Topical: Food, Pharmaceutical & Bioengineering Division

(425f) Transcriptomic and Proteomic Profiling of Ethanol-Tolerant Clostridium Thermocellum

Siguang Sui[1], Lin Tang Goh[2], Timothy J. Griffin[3] and Wei-Shou Hu[1], (1)Chemical Engineering and Materials Science, University of Minnesota, Minneapolis, MN, (2)Bioprocessing Technology Institute, Singapore, (3)Biochemistry, Molecular Biology & Biophysics, University of Minnesota, Minneapolis, MN

Clostridium thermocellum is a cellulolytic and ethanogenic organism that could convert cellulosic materials directly to bioethanol. However, its low ethanol tolerance has been an obstacle for the process industrialization. A strain tolerant of 6% (w/v) ethanol obtained by sequential transfer was capable of producing twice higher concentration of ethanol than wild type in fed-batch culture. Transcriptomic and proteomic analysis were performed on exponentially growing and stationary phase wild type and tolerant strain. Wild type and tolerant strain exhibited similar dynamics in transiting from growth to stationary phase in many gene classes including molecular chaperones, cellulosome, and ATP synthases. Whereas many genes underwent different responses between them, and these genes were involved in membrane transport of cellodextrin, oligopeptide, nitrogen and sulfate, maintaining redox homeostasis, such as ferredoxin, thioredoxin and rubredoxin, and hydrogen production. Combining transcriptomic data and iTRAQ-based quantitative proteomic results, genes and gene classes related to high ethanol tolerance feature were explored, which has shed light on the mechanism of ethanol tolerance of *Clostridium thermocellum*.

See more of this Session: Advances In Metabolic Engineering and Bioinformatics IA - Biofuels
See more of this Group/Topical: Food, Pharmaceutical & Bioengineering Division

(425g) Conversion of Proteins Into Biofuels by Engineering Nitrogen Flux

Yixin Huo[1], kwang Myung Cho[1] and James C. Liao[2], (1)Easel Biotechnologies, LLC, Los Angeles, CA, (2)Deparment of Chemical and Biomedical Engineering, University of California Los Angeles, Los Angeles, CA

Current biorefinery schemes are suboptimal for several reasons. For examples, all existing schemes, including sugar-based or cellulosic biorefining, lead to the accumulation of protein by-products, but there are no strategies to convert these by-products into liquid fuels. In additions, schemes based on algae have limited efficiency because cultures used for biofuel production must be starved so that they produce lipid feedstocks, resulting in less cell growth and less total CO_2 fixation. A heretofore unexplored solution to these limitations would be to use proteins as a feedstock. If proteins were deaminated and converted to fuel or chemicals, the reduced nitrogen could be recycled to achieve a nitrogen neutral biofuel production. More importantly, proteins are the major component of the photosynthesis apparatus, CO_2 fixation pathways and other biosynthetic and cell growth machinery. Proteins are the dominant fraction in fast-growing photosynthetic microorganisms and industrial fermentation residues. Thus, using proteins as feedstock might maximize growth and CO_2 fixation rates. However, current schemes aim to increase production of carbohydrates or lipids, rather than proteins. Proteins have not been used to synthesize fuels because of the difficulties of deaminating protein hydrolysates.

In this study, we apply metabolic engineering to generate*Escherichia coli* that can deaminate protein hydrolysates, enabling the cells to convert proteins to C4 and C5 alcohols at 56% of the theoretical yield. We accomplish this by introducing three exogenous transamination and deamination cycles, which provide an irreversible metabolic force that drives deamination reactions to completion. We show that *Saccharomyces cerevisiae*, *E. coli*, *Bacillus subtilis* and microalgae can be used as protein sources, producing up to 4,035 mg/l of alcohols from biomass containing ~22 g/l of amino acids. With this nitrogen-centric metabolic engineering strategy, all amino acids can be converted to their corresponding keto acids, and can in turn be further converted to fuels or chemicals, directly or through key metabolites such as acetyl-CoA or succinate. The compounds that could be produced from proteins include bulk chemicals, monomers and pharmaceutical intermediates, in addition to fuels. These results show the feasibility of using proteins for biorefineries, for which high-protein microalgae could be used as a feedstock with a possibility of maximizing algal growth and total CO_2 fixation.

For large-scale applications, the protein raw material could come from several sources. In the short term, waste proteins generated from the current fermentation, food processing and biofuel production industries could be used. In particular, genetically modified organisms used in fermentation cannot be disposed of as fertilizers or animal feed without additional treatment, and thus provide an excellent source material for protein-based biorefining. For long-term, large-scale applications, we envision using algal biomass, because proteins are the major component of fast-growing microalgae in open-pond cultures that are not artificially induced to accumulate lipids. Even with contamination by heterotrophic bacteria, the combined protein sources are still usable as raw material.

See more of this Session: Advances In Metabolic Engineering and Bioinformatics IA - Biofuels
See more of this Group/Topical: Food, Pharmaceutical & Bioengineering Division

(431a) Development of a Continuous Filter-Drier for Lab to Kilo-Lab Scale

David J. am Ende[1], David Pfisterer[1], Kevin Girard[1], Daniel Blackwood[2] and Ed Plocharczyk[3], (1) Chemical R&D, Pfizer, Inc, Groton, CT, (2)Pharmaceutical Development, Pfizer, Inc, Groton, CT, (3) Flow technology, D&M Continuous Solutions LLC, Greenwood, IN

Continuous isolation and drying is relatively rare in the pharmaceutical industry. There are several advantages to continuous filtration and drying provided the challenges of conveying and moving wet and dry solids can be overcome. These advantages include: 1) consistency of particle properties (eliminate batch to batch variability), 2) higher efficiency and reduced energy utilization, 3) smaller equipment size and portability. Highly consistent particle properties of API enable lower-cost drug product options such as direct compression operations. Direct compression and tableting operations require consistent API particle size and flow properties and high accuracy of drug-excipient blend compositions. Other motivating factors include the design of fit for purpose portable continuous equipment. The prototype filter drier system hardware design and system performance will be discussed.

See more of this Session: Applications of Continuous Processing In Manufacture of Drug Substance
See more of this Group/Topical: Food, Pharmaceutical & Bioengineering Division

(431b) Development of a Continuous Process for Tasisulam Sodium

Jennifer M. Groh, Timothy D. White, K. Derek Berglund and Martin D. Johnson, Chemical Product Research and Development, Eli Lilly and Company, Indianapolis, IN

The development of a synthetic route to tasisulam sodium (5-bromo-thiophene-2-sulfonic acid 2, 4-dichlorobenzoylamide sodium salt, hereafter referred to as tasisulam) utilizing continuous Schotten-Baumann reaction conditions is presented. This process was envisioned with the goal of creating a process with fewer unit operations utilizing the direct formation of the final sodium salt from a sulfonamide and acid chloride without isolation of the free acyl sulfonamide. Very tight release specifications were in place for the tasisulam API batch process and the challenges of meeting these requirements are detailed. Another processing goal was to minimize worker exposure to the cytotoxic API. This was accomplished by eliminating a solid isolation and reducing the total material in the system at any one time. Operating the Schotten-Baumann conditions in a continuous fashion gave two main advantages. It allows for inexpensive, dedicated lab hood equipment for API production, and it enables higher yield and better impurity rejection in the extraction, solvent exchange, and crystallization for this route. Batch development, continuous proof of concept, 50 g/day lab scale demonstration and 5 kg/day commercial scale runs will be presented.

See more of this Session: Applications of Continuous Processing In Manufacture of Drug Substance
See more of this Group/Topical: Food, Pharmaceutical & Bioengineering Division

(431c) API Attribute Control by Continuous Fast Crystallization/Precipitation

Dimitrios Zarkadas[1], Christopher Pridgen[2], Mourad Hamedi[1], Plamen Grigorov[3] and Michael Whittington[4], (1)Chemical Process Development & Commercialization, Merck and Co. Inc., Rahway, NJ, (2)Pharmaceutical Commercialization Technology, Merck and Co. Inc., Summit, NJ, (3)Chemical Process Development & Commercialization, Merck and Co. Inc., Union, NJ, (4)Chemical Process development & Commersialization, Merck and Co. Inc., Union, NJ

Increased bioavailabiltyand/or good API particle size and attribute control is needed to enhance orrender robust the pharmacokinetic profile of an increased number of currentdrugs and/or new molecular entities. Continuous fast crystallization/precipitationis a more effective method compared to batch crystallization/precipitation tocontrol API attributes and produce small particles for a number of reasons. Continuous fast crystallization offers higher mixing intensity and lower mixingtimes as well as a wider range of these two parameters. This results in increasedoperational flexibility. In addition, supersaturation is constant during acontinuous process thus producing a narrower PSD compared to batchcrystallization. Finally, the scale up factor from laboratory to commercialscale is two to three orders of magnitude lower for continuous crystallization.This paper describes a systematic approach for the implementation of continuousfast crystallization processes. Individual case studies showcase the advantagesoffered by continuous processing.

We identifiedthree main drivers for the implementation of continuous fast crystallizations.The first is manufacturing API with small particle size to enhancebioavailability. The second is particle size and/or other API attribute controlwithin a tight range. The third is the increased throughput and associated costreduction offered by continuous processes. We also identified a systematicapproach for the implementation of continuous fast crystallization and theirscale up from laboratory to pilot plant and commercial scale equipment. Thisapproach includes the determination of the fundamental kinetic constants of thesystem (mixing and induction time) and the subsequent selection of an appropriatemixer design for scale up.

The first case studyinvolves the isolation and attribute control of an amorphous API. Itsprecipitation ^{2}/g for 10 batches. Theequivalent average particle sizes are 220 and 170 nm, values indicating thegood PSD control offered by the continuous precipitation process.

The secondcase study involves a dissolution limited drug substance, which requiredmicronization to provide adequate exposure. However, we encountered severeclogging of the jet mill chamber during micronization, which severely limitedprocess throughput (3 days to micronize 12 kgs of drug substance). Theseproblems were overcome by altering the input to the micronization process. Wefound that producing API with a co-axial injection continuous crystallizationresulted in high API surface areas. The latter was proven considerably moreamenable to jet milling; the process throughput increased more than tenfold,from 200 g/hr to about 1.5-2.0 kg/kr.

See more of this Session: Applications of Continuous Processing In Manufacture of Drug Substance
See more of this Group/Topical: Food, Pharmaceutical & Bioengineering Division

(431d) Efficient and Cost Effective Dehydration of Pharmaceutical Streams by Pervaporation

Dimitrios Zarkadas[1], Ronnie Feng[1], Michael Wu[1], Plamen Grigorov[2], Azzeddine Lekhal[1] and Kevin Molter[1], (1)Chemical Process Development & Commercialization, Merck and Co. Inc., Rahway, NJ, (2) Chemical Process Development & Commercialization, Merck and Co. Inc., Union, NJ

Pervaporationis a continuous process that has been used successfully the last two decadesfor the dehydration of organic solvents at large scales and for improvingdistillation efficiency of azeotropic systems like ethanol-water. Similarly toother membrane processes, pervaporation offers several advantages. It oftenresults in lower costs, it is modular and therefore easy to scale up, offersmuch higher mass transfer rates per unit volume compared to conventional separationequipment and can effect separations difficult to achieve otherwise. Despitethe growing importance of pervaporation in the chemical industry, there are noreported systematic efforts to apply the technology in the dehydration ofpharmaceutical streams. The latter is important in several cases: reactionengineering (condensation reactions, moisture sensitive reactions, dehydrationof catalysts), crystallization (improve yield by water removal, generation ofsupersaturation) and solvent recovery. Dehydration of pharmaceutical streams iscurrently performed by atmospheric or vacuum distillation, which is inherentlyan inefficient process; large solvent volumes are required to drive off a smallamount of water and the separation often takes a long time which results inhigh operating costs and decreased throughput. In this paper we report ourongoing efforts to implement pervaporation as an alternative yet powerful toolfor dehydration applications in the pharmaceutical industry.

A systematicanalysis of frequently encountered dehydration problems in the pharmaceuticalindustry has revealed that pervaporation would be a better choice compared todistillation when the stream contains less than about 10% wt. water, forms anazeotrope with water or needs to be dehydrated to water contents less than0.5-1.0 % wt. In all other cases the selection between pervaporation anddistillation can be made on economic grounds. Two major operating costs havebeen identified for distillation: solvent (usage and disposal) and labor andoverhead costs associated with increased operating time. Pervaporation obviatesthe need to use solvent to remove water and therefore does not incur solventusage and disposal costs. If the membrane configuration is appropriatelydesigned, pervaporation can result in shorter operating times compared todistillation and therefore reduced operating costs. We have developed detailedperformance and economic comparison tools, which enable the design of processand cost efficient pervaporation processes.

The attractiveness ofpervaporation for dehydration applications is illustrated with severalexamples. These examples include the removal of water from a 2-MeTHF and anIpAc stream to improve yield during the subsequent isolation steps as well asthe improvements to a THF recovery process including a dehydration step. Theabove analysis and examples clearly illustrate that pervaporation can deliverthe required separation goal in a robust and more efficient manner compared todistillation, reduce solvent usage resulting in a greener process, reduceoperating time and overall result in a more cost efficient solution fordehydration applications in the pharmaceutical industry..

See more of this Session: Applications of Continuous Processing In Manufacture of Drug Substance
See more of this Group/Topical: Food, Pharmaceutical & Bioengineering Division

(431e) Failure Mode Analysis of Continuous Process with Process Models

Sze Wing (Candice) Wong, Chemical Process R&D, Eli Lilly and Company, Indianapolis, IN, Prashant B. Kokitkar, Engineering Technology Center, Eli Lilly and Company, Indianapolis, IN, K. Derek Berglund, Chemical Product Research and Development, Eli Lilly and Company, Indianapolis, IN and Martin Johnson, Chemical Process Research & Development, Eli Lilly &Co., Inc., Indianapolis, IN

Failure Mode Analysis of Continuous Process with Process Models

Sze-Wing Wong[1], Prashant Kokitkar, K. Derek Berglund, Martin D Johnson, and Shekhar K. Viswanath

Chemical Product R&D, Lilly Research Laboratories, Eli Lilly and Company, Indianapolis, IN 46285

Abstract: This presentation described the use of process models to drive process development of a continuous process at Eli Lilly and Company. In this case study, the continuous process is represented with two separate models: a reaction kinetics model and a counter-current extraction model. The reaction kinetics parameters were put into a three-in-series CSTR (Continuous Stirred Tank Reactor) model in which it can simulate different failure modes such as pump flow rate off-set or total pump failure. The resulting impurity profile of the reaction mixture becomes the input to the counter-current extraction model where the equilibrium data of all the components at each stages is known. With the combination of both process models, one can estimate how much process disturbances the system can handle. The simulation results could then be used to identify potential critical process parameters and also be used to set the process control strategy. This case study shows that the models can positively impact development of a continuous process even at early development phase, which is consistent with the quality by design principles.

[1] Corresponding author

E-mail address: wongsz@lilly.com

See more of this Session: Applications of Continuous Processing In Manufacture of Drug Substance
See more of this Group/Topical: Food, Pharmaceutical & Bioengineering Division

(431f) Novel Concepts for Continuous Purification of APIs Using Electro-Chromatography

Marie-G. Braunbruck[1], Heidrun Gruber-Woelfler[1], Peter Feenstra[1], Rene Laskowski[2], Hans-Joerg Bart[2] and Johannes G. Khinast[1], (1)Institute for Process and Particle Engineering, Graz University of Technology, Graz, Austria, (2)Lehrstuhl fr Thermische Verfahrenstechnik, Technische Universitt Kaiserslautern, Kaiserslautern, Germany

In pharmaceutical industry batch processing still dominates the manufacturing of the active pharmaceutical ingredients (APIs). One example is liquid chromatography, which is one of the basic separation methods for biosynthetic products but is still operated in batch mode. As continuous processes offer many advantages, approaches to develop continuous chromatographic processes have become increasingly important.
An effort in this area is the CAEC project, which deals with the development of a continuous annular electro-chromatography (CAEC) prototype [1]. This prototype combines the advantages of electrophoresis and preparative liquid chromatography in a continuous operation mode, thus leading to both, a high throughput and a high separation efficiency. Furthermore, the eluting analytes are monitored with an online detection system to ensure a continuous quality control.
We present the design, characterization and implementation of the stationary phase in this novel prototype. The stationary phase is based on a silica monolith which can be adjusted to the separation problem with functional groups, such as amino groups for ion exchange chromatography or alkyl chains for reversed phase chromatography [2]. In this way, not only the liquid phase but also the stationary phase can be adapted to specific problems which makes the CAEC universally applicable. The monolith is supported with silica particles that prevent the stationary phase from shrinking. This combination of particles and the monolithic structure increases the mechanical and chemical stability of the stationary phase while preserving a high porosity to achieve a high throughput.
The macroporous one-piece material was implemented in several moulds, such as a planar test cell that represents a fraction of the annular prototype and already allows a continuous separation via electro-chromatography. In this planar test cell hydrodynamics and heat transfer [3] were investigated and standard test systems could be successfully separated achieving theoretical plate numbers of over 20000 per meter and a throughput of up to 50 ml/h.
In addition, we present a simulation program that is implemented in MATLAB to support the design of the chromatographic experiments. The program is based on a stochastic (Monte Carlo) approach and algorithms for convection, diffusion and retention are implemented [4]. With this program chromatograms can be simulated based on the population distribution of

the analytes. Therefore, these simulations can help to determine the suitable stationary as well as mobile phase for a given separation problem. Moreover, the molecular interactions of the analytes with the stationary phase were studied using molecular dynamics (MD) simulations.

Overall, the outcomes of our approaches are powerful tools for future separation tasks in pharmaceutical industry and can be individually assigned to a broad spectrum of problems.

This work is supported by the 7th Research Framework Programme of the European Commission (Grant agreement number: NMP2-SL-2008-206707)

[1] https://caec.bci.tu-dortmund.de/

[2] Gruber-Woelfler, H.; Braunbruck, M.-G.; Feenstra, P.W.; Laskowski, R.; Bart, Khinast, J.G., Oral Presentation, 9th International Congress of Chemical and Process Engineering CHISA 2010, Prague, August, 2010

[3] Laskowski, R.; Bart, H.-J.; Gruber-Wölfler, H.; Feenstra, P.; Braunbruck, M.-G.; Khinast, J.; Hofmann, C.; Menges, G.; Werner, B.; Löb, P.; Hessel, V.:, 17th International Symposium on Electro- and Liquid Phase-separation Techniques (ITP), 2010, Baltimore, USA.

[4] McGuffin, V.L.;Wu, P.R.; Journal of Chromatography A, 722 (1996) 3

See more of this Session: Applications of Continuous Processing In Manufacture of Drug Substance

See more of this Group/Topical: Food, Pharmaceutical & Bioengineering Division

(432a) Taking PAT From the Laboratory to Commercial Manufacturing

Mark A. LaPack[1], Stephen B. Jeffery[2], Steven J. Doherty[3], Joseph R. Martinelli[1] and Christopher L. Burcham[4], (1)CPR&D Process Definition, Eli Lilly and Company, Indianapolis, IN, (2)Process Engineering, Eli Lilly and Company, Kinsale, Ireland, (3)Ms&t PAT, Eli Lilly and Company, Indianapolis, IN, (4)Chemical Product Research and Development; Particle Design Laboratory, Eli Lilly and Company, Indianapolis, IN

Measurements using PAT are applied to two main areas: 1) Gaining process understanding and 2) maintaining process control. The decision to use PAT for control applications is an outcome of the design and development of the process. The parameters to be controlled, the process design space, and control limits dictate the type of technology selected (pH, conductivity, mid-IR, mass spectrometry, etc.) and the user requirements (materials of construction, limit of detection, dynamic range, robustness, etc.).

An example will be presented where a dissolved gaseous byproduct further reacts to form an impurity that affects both product quality and yield. Based on process models, rationales were developed for monitoring either the reaction liquid or the headspace gas during the process designed to remove this gas. The focus of this presentation will be to describe the development of the PAT-based process control system, including discussions about technology selection, vendor selection, risk and contingencies, qualification, installation, and validation.

See more of this Session: Applications of PAT In a Manufacturing Setting
See more of this Group/Topical: Topical I: Comprehensive Quality by Design in Pharmaceutical Development and Manufacture

(432b) Combining Microwave Resonance Technology to Multivariate Data Analysis As a Novel PAT Tool to Improve Process Understanding In Fluid Bed Granulation

Vera Lourenco[1], Thorsten Herdling[2], Gabriele Reich[3], Jose Cardoso Menezes[1], Dirk Lochmann[2] and **Jens Schewitz**[2], (1)Institute for Biotechnology & Bioengineering, I.S.T., Lisbon, Portugal, (2)Quality Operations, PAT, Merck Serono, Darmstadt, Germany, (3)Institut for Pharmazie und Molekulare Biotechnologie, Heidelberg, Germany

A set of 192 fluid bed granulation batches at industrial scale was *in-line* monitored using microwave resonance technology (MRT) to determine moisture, temperature and density of the granules. The gathered data were analysed through multivariate data analysis techniques, such as multiway partial least squares (PLS) and principal component analysis (PCA) and multivariate batch control charts. The combination of all these techniques, along with a few *off-line* particle size measurements, led to a huge gain on the process knowledge and understanding. A seasonality effect could be put into evidence with its influence on the process ongoing and on the final granules particle size. Moreover it was demonstrated and quantified by means of a PLS regression the relation between the particle size and the MRT measurements, highlighting a potential ability of the MRT sensor to predict the final particle size of the granules.

This study contributed successfully to the continuous improvement of an industrial fluid bed granulation process, under the *Process Analytical Technology* (PAT) initiative to meet the *Quality by Design* (QbD) principles.

See more of this Session: Applications of PAT In a Manufacturing Setting
See more of this Group/Topical: Topical I: Comprehensive Quality by Design in Pharmaceutical Development and Manufacture

(432c) PAT Application and Validation of Online Particle Sizing Measurement for Controlling a Critical Quality Attribute

Kevin J. Bittorf, CMC Process & Alliance Management, Strategic Development Networks, Vertex Pharmaceuticals, Cambridge, MA

In accordance with ICH Q8/9/10 and the FDA QbD initiative, companies are adding more and more PAT equipment, installing more sophisticated control systems, and moving to real-time release. Throughout this process, many questions have come to light. How is PAT justified for controlling quality attributes? How is PAT incorporated into the overall control strategy? What is needed for to validate a PAT system? These questions will be answered in a case study of using PAT to monitor and control the critical quality attribute of particle size in a production setting.

See more of this Session: Applications of PAT In a Manufacturing Setting
See more of this Group/Topical: Topical I: Comprehensive Quality by Design in Pharmaceutical Development and Manufacture

(432d) Experimental and Theoretical Study of Lyophilization From a Packing of Vials: Using of Tunable Diode Laser Absorption Spectroscopy (TDLAS) for Process Monitoring and Model Calibration

Pavol Rajniak[1], Cassandra Mifkovich[2] and Leon Farber[2], (1)Pharmaceutical Commercialization Technology, Merck & Co., Inc., West Point, PA, (2)Merck & Co., Inc., West Point, PA

Objectives: 1/experimentally evaluate water sublimation and lyophilization of apharmaceutical formulation from a packing of vials at different conditions;2/monitor total sublimation and desorption rate using a tunable diode laser absorptionspectroscopy (TDLAS) sensor;3/evaluate distribution of heat transfer coefficients from gravimetric watersublimation studies; 4/develop mathematical models for water sublimation andprimary drying from a packing of vials; 5/evaluate cake resistance parametersfrom the primary drying sublimation rates monitored by the TDLAS; 6/perform aqualitative comparison between the SEM images of cakes manufactured atdifferent freezing conditions and cake properties evaluated by the model.

Methods: A simple gravimetric method was used to evaluatethe mass of water sublimed at different locations within a laboratory scale(FTS-Lyostar II) lyophilizer.Unidirectional mathematical model for ice sublimation from a single vial wasused for evaluation of heat transfer coefficients between the shelf and bottomof vials at different locations. Different cumulative distribution functionswere tested for calculation of distribution of evaluated heat transfercoefficients. A mathematical model combining the distribution of heat transfercoefficients with the unidirectional model of primary drying in a vial wasdeveloped and employed for calculation of sublimation rates from a packing ofvials. A simple optimization procedure was developed for evaluation of the cakeresistance parameters (pore size and skin resistance) by fitting the modelpredictions to the TDLAS mass flux data. SEM imaging was used for visualizationof the cake morphology.

Results: Weibull cumulative distribution function accuratelydescribes the distribution of heat transfer coefficients of an inhomogeneousshelf. Mathematical models for both, thepure ice sublimation and sublimation of water from a pharmaceutical(sucrose-based) formulation satisfactorily simulate the sublimation rates measuredexperimentally. A model assuming a homogeneous porous cake was unableto describe experimental observation of a lyophilization cycle withoutannealing. On the other hand, a model assuming the presence of a skin (skinresistance R_{skin} = 0.7) at the cakesurface followed by a porous structure with pore radii 70 microns very well describes experimentalobservations. A faster sublimation was measured using the TDLAS for a cyclewith annealing. The experimental mass flux can be satisfactorily predicted usinga cake model with a thinner skin at the surface (skin resistance R_{skin} = 0.1) followed by a porousstructure with pore radii 90 microns. SEM imaging nicely supports results oftheoretical evaluation showing a thicker skin followed by smaller pores forcakes from the cycle without annealing.

Conclusions: A multi-dimensional mathematical model oflyophilization from packing of vials in combination with the TDLAS probe is auseful tool for better understanding and faster development and scale-up oflyophilization processes.

See more of this Session: Applications of PAT In a Manufacturing Setting
See more of this Group/Topical: Topical I: Comprehensive Quality by Design in Pharmaceutical Development and Manufacture

(432e) Optical Coherence Tomography: A New PAT-Tool for Fast and Non-Destructive Analysis of Tablet Coating Quality

Daniel M. Koller[1], Christoph Schinwald[1], Otto Scheibelhofer[1], Michael Leitner[2] and Johannes G. Khinast[1], (1)Research Center Pharmaceutical Engineering GmbH, Graz, Austria, (2)RECENDT - Research Center for Non Destructive Testing GmbH, Linz, Austria

Optical coherence tomography (OCT) is a non-invasive imaging technique allowing fast and high-quality cross-sectional imaging of semi-transparent and turbid media. Biomedical research and diagnostics have been up to now the main driving forces for the reported applications and progress in OCT. The characteristics of OCT are the ability to provide high-resolution images and depth-resolved information also in strongly scattering media in a contact-free way. Hence OCT is well suited to image layered structures like retina, skin or multi-layered foils in the packaging industry. However, the application of OCT for the analysis and characterization of tablet coatings in the pharmaceutical sciences is still relatively new [1]. The image contrast is due to inhomogeneities in the refractive index of the sample material, and thus, OCT provides complementary information to other high resolution imaging techniques like terahertz imaging, X-ray computed tomography (CT) and magnetic resonance imaging (MRI).

In this work, high-speed spectral domain OCT was demonstrated as a new Process Analytical Process (PAT) tool for the characterization of pharmaceutical tablet coatings [2]. A Design of Experiments (DoE) was created to produce tablet batches with three coating process parameters. The coating thickness was determined via conventional testings with a µm-caliper or weight gain measurements and compared with OCT measurments. In addition, the results were related to investigations carried out with already established PAT tools like Near-Infrared (NIR) and Raman spectroscopy combined with qualitative and quantitative Multivariate Data Analysis (MVDA).

Finally, the influence of process parameters on the dissolution behavior as a critical quality attribute for the tablets was related to the disintegration times of the tablets for response surface modelling.

[1] Mauritz JMA, Morrisby RS, Hutton RS, Legge CH, Kaminski CF 2010. Imaging Pharmaceutical Tablets with Optical Coherence Tomography. Journal of Pharmaceutical Sciences 99(1):385-391.

[2] Koller DM, Hannesschläger G, Leitner M, Khinast JG. Non-Destructive Analysis of Tablet Coatings with Optical Coherence Tomography. European Journal of Pharmaceutical Sciences. 2011.

See more of this Session: Applications of PAT In a Manufacturing Setting

(432f) QbD-Driven PAT Applications for Precision Spray-Coating Development and Control

Charles Miller[1], Bruce T. Thompson[1], Fan Zhang-Plasket[1], Patrick T. Schilling[2], John Higgins[1], Gert Thurau[1], Mano Ramasamy[1] and Eric Ahuja[3], (1)Process Analytical Technologies, Merck and Company, Inc., West Point, PA, (2)Pharmaceutical Commercialization Technology, Merck & Co., Inc., West Point, PA, (3)Director Analytical Chemistry Development and Supply, Merck & Co, Inc, West Point, PA

Two of the core practices of Quality by Design (QbD), design space development and deployment of control strategies, can be greatly facilitated by Process Analytical Technologies (PAT). Ultimately, however, the actual usage of PAT methods for any QbD task must first be warranted by a careful and thorough risk assessment of the product, process and measurement system.

This work addresses one such case, where the product is an API-coated immediate release drug product that is manufactured by a precision spray coating process.

An initial risk assessment of the process indicated that the spray coating operation had considerable risk with respect to two critical quality attributes: content uniformity and composite assay. In response to this risk assessment, a PAT measurement system based on NIR diffuse reflectance spectroscopy was used for two different but equally critical tasks:

1) rapid tablet composition analysis during process development, to assist in design space development and, thus, mitigation of the content uniformity risk, and

2) at-line monitoring of coating growth during the spray coating process, to support the spray coating control strategy, and thus mitigate the composite assay risk

Details regarding the QbD approach, as well as the design and performance of the NIR methods used for development and control, will be discussed. In addition, a brief discussion of the hardware and software infrastructure required to deploy the at-line monitoring PAT method at the supply site will be made.

See more of this Session: Applications of PAT In a Manufacturing Setting
See more of this Group/Topical: Topical I: Comprehensive Quality by Design in Pharmaceutical Development and Manufacture

PROCESS ANALYTICAL TECHNOLOGY FOR RECOMBINANT PANDEMIC FLU VACCINES: VIRAL ULTRASTRUCTURE, AGGREGATION, AND BINDING

De-Hao Tsai,[1] Daniel I. Lipin,[2] Suvajyoti Guha,[1,3] Jeremy I. Feldblyum,[1,3] Kenneth D. Cole,[1] Kurt A. Brorson,[4] Michael R. Zachariah,[1,3] Michael J. Tarlov,[1] Anton P. J. Middelberg,[2] Leonard F. Pease III[1,5]

1. National Institute of Standards and Technology (NIST), Gaithersburg, MD 20899, USA
2. University of Queensland, Brisbane, Australia
3. University of Maryland, College Park, MD 20742, USA
4. Food and Drug Administration, While Oak, MD 10903, USA
5. University of Utah, Salt Lake City, UT 84112, USA

Abstract

Here we describe our progress in developing electrospray differential mobility analysis (ES-DMA) into a process analytical technology (PAT) for recombinant vaccines. The threat of pandemic flu (e.g. avian influenza H5N1, swine flu H1N1, etc.) remains a significant public concern. Recombinantly produced vaccines hold significant advantages over traditional means of growing vaccines including increased production rates, shorter times between viral identification and clinical treatment, and improved safety profiles. However, accelerated production methods also require advanced process control sensors (i.e. process analytical technologies) that can rapidly and accurately detect out-of-specification intermediates. ES-DMA has the potential to fulfill this need. Here we describe our efforts to use ES-DMA to (1) rapidly quantify the assembly state and integrity of recombinantly produced virus like particles (VLPs), (2) determine the viral aggregation state, and (3) quantify antibody-virus binding necessary to evaluate stoichiometry and vaccine efficacy.

Introduction

Recent threats of pandemic flu, including SARS, H5N1 avian influenza and the H1N1 swine flu of 2009, highlight the need for rapid vaccine selection and production.[1] Traditional manufacturing techniques require several months for vaccine stocks to mature, cannot be readily adapted to rapidly changing viral subtypes, and some individuals are allergic to egg based products.[1] Recent recombinant production strategies can overcome each of these challenges.[2-7] In recombinant strategies, cell stocks are induced to produce viral coat proteins that are genetically modified to express antigens specific to the flu virus. The proteins are then subsequently assembled into complete virus like particles (VLPs) before or after purification to remove adventitious (e.g. murine minute virus, Cache Valley virus) and endogenous (e.g. retrovirus) viruses in harmony with US Federal Code.[1,8] Similar recombinant strategies have been used to produce vaccines against the human papillomavirus (HPV), which is marketed by Merek as Gardasil.

To provide additional manufacturing flexibility, the US Food and Drug Administration (FDA) introduced the Quality by Design (QbD) and Process Analytical Technologies (PAT) initiatives. Though novel to many in the chemistry, biology, and biotechnology fields, these initiatives mirror traditional process control efforts well established in Chemical Engineering. Specifically, the PAT initiative calls for additional sensors at key locations within biomanufacturing processes to directly measure critical quality attributes related to the final product composition. Examples of relevance to VLP manufacture include sensors to determine the state of protein assembly, extent of protein and VLP aggregation, and interrogate VLP conjugation. When such sensors are integrated upstream from the final lot release, process control techniques can be used to adjust process settings (if allowed following QbD analysis), abort batches before they undergo very expensive post assembly separation, and to identify why processing goes awry in *post mortem* analysis. In so doing, PATs may increase the net process productivity and decrease the associated manufacturing costs.

Here we present an emerging technique, called electrospray differential mobility analysis (ES-DMA), that holds potential to act as a sensor for each of these applications. In ES-DMA, antibodies, antigens or antibody-antigen complexes in solution are electrosprayed to produce highly charged droplets.[9] These droplets then pass into a charge neutralization chamber, which dries the droplets and fixes the charge on the particles between +2 and -2.[10] Upon entry into the DMA, only positively charged particles are deflected toward a collection slit based on their charge-to-size ratio.[11] Because a vast majority of positively charged particles possess a +1 charge (>95% of 45 nm particles), essentially only particles of a particular size will traverse the collection slit at the exit of the DMA into a condensation particle counter (CPC).[12] Within the CPC, particles pass through a saturated butanol vapor and grow into droplets several microns in size, which can then be counted individually as they obscure light impinging on a photodetector.[9,13,14] The total number of particles passing through the detector is then summed and reported for a particular size. Scanning through a range of sizes assembles a complete size distribution. Peaks in the distribution correspond to individual proteins, protein assemblies (e.g. capsomers), virus & VLPs, or viral aggregates.[8,13-16]

We have previously reported the use of ES-DMA to analyze the size of viruses, VLPs and viral aggregates.[8,13-15] For example, VLPs assembled from solution can be heterogeneous in size or poorly assembled.[14] Figure 1 compares size distributions obtained using ES-DMA to distributions from TEM for VLPs with a 17 amino acid residue from avian influenza.[14] ES-DMA has superior resolution (subnanometer) from 5-500 nm.[17] Similarly, we have recently reported that ES-DMA can be used to rapidly interrogate the composition of the viral coat and that the size measured by ES-DMA most closely matches the edge-to-edge length.[8] We have also shown that temporal and thermal degradation can be tracked using the instrumentation.[8] Here we extend our analyses to viral degradation at low pH and detection of virus-antibody complexes, both of which are important in biomanufacturing environments because they are critical to removing adventitious viral contaminants and evaluating the humoral response induced by vaccines.

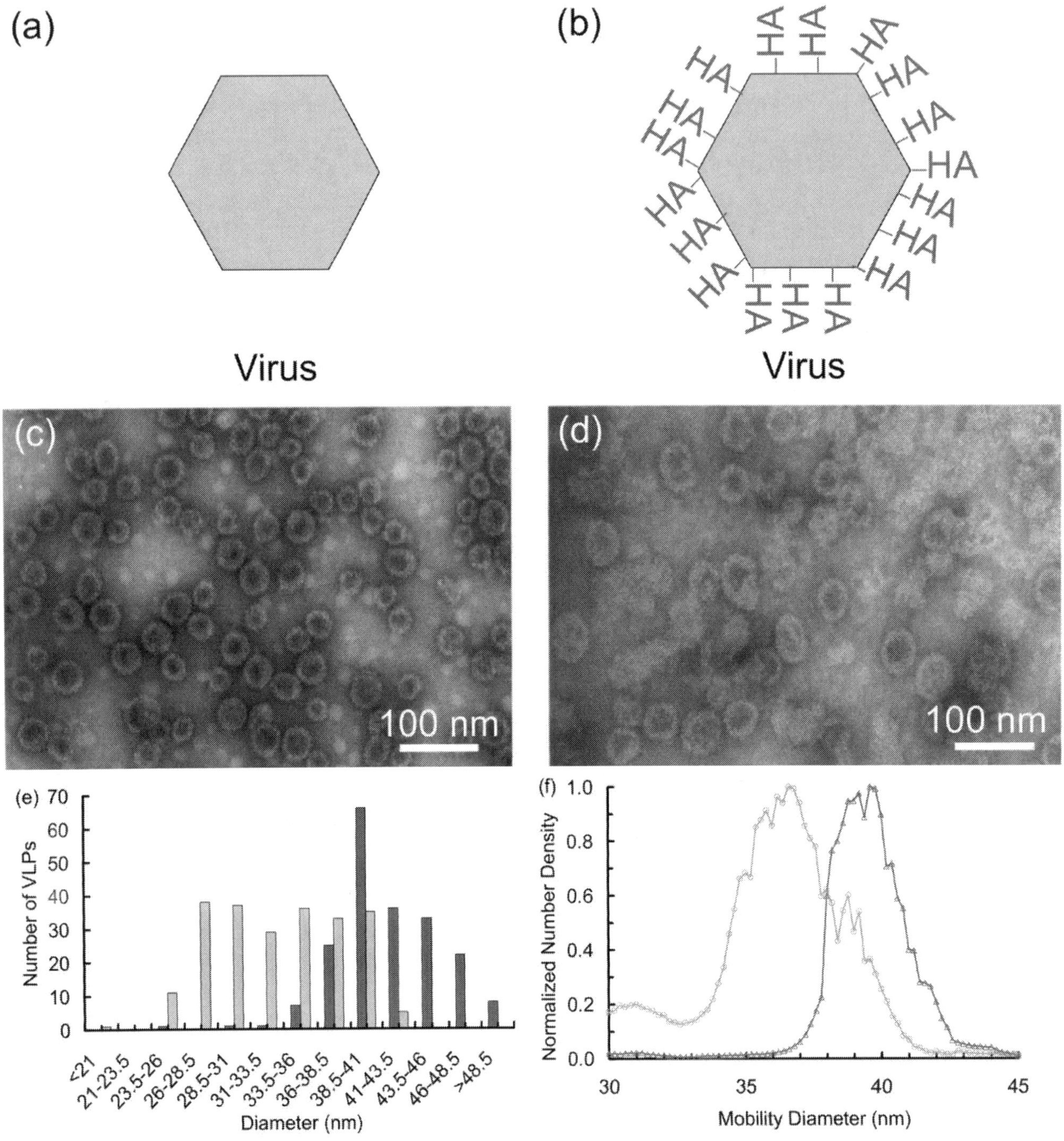

Figure 1. Representation of (a) bare polyoma VLP and (b) polyoma VLP with 17 amino acid reside expressed on external surface loops. TEM micrograph of (c) bare polyoma VLP and (d) polyoma VLP with 17 amino acid reside expressed on external surface loops. Size distributions obtained using (e) TEM and (f) ES-DMA comparing bare polyoma VLP (light bars or circles) and polyoma VLP with 17 aminoacid reside expressed on external surface loops (dark bars or triangles). Adapted from Pease, et al., Biotech. Bioeng., 102 (2009) 845-855.[14]

Materials and Methods

For analysis of acidic degradation, bacteriophage pp7 along with its host *Pseudomonas Aeruginosa* were obtained from the ATCC (Manassas, VA; accession numbers 15692-B4 and 15692). Stocks were prepared by ultracentrifugation using CsCl gradients, and intact phage were enumerated by plaque assays using their respective hosts as described elsewhere[15,18,19] to determine an initial titer of $6.9 \cdot 10^{12}$ pfu/mL. Infectivity loss experiments at low pH were performed by diluting the phage in 10 mmol/L tris at pH 7.5 into 100 mmol/L citric acid/sodium citrate at various pH levels (2.5 to 5.0 and 7.0) for 30 min before pH neutralization by increasing the sample volume by 20% with 3.0 mol/L tris, pH 9.0. Phage in these test articles were enumerated by a plaque assay. Dilutions of the virus stocks and pH treated samples were prepared and added to mid-log phase host and liquefied top agar, which was spread over tryptic soy agar plates. TEM images were obtained as described in detail elsewhere.[8]

For analysis of acidic degradation by ES-DMA, the phage were dialyzed into 10 mmol/L ammonium acetate at pH 7 to remove nonvolatile salts and limit sample conductivity to <0.2 S/m.[14] Studies (not reported) indicate that buffer exchange does not significantly impact phage infectivity. Acidic buffer solutions at pH's of 2.00±0.25, 3.00±0.25, 4.00±0.25 and 5.00±0.25 were prepared by titrating glacial acetic acid into 0.5-1.0 mL of an ammonium acetate solution until the desired pH was measured with an immersed pH probe (#525A, Orion). Ten microliters of the dialyzed virus samples were exposed to 190 μL of these acidified buffer solutions for 30 minutes before quenching the reaction with 190 μL of cold (4°C) ammonium hydroxide. Cold base counters temperature increases associated with mixing of non-ideal acid-base solutions.[20] Because the resulting neutralized test articles could not be electrosprayed due to excessive conductivity, the samples were again dialyzed into an approximately 3.1 mmol/L solution of ammonium acetate at neutral pH. Additional ammonium acetate buffer was added, if necessary, to bring all test article volumes (including controls) to 510±10 μL to achieve uniform levels of dilution.

The bacteriophage MS2 was selected for virus-antibody neutralization assays because it is a well studied biosafety level 1 bacteriophage and known antibodies were available. Phage (#15597-B1 from the ATCC) and E. Coli host (ATCC #15597) were propagated and purified as previously described.[13] MS2 samples were stored at 4°C or frozen at -80°C. A polyclonal antibody labeled P and a monoclonal antibody labeled G were obtained from Tetracore, Inc (Rockville, MD). Antibodies were stored at 4°C until use. The virus and antibodies formed complexes when combined in concentration ratios indicated in the figure captions and allowed to incubate for 4.0 h at ambient temperature before storing at 4°C. To determine the infectivity, incubations were serially diluted and active MS2 virus was measured by plaque assay on the host. Samples at a concentration of $3.45 \cdot 10^{13}$ virus particles/mL but as small as 30 μL were prepared for ES-DMA as follows. Samples were first dialyzed against 2 mmol/L ammonium acetate (neutral pH) using mini dialysis cups (#69590, Pierce, Inc.). Dialysis was necessary to reduce the concentration of nonvolatile salts that may encrust the surface of aerosolized viruses and virus-antibody complexes and to reduce solution conductivities ≤0.2 S/m.[9,14] Arrayed cups were floated on 50 mL of DI water to removed glycerol residues as per the manufacturer's instructions. Approximately 40 μL of sample was then added to each cup and nonvolatile salts allowed to dialyze for at least 3.5 hours at room temperature. The volume of the sample before and after dialysis, which varied from sample to sample, was measured. This dialysis step was previously found to not significantly diminish the infectivity of bacteriophages.[13] Immediately before insertion into the electrospray chamber the samples were diluted 20 fold to extend the amount of sample available for analysis without deleterious effect. Fetal bovine serum was also obtained and diluted in ammonium acetate to the concentration reported in the figure captions.

Solutions were electrosprayed with an Electrospray Aerosol Generator (TSI Inc., USA, #3480), which was operated with gas flow rates of 1.0 L/min of house air and 0.2 L/min of CO_2.[9] A vial containing the sample was placed under pressure, driving the sample solution through a capillary (25 µm in diameter, 24 cm in length) at approximately 66 nL/min. The capillary tip formed a tapered outlet at which voltages up to 3.5 kV were applied as needed until the exiting meniscus formed into a sharp Taylor cone and the measured current remained stable.[9] A stable cone was critical because other meniscus flow regimes fail to produce the narrow distribution of small droplet sizes necessary to achieve high quality, reproducible results. The highly charged droplets, formed at the tip of the Taylor cone, passed into a bipolar charger where the charge remaining on the droplets achieved a steady-state distribution dominated by +1, 0, -1 as described by Wiedensohler.[12] For example, charge neutralization of 45 nm particles (e.g. viruses, proteins, etc.) yields a charge distribution consisting of 61.5% with a net neutral charge, 16.1% with a single positive charge, 0.5% with a double positive charge, 21.1% with a single negative charge, 0.8% with a double negative charge and only 0.01% with charges outside this range.[12] These dry, charged particles (e.g. antibodies, viruses, antibody-virus conjugates) then transited the approximately 1.4 m of plastic Tygon tubing (1.6 cm diameter) connecting the exit of the ES to the entrance of the DMA. Due to adherence of the antibodies to the inside of the electrospray capillaries, quantifying the amount of free antibody before and after complexation to obtain a second supporting measure of antibody binding was not feasible. Adherence was confirmed by replacing the antibody containing sample by 2.0-20 mmol/L ammonium acetate (neutral pH) in the pressure chamber at the end of a run. Rapid sequential size distributions showed at least one peak corresponding to the free antibody, whose magnitude exhibited exponential decreases in consecutive runs as the antibody desorbed from the capillary wall.[21] Within the DMA, samples were separated based on their charge-to-size ratio.[14,22] The DMA (TSI Inc., USA, #3080) was operated with a sheath-to-sample flow rate ratio of 30:1.2 and voltages ranging from 0.0 kV to -10 kV were applied to the central electrode to classify the particles. The flow rate exiting the DMA was 1.0 L/min, which was supplemented by 0.5 L/min of HEPA filtered air as it entered the condensation particle counter (CPC). Within the CPC (TSI Inc., USA, #3025A), particles pass through a saturated butanol environment and grow into droplets several microns in size which can be counted individually as they obscure light impinging on a photodetector. Because a negative bias is applied to the DMA electrode, only positively charged particles are counted by the CPC. Data points were collected at a rate of three per minute with a step interval of 0.2 nm. The raw CPC count is divided by the fraction of positively charged particles described by Wiedensohler[12] to obtain the count for all particles regardless of their charge state when entering the DMA. The DMA voltage is converted into a spherically equivalent size by balancing the mobility of the particle against the mobility of the

DMA tuned by adjusting the voltage and flow rate as described elsewhere.[9,22] Conversion to size was performed assuming the particles to be spheres of diameter d with a Cunningham slip correction factor of $C_c = 1+Kn[\alpha+\beta \exp(-\gamma / Kn)]$, where $Kn = 2\lambda/d$, $\alpha = 1.257$, $\beta = 0.40$, $\gamma = 1.110$, and the gas mean free path at room temperature $\lambda = 66$ nm.[9] The mobility diameter corresponding to the ES-DMA peak maximum is reported as its size. The mobility diameter measured depends on the drag force experienced by the particle, which was proportion to the projected area along each of the particle's principle axes.[8,9,16,22-24]

Results and Discussion

Because we have previously demonstrated ES-DMA as a tool to characterize VLPs as influenza vaccines,[14] the extension of our efforts reported herein focuses on acidic degradation of viruses and antibody neutralization in preparation for future efforts with VLPs and other vaccine constructs.

Low pH Viral Degradation

We now address the degradation in acidic environments of phage pp7, which was chosen as a convenient biosafety level I model for small non-enveloped mammalian viruses.[15,18,19] Like many small viruses, pp7 consists of an RNA genome and an icosahedral capsid assembled from 180 identical copies of a single major capsid protein, having a molecular weight of 13.9 kDa.[15,18,19] (Additional proteins encoded by the pp7 genome probably contribute only marginally to the external capsid structure detected by the DMA due to their significantly lower concentration.) Figure 2 shows the size distributions of pp7 subjected to acid as a function pH. Two sets of peaks are evident, one for intact viruses and one for smaller species. The 23.4 nm peak maximum corresponds to the virus size we have measured previously for pp7.[8,13,15] The smaller peaks (<15 nm) represent free capsid proteins, capsomers, or capsomer fragments. The molecular weight correlation of Bacher, et al.,[10] identifies the primary peak (at 5.4 nm) as a dimer of the major capsid protein. Tars, et al., report that capsids of the viral genus *levivirus* often assemble as covalently linked dimers of the major capsid protein such that *levivirdae* capsids may often be regarded as assemblies of 90 dimers, termed capsomers, instead of 180 individual coat proteins commensurate with our observations.[25]

Several changes in the size distribution take place as pp7 is exposed to acid. First, the concentration of intact virus decreases substantially. Our finding of incomplete capsid degradation at pH 4.0 is in reasonable agreement with the previous work of Brorson, et al.,[26] who find incomplete degradation of the capsid of the xenotropic murine leukemia virus (X-MuLV), another small virus of biomanufacturing interest, at pH 3.8. Second, dimer peaks completely dominate the size distribution. Our data therefore support a mechanism for low pH inactivation where the intact pp7 virus particles disintegrate into dimers without accumulating intermediate species. Thus, ES-DMA can be used as a tool to evaluate virus inactivation mechanisms, which remain essential industry efforts to identify strategies to maximize virus safety.

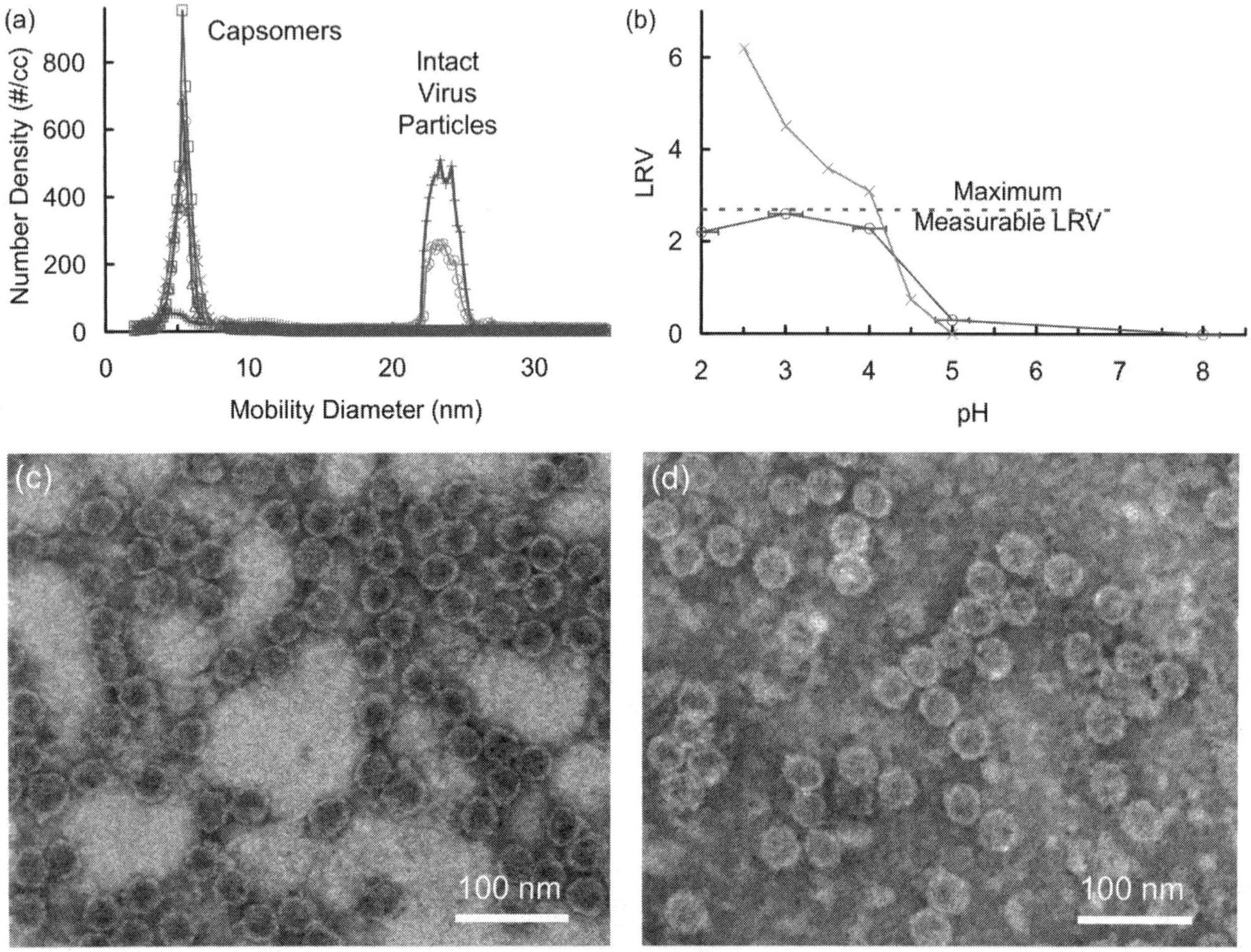

Figure 2. ES-DMA size distributions for (a) pp7 subjected to pH's of neutral (+), 5.0 (o), 4.0 (x), 3.0 (triangle), and 2.0 (square) for 30 minutes. (b) LRV versus pH for pp7 from ES-DMA (o) and plaque assay (x). TEM images of pp7 (c) untreated and (d) exposed to pH ~4.

We can quantify the loss of the intact virus particles using the logs of retained virus (LRV) formalism defined as $LRV = Log_{10}(N_{in}/N_{out}) = Log_{10}[(C_{in}V_{in})/(C_{out}V_{out})]$, where N_{in}, C_{in} and V_{in} are the number or moles, concentration, and volume of phage entering a process stream or at the beginning of a batch degradation process and N_{out}, C_{out} and V_{out} are the number or moles, concentration, and volume exiting the stream or at the termination a batch process.[8,27] The LRV represents a convenient means of representing concentration changes that can span several orders of magnitude in a unit operation. Larger LRV's represent greater viral clearance via removal or inactivation (i.e. a concentration decreases by three orders of magnitude from entrance to exit is described as changing by 3 log_{10} or an LRV of 3). Here we approximate the concentration ratio as the ratio of the counts under the peak (which ranges from 21.8 nm to 26.2 nm) of the control at neutral pH to the counts at the pH of interest. Figure 2b illustrates how the LRV measured by ES-DMA and plaque assays depends on pH. We find minimal changes in the LRV in the neutral range between pH 5 and 8, a large jump between pH 4 and 5, and a plateau below pH 4 for ES-DMA but not the plaque assay. This is because the plaque assay has a large linear range (up to 8 log_{10} in this experiment), whereas the dynamic range of ES-DMA is limited by its lower limit of detection ($\leq 10^9$ virus particles/mL).[27,28] Although ES-DMA boasts single particle counting like the plaque assay, the 66 nL/min flow through the electrospray is effectively diluted into a gas flow of 1.5 L/min, decreasing the apparent concentration by $\approx 2.3 \cdot 10^7$. Thus, a sample that begins at $6 \cdot 10^{11}$ viruses/mL, for example, is diluted by a factor of 50 in the preparation protocol, and then is further diluted in the instrument by $2.3 \cdot 10^7$ would be expected to sustain a maximum gas-phase count of 500 particles/cubic centimeter (see Figure 2a). Because the baseline has a number density of approximately unity, the maximum LRV measurable with this protocol and sample set is $\log_{10}(500/1) = 2.7$. This example implies that the plateau ≤pH 4 in Figure 2b is due to limitations of the measurement protocol and not degradation of the sample. The presence of the plateau indicates the uncertainty in these measurements for which one standard deviation at pH 2-4 equals 0.21 in $\log_{10}$ units. This uncertainty is comparable to that of the plaque assay, which has an uncertainty of approximately 0.3 $\log_{10}$ units.[27] We note in contrast that our method does not require long incubation times nor specific hosts. Indeed, the technique is fully functional even for viruses for which neither the genetic sequence nor the host has been fully characterized. Thus, within their respective dynamic ranges, curves derived from ES-DMA and the plaque assay essentially overlap within the linear range of the former. This argues that pp7 inactivation can be completely attributed to disintegration of the capsid into capsomers, rather than a more subtle effect such as denaturation of surface coat proteins.

Viral Neutralization

ES-DMA assays of nanoconjugate structure and stoichiometry rely on changes in the measured mobility size upon binding (see Fig. 3a).[9,24,29] Here we report on the virus-antibody conjugates formed by binding anti-MS2 antibodies (i.e. those specific to the bacteriophage MS2) to MS2. As antibodies recognize and bind to the viral coat, the antibody-virus complex forms, increasing the apparent size of the virus in the size distributions. Figure 3 shows that prior to binding the MS2 virus particle displays a symmetric monomodal peak at 23.2 nm in reasonable agreement with previous results of Cole, et al.,[13] and Hogan, et al.,[28] who find MS2 to be 23.0 nm and 24.1 nm, respectively. After binding Figure 3b the apparent size of the complex increases with larger solution concentrations of antibodies leading to further increases in size. Both the polyclonal P and monoclonal G antibodies increase the apparent size of the virus peak, though changes in size due to the former are more modest than those due to the latter. The apparent size of the virus complex as a function of the antibody concentration is summarized in Figure 3d using a dose response curve. Although some of the 2.2 nm shift may be due to one or more free antibodies clustering with the complex formed in solution as the droplet dried,[16,21,30] quantitative analysis (not shown) indicates that nonspecific aggregation alone cannot explain the size shifts observed, particularly for the G antibody, demonstrating that the ES-DMA cleanly distinguish antibodies that bind well from those that do not. We note in passing that these results support the MS2 specificity of both the P and G antibodies because nonspecific human antibodies mixed with MS2 showed no indication of complex formation in ES-DMA (data not shown). These results demonstrate the utility of size based discrimination of virus-antibody binding.

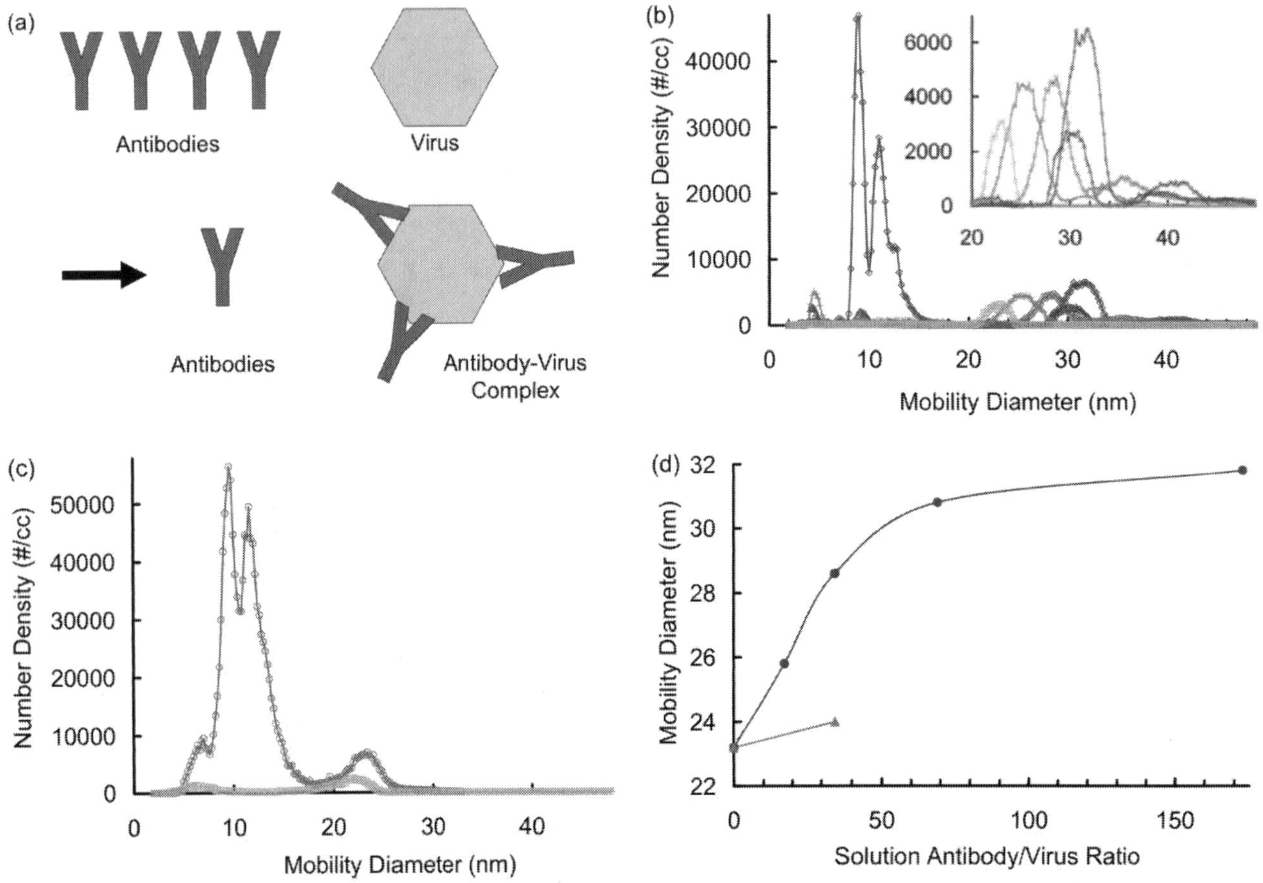

Figure 3. (a) Reaction scheme for antibody-antigen (viral) binding suggesting that the apparent size of the virus-antibody complex should exceed that of the bare virus. (b) ES-DMA size distributions of MS2 virus (squares) with monoclonal antibodies at molar solution concentration ratios of 173 (diamond), 69 (triangle), 35 (circle), and 17 (+) antibodies per virus. (c) ES-DMA size distributions of MS2 virus (squares) with 35 (circle) polyclonal antibodies/virus. (d) Summary dose-response curves of mobility size from ES-DMA versus solution antibodies-virus ratio for monoclonal (circle) and polyclonal antibodies (triangle).

A remarkable feature of the ES-DMA data for the G antibody is the presence of two peaks for the antibody-virus complexes as seen in the inset of Figure 3b. For example, at the highest G antibody concentration, peaks are observed at both 31 nm corresponding to one antibody bound virus and 40 nm corresponding to two antibody labeled virus particles. Similar bimodal distributions are observed at other concentrations. The initial virus sample (Fig. 3b) did not show any signs of virus dimerization, indicating the antibody to be responsible for formation of the two-virus complexes. We suspect that the antibody serves as a linker for the two viruses because each antibody possess two Fc regions, both of which can bind. The presence of bimodal distributions for the G antibody and their absence for the P antibody provides further evidence of the binding strength of the former. This result is particularly important because viral aggregation is one of the three primary ways in which an antibody can neutralize virus; ES-DMA holds the potential to rapidly and clearly distinguish this form of neutralization.

Although it is clear that increases in the apparent size of the virus peak reflect antibody binding, the exact correlation between size and stoichiometry remains uncertain. Two models are available to analyze the data, both of which approximate both antibody and viruses as spheres of diameter d_{ab} and d_v, respectively. Laschober, et al.,[29] assert based on volume conservation that the number of antibodies, n_{ab}, bound to a virus particle is equal to the volume of the thin spherical shell, $\pi(d_{av}^3-d_v^3)/6$, divided by the volume of an individual antibody, $\pi d_{ab}^3/6$,

$$n_{ab}=\frac{d_{av}^{3}-d_{v}^{3}}{d_{ab}^{3}} \quad . \tag{1}$$

A similar expression was used by Pease, et al., to approximate the surface binding of thiol terminated single stranded DNA on gold nanoparticles.[9] A second model more recently advanced by Pease, et al., draws upon a fundamental analysis of the forces acting on the complex within the DMA.[8,9,13-17,22-24,27,28,30-34] Briefly, the drag force acting on the complex within the DMA is proportional to the projected area of that complex when it is smaller than the mean free path of the carrier gas. By definition each complex has three principle axes.[22,35] A Brownian average of the projected area, A_i, orthogonal to each of these axes finds

$$d=\left[\frac{\sqrt{\pi}}{6}\sum_{i=1}^{3}A_i^{-\frac{1}{2}}\right]^{-1} \tag{2}$$

in the free molecular regime.[9,22] This equation has been successfully used in several recent articles to approximate the surface density of DNA on nanoparticles, size virus clusters, determine the number of quantum dots attached to bacteriophage λ, and evaluate the number of streptavidins on rod shaped nanoparticles, *inter alia*.[8,9,13-17,22-24,27,30-34]

Table 1 summarizes the predictions of the two equations for the MS2 data reported herein and data for human rhinovirus (HRV) serotype 2, as reported by Laschober, et al.[29] We include the latter with pause because the corresponding size distributions contain several peaks that were not clearly identified, perhaps due to degraded or impure virus samples. The uncertainty in stoichiometry using Eq. 1 based on an uncertainty in complex mobility diameter of ±0.5 nm is approximately ±4 antibodies as indicated by Laschober, et al.,[29] while the uncertainty using Eq. 2 remains less than ±1 antibody. In either case, the disagreement between the two models is significant. Laschober, et al., support their estimates by asserting that their construction is intuitively simple and that their predicted number of antibodies is approximately equal to the close packed limit of spherical antibodies on the viral surface in agreement with other estimates for HRV.[29] Proponents of the second model assert that it is more rigorous than the former because it is based on the fundamental physics underlying the measure of the complex's mobility within the DMA, the second equation can be derived from first principles whereas the first cannot, and the close packed limit is unlikely to be achieved in practice due to steric and electrostatic repulsion between neighboring antibodies.[11,22,36] Both models have been used previously in the literature and both make unphysical simplifying assumptions about the structure antibodies by asserting that they are spheres when it is well known that they have a Y-like structure. Final arbitration between the two models awaits more rigorous and advanced computational models—outside of the scope of this contribution—that account for the Y-like antibody structure, antibody orientation relative to icosahedral viral surface, intermolecular forces acting between antibodies and the virus particle as well as between neighboring antibodies, and advanced drag force calculations in the free molecular regime.

Table 1. Stoichiometry of Antibody-Virus Complexes

Sample (solution antibody:virus molar ratio)	Virus Size (nm)	Antibody Size (nm)	Antibody-Virus Complex Size (nm)	Complex Stoichiometry from Eq. 2	Complex Stoichiometry from Eq. 1
MS2 + G (17:1)	23.2	9.0	25.8	1.8	6.4
MS2 + G (35:1)	23.2	9.0	28.6	3.8	15.0
MS2 + G (69:1)	23.2	9.0	30.8	5.4	23.0
MS2 + G (173:1)	23.2	9.0	31.8	6.1	27.0
MS2 + P (35:1)	23.2	9.0	24.0	0.6	1.8
HRV2 + mAB-8F5 (30:1)	29.8	8.8	34.3	4.7	20.4
HRV2 + mAB-8F5 (60:1)	29.8	8.8	35.6	6.1	27.4
HRV2 + mAB-8F5 (90:1)	29.8	8.8	35.6	6.1	27.4
HRV2 + mAB-8F5 (150:1)	29.8	8.8	35.6	6.1	27.4

Binding of antibodies to the virus surface also neutralizes viral infectivity. To test the ability of the polyclonal P and monoclonal G antibodies to neutralize the infectivity of MS2, antibody-virus complexes were introduced to their *E. Coli* host. Fig. 4a shows both the mono and polyclonal antibodies substantially decrease the infectivity. As expected the G antibody was much more effective at decreasing the viral infectivity, to as little as 0.5% at the highest included concentration, than the polyclonal P due to its better binding stoichiometry. For both antibodies, increasing the solution concentration of the antibody leads to more effective neutralization in reasonable agreement with the dose-response curves above. Combining these results with stoichiometry determined by either equation shows that all reductions in infectivity fall on a single power-law fit. The uncertainty in the plaque assay remains commensurate with that above (though log-linear plots tend to emphasize uncertainty at smaller remaining infectivities). *This remarkable result indicates that decrements in activity or infectivity are primarily a function of binding stoichiometry and not antibody identity.* The infectivity decreases by one order of magnitude for every 3.2 or 13 antibodies bound, using Eq. 2 and 1, respectively. Furthermore, these results indicate that just four antibodies can decrease infectivity by 95% and seven antibodies decrease infectivity by 99% using Eq. 2. This implies that less than 8% of the 90 MS2 capsomers must be directly bound to achieve a 99% reduction in infectivity. In contrast, Eq. 1 requires 17 and 31 for 95% and 99% infectivity reduction, respectively. The dramatic difference in predicted neutralizing power emphasizes the need for more comprehensive modeling of binding and gas phase mobility.

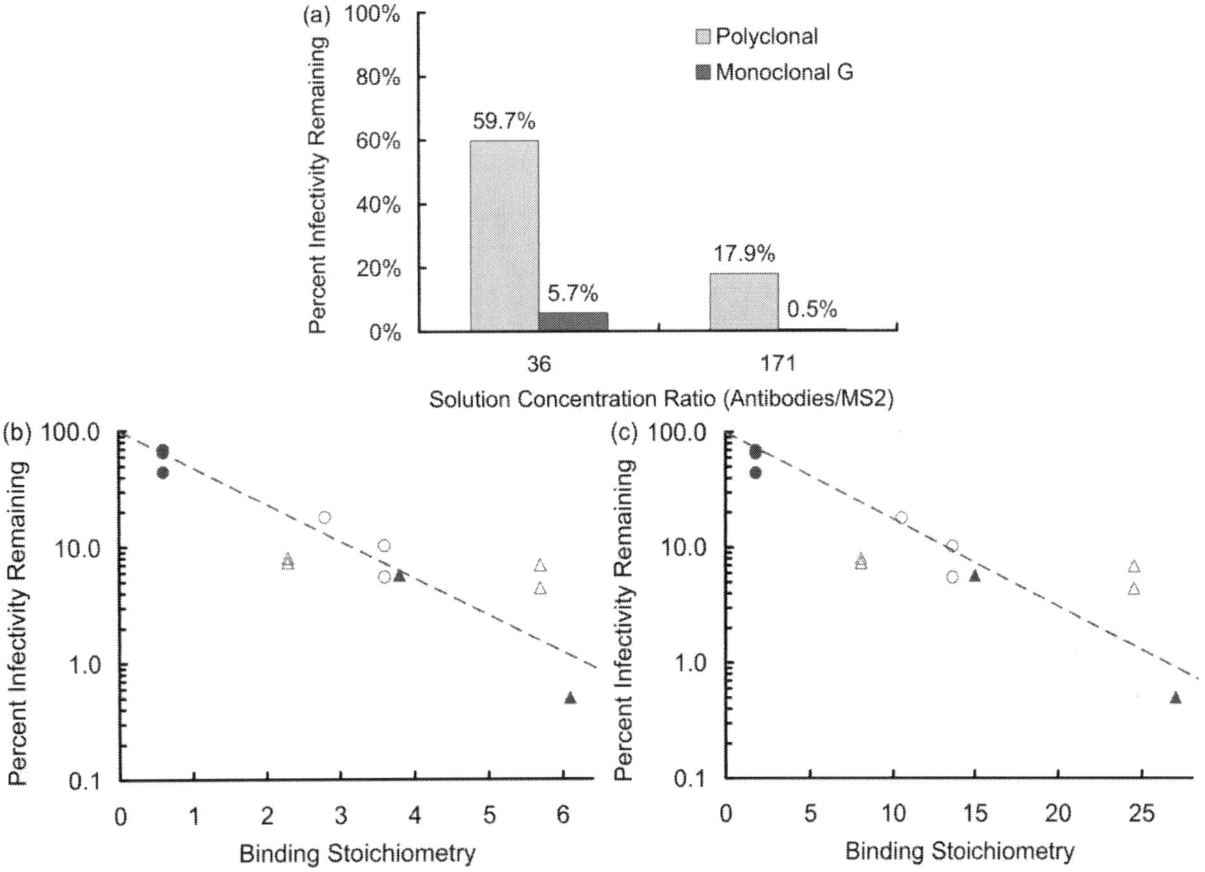

Figure 4. Percent infectivity remaining after antibody binding versus (a) solution concentration of neutralizing anti-MS2 polyclonal P and monoclonal G antibodies. Percent infectivity remaining, *PIR*, for both P (circles) and G (triangles) antibodies versus binding stoichiometry, n_{ab}, as determined by (b) Eq. 2 and (c) by Eq. 1 with close and open symbols representing directly measured and curvilinearly interpolated/extrapolated stoichiometries (from Figure 3d), respectively. The dashed line represents best fits including the point (0, 100%) of $PIR=10^{(2-0.315\, nab)}$ with $R^2 = 0.78$ for Eq. 2 and $PIR=10^{(2-0.075\, nab)}$ with $R^2 = 0.74$ for Eq. 1.

Finally, many neutralization assays do not take place in clean solutions containing only viruses and antibodies in nonvolatile ammonium acetate solutions, but occur in more challenging solutions and suspensions including patient serum. Figure 5 provides a size distribution for fetal bovine serum (fbs). Without dilution, the size distribution only increases in the range of 1-50 nm, but with increasing dilution, the peaks tail at progressively smaller sizes. With a twenty fold dilution, the serum background decays by 30-35 nm, allowing larger viruses to be evaluated using this neutralization assay. With 2000 fold dilution, the serum background decays by 12-13 nm allowing even smaller viruses such as pp7 to be analyzed using this assay. At this level of dilution, the large aggregates represented in Figure 5 are no longer formed but individual peaks for the albumins and globulins are observed.

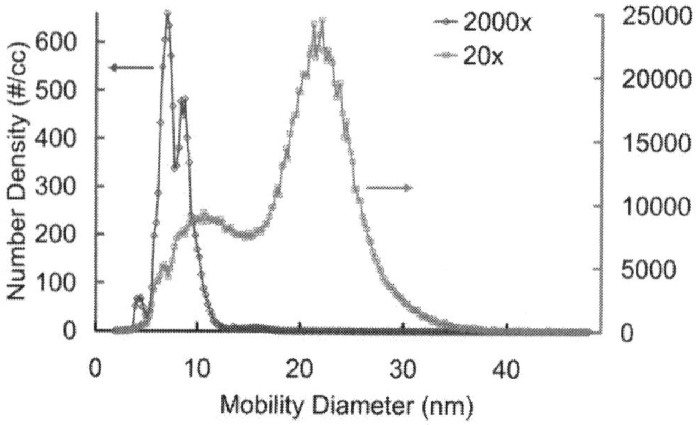

Figure 5. ES-DMA size distribution of positively charged molecular ions from fetal bovine serum diluted twenty (squares) and 2000 (diamonds) times in ammonium acetate buffer.

Conclusion

In summary, ES-DMA represents a powerful new technique to evaluate virus, VLP and vaccine properties and determine the relationship between their structure and properties. The technique is rapid, quantitative, and holds potential to quantitatively analyze binding stoichiometry and kinetics. Because of its design, one particle gives a single count regardless of its mass, enabling precise measurements even at the smallest sizes. In contrast to traditional biological assays, this method does not require long incubation times nor specific hosts. Indeed, the technique is fully functional even for viruses for which neither the genetic sequence nor the host has been fully characterized. These attributes make ES-DMA an attractive candidate with which to evaluate virus neutralization mechanisms. A highlight of the analysis above is that decrements in infectivity are primarily a function of antibody binding stoichiometry and not antibody identity. Future work will apply this technique to characterization of VLPs and adaptation of this technique to at-line analysis in biomanufacturing facilities.

Acknowledgements and Disclaimer

Reference to commercial equipment, supplies, or software neither implies its endorsement by the National Institute of Standards and Technology (NIST) or the Food and Drug Administration (FDA) nor that it is necessarily best suited for this purpose. The authors express appreciation to Tetracore (Rockville, MD) for their generous provision of antibodies specific to MS2. LFP recognizes provision of University of Utah startup funds supporting this endeavor.

References

1. Middelberg APJ, Rivera-Hernandez T, Wibowo N, et al. A microbial platform for rapid and low-cost virus-like particle and capsomere vaccines. *Vaccine.* 2011:doi:10.1016/j.vaccine.2011.1005.1075.

2. Hanslip SJ, Zaccai NR, Middelberg AP, Falconer RJ. Intrinsic fluorescence as an analytical probe of virus-like particle assembly and maturation. *Biochem Biophys Res Commun.* Oct 24 2008;375(3):351-355.
3. Hanslip SJ, Zaccai NR, Middelberg AP, Falconer RJ. Assembly of human papillomavirus type-16 virus-like particles: multifactorial study of assembly and competing aggregation. *Biotechnol Prog.* Mar-Apr 2006;22(2):554-560.
4. Pattenden LK, Middelberg AP, Niebert M, Lipin DI. Towards the preparative and large-scale precision manufacture of virus-like particles. *Trends Biotechnol.* Oct 2005;23(10):523-529.
5. Middelberg AP, Rivera-Hernandez T, Wibowo N, et al. A microbial platform for rapid and low-cost virus-like particle and capsomere vaccines. *Vaccine.* Jun 7 2011.
6. Liew MW, Rajendran A, Middelberg AP. Microbial production of virus-like particle vaccine protein at gram-per-litre levels. *J Biotechnol.* Oct 15 2010;150(2):224-231.
7. Ding Y, Chuan YP, He L, Middelberg AP. Modeling the competition between aggregation and self-assembly during virus-like particle processing. *Biotechnol Bioeng.* Oct 15 2010;107(3):550-560.
8. Pease LF, Tsai DH, Brorson KA, Guha S, Zachariah MR, Tarlov MJ. Physical characterization of icosahedral virus ultra structure, stability, and integrity using electrospray differential mobility analysis. *Anal Chem.* Feb 8 2011;83:1753-1759.
9. Pease LF, Tsai DH, Zangmeister RA, Zachariah MR, Tarlov MJ. Quantifying the surface coverage of conjugate molecules on functionalized nanoparticles. *J Phys Chem C.* Nov 22 2007;111(46):17155-17157.
10. Bacher G, Szymanski WW, Kaufman SL, Zollner P, Blaas D, Allmaier G. Charge-reduced nano electrospray ionization combined with differential mobility analysis of peptides, proteins, glycoproteins, noncovalent protein complexes and viruses. *J Mass Spectrom.* Sep 2001;36(9):1038-1052.
11. Knutson EO, Whitby KT. Aerosol classification by electrical mobility: apparatus, theory, and applications. *J. Aerosol. Sci.* 1975;6:443-451.
12. Wiedensohler A. An approximation of the bipolar charge-distribution for particles in the sub-micron size range. *J. Aerosol Sci.* Jun 1988;19(3):387-389.
13. Cole KD, Pease LF, Tsai DH, et al. Particle concentration measurement of virus samples using electrospray differential mobility analysis and quantitative amino acid analysis. *J Chromatogr A.* Jul 24 2009;1216(30):5715-5722.
14. Pease LF, Lipin DI, Tsai DH, et al. Quantitative characterization of virus-like particles by asymmetrical flow field flow fractionation, electrospray differential mobility analysis, and transmission electron microscopy. *Biotechnol Bioeng.* Feb 15 2009;102(3):845-855.
15. Lute S, Riordan W, Pease LF, et al. A consensus rating method for small virus-retentive filters. I. Method development. *PDA J Pharm Sci Technol.* Sep-Oct 2008;62(5):318-333.
16. Pease LF, Elliott JT, Tsai DH, Zachariah MR, Tarlov MJ. Determination of protein aggregation with differential mobility analysis: application to IgG antibody. *Biotechnol Bioeng.* Dec 15 2008;101(6):1214-1222.
17. Tsai DH, Zangmeister RA, Pease LF, Tarlov MJ, Zachariah MR. Gas-phase ion-mobility characterization of SAM-functionalized Au nanoparticles. *Langmuir.* Aug 19 2008;24(16):8483-8490.

18. Lute S, Aranha H, Tremblay D, et al. Characterization of coliphage PR772 and evaluation of its use for virus filter performance testing. *Appl Environ Microbiol.* Aug 2004;70(8):4864-4871.

19. Lute S, Bailey M, Combs J, Sukumar M, Brorson K. Phage passage after extended processing in small-virus-retentive filters. *Biotechnol Appl Biochem.* Jul 2007;47(Pt 3):141-151.

20. Smith JM, VanNess HS, Abbott MM. *Introduction to Chemical Engineering Thermodynamics.* 5th ed. New York: McGraw-Hill; 1996.

21. Kaufman SL, Skogen JW, Dorman FD, Zarrin F, Lewis KC. Macromolecule analysis based on electrophoretic mobility in air: globular proteins. *Anal Chem.* Jun 1 1996;68(11):1895-1904.

22. Pease LF, Tsai DH, Hertz JL, Zangmeister RA, Zachariah MR, Tarlov MJ. Packing and size determination of colloidal nanoclusters. *Langmuir.* Jul 6 2010;26(13):11384-11390.

23. Pease LF, Sorci M, Guha S, et al. Probing the nucleus model for oligomer formation during insulin amyloid fibrillogenesis. *Biophys J.* Dec 15 2010;99(12):3979-3985.

24. Pease LF, Feldblyum JI, Depaoli Lacaerda SH, et al. Structural analysis of soft multicomponent nanoparticle clusters. *ACS Nano.* Nov 23 2010;4(11):6982-6988.

25. Tars K, Fridborg K, Bundule M, Liljas L. The three-dimensional structure of bacteriophage PP7 from Pseudomonas aeruginosa at 3.7-A resolution. *Virology.* Jul 5 2000;272(2):331-337.

26. Brorson K, Krejci S, Lee K, Hamilton E, Stein K, Xu Y. Bracketed generic inactivation of rodent retroviruses by low pH treatment for monoclonal antibodies and recombinant proteins. *Biotechnol Bioeng.* May 5 2003;82(3):321-329.

27. Guha S, Pease LF, Brorson KA, Tarlov MJ, Zachariah MR. Evaluation of electrospray differential mobility analysis for virus particle analysis: potential applications for biomanufacturing. *J. Virolog. Meth.* 2011:submitted.

28. Hogan CJ, Kettleson EM, Ramaswami B, Chen DR, Biswas P. Charge reduced electrospray size spectrometry of mega- and gigadalton complexes: Whole viruses and virus fragments. *Anal Chem.* Feb 1 2006;78(3):844-852.

29. Laschober C, Wruss J, Blaas D, Szymanski WW, Allmaier G. Gas-phase electrophoretic molecular mobility analysis of size and stoichiometry of complexes of a common cold virus with antibody and soluble receptor molecules. *Anal Chem.* Mar 15 2008;80(6):2261-2264.

30. Pease LF. Optimizing the yield and selectivity of high purity nanoparticle clusters. *J. Nanopart. Res.* 2011;13 2157-2172.

31. Anumolu R, Gustafson JA, Magda JJ, Cappello J, Ghandehari H, Pease LF. Fabrication of highly uniform nanoparticles from recombinant silk-elastin-like protein polymers for therapeutic agent delivery. *ACS Nano.* Jul 5 2011;5:5374-5382.

32. Pease LF, Tsai DH, Fagan JA, et al. Length distribution of single-walled carbon nanotubes in aqueous suspension measured by electrospray differential mobility analysis. *Small.* Dec 2009;5(24):2894-2901.

33. Tsai DH, Pease LF, Zangmeister RA, Tarlov MJ, Zachariah MR. Aggregation kinetics of colloidal particles measured by gas-phase differential mobility analysis. *Langmuir.* Jan 6 2009;25(1):140-146.

34. Yim PB, Clarke ML, McKinstry M, et al. Quantitative characterization of quantum dot-labeled lambda phage for Escherichia coli detection. *Biotechnol Bioeng.* Dec 15 2009;104(6):1059-1067.

35. Happel J, Brenner H. *Low Reynolds number hydrodynamics.* The Hague: Kluwer Academic Publishers 1983.

36. Russel WB, Saville DA, Schowalter WR. *Colloidal Dispersons.* Cambridge, UK: Cambridge University Press; 1989.

(433a) Developing a Novel Preparation Method for Residual Biomass Polymer Blends

Jinwen Zhang, Composite Materials and Engineering Center, college of Engineering and Architecture, Washington State University, Pullman, WA

There is a growing interest in polymeric materials from renewable resources. Nevertheless, in facing a society where petroleum-based plastic products are almost ubiquitous, the challenge for the emerging biobased polymer materials is twofold: being cost effective and performance competitive while lowering the environmental footprint.

Soy protein meal, a residue after oil crush, has being explored as a new resource for plastics. However, owing to some obstacles such as water/moisture sensitivity, high melt viscosity, narrow process window and brittleness, neat soy protein (SP) plastics have little practical values. To overcome these disadvantages, blending SP with other biodegradable polymers has received increasing attention. Sugar beet pulp (SBP) is another agricultural residue which has received similar attention for plastic applications as SP. In the current study of SP or SBP polymer blends, SP and SBP are merely used as organic fillers in most cases. Although certain processing and end use properties can be improved with respect to that of neat SP and SBP plastics, the mechanical properties of the blends are often not satisfactory and even lower than that of both the neat polymer and the neat SP (or SBP) plastics.

In this work, we introduced a novel method to prepare polymer blends of residual biomass and biodegradable thermoplastic polymer. SP and SBP were used as plastic components during the compounding of the blends. The polymers used were poly(lactic acid) (PLA) and poly(butylene adipate-*co*-terephthalate) (PBAT), respectively. The mixing of the blends was carried out in a twin screw extruder and the test specimens were prepared by injection molding. Plasticization of SP and SBP by water during compounding played a critical role in determining SP (or SBP) to be a plastic component or filler in the mixing and hence the morphological structure of the resulting blends. When extra water was added, SP (or SBP) behaved like a melt during the compounding, consequently the resulting blends demonstrated significant different properties and morphological structures from those made of dry SP (or SBP). Because water evaporated during mixing and after drying, the resulting blends actually became in situ formed thermoplastic composites. Percolation network structure of SP (or SBP) was noted in some of these blends. Different effects of SP (or SBP) as a plastic component and filler on phase structure and mechanical properties of the resulting blends were investigated. Molecular weight changes of the thermoplastic PLA and PBAT after processing were examined. The change of SP molecular structure caused by melt processing was also evaluated by protein solubility test. Effects of various compatibilizers on the properties of the blends were also studied. The results suggest that when SP and SBP were used as plastic components in mixing, the resulting blends demonstrated an overall superior performance than when they are used as fillers.

(433b) Air-Templated Nanocellulose Reinforced Macroporous Foams

Alexander Bismarck, ICL, London, United Kingdom

In progres

See more of this Session: Biobased Materials I: General
See more of this Group/Topical: Forest and Plant Bioproducts Division

(433c) Polymerization of Lactide to Polylactic Acid and Co-Polymers of Polylactic Acid Using High Viscosity Kneader Reactors

Boyd T. Safrit and George E. Schlager, LIST USA, Inc., Charlotte, NC

Polymerization of lactide to polylactic acid (PLA) can be performed using conventional reactor technology such as stirred tank reactors, but the conversion and/or final molecular weight must be controlled to a lower level. At higher conversion and/or molecular weight, the reaction mass will become very viscous , which limits the ability of conventional reactor technology to provide adequate mixing, minimize mass transfer effects on reaction kinetics, remove exothermic heat of reaction and ensure proper heat transfer in order to eliminate hotspots/thermal degradation.

Kneader reactor technology has been used over 60 years in many high viscosity applications such as reactions and polymerization, devolatilization, and drying. This technology can handle the higher conversion and molecular weight polymerizations of lactide and other copolymers of lactide, while also providing the heat transfer required for proper temperature control. Using model kinetics and rheology data, a study was performed that shows the capability of kneader reactor technology for lactide polymerizations as well as other copolymers. Kneader reactor technology can also be used to remove the uncovered monomers from the polymer and expected results from the continuous operation of a polymerizer and finisher will be shown.

See more of this Session: Biobased Materials I: General
See more of this Group/Topical: Forest and Plant Bioproducts Division

(433d) Biodegradable Poly(hydroxyalkanoate) Foams

Amy Tsui, Qi Liao and Curtis W. Frank, Chemical Engineering, Stanford University, Stanford, CA

Poly(3-hydroxybutyrate-*co*-3-hydroxyvalerate) (PHBV) is a member of the bacterial polyester family of poly(3-hydroxyalkanoates) (PHAs) known for its ability for enzymatic degradation and good mechanical properties. The use of biodegradable polymers for producing sustainable materials has received increasing attention, and PHBV is an excellent candidate, shown to have similar properties to polypropylene (PP), known to be a tough and robust material. Specifically, PHBV could be a useful material for construction materials such as insulation foam. Current insulation foams are made primarily of polyurethane, polystyrene and polyethylene which are derived from non-renewable (petroleum) feedstock. Conventionally, foam structures in nonaqueous systems, such as polymers are stabilized simply by solidification due to vitrification, cross-linking, or crystallization (1). This requires that the polymer matrix have a high enough melt strength to retain cell structures until solidification occurs. However, processing of the PHBV into foam is limited by its low melt elongational viscosity and narrow thermal processing window. Low melt elongational viscosity contributes to cell coalescence and collapse, resulting in undesirable high density foam. The poor foamability of PHBV can be addressed via the addition of a nucleating agent, orotic acid.

Orotic acid (OA), a biodegradable and nontoxic chemical, was shown to be a very effective nucleating agent (NA) for crystallization in poly(3-hydroxybutyrate) (PHB) (2,3). Nucleating agents can also be used to promote bubble formation (4). With NAs, more nucleation sites are created for gas to diffuse into so bubbles are limited to a smaller, more uniform cell size (5), while also increasing cell density (6). Additional benefit to using nucleating agents is that they often act as plasticizers which reduce the processing temperature needed for melt extrusion (4). As NAs also promote crystallization, the addition of NAs may also reduce the time required for stabilization of polymeric foam structure.

To first evaluate the nucleating efficiency of OA in PHBV, films of PHBV and OA were made using the solvent casting method. The compositions ranged from 0 to 10 wt% OA. Polarized light optical microscopy (PLOM) showed that the OA particles were well-dispersed throughout the films, which is optimal for homogeneity of nucleation sites. Differential scanning calorimetry (DSC) indicated that 1 or 2 wt% OA resulted in the highest percent crystallinity. It was also seen that presence of OA resulted in more uniform crystal size, as indicated by the narrower full-width-half-maximum (FWHM) values of the melting endotherm. Further studies are being performed to evaluate the effect of OA content on crystallization kinetics.

Foaming of poly(3-hydroxybutyrate-co-3-hydroxyvalerate) (PHBV) with 5% hydroxyvalerate content (Tianan Biologic Materials Company, China) was performed using an activated azodicarbonamide (AZ) (Actafoam 780, Chemtura) chemical blowing agent that decomposed at 140-150°C, generating approximately 150 cc of nitrogen per gram of solid. After blending and drying the PHBV and blowing agent, materials were extruded through a ¾-inch single-screw extruder (L/D ratio 25:1, compression ratio 3:1, C.W. Brabender) equipped with a 2-inch horizontal flex-lip ribbon die. The extruder temperature profile was chosen based on the temperature requirements for melting the PHBV and decomposing the blowing agent, while preventing PHBV decomposition (7-9). Foaming occurred upon leaving the die. SEM images obtained on cryo-fractured surfaces of extruded foam samples showed varying cellular morphologies with changes in AZ content. Cell coalescence and collapse occurred when the amount of blowing agent AZ exceeded 1.5 parts per hundred (phr), suggesting that the PHBV melt strength does not withstand more increase in pressure posed by gas generation and expansion (10). Using the same procedure as for the PHBV foams, results from the addition of 1 or 2 wt% OA will be shown.

The biodegradable polyester PHBV is a promising material for replacing conventional petroleum-based polymers. However, the melt strength and thermal processing window of PHBV must be improved in order to develop an alternative insulation foam. Adding orotic acid as a nucleating agent for both crystallization and bubble formation could be a promising method for making such enhancements to the foamability of PHBV.

References

1. Thareja P; Ising BP; Kingston SJ; Velankar SS. Macromol. Rap. Comm. 2008, 29, 1329

2. Jacquel N; Tajima K; Nakamura N; Miyagawa T; Pan P; Inoue Y J. App. Polym. Sci. 2009, 114, 1287

3. Jacquel N; Tajima K; Nakamura N; Kawachi H; Pan P; Inoue Y J. App. Polym. Sci. 2010, 115, 709

4. Kim SG; Lee JWS; Park CB; Sain M J. App Polym. Sci. 118, 1691

5. Mills N Polymer Foams Handbook 2007, Chapter 3, 40

6. Rodrigue D; Gosselin R. The Effect of Nucleating Agents on PP Foam Morphology 2003, 169

7. G. Schmack; D. Jehnichen; R. Vogel; B. Tandler J. Polym. Sci. 2000, 38, 2841

8. M. Yamaguchi; K. Arakawa Euro. Polym. J. 2006, 42, 1479

9. R. Vogel; B. Tandler; D. Voigt; D. Jehnichen; L. Hausler; L. Peizsch; H. Brunig Macrom. Biosci. 2007, 7, 820

10. Q. Liao 2010 Ph.D. Thesis, Stanford University, Stanford, CA.

(433e) Carbon Fiber Reinforced BioPlastics: Present Status and Future Directions

Manju Misra, School of Engineering and the Department of Plant Agriculture, University of Guelph, Guelph, ON, Canada and **Amar K. Mohanty**, Department of Plant Agriculture & School of Engineering, Guelph, ON, Canada

In Progress

See more of this Session: Biobased Materials I: General
See more of this Group/Topical: Forest and Plant Bioproducts Division

(433f) Modification of Biomass Rheology: The Influence of Physical Chemistry

Joseph R. Samaniuk[1], C. Tim Scott[2], Thatcher W. Root[1] and Daniel J. Klingenberg[1], (1)Chemical and Biological Engineering, University of Wisconsin-Madison, Madison, WI, (2)USDA Forest Products Laboratory, Madison, WI

Chemicals derived from lignocellulosic biomass, including liquid fuels, are a sustainable alternative to petroleum derivatives. Many different processes have been proposed to convert lignocellulosic materials to liquid fuels but a common operation required in all of them is heating to high temperatures (110-230 ^{o}C). Reducing the concentration of water during this step decreases utility costs and decreases overall capital costs by reducing equipment size. Increasing the solids concentration however, results in rheological changes like a rapidly increasing yield stress. Mixing, pumping and pouring biomass slurries becomes difficult, if not impossible, above 20-30 wt% solids.

We have shown previously that the rheology of biomass can be controlled with the use of certain water soluble polymers (WSPs). AFM experiments have revealed that WSPs alter stresses in cellulose fiber suspensions by lubricating the surface of individual fibers. Using these additives, we have been able to mix and convey untreated biomass at solids concentrations up to 50 wt%. However, the efficacy of WSPs is reduced after certain treatments such as acid hydrolysis. How changes in the physical chemistry of biomass during treatment affect rheology modification with WSPs is unknown.

In this presentation our focus will be on the influence of the physical chemistry of both biomass and WSPs on rheological modifications. The physical chemistry of the biomass is altered by the treatment type, e.g. acid hydrolysis, and the treatment severity. Specific changes to the physical chemistry of the biomass are probed with fiber suspensions containing known chemistries, e.g. cellulose fibers vs. hemicellulose-rich fibers. The physical chemistry of the WSPs are altered by varying the type of polymer, the molecular weight, the degree of substitution, and by the addition of chemical additives. We will show how these variables influence not only the instantaneous rheological modification but also its behavior over time.

See more of this Session: Biobased Materials I: General
See more of this Group/Topical: Forest and Plant Bioproducts Division

Torrefied wood pellets production using natural biomass compounds

Venkata K Penmetsa and Philip H Steele*
Forest Products Department, Mississippi State University,
100 Blackjack Rd. Starkville, MS 39759. Tel. no: 662-325-8083
kpenmetsa@cfr.msstate.edu

The U.S. coal industry is highly interested in the potential for combusting torrefied wood as a co-feed with coal. Unlike biomass torrefied wood can be cofed with coal at a high percentage. Biomass combustion facilities cannot use the existing infrastructure and new biomass fed plants cost from $140 to $200 million to construct.

Torrefaction of biomass has attracted considerable attention as a means to reduce biomass weight by 70% through removal of water and the hemicellulosic fractions while retaining most of the original high-energy cellulose and lignin content. A further advantage of torrefied wood is that it has anti-microbial properties as well as limited water resistance such that microbial degradation is eliminated.

Unlike biomass torrefied wood can be co-fed with coal at a high percentage and requires no major infrastructure investment. One issue is the current inability to ship torrefied wood as coal is currently shipped. A waterproof pelletized product is required to allow bulk shipping via open railcar or by barge. Production of a waterproof torrefied wood pellets or briquettes has not been achieved to date.

Mississippi State University has developed a low-cost binder to produce a high-density pellet with high water resistance. The binder is produced from biomass and has a neutral pH of 6.2 such that the pellets do not produce corrosive gases during their combustion. No chemical cross linking agents, such as formaldehyde, are required making this binder very environmentally friendly. The binder is suitable to produce densified water repellent pellets with many different biomass feed stocks. The pellets have been subjected to crushing strength and water repellant test.

Pellet hardness was determined for each test specimen as described in ASTM standard D143 for small clear specimens. Each specimen was weighed immediately before and after the test and the density values were computed. All test results are given in Table 1.

The pellets were also tested in a water absorption test. The pellets were weighed before immersing in water and were fully immersed in water for 14 days. The pellets were weighed every 24 hours to monitor any increase in the weight of each pellet due to water absorption. The weight difference quantifies the amount of water absorbed inside the pellet pores. Figure 1 demonstrates the behavior of the pellets immersed in water over 14 days at various binder loadings as the weight percent of the total pellet weight. The 30%, 20% and 15% binder proportions showed very minor weight gains of water ranging from 0.05% to 0.6%.

These results indicate that at these high binder ratios the pellets were so densified that limited pore spaces are available for water to penetrate over time. These results on water weight gain over time indicate that binder loading magnitude was directly linked to water absorbance. The pellets with the lowest binder proportion had the highest water absorption values. Over 14 days the pellet produced with 2% binder had maximum weight gain of about 11%; the 5% binder loaded pellet had a water weight gain of 9.6%; that with 10% binder addition had a water weight gain of only 1.2%.

The study results showed that the tested binder produces a densified and water repellent bond for torrefied wood that is helpful in producing high initial strength with this strength substantially maintained following severe water exposure is more helpful in pelletizing torrefied wood. Both initial pellet compressive strength and strength following long-term water soaking the pellet strength is a relatively linear function of proportion of binder applied. However, even pellets with only 2% binder loading produced an initial strength of 764 kg/m^2 and 68% of this initial strength was maintained following the 14 days of water soaking followed by high-temperature drying to zero moisture content. While binder loadings of 30, 20 and 15% were tested and showed superior performance it is unlikely that these loadings will be economical due to the high amount of binder required. In addition, for production of a waterproof pellet at 10% loading appears to be more than adequate. Over the 14 days of soaking only 1.2% of water absorption was observed for this loading. In addition, nearly 80% of the initial compressive strength was maintained after water exposure and oven drying.

Table 1. Pellet weight (g) and compression strength of test pellets immediately after production compared to pellets water soaked for 14 days and oven dried to oven dry water content.

Spec. no.	Weight (g)	Binder composition (%)	Compression strength (kg/m^2)	Compression strength (kg/m^2) after 14 days in water
1	16.80	30	1095	960
2	16.92	20	987	846
3	16.76	15	934	800
4	16.90	10	905	718
5	16.39	5	841	630
6	16.42	2	764	519

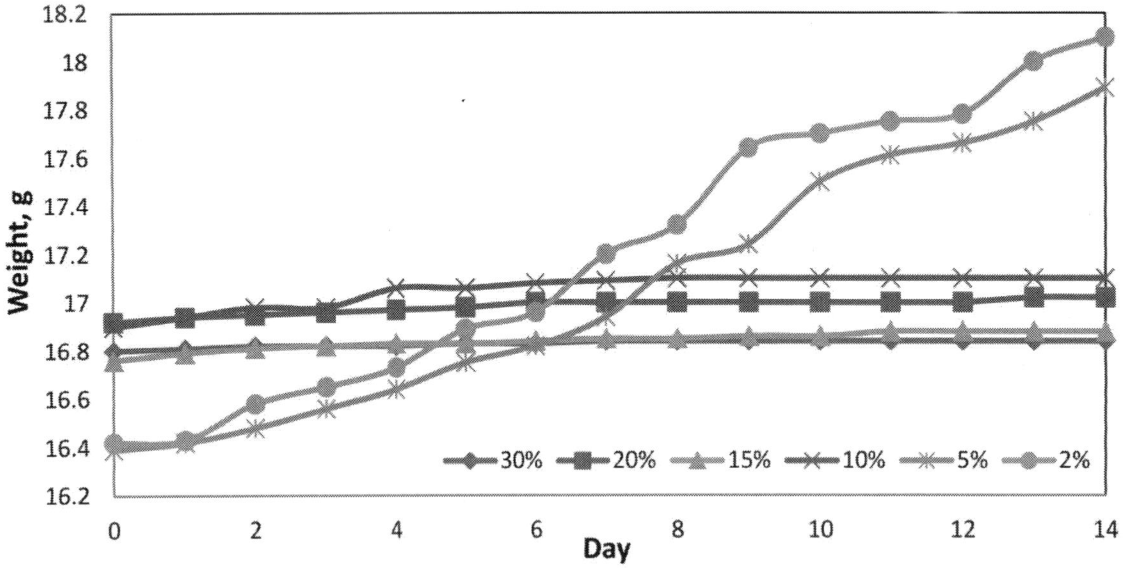

Figure 1: Daily water weight absorption values by pellet binder proportion for 14 days of soaking

For most applications it appears that a 2% loading will produce adequate initial compression strength at 764 kg/m^2. In addition, following water soaking 68% of the initial strength (519 kg/m^2) were maintained. It is not clear that the extra expense of producing a 10%, or higher, binder loading pellet would be cost effective versus the added cost of the additional binder.

References:

1. Fonseca Felfli, F., Luengo, C.A., Bezzon G. and Beaton Soler, P., 1998. *Bench unit for biomass residues torrefaction*, Biomass for Energy and Industry, Proceeding of Int. Conf., Würzburg, Germany, 8-11June, Ed. by Kopetz, Weber, Palz, Chartier and Ferrero, C.A.R.M.E.N., Rimpar Germany, 1593-1595.
2. Arcate, J.R., 2000. *New process for torrefied wood manufacturing,* Bioenergy update, vol 2, No. 4. (www.techtp.com/Bioenergy%20Update%20.pdf).
3. Wescott, J.M. and C.R. Frihart. 2004. *In* Proceedings of the 38th International Wood Composite Materials Symposium, Washington State University, Pullman, Washington. pp 199-206.

(437a) Mechanical Strain Decreases the Rate of Fibrin Gel Degradation by Plasmin

Arjun S. Adhikari, Armen Mekhdjian and Alexander R. Dunn, Chemical Engineering, Stanford University, Stanford, CA

Fibrin is a major component of blood clots and is formed by polymerization of fibrinogen. From a biomaterials standpoint, fibrin is of great interest because it can withstand strains of 200% and serves as an excellent tissue engineering scaffold. The mechanical properties of fibrin gels are likewise of fundamental medical importance: it is highly desirable to dissolve fibrin networks in the context of thrombosis, but mechanically robust clot formation is essential during wound healing. Fibrin networks are subject to strain both in vivo and in engineered biomaterials. However, very little is known about how mechanical stress may affect fibrinolysis by plasmin, the major enzyme responsible for clot dissolution. Here we describe micro-scale measurements that quantify the effect of mechanical strain on fibrin clot dissolution by plasmin. We find that 170% strain results in a dramatic, ~8-fold decrease in fibrinolysis rates. A poroelastic model of the fibrin network can account for the decrease in proteolysis rate that we observe. Our data and resulting model suggest that mechanical strain may be an important regulator of fibrinolysis in vivo.

See more of this Session: Biomaterial-Cell Interactions In Tissue Engineering
See more of this Group/Topical: Food, Pharmaceutical & Bioengineering Division

(437b) Engineering Synthetic ECM Analogs Using Elastin Mimetic Peptides

Dhaval Patel, School of Chemical and Biomolecular Engineering, Georgia Institute of Technology, Atlanta, GA, Rohan Menon, Department of Biomedical Engeering, Georgia Institute of Technology, Atlanta and Lakeshia Taite, Chemical and Biomolecular Engineering, Georgia Institute of Technology, Atlanta, GA

EngineeringSynthetic ECM Analogs Using Elastin Mimetic Peptides

Developing biomaterials whichprovide cellular cues for increased extra-cellular matrix (ECM) deposition hasrendered much research interest in the past few years. The ECM iscomprised of several cell secreted proteins which provide cellular cues formigration, proliferation, and differentiation. Furthermore, the ECM isresponsible for creating a suitable microenvironment for cell sustainabilitywithin a tissue. Current technology in biomaterial application limits theamount of ECM deposition resulting in decreased cell viability within thebiomaterial. Herein we engineered two different elastin mimetic peptides,EM-19 and EM-23, capable of inducing cells to produce a viable ECM once incorporatedinto a biomaterial. Elastin is a major ECM protein commonly found invascular tissue. Although highly hydrophobic, elastin strongly interactswith cells using the hexapeptide sequence VGVAPG triggering cells to inducevascular ECM formation. Our previous report suggests that the peptidesEM-19, AAKAAKVGVAPGAAKAAKK, and EM-23, AAKAAKVGVAPGRGDSAAKAAKK, are capable ofself-assembling within the deposited elastin matrix[1]. Thepresence of RGDS in EM-23 has also been shown to increase the ability of thepeptide to self-integrate into the elastin matrix while inducing more elastindeposition[1]. In this report, we covalently integrated eitherpeptide, 5 mol/ml, into a 10 % w/v poly(ethylene)glycol diacrylate (PEGDA) hydrogel with smooth muscle cells (SMCs). After48 hours incubation, we evaluated elastin production using a Fastin assay (Fig1). Our data suggests that either peptide has the capability of servingas ECM templates once incorporated into the hydrogel.

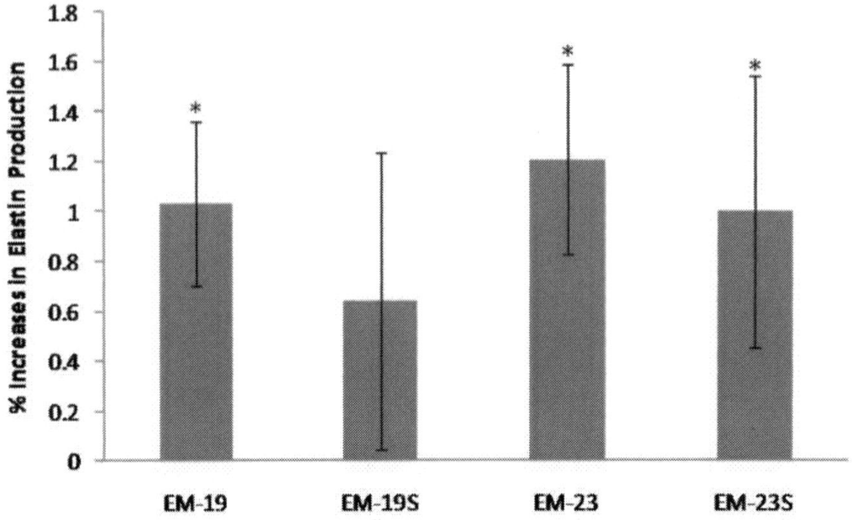

Fig 1. Percent increase for elastin productionin the presence of different elastin mimetic peptides within a 10 % w/v PEGDAhydrogel. Scrambled versions, where VGVAPG was scrambled to VGVPAG, ofEM-19 and EM-23 are denoted as EM-19S and EM-23S, respectively. Significantly increased elastin production was observed on all peptidesexcept for EM-19S in comparison to bare PEGDA; with EM-23 showing the highestability to induce elastin production. * denotes $p<0.05$ from PEGDA with atleast 6 samplings.

1. Patel, D., et.al., *Self-Assemblyof Elastin-Based Peptides into the ECM: the Importance of Integrins and theElastin Binding Protein in Elastic Fiber Assembly.* Biomacromolecules 2011 **12:** pg 432-440.

See more of this Session: Biomaterial-Cell Interactions In Tissue Engineering
See more of this Group/Topical: Food, Pharmaceutical & Bioengineering Division

(437c) Increased Schwann Cell Migration Speed and Directionality Occurs During Electrical Stimulation Through Polypyrrole Substrates

Leandro Forciniti, Chemical Engineering, University of Texas at Austin, Austin, TX, Jose Ybarra, Biology, University of Texas at Austin, Austin, TX, Muhammad H. Zaman, Biomedical Engineering, Boston University, Boston, MA and Christine Schmidt, Department of Chemical Engineering, University of Texas at Austin, Austin, TX

Current injury models suggest that Schwann cell migration and guidance are necessary for successful regeneration and synaptic reconnection after peripheral nerve injury. The ability of conducting polymers such as polypyrrole to exhibit chemical, contact, and electrical stimuli has led to much interest in their use for neural conduits. Despite this interest very little research has investigated the effect that electrical stimulation using polypyrrole has on Schwann cell behavior. Here we investigate the mechanism by which Schwann cells interact with polypyrrole in the presence of an electric field. We explored the effect that adsorption of different serum proteins on polypyrrole upon the application of an electric field has on Schwann cell migration and secretion of key biomolecules (i.e., NGF, BDNF, and CNTF). The results show an increase in average displacement with electrical stimulation, resulting in a net anodic migration. Moreover, indirect effect of protein adsorption due to the oxidation of the film upon the application of electrical stimulation was shown to have a larger effect on migration speed than on migration directionality. These results suggest that SC migration speed is governed by integrin mediated interactions with adsorbed proteins; whereas SC migration directionality is governed by electrically mediated phenomena. These results will prove invaluable in optimizing conducting polymers for their different biomedical applications such as nerve repair.

See more of this Session: Biomaterial-Cell Interactions In Tissue Engineering
See more of this Group/Topical: Food, Pharmaceutical & Bioengineering Division

(437d) Improving Cochlear Implant Performance by Directing Nerve Growth with Photopolymerized Micropatterns

Brad Tuft[1], Joseph Clarke[2], Scott White[3], Marlan Hansen[2] and C. Allan Guymon[3], (1)University of Iowa, Iowa City, IA, (2)Otolaryngology, University of Iowa, Iowa City, IA, (3)Chemical and Biochemical Engineering, University of Iowa, Iowa City, IA

Cochlear implants (CIs) provide auditory perception to individuals with severe to profound hearing impairments, however, CI performance has not significantly improved in recent years due to electrical current spread in the inner ear. A neural tissue engineering approach that guides nerve cell processes towards target electrodes may reduce the problematic current spread and enhance CI performance. Consequently, we are using the spatial and temporal control inherent to photopolymerization methodology to fabricate micropatterned methacrylate polymers that direct nerve cell growth based on substrate topographical cues. Micropatterned substrates are formed in a rapid, single-step reaction by selectively blocking light with glass-chrome, ronchi rule photomasks which have parallel line-space gratings from 2.5-100μm in width. The resultant pattern is a continuous series of parallel ridges and grooves at regular intervals and of various amplitudes that can be used for cellular contact guidance studies. Micro-feature depth is controlled and reproducibly generated from 220±40nm to 16±1.3μm by shuttering the light source at different time steps during the reaction. Regenerative growth of inner ear nerve cells orients to the direction of the micropattern and is strongly dependent on feature size. The ultimate goal of the research is to develop materials that elicit specific nerve cell behavior in order to improve the quality of life for CI patients.

See more of this Session: Biomaterial-Cell Interactions In Tissue Engineering
See more of this Group/Topical: Food, Pharmaceutical & Bioengineering Division

(437e) Fibril Alignment Induces Alignment of Sprouts From Endothelial Cell Spheroids In Fibrin Gel

Kristen T. Morin, Department of Biomedical Engineering, University of Minnesota, Minneapolis, MN and Robert T. Tranquillo, University of Minnesota- Twin Cities, Minneapolis, MN

The creation of a microvascular network within engineered tissue is an important step toward forming highly metabolic tissues (such as myocardium or liver) large enough for human use. It is critical that this network be aligned, as alignment provides natural inlet and outlet sides for perfusion. One potential method for creating such aligned microvasculature is to harness contact guidance to direct sprout outgrowth from endothelial spheroids. Human blood outgrowth endothelial cell (BOEC) spheroids were entrapped with neonatal human dermal fibroblasts in fibrin gel. The gels were aligned to varying degrees via magnetic fields or mechanically-constrained cell-induced gel compaction. After 7 days of culture, fibril alignment was measured with polarimetry, and spheroids were imaged with confocal microscopy. Sprout lengths and angles were measured in ImageJ, and an anisotropy index was defined (X component of sprout length:Y component of sprout length) to describe sprout alignment. In both alignment conditions, sprout alignment strongly correlated with fibril alignment, and sprout length correlated with both sprout and fibril alignment. Because magnetic fields provide a way to align gels without varying other factors, the data from the magnetically-aligned gels for the first time provide evidence that fibril alignment causes endothelial sprout alignment and that fibril alignment increases sprout lengths. Confounding factors (such as increased fiber density) preclude the same conclusion from the cell-aligned gel data; however those data show that the sprout alignment and lengthening effects can also be achieved using cell-induced alignment, a technique that is technically simpler and more relevant to tissue engineering.

See more of this Session: Biomaterial-Cell Interactions In Tissue Engineering
See more of this Group/Topical: Food, Pharmaceutical & Bioengineering Division

(437f) Dynamic Adhesion of Endothelial Progenitor Cells On Peptide-Coupled Hydrogels Under Flow

Wen Jun Seeto and Elizabeth A. Lipke, Department of Chemical Engineering, Auburn University, Auburn University, AL

Endothelial progenitor cells (EPCs) have the potential tobecome a reliable source of autologous cells for endothelialization ofintravascular devices and vascularization of tissue engineered constructs. Wehave characterized the rolling velocity and transient adhesion of EPCs ondifferent peptide-coupled hydrogels using a parallel plate flow chamber. Theseresults will be applied in the design of future biomaterial surfaces to enhanceendothelialization and improve EPC strength of adhesion under shear. EPCsare blood-derived cells and little is currently known about their response toshear stress. For these studies, umbilical cord blood outgrowthendothelial colony forming cells (ECFCs), one type of EPCs, were used. Toassess the specific interactions required for EFCFs to interact with materialcoatings under shear, copolymers poly(ethylene glycol)diacrylate (PEG-DA) and acryl-PEG-peptides were formed. Peptides included Arg-Gly-Asp-Ser (RGDS), Arg-Glu-Asp-Val (REDV), Tyr-Ile-Gly-Ser-Arg (YIGSR), and Arg-Gly-Glu-Ser (RGES). Shearexperiments were performed to examine EFCF rolling and adhesion on the hydrogelsurfaces. First, ECFCs were dissociated and suspended in flow media. Using aGlycotech parallel plate flow chamber, the ECFC cell suspension was then shearedover hydrogels containing peptides at shear rates of$20s^{-1}$, $40s^{-1}$, $80s^{-1}$, and $120s^{-1}$.Rolling velocity of ECFCs was shown to relate to shear rates and adhesionmaterial surface. Future studies will investigate the role of specific integrinbinding sites in ECFC rolling and capture under shear stress. Capture ofrolling ECFCs could be maximized by engineering biomaterials to incorporate theappropriate binding ligands. Our results provide a better understanding ofECFCs-material interactions under physiological shear stress and will aid inthe design of materials for stent coating and vascular grafts as well as forother intravascular applications.

See more of this Session: Biomaterial-Cell Interactions In Tissue Engineering
See more of this Group/Topical: Food, Pharmaceutical & Bioengineering Division

(437g) Cell Adhesion On Thermosensitive Bio-Inert Materials with a Tunable LCST

Kevin Ling[1], Erik Wischerhoff[2] and Lauren S. Anderson[1], (1)Chemical and Biomolecular Engineering, Lafayette College, Easton, PA, (2)Fraunhofer Institute for Applied Polymer Research, Potsdam-Golm, Germany

Smart, bioactive materials have recently emerged, capable of controlling or influencing biological processes at the molecular level. Thermoresponsive polymer brushes, such as those composed of poly (N-isopropyl acrylamide) (PNIPAM), allow for the gentle detachment of cells from their culture surface by simply changing the temperature below the lower critical solution temperature (LCST) of the polymer in water. At temperatures below the LCST, polymer chains are soluble and expand; above the LCST, chains are insoluble and collapse. Thus, these polymer brushes can switch wettability, between hydrophilic and hydrophobic states, altering cell adhesion.

In this work, we investigate changes in cellular phenotype in response to culture on thermosensitive bio-inert materials compared to tissue culture plastic (TCP). We present an alternative to PNIPAM: thermoresponsive polymer brushes with a tunable LCST prepared from random copolymers of 2-(2-methoxyethoxy)ethyl methacrylate (MEO_2MA) and oligo(ethylene glycol) methacrylate (OEGMA). We hypothesize that the advantage of MEO_2MA-OEGMA thermoresponsive brushes, namely the ability to modulate LCST, can be utilized to control cell adhesion. Changes in cell morphology and phenotype will be assessed compared to TCP and as a function of LCST.

To date, mouse fibroblasts were cultured on both thermoresponsive brushes and TCP for 48 hours and then harvested for gene expression analysis using real time RT-PCR. Genes regulating cell adhesion (FN1, Dusp2, RhoA), inflammation (IL-6), and apoptosis (Bcl-2, trp53) were evaluated. Cell morphology was assessed using phase contrast microscopy at 0, 6, 18, 24, 30, and 48 hours. No obvious morphological differences were observed in cells cultured on brushes or TCP. Gene expression analysis has revealed a small upregulation of all genes on thermobrushes relative to TCP. Specifically, fold changes ranged between 1-1.6 for all genes; future studies must determine if this change is biologically significant.

Thermoresponsive polymer brushes are biocompatible and non-cytotoxic. Bioengineering applications of modulating LCST to control cell adhesion include culture under serum free conditions and growth of cellular monolayers that can be released mechanically.

See more of this Session: Biomaterial-Cell Interactions In Tissue Engineering
See more of this Group/Topical: Food, Pharmaceutical & Bioengineering Division

(448a) Elucidating Human Transferrin-Receptor Interactions Using Linear Epitope Mapping

Pankaj Karande and Divya Chandra, Chemical and Biological Engineering, Rensselaer Polytechnic Institute, Troy, NY

Transferrin (Tf) is a carrier protein involved in the transport of iron (Fe3+). Tf binds to transmembrane transferrin receptors (TfRs) expressed on cell surfaces and transports Fe3+ by receptor mediated transcytosis (RMT). The Tf-TfR complex is an interesting system as it provides a natural biological pathway for delivery of Tf-bound cargo across cells and tissues. It has also been demonstrated that cancer cells over-expressing TfR can be selectively targeted with Tf-conjugated cytotoxic drugs. Understanding the interactions between Tf and TfR is therefore valuable for designing better therapeutic modalities for a broad range of physiological diseases. A few studies have highlighted putative amino acid residues involved in the formation of the Tf-TfR complex. The crystal structure of the N-lobe of Tf and the ectodomain of TfR has also been resolved independently. However, the crystal structure of the bound human Tf-TfR complex is still not available. We propose to use peptide microarrays for epitope mapping of the Tf-TfR complex to probe the interactions between Tf and TfR. Linear epitope mapping has been widely used as a technique for identifying antigenic regions in proteins in a fast, reliable high-throughput fashion. We have employed peptide microarrays of short peptides that are linear, continuous fragments of TfR to capture the interactions between TfR and its ligand. Microarrays printed with peptide fragments that span the entire sequence of TfR were tested against Tf to screen for regions on TfR that are involved in binding to Tf. Our results have confirmed prior findings but also highlighted the role of other important discontinuous regions on Tf involved in binding to TfR. We anticipate that our studies along with existing knowledge about the Tf-TfR system will further advance the understanding about this important transport and regulatory pathway.

See more of this Session: Drug Delivery I
See more of this Group/Topical: Food, Pharmaceutical & Bioengineering Division

(448b) Dendrimer-Based Nanotherapeutics for Targeted Post-Natal Therapy In Cerebral Palsy: Chemistry, Mechanism and In Vivo Efficacy

Rangaramanujam M. Kannan[1], Raghavendra Navath[2], Hui Dai[3], Bindu Balakrishnan[3], Roberto Romero[4] and Sujatha Kannan[5], (1)Chemical Engineering and Materials Science, Wayne State University, Detroit, MI, (2)Chemical Engineering, Wayne State University, Detroit, MI, (3)Pediatrics/ PRB, Wayne State University, Detroit, MI, (4)NICHD PRB/NIH, Wayne State University, Detroit, MI, (5)Pediatrics, NICHD PRB, Wayne State University, Detroit, MI

Neuroinflammation plays a key role in the pathogenesis of a wide range of neurodegenerative diseases such as cerebral palsy, autism, and Alzheimer's. Developing therapeutic approaches to treat inflammation in the central nervous system is a challenge, because: (1) inflammation and injury are often diffuse, precluding local delivery mechanisms (2) transport of nanoscale drug delivery vehicles, and even small drugs across the blood-brain-barrier is difficult to achieve (3) *postnatal* treatment of a *prenatal* injury to the brain is not expected to result in improvement in motor deficits. We hypothesized that a therapeutic strategy targeting activated microglia and astrocytes for sustained attenuation of ongoing neuroinflammation would improve outcomes in the post-natal period.

Maternal intrauterine inflammation resulting in microglial activation has been implicated in the development of periventricular leukomalacia and cerebral palsy. N-acetyl cysteine (NAC) is a drug that is currently being explored for the treatment of neuroinflammation in neonatal and perinatal applications. However, plasma binding of NAC significantly reduces the bioavailability requiring very high doses (100-300 mg/kg in animal models). Our translational research effort is focused on the preparation, in vitro, and in vivo evaluation of PAMAM dendrimer-based nanodevices for the delivery of therapeutics in general, and NAC in particular.

We have shown that PAMAM dendrimers, without any targeting ligands, can target neuroinflammation even upon intravenous administration. These dendrimers localize in activated microglia and astrocytes only in animals with cerebral palsy, but not in age-matched controls.

The nanodevices were evaluated extensively in a rabbit model of cerebral palsy. The biodistribution and efficacy of intravenously administered dendrimer-drug conjugates are compared with those of the free drug using a combination of tools. The efficacy is evaluated using a combination of neurobehavioral analysis, assessment of brain injury at the neuronal level, and at the DNA and RNA levels. Our studies show that dendrimer-drug conjugates are able to produce a dramatic improvement in motor deficits, in contrast to expectations that such improvements are not possible upon post-natal treatment. The dendrimer nanodevices were significantly better than even 100-fold higher free drug treatments, suggesting that these conjugates (~ 18000 Da) are able to cross the blood brain barrier and deliver the drug significantly better than free drug. Detailed mechanistic and efficacy studies will be presented.

See more of this Session: Drug Delivery I
See more of this Group/Topical: Food, Pharmaceutical & Bioengineering Division

(448c) Early T Cell Activation Events In the Context of Single Dose Vaccination with Polyanhydride Nanoparticles

Lucas Huntimer[1], **Kathleen A. Ross**[2], Amanda Ramer-Tait[1], Balaji Narasimhan[2] and Michael J. Wannemuehler[1], (1)Veterinary Microbiology and Preventive Medicine, Iowa State University, Ames, IA, (2)Chemical and Biological Engineering, Iowa State University, Ames, IA

Polyanhydrides based on sebacic acid (SA), 1,6-bis(*p*-carboxyphenoxy) hexane (CPH), and 1,8-bis(*p*-carboxyphenoxy)-3,6-dioxaoctane (CPTEG) have been shown to enhance protein stability with controlled release, a quality necessary for vaccine antigen delivery. Co-administration of soluble antigen with encapsulated F1-V antigen into polyanhydride nanoparticles leads to prolonged antibody titers of high avidity after a single dose administration that is protective in a *Yersinia pestis* challenge model at 40 weeks post vaccination.

Using transgenic T cells from OT I and OT II, in which the T cell receptors are specific for the model antigen ovalbumin (OVA), adoptively transferred into wild type C57BL/6-Thy1.1 recipients, antigen specific T cell expansion and contraction was identified in the secondary lymphoid tissues after vaccination. Draining lymph nodes and spleens of mice were excised at 1, 3, 5, 7 and 14 days post injection after a single soluble injection of antigen and T cell numbers as well as phenotype were quantified using flow cytometry. T cell proliferative cytokine IL-2 was also quantified using flow cytometry and ELISpot.

Data from single dose vaccinations with OVA in the transgenic antigen specific T cell (OT I and OT II) adoptive transfer model showed that T cell expansion peaked at 3 days in the draining lymph nodes and contraction of donor T cells follows at the later time points. IL-2 secretion correlates with the peak at 3 days post injection as well. Current studies aim to explore T cell expansion with the soluble antigen plus encapsulated antigen regimen established in previous work with *Y. pestis*. We hypothesize that the controlled release and immunostimulatory properties of polyanhydride nanoparticles encapsulating OVA will lead to an extended expansion of T cells in the draining lymph nodes, providing insight into the mechanisms behind the extended antibody titers.

See more of this Session: Drug Delivery I
See more of this Group/Topical: Food, Pharmaceutical & Bioengineering Division

(448d) Delivery of a Phospholipid Agent From a Silicone Hydrogel Contact Lens

William G. Pitt, Department of Chemical Enginering, Brigham Young University, Provo, UT, Yibei Zhao, Chemical Engineering, Brigham Young University, Provo, UT, Krystian X. Perez, Chemical and Biochemical Engineering, University of Iowa, Provo, UT, Jared L. Nelson, Global Research, CIBA VISION, Duluth, GA and John D. Pruitt, CIBA VISION, Duluth, GA

Silicone hydrogel contact lenses have been shown to quickly absorb dimyristoylphosphatidylcholine (DMPC) from an n-propanol solution, and then release it slowly into water or an artificial tear solution [1,2]. This paper presents an analysis of the transport phenomena associated with the uptake and release. To load a lens, it was carefully removed from the blister pack, lightly blotted on a lint-free tissue (KimWipe, Kimberly-Clark, Neenah, WI), and immersed in 5 mL of loading solution (0.15 % DMPC) for 30, 60 or 120 seconds. Then the lens was placed in a beaker of distilled deionized water (DD H_2O) at 35°C for 5 seconds, transferred to another beaker of DD H_2O for at least 5 seconds, rinsed with DD H_2O, and finally stored in a capped vial of DD H_2O for 2 to 20 minutes while the other lenses were prepared. To elute the DMPC from the lens, it was placed in water or artificial tear solution at 35°C and samples were taken at 0, 2, 4, 10, 24, 48 and 72 hrs.

Analysis of the data showed that amount of DMPC loaded into the lens increased with loading time and loading concentration. The DMPC eluted continuously from the lens during the 72 hours, but the elution rate decreased steadily with time. To determine if the elution might be diffusion controlled, the amount eluted was plotted versus the square root of time, and such a plot was linear for the first 10 hours, but deviated to slower elution at longer times. The first 10 hours of data was fit to a 1-dimensional model in Cartesian coordinates of diffusion from a semi-infinite solid with constant surface concentration. The model fit well and a diffusivity of 1.34×10^{-12} cm^2/s was estimated. This value is consistent with the reported diffusivities of other materials in silicone hydrogel contact lenses [3]. Given this diffusivity and a nominal lens thickness of 0.01 cm, the characteristic time constant of this system is $t = L^2/D_{AB}$ = 5200 hrs, where L is the half-thickness of the lens. Because this value is so much larger than 10 hrs (where deviation from the model occurs), the diffusion of DMPC from the lens must involve a more complex mechanism than the simple diffusion model. Subsequently we have applied more complicated models that include changing bulk concentrations during the elution, and the multiphase characteristic of silicone hydrogel contact lenses.

While these models may be of academic and perhaps esoteric interest to academicians and researchers who might design such lenses form DMPC delivery, the clinical significance of this research work is that DMPC can be easily loaded into a contact lens and then delivered slowly and consistently over the course of 10 hours. Since DMPC is a comfort product used in some types of eye drops designed to relieve "dry eye" irritation, the ability to deliver DMPC from a contact lens is of great interest to contact lens users who could obtain relief while wearing a contact lens.

References:

1. Pitt, W.G., Jack, D.R., Zhao, Y., Nelson, J.L., and Pruitt, J.D. "Loading and Release of a Phospholipid from Contact Lenses", *Optometry and Vision Science*, **88**(4), 502-506 (2011).

2. Pitt, W.G., Jack, D.R., Zhao, Y., Nelson, J.L., and Pruitt, J.D. "Transport of Phospholipid in Silicone Hydrogel Contact Lenses", *J. Biomat. Sci. Polym. Ed.*, **in press** (2011).

3. Kim, J., Conway, A. Chauhan, A., "Extended delivery of ophthalmic drugs by silicone hydrogel contact lenses" *Biomaterials*, **29**, 2259-2269 (2008).

See more of this Session: Drug Delivery I
See more of this Group/Topical: Food, Pharmaceutical & Bioengineering Division

(448e) Lytic Peptide-Polymer Conjugate with Enhanced Potency and Sensitivity to Tumor-Associated Protease

Ying Chau and Jieying Zhong, Chemical and Biomolecular Engineering, The Hong Kong University of Science and Technology, Hong Kong, Hong Kong

Lytic peptides are a group of cationic peptides capable of lysing cell membranes with quick action. They widely exist as natural antibiotic, and recently are being explored as anticancer agents for the unique property in overcoming drug resistance. However, there remain two major obstacles: 1) Relatively high local concentration and minimal length are required for effectiveness; 2) The lack of selectivity for cancer cells may cause serious side effect. To tackle these problems, we synthesized a polyvalent peptide-polymer conjugate, which significantly enhanced the anticancer potency of a short lytic peptide. The conjugate was further engineered with selectivity for cancer cells by targeting a tumor-associated protease.

Multiple copies of a hexopeptide KWKWKW ($(KW)_3$ for short) of mild membrane lysing activity were conjugated to 70kDa dextran with "click" chemistry to yield the polyvalent conjugate Dex-$(KW)_3$. All the reactions were monitored with NMR. The peptide loading was measured to be 0.05 mmol/g. The zeta potential of conjugate in pH 7.4 buffer was about 8 mV.The anticancer activity of polyvalent Dex-$(KW)_3$ was examined and compared with free $(KW)_3$ peptide. The potency of conjugate was significantly enhanced, with LC_{50} lower by three orders of magnitude than free peptide in two different cancer cell lines MCF-7 and PC-3. The same trend was observed in multi-drug resistant cancer cell line MES-SA/Dx5, suggesting the potential in overcoming MDR. In isothermal titration, the polyvalent conjugate showed stronger capability of membrane binding and disturbing than free peptide, which are attributed to the increased order and elevated local concentration of lytic peptides by polyvalent design.

To obtain a tumor targeted conjugate, an anionic quencher was fused to lytic peptide via a linker labile to MT1-MMP, a membrane protease actively involved in cancer progression. Since the positive charge mediates the first step of membrane binding and therefore critical to the activity of lytic peptides, the activity of conjugate should be neutralized until the removal of quenchers by MT1-MMP in tumor region. Preliminary data have shown the adjacent negative charge were capable of suppressing the cytotoxicity of conjugate. The protease specificity is being examined.

Novel lytic peptide-polymer conjugates with enhanced potency, tumor-targeted activity and improved pharmacokinetic properties were constructed. The work should contribute to the further development of an emerging class of promising anticancer agents.

REFERENCES

1. J. Zhong, Y. Chau, *Bioconjugate Chemistry,* 21, 2055-2064, 2010.

See more of this Session: Drug Delivery I
See more of this Group/Topical: Food, Pharmaceutical & Bioengineering Division

(448f) Development of Aptamer-Amphiphiles for Targeted Delivery to Colon Cancer Cells Expressing Fractalkine

Brett M. Waybrant[1], Elise A. Lamont[2], Timothy R. Pearce[3], Srinand Sreevatsan[2] and Efrosini Kokkoli[4], (1)Chemical Engineering and Materials Science, University of Minnesota, Minneapolis, MN, (2)Veterinary Population Medicine, University of Minnesota, St. Paul, MN, (3)Biomedical Engineering, University of Minnesota, Minneapolis, MN, (4)Department of Chemical Engineering and Materials Science, University of Minnesota, Minneapolis, MN

Delivering chemotherapeutic drugs directly to cancerous tissues would greatly improve patient outcomes and reduce debilitating side effects. To accomplish this, we developed an aptamer which binds to fractalkine with high affinity through the Systematic Evolution of Ligands by Exponential Enrichment, or SELEX process. Fractalkine is an exciting cell surface protein uniquely suited for targeted delivery. Fractalkine, expressed on tumor vasculature and certain cancers like colon cancer, is designed to capture leukocytes from circulation. The binding domain sits atop a mucin-like stalk extending the domain away from the cell surface allowing for maximum interaction with the circulating leukocytes or a targeted delivery vehicle. Aptamers are short single stranded nucleic acid sequences which bind to a target protein through hydrogen bonding, ring stacking, electrostatic and van der Waals forces. Aptamers have only recently been used as targeting moieties yet have many benefits over other ligands. Aptamers have binding affinities and specificities rivaling that of antibodies yet are a tenth the size, are easily obtained through automated DNA synthesis and have little to no immunogenicity.

The developed aptamer has high affinity for fractalkine as demonstrated through ELISA and cell based fluorescent assays. For use with liposome delivery vehicle, the aptamer was conjugated to a hydrophobic tail creating an aptamer-amphiphile which readily incorporates into the liposome lipid bilayer. The mild chemistry allows for conjugation of a wide range of hydrophobic molecules with little risk of degradation of the aptamer. Confocal microscopy and flow cytometry quantify the targeting capabilities of the aptamer functionalized liposomes to colon cancer cells expressing high levels of fractalkine while leaving healthy cells unaffected. This work demonstrates exciting possibilities for aptamer targeted delivery to fractalkine.

See more of this Session: Drug Delivery I
See more of this Group/Topical: Food, Pharmaceutical & Bioengineering Division

(448g) Molecularly Imprinted Extended-Wear Contact Lenses with Controlled Delivery of High Molecular Weight Comfort Molecules

Charles J. White, Matthew K. McBride and Mark E. Byrne, Biomimetic & Biohybrid Materials, Biomedical Devices, & Drug Delivery Laboratories, Department of Chemical Engineering, Auburn University, Auburn, AL

Up to 30 million contact lens wearers in the United States experience discomfort due to contact lens induced dry eye (CLIDE). Symptoms are typically treated through the application of eye drops containing macromolecular re-wetting agents, where a typical regime provides a cumulative mass delivery rate of 2.5 μg HPMC/day. Therapeutic soft contact lenses can alleviate CLIDE symptoms by slowly releasing comfort agent from the lens. We present the first extended wear silicone hydrogel contact lens engineered to controllably release of 120 KDa hydroxypropyl methylcellulose (HPMC) for extended periods. The lens was designed using lotrafilcon B, a commercial silicone hydrogel lens, and HPMC release rates were controlled by a molecular imprinting strategy. The release rate was tailored to deliver approximately 1,000 μ g of HPMC over a period of up to 60 days in a constant manner at a rate of 16 μ g/day under physiological flow rates. This rate is a 6 fold increase over the delivery from eye drops, and these lenses can release over the entire range of continuous wear, thereby maximizing comfort and convenience for the lens wearer. Release rates were significantly varied by controlling the imprinting effect. Adjusting the functional monomer to template ratio (M/T) to M/T values of 0, 0.2, 2.8, 3.4 corresponded to HPMC release durations of 10, 13, 23, and 53 days, respectively. These lenses had high optical quality and adequate mechanical properties for contact lens use. This work highlights the enormous potential of imprinting in the design and engineering of therapeutic contact lenses.

See more of this Session: Drug Delivery I
See more of this Group/Topical: Food, Pharmaceutical & Bioengineering Division

(453a) Removal of Selenite with Microbial Fuel Cells Utilizing Shewanella Oneidensis MR-1

Prithiviraj Chellamuthu, Department of Biological Sciences, University of Southern California, Los Angeles, CA, Mandy Ward, Department of Earth sciences, University of Southern California, Los Angeles, CA, Kenneth Nealson, Department of Earth Sciences, University of Southern California, Los Angeles, CA and Miriam Kronen, Department of Microbiology, University of Heidelberg, Heidelberg, Germany

Selenium is a trace element with important physiological roles in immune function and oxidative stress; however, at elevated levels it is known to be toxic to all forms of living organisms. Apart from naturally occurring selenium in rocks, selenium compounds are widely used in electronics, oil production and agricultural applications, which has resulted in the contamination of ground water. Selenite (+4 oxidation state) and selenate (+6 oxidation state) are the oxidized forms of selenium. Both are soluble, mobile and toxic in soil and water systems. Even at low levels of contamination these elements have deleterious effects on fauna, where bio-magnification plays a significant role in inducing toxicity. The EPA has set a limit of .05 parts per million (ppm) in drinking water. Hence remediation is important and a variety of methods have been proposed, including activated alumina, coagulation/filtration, reverse osmosis and bioremediation. Recent investigations have shown that bioremediation, which utilizes indigenous or engineered bacteria for the removal of selenium compounds from the soil and/or water systems could be a viable option.

Microbial fuel cell (MFC) systems utilize the catalytic ability of electrogenic bacteria, unique organisms that can convert energy from organic compounds to electricity through the respiration of solid electrodes. There has been much research into the applications of MFCs, including recent investigations that suggest that the cathode of a MFC can be used for bioremediation and the extraction of valuable metals. Bioremediation results from the ability of the bacteria to convert toxic compounds into more manageable, nontoxic forms. In a MFC, transformation of toxic compounds occurs when used as electron acceptors or oxidants in the cathode compartment. Researchers have shown the applicability of MFC bioremediation technology for the removal of various metals including copper, chromium, selenium and arsenic. In order to fully exploit the potential of this technology it is important to understand the electron transfer mechanism involved in this reduction/ remediation process.

Shewanella oneidensis MR-1 is a facultative anaerobic bacterium known for its versatility in using a variety of soluble and insoluble electron acceptors. This versatility stems from an array of multi-heme cytochromes which makes them an interesting candidate for bioremediation of toxic heavy metals such as uranium, selenium and chromium. Understanding respiratory electron transfer pathways will help in the design of engineered organisms for better removal of toxic contaminants from water. Using a MFC, we have identified the key cytochromes involved in the reduction of selenium under fuel cell conditions. We have also optimized the maximum concentrations for optimal removal rates without affecting the performance of the MFC. Using various electrochemical techniques, the operating parameters, such as internal resistance and maximum power output, have been determined. Our results indicate this technology could satisfy the stringent EPA water quality standards for removal of selenium.

See more of this Session: Fundamentals of Biomass Utilization
See more of this Group/Topical: Environmental Division

(453b) Reaction Mechanism of Catalytic Oxidative Depolymerization of Cellulosic Biomass

Hongfei Lin, Chemical and Materials Engineering, University of Nevada, Reno, NV

While biomass recalcitrance is one of the biggest challenges in biofuel industry, Nature has the key to unlock this puzzle. When a plant grows, the cell wall undergoes re-structuring. Recent biological studies found that hydroxyl radical (•OH), a very short-lived and strongly oxidative radical, is harnessed by the growing plant cells to loosen their cell walls by nonenzymatically cleaving the wall polysaccharides, although the physiological significance of this loosening mechanism is unclear. This nonenzymatic process sheds light on developing chemical oxidation methods for biomass deconstruction under mild reaction conditions. We are developing cost competitive catalytic biomass deconstruction strategies using aqueous phase partial oxidation (APPO) for releasing platform intermediates from solid biomass with a focus on the production of organic acid streams that can be readily upgraded and converted to commodity chemicals or infrastructure compatible liquid hydrocarbon fuels. The reaction mechanism of the APPO process has been investigated. A free radical route is presented to interpret the pathway of cleaving cellulosic biomass via the APPO process and the functionalities of oxidation catalysts are correlated to the reaction mechanism. By understanding the reaction mechanism, we are able to optimizing the catalytic system to realize high yield of the organic acids and to manipulate the selectivity of the APPO process.

See more of this Session: Fundamentals of Biomass Utilization
See more of this Group/Topical: Environmental Division

(453c) Infrared Spectroscopy and Thermo Gravimetric Analysis for Cellulose and Residue by Hydrothermal Decomposition

Lidong Geng and Zhongjun Zhu, Chinese Society for Environmental Sciences, Beijing, China

Abstract: Under the background of energy and resource shortage and the environment pollution, the utilization and research of biomass wastes as a kind of resource attracting worldwide attention. Hydrothermal reaction provides an effective method for the reutilization of biomass wastes. The disposal and reutilization of biomass wastes were realized by hydrothermal reaction in batch hydrothermal reactor. By using Fourier transform spectrum analysis, the microcosmic structure of micro crystallite cellulose and its products were compared. The pyrolysis behavior of cellulose was studied by thermo gravimetric analysis (TGA). The experimental results showed that the molecular structure of micro crystallite cellulose changed obviously and the decomposition of micro crystallite cellulose mainly occurred at the temperature range of 250-400°C. The weight loss(TG) and the rate of weight loss (DTG) shift to higher temperature region with the increase of heating-up rate. The thermal dynamics for biomass wastes was coincidence to the first order reaction dynamics by Coat-Redfern analysis. The FTIR and TGA analysis indicated that the hydrothermal decomposition of micro crystallite cellulose was composed by low temperature hydrolysis and high temperature pyrolysis.

See more of this Session: Fundamentals of Biomass Utilization
See more of this Group/Topical: Environmental Division

(453d) Landfill Gas Utilization In Gas Turbines: Experimental Results for Combustion of Landfill Impurities In Premixed Flames

Aydin Jalali[1], Mirmohammadyousef Motamedhashemi[2], Fokion N. Egolfopoulos[1] and Theodore Tsotsis[2], (1)Department of Aerospace and Mechanical Engineering, University of Southern California, Los Angeles, CA, (2)Mork Family Department of Chemical Engineering and Material Science, University of Southern California, Los Angeles, CA

Gas turbines and other conventional internal combustion engines can combust landfill gas (LFG), which is an important renewable energy source, to generate electricity. However, the corrosion and damage that can be caused by its impurities may reduce the operating life of these combustion engines when operating on LFG. One class of major impurities present in LFG is Siloxanes. In our research, the decomposition of Hexamethyldisiloxane (also referred to as L_2) present in LFG is experimentally investigated in order to provide a better knowledge of the way it is degraded during the combustion of LFG. Several experimental techniques were utilized in order to study premixed flat flames, and to gain a better understanding about the durability of L_2 in the flame environment and the kinetics of its decomposition. In our experiments, we have utilized the counterflow technique, which allows one to accurately probe and analyze the mechanisms leading to the formation of silica microparticles resulting from the decomposition of L_2. We also applied laser extinction techniques for the *in situ* monitoring of particle formation, as well as GC-MS techniques for measuring the kinetics of siloxane decomposition.

It was found that microcrystalline silica particles were generated during LFG combustion, leading to a rapid coverage of the surface of Ni/Cr sheets placed in the flame environment, and forming a light white layer of solid particles, whose surface was analyzed by atomic force microscopy (AFM). The size of these particles was then estimated using SEM/EDAX. The effect of L_2 concentration on its conversion along the flame was found for a number of different feed concentrations. Volume fractions of particles within the flame were also measured *via* the laser extinction method indicating a linear relationship between the concentration in the fuel and the corresponding volume of particles in the flue-gas. The temperature profile has been measured experimentally in order to identify the kinetics of burning of the siloxane compounds. This fundamental insight is important in terms of being able to accurately determine the maximum allowable siloxane content for biogas that is safe to use without leading to deposits and microparticle formation.

See more of this Session: Fundamentals of Biomass Utilization
See more of this Group/Topical: Environmental Division

(453e) New Modeling Framework of Enzymatic Hydrolysis of Cellulose

Yang Zhang, Deparment of Chemical Engineering, Michigan Technological University, Houghton, MI and Wen Zhou, Chemical Engineering, Michigan Technological University, Houghton, MI

Cellulose has long been recognized as a renewable carbohydrate source for human energy use. It is available in large quantities in plant biomass. In recent years, for both economic and environmental reasons, intense research efforts have been directed at the utilization of cellulose to produce short-chain soluble sugar oligomers, such as glucose or cellobiose, which are subsequently or concurrently metabolized by microorganisms to produce biofuels such as bioethanol. The hydrolytic conversion of cellulose into soluble, fermentable sugars is achieved by means of cellulase systems which contain three major types of enzymatic activities: (1) endoglucanases which cut beta(1,4) glucosidic bonds randomly at all internal bond sites of insoluble cellulose chains in solid substrates; (2) exoglucanases which cut bonds only at the ends of insoluble cellulose chains; and (3) b-glucosidases which hydrolyze soluble oligomer sugars into glucose. The endoglucanases and exoglucanases thus depolymerize long solid cellulose chains into short soluble sugars and this is believed to be the rate-limiting step of the cellulose hydrolysis process.

In this work, we develop a general framework for a realistic rate equation modeling of cellulose hydrolysis using noncomplexed cellulase. Our proposed formalism, for the first time, takes into account explicitly the time evolution of the random substrate morphology resulting from the hydrolytic cellulose chain fragmentation and solubilization. This is achieved by integrating novel geometrical concepts to quantitatively capture the time-dependent random morphology, together with the enzymatic chain fragmentation, into a coupled morphology-plus-kinetics rate equation approach. In addition, an innovative site number representation, based on tracking available numbers of beta(1,4) glucosidic bonds, of different site types, exposed to attacks by different enzyme types, is presented. This site number representation results in an ordinary differential equation (ODE) system, with a substantially reduced ODE system size, compared to earlier chain fragmentation kinetics approaches. This formalism enables us to quantitatively simulate both the hydrolytically evolving random substrate morphology and the profound, and heretofore neglected, morphology effects on the hydrolysis kinetics. By incorporating the evolving morphology on an equal footing with the hydrolytic chain fragmentation, our formalism provides a framework for the realistic modeling of the entire solubilization process, beyond the short-time limit and through near-complete hydrolytic conversion.

See more of this Session: Fundamentals of Biomass Utilization
See more of this Group/Topical: Environmental Division

(453f) Theoretical Analysis of Biodiesel Production Under High Temperature and Pressure Conditions

Fernando Olmos, Chemical and Biomolecular Engineering, UCLA, Los Angeles, CA and Vasilios Manousiouthakis, Chemical Engineering Department, University of California at Los Angeles, Los Angeles, CA

Typically, biodiesel is synthesized by the transesterification of vegetable oil in heterogeneous fashion, i. e. in multiple liquid phases since the reactants, methanol and vegetable oil, are immiscible. In this work, we propose that the synthesis of fatty acid methyl esters (FAME), which after purification become biodiesel, take place in a solution where methanol, vegetable oil, and FAME are all homogeneously dissolved. Carrying out the reaction scheme in this manner eliminates any mass transfer effects between the oil and methanol phases. In addition, the overall reaction rate is further enhanced by the gravity based removal of the glycerol final product from the reaction mixture. The proposed process is modeled as a multiphase reactor with three phases (vapor, FAME, glycerol) in equilibrium with each other. Reactions are only considered to be carried out in the FAME phase. The mathematical model constitutes a Differential-Algebraic-Equation (DAE) problem that is solved through the implementation of a hybrid Runge-Kutta/Newton-Raphson solver. Several temperature and pressure conditions are considered to explore the operational characteristics of this process. Comparisons with the original heterogeneous biodiesel production process are carried out.

See more of this Session: Fundamentals of Biomass Utilization
See more of this Group/Topical: Environmental Division

(454a) Multiparametric Characterization of Actively-Secreting B Cells Reveals Dissimilarities Between HIV-1 Specific Responses In Blood and GALT

Adebola O. Ogunniyi[1], Douglas Kwon[2], Navin Varadarajan[3], Bruce D. Walker[2] and J. Christopher Love[4], (1)Massachusetts Institute of Technology, Cambridge, MA, (2)AIDS Research, Ragon Institute of Massachusetts General Hospital, Massachusetts Institute of Technology, and Harvard, Charlestown, MA, (3)Dept. of Chemical & Biomolecular Engineering, University of Houston, Houston, TX, (4) Department of Chemical Engineering, The David H. Koch Institute for Integrative Cancer Research, Massachusetts Institute of Technology, Cambridge, MA

Traditional strategies for producing antibody-based vaccines have failed against the human immunodeficiency virus type 1 (HIV-1), hence, more deliberate approaches at immunogen design are necessary if a protective therapy is to be developed. Some persons infected by the virus are known to develop potent neutralizing antibody responses, but how these responses develop is still poorly understood. Furthermore, the identification of effective antibodies from such individuals remains challenging. Existing analyses of B cells from infected individuals have focused largely on the dormant memory B cell population, ignoring contributions from actively-secreting B cells (ASCs). In this study, we profile the repertoire of activated B cells in multiple HIV-infected patients for surface-expressed proteins, immunoglobulin isotype, and the ability of secreted antibodies to bind HIV-1 BaL, a tier 1 HIV-1 strain. Using microengraving technology, blood and tissue samples from multiple patients were analyzed. This talk will describe a comparative analysis of the HIV-1 specific B cells present in the different compartments.

See more of this Session: High Throughput Technologies
See more of this Group/Topical: Food, Pharmaceutical & Bioengineering Division

(454b) Quantitative Proteomic Discovery of the Epigenetic Mechanisms That Control Human Cytomegalovirus Infection

Peter A. DiMaggio Jr.[1], Christine M. O'Connor[2], Thomas Shenk[1] and Benjamin A. Garcia[1], (1) Department of Molecular Biology, Princeton University, Princeton, NJ, (2)Department of Molecular Genetics, Lerner Research Institute, Cleveland, OH

Human cytomegalovirus (HCMV) is a large, double strand DNA virus that is a member of the herpesvirus family. HCMV establishes a life-long, persistent infection in its host that remains asymptomatic in healthy individuals. However, the virus can cause severe morbidity and mortality upon reactivation in immunocompromised individuals. To date, no preventative HMCV vaccinations exist and treatment options suffer from low bioavailability, toxicity, and viral resistance. Several studies have shown that HCMV viral DNA associates with host histones and undergoes chromatinization upon entering the nucleus of infected cells and that changes in chromatin structure are strongly correlated with changes in viral gene expression. Thus, an understanding of the epigenetic mechanisms driving viral gene expression and replication would benefit the design of novel therapeutics.

In this work, we have utilized liquid chromatography tandem mass spectrometry (LC-MS/MS) to identify and quantify the dynamic histone post-translational modifications (PTMs) expressed during HCMV lytic infection (TB40/E BAC-derived clinical strain) in primary fibroblasts. Specifically, we examined the changes in histone modifications over a 96 hour time course to provide a sufficient sampling of snap shots corresponding to the immediate early (IE), early (E), and late (L) stages of viral infection. Over 100 Bottom Up LC-MS/MS experiments were conducted and quantitative analyses of more than 60 modified histone H3 and H4 isoforms during the time course of infection revealed significant up- and down-regulation of several histone post-translational modifications, including a substantial increase in histone H3 K79 dimethylation (H3K79me2). Previous studies have associated K79 methylation with the recruitment of DNA damage repair machinery. Consistent with this observation, HCMV infected cells had increased levels of chromosomal damage relative to mock infected cells. Additionally, we found that the RNA levels of DOT1L, the methyltransferase specific for "writing" H3K79me2, had also increased over the course of HCMV infection.

Based on these findings, we investigated the functional role of this H3K79me2 modification on HCMV growth by shRNA knock-down of DOT1L. Two independent DOT1L knock-down stable cell lines expressing different shRNA sequences were generated in primary fibroblasts. We confirmed the knock-down of DOT1L via both RNA analysis and LC-MS/MS. We then assessed viral replication in these DOT1L knock-down cells and found that HCMV growth was reduced. To determine the stage in viral growth that this post-translational histone modification was predominantly impacting, we examined viral protein expression/localization as well as viral DNA replication. Although viral proteins were expressed and localized similarly across the stable cell lines, we found that viral DNA replication in the knock-down cells was impaired when compared to the scramble knock-down cells. Current experiments include genome-wide microarray, ChIP-Seq and heavy methyl SILAC experiments to further characterize the systems-level function of the histone PTMs during HCMV infection.

See more of this Session: High Throughput Technologies
See more of this Group/Topical: Food, Pharmaceutical & Bioengineering Division

(454c) Parallel Analysis of Transcription Factors

Betul Bilgin and S. Patrick Walton, Chemical Engineering, Michigan State University, East Lansing, MI

Gene expression is regulated by transcription factors (TFs), proteins that bind specific sites of chromosomal DNA to enact the response to intracellular and extracellular stimuli. TFs work alone and in concert with other TFs and other proteins to generate the phenotypes that are observed at the cellular level. Given that there are roughly 3,000 human TFs, the potential network of responses and interactions is considerable. However, the most common means for describing activation of a pathway by a stimulus is the measurement of the activation of a single TF. This limited focus potentially ignores important events that contribute to the cellular response. Thus, the measurement of multiple TFs in parallel would be advantageous.

Several methods to analyze multiple TF s in parallel exist. However, these are limited by the number of TFs that can be measured in one assay, relatively poor sensitivity, and limited flexibility. We have sought to develop a new, scalable, flexible, and sensitive approach to analysis of TF levels. Based on magnetic bead affinity separation, we have devised an approach to measure TF levels, beginning with human TFs, NF-kB and Ap1, in parallel with ~10-fold improved sensitivity over existing approaches. Further results and challenges will be described.

See more of this Session: High Throughput Technologies
See more of this Group/Topical: Food, Pharmaceutical & Bioengineering Division

(454d) Microfluidics-Based Quantitative Screening of C. Elegans with Adaptive Algorithm

Hyewon Lee[1], Matthew Crane[2] and Hang Lu[1], (1)School of Chemical & Biomolecular Engineering, Georgia Institute of Technology, Atlanta, GA, (2)Interdisciplinary Program in Bioengineering, Georgia Institute of Technology, Atlanta, GA

Forward genetic screening based on mutagenesis is a powerful tool to identify components in pathways in model organisms such as the nematode *C. elegans*. This screening usually requires imaging a large number of animals to reach saturation and/or to overcome the rare chance to find mutants. Conventional manual microscopy is usually labor-intensive and time-consuming. Recent development of microfluidic systems has allowed for increases in the throughput of experiments with minimized human intervention. However, most devices are so complicated that their fabrication is hard, especially for non-experts. Here, we present a simple single-layer microfluidic device for imaging and sorting *C. elegans*. Worms are loaded by pressure-driven flow and positioned in the imaging zone by controlling side valves. Once worms are loaded, they are immobilized by low temperature of side cooling channel. After complete immobilization, worms are imaged and then they can be sorted as either wild-type or mutant. This operation is semi-automated by developed program in MATLAB. This yields a throughput of 100-500 worms/hr, 1-2 orders of magnitude higher than conventional approach. In addition to microfluidics, here we present the simple algorithm that finds target mutants. Target mutants are sorted usually based on the changes of certain morphological markers, such as GFP transcriptional markers. This selection in some scenarios is straightforward and many genetic screens are already performed with qualitative analysis of gene expressions. However, some changes are too subtle to be identified by eyes even at high magnification. Furthermore, since gene expressions may be heterogeneous due to stochasticity, and certain environmental fluctuations such as culture temperature and bacterial food conditions can change gene expressions from day to day, a simple threshold of GFP intensity is insufficient in selecting putative mutants with an acceptable small false positive rate. To look for mutants with subtle changes, we used the adaptive quantification algorithm. First, fifty mutagenized worms are imaged, and assuming the rate of finding relevant mutant is low (e.g. below 2%), the minimum value of target reporter expression in this group is set as a screening criterion. Putative mutant worms having lower expression level compared to the sorting criterion were sorted. The sorting criterion is updated after each 50 additional worms such that a reasonable sorting rate is maintained. This algorithm is easy to implement and minimize efforts for verifying selected targets. We use this system and the algorithm to quantify the regulation of tryptophan hydroxylase, a key enzyme for serotonin synthesis to understand how environmental signal can be integrated for various metabolic adjustments. Using this high-throughput and quantitative method, we found several putative mutants, which would have been difficult and taken longer to find.

See more of this Session: High Throughput Technologies
See more of this Group/Topical: Food, Pharmaceutical & Bioengineering Division

(454e) Temporal Gene Expression Profiling In Live Cell Array: Application In Understanding Mesenchymal Stem Cell Differentiation towards Myogenic Lineage

Janhavi Moharil, Pedro Lei, Jun Tian and Stelios Andreadis, Chemical and Biological Engineering, University at Buffalo-SUNY, Amherst, NY

Livecell microarrays offer a nondestructive way to monitor gene expression unlikemost other high throughput methods such as RT-PCR or cDNA microarrays. However theysuffer from spot-to-spot variations in signal because of differences in virustiter and transduction efficiencies. To overcome this limitation, a noveldual-promoter lentiviral vector (LVDP) carrying human PGK promoter driving theexpression of DsRed serving as an internal control and the promoter of interestdriving the expression of ZsGreen was designed in our laboratory. Herein, wedemonstrate the application of LVDP in high throughput monitoring of real timegene expression during myogenic differentiation of human Bone Marrow derivedMesenchymal Stem Cells (hBM-MSCs). Specifically we cloned a library of 14promoters and 18 response elements upstream of ZsGreen in the LVDP andmonitored the promoter activity using continuous fluorescence imaging. Weobtained the dynamic expression profiles of these promoters and responseelements under growth and differentiation conditions for a period of severaldays (Fig. 1). The gene expression dynamics revealed that treatment withTGF-β and Heparin stimulated a 2 to 4-fold increase of many early myogenicmarkers such as α-SMA, SM22, and Myocardin with a $t_{1/2}$ of ~24 hourswhile late markers such as SM-MHC and Calponin exhibited only 1.2-1.5-foldincrease, suggesting partial myogenic differentiation. Interestingly, thecombination of TGF-β and BMP4 showed higher expression of SM-MHC at 7 dayspost treatment, suggesting a synergistic effect of these two factors onmyogenic differentiation. The promoter/RE activity data correlated well withmRNA expression as assessed by RT-PCR and with protein levels as demonstratedby immunostaining. Collectively our data demonstrates a novel method tocapture real time gene expression dynamics during myogenic differentiation inlive cells. In contrast to standard gene expression monitoring methods, our resultsprovide rich dynamic information of geneexpression over a period of several days.We are currently using this set of dynamic datain mathematical models to decipher the generegulatory networks at work during stem cell differentiation.

Figure 1: Temporal gene expression profiles. The heatmap of gene expression dynamics on treatment of hBM-MSC with TGF-β and Heparin is shown. The x axis represents the time in hours and the y axis shows the response elements or promoters. Dendrogram is computed by hierarchical clustering.

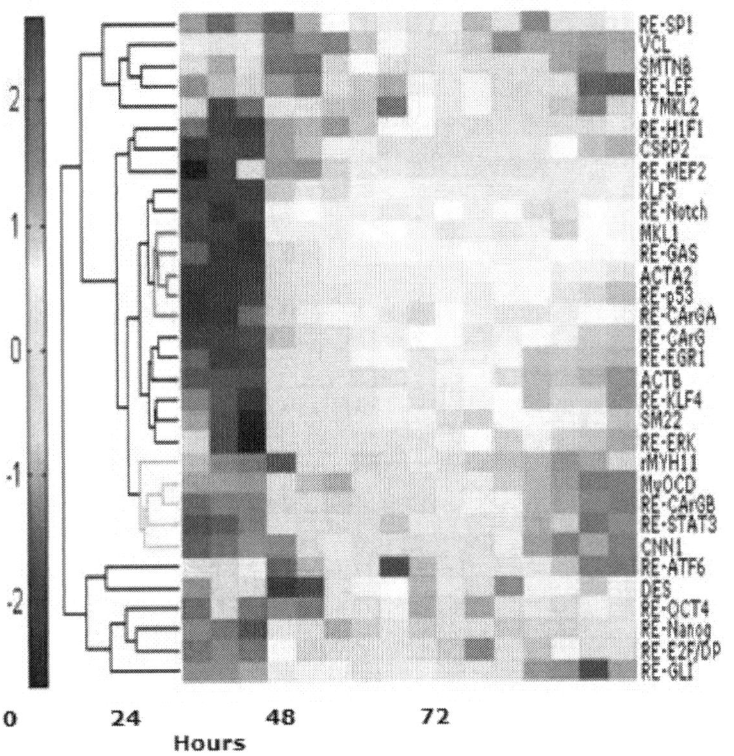

See more of this Session: High Throughput Technologies
See more of this Group/Topical: Food, Pharmaceutical & Bioengineering Division

(454f) Use of 3D Free Form Fabrication for Design of Functional Human Tissues

Gurtej Singh[1], J. P. Trasatti[2], Vivian Lee[3], Guohao Dai[3] and Pankaj Karande[1], (1)Chemical and Biological Engineering, Rensselaer Polytechnic Institute, Troy, NY, (2)Chemistry, Rensselaer Polytechnic Institute, Troy, NY, (3)Biomedical Engineering, Rensselaer Polytechnic Institute, Troy, NY

Skin forms a first line of defense against pathogens, environmental insults and trauma. In addition to being a formidable physical barrier, skin hosts a highly sophisticated and robust network of immune surveillance agents. The high accessibility of skin and its inherent immunological activity make it an attractive interface for vaccination. Transdermal vaccination is especially advantageous since it can induce cellular (cytotoxic T cell based) immunity as well as humoral (B cell or antibody based) immunity to ensure adequate protection against an infection. Contemporary literature in skin immunobiology has highlighted key immune components and their roles in the cutaneous immune network. To develop effective transdermal immunotherapeutics, a physiologically relevant, multi-compartmental, integrated immune model of the skin is needed. This study describes the design of an in vitro, multi-compartmental, three-dimensional, Human Immunocompetent Skin Tissue Experimental Model (HISTEM) that adequately captures the innate immune response in the dermal/epidermal compartment as well as the adaptive immune response in the skin draining lymph nodes. A novel cell-printing technology called three-dimensional free form fabrication (3DFFF) is employed for the design of HISTEM. 3DFFF is a powerful and flexible technology that allows printing of multi-cellular, multi-layer skin tissue structures in a high-throughput manner. HISTEM has been tested for its ability to initiate innate and adaptive immune responses when challenged with known irritants, allergens and antigens. The long term goal of this study is to build upon the successful development of HISTEM, and focus on rational design of novel transdermal vaccines and immunomodulators. Skin-patch based transcutaneous vaccines are expected to make a significant impact on mass vaccination programs as well as limit the accidental spread of infections among healthcare providers through needle-sticks. Beyond use as prophylactics against systemic infections, transdermal immunotherapeutics have the potential as novel interventions in genetic and acquired immunological disorders of the skin, cutaneous malignancies and immunosuppressive therapies for patients receiving skin transplants.

See more of this Session: High Throughput Technologies
See more of this Group/Topical: Food, Pharmaceutical & Bioengineering Division

(467a) Modeling the Residence Time Distribution In a Batch Fermentor: Comparison of CFD Prediction with Experiment

Abhishek Dutta[1], **Rita Lencastre Fernandes**[2], Anna-Lena Heins[3], Anna Eliasson Lantz[3], Anker D. Jensen[2], Krist V. Gernaey[2] and Ingmar Nopens[1], (1)Biomath, University of Ghent, Ghent, Belgium, (2) Department of Chemical and Biochemical Engineering, Technical University of Denmark, Kgs. Lyngby, Denmark, (3)Department of Systems Biology, Technical University of Denmark, Kgs. Lyngby, Denmark

Continuousstirred tank reactors (CSTR) are ubiquitously used as fermenters in the biotechnologicalproduction of enzymes, pharmaceuticals, metabolites, antibodies, etc. They arealso used in the bioprocess industry for the production of bio-ethanol, bio-hydrogenand even in mammalian cell culture. In most cases, a scale-up of these processesesis necessary to yield commercial success and viability. As with most scaling-upoperations, mixing is also a crucial parameter in a biological process and isoften not ideal when operating at industrial scale. Poor mixing in a CSTR may becaused by gas by-passing and back-mixing phenomena leading to the formation of deadzones where, in the case of a bioreactor, cells will experience a limited supplyof nutrients and oxygen. In large-scale bioreactors gradients of e.g.substrate, oxygen and pH arise as a result of non-ideal mixing, resulting in a spatiallyheterogeneous extracellular environment. This fact has been suggested to be themain reason behind lower product yields, but higher cell viability, in comparisonto lab scale cultivation (Enfors et al., 2001)

Understanding bioreactor hydrodynamics in general, andnon-ideal mixing in particular, is a key problem to understand the inherent bioprocess.The Residence Time Distribution (RTD) is one such way to characterize thenon-ideal mixing in a CSTR (Choi et al., 2004). Toestimate the RTD, a glucose tracer study has been performed experimentally in aSartorius CPlus batch bioreactor, havinga working capacity of 10 litres, and the results werecompared using Computational Fluid Dynamics (CFD) simulations. Several caseswith varying impeller speed have been considered to study the effect of mixingphenomena on the RTD. The mean and variance of the distribution are analyzed forthe extent of the deviation of the reactor behavior from its ideal state.Furthermore, the sensitivity of the sliding mesh is tested by varying theposition of the interface zone. This would eventually contribute to building anoptimized model of the bioreactor, and will be used for integration withbiological models aiming at improving performance prediction in the cases ofpoorly-mixed reactors.

<u>Reference</u>

[1]Enfors, S.O., Jahic, M., Rozkov, A., Xu, B., Hecker, M., Jrgen, B. et al. (2001). Physiologicalresponses to mixing in large scale bioreactors. J. Biotechnol., 85, 175-185.

[2] Choi, B.S., Wan, B., Philyaw, S.,Dhanasekharan, K., Ring, T.A. (2004). Residence Time Distribution in a stirredtank: comparison of CFD predictions with experiment. Ind. Eng. Chem. Res., 43,6548-6556.

See more of this Session: Mixing In the Production of Biopharmaceuticals
See more of this Group/Topical: North American Mixing Forum

(467b) Mixing It up: Probing Cell Culture Bioreactors Using CFD Modeling

E. Todd Sorensen[1], Pratik Jaluria[1], Luke Munholand[2] and Derek Adams[1], (1)Upstream Development, Alexion Pharmaceuticals, Cheshire, CT, (2)Fluent Inc., Lebanon, NH

Process scale-up in the biopharmaceutical industry is complex and empirical, because of the presence of living cells used to manufacture the desired product. Experimentation is increasingly difficult as the scale increases, due to regulatory and cost constraints, so there is a heavy reliance on scaling calculations and small-scale experiments. Simplified mixing experiments are often conducted in conjunction with standard bulk calculations to derive physical characteristics of each scale, such as volume-to-volume-per-minute (VVM), Power to Volume (P/V), Tip Speed, mass transfer coefficient (k_La), maximum shear stress, and recirculation time.

We scaled up one of our processes from the 2L to the 10,000L bioreactor initially by maintaining the VVM and P/V ratios, which resulted in a 30% drop in Maximum Viable Cell Density (VCD) with a 35% drop in productivity from the 2L to the 10,000L scale. Since we have chosen the 2L bioreactor to be our main development scale we wanted to increase our understanding about the differences between the 2L and 10,000L scales. In order to assist us, we are beginning to use Computational Fluid Dynamics (CFD) as an additional tool, allowing us to explore the internal fluid dynamics within each tank with more spatial and temporal detail. Our relatively simple CFD models are analyzed against directly-comparable physical experiments (where possible) to validate the results.

Thus far, we have determined that mixing times of liquids at large scale were 8 times longer and exhibited relatively long-lived regions of high concentrations of added components. This result is important for our processes, where significant quantities of non-physiological solutions (e.g., concentrated nutrient feeds and base) are routinely added to the reactor. We believe that the slower mixing in the large-scale bioreactors permits zones of different concentrations of components ("mini – environments") to exist when liquids are added to the bioreactor. These "mini-environments" certainly contain pH gradients, may exhibit hypoxic characteristics, likely have altered Carbon Dioxide levels, and/or contain elevated Osmotic environments. This indicates that scaling biological processes may require more than just maintaining bulk physical characteristics but instead rely on scaling mini-environments to allow for adequate gas exchange and liquid-liquid mixing. Overall, this work highlights our initial phase of development to enhance biological process scale-up using customizable CFD software.

See more of this Session: Mixing In the Production of Biopharmaceuticals
See more of this Group/Topical: North American Mixing Forum

EFFECT OF DIFFERENT AGITATION SYSTEMS AND PROTOCOLS ON THE GROWTH OF *NICOTIANA TABACUM* CELLS: COMPARISON OF GROWTH KINETICS IN ERLENMEYER FLASKS, CONVENTIONAL STIRRED TANKS, AND ECCENTRICALLY AGITATED TANK BIOREACTORS

Josefina Castillo-Reyna[1] and Mario M. Alvarez[2]

[1] *Departamento de Ingeniería Química y Biotecnología, Tecnológico de Monterrey at Estado de México, Estado de México, México*

[2] *Centro de Biotecnología-FEMSA, Tecnológico de Monterrey at Monterrey, Monterrey, NL. México*

Erlenmeyer flasks and stirred tank bioreactors are the most commonly used bioreactors for the lab culture of plant cells. We compare the performance of these systems against and novel eccentrically agitated stirred tank bioreactor for the particular case of culture of *Nicotiana tabacum* cells. We measured the specific growth rate, total cell density, cell viability and integrity/morphology over time of tobacco cells in 300 mL Erlenmeyer flasks, and 1L conventional and eccentrically stirred bioreactors. We observed that, under some agitation regimes, concentrically agitated systems inflicted damage to the tobacco cells, particularly at high cell densities. At these conditions, higher specific growth rates and cell densities were observed in Erlenmeyer flasks and eccentrically agitated bioreactors. Our results suggest eccentric bioreactors offer with a more benign and scalable growth environment for the culture of shear sensitive plant cells.

Key words: tobacco cells, stirred tank bioreactor, plant cell culture, mechanical stress, mass transfer

INTRODUCTION

The culture of plant cells in bioreactors has been suggested as a promising platform to produce high added-value bioproducts such as recombinant enzymes or specialized biopharmaceuticals. Plant cell culture requires of relatively inexpensive culture media, particularly as compared to animal cell culture. In addition, molecular techniques to transform plant cells, for example *Nicotiana tabacum*, have achieved a good degree of maturity: for example, the expression of molecules of high value in health (Magnuson et al., 1998; James et al., 2000; Hong et al., 2002; Judge et al., 2004). At laboratory scale, plant cells are routinely cultured in Erlenmeyer flasks. At an intermediate scale, airlift reactors for low to moderate cell densities, and vessels with paddle–type or helical-ribbon impellers for high cell density systems have been suggested as reactors for plant cell culture (Doran, 1999; Doran, 2000).

Research work on the area of reactor engineering for plant cells is limited. In general, suspension cell culture techniques in bioreactors are restricted to Erlenmeyer flasks, conventional stirred tanks, and air-lift bioreactors. In particular, limited kinetic data is available on the culture of tobacco cells in instrumented bioreactors.

A number of reviews have dealt with the choice of bioreactors for plant cell suspension culture (Doran *et al.*, 1993) and most emphasis has been placed on modifications of the conventional STR, with bubble aeration, employing a variety of impeller designs. However, equipment normally employed for microbial cultivation may not be immediately applicable to plant cells, due to differences between the characteristics of both individual cells and suspensions of the respective systems (Kieran *et al.*, 1997) that have direct implications for the design and scale up of suspension culture systems. Plant cells are large, and when they are exposed to turbulent shear fields where the eddy size approaches the cell size, the cells can be exposed to a twisting motion that can damage them. Lower levels of shear appear to affect cell surface receptors and nutrient transport. Reactors with high shear must be avoided. Stirred tanks designed for the culture of bacteria are not good choices, but modified stirred tanks can be suitable (Paul *et al*., 2004).

Here, we study the effect of three different agitation conditions on the specific growth rate, mortality rate, morphology, and viability of *Nicotiana tabacum* cells cultured in concentrically and eccentrically agitated instrumented bioreactors (1 L effective volume). Growth results from Erlenmeyer flask culture are also presented for comparison.

MATERIALS AND METHODS

Biological Material and culture media

Tobacco cells *Nicotiana tabacum* 1 (NT-1) were used in this project. Cells were obtained from *calli* kindly donated by Dr. Paula G. Miller from the Chemical and Biomolecular Engineering Department at Cornell University, Ithaca, N.Y. Tobacco cell suspension culture and *calli* were maintained in **NT1** medium containing 4.3 g Murashig & Skoog basal salts, 30 g sucrose, 50 ml MES (stock 20x), 10 ml B1-Inositol (stock 100x), 3 ml Millers I (stock) and 2.2 ml 2,4-D (stock) in deionized water and adjust pH to 5.7 with KOH. For solid medium include 8 g of agar (Mayo *et al*., 2006).

NT1 Cell Growth and maintenance in solid media

Tobacco cell suspension culture and *calli* were maintained in **NT1** medium containing 4.3 g Murashig and Skoog basal salts, 30 g sucrose, 50 ml MES (stock 20x), 10 ml B1-Inositol (stock 100x), 3 ml Millers I (stock) and 2.2 ml 2,4-D (stock) in deionized water. The pH of the medium was adjusted to 5.7 with KOH. For solid medium, 8 g of agar were added (Mayo *et al*., 2006) to this basic formulation. *Calli* were grown in Petri dishes covered with aluminum foil, placed in the incubator to 25° C and kept in darkness (Loyola-Vargas & Vazquez-Flota, 2006). After three incubation weeks, a re-culture was done. Antibiotic solutions were prepared and sterilized by filtration. They should be added after autoclaving when the medium has cooled to 50 °C (Mayo *et al*., 2006).

Cell culture experiments in flasks

Culture experiments of tobacco cells were conducted in Erlenmeyer flasks under different inoculums concentrations (1 g/L and 5 g/L). Erlenmeyer flasks of 1 L of capacity with 500 ml NT1 medium were used. Flasks were inoculated by adding 1 g/L or 5 g/L of *callus* previously established in agar plates (Mayo *et al*., 2006). Erlenmeyer flasks were inoculated with 12-16 days old suspension cultures. All flasks were covered with aluminum foil to keep them in darkness. Flasks were placed in an orbital shaker at 150 rpm and a temperature from 25 to 27°C (similar than in Loyola and Vazquez, 2006). A medium change was performed weekly. Flasks were sampled for analytical measurements by taking a volume of 5-ml daily. Specific growth rates at the exponential phase, maximum cell densities, and mortality rates were calculated for each Erlenmeyer experiment.

Cell culture in bioreactors

Culture experiments were conducted in 350 mL instrumented concentric stirred tank bioreactors (from Infors®, Switzerland) under 100, 200, and 300 RPM. Initial cell concentrations of 5 g/L were used. Additional culture experiments were performed in a 2L instrumented bioreactor (from Applikon®, Netherlands). The reactor multiport stainless steel head plate was modified according to a design by

Alvarez *et al.* (2008) to allow for concentric and eccentric agitation. Temperature was set at 26°±1°C in all bioreactor experiments.

Evaluation of cell biomass, viability and integrity

Cell biomass was determined by Fresh Cell Weight and dry cell weight evaluation from well mixed samples taken at different culture times (see for example Loyola & Vazquez, 2006). Due to the small sample size, freeze drying was used to dry. Packed Volume was evaluated by taking 5 ml of sample in a 15-ml conic centrifuge tube. Viability determinations were made daily through a staining using Fluorescein Diacetate (FDA) and Propidium Iodide (PI) Techniques. A drop of sample was placed in a slide and observed in 10X at the microscope. Cell counting was done by triplicate in different fields. An average was obtained and the results were expressed in percentage of viability (Loyola & Vazquez, 2006). For FDA observations filter no. 46 carl zeiss was used. Propidium Iodide is an intercalating agent and a fluorescent molecule used to stain DNA.

Carbohidrate determination

Sucrose, Glucose and Fructose concentrations were monitored during cell culture by HPLC from 5 mL supernatant samples. Carbohydrate quantification was performed in HPLC Waters model W1515 with a refractive index detector model W2414, using a SGE EXIL AMINO column in an isocratic run. The chromatographic conditions were: flow of 1 mL/min; mobile phase of Acetonitrile-Water 80: 20; column temperature of 30°C; detector Temperature of 30°C; and injection Volumen: 20 µL. Calibration Curves were made at several concentrations: 2,4,6,8 and 10 mg/mL for each carbohydrate. The standards used for calibration curves were : D-(-)- fructose > 99% (Sigma CAS 57-48-7) ; D-(+)-glucose > 99.5% (Sigma CAS 50-99-7) ; and sucrose >99.5% (Fluka CAS 57-50-1).

RESULTS AND DISCUSSION

Growth kinetics of tobacco cells

The growth behaviour of tobacco cells during exponential phase, both in Erlenmeyer flasks or instrumented bioreactors can be easily characterized using simple first order models. Figure 1 present a typical growth curve based on total biomass density (v.gr. g of fresh cell weight (FCW/L) and g of dry cell weight (DCW/L)) from tobacco cells cultured in Erlenmeyer flasks.

The curve exhibit a typical behavior, with a relatively large lag phase, and an exponential growth period followed by a slow decay in total biomass. Specific growth rates at the exponential phase, maximum cell densities, and mortality rates were calculated for each experiment. For the exponential growth phase, a simple first order model with respect to biomass (v.gr. $d[X]/dt= \mu [X]$) fits the data with correlation coefficients greater than $r^2 = 0.95$. In this equation, [X] is the total biomass (the model holds for both wet and dry cell biomass data), t is time and μ is the specific cell growth rate. Similarly, for the biomass decay phase, a model of the type $d[X]/dt= - k_m [X]$ describes the observed behavior. Therefore, both exponential growth and cell death can be adequately and simply described by only one parameter, μ and k_m respectively.

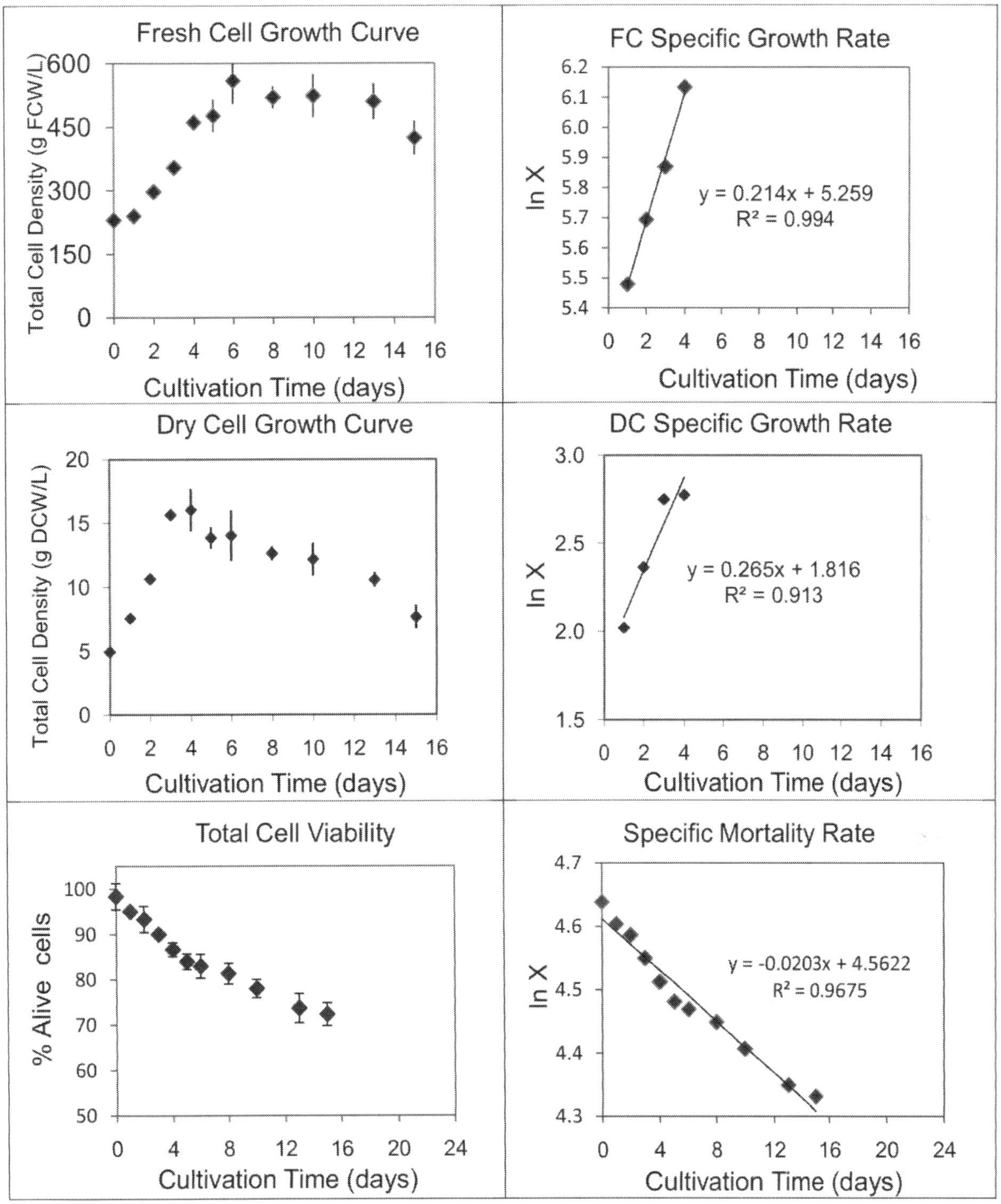

Figure 1. Growth curve of NT1 cells in 1 L flasks with an *inoculum* of 5 g dry cell/L (in orbital shaker at 26° C in darkness). Samples were taken every third day. Results corresponding to the average of 3 independent experiments are shown. Error bars indicate one standard deviation.

Growth of tobacco cells in Erlenmeyer flasks

We found that growth kinetic parameters do depend on inoculums concentration.In experiments conducted 1 g/L of inoculum, growth rates of tobacco cell fluctuate between 0.107 and 0.115 day^{-1}. In contrast, in cultures inoculated at 5 g/L, growth rates were estimated between 0.20 and 0.26 day^{-1}. Agitation conditions in Erlenmeyer flasks promote cell aggregation. Our experiments suggest that this aggregation, and probably a more efficient cell signalling/communication associated with it, is responsible from faster growth. We did not find evidence of cell fragmentation or damaged induced by the agitation conditions under Erlenmeyer agitation. In general, cell viability remained high (above 70%) for the first week of culture and was not influenced by initial cell concentration. Similarly, the mortality rate averaged 0.02 day^{-1} in experiments at the two different inoculums conditions.

Experiments in concentrically agitated bioreactors

Additional experiments in instrumented bioreactors were performed in 300 mL INFORS® bioreactors. The rational for these series of experiments was to simultaneously test different agitation conditions maintaining all the rest of the relevant operational variables the same. This experiment was of great importance to directly observe the effect of shear stress caused by high agitation speeds and measure its consequences on the growth kinetic parameters, cell morphology and aggregation, and yields of the culture. Figure 2 show results from experiments performed in Infors® bioreactors. DCW growth curves, (b) Total Cell Viability, (c) specific rates for growth, and (d) mortality results are presented. Runs at three different agitation values (100, 200 and 300 RPMs) are compared.

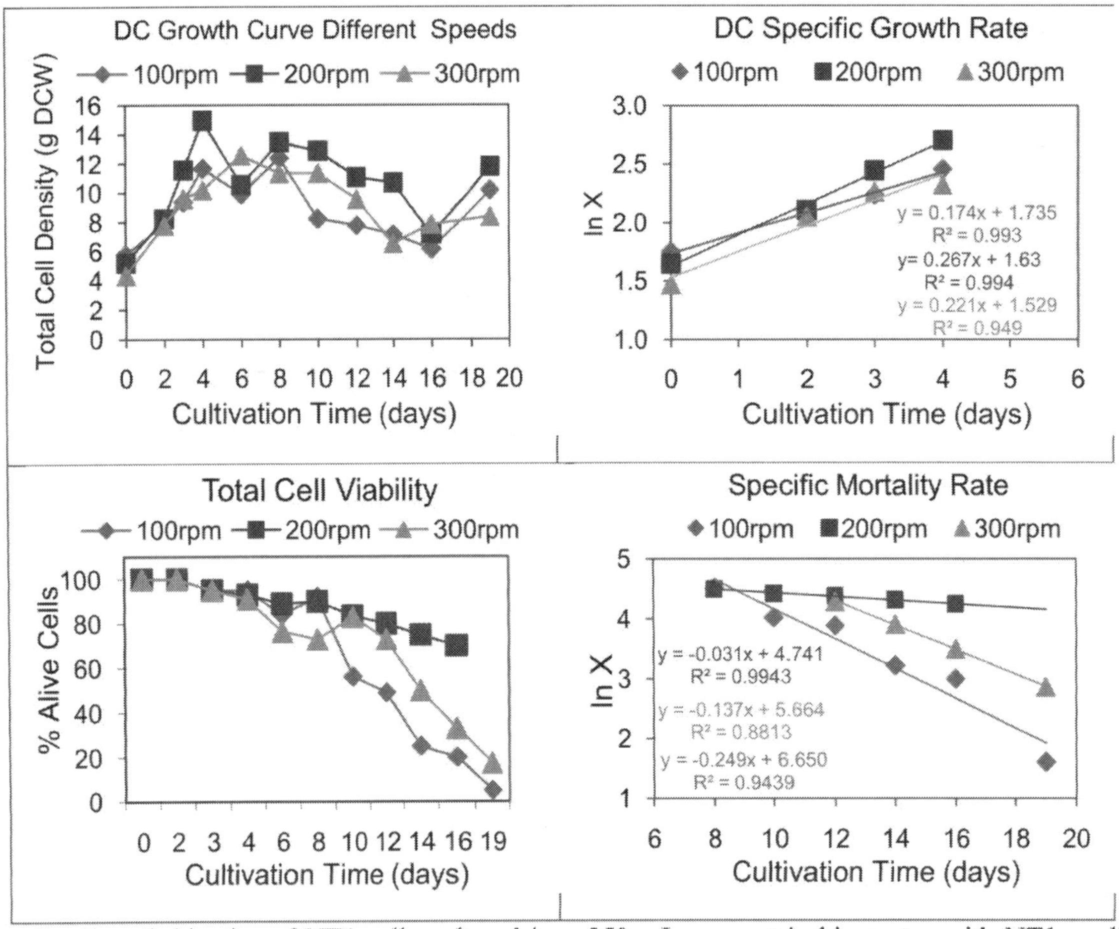

Figure 2. Growth kinetics of NT1 cells cultured in a 350 mL concentric bioreactor with NT1 medium at 3 different agitation speeds: 100, 200 and 300 rpm and 26° C in darkness. The initial *inoculum* was approximately 5 g dry cell /L.

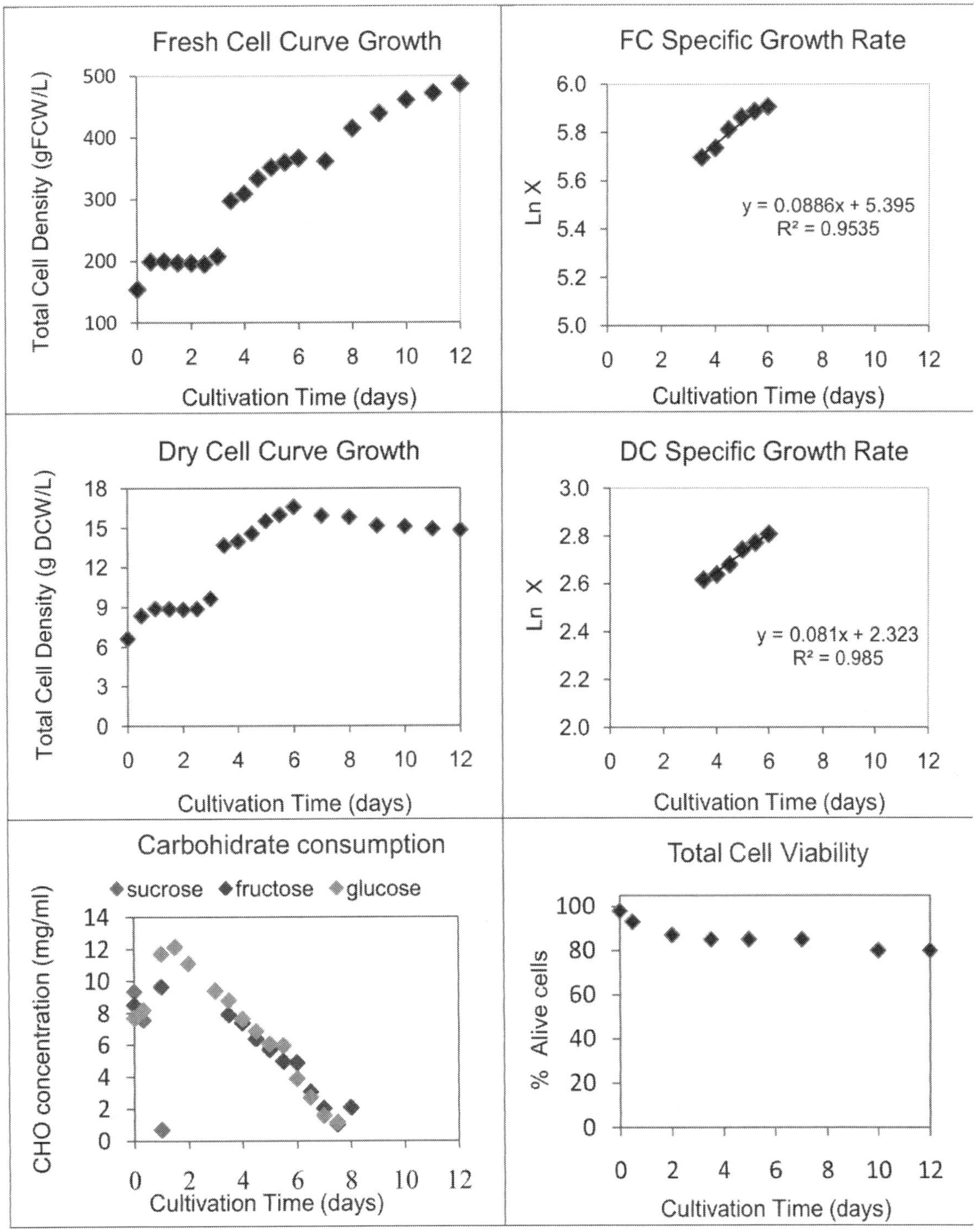

Figure 3. Growth curve (6) of NT1 cells in Applikon® bioreactor at 50 rpm for the first 4 days and 300 rpm until the end of the experiment, and 26° C in darkness. The *inoculum* was approximately 6.6 g dry cell /L.

Overall, better growth characteristics are observed at 200 RPM (i.e. a higher maximum cell density is achieved and higher cell counts are maintained throughout the evolution of the culture experiment. The specific growth rate observed at 100 RPM (μ =0.174 day^{-1}), is lower than those observed at higher speeds. The highest specific growth rate is observed when agitation is set at 200 RPM (μ=0.26 h^{-1}). This value is consistent to the highest specific growth rate observed in Erlenmeyer flasks at high inoculum concentrations. Not only the specific growth rate is maximized at 200 RPM, also the cell death rate is the lowest for the 200 RPM experiment, as illustrated by Figure 2. For experiments conducted at 100 rpm, the average maximum cell yield was 12.16 g DCW/L, an intermediate value when compared with the achieved at 200 and 300 rpm. The best yield obtained was 14.33 g DCW/L, corresponding to experiments at 200 rpm. In all cases, the maximum cell concentration was achieved between day 3 and 4. Regarding specific growth rates and doubling times, the highest growth rate (and therefore the minimum doubling time) was obtained in experiment at 200 rpm (0.294 days^{-1} and D_T= 2.35). The average specific growth rate during exponential phase for experiments at 100 and 300 rpm was practically identical, 0.23 and 0.24 days^{-1} respectively. In all experiments, cell viability remained above 80% for at least the first 8 days of culture. Remarkably, in experiments at 200 RPM, cell viability remained higher than 80% until day 14. In contrast, for the experiments at 100 and 300 rpm an abrupt decay in viability was observed after day 8, with viability values of 20 or 30% at day 19.

These results suggest that the most effective agitation speed to grow *N. tabacum* cell suspensions in Infors® bioreactors is 200 rpm. Results suggest that at 100 RPM, the culture growth is limited by oxygen mass transfer, and agitation intensity is not enough to sustain faster growth and higher cell densities. At 300 RPM, the agitation intensity is such that it induces cell damage, diminishing the proliferation rate and increasing the mortality rate. Vigorous agitation has been proposed as the reason for some plant cell lines to fail to grow optimally in stirred tank bioreactors. Plant cells have been regarded as being sensitive to shear stress because of their rigid cell wall, relatively large size and extensive vacuole volume. This hypothesis is supported by photographic evidence, taken at different cultivation times during the experiments at different agitation rates (not shown).

Experiments in eccentrically agitated bioreactors

Additional experiments were conducted at different *inoculum* concentrations and using different operational strategies in stirred tanks agitated with inclined disc impellers mounted in eccentric shafts. Eccentric bioreactor configurations have been referred as adequate for proper mixing at low Reynolds regimes, therefore avoiding cellular damage induced by intense agitation (Alvarez et al., 2002; Sánchez-Cervantes et al., 2006). Therefore, adequate cell densities and high growth rates were expected for these culture experiments. However, in experiments at low inoculums concentrations, namely 1 g/L, lower specific growth rate than in Erlenmeyer flasks were observed. Best results were obtained with high inoculums concentrations (5-6 g/L), with observed specific growth rates similar to those observed in Erlenmeyer flasks (Figure 3).

Two stage protocols in eccentrically agitated bioreactors were also assayed. In high inoculums experiments, cells were cultured under eccentric agitation at low speed, namely 50 RPM, for several days (during exponential growth) to later increase agitation speed to 300 RPM. Attending the scaling-up criteria of maximum tip speed velocity, this agitation condition would be comparable to more than 360 RPM in the 350 ml concentric bioreactors referred previously. However, no cell damage or increased mortality rate was observed in cell density high density cultures in eccentric bioreactors. Indeed, viabilities above 80% were sustained for twelve days of culture (Figure 3).

CONCLUSIONS

Our experimental results suggest than eccentrically agitated systems can be used as a cell culture system for shear sensitive plant cells. Similar specific growth rates were observed in highly inoculated eccentrically agitated bioreactors and Erlenmeyer flasks of similar scale (1-2 L). However, in smaller volume concentrically agitated bioreactors, mortality rates, cell viability and yields are a strong function of agitation rate. Overall yields comparable to those observed in Erlenmeyer flasks were observed only in a narrow window of agitation speeds, which complicates culture scalability.

REFERENCES

1. Alvarez, M. M., Arratia, P. E., & Muzzio, F. J. (2002). Laminar mixing in eccentric stirred tank systems. *Canadian Journal of Chemical Engineering, 80*(4), 546-557.
2. Doran, P. M. 2000. Foreign protein production in plant tissue cultures. Biotechnology. **11**:199–204.
3. Doran, P.M. 1993. Design of reactors for plant cells and organs. Advances in Biochemical Engineering Biotechnology. **48**: 117-168.
4. Doran, P.M. 1999. Design of mixing systems for Plant cell suspensions in stirred Reactors. Biotechnology Progress. **15**: 319-355.
5. Hong, S. Y., Kwon, T. H., Lee, J. H., Jang, Y. S. & Yang, M.S. 2002. Production of biologically active hG-CSF by transgenic plant cell suspension culture. Enzyme and Microbial Technology **30**: 763–767. Elsevier. South Korea.
6. James, E.A., Wang, C., Wang, Z., Reeves, R., Shin, J. H., Magnuson, N. S., & Lee, J. M. 2000. Production and characterization of biologically active human GM-CSF secreted by genetically modified plant cells. Protein Expression & Purification. **19**:131–138.
7. Joshi, J.B., Elias, C.B. & Patole, M.S., 1996. Role of hydrodynamic shear in the cultivation of animal, plant and microbial cells. Chemical Engineering Journal. **62**: 121–141.
8. Judge, N., A., Mason, H., & O'Brien, A. D. 2004. Plant Cell-Based Intimin Vaccine Given Orally to Mice Primed with Intimin Reduces Time of *Escherichia coli* O157:H7 Shedding in Feces. Infection and Immunity. **72**: 168-175. American Society for Microbiology.
9. Kieran P.M., MacLoughlin, P.F., & Malone D.M. 1997.Plant cell suspension cultures: some engineering considerations. Journal of Biotechnology. Vol. **59**: 39–52. Elsevier.
10. Loyola-Vargas, V.M. & Vázquez-Flota, F. 2006. Plant Cell Culture Protocols. Second Edition. Methods in Molecular Biology. Humana Press Inc.
11. Magnuson, N. S., Linzmaier, P.M., Reeves, R., An, G., Hay Glass, K., & Lee, J.M. 1998. Secretion of biologically active human interleukin-2 and interleukin-4 from genetically modified tobacco cells in suspension culture. Protein Expression & Purification. **13**:45–52.
12. Mayo, K.J., Gonzales, B.J. & Mason, H.S. 2006. Genetic transformation of tobacco NT1 cells with *Agrobacterium tumefaciens*. Nature Protocols. **1**: (3) 1105-1111.
13. Paul, L. E., Atiemo-Obeng, V.A.& Kresta, S.M. 2004. Handbook of Industrial Mixing: Science and Practice. John Wiley & Sons, Inc. Sponsored by the North American Mixing Forum.
14. Sanchez-Cervantes I. M., Lacombe J., Muzzio F.J., Alvarez M.M. 2006. Novel bioreactor design for the culture of suspended mammalian cells. Chemical Engineering Science. Volume 61, Issue 24, December 2006, pages 8075-8084. Part I: Mixing characterization.

(467d) Bioreactor Mass Transfer At Large Scale

Sperry Brown, Amgen, Inc., Thousand Oaks, CA

Mass transfer in bioreactors has been studied at large scale in Amgen's pilot plant. Implications of the relationships between sparging, agitation on mass transfer and reactor performance will be discussed.

See more of this Session: Mixing In the Production of Biopharmaceuticals
See more of this Group/Topical: North American Mixing Forum

(467e) Development of a Method to Measure Bioreactor Mass Transfer

Gregory Frank, Amgen, Thousand Oaks, CA

Determination of bioreactor mass transfer involves a number of decisions, including what conditions, media, experimental design and many other. The goal of developing a protocol is to have a rapid, low cost, and robust methodology that provides mass transfer data that's representative of conditions experienced in actual cell culture.

See more of this Session: Mixing In the Production of Biopharmaceuticals
See more of this Group/Topical: North American Mixing Forum

Experimental measurement and CFD modelling of bubble size distribution in an aerated stirred tank

Zbyněk Kálal, Milan Jahoda, Institute of Chemical Technology, Prague, Czech Republic

Summary

The main topic of this study is experimental measurement and modelling of bubble size distribution in an aerated stirred vessel using CFD (Computational Fluid Dynamics) method. The studied system consisted of a cylindrical vessel of diameter $T = 0.29$ m, which was equipped with six-bladed Rushton turbine and filled with tap water to the height $H = T$. The secondary phase was introduced through a ring sparger situated under the impeller. Geometry and mesh of the computational domain were created using ANSYS DesignModeler and ANSYS Meshing software, calculations were carried out in CFD solver CFX 12.1 using Euler-Euler multiphase model together with k- turbulent model. The rotating impeller was modelled via MRF (Multiple Reference Frames) method and coalescence and breakage of bubbles using homogeneous MUSIG (Multiple Size Group) model with 13 classes of bubble size. Regarding the fact that these processes of particle size change represent from physical point of view very complicated and to a large extent stochastic phenomena (similarly as the very turbulent flow), the applicability of population balance models is presently very limited and numerical results should be always compared and fitted against experimental data.

Introduction

Presently the one-phase systems are well explored both experimentally and using simulations. Detailed description of dynamic behaviour of multiphase systems and determination of hydrodynamical variables is a difficult task because of complexity of interaction between phases. The presence of dispersed phase influences the dynamic of disperging phase. Interactions between phases include various mechanisms of momentum exchange like drag, lift and virtual mass force. However, in stirred tanks only contribution of drag force is significant and the others can be neglected during calculations [1].

The accuracy of results is usually affected by the particle size distribution modelling of the secondary phase, whether it is solid particles, droplets or bubbles. The size change can be caused by combination of different phenomena such as nucleation, growth, dissolution, aggregation (coalescence) or breakup of the particle. To capture this process it is necessary to use additional equation describing changes in particle size distribution together with transport or enthalpic equations for multiphase flow. This equation is generally referred to as population balance equation (PBE) and is used to describe phenomena such as crystallization, precipitation, granulation, fluid polymerization or even coalescence and breakage of bubbles in bubble columns and stirred tanks.

To simplify and speed up the calculation, constant size of dispersed (gas) phase is often assumed during mathematical modelling of liquid-gas system. This idea, however, does not fully capture the real situation, since via bubble-bubble and bubble-liquid interactions coalescence and breakage occur. So the constant bubble size model cannot correctly describe interfacial forces and other interactions between phases as heat and mass transfer, because their magnitude is dependent just on the bubble size.

The aim of this text is to give basic theoretical description of particle coalescence and breakup and possibilities of their modelling using CFD method in an aerated stirred vessel.

Theoretical part

Population balance equation and its solution

In general, PBE is a balance equation of the number density probability of some particle property (usually size). It defines how population of particles develop over time. Neglecting growth and nucleation term the general PBE can be written as follows

$$\frac{\partial}{\partial t}[n(V,t)]+\nabla\cdot[\vec{U}n(V,t)]=B_{ag}(n,V,t)-D_{ag}(n,V,t)+B_{br}(n,V,t)-D_{br}(n,V,t) \tag{1}$$

where $n(V,t)$ is the bubble number density function. It is defined as the number of bubbles with volume between V and $V+dV$ per unit volume and per class width dV. Thus, integration of $n(V,t)$ over all bubble volumes results in the total number of bubbles per volume. $B_{ag}(n,V,t)$, $B_{br}(n,V,t)$ express birth of bubbles of size V to $V+dV$ due to coalescence and breakage, $D_{ag}(n,V,t)$, $D_{br}(n,V,t)$ corresponding particle death, see following equations

$$B_{ag}(n,V,t)=\frac{1}{2}\int_0^V a(V-V',V')n(V-V',t)n(V',t)dV' \tag{2}$$

$$D_{ag}(n,V,t)=\int_0^\infty a(V,V')n(V,t)n(V',t)dV' \tag{3}$$

$$B_{br}(n,V,t)=\int_V^\infty g(V')\beta(V\mid V')n(V',t)dV' \tag{4}$$

$$D_{br}(n,V,t)=g(V)n(V,t) \tag{5}$$

Here $a(V-V',V')$ is the coalescence rate between bubbles of size $V-V'$ and V', $g(V')$ breakup rate (frequency) of bubbles of size V', i.e. the fraction of particles of volume V' breaking per unit time, and $\beta(V\mid V')$ probability density function of particles breaking from size V' to a particle of size V.

From the former equations it is obvious that PBE is a complicated integro-partial differential equation. As a consequence, it is possible to solve it analytically only in rare cases. Therefore various numerical methods are used to deal with it, e.g. Monte Carlo method, Method of Classes - CM, Standard Method of Moments - SMM or Quadratic Method of Moments - QMOM. Presently the best explored method is the CM, which is based on discretizing of continuous size range of the particle population in a finite number of size intervals (usually called bins). Each of these intervals is represented by one specific value and each particle can interact with particles in other groups. The PBE for class i can be written as

$$\frac{\partial}{\partial t}(\rho_g n_i)+\nabla\cdot(\rho_g\vec{U}_g n_i)=\rho_g(B_{ag,i}-D_{ag,i}+B_{br,i}-D_{br,i}) \tag{6}$$

where n_i represents the number of bubbles of given class per unit volume

$$n_i=\int_{V_i}^{V_{i+1}} n(V,t)dV \tag{7}$$

As mentioned earlier, every class is represented by a specific value of particle size V_i. However, due to coalescence and breakage phenomena also particles, the volume of which V lies somewhere between the values of two neighbouring classes V_i and V_{i+1}, can be produced. This problem is solved by splitting the particle into two neighbouring classes. More information about the CM including detailed derivation and discretization of source terms can be found in [2].

Method of Classes is used especially in cases when the size of largest and smallest particle are known and this range is not too big, so that it is possible to discretize the population with relatively small number of intervals. To gain sufficiently exact results it is recommended to use at least 12-15 classes. The main advantage of this attitude is direct computation of particle size distribution, although at the expense of significant computational demands, because one transport equation has to be solved for every bubble class.

Coalescence and breakup models

The modelling of bubble, drop or solid particle coalescence and breakup is very complex and it is based on the knowledge of coalescence and breakup rates and daughter size distribution (see equations 2-5).

Coalescence between bubbles occurs in three steps. Firstly, the bubbles collide together and trap a small amount of liquid between them. Subsequently, this liquid drains till the liquid film reaches its critical thickness. Finally, the film ruptures and bubbles aggregate. In reality, the duration of collisions is limited and coalescence will occur only if the interaction time is sufficiently long enough for the film to drain out to critical thickness, therefore the concept of collision efficiency is introduced. The resulting coalescence rate is then determined as a product of collision frequency and collision efficiency.

Various mechanisms of coalescence are known presently. Buoyancy-driven collisions result from the difference in rise velocities of bubbles having different sizes and thus different buoyancy force acting on them. Furthermore, during the rise of a bubble through a liquid some of this liquid is carried up and accelerated behind this bubble, which is known as a wake. When another bubble enters the wake region of the leading bubble, it will accelerate and may coalesce with the preceding one. However, all these mechanisms are usually neglected, and in the stirred tanks with high level of turbulence especially, in comparison with turbulent-induced coalescence, which result from random motion of bubbles due to turbulent fluctuations.

The random motion of fluid particles in a turbulent flow is usually assumed to be similar to the random movement of gas molecules, thus most of the present models for coalescence were derived analogously to classical kinetic theory of gases, where collisions between molecules are considered. Thus, various coalescence models show similar trends, e.g. a monotonous increase in the coalescence rate with increase in the particle diameter.

One of the most widely used coalescence model is the one given by Prince and Blanch [3]. The collision frequency ω_{ag} and collision efficiency P_{ag} are expressed as following

$$\omega_{ag}(V_i, V_j) = 0.356\pi(d_i + d_j)^2 \varepsilon^{1/3}\left(d_i^{2/3} + d_j^{2/3}\right)^{1/2} \tag{8}$$

$$P_{ag}(V_i, V_j) = \exp\left(-\frac{\rho_l^{1/2} r_{ij}^{5/6} \varepsilon^{1/3}}{4\sigma^{1/2}} \ln\frac{h_0}{h_f}\right) \tag{9}$$

where equivalent radius r_{ij} is

$$r_{ij} = \left(\frac{1}{2}\left(\frac{1}{r_i} + \frac{1}{r_j}\right)\right)^{-1} \tag{10}$$

and h_0, h_f are initial and critical liquid film thickness separating the two coalescing particles (10^{-4} a 10^{-8} m).

Bubble breakage generally depends on the balance between external stresses that disrupt the bubble, and viscous stresses, represented by surface tension σ, that resist the bubble deformation. Most of the published breakage models are based on similar assumptions: 1) breakup is caused by turbulent eddies of size λ hitting the surface of bubbles with diameter d; 2) for the breakup to occur, the eddies must have energy higher than the surface energy of the bubble and simultaneously they should be smaller or equal to the size of the bubble; 3) eddies bigger than the bubble are supposed just to transport this bubble. Following these arguments, breakup frequency is given, similarly as in the case of coalescence, by a collision rate between particles and turbulent eddies multiplied by a collision (or breakage) efficiency. Because of the second and third assumptions, the breakage rate of particles should decrease with their decreasing size. For very small bubbles the breakup rate is close to zero, because the eddies which are even smaller than the bubbles do not have enough energy to break them and larger eddies only transport these bubbles.

The main difference among the available models is in their prediction of daughter size distribution (DSD), which represents the fraction of bubbles of size V' that break into bubbles of the size V. Some of the models assume a uniform distribution [3]. Others has a shape of truncated normal distribution which is centered at the half of the particle size [4]. It means, that these models prefer equal-sized breakage. In contrast, some other models favours unequal breakup [5], i.e. a division of the mother particle in one small and one big daughter particle, which is motivated by the fact that the unequal-sized breakup is less energetically demanding than the equal-sized one and thus should be more probable.

Luo and Svendsen based their model on assumption of isotropic turbulence and considered only binary breakup. The significant difference between this model and its predecesssors is that it gives not only overall breakage rate, but also partial breakage rate, i.e. the breakage rate for a particle of size V' splitting into a particle of size V and its complement. This partial breakup rate is expressed as

$$g(V,V') = C_B 0.9238 \frac{\varepsilon^{1/3}}{d^{2/3}}(1-\alpha_g)\int_{\xi_{min}}^{1} \frac{(1+\xi)^2}{\xi^{11/3}} \exp\left(-\frac{12 c_f \sigma}{\beta \rho_l \varepsilon^{2/3} d^{5/3} \xi^{11/3}}\right) d\xi \qquad (11)$$

where σ is the surface tension between phases, ξ dimensionless eddy size, c_f change in the surface area, β is a constant equal to 2.05 and C_B a calibration coefficient.

$$c_f = f^{2/3} + (1-f)^{2/3} - 1, \quad f = \frac{V}{V'}, \quad \xi_{min} = 11.4\frac{\eta}{d}, \quad \eta = \left[\frac{\left(\frac{\mu_l}{\rho_l}\right)^3}{\varepsilon}\right]^{1/4} \qquad (12)$$

where η is Kolmogorov micro-scale. The total breakup frequency of particles of size V' is then computed as an integral over all possible daughter particle sizes.

$$g(V') = \frac{1}{2}\int_0^1 g(V,V')df = \int_0^{0.5} g(V,V')df \qquad (13)$$

One half before the integral is introduced because during integration every breakup is included actually twice. The integral is symmetrical around the value 0.5, so that a change of the upper limit of integration to this value and removing the one half from in front of the integral would yield the same result. A great plus of this model is that the DSD is derived directly from the model for breakage rate and found by normalizing the partial breakage rate by the overall breakage rate.

$$\beta(V \mid V') = \frac{1}{V'}\frac{g(V,V')}{g(V')} = \frac{2\int\limits_{\xi_{min}}^{1} \frac{(1+\xi)^2}{\xi^{11/3}} \exp\left(-\frac{12 c_f \sigma}{\beta \rho_l \varepsilon^{2/3} d^{5/3} \xi^{11/3}}\right) d\xi}{V' \int\limits_{0}^{1}\int\limits_{\xi_{min}}^{1} \frac{(1+\xi)^2}{\xi^{11/3}} \exp\left(-\frac{12 c_f \sigma}{\beta \rho_l \varepsilon^{2/3} d^{5/3} \xi^{11/3}}\right) d\xi df} \quad (14)$$

Experimental part

Experiments were carried out in a cylindrical vessel with inner diameter $T = 0.29$ m, which was filled with tap water to a height $H = T$ (Fig. 1). Four baffles of width $b = T/10$ and thickness of 10 mm were placed evenly around the perimeter of vessel wall. The tank was equipped with a standard six-bladed Rushton turbine of diameter $D = T/3$, the distance of which from the bottom was also $C = T/3$. It was mounted on a central shaft rotating with constant speed of 300 rpm. Halfway between the vessel bottom and lower edge of the impeller there was located a ring sparger, on the upper side of which there was six point outlets with a diameter of 1.5 mm through which air was introduced in the system. Volumetric gas flow rate was equal to 0.2 vvm.

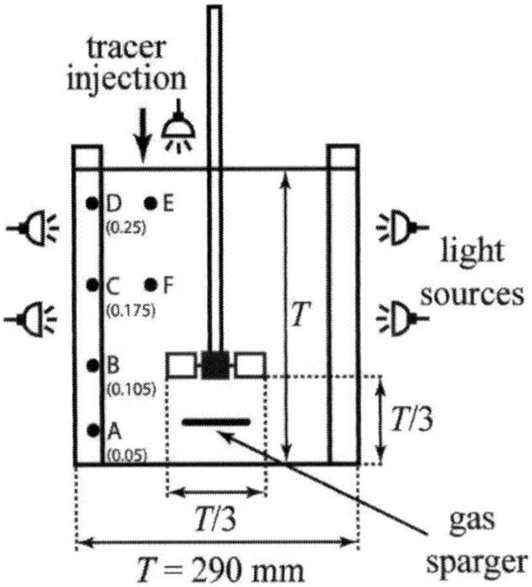

Fig. 1: Geometry of the stirred tank with six points for bubble size distribution measurement (A-F); the numbers in parentheses refer to vertical distance from the vessel bottom [m]

The determination of bubble size distribution (BSD) was accomplished using photographic analysis. Firstly, the tank was placed in a rectangular, glass vessel also filled with water, which prevented the distortion of the view when looking through rounded wall. The photographing and consequent evaluating of dense dispersions is generally very difficult due to overlapping of bubbles. Therefore, the tank was placed in a darkened room and a set of lights was placed on both sides of the tank and above the free liquid surface. These lights enlightened just the plane where bubble snapshots were taken, so that the unwanted background was significantly suppressed and identification of bubbles situated in the plane was facilitated.

The BSD was investigated at six points (see Fig. 1), which were all located in the vertical plane between baffles. Positions A-D were placed in the vicinity of the wall (approximately in the distance of two centimeters), points E and F then halfway between vessel wall and impeller shaft. During taking photographs a needle of known dimensions was placed in the desired position and served as a point at which the image was focused and simultaneously as a scale in the process of evaluating of photographs. Since the fluid moves in the mixed tank at considerable speed, it was necessary to use a low value of the exposure time, 1/1250 s, to gain sharp contours of the bubbles. Furthermore, appropriate depth of field was adjusted so as to obtain sharp outlines only of those bubbles that were found in the plane of interest. At each point about 50 photographs was taken. The captured bubbles were manually identified and their size was evaluated via image analysis in NIS-Elements software. The equivalent diameter of a spherical bubble was determined from the area of the circle which was set equal to the area of the bubble evaluated using image analysis.

For the simulations it was also needed to get the size of bubbles at the gas inlet because this parameter represents one of the boundary conditions of the numerical model. However, even at low gas flow rates gas jetting occurs, which prevents the identification of bubble size. Therefore, the flow rate was reduced to as low a value that at gas sparger separated bubbles were formed. Their size was used as input parameter of the model.

Regarding the fact that local gas hold-up measurement is very complicated and requires sophisticated equipment such as CT [6], only global gas hold-up was measured on the principle of liquid height elevation. Because the liquid surface in the tank fluctuate during stirring and so does not allow simple reading of the elevation, a sequence of photographs of the surface was acquired under ungassed and gassed condition. Using image analysis the averaged position of the liquid surface was determined for each condition and according to the difference in water level the gas hold-up was calculated.

Computational part

Geometry of the studied system was created in the preprocessor DesignModeler. The domain was classically divided into rotating part containing impeller and stationary part. For mesh generating another preprocessor was used Meshing. The final mesh contained 185 000 cells (Fig. 2).

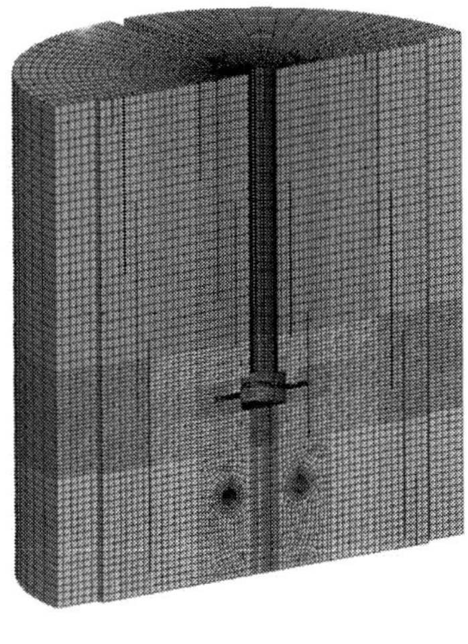

Fig. 2: Computational mesh

CFD simulations were carried out in software CFX 12.1. For numerical modelling of multiphase flow the most complex Eulerian model was chosen, so that for each phase a set of transport equations is solved. Because the system was in fully turbulent regime ($\mathrm{Re} = 4.49 \cdot 10^4$), *k-* turbulent model was used. The impeller was modelled via the MRF method. The upper part of the domain was set as a degassing boundary, which prevents water to leave the domain, but allows air to escape. The system was considered as isothermal and due to negligible solubility of air components transport between phases was omitted. On the basis of experiments the bubble diameter at the inlet was set to 8 mm. The PBE was solved using CM with 13 bubble groups in range from 0.1 to 16 mm, so that thirteen additional equations was necessary to solve. Coalescence kernel was calculated using Prince & Blanch model, breakage term then using Luo s model. The drag force was approximated according to the correlation given by Ishii and Zuber [7], other interaction forces were neglected.

Results

Experiments

At each investigated point 250-500 bubbles was marked and evaluated. The maximum bubble size observed in the system was around 6 mm, whereas the smallest detectable bubbles were of 0.1 mm in diameter. Most probably smaller bubbles existed, but due to limitations of the camera resolution and human eye during manual evaluation of taken photographs these were not incorporated. However, these bubbles contain only a very low fraction of gas volume and interfacial area. More than 90 % of both gas volume and interfacial area is stored in bubbles bigger than 1 mm. The average diameter varied from 0.5 to 1.2 mm and around 60 % of detected bubbles had diameter smaller than 1 mm.

To view bubble size distribution a quantity called Sauter mean diameter (SMD) d_{32} is usually used, which characterizes an average particle size. It is especially important in calculations where the active surface area is crucial.

$$d_{32} = \frac{\sum_i n_i d_i^3}{\sum_i n_i d_i^2} \tag{15}$$

The lowest SMD was detected in the impeller discharge stream where the turbulent dissipation energy has the largest value, which causes intensive breakup of bubbles (Fig. 3). The bubble diameter then slowly increases from the impeller towards the liquid surface and from the vessel wall towards the shaft because in the volume above impeller plane coalescence prevails. The smallest SMD was observed in position A. This was caused by the fact that the system was working in loading regime when vast majority of bubbles follows the upper recirculation loops and only very small bubbles are captured by the lower loops. These are then trapped here, coalesce and slip upward along the wall. On the contrary, if the system works in complete dispersion regime, the SMD in this position is considerably higher than in position B, because large bubbles are formed from bigger bubbles entrapped in the lower recirculation loops. The average diameter of bubbles entering the system was determined to be equal to 8 mm.

The gas hold-up in the system was found to be equal to 0.7 vol.%, which corresponds well with the correlation for systems with one Rushton turbine given by van't Riet [8].

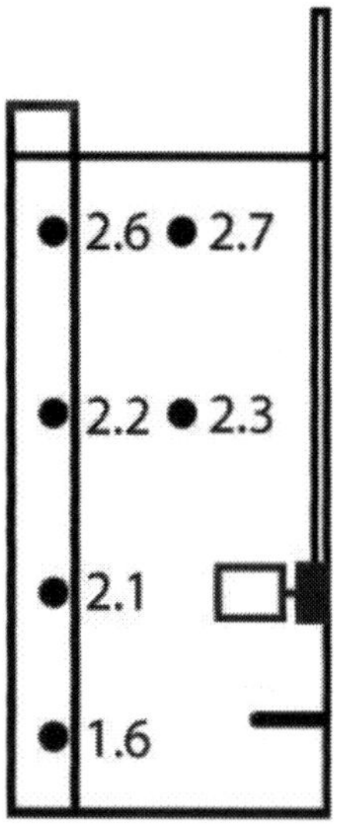

Fig. 3: Measured SMD in the tank [mm]

CFD simulations

As mentioned earlier, the bubbles were divided in 13 classes ranging from 0.1 to 16 mm in diameter. The breakup model by Luo was used. However, this model showed very strong underestimation of the SMD in the domain, especially in the impeller discharge stream. Practically all bubbles here were completely smashed in very small particles with diameter close to the size of the smallest class, i.e. 0.1 mm. This is caused most probably due to U-shaped daughter size distribution of the model. In Fig. 4 there are depicted dimensionless DSD curves for three different turbulent dissipation rates and constant bubble diameter and gas volume fraction.

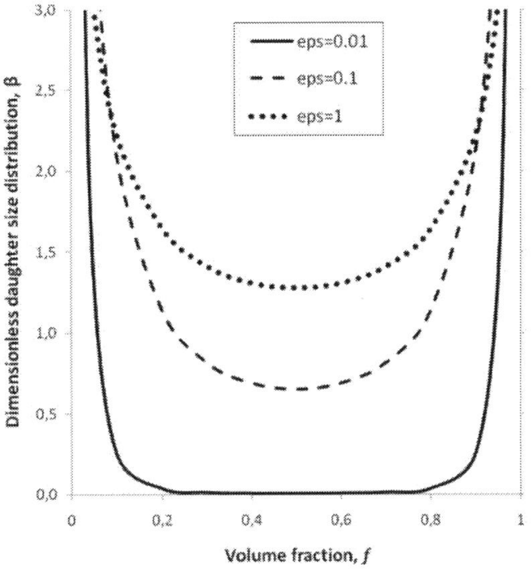

Fig. 4: Evolution of the dimensionless daughter size distribution of the Luo and Svendsen model with the change of turbulent dissipation energy (d=0.008 m, =0.01)

From the Fig. 4 it is obvious, that the model greatly favours the formation of a very small daughter particle and its complement, which stays after the breakup process nearly as big as the original mother particle. Considering the fact that the entering particles have diameter equal to 8 mm, thus being very unstable in turbulent flow of the system and having high breakup frequency, large number of small particles is with high probability gradually peeled from these bubbles. The breaking mother particle produces a lot of very small particles while still maintaining a high breakup rate. In the end, a great number of particles of the smallest size is created and the resulting SMD is underpredicted. The situation is depicted in Fig. 5.

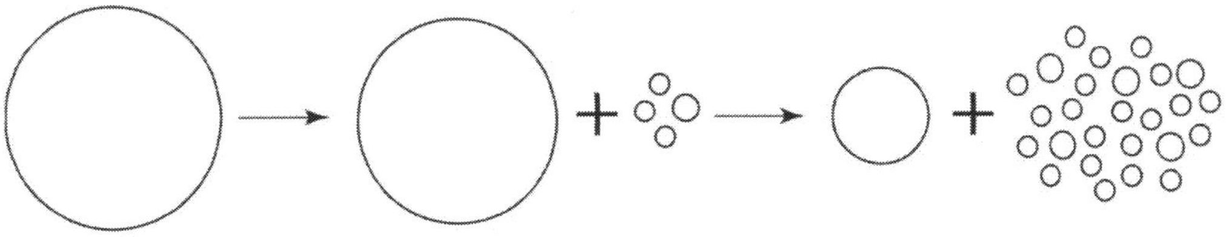

Fig. 5: Gradual breakup of a big mother particle

To remedy this issue either breakup rate must be decreased by adjusting an appropriate calibration coefficient of the breakup model, or the size of the smallest bubble group must be increased which results in formation of bigger bubbles by the breakup. However, even tenfold decrease of the breakup rate brought only a partial improvement of the resulting SMD which remained significantly lower than the experimental values (Fig. 6a). Much better agreement between simulations and experiments was achieved by setting the size of the smallest bubble class equal to 1.5 mm (see Fig. 6b). Nevertheless, this approach is not optimal either. Neglecting of particles smaller than is the size of the smallest class change the bubble size distribution because small bubbles are excluded. On the other hand, keeping in mind that a significant majority of bubble volume and interfacial area is stored in bigger bubbles, this method could be used as a good approximation for e.g. mass transfer simulations. A combination of adjusting the calibration coefficient of the breakup model and the size value of the lowest class seems to be the most powerful way how to handle the model given by Luo, but still it requires fitting against experimental data and its real predictivity is questionable. Use of different breakup models ([9],[10]) with DSD less emphasizing the formation of very small bubbles should be considered.

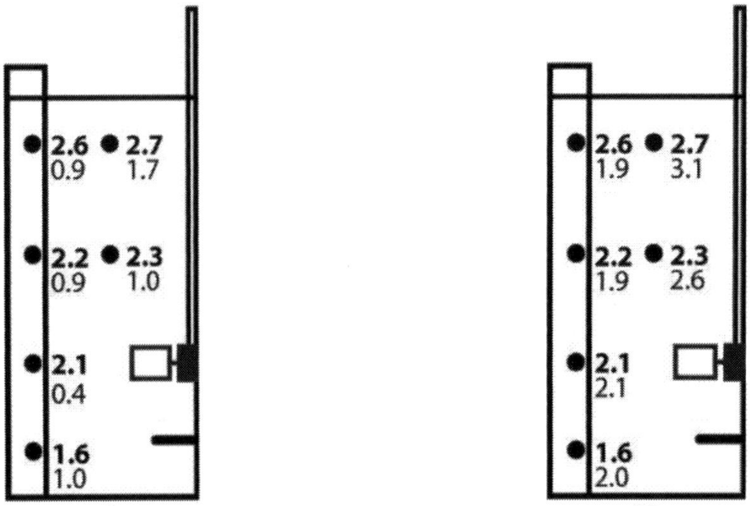

a) **b)**

Fig. 6: Sauter mean diameter [mm] (1. line - experiments, 2. line - simulations)
a) The breakup rate equal to $0.1 \cdot g(V')$
b) The smallest bubble size $d_{b,\min} = 1.5$ mm

The gas hold-up predicted by Ishii-Zuber correlation was about 30 % lower than the value gained from the experiments. Therefore, turbulent drag correction should be included in the simulations. This correction is based on introducing of effective viscosity (sum of molecular viscosity and a term proportional to turbulent viscosity) in the computation of Reynolds number which enters calculation of the drag coefficient [11].

Conclusion

The bubble size distribution was measured in an air-water stirred tank using photographic analysis and these data were used for comparison with numerical results. For computations the breakup term by Luo and coalescence term by Prince and Blanch were used. However, this combination showed great underestimation of the predicted Sauter mean diameter. This is probably caused by the U-shaped daughter size distribution of the breakup model which significantly favours formation of very small particles, thus decreasing the value of the SMD. Therefore, fitting against experimental data is needed. This can be achieved by either decreasing of the breakup rate (alternatively increasing of coalescence rate) or adjusting of the bubble size of the smallest class. The latter attitude distorts the BSD because no very small bubbles (smaller than the lowest class) are formed, but since a great majority of bubble volume and interfacial area is stored in bigger bubbles, this method could be used as a good approximation for e.g. mass transfer simulations. Generally, the available breakup models are derived and validated only for certain flow conditions and their transferability is very limited, so that further research and development of these models is necessary.

Acknowledgements
This project has been supported by the Czech Science Foundation (Grant: 104/09/1290 and 104/08/H055) and by the Czech Ministry of Education (Grant: MSM 6046137306).

References

1. Scargiali F., D Orazio A., Grisafi F., Brucato A. (2007), "Modelling and simulation of gas-liquid hydrodynamics in mechanically stirred tanks", *Chemical Engineering Research and Design 85*, pp. 637-646
2. Kumar S., Ramkrishna D. (1996), "On the solution of population balance equations by discretization-I. A fixed pivot technique", *Chemical Engineering Science 51*, pp. 1311-1332
3. Prince M.J., Blanch H.W. (1990), "Bubble coalescence and break-up in air-sparged bubble columns", *AIChE Journal 36*, pp. 1485-1499
4. Martínez-Bazán C., Montanes J.L., Lasheras J.C. (1999), "On the breakup of an air bubble injected into fully developed turbulent flow. Part 1. Breakup frequency", *Journal of Fluid Mechanics 401*, pp. 157-182
5. Luo H., Svendsen H.F. (1996), "Theoretical model for drop and bubble breakup in turbulent dispersions", *AIChE Journal 42*, pp. 1225-1233
6. Khopkar A.R., Rammohan A.R., Ranade V.V., Dudukovic M.P. (2005), "Gas-liquid flow generated by a Rushton turbine in stirred vessel: CARPT/CT measurements and CFD simulations", *Chemical Engineering Science 60*, pp. 2215-2229
7. Ishii M., Zuber N. (1979), "Drag coefficient and relative velocity in bubbly, droplet or particulate flows", *AIChE Journal 25*, pp. 843-855
8. van't Riet K. (1975), "Turbine agitator hydrodynamics and dispersion performance", *PhD Thesis*, University of Delft
9. Lehr F., Millies M., Mewes D. (2002), "Bubble-size distributions and flow fields in bubble columns", *AIChE Journal 48*, pp. 2426-2443
10. Laakkonen M., Moilanen P., Alopaeus V., Aittamaa J. (2007), "Modelling local bubble size distributions in agitated vessels", *Chemical Engineering Science 62*, pp. 721-740
11. Bakker A., van den Akker H.E.A. (1994), "A computational model for the gas-liquid flow in stirred reactors", *Chemical Engineering Research and Design 72*, pp. 573-582

(477a) Improved Thermostability of AEH by Combining B-FIT Analysis and Structure-Guided Consensus Method

Janna Blum[1], M. Daniel Ricketts[2] and Andreas S. Bommarius[1], (1)School of Chemical and Biomolecular Engineering, Georgia Institute of Technology, Atlanta, GA, (2)School of Chemistry and Biochemistry, Georgia Institute of Technology, Atlanta, GA

alpha-amino ester hydrolases (AEHs) catalyze the synthesis and hydrolysis of α-amino β-lactam antibiotics. The AEH enzymes have been shown to feature excellent synthetic capability but suffer from poor thermostability. AEH from *Xanthomonas campestris* exhibits an optimal temperature of 25°C, observed half-life of 5 minutes at 30°C and a T_{50}^{30}, the temperature at which the half-life is 30 minute, of 27°C. To improve the thermostability of the AEH, a modified structure-guided consensus model of seven homologous enzymes was generated along with analysis of the B-values from the available crystal structures of AEH from *Xanthomonas citri.* A family of stabilized variants was created including a consensus-driven triple variant, A275P/N186D/V622I. In the third round, independent NNK saturation of two high B-factor sites, K34 and E143, on the triple variant resulted in our best variant, the quadruple mutant E143H/A275P/N186D/V622I, with a T_{50}^{30} value of 34°C (7°C improvement) and 1.3-fold activity compared to wild-type.

See more of this Session: Protein Engineering - IV More Applications
See more of this Group/Topical: Food, Pharmaceutical & Bioengineering Division

(477b) USING the UNCONVENTIONAL NUCLEOTIDE Binding-Site for Affinity Purification of Antibodies

Nathan J. Alves[1], Michael W. Handlogten[1], Jonathan D. Ashley[1], Tanyel Kiziltepe[2] and **Basar Bilgicer**[1], (1)Chemical and Biomolecular Engineering, University of Notre Dame, Notre Dame, IN, (2) Advanced Diagnostics and Therapeutics / Chemical and Biomolecular Engineering, University of Notre Dame, Notre Dame, IN

This talk will describe a novel method to purify monoclonal and polyclonal antibodies (immunoglobulins) from cell lysates and ascites fluids using affinity chromatography where a small molecule (nucleotide analog) presented on the immobile phase binds to the unconventional nucleotide binding site that is present on all immunoglubulins. We targeted the nucleotide binding site present on each Fab arm of imminoglobulins with a small molecule that was discovered using computational methods for *in silico* screening. It has a K_d between 10 mM to 1 mM for this site on various IgG & IgE antibodies we tested originating from various species. Rituximab was used for proof of concept experiments. Antibody capture was accomplished by injection of samples while running equilibration buffer (50 mM sodium phosphate pH 7.0) and elution of the antibody was achieved by running a gradient of mild elution buffer (2M NaCl in 50 mM phosphate pH 7.0). Column's capture efficiency was >95%, with a purified antibody of >98% yield. This small molecule affinity purification method is further tested for a number of necessary parameters such as: injection concentrations, volumes, wash/bind time, elution gradient, antibody/protein-contaminant combinations, and effects of injection buffer. This method provides a superior alternative to the protein-A affinity purification method that is widely used for purification of humanized and chimeric antibodies.

See more of this Session: Protein Engineering - IV More Applications
See more of this Group/Topical: Food, Pharmaceutical & Bioengineering Division

Construction of Intein-Mediated Bioactive Protein Hydrogel

Zhilei Chen and Miguel Ramirez, Chemical Engineering, Texas A&M University, College Station, TX

Hydrogels made entirely of proteins/peptides or hybrid proteins/polymers have found applications as tissue engineering scaffolds, drug delivery carriers, controlled release depots and as key components of biosensing and bioanalytical devices. Hydrogels formed through a self-assembly process have the potential to produce precisely defined, hierarchical 3D structures – a particularly desirable trait when assembling enzymes/catalysts for cascade reactions. Taking advantage of intein-mediated protein ligation, we have constructed a protein hydrogel with the potential to self-assemble in situ for therapeutic applications. The building blocks of our protein hydrogels are two protein triblock copolymers that are expressed separately in E.coli. Each copolymer contains a bioactive protein flanked by one half of a naturally split intein and a crosslinker comprising multimeric protein subunit. Upon mixing of these two protein copolymers, the split intein in each copolymer covalently links the crosslinker/bioactive protein from the two triblock copolymers, resulting in the formation of a protein polymer molecule capable of selfassembling into a hydrogel. The gelation process is rapid upon mixing of the protein copolymers. This hydrogel is high stable in solution and facilitates high loading of enzyme. The described protein hydrogel technology enables precise molecular control of the constituent protein elements through manipulation of the genetic code and could provide scaffolds for industrial biocatalysis and tissue engineering applications.

(477d) Engineering a Reversible and High-Affinity Scaffold for Protein Purification

Fang Liu and Wilfred Chen, Chemical Engineering, University of Delaware, Newark, DE

While protein purification has long been dominated by standard chromatography, the relatively high cost and complex scale-up have promoted the development of alternative non-chromatographic separation methods. Here we developed a new non-chromatographic affinity method for the purification of secreted proteins. The approach is to engineer *E. coli* cells to secrete target proteins that are fused with an affinity tag. Direct purification and recovery can be achieved using a thermo-responsive elastin-like protein (ELP) scaffold containing the capturing ligands. Naturally occurring cohesin-dockerin pairs, which are high-affinity protein complex responsible for the formation of cellulosome in anaerobic bacteria, were used as the model. By exploiting the highly specific interaction between the dockerin and cohesin domain and the reversible aggregation property of ELP, highly purified and active dockerin-tagged proteins were recovered directly from crude cell extracts in a single purification step with yields achieving over 90%. Incorporation of a self-cleaving intein domain enabled rapid removal of the affinity tag from the target proteins. This method offers great flexibility as a wide range of affinity tags and ligands can be used.

See more of this Session: Protein Engineering - IV More Applications
See more of this Group/Topical: Food, Pharmaceutical & Bioengineering Division

(477e) Azide-Bearing Amino Acid Incorporation Into Recombinant Proteins In E. Coli

Diya M. Abdeljabbar and A. James Link, Chemical & Biological Engineering, Princeton University, Princeton, NJ

Unnatural amino acids can be incorporated into recombinant proteins *in vivo* through the use of engineered aminoacyl-tRNA synthetases (aaRSs). In studies conducted to screen methionyl-tRNA synthetase (MetRS) mutants that allow for the incorporation of the unnatural amino acid azidonorleucine (ANL), a MetRS variant with a leucine to glycine (L13G) mutation in its methionine binding pocket was identified. Methionine auxotrophic and prototrophic bacteria were engineered by replacing their genomic MetRS allele, *metG*, with the mutant synthetase gene, *metG**, which encodes for the MetRS L13G variant. The resulting engineered strains were able to replace methionine with ANL at high levels. Both engineered auxotrophic and prototrophic strains produced more ANL substituted protein than the wild-type strains with plasmid-borne copies of *metG** in addition to genomic copies of native *metG*. Despite the fact that ANL was incorporated at lower levels in the prototrophic strain, more functional recombinant protein was produced in this strain versus its auxotrophic counterpart. However, both strains were able to produce recombinant proteins with sufficient azide chemical handles in order to perform click chemistry. The incorporation of azide-bearing amino acids can be utilized for various applications using the engineered MetRS L13G.

See more of this Session: Protein Engineering - IV More Applications
See more of this Group/Topical: Food, Pharmaceutical & Bioengineering Division

(477f) Site-Specific Labeling of Transcription Factors Using Non-Canonical Amino Acids and Ring Strain-Promoted [2+3] Cycloaddition Via Copper-Free Click Chemistry

Younghoon Kim[1], Sunghoon Kim[2], John A. Katzenellenbogen[2] and Charles M. Schroeder[1], (1) Department of Chemical and Biomolecular Engineering, University of Illinois at Urbana-Champaign, Urbana, IL, (2)Department of Chemistry, University of Illinois at Urbana-Champaign, Urbana, IL

In this work, we demonstrate labeling of transcription factor proteins by incorporation of non-natural amino acids containing bioorthogonal chemical functionalites (azide or alkyne) followed by copper-free click chemistry. The ring strain promoted [3 + 2] dipolar cycloaddition of dibenzocyclooctyne with azides, known as "copper-free click chemistry", provides rapid and site-specific labeling of zinc finger proteins (ZFPs) as transcription factors (TFs). Following expression, purification and labeling, TFs retain their sequence-specific DNA-binding activity under native conditions, whereas the activity of TFs labeled by copper-mediated click chemistry was ablated. Copper-free click chemistry exhibited high reaction yields, even at low temperatures or in the presence of high salt concentrations, which are generally unfavorable reaction conditions for common methods involving hydrolysis-mediated protein modification. Residue-specific replacement of methionine by azidohomoalanine was used to achieve mono- or multi-functionalization of engineered or "synthetic" transcription factor proteins. As a model transcription factor, we engineered a zinc-finger protein to contain a single methionine residue at the N-terminus. In addition, we constructed a chimeric TF protein by fusing a protein tag containing multiple methionine residues at the N-terminus and a zinc-finger domain without methionine sites at the C-terminus. In this way, the engineered protein contained multiple sites for chemical labeling, and the chimeric protein was assembled and assayed for functionality after the protein modification. For all engineered TF proteins, the copper-free click reaction afforded good yields as determined by spectroscopic and fluorescence measurements, which was independent of the protein purification buffer. Finally, using mass spectrometry analysis, we determined the number of solvent exposed methionine sites on TF proteins. Overall, we anticipate that copper-free click chemistry mediated by non-natural amino acid incorporation will provide a robust and viable route for site-specific labeling of metal-ion containing proteins, which comprise a universal class of transcription factor proteins.

See more of this Session: Protein Engineering - IV More Applications
See more of this Group/Topical: Food, Pharmaceutical & Bioengineering Division

(477g) Engineering a Ligand-Free Molecular Switch In Bacterial Flagellar Motor

Pushkar Lele, Vedavalli Nathan and Howard C. Berg, Department of Molecular and Cellular Biology, Harvard University, Cambridge, MA

E. coli, a well characterized gram-negative bacterium, performs a biased-random walk by rotating its flagella alternately clockwise and counter-clockwise, using tiny reversible electric motors (45 nm dia). Changes in direction of rotation are aided by a molecular switch activated by a signaling molecule (ligand), which allows the bacterium to migrate over relatively large distances in response to attractants or repellents. This phenomenon, known as chemotaxis, is of considerable interest to researchers because it is a model system for understanding the principles of information processing and decision making that control vital functions of living organisms, including motility, gene expression, cytoskeleton dynamics and growth. However, at the heart of chemotaxis in bacteria lies the molecular switch, which is not well understood. The switch enables a change in the direction of rotation of the flagellum, thereby allowing the bacterium to move in a preferred direction. I will discuss our recent efforts on developing a ligand-free molecular switch in the flagellar motor. I will present observations of switching dynamics of this particular switch as well as describe the measurements of protein-turnover of a key component in the switch. These results reveal important aspects of protein sub-unit interactions and are consistent with the notion of stochasticity in conformations in multi-protein complexes.

See more of this Session: Protein Engineering - IV More Applications
See more of this Group/Topical: Food, Pharmaceutical & Bioengineering Division

CLASSICAL AND CHEMICAL IMAGING APPLIED TO FOOD "GRAINS" PROCESSING AND CONTROL: A CRITICAL REVIEW

Prof. Giuseppe Bonifazi
Sapienza – Università di Roma, Rome, Italy

Abstract

Many natural and/or artificial food products are usually present on the market as **P***articulate* **S***olids* (PS) or "grains". Their quality is defined according to the specific attributes of the single grain and/or the grains as a bulk system. PS characterization, inspection and quality control, inside a specific process and/or handling actions, are usually based on the collection of some specific particles physical-chemical attributes. These attributes represent the key factor to take into account when hardware and software (HW&SW) integrated architectures have to be developed to reach the previous mentioned goals. In this context the possibilities offered by classical **I***mage* **P***rocessing* and **A***nalysis* (IP&A) and specialized **H***yper*Spectral (or **C***hemical*) **I***maging* (HSI) based architectures, and related detection-control logics, able to play both *off-* and/or *on-line* are presented and critically evaluated in order to fulfill the previous mentioned goals. Some of the effects related to the adoption of different hardware configurations and different logics to perform the analysis and the selection of different algorithms according to the different characterization, inspection and quality control actions are presented.

(494b) A New Drug Screening Method for Chinese Herbal Medicines As Enhancers for Embryonic Stem Cell Proliferation

Ding Li, William G. Lowrie Department of Chemical and Biomolecular Engineering, The Ohio State University, Columbus, OH, Jufang Wang, School of Biosciences & Bioengineering, South China University of Technology, Guangzhou, China and Shang-Tian Yang, Department of Chemical and Biomolecular Engineering, Ohio State University, Columbus, OH

In this work, we developed a new drug screening method based on three-dimensional (3-D) cell culture for Chinese herbal medicines and for the first time discovered the potential application of a traditional herbal medicine as a new proliferation enhancer of embryonic stem cell culture through this method. This new screening method used nonwoven polyethylene terephthalate (PET) fibers as scaffolds for the 3-D culture of cells expressing enhanced green fluorescent protein (EGFP) in the medium with herbal medicine extracts. The extracts of three traditional Chinese herbal medicines, *Epimedium dacidii Franch* (EDF), *Ganoderma lucidum* spore (GLS), and *Ginkgo biloba* (GB), were selected as the screening targets and their effects on the proliferations of three cell lines (i.e., mouse embryonic stem (mES) D3 cells, colon cancer HT-29 cells, and breast cancer MCF-7 cells) were evaluated by measuring EGFP fluorescence intensities at a certain wave length. For each herbal medicine, the extracts were dissolved in DMEM cell culture medium at the concentrations of 0.01%, 0.1%, and 1% (w/v). Cell culture in the medium containing deoxycholic acid (DCA) (0.1 mM, 1 mM, and 10 mM) was used as the positive control for screening. The cell morphologies under different culture conditions were also characterized via scanning electron microscopy (SEM). Results showed that the growth of all three cell lines were significantly inhibited in the medium with 1% of herbal medicine exacts, which indicated that all three herbal medicines demonstrated strong cytotoxicity at 1% concentration regardless of the cell type. At a concentration of 0.1%, the herbal medicines showed different effects on cell proliferation depending on the cell type: GLS displayed a promotional effect on mES D3 cell proliferation but a reverse effect on the two cancer cell lines, while both EDF and GB inhibited the proliferation of mES D3 cells but showed no apparent impact on the two cancer cells. Those results were also supported by cell morphologies on SEM images. Based on the above results, we selected GLS as a new proliferation enhancer and for the first time used DMEM medium containing 0.1% GLS for stem cell culture. The new medium enhanced the specific growth rate and the maximum cell density of mES D3 cells by 28.6% and 29.1%, respectively, compared with the cell culture in regular DMEM medium. In summary, this 3-D cell culture-based screening method is feasible and efficient in evaluating the effects of herbal medicines on cell growth. GLS selected by this method shows its potential applications in mass production of embryonic stem cells and for treating cancers.

See more of this Session: Advances In Food and Bioprocess Engineering
See more of this Group/Topical: Food, Pharmaceutical & Bioengineering Division

(494c) Effects of High Pulsed Magnetic Field On Enzymes and Microorganisms In Strawberry Juice

Kaijun Xiao[1], Lan-ping Luo[1], Jiang-tao Jin[1], Bi-sheng Zheng[1] and Yuan Ren[2], (1)College of Light Industry and Food Science, South China University of Technology, Guangzhou, China, (2)College of Environmental Science and Engineerin, South China University of Technology, Guangzhou, China

This work is to evaluate the effects of high pulsed magnetic field (PMF) on enzymes, microorganisms and nutrient ingredient in strawberry juice, and discuss the mechanisms of PMF for sterilization and enzyme inactivation. Different HPMFs with the field strengths ranging from 9.3 to 19.0 tesla (T) for 3 to 12 pulses were applied. A maximum inactivation for polyphenol oxidase (PPO) was achieved by PMF at 17.3 T for 9 pulses, while for peroxidase (POD) was at 17.3 T for 12 pulses that 60 % of the enzyme was inactivated. Total bacterial number in strawberry juice reduced less than 100 cfu/mL without nutrition destroying at 17.3 T for 9 pulses or more, among them molds and yeasts were totally killed. In order to explore the mechanism of inactivating enzyme by high PMF, circular dichroism (CD) spectra and the fluorescence chromatography were applied to analyze the enzyme structure. They proved that alteration of the tertiary structure is the reason for the inactivation of POD under high PMF. Fluorescence microscope observation also proved that the damaged *Aspergillus* genus was increased with the increasing of magnetic field strength.

See more of this Session: Advances In Food and Bioprocess Engineering
See more of this Group/Topical: Food, Pharmaceutical & Bioengineering Division

(494d) Compressed Gas Fractionation of High Value Lipids From Plants

Steve J. Tallon, Industrial Research Ltd, Lower Hutt, New Zealand

Plants contain a range of interesting lipophyllic compounds with potential dietary or topical health benefits. These include antioxidants, phospholipids important to cell function, and polyunsaturated lipids. This paper describes a compressed gas extraction process using dimethyl ether to extract polar lipids and long-chain PUFA from plant materials including New Zealand ferns, and marine algae. A fractionation process using supercritical fluid CO_2 can be further used to isolate individual compounds.

See more of this Session: Advances In Food and Bioprocess Engineering
See more of this Group/Topical: Food, Pharmaceutical & Bioengineering Division

(494e) Furan Formation During UV-Treatment of Fruit Juices

Mahesh Bule[1], Kiran Desai[1], Brian Parisi[2], **Satish J. Parulekar**[3], Peter Slade[2], Rekha Singhal[1] and Alfredo Rodriguez[2], (1)Food Engineering and Technology Department, Institute of Chemical Technology, Mumbai, India, (2)National Center for Food Safety and Technology, Illinois Institute of Technology, Summit-Argo, IL, (3)Chemical and Biological Engineering, Illinois Institute of Technology, Chicago, IL

Furan is a suspected human carcinogen and reported to be produced by thermal or ultraviolet light (UV-C) treatment of various foods, such as fruit juices and sugar solutions. This study explores the formation of furan during UV-treatment of high fructose corn syrup (HFCS) solution and simulated juices containing it and validates the results with two real juices, apple juice and apple cider. Studies performed on individual components of fruit juices revealed fructose to be the main constituent in fruit juices that is responsible for furan formation during UV-treatment. Furan formation was promoted at acidic pH, yet was suppressed in the presence of ascorbic acid. Furan synthesis is strongly influenced by absorbance of the treated solution, with increasing absorption coefficient leading to reduced furan formation. Formation of furan in fruit juices sweetened with HFCS when subjected to UV treatment has been firmly established.

See more of this Session: Advances In Food and Bioprocess Engineering
See more of this Group/Topical: Food, Pharmaceutical & Bioengineering Division

SUPERCRITICAL CARBON DIOXIDE INACTIVATION OF *LISTERIA MONOCYTOGENES* SPIKED ON DRY CURED HAM PIG SKIN

Giovanna Ferrentino, Sara Balzan, Sara Spilimbergo, University of Trento, Italy

Abstract

Supercritical carbon dioxide (SC-CO_2) treatment has been largely demonstrated to be a promising low temperature preservation method for liquid foods. Due to the high potentials of the process -it induces microbial inactivation avoiding undesirable alterations of quality and nutritional attributes- the research has been recently turned into the applications of SC-CO_2 treatment to solid foods.

In this concern, a feasibility study of SC-CO_2 treatment for the inactivation of *Listeria monocytogenes* (ATCC19115) inoculated on dry cured ham pig skin is presented.

The sample was treated in a multi-batch lab scale apparatus at 12.0 MPa and the inactivation kinetics determined as a function of temperature (35÷50°C), treatment time (5÷30 minutes) and microbial load (10^7 cfu·g^{-1} and 10^3 cfu·g^{-1}). Additionally, color modifications during the process were detected in terms of reflectance spectra and L^*, a^*, b^* parameters thanks to an innovative on-line spectroscopic technique.

The results show that a treatment at 50°C, 12.0 MPa for 15 minutes induces total inactivation of *L. monocytogenes* with an initial load of 10^7 cfu·g^{-1}, while less severe processing conditions (45°C, 12.0 MPa 5 minutes of treatment) are sufficient if the initial load is decreased down to 10^3 cfu·g^{-1}; furthermore SC-CO_2 process does not seem to induce significant changes in the color of the sample.

In conclusion, the present study demonstrates the efficiency of the process, suggesting SC-CO_2 as an innovative technology for the pasteurization/sterilization of solid food products.

Introduction

During the last two decades consumers' preferences for ready to eat (RTE) products have increased considerably due to the modern lifestyles that require quick, easy and cheap prepared meals. Among RTE foods, cured hams are the most consumed products around the world. However, the manufacture of this product increases the risk of contamination, mainly during slicing and packaging. Among pathogenic microorganisms, *Listeria monocytogenes* constitutes the major concern for this kind of product, as it is a wide spread environmental microorganism, it is difficult to eradicate from the product environment and is able to grow during refrigeration and storage. Dry cured ham presents chemical intrinsic characteristics such as low water activity (usually lower than 0.92), high salt concentration (higher than 4%) and long ripening time (up to 12 months) which do not support the growth of *L. monocytogenes* at refrigerated temperature during the first processing phase [1]. However the product can be exposed to cross recontamination at the end of ripening, during or after deboning, slicing, and packaging in modified atmosphere or under vacuum [2]. The high severity of listeriosis and the high mortality rate -around 20%- makes the elimination or at least the reduction of *L. monocytogenes* a compulsory step to put the product into the market.

Different methods have been applied as treatments to prevent, eliminate, or reduce the presence of *L. monocytogenes* or other pathogens in "ready to eat" foods [3, 4, 5]. These treatments can prevent growth or reduce the presence of *L. monocytogenes* and other pathogens but, with the exception of heat treatments, only rarely they manage to achieve total inactivation. In particular, thermal treatment can induce negative and undesirable modifications of the quality attributes of the product.

In this regard, SC-CO_2 processing represents an attractive new preservation technology as it potentially meets the consumers' demands for a product with a high nutritional and organoleptic quality and, at the same time, a high microbial stability and hygienic safety.

The use of compressed CO_2, which is regarded in the United States by the FDA (Food and Drug Administration) as a GRAS (Generally Recognized As Safe) substance, has been investigated intensively as bactericidal agent in low-temperature and wide-target alternative processes to conventional ones on liquid substrates [6, 7, 8, 9, 10]. Few papers concern the effect of SC-CO_2 processing on solid substrates due to the complexity of the solid matrix, which can make CO_2 bactericidal action more difficult, and to the lack of information about the inactivation mechanism [11, 12]. As for liquids, the technology is attending to be promising also for high-value solid products with high profitability [13].

To this concern, the present work focuses on the applicability of SC-CO_2 treatment for the inactivation of *L. monocytogenes* spiked on dry cured ham surface. The objectives of the study are: 1. to optimize the process conditions (temperature and time) to obtain total inactivation of *L. monocytogenes*; 2. to investigate the process conditions assuring total inactivation of the microorganism at a microbial concentration spiked on the ham surface which mimics the natural contamination of the product; 3. to monitor the color of the sample during as well as before and after the entire process thanks to an innovative on-line spectroscopic technique.

Materials and Methods

Microbial suspension

The microbial strain used for the experiments was *Listeria monocytogenes* ATCC19115. Microbial culture was grown in 10 mL Brain Heart Infusion (BHI) broth at 37°C overnight. Then, the culture was transferred to a 200 mL flask of BHIB and grown at 37°C overnight. Cell growth was carried out in mixed conditions (220 rpm) in a shaking incubator. The microbial suspension was centrifuged at 6000 rpm for 10 minutes at 4°C, the surnatant removed and the pellet resuspended in 25 mL of phosphate buffer solution (PBS) to reach a final concentration of about 10^9 cfu/mL (colony forming unit/mL).

Sample preparation

Slices of dry cured ham pig skin, purchased from a local market, were cut in 2 grams pieces with a surface of about 2 cm^2. Each piece was spiked with 50 μL of *L. monocytogenes* suspension to reach the final desired concentration of microbial cells. The inoculum was spread over the surface until its absorption. Two inocula (10^7 and 10^3 $cfu \cdot g^{-1}$) were spiked on the sample to carry out experiments at different initial microbial concentrations. Experiments at decreased concentration were performed to mimic the potential contamination of *Listeria* species occurred on the surface of the product.

High pressure carbon dioxide apparatus

SC-CO_2 treatment was carried out in a multi-batch apparatus (Figure 1). The system consists of 10 identical reactors with an internal volume of 15 ml connected in parallel, so that each experimental run provides a set of experimental data taken at identical process conditions but different treatment times.

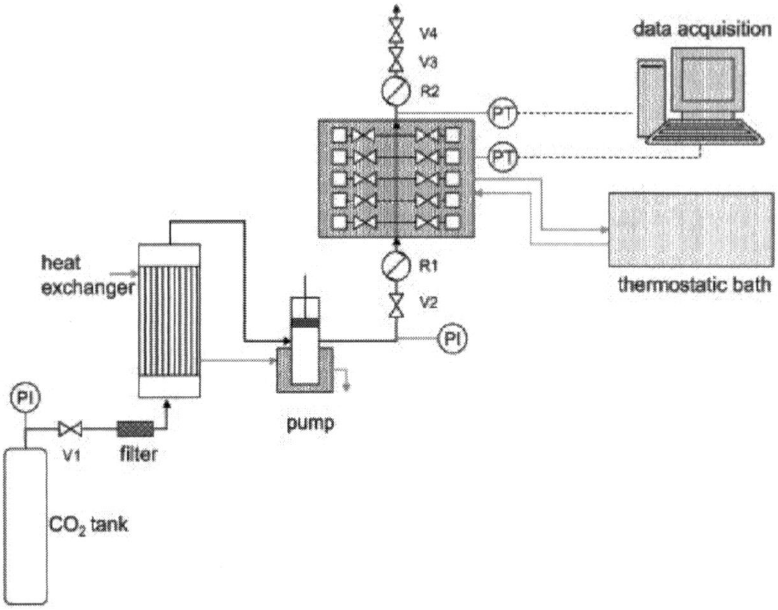

Figure 1. Schematic of the multi-batch apparatus. V-1÷V-4: valves; R1, R2: Electrical resistance; PT: pressure transducer; PI: pressure manometer.

Each reactor is connected to an on-off valve that can be used to depressurize it independently from the others. The ten reactors are submerged in a single temperature-controlled water bath so that the temperature is uniform in all of them. Liquid CO_2 (Messer, Carbon dioxide 4.0, purity 99.990%) is fed into the reactors by a volumetric pump (LEWA, mod. LCD1/M910s) with a maximum flow rate of 11 L/h after passing through a cooling bath containing a solution of water (50%) and antifreeze (50%) to reach low temperatures (approximately -3.0°C). The apparatus is provided with a pressure transducer while one cover lid of the ten reactors is equipped with a fixed temperature probe (Pt 100 Ω). The operating parameters (temperature and pressure) are continuously recorded by a real time acquisition data system (NATIONAL INSTRUMENTS, field point FP-1000 RS 232/RS 485) and monitored by a specific software (LabVIEW ™ 5.0).

Process conditions

Spiked samples were transferred into the reactors and treated at a constant CO_2 pressure of 12.0 MPa and different operative conditions of temperature (35, 40, 45 and 50°C), and treatment times (5, 15 and 30 min). Previous studies indicated that a pressure around 10.0 MPa was an optimal process condition to induce a pasteurizing effect on solid food matrixes [14].

Enumeration of survivals

Two grams of dry cured ham pig skin were homogenized (1:3) with PBS in a Stomacher 400 (International P. B. I., Milano, Italy) at 230 rpm for 2 minutes. The homogenate was serially diluted in PBS for the enumeration of *L. monocytogenes* colonies by plate count technique. Depending on the expected counts the adequate decimal dilution was spread manually on the selective media O. A. Listeria Agar (Liofilchem s. r. l., Teramo, Italy) petri dishes. The plates were incubated at 37°C for 48 h.

The microbial concentration of *L. monocytogenes* was reported as the Log of the microorganisms (*N*) versus time, where *N* ($cfu \cdot g^{-1}$) is the number of cells in the samples. Three independent experiments have been carried out for each single experimental condition (temperature, pressure, time). The results are calculated as the mean value of three replications. Standard deviations were shown by error bars.

On-line color monitoring

The color appearance of the sample during the treatment was on-line monitored with an innovative spectroscopic apparatus (Figure 2). The principal element of the system is a fiber optic probe (Ocean Optics Inc., Dunedin, FL) connected to a spectrophotometer (S2000 CCD, Ocean Optics Inc., Dunedin, FL) which transmits the light from a halogen lamp to the sample and acquires the reflected light.

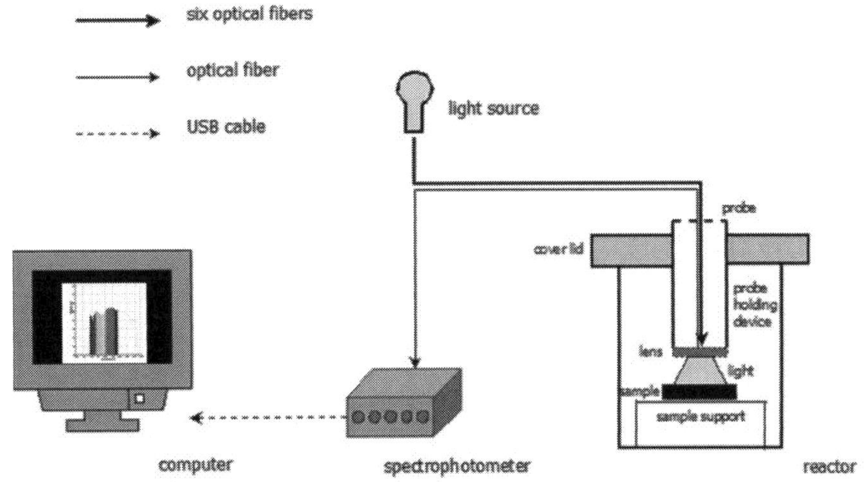

Figure 2. On-line color monitoring apparatus.

The probe is fixed vertically on the cover lid of a CO_2 pressurized reactor with an internal volume of 310 ml. The reflectance spectrum is continuously acquired and monitored during the supercritical CO_2 process by a specific software (Spectra Suite®, Ocean Optics, Dunedin, FL) providing L* (lightness), a* (redness) and b* (yellowness) color coordinates. Color measurements were performed in triplicate and mean values and standard deviations evaluated.

Results

Effect of temperature and treatment time

The inactivation kinetics of *L. monocytogenes* spiked on dry cured ham surface at different temperature (35, 40, 45, 50°C) as a function of the treatment time are reported in Figure 3. The results clearly demonstrate that the temperature plays an important role in increasing the inactivation rate: total inactivation is obtained at 45°C after 30 min or 50°C after only 15 minutes of treatment. At lower temperature (35 and 40°C), the inactivation level (-4 and 5 log reductions respectively) attained after 5 min of treatment remains almost constant although the time is increased to 30 min.

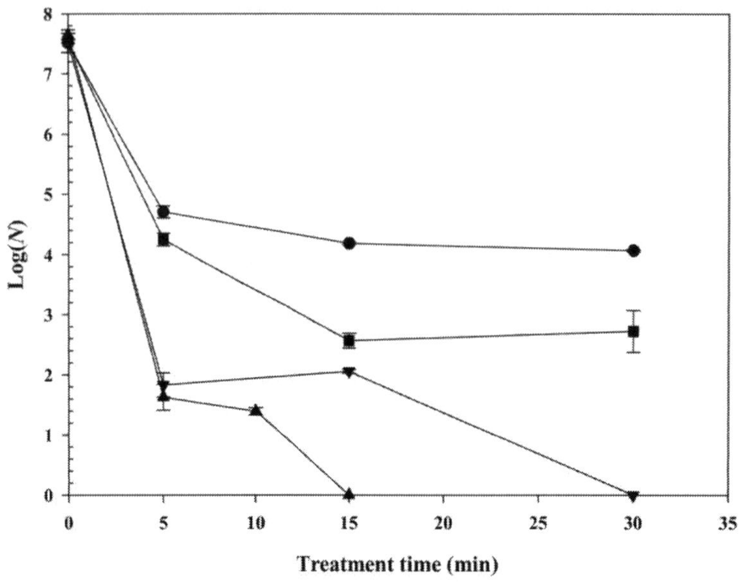

Figure 3. Inactivation kinetics of *L. monocytogenes* spiked on dry cured ham surface at 12.0 MPa as a function of treatment time and temperature: ●, 35°C; ■, 40°C; ▼, 45°C; ▲, 50°C.

Effect of the initial microbial concentration

Table 1 shows the inactivation of *L. monocytogenes* spiked on the surface of dry cured ham at different microbial concentrations (7 and 3 Log(cfu·g^{-1})). SC-CO_2 treatment was performed at 12.0 MPa, 45°C and the treatment time varied from 5 to 30 min. The experimental results demonstrate that the inactivation rate is a function of the initial microbial load spread on the sample: lowering the initial microbial load a faster inactivation is obtained. Total inactivation is achieved after 30 and 5 min of treatment starting from respectively 7 and 3 Log(cfu·g^{-1}) of the initial microbial load spiked on the sample.

It is worth noticing that this finding is of relevant importance considering that a concentration of 3 Log(cfu·g^{-1}) mimics the natural contamination of *Listeria* species usually detected on the surface of the dry cured ham during the processing.

Table 1. Concentration of *L. monocytogenes* spiked on dry cured ham surface at different microbial loads spiked on the samples (Treatment conditions: 12.0 MPa, 45°C).

Treatment time (min)	**Sample 1 LogN (cfu·g^{-1})**	**Sample 2 LogN (cfu·g^{-1})**
0	7.30±0.18	3.1±0.43
5	2.82±0.12	N.D.*
15	1.83±0.20	-
30	N.D.*	-

*N.D.: not detected

Effect of the process on color parameters

The reflectance spectra of dry cured ham surface treated at 12.0 MPa, 50°C for 15 minutes are reported in Figure 4. The signals were acquired before and after the treatment and on-line to monitor possible color changes during the process. The intensity of the reflectance curve of the sample after the treatment is not significantly different compared to the one acquired before. Both spectra show a high reflectance intensity in the 550-650 nm wavelength region corresponding to the red/orange color range. The on – line reflectance spectra acquired at different times (1, 5 and 15 minutes) are all overlapped showing no defined peaks. The similarity of the reflectance profiles over the entire wavelength range indicates the stability of the color of the sample during the treatment.

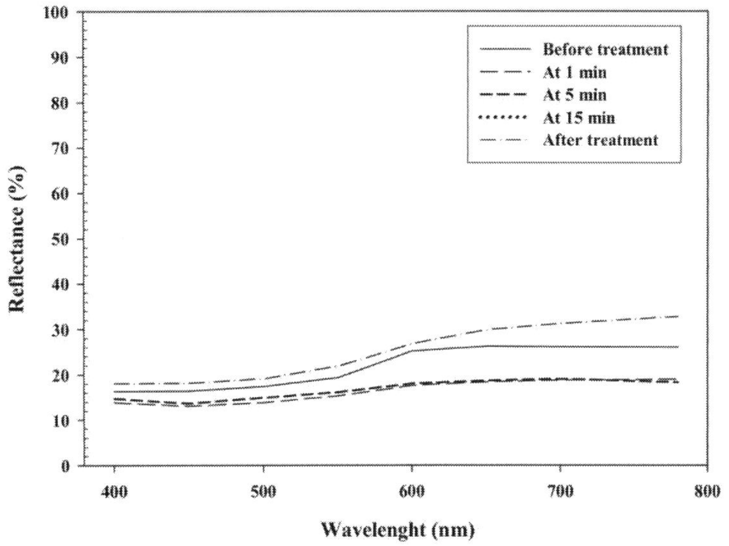

Figure 4. Reflectance spectra of dry cured ham surface acquired before, after the treatment and on – line at 1, 5, 10, and 15 min during the SC-CO_2.

The colorimetric parameters on-line acquired during the treatment are reported in Table 2. The experimental data show no significant changes in L*, a* and b* values over the entire process length.

Table 2. On-line acquisition of color parameters for dry cured ham surface during the SC-CO_2.

Treatment time (min)	L*	a*	b*
1	46.6±1.5	2.8±0.8	1.3±0.8
5	45.5±2.8	3.0±0.9	1.63±0.8
15	46.8±1.0	3.4±1.4	1.93±0.9

The color coordinates were also measured before and after the treatment: L* varied from 54.1 to 55.1, a* from 4.5 to 6.4, while b* from 5.5 to 7.2. The results reflect the reflectance profiles reported in Figure 3. The high a* and b* values can be probably associated to the presence of residual CO_2 in the high pressure vessel after the decompression of the system.

Conclusions

The present study clearly demonstrates the potential of the SC-CO_2 treatment for the inactivation of *L. monocytogenes* spiked on dry cured ham surface. The experimental findings show that: 1. temperature and treatment time influence the inactivation rate of *L. monocytogenes*: total inactivation is obtained at 45°C, 30 min or 50°C, 15 min with an extremely large contamination of the samples (about 10^7 cfu·g^{-1}); 2. decreasing the degree of contamination down to 10^3 cfu·g^{-1}, in order to mimic the real contamination of the product, just 5 minutes of treatment are sufficient to reach total inactivation at 45°C; 3. color measurements during and after the treatment reveal no significant changes of the visual appearance of the sample; 4. the results demonstrate that the SC-CO_2 process allows to comply with the concept of "absence of risk" and with the limit of *L. monocytogenes* detection required for human safety.

Acknowledgements

The research leading to these results has received funding from the European Community's Seventh Framework Program (FP7/2007-2013) under grant agreement no. 245280, also known under the acronym PRESERF.

References

1. International Commission on Microbiological Specifications for Foods (1996.), "Microorganisms in Foods 5. Microbiological Specifications of Food Pathogens," ed. London, Blackie Academic & Professional Press.
2. Tompkin, R. B. (2002), "Control of *Listeria monocytogenes* in the food processing environment," Journal of Food Protection 65:709–725.
3. Porto-Fett A. C. S., Call J. E. and J. B. Luchansky (2008), "Validation of commercial process for inactivation of *Escherichia coli* O157H7, *Salmonella Typhimurium*, and *Listeria monocytogenes* ion the surface of whole muscle beef jerky," Journal of Food Protection 71:918–926.
4. Shabala L., Lee S. H., Cannesson P. and T. Ross (2008), "Acid and NaCl limits to growth of *Listeria monocytogenes* and influence of sequence of inimical acid and NaCl levels on inactivation kinetics," Journal of Food Protection 71:1169–1177.
5. Gola S., Frustoli M., Rovere P. and L. Miglioli (2003), "Inattivazione di *Listeria monocytogenes* in prosciutto crudo trattato con la pressione idrostatica. – Inactivation of *Listeria. Monocytogenes* in dry cured ham by hydrostatic pressure treatment," Ind Conserve 78: 441–449.
6. Daniels J. A., Krishnamurthi R. and S .S. H. Rizvi (1985), "A review of effects of carbon dioxide on microbial growth and food quality," Journal of Food Protection 48: 532-537.
7. Cheftel J. C. (1995), "High-pressure, microbial inactivation and food preservation," Food Science and Technology International 1: 75-90.
8. Spilimbergo S. and A. Bertucco (2003), "Non-thermal bacteria inactivation with dense CO_2," Biotechnology Bioengineering 84: 627-638.
9. Damar S. and M. O. Balaban (2006), "Review of dense phase CO_2 technology: Microbial and enzyme inactivation, and effects on food quality," Journal of Food Science 71: R1-R11.
10. Zhang J., Davis T. A., Matthews M. A., Drews M. J., LaBerge M. and Y. H. H. An (2006),

"Sterilization using high-pressure carbon dioxide," Journal of Supercritical Fluids 38: 354-372.
11. Mazzoni A. M., Sharma R. R, Demerci A. and G. R. Ziegler (2001), "Supercritical carbon dioxide treatment to inactivate aerobic microorganisms on alfalfa seeds," Journal of Food Safety 21: 215 – 223.
12. Calvo, L. and E. Torres (2010), "Microbial inactivation of paprika using high-pressure CO_2," Journal of Supercritical Fluids 52: 134 – 141.
13. Ferrentino G. and S. Spilimbergo (2011), "High pressure carbon dioxide pasteurization of solid foods: current knowledge and future outlooks," Trends in Food Science and Technology 22: 427 – 441.
14. Choi Y. M., Bae Y. Y., Kim K. H., Kim B. C. and M.S. Rhee (2009), "Effects of supercritical carbon dioxide treatment against generic *Escherichia coli*, *Listeria monocytogenes*, *Salmonella typhimurium*, and *E. coli* O157:H7 in marinades and marinated pork," Meat Science 82: 419 – 424.

(494g) The Use of Real-Time PCR to Determine the Growth Kinetics of Fungal Biomass On Solid Food At Different Water Activities

J.M.R. Apollo Arquiza and Jean B. Hunter, Biological and Environmental Engineering, Cornell University, Ithaca, NY

Fungal growth on solid foods usually make them unfit for human consumption but certain specialty foods require fungi to produce their characteristic properties. In either case, a reliable way of measuring biomass is needed in order to monitor fungal growth rates on these substrates and determine how they are affected by various factors such as water activity. In this study, the use of the species-specific real-time polymerase chain reaction (rt-PCR) to quantify the DNA of a target fungus and then relate this to its biomass was investigated. The developed procedure was utilized to determine the growth kinetic parameters of fungi growing on solid food. The advantage of this method is that fungal growth can be measured even with extensive substrate infiltration and that the growth rates of individual species in the presence of other fungi can be also be obtained. In this method, the ratio of the amount of DNA to biomass is a crucial parameter for converting the quantified DNA to biomass so the effects of the fungal species and age of colony on this ratio were studied. The test fungi used were Aspergillus niger and Penicillium chrysogenum which are species commonly involved in food spoilage. The kinetic parameters of the growth of these fungi on food were measured at different water activities. This method would be useful in determining the minimum water activity that will allow the growth of specific fungi on foods.

See more of this Session: Advances In Food and Bioprocess Engineering
See more of this Group/Topical: Food, Pharmaceutical & Bioengineering Division

(495a) Designing Novel Cellulase Systems Through Agent Based Modeling and Global Sensitivity Analysis

Advait Apte, Biological Systems Engineering, Virginia Tech, Blacksburg, VA, Ryan S. Senger, Biological Systems Engineering Department, Virginia Tech, Blacksburg, VA and Stephen S. Fong, Department of Chemical and Life Science Engineering, Virginia Commonwealth University, Richmond, VA

The efficiency of enzymatic digestion of lignocellulosic substrates is an important area of research for liquid biofuels produced through fermentation. Given the large research activity in this area, fewer research efforts have been directed towards directly comparing the different cellulolytic enzymatic systems available to microorganisms of interest. Specifically, these are (i) cellulase enzymes supported by a cellulosome and (ii) secreted cellulases. There is a growing need to assess the competing mechanisms of these enzyme systems, independent of the underlying organisms, to compare their efficiency in hydrolyzing cellulose. In this study, we have used agent based modeling (ABM) to build a computational model of complexed (cellulosome) and non-complexed (secreted enzyme) cellulose hydrolysis. Statistical analyses show that complexed cellulose hydrolysis involving an extracellular cellulosome is advantageous to non-complexed cellulose hydrolysis. Additional simulations carried-out to measure the specific role of different cellulases suggest positive correlation between the concentration of endoglucanases and efficiency of cellulose hydrolysis. These simulations were further supplemented by a global sensitivity analysis of the system. From these results, we have derived a series of rules for metabolic engineers and synthetic biologists to facilitate choosing effective cellulase systems or designing new ones based on the lignocellulosic substrate available.

See more of this Session: Advances In Metabolic Engineering and Bioinformatics IB - Biofuels
See more of this Group/Topical: Food, Pharmaceutical & Bioengineering Division

(495b) Improving Exogenous Sugar Utilization by Engineered Transporters

Eric M. Young, Chemical Engineering, University of Texas at Austin, Austin, TX and Hal Alper, Chemical Engineering, The University of Texas at Austin, Austin, TX

Metabolic engineering is a powerful platform for controlling cellular production of biomolecules. However, classical metabolic engineering tools do not address molecular transport across the cellular membrane – the key first step in many pathways. Molecular transport is rarely an issue when an organism utilizes its preferred carbon source, yet limitation is clearly evident when altering the carbon source preferences of an organism. As an example, molecular transport limits pentose utilization in *Saccharomyces cerevisiae*, an important problem in lignocellulosic biomass conversion to biofuels. Due to the lack of specific D-xylose and L-arabinose transporters, pentose uptake occurs via hexose transporters. Metabolic flux is limited by low pentose affinity and hexose inhibition. Here, we present evidence that heterologous proteins may be engineered using directed evolution to improve *S. cerevisiae* pentose growth characteristics and ultimately biofuels production. This study provides interesting insight into the possibility of molecular transporter engineering to improve metabolic pathways.

See more of this Session: Advances In Metabolic Engineering and Bioinformatics IB - Biofuels
See more of this Group/Topical: Food, Pharmaceutical & Bioengineering Division

(495c) Metabolic Engineering of Saccharomyces Cerevisiae for Improved Xylose Utilization

Hang Zhou, Jingsheng Cheng and Gregory Stephanopoulos, Chemical Engineering, Massachusetts Institute of Technology, Cambridge, MA

Xylose is the major pentose and the second most abundant sugar in lignocellulosic feedstocks. Therefore, the efficient utilization of xylose is required for the cost-effective bioconversion of lignocellulose. Metabolic engineering approaches have been extensively exploited on the yeast *S. cerevisiae* for improved xylose utilization. However, the bioconversion of pentoses to ethanol still presents a considerable economic and technical challenge.

In this study, we started with the rational genetic engineering of a laboratory S. cerevisiae strain to express the xylose metabolic pathway, including the xylose isomerase (XI) and the xylulokinase (XK). The non-oxidative pentose phosphate pathway (PPP) was also overexpressed to facilitate pentose assimilation. The resulting strains, H131-XYLA31, exhibited slow but significant aerobic growth (μmax=0.031±0.022 h^{-1}), establishing a baseline for further advancement.

The engineered strain mentioned above was then used to initiate a three-staged evolutionary engineering, through aerobic and oxygen-limited sequential batch cultivation followed by xylose-limited anaerobic chemostat cultivation. Continuous improvement was observed during adaption of the H131-XYLA31. The finally isolated strain, H131E8-XYLA31, displayed a significantly increased anaerobic growth rate (0.120±0.004 h^{-1}) and xylose consumption rate (0.916 $g{\cdot}g^{-1}h^{-1}$) compared to its parent strain.

Upon successful evolutionary engineering, the H131E8-XYLA31 was further modified by complementing the auxotrophic markers arg4 and leu2, resulting in H153E10-XYLA31 with greatly boosted aerobic growth. Moreover, adding the anaerobic growth factors ergosterol and Tween 80 to the medium enabled a maximum anaerobic growth rate of 0.199 h-1 and a specific xylose consumption rate of 1.647g·g-1h-1 in batch fermentation, 65% and 37% higher than those of the best reported xylose-fermenting strain RWB 218, respectively.

In order to identify the genotypes responsible for quick xylose utilization in the evolutionarily engineered strain, an inverse metabolic engineering approach was applied to the evolved strain, revealing the tandem duplication of *XYLA*. The structure of *XYLA* integration, coupled with qPCR, DNA/RNA blotting, and enzyme activity assay results, suggests that the high expression level of XI is a major recombination event during the evolution and is necessary for efficient xylose assimilation.

Moreover, mutations on *GRE3* that potentially abolish XR activity were discovered in the evolutionarily engineered strains, supporting the hypothesis that the elimination of XR activity was favorable for xylose utilization by reducing the production of xylitol and hence alleviating the inhibition of XI activity caused by xylitol.

In this study, we successfully applied rational and combinatorial metabolic engineering approaches for both constructing rapid xylose-fermenting strains and identifying novel genetic characteristics. The work serves as a practical milestone in lignocellulose conversions. The constructed strains can be used as hosts for further advancement, and the metabolic engineering techniques have proven to be effective tools for future strain evolution.

See more of this Session: Advances In Metabolic Engineering and Bioinformatics IB - Biofuels
See more of this Group/Topical: Food, Pharmaceutical & Bioengineering Division

(495d) Elementary Mode Analysis of Ralstonia Eutropha Metabolism for the Production of Useful Metabolites

Gilsinia Lopez and Friedrich Srienc, Department of Chemical Engineering and Materials Science, and BioTechnology Insititute, University of Minnesota, Minneapolis/St.Paul, MN

Ralstonia eutropha H16 is a Gram-negative, facultative chemolithoautotrophic bacterium with the capability to synthesize many useful metabolites. One of the keys to the organism's lifestyle is its ability to use—alternatively or concomitantly—both organic compounds and molecular H2 as sources of energy. In the absence of environmental O2, R. eutropha H16 can switch to anaerobic respiration; a complete denitrification pathway allows the organism to exploit alternative electron acceptors such as NO3- and NO2-. Correlated with the flexible bioenergetics of the organism is its capacity to shift between heterotrophic and autotrophic growth modes. It fixes CO2 via the Calvin-Benson-Bassham (CBB) cycle *and* produces several useful metabolites like Poly(3-hydroxybutyric acid), isobutanol , 2,3-butanediol and ethanol. It can be easily manipulated genetically for the creation of knock out mutations, for integration of genes into the chromosome and it can maintain plasmids with appropriate selection techniques. To quantitatively evaluate the capabilities of the metabolism we have set up a model of the central metabolism that is based on known pathways of the organism. This model includes heterotrophic as well chemolithoautotrophic growth modes. The metabolic model consists of 21 reversible reactions, 51 irreversible reactions, 56 internal metabolites and 16 external metabolites exchanged through the cell membrane. Elementary Mode analysis on this model identified 55,585 modes for the litho-autotrophic growth mode and 61521 modes for the heterotrophic mode on fructose as carbon and energy source. We have used the results from this analysis to predict key genetic alterations in the metabolism that would direct the metabolic flux towards metabolites of interest including basic and advanced bio fuels as well as Poly(3-hydroxybutyric acid) and other secondary metabolites. The experimental realization of predicted metabolic networks is currently underway.

See more of this Session: Advances In Metabolic Engineering and Bioinformatics IB - Biofuels
See more of this Group/Topical: Food, Pharmaceutical & Bioengineering Division

Analysis of growth of *Thermus thermophilus* HB8 using mini-bioreactors and defined growth medium without yeast extract

Jing Lu and Maciek R. Antoniewicz

Department of Chemical Engineering, University of Delaware

150 Academy St, Newark, DE 19716

ABSTRACT

The growth of *Thermus thermophilus* HB8, a thermophilic bacterium, was characterized in batch cultures under defined experimental growth conditions. First, the defined growth medium was optimized to maximize the growth rate without the need to add yeast extract to the medium. After that, growth rates were measured under aerobic growth conditions for temperatures ranging between 50 C and 90 C with glucose as the only carbon source. The optimum growth temperature was determined to be 85 C. The maximum growth rate of *T. thermophilus* was 0.28 1/hr, which corresponded to a doubling time of about 2.5 hours. In our growth experiments, *T. thermophilus* was cultured in custom designed mini-bioreactors at 10-mL scale that were constructed using Hungate tubes and septum caps. We used uniformly 13C-labeled glucose [U-13C]glucose as the carbon source in our experiments and measured [13C]CO2 production in the offgas by an online mass spectrometer (at m/z 45). With this setup we followed the growth of cells in real-time. In addition to measuring CO2 production, we also measured glucose concentration and optical density at 600 nm (OD600) during the cultures. From all of these measurements, we calculated growth rates for the exponential growth phase. We are currently investigating the potential of using *T. thermophilus* for the production of various biofuels.

(495f) Comparative Genomic Analysis of a High n-Butanol Tolerant and Producing Mutant Strain of Clostridium

Jingbo Zhao, William G. Lowrie Department of Chemical and Biomolecular Engineering, The Ohio State University, Columbus, OH and Shang-Tian Yang, Department of Chemical and Biomolecular Engineering, Ohio State University, Columbus, OH

n-Butanol is an important industrial chemical and also has been proposed as an alternative biofuel superior to ethanol. Biological production of *n*-butanol is promising because it can use renewable biomass as feedstock and is environmentally friendly. Unfortunately, *n*-butanol is more toxic than ethanol, which leads to the low final *n*-butanol titer and consequently high recovery cost. To improve *n*-butanol tolerance and final titer, a parent *Clostridium* strain commonly used in acetone-butanol-ethanol (ABE) fermentation was evolutionarily engineered to finally derive a mutant strain that produced 21 g/L *n*-butanol in free cell fermentation and 28 g/L in immobilized cell fermentation. Besides dramatically improved *n*-butanol tolerance and production, compared with the parental strain, the mutant also showed improved autolysis resistance, robust adjustment of membrane fatty acid composition, more complete substrate utilization, and remarkable proteomic change. To further investigate the molecular basis for the acquired traits of the mutant, the complete genomes of the parent and mutant strains were sequenced via the Illumina Hiseq2000 and comparative genomic analysis was conducted. SNP and short indel analyses revealed many variations which might have contributed to the characteristics of the hyper-*n*-butanol tolerant and producing mutant strain. The results can be used in further strain improvement via inverse metabolic engineering and evolutionary engineering.

See more of this Session: Advances In Metabolic Engineering and Bioinformatics IB - Biofuels
See more of this Group/Topical: Food, Pharmaceutical & Bioengineering Division

(495g) Probing Carbon Flow Patterns in a Photosynthetically Efficient Marine Diatom Toward Biofuel Production

Yuting Zheng and **Ganesh Sriram**, Chemical and Biomolecular Engineering, University of Maryland, College Park, MD

Diatoms are unicellular, photosynthetic, marine algae that constitute a major component of phytoplankton. The abundant lipid content of diatoms motivates their use in the manufacture of lipid-derived biofuels such as biodiesel. The recent sequencing of diatom genomes has led to the hypothesis that diatom biochemistry is unusual as compared to that of other algae or plants. Specifically, the observation that diatoms exhibit high photosynthetic productivity (>20% of global photosynthetic productivity and molecular oxygen output) despite surviving in CO_2-depleted environments points to the operation of an efficient photosynthetic mechanism in them. However, a quantitative investigation of diatom biochemistry in terms of metabolic flux measurements through central carbon metabolism has not been reported in the literature. Performing this investigation will have implications toward engineering diatoms for improved CO_2 fixation and lipid production and ultimately toward engineering synthetic CO_2-sequestering devices.

Towards this objective we performed isotope-assisted metabolic flux analysis (MFA) on the model diatom *Phaeodactylum tricornutum*. This organism is an ideal choice for studying diatom biochemistry because both a sequenced genome and reverse genetics tools are available for it. In steady state isotope labeling experiments performed under constant light, we supplied *P. tricornutum* cultures with combinations of several $^{13}C/^{12}C$ carbon sources including CO_2 and glucose. We measured isotopomer abundances of biomass components such as proteinogenic amino acids by mass spectrometry, and employed metabolic network modeling to convert the isotopomer abundances to metabolic fluxes. In this presentation we will discuss how the isotopomer data and flux results from these experiments suggest atypical and previously unexpected metabolic flux patterns, such as the path traced by CO_2 through central carbon metabolism to lipids. We anticipate that the knowledge generated from this investigation to be pivotal towards developing metabolic engineering strategies for *P. tricornutum* and other diatoms.

See more of this Session: Advances In Metabolic Engineering and Bioinformatics IB - Biofuels
See more of this Group/Topical: Food, Pharmaceutical & Bioengineering Division

(497a) Transferring Monitoring Models Between Different Scales Through Multivariate Statistical Techniques

Emanuele Tomba[1], Pierantonio Facco[1], Fabrizio Bezzo[1], Salvador Garca-Muoz[2] and Massimiliano Barolo[1], (1)DIPIC - Department of Chemical Engineering Principles and Practice, University of Padova, Padova, Italy, (2)Process Modeling & Engineering Technology, Pfizer Worldwide Research & Development, Groton, CT

The challenge of transferring a process monitoring model across different scales is a different problem than the one addressed in previous studies that have focused on transferring operating procedures looking to produce the same grade of product [1, 2, 3, 4].

In the development of a new pharmaceutical product, it is of paramount importance to shorten the launch time of the product, and for new compounds it may be that this is the first time the product is manufactured at pilot scale. At the pilot scale, an extended experimental campaign can be carried out to gain process understanding and disclose potential pitfalls in the process equipment, operating conditions or control configuration. This experimental campaign eventually leads to the definition of a process equipment layout as well as a set of proper ("normal") process operating conditions that can guarantee the required product quality with acceptable variability. If the experimental work at the pilot scale is extended to produce a data set where common cause variability can be observed, then a multivariate statistical process monitoring model can be designed and used to detect (and possibly diagnose) process faults and malfunctions at the pilot-plant scale in real time, thus gaining further process understanding.

When the product is transferred to the commercial manufacturing scale, it is highly desired to have an effective process monitoring framework available as quickly as possible. Carrying out an extended experimental campaign to produce data (where the common cause variability can be observed) is not a viable option at this scale, and therefore the issue arises on whether the monitoring model developed at the pilot-plant scale can be transferred up to the production scale to monitor the manufacturing process until a sufficient amount of data have been collected to design a full production-scale monitoring model.

This work addresses the above issue. Some approaches will be presented that use multivariate statistical techniques [1, 5, 6] and enable transferring a process monitoring model between different scales. The approaches are adaptive, in that initially they rely heavily on the wealth of information embedded in the pilot-scale data, but the relevance of these data is progressively discounted as long as more production-scale data become available. The proposed approaches are tested and critically evaluated using real-world data coming from a spray-drying process [7], and prove to be very effective at monitoring normal conditions and quickly detecting process faults at the production scale using data from the pilot scale. The presented methods will accelerate the implementation of continuous quality assurance programs in pharmaceutical manufacturing by transferring a greater portion of the knowledge generated at the pilot scale.

References

[1] Garcìa-Muñoz, S., J. F. MacGregor, and T. Kourti (2005). Product transfer between sites using Joint-Y PLS. *Chemom. Intell. Lab. Sys.*, **79**, 101-114.

[2] Jaeckle, C. M., and J. F. MacGregor (2000). product transfer between plants using historical process data. *AIChE J.*, **46**, 1989-1997.

[3] Jaeckle, C. M., and J. F. MacGregor (2000). Industrial application of product design through the inversion of latent variable models. *Chemom. Intell. Lab. Sys.*, **50**, 199-210

[4] Jaeckle, C. M., and J. F. MacGregor (1998). Product design through multivariate statistical analysis of process data. *AIChE J.*, **44**, 1105-1118.

[5] Jackson, J. E. (1991). A user's guide to principal components. John Wiley & Sons Inc., New York (U.S.A.).

[6] Geladi, P. and R. Kowalski (1986). Partial least squares regression: a tutorial. *Anal. Chim. Acta*, **185**, 1-17.

[7] Dobry E.D., D.M. Settell, J.M. Baumann, R.J. Ray, L.J. Graham and R.A. Beyerinck (2009). A model-based methodology for spray-drying process development. *J. Pharm. Innov.*, **4**, 133-142.

See more of this Session: Applications of Multivariate Data Analysis In Pharmaceutical Process Development: Traditional and Advanced Multivariate Methods
See more of this Group/Topical: Topical I: Comprehensive Quality by Design in Pharmaceutical Development and Manufacture

(497b) Implementation of Genetic Algorithms In the Generation of High-Order Statistical Models

Brendan C. Mack, Nathan Domagalski, Daniel Hallow, Michael Fenster, Lindsay Hobson and Jose Tabora, Bristol-Myers Squibb Company, New Brunswick, NJ

To support the quality by design (QbD) philosophy, it is critical to understand the behavior of chemical reactions used for API manufacturing. The parameter response space may be mapped experimentally and further explored by mechanistic or statistical modeling. In this work, the formation of a pharmaceutical intermediate was studied via a series of fraction factorial experimental designs. The results were then used to construct statistical models for several responses, including starting material conversion, desired product formation, and key impurity formation. The models were constructed using a genetic algorithm to select optimal variables from all the possible linear, collinear and quadratic terms. Although the models demonstrated good agreement with many of the experimental responses, product formation was not accurately predicted suggesting either experimental error of inadequate representation from a quadratic statistical model. To improve the models different approaches were explored to increase the order of the statistical model. The resulting models had much better predictive power than the quadratic models. This talk will discuss the systematic construction of high order statistical models and their implementation in the QbD paradigm.

See more of this Session: Applications of Multivariate Data Analysis In Pharmaceutical Process Development: Traditional and Advanced Multivariate Methods
See more of this Group/Topical: Topical I: Comprehensive Quality by Design in Pharmaceutical Development and Manufacture

(497c) Understanding the Impact of the Properties of Raw Materials Onto Product Quality Using JR-PLS

Salvador Garca-Muoz, Process Modeling & Engineering Technology, Pfizer Worldwide Research & Development, Groton, CT

Multiple proposals are found in literature to analyze the impact of raw materials onto product using data analysis. From the methods proposed, the L-PLS is the only method that not only considers the ratio in which ingredients are mixed, but also the physical properties of the lots of raw material. This method however is limited to analyze data from materials of the same nature such that all the physical properties apply to all materials. This is not the case when a product includes materials that need to undergo different physcial/chemical tests due to theiir nature. Modeing this scenario using the L-PLS would include blocks of missing data with non-overlapping rows or columns.

A new method is presented to handle this situations, the Joint-R PLS (JRPLS) combines the JYPLS and the LPLS methods providing a plattform to model the impact of the physical/chemical properties of multiple ingredients of different natures onto the final quality attributes of the product. The use and advantage of the model is shown in the troubleshooting of multiple scenarios where a complex relation of physical/chemical properties across multiple ingredients was identified as the root-cause of a situation being investigated.

See more of this Session: Applications of Multivariate Data Analysis In Pharmaceutical Process Development: Traditional and Advanced Multivariate Methods
See more of this Group/Topical: Topical I: Comprehensive Quality by Design in Pharmaceutical Development and Manufacture

(497d) De-Risking Scale-up of a High Shear Wet Granulation Process Using Latent Variable Modeling and near Infrared Spectroscopy

Koji Muteki, Ken Yamamoto, George L. Reid and Mahesh K. Krishnan, Pfizer Inc, Groton, CT

In the development of wet granulated drug products, two primary sources of variance (disturbance) include the operational scale of the high shear wet granulation (HSWG) process and Active Pharmaceutical Ingredient (API) lot-to-lot variability, particularly for formulations containing a high drug load. This paper presents a novel Process Analytical Technology (PAT) strategy using Latent Variable Modeling (LVM) with Near Infrared (NIR) spectroscopy to reduce risk in scale-up operations of the HSWG process while simultaneously accounting for API lot-to-lot variability, even with limited manufacturing history. The process involves building a Partial Least Square (PLS) model among the API material properties, the HSWG process parameters and the NIR end points of the HSWG process based on small scale DOE (Design of Experiment) batches. The PLS model is then used in an optimization framework to find suitable small scale mechanical process parameters (impeller/chopper speed) that approximate a previous large scale operation so as to keep the NIR end point of large scale operation constant. Prior to making additional large scale batches with a new lot of API, NIR end points of large-scale HSWG with the new API lot are predicted based on the PLS model developed from the small-scale operation. If the predicted NIR end point for the HSWG using the new API lot is not within the target region, the risks associated with the scale-up operation can then be significantly reduced by modifying other HSWG process parameters such as the total amount of water added or total granulation time to achieve the target region. A case study is presented that demonstrates the effectiveness of this methodology during development and scale-up of a drug product manufactured using a HSWG process.

See more of this Session: Applications of Multivariate Data Analysis In Pharmaceutical Process Development: Traditional and Advanced Multivariate Methods
See more of this Group/Topical: Topical I: Comprehensive Quality by Design in Pharmaceutical Development and Manufacture

(497e) Diverse Uses of Latent Variable Methods In Vaccines and Biologics Manufacturing

Charles Miller, Louis Obando and John Higgins, Manufacturing Division, Process Analytical Technologies (PAT), Merck and Company, Inc., West Point, PA

It is generally understood that latent variable multivariate modeling methods have great value-add potential in vaccines and biologics enterprises. What is relatively less-known, however, is the diversity of tasks that these tools can support in such applications. Although the fundamental math and statistics behind these tools are fixed, each of these different tasks require unique considerations with respect to several attributes, including data alignment/integration, data preprocessing and scaling, model validation, computing hardware, and visualization of results.

This paper will use recent case studies to illustrate the various ways that latent variable methods support current vaccines and biologics efforts at Merck, which include QbD initiatives. These case studies will include retrospective analyses of existing operations for troubleshooting, exploratory modeling of processes for continuous improvement efforts, multi-scale models for scale-down model qualification, and consistency monitoring of manufacturing operations.

See more of this Session: Applications of Multivariate Data Analysis In Pharmaceutical Process Development: Traditional and Advanced Multivariate Methods
See more of this Group/Topical: Topical I: Comprehensive Quality by Design in Pharmaceutical Development and Manufacture

(497f) Insights Into Lactate Metabolism Through Multivariate Analysis of Cell Culture Bioprocess Data

Huong Le[1], Santosh Kabbur[2], Ziran Sun[3], Luciano Pollastrini[3], Kevin Johnson[3], Keri Mills[3], George Karypis[2] and Wei-Shou Hu[1], (1)Chemical Engineering and Materials Science, University of Minnesota, Minneapolis, MN, (2)Computer Science and Engineering, University of Minnesota, Minneapolis, MN, (3)Manufacturing Sciences and Technical Services, Genentech, Inc., Vacaville, CA

Multivariate data analysis has recently emerged as a critical tool to gain fundamental understandings of process characteristics, and thus assisting the search for possible intervention means to improve process performance. The time dynamics of 150 process parameters acquired from 243 production runs from the Genentech's Vacaville manufacturing facility were investigated in this study. Two complementary multivariate approaches, kernel-based support vector regression (SVR) and partial least square (PLS) regression, were used to predict process outcome indicated by the final antibody concentration and the final lactate concentration. Both productivity and cellular metabolic state can be predicted accurately in the early stage of the inoculum train, suggesting that the history of the culture exerts critical impact on the final process outcome. Furthermore, specific lactate production rate, viable cell density, and viability were identified as key parameters during the inoculum train contributing to the final process performance. In the production stage, most critical parameters were also related to lactate metabolism and cell growth, suggesting the important role of these two aspects of process characteristics. Interestingly, lactate consumption appeared to be a prominent factor in determining final process outcome, and was further investigated. This study represents an important step towards implementing PAT and QbD principles to enhance process understanding and facilitate dynamic control for improved process robustness.

See more of this Session: Applications of Multivariate Data Analysis In Pharmaceutical Process Development: Traditional and Advanced Multivariate Methods
See more of this Group/Topical: Topical I: Comprehensive Quality by Design in Pharmaceutical Development and Manufacture

(506a) Continuous Production of Stable Drug Nanoparticle Suspensions Via Wet Stirred Media Milling

Ecevit Bilgili, Chemical, Biological and Pharmaceutical Engineering, New Jersey Institute of Technology, Newark, NJ and Afolawemi Afolabi, Chemical, Biological, and Pharmaceutical Engineering, New Jersey Institute of Technology, Newark, NJ

One way to improve the bioavailability of poorly water-soluble compounds is to reduce particle size of drug crystals down to nanoscale via wet stirred media milling. To prevent aggregation of nanoparticles, polymers and surfactants are dissolved in water acting as stabilizers via adsorption onto the drug crystals. The nanoparticle suspensions can be either filtered/sterilized and packaged (oral liquid, nasal, inhalation, parenteral) or can be converted into a solid dosage forms (capsules, tablets) upon various spraying processes such as spray drying, fluidized bed spray drying or fluidized bed coating/granulation.

In the last two decades, ample experimental data were generated in the area of nanoparticle suspension production via media milling. Recirculation mode of milling operation, where the nanoparticle suspension in a holding tank continuously circulates through the stirred media mill, has been commonly used in lab, pilot, and commercial scales. Although the circulation is continuous, the recirculation operation mode is overall a batch operation, requiring significant number of batches for a large-volume pharmaceutical product. To the best knowledge of the authors, a detailed engineering analysis of the continuous wet stirred media milling of drugs does not exist in the open literature. We present a comparative analysis of recirculation mode and continuous mode by studying the wet milling of Griseofulvin as a model drug in a stirred media mill. A suitable set of operating and formulation parameters was obtained by conducting milling experiments in the recirculation mode. The knowledge generated from recirculation milling guided the experiments with continuous milling mode, where drug suspension was passed through the stirred media mill multiple times. This multi-pass continuous operation with a single mill closely mimics the operation of a number of mills in series. Impact of various processing parameters on the mean residence time and the process dynamics has been studied with the hope of optimizing the continuous milling process. Our results suggest that continuous milling of drugs in a series of few mills leads to the production of stable nanoparticle suspensions. Continuous mode can produce as small particles as recirculation mode given sufficient residence time in a mill or minimum number of mills in series.

See more of this Session: Applications of Continuous Processing In Manufacture of Drug Product
See more of this Group/Topical: Food, Pharmaceutical & Bioengineering Division

(506b) A Plant-Wide Dynamic Model of a Continuous Pharmaceutical Process

Brahim Benyahia, Richard Lakerveld and Paul I. Barton, Chemical Engineering, Massachusetts Institute of Technology, Cambridge, MA

The pharmaceutical industry is undergoing rapid changes. Economic, environmental, health and safety requirements in conjunction with the regulatory environment are driving the sector towards more efficient processes. This requires innovation and cutting edge scientific, engineering knowledge and quality management systems. As a result, shifting from the traditional batch-wise to continuous pharmaceutical manufacturing (CPM) becomes a relevant alternative. In addition, the use of new process analytical technology (PAT), particularly on line, enables engineers to get better understanding and control of the underlying physical and chemical phenomena and makes it possible to build efficient CPM plants.

In this work, a plant-wide dynamic model of a continuous pharmaceutical pilot plant, including upstream and downstream units, is presented. This model is used to study the effect of design parameters such as purge ratio, reactor and crystallizer volumes on yield and purity. The objective is to minimize off-spec tablets production, particularly during the start up. The main specification used are the impurities levels. Impurities can arise out of raw materials, reagents, intermediates used in synthesis or bio-synthesis, by-products, degradation products, residual solvents or isomers. These impurities can affect safety and also efficacy of the drug. Therefore, a high level of purity of drug substance and the control of the impurities profile are important criteria in manufacturing of a safe and efficient drug. In the present work, all raw materials, solvent and excepients are considered to be pure, and only impurities that are from reactions (i.e. by-products, isomers) and residual reagents, intermediates and solvents are considered.

See more of this Session: Applications of Continuous Processing In Manufacture of Drug Product
See more of this Group/Topical: Food, Pharmaceutical & Bioengineering Division

(506c) NIR Spectroscopy with Multivariate Statistical Process Control for Selected Pharmaceutical Processes

Nicolas Heigl[1], Patrik Wahl[1], Otto Scheibelhofer[1], Daniel Koller[1], Merten Schlingmann[2], Franz Reiter[2], Gerhard Straka[2] and Johannes G. Khinast[3], (1)Research Center Pharmaceutical Engineering GmbH, Graz, Austria, (2)G.L. Pharma GmbH, Lannach, Austria, (3)Institute for Process and Particle Engineering, Graz University of Technology, Graz, Austria

Process analytical technology (PAT) is moving from research to the pharmaceutical industry and thereby providing adjuvant tools to the companies, especially for the continuous process monitoring. Thus a detailed understanding of the critical process parameters (CPPs) can be achieved following the guideline: "Quality cannot be tested into products; it has to be built in by design" [1, 2]. Here, we present two industrial use cases of near-infrared (NIR) spectroscopy with multivariate data analysis (MVDA) as a PAT tool in collaboration with G.L. Pharma GmbH.

Hot-melt extrusion is a novel method to embed drugs with a poor solubility in a polymer or wax matrix and thereby manufacturing retarded or delayed release forms. In order to determine the content homogeneity of the active pharmaceutical ingredient (API) in the extrudates, a Dynisco NIR probe was mounted at the extrusion die. Multivariate statistical process control (MSPC) was used to follow the drug content distributions within the sample matrix and for sample uniformity analysis. Furthermore, considerable spectral fluctuations indicated non-uniformly distributed ingredients due to a minimum feed rate of the implemented blender.

Fluid bed granulation is generally applied to gain uniform granules which are superior in terms of further processing, e.g. compaction and mixing, to bulk powders. The NIR spectral region shows strong water absorption signals and can therefore be used for the residual water determination. Specific NIR water signals can then be correlated with classical process data like outlet air humidity and core bed temperature. In this case the NIR spectral data was collected using a Sentronic SentroPAT FO process spectrometer which operates in the spectral region between 1100 nm and 2200 nm. The NIR probe was located at the sampler position of the granulator, about 30 cm above the ground plate.Here a batch analysis via NIR spectroscopy revealed a potential reduction of the process time by around 40 %.

[1] ICH, ICH Q8(R2), Pharmaceutical Development Part I: Pharmaceutical development

[2] ICH, ICH Q8(R2), Pharmaceutical Development Part II: Annex to pharmaceutical development

See more of this Session: Applications of Continuous Processing In Manufacture of Drug Product

(506d) Optimizing Continuous Powder Mixing Processes Using Periodic Section Modeling

Yijie Gao, Chem and BioChem Engineering, Rutgers University, Piscataway, NJ, Fernando. J Muzzio, Chemical and Biochemical Engineering, Rutgers University, Piscataway, NJ and Marianthi G. Ierapetritou, Department of Chemical and Biochemical Engineering, Rutgers, The State University of New Jersey, Piscataway, NJ

Optimizingcontinuous powder mixing processes using periodic section modeling

Yijie Gao, Fernando Muzzio andMarianthi G. Ierapetritou

Dept. of Chemical and Biochemical Engineering, RutgersUniversity, Piscataway, NJ 08854

Continuouspowder mixing is considered as an advantageous choice to batch mixing forbetter controllability and easier scale-up of production rate. Many works havebeen performed on the performance improvement of the continuous mixing process.In this study, the effects of operating conditions are investigated. Using theperiodic section modeling developed in our previous study [1],we investigated the effects on cross-sectional mixing, and on axial mixing and motion.The variance decay rate of the batch-like mixing in the cross-sectionaldirections (k_b), as wellas the mean particle velocity in the axial direction (v_x), is used as keyindices of performance in the continuous mixing process.

We consider mixingat different fill levels and blade speeds in order to characterize theireffects on powder mixing. Mixing cases with materials of different particlesize, particle density and cohesiveness are also compared. Results show similarcontours for both k_b and $v_x$among the studied cases, indicating that the optimal operation obtained incontinuous mixing of one mixing case can be directly used as the optimalcondition at another case. Based on the characteristics of contours, it is alsonoticeable that both variance decay rate $k_b$and axial velocity $v_x$increase with the blade speed. This leads to a trade-off on the improvement ofthe overall cross-sectional mixing efficiency. To resolve this issue, twooperating parameters are changed simultaneously, i.e. increase of blade speedand decrease of blade angle, or increase of both blade speed and weir height.While variance decay rate $k_b$increases proportionally with blade speed, an approximately constant axialvelocity $v_x$is obtained using this strategy. Results show dramatic improvement of theoverall cross-sectional mixing efficiency, and provide a promising directiontowards optimizing continuous powder mixing design and operating conditions.

Reference

1. GaoY, Ierapetritou MG, Muzzio FJ. Periodic section modeling of convectivecontinuous powder mixing processes. *AIChEJournal*. 2011; In Press: DOI: 10.1002/aic.12348.

See more of this Session: Applications of Continuous Processing In Manufacture of Drug Product
See more of this Group/Topical: Food, Pharmaceutical & Bioengineering Division

(506e) A Powder Mixing Model for a Small-Scale Continuous Mixing/Direct Compression Manufacturing Process

Daniel O. Blackwood[1], Bruno C. Hancock[1], Mark Polizzi[1], Bruce C. MacDonald[1], Hugh J. Clarke[2], Howard W. Ward II[2] and Yang A. Liu[2], (1)Pharmaceutical Development, Pfizer Worldwide Research and Development, Groton, CT, (2)Analytical Development, Pfizer Worldwide Research and Development, Groton, CT

Abstract will be uploaded at a later date.

See more of this Session: Applications of Continuous Processing In Manufacture of Drug Product
See more of this Group/Topical: Food, Pharmaceutical & Bioengineering Division

(506f) Model-Based Control of An Integrated Continuous Pharmaceutical Manufacturing Process

Rohit Ramachandran[1], Anwesha Chaudhury[2] and Marianthi G. Ierapetritou[1], (1)Department of Chemical and Biochemical Engineering, Rutgers, The State University of New Jersey, Piscataway, NJ, (2)Chemical and Biochemical Engineering, Rutgers, The State University of New Jersey, Piscataway, NJ

The pharmaceutical industry is a tightly regulated industry where all production must comply with good manufacturing practices (GMP) and quality requirements should be strictly satisfied. Historically, manufacturing in the pharmaceutical industry has been carried out in batch configuration which potentially results in expensive, inefficient and poorly controlled processes. Recently, both pharmaceutical industries and regulatory authorities have recognized that continuous manufacturing has significant potential to improve product quality. The shift from batch to continuous processing would necessitate the simulation of a complex process plant with interconnected unit operations, recycle and bypass streams and heat integration. This significantly alters the plant-wide process dynamics and increase the overall complexity of the integrated process. The interactions between process units may lead to poor performance of the decentralized unit-operation level control systems which are currently studied and found in the literature. Therefore a robust control strategy of the integrated process is required to support the overall plant-wide control strategy that needs to be adopted. The lack of an efficient control strategy for the integrated process remains a bottleneck in the overall implementation of a developing an efficient process control framework for the pharmaceutical manufacturing process.

The overall objective of this study is to develop a robust regulatory and advanced control framework to maintain critical quality attributes (CQA) of tablets (or intermediate products) at specified set-points given process disturbances, for the integrated pharmaceutical process consisting of several inter-connected unit operations. The focus will be on the manufacturing of oral solid dosage drugs which consist of approximately 85% of the entire pharmaceutical production. A typical manufacturing process for a powder based product (e.g., tablets and most capsules) is via continuous direct compaction and involves multiple processing steps, which are powder feeding, powder blending and tablet compaction. Tablet dissolution is also considered as a product model to relate tablet properties to the final dissolution rate which is required by the regulatory authorities to be tightly controlled. Simulations show that that at very mild interactions, a regulatory control strategy is able to maintain set-points at desired values. However, at moderate to high process interactions, oscillatory behavior of controlled variables is seen. Advanced control strategies (e.g. decouplers and model predictive control) were able to mitigate these deviations from set-point.

See more of this Session: Applications of Continuous Processing In Manufacture of Drug Product

(524a) The Effects of Hydrophobically Modified Chitosan On Liposomes

Rubo Zheng[1], Jingjing Zhan[1], Matthew B. Dowling[2], Srinivasa R. Raghavan[3], Joy St. Dennis[1], Pradeep Venkataraman[1], Jibao He[4] and Vijay T. John[1], (1)Department of Chemical and Biomolecular Engineering, Tulane University, New Orleans, LA, (2)University of Maryland, College Park, MD, (3) Department of Chemical and Biomolecular Engineering, University of Maryland, College Park, MD, (4) Central Instrumental Facilities, Tulane University, New Orleans, LA

We present results of work showing that chitosans that are hydrophobically modified with long alkyl groups have a protective influence on liposomes, allowing enhanced circulation times and sustained drug delivery. The alkyl groups partition into the lipid bilayers allowing the chitosan to form a coating that stabilizes the liposome. Such coated liposomes have extended circulation times and are protected from degradation by serum enzymes. We show the results of such coatings for spherical phospholipid liposomes and for a class of novel tubular liposomes obtained through the addition of sphingolipids (ceramides) to the phospholipid bilayer. Details of liposome structure, dynamics, and morphology are characterized through cryoelectron microscopy and high resolution NMR. As the concentration of hydrophobically modified chitosan is increased, the system transitions from a liquid to a gel where the lipsomes act as nodes in a network structure. The transition is understood through detailed rheological characterization. Results are presented on drug release from such modified liposomes as correlated with liposome structure and system viscoelastic characteristics.

See more of this Session: Drug Delivery II
See more of this Group/Topical: Food, Pharmaceutical & Bioengineering Division

(524b) Combating Pseudomonas Aeruginosa Biofilm Infections Using Synergistic Drug Therapies

Stacy Ross, Pharmaceutical Sciences and Experimental Therapeutics, The University of Iowa, Iowa City, IA, Annie Kock, Chemical and Biochemical Engineering, The University of Iowa, Iowa City, IA, Lucy Sanchez, Chemistry, The University of Iowa, Iowa City, IA and **Jennifer Fiegel**, Pharmaceutics and Translational Theraputics; Chemical and Biochemical Engineering, The University of Iowa, Iowa City, IA

Bacterial biofilms are formed when planktonic bacteria attach to a surface and create a polysaccharide matrix for protection. Within the body, bacteria in biofilms are less susceptible to antibiotic therapy and the immune system, causing current antibiotic treatments to be ineffective. Combinations of antibiotic drug compound have also been unable to eradicate biofilms. A new approach has focused on the use of dispersion compounds to entice the bacteria out of the biofilm, potentially enabling traditional antibiotics to kill the population of free-swimming bacteria. Therefore, a combination therapy of both antibiotics and dispersion compounds may provide a synergistic effect in breaking up and eradicating biofilms. It is our goal to develop dry powder aerosols to co-deliver dispersion and antibiotic compounds against pulmonary *Pseudomonas aeruginosa* biofilms. This bacterium causes chronic lung infections in various patient populations, including patients with cystic fibrosis, patients on mechanical ventilation, and immunocompromised patients.

Dry powders were formulated via spray drying to contain antibiotic, dispersion compound, and L-leucine to improve the flow properties of the powders. Key formulation and process parameters were varied to optimize the physical properties of the spray dried powders, including solids concentration in the initial solution, solution pH, liquid flow rate, atomizer air flow rate, and spray dryer inlet temperature. Optimized dry powder aerosols were formulated with up to 90% drug and had aerodynamic properties appropriate for deposition across the tracheobronchial and alveolar regions (mass median aerodynamic diameter = 3.0 – 5.3 μm).

Screening studies were conducted to evaluate the ability of a single dose co-treatment of antibiotic and dispersion compounds to eradicate *in vitro* cultures of a biofilm-forming mucoid strain of *Pseudomonas aeruginosa* (BAA-47; ATCC). Biofilms were exposed to various combinations of 6 antibiotic and 6 dispersion compounds. Twenty four hours after treatment, biofilms were stained with BacLight Live/Dead syto9 and propidium iodide stains, visualized via confocal microscopy, and quantified using COMSTAT. Dispersion compounds tested individually resulted in biofilms growth or no statistical change in growth. Synergy was found with four antibiotics and various dispersion compounds in reducing the total percent live bacteria remaining after treatment. Overall, results from these studies suggest that fine-tuning of antibiotic and dispersion compound concentrations may allow lower doses of antibiotic to be delivered to eradicate the biofilms.

See more of this Session: Drug Delivery II
See more of this Group/Topical: Food, Pharmaceutical & Bioengineering Division

(524c) Drug Delivery Using Stimuli-Responsive Polymer Gel Spheres

Pravin Ninawe and **Satish J. Parulekar**, Chemical and Biological Engineering, Illinois Institute of Technology, Chicago, IL

Polymeric gels that undergo deformation upon appropriate changes in pH or temperature have considerable promise as drug delivery vehicles. Uptake of drug macromolecules into swelling and non-swelling gel spheres and release of drug macromolecules from deswelling and non-deforming gel spheres into a target fluid are investigated here. A mathematical model for gel-solution composite, a composite of a distributed parameter system (gel spheres) and a lumped parameter system (surrounding solution), is developed. The polymer network displacement in swelling/deswelling gels is described by a stress diffusion coupling model. The analytical solution for network displacement is used to predict solvent intake by swelling gels, solvent efflux from deswelling gels, and changes in pressure, porosity and effective drug diffusivity resulting from network displacement. These in turn influence drug uptake during and after gel swelling and drug release from gel during and after gel deswelling. Numerical results illustrate benefits of gel swelling for drug loading and merits of different modes of drug release. Comparisons are made, as concerns drug uptake and drug release, with gels not subject to deformation.

See more of this Session: Drug Delivery II
See more of this Group/Topical: Food, Pharmaceutical & Bioengineering Division

(524d) Synthesis of Polyanhydride Particles As An Antibiotic Delivery Platform Against Intracellular Pathogens

Yashdeep Phanse[1], **Timothy Brenza**[2], Brenda Carrillo-Conde[2], Balaji Narasimhan[2] and Bryan H. Bellaire[1], (1)Vet Microbiology & Preventive Medicine, Iowa State University, Ames, IA, (2)Chemical and Biological Engineering, Iowa State University, Ames, IA

Polyanhydride particles elicit unique cellular responsesfrom immune cells by stimulating particle internalization and by directingintracellular trafficking. Through encapsulating antibiotics within the polyanhydride particles we can achieve sustained release ofthe antimicrobial compounds. The chemistry of the polymer and particle size canbe altered to influence the release of the antibiotic and the fate of theparticle within target cells can be optimized. Many microbial pathogens protectthemselves against immune defenses by residing and replicating within host cells.The sequestered environment within cells not only protects the pathogen fromkilling by extracellular serum proteins, but also from being exposed to highconcentrations of antibiotics. One such pathogen, *Brucellaabortus*,survives and replicates within macrophages. We encapsulated doxycyclinein polyanhydride nanoparticles based on sebacic acid (SA) and 1,6-bis(*p*-carboxyphenoxy)hexane(CPH). Results have demonstrated that encapsulation of antibiotics enablesgreater antimicrobial activity by improving intracellular delivery andsustaining continual antibiotic release. The overall effective concentration fortreatment was significantly lower compared to current standards of treatmentwith soluble antibiotics.

See more of this Session: Drug Delivery II
See more of this Group/Topical: Food, Pharmaceutical & Bioengineering Division

(524e) Discovery of Tight Junction Disrupting Peptides for Drug Delivery to the Brain

J. P. Trasatti, Chemistry, Rensselaer Polytechnic Institute, Troy, NY and Pankaj Karande, Chemical and Biological Engineering, Rensselaer Polytechnic Institute, Troy, NY

Currently, patients suffering from brain related disorders such as Alzheimer's, Parkinson's, Huntington's, epilepsy, stroke, brain tumors and sclerosis show very poor clinical outcome. Advances in basic neurobiology, medicinal chemistry, and protein and enzyme engineering have led to the discovery of a large number of potential therapeutics for treatment of brain related disorders. However, most candidate therapeutics will never make it to the clinic because their physico-chemical properties prevent them from crossing the blood brain barrier (BBB). We plan on addressing this important challenge by designing peptides that can selectively and reversibly permeabilize the BBB to therapeutic molecules. Specifically, we have discovered peptides that bind to claudin 5 an important transmembrane protein involved in the formation of tight junctions of the BBB. Claudin 5 binding peptides disrupt tight junctions of the BBB that are responsible for regulating transport across the BBB. Peptides are inexpensive as compared to fusion-protein or antibody-based therapeutics and are expected to show low toxicity and immune rejection. Further, our proposed strategy does not require chemical modification of the active drug, and can be used for delivery of promising candidates that have been discovered previously or are used in clinical settings with limited efficacy. If successful, this proposal will also provide a new lease on drugs that have failed due to poor brain penetration.

See more of this Session: Drug Delivery II
See more of this Group/Topical: Food, Pharmaceutical & Bioengineering Division

(524f) Lung Targeting with Triggered Release Using Gel Microparticles with Encapsulated Nanoparticles

Nathalie M. Pinkerton[1], Lei Shi[1], Jiandi Wan[2], Bryan Benson[1], Robert K. Prud'homme[1], Howard A. Stone[2] and Patrick J. Sinko[3], (1)Chemical and Biological Engineering, Princeton University, Princeton, NJ, (2)Mechanical and Aerospace Engineering, Princeton University, Princeton, NJ, (3)Department of Pharmaceutics, Rutgers University, Piscataway, NJ

The American Cancer Society estimated that in 2009, 1,479,000 new cancer cases would be diagnosed in the United States of which 219,000 would be lung and bronchus related. Although only the second most prevalent type of cancer, lung cancer is the most lethal accounting for a projected 159,000 deaths in the United States. National Cancer Institute assesses that results of standard treatment are generally poor with only a 15 percent 5-year survival rate for combined cancer stages. Challenges facing the current chemotherapy drugs include, hydrophobicity of therapeutic agents, excessive systemic toxicity in healthy tissues and inability to specifically target lung tumors. The drug delivery system described herein aims to overcome these challenges by selectively targeting the lung to deliver hydrophobic anti-cancer drugs and enabling triggered drug release.

The design of the human lung affords unique targeting options. Delivery via inhalation of anti-cancer drugs has been explored; however, low absorption and poor lung distribution of drugs has limited this avenue. A more promising approach involves passive targeting of the lung via the venous blood stream. The lung receives the entire venous blood supply from the heart and passes it through the intricate capillary beds on the alveoli. Large particles in the venous blood are thus trapped in these capillary beds. This filtering phenomenon can be used to selectively deliver particles to the lung. Such delivery methods have been safely employed in pulmonary perfusion diagnostic agents; however the use of this novel delivery route for chemotherapeutic drugs has not been appreciated or utilized by the drug delivery community.

The lung-specific delivery system employs both passive and active targeting to intravenously deliver anti-cancer drugs to tumor cells. The first layer of the delivery system is a poly(ethylene glycol) based gel microparticle (GMP) designed to take advantage of the venous lung filtration pathway and passively accumulate in the lungs after intravenous injection into the body. The second layer of the delivery system consists of nanoparticles (NP) embedded in the GMP matrix. The NPs are loaded with anti-cancer drugs and their surfaces can be decorated with ligands to actively target cancer cells once released from the gel matrix.

The size and modulus of the GMPs affect the targeting, accumulation and clearance from the lungs. Once deposited, triggered drug release from the GMP is controlled by two mechanisms. First, the GMP matrix is synthesized with crosslink junctions with tunable degradation kinetics. The construct involve PEG-di-(lactide-co-acrylate) chains where the number of lactide repeats controls the rate of hydrolysis. The hydrolysis of the network controls the release of the NPs *from the network.* In addition, a second level of release is provided by release of pro-drug conjugates *from the NP core*. Conjugation of hydrophobic lipids to the active drug via hydrolytically unstable acetal or orthoester linkages enables release of the drug from the NP core. This dual level of control of release enables precise tailoring of drug release in the lung. This dual-delivery system affords the ability to sustain a high concentration of anti-cancer drugs in the lungs while minimizing the systemic exposure and accordingly reducing the side effects.

See more of this Session: Drug Delivery II
See more of this Group/Topical: Food, Pharmaceutical & Bioengineering Division

(524g) Mechanistic Studies of Self Emulsifying Drug Delivery Systems for the Oral Delivery of Lipophilic Drugs

Fulden Buyukozturk[1], James Benneyan[2] and Rebecca L. Carrier[1], (1)Chemical Engineering, Northeastern University, Boston, MA, (2)Mechanical and Industrial Engineering, Northeastern University, Boston, MA

Oral drug delivery is a $35 billion dollar industry [1] and the most preferred way of drug administration. However, the oral route is not possible for 50% of currently marketed drug compounds due to their low solubility in water. Forty percent of prospective drugs coming out of discovery pipelines are lipophilic compounds, which leads to difficulty in achieving acceptable oral bioavailability. Lipid based drug delivery systems, and in particular self-emulsifying drug delivery systems (SEDDS), show great potential for enhancing oral bioavailability of lipophilic drugs, as well as offering the advantage of minimal processing and inherent stability, but have not been broadly applied, largely due to lack of general formulation guidance and lack of knowledge of how these systems function to enhance bioavailability [2, 3]. SEDDS are oil in water emulsions that typically have droplet sizes ranging from a few nanometers to hundreds of nanometers and consist minimally of oil, surfactant, and the drug to be delivered dissolved in oil (Figure 1). SEDDS are spontaneously formed in the gently mixed aqueous gastrointestinal (GI) environment [4].

To gain a better understanding of the dependence of emulsion drug delivery function in the GI tract on formulation composition, a quantitative study of properties central to emulsion function was conducted utilizing representative self-emulsifying formulations based on a 3^3 factorial experimental design. Formulations have been studied as follows. Oils from three different structural classes (long chain triglyceride (Soybean oil), medium chain trigylcerides (Neobee M5), propylene glycol dicaprylate/ dicaprate (Captex 200)) and surfactants with hydrophilic-lipophilic balance (HLB) values ranging from 10-15 (Cremophor EL, Tween 80, a mixture of Capmul MCM and Labrasol) were combined at three different oil-to-surfactant weight ratios (9:1, 5:1, 1:1). A model drug, Naproxen, was included in formulations to test the influence of drug incorporation on emulsion properties. One of the proposed mechanisms by which SEDDS are believed to influence oral bioavailability of water insoluble drugs is by diverting drug absorption to the lymphatic pathway. Another major mechanism is the formation of complex colloidal structures due to digestion of the oil component of the formulation in the intestine, which alters the overall drug solubility. Upon reaching the intestine, SEDDS begin to be hydrolyzed by lipase. Upon inception of digestion, digestion products are released from the surface of SEDDS emulsion droplets into the aqueous medium, forming complex colloidal structures. Meanwhile, drug compounds are released from SEDDS into intestinal fluid, and a portion of released drug further partitions into colloidal structures.

In this study, the influence of formulation composition on lymphatic drug transport was assessed in vitro by quantifying secretion of Apolipoprotein B (Apo B), a lipoprotein secretion marker, which is proportional to the increase of lymphatic drug transport, in intestinal epithelial Caco-2 cells exposed to formulations. Lipophilic drug associates with the triglyceride lipid core of the lipoproteins, and it has been suggested that amount of drug transported through the lymphatic pathway is proportional to the amount of lipoprotein secreted by intestinal cells. In vitro tests for Apo B secretion require 20 hour incubation of formulations with Caco-2 cells. Therefore, along with the Apo B test, MTT toxicity tests were carried out to assess the possible cytotoxicty of formulations due to the long incubation period. Low dilution formulations in biorelevant media (1:100, 1:200, 1:500) resulted in low levels of cell viability. 1:1000 dilution formulations, on the other hand, resulted in reasonably high viabilities; only formulations containing Captex 200 had low (30%) viabilities. 1:1000 dilution formulations therefore were used for Apo B secretion tests. Quantification of Apo B secretion with an enzyme-linked immunosorbent assay suggested the influences of certain formulation parameters on lymphatic transport. Formulations of Cremophor EL at high surfactant ratios showed a trend of inhibiting ApoB secretion compared to negative control of biorelevant media. In general, results did not indicate enhancement of lymphatic transport by the formulations tested.

Secondly, in an effort to perform real time quantitative tracking of drug partitioning and drug release into colloidal phases during in vitro digestion, electron paramagnetic resonance (EPR) was used to track a model drug (spin probe: Tempol Benzoate). In vitro lipolysis experiments were conducted in fasted state biorelevant simulated intestinal fluid. Formulations including Tempol Benzoate were added to 20 ml digestion buffer at 1:100 ratio, and digestion was started with the addition of 2.2 ml pancreatin extract. Samples were collected at specific time intervals (0, 5, 20, 50 minutes), and the digestion process was terminated by the addition of enzyme inhibitior, 4-BPB (4-bromophenacyl bromide). Samples were analyzed with EPR. Differences in peak to peak distances (a_N) and peak widths among EPR spectra give information related to probe distribution among different phases (oil vs. vesicles and micelles vs. water). For the formulations studied, results indicate a release of the probe from oil phase into biorelevant micellar phase over 15 minutes in the absence of digestive enzymes and an increased partitioning of the probe into the micellar phase over a 30 minute time period in the presence of digestion enzymes. Correlation of quantitative results on drug partitioning and lymphatic transport with formulation design parameters will be ultimately incorporated in an overall mechanistic mathematical model to predict and optimize oral absorption of a drug administered with SEDDS.

References

[1] Drug Deliver Markets. Edition 2 Volume I: Oral Delivery, Jan. 2007

[2] A.J. Humberstone, W.N. Charman, Lipid-based vehicles for the oral delivery of poorly water soluble

drugs. Adv. Drug Deliv. Rev 25(1) (1997) 103-128.

[3] R. Neslihan Gursoy, S. Benita, Self-emulsifying drug delivery systems (SEDDS) for improved oral delivery of lipophilic drugs. Biomed. Pharmacother. 58(3) (2004) 173-182.

[4] J. Yamamoto, H. Tanaka, Transparent nematic phase in a liquid-crystal-based microemulsion. Nature 409(6818) (2001) 321-

See more of this Session: Drug Delivery II
See more of this Group/Topical: Food, Pharmaceutical & Bioengineering Division

(527a) Formation of Itraconazole-Malic Acid Cocrystals In a CO2-Expanded Liquid

Courtney A. Ober and Ram B. Gupta, Chemical Engineering, Auburn University, Auburn, AL

Itraconazole, an antifungal drug, has poor aqueous solubility. To improve the dissolution rate, cocrystals of itraconazole with l-malic acid were prepared in tetrahydrofuran using supercritical carbon dioxide as an antisolvent. In this novel process, the drug (itraconazole) and cocrystal former (l-malic acid) are dissolved in a liquid solvent (tetrahydrofuran). Cocrystals of the two components are precipitated by pressurization with CO_2. The liquid solvent is then removed and the cocrystals dried with excess supercritical CO_2. These cocrystals were compared to cocrystals produced by using n-heptane, a traditional liquid antisolvent. The cocrystals produced by each method were characterized for crystallinity by X-ray diffraction (XRD), thermal behavior by differential scanning calorimetry (DSC), composition by high pressure liquid chromatography (HPLC), and size and surface morphology by scanning electron microscopy (SEM). Dissolution studies were also conducted for the cocrystals.

The XRD results indicate that cocrystals of itraconazole with l-malic acid can be produced using either supercritical CO_2 or a traditional liquid solvent, such as n-heptane. Both XRD and DSC confirm the presence of an itraconazole-malic acid cocrystal and verify the absence of either of the pure components. HPLC confirmed that the composition of the cocrystals matched that previously reported in the literature. Although the cocrystals produced by CXLCF were slightly larger than those precipitated with n-heptane, their microporous structure allowed for improved dissolution. Through use of XRD, DSC, HPLC, SEM, and dissolution studies, it is shown that the proposed cocrystallization is a viable method for producing pharmaceutical cocrystals of poorly water-soluble drugs to improved dissolution and reduce organic solvent content.

See more of this Session: Engineered Composite Particulate Systems for Pharmaceutical Active Ingredient Delivery
See more of this Group/Topical: Particle Technology Forum

(527b) Preparation of Calcium Stearate Pellets by Extrusion/Spheronization: Effect of Drying Technique

Simone Schrank[1], Aden Hodzic[2], Ben Glasser[3], Andreas Zimmer[1], Johannes G. Khinast[4] and Eva Roblegg[1], (1)IPS/Department of Pharmaceutical Technology, University of Graz, Graz, Austria, (2) Research Center Pharmaceutical Engineering (RCPE), Graz, Austria, (3)Rutgers University, Piscataway, NJ, (4)Institute for Process and Particle Engineering, Graz University of Technology, Graz, Austria

Pharmaceutical pellets intended for oral dosing are frequently prepared via the wet extrusion/ spheronization technique. The last "active" step of this process is drying. Although the drying method and/or the drying conditions applied are considered to highly influence the final pellet characteristics, investigations in this field are rare. Some studies are reported, mostly evaluating the moisture removal from pellets containing microcrystalline cellulose (MCC) [e.g.: 1].

The current study focuses on the drying behavior of calcium stearate (CS) pellets containing ibuprofen as model drug. These matrixpellets were shown to be spherical in shape and mechanically stable before [2]. After spheronization three different drying methods were applied; i) desiccation (D) over silica gel at ambient conditions, ii) drying in a fluid bed (FB) apparatus at an inlet air temperature of 40 °C and iii) lyophilization after freezing the pellets in liquid nitrogen (-196 °C). It was ensured via differential scanning calorimetry (DSC) that no interactions between the primary powders occurred during liquid removal. Furthermore, it was proved by X-ray measurements that neither CS nor ibuprofen underwent changes in their crystal stated during preparation.

The drying profile of the pellets dried via desiccation revealed a non-linear shape, where the rate of drying decreased throughout the process. This is mainly attributed to the shrinking of pellets by 13% during liquid removal causing changes in the surface area. CS was shown to swell in contact with the granulation liquid (i.e., 50% ethanol) and is likely to shrink again during drying. The shrinking of pellets during FB drying was less pronounced due to significantly lower drying times (i.e., by 8%). During lyophilization shrinking was largely suppressed since no liquid was involved resulting in minimal capillary flow. The differences in the shrinking behavior correlated with the pellet size distributions. The shape was not influenced by the drying conditions and the aspect ratio was below 1.2 for all batches.

Drug release studies demonstrated that pellets dried via desiccation liberated the lowest and pellets dried via lyophilization released the highest amount of ibuprofen. However, kinetics were best described by the Higuchi approach (i.e., diffusion based on Fick´s law) for all investigated batches. Differences in drug liberation behavior can occur due to variations in the spatial drug distribution throughout the pellets. It was shown before that for high drying rates (i.e., FB drying in the current study) dissolved components are likely to be transported towards the outer shell [3] where they recrystallize. Therefore, the ibuprofen distribution along the pellet cross sections was investigated using Raman mapping. However, ibuprofen was rather homogeneously distributed throughout all pellets despite the different ways of liquid removal. Although ibuprofen is highly soluble in the granulation fluid (i.e., 48 g/l), only small amounts of drug were expected to be dissolved during pellet preparation. Moreover, it is known from literature that ibuprofen causes phase separation of 50% ethanol at 40 °C [4] (i.e., the applied FB drying temperature). Subsequently, the dissolved ibuprofen crystallized in the water phase due to its poor water solubility [5]. Thus, the active pharmaceutical ingredient was not transported towards the external surface. Another reason for the different dissolution rates may be differences in structure and/or morphology. The crushing strength decreased in the order D, FB and L where FB and L showed similar results. This can be explained by the minimal extent of shrinking during L resulting in different morphological characteristics. During FB drying pellets were exposed to mechanical load. Thereby, cracks were likely to be formed which in turn lead to pellet breakage at lower forces. Moreover, it was found from the He-pycnometer measurements that during FB drying closed pores were more pronounced to be formed. Consequently, the variations in drug release characteristics can be explained by drying induced changes in the pellet morphology.

[1] B. Song, S.L. Rough, D.I. Wilson, Int. J. Pharm. 332 (2007), 38-44.

[2] E. Roblegg, S. Ulbing, S. Zeissmann, A. Zimmer, Eur. J. Pharm. Biopharm. 75 (2010), 56-62.

[3] A. Lekhal, B.J. Glasser, J. G. Khinast, Chem. Eng. Sci. 59 (2004), 1063-1077.

[4] J. Manrique, F. Martinez, Lat. Am. J. Pharm. 26 (2007), 344-354.

[5] M. Jbilou, A. Ettabia, A.M. Guyot-Hermann, J.C. Guyot, Drug Dev. Ind. Pharm. 25 (1999), 297-305.

See more of this Session: Engineered Composite Particulate Systems for Pharmaceutical Active Ingredient Delivery
See more of this Group/Topical: Particle Technology Forum

AUTHOR INDEX

AUTHOR INDEX

AUTHOR INDEX

AUTHOR INDEX

AUTHOR INDEX

AUTHOR INDEX

AUTHOR INDEX

AUTHOR INDEX

AUTHOR INDEX

AUTHOR INDEX

AUTHOR INDEX

AUTHOR INDEX

AUTHOR INDEX

AUTHOR INDEX

AUTHOR INDEX

AUTHOR INDEX

AUTHOR INDEX

AUTHOR INDEX

AUTHOR INDEX

Food, Pharmaceutical and Bioengineering Division

Core Programming Topic at the 2011 AIChE Annual Meeting

Minneapolis, Minnesota, USA
16-21 October 2011

Volume 2 of 2

ISBN: 978-1-61839-735-5

Printed from e-media with permission by:

Curran Associates, Inc.
57 Morehouse Lane
Red Hook, NY 12571

Some format issues inherent in the e-media version may also appear in this print version.

Printed by Curran Associates, Inc. (2012)

For permission requests, please contact AIChE
at the address below.

AIChE
3 Park Avenue
New York, NY 10016-5991

Phone: (203) 702-7660
Fax: (203) 775-5177

www.aiche.org

Additional copies of this publication are available from:

Curran Associates, Inc.
57 Morehouse Lane
Red Hook, NY 12571 USA
Phone: 845-758-0400
Fax: 845-758-2634
Email: curran@proceedings.com
Web: www.proceedings.com

TABLE OF CONTENTS

Volume 1

Volume 2

(527c) Particle Engineering Via Dry Coating: Development of a Novel Material Sparing Technology for Pharmaceutical Powders

Rajesh N. Dave, Chemical Engineering, New Jersey Institute of Technology, Newark, NJ, Lauren Beach, Otto H. York Department of Chemical, Biochemical and Pharmaceutical Engineering, New Jersey Institute of Technology, Newark, NJ, James V. Scicolone, New Jersey Center for Engineered Particulates, New Jersey Institute of Technology, Newark, NJ and Lakxmi Gurumurthy, Department of Chemical, Biological and Pharmaceutical Engineering, New Jersey Center for Engineered Particulates, Newark, NJ

Typically, most Active Pharmaceutical Ingredients (APIs) are extremely cohesive which adversely affects manufacturing processes such as feeding, blending, tabletting and leads to inferior quality of the product. In this work, different pharmaceutical powders (several APIs and excipients) of varying size range are considered for flow and packing density improvement via dry coating based surface modification. The main focus of this presentation is on developing a novel material sparing method that utilizes only small amount of API powders. The idea is to generate information during early API development to assess if the API requires flow improvement and to what extent the flow improvement can be achieved via dry coating. This methodology is validated via multi-faceted characterization, as well as comparison with other batch as well as continuous dry coating methods. Morphology, size and size distribution of these powders are characterized by Scanning Electron Microscopy (SEM) and laser diffraction (Sympatec Helos/Rodos) respectively. Flow characterization is performed using angle of repose (AoR), bulk and tapped density, critical outlet diameter via flodex, Vibration Packed Density (VPD) and shear testing (Freeman FT4). This allows for developing guidelines for the applicability of the novel technology as well as the dry coating approach. The results presented are expected to have an impact on the formulation development of poorly flowing APIs.

See more of this Session: Engineered Composite Particulate Systems for Pharmaceutical Active Ingredient Delivery
See more of this Group/Topical: Particle Technology Forum

(527d) Microstructural Characterization and Dissolution Behavior of Drug/Semicrystalline Polymer Systems

Qing Zhu[1], Lynne Taylor[2], Michael T. Harris[1] and Hsin-Yun Hsu[1], (1)Chemical Engineering, Purdue University, West Lafayette, IN, (2)Industrial and Physical Pharmacy, Purdue University, West Lafayette, IN

Time-resolved simultaneous small-angle X-ray scattering/wide-angle X-ray scattering (SAXS/WAXS) and second harmonic generation (SHG) microscopy has been employed to study the crystallization kinetics and structural evolution of drug/polymer systems. It was found that the drug formed larger crystallites at higher crystallization temperatures, and the overall crystallization kinetics seems slower. The dissolution study showed that the samples crystallized at higher temperatures shows much slower dissolution rate with different dissolution mechanism, compared to that prepared at lower temperatures, which is consistent with the SHG microscopy results.

See more of this Session: Engineered Composite Particulate Systems for Pharmaceutical Active Ingredient Delivery
See more of this Group/Topical: Particle Technology Forum

(527e) *In Vivo* Cellular Uptake of PEGylated Nanoparticles

Xi Zhan, Kenny K. Tran and Hong Shen, Department of Chemical Engineering, University of Washington, Seattle, WA

Polyethylene glycol (PEG) has been broadly utilized as a strategy in nanoparticle surface functionalization to prolong circulation and reduce nonspecific distribution in drug delivery systems. Modification of nanoparticles with PEG prevents protein adsorption and opsonization to delay non-specific phagocytosis. Previously we have shown PEG-mediated non-specific cellular uptake is cell-type dependent in a range of cell lines. In this study, we examined the *in vivo* uptake of polystyrene nanoparticles with varying PEG density on their surface. Unmodified nanoparticles showed high uptake by different cell types while highly PEGylated particles reduced internalization. Nanoparticles with low and intermediate PEG density showed cell-type dependent internalization. Internalization by phagocytes was highly affected by PEG density while epithelial cells internalized particles independent of PEG density.

See more of this Session: Engineered Composite Particulate Systems for Pharmaceutical Active Ingredient Delivery
See more of this Group/Topical: Particle Technology Forum

On single step assembly of uniform core-shell microparticles with tunable controlled release functionalities

W. **Liu**, W.D. **Wu**, C. **Selomulya**, X. D. **Chen***
Department of Chemical Engineering, Monash University VIC 3800, Australia
Department of Chemical and Biochemical Engineering, Xiamen University, 361005 Fujian Province, P.R. China
E-mail for correspondence: cordelia.selomulya@monash.edu

Core/shell microparticles are of interest for various applications including pharmaceutical engineering, due to the possibility of fine-tuning individual functionalities of the different parts. Previously, core-shell structured microparticles have been produced using concentric nozzles to deliver the core and shell components separately, followed by chemical consolidation of the droplets and further drying steps. To the best of our knowledge, no core-shell structure formation from a homogeneous precursor via spray drying using a single nozzle (orifice) has ever been reported. Furthermore, ordinary lab-scale spray dryers usually produce particles of various characteristics, e.g. size, morphology, structure within the same batch. The lack of reproducibility renders it difficult to correlate the properties of particles formed with the nature of the precursors or other process parameters.

Here we report a new approach for continuous and scalable production of core/shell microparticles, using acrylic nanoparticles (Eudragit® RS) as the shell, and hydrolized tetraethyl orthosilicate (TEOS) forming the silica core upon the single-step assembly. The method is capable of forming monodisperse microparticles from homogeneous precursors without the use of any organic solvents. Precise control of particle uniformity is realized via the use of a micro-fluidic spray drying technique, which has been previously demonstrated to generate monodisperse chitosan microparticles of specific sizes [1] and polymeric microparticles for controlled drug release applications [2, 3].

Evaporation induced self-assembly attained distinct phase separation among the different components during drying and resulted in the core-shell spatial configuration (**Figure 1**), while a precise control over particle uniformity was accomplished via a microfluidic jet spray dryer. Direct control over shell thickness can be achieved from the ratio of the core and shell ingredients in the precursors. A fluorescent compound, Rhodamine B, was used as a highly water-soluble model component to investigate the controlled release properties of these microencapsulates, with the release behaviors shown to be significantly dependent upon their architectures.

References:

1. Liu, W., Wu, W., Selomulya, C., Chen, X. D., 2011, Uniform chitosan microparticles prepared by a novel spray-drying technique, *International Journal of Chemical Engineering*, doi:10.1155/2011/267218, Article ID 267218.
2. Liu, W., Wu, W., Selomulya, C., Chen, X. D., 2011, A single step assembly of uniform microparticles for controlled release applications, *Soft Matter*, 7, 3323-3330.
3. Wu, D. W., Liu, W., Selomulya, C., and Chen, X.D., 2011, Synthesis of uniform silica microencapsulates via spray drying: effects of additives on particle size, morphology and in vitro drug release, (2011), Soft Matter. doi:10.1039/C1SM05879G (*in press*).

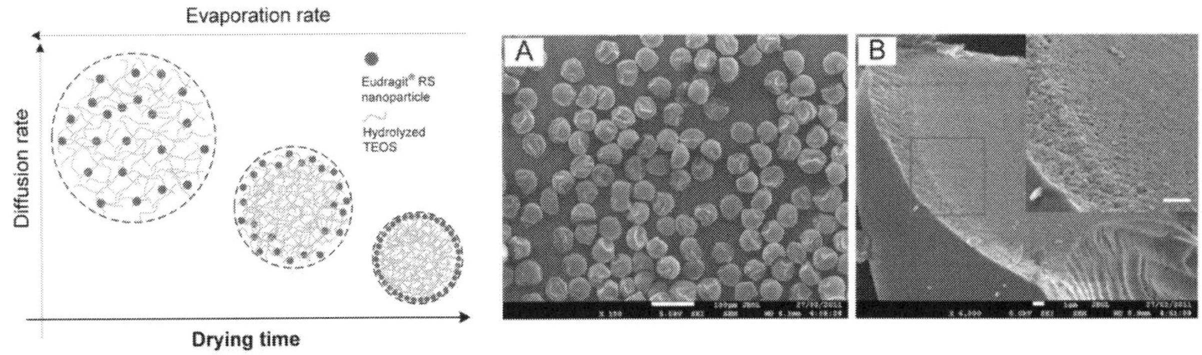

Figure 1 A schematic diagram of particle formation due to evaporation induced self-assembly during spray drying, with FESEM images of (A) monodisperse spray dried microparticles with average diameter of 55 μm, with (B) cross-section displaying (silica) core – (polymeric) shell configuration (inset scale bar: 1μm).

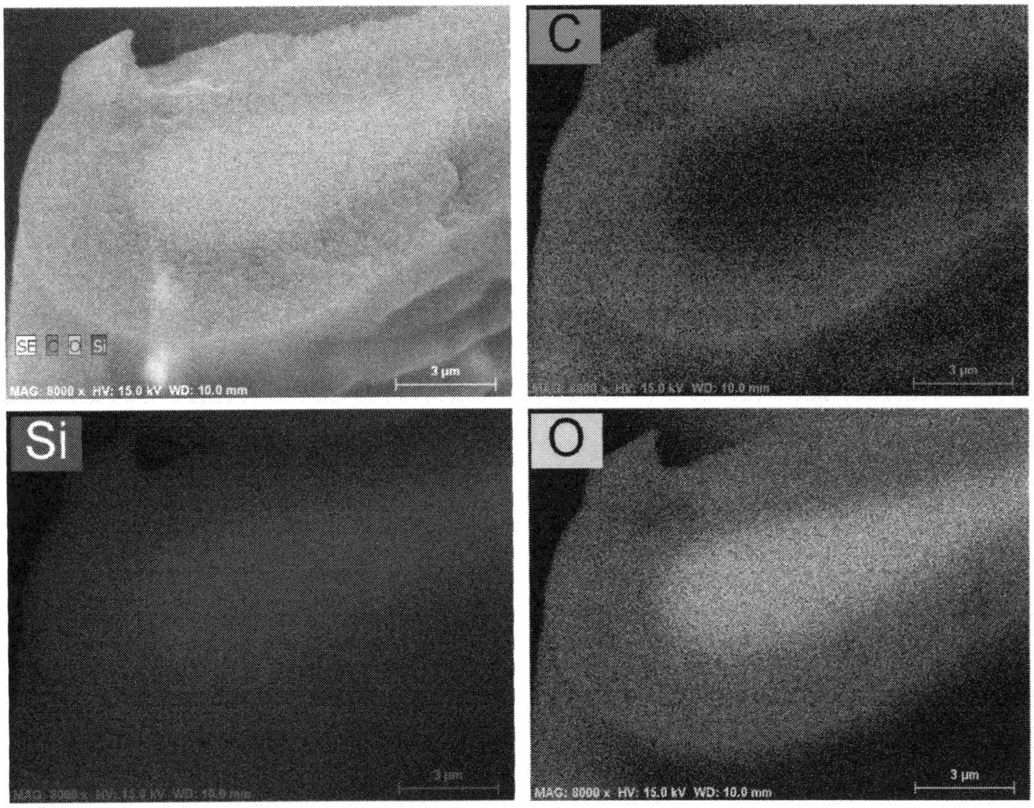

Figure 2 Elemental distribution maps of the cross section of microparticles composed of silica and polymer in 1 : 1 ratio (wt). The elemental "C" originated from Eudragit® RS is mostly present on the outer layer of the particles, indicating the formation of mostly polymeric shell, whereas the elementals "Si" and "O" are concentrated on the inner particles, implying silica as the main core component.

Spray Drying of Uniform Silica Microencapsulates: Implications for Controlled Release

Winston Wu[1], Wenjie Liu[1], Xiao Dong Chen[1,2], and **Cordelia Selomulya[1]**
[1] Department of Chemical Engineering, Monash University, Clayton VIC 3800 Australia
[2] Department of Chemical and Biochemical Engineering, College of Chemistry and Chemical Engineering, Xiamen University, 361005 Fujian Province, P.R. China

Extended Abstract

Spray drying is a scalable route for particle formation with easy product recovery. Conventional spray drying produces broad distributions of particles with various morphologies in a single batch, due to non-uniform distribution of droplets / spray trajectories and different residence times [1]. This renders any interpretation of particle behaviour in relation to their physico-chemical properties virtually impossible. Here, uniform silica microparticles encapsulating vitamin B_{12}, which was homogenously distributed within each particle, were synthesised in a single step via a micro-fluidic-jet-spray-dryer (Figure 1). The dryer utilised a micro-fluidic-aerosol-nozzle for continuous generation of monodisperse droplets from complex precursors [2]. The uniformity of the particles allowed direct correlations between their properties and the release behaviour of vitamin B_{12} to be investigated, without the complications of wide size distribution or non-uniform shapes. The effects of additives in the forms of lactose and Na-alginate to the properties of these microparticles were demonstrated. Spherical particles with relatively smooth surface were obtained with lactose addition, while incorporation of Na-alginate resulted in noticeably increasing surface roughness with higher amount added (Figure 2). Lactose addition also accelerated the release of the encapsulated vitamin B_{12}, due to the relatively fast lactose dissolution that allowed buffer to penetrate into the matrix to facilitate diffusion and silica erosion. On the contrary, Na-alginate slowed down the release considerably by serving as an additional barrier to decelerate silica matrix erosion. Release kinetics data indicated diffusion as the main release mechanism independent of the additives. The release profiles from different compositions of the synthesized particles demonstrated good agreement with computational results, highlighting the reproducibility of particles produced with this technique and the ability to modulate the release behaviour directly from the precursor compositions [3].

References:

1. Woo, M.W., Rogers, S., Lin, S.X.Q., Selomulya, C., Chen, X. D., 2011, Numerical probing of a low velocity concurrent pilot scale spray drying tower for mono-disperse particle production - unusual characteristics and possible improvements, *Chemical Engineering and Processing: Process Intensification*, doi:10.1016/j.cep.2011.02.007 (*in press*).
2. Wu, W., Amelia, R., Hao, N, Selomulya, C., Zhao, D., Chiu, Y., Chen, X. D., 2011, Assembly of Uniform Photoluminescent Microcomposites Using a Novel Micro-Fluidic-Jet-Spray-Dryer, *AIChE Journal*, doi: 10.1002/aic.12489 (*in press*).
3. Wu, W., Liu, W., Selomulya, C., and Chen, X. D., 2011, Synthesis of Uniform Silica Microencapsulates via spray drying: Effects of Additives on Particle Size, Morphology and In Vitro Drug Release, *Soft Matter*, doi:10.1039/C1SM05879G (*in press*).

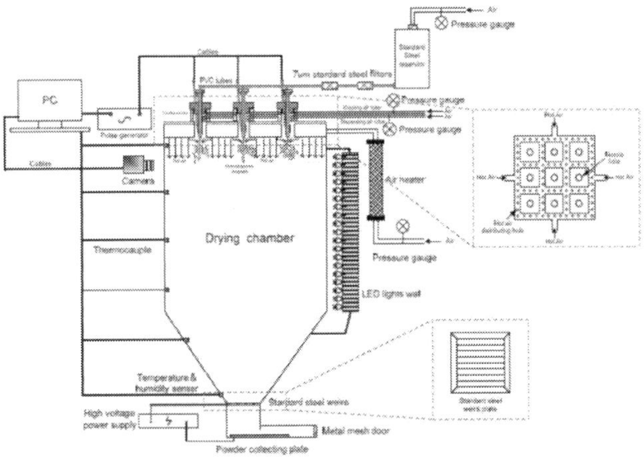

Figure 1 – Schematic of the micro-fluidic-jet-spray-dryer

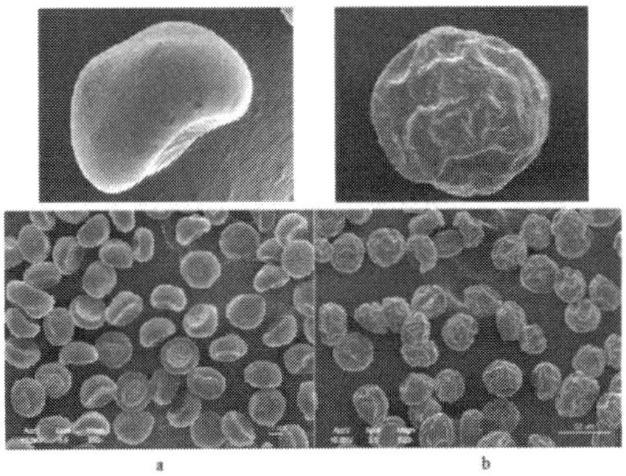

Figure 2 – Uniform silica microencapsulates with
(a) 0.5%w/v Na-alginate; (b) 1.0%w/v Na-alginate

(529a) Integrated Biogeochemical Modeling for Sustained Uranium Removal

Jiao Zhao and Radhakrishna Mahadevan, Department of Chemical Engineering and Applied Chemistry, University of Toronto, Toronto, ON, Canada

Improper management or accidental leakage ofhazardous wastes may lead to contaminated air, soil and water, which is notonly hazardous to public health but also to the ecosystem. For example, uranium contamination is aserious concern at several sites in theU.S. The soluble hexavalent uranium [U(VI)] can migrate to groundwater insurrounding areas, posing a problem for the drinking water safety. Whileevidences have shown that the reduction of soluble U(VI) to insoluble tetravalenturanium [U(IV)] is mainly a biotic process (microbial dissimilatory metalreduction), some bench-scale experiments indicate that U(VI) is readily reducedby some abiotic process (e.g. natural reductants such as ferrous iron and hydrogensulfide). Since experimental investigation of biotic and abiotic reduction wasconducted under different environmental conditions, it is necessary to developmodels that combine both biotic and abiotic factors for predicting establishmentof sustained uranium removal at actual environmental interfaces.

In this work, we developed a comprehensive 3-Dreactive-transport model for a recent column experiment (Moon et al., 2010)packed with the sediment from Old Rifle, CO. Groundwater from the Old Riflesite (which contains ca. 9 mM of sulfate but was amended with 20 μM U(VI)) waspumped upflow through the column at a rate of 0.035 ml/min. The biotic processsuch as genome-scale and kinetic models for microbial communities (*Geobacter*,*Rhodoferax* and *Desulfovibrio*) and electron capacitance for *Geobacter*(Zhao et al., 2010) has been coupled to the abiotic process in HYDROGEOCHEM(Yeh et al., 1998), where the finite element method was used to iterativelysolve governing equations by discretizing the column with 9 elements and 36nodes (Fig. 1A). The abiotic reduction of U(VI) is based on two publishedmechanisms -- uranium-hydroxyl species are the ones being reduced by sulfide(Hua et al., 2006) and the surface-adsorbed uranium species can be reduced bysurface-bound Fe(II) (Rosso et al., 2006 and Liger et al., 1999).

The model predicts that the biotic process plays amajor role in the U(VI) reduction under actual sediments (Fig.1B), which isconsistent with field-experiments (Anderson et al., 2002). Due to the mechanismof electron capacitance through which the reduced cytochromes can transferelectrons directly to oxidized metal ions with high rates (Wang et al., 2008),the uranium bioremediation process is still efficient and effective even whenthe sediments has been dominated by sulfate-reducing bacteria after 40 days ofacetate amendment. This simulation supports the recent findings (Williamset al., 2011) suggesting that following prolongedacetate addition, the small fraction of planktonic *Geobacter* species maystill be largely responsible for the vast bulk of U(VI) immobilization.Further, the model provides an explanation for the insignificant effect exertedby the abiotic process on the U(VI) reduction. Due to the dominant uranium-carbonatespecies present in the carbonate-containing system, the calculated total concentrationsof uranium-hydroxyl species are very low, leading to an extremely slow sulfide-drivenU(VI) reduction. Similarly, the low concentration of the surface-bound U(VI) (Moonet al., 2007) results in the low rate of surface catalysis of U(VI) reductionby Fe(II).

The model results suggest that themicrobially-generated hydrogen sulfide by sulfate-reducing bacterium may behelpful for the U(VI) reduction if the carbonate concentration can be reducedin the column amended with 20 μM U(VI). Fig. 2 indicates that by reducing theinfluent bicarbonate concentration from 8 mM to 3.5 mM, the total concentrationof uranium-hydroxyl species is significantly increased, generating a morethorough removal of the soluble U(VI) for long-term uranium immobilization. Itshould be noted that such abiotic process is dependent upon certaincombinations of experimental conditions, including the concentrations of U(VI),sulfide, bicarbonate and Ca(II). For example, if U(VI) concentration is low(e.g. typical values for Rifle groundwater), it might be possible that thebicarbonate amendment is beneficial to U(VI) reduction, since it will improvethe U(VI) desorption process, and hence increase the probability of the bioticreactions in groundwater.

Thecomprehensive model may provide auseful tool for investigating mechanisms and dynamics of abiotic and bioticinteractions under actual environmental conditions and subsequently for designingan optimal strategy for sustained uranium removal.

Reference

AndersonRT, Vrionis HA, Ortiz-Bernad I, Resch CT, Long PE, Dayvault R, Karp K, MarutzkyS, Metzler DR, Peacock A, White DC, Lowe M, Lovley DR. 2003. Stimulating the insitu activity of Geobacter species to remove uranium from the groundwater of auranium-contaminated aquifer. Appl Environ Microbiol 69(10):5884C5891.

Hua B, XuH, Terry J, Deng B. 2006. Kinetics of uranium(VI) reduction by hydrogen sulfidein anoxic aqueous systems. Environ. Sci. Technol. 40, 4666-4671

Liger E,Charlet L, Cappellen V P. 1999. Surface catalysis ofuranium (VI) reduction byiron(II). Geochim. Cosmochim. Acta 63,2939C2955.

Moon H,Komlos J, Jaffe PR. 2007. Uranium reoxidation in previously bioreduced sedimentby dissolved oxygen and nitrate. Environ Sci Technol 41:4587C4592.

Moon H,McGuinness L, Kukkadapu RK, Peacock AD, Komlos J, Kerkhof LJ, Long PE, JaffePR. 2010. Microbial reduction of uranium under iron- and sulfate-reducingconditions: Effect of amended goethite on microbial community composition anddynamics. Water Res 44:4015C4028.

Rosso KM. 2006. Mechanims and dynamics of abiotic and biotic interactions atenvironmental interfaces. ERSD Annual Report.

Wang Z,Liu C, Wang X, Marshall MJ, Zachara JM, Rosso KM, Dupuis M, Fredrickson JK,Heald S, Shi L. 2008. Kinetics of reduction of Fe(III) complexes by outermembrane cytochromes MtrC and OmcA of Shewanella oneidensis MR-1. Appl EnvironMicrobiol 74(21):6746C6755.

WilliamsKH, Long PE, Davis JA, Steefel CI, Wilkins MJ, N'Guessan AL, Yang L, NewcomerD, Spane FA, Kerkhof LJ, McGuinness L, Dayvault R, Lovely DR. 2011. Acetateavailability and its influence on sustainable bioremediation ofuranium-contaminated groundwater. Geomicro J. In press.

Yeh GT,Salvage KM, Gwo JP, Zachara JM, Szecsody JE. 1998. HYDROBIOGEOCHEM: A CoupledModel of HYDROlogic Transport and Mixed BIOGEOCHEMical Kinetic/EquilibriumReactions in Saturated-Unsaturated Media. Oak Ridge National Laboratory, Centerfor Computational Sciences, ORNL/TM-13668, Oak Ridge, TN37831.

Zhao J,Scheibe TD, Fang Y, Lovley DR, Mahadevan R. 2010. Modeling and sensitivityanalysis of electron capacitance for geobacter in sediment environments. JContam Hydrol 112(1C4):30-44.

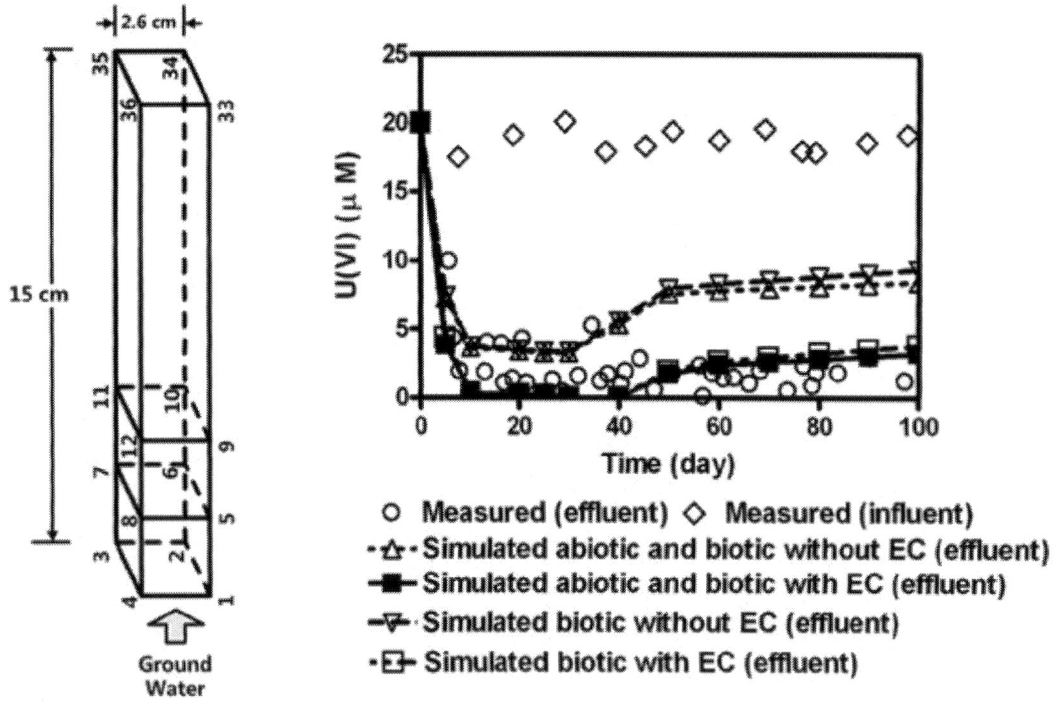

(A) (B)

Fig. 1. Experimental3-D domain for the flow-through column (A) and the effect of abiotic and bioticprocess on U(VI) reduction (B). EC C Electron Capacitance for *Geobacter*.

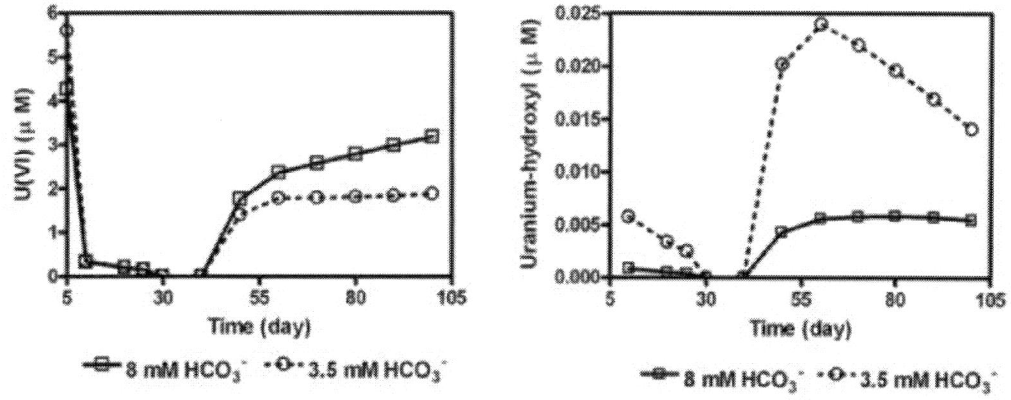

Fig. 2.Effect of carbonate concentration on total concentration of U(VI) species (A)and U(VI)-hydroxyl(B) in effluent.

See more of this Session: Fundamentals of Environmental Biotechnology
See more of this Group/Topical: Environmental Division

(529b) The Role of Geobacterial Cytochromes In Uranium Reduction

Bhushan P. Awate, Chemical engineering and Materials science, Michigan State University, East Lansing, MI, Gemma Reguera, Microbiology and Molecular Genetics, Michigan State University, East Lansing, MI and R. Mark Worden, Chemical Engineering and Materials science, Michigan State University, East Lansing, MI

Anthropogenic use of uranium has led to environmental contamination. Leaching of soluble uranium into the groundwater can lead to widespread human exposure and serious health hazards. One strategy to arrest groundwater contamination is to augment uranium precipitation via subsurface bacteria. *Geobacter* species have been shown to electrochemically reduce uranium from a soluble form U(VI) to an insoluble form U(IV). *Geobacter* cells use electron-carrying proteins to transfer electrons generated during respiration from the cytoplasm, through the cell membrane, to electron acceptors (mostly metal oxides) located outside the cell. Although *Geobacter's*'role in uranium reduction is widely recognized, the electron-transfer mechanism(s) used by these bacteria is not well understood. One distinctive feature of *Geobacter* species is the presence of more than 100 *c*-type cytochromes. These cytochromes are postulated to function as a heme network that can temporarily store electron produced during respiration and then shuttle the electrons to external electron acceptors, possibly with the involvement of conductive pili. Elucidation of *Geobacter*'s electron transfer mechanisms is particularly challenging due to the inherent complexity of *Geobacter's* heme network and the possibility of pleiotropic effects following gene knockouts in whole cells. In this study, we describe a biomimetic strategy, in which molecular architectures that mimic the structure and composition of *Geobacter*'s heme network are assembled on an electrode, and the resulting biomimetic interface is electrochemically characterized. This "bottom-up" approach enables hypotheses about structure-function relationships among key components of the electron transfer system to be tested. To illustrate the concept, a biomimetic interface was assembled that contained the periplasmic cytochrome PpcA, which is the only cytochrome that is conserved in all the members of the family *Geobacteraceae*. Recombinant PpcA from *Geobacter sulfurreducens* was overexpressed in *Escherichia coli*, purified, and immobilized on an alkanethiol self assembled monolayer bound to a gold electrode. Redox activity of the PpcA-containing interface was characterized using cyclic voltammetry, amperometry and chronocoulometry. The relative rates of PpcA-mediated reduction of soluble iron and uranium species were determined. The results are discussed in terms of the PpcA's possible role in groundwater uranium fixation.

See more of this Session: Fundamentals of Environmental Biotechnology
See more of this Group/Topical: Environmental Division

(529c) Evaluation of Non-Growth Biokinetic Models for Chromate Reduction with *Shewanellae*

Lewis Hsu[1], Kenneth H. Nealson[2] and Massoud Pirbazari[1], (1)Sonny Astani Department of Civil and Environmental Engineering, University of Southern California, Los Angeles, CA, (2)Department of Earth Sciences, University of Southern California, Los Angeles, CA

Hexavalent chromium is a carcinogenic pollutant oftenassociated with leather tanning and electroplating industries. Additionalapplications have included its use as a corrosion inhibitor in cooling waterand as a colorant in dye and pigment compounds. Its widespread use inindustrial application has often been met with poor waste management practices,leading to its release into the natural environment.

Bioremediation has been proposed as a viable remediationtool to reduce hexavalent chromium to its trivalent state. The rationale behindthis approach is that while the hexavalent form is very soluble and mobile inthe environment, the trivalent form is known to form insoluble precipitatesunder most environmental conditions. Such a transformation results in chromium immobility.Moreover, by utilizing microbes encountered in the subsurface environment, thepotential of designing an *in situ* treatment approach is much moreappealing when compared to more invasive *ex situ* approaches.

In particular, *Shewanella* species have beenproposed as model organisms capable of hexavalent chromium reduction. Thewidespread abundance of *Shewanella* in nature combined with the breadthof genomic and physiological knowledge make it an ideal organism to study inorder to understand mechanisms of chromium reduction in the environment. Tothis end, several *Shewanella* species were evaluated for their ability toreduce chromium under anaerobic conditions in batch reactors.

Data generated from these experiments was used toestimate biokinetic parameters for models proposed in literature for remediationof other types of pollutants. In this manner, we have evaluated the applicationof these models for hexavalent chromium reduction. A transformation capacity modelwas selected based on its agreement with experimental data. Furthermore, thesensitivity of this model to changes in biokinetic parameters was determined,as well as the effects of pH, temperature, and electron donor. The data generatedfrom these studies suggests that biomass, pH, and electron donor are among the keyparameters to consider when using designing bioremediation based treatmentsystems.

See more of this Session: Fundamentals of Environmental Biotechnology
See more of this Group/Topical: Environmental Division

Biotransformation of Coffee Pulp/Husk by Ligniculous Filamentous Fungi

María del Pilar Sierra-Gómez[1], *and Patricia Ortiz-Bermúdez*[1]
[1]*Department of Chemical Engineering, University of Puerto Rico Mayagüez Campus*
E-mails: maria.sierra@upr.edu, patricia.ortiz3@upr.edu

Abstract - Agro-industrial residues have been increasingly proposed as substrates in several bioprocesses. This is mainly because the recycling of these residues contributes to solve prevailing pollution waste management problems. Coffee pulp is the main byproduct in the wet processing of coffee; one of Puerto Rico's important agricultural commodities. Coffee pulp is close to 94% of the total weight of the coffee cherry, therefore innovative and environmentally friendly methods to reutilize this waste are necessary. The bioavailability of coffee pulp is restricted by the presence of recalcitrant and toxic compounds such as lignin, caffeine, and tannins. Thus, we propose the utilization of ligniculous filamentous fungi to detoxify coffee pulp, and convert it into an amenable source of carbohydrates and protein.

We have screened twelve strains of ligniculous filamentous fungi for caffeine tolerance. These strains were inoculated in caffeine concentrations of 0.2, 0.5, 1.0 and 4.0%. Caffeine degradation by candidate strains was assessed in liquid medium with defined amounts of pure caffeine. Experiments were conducted on minimal nutrient conditions at pH 4.5-5.5 and 60-70% moisture. Identification of caffeine reduction and its metabolites was done by HPLC analysis. Solid-state fermentations (SSFs) were carried out to assess the activity of ligniculous filamentous fungi on the coffee pulp. Chemical analysis of the degraded material was conducted to assess whether caffeine detoxification has occurred under SSF conditions. These analyses determined the extent of detoxification carried out by candidate strains on the actual coffee pulp.

Our preliminary results indicate that *Alternaria alternata* showed an unreported tolerance to caffeine of up to 1.0% on plates in rich medium. Experiments under defined liquid medium conditions with caffeine as a sole nitrogen source revealed that *A. alternata* detoxified up to 50% of the caffeine within 70 days. It is noteworthy that reduction of caffeine was also observed in experiments using caffeine as a sole carbon and nitrogen source although the percentage of reduction of caffeine was less. *Aspergillus tamarii*, a well-known caffeine detoxifier was grown in parallel with and its performance was similar to that of *A. alternata* under our culture conditions. SSF experiments demonstrated that coffee pulp supported the growth of *A. alternata,* preferentially at the larger particle size (i.e. #14 sieve supported growth better than #30 sieve). Preliminary extraction of caffeine from treated coffee pulp suggests that reduction of caffeine is possible under the SSF conditions we tested. Experiments are currently underway to assess caffeine detoxification of coffee pulp quantitatively, using *A. alternata* and *A. tamarii* as model strains.

(529e) Detecting Biological Networks In Clostridium Thermocellum Using Computational Approaches

Wen Zhou and Vinay Patil, Chemical Engineering, Michigan Technological University, Houghton, MI

Lignocellulosic plant biomass has long been recognized as a potential low-cost and sustainable source of mixed sugars for production of biofuels and other value-added chemicals. Lignocellulosic biomass comes from plant cell wall which is a composite structure with crystalline cellulose, hydrated hemicellulose, and lignin as major components. Corn stover, switchgrass, miscanthus, woodchips and the byproducts of lawn and tree maintenance are some of the more popular cellulosic materials for biofuel production. Production of biofuels from lignocellulose has the advantage of abundant and diverse raw material compared to sources like corn and cane sugars, but requires a greater amount of processing to make the sugar monomers available to the microorganisms that are typically used to produce ethanol by fermentation. Plants have evolved superb mechanisms for resisting assault on their cell wall structural sugars from the microbial and animal kingdoms, collectively known as biomass recalcitrance. Current technologies for the biological conversion of lignocellulosic biomass generally include pretreatment of biomass, hydrolysis of polysaccharides, and fermentation of C5 and C6 sugars.

Clostridium thermocellum is an anaerobic, thermophilic bacterium. C. thermocellum has garnered great research interest due to its cellulolytic and ethanologenic abilities, being capable of directly converting a cellulosic substrate into ethanol. This makes it useful in converting biomass into a usable energy source. The degradation of the cellulose is carried out in the bacterium by a large extracellular cellulase system called a cellulosome, which contains nearly 20 catalytic subunits. The cellulase system of the bacterium significantly differs from fungal cellulases due to its high activity on crystalline cellulose. In this work, I will present our recent work on understanding and improving the process of biologically converting biomass into biofuels through computational systems biology approaches. Specifically, I will focus my presentation on elucidating of a number of key regulatory pathways in cellulolytic organisms relevant to bioenergy production.

See more of this Session: Fundamentals of Environmental Biotechnology
See more of this Group/Topical: Environmental Division

(529f) Investigating the Mechanisms of Arsenic Removal by Microbial Layer In a Bio-Sand Filter

Hannah Gray[1], Charlotte Chan[1], **Varadarajan Ravindran**[2], Lewis Hsu[1] and Massoud Pirbazari[1], (1) Sonny Astani Department of Civil and Environmental Engineering, University of Southern California, Los Angeles, CA, (2)University of Southern California, Los Angeles, CA

The present study addresses the enormous global demand for safe drinking water in developing countries, and focuses on the removal of arsenic and bacterial pathogens from potable water supplies using a bio-sand filter. The bio-sand filter was constructed of a transparent polyethyl resin to facilitate visual observation of microbial activity and reactor flows. The filter medium essentially consisted of a supporting gravel layer, large sand layer, and biofilm layer containing a diverse microbial population, acclimated to a natural surface water source. Our work showed that the bio-sand filter r in-tandem with a pre-treatment column of iron-oxide-coated sand were effective in eliminating these contaminants. The present work focuses on determining the actual mechanisms of arsenic removal within the filter medium. In this study, the previous filter design was adapted to include a layer of iron-oxide-coated sand. Four sampling ports were strategically located to separately quantify arsenic removals in the biofilm and iron-oxide-coated-sand layers. Both sodium arsenate and sodium arsenite were spiked in the filter influent to determine whether arsenic removal was attributable to redox reactions.

The study is also directed at determining the actual mechanisms of arsenic removal within the filter, and the microbial strains primarily responsible for arsenic removal. Four sampling ports were strategically located in the bio-sand filter to separately quantify arsenic removals in the biofilm and iron-oxide-coated-sand layers. Both sodium arsenate and sodium arsenite were spiked in the filter influent to determine whether arsenic removal was attributable to redox reactions. The total arsenic removal was determined using the inductively coupled plasma –mass spectrometry (ICP-MS) technique. The bacterial removals were determined by the If arsenic removal occurred within the biofilm layer, the implications would be important, highlighting the ability of bacteria to store or process harmful forms of arsenic. Samples of schmutzdecke (biofilm layer) were obtained from the filter for microbial community analysis. Identification of dominant bacterial strains was accomplished by DNA isolation, PCR techniques, and DGGE fingerprinting. The DNA materials were isolated from these bio-layer samples, and the bacterial rRNA sequences were amplified by polymerase chain reaction (PCR) techniques. The PCR products were subsequently separated by denaturing gradient gel electrophoresis (DGEE), and sequenced for fingerprinting identification of dominant bacterial strains, before and after exposure to arsenic. Similar samples were also collected for scanning electron microscopy analysis to morphologically distinguish the microbial species associated with arsenic removal. Elemental analysis was undertaken to determine the locations of any immobilized arsenic species. The removals of bacterial pathogens (coliform used as surrogates for bacteria) in the bio-sand filter easily exceeded 5-logs (99.999%). The presentation would discuss the removals of arsenic and bacteria in the bio-sand filter at various depths of the filter. More importantly, it would address the mechanisms and interactions and mechanisms of arsenic removal in the bioactive filter.

See more of this Session: Fundamentals of Environmental Biotechnology
See more of this Group/Topical: Environmental Division

(535a) Intracellular Trafficking of Toll-Like Receptor 9 Ligands

Helen C. Chen, Chemical Engineering, University of Washington, Seattle, Seattle, WA and Hong Shen, Department of Chemical Engineering, University of Washington, Seattle, WA

The innate immune system relies on innate immune cells to survey their environment for pathogens and to mount responses through the activation of pattern recognition receptors. To activate intracellular pattern recognition receptors such as toll-like receptor 9 (TLR9), innate immune cells must first internalize exogenous material. However, complex cellular processes control the intracellular fate of internalized ligands, and ultimately, the cytokine secretion profiles. In this study, we aim to (1) examine the trafficking of TLR9 ligands, and (2) identify the possible intracellular mechanisms which affect the trafficking and immunogenicity of TLR9 ligands.

See more of this Session: Intracellular Processes I
See more of this Group/Topical: Food, Pharmaceutical & Bioengineering Division

(535b) Toward a Quantitative and Predictive Model of Growth for An RNA Virus

Collin M. Timm and John Yin, Chemical and Biological Engineering, University of Wisconsin-Madison, Madison, WI

Viral infections are the cause of many human diseases such as the common cold, AIDS and even some cancers. In order to fight and prevent virally caused diseases we must understand the viral replication process. Our approach is to use kinetic models based on known viral mechanisms to predict viral growth. A model that can predict virus growth can be used to design virus systems to generate vaccines, to suggest anti-viral strategies, and to engineer viruses that selectively kill cancerous cells while avoiding healthy tissue. The viral system we use is the well studied vesicular stomatitis virus (VSV). VSV has a negative sense RNA genome which encodes five viral proteins. Upon introduction of the viral genome into the cell, the viral polymerase begins transcribing the five genes sequentially with a measureable attenuation at each gene junction. The attenuation leads to viral gene expression control from a single promoter, with the genes closer to the promoter being expressed higher than the next gene in the sequence. As the mRNAs accumulate and are translated, the viral polymerase is transformed from a transcriptase to a replicase by interaction with produced viral proteins. The replicase generates full length anti-genome which is not translated but used as a template to produce more viral genomes. The genomes associate with viral proteins and produce infectious viral particles.

While the general mechanisms of the VSV replication cycle are well studied, the relative rates of the processes are unknown. In order to determine some of the unknown rates we have developed an assay to measure absolute numbers per cell of viral mRNAs, genomes and anti-genomes during the infection process. The kinetic data for viral RNA is used with a previously published model to predict kinetic parameters. In order to provide a rich data set for parameter estimation and model refinement, the initial condition of the viral infection was varied from a low virus to cell ratio to a high virus to cell ratio. A low virus to cell ratio represents the initiation of infection in an organism, while a high ratio represents the conditions that would be seen in an infection spreading through tissue. Interestingly, while the published model was able to predict viral growth it underestimated the levels of viral RNA by orders of magnitude. These results suggest unknown and important mechanisms in the viral infection process which must be included in the predictive model. This research will lead to development of a true predictive model of VSV growth which can be extended to other virus systems for the design of vaccines, anti-viral strategies, and anti-cancer strategies.

See more of this Session: Intracellular Processes I
See more of this Group/Topical: Food, Pharmaceutical & Bioengineering Division

(535c) Quantifying Metabolism of Differentiated Adipocytes Using 13C-Metabolic Flux Analysis

Maciek R. Antoniewicz[1], Scott B. Crown[2] and Nick Marze[2], (1)Chemical Engineering, University of Delaware, Newark, DE, (2)Chemical Engineering, University of Delaware, Newark

Obesity and type II diabetes are increasing health epidemics in the United States and around the world and are characterized by dysregulation of lipid metabolism. A more complete understanding of *de novo* lipid synthesis at the metabolic level is required to develop novel therapeutics for the treatment of these disorders. Towards this end, we are implementing a systems biology approach using metabolic flux analysis and stable isotope ^{13}C-tracers to determine the active reaction pathways in an adipocyte cell line 3T3-L1. Differentiated white adipocytes were incubated with [U-^{13}C]glucose, [1,2-^{13}C]glucose, [U-^{13}C]glutamine, and [5-^{13}C]glutamine and the incorporation of ^{13}C-atoms into metabolites from each tracer was analyzed with gas chromatography mass spectrometry (GC-MS). Metabolic fluxes were then determined using Metran software based on the EMU framework.

[U-^{13}C]glucose and [1,2-^{13}C]glucose did not reach isotopic steady-state within 24 hours. We speculate that an initially unlabeled metabolite pool is exchanging with central metabolism and slowly becomes enriched. [U-^{13}C]glutamine and [5-^{13}C]glutamine achieved isotopic steady-state in intracellular metabolites within 6 hours. For glutamine tracers, the mass isotopomer distribution of citrate indicated activity of reductive carboxylation in which reverse citric acid cycle flux occurs from a-ketoglutarate to citrate. This result provided evidence for isocitrate dehydrogenase as a potential modulator of adipocyte metabolism. Fatty acid profiling and mass isotopomer analysis of the differentiated adipocytes was conducted. Fatty acids varied from 14 to 18 carbons in chain length, included even and odd chain length, and were saturated and monounsaturated. For glutamine tracers, isotopomer spectral analysis (ISA) was conducted on the mass isotopomer distributions for fatty acids. ISA suggested that odd-chain fatty acid synthesis proceeded from elongation of a three-carbon precursor rather than by alpha-oxidation as was previously assumed. Together, these results provide valuable new information regarding *de novo* lipogenesis in adipocytes that can be used to identify new drug targets to combat excessive fat accumulation.

See more of this Session: Intracellular Processes I
See more of this Group/Topical: Food, Pharmaceutical & Bioengineering Division

(535d) Systems-Level and Subcellular Resolution Dynamic Analysis of Trophectodermal Differentiation In Human Embryonic Stem Cells

Prasenjit Sarkar[1], Shan M. Randall[2], Timothy S. Collier[2], David C. Muddiman[2] and **Balaji Rao**[3], (1) Chemical and Biomolecular Engineering, North Carolina State University, Raleigh, NC, (2)Chemistry, North Carolina State University, Raleigh, NC, (3)Department of Chemical and Biomolecular Engineering, North Carolina State University, Raleigh, NC

The pluripotency of human embryonic stem cells (hESCs) is regulated by a complex network of interacting pathways. Here we present a systems-level quantitative analysis of the molecular mechanisms regulating differentiation of hESCs along the trophectodermal lineage. The subcellular localization of proteins within a pathway, and their relative abundance govern the influence of a specific pathway on hESC fate. We developed a discontinuous sucrose gradient based protocol for simultaneous isolation of the nuclear, membrane and cytosolic subcellular fractions from undifferentiated and differentiating hESCs and conducted proteomic analysis on each of these three fractions. Further, we also used the method of spectral counting to estimate the relative abundance of all proteins identified in the cytosolic fraction. The integration of protein identifications with subcellular localization data and estimates of protein abundance allowed us to uncover several potential regulatory mechanisms pertaining to both hESC pluripotency and trophectodermal differentiation.

The hESC microenvironment plays a significant role in the regulation of signaling pathways in hESCs. The complex microenvironment of hESCs is determined in part by endogenous factors that are secreted by hESCs. However, experimental characterization of the hESC secretome is challenging. Since secreted proteins are synthesized in the endoplasmic reticulum (ER), we quantitatively characterized the microenvironment of hESCs as well as cells undergoing differentiation through proteomic analysis of the ER fractions. Interestingly, the list of secreted proteins identified using this approach is consistent with a previous proteomic analysis of hESC conditioned medium.

Taken together along with real-time PCR data, our proteomic analysis provides a quantitative understanding of the molecular processes regulating trophectodermal differentiation in hESCs.

See more of this Session: Intracellular Processes I
See more of this Group/Topical: Food, Pharmaceutical & Bioengineering Division

(535e) Regulation of Mesenchymal Stem Cell Myogenic Differentiation by Cell-Cell Adhesion: The Role of Cadherins In Differentiation

Stella Alimperti, Chemical and Biological Engineering, State University of New York at Buffalo, Amherst, NY and Stelios T. Andreadis, Department of Chemical and Biological Engineering, University at Buffalo, the State University of New York, Amherst, NY

Regulationof mesenchymal stem cell myogenicdifferentiation by cell-cell adhesion: the role of cadherinsin differentiation

Stella Alimperti[1] and Stelios Andreadis[1,2]

[1] Department of Chemical and Biological Engineering,University at Buffalo, The State University of New York, Amherst, 14260; [2]Center of Excellence in Bioinformatics and LifeSciences, Buffalo, NY 14203

During mesenchymal stem cell (MSC) differentiation towardmyogenic lineage, early (aSMA), intermediate (calponin) and late (myosin heavy chain,MHC) markers are expressed. It is well known that MSC can be coaxed to the myogenic lineage usingsoluble factors such as TGF-β1 but cell-cell communication has not been examinedin the context of stem cell differentiation. We report that confluencypromoted myogenic differentiation of MSC even underconditions that suppressed differentiation e.g. in the presence of bFGF. Specifically, cell-cell contact increased expressionof myogenic markers such as αSMA, calponinand smoothelin. Since cell-cell adhesion is mediatedby cadherins, we examined how two cadherinsnamely, N-cadherin or OB-cadherin may affectcell-cell contact mediated MSC differentiation. Here we report that OB-cadherinplays an important role in myogenic differentiation. Gene knockdown studiesrevealed that OB-cadherin mediated cell proliferation, aSMA expression and smooth muscle cell contraction. The relationshipbetween OB-cadherin and N-cadherin was also examined as well as intracellular signalingpathways and transcription factors such as myocardin and MRTF that may bemediating the signals of cadherin engagement. These results are novel as they identifycell-cell contact as a potent inducer of myogenicdifferentiation through cadherins and theirintracellular mediators.

See more of this Session: Intracellular Processes I
See more of this Group/Topical: Food, Pharmaceutical & Bioengineering Division

(535f) Efflux and Metabolism: Elucidating the Mechanisms of Metabolite Overflow and Secretion

Lon Chubiz, Chemical and Biomolecular Engineering, University of Illinois, Urbana, IL and **Christopher V. Rao**, Chemical and Biomolecular Engineering, University of Illinois, Urbana Champaign, Urbana, IL

Bacteria such as *Escherichia coli* have evolved a complex repertoire of mechanisms for limiting the uptake of toxic compounds from their environment. These include altering the structure of the outer membrane, expressing efflux transporters, and redirecting metabolic flux. While these mechanisms have traditionally been implicated in intrinsic antibiotic resistance, we have recently shown that they also limit the accumulation of metabolic intermediates. For example, 4-hydroxybenzoate, 2,3-dihydroxybenzoate and anthranilate, involved in ubiquinone, enterobactin and tryptophan biosynthesis, respectively, induce the same resistance and tolerance mechanisms as do antibiotics such as salicylic acid. In addition, we have identified a number of efflux transporters that secrete sugars from the cell. These results suggest that many of the antibiotic resistance mechanisms intrinsic to bacteria are also utilized to prevent the accumulation of toxic metabolites and potentially to function as relief valves for metabolite overflow.

In this work, we will discuss our ongoing efforts to elucidate the mechanisms of intrinsic antibiotic resistance in *E. coli* and its role in metabolism. We will also discuss how these finding may yield new approaches and targets for metabolic engineering.

See more of this Session: Intracellular Processes I
See more of this Group/Topical: Food, Pharmaceutical & Bioengineering Division

(535g) Single-Cell Analysis of Cellular Quality Control In S. Cerevisiae: How Low' Can You Go?

Carissa L. Young[1], Theresa Yuraszeck[2], Jeffrey Caplan[3], Kirk J. Czymmek[3], Francis J. Doyle III[2] and Anne Skaja Robinson[4], (1)Chemical Engineering, University of Delaware, Newark, DE, (2)Chemical Engineering, University of California, Santa Barbara, CA, (3)DBI BioImaging Center, Delaware Biotechnology Institute, Newark, DE, (4)Department of Chemical Engineering, University of Delaware, Newark, DE

Within the secretory pathway of eukaryotic cells, the endoplasmicreticulum (ER) is responsible for maintaining the fidelity of protein synthesisand maturation. A variety of insults including nutrient deprivation, pathogenicinfection, and chemical treatment, collectively termed ER stress', inducequality control mechanisms to recover cell homeostasis. ER associateddegradation (ERAD), unfolded protein response (UPR), and autophagy are qualitycontrol pathways that occur at various timescales, encompass variations in thespatial organization of multiple organelles, and alter select proteinconcentrations and intracellular localization. Surprisingly, all three pathwaysare activated in several neurodegenerative(e.g. Alzheimer, Parkinson, Sclerosis, Huntington, etc.) and hereditary diseases (e.g. Marinesco-Sjgrensyndrome, Woozy mouse, nonpolyposiscolorectal cancer-associated small-bowel cancer, prostate cancer, etc.).In all cases, as a result of ER stress, there is evidence of atypical, intracellularprotein distribution during disease manifestation. Yet, how the accumulation ofdisease-specific proteins is involved in compromising quality control remains elusive.

Byimplementing DNA recombinant systems combined with high-resolution imagingtechniques, we have determined that protein redistribution, resultant spatialeffects, and organelle modifications are a consequence of the cell's responseto ER stress in yeast, *S. cerevisiae*.In pursuit of a thorough analysis of protein redistribution at the subcellularlevel, multiple yeastexpression cassettes have been created to test the effects of codon-optimizedfluorescent variants, small epitope tags, polylinkerlength for N and C terminal tags, and the inclusion of essential retrieval sequencesfor ER luminal chaperones and foldases. A photoconvertible GFP variant (i.e. mEos2) and six-residue tetracysteine motif required for FlAsH (fluorescein arsenical helix binder)-based technology was used to investigatediscrete subpopulations of tagged proteins using live-cell imaging methods, correlativemicroscopy, and super-resolution techniques. Using fluorescent protein (FP)variants as probes, we have expressed endogenous proteins with FP fusions inorder to continuously monitor protein trafficking; analyzed localization effectsof proteins involved in ERAD; examined organelle dynamics under variousenvironmental conditions to initiate ER quality control (ERQC) and co-localizationwith cytoskeleton structures, actin and β-tubulin; while confirming theexistence of cellular variability during UPR activation and its directcorrelation to age at the level of single-cell analysis.

Itis now evident that endogenous proteins involved in quality controlredistribute within the cell, specifically the endoplasmic reticulum (ER), inorder to perform essential functions that maintain cell homeostasis. Usingnovel techniques to image cells such as correlative microscopy, a combinationof confocal light microscopy (CLM) and transmission electron microscopy (TEM),as well as Focused Ion Beam (FIB) microscopy, we have developed entire three-dimensionalorganelle reconstructions of yeast cells at electron microscope resolution. Interestingly,our microarray analysis and q-PCR validation have established noveldown-regulation of selective genes in *S. cerevisae* as a consequence of ER stress. Subsequently,we investigated the effects of cellular disturbances inducing ER stress on a globallevel by examining select proteins in various organelle targets and elucidatingthe dramatic changes in organelle morphology.

See more of this Session: Intracellular Processes I
See more of this Group/Topical: Food, Pharmaceutical & Bioengineering Division

(540a) The Free Energy of a Salt Bridge From Simulations

Andrew D. White and Shaoyi Jiang, Chemical Engineering, University of Washington, Seattle, WA

Understanding when and how salt bridges are stabilizing is still an open question. Researchers have argued that in the core of a protein where there is no charge screening, salt bridges will stabilize proteins more than on the surface of proteins. However, mutagenesis studies have shown that to not always be the case. One key question is how well the different ionic amino acids(K,E,R,D) pair without any influence from the protein and how their solvation affects this. We used state of the art free energy techniques, replica exchange metadynamics, to quantify both the free energy of forming salt bridges between the different amino acids as well as the free energy of desolvating them. Combining these two terms into a thermodynamic cycle gives insight into how the local environment affects pairs. Specifically, we've found that the desolvation is as important, and more important in some cases, than the formation of salt bridges. The differences in free energy of salt bridges can be explained by a combination of the differences in geometry and solvation. Additionally, we've examined pairs on hundreds of proteins from the protein data bank (PDB) in order to compare our results with what is observed in Nature. Our results match the ranking of the salt bridges the PDB. Finally, we've created a probability based free energy surface from the PDB data and compared it with our simulation based free energy surface.

See more of this Session: Molecular Modeling of Biophysical Processes I - Molecular Binding and Protein Structure and Dynamics
See more of this Group/Topical: Food, Pharmaceutical & Bioengineering Division

Forced dissociation of tight-binding cucurbituril-guest systems.

Simulation studies of rupture force, mechanical stress, and binding affinity.

Camilo Velez-Vega and Michael K. Gilson

Cucurbiturils are ring-shaped molecules (also known as CB[n], where n is the number of glycouril monomers comprising the ring) that have been successfully used as molecular recognition hosts, being able to bind cationic guests with high affinities. Moreover, given their interesting binding properties and fairly easy synthesis, CB[n]s have numerous other applications which include their use as drug carriers, catalysts for supramolecular chemistry, and as basic units for the formation of polyrotaxanes. Given their small size and diverse intra and inter molecular interactions, CB[n]-guest complexes are also optimal test beds for innovative in-silico approaches pertaining to the modeling of biophysical processes.

Atomistic stress analysis and rupture force and binding affinity calculations were performed on two high affinity cucurbituril-guest systems, based upon simulations of forced dissociation processes. The unbinding processes were modeled using constant velocity pulling along different routes as in atomic force microscopy (AFM) experiments, and some of the resulting paths were further equilibrated via umbrella sampling simulations. Length-force profiles computed with a low spring constant typical of experimental AFM conditions yielded good agreement with available measured rupture forces. Simulations with a high spring constant, on the other hand, were found to be more suitable for computing binding free energies via a novel expression that yielded converged results quite efficiently. The computed affinities of the two systems are ranked correctly, though they still deviate from those obtained via prior isothermal titration calorimetry experiments. For these complexes, the force profiles were found to depend critically upon how the artificial pulling bond was attached to the host and the guest molecules. For instance, serious irreversibility occurred when the pulling force was attached to a flexible part of the guest molecule that was not directly engaged in the high-force transition state. Furthermore, force profiles were also found to be considerably influenced by the pulling path allowed by the selected attachment points. Interestingly, the binding affinities calculated from the irreversible and the more reversible force profiles are similar, indicating that for these complexes, the deviations caused by irreversible pulling were masked upon calculation of the binding affinities. However, for the general case, the accuracy of the binding affinities obtained from pulling simulations may improve by use of a synergistic approach in which the conventional practice of sampling diverse paths is complemented by the identification and prevention of irreversibilities arising from "soft-spring" effects due to the choice of attachment points.

On the other hand, mechanical stress calculations for these host-guest complexes proved to be informative regarding the key interactions controlling their dissociation processes. In particular, we observed that the unbinding transition of the first complex is characterized by a stepwise process, which is initially dominated by electrostatic interactions between the guest's ammoniums and the host's carbonyl groups, and subsequently limited by the extraction of a bulky moiety, comprising the center of the guest; this last step requires some bond stretching at the cucurbituril's extraction portal. Conversely, the dissociation of a second complex with a more slender guest is mainly driven by successive electrostatic interactions between the different guest's ammoniums and the host's carbonyl groups. In general, stress calculations are valuable tools for providing details on interaction-specific contributions to the transition mechanisms that would otherwise be difficult to discern.

Lastly, we report our recent efforts on the efficient and accurate identification of binding free energies and mechanisms of protein-ligand complexes, via optimized forced dissociation calculations.

(540c) Computational Discovery and Characterization of Peptide Chains That Recognize the tRNALys3 Anticodon Loop

Xingqing Xiao[1], Carol K. Hall[1] and Paul Agris[2], (1)Chemical and Biomolecular Engineering, North Carolina State University, Raleigh, NC, (2)Department of Biological Sciences, University of Albany, State University of New York, Albany, NY

HIV (human immunodeficiency virus) and AIDS (acquired immunodeficiency syndrome) are responsible for the deaths of over 25 million people. Our efforts are aimed at discovering potential therapies capable of breaking the reverse transcription of HIV. Since HIV replication depends strongly on the $tRNA^{Lys3}$ primer, Agris and coworkers have been developing peptide sequences to bind to primer $tRNA^{Lys3}$ with a higher affinity and specificity than viral RNA. A short promising peptide sequence, 15 amino acids in length — *RVTHHAFLGAHRTVG*, was synthesized and tested in their lab. Our goal in this work is to use simulation to: (1) understand the mechanism by which this sequence recognizes and binds to $tRNA^{Lys3}$ and (2) design additional sequences with superior affinity and specificity. The AMBER package was used to simulate the binding behavior of $tRNA^{Lys3}$ and the 15-amino acid peptide at the atomic level. Our results revealed that the arginine, which is positively charged, on the peptide chain, dominates the binding of the peptide sequence due to its interaction with the negatively-charged phosphate backbone on the tRNA. However this binding is general and does not enhance the specificity that is necessary for $tRNA^{Lys3}$ recognition. In contrast the hydrophilic histidine binds selectively to the loop of $tRNA^{Lys3}$, conferring specific recognition. The phenylalanine contributes positively to both binding formation and peptide recognition. The loop on $tRNA^{Lys3}$ plays an important energetic role in both the binding affinity and specificity. By calculating and analyzing the binding free energy, we find that the van der Waals interaction is the key to recognizing specific molecules, while the electrostatic interaction essentially determines the strength of the binding affinity. We are developing an algorithm that combines self-consistent mean field (SCMF) with a Monte Carlo (MC) search algorithm in sequence space to determine optimal peptide sequences which can then be tested for binding affinity and specificity with $tRNA^{Lys3}$ in the Agris lab.

See more of this Session: Molecular Modeling of Biophysical Processes I - Molecular Binding and Protein Structure and Dynamics
See more of this Group/Topical: Food, Pharmaceutical & Bioengineering Division

(540d) Effect of Local Sequence On the Genomic Positioning of Nucleosomes

Gordon S. Freeman[1], Daniel M. Hinckley[1], Vanessa Ortiz[2], Gregory N. Toepperwein[1] and Juan J. de Pablo[1], (1)Chemical and Biological Engineering, University of Wisconsin-Madison, Madison, WI, (2) Chemical Engineering, Columbia University, New York, NY

DNA encodes the genetic information necessary for most living organisms. While cellular functions require this information be accessible, the genome (human genome ~ 2 meters in length) must also be highly compacted in order to fit within the nucleus of the cell (~10 μ m). In eukaryotes, the compact form of the genome is the chromosome. At the lowest level of organization, the chromosome consists of the nucleosome in which approximately 147 base-pairs of DNA are wrapped around a core histone protein complex. Adjacent nucleosomes are then separated by short nucleosome-free linker regions of DNA.

The mechanism of nucleosome positioning on the genome is not well understood with competing viewpoints being presented in the literature[1,2]. It is known that certain sequence motifs have a higher affinity for nucleosome formation both *in vitro* and *in vivo* compared to others[3,4]. However, whether this local affinity, or lack thereof, dominates nucleosome positioning *in vivo* and thus constitutes a "code" for nucleosome positioning is unclear; recent data suggests that the location of nucleosomes on the genome is largely statistical in nature with sequence preference playing a minor role[5]. These emerging viewpoints have been formulated in the context of experimental data; molecular models capable of describing nucleosome positioning are not yet available.

In this work, we employ a coarse grain model for DNA[6,7] and the histone protein complex in combination with a novel molecular simulation technique to study the effect of dinucleotide base steps on the preferred orientation of DNA wrapped around the histone complex. Recently, this coarse-grain model was used to determine the work required to bend different DNA sequences to the curvature required by nucleosomes[8]. We study sequences known to show high affinity for nucleosome formation as well as sequences that do not. We examine the orientational preference of these sequences and identify those motifs that enhance the stability of the nucleosome structure. In addition, to address the question of whether sequence alone, in the absence of DNA binding factors, can drive the positioning of nucleosomes, we determine the free energy surface associated with the position of a histone complex along a strand of DNA with a region in which nucleosome formation is not observed *in vivo* (e.g. a protein binding site) and an adjacent region that favors nucleosome formation.

GSF and DMH are supported by an NHCRI training grant to the Genomic Sciences Training Program, T32HC002760.

[1]Kaplan *et al.*, *Nature Structural & Molecular Biology* **17**, 918 (2010).

[2]Zhang *et al.*, *Nature Structural & Molecular Biology* **17**, 920 (2010).

[3]Segal *et al.*, *Nature* **442**, 772 (2006).

[4]Kaplan *et al.*, *Nature* **458**, 362 (2009).

[5]Zhang *et al.*, *Nature Structural & Molecular Biology* **16**, 847 (2009).

[6]Knotts, *et al.*, *Journal of Chemical Physics* **126**, 084901 (2007).

[7]Sambriski *et al.*, *Biophysical Journal* **96**, 1675 (2009).

[8]Ortiz *et al.*, *Physical Review Letters*, in press.

See more of this Session: Molecular Modeling of Biophysical Processes I - Molecular Binding and Protein Structure and Dynamics
See more of this Group/Topical: Food, Pharmaceutical & Bioengineering Division

(540e) Molecular Architecture of a Late 40S Ribosome Assembly Intermediate

Harish Vashisth and Charles L. Brooks III, Chemistry and Biophysics, University of Michigan, Ann Arbor, MI

Ribosomes are self-assembling macromolecular machines which catalyze the synthesis of proteins in all cells, and are made up of two subunits that differ in size due to the presence of RNA and proteins in different ratio. Although significant structural details of the organization of mature ribosomes in prokaryotes and eukaryotes have been elucidated in the past decade, ribosome biogenesis and assembly processes are poorly understood. Ribosome assembly in eukaryotes requires at least 200 essential transiently associating assembly factor proteins that facilitate the RNA folding and recruitment of ribosomal proteins. We have determined the structure of a late cytoplasmic 40S ribosome assembly intermediate from *Saccharomyces Cerevisiae* with the help of Cryo-electron microscopy and multiscale molecular simulation techniques. We have collected maps of the immature 40S particles and its components at various resolutions, and apply molecular dynamics based flexible fitting (MDFF) to fit an initial atomic model in to the cryo-EM maps of immature 40S particles with minimal structural perturbation. We verify that the largest conformational change observed in RNA can be robustly generated with an independent technique called normal mode flexible fitting (NMFF). We find that the pre-40s particles in addition to ribosomal RNA and proteins also contain 7 types of assembly factors, the joint activity of which blocks the translation initiation pathway in immature 40S subunits.

See more of this Session: Molecular Modeling of Biophysical Processes I - Molecular Binding and Protein Structure and Dynamics
See more of this Group/Topical: Food, Pharmaceutical & Bioengineering Division

(540f) Molecular Insight Into the Inhibition Effect of (-)-Epigallocatechin-3-Gallate On the Conformational Transition of Amyloid Peptide 42

Fufeng Liu, Xiaoyan Dong and Yan Sun, Department of Biochemical Engineering, School of Chemical Engineering and Technology, Tianjin University, Tianjin, China

Alzheimer's disease is the most fatal neurodegenerative disorder which is caused by the disfolding and aggregation of amyloid-β peptides. Much experimental evidence indicates that (-)-epigallocatechin-3-gallate (EGCG) inhibits the fibrillogenesis of amyloid-β peptide 42 ($A\beta_{42}$) and alleviates its associated cytotoxicity. However, a detailed knowledge of the molecular mechanism of the inhibition effect of EGCG on the conformational transition of $A\beta_{42}$ remains unclear due to the limitations of current experimental techniques. In this work, molecular dynamics simulations and molecular mechanics-Poisson-Boltzmann surface area (MM-PBSA) analysis were coupled to address the issue. It is found that the preferential interaction of EGCG is the origin of its inhibition effects. Namely, EGCG molecules expel the water from the surface of $A\beta_{42}$, cluster with each other and interact directly with the peptide. The results of free energy decomposition calculated by MM-PBSA indicate that nonpolar interactions contribute more than 70% to the binding free energy of the EGCG-$A\beta_{42}$ complex, while polar interactions (i.e., hydrogen bonding) play a minor role. It is identified that there are 12 important residues of $A\beta_{42}$ that strongly interact with EGCG (i.e., Phe4, Arg5, Phe19, Phe20, Glu22, Val24, Gly29, Leu34-Gly37, and Ile41). Furthermore, it is observed that nonpolar interactions are provided by the side chains of hydrophobic residues (Phe, Met and Ile), while polar interactions are mainly formed by some of the main chains of $A\beta_{42}$, of which the main chains of Gly29 and Gly37 contributes greatly. The work shed light at the atomic level on the inhibition effect of EGCG on the conformational transition of $A\beta_{42}$, and the findings are considered critical for the design of small-molecule inhibitors of aggregation with therapeutic potential in Alzheimer's disease.

See more of this Session: Molecular Modeling of Biophysical Processes I - Molecular Binding and Protein Structure and Dynamics
See more of this Group/Topical: Food, Pharmaceutical & Bioengineering Division

(540g) Enhanced Meta-Analysis of Acetylcholine Binding Protein Structures Reveals Conformational Signatures of Agonism In Nicotinic Receptors

Spencer T. Stober and Cameron F. Abrams, Department of Chemical and Biological Engineering, Drexel University, Philadelphia, PA

The soluble acetylcholine binding protein (AChBP)has served for many years as a structural proxy for members of the superfamilyof pentameric ligand-gated ion channels (LGICs),which are integral to many biological functions across an array ofspecies. Unfortunately, it has so farproven difficult to extract detailed conformational signatures unique to agonism and antagonism of LGIC's from the large set of AChBP crystal structures in both apoand ligand-bound states, primarily because AChBPconformations are fairly invariant across this set (mean RMSD < 1.5),resulting in alignment artifacts that make comparing structures for thepurposes of identifying conformational changes problematic. To address this, we have undertaken asystematic approach to elucidate conformational signatures displayed by AChBP that differentiate apo/antagonist-boundfrom agonist-bound states. Our methodside-steps alignment artifacts by using statistical inference based on inter-α-carbondistances from both crystallographic states and conformations sampled duringlong molecular dynamics (MD) simulations. Using this method in conjunction with MD simulation of an agonist-boundstructure, we observed that binding of agonists of the nicotinic acetylcholinereceptor (nAChR) to AChBPelicits clockwise rotation of the inner β-sheet with respect to the outerβ-sheet, causing tilting of the cys-loop. This motion is similar to that suggested bycomparing conformations of α and non-α subunits in the heteromeric nAChR, and ispotentially important in transmission of the gating signal to the transmembrane domain of a LGIC (Unwin,J. Mol. Biol. 2005;346:967-989). Themethod is also successful in discriminating partial from full nAChR-agonists, and supports the hypothesis that aparticularly controversial ligand, lobeline, is infact an LGIC antagonist.

See more of this Session: Molecular Modeling of Biophysical Processes I - Molecular Binding and Protein Structure and Dynamics
See more of this Group/Topical: Food, Pharmaceutical & Bioengineering Division

(562a) Re-Engineering Multicloning Sites for Function and Convenience

Nathan Crook, Chemical Engineering, University of Texas at Austin, Austin, TX, Elizabeth S. Freeman, Department of Biology, Washington University in St. Louis, St. Louis, MO and Hal Alper, Chemical Engineering, The University of Texas at Austin, Austin, TX

Metabolic engineering efforts require well-tuned expression of heterologous genes. Often, this task is accomplished through the use of plasmids and synthetic genetic circuits intended to express controlled levels of any gene of interest. These expression cassettes usually contain pre-defined multiple cloning sites (MCSs) which enable the facile expression and cloning of recombinant genes. In all synthetic biology applications, multiple cloning sites are thought to be benign, non-interacting elements that exist for mere convenience. However, a promoter element is usually placed upstream of the MCS, resulting in additions to the 5' untranslated region which may influence downstream gene expression in eukaryotes. Here, we demonstrate that mRNA secondary structure in common MCSs can significantly inhibit protein translation in multiple transcriptional contexts, especially for short, codon-optimized genes. Furthermore, we show that this inhibition is highly dependent on cloning location, resulting in drastic, nonmonotonic changes in protein production as different restriction sites are used. To mitigate this effect, we developed a model of structure-based translation inhibition which can predict the effects of mRNA secondary structure on protein expression in eukaryotes, in particular, the yeast *Saccharomyces cerevisiae*. This model enables the rational re-engineering of multicloning sites to allow consistently high levels of translation regardless of gene context. We demonstrate this utility through experimental validation of these improved genetic elements. Several novel multicloning sites were developed for use in conjunction with a wide range of promoters which show significant improvement over commonly used constructs, validating the predictive ability of our model. Moreover, these experiments indicate that rational, context-specific optimization of biological parts represents both an attainable and desirable goal for improvement of biosynthetic parameters. Finally, these results demonstrate that synthetic biology parts interact, and thus cannot be designed independently.

See more of this Session: Advances In Metabolic Engineering and Bioinformatics II - Computational
See more of this Group/Topical: Food, Pharmaceutical & Bioengineering Division

(562b) MetRxn: Reaction and Metabolite Standardization and Congruency Across Databases and Genome-Scale Metabolic Models

Patrick Suthers, Chemical Engineering, The Pennsylvania State University, University Park, PA, Akhil Kumar, Computer Science and Engineering, The Pennsylvania State University, University Park, PA and Costas D. Maranas, Department of Chemical Engineering, University Park, PA

The ever-accelerating pace of DNA sequencing and annotation information generation is spearheading the global inventorying of metabolic functions across all kingdoms of life. Increasingly, metabolite and reaction information is organized in the form of community, organism, or even tissue-specific genome-scale metabolic reconstructions. These reconstructions account for reaction stoichiometry and directionality, gene to protein to reaction associations, organelle reaction localization, transporter information, transcriptional regulation and biomass composition. Already over 40 genome-scale models are available for eukaryotic, prokaryotic and archaeal species and are becoming indispensable for computationally driving engineering interventions in microbial strains for targeted overproductions, elucidating the organizing principles of metabolism and even pinpointing drug targets. A key barrier to the pace of extraction of metabolic knowledge from data is our inability to directly make use of metabolite/reaction information from databases (e.g., BRENDA, KEGG, BioCyc, UM-BBD, PubChem, ChEBI, Reactome.org, Rhea, etc.) or other metabolic models due to incompatibilities of representation, duplications and errors. Therefore, the inadvertent inclusion of multiple replicates of the same metabolite, stoichiometrically inconsistent and/or elementally/charge unbalanced reactions can lead to erroneous model predictions and missed opportunities to reveal (synthetic) lethal gene deletions, repair network gaps and quantify metabolic flows. There have already been a number of efforts aimed at addressing some of these limitations such as the Rhea database and Model SEED. Motivated by this challenge we recently developed the MetRxn knowledgebase that integrates, using internally consistent descriptions, metabolite and reaction information from 6 databases and 34 metabolic models Metabolite and reaction data was first downloaded from BRENDA, KEGG and BioCyc using a variety of methods based on protocols such as SOAP, FTP and HTTP. We subsequently pre-processed the data into flat files that were imported into MetRxn. All original information pertaining to metabolite name, abbreviations, metabolite geometry, related reactions, catalyzing enzyme and organism name, gene-protein-reaction associations, and compartmentalization was retained. For all 34 genome-scale models ancillary information culled form the corresponding publications was also imported. The "raw data" from both databases and models was unified using standard SQL scripts on a MySQL server. We used Marvin (Chemaxon) to analyze all 231,085 raw metabolite entries containing structural information (out of a total of 322,936 entries). Metabolite atom bond connectivity was calculated at a fixed pH of 7.2 and converted into standard Isomeric SMILES format. Metabolites were also annotated with Canonical SMILES using the OpenBabel Interface from ChemSpider. Metabolites with missing structural information were re-visited during the reaction reconciliation step. After generating the initial metabolite associations, we identified reaction overlaps using the reaction synonyms and reaction strings along with the metabolite SMILES representations. During this step, reactions were flagged as single-compartment or two-compartment (i.e., transport reactions). Using the corrected metabolite elemental composition and protonation states, reactions are evaluated for charge and elementally balance. We used a linear optimization program to charge and elementally balance all reactions. MetRxn contains over 62,000 unique metabolites and 56,000 unique reactions. In this talk, we present the web-based access to MetRxn, describing updates in its increased scope of standardized metabolic models. Specifically, we explore the impact standardization has on genome-scale models and resulting flux balance analysis of the models. We also explore some interesting observations with model content overlaps and contrast these to phylogenetic trees. We also discuss SBML export and a web-based API for integration with external applications.

See more of this Session: Advances In Metabolic Engineering and Bioinformatics II - Computational
See more of this Group/Topical: Food, Pharmaceutical & Bioengineering Division

(562c) Modeling Metabolic Variation In Mutant Library Selections for Sugar Utilization

Joost Groot and Ryan Gill, Chemical and Biological Engineering, University of Colorado, Boulder, CO

Selections are a powerful tool to obtain microbial strainswith superior traits like improved cellulosic sugar utilization.However, the genetic mechanisms behind selections are complex and only partiallyunderstood at best. Better understanding of how genetic variation confersgrowth, fitness effects will allow for improved design of selections. Here theapplications and limitations of Elementary Flux Mode (EFM) modeling of genomicmutant library growth phenotypes will be discussed for *E.coli* selectionson cellulosic sugars.

Trackable Multiplex Recombineering (TRMR) creates acollection of mutants each with either a specific gene overexpressed or knocked-down[1].This genetically engineered variance speeds up the discovery and identificationof mutants with high growth/fitness thereby indicating mutations/genetic mechanisms that dominate a selection.

Elementary Flux Mode (EFM) Analysis allows for an evaluationof the metabolic pathways of a cell. A recent EFM method[2] allows formodeling changes in fluxes in mutants with altered expression of metabolicgenes. The method is adapted to assess variability in growth phenotypes in a mutantlibrary. An evaluation of growth (yield) variability is given for differentselection conditions (e.g. glucose, xylose growth) and an extension to modelmutant fitness of a TRMR library is proposed.

Sugars like glucose and xylose are metabolized via differentpathways. Quantifying the variation in mutant selections based on metabolicpathway utilization can help assess its contribution to divergence in selectionsand could aid in relating its effect to variation in regulation and (often unanswered) fundamental issues like rate versus yield strategies, the relation betweengrowth and fitness.

References

1 Warner, J.R., Gill, R.T. et al (2010) Nature Biotechnology28: 856862

2 Zhao, Q. and Kurata, H. (2009) Bioinformatics25(13):1702-1708

See more of this Session: Advances In Metabolic Engineering and Bioinformatics II - Computational
See more of this Group/Topical: Food, Pharmaceutical & Bioengineering Division

(562d) Design Principles for 13C-Metabolic Flux Analysis Using Elementary Metabolite Unit Basis Vectors (EMU-BV)

Maciek R. Antoniewicz, Chemical Engineering, University of Delaware, Newark, DE and Scott B. Crown, Chemical Engineering, University of Delaware, Newark

Metabolic flux analysis (MFA) is a powerful tool for elucidating *in vivo* fluxes in microbial and mammalian systems. A key step in ^{13}C-MFA is the selection of an appropriate isotopic tracer to observe fluxes in a proposed network model. Despite the importance of MFA in metabolic engineering and beyond, current approaches for tracer experiment design are still largely based on trial-and-error. The lack of a rational methodology for selecting isotopic tracers prevents MFA from achieving its full potential. Here, we introduce a new technique for tracer experiment design based on the concept of elementary metabolite unit basis vectors (EMU-BV). We demonstrate that any metabolite in a network model can be described as a linear combination of so-called EMU basis vectors, where the corresponding coefficients indicate the fractional contribution of the EMU-BV to the product metabolite. The strength of this approach is the complete decoupling of substrate labeling, i.e. the EMU-BV, from the dependence on free fluxes, i.e. the coefficients. In this work, we demonstrate that flux observability depends inherently on the number of independent EMU-BV and the sensitivities of coefficients with respect to free fluxes. Specifically, the number of independent EMU-BV places hard limits on how many free fluxes in a model can be determined. This constraint is used as a guide for selecting feasible substrate labeling. In three example models, we demonstrate that by maximizing the number of independent EMU-BV the observability of a system is optimized. Inspection of sensitivities of the coefficients with respect to free fluxes provides additional constraints for a proper selection of tracers. In summary, this work provides a fresh perspective on an important topic in metabolic engineering and gives practical guidelines and design principles for *a priori* selection of isotopic tracers for ^{13}C-MFA studies.

See more of this Session: Advances In Metabolic Engineering and Bioinformatics II - Computational
See more of this Group/Topical: Food, Pharmaceutical & Bioengineering Division

(562e) Improving the Efficiency of 13C Isotopic Instationary Metabolic Flux Analysis

Xiaofeng Zhang, Department of Chemical and Biomolecular Engineering, University of Maryland, College Park, Greenbelt, MD, Ian MacIntire, Department of Chemical and Biomolecular Engineering, University of Maryland, College Park, College Park, MD and Ganesh Sriram, Chemical and Biomolecular Engineering, University of Maryland, College Park, MD

Isotope-assisted metabolic flux analysis (MFA) is a powerful experimental tool for quantifying metabolism and a cornerstone of metabolic engineering. This technique can be classified into two categories based on how sampling is performed: steady state MFA, which involves sampling at an isotopic steady state or at least a pseudo-steady state, and instationary MFA, which involves multiple, transient sampling before steady state is attained. Compared with steady state MFA, instationary MFA can lead to shorter experimental times and is even indispensable for quantifying fluxes in certain metabolic scenarios such as photosynthetic CO_2 assimilation. However, this technique is relatively less popular as compared with steady state MFA due to two major difficulties associated with implementing it. One is the experimental requirement of obtaining precise measurements of pool sizes, or metabolite concentrations, which is unnecessary in steady state MFA. The second, computational hurdle is that instead of algebraic equations, systems of ordinary differential equations (ODEs) need to be solved repeatedly in instationary MFA, which results in huge computational cost. In this presentation we will introduce two ways to reduce this computational burden. Firstly, we will introduce a novel analytical method to solve the ODE systems in instationary MFA. This analytical method requires significantly less computational time over the commonly used numerical method when applied to many common metabolic networks. Additionally, we present a new metabolic network decomposition method that is similar to the concept of 'bondomers'. While the concept of 'bondomers' can only be applied to experiments that employ a mixture of uniformly labeled and naturally abundant carbon sources, the decomposition method presented here can be employed for several more labeling experiment scenarios. We anticipate that the combination of these two improvements will make the solution to the mathematical models of instationary MFA more efficient and thus extend the application of instationary MFA to larger metabolic networks.

See more of this Session: Advances In Metabolic Engineering and Bioinformatics II - Computational
See more of this Group/Topical: Food, Pharmaceutical & Bioengineering Division

(562f) Cost Efficiency Model for the Network-Wide Quantitative Determination of Enzyme Activities

Hong Yang and Igor G.L. Libourel, Department of Plant Biology, BioTechnology Institute, University of Minnesota, St. Paul, MN

Flux Balance Analysis (FBA) has become the most commonly applied method to evaluate internal fluxes from extracellular flux measurements and an inferred objective function that is thought to be enforced by evolution. However, FBA results in a high-dimensional solution space instead of a unique flux distribution. Kinetic models on the other hand do return a unique flux solution, but their development has been hindered by difficulties in collecting adequate enzyme activity data. We present an enzyme cost efficiency model designed to simultaneously determine all enzyme activities from enzyme concentration measurements. The premise of this model is the economical usage of enzymes, which implies a minimal investment in protein cost for a given flux demand. The cost efficiency model corresponds to a constrained optimization problem that minimizes the distance between measured and simulated enzyme concentrations and maximizes the biomass production. A bidirectional mass action model links enzyme activities and concentrations to the flux distribution. The forward and reverse rate constants were constrained by the equilibrium rate constants which reduces the number of activity variables by a factor of two, whilst enforcing thermodynamic consistency. The parameter fitting was solved as a double non-linear inverse problem. Enzyme activities were fitted in an outer loop and enzyme concentrations were optimized in an inner loop. This cost efficiency model presents a novel optimization framework that enables the formulation of kinetic models for large metabolic networks suitable for rational design of metabolism.

See more of this Session: Advances In Metabolic Engineering and Bioinformatics II - Computational
See more of this Group/Topical: Food, Pharmaceutical & Bioengineering Division

(562g) Combining Experimental Data with Sampling Know-How: Large-Scale Kinetic Modeling of Chlamydomonas Reinhardtii

Jon Bollinger and Mariajose Castellanos, Chemical and Biochemical Engineering, UMBC, Baltimore, MD

Genome-scale kinetic metabolic modeling is considered challenging and even infeasible because of poorly known kinetics and the presumed computational limitations associated with solving systems of nonlinear differential equations; we present an approach that constructs truly kinetic models using experimentally-derived information while taking cues from sampling-based methodologies (such as Metabolic Control Analysis and Ensemble Modeling). Recognizing that kinetic parameters are inherently uncertain (e.g. *in vitro* vs. *in vivo* enzyme behavior) and intracellular concentrations typically unknown, especially for less-characterized species, parameters are allowed to fluctuate randomly in order to create ensembles of kinetic models that are not at all dependent on flux-derived steady-states. Rather, the resulting model ensembles are allowed to develop to their respective steady-states, which represent the possible cell behaviors allowable within the bounds of known thermodynamics, the known spaces of parametrized enzyme kinetics, and cell biology. The ensembles solve without extreme time-burden. We applied this methodology to the model algae *Chlamydomonas reinhardtii*, of interest to groups working toward improved yields of biofuels and other value-added products. The model system comprised 4 organeller compartments and 469 reactions and incorporated known inhibition kinetics for highly regulated cycles (e.g. Calvin, TCA) and other control points. Ensembles comprised 500-1000 models, and reproduced known trends in energy states, uptake/evolution of extracellular species, and intracellular concentration shifts under nutrient deprivation and genomic edits. Subsequent studies focused genomic changes affecting the Calvin cycle enzymes toward increased biomass accumulation, per the interest of experimentalist collaborators.

See more of this Session: Advances In Metabolic Engineering and Bioinformatics II - Computational
See more of this Group/Topical: Food, Pharmaceutical & Bioengineering Division

(571a) Evaluation of Escherichia Coli Proteins That Burden Recombinant Protein Purification On Inexpensive, Non-Affinity-Based Resins

Patrick R. Bartlow[1], Neha Tiwari[2], Robert R. Beitle[2] and Mohammad M. Ataai[3], (1)Bioengineering, University of Pittsburgh, Pittsburgh, PA, (2)Ralph E. Martin Department of Chemical Engineering, University of Arkansas, Fayetteville, AR, (3)Chemical and Petroleum Engineering, University of Pittsburgh, Pittsburgh, PA

Escherichia coli has long been a favored host for rapid, scalable expression of recombinant proteins for commercial or academic use. Upstream advantages must be coupled with robust downstream processes, however, to maximize its economic benefits. While affinity chromatography is unrivaled in its selectivity and resolving capabilities, it comes with significant cost. This presentation reports on preliminary efforts to integrate downstream processing with upstream host design by evaluating co-eluting host proteins that most severely burden purification by non-affinity-based column processes. Phosphoenolpyruvate carboxykinase and peptidase D were significant contaminants during serial purification of GFPuv by hydrophobic interaction and anion exchange chromatography. With the intent to improve downstream efficiency with cheaper resins via host modifications, implications for genetic knockout or site-directed mutagenesis that diminishes contaminant retention are discussed for these and other identified host proteins.

See more of this Session: Bioseparations / Downstream Processing of Biopharmaceuticals
See more of this Group/Topical: Food, Pharmaceutical & Bioengineering Division

Proteomics approach for predicting separation behavior of a mixture of proteins during downstream purification

Ryan K. Swanson[1], Ruo Xu[2], Dan S. Nettleton[2] and Charles E. Glatz[1]

[1]Department of Chemical and Biological Engineering, Iowa State University, Ames, IA 50011

[2]Department of Statistics, Iowa State University, Ames, IA 50011

Recombinant protein production has many applications spanning a wide range of scientific fields, with downstream purification being the most costly part of the overall process. One reason for this stems from a lack of knowledge of the selected host cell protein's (HCP) separation behavior during downstream purification. The process of selecting the downstream purification method(s) as well as the host cell can benefit from an accurate prediction of the HCP separation behavior, thereby reducing the resources needed to investigate both. The HCP separation behavior was characterized using a method where three of the most common protein properties involved in downstream purification were obtained by aqueous two-phase partitioning or ATPS (surface hydrophobicity) and 2-dimensional electrophoresis or 2DE (pI or isoelectric point and molecular weight). Once the proteins had been characterized or "mapped-out" by the three properties, a multivariate model was generated with each characterization property used as a predictor variable and separation behavior as the response variables. The accuracy of predicting a proteins separation behavior using this 3-dimensional characterization method (ATPS + 2DE) had been previously verified for ion exchange chromatography using a set of known model proteins. Here results for hydrophobic interaction chromatography will be added, again using a set of known proteins. This talk will additionally discuss the extension of alternative statistical models for predicting the separation behavior of selected downstream purification methods from 3D characterization of a complex HCP extract (transgenic corn).

Figure 1 shows the chromatogram that resulted from the cation exchange (CEC) separation (selected downstream purification method) on a mixture of native corn germ-rich fraction proteins extracted at pH 7 (selected complex HCP extract).

Figure 2 (a) shows the corn germ extract 2DE gel scan image and (b) shows the 3D scatter plot (or "map") of the three characterization properties for each protein. 2DE gels and 3D scatter plots were also prepared for each of the 6 CEC fractions shown in Figure 1 (Wash, 10% B, 20% B, 40% B, 60% B, 100% B) as well as for both phase solutions that resulted from ATPS partitioning of the corn germ extract.

Table 1 shows the overall root mean squared error (RMSE) value for the multivariate model that was developed using all of the characterized corn germ proteins (497). The low RMSE value (0.0406) demonstrates the multivariate model's ability to accurately predict the separation behavior of a complex mixture of proteins during a selected downstream purification technique. Also shown are the variable importance value (a reflection of the importance of each predictor variable towards predicting the correct value of the response variables, or VIV) results and the values conform to what is expected based on intuition when dealing with CEC separation, i.e. that pI has the highest VIV followed by MW and surface hydrophobicity.

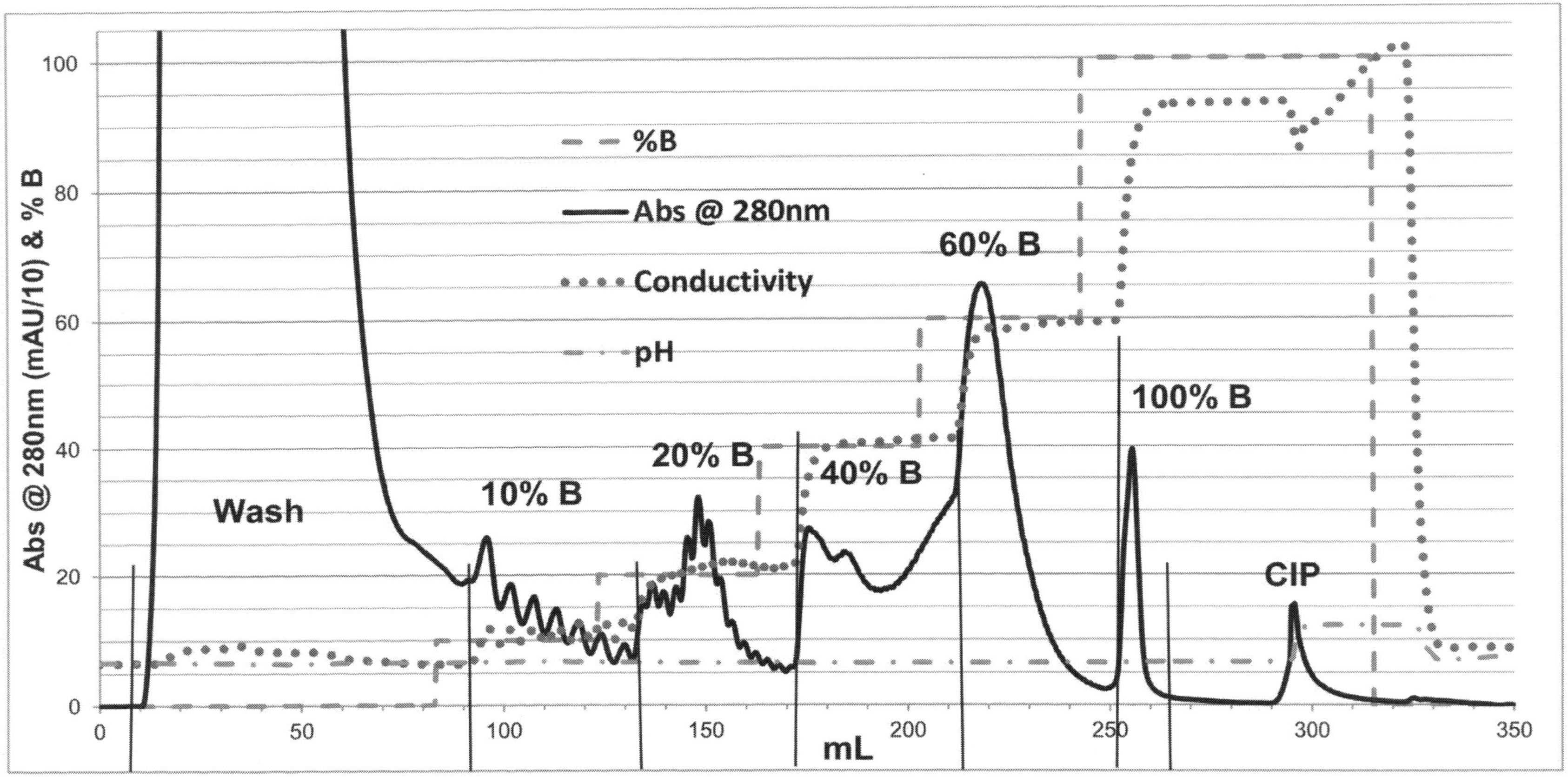

Figure 1: Chromatogram resulting from CEC separation of corn germ rich fraction extracted at pH 7.0. The vertical lines represent the points where the fraction volumes were pooled to represent each elution fraction and to capture each elution step's absorbance peak. The y-axis is scaled to mAU/10 (milli-absorbance units/10) and corresponds to %B concentrations as well (B = elution buffer: 50 mM $NaPO_4$ + 1 M NaCl @ pH 7).

a)

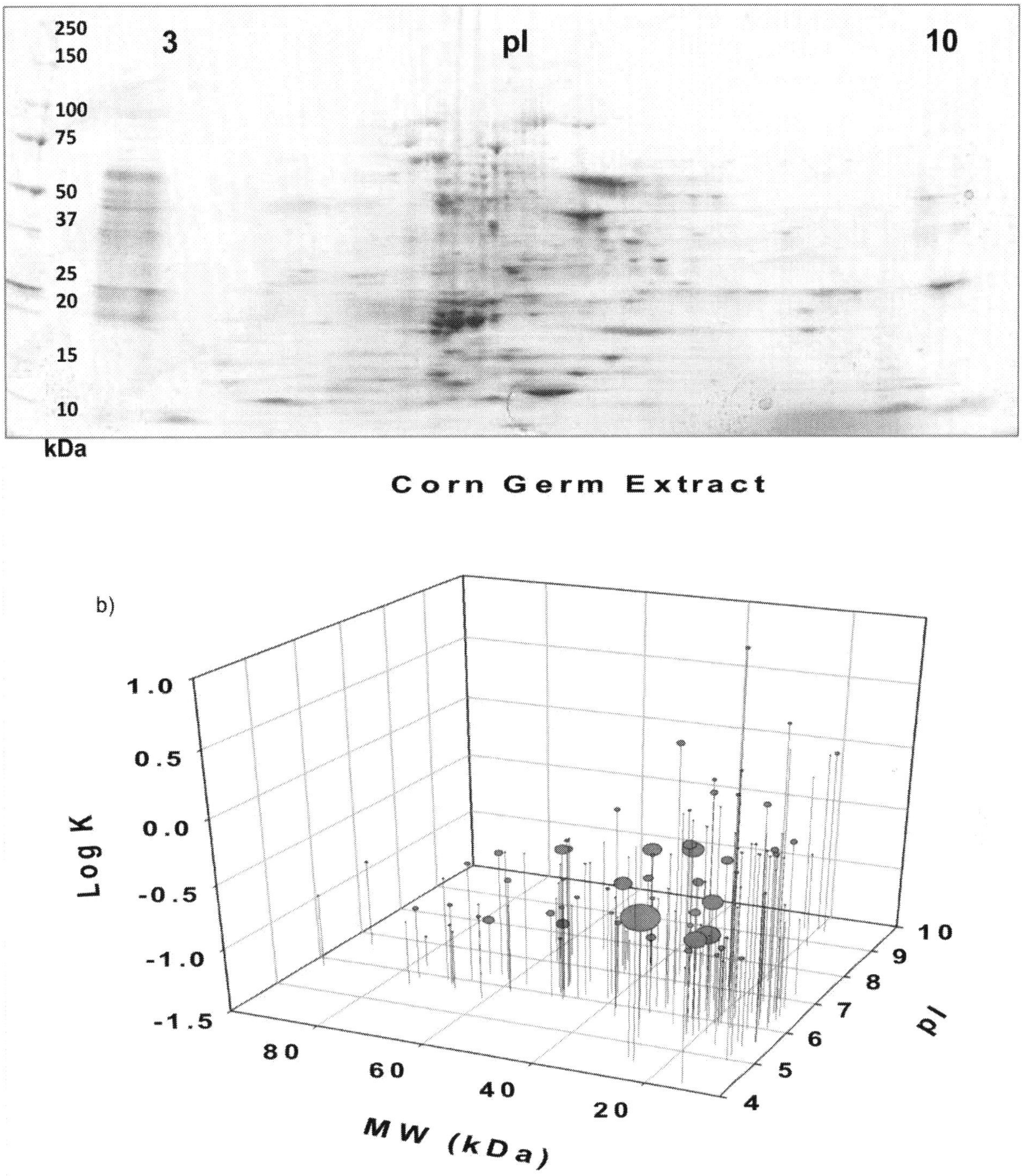

Figure 2: a) Corn germ extract 2DE gel scanned image with the molecular weight standard markers shown on the left in kDa and the pI range, which was assumed to be linear, shown across the top b) 3D scatter plot developed from the 2DE gels where each proteins pI is plotted on the x-axis, MW on the y-axis and Log K on the z-axis: sphere size represents % of total protein present in corn germ extract (1/10th scale). The spots shown all are $\pm$30% of mass balance & $\geq$0.1% of the total corn germ protein extract.

	Predictor Variable Importance Values			
Model Description	**pI**	**MW**	**Log K**	**RMSE**
497 characterized corn germ proteins	0.66	0.21	0.13	0.0406

Table 1: Variable Importance Value (VIV) for each of the three characterization (predictor) variables. Also shown is the overall root mean squared error (RMSE) result calculated for the multivariate model that was developed using all of the characterized corn germ proteins (497). RMSE value units are molarity of salt to elute.

(571c) A Mathematical Model for Protein Purification with Affinity Membranes

Simone Dimartino, Cristiana Boi and **Giulio C. Sarti**, University of Bologna, Bologna, Italy

Affinity chromatography represents one of the most important and widely used unit operations in the biotechnology industry. However, bead-based columns suffer from several limitations such as high pressure drop, slow mass transfer through the diffusive pores and strong dependence of the binding capacity on flow rate. One possible alternative to overcome these drawbacks is represented by convective media columns packed with affinity membranes. In order to promote their application in large scale processes, it is imperative to develop a reliable simulation tool able to describe the process performance in a predictive way.

In this work we present a mathematical model for the purification of immunoglobulin G in columns packed with affinity membranes, and discuss the enhanced performance versus packed bead columns. Such chromatographic model is based on species mass balance equation over the convective medium, coupled with a suitable kinetic equation which represents the interaction between the IgG target molecule and the ligand immobilized on the porous support. A detailed analysis of the experimental data indicates that a bi-Langmuir binding kinetics is essential for a correct process description up to the saturation of the stationary phase.

The proposed model has been validated with experimental data obtained in chromatographic cycles performed with affinity membranes. The stationary phases studied derive from the functionalization of membranes with natural and synthetic affinity ligands that show high specificity towards IgG.

Model simulations are in good agreement with all of the experimental affinity cycles, demonstrating the accuracy of the model to describe the transport phenomena in the column and the adsorption binding mechanism.

The scale-up of the improved affinity materials for industrial applications is also addressed.

See more of this Session: Bioseparations / Downstream Processing of Biopharmaceuticals
See more of this Group/Topical: Food, Pharmaceutical & Bioengineering Division

(571d) ELP-ZZ Fusions for the Purification of Immunoglobulins by Affinity Precipitation

Bhawna Madan[1], Garima Chaudhary[2], Steven M. Cramer[3] and Wilfred Chen[1], (1)Chemical Engineering, University of Delaware, Newark, DE, (2)Chemical Engineering, University California Riverside, Riverside, CA, (3)Howard P Isermann Department of Chemical and Biological Engineering, Rensselaer Polytechnic Institute, Troy, NY

The increase in demand of monoclonal antibodies for therapeutic applications along with the high manufacturing cost has made it necessary to evaluate better process options and technologies. Downstream purification of antibodies, which typically accounts for 50-80% of the total manufacturing cost, is a prime cost-cutting target. Affinity precipitation is an attractive alternative to traditional chromatographic methods by affording effective purification using a simple environmental trigger. In the current study, a Fc-binding zz domain fused to the C-terminus of an elastin like protein [ELP] was used as a new capture scaffold. The tunable hydrophobic interaction exhibited by the stimuli-responsive ELP was exploited as a non-covalent method to reversibly capture antibodies based on interaction with the zz domain. Up to 96% IgG capturing was achieved in a single precipitation step, while over 95% of the captured IgG was recovered after elution. The ELP-zz capture scaffold can be reused three times without any loss of binding efficiency. More importantly, one-step purification was also demonstrated for secreted IgG directly from hybridoma supernatant. Utility of this new technology for industrial IgG purification will be discussed as well.

See more of this Session: Bioseparations / Downstream Processing of Biopharmaceuticals
See more of this Group/Topical: Food, Pharmaceutical & Bioengineering Division

(571e) Reversible Oligomerization and Irreversible Coagulation of Monoclonal Antibodies In Downstream Processing

Paolo Arosio, Simonetta Rima, Baptiste Jaquet, Thomas Mller-Spth and Massimo Morbidelli, Chemistry and Applied Biosciences, ETH Zrich, Zurich, Switzerland

Therapeutic proteins represent a fast growing class of drugs in the pharmaceutical market. Achieving a better understanding of the mechanisms responsible for protein aggregation is crucial for successful commercialization of protein drugs. In this work, we investigate the aggregation stability of several mAbs in conditions relevant for downstream processing. The time evolution of the aggregates is analyzed by a combination of dynamic light scattering, size exclusion chromatography and field flow fractionation. Secondary structure changes are monitored by circular dichroism and thioflavin binding.

The results get insights into the intricate combined effect of pH, salt, kind of salt, temperature and antibody class on aggregation stability. Depending on operating conditions either reversible oligomerization or aggregation into insoluble, high molecular weight aggregates may occur. The experimental results are rationalized by providing a mechanistic description of both processes which connects the aggregation kinetics to the protein conformational stability and colloidal stability.

Particularly, an anion specific effect in inducing aggregation is observed according to a ranking which follows the Hofmeister series with the exception of the sulphate anion. An explanation of the observed anion effect at the molecular level is proposed, which could also be extent to the stabilizing/de-stabilizing effect of co-solutes on protein stability.

References:

P. Arosio, G. Barolo, T. Müller-Spath, H. Wu, and M. Morbidelli, *Pharmaceutical Research*; **2011**, *in press*

See more of this Session: Bioseparations / Downstream Processing of Biopharmaceuticals
See more of this Group/Topical: Food, Pharmaceutical & Bioengineering Division

(571f) Effect of Feed Solution Properties On Viral Filtration Performance

Joshua Walker, Process Biochemistry / Purification Scale up, Biogen Idec, Cambridge, MA

Effect of feed solution properties on viral filtration performance

Viral safety is an essential component of biopharmaceutical products derived from cell lines of animal origin. One aspect of viral safety is the incorporation of viral clearance capability into the manufacturing process. For example, nano-filtration is commonly employed to provide removal of a variety of large (e.g., retrovirus) and small (e.g., parvovirus) virus types. Parvovirus's have a diameter of 18-26 nm, whereas a typical IgG antibody has a hydrodynamic diameter of 8-12 nm. In order to achieve robust removal of these viruses while maintaining product throughput and minimal filtration area, viral filters are required to have a very narrow pore size distribution and are often sensitive to operating conditions. Therefore, careful optimization of operating parameters will ensure a robust, economical and scalable virus removal step.

Scaled-down systems were utilized to model and optimize the performance of a variety of commercial viral filters. Studies were performed evaluating the individual and combined effects of protein concentration, pH, ionic strength, and impurity levels on filter capacity, flux performance, and product recovery. In addition, these results were used to determine the optimal location of a viral filtration step in the purification process.

See more of this Session: Bioseparations / Downstream Processing of Biopharmaceuticals
See more of this Group/Topical: Food, Pharmaceutical & Bioengineering Division

(571g) "Kinetically Limited" Differential Centrifugation As An Inexpensive and Readily Available Alternative to Centrifugal Elutriation

Byung-Doo Lee[1], Jinwang Tan[1], Luis Polo-Parada[2] and **Shramik Sengupta**[1], (1)Biological Engineering, University of Missouri, Columbia, MO, (2)Medical Pharmacology & Physiology, University of Missouri, Columbia, MO

When one desires to separate two species with similar densities, but differing sedimentation velocities (due to differences in size), Centrifugal Elutriation, a method requiring specialized equipment, is the usual method of choice. We present here a method for achieving similar separations that can be realized using standard benchtop centrifuges. It involves loading the seperands as a layer on top of a denser buffer of a specified length, and running the benchtop centrifugation process for a calculated amount of time that ensures that *all* of the faster moving species are collected at the bottom, while *all* of the slower moving species remain in the buffer. We demonstrate the use of this procedure to isolate bacteria from Blood Culture Broth (a mixture of bacterial growth media, blood and bacteria).

See more of this Session: Bioseparations / Downstream Processing of Biopharmaceuticals
See more of this Group/Topical: Food, Pharmaceutical & Bioengineering Division

(584a) Triggered Binding Activity of Anti-HER2/ Neu Lipid Vesicles Encapsulating Doxorubicin Enhances Tumor Control In Vivo While Preserving the Potential for Low Toxicities

Amey Bandekar, Othmer-Jacobs Department of Chemical and Biological Engineering, Polytechnic Institute of NYU, Brooklyn, NY and Stavroula Sofou, Biomedical Engineering, and Chemical and Biochemical Engineering, Rutgers University, Piscataway, NJ

Ligand-mediated uptake of nanometer-sized drug carriers by cancer cells is usually necessary, since extravasated non-targeted nanocarriers into the tumor interstitium may not associate extensively and fast with the cancer cells that constitute the tumor. Typically, though, in cancer cells the extracellular leaflet of their plasma membrane is not characterized by unique molecular targets but by overexpression of antigens that are relatively downregulated in healthy cells. Consequently, during circulation in the blood, conventionally targeted drug delivery carriers with "exposed" targeting ligands would specifically bind to receptors expressed by non-cancerous cells ultimately increasing toxicities. Toxicities may also increase due to uptake of carriers by reticuloendothelial cells identifying the targeting ligands during circulation of carriers.

To enable selective cell-kill, we designed functionalized lipid vesicles with pH-triggered heterogeneous membranes and encapsulated doxorubicin that exhibit tunable surface topography. These vesicles "hide" (mask) the targeting ligands from their surface during circulation in the blood, and only progressively "expose" these ligands as they gradually penetrate deeper into the tumor interstitium, where after endocytosis they burst-release their contents. The stimulus to activate the binding reactivity is the pH gradient between the blood stream (pH 7.4–7.0) and the increasingly acidic pH inside the tumor interstitium (pH 6.7–6.0). Doxorubicin release is activated at the endosomal pH 5.5–5.0. *In vitro*, these targeted vesicles exhibit superior killing efficacy of cancer cells - with lowering the extracellular pH from 7.4 to 6.0 - compared to established liposome formulations. *In vivo*, in a murine vascularized breast cancer model, these antiHER2/neu-short-peptide vesicles loaded with doxorubicin exhibit superior tumor control compared to established formulations of liposomal doxorubicin labeled with identical antiHER2/neu peptides. Our findings demonstrate the potential of targeted lipid vesicles with triggered binding reactivity to increase the therapeutic efficacy of targeted immunochemotherapy while keeping toxicities and potential immunoreactivity to a minimum.

See more of this Session: Drug Delivery III
See more of this Group/Topical: Food, Pharmaceutical & Bioengineering Division

(584c) Sustained In Vivo Release From Imprinted Daily Wear Contact Lenses

Arianna Tieppo[1], Charles J. White[1], Amanda C. Paine[1], Meredith L. Voyles[2], Matthew K. McBride[1] and Mark E. Byrne[1], (1)Biomimetic & Biohybrid Materials, Biomedical Devices, & Drug Delivery Laboratories, Department of Chemical Engineering, Auburn University, Auburn, AL, (2)Clinical Sciences, Ophthalmology, College of Veterinary Medicine, Auburn University, Auburn, AL

Ocular tear flow poses a significant challenge to effective and efficient delivery of drugs to the eye. Typically, less than 1-7% of drug delivered through eye drop formulations is absorbed. However, eye drops and ointments account for 90% of ophthalmological formulations on the market today. Thus, there exists a considerable unmet need and strong motivation for more efficacious delivery of ocular therapeutics. Molecular imprinting is an effective technique to create macromolecular memory within a polymer, allowing greater control of drug transport from hydrogels. Imprinted lenses can deliver a constant drug concentration at a high effectiveness level for extended periods leading to significantly increased duration of action. The enormous advantage of using imprinting mechanisms is that the release rate, and thus the concentration reached, can be altered by changing the engineering design of such systems.

In this work, we demonstrate the successful *in vivo* extended release of a small molecular weight therapeutic, ketotifen fumarate (MW=425), from molecularly imprinted, therapeutic contact lenses. This is the first time that a steady, effective concentration of drug is maintained in the tear film from a contact lens for an extended period of time for the entire duration of lens wear. Poly(HEMA-co-AA-co-AM-co-NVP-co-PEG200DMA) soft contact lenses were prepared (100±5μm thickness, diameter 11.8 mm, power zero), and a constant tear film concentration of 170±30 μ g/mL was measured for up to 26 hours in a New Zealand white rabbit model. The results showed a dramatic increase in ketotifen mean residence time (MRT) and bioavailability compared to topical drop therapy and drug soaked lenses. The MRT for imprinted lenses was 12.47±3.99 hours, almost 37.0 fold greater than 0.035% eye drops (Zaditor®). Furthermore, the bioavailability ($AUC_{0\text{-}26hrs}$) was 127 fold greater for imprinted lenses than topical eye drops. The results indicate that molecular imprinting provides an exciting rational engineering strategy for sustained release. It is clear from this work that imprinted lenses are very promising combination devices and are much more effective and efficient delivery devices than eye drops.

See more of this Session: Drug Delivery III
See more of this Group/Topical: Food, Pharmaceutical & Bioengineering Division

(584e) Hemodynamics-Driven Design of Carriers for Imaging and Drug Delivery In Atherosclerosis

Alex Thompson[1], Katawut Namdee[2], Supriya Mocherla[1], Eric Mastria[3], Diane Bouis[4] and Omolola Eniola-Adefeso[1], (1)Chemical Engineering, University of Michigan, Ann Arbor, MI, (2)Biomedical Engineering, University of Michigan, Ann Arbor, MI, (3)Biomedical Engineering/Medical School, Duke University, Durham, NC, (4)Internal Medicine, University of Michigan, Ann Arbor, MI

@font-face { "Calibri"; }p.MsoNormal, li.MsoNormal, div.MsoNormal { margin: 0in 0in 10pt; line-height: 115%; font-size: 11pt; ""; }div.Section1 { page: Section1; }

The development of vascular-targeted carriers (VTC) for the delivery of therapeutics could greatly improve the treatment of many human diseases. Spherical nanoparticles are commonly used as potential drug carriers due to their ability to easily navigate microvasculature and ease of fabrication/drug loading. However, recent literature has shown that spherical nanoparticles do not efficiently marginate and adhere to the vascular wall in medium/large blood vessels (M/LBV) compared to larger micron sized particles.[1] Therefore spherical nanoparticles may not be effective in treating diseases which occur in M/LBV, such as atherosclerosis. Recently, particle shape has received attention as a parameter that can be used to improve the performance of VTCs. The purpose of this study is to examine the hemodynamics of rod-shaped particles of different aspect ratios relative to spheres of equal volume both *in vitro* in human blood flow under physiological conditions relevant in M/LBVs and *in vivo* in a mouse model of atherosclerosis.

Materials and Methods: Polystyrene rod-shaped particles were fabricated via a polymer film stretch method previously described in literature.[2] Selectin-targeted (via Sialyl Lewis ^{a}) spheres and rods at a fixed concentration were mixed with RBC and saline flow buffer, and allowed to flow over a layer of IL-1β activated human umbilical vein endothelial cells (HUVEC) using a parallel plate flow chamber. The number of adherent particles under various flow conditions was imaged/quantified using brightfield and fluorescence microscopy. The particles explored ranged in diameter from 0.5 – 5 μ m (equivalent spherical diameter, ESD, for rods).

Results and Discussion: With *in vitro* assays we find that rod shaped particles with ESD >2 μm have a higher binding efficiency to HUVEC monolayer at the wall from bulk human blood flow than spheres of equivalent volume and targeting ligand site density. The folds increase in the adhesion of rods over spheres increased both with increasing blood shear rate and increasing particle aspect ratio. Overall, the observed flow adhesion pattern for micron-sized rods was due to a combination of both a higher affinity (localization) of rods for the vascular wall, as well as an increased ability of rods to resist detachment from the vascular wall due to its streamlined shape. Interestingly, there appears to be a minimum major axis length requirement for rods with ESD <2μm to display a significant increase in flow adhesion over their equivalent spheres. Result from *in vivo* assays in mice show that rods have significantly higher localization to plaque surface in the aorta compare to nanospheres.

Conclusions: Our study shows that shape can be a very useful and tunable parameter in the design of vascular-targeted carriers for imaging and treatment of atherosclerosis and other diseases of M/LBVs.

References:

[1] Charoenphol et al. **Biomaterials** (2010) vol 31, 1392-1402.

[2] Ho et al. **Colloid Polym Sci** (1993) vol 271 n 5, 469-479.

See more of this Session: Drug Delivery III
See more of this Group/Topical: Food, Pharmaceutical & Bioengineering Division

(584f) Impact of Food-Associated Stimuli On Barrier Properties of Gastrointestinal Mucus to Particle Transport

Hasan M. Yildiz and Rebecca Carrier, Chemical Engineering, Northeastern University, Boston, MA

Impact of Food-associated Stimuli on Barrier Properties of Gastrointestinal Mucus to Particle Transport

Hasan Yildiz, yildiz.h@husky.neu.edu

Rebecca L. Carrier, rebecca@coe.neu.edu

Introduction:

Mucus is composed of a highly complex network including glycoproteins, lipids, cellular and serum macromolecules. Orally delivered drugs and nutrients must diffuse through mucus in order to enter the circulatory system. The transport of drug vectors through the mucus barriers has shown significant effects to enhance drug delivery systems, however the transport of these vectors and the barrier properties of mucus are poorly characterized. The main objectives of this study were to analyze the barrier properties of mucus and understand the mechanisms of effects of physiologically relevant stimuli: Lipids, pH and [Ca^{+2}] on particle transport through the gastrointestinal mucus barrier. All these factors are contributed with gastrointestinal changes regarding food intake and they can also be easily controlled.

Materials and Methods:

Native porcine intestinal mucus was collected from porcine jejunum within 2 h of slaughter. 200 nm amine-, carboxylate-, and sulfate-modified microsphere solutions (2 % solids in distilled water with 2 mM azide) were diluted in media: fed state levels of maleate buffer, bile salts-maleate buffer and fed state lipids in digested states. All these media were prepared at a relevant ph (3.5, 5.5, 6.5), physiological [Ca^{+2}] concentrations (5, 10, 20 mM). 10 µl of diluted particle solutions (0.0025 % wt/vol) was added to 200 µl of native mucus for a final particle concentration of 1.25 x 10^{-4} % wt/vol. Particle diffusion was measured by tracking the positions of microspheres using an Olympus IX51 microscope at 40 X magnification and videomicroscopy. 10 µ l of diluted particle solution was added to 200 µ l of native mucus. Samples were covered and left at room temperature for 2 h prior to microscopy. Five videos were taken from each individual preparation for a total 100 particle trajectories. The coordinates of particle centroids were used to calculate time-averaged mean squared displacements: MSD=[x(t+r)-x

$(t)]^2+[y(t+r)-y(t)]^2$, where t is the time scale and x, y are the coordinates. Two dimensional time-independent diffusion coefficients (D_0) were determined by plotting MSD versus time scale to the anomalous subdiffusive transport relation: $MSD=4D_0T^{\alpha}$, where α is the anomalous exponent, which is the degree of particle transport obstruction.

Results:

Lipid content associated with fed state intestinal contents significantly decreased the transport rate of microspheres through gastrointestinal mucus. The affinity of pig gastric mucus glycoprotein $[Ca^{+2}]$ was examined, and it was found that a higher concentration of $[Ca^{+2}]$ resulted in the reduction of nanoparticle transport rate. This is likely due to binding of $[Ca^{+2}]$ to mucin glycoproteins and thus enhanced cross-linking of the mucus gel network. Elevation of pH affects mucus viscoelasticity, leading to less hindered particle transport.

Conclusions:

Gastrointestinal mucus can pose a significant barrier to drug carrier transport dependent on medium properties. Promising results provide the influence of lipids, $[Ca^{+2}]$ and pH support the significance of food contents on mucus barrier properties and also the use of naturally changing factors in the GI tract, lipids, Ca^{+2} and pH, to control drug carrier or pathogen transport.

See more of this Session: Drug Delivery III
See more of this Group/Topical: Food, Pharmaceutical & Bioengineering Division

(584g) Diffusion Coefficient Prediction Model for Sustained Release From Silk Films

Daniel J. Hines, Chemical and Biological Engineering, Tufts University, Medford, MA and David L Kaplan, Biomedical Engineering, Tufts University, Medford, MA

The overall goal of the research project is to develop a prediction model set that can accurately estimate the diffusion coefficient for small molecule release from silk films by accounting for the physiochemical properties of the chemical. This type of a predictive model would benefit the design of silk fibroin sustained release devices for drug delivery by simulating release and reducing experimentation. Also, a robust predictive model will help to identify material properties that can be manipulated as design variables to achieve a target release profile. In order to develop this predictive model set the diffusional release mechanism was confirmed by fitting an analytical diffusion model to the experimental release data. A group of synthetic dyes varying in molecular weight, solubility, partition coefficient, pK_a, and pK_b were chosen for release from silk fibroin films to construct the predictive model. Synthetic dyes were chosen for the experimental design because they are relatively stable, their release into solution can be easily detected by spectroscopy, and they exhibit a wide range of physiochemical factors that can impact sustained release kinetics. Also there exist many classes of synthetic dyes that are derived from the same precursors and share similar structures. This allows the information to be approached as classes of pseudo drug libraries. Once several diffusion coefficient estimates were obtained for the dyes, empirical relationships were constructed between the diffusion coefficient estimates and the physiochemical properties of the dyes with multivariate regression analysis. Cross validation of the predictive model with a separate data was used to judge its interpolative and extrapolative predictive power in simulation. The developed model set can be updated for material properties, so that release kinetic relationships for more complex systems can be determined, in order to expand the predictive capabilities of the model set.

See more of this Session: Drug Delivery III
See more of this Group/Topical: Food, Pharmaceutical & Bioengineering Division

(593a) The Mammalian MAPK/ERK Pathway Exhibits Properties of a Negative Feedback Amplifier

Oliver Sturm[1], Richard Orton[1], Joan Grindlay[2], **Marc R. Birtwistle**[3], Vladislav Vyshemirsky[1], David Gilbert[1], Muffy Calder[1], Andrew Pitt[4], Boris N. Kholodenko[3] and Walter Kolch[3], (1)Computing Science, University of Glasgow, (2)Beatson Institute for Cancer Research, (3)Systems Biology, University College Dublin, Belfield, Ireland, (4)Sir Henry Wellcome Functional Genomics Facility, University of Glasgow

Three-tiered kinase modules, such as the Raf–MEK (mitogen-activated or extracellular signal–regulated protein kinase kinase)–ERK (extracellular signal–regulated kinase) mitogen-activated protein kinase pathway, are widespread in biology, suggesting that this structure conveys evolutionarily advantageous properties. We show that the three-tiered kinase amplifier module combined with negative feedback recapitulates the design principles of a negative feedback amplifier (NFA), which is used in electronic circuits to confer robustness, output stabilization, and linearization of nonlinear signal amplification. We used mathematical modeling and experimental validation to demonstrate that the ERK pathway has properties of an NFA that (i) converts intrinsic switch-like activation kinetics into graded linear responses, (ii) conveys robustness to changes in rates of reactions within the NFA module, and (iii) stabilizes outputs in response to drug-induced perturbations of the amplifier. These properties determine biological behavior, including activation kinetics and the response to drugs.

See more of this Session: Intracellular Processes II
See more of this Group/Topical: Food, Pharmaceutical & Bioengineering Division

(593b) Quantifying Cross-Talk Among Interferon-γ, Interleukin-12 and Tumor Necrosis Factor Signaling Pathways within a T_H1 Cell Model: A Model-Based Inference Approach

David J. Klinke, Emily Chambers and Ning Cheng, Chemical Engineering, West Virginia University, Morgantown, WV

T helper cells integrate biochemical cues present within the tissue microenvironment to orchestrate immunity via production of cytokines. Prior discoveries reveal a qualitative understanding of how T helper cells process this biochemical information, as frequently depicted as signaling cartoons. However, the lack of methods for quantifying how well these signaling cartoons apply to a particular cell type presents a major hurdle for translating our knowledge of immunity across systems. The objective of this study was to illustrate how model-based inference methods, in conjunction with quantitative cytometry-based methods, can be used to reason about the relative contributions of different putative branches within a signaling network. A cellular model of mouse T_H1 cells was used to quantify the functional response to Interleukin-12 (IL-12), a key cytokine that links innate to adaptive immunity. Our results demonstrate that the response of T helper cells to IL-12, as modeled by a mouse IL-12-dependent T_H1 cell line, is a dynamic nonlinear process that reflects a hysteresis in response and engages a positive feedback mechanism via direct activation of STAT1. The hysteresis in the dose-response curve to IL-12 creates a transient "memory" by sustaining cytokine secretion following the withdraw of the stimuli. In summary, this combined experimental and computational approach illustrates how model-based inference can be used to obtain greater fidelity in understanding how cells process and act upon biochemical cues present within the tissue microenvironment.

See more of this Session: Intracellular Processes II
See more of this Group/Topical: Food, Pharmaceutical & Bioengineering Division

(593c) Reorientation During Cell Migration Occurs by a PI3K-Dependent Steering Mechanism

Erik S. Welf, Chemical and Biomolecular Engineering, North Carolina State University, Raleigh, NC and Jason Haugh, Chemical & Biomolecular Engineering, North Carolina State University, Raleigh, NC

Cell migration requires coordination between intracellular signaling and extracellular cues to orchestrate the cellular processes necessary for efficient cell movement. Further, the spatial heterogeneity and stochasticity of cell signaling events involved in migration require live-cell, spatially-resolved measurements of cell signaling activity along with specialized analytical techniques to determine relationships between dynamic signaling events in a constantly-changing cell. In this work, we use total internal reflection fluorescence (TIRF) microscopy to image membrane-localized signaling events during fibroblast migration and computational techniques to analyze the dynamic localization of these signaling events. Signaling through the phosphoinositide 3-kinase (PI3K) pathway, which occurs primarily in protruding regions of the cell, is thought to affect cell motility, yet its exact roles in random and chemotactic migration remain controversial. By mapping the spatiotemporal dynamics of cell protrusion/retraction, localization of PI3K signaling, and cell morphology during migration, we show that fibroblasts reorient polarity through a PI3K-dependent mechanism involving the branching and pivoting of protrusions. Surprisingly, maximal PI3K signaling occurs just after local protrusion, and inhibition of PI3K signaling does not affect migration speed. During chemotaxis, we find that fibroblasts use this branch-and-pivot steering mechanism to move up a concentration gradient via preferential signaling and protrusion in the branch experiencing the highest concentration of chemoattractant.

See more of this Session: Intracellular Processes II
See more of this Group/Topical: Food, Pharmaceutical & Bioengineering Division

(593d) A Mechanism Leading to Imbalanced EGFR Oncogenic Signaling Common to Lung and Brain Cancers with Distinct EGFR-Activating Mutations

Janine M. Hall, Christopher M. Furcht, Andrs Muoz Rojas and **Matthew J. Lazzara**, Chemical and Biomolecular Engineering, University of Pennsylvania, Philadelphia, PA

Elevated expression of the epidermal growth factor receptor (EGFR) is observed in many cancers, but efforts to treat such cancers with EGFR-targeted therapeutics have resulted in limited success. Surprisingly, EGFR mutations resulting in constitutive receptor activity tend to promote cellular sensitivity to EGFR-targeted inhibitors, an effect which has been suggested to result from a general augmentation of both pro- and anti-apoptotic signaling activity downstream of the receptor in which pro-apoptotic pathways are more sustained in response to receptor inhibition. Previous work in our lab demonstrated that the EGFR-activating mutations in exons 19 and 21 which arise in non-small cell lung cancer (NSCLC) lead to enhanced cellular sensitivity to the EGFR kinase inhibitors gefitinib and erlotinib in part because of an impairment in the activation of the extracellular regulated kinase (ERK) due to altered function of the protein tyrosine phosphatase SHP2. This mechanism appears to involve the sequestration of biochemically active SHP2 with the endocytosis-impaired EGFR mutants at the plasma membrane in a manner which prevents the participation of SHP2 in the activation of ERK. In the present study, we sought to determine the extent to which a similar mechanism might exist in the context of elevated expression of the structurally distinct, constitutively active EGFRvIII mutant (which results from an in-frame deletion of exons 2-7) in glioblastoma multiforme (GBM) cells. Using a panel of GBM cell lines expressing three different levels of EGFRvIII or a kinase-dead counterpart, we found that the magnitude of EGF-mediated activation of ERK decreased sharply with increasing EGFRvIII expression levels. In fact, for cells with the highest expression of EGFRvIII (two million receptors per cell), treatment with 10 ng/mL EGF failed to induce detectable ERK phosphorylation. Also consistent with our findings in NSCLC cells, we found that increasing expression of EGFRvIII reduced the rate of ligand-mediated receptor internalization. Immunofluorescence imaging revealed that ligand-induced redistribution of SHP2 from the cell center outward was least pronounced for cells with the highest EGFRvIII expression levels, where SHP2 was basally distributed throughout the cell. Immunoprecipitation of SHP2 from cell lysates revealed that basal association of SHP2 with the adapter protein GAB1 and EGFR increased with increasing EGFRvIII expression. Taken together, the preceding data suggest that sufficiently high EGFRvIII expression may result in sequestration of SHP2 with adapter proteins and the receptor at the cell periphery in a complex which may lead to an impaired ability to promote ERK activity in response to EGFR ligands. We further explored this model in the panel of GBM cell lines through stable knockdown of SHP2 using RNA interference. SHP2 knockdown led to decreased ERK phosphorylation in all cell lines, but the magnitude of this effect decreased with increasing EGFRvIII expression. Consistent with this observation, we also found that SHP2 knockdown inhibited rates of cell proliferation least significantly in cells with the highest EGFRvIII expression. Thus, our data confirm the hypothesis that as EGFRvIII expression increases the extent to which SHP2 participates in ERK activation is diminished in a manner that reduces overall ERK activity levels and alters cellular proliferation rates. Our findings further demonstrate that structurally distinct EGFR-activating mutations may result in qualitatively similar perturbations to downstream signaling leading to the surprising impairment of receptor-mediated ERK activation. Given that elevated expression of these mutant receptors generally promotes cellular response to EGFR inhibitors, this work identifies SHP2 and ERK as potentially critical nodes in the signaling network downstream of EGFR for co-inhibition in settings where EGFR inhibitors are not effective as single agents.

See more of this Session: Intracellular Processes II
See more of this Group/Topical: Food, Pharmaceutical & Bioengineering Division

(593e) Mitochondrial Dysfunction In Hepatic Lipotoxicity

Robert Egnatchik and Jamey Young, Chemical and Biomolecular Engineering, Vanderbilt University, Nashville, TN

Excess fat accumulation in the liver results in nonalcoholic fatty liver disease (NAFLD), which is closely associated with obesity and insulin resistance. NAFLD is present in up to 30% of the American population, and can progress to a more severe form of liver dysfunction known as nonalcoholic steatohepatitis (NASH). Hepatocyte apoptosis is a prominent feature of NASH and correlates with disease severity. Prior *in vitro* studies have demonstrated that saturated fatty acids (SFAs) are potent inducers of apoptosis, whereas monounsaturated fatty acids (MUFAs) predominantly induce triglyceride accumulation without triggering apoptosis. SFAs can induce both endoplasmic reticulum (ER) stress and oxidative stress, both of which appear to play a causal role in triggering apoptosis. We have previously shown that SFAs alter central carbon metabolism by reducing glucose uptake and lactate production while increasing glutamine oxidation, and we hypothesize that this uncoupling of glycolytic and mitochondrial metabolism is linked to increased accumulation of reactive oxygen species (ROS) in SFA-treated cells.

Recently, we have investigated a possible mechanism in which calcium translocation from ER to mitochondria leads to overstimulation of oxidative metabolism, resulting in ROS accumulation and lipoapoptotic cell death. Our approach to study SFA toxicity uses metabolic inhibitors, calcium chelators, antioxidants, and a [U-$^{13}C_5$]glutamine isotopic tracer to quantify phenotypic changes in H4IIEC3 rat hepatoma cells treated with 200-400 μ M palmitate. First, to confirm that ROS is critical in promoting lipoapoptosis, the antioxidant n-acetyl cysteine (NAC) was added to cells treated with palmitate, which resulted in an approximately 50% increase in 24-hour viability. Therefore, the overproduction of ROS is a critical step in initiating palmitate lipoapoptosis in hepatic cells and is a potential therapeutic target. Next, we hypothesized that reducing mitochondrial metabolism would result in reduced ROS production and increased viability. Through co-treatment with the mitochondrial antagonist phenformin, a complete reversal of palmitate-induced ROS was achieved. Since phenformin is known to stimulate glucose uptake, we used insulin combined with palmitate to examine the effects of increased glucose uptake. Insulin co-treatment enhanced glucose metabolism, but unlike phenformin resulted in enhanced oxidative stress and higher levels of cell death in SFA-treated cells. Therefore, increased mitochondrial metabolism and not decreased glucose uptake is responsible for increased oxidative stress in hepatic lipoapoptosis.

To explore how lipotoxic SFA treatments increase mitochondrial metabolism, we hypothesized a mechanism in which ER stress-induced calcium release overstimulates mitochondrial function. To limit intracellular calcium trafficking, the membrane permeable calcium-specific chelator BAPTA was added to palmitate-treated cells. Addition of BAPTA normalized SFA-induced oxidative stress and increased cell viability. This supports our hypothesis that calcium has a role in promoting mitochondrial metabolism, ROS accumulation, and apoptosis resulting from SFA overexposure.

In summary, our recent experiments using intracellular calcium chelators (BAPTA) and mitochondrial antagonists (Phenformin) to rescue viability while reducing ROS implicate a critical role for ER calcium release and mitochondrial overactivation in promoting oxidative stress and lipoapoptosis in SFA-treated hepatic cells. Our presentation will highlight these results and discuss how mitochondrial metabolism and oxidative stress represent promising targets for treating NASH and lipotoxicity.

See more of this Session: Intracellular Processes II
See more of this Group/Topical: Food, Pharmaceutical & Bioengineering Division

(593f) Multisite Phosphorylation Underlies Ultrasensitive Response of CDK Inhibitor Sic1 In Yeast Cell Cycle Regulation

S. Marjan Varedi K.[1], Hangtian Song[2], Wendi A. Hale[2], Kristina Hakansson[2] and Xiaoxia Nina Lin[1], (1)Chemical Engineering, University of Michigan - Ann Arbor, Ann Arbor, MI, (2)Chemistry, University of Michigan - Ann Arbor, Ann Arbor, MI

Sic1 is a cyclin dependent kinase (CDK) inhibitor in the budding yeast *Saccharomyces cerevisiae.* This protein inhibits the Clb5,6-Cdc28 kinase required for DNA replication in the S phase and is believed to contribute to precise timing of the G1 to S transition by undergoing switch-like multisite phosphorylation-mediated degradation. During the late G1 phase, as the activity of Cln2-Cdc28 kinase complex increases, Sic1 becomes phosphorylated at up to nine sites. The phosphorylated Sic1 is recruited by the Cdc4 protein for proteolysis which eventually leads to Clb5,6-Cdc28 kinase release and DNA replication. Computational studies suggest that multisite phosphorylation could lead to highly non-linear degradation of Sic1 and thus a switch-like entry into the S phase. Here, we demonstrate experimentally that the *in vitro* binding of phosphorylated Sic1 to Cdc4 in response to the increasing activity of Cln2-Cdc28 kinase is indeed ultrasensitive (Hill coefficient $n_H \sim 3.2$) and the ultrasensitivity decreases as the number of phosphorylable sites decreases to five ($n_H \sim 1.0$). In our steady-state binding assays, mutant Sic1 proteins with less than five phosphorylable sites do not bind to Cdc4 and the Cdc4-bound fraction of Sic1 increases as the number of phosphorylable sites increases. In parallel, mass spectrometry analysis showes that the wild-type Sic1 is phosphorylated mostly on less than five sites in the initial lag phase of the binding curve. Interestingly, the distribution shifts dramatically toward six- and higher orders of phosphorylated states as the concentration of Cln2-Cdc28 increases. This study highlights the potential role of multisite protein phosphorylation in converting a graded input (Cln2-Cdc28 activity) to an ultrasensitive output (binding to Cdc4) and thereby generating a switch-like cellular behavior such as the G1/S transition.

See more of this Session: Intracellular Processes II
See more of this Group/Topical: Food, Pharmaceutical & Bioengineering Division

(593g) Interpretation of Dynamic Signals: How Genes Respond When the Signal Changes

Gregory T. Reeves[1], Nathanie Trisnadi[2], Thai Truong[2], Marcos Nahmad[2], Sophie Katz[2] and Angelike Stathopoulos[2], (1)Chemical and Biomolecular Engineering, North Carolina State University, Raleigh, NC, (2)Biology, California Institute of Technology, Pasadena, CA

Tumorigenesis, stem cell maintenance and differentiation, and tissue patterning are all inherently multicellular phenomena that rely heavily on cell-cell communication processes. However, most engineering approaches to these problems have focused on single-celled organisms, or, at best, aggregates of single-celled organisms. Bona-fide multicellular models organisms are also necessary to fully understand cell-cell signaling in its native context. Therefore, to uncover the fundamental physical and biological principles behind cell-cell signaling, we study tissue patterning in the early embryo of the fruit fly *Drosophila melanogaster*. In particular, we are studying how a spatially-distributed chemical signal patterns the dorsal-ventral axis of the embryo in the first several hours of development. This chemical signal is a transcription factor called Dorsal, and the nuclei in the early embryo respond to Dorsal in a concentration-dependent fashion: their differentiation state is determined by the amount of Dorsal they are exposed to [1]. Recent work has shown that the strength of this signal varies appreciably in time, raising the question of how nuclei can reliably read a constantly-changing signal and differentiate accordingly [2-4]. Detailed studies of gene expression patterns reveal that nuclei respond to Dorsal in a real-time fashion, changing along with the changing signal. This finding has implications for understanding the dynamics of gene expression when exposed to transient stimuli in the context of a multicellular organism. One important implication is the ability of the embryo to respond to errors in gene expression patterns, in that if patterns are initially specified incorrectly, enough time and plasticity are available for correction. Endnotes: [1] Roth, S.; Stein, D. & Nüsslein-Volhard, C. (1989). *Cell*, **59**: 1189-1202. [2] Liberman, L. M.; Reeves, G. T. & Stathopoulos, A. (2009). *Proc Natl Acad Sci U S A*, **106**: 22317-22322. [3] Delotto, R.; Delotto, Y.; Steward, R. & Lippincott-Schwartz, J. (2007). *Development*, **134**: 4233-4241. [4] Reeves, G. T., et al. (2011). In preparation.

See more of this Session: Intracellular Processes II
See more of this Group/Topical: Food, Pharmaceutical & Bioengineering Division

(595a) In-Silico Design of Experiments As a Tool for Nonlinear Sensitivity Analysis of Detailed Models

Alexandros Kiparissides[1], A. Mantalaris[1], Stratos Pistikopoulos[1] and **Christos Georgakis**[2], (1)Centre for Process Systems Engineering, Department of Chemical Engineering, Imperial College, London, United Kingdom, (2)Systems Research Institute and Chemical and Biological Engineering, Tufts University, Medford, MA

Mathematical modellingprovides a systematic way to study quantitatively the behaviour of complexsystems composed of numerous components and some of which might be described atdifferent levels of detail. Engineers and scientists who postulate such modelsinvariably chose to include more rather than less detail is their descriptionof the physical, chemical or/and biological system of interest. The inclusionof more and more details in such models increases the number of parameters thatneed be estimated from experimental data. The estimation of an increased numberof parameters increases the number of experimental data needed. However, evenif this requirement is met, the estimation of parameters associated withinsignificant model details is extremely difficult because of their minimalimpact on the model predictions being matched to the experimental data. Thisdifficulty cannot be easily removed because we do not know *a priori*which model components are insignificant in the region of present interest.Even in the case where the values of the model parameters are known, it is ofinterest to investigate the sensitivity of the model predictions on the valuesof the input variables of the process. Exhaustive simulations, especially ofvery complex models, do not provide an effective answer to the problem becauseof the computational cost. Knowing which parameters and to what extenthave a significant effect on the model output allows for their accurateestimation through tailor-made experiments (Kiparissides et al 2011).Parameters with negligible sensitivity, quantified by a Sensitivity Index (SI),can be fixed at their literature or their initially assumed value as they donot affect the model predictions

Existing model sensitivity methodsare commonly classified in three broad categories: screening, local and globalmethods (Saltelli, 2000). For the general case of non-linear ODE models,commonly met in engineering applications, global methods and variance basedglobal methods in particular have the edge over their counterparts as discussedpreviously (Chan et al 1997, Saltelli et al 2000, Kiparissides et al 2009). Briefly, the ability to estimate higher-level indices, which quantify theeffect on the output of parameter-parameter interactions, gives global methodsthe edge over their local counterparts. Alas, global methods usually requirethe estimation of a large, often prohibitive, number of model evaluations.

Here we propose touse, in a different context, a methodology developed more than fifty years ago.It is called "Design of Experiments" and has become a classic experimental toolin understanding how different operating conditions affect the performance of aprocess (see for example Box and Draper

(1987)). Herein, we will apply this techniqueon the mathematical model of the process instead of on an experimental set-up, hencethe use of the term *in silico*. Similar to its original use in extractingthe maximum amount of information from the process with the minimum number ofproperly designed experiments, we will use it here so that we can obtain the maximuminformation on the sensitivity of the model to its parameters with the minimum numberof simulations of the full model. This parallel use of the design ofexperiments concepts and tools has also been used to understand the operabilitycharacteristics of continuous chemical processes by Li and Georgakis (2010),where it was called design of selective calculations.

We utilise a DAEmodel that describes in substantial detail hybridoma cell growth and proliferationin batch cultures (Bang and Jarford 2000). We show that the use of DoE for thegeneration of a response surface model (RSM) approximation of the full model canlead to the same quality of sensitivity information as the established variancebased GSA methods at a significantly lower computational cost. We show that theexplicit information in the RSM model provides the needed sensitivityinformation with respect to the parameters of the detailed model. Besides theirrelative absolute magnitude, the RSM model parameters provide additionalinformation that previous methods did not avail. The results of the proposedapproach are in full agreement with those of the Sobol' method, one of the mostwidely used previously, and this is done at a fraction of the computationalcost.

The proposedapproach offers directional and interactional information about the full model ata minimal computational cost. Moreover, we show that by estimating an approximateRSM model, minimal information is lost in terms of parameter significance and sensitivity.This is highlighted by comparing the results of the Sobol' method when appliedto the full model to the results of the Sobol' method when applied to the estimatedRSM model.

References:

1. Box G.E.P., Draper N.R. (1987) Empirical Model Building and Response Surfaces. Wiley Press.
2. Chan K., Saltelli A., Tarantola S. (1997) Sensitivity Analysis of the model output: variance-based methods make the difference. Proceedings of the 1997 Winter Simulation Conference.
3. Jang J.D., Barford J.P. (2000) An unstructured kinetic model of macromolecular metabolism in batch and fed-batch cultures of hybridoma cells producing monoclonal antibodies. Biochem. Eng. Journal, 4, 153-168.
4. Kiparissides A., Kucherenko S., Mantalaris A., Pisitikopoulos E.N. (2009) Global Sensitivity Analysis Challenges in biological systems modeling. Ind. Eng. Chem. Res., 48 (15), pp 7168-7180.
5. Kiparissides A., Koutinas M., Kontoravdi C., Mantalaris A., Pistikopoulos E.N. (2011) °Closing the loop in biological systems modeling From the in silico to the in vitro. Automatica, doi:10.1016/j.automatica.2011.01.013.
6. Li, L., Georgakis, C. (2010), On the Calculation of Operability Sets of Nonlinear High-Dimensional Processes Ind. & Eng. Chem. Res. 49, (17) pp.8035-8047
7. Saltelli A., Chan K., Scott E.M. (2000) Sensitivity Analysis, Wiley Press.

(595b) In Silico Prediction of Cancer GI50

E. A. Whitebay, J. D. Ramsey, B. J. Neely and K. A. M. Gasem, School of Chemical Engineering, Oklahoma State University, Stillwater, OK

Current methods of cancer treatment have improved vastly the survivability of humans from many forms of cancer. Traditional chemotherapy remains a key treatment method, offering distinct advantages over other treatment options, especially in the treatment of metastasized tumors. Much effort has been devoted to identifying new, more effective chemotherapeutic agents. Currently, the National Cancer Institute (NCI) analyzes the potential anticancer activity of over 3,000 compounds per year. However, less than one percent of these drugs will be marketed, mostly due to poor anticancer activity and toxicity issues.

To provide a faster and more accurate methodology for identification of potential anticancer agents, we have developed *in silico* QSAR (quantitative structure-activity relationship) models to predict the concentration of chemical required to reduce the cancer growth rate by 50% (GI_{50}) for the breast cancer cell line MCF-7. These screening models use nonlinear neural networks to provide *a priori* predictions of GI_{50} for potential anticancer agents with an overall root-mean square error of 0.49 Log (GI_{50}). As a screening tool capable of greatly reducing time and expense of new drug development, the model is employed to remove unqualified candidate anticancer compounds, while identifying and retaining highly qualified candidates for experimental trials. Further, this methodology facilitates a greater understanding of the relationship between molecular structure and growth inhibition potential.

See more of this Session: Mathematical and Computational Biosystems Engineering
See more of this Group/Topical: Computing and Systems Technology Division

(595c) A Mesoscopic Membrane Model for the Investigation of Mechanisms Underlying Caveolae-Mediated Endocytosis

Belinda S. Akpa[1], Richard D. Minshall[2], Lewis E Wedgewood[3] and Ludwig C Nitsche[3], (1)Chemical Engineering and Bioengineering, University of Illinois at Chicago, Chicago, IL, (2)Pharmacology and Anesthesiology, University of Illinois at Chicago, Chicago, IL, (3)Chemical Engineering, University of Illinois at Chicago, Chicago, IL

Caveolae, small, flask-shaped invaginations of mammalian plasma membranes, are ubiquitous features of endothelial cells. They comprise about 15% of the cell volume and account for >95% of the plasmalemmal vesicles within the cells. Caveolae are implicated as essential vesicle carriers mediating endocytosis in endothelial cells. In this work, we report the development of a computational model which can address how changes in caveolin-1 oligomerization and key signaling reactions impact upon membrane invagination, budding, and internalization by (i) controlling the stability of caveolin polymers in the plasma membrane and (ii) their ability to facilitate changes in membrane curvature.

A novel mesoscopic, particulate model is developed for cell membranes in which local directionality (surface normal vector) is extracted from the relative arrangement of isotropic particles rather than being embedded in the more complex (i.e., rod-like or multi-bead) internal structure of lipid particles frequently employed in coarse-grained or atomistic simulations. Based on the normal vector, an anisotropic force law is applied having three parts: (1) A pairwise additive, two-dimensional Lennard-Jones force preserves in-plane cohesion. (2) A normal restoring force penalizes excursions from the plane and maintains a spontaneous curvature, thereby imparting bending elasticity. (3) An anisotropic Brownian force confers in-plane fluidity. This approach is contrasted with previous models in the dependence of the normal restoring force on the mean and Gaussian curvatures.

The membrane model is to be combined with a previously developed model of the Cav-1 protein coat characteristic of caveolae in endothelial cells. This Brownian dynamics simulation, adapted from polymer kinetic theory, represents the protein chains as bead-spring assemblies and tracks their stochastic motion as forces that are transmitted to the membrane. By combining Brownian dynamics of Cav-1 oligomers anchored to the membrane with the elastic membrane response, a model of caveolin-mediated membrane deformation will be obtained.

See more of this Session: Mathematical and Computational Biosystems Engineering
See more of this Group/Topical: Computing and Systems Technology Division

(595d) Modeling Heterogeneity In Populations of Self-Renewing Human Embryonic Stem Cells

Jincheng Wu and Emmanuel (Manolis) S. Tzanakakis, Chemical and Biological Engineering, State University of New York at Buffalo, Buffalo, NY

Despite the great potential of human pluripotent stem cells as a source of therapeutics for regenerative medicine, their efficient expansion and differentiation remains challenging. To that end, cell population modeling can contribute to the optimization and standardization of conditions for the efficient maintenance and differentiation of stem cells. In this work, we developed a population balance equation (PBE) model to describe and predict the dynamics of human embryonic stem cell (hESC) self-renewal. Intrinsic and extrinsic sources of stem cell population heterogeneity were considered. The model is based on experimental data obtained in our laboratory.

A PBE model was constructed for the heterogeneous self-renewal of hESCs with a physiological state vector comprising elements pertinent to hESC proliferation and the expression level of pluripotency markers. These variables were measured experimentally for a variety of hESC maintenance conditions and their rates of change were represented by ordinary and stochastic differential equations. Solutions to the model were obtained primarily by Monte Carlo methods.

Model parameters pertinent to cell growth, division and marker expression were calculated by comparing simulation results with experimental data. Sensitivity analysis was also performed to gauge the effect of parameter perturbation on hESC heterogeneity. Additional experiments for the self-renewal of hESCs under different conditions were carried out to validate the model. The results correlate well with the flow cytometry data for hESC populations. The contribution of intrinsic sources considered here to hESC heterogeneity was found to be almost 40%. Finally, the PBE model successfully depicts both the proliferation of hESCs and maintenance of their pluripotency in a heterogeneous population. Current efforts concentrate on expanding the model to include additional sources of heterogeneity stemming from hESC differentiation.

The quantitative framework developed can complement and accelerate investigations utilizing stem cell populations by reducing costly and labor-intensive experimental procedures. Moreover, the model's predictive power can be useful in the design and operation of processes for the large-scale expansion and differentiation of human pluripotent stem cells to therapeutically useful cells.

See more of this Session: Mathematical and Computational Biosystems Engineering
See more of this Group/Topical: Computing and Systems Technology Division

(595e) Systematic Engineering Approach to Development and Identification of Physiologically-Based Pharmacokinetic Models

Cierra Hall[1], Martina Heitzig[2], Grkan Sin[2], Rafiqul Gani[2] and Andreas Linninger[1], (1)Laboratory for Product and Process Design, University of Illinois at Chicago, Chicago, IL, (2)Department of Chemical and Biochemical Engineering, Technical University of Denmark, Kgs. Lyngby, Denmark

Whole body physiologically-based pharmacokinetic (PBPK) modeling attempts to model the uptake and distribution of a drug inside the body. Traditionally, pharmacokinetic models have been developed by simple curve fitting of experimental data without considering the mechanistic and physics of the system. The drawback of such models is their lack of physical insights and the pure extrapolation features to untested conditions, to other drugs as well as to individuals of the same species or different species. Lately, PBPK models, which are based on first principles, have received increasing attention. In this work, a first principles engineering approach to PBPK modeling that is based on conservation equations and advanced scaling laws is proposed. The developed global pharmacokinetic models are based on biochemical, anatomical and physiological data.

The resulting PBPK models are able to extrapolate and scale between species and individuals because they are based on actual occurring physical mechanisms and allow the application of more advanced scaling laws than just considering the differing masses of the individuals. Due to their improved scaling qualities, first principles PBPK models have the potential to reduce the number of undesirable as well as time- and cost-intensive animal experiments during the drug development phase. Further, the development of first principle-based PBPK models leads to a gain of insight in the occurring processes during drug uptake by an organism. Having these models at hand moreover allows the determination of accurate and optimal drug dosage for desired therapeutic effects.

However, the development of first principles PBPK models, their identification, and discrimination between different candidate models is a non-trivial task inherent with a number of challenges which are related to identifying the occurring mechanisms inside the body and the appropriate degrees of detail of the models with respect to the modeling goal and the available experimental data:

Firstly, a large number of alternative PBPK models can be generated depending on the degree of detail considered and the assumptions made about the different considered mechanisms in the body. Further, first principle PBPK models are usually complex models which contain a large number of unknown parameters that are not always all identifiable by the limited experimental data accessible. A strategy needs to be found to address the problem of discriminating between the different candidate models and identifying their parameters in order to find the model that is best supported by the experimental data available. In this context, design of experiments for model discrimination and parameter estimation plays an important role. If the models cannot be discriminated based on the available data, or model parameters that are important for the desired model outputs are not identifiable by the available data, methods for optimal design of experiments need to be applied. Design of experiments is especially important in pharmacokinetic modeling because in order to obtain the required datapoints animals need to be sacrificed.

In this work, a systematic methodology for the development, analysis, identification (including sensitivity analysis, identifiability analysis and parameter estimation) and discrimination between different pharmacokinetic candidate models is proposed. The methodology is implemented into a computer-aided modeling framework. The benefit of the proposed systematic approach is an improvement of model quality and an increase of efficiency of the modeler as well as the experimentalist and thereby a reduction of time and resources required for model development, identification, discrimination and application. This is achieved by designing the structure of the computer-aided modeling framework such that it can handle the work-flows and data-flows associated with the different modeling tasks, combining state-of-the-art modeling techniques for the different work-flow steps as well as supporting model-documentation and model reuse.

The proposed modeling framework for PBPK modeling is highlighted through a case study which deals with the development of a PBPK model for a rat based on available experimental data.

See more of this Session: Mathematical and Computational Biosystems Engineering
See more of this Group/Topical: Computing and Systems Technology Division

(595f) A Compartmentalized Sepsis Model Based On Neutrophil Kinetics As a Decision Supporting Tool of Therapeutic Intervention

Sang Ok Song[1], Justin S. Hogg[2], Robert S. Parker[3] and Gilles Clermont[1], (1)Critical Care Medicine, University of Pittsburgh, Pittsburgh, PA, (2)Computational Biology, University of Pittsburgh, Pittsburgh, PA, (3)Chemical and Petroleum Engineering, University of Pittsburgh, Pittsburgh, PA

The complex pathophysiological interaction network of sepsis and its systemic nature involve many inflammatory mediators and a number of cell types, tissues and organs. Mechanisms driving the high mortality and morbidity observed in sepsis are therefore difficult to intuit and elucidate. Consequently, diagnostic, prognostic and therapeutic options have been limited due to the high variability and non-specificity of symptoms. As a broad-spectrum immune modulating therapy, an extracorporeal blood purification by hemoadsorption (HA) device using CytoSorb^TM beads (CytoSorbents, Monmouth Junction, NJ) has shown the capacity to alter circulating cytokine levels and improve survival in experimental endotoxemia and cecal ligation and puncture (CLP) induced sepsis in animals. However, the exact mechanisms by which the HA treatment resulted in reduced organ injury and improved outcome remain unclear.

Balancing the complexity of the inflammatory response, the insufficient knowledge of mechanistic details, and the relative paucity of data, we developed a mathematical model having a conceptual abstraction of the system behaviors across multiple tissue compartments. We seek to account for available experimental evidence and provide useful perspective on the pathophysiologic mechanisms of sepsis. The model focuses on the systemic behavior of neutrophils in response to infectious challenges by phenotyping circulatory neutrophils as they respond to inflammatory signals. The network components and interactions were assembled based on qualitative domain knowledge of the acute inflammatory response, including multiple phenotypes of neutrophils and major molecular effectors in distinct compartments (blood, peritoneum, and lung). In order to capture the impaired recruitment of neutrophils, which is one of the key pathophysiologic features of severe sepsis, several lines of evidence about the mechanisms influencing neutrophil migration in sepsis were incorporated in the model. Based on prior and concurrent experimental data, and our previous simulation results, we hypothesized that the HA device adsorbs circulating neutrophils as well as pro- and anti-inflammatory mediators (e.g., IL6, IL10, TNF, IL1, IL8) from the circulation.

Serial cytokine/chemokine measurements and neutrophil counts in distinct compartments were measured in rats that underwent CLP with sham (no HA treatment) or an HA intervention of 4 hour duration 18 hours after the induction of CLP. The experimental data sets were categorized into four subgroups (HA survival, HA death, Sham survival, Sham death) and the resulting four objective functions were constructed to constrain the model behaviors separately.

We computed quantitative predictions over an ensemble of parameter sets statistically drawn from all sets that are consistent with the available experimental data. An adaptive Markov Chain Monte Carlo where the step scales are periodically optimized based on the Hessian matrix of the cost function was used to efficiently sample the plausible parameter sets. Analysis of the parameter ensembles sampled from different target distributions provided useful insights on the pathophysiologic mechanisms represented by parameter subsets strongly associated with an adverse outcome and hypothesized the effects of HA on the dynamics of the system. In addition, model simulations predict a subset of animals (in their parametric description) most likely to benefit from the HA intervention. Other factors such as treatment timing and duration are discussed based on different plausible hypotheses. The mathematical model constructed here provides a platform for generating and testing hypotheses in silico, as well as motivating further translational (experimental) studies to advance our understanding of the complex biological response to inflammation and sepsis.

See more of this Session: Mathematical and Computational Biosystems Engineering
See more of this Group/Topical: Computing and Systems Technology Division

(595g) A Lattice-Based Computational Model of Microscopic Bone Resorption In Cortical Bone Multicellular Units

Junhwan Jeon[1], Pascal R. Buenzli[2], Peter Pivonka[2], David W. Smith[2] and Peter T. Cummings[1], (1) Deptartment of Chemical and Biomolecular Engineering, Vanderbilt University, Nashville, TN, (2) Engineering Computational Biology Group, The University of Western Australia, Australia

Bone resorption by osteoclasts is an essential phase in the bone remodelling cycle as it creates the physiological conditions for subsequent bone formation. While several properties of osteoclastic bone resorption in cortical Bone Multicellular Units (BMU) have been assessed experimentally, the precise spatio-temporal dynamics, movement pattern, apoptotic state (single nuclei vs. whole cell) of the osteoclasts remain to be elucidated. Furthermore, the individual effects that these behaviours confer on the shape and extent of the resorption cavity are unclear. In this work, we develop a lattice-based computational model focused on bone resorption in cortical BMUs to address these questions. Our model takes into account the interaction of osteoclasts with the bone matrix, the interaction of osteoclasts with each other, the production of osteoclasts from the tip of a growing blood vessel, apoptosis time of osteoclasts, and the renewal of osteoclasts' nuclei by cell fusion. All these features are shown to strongly affect the geometrical properties of the developing resorption cavity including shape, size, and progression rate, but also the resorption pattern, lifespan and average resorption rate of individual osteoclasts, as well as their spatial distribution in the BMU. In particular, our numerical simulation results indicate that the growth of a blood vessel and the continual renewal of the osteoclasts' nuclei are essential physiological features for functional resorption by cortical BMUs, without which balanced bone renewal may be compromised.

See more of this Session: Mathematical and Computational Biosystems Engineering
See more of this Group/Topical: Computing and Systems Technology Division

(595h) Optimal Control of Fed-Batch Fermentation Process Using Modified Iterative Dynamic Programming

Zheng Li and **M. Nazmul Karim**, Department of Chemical Engineering, Texas Tech University, Lubbock, TX

This article proposes a feeding strategy based on a metabolic networks model to optimize ethanol production using *Escherichia coli* KO11. In metabolic engineering, growth behavior and product secretion from microorganisms can be predicted using constraint-based analysis of metabolic networks. In this study, a new approach of using stoichiometric model combined with kinetic model allows for precise prediction of ethanol production and formation of other byproducts under specific fermentation environment. An optimal feeding strategy and minimization of total fermentation time are accomplished by iterative Dynamic Programming approach, which maximizes ethanol production and also minimizes toxic byproduct formation. The switching time from batch to fed- batch is also taken as variable to be optimized. All variables are computed by using iterative Dynamic Programming utilizing a stoichiometric model. Iterative dynamic programming is modified to increase computational efficiency. Simulation results are promising.

See more of this Session: Mathematical and Computational Biosystems Engineering
See more of this Group/Topical: Computing and Systems Technology Division

(597a) Estimating the Free Energy of the Association of Amyloid Fibril Forming Peptides

Diwakar Shukla, Chemical Engineering, Massachusetts Institute of Technology, Cambridge, MA and Bernhardt L. Trout, Massachusetts Institute of Technology, Cambridge, MA

In order to computationally estimate the free energy of association between proteins, association of two proteins in solution have to simulated. However, the required computational time for such a system would be huge. Therefore, association of small amyloid peptides is used as a model system to identify the pathways for their association. These peptides are involved in the contact between the associating proteins and form aggregates which are similar to the protein aggregates. The mechanism of addition of a soluble unstructured monomer to a preformed ordered amyloid fibril is a complex process. On the basis of the kinetics of monomer disassociation of Aβ(1-40) from the amy- loid fibril, it has been suggested that deposition is a multistep process involving a rapid reversible association of the unstructured monomer to the fibril surface (docking) followed by a slower conformational rearrangement leading to the incorporation onto the underlying fibril lattice (locking). Several computational studies have reported many aspects of the dock-lock mechanism for amyloid formation. Similarly, there are various recent studies which have investigated the thermodynamics of this process. However, these studies are limited in accuracy and detail. We have used technqiues such as Temperate accelerated MD, String method in collective variables, and Voronoi-Tesselated Milestoning to obtain the free energy of the peptide association. The advantage of using string method as compared to other computational techniques is that the computational time does not scale as the number of order parameter. We have also investigated the effect of small molecules on the free energy of peptide association.

See more of this Session: Molecular Modeling of Biophysical Processes II - Membranes and Biological Polymers
See more of this Group/Topical: Food, Pharmaceutical & Bioengineering Division

(597b) The Role of Aromatic Residues In the Catalytic Domain Tunnel of Hypocrea Jecorina Cellobiohydrolase Cel6A

Christina M. Payne[1], Deanne W. Sammond[1], Michael E. Himmel[1], Michael F. Crowley[1] and Gregg T. Beckham[2], (1)Biosciences Center, National Renewable Energy Laboratory, Golden, CO, (2)National Bioenergy Center, National Renewable Energy Laboratory, Golden, CO

The glycoside hydrolase Family 6 cellobiohydrolase (Cel6A) from *Hypocrea jecorina* – formerly *Trichoderma reesei* – plays an important role in the enzymatic cocktail used to breakdown cellulosic feedstock to soluble sugars. The catalytic domain of this enzyme includes an enclosed tunnel consisting of four structurally-defined binding sites through which cellulose threads and is hydrolyzed to cellobiose or glucose. Along the length of the tunnel, aromatic residues assist in glucose ring binding via stacking interactions. We use molecular simulation and free energy calculations – specifically, thermodynamic integration – to provide insight into how mutations to the aromatic residues in the active site tunnel affect enzymatic processivity, protein structure, and the conformation of the catalytically-active complex. As aromatic residues are ubiquitously found in the tunnels of carbohydrate-active enzymes, these results have implications for understanding protein-ligand binding for general carbohydrate turnover in the biosphere and in protein engineering for enhanced conversion rates.

See more of this Session: Molecular Modeling of Biophysical Processes II - Membranes and Biological Polymers
See more of this Group/Topical: Food, Pharmaceutical & Bioengineering Division

(597c) Ultrasonic Stimulation of Chondrocytes: Harmonic Analysis Elucidates Frequency Dependent Bioeffects

Tobias Louw[1], Sanjukta Guhathakurta[2], Anuradha Subramanian[2] and Hendrik Viljoen[2], (1)University of Nebraska, Lincoln, NE, (2)Chemical and Biomolecular Engineering, University of Nebraska, Lincoln, NE

Therapeutic ultrasound has long been used to mediate the healing of fractures. Recently, ultrasonics has been applied to bioreactors to enhance the properties of synthesized tissue. It has been found that cartilaginous tissue in particular benefits from ultrasound stimulation, increasing cellular viability and the production of type II collagen in the extracellular matrix. The exact mechanism responsible for the beneficial effects of ultrasound is as yet unknown, although many new theories are in development. Focus has mostly been placed on cavitation effects, which would harm cells while increasing cellular membrane permeability.

We provide an alternative theory based on harmonic acoustic analysis in which we investigate the gain and phase of the cellular and nucleic membrane in relation to the incident acoustic wave. We model the cell using linear acoustic theory for fluids and solids. An elegant solution is obtained by assuming harmonic time dependence and expanding the problem as an infinite series of Legendre polynomials. The solution is both accurate and computationally inexpensive, allowing us to compare the effects of the acoustic wave over a wide range of frequencies. This comparison reveals the fundamental resonance frequency of the cell as well as the associated phase shifts. We hypothesize that the cell membrane wall thickness varies harmonically along with the flow of nutrients into and out of the cell. Above the resonance frequency, this phase shift aids the flow of nutrients into the cell, while preventing nutrients from exiting. This theory is validated by experimental results showing increased concentrations of nutrients in chondrocyte cells exposed to ultrasonic stimulation.

See more of this Session: Molecular Modeling of Biophysical Processes II - Membranes and Biological Polymers
See more of this Group/Topical: Food, Pharmaceutical & Bioengineering Division

(597d) Exploring Virus Maturation Pathways Through Computer Simulations

Eric R. May, Chemistry & Biophysics, University of Michigan, Ann Arbor, MI and Charles L. Brooks III, Department of Chemistry and Biophysics Program, University of Michigan, Ann Arbor, MI

During the maturation process of the bacteriophage HK97 the capsid swells to a larger size and also changes shape by becoming more faceted (and less spherical). The HK97 capsid consists of 420 protein subunits, the dimension of the capsid is ~50 nm and the time scale for maturation is on the order of hours; studying the maturation phenomenon from a molecular scale clearly necessitates a multiscale/enhanced sampling approach. As structures exist for both the immature and mature virus capsid, we have constructed structural pathways connecting the two end states. This is done by constructing an initial "naive" pathway and then refining this pathway using a string method to evolve the pathway to a minimum energy pathway (MEP). From the MEP we can compute the free energy profile and other physical properties. From in-vitro studies reducing the system pH enhances formation of the mature form and we will be examine how a low pH versus neutral pH environment will affect the thermodynamic and kinetic properties of the system. Lastly, I will cover our attempts at understanding if icosahedral symmetry breaking occurs during maturation and if it facilitates formation of the mature state. This is accomplished by employing coarse-grained simulations of the entire capsid with a potential energy function funneled toward the mature state.

See more of this Session: Molecular Modeling of Biophysical Processes II - Membranes and Biological Polymers
See more of this Group/Topical: Food, Pharmaceutical & Bioengineering Division

(597e) Quantification of Sugar Binding Affinity and Study of Proton Translocation In Lactose Permease of *Escherichia Coli*

Pushkar Y. Pendse, Department of Chemical and Biomolecular Engineering, University of Maryland, College Park, College Park, MD and Jeffery Klauda, Department of Chemical and Biomolecular Engineering, University of Maryland, College Park, MD

In biological cells, molecular traffic in and out of the cell is mainly controlled by membrane transport proteins. The Major Facilitator Superfamily (MFS) is an important class of membrane transporters whose members are found in almost all types of organisms and are very diverse in terms of substrate transport. Lactose Permease (LacY) of *E.coli*, a secondary active transporter that transports various sugar molecules across the plasma membrane by proton symport mechanism, is studied as a model for the MFS proteins.

Although LacY binds to different anomeric states of a disaccharide with the same specificity, binding affinity of the α-anomer is greater than β (Sahin-Toth et al., *Biochem*, 2000). Only a subtle change in the substrate stereochemistry results in a significant change in its binding affinity to LacY. To investigate this anomeric binding phenomenon, free energies of binding of αα-, αβ-, and ββ-$(\mathrm{Gal}p)_2$ to LacY are calculated using the alchemical free energy perturbation (FEP) method (Mobley et al., *JCP*, 2006). Since free energy is a state function, the double decoupling scheme was used to construct an alchemical thermodynamic cycle between a sugar molecule bound to LacY and a solvated sugar molecule. For FEP calculations, sampling windows are generated from independent molecular dynamics (MD) trajectories for all the alchemical transformation steps using the CHARMM simulation package. The sampling data is processed with the weighted histogram analysis method (WHAM) to get the ΔG values. Since the sugar molecules bind to LacY in multiple binding conformations, multiple independent MD simulations were performed to get all the binding modes. The initial protein conformations were obtained from the x-ray crystal structure (1PV7) and our model for the outward-facing state (Pendse et al., JMB, 2010) to get the conformations in the inward- and outward-facing states of LacY, respectively. To validate our methods, ΔG_{bind} for p-nitrophenyl α-D-galactopyranoside (NPG) from our simulations was compared and agrees favorably with the values obtained from the isothermal calorimetry experiments (Nie et al., *JBC*, 2006). The binding of the α-anomer of a sugar was determined to be more favorable compared to the β-anomer. The dissimilarity in binding affinities was found to be a result of differences in the binding conformations for different anomeric states.

It has been established that protonation of LacY, in other words binding of a proton, is a necessary precursor to sugar binding. LacY then couples translocation of protons with the transport of sugar molecules but the mechanism of this coupling is unclear. The hybrid quantum mechanics/ molecular mechanics (QM/MM) approach is used to study the proton translocation. The self consistent charge-density functional tight binding (SCC-DFTB) method is used for the QM calculations. The generalized hybrid orbital (GHO) method is used to treat the transition between the QM and MM regions across the covalent bonds. The QM/MM calculations are used to understand the key protein residues involved in proton translocation and the translocation pathway.

See more of this Session: Molecular Modeling of Biophysical Processes II - Membranes and Biological Polymers
See more of this Group/Topical: Food, Pharmaceutical & Bioengineering Division

(597f) Avoiding Macrophages: Design and Rationale Behind a 'marker of Self' Peptide

Diego A. Pantano, Chemical and Biomolecular Engineering, University of Pennsylvania, Philadelphia, PA, Michael L. Klein, Institute for Computational Molecular Science and Department of Chemistry, Temple University, Philadelphia, PA and Dennis E. Discher, Chemical & Biomolecular Engineering, University of Pennsylvania, Philadelphia, PA

Macrophages are cells immediately available to combat wide range of pathogens without requiring prior exposure. They respond upon contact in a two-step process. In the first step, the macrophage is activated by contact to anything, even your own cells. In the second stage, and just before a macrophage engulfs its target, it checks for identification, that includes the 'marker of self' protein CD47, a widely expressed membrane protein. If CD47 is present, it causes the macrophage to disengage, making the activated phagocytosis process less efficient. Attachment of "self" CD47 on avidin-beads has shown a reduction in the rate of the phagocytosis by the macrophage.

Expression of CD47's functional domain is technically challenging, which encouraged us to search for a smaller peptide showing phagocytosis inhibition. We employed molecular dynamics simulations to understand the binding process and design a useful peptide based on the crystallographic structure of CD47/SIRP-α complex. We tested different peptides, each of them represent different strategies aiming to preserve the binding region structure. We found that some of these strategies maintained the structure better than others. Affinity experiments agreed with our predictions, and suggest the real possibility of more biocompatible materials with a small peptide.

See more of this Session: Molecular Modeling of Biophysical Processes II - Membranes and Biological Polymers
See more of this Group/Topical: Food, Pharmaceutical & Bioengineering Division

(597g) Interaction Between PEG and Conjugated Insulin: a Molecular Dynamics Simulation and Experimental Observation

Cheng Yang, Chemical Engnieering, Institute of Biochemcial Engineering, Beijing, China, Rui Wang, Department of Chemical Engineering, Institute of Biochemcial Engineering, Beijing, China, Diannan Lu, Chemical Engineering, Institute of Biochemcial Engineering, Beijing, China and Zheng Liu, Chemical Engineering, Tsinghua University, Beijing, China

An all-atom level molecular dynamics simulation using an annealing procedure was proposed and applied to examine the interaction between PEG and conjugated protein, an established method with proven effectiveness in protein stabilization. For the present study, insulin was used as the model protein, PEG with different chain lengths and topological structures were applied for the protein PEGylation. It was shown that an entangled structure including PEG and insulin can be formed. The driving force of this complex was proven to be hydrophobic interaction between PEG and the conjugated insulin, at the same time, a large number of hydrogen bonds was detected between PEG and surrounding water. An increased molecular size and decreased solvent accessible surface area of protein are observed, verifying the prolonged circulation time *in vivo* and the improved resistance against proteolysis and immunogenic side effects of protein-PEG conjugate. The enhanced stability of protein-PEG conjugate is also validated at accelerated denaturation environment *in silico*. Moreover, we also dictated there will be an optimum length for conjugated PEG, and then the protecting mechanism of linear, branch, and brush PEG was elucidated. The simulation results were validated by bioconjugation experiments between PEG and insulin. The method above provided a molecular insight into the conformation transformation of PEG-protein conjugates at all-atom level and offered an enabling tool for the molecular design of PEGylated protein pharmaceuticals.

See more of this Session: Molecular Modeling of Biophysical Processes II - Membranes and Biological Polymers
See more of this Group/Topical: Food, Pharmaceutical & Bioengineering Division

(607a) CMC Reviewer Expectations for ANDAs Using QbD Approaches

Robert L. Iser, Division of Chemistry, FDA/CDER/Office of Generic Drugs, Rockville, MD

There has been an increase in science and risk-based approaches under the Quality by Design paradigm in new drug submissions to the FDA. There are particular challenges for generic drug manufacturers with respect to using QbD approaches, especially with more complex dosage forms, such as modified release (MR) products. In an effort to increase transparency with the generic industry and the pharmaceutical industry at large, this presentation will provide insight into CMC reviewer expectations when evaluating pharmaceutical development studies, the applicant's process and product understanding, use of risk assessment tools, and proposed control strategies.

See more of this Session: QbD for Modified Release Dosage Forms
See more of this Group/Topical: Topical I: Comprehensive Quality by Design in Pharmaceutical Development and Manufacture

MECHANISTIC MODELING AND COMPUTER SOFTWARE FOR QUALITY BY DESIGN IN FORMULATION OF MODIFIED RELEASE DOSAGE FORMS

Xiao Yu Wu[1] *and Yousheng Zhou*[2]
[1]*Leslie Dan Faculty of Pharmacy, University of Toronto,* [2]*AP-CAD Inc., Toronto, ON, Canada*

Introduction

Modified release dosage forms (MRDFs) can be designed and manufactured in various shapes and structures using different materials. Depending on the design, formulation and properties of drug and excipients (e.g. polymers), drug release from MRDFs can undergo varying mechanisms. In addition, the external conditions such as mixing degree, pH and volume of the releasing medium, may play an important role in the release kinetics of the drug. Therefore, thorough understanding of drug release mechanism of an MRDF and establishing a mechanistic model with consideration of important factors are essential for quality by design (QbD) in the development of MRDFs with target performance.

This presentation first describes the role of mathematic modeling and computer simulation in QbD, reviews the difference between mechanistic modeling and statistical modeling, and then discusses how mechanistic modeling and computer software can be employed for QbD in formulation of MDRFs using actual MDRFs as examples. Various multiparticulate systems are analyzed including matrix beads/pellets and minitablets with or without membrane coating, drug granules coated with polymer membrane, sugar pellets coated with drug layer and polymer coating. Prediction of *in vivo* performance from *in vitro* release profile and design of a MRDF from targeted *in vivo* plasma concentration profile will also be demonstrated.

The Role of Mechanistic Modeling and Computer Simulation in QbD for MRDFs

Although the concept of QbD in pharmaceutical product development and manufacturing has been proposed since 2004 by the FDA, mathematical tools for achieving QbD goals in MRDF design have not yet been fully developed. As illustrated in Figure 1, to implement QbD for development of pharmaceutical products, such as MRDFs, several steps are essential. Firstly we need to gain scientific understanding of how formulation and processing factors affect product quality and performance by theoretical analysis and experimental testing. Secondly we need a mathematic model with predictive power that is able to describe the relationship between formulation/process variables and product performance characteristics. Because of comlexicity of such relationship, sophisticated computer programming is required to make quantitative prediction. With the assistance of computer simulation, we are then able to design the formulation and process to assure product quality and desired attributes. Finally the performance of the products has to be confirmed by clinical studies. If the clinical results are satisfactory, the product development will be progressed to the next stage. Otherwise, the parameters in the model for the formulation and/or manufacturing will be adjusted. Modification of the model may also be required to take into consideration of new knowledge gained from the clinical studies. Therefore, mathematical modeling and computer simulation are useful tools for implementation of QbD of MRDFs.

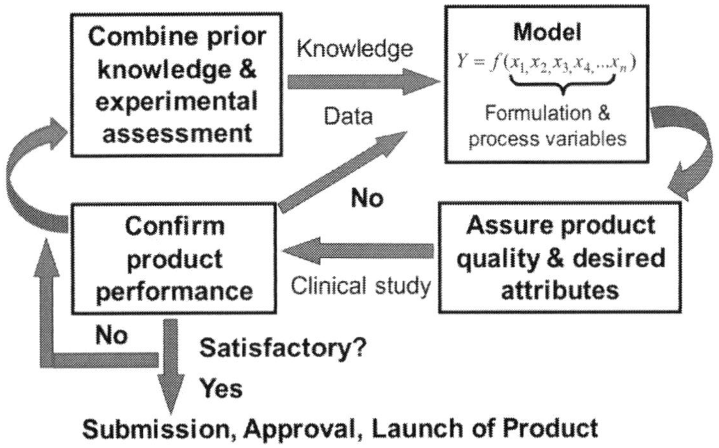

Figure 1. Schematic illustration of essential steps for implementation of QbD

Mechanistic Models versus Statistical Models

A variety of mathematical models and modeling methods can be employed to describe the relationship between formulation/process variables and product quality. Generally these models can be classified into two categories: (1) mechanistic models and (2) statistical models. The two types of models require different methodologies and input data sets and both have pros and cons as illustrated in Figure 2. For complex systems involving multiple variables simultaneously with unclear mechanistic relationships, statistical models exhibit advantages over mechanistic models. However, these models are post-experience models established by empirical correlation or statistical analysis and require a large amount of experimental data. To design MRDFs scientifically, mechanistic models are necessary as they can correlate excipient properties, drug properties, formulation composition, dosage form structure with the quality and performance of MRDFs based on underlying mechanisms and science.

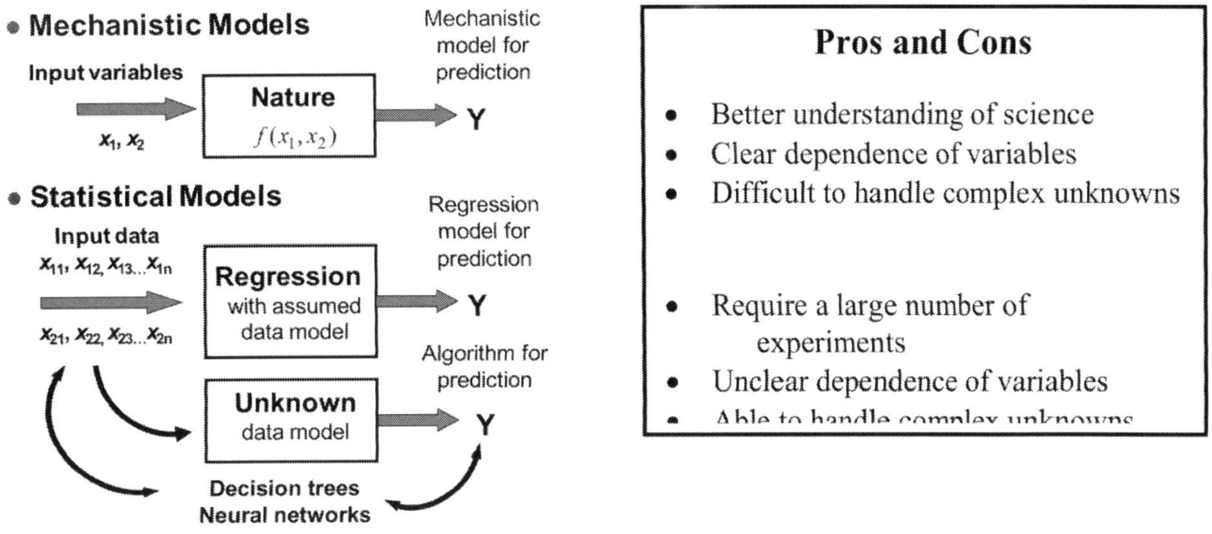

Figure 2. Comparison of mechanistic models and statistical models

Mechanistic Model-based Computer-Aided QbD of MRDFs

To build mechanistic models for MRDFs, we must first understand thoroughly the drug of interest (i.e., its physicochemical properties, pharmacology, pharmacokinetics, etc.), the drug delivery system and drug release mechanisms. Depending on the formulation and structure of MRDFs and drug properties, drug release kinetics may be controlled by drug diffusion, dissolution, polymer swelling/erosion, ion exchange, osmotic pressure, or a combination of several of these mechanisms. After having identified a rate-limiting release mechanism by theoretical analysis and preliminary experiments, we can derive a mechanistic model for the given drug and the delivery system. The model parameters, e.g. diffusion coefficient and partition coefficient, can be obtained by fitting experimental data of drug release with the model or determined by independent experiments or theoretical calculations (1-6). The experiment-verified parameters and model can then be applied to make predictions, to answer what-if questions, and ultimately to design MRDFs with desired quality attributes. In this presentation, we demonstrate the application of mechanistic model-based computer-aided QbD in typical polymer-coated multiparticulate dosage forms.

Multi-particulate systems such as ensembles of coated matrix beads/pellets or drug granules are extensively used for both systemic and local drug delivery. The multi-unit dosage forms are advantageous over single-unit ones in certain applications due to their flexibility in dosing and administration and low risk of dose dumping. In some formulations drug granules made of drug crystals are coated with film-forming polymers, and in some formulations a drug layer is sprayed onto sugar cores and then coated with a barrier polymer.

Application of Mechanistic Model-Based Computer-Aided QbD to Eudragit® Polymer-Coated Drug Granules

For a given drug, two formulation variables of coated drug granules, i.e., permeability and thickness of the coating, can be tailored to achieve target release profiles. EUDRAGIT® RL and RS polymers are acrylic polymers with different permeability due to a difference in the ionic monomer content which is higher in RL than in RS. They are widely used coating materials and can be mixed at any ratio to obtain desired permeability. In addition, one may apply different coating thickness of the polymers at a fixed RL/RS ratio to achieve desirable release profiles. However, it is quite challenging even for an experienced formulation scientist to decide a right thickness and RL/RS ratio to match a target curve for a given drug without multiple trial-and-error attempts.

Here we use theophylline as a model drug to demonstrate how mechanistic models and computer software can assist the design of EUDRAGIT® RL/RS coated drug granules. Based on the dissolution data in literature (7), the effective permeability of theophylline across coatings with various EUDRAGIT® RL/RS ratios was evaluated by using the optimization module in AP-CAD® software. The model-predicted release curves were then computed by using the obtained permeability, granule diameter of 0.55 mm, drug solubility $C_s \approx 0.01$ g/cm^3 and initial drug loading $C_o \approx 1$ g/cm^3 (8). Figure 3 is a typical plot of model prediction compared with the test data. The correlation between theophylline permeability and the RL/RS ratio was used to predict release kinetics of granules with any given EUDRAGIT® RL/RS ratios. Figure 4 illustrates the effect of EUDRAGIT® RL/RS ratio on the release profiles for a fixed weight gain of 12%. Figure 5 portrays how weight gain of coating influences drug release profiles of granules at a given RL/RS ratio of 1:7. These examples demonstrated that formulation scientists can evaluate quickly different formulation options (i.e., RL/RS ratio) and physical parameters (e.g. coating thickness) to achieve target curves using this computational tool.

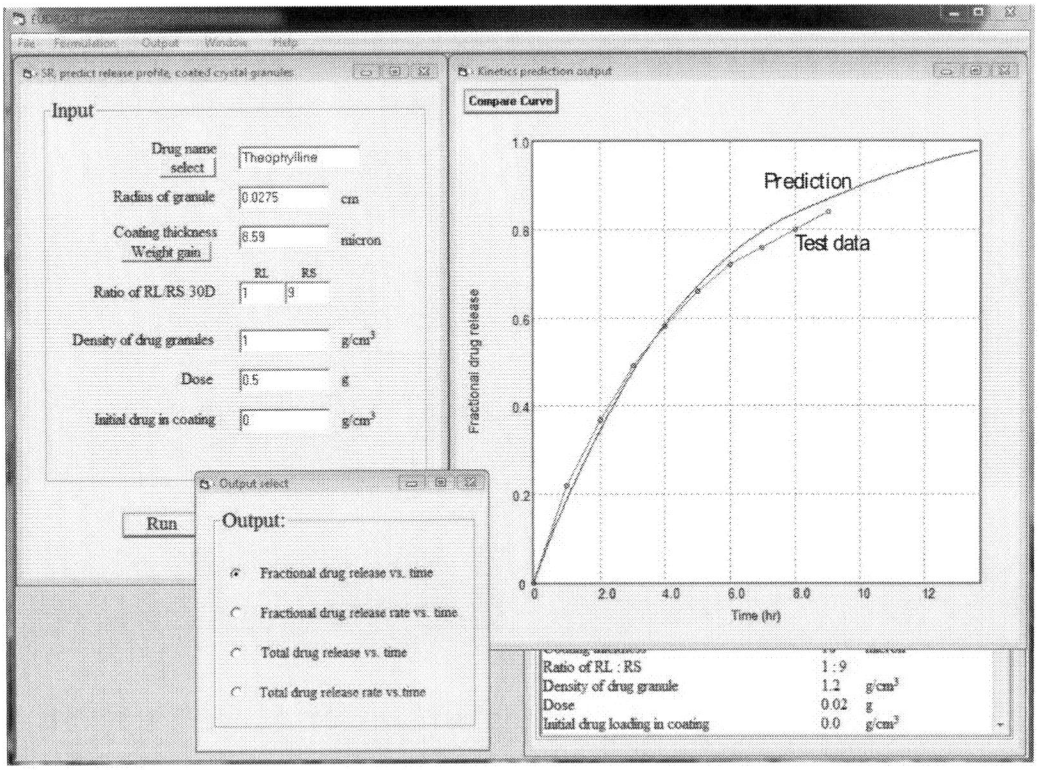

Figure 3. Comparison of model-predicted release profile with test data of theophylline granules

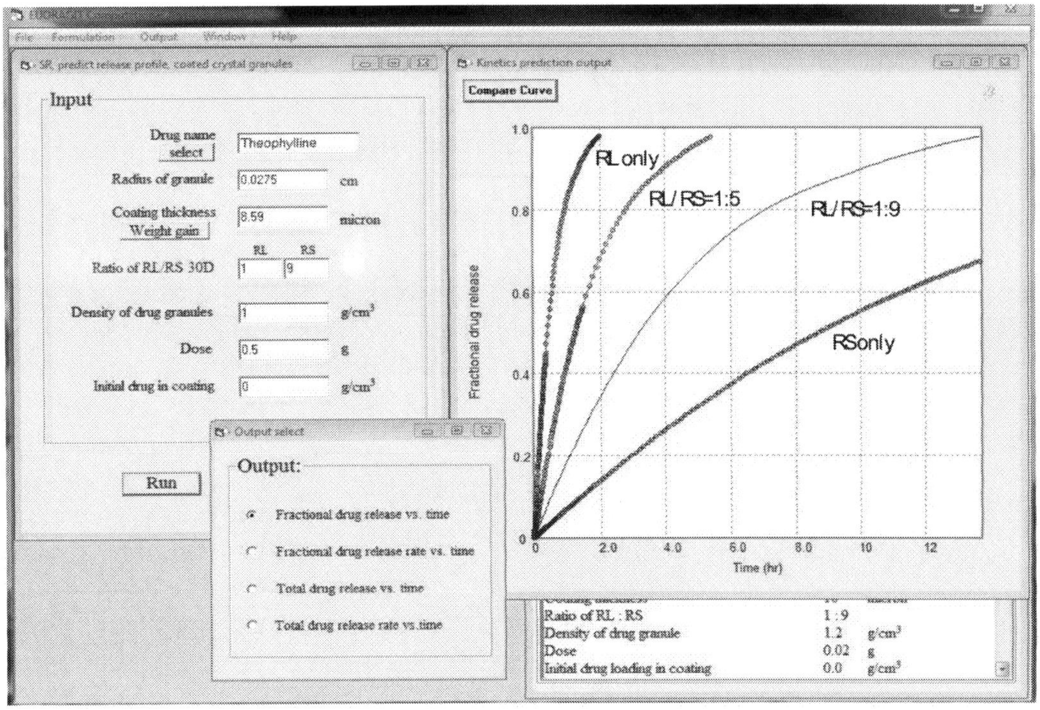

Figure 4. Effect of Eudragit® RL/RS ratio on theophylline release kinetics.

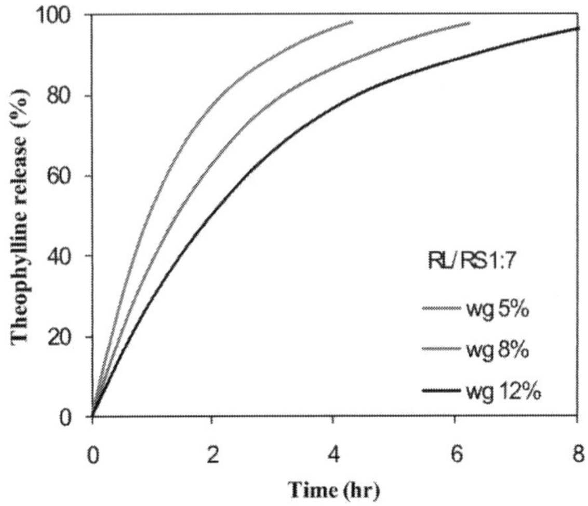

Figure 5. Effect of weight gain of coating on release profiles of theophylline granules coated with a Eudragit® RL/RS ratio of 1:7

Application of Mechanistic Model-Based Computer-Aided QbD to Surelease®-coated Drug Sugar Beads

Surelease® is an aqueous dispersion of ethylcellulose which is widely used as a barrier membrane in MRDFs such as coated beads and tablets. Since ethylcellulose is hydrophobic, a hydrophilic polymer (so called pore former), e.g. hydroxypropylmethylcellulose (HPMC) is often added in the dispersion to induce water-filled channels thus increasing drug release rate. Therefore, by adjusting the HPMC content in the coating, as well as the coating thickness, we are able to obtain different drug release profiles. A similar procedure of model building and verification presented above was used to study Surelease® coated metoclopremide HCl beads. Unlike the theophylline granules, this multiparticulate system has different structure and design. The drug was first sprayed onto sugar beads (nonpareil beads) and then coated with Surlease® E-7-7050 with or without pore former HPMC-E15 (9). A mechanistic mode was thus built with consideration of such structure and design.

Figure 6 portrays a typical plot of model-predicted release profile of metoclopremide HCl on sugar beads coated with Surelease® E-7-7050 and 5% pore former HPMC-E15 compared with test data (9) (the curve with circles). The model parameters were estimated from experimental data from Ref. 9 and the line was predicted by computer simulation using AP-CAD® software package for coated multiparticulate dosage forms. The verified model parameters, mathematical model and computational procedures were then used to analyze formulation and design factors. The influence of concentration of the pore former on the release kinetics is revealed in Figure 7. The impact of weight gain after coating (a measure of coating thickness) on drug release profiles at a given concentration of pore former (5%) is shown in Figure 8.

The above results indicate that, using the computer simulation, we can predict drug release profiles for given pore former concentration and coating thickness. On the other hand, we can identify an optimal pore former concentration and/or coating thickness for a desired release profile. This computational procedure can be combined with convolution method and drug pharmacokinetic data to predict plasma concentration profile of the drug; or with deconvolution method to design a coating formulation and thickness for a target plasma concentration profile (10).

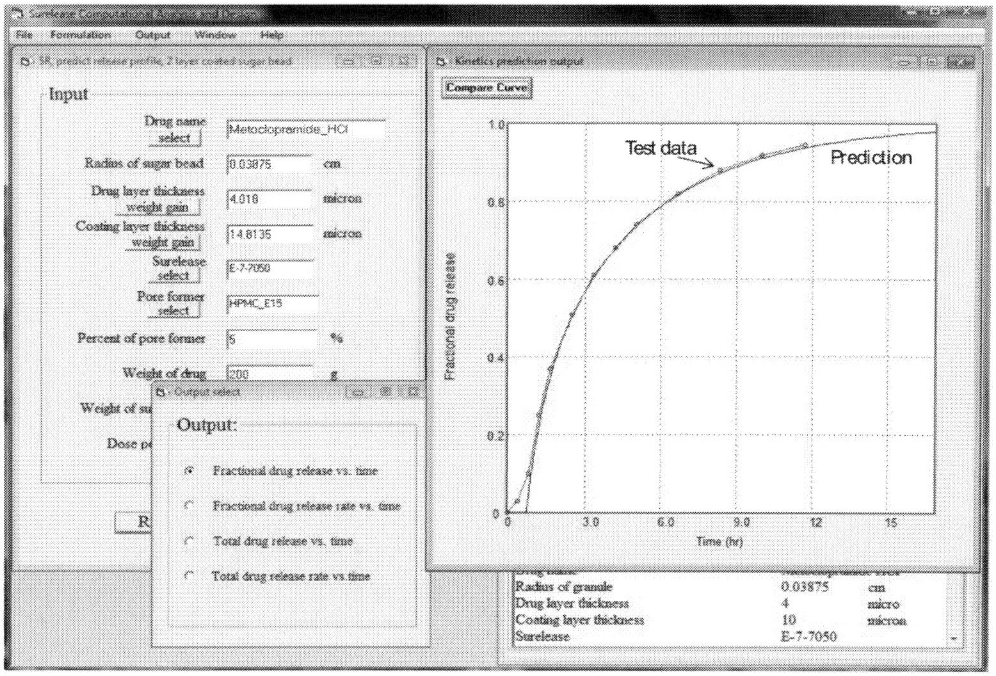

Figure 6. Comparison of model-predicted release profile with test data of metoclopremide HCl pellets

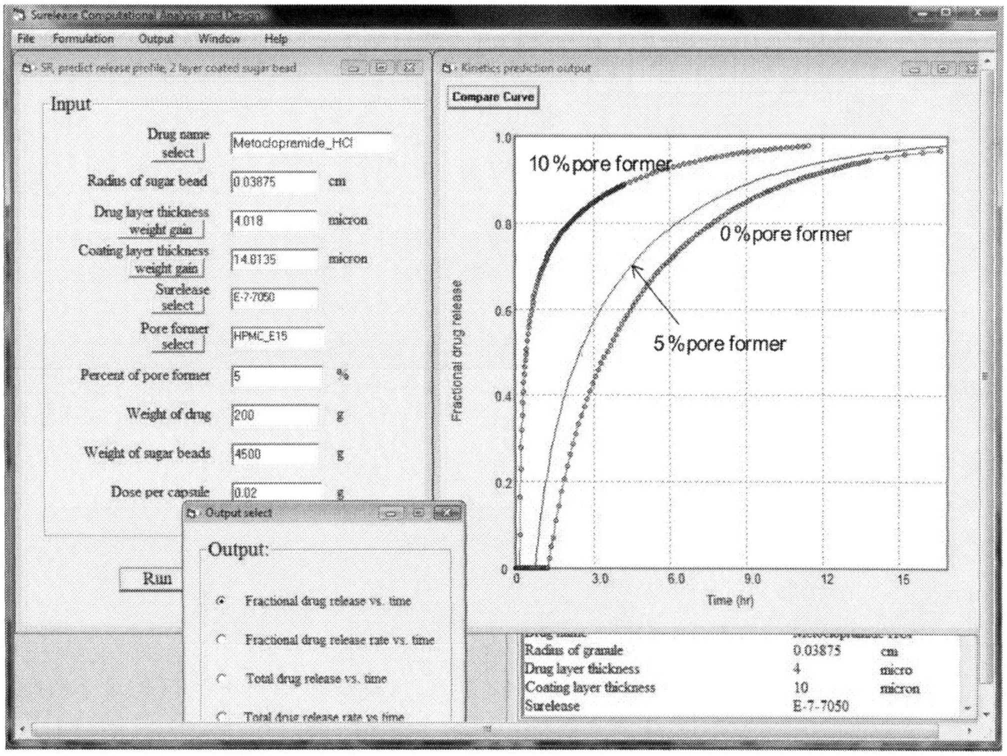

Figure 7. Effect of concentration of pore former HPMC-E15 on release profile of metoclopremide HCl pellets coated with Surlease® E-7-7050

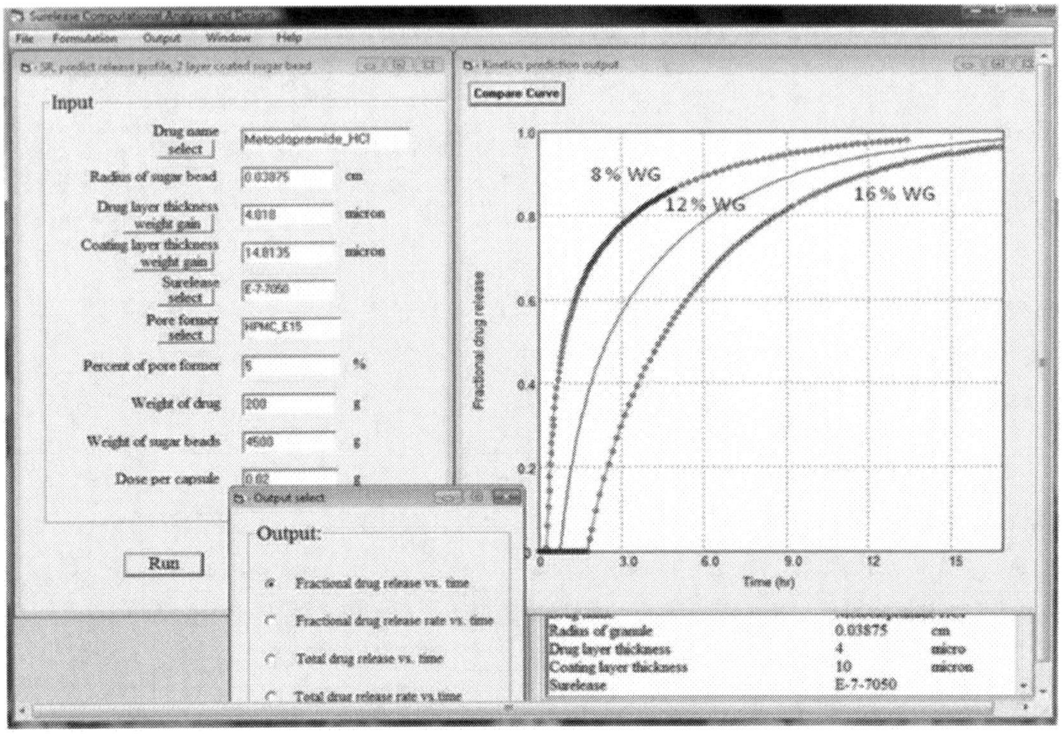

Figure 8 Effect of coating weight gain (a measure of coating thickness) on release profile of metoclopremide HCl pellets coated with Surelease® E-7-7050 and 5% pore former HPMC-E15

Conclusions

Mechanistic models and computer software are useful tools for QbD in formulations of MRDFs, as evident by the examples of EUDRAGIT® RL/RS 30D coated theophylline granules and Surelease® coated metoclopremide HCl sugar beads. This work has demonstrated that desired release profiles of various chemical entities can be achieved by altering physical parameters such as coating thickness (weight gain), and formulation parameters, e.g. EUDRAGIT® RL/RS ratio, or Surelease® and HPMC concentrations. The *in sillico* experiments can be accomplished in hours instead of days or weeks thus enabling scientists to explore greater formulation options and to define design space quickly.

Reference

1. Zhou, Y., Chu, J. S., Li, J. X., and Wu, X. Y. (2010), Theoretical analysis of release kinetics of coated tablets containing constant and non-constant drug reservoirs. *International Journal of Pharmaceutics, 385*(1-2), 98-103.
2. Zhou, Y., Chu, J. S., Zhou, T., and Wu, X. Y. (2005), Modeling of dispersed-drug release from two-dimensional matrix tablets. *Biomaterials, 26*(8), 945-952.
3. Zhou, Y., Chu, J. S., and Wu, X. Y. (2004), Theoretical analysis of drug release into a finite medium from sphere ensembles with various size and concentration distributions. *European Journal of Pharmaceutical Sciences, 22*(4), 251-259.
4. Wu, X. Y., Zhou, Y. (1999), Studies of diffusional release of a dispersed solute from polymeric matrixes by finite element method. *Journal of Pharmaceutical Sciences, 88*(10), 1050-1057.
5. Wu, X. Y., Zhou, Y. (1998), Finite element analysis of diffusional drug release from complex matrix systems. II. factors influencing release kinetics. *Journal of Controlled Release, 51*(1), 57-71.
6. Zhou, Y., Wu, X. Y. (1997), Finite element analysis of diffusional drug release from complex matrix systems. I. complex geometries and composite structures. *Journal of Controlled Release, 49*(2-3), 277-288.
7. Sinchaipanid, N., Chitropas, P., and Mitrevej, A. (2004), Influences of Layering on Theophylline Pellet Characteristics, *Pharmaceutical Development and Technology*, 9(2), 163–170.
8. EUDRAGIT Application Guidelines, (2008) 10th edition.
9. Sadeghi, F., Ford, J. L., Rubinstein, M. H. and Rajabi-siahboomi, A. R., (2000), Comparative study of drug release from pellets coated with HPMC or Surelease, *Drug Development and Industrial Pharmacy*, 26(6) 651-660.
10. Dutta, S., Qiu, Y., Samara, E., Gao, G., Granneman, G. (2005), Once-a-day extended-release dosage form of divalproex sodium III: Development and validation of a level A in vitro–in vivo correlation (IVIVC), *Journal of Pharmaceutical Sciences*, *94*(9), 1949-56.

(607c) Mechanistic Investigation of Drug Release From Asymmetric Membrane Tablets

Mary T. am Ende, Pharmaceutical Development, Pfizer Worldwide Research and Development, Groton, CT

An asymmetric membrane (AM) tablet was developed for a soluble model compound to study the in vitro drug release mechanisms in challenge conditions, including osmotic gradients, concentration gradients, and under potential coating failure modes. Porous, semipermable membrane integrity may be compromised by a high fat meal or by the presence of a defect in the coating that could cause a safety concern about dose-dumping. Drug release from the AM system studied was determined to be robust to varying and extreme challenge conditions. The conditions investigated included varying pH, agitation rate, media osmotic pressure, media saturated with drug to eliminate the concentration gradient, simulated high fat meal, and intentionally placed film coating defects. Osmotic and diffusional shut off experiments suggest that the mechanism governing drug release is a combination of osmotic and diffusional at approximately 90-95% and 10-5%, respectively. In addition, the coating failure mode studies revealed this formulation and design is not significantly affected by a high fat meal or by an intentionally placed defect in the film coating, and more specifically, did not result in a burst of drug release.

See more of this Session: QbD for Modified Release Dosage Forms
See more of this Group/Topical: Topical I: Comprehensive Quality by Design in Pharmaceutical Development and Manufacture

(607d) QbD for Generic MR Drug Products

Yue (Helen) Teng, Office of Generic Drugs, FDA, Rockville, MD

The Office of Generic Drugs ensures the therapeutic equivalence and quality of generic drug products. This is based principally upon a regulatory paradigm that requires pharmaceutically equivalent drug products containing the same active ingredient, dosage form, and strength (among other things), and being bioequivalent to the brand name product. Also under this paradigm, the batch to batch quality of generic drug products is ensured based upon batch release testing according to specifications. While this approach has been used with great success toward bringing the availability of high quality of generic drug products, the vast majority of pharmaceuticals approved using this framework have been solution and immediate release oral products, which are inherently simple in design. Drug products however, have increased in design complexity, to encompass modified (oral) release products, among others complex dosage forms. As these pharmaceuticals have increased in their design complexity, approaches adopted to ensure the equivalence and quality of generic drug product pharmaceuticals, must likewise evolve.

FDA's Pharmaceutical cGMPs for the 21st Century: A Risk-Based Approach and Process Analytical Technology Initiatives are meant to enhance and modernize the regulation of pharmaceutical manufacturing and product quality. As a result of these initiatives and the trend of increasingly complex pharmaceutical products, OGD/GPhA has developed a mock example of pharmaceutical development report for a modified release product. This report illustrates how ANDA applicants can move toward implementation of quality by design (QbD). Through establishing quality target product profile (QTPP), identifying critical quality attributes (CQAs), performing risk assessment, conducting design of experiment, and setting control strategies, quality of generic modified release products is ensured.

See more of this Session: QbD for Modified Release Dosage Forms
See more of this Group/Topical: Topical I: Comprehensive Quality by Design in Pharmaceutical Development and Manufacture

FLUID BED COATING PROCESS FOR MODIFIED RELEASE APPLICATIONS

Orapin Rubino, Ph.D., Glatt Air Techniques, Inc., Ramsey, New Jersey

Biography

Orapin Rubino is Director, Formulation and Product Development, Pharmaceutical Services Division at Glatt Air Techniques, Inc. Her experience focus on development of solid dosage forms (sustained and modified release formulations, granulation, direct pelletization, pellet coating, drying, encapsulation and tableting processes). She received her B.S. degree in Pharmacy from Thailand and Ph.D. degree in Pharmaceutics from UNC Chapel Hill. She then spent two years as a postdoctoral research associate in Protein Chemistry (physical biochemistry) at Duke University, NC. Prior to joining Glatt in 1998, she worked as a Senior Process Scientist and Development Scientist at Dade Behring (formerly DuPont Medical Products) in Delaware for about 5 years. Her responsibilities at DuPont included formulation development, scale up and technical support for granulation and tableting of clinical diagnostic reagents.

Abstract

Wurster process (bottom spray fluid bed coating process) is a process of choice for coating of multiparticulates to produce modified or controlled release dosage forms. The unique features of the Wurster process include air distribution plates (up bed and down bed plates) and partition that organize fluidization of particles through the "coating zone". The nozzle is mounted at the bottom of a product container. The short distance between coating materials and particles during the coating process and the concurrent spray (same direction of particle flow and spray droplets) minimize spray drying. This principle contributes to high coating uniformity and coating efficiency of the Wurster process. This process can be used to produce pellets by drug layering process, follow by polymer coating to obtain modified or controlled drug release.

The fluidization pattern of particles is controlled by air distribution plates, partition height, and process air volume. Coating process parameters that mainly control the process include inlet air temperature, atomization air pressure/volume, and spray rate. Formulation aspect such as composition and solid content of the coating preparation can affect the drug release from the pellets.

Fluid Bed Coating

Diagram: Bottom Spray Coater (Wurster Process)

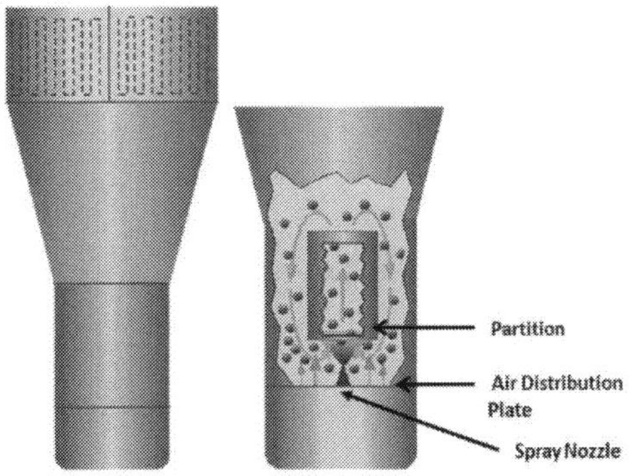

The principle and advantages of the Wurster process will be presented. Equipment, coating process parameters and scale up considerations will be discussed.

(607f) Implementation of QbD In the Development of a CR Dosage Form

Nancy E. Sever, Global Formulation Sciences, Abbott Laboratories, Abbott Park, IL, **Michael Brian Mackaplow**, Global Formulation Sciences- Solids, Abbott, Abbott Park, IL and Didier LeFebvre, Abbott, Abbott Park, IL

Use of QbD in characterization and development of a modified release dosage form will be discussed. Special emphasis will be placed on the early stages of development and the utilization of material sparing, strategic experimental design to ascertain the critical parameters affecting the performance of the dosage unit. Lab to Pilot to Commercial scale transfer will be discussed in terms of a QbD based approach. In addition to formulation development, opportunities for QbD in analytical method development will also be presented.

Additional information will be provided upon legal approval - End of May.

See more of this Session: QbD for Modified Release Dosage Forms
See more of this Group/Topical: Topical I: Comprehensive Quality by Design in Pharmaceutical Development and Manufacture

(607g) Control Strategy for Manufacturing of Modified Release Capsules: A Regulatory Perspective

Zhigang Sun, Office of Generic Drugs, OPS, CDER, FDA, Rockville, MD

Compared to simple dosage forms, there will be more challenges for modified release dosage forms to assure high drug product quality as well as manufacturing consistency due to its complexity of formulation design and process development. Quality by design (QbD) provides an effective tool to systematically evaluate, understand, and refine the formulation and manufacturing process to ensure predefined product quality. In this presentation, the modified release capsule manufacturing process will be used as an example to illustrate how to implement the QbD principles to enhance process understanding and to improve process controls in order to provide assurance of reproducible commercial manufacturing for consistent product quality. The purpose of this talk is to provide a regulatory perspective with regard to the QbD based control strategy for manufacturing of modified release drug products.

See more of this Session: QbD for Modified Release Dosage Forms
See more of this Group/Topical: Topical I: Comprehensive Quality by Design in Pharmaceutical Development and Manufacture

(623a) A Systems Approach to Decipher Stress Response Networks In Mycobacteria

Arun N. Prasanna, Thawfeek Mohammed Varusai and **Sarika Mehra**, Department of Chemical Engineering, Indian Institute of Technology, Bombay, Mumbai, India

The repertoire of Mycobacteria includes both pathogenic and non-pathogenic bacteria. Despite a century of research, the mechanism of survival of mycobacteria within the host and the host-pathogen interaction has not been understood completely. Inside the macrophage, the bacteria encounter many environmental stresses such as acidic pH and reactive oxygen. A complex network of sigma factors and two-component systems with extensive cross-talk is involved in the survival of the bacteria. In response to each stress, a variety of proteins are induced that help mycobacteria adapt to the new environment.

The objective of this work is to characterize this regulatory network in response to single and multiple stresses using *M. smegmatis* as a model system. Both exponentially growing cells and cells from stationary phase were subjected to different kind of stresses, including acid shock, starvation, heat shock and ethanol stress. Cell viability was measured after a defined period of exposure. The stresses are applied either alone or in combinations. Surprisingly, in certain cases, viability increases in the presence of multiple stresses, as opposed to a single stress alone. It is hypothesized that the presence of a stress helps the bacteria to adapt to the other stress condition. To identify the molecular basis of this adaptation, transcriptional profile of key sigma factors is measured.

Based on the current knowledge of the stress response transcriptional networks in Mycobacteria, a Boolean model is developed considering six types of stresses – hypoxia, heat shock, surface stress, nutrient stress, oxidative stress and cold shock. The model is simulated to evaluate the ability of this network to differentiate between stresses. The model suggests that in addition to induction of a sigma factor, the concentration levels distinguish between multiple stress combinations. Further, a robustness analysis is performed to identify the key nodes of the network.

See more of this Session: Poster Session: Bioengineering
See more of this Group/Topical: Food, Pharmaceutical & Bioengineering Division

(623aa) Determination of Progerin Lamin A Phase Association and Orientation to Phospholipid Membranes

Agnieszka Kalinowski[1], **Peter N. Yaron**[2], Mathias Lsche[3] and Kris N. Dahl[1], (1)Biomedical Engineering, Carnegie Mellon University, Pittsburgh, PA, (2)Department of Chemical Engineering, Carnegie Mellon University, Pittsburgh, PA, (3)Physics, Carnegie Mellon University, Pittsburgh, PA

Mutations in lamin A cause several disease states, called laminopathies. One of these, Hutchinson Gilford Progeria Syndrome (HGPS), produces a lamin A variant, progerin, that is improperly processed and accumulates at the nuclear membrane, triggering changes in gene expression, heterochromatin organization, as well as mechanical and morphological properties of the nucleus. Accumulation results in the formation of fibrous microdomains that alter nuclear mechanical behavior and overall cellular function. The molecular details of how the association of progerin lamin A with the nuclear membrane leads to the observed disease pathology is not known.

Membrane affinity of lamin A is regulated by a series of post-translational modifications that attach hydrophobic farnesyl and carboxy-methyl groups to the C-terminal cysteine group. Previous work has established that farnesylation alone is not sufficient for membrane association and implies other interactions contribute to the enhanced association of progerin lamin A. Enhanced membrane association is thought to be contributed by a terminal carboxy-methylation and polybasic stretch of histadines brought closer to the membrane surface by a 50 amino acid deletion (delta 50) that occurs on progerin lamin. This causes an accumulation of the delta 50 lamin A protein at the nuclear membrane in comparison to normal wild-type lamin A, as demonstrated in cells from HGPS patients with fluorescence and electron microscopy. We believe increased membrane retention enhances the aggregation of the lamin proteins into microdomains by preferential protein conformation and orientation at the membrane to maximize binding and simultaneously alters local membrane structure affecting downstream cellular signaling. We aim to determine the changes in protein conformation and orientation relative to the membrane surface caused by each membrane active moiety and change in membrane structure due to enhanced protein retention. To study lipid phase association of progerin lamin A we will use Giant Plasma Membrane Vesicles (GPMVs) Giant Unilammelar Vesicles (GUVs) and Langmuir Monolayers (LM) coupled with fluorophore labelled protein variants. To study protein orientation at the membrane interface we will use a combination of tethered bilayer membranes (tBLMs) and Langmuir monolayers.

See more of this Session: Poster Session: Bioengineering
See more of this Group/Topical: Food, Pharmaceutical & Bioengineering Division

(623ab) CD47 Inhibit the Immune System Response Against Nanoparticles In Vivo

Pia L. Rodriguez and Dennis E. Discher, Chemical and Biomolecular Engineering, University of Pennsylvania, Philadelphia, PA

A major challenge for injecting particles or implanting biomaterials into the body is that they activate immune cells such as macrophages, the cells that normally function to clear invading pathogens. Interestingly, macrophages have a surface receptor mechanism which prevents them from phagocytosing our own cells. During initial macrophage engulfment, macrophages recognize foreign and self targets because they both have antibodies or plasma complement proteins on their surface. However, before the macrophage engulfs the target, cells are checked for the presence of the surface protein CD47. CD47 is recognized by the receptor SIRPa on macrophages which inhibits phagocytosis, and when we attach a 100 aa extracellular component of CD47 to micro- and nano-sized particle, we demonstrate, *in vitro*, that this sufficient to inhibit phagocytosis of these small particles. Now we are focusing our efforts on understanding how these results are consistent with *in vivo* systems. In vivo, we are currently testing whether nanoparticles that have surface immobilized hCD47 will not be phagocytosed in nobese diabetic (NOD) mice, because we tested that hCD47 actually bind to this mSirpa strain. We will detail the dependence on protein densities and the length scale dependence for this fundamental facet of immunocompatability.

See more of this Session: Poster Session: Bioengineering
See more of this Group/Topical: Food, Pharmaceutical & Bioengineering Division

(623ac) Modeling of the Transcriptional Regulation of Tryptophan Hydroxylase In C. Elegans

Hyewon Lee and Hang Lu, School of Chemical & Biomolecular Engineering, Georgia Institute of Technology, Atlanta, GA

Organisms respond to environmental conditions and modulate their physiological outcomes. These responses vary from individual to individual even in isogenic populations. Many previous studies described physiological consequences such as alternative cell fate decision as a result of gene regulatory networks. However, it is still not well understood how genes integrate intrinsic and environmental signals. Here, we analyzed transcriptional regulation of tryptophan hydroxylase (*tph-1*) in *C. elegans* using mathematical modeling. We are interested in *tph-1* because it encodes a key enzyme for serotonin synthesis and its expression is regulated by several environmental factors including food and temperature. Furthermore, changes in *tph-1* expression level can modulate physiological consequences like aging, fat storage, and egg laying. We hypothesize that regulatory mechanism from environmental perception to physiology can be revealed by analyzing transcriptional regulation of *tph-1*. In the model, *tph-1* expression of wild-type and that of *unc-43* gain-of-function mutant were compared. CaMKII encoded by *unc-43* is one of the major effectors of *tph-1* expression in ADF neurons. It has been known that *unc-43(gf)* increased the level of *tph-1* expression but the detailed mechanism of its regulation is yet to be clarified. The promoter activity of *tph-1* was analyzed by characterizing transcriptional activities. There are two known transcriptional factors in *tph-1* regulation: one inhibitor and one activator. These transcriptional factor activities are regulated by phosphorylation from signal transduction network (kinase cascade). Based on modeling results, we can confirm the fact that the promoter of *tph-1* has high activity in stressful condition. After developing a deterministic model, we applied the Gillespie algorithm to include stochastic effects on *tph-1* transcription. The distribution of *tph-1* expression of wild-type and that of *unc-43 (gf)* mutant were compared. Simulation results agree with experimental results, showing that mutant population has a large deviation compared to wild-type even considered elevated *tph-1* expression on average. This result may reflect higher variability under stressful condition in environment.

See more of this Session: Poster Session: Bioengineering
See more of this Group/Topical: Food, Pharmaceutical & Bioengineering Division

(623ad) On Damped Wave Diffusion of Oxygen In Iselts of Langergahns: Part-VI Comparison of Parabolic and Hyperbolic Models In a Finite Sphere for Pulse Boundary Condition

Kal Renganathan Sharma, Department of Chemical Engineering, Prairie View A & M University, Prairie View, TX

Oxygen availability becomes limited in some regions of the tissue. The metabolic rate in the cells and demand for oxygen is greater than the oxygen that has diffused to that region. It becomes a diffusion limited process. Due to this phenomena growth of multicellular systems over 100 μ m does not happen. A condition called *hypoxia* has been observed in Brockman bodies in fish. Oxygen supply in addition to diffusion also comes about by the circulatory system and by hemoglobin molecule. Oxygen is carried in the blood by convection to capillaries by the circulatory system. Islets of Langerhans (Figure 1.0) are spheroidal aggregates of cells that are located in the pancreas (2). They secrete harmones that are involved in glucose metabolism particulary insulin. Transplantation of isolated cells is a promising treatment for some forms of Type I diabetes. Islets removed from the pancreas are devoid of their internal vacularization. The metabolic requirement of the cells require oxygen to diffuse from the external envirnonment and through the oxygen-consuming islet tissue. The oxygen supply is a critical limiting factor for the functionality and feasibility of islets that are encapsulated, placed in devices for *implantation*, cultured, used in aneorobic conditions. Theoretical models are needed to describe the oxygen diffusion. The parameters of the model require knowledge of the consumption rate of oxygen, oxygen solubility, effective diffusion coefficient to oxygen in the tissue. Simultaneous diffusion and reaction model was developed by Colton (3). This study is a continuation in the series of studies undertaken. A finite sphere is considered. The transient concentration profile is obtained in both the asymptotic limits of zeroth order and first order limits of the Michaelis and Menten kinetics. The pulse boundary condition is studied. Both parabolic Fick diffusion and hyperbolic damped wave diffusion and relaxation are considered. The mass transfer coefficients are calculated. Sharma number effects in artificial pancreas is discussed.

See more of this Session: Poster Session: Bioengineering
See more of this Group/Topical: Food, Pharmaceutical & Bioengineering Division

(623ae) Experimental Analysis and Model-Based Optimization of Microalgae Growth in Photo-Bioreactors Using Flue Gas

Lian He, Enery, Environmental and Chemical Engineering Department, Washington University in St. Louis, St Louis, MO, Gang Wu, Enery, Environmental and Chemical Engineering Department, Washington University in St. Louis, Saint Louis, MO and Yinjie Tang, Department of Energy, Environmental and Chemical Engineering, Washington University, St Louis, MO

Three strains of microalgae, *Chlorella sp.*, *Tetraselmis suecica* and *Synechocystis* PCC 6803, were grown with the flue gas generated by natural gas combustion. Continuous exposure of cultures to flue gas significantly reduced all algal growth, while *Chlorella sp.* showed the highest tolerance to high concentration CO_2. Two methods can be used to overcome flue gas inhibitory effect on biomass growth: reduce in-flow CO_2 concentration or apply on-off mode for flue gas treatment. To design efficient strategies for flue gas utilization in algal photo-bioreactors, we employed a Monod-based kinetic model to describe phototrophic biomass growth in response to CO_2 control dynamics. Via static optimization, the model designed the control dynamics of in-flow CO_2 at low concentration to support the maximal growth rate by gradually increasing CO_2 concentration. On the other hand, we also tested a much simpler, but industrial feasible, on-off control mode for direct flue gas treatment. Based on the reported algal kinetic parameters (e.g., maximal specific growth rate and inhibition constant), the model showed that gas-on (~10 seconds) and gas-off (5~9 minutes) had more than 90% of theoretically optimal biomass growth rate (as achieved by optimal control of dynamic in-flow CO_2 profiles). The models developed in this study could be useful for rational design of algal bioprocess using high concentration flue gases.

See more of this Session: Poster Session: Bioengineering
See more of this Group/Topical: Food, Pharmaceutical & Bioengineering Division

(623af) Solvent-Enhanced Transformations of Steroids by Beauveria Bassiana

Richard Gonzlez[1], Felipe Nicolau[1], Horacio F. Olivo[2] and Tonya L. Peeples[1], (1)Chemical and Biochemical Engineering, University of Iowa, Iowa City, IA, (2)Division of Medicinal and Natural Products Chemistry College of Pharmacy, University of Iowa, Iowa City, IA

The filamentous fungus *Beauveria bassiana* has the capacity to enhance the yield and hydroxylation selectivity of steroid biotransformation. Organic solvents serve as carbon sources during growth of *B. bassiana* and have been shown to induce the expression of oxidative enzymes. Expression varies with the alkane compound. Current research efforts apply organic solvents to enhance the conversion of androst-4-ene-3,17-dione to desired steroid products. The long-term goal is to standardize and control the biocatalysis capacity of *B. bassiana* in the biotransformation of a variety of steroids under different operational parameters. The selection of suitable n-alkane solvent for optimal yield and selectivity, and the determination of the optimal pH and reaction times will be presented. This research will provide new insight on the synthesis of steroids that are challenging to obtain by conventional chemistry. It will also offer enhancement opportunities for industrial applications utilizing steroid molecules.

See more of this Session: Poster Session: Bioengineering
See more of this Group/Topical: Food, Pharmaceutical & Bioengineering Division

(623ag) On Damped Wave Diffusion of Oxygen In Iselts of Langergahns: Part-VIII Comparison of Parabolic and Hyperbolic Models In a Finite Sphere Subject to Periodic Boundary Condition

Kal Renganathan Sharma, Department of Chemical Engineering, Prairie View A & M University, Prairie View, TX

Oxygen availability becomes limited in some regions of the tissue. The metabolic rate in the cells and demand for oxygen is greater than the oxygen that has diffused to that region. It becomes a diffusion limited process. Due to this phenomena growth of multicellular systems over 100 μ m does not happen. A condition called *hypoxia* has been observed in Brockman bodies in fish. Oxygen supply in addition to diffusion also comes about by the circulatory system and by hemoglobin molecule. Oxygen is carried in the blood by convection to capillaries by the circulatory system. Islets of Langerhans (Figure 1.0) are spheroidal aggregates of cells that are located in the pancreas (2). They secrete harmones that are involved in glucose metabolism particulary insulin. Transplantation of isolated cells is a promising treatment for some forms of Type I diabetes. Islets removed from the pancreas are devoid of their internal vacularization. The metabolic requirement of the cells require oxygen to diffuse from the external envirnonment and through the oxygen-consuming islet tissue. The oxygen supply is a critical limiting factor for the functionality and feasibility of islets that are encapsulated, placed in devices for *implantation*, cultured, used in aneorobic conditions. Theoretical models are needed to describe the oxygen diffusion. The parameters of the model require knowledge of the consumption rate of oxygen, oxygen solubility, effective diffusion coefficient to oxygen in the tissue. Simultaneous diffusion and reaction model was developed by Colton (3). This study is a continuation in the series of studies undertaken. A finite sphere is considered. The transient concentration profile is obtained in both the asymptotic limits of zeroth order and first order limits of the Michaelis and Menten kinetics. The periodic boundary condition is studied. Both parabolic Fick diffusion and hyperbolic damped wave diffusion and relaxation are considered. The mass transfer coefficients are calculated. Sharma number effects in artificial pancreas is discussed. The time taken to reach steady state in the hyperbolic models are calculated. Two term and three term approximations of the transcedental equation in the boundary condition was evaluated. The use of wave term in the boundary condition and independently the use of the final condition in time renders the occurences of "overshoot" more a mathematical artefact rather than depiction of reality. The method of complex concentration is found to be useful in evaluating the phase lag and frequency of oscillations. Experimental apparatus may be constructed based on this study to measure relaxation times and binary diffusivity under transient conditions.

See more of this Session: Poster Session: Bioengineering
See more of this Group/Topical: Food, Pharmaceutical & Bioengineering Division

(623ah) Toxicity, Transport, and Accumulation of Hydrocarbons In Saccharomyces Cerevisiae

Hua Ling, Binbin Chen, Hongxin Zhao, Pei Yu Lim, Hazarki Yaohari, Susanna Leong and **Matthew Wook Chang**, School of Chemical and Biomedical Engineering, Nanyang Technological University, Singaproe, Singapore

Despite the widespread presence of hydrocarbons, including alkanes, in the environment, there is a lack of understanding of interactions between hydrocarbons and biological systems. In this study, we aimed to elucidate molecular mechanisms underlying toxicity, transport, and accumulation of alkanes in *Saccharomyces cerevisiae*. Our results show that medium chain alkanes significantly decreased the number of the viable cells, and the killing percentage reached 72.7 after 48-hour incubation. Despite this toxicity, we observed no apparent damage on the cell surface using Field Emission Scanning Electron Microscope (FESEM). Interestingly, upon exposure to alkanes, intracellular medium chain alkanes were detected in *S. cerevisiae*, implying that some alkanes may be diffused and/or transported into *S. cerevisiae*. Moreover, from microarray-based transcriptome analysis, we identified the genes potentially associated with alkane toxicity and transport in *S. cerevisiae*. This study provides new insights into molecular and cellular mechanisms underlying the toxicity, transport, and accumulation of alkanes in *S. cerevisiae*.

See more of this Session: Poster Session: Bioengineering
See more of this Group/Topical: Food, Pharmaceutical & Bioengineering Division

(623ai) Adiponectin Levels Mediate Recovery of Renal Function In a Novel Model of Podocyte Ablation

Joseph M. Rutkowski and Philipp E. Scherer, Touchstone Diabetes Center - Department of Internal Medicine, UT Southwestern Medical Center, Dallas, TX

Adiponectin levels are altered in renal pathologies: low levels are associated with metabolic dysfunction and type II diabetes while heightened levels have been measured in chronic kidney disease. To explore the effects of adiponectin on kidney disease progression, a novel murine model of podocyte apoptosis was created using a PODocin promoter driven cassette for Apoptosis Through Targeted Activation of Caspase 8 (POD-ATTAC). Upon induction by a low molecular weight dimerizing agent, the ATTAC construct induces caspase-mediated apoptosis in a cell-specific manner. These mice not only demonstrated significant kidney damage to mimic human renal disease, but also restored podocyte filtration function over time in a dimerizer dose-dependent manner. Adiponectin knockout POD-ATTAC mice developed massive, irreversible albuminuria and renal failure over time, while adiponectin over-expressing (OE) POD-ATTAC mice demonstrated rapid recovery from heightened albuminuria. Renal histology in OE kidneys following recovery showed improved overall renal health with reduced interstitial fibrosis suggesting protective effects of adiponectin throughout the kidney. Thus, despite similar number of podocytes ablated as compared to wildtype POD-ATTAC mice, these findings demonstrate that increased adiponectin levels in OE mice prevent damage to other renal cell types and progression to renal failure. The POD-ATTAC model provides an excellent platform for further studies of podocyte regeneration and potential roles of adiponectin on renal function in chronic kidney disease.

See more of this Session: Poster Session: Bioengineering
See more of this Group/Topical: Food, Pharmaceutical & Bioengineering Division

(623aj) Assaying Secreted Titer of Heterologous Proteins In Salmonella

Kevin James Metcalf and Danielle Tullman-Ercek, Chemical and Biomolecular Engineering, University of California, Berkeley, Berkeley, CA

Inefficient bacterial secretion causes most heterologous protein to be expressed as inclusion bodies. Additionally, insoluble aggregation and host toxicity limit titer. We expect to achieve high titer of soluble heterologous proteins in bacterial culture by engineering the type III secretion system (T3SS) of *Salmonella typhimurium* for efficient peptide secretion to the extracellular fluid. The T3SS is a transmembrane pump found in many Gram-negative pathogenic bacteria. It translocates peptides from the cytoplasm across both the inner and outer membranes of the bacterium and the cell membrane of the host cell. Spider silk monomers have been successfully secreted into the extracellular fluid in the absence of host cells using the wild-type T3SS. However, reported titers are several orders of magnitude below those achieved through routine industrial bacterial fermentation and secretion efficiency is less than 20%. We report the development of a high-throughput fluorescent assay for secretion efficiency quantification using flow cytometry. Single cell analysis is achieved through the expression of a secretion-tagged transcription factor that represses gfp production. Such a screening method will enable the rapid, high-throughput quantification of secretion efficiency and assist in the development of a T3SS-based prokaryotic protein expression system.

See more of this Session: Poster Session: Bioengineering
See more of this Group/Topical: Food, Pharmaceutical & Bioengineering Division

(623ak) Metabolic Engineering of Escherichia Coli for Dihomomethionine Production

Aram Kang, Wendy Chen, Susanna Su Jan Leong, Chi Bun Ching and **Matthew Wook Chang**, School of Chemical and Biomedical Engineering, Nanyang Technological University, Singaproe, Singapore

This study aimed to metabolically engineer *Escherichia coli* to produce dihomomethionine (DHM). Towards this aim, genes involved in DHM biosynthesis pathways in plant cells were introduced to *E. coli*. Utilization of *E. coli* as an engineering host has two potential advantages in this study. First, *E. coli* has a small number of methionine-consuming metabolic pathways, of which fluxes could be more easily directed to methionine chain-elongation pathways. Second, *E. coli* does not have cellular compartments, which might increase the synthesis efficiency of α-keto acids. In this work, production of DHM was detected and quantitated in *E. coli* by mass spectrometry. To optimize DHM production level in *E. coli*, three strategies were applied. First, optimal strength of promoters was determined by steady-state modeling, and various combinations of promoters were experimentally tested. Second, methionine-metabolizing pathways were sequentially disrupted by gene-knockout mutagenesis. Third, metabolic fluxes in the knockout mutants were modified by in silico flux balance analysis. Furthermore, culture conditions including temperature and medium compositions were optimized. In particular, effects of methionine supplementation on the production yield were investigated. These results represent the first report of DHM production in microbes and provide new insights into the feasibility of DHM production in microbial hosts.

See more of this Session: Poster Session: Bioengineering
See more of this Group/Topical: Food, Pharmaceutical & Bioengineering Division

(623al) Engineering E. Coli for Improved Efflux of Fuel Molecules

Susanna Su Jan Leong, Jee Loon Foo, Matthew Wook Chang and Chi Bun Ching, School of Chemical and Biomedical Engineering, Nanyang Technological University, Singaproe, Singapore

Engineering *E. coli* for improved efflux of fuel molecules

Foo Jee Loon, Matthew Wook Chang, Chi Bun Ching, Susanna Su Jan Leong

School of Chemical and Biomedical Engineering, Nanyang Technological University

62 Nanyang Drive, Singapore 637459, Singapore

Escherichia coli is the production host used for the production of many biofuels. However, many organic solvents are toxic to bacteria and the tolerance of bacteria to these solvents has been found to involve exportation of the molecules out of the cells through efflux pump systems. In the context of biofuel production, these efflux pumps could be employed to confer greater solvent tolerance of the production host as well as promote secretion of the produced biofuel out into the external medium, thus increase the efficiency and reduce cost of biofuel production.

Of the few efflux pump systems identified in *E. coli*, the AcrAB-TolC is probably the best characterised. In order to facilitate the use of the AcrAB-TolC system for higher efflux efficiency, we have performed directed evolution on the AcrB inner membrane transporter for increased efflux of organic molecules, in this case, n-octane. Mutant libraries of AcrB were generated through random mutagenesis and variants with accelerated extrusion of intracellular n-octane were isolated after enrichment screening based on the n-octane tolerance of *E. coli* transformed with the mutant genes. The engineered AcrB mutants could be utilized in *E. coli* for more efficient biofuel production. Additionally, the information gathered from the mutants would enable us to further our understanding of the pump and aid rational design of the efflux system in the future.

See more of this Session: Poster Session: Bioengineering
See more of this Group/Topical: Food, Pharmaceutical & Bioengineering Division

(623am) Method for Using Peptide Labeling In Metabolic Flux Analysis

Dominic Mandy[1], Hong Yang[1], Douglas K. Allen[2] and Igor GL Libourel[1], (1)Department of Plant Biology, University of Minnesota, St. Paul, MN, (2)USDA, Saint Louis, MO

Isotope labeling-based metabolic flux analysis estimates internal metabolic fluxes by fitting predicted labeling patterns to experimental labeling data. Labeling data can be acquired through isotope discriminating techniques such as Nuclear Magnetic Resonance (NMR) spectroscopy or Mass Spectrometry (MS). The relative isotopomer distributions of metabolites can be calculated directly for a given set of flux values and substrate label using the EMU framework. Most of such labeling measurements are derived from amino acids recovered from protein hydrolysates. Here we present a method to directly use complete peptides instead. The aim of this new approach is to retain the compartmental and temporal information associated with each peptide. Because the mass isotopomer distribution (MID) of each peptide is produced by the Cauchy Product of the MIDs of its individual amino acids, the MID of a peptide can be readily determined.

See more of this Session: Poster Session: Bioengineering
See more of this Group/Topical: Food, Pharmaceutical & Bioengineering Division

(623an) Do Bacteria Optimize Their Contact Area with Surfaces?

Mehdi Kargar, Jeffrey R. Saucke, Amrinder S. Nain and **Bahareh Behkam**, Mechanical Engineering, Virginia Tech, Blacksburg, VA

Using our previously developed Spinnert based Tunable Engineered Parameters (STEP) technique, polystyrene (PS) fibers ($2R_f$: 300 nm to 600 nm, length: 10 mm, $2D_f$: 0 nm to microns) were deposited in aligned configuration on both glass and PS substrates. Pseudomonas aeruginosa (PAO1), a common human pathogen was used as the model organism in this research. Static retention assays were conducted by placing triplicates of each sample in a suspension of bacteria in phosphate buffered saline (PBS) with starting optical density, $OD_{600}nm = 0.3$ for 2 hours at 37 °C. Five most predominant modes of interactions between the bacteria and parallel nanofibers were identified. Scanning electron microscope imaging was utilized to quantify line density (number of bacteria/fiber length) of bacteria attachment for all five possible cases. Comparing results brings this fact in focus that for surfaces with polystyrene nanofibrous features; more hydrophobic substrate promotes the modes in which bacteria prefer unaligned condition. Therefore, we conclude that favorable conditions for bacteria attachment are dependent upon two parameters: topography and hydrophobicity. It has been demonstrated that bacterial cell do not always maximize their contact area with substrate and that surface energy affect their preferred mode of interaction and attachment density. We plan to use these findings towards identifying the optimum nanofiber dimension and spacing for minimizing bacterial adhesion and biofilm formation. Finding from this work can also be used towards developing more effective nanofibrous tissue scaffolds which need to promote adhesion of mammalian cells and minimize bacterial adhesion.

See more of this Session: Poster Session: Bioengineering
See more of this Group/Topical: Food, Pharmaceutical & Bioengineering Division

(623ao) Engineering S. Cerevisiae for Production of Medium Chain Hydrocarbons

Susanna Su Jan Leong, Pei Yu Lim, Hazarki Yaohari, Hongxin Zhao, Bin Bin Chen, Hua Ling and Matthew Wook Chang, School of Chemical and Biomedical Engineering, Nanyang Technological University, Singaproe, Singapore

Engineering *S. cerevisiae* for production of medium chain hydrocarbons

Pei Yu Lim, Hazarki Yaohari, Hongxin Zhao, Bin Bin Chen, Hua Ling, Matthew Wook Chang, Susanna Su Jan Leong

School of Chemical and Biomedical Engineering, Nanyang Technological University,

62 Nanyang Drive, Singapore 637459, Singapore

Biofuel production in microbial platforms using renewable sugar feedstocks is a promising strategy that can potentially address shortage of non-renewable petroleum fuels, which has become a global problem today. The advancement of genetic and metabolic engineering tools has opened the way for rational engineering of microbes to produce a wide range of target products including fuel molecules. In this study, we report the engineering of *S. cerevisiae* for the production of saturated hydrocarbons, which are essential components of petroleum fuels. We constructed a pathway in *S. cerevisiae* that demonstrated efficient conversion of medium chain fatty acids into alkanes *via* two steps (i.e. reduction followed by decarbonylation) with uncompromised cell viability. Protein engineering efforts are also underway to further improve conversion yield to attain economic viable titers of the end product.

See more of this Session: Poster Session: Bioengineering
See more of this Group/Topical: Food, Pharmaceutical & Bioengineering Division

(623ap) Collagen Binding Domain Fusion Proteins for Therapeutics: Fermentation and Isolation

Hannah Moreland[1], McKinzie Fruchtl[2], Robert R. Beitle[2] and Joshua Sakon[3], (1)Department of Biological and Agricultural Engineering, University of Arkansas, Fayetteville, AR, (2)Ralph E. Martin Department of Chemical Engineering, University of Arkansas, Fayetteville, AR, (3)Department of Chemistry and Biochemistry, University of Arkansas, Fayetteville, AR

Interest in the production of recombinant proteins consisting of collagen binding domain (CBD) fused to a bioactive material has increased due to the targeting / attachment capabilities of CBD. For example, CBD fusions can be applied to reversing of bone density loss and repair of eardrum, specifically, by choosing an appropriate fusion partner (parathyroid hormone or epidermal growth factor). The production of CBD fusions was examined using batch and fed batch culturing of *Escherichia coli* to express the fusion proteins, and affinity chromatography to isolate the final product. Different medium formulations and induction strategies were tested in order to develop a production strategy lacking yeast extract or other difficult-to-validate materials, and various methods of recovery were verified to be effective. Data indicated that high productivity of these novel CBD fusion proteins can be achieved for use during cancer treatment and tympanoplasty, respectively.

See more of this Session: Poster Session: Bioengineering
See more of this Group/Topical: Food, Pharmaceutical & Bioengineering Division

(623aq) Effect of Surface Functional Groups On Interactions of Engineered Nanomaterials with a Lipid Bilayer

Alexander Negoda, Chemical Engineering and Material Science, Michigan State University, East Lansing, MI, Gregory Baker, Chemistry, Michigan State University, East Lansing, MI and Robert M. Worden, Chemical Engineering & Materials Science, Michigan State University, East Lansing, MI

Engineered nanomaterials (ENM) are being introduced into a wide range of commercial products. Recent studies have shown that ENM can induce health risks. ENM could induce toxicity in a variety of ways, several of which involve molecular interactions with cell membranes, including creating pores in the membrane and entering cells or organelles through the membrane. However, the mechanisms underlying ENM's molecular interactions with cell membranes are poorly understood. This paper describes electrophysiological studies to explore the effect of surface functional groups on interactions between polymeric nanoparticles and synthetic BLM that mimic cell membranes. Polypropargyl glycolide (PGL) nanoparticles decorated with different surface functional groups (PEG, OH, COOH) were fabricated. The sizes and aggregation rates of the ENM were then determined using light scattering. Synthetic diphytanoyl phosphocholine BLM were then challenged with the ENM, and resulting current spikes and integral conductance were measured for each particle type. PGL nanoparticles having PEG and OH surface groups resulted in current spikes, while nanoparticles having COOH groups generated both current spikes and integral conductance. Frequency of current spikes having a relatively high conductance (>100 pS), as well as total charge transferred through the BLM, increased in the order PEG<OH<COOH. Ion selectivity of the ENM-generated conductance was investigated by measuring transmembrane currents at various applied potentials in the presence of a 10:1 ion gradient, before and after reversing the applied potential. The results showed no evidence that the conductance triggered by the surface-functionalized polymeric ENM nanoparticles was ion selective.

See more of this Session: Poster Session: Bioengineering
See more of this Group/Topical: Food, Pharmaceutical & Bioengineering Division

EFFECT OF THE ELECTROSTATIC POTENTIAL ON THE INTERNALIZATION MECHANISM OF CELL PENETRATING PEPTIDES

*Karen A. Flores[1], J. Cristian Salgado[*1,2], Gerald Zapata-Torres[3,2], Ziomara P. Gerdtzen[4,2], María-Julieta Gonzalez[5], Marcela A. Hermoso[*6]*

[1] Laboratory of Process Modeling and Distributed Computing, Department of Chemical Engineering and Biotechnology, University of Chile, Santiago, Chile. [2] Millennium Institute for Cell Dynamics and Biotechnology: a Centre for Systems Biology, University of Chile, Santiago, Chile. [3] Department of Inorganic and Analytical Chemistry, Faculty of Chemical and Pharmaceutical Sciences, University of Chile, Santiago, Chile. [4] Centre for Biochemical Engineering and Biotechnology, Department of Chemical Engineering and Biotechnology, University of Chile, Santiago, Chile. [5] Laboratory of Cell Biology, Cell and Molecular Program, Faculty of Medicine, University of Chile, Santiago, Chile. [6] Laboratory of Innate Immunology, Institute of Biomedical Sciences, Faculty of Medicine, University of Chile, Santiago, Chile. [*]Corresponding authors: Both authors share senior authorship.

Background

Cell Penetrating Peptides (CPPs) can internalize and transport molecules that are unable to cross the plasma membrane by themselves, such as drugs [1], peptides [2], proteins [3], peptide nucleic acids [4] and nanoparticles [5]. The increased therapeutic interest on these peptides focuses in their ability to facilitate the internalization of biologically active molecules [6]. The understanding of the internalization mechanisms used by these peptides contributes to the design of new CPPs with improved characteristics.

There is evidence for several routes of entry used by CPPs, both dependent and independent of the endosomal pathway [7]. Due to the acidification of the endosome upon endocytosis and its potential for chemical alteration of the transported drug it is desirable that CPPs enter directly to the cytoplasm. The transduction mechanism is based on the formation of nucleation zones in restricted areas of the cell membrane, leading to a rapid internalization of CPPs. This mechanism has been described for the internalization of oligo-arginine R9 and TAT peptides in several cell types [8].

Proteins in the Toll-like receptor TLR4 signaling pathway are potential targets for therapeutic interventions, where blocking peptides could act as antagonists modulating harmful or chronic inflammatory responses. TLRs share a cytoplasmic domain or TIR (TLR receptor domain and Interleukin 1, IL-1R), and their activation induces the recruitment of adapter proteins such as TIRAP. A peptide containing the BB loop of TIRAP (highly conserved among TIR proteins) could act as a specific inhibitor of TIRAP, competing for the same interaction site of TLR4 [9]. This motivates the design of new blocking peptides for the TLR signaling pathway, capable of crossing the plasmatic membrane without the need of CPPs.

It has been reported that transduction is the internalization mechanism used by the arginine oligopeptide R9 and the arginine rich TAT peptide (pepTAT) [8]. The TIRAP peptide (pepTIRAP), which also has a high content of cationic clusters of arginine and lysine residues, internalizes directly into the cytoplasm without altering its structure [10]. This suggests that the presence of these cationic clusters is important for the internalization of these peptides.

The aim of this work is to contribute to the understanding of the internalization mechanism used by pepTIRAP. To achieve this, three-dimensional models of pepTIRAP, pepTAT and $pepTIRAP^{ALA}$ were developed in order to study their structural and electrostatic potential features. These properties were correlated with experimental internalization results for pepTIRAP and $pepTIRAP^{ALA}$ using a novel mathematical model in order to quantify and explain the internalization of these peptides.

Methodology

Molecular Modeling

An exhaustive search of similar sequences for pepTIRAP, $pepTIRAP^{ALA}$ and pepTAT and their alignment was carried out using (1) FUGUE (Version 2.0) [11], which uses the HOMSTRAD database and (2) the Protein Data Bank database (PDB) [12]. The structures were built using MODELLER, satisfying the spatial constraints imposed by templates [13]. The VADAR online server [14] was used to detect possible structural errors, and errors were corrected with MODELLER. The final validation of the structures was performed using VADAR and MolProbity servers [15].

The online server PDB2PQR was used to preprocess the coordinate files generated by MODELLER considering a PARSE force field and pH 7 [16]. The calculation of the electrostatic potential field (EPF) for each peptide was performed using the APBS software [17].

Experimental methods

For all culture experiments, HeLa cells were seeded in 48-well plates and incubated at 37 °C, 5 % CO_2 in DMEM containing 10% FCS. penicillin and streptomycin. Cell viability was determined by staining with propidium iodide (PI), an indicator of late apoptosis, using flow cytometry. The threshold of autofluorescence was set to 0.7 %.

Biotinylated pepTIRAP and $pepTIRAP^{ALA}$ used in this study were provided by Dr. Walter Low [10]. To asess viability upon treatment cultures treated either with PBS or with Staurosporine 0-2 µM , were carried out. To study the effect of peptide concentration on viability, cell were exposed to pepTIRAP and $pepTIRAP^{ALA}$ in PBS 0-100 µM for 16 hours and analyzed by flow cytometry.

To evaluate peptide internalization cells were incubated with peptide at 0 and 40 µM for 2 hours. PepTIRAP and $pepTIRAP^{ALA}$ internalization was quantified by flow cytometry using streptavidin conjugated to r-phycoerythrin.

A mathematical model was also proposed and used to discriminate the fluorescence of streptavidin-R-PE associated to: biotinylated peptide internalized and located on the cell membrane, specific binding to biotinylated cytoplasmatic compounds, and non specific associations to the extracellular matrix.

Results

The structural models obtained have high structural similarities among them with a predominantly α-helical secondary structure that involves more than 69\% of the amino acids. Regarding the electrostatic surface potential of pepTAT and pepTIRAP, a predominantly positive EPF is found on the surface of both peptides. The lack of cationic clusters in $pepTIRAP^{ALA}$ produces a more balanced distribution of its EPF compared to pepTAT and pepTIRAP. The main effect of the replacement of cationic clusters is the alteration on the characteristics of the peptides electrostatic potential. We postulate that due to its predominantly positive EPF, pepTIRAP interacts with molecules on the cell membrane, facilitating its internalization. The characteristics of the EPF in $pepTIRAP^{ALA}$ result in weaker interactions with the cell membrane that decrease its internalization.

For HeLa cells after incubation with pepTIRAP and $pepTIRAP^{ALA}$, an increase in propidium iodide (PI) fluorescence (cell death) is observed as pepTIRAP concentration is increased. This change is more abrupt over 40 μM. This is not the case for $pepTIRAP^{ALA}$ where no changes in fluorescence are observed. The increment of cell death observed at 40 μM of pepTIRAP, is related to a change in the internalization mechanism used.

To determine and quantify peptide internalization, pepTIRAP and $pepTIRAP^{ALA}$ were labeled with biotin, which specifically binds to streptavidin-R-PE (STR-R-PE). The degree of internalization is determined by flow cytometry. The total fluorescence measured corresponds to the sum of non-specific interactions of STR-R-PE to non-biotinylated compounds on the cell membrane, specific binding to biotinylated peptides located outside the cell, to internalized biotinylated peptides, and other endogenously biotinylated proteins. Non-specific associations of STR-R-PE to membrane bound compounds were determined experimentally. Experimental fluorescence measurements after incubation with pepTIRAP 40 μM are shown in Figure 1 A and B. Figure 1 A shows results for cells without fixation-permeabilization treatment (NFP), where STR-R-PE is unable to enter the cells and only binds to peptides associated to the extracellular membrane and the extracellular matrix. Experimental results for cell with fixation-permeabilization treatment (FP) are shown in Figure 1 B. STR-R-PE can enter the cells and bind specifically to biotinylated peptides and endogenously biotinylated compounds. An increase in fluorescence is observed with increasing concentrations of STR-R-PE as it binds to pepTIRAP and other extracellular compounds. There is an heterogeneous fluorescence distribution for events above the autofluorescence threshold, suggesting a diverse amount of peptide located on the cell membrane among the population. The increase in fluorescence with STR-R-PE in Figure 1 B is higher than the one in Figure 1 A, since STR-R-PE is detecting in addition to extracellular interactions, intracellular biotinylated compounds. Heterogeneity in fluorescence distribution for events above the autofluorescence threshold is similar as of non-permeabilized cells. Fluorescence distribution heterogeneity can be attributed to differences in the degree of pepTIRAP internalization. Figure 1 C and D show experimental fluorescence measurements after incubation with pepTIRAP^{ALA}. The increase in fluorescence with increasing concentrations of STR-R-PE for non-permeabilized cells, is also observed as for the case of pepTIRAP (Figure 1 C). However, the lower percentage of the population above the autofluorescence threshold and the narrower distribution range indicates a weaker interaction of pepTIRAP^{ALA} with the cellular membrane compared to pepTIRAP. Experimental results for permeabilized cells incubated with p pepTIRAP^{ALA} in Figure 1 D show lower increase in the percentage of events above the autofluorescence threshold compared to pepTIRAP, indicating a lower internalization of pepTIRAP^{ALA} compared to pepTIRAP.

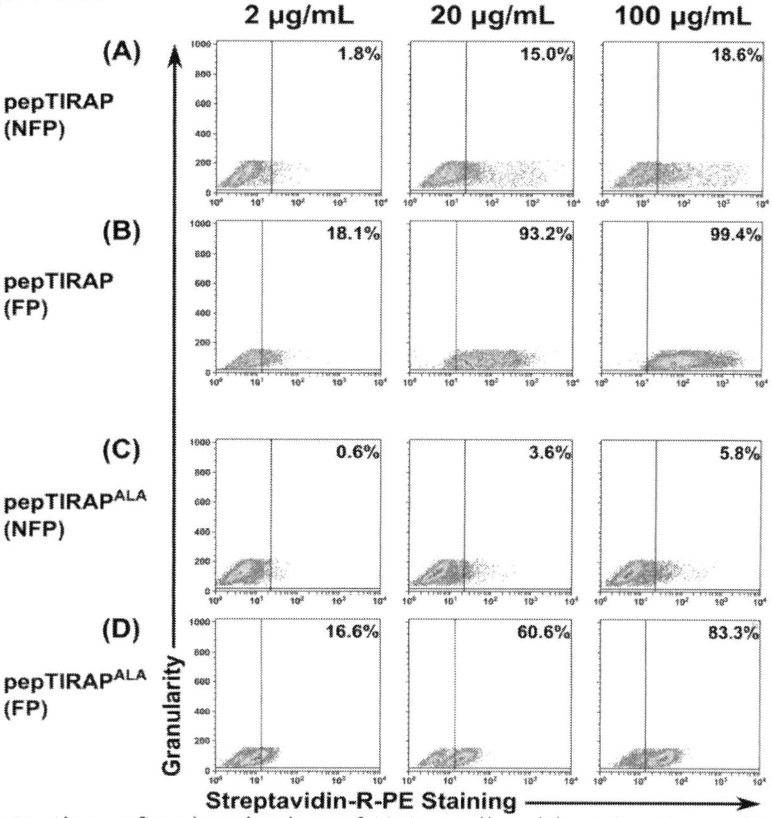

Figure 1: Peptide detection, after incubation of HeLa cells with 40 µM pepTIRAP and pepTIRAP^{ALA} for 2 hours. Granularity vs. Streptavidin-R-PE staining dot plots for Streptavidin-R-PE concentrations of 2, 20 and 100 µg/mL. Incubation of non-permeabilized (A) and permeabilized cells (B) with pepTIRAP; Incubation of non-permeabilized (C) and permeabilized cells (D) with pepTIRAP^{ALA}. The vertical line marks the autofluorescence threshold for viable cells.

In order to analyze the mean fluorescence intensity (MFI) and the contribution of each of its components to the total measured fluorescence, a mathematical model was proposed and used to quantify fluorescence contributions due to internalization, specific and non-specific interactions from flow cytometry measurements (data not shown). The model considers both specific and non-specific interactions for STR-R-PE and allows the quantification of biotinylated peptides located inside and outside the cell. The total fluorescence measured is the result of specific (MFI_S) and non-specific (MFI_{NS}) interactions. To build the model we consider non-specific associations to be directly proportional to STR-R-PE concentration, and that specific associations follow a saturation curve.

The maximum fluorescence calculated for pepTIRAP located outside the plasma membrane is almost 40-fold higher than that for pepTIRAP^{ALA}. In the case of internalized peptide, this difference is even higher. Internalized pepTIRAP is more than a hundred times higher than internalized pepTIRAP^{ALA}. According to these results, pepTIRAP is highly internalized in HeLa cells after 2 hours of incubation. In contrast pepTIRAP^{ALA} binds poorly and scarcely crosses the plasma membrane. The maximum fluorescence associated to internalized pepTIRAP is 3-fold the fluorescence associated to pepTIRAP located on the membrane. In spite the fact that pepTIRAP is highly internalized, the amount of peptide bound to the membrane is considerable, suggesting that pepTIRAP can establish strong interactions with the cell membrane.

Conclusions

Molecular modeling studies showed that the replacement of cationic amino acids of a peptide derived from the TIRAP protein (pepTIRAP) with alanine (pepTIRAP^{ALA}), resulted in a decrease in the positive electrostatic potential on the peptide, even when its three dimensional structure remains mostly unaltered. Cell viability is reduced as pepTIRAP concentration increases, which is not observed for pepTIRAP^{ALA}. Flow cytometry results show that pepTIRAP enters the cell, while pepTIRAP^{ALA} is not considerably internalized. In addition, at the same concentration pepTIRAP binds to the membrane to a greater extent than pepTIRAP^{ALA}.

A mathematical model was developed for quantifying the amount of peptide internalized and the amount bound to the membrane from flow cytometry measurements. Results show that the amount of pepTIRAP internalized is in the same order of magnitude than the amount found bound to the membrane, and it is significantly higher than the amount of pepTIRAP^{ALA} that enters the cell.

We believe that the difference in internalization between pepTIRAP and pepTIRAP^{ALA} is due mainly to differences in their electrostatic potential, suggesting that a mechanism dependent on the electrostatic interactions between the peptide and molecules on the membrane triggers peptide internalization.

Acknowledgements

This work was partially supported by the Millennium Scientific Initiative ICM project P05-001-F, FONDECYT Research Initiation Grants 11080016 and 11090268, FONDECYT Grants 1080006 and 1080290.

References

1. Nori, A., K. D. Jensen, M. Tijerina, P. Kopeckova, and J. Kopecek (2003) "Tat-conjugated synthetic macromolecules facilitate cytoplasmic drug delivery to human ovarian carcinoma cells". Bioconjug. Chem. 14: 44–50.

2. Choi, M., S. Rolle, M.Wellner, M. C.Cardoso, C. Scheidereit, F. C. Luft, and R. Kettritz (2003) "Inhibition of NF-kB by a TAT-NEMO-binding domain peptide accelerates constitutive apoptosis and abrogatesLPS-delayed neutrophil apoptosis". Blood 102: 2259–2267.

3. Fawell, S., J. Seery, Y. Daikh, C. Moore, L. L. Chen, B. Pepinsky, and J. Barsoum (1994) "Tat-mediated delivery of heterologous proteins into cells". Proc. Natl. Acad. Sci. U. S. A. 91: 664–668.
4. Turner, J. J., G. D. Ivanova, B. Verbeure, D. Williams, A. A. Arzumanov, S. Abes, B. Lebleu, and M. J. Gait (2005) "Cell-penetrating peptide conjugates of peptide nucleic acids (PNA) as inhibitors of HIV-1 Tat-dependent trans-activation in cells". Nucleic Acids Res. 33: 6837–6849.
5. Lewin, M., N. Carlesso, C. H. Tung, X. W. Tang, D. Cory, D. T. Scadden, and R. Weissleder (2000) "Tat peptide-derivatized magnetic nanoparticles allow in vivo tracking and recovery of progenitor cells". Nat. Biotechnol. 18: 410–414.
6. Yoshikawa, T., T. Sugita, Y. Mukai, Y. Abe, S. Nakagawa, H. Kamada, S. Tsunoda, and Y. Tsutsumi (2009) "The augmentation of intracellular delivery of peptide therapeutics by artificial protein transduction domains". Biomaterials 30: 3318–3323.
7. Deshayes, S., M. C. Morris, G. Divita, and F. Heitz (2005) "Cell-penetrating peptides: tools for intracellular delivery of therapeutics". Cell. Mol. Life Sci. 62: 1839–1849.
8. Duchardt, F., M. Fotin-Mleczek, H. Schwarz, R. Fischer, and R. Brock (2007) "A comprehensive model for the cellular uptake of cationic cell-penetrating peptides". Traffic 8: 848–866.
9. Horng, T., G. M. Barton, and R. Medzhitov (2001) "TIRAP: an adapter molecule in the Toll signaling pathway". Nat. Immunol. 2: 835–841.
10. Low, W., A. Mortlock, L. Petrovska, T. Dottorini, G. Dougan, and A. Crisanti (2007) "Functional cell permeable motifs within medically relevant proteins". J. Biotechnol. 129: 555–564.
11. Shi, J., T. L. Blundell, and K. Mizuguchi (2001) "FUGUE: sequence-structure homology recognition using environment-specific substitution tables and structure-dependent gap penalties". J. Mol. Biol. 310: 243–257.
12. Bernstein, F. C., T. F. Koetzle, G. J. Williams, J. Meyer, E. F., M. D. Brice, J. R. Rodgers, O. Kennard, T. Shimanouchi, and M. Tasumi (1977) "The Protein Data Bank: a computer-based archival file for macromolecular structures". J. Mol. Biol. 112: 535–542.
13. Eswar, N., B. Webb, M. A. Marti-Renom, M. Madhusudhan, D. Eramian, M.-y. Shen, U. Pieper, and A. Sali (2002) "Current Protocols in Bioinformatics: Comparative Protein Structure Modeling Using Modeller". John Wiley & Sons, Inc.
14. Willard, L., A. Ranjan, H. Zhang, H. Monzavi, R. F. Boyko, B. D. Sykes, and D. S. Wishart (2003) "VADAR: a web server for quantitative evaluation of protein structure quality". Nucleic Acids Res. 31: 3316–3319.
15. Davis, I. W., A. Leaver-Fay, V. B. Chen, J. N. Block, G. J. Kapral, X. Wang, L. W. Murray, r. Arendall, W. B., J. Snoeyink, J. S. Richardson, and D. C. Richardson (2007) "MolProbity: all-atom contacts and structure validation for proteins and nucleic acids". Nucleic Acids Res. 35: W375–383.
16. Dolinsky, T. J., J. E. Nielsen, J. A. McCammon, and N. A. Baker (2004) "PDB2PQR: an automated pipeline for the setup of Poisson-Boltzmann electrostatics calculations". Nucleic Acids Res. 32: W665–W667.
17. Baker, N. A., D. Sept, S. Joseph, M. J. Holst, and J. A. McCammon (2001) "Electrostatics of nanosystems: application to microtubules and the ribosome". Proc. Natl. Acad. Sci. U. S. A. 98: 10037–10041.

(623as) Enhancement of Activity of Cross-Linked Enzyme Aggregates by a Novel Sugar-Assisted Precipitation Strategy

Mengfan Wang[1], **Wei Qi**[2], Rongxin Su[1] and Zhimin He[2], (1)Chemical Engineering Research Center, School of Chemical Engineering and Technology, Tianjin University, Tianjin, China, (2)State Key Laboratory of Chemical Engineering,Chemical Engineering Research Center, School of Chemical Engineering and Technology, Tianjin University, Tianjin, China

Enhancement of activity of cross-linked enzyme aggregates by anovel sugar-assisted precipitation strategy

Mengfan Wang[2],Wei Qi[1], Rongxin Su[2]and Zhimin He[1],

1) State Key Laboratory of Chemical Engineering, School ofChemical Engineering and Technology, Tianjin University, Tianjin, China. 2)Chemical Engineering Research Center, School of Chemical Engineering andTechnology, Tianjin University, Tianjin, China.

*CorrespondingAuthor's E-mail: qiwei@tju.edu.cn

Using enzymes as biocatalysts in chemicalreactions has numerous advantages including selectivity, specificity, mildreaction conditions, stereoselectivity with pure enantiomers as products, andenvironmental friendly. However, the enzymes are sensitive and fragile totemperature, system pH and reaction medium. Recently, cross-linked enzymeaggregates (CLEAs) were developed as an efficient approach to obtainimmobilized enzymes without the use of any pre-existing carriers [Sheldon, 2005]. CLEAs arestable over a wide pH and temperature ranges, tolerant to organic medium,reusable, and cost-effective. The preparation of CLEAs includes two steps:precipitation (usually using water-miscible organic solvents) andcross-linking. However, the water-miscible organic solvents generally inducethe denaturation of enzyme [Bellezza,2009], which leads to the activityloss of enzyme and finally lowers the activity yield of resulting CLEAs.

It is well known that sugar is an idealstabilizer of various proteins in natural dehydrated environments as well asduring lyophilization process. Therefore, in this work, glucose, sucrose, ortrehalose was firstly introduced in the precipitation step to prepare thesugar-assisted CLEAs. For sugar-assisted CLEAs of penicillin G acylase (PGA),79.3% of the activity yield was obtained by adding 20% trehalose and usingacetone as the precipitant, which was 30% activity higher than that of CLEAsprepared without sugar. With the increasing of trehalose concentration, thefinal activity of trehalose-assisted CLEAs was increased and the aggregatessize became larger according to scanning electron microscope (SEM) images.Confocal laser scanning microscopy (CLSM) and fourier transform infraredspectroscopy (FTIR) studies showed that the polar microenvironment and thesecondary structure of native PGA were preserved to some extent when CLEAs wereprepared with this novel sugar-assisted strategy (as **Figure 1 A-C** showed) .The kinetic studies also indicated the enhanced affinity to substrate, higherresist-inhibition capacity and catalysis efficiency of sugar-assisted CLEAs.Moreover, sugar-assisted CLEAs were significantly more stable when incubated athigh temperature for a long time. The strategy may provide a feasible andefficient solution to improve the properties of CLEAs so as to broaden theirapplication in biocatalysis and biotransformation.

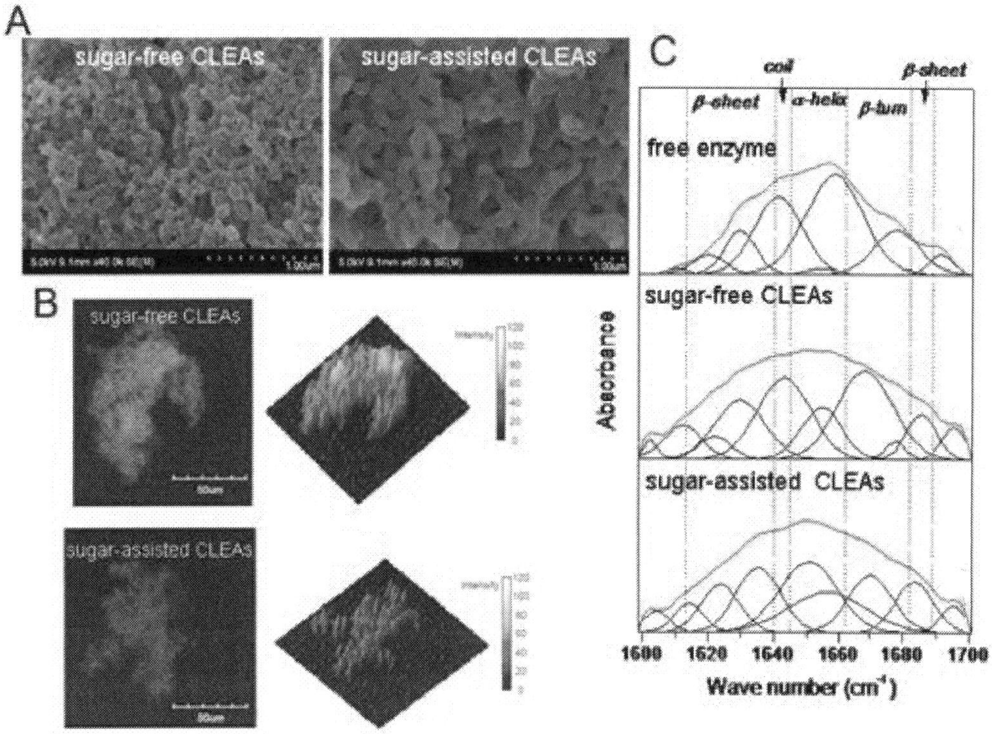

Figure 1(A) SEM, (B) CLSM, (C) FTIR of sugar-free and sugar-assisted CLEAs.

Acknowledgement

The authors acknowledge the financial supports receivedfrom the Program for New Century Excellent Talents in Chinese University(NCET-08-0386), the Key Project of Chinese Ministry of Education (108031), theNatural Science Foundation of China (20976125, 20806057, 31071509) and Tianjin(10JCYBJC05100), and the Program of Introducing Talents of Discipline toUniversities of China (B06006).

References

(1) Sheldon R.A.,Schoevaart R. and Van Langen L.M. Cross-linked enzyme aggregates (CLEAs): Anovel and versatile method for enzyme immobilization (a review), Biocatalysisand Biotransformation, 23, 141-147, 2005.

(2) Bellezza F.,Cipiciani A., Cinelli S. and Onori G. Influence of alcohols and osmolytes onthermal stability and catalytic activity of myoglobin: Co-solvent clusteringeffects, Chemical Physics Letters, 482, 139-142, 2009.

See more of this Session: Poster Session: Bioengineering
See more of this Group/Topical: Food, Pharmaceutical & Bioengineering Division

(623at) Inhibition and Deconstruction of Insulin Amyloid Fibrillation

Rui Liu, **Rongxin Su**, Wei Qi and Zhimin He, School of Chemical Engineering and Technology, Tianjin University, Tianjin, China

Amyloid fibrils are associated with the pathologies of more than 20 human diseases, including Alzheimer's, type II diabetes, Parkinson's and Huntingtons's [Dobson, 2003; Murphy and Kendrick, 2007]. Moreover, the ability of proteins to form amyloid structures poses problems in its production, storage and delivery[Manning et al., 2010]. Especially for protein drugs like insulin, it is extremely dangerous to undergo amyloid fibrillation. Therefore, it is critical to find new strategies for destroying amyloid fibrils and developing potential inhibitors.

This paper will present our work on the inhibition and deconstruction of insulin amyloid fibrils. we firstly used NH_3-H_2O_2 two-step method to monitor the cleavage and regeneration of disulfide bonds between insulin A-chain and B-chain, and investigated the inhibition effects on insulin aggregation by small polyphenolic molecules and laser irradiation. The main ideas and conclusions are summarized as follows.

(1) The NH_3-H_2O_2 two-step method was investigated to destroy insulin amyloid fibrils and help its conformations back to the native state. ThT fluorescence results showed that insulin fibrils were completely soluble in 25% $NH_3 \cdot H_2O$ after 2 h incubation. The secondary structure contents (α-helix, β-sheet, random coil) of insulin monomers and $NH_3 \cdot H_2O$-treated insulin fibrils predicted by K2D method [Whitmore and Wallace, 2008] were 0.80, 0.01, 0.19 and 0.79, 0.00, 0.21, respectively, indicating conformational similarities between them. The HPLC peak of insulin monomer moves to lower retention time with the increase of $NH_3 \cdot H_2O$ concentrations. Furthermore, the disulfide bonds between insulin A-chain and B-chain were regenerated by hydrogen peroxide oxidation of sulfydryls.

(2) The polyphenol was studied to inhibit insulin amyloid fibrillation. In the absence of 0.1 M NaCl, the inhibition effects of polyphenols were concentration-dependent. ThT fluorescence results showed that there was no significant decrease in insulin fibrils by adding 50 mM and 100 mM resveratrol and trehalose, while a significant inhibition effect on insulin aggregates was shown with the lag time ~61.9 h by adding low concentration procyanidine (10 mM). The results indicated that procyanidine could provent insulin amyloid fibrillation by inhibiting nucleation. In the presence of 0.1 M NaCl, the kinetics of insulin amyloid fibrillation incubated was similar to that without 0.1 M NaCl, but have shorter lag time. In sum, the inhibition ability of procyanidine, resveratrol and trehalose decreased in turn. They may affect the process of insulin fibrillation by inhibiting insulin nucleation, preventing growth of protofilament and with no significant inhibition effects, respectively.

(3) A fibrillation inhibition experiment using laser beam irradiation is introduced. During heat-induced aggregation, the size distribution of two insulin solutions obtained by online and offline dynamic light scattering were different. The laser-on insulin in the presence of 0.1 M NaCl exhibited fewer fibrils than the laser-off insulin, whereas no insulin fibril under laser irradiation was observed in the absence of 0.1 M NaCl for 45 h incubation. According to the CD results, the laser-irradiated insulin solution maintained mainly an α-helical conformation, but the laser-off insulin solution formed bulk fibrils followed by a significant increase in β-sheet content for 106 h incubation. These findings provide an inhibition method for insulin amyloid fibrillation using the laser irradiation and demonstrate that the online long-time laser measurements should be carefully used in the study of amyloid proteins because they may change the original results.

References

1. Dobson C. M. Protein folding and misfolding, *Nature*, 426, 884-890, 2003
2. Murphy R. M. and Kendrick B. S. Protein misfolding and aggregation, *Biotechnology Progress*, 23, 548-552, 2007
3. Manning M. C., Chou D. K., Murphy B. M., Payne R. W. and Katayama D. S. Stability of protein pharmaceuticals: an update, *Pharmaceutical Research*, 27, 544-575, 2010
4. Whitmore L. and Wallace B. A. Protein secondary structure analyses from circular dichroism spectroscopy: Methods and reference databases, *Biopolymers*, 89, 392-400, 2008

See more of this Session: Poster Session: Bioengineering
See more of this Group/Topical: Food, Pharmaceutical & Bioengineering Division

(623au) Identification of Physically Unlinked Determinants of Complex Phenotypes

Yuqi Guo, Maria Priscila Almario and Katy Kao, Chemical Engineering, Texas A&M University, College Station, TX

Identification of physically unlinked determinants of complex phenotypes

The use of microbial systems for the production of valued compounds and fuels provide a potentially greener and sustainable route for chemical production. However, one of the challenges in microbial biofuel and biorenewables production is the low tolerance of the microbial hosts to toxic compounds in the feedstock and products. Unfortunately, the engineering of microorganisms for increased robustness is currently restricted by the lack of knowledge of the molecular mechanisms involved. Here, we focus on addressing this challenge using combinations of deletion and overexpression studies in *E. coli* to identify the genetic determinants involved in butanol tolerance.

See more of this Session: Poster Session: Bioengineering
See more of this Group/Topical: Food, Pharmaceutical & Bioengineering Division

(623av) Effects of Inoculation Methods On Fumaric Acid Fermentation by Rhizopus Oryzae

Kun Zhang, Chemical and Biomolecular Engineering, The Ohio State University, Columbus, OH and Shang-Tian Yang, Department of Chemical and Biomolecular Engineering, Ohio State University, Columbus, OH

The production of fumaric acid from renewable resources by fungal fermentation has received much attention due to its extensive applications in polymer, medicine and food industries and the rapidly increased production cost of petroleum-based chemicals. However, current fungal fermentation process suffers from low product yield and productivity mainly due to difficulties in controlling cell morphology and mass transfer during fermentation in conventional stirred-tank bioreactors.

The production process of fumaric acid by *Rhizopus oryzae* involves two steps: seed culture and acid production. By using soybean meal hydrolysate as nitrogen source, cell morphology of dispersed tiny mycelia clumps was obtained in seed culture phase with enhanced fumaric acid production. However, during the inoculation process, the mycelia clumps concentration in the inoculums varies from batch to batch, causing inconsistent fermentation performance. Also, different inoculation methods, such as the inclusion of residual seed culture broth (RSCB) in the inoculums, can significantly affect the final results. Therefore, in this study, with the purpose of finding the best inoculation method for fumaric acid production, the effects of different inoculation methods and some key variables such as inoculation size were investigated. Comparing to the method in which RSCB was removed from the inoculums, the inclusion of certain amount of RSCB was beneficial for the fermentation. The relationships of the cell concentration and RSCB content in the inoculums with the fumaric acid production were also studied. Since the properties of RSCB such as the residual nitrogen content was directly related to the seed age. The effects of seed age on the fumaric acid fermentation were studied. In addition, a low pH around 2 in the final seed culture broth was observed. Since $CaCO_3$ is commonly used as the neutralizing agent in the fermentation medium to maintain pH ~5, NaOH was first added to adjust the pH close to 5 in order to lower the bubble production from the reactions between $CaCO_3$ and acids. However, fermentation results showed a negative effect of such pH adjustment and that direct addition of $CaCO_3$ with the sudden increase of dissolved $CO\text{-}_2$ concentration in the medium was favorable to the fumaric acid fermentation. By studying the effects of these inoculation methods and some key variables, it is expected to achieve a higher fumaric acid production with consistent fermentation performance.

See more of this Session: Poster Session: Bioengineering
See more of this Group/Topical: Food, Pharmaceutical & Bioengineering Division

(623aw) Structural Elucidation of Unknown Alkaloids Present In Catharanthus Roseus by Precursor Ion Fingerprinting

Carlos Daniel Flores Espronceda, Chemical and Biological Engineering, Iowa State University, Ames, IA, Guy W. Sander, Chemical Engineering, University of Minnesota Duluth, Duluth, MN, M. Ann Perera, W.M. Keck Metabolomics Research Laboratory, Iowa State University, Ames, IA and Jacqueline V. Shanks, NSF Engineering Research Center for Biorenewable Chemicals (CBiRC), Iowa State University, Ames, IA

Alkaloids are a group of pharmacologically active compounds present in the secondary metabolism of different organisms. Two examples are vincristine and vinblastine, two alkaloids extracted from *C. roseus* and extensively used in cancer therapy. These molecules are characterized by having a complex structure with several chiral centers; this renders their synthetic production impossible. Since secondary metabolites in plants tend to be present in a relatively low concentration, research has been performed on the use of metabolic engineering to increase the yield of these compounds. One limitation of this technology is the lack of complete knowledge of the metabolic pathway that leads to the biosynthesis of these alkaloids. In order to elucidate the pathway, it is necessary to identify the enzymes and intermediates involved in the series of reactions. The focus of this work is to reveal and confirm the structure of some intermediates present downstream of Tabersonine, a precursor for the alkaloids of interests. The compounds were purified using High Performance Liquid Chromatography and their fragment ions were characterized using Liquid Chromatography/Mass Spectrometry and Fourier Transform Ion Cyclotron Resonance. Among those intermediates verifieded are 16-hydroxytabersonine and 6,7-dehydroechitovenine.

See more of this Session: Poster Session: Bioengineering
See more of this Group/Topical: Food, Pharmaceutical & Bioengineering Division

(623ax) Design and Use of Oligonucleotide Microarray Probes for Detection of Food-Borne Pathogenic Salmonella Enterica Serotypes

Hwa Hui Shin, Byeong Hee Hwang and Hyung Joon Cha, Department of Chemical Engineering, Pohang University of Science and Technology, Pohang, South Korea

Efficient detection of food-borne pathogenic bacteria is very important to reduce danger of severe communicable diseases. Thus, there have been many trials to exploit fast, reliable, and sensitive detection and diagnosis system. We have also developed the oligonucleotide microarray system for 11 pathogenic bacteria including *Salmonella* species based on the detection principle of 16S rRNA information. It was quite powerful but we found some difficulty in distinguish very closed bacterial species such as *Salmonella enterica* serotypes Enteritidis, Choleraesuis, and Typhimurium. Salmonellosis should be prevented because *Salmonella* is one of frequently spread notorious communicable pathogens. In the current work, we developed a new microarray system to distinguish *Salmonella enterica* serotypes.

See more of this Session: Poster Session: Bioengineering
See more of this Group/Topical: Food, Pharmaceutical & Bioengineering Division

(623ay) Development of Brushite Bone Cement Using Recombinant Mussel Adhesive Protein

So Yeong Bahn, School of Interdisciplinary Bioscience and Bioengineering, Pohang University of Science and Technology, Pohang, South Korea, Yoo Seong Choi, Department of Chemical Engineering, Chungnam National University, Daejeon, South Korea and Hyung Joon Cha, Department of Chemical Engineering, Pohang University of Science and Technology, Pohang, South Korea

Increased cases of skeletal defects have led requirements of bone graft. As an alternative to autograft and allograft, a wide range of materials are considered in replacement of local bone loss. Calcium phosphate materials, which have similar composition to bone, are the most attractive as promising bone substitutes because of their biological advantages. Especially, brushite-based calcium phosphate cements are paid attention with interest of their fast resorption rates. However, the low mechanical strength limits their use to only non load-bearing applications. Therefore, to improve mechanical strength of brushite-based calcium phosphate bone cement, we used recombinant mussel adhesive protein (MAP). Recombinant MAPs are biocompatible and biodegradable mussel-derived proteins for underwater adhesion. These harmless bioadhesives were considered as bone cement additives. In the present work, we investigated effects of recombinant MAP addition on formation of brushite-based calcium phosphate cement.

See more of this Session: Poster Session: Bioengineering
See more of this Group/Topical: Food, Pharmaceutical & Bioengineering Division

(623az) Recombinant Shematrin: Silk-Like Protein From the Ocean

Dooyup Jung[1], Yun Jung Yang[1], Dong Gyun Kang[1], Yoo Seong Choi[2] and Hyung Joon Cha[1], (1) Department of Chemical Engineering, Pohang University of Science and Technology, Pohang, South Korea, (2)Department of Chemical Engineering, Chungnam National University, Daejeon, South Korea

Shematrins are glycine-rich proteins originally derived from the pearl oyster *Pinctada fucata*, and they are suggested to be given as organic matrix proteins for calcification. We noticed that shematrins contain repetitive silk-like peptides, which are similar with those of spiders and silkworms. Especially, we chose shematrin-5, which has Gly-Gly-(Lys/Tyr) repeat, as our target. This domain is thought to build up 3_1-helix secondary structure and thus, it might show remarkable extensibility when formulated by appropriate methods such as wet spinning and film casting. In the present work, we constructed recombinant *Escherichia coli* system carrying the codon-optimized gene of hexahistidine-fused shematrin-5. We successfully expressed and purified recombinant shematrin-5 silk-like protein for wet-spun fiber formation and measurement of its mechanical properties. We expect that production and formulation of recombinant shematrin may give a source for researchers to study its role in biomineralization process and mechanical functions in nature. Moreover, we expect to use this marine silk in various application areas including medical and textile industries.

See more of this Session: Poster Session: Bioengineering
See more of this Group/Topical: Food, Pharmaceutical & Bioengineering Division

(623b) Systems Biology of Glucose Metabolism In Humanized Yeast

Goutham Vemuri, Quantitative Systems - Biology Department of Chemical and Biological Engineering, Chalmers University, Goteborg, Denmark

One of the most remarkable characteristics of living systems is their ability to respond to stimuli. Sensing stimuli, transmitting the information to invoke appropriate response is largely carried out by specific signal transduction pathways, which are activated/deactivated by reversible phosphorylation. In this presentation, I will demonstrate how the information flux can be engineered to influence metabolic flux.

I will present the progress we made in our understanding of the signal transduction pathways that control and coordinate metabolism with growth and cellular processes. Specifically, I will focus on the signaling pathway that is controlled by Snf1 kinase, a highly conserved global regulator. This kinase controls energy homeostasis, regulates glucose metabolism and lipid synthesis as well as coordinates these processes with cell growth. Indeed, the homolog of this kinase is the target for diabetes drugs in the market today. The metabolic pathways as well as their regulation are largely conserved between humans and yeast, enabling this microorganism to serve as a prototype to gain deeper insight into metabolic malfunction in humans. In yeast, we determined the activation of Snf1, its interaction with other key kinases and the evolution of this interaction network as well as the consequences of its replacement with its mammalian counterpart. We discovered that the mammalian kinase was less efficient in regulating biosynthesis of amino acid and lipid biosynthesis in yeast. I will end with a perspective on the implications of our findings on engineering signaling pathways to manipulate metabolic flux.

See more of this Session: Poster Session: Bioengineering
See more of this Group/Topical: Food, Pharmaceutical & Bioengineering Division

(623ba) Functional Nanofibers Fabricated by Mussel Adhesive Proteins for Tissue Engineering

Bum Jin Kim, School of Interdisciplinary Bioengineering and Bioscience, Pohang University of Science and Technology, Pohang, South Korea, **Yoo Seong Choi**, Department of Chemical Engineering, Chungnam National University, Daejeon, South Korea and Hyung Joon Cha, Department of Chemical Engineering, Pohang University of Science and Technology, Pohang, South Korea

Various bioactive molecules have been introduced on the surface of nanofibers to design biomimetic tissue engineering scaffolds. For this achievement, generally, typical conjugation chemistry and physical adsorption procedures have been used after surface activation treatment steps. However, these strategies have limitations such as requirement of multiple steps, low stability or efficiency, and potential remaining toxic substances. Previously, we suggested that recombinant mussel adhesive proteins (MAPs) which have a great cell adhesion ability and biocompatibility can be the potential biomaterials for tissue engineering. In the present work, through the typical electrospinning procedure, we fabricated novel composite nanofibers using polycaprolactone (PCL) and MAPs by varying their mixing ratios. Using the FT-IR, XPS, and contact angle analyses, we indentified well-exposition of MAPs on the surface of composite nanofibers to provide the friendly environment for cell-nanofiber interaction. Mechanically, MAP-contained nanofibers showed the stiff and brittle properties but the tensile strength was significantly increased by addition of MAPs. Through in vitro cell culture experiments, we found that cell adhesion and proliferation were also highly enhanced on the PCL/MAP nanofibers compared to sole PCL nanofibers. Also, we investigated the use of PCL/MAP nanoifibers as coating platform for many biomolecule types by simple coating without any chemical treatments. Collectively, our newly fabricated MAP-based nanofiber can be used as functional tissue engineering scaffolds.

See more of this Session: Poster Session: Bioengineering
See more of this Group/Topical: Food, Pharmaceutical & Bioengineering Division

(623bb) Marine Silk Protein From Sea Anemone

Yun Jung Yang[1], Yoo Seong Choi[2], **Dooyup Jung**[1] and Hyung Joon Cha[1], (1)Department of Chemical Engineering, Pohang University of Science and Technology, Pohang, South Korea, (2)Department of Chemical Engineering, Chungnam National University, Daejeon, South Korea

The extraordinary biocompatibility of silk protein is the strongest factor of being used as a novel texture for centuries. With this benefit, usage of silk reaches to the material industry and medical area. Property of silk protein is organism-dependent. To be specific, silk from spider is stronger and tougher than silkworm. The researches for silk protein have been mainly focused on spider and silkworm. In the present work, we investigated new silk-bearing marine organisms to figure out flesh property and possibility of novel silk protein to apply industries. We designed and constructed new silk-like protein (designated to 'aneroin') from sea anemone. With simple heat shock and acid purification process, we obtained recombinant aneroin successfully. We spun about 50-100 mm of aneroin with no defects by wet spinning. In addition, electrospinning of aneroin was also conducted.

See more of this Session: Poster Session: Bioengineering
See more of this Group/Topical: Food, Pharmaceutical & Bioengineering Division

(623bc) Development of Lectin Immunoassay for Detection of Glyco-Chain Conjugated with Biomarker

Yusuke Emori[1], Yuji Ohigashi[1], Yoichi Kumada[2] and Michimasa Kishimoto[2], (1)Department of Chemistry and Materials Technology, Kyoto Institute of Technology, Kyoto, Japan, (2)Department of Biomolecular Engineering, Kyoto Institute of Technology, Kyoto, Japan

Recently, ithas been clarified that various glycol-chains with special structures areconjugated with protein biomarkers according to appearance and/or proceeding ofserious diseases in human body. Therefore, sensitive detection methods for notonly amount of biomarker itself, but also its glycol-chain structure are highlyrequisite in the field of clinical diagnosis. In this study, we developed anovel lectin immunoassay method that special glyco-form structures conjugatedcould be specifically detected with high sensitivity and S/N ratio usingsingle-chain Fv (scFv) antibody and HRP-labeled lectin. Polystyrene-bindingpeptide (PS-tag: RIIIRRIRR) which was originally isolated in our researchgroup, was genetically-fused at the C-terminal region of scFv (30kDa) forsite-specific attachment with higher density and maintenance of higherantigen-binding activity, compared with the conventional whole antibody (wholeAb, 150kDa). PS-tag-fused scFv (scFv-PS) was over-expressed in *E. coli*, and it was site-specificallyimmobilized on the surface of polystyrene (PS) plate. When the scFv-immobilizedPS plate was used in sandwich ELISA using carcinoembryonic antigen (CEA) as amodel glycobiomarker, more than 50-fold higher sensitivity was detected,compared with the conventional PS plate immobilized with whole Ab. Lectinimmunoassay using the scFv-immobilized PS plate and 12 types of HRP-labeldlectins allowed detection of signals on the basis of the specific interactionbetween lectin and glyco-chain conjugated. Much higher signal / noise (S/N) ratioswere obtained from scFv-immobilized PS plate in the lectin immunoassay, whilehigher background signals detected from whole Ab-immobilized PS platesignificantly decreased S/N ratios, because lectins strongly bound to the wholeAb which was also a kinds of glycoproteins. Thus, the lectin immunoassay usingscFv-immobilized PS plate developed in this study is considerably useful fordetection of a variety of glyco-forms conjugated with protein biomarkers.

See more of this Session: Poster Session: Bioengineering
See more of this Group/Topical: Food, Pharmaceutical & Bioengineering Division

(623bd) A Mussel Adhesive Protein Fused with the BC Domain of Protein A Is a Functional Linker Material That Efficiently Immobilizes Antibodies Onto Diverse Surfaces

Chang Sup Kim[1], **Yoo Seong Choi**[2] and Hyung Joon Cha[1], (1)Department of Chemical Engineering, Pohang University of Science and Technology, Pohang, South Korea, (2)Department of Chemical Engineering, Chungnam National University, Daejeon, South Korea

The efficient immobilization of antibodies onto solid surfaces is a vital factor for the sensitivity and specificity of various immunoassays and immunosensors. In the present work, we designed and produced a novel linker protein, BC-MAP, in *Escherichia coli* by genetically fusing mussel adhesive protein (MAP) with two domains (B and C) of protein A (antibody binding protein) for efficient antibody immobilization on diverse surfaces. Through direct surface coating analyses, we found that BC-MAP successfully coated diverse surfaces including glass, polymers, and metals, but the BC domain alone did not. Importantly, antibodies were efficiently immobilized on BC-MAP-coated surfaces, and the immobilized antibodies interacted selectively with their corresponding antigen. Quartz crystal microbalance analyses showed that BC-MAP has excellent antibody-binding ability compared to BC protein on gold surfaces. These results demonstrate that the MAP domain, with uniquely strong underwater adhesive properties, plays a role in the direct and efficient coating of BC-MAP molecules onto diverse surfaces lacking any additional surface treatment, and the BC domain of BC-MAP contributes to the selective and oriented immobilization of antibodies on BC-MAP-coated surfaces. Thus, our BC-MAP fusion protein could be a valuable novel linker material for the facile and efficient immobilization of antibodies onto diverse solid supports.

See more of this Session: Poster Session: Bioengineering
See more of this Group/Topical: Food, Pharmaceutical & Bioengineering Division

(623be) Production of Islet Cell Progeny From Human Pluripotent Stem Cells In Microcarrier Cultures

Jincheng Wu, Yongjia Fan, Lye T. Lock and Emmanuel (Manolis) S. Tzanakakis, Chemical and Biological Engineering, State University of New York at Buffalo, Buffalo, NY

Human pluripotent stem cells (hPSCs) can serve as inexhaustible sources of islet cells for pancreatic tissue engineering and diabetes therapies. Essential for the success of stem cell-based diabetes applications is the development of efficient differentiation protocols and bioprocesses for converting large quantities of hPSCs into genetically normal and functional insulin-producing cells. Here, we employed multiple human embryonic stem cell (hESC) lines to develop a protocol for guiding stem cells along pancreatic cell lineages. Furthermore, our differentiation approach was adapted to the scalable culture of hPSCs in stirred-suspension microcarrier bioreactors.

A stage-wise differentiation strategy was established to sequentially differentiate hESCs into definitive endoderm, primitive gut tube, posterior foregut and pancreatic islet cells. Differentiation was induced in serum-free media with physiological factors involved in embryonic pancreas development while formation of embryoid bodies was avoided. The differentiating hPSC populations were analyzed by quantitative PCR, immunostaining and flow cytometry and ELISA (for hormone secretion). Microcarrier bioreactor cultures of hPSCs were carried out and analyzed (e.g. cell concentration, viability, lactate dehydrogenase activity etc.) as we described[1,2].

Human PSC monolayers were coaxed to definitive endoderm cells co-expressing FOXA2 and SOX17 as we reported[1]. Subsequent incubation with retinoic acid and FGF7 led to the emergence of HNF1B^{+}/ HNF4A^{+} cells after 4-5 days while the first PDX1^{+}/HNF6^{+} cells appeared on day 8. Additional factors were also explored for curtailing cell apoptosis. The fraction of cells transitioning between successive stages was above 70% as assessed by flow cytometry for the co-expression of relevant markers. These cells gave rise to islet cell-like progeny including cells expressing PDX1, NGN3 and NKX6.1. Cells were also immunopositive for C-peptide and insulin while release of these antigens was detected in response to glucose and secretagogues.

The differentiation scheme was also translated to stirred-suspension microcarrier cultures of hPSCs. Cells were expanded on microcarrier bioreactors under non-differentiating conditions as described[1]. Then, the differentiation protocol developed in static cultures was applied. Separate experiments were performed to select bioreactor culture conditions (e.g., the initial cell concentration, cell/microcarrier ratio, agitation rate) which allow the propagation of cells without hindering their pancreatogenic commitment. Cells on microcarriers transitioned through the same stages exhibiting appropriate markers leading to a population with 40% PFG cells. Cells further gave rise to islet-like cells expressing PDX1,

NKX2.2, NKX6.1 and GLUT2. Over 8% of hPSC-derived cells actively transcribed insulin. These cells were also immunoreactive to insulin and C-peptide. Current efforts focus on increasing the yield of insulin-producing cells in the differentiation of hPSCs and assessing their function in vivo.

A protocol was established for the efficient generation of pancreatic islet progeny from hPSCs using physiological factors in the absence of serum and without the formation of embryoid bodies. Moreover, this differentiation strategy was successfully combined with the microcarrier bioreactor culture of hPSCs. This is an important step toward the establishment of bioprocesses for the generation of functional islet cells for pancreatic tissue engineering and diabetes therapies.

Acknowledgements: Support for this project was provided by a grant (C024355) to E.S.T. from the New York State Stem Cell Trust (NYSTEM).

References:

1. Lock, L.T. et al. Tissue Eng. Part A 2009, 15:2051-63.

2. Kehoe, D. et al. Tissue Eng. Part A. 2010, 16:405-21.

See more of this Session: Poster Session: Bioengineering
See more of this Group/Topical: Food, Pharmaceutical & Bioengineering Division

(623bf) Complete Cell-Free Conversion of Glucose to Hydrogen

Joseph A. Rollin and Percival Zhang, Biological Systems Engineering, Virginia Polytechnic Institute and State University (Virginia Tech), Blacksburg, VA

Biohydrogen has great potential as a future transportation fuel and commodity chemical. Cell-free production of hydrogen offers numerous engineering advantages, such as increased reaction rates, lack of competing pathways, and minimization of biocatalyst cost per unit product. Previous studies from our group have demonstrated the conversion of starch and cellulosic substrates to hydrogen, using such a cell-free synthetic pathway biotransformation (1,2). In this study, we broaden the substrate range to include monomeric glucose without the use of ATP, increase the reaction rate by an order of magnitude, and demonstrate complete conversion for the first time.

1. Y.-H.P. Zhang, B.R. Evans, J.R. Mielenz, R.C. Hopkins, M.W.W. Adams, PLoS One, 2 (2007) e456.
2. X. Ye, Y. Wang, R.C. Hopkins, M.W.W. Adams, B.R. Evans, J.R. Mielenz, Y.-H.P. Zhang, ChemSusChem, 2 (2009) 149-152.

See more of this Session: Poster Session: Bioengineering
See more of this Group/Topical: Food, Pharmaceutical & Bioengineering Division

A SYSTEMIC MODEL FOR THE RECEPTOR COMPLEX NRC/MASC BASED ON ITS INFERRED INTERPROTEIC NETWORK

Jaime A. Campos[1,3], Ziomara P. Gerdtzen[†1,2], J. Cristian Salgado[†1,2,3]

[1] Department of Chemical Engineering and Biotechnology, University of Chile, Av. Beauchef 850, Santiago 837-0448, Chile
[2] Millennium Institute for Cell Dynamics and Biotechnology
[3] Laboratory for Process Modeling and Distributed Computing
[†] Equal contribution authors

Background

Theoretical and experimental studies suggest that there is a relationship between glutamate transmission and the triggering of cognitive diseases such as schizophrenia [1, 2]. The importance of the receptor complexes of glutamate lays on their role in the reception of the neuronal code, in the plasticity of the synapses and in the internal structure of the postsynaptic terminal (PST) [1, 4]. Two characteristics of the glutamate receptors make them interesting targets of study [5]: i) They are connected to a large number of proteins, configuring modulatory complexes and initiating signaling pathways. ii) Proteomic studies of the postsynaptic density (PSP) have shown that there is a relationship between proteins belonging to receptors complexes and the triggering of cognitive diseases. All this characteristics make the study of glutamate receptor complex fundamental for understanding the functionality and robustness of the synaptic plasticity phenomena, and it offers a new approach for studying the triggering of cognitive diseases.
Previous studies have identified and classified 1124 proteins belonging to the post synaptic density (PSD) [5- 7]. Studies of the functional composition of PSD revealed its specialization through the enrichment with protein domains such as: scaffolding and phosphorylation, membrane localization, domains associated with Ca2+ and GTP-dependent signalling [8, 9]. The glutamate receptor family comprises part of the PSD. The N-methyl-D -aspartate (NMDA) ionotropic receptor along with the metabotropic receptor (mGluR) activate signalling and plasticity mechanisms. On the other hand, the a-amino-3-hydroxy-5-methylisoxazole-4-propionate(AMPA) receptor transmits the depolarisation of the membrane.
The role of this complex in synaptic function, behaviour and cognitive disease, and its role in the triggering of mental diseases has been studied [8–10]. Almost 25% of the MASC complex proteins were identified as fundamental for normal synaptic and behaviour plasticity. In addition, clusters of proteins that share a functional profile were identified using interproteic interactions (PPIs), resulting in a three layer structure for the system, shown in Figure 1.
The identification of interproteic interactions by traditional experimental methods usually has low confidence levels and it is low-throughput [3, 11, 12]. The use of

different classification algorithms has been proposed in order to identify new interactions to overcome the low confidence issue [11].

In this work we present the integration of the two methodologies described above: the modular approach for identifying the functional structure of the receptor complex, and Naïve-Bayes for the inference of a putative interproteic interactions network. Through the inference of a protein interaction network it is possible to perform a new systemic study of the receptor complex with information not considered before and with a high confidence level, further increasing the knowledge of the complex network's internal relationships. We aim to relate the specific structure and functionality of the modules with cognitive diseases and alterations in the synaptic plasticity. The methodology presented can be applied for the inference interproteic interactions and network structures in other protein networks of interest.

Methodology

The general methodology used in this work consists of three consecutive stages:

1. *The protein-protein interactions inference stage:* Different data sets were implemented through the Naïve-Bayes algorithm: the orthologic interaction classes in [13] and microarray data sets [15 - 18]. This classification algorithm allowed us to infer protein-protein interactions (PPIs) that had not been detected empirically. For each interaction the posterior odds (O_{Post}) were calculated and used to select the PPIs with at least 50% chance of being a true positive. This putative PPI network was merged with other two inferred networks previously reported [13,14] delivering a novel network, defined as the consensus network.
2. *The receptor complex clustering stage*: Interactions between proteins that belong to the NRC/MASC complex were selected and a series number of mapping processes were performed using the consensus network previously obtained. In order to obtain comparable results the methodology applied in [8] was used. This group of interactions were merged with the PPIs previously reported into a dual network. The modules of receptor complex were identified through the application of a clustering algorithm over the PPIs. A second clustering algorithm and manual rearrangement of proteins was applied over the four biggest clusters.
3. *Modules characterisation and functional structure definition stage*: The last stage of this work consisted in the study of the modules, focusing on the particular characteristics of each cluster and their functional properties in terms of the interaction between belonging elements. In addition, we studied the relationships between modules and their association to the triggering of different mental disorders, and the relevance of these modules in impairing synaptic and behavioural plasticity.

Results

The consensus network restricted to the core functional elements of the complex, has a similar number of proteins as the network presented by Pocklington et al., which was defined only through the interactions reported in literature. In addition, there is a higher number of interactions based on literature mining than putative interactions.
The number of proteins of the dual network represent 76% of the total NRC/MASC complex, where the number of proteins integrated by this work corresponds to 21% of the total number of proteins in complex. One of the most important features of the consensus network is the increase of the number of PPIs with respect to the PPIs documented experimentally. As shown in Results, the putative network and the literature based share only 6% of their PPIs, and there is an increase of 60% in the number of PPIs in the putative network with respect to the literature based one. A bigger and denser network was generated, with an average connectivity of 5.6 interactions per protein. Due this integration of proteins and interactions, a new set of modules and therefore, a new functional structure arises, shown in Figure 1.
New additions to the literature based network improved the percentage of proteins with specific functional families, such as Channels and receptors, Cell adhesion and cytoskeletal, Signalling molecules and enzymes and Synaptic vesicles/protein transport, which increased almost 30% with respect to the total proteins from the MASC complex. This represents an important advance for the definition of the modular structure and functional organisation of the complex, as these key families are the ones that define the internal structure of the NRC/MASC complex.

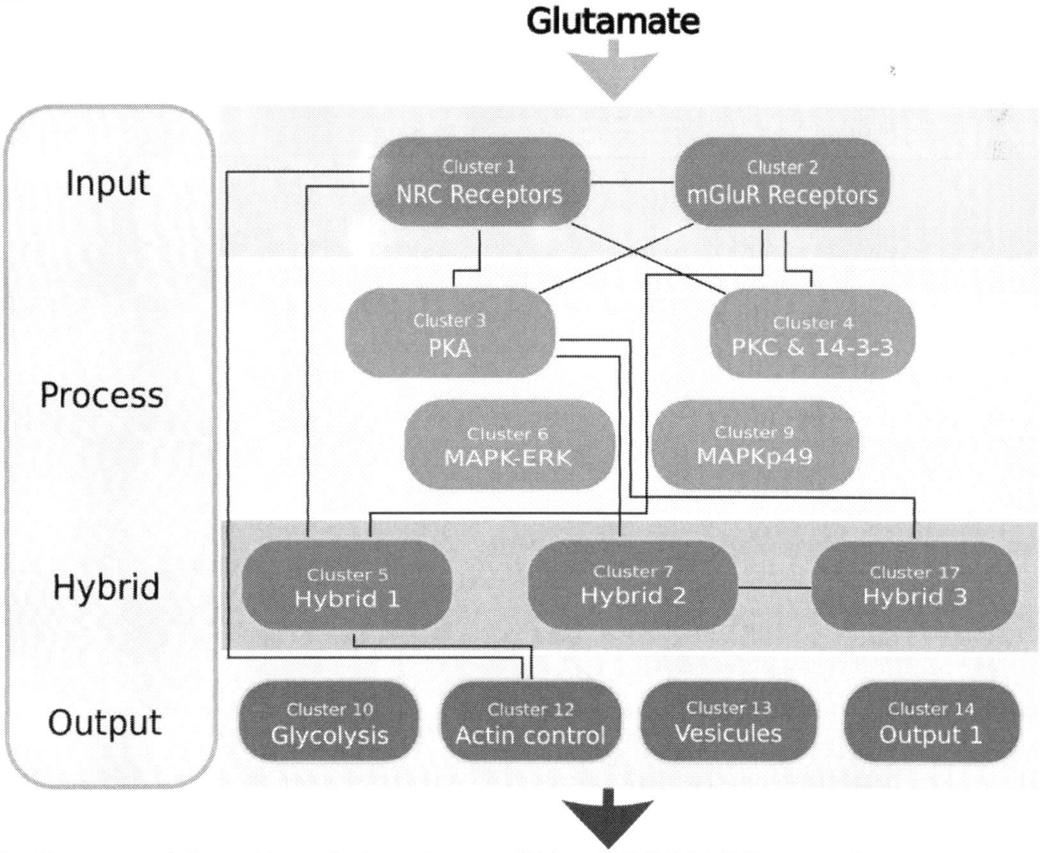

Figure 1: Proposed functional structure of the NRC/MASC complex.

Conclusions

The methodology we have developed in this work delivered relevant information for the study of the NRC/MASC complex: the modules obtained through the clustering algorithm, the functional profiles for each cluster and the particular relationships between those modules. This led to a novel four-layered structure for the NRC/MASC complex. The internal sub-structures of the complex as well as the localisation of proteins for the different phenotypic annotations in the complex were identified. All this information was integrated in a functional structure, which extends the previously reported functional model. The proposed model recovers and provides deeper insight into the core elements of the previously reported structure.

The methodology presented in this work is suitable for the study of any particular network, whose components have been experimentally identified. The capacity to identify new and relevant PPIs and integrate them into the network is a valuable tool. We expect that this work guides future experimental studies in order to unveil the underlying structure of the NRC/MASC complex and verify the functional characteristics of its modules.

Acknowledgements

This work was supported by FONDECYT Initiation grants 11080016 and 11090268.

References

1. Goff D.C., Coyle J.T., (2001)"The emerging role of glutamate in the pathophysiology and treatment of schizophrenia". Am J Psychiatry 158(9):1367–77.
2. Lang U.E., Puls I., Muller D.J., Strutz-Seebohm N., Gallinat J. (2007) "Molecular mechanisms of schizophrenia". Cell Physiol Biochem 20(6):687–702.
3. Choudhary J., Grant S.G. (2004) "Proteomics in postgenomic neuroscience: the end of the beginning". Nat Neurosci, 7(5):440–5.
4. Grant S.G., O'Dell T.J. (2001) "Multiprotein complex signaling and the plasticity problem". Curr Opin Neurobiol, 11(3):363–8.
5. Husi H., Grant S.G. (2001) "Isolation of 2000-kDa complexes of N-methyl-D-aspartate receptor and postsynaptic density 95 from mouse brain". J Neurochem, 77:281–91.
6. Collins M.O., Husi H., Yu L., Brandon J.M., Anderson C.N., Blackstock W.P., Choudhary J.S., Grant S.G. () "Molecular characterization and comparison of the components and multiprotein complexes in the postsynaptic proteome". J Neurochem, 97 Suppl 1:16–23.
7. Farr C.D., Gafken P.R., Norbeck A.D., Doneanu C.E., Stapels M.D., Barofsky D.F., Minami M., Saugstad J.A. (2004) "Proteomic analysis of native

metabotropic glutamate receptor 5 protein complexes reveals novel molecular constituents". J Neurochem, 91(2):438–50.
8. Pocklington A.J., Cumiskey M., Armstrong J.D., Grant S.G. (2006) "The proteomes of neurotransmitter receptor complexes form modular networks with distributed functionality underlying plasticity and behaviour". Mol Syst Biol, 2:2006.0023.
9. Pocklington A.J., Armstrong J.D., Grant S.G. (2006) "Organization of brain complexity–synapse proteome form and function". Brief Funct Genomic Proteomic, 5:66–73.
10. Armstrong J.D., Pocklington A.J., Cumiskey M.A., Grant S.G. (2006) "Reconstructing protein complexes: from proteomics to systems biology". Proteomics, 6(17):4724–31.
11. Jansen R., Yu H., Greenbaum D., Kluger Y., Krogan N.J., Chung S., Emili A., Snyder M., Greenblatt J.F., Gerstein M. (2003) "A Bayesian networks approach for predicting protein-protein interactions from genomic data". Science, 302(5644):449–53.
12. Qi Y., Bar-Joseph Z., Klein-Seetharaman J. (2006) "Evaluation of different biological data and computational classification methods for use in protein interaction prediction". Proteins, 63(3):490–500.
13. Rhodes D.R., Tomlins S.A., Varambally S., Mahavisno V., Barrette T., Kalyana-Sundaram S., Ghosh D., Pandey A., Chinnaiyan A.M. (2005) "Probabilistic model of the human protein-protein interaction network". Nat Biotechnol, 23(8):951–9.
14. Xia K., Dong D., Han J.D. (2006) "IntNetDB v1.0: an integrated protein-protein interaction network database generated by a probabilistic model". BMC Bioinformatics, 7:508.
15. Rosenwald A., Wright G., Chan W.C., Connors J.M., Campo E., Fisher R.I., Gascoyne R.D., Muller-Hermelink H.K., Smeland E.B., Giltnane J.M., Hurt E.M. (2002) "The Use of Molecular Profiling to Predict Survival after Chemotherapy for Diffuse Large-B-Cell Lymphoma". N Engl J Med, 346(25):1937–1947.
16. Segal N.H., Pavlidis P., Noble W.S., Antonescu C.R., Viale A., Wesley U.V., Busam K., Gallardo H., DeSantis D., Brennan M.F., Cordon-Cardo C., Wolchok J.D., Houghton A.N. (2003) "Classification of Clear-Cell Sarcoma as a Subtype of Melanoma by Genomic Profiling". J Clin Oncol, 21(9):1775–1781.
17. Su A.I., Welsh J.B., Sapinoso L.M., Kern S.G., Dimitrov P., Lapp H., Schultz P.G., Powell S.M., Moskaluk C.A., Frierson J. Henry F., Hampton G.M. (2001) "Molecular Classification of Human Carcinomas by Use of Gene Expression Signatures". Cancer Res, 61(20):7388–7393.
18. van 't Veer L., Dai H., van de Vijver M., He Y., Hart A. (2002) "Gene expression profiling predicts clinical outcome of breast cancer". Nature, 415(6871):530–536.

(623bh) Cellulase Recycling and Reuse: Comparison of Immobilization, Ultrafiltration and Re-Adsorption

Rongxin Su, Wei Qi and Zhimin He, School of Chemical Engineering and Technology, Tianjin University, Tianjin, China

*Corresponding Author's E-mail: surx@tju.edu.cn

Until now, the enzyme cost is still the dominating economic barrier to produce ethanol from lignocellulosic biomass. Therefore, lower enzyme cost is being pursued for the industrialization of cellulosic bioethanol. The recycling of enzymes has been explored as an effective way of reducing the high cost of enzymes [1, 2]. After enzymatic hydrolysis of lignocellulosic substrates, some enzymes remain free in the hydrolysate, while others are bound to the residual substrates. In our lab, we compare several strategies developed to recover and reuse the enzymes, including ultrafiltration, readsorption onto fresh substrates, and enzyme immobilization.

(1) Immobilization. Immobilization of enzymes is an important approach for enzyme recycling. Among cellulases, b-glucosidase is frequently used to supplement cellulase preparations for hydrolysis of cellobiose to glucose. Compared to free cellobiase, immobilized cellobiase had improved thermal stability and mechanical strength. It could be reused several times in a continuous reaction packed column. Moreover, immobilized cellobiase is readily separated from the substrate and products, which can simplify subsequent purification processes. However, few studies have used lignocellulosic biomass as substrates and none have been done at high solids loadings. The use of immobilized cellulases for lignocellulose hydrolysis is still a technically difficult task because immobilization may hinder enzyme access to insoluble cellulose. Multi-enzymes complex that are required for hydrolysis of lignocellulose also increase the difficulty in immobilization process.

(2) Ultrafiltration. Ultrafiltration has been proven to be capable of recovering cellulases and b-glucosidase, and it has the capacity to continuously remove sugars and other small compounds that may potentially inhibit enzyme activity. For this reason, a membrane bioreactor (MBR) may provide some advantages in performing enzymatic hydrolysis. The MBR combines product inhibitor removal and cellulase reuse in one device, which facilitates operation and reduces equipment costs by eliminating an extra reactor. Another advantage of MBR is that the enzymatic hydrolysis rate increases because product inhibition is lowered compared with that in a conventional batch reactor. In the fed-batch MBR, fresh substrate was supplemented five times at the rate of 10 g cellulose per liter each time, resulting in a total cellulose loading of 7%. After 60 h, the enzyme utilization efficiency was 3.13-fold of MBR. An additional process, like reverse osmosis, is required on the downstream processes to concentrate the sugar products.

(3) Re-adsorption. Cellulases have relatively high stability and high affinity for cellulose. After hydrolysis, most cellulases are free in the hydrolysate. Therefore, free cellulases can be recovered for further use by readsorption onto fresh substrates and subsequent microfiltration. One of the most important factors that limit enzyme recovering in the readsorption process is the adsorption of cellulases to lignin. To reduce the nonproductive adsorption, nonionic surfactants, such as Tween 80, were added to increase free enzymes in the supernatant. In this case, the enzymes (e.g. Spezyme CP) can be used in four successive rounds of enzyme recycling, achieving an above 80% hydrolysis yield after the fourth round. Compared to ultrafiltration, enzyme recycling by readsorption is unable to recover b-glucosidase because it does not typically bind to the cellulosic substrate. Therefore, the supplementation of fresh b-glucosidase is commonly necessary to subsequent rounds of hydrolysis.

Acknowledgement

The authors acknowledge the financial supports received from the Program for New Century Excellent Talents in Chinese University (NCET-08-0386), the Natural Science Foundation of China (20976130, 20806057).

References

[1] Steele E, Raj S, Nghiem J, Stowers M (2005) Enzyme recovery and recycling following hydrolysis of ammonia fiber explosion-treated corn stover. Appl Biochem Biotechnol 121:901-910.

[2] Tu MB, Chandra RP, Saddler JN (2007) Evaluating the distribution of cellulases and the recycling of free cellulases during the hydrolysis of lignocellulosic substrates. Biotechnol Prog 23:398-406.

See more of this Session: Poster Session: Bioengineering
See more of this Group/Topical: Food, Pharmaceutical & Bioengineering Division

(623bi) Application of Hydroxyproline-O-Glycosylation Code for Enhanced Recombinant Protein Production In Plants and for Reconstruction of Plant Cell Wall for Improved Biomass Processability

Jianfeng Xu, Arkansas Biosciences Institute, Arkansas State University, Jonesboro, AR, Xumeng Ge, Arkansas Biosciences Institute, Arkansas State University, Jonsboro, AR and Marcia Kieliszewski, Chemistry and Biochemistry, Ohio University, Athens, OH

Hydroxyproline-O-glycosylation involves post-translational hydroxylation of proline to hydroxyproline (Hyp) and subsequent glycosylation, a modification that is unique to plants and green algae. Our earlier work with synthetic genes encoding various Hyp-rich glycoproteins (HRGPs) expressed in plant cells elucidated a Hyp-*O*-glycosylation "code", that is a peptide sequence directs the Hyp-*O*-glycosylation; specifically contiguous Hyp residues, as in X-Hyp-Hyp are sites of oligoarabinosylation; in contrast, clustered non-contiguous Hyp residues, as in X-Hyp-X-Hyp repeats are mainly sites of highly branched arabinogalactan polysaccharide addition. Where X is often Ser or Ala. These results demonstrated the feasibility of Hyp-*O*-based glycoprotein design in plants and triggered the following applications: 1) by introducing a Hyp-O-glycosylation tag to recombinant proteins expressed in tobacco BY-2 cells, we dramatically enhanced the production of secreted proteins up to 1500 fold; 2) using the same approach, we significantly improved the yields of recombinant proteins transiently expressed in *Nicotiana benthamiana*; 3) by engineering specific Hyp-*O*-based "designer" biopolymers into plants, we could reconstruct the plant cell walls for improved biomass processability.

See more of this Session: Poster Session: Bioengineering
See more of this Group/Topical: Food, Pharmaceutical & Bioengineering Division

Effect of UV Curing on Peptide Antimicrobials and Considerations for Incorporation in Polymeric Coatings

Fatima Alim, Jennifer A. Neff, Allvivo Vascular, Inc., Lake Forest, CA

Statement of Purpose

Increases in hospital acquired infection rates and their associated costs have lead to a demand for surface active antimicrobial coatings. Numerous strategies exist to impart antimicrobial activity on the surface of medical devices. Commonly used approaches involve impregnation of an antimicrobial agent directly in the material used to form the device, application directly onto the surface of the device, or encapsulation in a polymeric surface coating. In the latter case, UV curing is a preferred method for bonding polymer coatings to device surfaces due to quick processing speeds and low manufacturing costs. Despite the high use of UV, studies that describe its effects on activity of antimicrobials are lacking.

There is a great deal of interest in the use of antimicrobial peptides for medical device coatings because they are not likely to promote the development of antibiotic resistant bacteria. However, when attempting to incorporate an antimicrobial peptide, one is faced with multiple interactions that occur from UV exposure that may reduce or completely eliminate an antimicrobial peptide's activity. These interactions narrow the coating formulation and application approaches that can be effectively utilized and more information is needed to understand how UV and UV initiated crosslinking reactions within a coating layer affect peptide activity.

In this study, coatings comprised of the lantibiotic, nisin, incorporated in a polymer gel composed of Pluronic F127 (F127) were evaluated. Nisin displays excellent antimicrobial activity against gram positive bacteria including *Staphylococcus epidermidis* .[1] This study provides insight into the affect of UV exposure on the activity of two natural variants of nisin, nisin A and nisin Z, taking into account the effect of peptide concentration, UV intensity, and UV duration. This study also examines the affect of UV on the antimicrobial activity of peptide- polymer combination coatings.

Methods

High purity nisin (≥95%, Handary) was dissolved in 0.02M degassed HCl at various concentrations. Solutions were then UV treated in UV grade quartz tubes at different intensities and durations using a fusion electrode-less microwave excited UV system. Two UV intensities were chosen for comparison, 277 mW/cm^2 (Intensity A) and 554 mW/cm^2 (Intensity B). Solutions were diluted to 1 mg/ml and antimicrobial activity assessed by the ZOI assay. Petri dishes were overlay with 25ml agar enriched with trypticase soy broth (TSB). A liquid culture of *Staphylococcus Epidermidis* (ATCC, P5984) that was prepared by incubating a single colony in TSB was then streaked onto the agar plates. Wells were bore into each quadrant of the plate and 30 μl aliquots of solution added to each well. Plates were incubated at 37°C for 18 hours. ZOI diameters were measured and used for calculating ZOI areas. The ZOI assay provides a quantitative measurement of changes in antimicrobial activity where reduction in ZOI correlates with a reduction in antimicrobial activity.

To determine the effect of UV on nisin A's activity in the presence of F127, two sets of solutions with varying concentrations of F127 (0, 0.5, 1, 5, 10 and 20% w/v) and 2 mg/ml nisin A were prepared. An additional formulation containing 20% F127 and 20 mg/ml nisin A was also prepared. One set of each formulation was treated with UV at intensity B for 1 min and the other served as a no UV control. These solutions were subjected to the ZOI assay as described above and the turbidometric assay as follows. A liquid culture of *S. Epidermidis* was diluted to 10^7 colony forming units/ml. Aliquots of 190 µl of the bacteria culture were added to each well of a 96 well plate. Sample types were diluted and 10 µl aliquots were subsequently added in triplicates to each well to obtain 5, 10, 20, 40 or 100 µg Nisin/ml Staph. E., where these concentrations are above the minimum inhibitory concentration (MIC), (4.2 µg/ml).[1] The 96-well plate was incubated at 37°C for 18 hours and the turbidity of the wells measured spectrophotometrically at 600 nm. Any absorbance obtained indicates growth of *S. Epidermidis* and hence a loss in nisin A antimicrobial activity.

Results

Nisin A's antimicrobial activity decreased linearly with increasing UV duration (Figure 1). For the doses and times evaluated here, the effect of UV on nisin A activity depended primarily on the total dose of UV delivered. There was a slight trend toward greater retention of activity with the application of higher intensity UV for shorter times vs lower intensities for longer times.

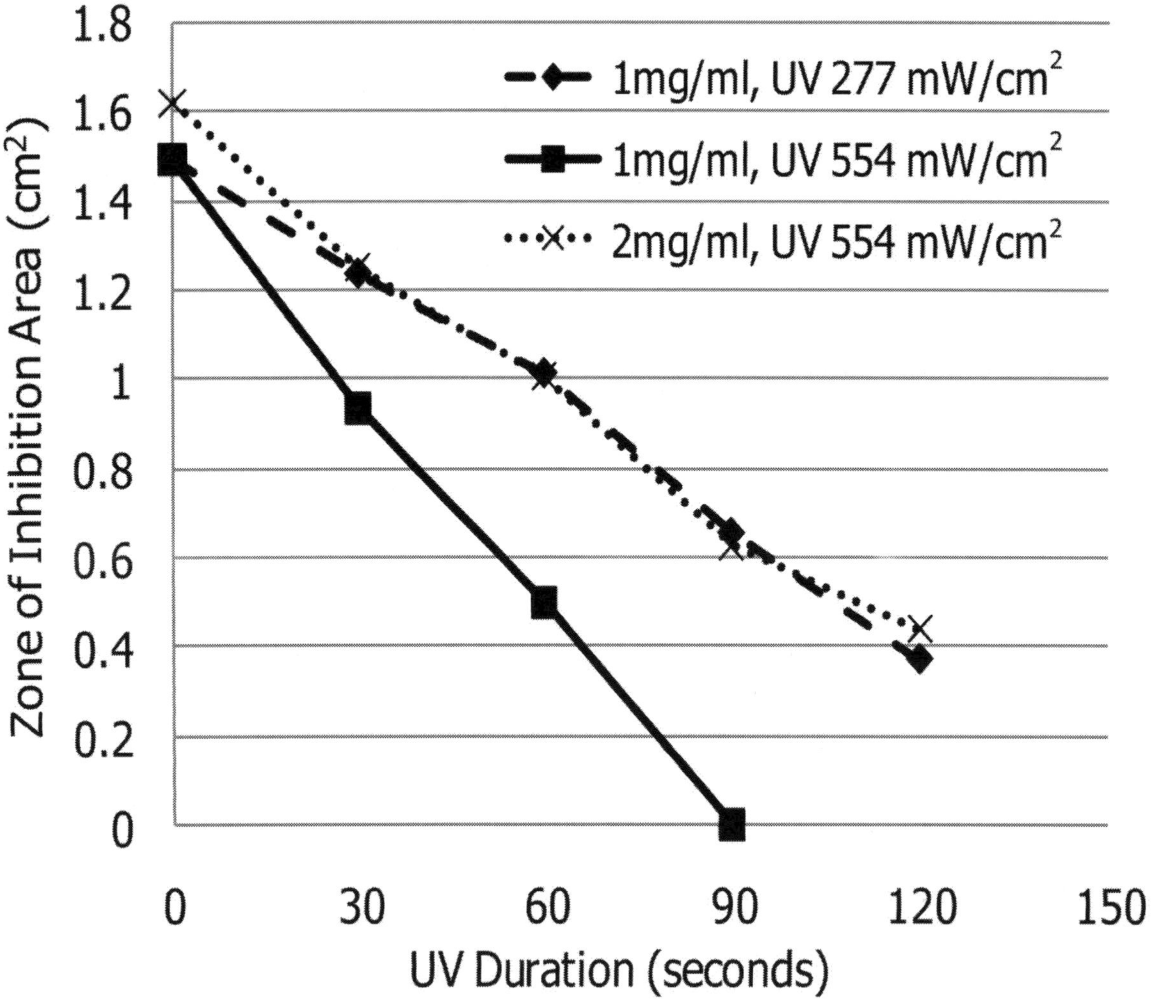

Figure 1: Nisin A UV stability as determined by ZOI assay. (n=3, standard deviations less than 0.1)

Nisin Z displayed poor stability to UV relative to nisin A. Even at high concentrations, UV treated nisin Z did not yield a measurable ZOI. After UV exposure at the lowest dose, Nisin Z lost more than 90% activity. Nisin A and nisin Z are nearly identical in structure and differ only in the type of amino acid at position 27 (His in nisin A and Asn in nisin Z). The two variants are also equally distributed among nisin-producing *Lactococcus lactis* strains and are generally thought to have similar activities and chemical stability with the only difference being that Nisin A displays greater solubility at low pH[2, 3]. To our knowledge, this is the first report that nisin A is substantially more stable to UV exposure than nisin Z.

The percent reduction in the activity of pure nisin A upon UV exposure was found to depend on the peptide's concentration (Figure 2). Nisin A was more stable to UV when exposed at higher concentrations. A similar trend was obtained with commonly used nisin products such as Nisaplin® that contain very low weight percentages of nisin (2.5% nisin, 22.5% denatured milk solids)(data not shown).

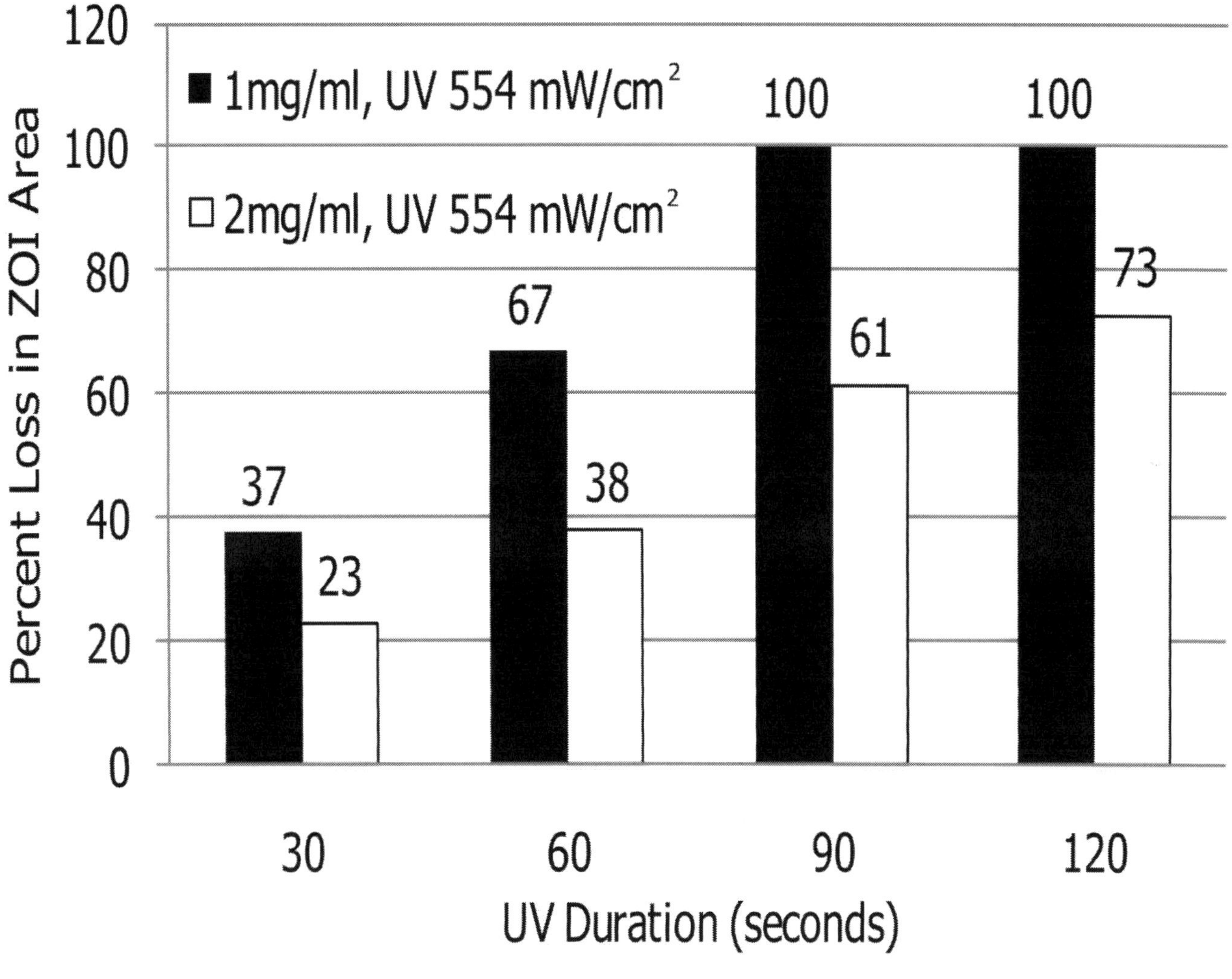

Figure 2: Percent loss in ZOI area for nisin A exposed to UV relative to untreated nisin A.

In the secondary study, the effect of UV exposure on nisin A in the presence of varying concentrations of F127 was evaluated. Using a standard ZOI assay, it was found that control samples containing mixtures of nisin with F127 produced ZOIs that were comparable to those obtained for nisin alone. Samples containing a mixture of F127 and nisin that were treated with UV resulted in smaller ZOIs with increasing F127 concentration. Control samples of UV treated nisin, showed a loss in ZOI area relative to untreated nisin. However, the difference was not as great as that observed for solutions containing high concentrations of F127, indicating that the loss of activity observed in the mixed samples was due to an interaction with F127. Prior NMR studies indicate that some crosslinking between Pluronic® copolymers occurs with irradiation. Based on the results from the agar diffunsion assay, it was not clear if the nisin was entrapped within F127 structures that impaired its diffusion into the agar or possibly forming chemical crosslinks with F127 upon UV exposure.

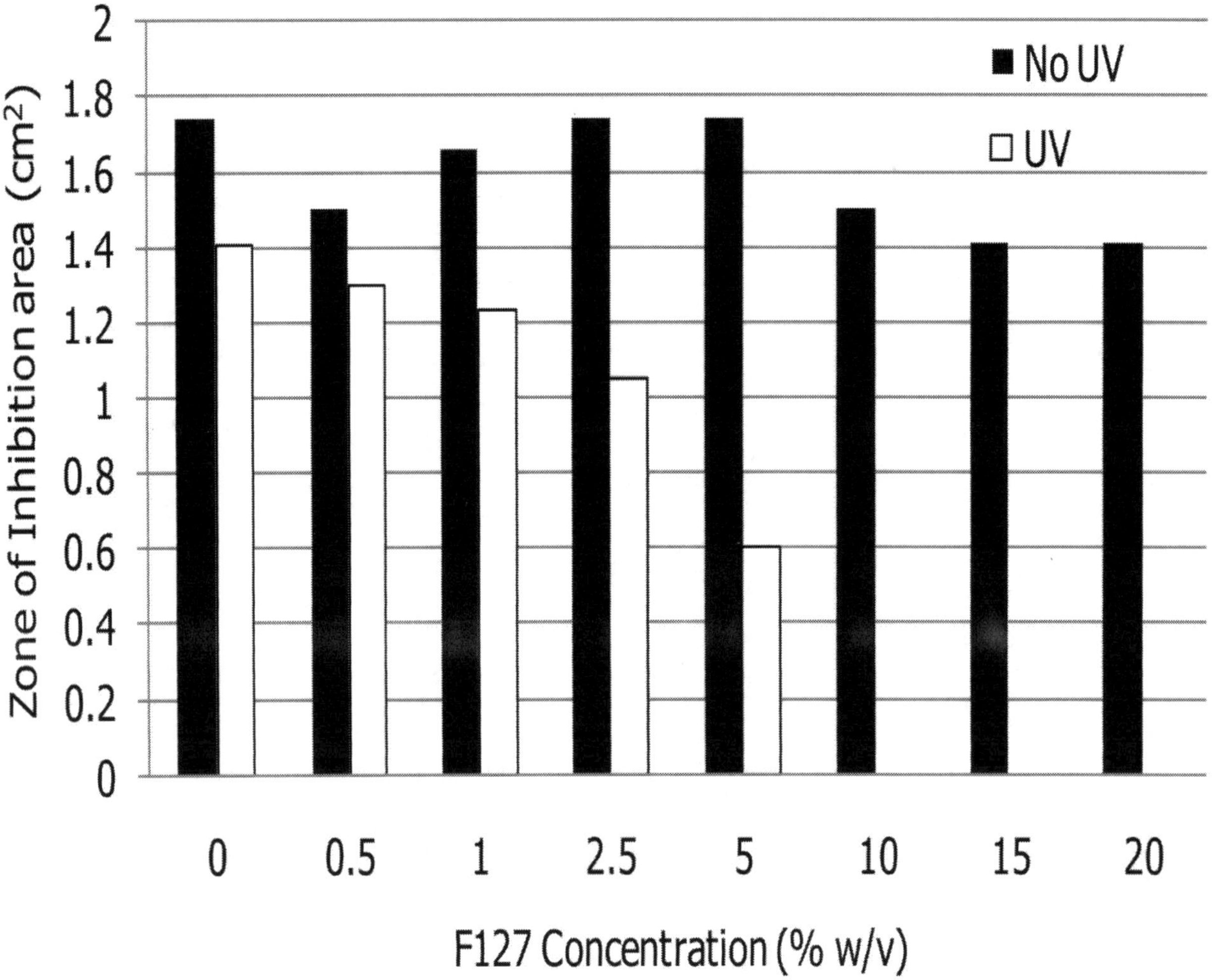

Figure 3: Effect of F127 concentration on entrapment of Nisin A. (n=3, standard deviations less than 0.1)

In an effort to better understand the system, a similar experiment was conducted using a turbidometric assay (Table 1). UV treated mixtures of 2 mg/ml nisin A and F127 showed increasing

MIC with increasing F127 concentration. At concentrations above 10% F127, the resulting MIC was above 100 μg/ml. In contrast, mixtures containing 20 mg/ml nisin and 20% F127 retained an MIC below 5 μg/ml upon UV exposure, indicating almost no loss in activity. This result is consistent with the prior finding that increasing nisin concentration results in higher stability to UV exposure. It also appears that at the higher nisin concentration of 20 mg/mL, it is possible to incorporate higher concentrations of F127 in conjunction with UV exposure without substantial losses in activity.

Table 1. Turbidity of bacteria cultures treated with solutions of nisin and Pluronic® F127. Solutions containing either 2 mg/ml or 20 mg/ml nisin and various concentrations of Pluronic F127 were treated with UV, diluted, and subsequently added to *S. Epidermidis* cultures. Samples marked with X were turbid.

Nisin Concentration Added to Bacteria Culture (μg /ml *S. Epidermidis*)	2 mg/ml Nisin A								20 mg/ml Nisin A	
	% F127								% F127	
	0	0.5	1	2	5	10	15	20	0	20
5	X	X	X	X	X	X	X	X		
10				X	X	X	X	X		
20				X	X	X	X	X		
40					X	X	X	X		
100						X	X	X		

Conclusions

Exposure of nisin A to UV results in a loss of activity that depends primarily on the total dose of UV delivered. Although nisin A and nisin Z are nearly identical in structure, nisin A displayed substantially higher stability in the presence of UV compared with nisin Z. Nisin Z looses activity rapidly upon UV exposure and may not be suitable for applications involving UV processing. UV treatment of mixtures containing 2 mg/ml nisin A and F127 resulted in increasing loss of activity with increasing F127 concentration. However, for gels containing 20 mg/ml nisin, the loss of activity appears to be insignificant, even when using higher concentrations of F127. This indicates that it is feasible to apply UV to produce nisin containing gels that retain potent activity.

References

1. Mota-Meira, M., et al., "MICs of mutacin B-Ny266, nisin A, vancomycin, and oxacillin against bacterial pathogens." *Antimicrob Agents Chemother*, 44 (1), pp 24-29 (Jan. 2000).
2. Rollema, H. S., et al., "Improvement of solubility and stability of the antimicrobial peptide nisin by protein engineering." *Appl Environ Microbiol,* 61 (8), 2873-8 (Aug. 2000).
3. de Vos, W. M., et al., "Properties of nisin Z and distribution of its gene, nisZ, in Lactococcus lactis." *Appl Environ Microbiol,* 59 (1), 213-8. (Jan 1993).

(623bk) Production of Lignocellulases by Trichoderma Viride In Free-Cell and Immobilized-Cell Fermentations: A Comparative Study of the Secretome

Tian-Qing Lan[1], Shang-Tian Yang[2] and Dong Wei[1], (1)School of Light Industry and Food Sciences, South China University of Technology, Guangzhou, China, (2)Chemical and Biomolecular Engineering, The Ohio State University, Columbus, OH

Abstract: Lignocellulases play an important role in the development of bioethanol. *Trichoderma viride* has been widely used to produce lignocellulases in submerged fermentation. However, very little is known about the effects of the fermentation conditions on lignocellulases production and secretion. In this study, production of lignocellulases and secretome of *Trichoderma viride* in submerged fermentations with free cells and immobilized cells in a rotating fibrous-bed bioreactor (RFBB) were investigated and compared. In general, the enzyme activities of FPase and cellobiase and the saccharification yield of sugarcane bagasse using the fermentation broth produced in the immobilized-cell fermentation were respectively 35.4%, 125.2% and 55.3% higher than those from the free-cell fermentation. The secretomes or the protein profiles in these fermentation broths were analyzed with two-dimensional gel electrophoresis and MALDI-TOF-TOF mass spectrometry, which identified 24 protein spots with significantly different expression/secretion levels. Among them, cellobiohydrolase CBH Ⅱ (MW: 39.2 kDa, pI: 4.83), endoglucanase EG Ⅱ (MW: 44.9 kDa, pI: 4.97) and α-amylase (MW: 51.0 kDa, pI: 4.65) were highly expressed and secreted in the immobilized-cell fermentation, while free-cell fermentation gave more cellobiohydrolase CBH Ⅰ (MW: 47.1 kDa, pI: 4.29), CBH Ⅰ (MW: 52.8 kDa, pI: 4.33), CBH Ⅰ (MW: 54.6 kDa, pI: 4.81), and endoglucanase EG Ⅳ (MW: 35.9 kDa, pI: 5.29). Swollenin (MW: 53.1 kDa, pI: 4.8) was also identified in both fermentations, but at different pI locations on the gel. These results suggested that different secretomes might have contributed to the higher saccharification efficiency of the fermentation broth produced in the immobilized-cell fermentation.

See more of this Session: Poster Session: Bioengineering
See more of this Group/Topical: Food, Pharmaceutical & Bioengineering Division

(623bl) Syngas Fermentation In Trickle Bed Reactor Using "Clostridium Ragsdalei" Strain P11

Mamatha Devarapalli[1], **Hasan K. Atiyeh**[1], Randy S. Lewis[2] and Raymond L. Huhnke[1], (1)Biosystems and Agricultural Engineering, Oklahoma State University, Stillwater, OK, (2)Department of Chemical Engineering, Brigham Young University, Provo, UT

Syngas fermentation involves complex biochemical reactions of gaseous substrates to produce liquid products such as ethanol, butanol and acetic acid using special group of acetogenic microorganisms. Ethanol productivity and syngas conversion efficiency are severely affected by the low solubility of CO and H_2, the primary components of syngas, in the fermentation media. One way to improve ethanol productivity and syngas conversion efficiency is to increase the rate at which CO and H_2 are transferred into the fermentation media. A trickle bed reactor (TBR) can enhance gas-liquid mass transfer efficiency by minimizing the liquid resistance to mass transfer because of a very thin liquid film that is in contact with the gas. Results on mass transfer characteristics of a 1L TBR with two packing materials and syngas fermentations using "*Clostridium ragsdalei*" strain P11 will be presented. A counter-current flow of liquid and gas configuration in the TBR were used. The optimal conditions based on the highest gas-liquid mass transfer coefficients for CO and H_2 were considered in syngas fermentation experiments. A commercial syngas mixture containing 40% CO, 30% CO_2, 30% H_2, by volume, was fed to the TBR at 37°C. Cell growth rates, ethanol and acetic acid production rates, CO and H_2 consumption rates were determined.

See more of this Session: Poster Session: Bioengineering
See more of this Group/Topical: Food, Pharmaceutical & Bioengineering Division

(623bm) Effect of Propionyl-CoA Synthetase From Ralstonia Solanacearum by LC-MS/MS Quantification of Acyl-CoA Thioesters In Escherichia Coli

John W. Armando and Blaine A. Pfeifer, Chemical and Biological Engineering, Tufts University, Medford, MA

Metabolomics has become an increasingly important methodology for analyzing perturbations in cellular systems along with the more established proteomics and genomics tools currently available today. More specifically, targeted metabolite analysis allows direct quantification of small molecules of interest, and can give a snapshot of dynamic metabolic flux. This paper utilized quantitative LC-MS/MS to profile the short chain acyl-CoA levels of several engineered *Escherichia coli* constructed for heterologous polyketide production. It was observed upon feeding propionate, the engineered *E. coli* strains had increases in both propionyl- and methylmalonyl-CoA of ~6 to 30 fold and ~3.7 to 6.8 fold. The observed increases in acyl-CoA levels reflect the genetic modifications utilized for improved polyketide production. To further improve the levels of available acyl-CoA molecules, a flexible propionyl-CoA synthetase gene from *Ralstonia solanacearum* (*prpE-RS*) was subcloned and expressed in the engineered strain BAP1 (Watanabe et al 2004, Pfeifer et al 2000)). Watanabe et al. demonstrated substrate flexibility of prpE-RS in vitro, which should produce an increase in propionyl-, acetyl-, and butyryl-CoA when expressed in *Escherichia coli*. Induction of *prpE-RS* resulted in ~1.5, 15, and 8.5 fold increases in acetyl-, butyryl-, and propionyl-CoA respectively, when fed with the corresponding substrate. When compared to the empty vector control, no significant increases in acyl-CoA levels were observed, indicating that the substrate flexibility observed may be a result of the native T7 controlled PrpE rather than the heterologously expressed PrpE-*RS*. Additionally, the propionate transporter ATOAD was expressed with PrpE-*RS* resulting in a ~1.44, and 1.34 fold increase in butyryl and acetyl-CoA. No significant increase in propionyl-CoA was observed when ATOAD was expressed with PrpE-*RS*. As a result, the introduction of the flexible PrpE-*RS* and propionate transporter ATOAD did not significantly improve the acyl-CoA levels in *Escherichia coli*, but it was observed that the native prpE overexpressed in our engineered strains may have an intrinsic substrate flexibility resulting in increased acetyl-, propionyl-, and butyryl-CoA levels upon over-expression providing a more robust platform for polyketide production.

See more of this Session: Poster Session: Bioengineering
See more of this Group/Topical: Food, Pharmaceutical & Bioengineering Division

(623bn) Developing a Miniaturized Assay for Label-Free, High-Throughput Biomolecule Analysis

Aaron M. Joffe, Thuyoanh Truong, William S. Ferguson and **Roger C. Lo**, Department of Chemical Engineering, California State University, Long Beach, Long Beach, CA

Microfluidics technology has been widely applied to the miniaturization of analytical methods and chemical and biological processes because it has many advantages, such as significant reduction in analysis time, much lower waste production, and enhanced system performance and functionality by integrating different components onto microfluidic devices. These applications are usually called micro total analysis systems (μTAS) or lab on a chip (LOC). Among all the detection methods developed for microfluidic devices, laser-induced fluorescence (LIF) still remains the most popular because of its high sensitivity and compatibility with microfluidic devices. However, conventional fluorescence-based detection requires the analytes to be self-fluorescent or to be labeled with a fluorescence tag and thus results in limited applicability. Another issue of fluorescence-based detection is the analyte labeling process, which is not always straightforward and may cause undesired results, especially for biomolecules like proteins. UV absorbance-based detection has been a good alternative because it's a label-free, nondestructive technique and has been widely used in many commercially available capillary electrophoresis and chromatography systems. However, the integration of UV absorbance-based detection into microfluidic devices is not always straightforward due to the short optical pathlength (usually between 5 and 50 μm) of typical microchannel designs. Additional difficulty comes from the need of UV-transparent material, usually quartz or fused silica, to construct microfluidic devices compatible with UV absorbance-based detection. This results in a significant increase in the cost and complexity of the device fabrication process.

To address these problems, we have designed and fabricated a new high-aspect-ratio microfluidic chip using a commercially available UV-curable glue and simple soft lithographic technique, and integrated it with a two-dimensional UV-absorbance imaging detector. The UV imaging detection technique we use here allows simultaneous detection of analytes in multiple microchannels on a single microchip with a high level of spatial and temporal resolution. As a proof of concept, we used this system to perform microchip DNA gel electrophoresis without labeling. This system can be extended to other biomolecule assays by changing the chip design. After optimization, we will use this system to perform a miniaturized, label-free protein immunoblot to validate the sexually dimorphic expression of two epigenetic protein biomarkers, H3K9/14Ac and H3K9Me3, from neonatal mouse brains.

See more of this Session: Poster Session: Bioengineering
See more of this Group/Topical: Food, Pharmaceutical & Bioengineering Division

(623bo) Analysis of Microbial Population Dynamics Using the Visualizing Evolution In Real Time (VERT) System

James D. Winkler and Katy Kao, Chemical Engineering, Texas A&M University, College Station, TX

Adaptive evolution has been harnessed to improve industrially relevant strain qualities such as inhibitor tolerance, product titer, and product yield for decades. Despite the importance of this tool in industry and academia, many questions remain concerning how new adaptive mutants compete within a population. Sampling populations for adaptive mutants requires lengthy analysis of isolated mutants and provides only indirect information concerning the relative proportion of each mutant within the entire population. However, even in light of these difficulties, much progress has been made in experimental and theoretical analysis of evolutionary dynamics.

Unlike previous attempts to quantify or model population dynamics in response to an external stressor, the Visualizing Evolution in Real Time (VERT) system is capable of tracking fluorescently labeled subpopulations during adaptive evolution in the laboratory. We exploit this unique feature of VERT to compare the population dynamics of three evolving microbial species in order to provide new insight into process of adaptive evolution and quantify parameters of interest, such as the number of possible adaptive mutations in a given environment. Based on the observed population dynamics data, an algorithm capable of autonomously detecting adaptive mutants within a population is also developed and validated against experimental data. Together, these results represent an important advance in the understanding of laboratory adaptive evolution.

See more of this Session: Poster Session: Bioengineering
See more of this Group/Topical: Food, Pharmaceutical & Bioengineering Division

(623bp) Metabolic Engineering of P. Freudenrichii for Enhanced Propionic Acid Production

Ehab Mohamed Ammar, William G. Lowrie Department of Chemical and Biomolecular Engineering, Ohio State University, Columbus, OH and Shang-Tian Yang, Department of Chemical and Biomolecular Engineering, Ohio State University, Columbus, OH

Propionic acid is widely used as a chemical intermediate in various manufacturing processes, and its salts are common food preservatives. It is mainly produced via petrochemicals, but there is increasing interest in its production by fermentation these days. Unfortunately, fermentative production provides low yields due to the formation of other organic acids, mainly acetic, as byproducts. In addition, the accumulation of propionic acid in the medium is inhibitory to cells. In this work, the effect of up-regulating some genes in the propionate biosynthesis pathway to shift the carbon flux towards propionate was studied. Higher propionate yields from sugar fermented were obtained. Surprisingly, acetate yields increased assumingly to provide more NADH to support propionate formation. Consequently, up-regulating the propionate pathway is better accompanied by down-regulating the acetate pathway. We anticipate that such approach in combination with process engineering can facilitate the development of a fermentation process for the economic production of propionic acid from sugars and other renewable feedstocks.

See more of this Session: Poster Session: Bioengineering
See more of this Group/Topical: Food, Pharmaceutical & Bioengineering Division

(623br) Computational and Thermodynamic Characterization of a Mostly Disordered Protein and An Aging Related Mutant

Agnieszka M. Kalinowski[1], Zhao Qin[2], Markus J. Buehler[2] and **Kris Noel Dahl**[3], (1)Biomedical Engineering, Carnegie Mellon University, Pittsburgh, PA, (2)Civil and Environmental Engineering, MIT, (3)Chemical Engineering, Carnegie Mellon University, Pittsburgh, PA

Hutchinson-Gilford progeria syndrome (HGPS) is a premature aging syndrome caused by the expression and accumulation of a mutant lamin A protein, progerin. We report experimental and computational results that characterize the thermodynamic stability of the lamin A and progerin tail domains, which are mostly disordered. We study protein unfolding using tryptophan fluorescence and circular dichroism of purified, recombinant wild type (wt) and progerin tail domains. Both methods indicate that the progerin tail has a higher transition melting temperature than its wt counterpart. Replica Exchange Molecular Dynamics (REMD) simulations predict the progerin tail to be a more compact structure. Simulated protein unfolding confirms our experimental results; the predicted energy to unfold the progerin tail is higher than the wt lamin A tail. Structural simulations of the wt lamin tail show that the main site for proteolysis during normal posttranslational modification of lamin A is an unfolded region located at the periphery of all the predicted protein configurations, suggesting ample accessibility to proteolysis. Progerin lacks this region for posttranslational proteolysis, but its further compactness suggests a general resistance to turnover and more stable filament structures, potentially leading to the toxic accumulation observed in disease and during normal aging. We believe that this combination of thermodynamic characterization and REMD simulations is a useful method of studying structure of disordered proteins.

See more of this Session: Poster Session: Bioengineering
See more of this Group/Topical: Food, Pharmaceutical & Bioengineering Division

(623bs) Characterization of Short Chain Alcohol Dehydrogenases From Lactobacillus Brevis

Shawn Pugh and David R. Nielsen, Chemical Engineering, Arizona State University, Tempe, AZ

The substrate specificity of several previously uncharacterized short chain alcohol dehydrogenases from *Lactobacillus brevis* was explored. Six putative enzymes were cloned and overexpressed in *E. coli*. Protein BLAST (Basic Local Alignment Search Tool) analysis was done to determine possible substrates for which each protein may possess activity. The catalytic activity of LVIS_0108 on D-gluconate was confirmed and its dual cofactor (NAD^+ and $NADP^+$) specificity was investigated. In addition, the gene LVIS_0347, identified by Hummel *et al.* as the (R)-stereospecific alcohol dehydrogenase (RADH), was found to possess broad substrate activity on several short chain aldehydes and ketones, previously uncharacterized to perform such activity. Catalytically efficient, broad substrate alcohol dehydrogenases are especially useful for biofuel synthesis pathways. To demonstrate its relevance to the biofuels industry and its utility as a broad substrate alcohol dehydrogenase, a detailed catalytic characterization was performed to further investigate the activity of LVIS_0347. The remaining four putative enzymes remain under investigation as to their substrate specificity and functionality in *L. brevis*.

See more of this Session: Poster Session: Bioengineering
See more of this Group/Topical: Food, Pharmaceutical & Bioengineering Division

(623bt) Integrating Functional Genomics Tools to Survey Retrovirus Production In Human Cells

Ana Filipa Rodrigues[1], Ana Teixeira[1], Nuno Carinhas[1], Vanessa S. Bandeira[1], Paula M. Alves[1], Rui Oliveira[2], Wei-Shou Hu[3] and Ana S. Coroadinha[1], (1)Animal Cell Technology Unit, IBET/ITQB, Oeiras, Portugal, (2)Departamento de Qumica, REQUIMTE/CQFB, Universidade Nova de Lisboa, 2829-516 Caparica, Portugal, (3)Chemical Engineering and Materials Science, University of Minnesota, Minneapolis, MN

Retrovirus derived particles found valuable biotechnological application in Vaccinology and Gene

therapy, although challenging from the manufacturing view-point, due to the low titers and high content of

contaminant defective particles. Additionally, retroviruses such as HIV are a major world-wide health

concern with the cell-virus interaction dynamics poorly understood. Thus, an integrated overview on the

cellular determinants of retrovirus production in human cells can provide a valuable starting point and a

potential tool to address these and other issues.

In this work, we analyzed transcriptome changes between retrovirus producers and the corresponding

parental cells using two different human cell lines, and took a pathway analysis approach to identify

biochemical networks overrepresented in the virus production state. Nearly 200 pathways were identified,

majorly represented by signaling and metabolism. Retrovirus production induced transcriptome changes

associated with apoptotic and cell survival pathways suggesting a balance between both networks, a typical

feature of virus infection scenario. Fatty acid metabolism was among the most prominently up regulated

pathways, along with sphingolipid and phospholipid metabolism, suggesting increased membrane recycling dynamics for viral replication. Amino acid degradation was notably up-regulated as well. Virus production appears to result in higher energy demands, evidenced by the strong up-regulation of the oxidative phosphorylation and electron transport chain. Oxidative stress, pyrimidine metabolism and protein synthesis and post-translation modification were also highlighted.

Based on the pathway analysis output, we investigated particular transcriptome-fluxome correlations. Amino acid catabolism profile, one of the most significantly enriched pathways in terms of gene expression, was analyzed by HPLC and found to be considerable increased, up to 2 fold, in virus producer cells. Other suggested pathways are currently under study, some of which further substantiate pathways analysis results, including oxidative stress metabolism and lipid biosynthesis.

In parallel, the integration of transcriptomic and metabolomic data is being used to feed a genomescale constraint-based reconstruction of the metabolic network, in which a lumped reaction comprising all building blocks for virus production was included. Such flux-balance analysis based model should help to elucidate the cell fluxome alterations induced by a virus productive state and the rules to either enforce or inhibit such state. This work has a direct application in the field of retrovirus manufacture for Vaccinology and Gene Therapy, providing a framework to guide the rational design of process engineering and cell line development. It can also be relevant basic virology providing further insights on the cell-virus interaction and disclosing potential therapeutic targets to inhibit viral replication.

See more of this Session: Poster Session: Bioengineering

(623c) Inclusive Design of Flux-Determining 13C NMR Experiments

Jonathan Meade, Chemical Engineering, Carnegie Mellon, Pittsburgh, PA, Saleem A. Khan, Microbiology and Molecular Genetics, University of Pittsburgh, Pittsburgh, PA, Mohammad M. Ataai, Chemical and Petroleum Engineering, University of Pittsburgh, Pittsburgh, PA and Michael M. Domach, Chemical Engineering, Carnegie Mellon University, Pittsburgh, PA

NMR is supposed to fill in the flux value gaps that result when extracellular measurements are insufficient for contending with the degrees of freedom and redundancy present in a metabolic network. Many methods use several analytes and randomly search for a flux set that minimizes the error between the calculated and experimental NMR data. While this approach can be effective, one problem that this presentation will introduce is multiple solutions have been found for the same cell/mutant under similar growth conditions. This is problematic because NMR is supposed to be the resolver of flux possibilities; hence, finding different answers in the literature for the same cell/mutant is unsettling. We suggest that this problem can originate due to (i) coupling flux constraints to iterative error minimization rather than making full use of MFA capabilities, (ii) assuming no error in external measurements, (iii) failing to distinguish between gross error minimization and contrast between the alternate flux set solutions, (iv) extrapolating wild-type physiology to mutants, and (v) not recognizing spectrum uniqueness in terms of fluxes. This presentation will describe a maximally inclusive formulation for the NMR experimental design problem. The flux bounds and thus the possible flux sets are developed based on factors (i) – (iv). Label design is then optimized for an analyte or set. A test for whether the relative and nondiscarded information in spectra will lead to a unique flux distribution is then performed. Finally, for experimental design and then problem closure, we will show that for some problems, quasi-linearity exists where the isotopomer distribution vectors for analytes at flux extreme points can be combined in a linear fashion, which then makes the NMR data-to-flux problem a directed and bounded flux-determining problem. The example used to illustrate the approach and computations will be determining the fluxes in an E. coli mutant that has been shown to provide increased yield of therapeutic DNA. The yield evidence illustrates that the mutation "works", but understanding exactly why in terms of fluxes is unresolved because two flux solutions have been reported for the mutant by NMR, where only one is consistent with flux optimization analysis. Thus, this case study has practical and basic utility.

See more of this Session: Poster Session: Bioengineering
See more of this Group/Topical: Food, Pharmaceutical & Bioengineering Division

PERVAPORATIVE SEPARATION OF BIOETHANOL USING A COMPOSITE HOLLOW FIBER MEMBRANE

Hong-Joo Lee[1], Eun-Jin Cho[2], Seung-Gon Wi[2] and Hyeun-Jong Bae[2],
[1]Dept. Bioenergy Sci. Technol, Chonnam National University, Gwangju, South Korea,
[2]Bio-energy Research Institute, Chonnam National University, Gwangju, South Korea

Abstract

In this study, a composite hollow fiber membrane with a thin PDMS (Polydimethylsiloxane) active layer on the PEI (polyetherimide) macroporous structures was used for pervaporative separation of ethanol from fermentation broth. Pervaporation was carried out in model solutions of ethanol/water binary mixtures with a wide concentration range to investigate the influence of temperature and concentration on the process performances. The ethanol flux values increased linearly due to the increasing driving force of the concentration difference, while the separation factor sharply decreased in dilute solutions. The influence of temperature on permeation performance showed that the increase of ethanol flux with temperature was increased. In this study, bioethanol produced from fermentation using *Saccharomyces cerevisiae* was used as feed for pervaporation of the composite hollow fiber module. The pervaporative ethanol separation from the fermentation broth did not show a significant concentration polarization, nor a fouling effect on the hollow fiber membranes. A higher ethanol flux was observed in bioethanol pervaporation, than in the model solution, which can be understood in terms of the hydrophobic surface of the PDMS membrane, and rejection due to the presence of materials in the fermentation broth.

Introduction

Please Concerns over the environment with respect to the depletion of fossil fuels have increased interest in technologies related to renewable energy sources. Ethanol is one of the most important renewable fuels, with a potential to reduce negative environmental impacts [1,2]. In most cases, however, distillation is not an economical process for recovery due to the low concentration of alcohol and high concentration of solids in the fermentation broth. Among various separation processes, pervaporation is considered to be one of the most promising approaches in recovery of alcohols from fermentation broths since it is simple, nontoxic to fermenting microorganisms, and potentially less expensive than distillation [3-5]. In addition, pervaporation offers new possibilities for the use of fermentation, allowing removal of the ethanol produced by reducing the natural inhibition of cell growth caused by high ethanol concentrations. Applications of pervaporation to a variety of areas have already been suggested, such as dehydration of organic solvents, removal of dilute organic compounds from the aqueous phase, and separation of organic-organic mixtures without significant limitations [6,7]. When a membrane is in contact with a liquid mixture, one of the components can be preferentially removed from the mixture due to its higher affinity and/or diffusivity in the membrane.

Nowadays, development of hollow fiber modules for pervaporation applications has recently attracted interest due to its high packing density, self-supporting structure, and economical superiority [8-10]. Composite membranes, where a thin separation layer is supported on a highly porous substrate, give a sufficiently high permeation flux. It is accepted that the porous support provides mechanical strength and the separation layer determines the process performance of the composite membrane in terms of flux and separation factor [11-13]. Among the PDMS (Polydimethylsiloxane) membrane with polymeric supports, only a few results using the composite PDMS/PEI(polyetherimide) membranes for the pervaporation applications were reported and the composite membranes were the sheet type with small area for now as far as we know [14-16].

In this study, a composite hollow fiber membrane consisting of PDMS as a thin active layer was examined in pervaporation separations. In order to determine optimal pervaporation operating conditions and investigate the influence of temperature and feed concentration, model solutions of ethanol/water with a wide concentration range were used as feed in the study. In addition, the pervaporative separation of bioethanol from fermentation broth was performed with different concentration of bioethanol.

Experimental

Fermentation was carried out in a FMT-05 fermenter (Fermentec, Korea), using a medium containing 5 g/L yeast extract, 5 g/L peptone, 3.3 g/L potassium phosphate, 0.2 g/L ammonium sulfate, 0.4 g/L magnesium sulfate, and 100 g/L glucose. For the inoculum, a spore suspension of *Saccharomyces cerevisiae* KCTC 7906 was cultured on the yeast extract, peptone, and glucose at 30°C for 36 hours. Three hundred mL of the actively growing culture was added to the fermenter and kept under anaerobic conditions before and after inoculation. After 24–48 hour s of fermentation at 30 °C and an agitation speed of 300 rpm, the fermentation broth was filtered through a Millipore mixed cellulose esters membrane filter paper.

The composite hollow fiber membrane module was kindly supplied by Airrane, Korea. It consisted of an active layer of PDMS on a polyetherimide support. The effective area of the composite hollow fiber membrane was 903 cm^2. Outer and inner diameter of a fiber was 920 μm and 710 μm, respectively (thickness of a fiber: 105 μm). The hollow fiber membrane module was used for the pervaporation of synthetic solutions of a binary mixture of ethanol/water. One liter of the solutions in a conical flask was circulated at the constant flow rate of 1.8 L/min in the experimenrtal setup of the pervaporative separation shown in Fig. 1. The flask was sealed, and a condenser was used to condense volatiles. The feed was kept at atmospheric pressure, whereas the permeate pressure was maintained 5 mmHg using a vacuum pump in all experiments.

A temperature-controlled water bath ranging from 20 to 60 °C was used to heat the retentate to the desired temperature in the pervaporative separation from synthetic solutions. Different concentrations of the synthetic solutions were used as feed to investigate the influence of concentration on the pervaporation performance. The filtered fermentation broth was used as the feed for the bioethanol pervaporation separation. After 2 h circulation, the product was sampled periodically in a cold trap immersed in liquid nitrogen. The pervaporation performance was investigated in terms of ethanol flux (J_p) and ethanol separation factor (α) as follows:

$$J_p = \frac{W}{A_m t} \qquad (1)$$

$$\alpha = \frac{y/(1-y)}{x/(1-x)} \qquad (2)$$

where W is the weight of the condensate (g), A_m the membrane area (m^2), t the duration time (*h*), *x* and *y*, weight fraction of components in feed and permeate, respectively. Ethanol concentration was determined using an HPLC equipped with an RI detector (Waters 2414, USA) using a

Rezex RPM column (4.6x300 mm; Phenomenex, USA). HPLC-grade water was supplied at a flow rate of 0.6 mL/min as a mobile phase, while the temperature was controlled at 85 ^{o}C.

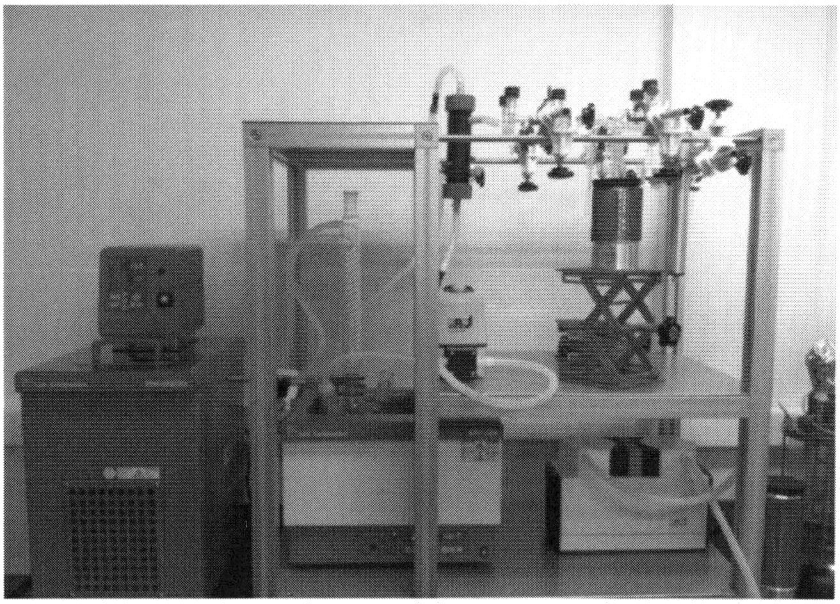

Fig. 1. Experimental setup of the pervaporative separation.

Results and Discussion

Pervaporation membrane performance in model solutions

In this study, the feed concentration was increased from 1.0% to 10.0% in order to observe the influence of feed concentration on the pervaporation performance estimated by ethanol flux and the separation factor. The recovered ethanol concentration increased linearly by up to 4%, showing that the ethanol concentration in the product was 30%. At concentrations higher than 4%, the ethanol concentration in the product showed similar values. Solution-diffusion is the generally accepted mechanism of mass transport through non-porous membranes. In a pervaporative separation, mass transport through a membrane is a rate-controlling process, which is generally governed by Fick's first law. Therefore, the ethanol flux is proportional to the feed concentration, and the driving force is the difference between the feed and permeate concentrations [6]. It has been suggested that the higher ethanol flux at higher feed concentration is due to a higher driving force of the pervaporation of ethanol through the PDMS membrane [17].

This study showed that the ethanol flux and separation factor have a strong dependence on feed concentration. The ethanol flux showed similar values at around 20 g/m^{2}h at concentrations lower than 2.0%. The ethanol flux increased linearly at concentrations higher than 2.0%, resulting in increased ethanol flux, up to 80 g/m^{2}h at a feed concentration of 10.0%, as shown in Fig. 2. The total flux in the PDMS/PEI composite hollow fiber membrane was measured to about 250 g/m^{2}h. In general, membrane performance shows a trade-off between the flux and separation factor, that is, the separation factor decreases as the flux increases. The separation factor decreased linearly at concentrations lower than 6.0%, and the resulting values did not decrease significantly at concentrations higher than 6.0% [18].

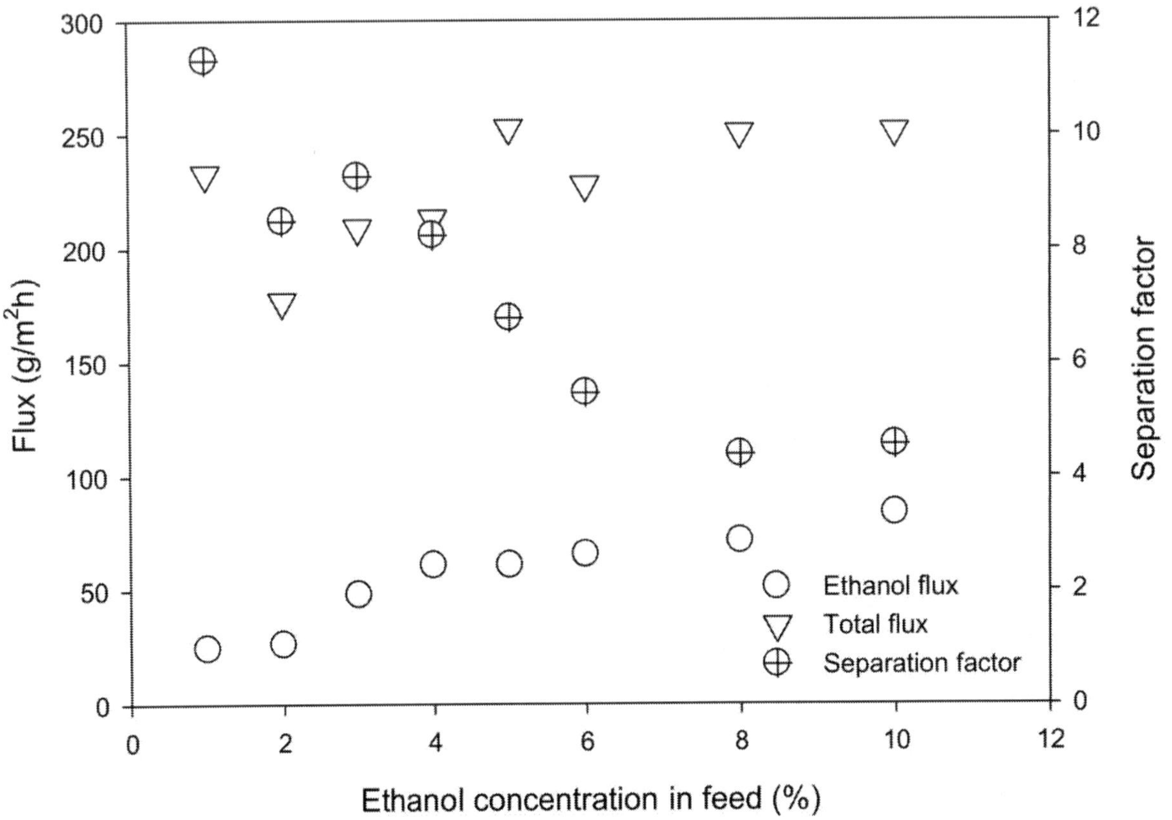

Fig. 2. Influence of ethanol concentration on the pervaporative separation of ethanol.

In pervaporation, temperature affects solubility and diffusivity in the membrane structure, as well as the permeation flux [19]. The influence of temperature on pervaporation performance was investigated in the pervaporative separation at a feed concentration of 5.0%. Table 1 shows the influence of temperature on the pervaporation performance in the composite hollow fiber membrane. As shown in the table, the ethanol concentration in the product decreased from 27% at the temperature of 20 °C to 21% at 60 °C. In addition, ethanol flux value of 67 g/m^2h was estimated at a temperature of 60 °C, which is 2.5 times higher than that at 20 °C. It is suggested that increased flux at high temperature is related to the increased swelling capacity and water desorption due to weak adsorption capacity on the hydrophobic membrane surface [18]. As for the separation factor, the estimated value decreased from 7 to 5 as the temperature increased from 20 °C to 60 °C.

The temperature dependence of the permeation flux can be expressed by an Arrhenius-type relationship:

$$J_p = A_p \exp(-\frac{E_a}{RT}) \quad (3)$$

where J_p is the permeation flux, A_p the pre-exponential factor, R the ideal gas constant, T the temperature and E_p the apparent activation energy of permeation. The apparent activation energy accounts for the effect of temperature, which characterizes the temperature dependence of membrane permeability. The apparent activation energy of ethanol permeation was calculated to be 19.3 kJ/mol. In the relationship between water flux and operating temperature, the apparent activation energy of water permeation was calculated to be 23.3 kJ/mol. The higher activation energy value of water flux indicates that as the temperature increased, water flux increased more than ethanol flux did.

Table 1. Influence of temperature on the pervaporative separation of ethanol

Temp (°C)	Product Conc (%)	Ethanol Flux (g/m^2h)	Separation factor	Total Flux (g/m^2h)
60	21.3	67.4	5.1	302.879
50	23.0	41.2	5.7	196.124
40	32.0	78.0	8.9	236.766
30	25.3	34.5	6.4	130.122
20	26.9	26.7	7.0	98.450

Pervaporative separation of bioethanol from fermentation broth

In this study, bioethanol was produced through fermentation, and the pervaporative separation was executed at the temperature of 40 °C from the fermentation broth. Three repeated experiments executed with different feed concentrations showed that the bioethanol recovery and ethanol flux showed higher values at high bioethanol concentration. At low concentrations (1.19 and 1.53%), ethanol concentration in the product was below 10%. Meanwhile, ethanol concentration in the product increased up to 29% at the feed concentration of 4.57%. Ethanol flux was estimated to 18 g/m^2hr at low ethanol concentrations in feed, and 70 g/m^2hr at high ethanol concentrations. Similar separation factor values of 7~8 were observed throughout all experiments.

Table 2 compares the pervaporation results for the fermentation broth and ethanol/water synthetic binary system. It is clearly shown that pervaporation resulting in the fermentation broth did not decrease the membrane performances. In the pervaporation of fermentation broth, even the ethanol flux and separation factor values were higher compared to the model solution (5.0%) due to the rejection increase in the fermentation broth and the hydrophobic surface of PDMS membrane [20].

Table 2. Pervaporative separation of bioethanol from fermentation broth

Run		Ethanol concentration in feed (%)	Ethanol concentration in product (%)	Flux (g/m^2hr)	Separation factor
Fermentation broth	Run 1	1.2	7.7	18.4	6.9
	Run 2	1.5	9.8	17.74	7.0
	Run 3	4.6	28.7	69.9	8.4
Synthetic solution		1.0	10.3	24.5	11.3
		5.0	26.2	60.9	6.8

Concluding Remark

In this study, a composite hollow fiber membrane consisting of the active thin layer of PDMS on a polyetherimide support was used for the pervaporative separation of ethanol from ethanol fermentation broth and model solutions of water and ethanol mixtures. The influence of feed concentration and temperature was investigated in the model solutions of water/ethanol with a concentration range of between 1.0 and 10.0% were supplied as feed. Ethanol flux and separation factor each showed a strong dependence on feed concentration. The ethanol flux values increased linearly due to the increase in the driving force of the concentration difference, while the separation factor values sharply decreased in dilute solutions. The ethanol flux through the composite hollow fiber membranes increased as the temperature increased.

Different concentrations of bioethanol, produced from fermentation using *Saccharomyces cerevisiae*, were used as feed for the pervaporative separation with the composite hollow fiber membrane. In three repeated experiments with different concentrations in feed, the recovered ethanol concentration and ethanol flux showed high values at high bioethanol concentrations. In addition, comparison of the pervaporation results between the fermentation broth and synthetic ethanol/water solution clearly showed that the fermentation broth did not foul the membrane, nor give significant concentration polarization.

Acknowledgements
This work was supported by Priority Research Centers Program through the National Research Foundation of Korea (NRF) funded by the Ministry of Education, Science and Technology (Project No. 2010-0020141) and from the World Class University project of the Ministry of Science and Technology of Korea (R31-2009-000-20025-0) to H.-J. Bae.

References

1. Cardona, C.A., Sanchez, Ó.J. (2007), "Fuel Ethanol Production: Process Design Trends and Integration Opportunities-Review," *Bioresource Technol.* 98, pp. 2415-2457.
2. Huang, H.J., Ramaswamy, S., Tshirner, U.W., Ramarao, B.V. (2008), "A Review of Separation Technologies in Current and Future Biorefineries," *Sep. Purif. Technol.* 62, pp. 1-21.
3. Wu, Y., Xiao, Z., Huang, W., Zhong, Y. (2005), "Mass Transfer in Pervaporation of Active Fermentation Broth with a Composite PDMS Membrane," Sep. Purif. Technol. 42, pp. 47-53.
4. Mohammadi, T., Aroujalian, A., Bakhshi, A. (2005), "Pervaporation of Dilute Alcoholic Mixtures Using PDMS membrane," *Chem. Eng. Sci.* 60, pp. 1875-1880.
5. Qureshi, N., Meagher, M.M., Huang, J., Hutkins, R.W. (2001), "Acetone Butanol Ethanol (ABE) Recovery by Pervaporation Using Silicate-silicone Composite Membrane from Fed-batch Reactor of *Clostridium acetobutylicum*," *J. Membr. Sci.* 187, pp. 93-102.
6. Shao, P., Huang, R.Y.M. (2007), "Polymeric Membrane Pervaporation-Review," *J. Membr. Sci.* 287, pp. 162-179.
7. Fleming, H.L., Slater, C.S. (1992), "Pervaporation," in *Membrane Handbook*, W.S. Winston Ho, K.K. Sirkar ed., Van Nostrand Reinhold, New York.
8. Zhan, X., Li, J., Huang, J., Chen, C. (2010), "Enhanced Pervaporation Performance of Multi-layer PDMS/PVDF Composite Membrane for Ethanol Recovery from Aqueous Solution," *Appl. Biochem. Biotechnol.* 160, pp. 632-642.
9. Shi E., Huang W., Xiao Z., Li, D., Tang, M. (2007), "Influence of Binding Interface between Active and Support Layers in Composite PDMS Membranes on Permeation Performance," *J. Appl. Polym. Sci.* 104, pp. 2468-2477.
10. Chang, C.L., Chang, M.S. (2004), "Preparation of Multi-layer Silicone/PVDF Composite Membranes for Pervaporation of Ethanol Aqueous Solutions," *J. Membr. Sci.* 238, pp. 117-122.
11. Liu, G., Xiangli, F., Wei, W., Liu, S., Jin, W. (2011), "Improved Performance of PDMS/ceramic Composite Pervaporation Membranes by ZSM-5 Homogeneously Dispersed in PDMS via a Surface Graft/coating Approach," *Chem. Eng. J.* In Press.
12. Wei, W., Xia S., Liu, G., Dong, X., Jin, W., Xu, N. (2011), "Effects of Polydimethylsiloxane

(PDMS) Molecular weight on Performance of PDMS/ceramic Composite Membranes," *J. Membr. Sci.* 375, pp. 334-344.

13. Xiangli, F., Chen, Y., Jin, W., Xu, N. (2007), "Polydimethylsiloxane (PDMS)/ceramic Composite Membrane with High Flux for Pervaporation of Ethanol-water Mixtures," *Ind. Eng. Chem. Res.* 46, pp. 2224-2230.
14. Galindo, M.O., Clar, A.I., Miranda, I.A., Greus, A.R. (2001), "Characterization of Poly(dimethylsiloxane)–poly(methyl hydrogen siloxane) Composite Membrane for Organic Water Pervaporation Separation," *J. Appl. Polym. Sci.* 81, pp. 546-556.
15. Zhao, C., Li, J., Chen, J., Qi, R., Luan, Z. (2009), "Separation of Sulfur/gasoline Mixture with Polydimethylsiloxane/polyetherimide Composite Membranes by Pervaporaiton." *Chinese J. Chem. Eng.* 17, pp. 707-710.
16. Zhao, C., Li, J., Qi, R., Chen, J., Luan, Z. (2008), "Pervaporation Separation of n-heptane/sulfur Species Mixtures with Polydimethylsiloxane Membranes," *Sep. Purif. Technol.* 63, pp. 220-225.
17. Thongsukmak, A., Sirkar, K.K. (2007), "Pervaporation Membranes Highly Selective for Solvents Present in Fermentation Broths," *J. Membr. Sci.* 302, pp. 45-58.
18. Quresh, N., Meagher, M.M., Hutkins, R.W. (1999), "Recovery of Butanol from Model Solutions and Fermentation Broth Using a Silicate/silicone Membrane," *J. Membr. Sci.* 158, pp. 115-125.
19. Fouad, E.A., Feng, X. (2009), "Pervaporative Separation of n-butanol from Dilute Aqueous Solutions Using Silicate-filled Poly(dimethyl siloxane) Membranes," *J. Membr. Sci.* 339, pp. 120-125.
20. Chovau, S., Gaykawad, S., Straathof, A.J.J., Van der Bruggen, B. (2011), "Influence of Fermentation By-products on the Purification of Ethanol from Water Using Pervaporation." *Bioresource Technol.* 102, pp. 1669-1674.

(623e) MicrobesFlux: a Web Based Platform for Reconstruction and Analysis of Genome-Scale Metabolic Models

Xueyang Feng[1], You Xu[2], Yixin Chen[2] and Yinjie J. Tang[1], (1)Department of Energy, Environmental and Chemical Engineering, Washington University, St Louis, MO, (2)Department of Computer Science and Engineering, Washington University, St. Louis, MO

Arising interests have focused on metabolic flux analysis of genome-scale networks in microorganisms. Fluxomics approach empowers systems analysis of cell-wide functions and regulations. Although such *in silico* analysis of metabolic networks has been applied in diverse microorganisms, the model development still lags behind the pace of high throughput genome sequencing. To bridge the gap between the genome annotation and the functional characterization, it is necessary for biologists to have convenient fluxomics tools for reconstruction of genome-scale models. Here, we present MicrobesFlux, a web-based platform which performs constraint-based flux balance analysis under both steady and dynamic metabolic statuses. MicrobesFlux contains five features as follows. First, it has a user-friendly interface to import annotated metabolic network from KEGG database. Second, it allows genome reconstructions (e.g. gene knock-out or introducing heterologous pathways). Third, it automatically creates a constraint-based flux balance model to quantify the intracellular flux distributions. Fourth, it can perform dynamic flux analysis via static optimization approach. Fifth, it facilitates the optimization of large-scale metabolic models via cloud computation methods.

See more of this Session: Poster Session: Bioengineering
See more of this Group/Topical: Food, Pharmaceutical & Bioengineering Division

(623f) Mathematical Modeling of the Amensal Relation Between Pseudomonas Aeruginosa and Micrococcus Luteus - Experimental and Simulation Studies

Baskaran Sivaprakash, Chemical Engineering, Annamalai University, Chidambaram, India, Singaram Jayalakshmi, Marine Biology, Annamalai University, Chidambaram, India and Thirugnanam Karunanithi, Engineering, Anna University, Coimbatore

Large numbers of microbial operations in industries employ mixed populations and in natural systems mixed culture of microorganisms are the rule rather than the exception. The mixed species interact in complex ways. Amensalism is a negative kind of interaction in which the growth of one species is inhibited by the presence of another, whose growth is unaltered due to interaction. The experimental studies (batch conditions) on the pure and mixed culture operations of the bacterial system *Pseudomonas aeruginosa* and *Micrococcus luteus* proved the existence of amensal interaction against luteus species. The experiments were carried out with various initial substrate concentrations (0, 100, 200, 300 and 400 mg/ml) to study its influence on the pure and mixed culture growth. Suitable models were developed to predict the pure and associated growth rates based on logistic equations. The k values in the logistic equation remained constant with respect to the initial substrate dosage for both the species (pseudomonas species: 0.2417 h^{-1}; micrococcus species: 0.2066 h^{-1}). However the other constant β showed decreasing trend (cubic polynomial) with increasing glucose concentrations for both the species. The interaction effect of species 2 on 1 is negligible whereas the interaction function $f_{21}(x_1, x_2)$ for the second by the first species is found to be second order. The interaction parameter a_{21} varies as a cubic polynomial with respect to the initial substrate concentration. Extension of the batch case equations to continuous operation was carried out using Chemostat model. The simulation studies yielded the range of dilution rates to be maintained for coexistence of both the species for the various initial concentrations and the respective steady state microbial concentrations.

See more of this Session: Poster Session: Bioengineering
See more of this Group/Topical: Food, Pharmaceutical & Bioengineering Division

(623g) Host Selection for Synthetic Pathways Using a Genome-Scale Model Database

Hadi Nazem-Bokaee, Biological Systems Engineering, Virginia Tech, Blacksburg, VA and Ryan S. Senger, Biological Systems Engineering Department, Virginia Tech, Blacksburg, VA

The number of available genome-scale models is growing exponentially, and the standards and formats in which they are published remain diverse. This has made it difficult for the computational biologist to compare several genome-scale models side-by-side. One notable application for this analysis is host selection for a novel synthetic pathway based on the metabolic capabilities of several organisms. To do this, genome-scale model information must be organized and stored in not only a retrievable but also an easily modifiable format. Several databases have been developed over the past decade for storing and using the genomic, enzymatic, metabolic pathway, and expression/phenotypic data. During this time, genome-scale models have become established as a systems-level tool for integrating this information. Thus, newer databases, such as the BiGG database, The SEED, MEMOSys (and others soon to be published) have been created to house metabolic network reconstructions. Here, we present the first database of genome-scale models that allows the user to incorporate user-defined synthetic pathways to any model and retrieve a standard or modified model in Systems Biology Markup Language (SBML) format. This database is of special interest to metabolic engineers and systems biologists since it can (i) act as a platform for exporting existing genome-scale models, with thermodynamic and pKa data, and modifying models with updated biochemical and phenotypic data, (ii) it can be used to compare reactions, genes, and compounds contained in different organisms and give insights into the similarities and discrepancies of existing models, and (iii) it can be used for synthetic pathway incorporation and prediction of potential pathway usage. Over 20 genome-scale models have been added to the database to date. Here, we will present a novel application of host selection for optimal expression of a given synthetic pathway to demonstrate the usefulness of the database approach to this important problem.

See more of this Session: Poster Session: Bioengineering
See more of this Group/Topical: Food, Pharmaceutical & Bioengineering Division

(623h) Reverse Engineering the "Small-World" Gene Networks

Prateek Gupta and **Sanjeev Garg**, Chemical Engineering, IIT Kanpur, Kanpur, Uttar Pradesh, India

The cDNA microarray data from experiments is used to arrange genes according to similarity in the gene expression ratios. This arrangement is in the form of subgroup or clusters, which are generally much smaller in number compared to the number of genes they represent, using various mathematical techniques. The average expression profiles of clusters generated can be used to model the genetic network with the help of reverse engineering techniques. The gene network gives an idea of existence and the level of interactions in-between the gene clusters, which can be put to use for many potential applications.

In this work, a robust clustering algorithm was improved upon and implemented using genetic algorithms with the jumping gene and siRNA adaptations. A multi-objective formulation of the algorithm was able to produce biologically relevant clusters for test data, and the work is in progress to get the results on real life data. Using the average cluster profiles, the reverse engineering of the gene networks was done using graph theoretical approaches. The approach exploits the fact that gene networks follow the 'small world phenomenon'. The network was build up, starting from a single node, using fewer objectives and more constraints. This enabled us to solve a smaller dimensionality problem and solve the larger problem with a faster convergence rate. The following two models were proposed and implemented

1. Step by step optimization – uses the results of the n node problem as an initial seed to the GA of n +1 node problem
2. Step by step optimization with fixed previous topology – fixes the topology generated at the n node problem and the GA only optimizes the interaction of the $(n+1)^{th}$ node, thus reducing number of variables

A set of Pareto optimal solutions were generated for each of the two cases which were in coherence with the available biological information.

See more of this Session: Poster Session: Bioengineering
See more of this Group/Topical: Food, Pharmaceutical & Bioengineering Division

(623i) Growth of *B. Methanolicus* MGA3 In Seawater-Based Media In a Bioreactor

Ankit Patel, **James Yeary** and Claire F. Komives, Chemical & Materials Engineering, San Jose State University, San Jose, CA

Bacillus methanolicus is a thermotolerant bacterium that utilizes methanol as a sole carbon and energy source. The production of lysine and glutamic acid have been demonstrated with good yields in bioreactor processes. Previous work showed that B. methanolicus PB1 can be adapted to media made from artificial seawater instead of pure water. The use of seawater is advantageous for potential large scale processes that are near a seawater source. This work demonstrated the adaptation of B. methanolicus MGA3 to the same artificial seawater (ASW) media and further demonstrated that the bacteria can be grown in a bioreactor with ASW media. The presentation will describe the strategy of adapting the cells to the high salt media and growth in the bioreactor, and also show how the yield of cells on methanol compares in both the standard water-based media and ASW media, based on chemostat culture. The potential for ASW media to be used in large scale culture will be discussed.

See more of this Session: Poster Session: Bioengineering
See more of this Group/Topical: Food, Pharmaceutical & Bioengineering Division

(623j) Confinement Can Destabilize Alpha-Helix Folding Proteins by Stabilizing the Beta Structures

Leili Javidpour, Department of Physics, Sharif University of Technology, Tehran, Iran and Muhammad Sahimi, University of Southern California, Los Angeles, CA

Protein folding in confined media has attracted wide attention over the past decade due to its importance to both in-vivo and in-vitro applications. It is generally believed that protein stability increases by decreasing the size of the confining medium, if its interaction with the confining walls is repulsive, and that maximum folding temperature occurs for a pore size only slightly larger than the smallest dimension of protein folded state. However, protein stability at pore sizes very close to the size of the folded state has not received enough attention. Unfortunately, an in depth study of this problem at atomistic level is not possible by the present experimental apparatuses. Using 0.3 millisecond-long molecular dynamics simulations, we show that proteins with alpha-helix native state can be destabilized at high confinement due to entropic stabilization of the protein states that contain the beta structures. In contradiction to the present theoretical explanations, which do not consider entropy of the misfolded states, we find that the folding temperature maximum occurs at larger pore sizes for smaller alpha helices. These results shed light on many recent experimental observations that could not be explained by the present theories.

Our work demonstrates the importance of entropic effects of proteins misfolded states in highly confined environments. The results support the concept of passive effect of GroEL on protein folding by preventing it from aggregation in crowded environment of cells, and provide deeper clues to the \alpha-\beta conformational transition, believed to contribute to Alzheimer's and Parkinson's diseases. The strategy of protein and enzyme stabilization in confined media may also have to be revisited in the case of high confinement.

See more of this Session: Poster Session: Bioengineering
See more of this Group/Topical: Food, Pharmaceutical & Bioengineering Division

(623l) Micelle Membrane Interactions Investigated with Coarse-Grained Molecular Dynamics

Sharon M. Loverde[1], Wataru Shinoda[2], Dennis E. Discher[1] and Michael L. Klein[3], (1)Chemical & Biomolecular Engineering, University of Pennsylvania, Philadelphia, PA, (2)National Institute of Advanced Industrial Science and Technology, Tsukuba, Japan, (3)Institute for Computational Molecular Science and Department of Chemistry, Temple University, Philadelphia, PA

Interactions of spherical micelles with lipid membranes are of fundamental interest for applications of delivery. Here, we address the interactions of a PEO-PCL spherical micelle with a dimyristoylphosphatidylcholine (DMPC) lipid membrane. We utilize a model for poly(ethylene oxide) - poly(caprolactone) (PEO-PCL) that is consistent with the experimentally found phase behavior, and a model for DMPC that has been well parameterized by the Klein group. In order to parameterize the cross interactions we examine mixtures of PEG-lipid and DMPC lipid in the monolayer to ensure the correct mixing and surface tension, in agreement with known experimental measurements. We next examine the potential of mean force between the micelle and membrane.

See more of this Session: Poster Session: Bioengineering
See more of this Group/Topical: Food, Pharmaceutical & Bioengineering Division

(623m) Engineering the Robustness of Industrial Microbes Through Introducing Metabolic Redundancy

Yin Li, Linjiang Zhu and Yanping Zhang, Institute of Microbiology, Chinese Academy of Sciences, Beijing, China

Robustness refers to the ability to maintain phenotypic stability when facing diverse perturbations. It has been shown that microbial robustness coping with environmental stresses can be improved through overexpression of chaperone proteins. We show that the stress resistance of lactic acid bacteria (*Lactococcus lactis* and *Lactobacillus sanfranciscensis*) can be improved by introducing glutathione (GSH), a tripeptide which cannot be produced by these bacteria. We further show that introducing the GSH biosynthetic pathway into *Clostridium acidobutylicum* can improve the aero- and solvent-resistance of the host. Lacking the native GSH biosynthetic ability in these Gram-positive bacteria suggests that GSH is not primarily needed for their physiology, thus can be considered as redundant for their metabolism. Our results demonstrated that introducing this metabolic redundancy can improve the robustness of the host. Therefore, strategies of exploiting and engineering the functional metabolic redundancy may help microbes to better adapt themselves to the industrial environments.

See more of this Session: Poster Session: Bioengineering
See more of this Group/Topical: Food, Pharmaceutical & Bioengineering Division

(623n) Classification of Ketoacyl Synthases by Their Primary and Tertiary Structures

Yingfei Chen, David C. Cantu and Peter J. Reilly, Chemical and Biological Engineering, Iowa State University, Ames, IA

Ketoacyl synthases (KSs) are enzymes that catalyze the condensation of malonyl-acyl carrier protein (ACP) or malonyl-CoA with acyl-ACP or acyl-CoA units. This condensation reaction adds two carbon atoms to the acyl chain, and it is a key step in the fatty acid synthesis cycle. Our classification of KSs according to their primary and tertiary structures provides some insights into this enzyme group, as the KSs in each family have the same catalytic residues, mechanisms, and similar tertiary structures. Phylogenetic analysis of all families has been done to separate them into subfamilies for further study. At present, there are five KS families. KS1 members, overwhelmingly named β-ketoacyl-ACP III or its variants, are produced predominantly by bacteria. Members of KS2 are mainly produced by plants, and they are usually long-chain fatty acid elongases/condensing enzymes and 3-ketoacyl-CoA synthases. KS3, a very large family, is composed of bacterial and eukaryotic β-ketoacyl-ACP synthases I and II, often found in multidomain fatty acid and polyketide synthases. Most of the chalcone synthases, stilbene synthases, and naringenin-chalcone synthases in KS4 are from eukaryota. KS5 members are all from eukaryota, most are produced by animals, and they are mainly very long chain fatty acid elongases. KS sequences, tertiary structures, and family classifications are available on the continuously updated ThYme (Thioester-active Enzyme) database.

See more of this Session: Poster Session: Bioengineering
See more of this Group/Topical: Food, Pharmaceutical & Bioengineering Division

(623o) Supercooling: An Alternative Biopreservation Scheme

O. Berk Usta[1], Basak Uygun[2], Korkut Uygun[3] and Yeonhee Kim[1], (1)Surgery, Massachusetts General Hospital / Harvard Medical School, Boston, MA, (2)Center for Engineering in Medicine, Massachusetts General Hospital, Harvard Medical School, Shriners Hospital for Children, Boston, MA, (3)Center for Engineering in Medicine, MGH/Harvard Medical School, Boston, MA

The objective of this study is to extend the viable preservation time of hepatic tissues by sub-zero non-freezing (SZNF) storage in a supercooled preservation medium. The hypothesis here is that the preserved liver viability is correlated strongly to metabolic activity. In this session I will discuss our efforts on this alternative preservation scheme with results from both in-vitro and invivo studies.

See more of this Session: Poster Session: Bioengineering
See more of this Group/Topical: Food, Pharmaceutical & Bioengineering Division

(623p) Control of Enzyme Activity by Introduction of Molecular Recognition Moiety

Yuhei Oshiba[1], Takanori Tamaki[1], Hidenori Ohashi[1], Taichi Ito[2], Hidehiko Hirakawa[3], Satoshi Yamaguchi[3], Teruyuki Nagamune[4] and Takeo Yamaguchi[1], (1)Chemical Resources Laboratory, Tokyo Institute of Technology, Yokohama, Japan, (2)Center for Disease Biology and Integrative Medicine, The University of Tokyo, Tokyo, Japan, (3)Department of Chemistry and Biotechnology, School of Engineering, The University of Tokyo, Tokyo, Japan, (4)Department of Chemistry and Biotechnology, Department of Bioengineering, School of Engineering, The University of Tokyo, Tokyo, Japan

Regulation of enzyme activity by biding a ligand or protein, such as allosteric enzyme and signal transduction, is common in vivo. This enzyme function is attractive from a view of application for bio reactor, DDS, and bio sensor. Some research groups developed conjugation of enzyme and stimuli responsive polymer to control enzyme activity or specific binding by stimulus such as temperature, pH, and light for affinity separation, microfluidic protein analysis and capture[1-3]. However, in order to control the enzyme activity by molecular recognition, the problem is that how to modify the surface of the enzyme to create the binding site for another ligand and how to control the activity of the enzyme when the ligand is recognized. There are some previous studies controlling enzyme activity by genetic modification[4] or antigen-antibody complex reaction[5]. These approaches utilize conformational change of enzyme with genetic modification or protein-ligand interaction. However, the reaction mechanism is complicated and is mostly difficult to know, thus we must repeat the genetic modification process and change the position of binding site by trial and error approach. The approach for making molecular recognition enzyme in previous researches also lacks versatility.

Our approach to develop novel functional enzyme is that molecular recognition moiety is conjugated only near active pocket of the mutant enzyme. In this research, Cytochrome P450cam was used as enzyme and biotin with spacer arm was used as molecular recognition moiety. This enzyme regulates its activity by utilizing avidin-biotin affinity interaction as the following. In the absence of avidin, there is no steric hindrance and the enzymatic reaction proceeds normally. On the other hand, in the presence of avidin, the avidin that binds to biotin covers the active pocket of the enzyme, then enzyme reaction does not proceed by the inhibition of substrate diffusion. Therefore, the enzyme can control activity itself in response to specific molecule. The advantage of this approach is that we can design the required enzymatic function by changing both enzyme and molecular recognition moiety.

In this study, we firstly fabricated genetically modified P450cam that has one thiol group near its active pocket, abbreviated as 3mD. This mutant has a property that thiol-reacticve pegylated biotin is able to be conjugated only at targeted position. Then, conjugation of 3mD and pegylated biotin with different length of spacer arm was performed. Finally, the concept of this study was proven and the effect of the length of spacer arm on the avidin signal response was also examined.

The activity of 3mD was retained compared to that of wild type P450cam. Furthermore, 3mD kept their tertial structure after genetic modification by measuring absorption peak of ferrous CO-complexed state [6]. This shows that 3mD maintained enough activity without any denaturation. The number of thiol group on the surface of 3mD were investigated. From thiol group determination, about 0.7 mol of thiol groups existed in 1 mol of 3mD, while thiol group on the surface of wild type P450cam was not detected. This result is close to theoretical value of 3mD. These results suggest that 3mD was prepared intendedly.

Then, three kinds of pegylated biotins that have different length of spacer arm were conjugated to 3mD. The length of spacer arms are 39 atoms (biotin-PEG_{11}), 12 atoms (biotin-PEG_2), and 2 atoms (biotin-$Alkyl_2$) respectively. This conjugation was confirmed by determining the amount of thiol groups before and after biotinylation and about 70% of thiol groups were successfully biotinylated. Finally, we evaluated each activity of biotinylated 3mD before and after avidin binding. In the case of biotin-PEG_{11}-3mD, which has the longest spacer arm, the activity did not change when avidin binds to biotin. On the other hand, the activity of 3mD with shorter spacer arm, especially biotin-$Alkyl_2$-3mD, dramatically decreased to 60% after avidin binding. The diameter of the active pocket is around 17 angstrom, which is almost the same as the length of spacer arm in biotin-$Alkyl_2$-3mD. The conjugation site exists on the circumference of this pocket. As for biotin-PEG_{11}-3mD, bound avidin was too far from the active pocket of 3mD to cover the pocket. Meanwhile, as the length of spacer arm is shorter, bound avidin get closer to the active pocket, and the avidin appropriately covered the active pocket in biotin-$Alkyl_2$-3mD. We can conclude that we successfully developed novel functional enzyme and optimized its nanostructure.

References

[1] A. S. Hoffman, *Clinical Chemistry* **46**, 1478 (2000).

[2] A. S. Hoffman et al., *Journal of Biomedical Materials Research* **52**, 577 (2000).

[3] C. D. H. Alarcon, S. Pennadam, C. Alexander, *Chemical Society Reviews* **34**, 276 (2005).

[4] F. T. K. Lau, A. R. Fersht, *Nature* **326**, 811 (1987).

[5] C. A. Brennan, K. Christianson, M. A. Lafleur, W. Mandecki, *Proceedings of the National Academy of Sciences of the United States of America* **92**, 5783 (1995).

[6] S. A. Martinis et al., *Biochemistry* **35**, 14530 (1996).

See more of this Session: Poster Session: Bioengineering

ENGINEERING GALACTOSE METABOLISM FOR IMPROVED CELL GROWTH OF CHO CELLS IN CULTURE

Natalia Eugenia Jiménez[1,2], Camila Alejandra Wilkens[1,2], and Ziomara Patricia Gerdtzen[1,2]

[1] Centre for Biochemical Engineering and Biotechnology, Department of Chemical Engineering and Biotechnology, University of Chile, Santiago, Chile
[2] Millenium Institute for Cell Dynamics and Biotechnology: a Centre for Systems Biology, University of Chile, Santiago, Chile.

Background

Chinese hamster ovary (CHO) cells are one of the main hosts for industrial production of therapeutic proteins, owing to well-characterized technologies for gene transfection, amplification, and selection of high-producer clones. This has motivated the search for different strategies for the improvement of their specific productivity being one of the key points for this approaches the reduction of metabolic end-products like lactate and ammonia.

The use of alternative carbon sources have been studied for the reduction of lactate in mammalian cell culture, as they are metabolized more slowly than glucose. This reduction in carbon uptake leads to lower production of metabolic end-products. Unfortunately, cultures grown in fructose and galactose are unable to achieve high cell growth rates, so they are not used as sole carbon sources. In a previous work we studied cultures in a mixture of glucose and galactose, and observed that glycolitic fluxes during glucose consumption were very high, causing pyruvate accumulation. Increased lactate production was linked to this high flux of pyruvate production [1]. During galactose consumption the flux towards pyruvate production was decreased compared to the culture in glucose, causing a metabolic shift towards lactate consumption to supply pyruvate for energy production.

A previous work showed that when CHO cells were transfected with a fructose transporter they were able to achieve higher cell densities than the wild type [2]. Other studies have observed that over-expression of limiting step enzymes lead to increased flux in the metabolic pathways associated [3,4]. For instance, over-expression of the PYC gene and MDH II enhanced energy metabolism enabling higher fluxes through the TCA cycle.

In the following work we propose to improve the metabolism of CHO cells growing in galactose by introducing multiple changes in key steps of galactose metabolism. We determined that the limiting steps of galactose metabolism were transport and phosphorilation of galactose by galactokinase.

Methodology

t-PA producing CHO TF 70R cells were transfected with genes involved in galactose metabolism, to enhance cell growth by improving galactose metabolism. Cells were transfected with the Galactokinase (GALK1) gene from *Mus musculus*, using Lipofectamine and selected with G418. Experiments were performed to analyze cell proliferation, carbon source consumption lactate production for the obtained pooled clones, as a preliminary tool to determine if the over-expression of this protein has a positive effect.

Based on these results, cells were selected by limiting dilution in order to obtain clones with lower lactate production and higher growth rates.

Transfected and selected cells were grown in protein free media with 2% Fetal Bovine Serum, supplemented with glucose and galactose to define two different culture conditions with a final concentration of 20 mM of hexose: 6 mM glucose and 14 mM galactose (GalKGG6/14); and 20 mM galactose (Gal20). Extracellular metabolites were measured.

Results

In combined carbon source experiments, two phases can be distinguished. During the first phase, cells produce biomass, tPA, lactate and alanine. When glucose is depleted wild type GG6/14 culture enters a second metabolic stage where no significant growth is observed and galactose is consumed along with extracellular lactate and alanine which is consistent with lactate being used as a supplementary carbon source to support energy metabolism.

The positive effect of the over-expression of galactokinase (GALK1) are shown in their growth curves. After transfection cultures exhibit a higher growth rate in galactose due to their ability of metabolize this sugar at a increased rate, indicating that the introduction of these gene does not increase galactose uptake in a way that would generate secondary metabolites. These cultures also exhibit lower lactate accumulation. Control cells are unable to grow on media with galactose as the only carbon source. Transfected cells were able to maintain high viability during 100 hours in this media.

Selected clones show a similar behavior but with enhanced characteristics like growth rate and maximum cell density; in addition the effect of metabolic stress associated to the over-expression of this protein was shown by growing cells using glucose as a unique carbon source.

The over-expression of the Slc2a8 (GLUT8) Galactose transporter could enhance the positive effects of the introduction of the galactokinase gene by increasing the availability of substrate at an intracelullar level. Before transfection, mutation of a dileucine motif to alanine is required to allow the expression of this transporter in the plasmatic membrane [5]. For that purpose an strategy using PCR with primers that include this substitutions was performed and used to transform cells in culture. Selection and further analysis are currently undergoing.

Conclusions

Metabolism can be modified through cell engineering to increase cell growth while maintaining low lactate production rates, enabling cells to utilize alternative carbon sources. Specifically this can be achieved with galactose.

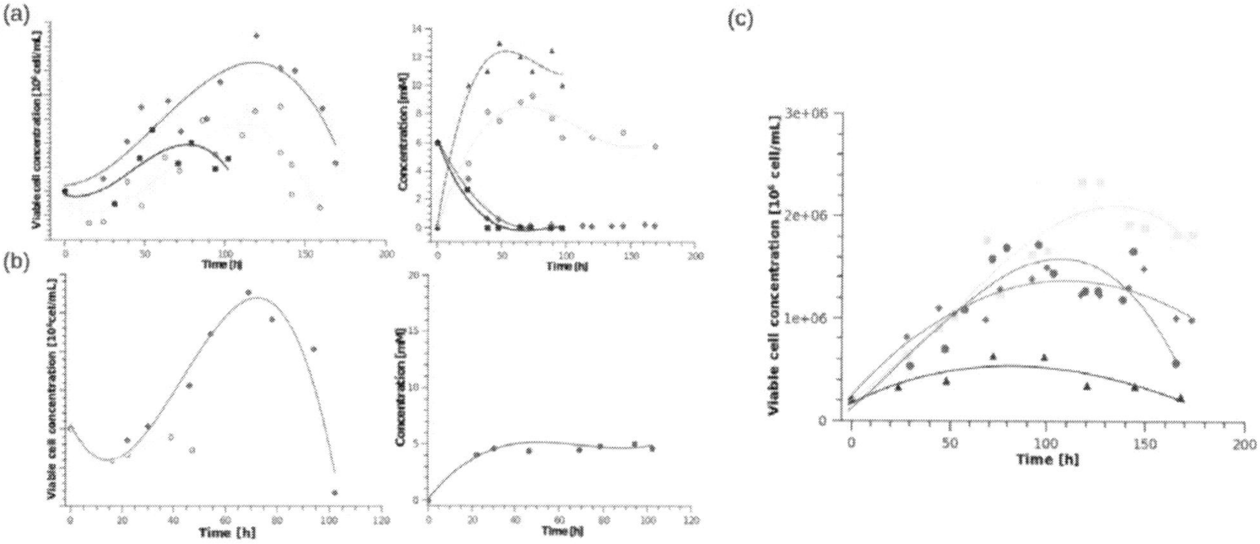

Figure 1: Cell density, glucose and lactate concentration in GG6/14 (a) and Gal20 media (b). (•) CHO tPA,( ) CHO tPA-pcDNA3.1(+), (◆) CHO tPA-GALK1 (pool) , glucose (■,■), lactate (,■). Cell density of selected clones (c), growth in glucose media (▲), growth in GG6/14 (, ●,◆).

Acknowledgements

This work was supported by FONDECYT Initiation Grant 11090268.

References

1. Wilkens C.A., Altamirano C., Gerdtzen Z.P. (2011) "Comparative metabolic analysis of lactate for CHO cells in glucose and galactose" *Biotechnology and Bioprocess Engineering* (In press)
2. Wlaschin K, Hu WS (2007) "Engineering cell metabolism for high density cell culture via manipulation of sugar transport". *Journal of Biotechnology*, 13:1023-1027
3. Irani, N., Wirth, M., van Den Heuvel, J., Wagner, R., (1999) "Improvement of the primary metabolism of cell cultures by introducing a new cytoplasmic pyruvate carboxylase reaction" *Biotechnol Bioeng.*, 66(4):238-46
4. Chong, W.P., Reddy, S.G., Yusufi, F.N., Lee, D.Y., Wong, N.S., Heng, C.K., Yap, M.G., Ho, Y.S., (2010) "Metabolomics-driven approach for the improvement of Chinese hamster ovary cell growth: overexpression of malate dehydrogenase II." *J Biotechnol*, 147(2):116-21
5. Diril M.K., Schmidt S. Krauß M., Gawlik V., Joost H-G., Schürmann A., Haucke V., Augustin R., (2009) "Lysosomal localization of GLUT8 in the testis – the EXXXLL motif of GLUT8 is sufficient for its intracellular sorting via AP1- and AP2-mediated interaction" *FEBS Journal* 276: 3729–3743

(623s) Shear Stress and Compression Induced Changes In Genome Rheology and Reorganization

Elizabeth A. Booth-Gauthier[1], Turi A. Alcoser[2], Ge Yang[2] and **Kris Noel Dahl**[1], (1)Chemical Engineering, Carnegie Mellon University, Pittsburgh, PA, (2)Biomedical Engineering, Carnegie Mellon University, Pittsburgh, PA

Many cell types respond to applied force by altering gene and protein expression, but the mechanisms of this response have not been fully elucidated. Endothelial cells line blood vessels and respond to shear stress. Osteocytes are present in bone and respond to compression. We examined reorganization of genes in the nucleus of HeLa and HUVEC (endothelial) cells under shear stress and in HeLa and SAOS-2 (osteogenic) cells under compression by tracking fluorescent intranuclear markers. By studying large scale nuclear response under physiological levels of shear stress and compression, we are able to show that there is both a shear stress and compression induced change in nuclear reorganization. We observe a shift in nuclear response to force at ~30 minutes after the application of forces from increased random reorganization to a directed reorganization. Also, we observe that nuclei stiffened by overexpressing a mutant nuclear structural protein, progerin, show little response to shear, potentially from altered mechanotransductive response. Although not gene specific, these large scale rheological responses across the nucleus suggest that there is nuclear reorganization that could correspond to substantial changes in gene expression.

See more of this Session: Poster Session: Bioengineering
See more of this Group/Topical: Food, Pharmaceutical & Bioengineering Division

(623t) Engineering Zinc Finger Proteins Using Accessory Binding Modules

Younghoon Kim, Nathan Yee, Peiyi Wang, Kathryn Trenshaw and Charles M. Schroeder, Department of Chemical and Biomolecular Engineering, University of Illinois at Urbana-Champaign, Urbana, IL

We report on the engineering of zinc finger proteins (ZFPs) as synthetic transcription factors by appending accessory modules to enhance protein binding affinity to target DNA. Over the last few years, transcription factor engineering has provided a valuable approach to genetically engineer a wide variety of cell types. ZFP technology is now being used to directly edit chromosomal DNA by creating synthetic nucleases. Recently, this approach has been used to produce genetically enhanced crops and improved biophama products and to cure genetic disease. The central feature of this technology is generation of synthetic transcription factors that bind to "target" DNA sequences with high specificity. In this way, modular assembly provides a convenient approach to synthesize target-specific ZFPs by generating modular proteins consisting of tandem repeats of well-characterized zinc fingers through standard molecular cloning. Oligomerized pool engineering (OPEN) and context-dependent assembly (CoDA) are alternative, albeit more complex protocols for creating ZFPs.

In this work, we describe the direct engineering of two- and three-finger ZFPs to improve DNA binding site recognition specificity and affinity. We generated chimeric ZFPs containing the accessory module PAR from the yeast transcriptional activator ADR1, a zinc finger transcription factor from Saccharomyces cerevisiae. We observed that addition of the PAR binding module increased the binding affinity for chimeric ZFPs to target DNA sequences without affecting the selectivity for two-finger ZFPs. Two-finger ZFPs are generally not used for targeted DNA binding, because the functionality of ZFPs containing only two modules results in a loss in binding affinity and selectivity. Here, we show that ZFPs consisting of only two-finger modules can be successfully engineered to target specific DNA sequences by appending accessory binding modules. Based on this work, we anticipate that these chimeric ZFPs will enable new applications involving genetic modification. When combined with the CoDA protocol for ZFP assembly, pre-screened two-finger ZFPs can be used as simple "unit building blocks", thereby enabling enhanced design and engineering of target-specific and tight binding synthetic transcription factors.

See more of this Session: Poster Session: Bioengineering
See more of this Group/Topical: Food, Pharmaceutical & Bioengineering Division

(623u) Monte Carlo Methodologies for the Structural Characterization of Globular Proteins

Yingying Jiang, Chemical Engineering, University of Texas at Austin, Austin, TX

Using MC methods similar to those used to characterize the cavity size distribution in synthetic polymers, new statistical metrics have been defined for the structural characterization of globular proteins. Some of these metrics include volume, surface area, asymmetry ratio, interior cavity size distribution, and the identification of percolation channels. Proteins that have been studied include hemoglobin, myoglobin, lysozyme, and aquaporins. An analysis of cavity statistics provides an efficient method to quantify local properties such as packing density and transport pathways. The average cavity sizes of studied proteins range from 6.0 to 7.0 Å.

See more of this Session: Poster Session: Bioengineering
See more of this Group/Topical: Food, Pharmaceutical & Bioengineering Division

(623v) Fungal Processes for Direct Bioconversion of Cellulose to Hydrocarbons

Kristopher A. Hunt[1], Natasha Mallette[1], Ross P. Carlson[1], Mitchell Smooke[2], Gary Strobel[3], Scott Strobel[4] and Brent Peyton[5], (1)Chemical and Biological Engineering, Montana State University, Bozeman, MT, (2)School of Engineering, Yale University, New Haven, CT, (3)Plant Sciences & Plant Pathology, Montana State University, Bozeman, MT, (4)Molecular Biophysics and Biochemistry, Yale University, New Haven, CT, (5)Thermal Biology Institute of Chemical and Biological Engineering, Montana State Univeristy, Bozeman, MT

As the scarcity of renewable feedstocks intensifies, it will become more important to utilize waste streams such as cellulosic biomass. An endophytic fungus, *Ascocoryne sarcoides* NRRL50072, has been shown to degrade cellulose and produce a hydrocarbon mixture termed "mycodiesel." The aims of this model cellulose-to-hydrocarbons system are i)identify key metabolic pathways for production of hydrocarbons using elementary flux mode analysis, ii)validate the model using experimental optimization data, and iii)determine optimal conditions for expression of those pathways or ideal deletion sets to maximize production of hydrocarbon. Additionally, combustion accounts for 2/3 of petroleum usage, and there is a potential for improvement in thermal efficiency by 25% to 50% through investments in fuel/engine concepts, with 7.3 billion barrels of oil per year ($550B), there is a potential savings of 3 million barrels of oil per day (out of 20M). Given the long-term use of liquid and gaseous fuels, it is clear that a coordinated experimental and computational program is needed to model combustion in both laminar and turbulent systems with the goals of enhancing efficiency and reducing pollutant formation. This work has begun developing such a program using fuel mixtures relevant to the fungal synthesized mycodiesel.

See more of this Session: Poster Session: Bioengineering
See more of this Group/Topical: Food, Pharmaceutical & Bioengineering Division

(623w) Protein Engineering for High-Throughput Screening and Protease Dynamic Analysis

Jingshu Zhu[1], Matthew A. Saunders[1], Larry A. Sklar[2], Bruce S. Edwards[2] and Steven W. Graves[1], (1) Center for Biomedical Engineering, The University of New Mexico, Albuquerque, NM, (2)Center of Molecular Studies & Dept. of Pathology, The University of New Mexico, Albuquerque, NM

Designing simple and sensitive enzymatic analysis methods has been of great interest for enzymatic kinetic assay and high-throughput screening (HTS). The need to measure multiple protease activities simultaneously has resulted in the invention of multiplex microsphere based protease assays. Previously, we have described the development of a microsphere based assay for the proteases Factor Xa, *Bacillus anthracis* Lethal Factor (LF), and *Clostridium botulinum* type A light chain (BoNT/A LC) via high throughput flow cytometry. The selection of potential inhibitors to these proteases is of great importance due to their key roles in the lethality of anthrax and the toxicity of BoNT. This screening platform uses suspension microsphere arrays coated with streptavidin and multiplexed via different levels of intrinsic microsphere fluorescence. The fluorescent fusion protein, which contains a biotinylation tag and a GFP domain separated by the full-length protein substrate, is bound to the microsphere and protease activity is detected via loss of fluorescence from the microsphere surface by flow cytometry. Here we present the development of a multiplex substrate set that has been engineered for use with three proteases (BoNT/A, F LCs and LF) simultaneously. These substrates have been attached to a four-plex microsphere set and mixed with the proteases that were arrayed in 1536-well plates containing the >1100 compounds of the Prestwick library. We have demonstrated multiplex microsphere based protease assays analyzed by flow cytometry and high throughput flow cytometry based screening that discovered several interesting inhibitors to the proteases. We will also present confirmatory assays and dose response curves for the identified compounds. The microsphere based flow cytometry assay is a promising approach with several advantages including: low substrate costs, small reaction volumes, homogenous assay format, and use of full length substrates to detect all pertinent interactions; however, it has limitations in collection of traditional kinetic data. Therefore, we have developed several novel methods to study the protease/substrate kinetics that enable determination of key kinetic constants of protease activities. One such method uses fluorescence resonance energy transfer (FRET) between a full length protease substrate and a fluorescently labeled streptavidin protein to determine the solution based enzymatic activities. In this case, we used biotinylation tag and GFP labeled protein substrates connected to streptavidin-Cy3 conjugate as the cleavage targets for proteases in solution. Based on the FRET from GFP to Cy3, the fluorescence intensity change was recorded on a fluorometer or a plate reader. Additionally, the development of another microsphere assay based on fusion protein substrates and engineered biomimetic lipid bilayer microsphere surfaces will be presented. It is capable of screening several different matrix metalloproteases (MMPs) activities which are suggested to require substrates in lipid bilayers. In this study, a MMP substrate was designed with a GFP domain and a membrane insertion active tag. In summary, we present several new protease assays that utilize custom engineered fusion proteins that enable measurement of both surface based and full-length solution based protease assays. As such, these fusion proteins enable high-throughput inhibitor screening and detailed kinetic studies.

See more of this Session: Poster Session: Bioengineering
See more of this Group/Topical: Food, Pharmaceutical & Bioengineering Division

(623x) Non-Chromatographic Protein Purification Via Mini-Intein Cleavage

Najla Valdes, Miguel Ramirez and Zhilei Chen, Chemical Engineering, Texas A&M University, College Station, TX

Current affinity tag-based protein purification approaches are not ideal for biotechnological applications. We report the development of a column-free, protease-free protein purification technology that takes advantage of an engineered pH-sensitive split-intein and a stimulus-responsive elastin-like-polypeptide (ELP) tag. This engineered split-intein catalyzes a single C-terminal cleavage reaction in the presence of reducing agent. The ELP is fused to the N-terminal fragment of the split intein and the protein of interest (POI) to the C-terminal fragment of the intein. These two fusion proteins are expressed individually in E. coli and mixed after cell lysis. In the absence of reducing agent, the N- and C-inteins associate strongly with each other without cleavage, physically connecting the ELP to POI. High salt conditions trigger ELP-mediated phase separation of the POI. Next, the precipitate containing the POI is resuspended in a low-salt, reducing buffer to resolubilize the aggregate and induce intein cleavage, releasing the POI from the ELP-intein complex. Finally, the resulting mixture is subjected to high salt conditions again to precipitate out the undesired ELP-intein complex from the solution, leaving only purified POI in the solution phase. This protein purification technology is simple, convenient, and highly efficient and should facilitate cost-effective purification of proteins on a large scale.

See more of this Session: Poster Session: Bioengineering
See more of this Group/Topical: Food, Pharmaceutical & Bioengineering Division

(623y) Raft Registration Across Bilayers In a Molecularly Detailed Model

Diego A. Pantano, Chemical and Biomolecular Engineering, University of Pennsylvania, Philadelphia, PA, Preston B. Moore, University of the Sciences in Philadelphia, Philadelphia, PA, Michael L. Klein, Institute for Computational Molecular Science and Department of Chemistry, Temple University, Philadelphia, PA and Dennis E. Discher, Chemical & Biomolecular Engineering, University of Pennsylvania, Philadelphia, PA

Membrane lateral heterogeneity is accepted as a requirement for the function of biological membranes. Coupling between bilayer leaflets plays an important role in biomembrane function inducing and registering rafts across leaflets for various cellular signals. Nevertheless registration mechanisms remain undetermined and several suggestions have been made: including chain interdigitation, cholesterol translocation in the case of lipids, interfacial energy, electrostatic coupling, van der Waals interactions and composition-curvature coupling. Recent experiments with mixtures of polymer amphiphiles that exhibit domain registration upon ligand-induced segregation suggest a general mechanism. We used coarse grained molecular dynamics (CGMD) simulations to show that raft registration arises spontaneously in bilayers upon segregation of a calcium or ligand-induced ordered phase from a liquid disordered phase. Thickness mismatching between phases induces a localized curvature change of the liquid phase which guides distant domains together and ultimately stabilizes the registered state. The absence of explicit charge in the model and the fact that domain size modulates the strength of transmembrane coupling demonstrate that collective interactions are sufficient to induce raft registration.

See more of this Session: Poster Session: Bioengineering
See more of this Group/Topical: Food, Pharmaceutical & Bioengineering Division

(623z) A Kinetic Model for the Storage of Iron In Ferritin

J. Cristian Salgado[1], Alvaro Olivera-Nappa[2], Ziomara P. Gerdtzen[2], Victoria Tapia[3], Elizabeth Theil[4], Carlos Conca[5] and Marco T. Nuez[3], (1)Department of Chemical Engineering and Biotechnology, University of Chile, Millennium Institute for Cell Dynamics and Biotechnology: a Centre for Systems Biology, Santiago, Chile, (2)Centre for Biochemical Engineering and Biotechnology, Department of Chemical Engineering and Biotechnology, Millennium Institute for Cell Dynamics and Biotechnology, University of Chile, Santiago, Chile, (3)Department of Biology, Millennium Institute for Cell Dynamics and Biotechnology: a Centre for Systems Biology, University of Chile, Santiago, Chile, (4)Department of Nutritional Sciences and Toxicology, University of California, Berkeley, CA, (5)Department of Mathematical Engineering, Centre for Mathematical Modeling, Millennium Institute for Cell Dynamics and Biotechnology: a Centre for Systems Biology, University of Chile, Santiago, Chile

Ferritin plays a key role in the regulation of iron cellular levels, acting as the main intracellular iron storage protein. Studying the mechanism used by ferritin to store iron will lead to a better understanding of cellular iron homeostasis. We propose a kinetic model for iron storage in ferritin based on the main reactions for ferritin and iron, considering a heterogeneous population of ferritin species differentiated by the amount of iron stored in them. The dynamics of iron storage in ferritin during the process of intestinal iron absorption was studied simulating a set of differential equations based on the kinetic model. Kinetic parameters and enzymatic constants for the model were directly taken or calculated from published works. The simulation of the model generated results in agreement with experimental data. In particular, the model was able to recover the distribution pattern of iron into ferritin protein nanocages with different iron content experimentally observed, validating in this way the kinetic model and its parameters. Results highlight the function of ferritin as a controller of the cytoplasmic labile iron pool. Simulations show that ferritin stabilizes the cytosolic iron pool for a wide range of total iron concentrations, including the case when the system is subject of large iron perturbations, as well as iron starvation. According to our results the labile iron pool stabilization process occurs very fast, suggesting that ferritin is located in a first line of defense protecting the cell from excess or starvation iron conditions. This initial rapid response would be follow by slower transcriptional regulation mechanisms that promote long term adaptation to iron stress conditions. Finally, this results suggest that iron buffering and labile iron pool stabilization are possibly the main functions of ferritin in the cell.

This work was partially supported by the Millennium Scientific Initiative ICMproject P05-001-F, FONDECYT Research Initiation Grant 11080016 andFONDECYT Projects 1070840 and 1050048. AO-N was supported by an ICDBpost-doctoral fellowship

See more of this Session: Poster Session: Bioengineering
See more of this Group/Topical: Food, Pharmaceutical & Bioengineering Division

() Tissue Engineering the Retina: Soft Lithography-Patterned Microchannel PLGA Scaffolds Guide the Morphogenesis of Dissociated Newborn Mouse Retinal Cells and Human Embryonic Stem Cell-Derived Retinal Cultures

Andrew C. McUsic, Department of Bioengineering, University of Washington, Seattle, WA and Thomas A. Reh, Department of Biological Structure, University of Washington, Seattle, WA

While embryonic stem cell-derived photoreceptors promise a cell source for enhanced in vitro models of retinal degenerative diseases, the more differentiated characteristics of retinal cells do not develop in the dissociated cultures typical of ES cells. Therefore, we have attempted to reconstruct organized retinal tissue from dissociated retinal cells into an array of aligned units that more faithfully mimics the retina. We report the first attempt to solvent-process poly(lactic-co-glycolic acid) into a microchannel scaffold format to achieve this geometric constraint. We compared the effect of PLGA concentration on channel morphology and, along with other culture conditions, on the infiltration of dissociated newborn mouse retinal cells into the channels. Culturing scaffolds at the gas-liquid interface and with low serum significantly increased infiltrated cell viability compared to submerged cultures, resulting in 12.3 +/- 1.70 cells per channel after three days and 10.7 +/- 0.80 cells per channel after seven days. Rod photoreceptors and Müller glia aligned processes parallel to the microchannel walls. Otx2+ and Pax6+ subpopulations recapitulated lamination behavior. We constructed scaffold/retinal pigment epithelium explants and observed rods extending rhodopsin-positive processes toward RPE cells. Finally, human embryonic stem cell-derived photoreceptors exhibited infiltration and morphological characteristics similar to mouse retinal cells inside the scaffolds. These findings further efforts to generate tissue-level retinal models from dissociated cells for use in a variety of applications.

See more of this Session: Poster Session: Bioengineering
See more of this Group/Topical: Food, Pharmaceutical & Bioengineering Division

() Redesigning E. Coli Metabolism for Obligate Anaerobic Production of Biofuels

Cong Trinh, Department of Chemical and Biomolecular Engineering, University of Tennessee, Knoxville, Knoxville, TN

The biological conversion of lignocellulosic biomass into biofuels has emerged as a promising route to sustainable energy. Fermentation is the most efficient route to synthesize biofuels that are reduced relative to fermentative sugars. However, most pathways that synthesize advanced biofuels are not obligately anaerobic. Thus, if we could redesign cell metabolism to use advanced biofuel-producing pathways as heterologous, obligate anaerobic pathways, we would achieve higher product yields and make large-scale production feasible and cost-effective because supply and precise control of oxygen are not required. To demonstrate the approach, elementary mode analysis was applied to redesign efficient *Escherichia coli* cells optimized for biofuels production such as short, medium, and long chain length alcohols from fermentative sugars under anaerobic conditions. The analysis decomposed the *E. coli* metabolic network into unique and elementary pathways from which only the optimal routes for alcohol production were selected. The designed cells were constrained to operate according to these selected pathways through multiple gene deletion, addition, and over-expression. In addition, the cells were specifically designed to tightly couple both biomass and alcohol production, which represented a unique property exploited for molecular pathway evolution to improve alcohol production and tolerance. Here we present the design, construction, and characterization of some of designed cells, and evaluate their performances with model prediction.

See more of this Session: Poster Session: Bioengineering
See more of this Group/Topical: Food, Pharmaceutical & Bioengineering Division

(625aa) Parallel Decomposition Strategies for Estimation of a Spatially Distributed, Nonlinear, Childhood Infectious Disease Model

Daniel Word[1], Jean-Paul Watson[2], David L. Woodruff[3] and Carl D. Laird[1], (1)Chemical Engineering, Texas A&M University, College Station, TX, (2)Sandia National Laboratories, Albuquerque, NM, (3) Graduate School of Management, University of California, Davis, Davis, CA

Understanding the factors that influence the dynamics of childhood infectious diseases is important for developing effective methods of controlling and reducing disease spread. The development of reliable, accurate mathematical models has been useful to help identify and understand factors that contribute to disease dynamics, but the accuracy of these models depend upon appropriate model parameter values which are unknown and must be estimated from available data.

Previously, we have shown efficient techniques to estimate seasonal transmission parameters using discrete-time and continuous-time infectious disease models. These models are formulated using differential equations and, using the simultaneous discretization approach, give rise to large-scale nonlinear optimization problems for estimating transmission parameters in large cities where diseases are endemic. However, in smaller cities fadeout is often observed making estimation much more difficult. For example, it has been shown that in urban areas larger than around 300,000 people, without the influence of vaccination, measles is self-sustaining. Since the population in many areas is significantly below this critical community size, stochastic fadeout is commonly seen and can cause extended periods of time with no observed cases. Cases are observed again only after an infection is imported from outside the community. To reasonably estimate disease transmission parameters in these smaller communities, and to better understand the factors behind this spatial coupling, the importation of new cases between cities must be considered. This can be done by including a spatial component in the model that includes information about the geographic location and population of communities. However, parameter estimations using long time horizons and spatial components with many cities create nonlinear estimation problems that can be intractably large for general estimation approaches.

In this work, we demonstrate an effective solution approach for estimating transmission parameters in a continuous time, spatially distributed infectious disease model using both progressive hedging and internal linear decomposition. These decomposition techniques enable parallel solution of these large-scale nonlinear optimization problems. We demonstrate our approach using measles incidence data from 60 cities in England and Wales, and measles incidence data from the 72 provinces of Thailand. This solution approach allows efficient solution for large problems and shows good scalability with problem size.

See more of this Session: Poster Session: Engineering Fundamentals In Life Science

(625ab) Decoupling Adhesion From Cohesion In Biofilm-Substrate Composites

Hallie Zeller, Department of Mechanical Engineering, Lafayette College, Easton, PA and **James K. Ferri**, Department of Chemical and Biomolecular Engineering, Lafayette College, Easton, PA

Biofilm formation is enhanced by bacteria through their production extracellular polymeric substances (EPS). Therefore, the mechanical properties, particularly the elastic modulus and the failure stress of these biocomposites, coupled with the adhesion energy between the biofilm and substrate are important in quantitative prediction of microbe fate in environmental applications.

Recently, we introduced a technique allowing fabrication and characterization of freestanding asymmetric membranes at fluid-fluid interfaces. Using this technique, biofilms of the Escherichia coli mutant D21g were synthesized by adsorption and complex formation of bacteria at the air-water interface of a pendant drop. By comparing the shape of the drop during dilation of its surface with continuum models, the elasticity (i.e cohesive interactions) in the biofilm can be quantified.

For biofilms grown on an elastic solid substrate, the problem break-up and release is a more complex balance between cohesive interactions in the film and the adhesion energy between the film and substrate. By measuring planar biaxial deformation characteristics of the composite biofilm-substrate system, the adhesion between biofilm and substrate can be decoupled from the elastic contributions of each. Here, we present biaxial mechanical data, ie.e force-displacement and digital image correlation, for d21g biofilms grown on polydimethylsiloxane (PDMS). Although biofilm intrinsic elasticity (cohesion) depends on age, growth phase, and other environmental conditions, the data show that interfacial effects (i.e. adhesion) dominate the composite system deformation in most cases studied. More importantly, this work describes how adhesion energy can be decoupled from cohesive interactions and gives the basis for a technique which can be used to study the impact of cleaning agents or therapies design to mitigate biofilm adhesion.

See more of this Session: Poster Session: Engineering Fundamentals In Life Science
See more of this Group/Topical: Food, Pharmaceutical & Bioengineering Division

(625b) An Automated Tissue Digester for Pancreatic Islet Production

Zhongliang Lu[1], Paul W. Todd[2] and Thomas Richard Hanley[1], (1)Chemical Engineering Department, Auburn University, Auburn, AL, (2)Space Hardware Optimization Technology, Inc, Greenville, IN

Pancreatic digestion is the primary step in preparing islets of Langerhans for clinical transplantation for restoring euglycemia in Type-1 Diabetes patients. In 1988, C. Ricordi et al. developed a semi-automated method in which pancreas fragments and glass balls were placed into a cylindrical vessel to form a closed-loop recirculation and dissociation system, operated by intermittent manual shaking of the vessel. However, the yield of islets from this technique tends to be variable and depends on several parameters. Collagenase-containing enzyme blends used for releasing islets from the exocrine tissue also produce variable results and can cause damage to islets. Operator function (shaking the vessel by hand) also produces variability from cases to case. An automated digestion unit capable of producing reproducible operating parameters is needed for the consistent preparation of islets.

This research focuses on the incorporation of an automated horizontally rotating digester consisting of a rotating outer cylinder with hemispherical baffles and a counter-rotating cylindrical core with hemispherical flow diverters. The specified spacing between baffles and diverters is one variable under investigation, along with the selection of rotation speeds. This configuration is designed to enhance the turbulent effect and maximize contact area between tissue fragments and walls. Inlet and outlet rotary unions, drive pulleys, the core rotator diameter, and the shell rotator length, speed of associated pumps, and operation of valves are variables that can be adjusted for optimization studies. In the digestion process, the rotation of the inner core is adjusted for optimum dissociation, and independently the exterior wall rotation (in the opposite direction) is optimized to maintain particles in suspension without centrifuging them. As particle size distribution is diminished during digestion, rotation speeds will need to be changed, so a particle tracking velocimeter (PTV) is applied as the imaging system in order to monitor the tissue particles. In addition, calculations of relevant parameters including Reynolds number (Re), shear rate, energy dissipation rate per volume and Kolmogoroff length have performed. These calculations have been made to compare the Ricordi chamber, a (Rushton) stirred tank and the novel digester. This algebraic operation has been followed by a homogeneous single-phase computational fluid dynamics (CFD) model developed for simulating the flow patterns in the digester. The validation of the model is performed through the comparison of numerical results with important experimental data. The designed outcome is a predictable, controllable fluid shear environment, optimized differential rotation mode, adjustable vessel volume, and larger collision contact area. These characteristics can be applied to an automatic mode instead of manual operation thereby enhancing the predictability of pancreatic islet preparation.

See more of this Session: Poster Session: Engineering Fundamentals In Life Science
See more of this Group/Topical: Food, Pharmaceutical & Bioengineering Division

(625c) Mathematical Optimization of Cryoprotectant Addition and Removal Procedures for the Purpose of Adherent Cell Vitrification

Allyson Fry, Danielle McClanahan and Adam Z. Higgins, School of Chemical, Biological, and Environmental Engineering, Oregon State University, Corvallis, OR

Long-term storage ofadherent cells is necessary for off-the-shelf availability of tissue engineeredconstructs, cell-based biosensors, and pharmaceutical testing materials.Cryopreservation in an ice-free glassy state (i.e., vitrification) is apromising approach for the preservation of adherent cells and tissue becauseformation of ice crystals is known to be damaging. To achieve vitrification,high concentrations of cryoprotectant additives (CPAs) are required. However,the addition and removal of CPAs can lead to cell damage due to osmoticallyinduced cell volume changes and chemical toxicity, effects that becomes moreprominent with increasing CPA concentration. Preventing toxicity is asignificant challenge when designing vitrification procedures, yet toxicityminimization is not addressed in traditional multi-step CPA addition andremoval procedures. We report a new mathematical optimization strategy for CPAaddition and removal procedures which involves minimization of a toxicity costfunctional. Membrane permeability parameters, osmotic tolerance limits and atoxicity cost functional are required as inputs to our optimization algorithm.Permeability parameters for endothelial monolayers were determined previouslyusing a fluorescence quenching technique. Osmotic tolerance limits weredetermined by exposing cell to a range of anisotonic solutions and assessingviability with a resazurin based fluorescence assay. The 95% viability windowcorresponded to 20-200% of the isotonic cell water volume. To determinetoxicity rates, cells were exposed to 1, 3, 5 and 7 molal CPA solutions forvarious time intervals, followed by estimation of viability as describedabove. To ensure that decreased viability was due to toxicity and not osmoticdamage, multi-step addition and removal procedures were used. As expected,damage due to toxicity increased with exposure time and with CPAconcentration. Toxicity rates were determined by fitting the viability datausing a first order rate model. The concentration-dependence of the toxicityrates was approximated using a power law model, which was then used to definethe toxicity cost functional. Two step procedures for addition and removal ofglycerol at 37°C were predicted using our iterative optimization strategy. Theresulting toxicity-minimized addition procedure included an initial step inwhich cells were induced to swell during exposure to a relatively low glycerolconcentration (less than 1 molal), followed by shrinkage in the second stepwhen the cells were exposed to the goal concentration of 17 molal glycerol.Using endothelial cell monlayers, we compared the toxicity minimized procedureto single-step and traditional multi-step procedures. In all cases, carryingout addition and removal of glycerol at 37°C resulted in low cell yields, withthe highest yield attained using the toxicity-minimized procedure (<40%).In an attempt to improve viability, we performed the 17 molal addition step onice. Cells were essentially nonviable for single-step (<1% viability) andmulti-step procedures (3.90.5%), however the toxicity-minimized procedureexhibited significantly higher viability (8015%). Our results demonstratethat the toxicity minimization strategy yields improved viability afterexposure to vitrifiable glycerol solutions compared with conventional multistepCPA addition and removal methods.

See more of this Session: Poster Session: Engineering Fundamentals In Life Science
See more of this Group/Topical: Food, Pharmaceutical & Bioengineering Division

Modeling of the Interaction between the Serotonin Regulation Pathway and the Stress Response, mediated by the Hypothalamic-Pituitary-Adrenal Axis, using a Boolean Approximation: a Novel Study for Depression.

Oscar Moreno-Ramos[1,2], María Claudia Lattig[1], Andrés Fernando González[2]

[1.] Laboratorio de Genética Humana, Departamento de Microbiología, Universidad de los Andes
[2.] Grupo de Diseño de Productos y Procesos, Departamento de Ingeniería Química, Universidad de los Andes

Depression shows a high prevalence in the overall population and is a major public health problem worldwide (WHO, 2004). According to different studies performed in twins, a heritability of approximately 40% is observed and it increases if another member of the family suffers from other types of affective disorders. Thus, it has been hypothesized that depression has a strong genetic background, and genes in the serotonin metabolic pathway have been implicated in the occurrence of the disease.

Environmental factors such as stressor events are also strongly associated with depression. The principal aim of this work was to model the interaction between the serotoninergic pathway and the stress response pathway mediated by the hypothalamic-pituitary-adrenal axis using a Boolean approximation. For this, we constructed a network with 41 nodes by a bibliographic mining. Then we also determined which genes or metabolites, when knocked out in-silico, cause the highest impact on the model stability calculating a Hamming distance metric.

Furthermore, in this model we simulate the impact antidepressants have on the biological pathway. The results obtained indicate that the network is stable with or without stress, but that stability is reached faster when the pathway is run without stress. Also, we observed that the biological model generated predicts steady states (attractors) for each of the different runs performed. Those attractors change in shape, especially when antidepressants were also included in the simulation. We also found that when the model involves both the serotoninergic pathway and stress, antidepressants (SSRIs) do not have the expected effect of modulate a constant serotonin.

This work also predicted that the genes that cause the highest impact on this model stability were CREB, BDNF (which in different association studies, has been found to be associated to major depression disorder) and TRkB; followed by the genes and metabolites that are related to the serotonin synthesis.

EFFECTS OF CONVECTIVE FLOW AND EXTRACELLULAR INHIBITORS ON EGFR SIGNALING – MATHEMATICAL MODELING

Marek Nebyla, Institute of Chemical Technology Prague, Prague, Czech Republic
Michal Přibyl, Institute of Chemical Technology Prague, Prague, Czech Republic

Abstract

We study effects of convective flow on velocity of signal transmission in epithelial tissue. It has been found that the convective transport is necessary in many developmental processes such as vascular and lymphatic system formation. Here, we focus on a one-layer epithelium system. We consider positive feedback loop ligand transport -> ligand-receptor binding -> intracellular processing -> ligand release. When a combination of diffusion and convective transport is considered in the extracellular space, we can see different regimes of ligand propagation or withdrawal at the epithelium. Parametrical studies dealing with effects of strength of ligand-receptor binding, geometric parameters, importance of ligand-receptor endocytosis or relative strength of ligand formation on signal propagation will be shown. We have developed three mathematical models that differ according to flow character above the epithelium: (i) plug flow along the epithelium, (ii) simplified plug flow model that exploits the thin fin approximation, (iii) laminar flow along the epithelium. We also found exact solutions of the models (i) and (ii) in asymptotic cases. All models were analyzed numerically (Comsol). When results of these models were compared, we realized that the simplified model (ii), which is spatially one-dimensional, describes behavior of the system with sufficient precision in most of physically relevant regimes. Hence, comprehensive continuation analysis of the model was carried out (AUTO-07p).

Further, we extended the model for inhibition of EGFR signaling by extracellular inhibitors. The work was motivated by problems related to glioblastoma treatment. There is a blood-brain barrier that surrounds blood capillaries in brain. It often makes impossible to delivery chemotherapeutics in tumor. Now, convection enhanced transport delivery of therapeutics can be used in treatment. Here, we consider constant flux of EGFR inhibitor (EGFR antibodies) through the barrier to astrocyte tumor. Effects of intensity of inhibitor flux and oriented convection transport will be presented.

Introduction

Intracellular signalling induced by growth factors can result in the production and extracellular secretion of these factors into the extracellular medium. This process produces a positive feedback loop between ligand binding and ligand release. This feedback can generate signal waves travelling at a constant speed when coupled with intracellular processes. Here, we model the influence of convective flow and model parameters on the velocity of signal propagation in a single-layer epithelium.

Cell communication in particular cases can be mediated by the epidermal growth factor receptor (EGFR) and its extracellular ligands. EGFR is a transmembrane glycoprotein serving as receptor for epidermal growth receptor, transforming growth factor α and the others. Together with another three proteins of similar structure and function, it is a member of the c-erb family of receptor tyrosine kinases. EGFR is important due to its role in the signalling pathways leading to signal transduction cascades. These cascades lead to DNA synthesis and cell proliferation [1, 2].

Many growth factors and cytokines binding to EGFR are released by intracellular or surface proteases [3]. The ligand-releasing enzymes can be activated by the signalling pathways that are stimulated by the ligands that they release. Hence, a positive feedback loop can be established when a cell effectively responds to and recaptures the secreted ligands [3, 4, 5]. Cells can reversibly bind the secreted ligand. The ligand-receptor complexes on the cell surface stimulate the intracellular processes leading to further ligand release. The ligand-receptor complexes can also be internalized into the cell by endocytosis (Figure 1) [6]. Here, we focus on such positive feedback loops. We consider epithelium formed by a single layer of cells covered with surface EGFRs.

In healthy cells, the expression of EGFR ranges from 40 000 to 100 000 receptors per cell [7]. However, EGFR is over-expressed in the majority of solid tumours e.g., breast cancer, head-and-neck cancer, colon cancer etc [8]. Such over-expression can lead to the permanent activation of downstream signalling pathways, which results in uncontrolled cell division with invasiveness characteristics [1, 9]. This permanent activation can be reduced by using extracellular and intracellular inhibitors. These inhibitors can be used in cancer treatment. Extracellular inhibitors are monoclonal antibodies that block ligand binding to the extracellular domain of EGFR. Intracellular inhibitors are small molecules that inhibit kinase activity of EGFR. Without this activity, downstream signalling pathways cannot be activated. In our work, we extended our model for the influence of extracellular inhibitors on signal propagation.

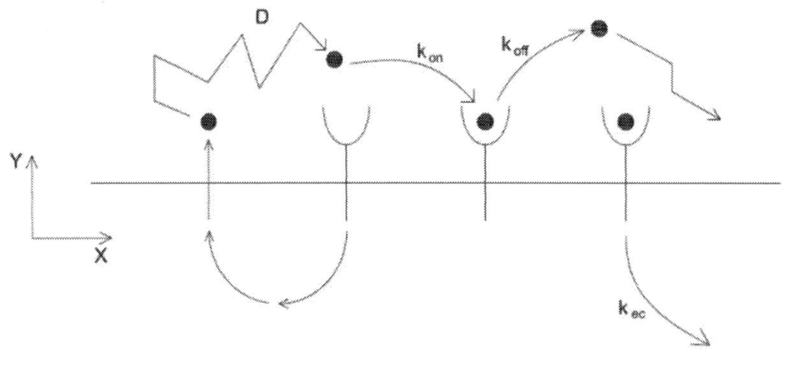

Figure 1: Receptor-mediated processes in the positive feedback loop: ligand release, extracellular transport, reversible binding, and endocytosis. D - ligand difusivity; k_{on} - ligand-receptor association constant; k_{off} - complex dissociation constant; k_{ec} - ligand-induced internalization rate constant

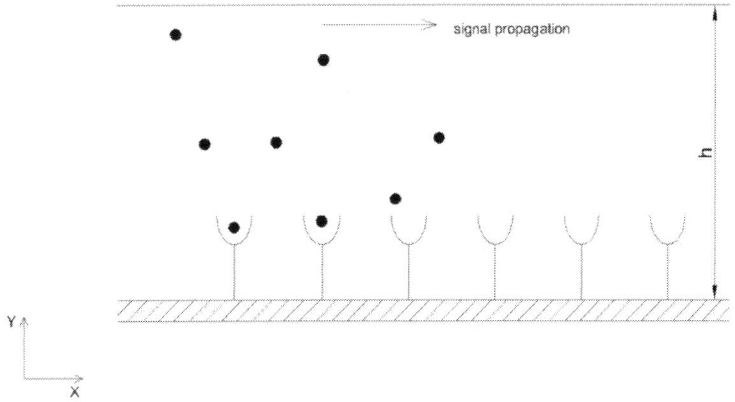

Figure 2: The geometry of cell communication in the model of positive feedback loop

Model Formulation

Equations and Scaling

We consider an infinite layer of cells covered with an extracellular medium of height h (Figure 2). Our model describes the coupled dynamics of secreted ligands (S), surface EGFR receptors (R), ligand-receptor complexes (C), and ligand-releasing proteases (P). Ligands reversibly bind to the receptors, the ligand-receptor complexes are degraded through ligand-mediated endocytosis. Q_r denotes the rate of EGFR synthesis, g_p and g_r are linear gains in the protease and ligand production, respectively. The balance equations and boundary conditions describing coupled dynamics are follows:

$$\frac{\partial S}{\partial t} = -v_x \frac{\partial S}{\partial x} + D\left(\frac{\partial^2 S}{\partial x^2} + \frac{\partial^2 S}{\partial y^2}\right), \tag{1}$$

$$\frac{\partial R}{\partial t} = -k_{on} R\overline{S} + k_{off} C - k_{er} R + Q_r, \tag{2}$$

$$\frac{\partial C}{\partial t} = k_{on} R\overline{S} - \left(k_{off} + k_{ec}\right) C, \tag{3}$$

$$\frac{\partial P}{\partial t} = -k_p P + g_p \sigma(C), \tag{4}$$

$$-D\left.\frac{\partial S}{\partial y}\right|_{y=0} = k_{off} C - k_{on} RS + g_r P \qquad \left.\frac{\partial S}{\partial y}\right|_{y=h} = 0. \tag{5}$$

Where t is time, x is spatial coordinate parallel to cellular layer, y is the spatial coordinate perpendicular to the cellular layer, and $\overline{S}$ denotes the ligand concentration at the cellular layer $\overline{S}(x,t) \equiv S(x,0,t)$. D is diffusivity of ligand, k_{ec} is ligand induced internalization rate constant, k_{er} is constitutive internalization rate constant, k_{on} denotes receptor/ligand association constant, k_p is protease degradation rate constant and $R_0 = Q_r / k_{er}$ means number of receptors per cell surface in absence of ligand. There is no convection considered in the *y*-direction.

In our model, ligand generation is controlled by the activity of the protease and has zero order with respect to the amount of cell-associated ligand precursor. The ligand-receptor complexes are internalized in process with rate constant k_{ec}. This process has zero order. The synthesis of the protease, $\sigma(C)$, is described by the sigmoidal function of the amount of occupied receptors [6].

Following transformations were used to convert the model into the dimensionless form:

$$\tilde{t} \equiv tk_p, \quad \tilde{x} \equiv x/L_x, \quad \tilde{S} \equiv S/S_0$$
$$\tilde{C} \equiv C/C_0, \quad \tilde{P} \equiv P/P_0, \quad \tilde{R} \equiv R/R_0, \tag{6}$$

where

$$S_0 = \frac{(k_{off} + k_{ec})g_r g_p k_{er}}{k_p k_{on} k_{ec} Q_r}, \quad C_0 = \frac{g_r g_p}{k_p k_{ec}}, \quad P_0 = \frac{g_p}{k_p}, \quad R_0 = \frac{Qr}{k_{er}}, \quad L_x = \frac{Dk_{er}}{Q_r k_{on}}. \tag{7}$$

The time scale and vertical length scale are set by the dynamics of protease degradation and by the height of the medium, respectively. After rescaling, the model has following form:

$$\tilde{\tau}_S \frac{\partial \tilde{S}}{\partial \tilde{t}} = -\tilde{v}_x \frac{\partial \tilde{S}}{\partial \tilde{x}} + \frac{\partial^2 \tilde{S}}{\partial \tilde{x}^2} + \frac{1}{\alpha^2}\frac{\partial^2 \tilde{S}}{\partial \tilde{y}^2}, \tag{8}$$

$$\tilde{\tau}_R \frac{\partial \tilde{R}}{\partial \tilde{t}} = 1 - \tilde{R} + \gamma\left[(1-\beta)\tilde{C} - \tilde{S}\tilde{R}\right], \tag{9}$$

$$\tilde{\tau}_C \frac{\partial \tilde{C}}{\partial \tilde{t}} = \tilde{S}\tilde{R} - \tilde{C}, \tag{10}$$

$$\frac{\partial \tilde{P}}{\partial \tilde{t}} = -\tilde{P} + \sigma(\tilde{C}), \tag{11}$$

$$-\left.\frac{\partial \tilde{S}}{\partial \tilde{y}}\right|_{\tilde{y}=0} = \alpha\left[(1-\beta)\tilde{C} - \tilde{S}\tilde{R} + \beta\tilde{P}\right] \qquad -\left.\frac{\partial \tilde{S}}{\partial \tilde{y}}\right|_{\tilde{y}=1} = 0. \tag{12}$$

There are six dimensionless parameters in the rescaled model:

$$\alpha = \frac{hQ_r k_{on}}{Dk_{er}}, \quad \beta = \frac{k_{ec}}{k_{off} + k_{ec}}, \quad \gamma = \frac{g_r g_p (k_{ec} + k_{off})}{k_p k_{ec} Q_{er}}, \tag{13}$$

$$\tau_S = \frac{k_{er}^2 D k_p}{Q_R^2 k_{on}^2}, \qquad \tau_C = \frac{k_p}{k_{off} + k_{ec}}, \qquad \tau_R = \frac{k_P}{k_{er}}.$$

For the case of plug flow, dimensionless velocity $\tilde{v}_x$ is defined as $\tilde{v}_x = \frac{v_x k_{er}}{Q_R k_{on}}$. Model with laminar flow along the epithelium has dimensionless velocity $\tilde{v}_x$ defined as $\tilde{v}_x = 6\frac{ZL_x}{D}(\tilde{y} - \tilde{y}^2)$, where Z is average velocity. Third model for the case of simplified plug flow was obtained after thin-layer approximation for ligand. The equation (8) has now form

$$\tilde{\tau}_S \frac{\partial \tilde{S}}{\partial \tilde{t}} = -\tilde{v}_x \frac{\partial \tilde{S}}{\partial \tilde{x}} + \frac{\partial^2 \tilde{S}}{\partial \tilde{x}^2} + \frac{1}{\alpha}\left[(1-\beta)\tilde{C} - \tilde{S}\tilde{R} + \beta\tilde{P}\right]. \qquad (14)$$

When results of these models were compared, we realized that the simplified model with plug flow, which is spatially one-dimensional, describes behavior of the system with sufficient precision in most of physically relevant regimes. Hence, we used this model for all future simulations and computations.

The model system now consists of equations 14 and 9-12. These equations were used in the dynamical modelling of the system with one-dimensional geometry. For purposes of continuations of heteroclinic orbits, the mathematical model was transformed into the moving coordinates system:

$$\frac{d^2\tilde{S}}{d\xi^2} = (\tilde{v}_x - \tau_s U)\frac{d\tilde{S}}{d\xi} - \frac{1}{\alpha}\left[(1-\beta)\tilde{C} - \tilde{S}\tilde{R} + \beta\tilde{P}\right], \qquad (15)$$

$$\frac{d\tilde{R}}{d\xi} = -\frac{1}{U\tau_R} + \frac{\tilde{R}}{U\tau_R} - \frac{\gamma}{U\tau_R}\left[(1-\beta)\tilde{C} - \tilde{S}\tilde{R}\right], \qquad (16)$$

$$\frac{d\tilde{C}}{d\xi} = -\frac{1}{U\tau_C}\left(\tilde{C} - \tilde{S}\tilde{R}\right), \qquad (17)$$

$$\frac{d\tilde{P}}{d\xi} = \frac{\tilde{P}}{U} + \frac{H}{U}\left(\tilde{C} - \tilde{C}_0\right), \qquad (18)$$

where U is the dimensionless velocity of wave propagation and ξ denotes positional coordinate and it is defined as $\xi = \tilde{x} - U\tilde{t}$. Sigmoidal function is approximated with the Heaviside nonlinearity: $\sigma(\tilde{C}) = H(\tilde{C} - \tilde{C}_0)$. The value of $\tilde{C}_0$ defines the threshold of the nonlinearity. In numerical computations, we approximated the Heaviside function with a modified hyperbolic tangent function.

In model with inhibitor, there are two new variables: (A) denotes concentration of extracellular inhibitors and (B) is an amount of inhibitor-receptor complexes. Model is extended for balance equations of inhibitors and of inhibitor-receptor complexes and balance equation of receptors is modified. New and modified equations and boundary conditions are follows:

$$\frac{\partial A}{\partial t} = -v_x\frac{\partial A}{\partial x} + D_A\left(\frac{\partial^2 A}{\partial x^2} + \frac{\partial^2 A}{\partial y^2}\right), \tag{19}$$

$$\frac{\partial R}{\partial t} = -k_{on}R\bar{S} + k_{off}C - k_{on,A}AR + k_{off,A}B - k_{er}R + Q_r, \tag{20}$$

$$\frac{\partial B}{\partial t} = k_{on,A}AR - (k_{off,A} + k_{eB})B, \tag{21}$$

$$-D\frac{\partial A}{\partial y}\bigg|_{y=0} = k_{off,A}B - k_{on,A}AR \qquad \frac{\partial S}{\partial y}\bigg|_{y=h} = g_A(t). \tag{22}$$

D_A denotes inhibitor diffusivity, $k_{on,A}$ is inhibitor-receptor association constant, $k_{off,A}$ is complex receptor-inhibitor dissociation constant, k_{eB} denotes inhibitor induced internalization rate constant and $g_A(t)$ is inhibitor flux into extracellular medium.

After rescaling, the model with inhibitor has following form:

$$\tilde{\tau}_S\frac{\partial \tilde{S}}{\partial \tilde{t}} = -\tilde{v}_x\frac{\partial \tilde{S}}{\partial \tilde{x}} + \frac{\partial^2 \tilde{S}}{\partial \tilde{x}^2} + \frac{1}{\alpha^2}\frac{\partial^2 \tilde{S}}{\partial \tilde{y}^2}, \tag{23}$$

$$\tilde{\tau}_S\frac{\partial \tilde{A}}{\partial \tilde{t}} = -\tilde{v}_x\frac{\partial \tilde{S}}{\partial \tilde{x}} + \varepsilon_A\frac{\partial^2 \tilde{A}}{\partial \tilde{x}^2} + \frac{\varepsilon_A}{\alpha^2}\frac{\partial^2 \tilde{A}}{\partial \tilde{y}^2}, \tag{24}$$

$$\tilde{\tau}_R\frac{\partial \tilde{R}}{\partial \tilde{t}} = 1 - \tilde{R} + \gamma\left[(1-\beta)\tilde{C} - \tilde{S}\tilde{R}\right] + (1-\beta_A)\tilde{B} - \tilde{A}\tilde{R}, \tag{25}$$

$$\tilde{\tau}_C\frac{\partial \tilde{C}}{\partial \tilde{t}} = \tilde{S}\tilde{R} - \tilde{C}, \tag{26}$$

$$\frac{\partial \tilde{P}}{\partial \tilde{t}} = -\tilde{P} + \sigma(\tilde{C}), \tag{27}$$

$$\tilde{\tau}_B\frac{\partial \tilde{B}}{\partial \tilde{t}} = \tilde{A}\tilde{R} - \tilde{B}, \tag{28}$$

$$-\frac{\partial \tilde{S}}{\partial \tilde{y}}\bigg|_{\tilde{y}=0} = \alpha\left[(1-\beta)\tilde{C} - \tilde{S}\tilde{R} + \beta\tilde{P}\right] \qquad -\frac{\partial \tilde{S}}{\partial \tilde{y}}\bigg|_{\tilde{y}=1} = 0, \tag{29}$$

$$-\frac{\partial \tilde{A}}{\partial \tilde{y}}\bigg|_{\tilde{y}=0} = \frac{\alpha}{\varepsilon_A}\varphi_A\left[(1-\beta_A)\tilde{B} - \tilde{A}\tilde{R}\right] \qquad -\frac{\partial \tilde{A}}{\partial \tilde{y}}\bigg|_{\tilde{y}=1} = -\psi_A, \tag{30}$$

with these new scaling factors and dimensionless parameters:

$$B_0 = \frac{Q_r}{k_{off,A} + k_{eB}}, \quad A_0 = \frac{k_{er}}{k_{on,A}}, \quad \varepsilon_A = \frac{D_A}{D}, \quad \beta_A = \frac{k_{eB}}{k_{off,A} + k_{eB}},$$
$$\varphi_A = \frac{k_{on,A}}{k_{on}}, \ \psi_A = \frac{g_A h k_{on,A}}{D_A k_{er}}, \ \bar{\tau}_B = \frac{k_p}{k_{off,A} + k_{eB}}. \tag{31}$$

Transformation into the moving coordinates has analogous form as model without inhibitor.

Steady States

Our mathematical models have two uniform steady states. The first steady state corresponds to the "off" state of the positive feedback loop, with the zero rate of ligand release. The second steady state corresponds to the "on" state of the positive feedback loop, with nonzero amount of ligand-receptor complexes and appreciable rate of ligand release.

Parametric study

We conducted series of parametric studies focused on effects of the dimensionless model parameters on the signal propagation velocity. Examined parameters for the model without inhibitor were α, β, γ, τ_S, τ_C, τ_R, $\tilde{v}_x$ and threshold $\tilde{C}_0$, and for the model with inhibitor they were the same plus ε_A, β_A, ϕ_A, ψ_A, τ_B. First, we carried out the dynamical simulations of wave propagation using computer program Comsol Multiphysics 3.5. Results from these simulations were used as starting orbit connecting the two steady states for continuation program AUTO-07p. In this program, the parametric studies for three convective velocities were carried out. Velocities used were -0.5, 0 and 1, respectively. In the parametric studies, one parameter was changed, while the others remained constant.

Results

Shape of the concentration curves of ligands, free receptors, ligand-receptor complexes, protease, inhibitors and inhibitor-receptor complexes in space in particular time is in the Figure 3. These so-called waves are moving with time from the steady state, which corresponds to "on" state of the positive feedback loop, to the steady state corresponding to the "off" state of the positive feedback loop.

Results were not fully complete at the time of Presentation Record upload deadline. More at poster session in Exhibit Hall B (Minneapolis Convention Center) at Wednesday, October 19.

Conclusion

We studied effects of convective flow on velocity of signal transmission in epithelial tissue. Our model accounts for coupled dynamics of ligand release, transport, binding and complex endocytosis. This model provides information about signal propagation velocity and about concentration of ligands, protease, receptors and complexes in time and space.

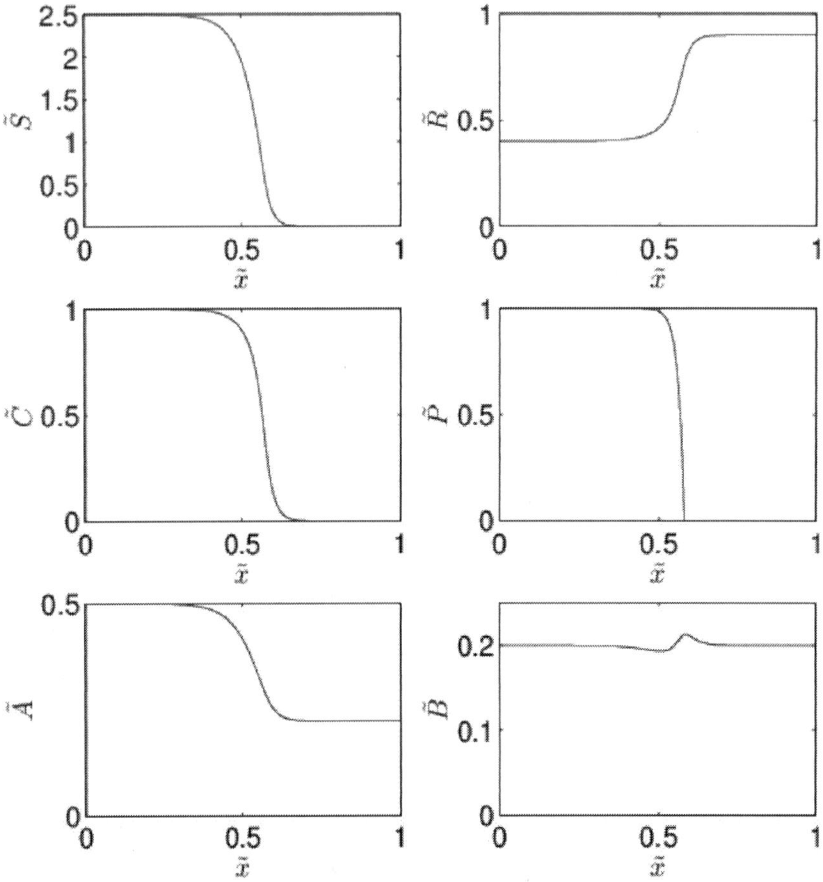

Figure 3: Shape of the concentration curves in particular time for ligands, free receptors, ligand-receptor complexes, protease, inhibitors and inhibitor-receptor complexes, respectively. The parameters used are $\alpha = 1$, $\beta = 0.5$, $\beta_A = 0.5$, $\gamma = 1$, $\varepsilon_A = 1$, $\phi_A = 1$, $\psi_A = 0.1$, $\tau_S = 0.002$, $\tau_C = 0.1$, $\tau_R = 0.5$, $\tau_B = 0.1$ and $\widetilde{C}_0 = 0.3$. Convective velocity was $\widetilde{v}_x = 0$.

This model was extended for influence of inhibition of EGFR signaling by extracellular inhibitors.

We compared three models with different type of flow along the epithelium and results suggest that simplified plug flow model that exploits the thin fin approximation describes behavior of the system with sufficient precision in most of physically relevant regimes.

Parametric studies for main model parameters in 1D geometry were carried out. Each parametric study was made for three different convective velocities. Results suggested that in most cases the dependence of propagation velocity on the examined parameter is for all three convective velocities similar. There are some exceptions.

Presented work can contribute to the development of artificial tissues, biomimetic systems and a better understanding of developmental processes in epithelia.

Acknowledgement

The authors thank for the support of the research by the grant of the Czech Grant Agency 104/08/H055, by grants MSMT KONTAKT ME10036, MSMT 604617306, and specific university research (MSMT No 21/2011).

References

1. Herbst, R. S., Review of epidermal growth factor receptor biology, International Journal of Radiation Oncology Biology Physics, 59 (2): 21-26 Suppl. S 2004
2. http://en.wikipedia.org/wiki/Epidermal_growth_factor_receptor
3. Dello, S., Rovida, E., Transduction of cell surface regulatory molecules via ectodomain shedding, Biological Chemistry, 383 (1): 69-83 Jan 2002
4. Carpenter, G., Employment of the epidermal growth factor receptor in growth factor-independent signaling pathways, Journal of Cell Biology, 146: 697-702, 1999
5. Gschwind, S., Zwick, E., Prenzel, N., Leserer, M., Ullrich, A., Cell communication networks: epidermal growth factors receptor transactivation as a paradigm for interreceptor signal transmission, Oncogene, 20: 1594-1600
6. Přibyl, M., Muratov, C. B., Shvartsman,. S. Y., Long-range signal transmission in autocrine relays, Biophysical Journal, 84(6): 3624 – 3635, 2003
7. Carpenter, G; Cohen, S, Epidermal Growth-Factor, Annual Review of Biochemistry, 48: 193-216 1979
8. Herbst, R. S.; Langer, C. J., Epidermal growth factor receptors as a target for cancer treatment: The emerging role of IMC-C225 in the treatment of lung and head and neck cancers, Seminars in Oncology, 29 (1): 27-36 Suppl. 4 FEB 2002
9. Ethier, S. P., Signal transduction pathways: The molecular basis for targeted therapies, Seminars in Radiation Oncology, 12 (3): 3-10 Suppl. 2 Jul 2002
10. Deen, W. M., Analysis of transport phenomena, Oxford University Press, New York, 1998

(625f) Implications of Ultradian Rhythms In Glucocorticoid Secretion

Jeremy D. Scheff[1], Alyssa K. Kosmides[1], Steven E. Calvano[2], Stephen F. Lowry[2] and Ioannis P. Androulakis[3], (1)Biomedical Engineering, Rutgers University, Piscataway, NJ, (2)Department of Surgery, UMDNJ-Robert Wood Johnson Medical School, New Brunswick, NJ, (3)Biomedical Engineering, Chemical & Biochemical Engineering, Rutgers University, Piscataway, NJ

Glucocorticoids (GCs) are a class of hormones which activate signal transduction pathways leading to the transcription of a broad range of genes, including those conferring anti-inflammatory effects. In humans, the GC cortisol is produced in the adrenal gland and secreted into systemic circulation in a series of discrete bursts. This discrete release pattern produces significant ultradian rhythms (oscillatory period ~1hr) in plasma cortisol concentration, and circadian variability in the size of the bursts leads to the characteristic circadian rhythms in cortisol levels. It has recently been observed that cells exposed to pulsatile levels of GCs develop a distinct phenotype relative to cells given constant cortisol treatment, even when the total GC exposure is equal (McMaster, Jangani et al. 2011). We applied computational systems biology techniques towards understanding both the mechanistic origins of this observed differential expression and translational implications of the importance of GC pulsatility in vivo.

To mathematically investigate the effect of ultradian GC pulsatility on transcription of GC-responsive genes, we combined a model of rhythmic GC production (Walker, Terry et al. 2010) with a pharmacodynamic model of GC action (Yao, DuBois et al. 2006). The hypothalamic-pituitary-adrenal (HPA) axis is modeled so that the negative feedback loop between GC secretion and adrenocorticotropic hormone (ACTH) production produces oscillations in these hormones. GCs are released from the adrenal gland into circulation. The glucocorticoid receptor (GR) is a cytosolic receptor, so it interacts with GCs that have crossed the cell membrane into the cytoplasm. Then, the activated GR translocates into the nucleus where it regulates gene transcription. This combined model consists of 7 equations and 14 parameters.

The nonlinear binding kinetics between GC and GR can fundamentally drive differential expression between the constant and pulsatile simulations shown in **Figure 1**. In other words, the peak value of the ultradian rhythm is in the region where GR is becoming saturated and the system gets diminishing returns for increasing GC levels. This type of homeostatic behavior is important in maintaining the responsiveness of the HPA axis by allowing for the rapid production of high levels of GCs and the responsiveness of the GC signaling pathway in peripheral cell, as a high constant level of GC exposure typically leads to tolerance.

The model of ultradian GC production allows for the assessment of the effects of dysregulated HPA axis function. In particular, we explore the implications of altering the feedback loop from GC to ACTH, which is ultimately the driving force behind the oscillatory hormone release. This is done in two different ways, by perturbing (1) the timing of signal transduction from GC to ACTH and (2) the strength of feedback from GC to ACTH.

When the time delay for the effect of GC on ACTH production is decreased, the frequency of ultradian rhythms increases. This means that there is a shorter time window for both the secretion and elimination of GCs, culminating in blunted rhythms. GC levels to rise and fall, which results in lower amplitude ultradian oscillations. (Veldhuis, Iranmanesh et al. 1989) reports that there is a positive relationship between the amount of GC released in each burst and the time between bursts. When the time delay is left constant and the parameter representing the strength of feedback from GC to ACTH is decreased, the ultradian amplitude is lowered without a change in frequency. In both cases, a flatter, chronically elevated hormone profile as the time delay is decreased leads to the loss of the downstream responses to pulsatility described above.

The work presented here has important translational implications. As exogenous GC therapy is applied clinically without considering pulsatility, this modeling work allows for comparing modes of therapeutic delivery of GCs. Under stress, endogenous rhythmicity in hormone release is often diminished or lost, and the downstream transcriptional implications of these changes can be assessed by this model.

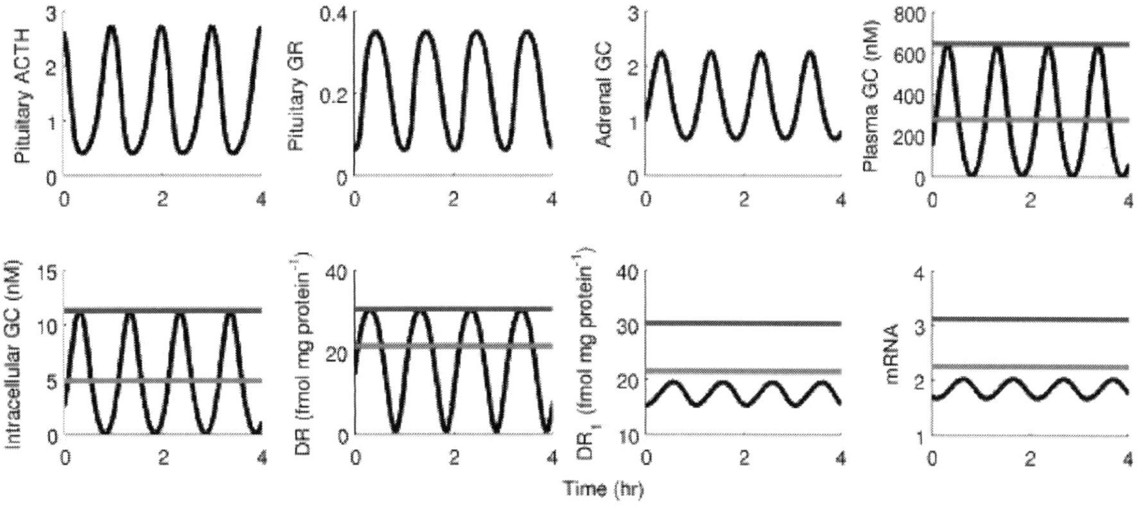

***Figure 1**: Pulsatile production of glucocorticoids (GC) arises from the HPA axis through interactions between ACTH and glucocorticoid receptor (GR) binding in the pituitary, culminating in the release of adrenal GC. Then, the signal propagates from intracellular GC (GC) concentration, to the activated GC-GR complex (DR), and translocation of the complex into the nucleus (DR1) where it regulates the transcription of GC-responsive genes (mRNA). Three simulations are shown: pulsatile (black), constant with the same exposure as pulsatile (red), and constant with the maximum concentration of pulsatile*

(blue). Gene pulsing is observed in response to ultradian rhythms, and both of the constant GC simulations produce higher levels of gene transcripts.

References

- McMaster, A., M. Jangani, et al. (2011). "Ultradian cortisol pulsatility encodes a distinct, biologically important signal." PLoS One 6(1): e15766.
- Veldhuis, J. D., A. Iranmanesh, et al. (1989). "Amplitude modulation of a burstlike mode of cortisol secretion subserves the circadian glucocorticoid rhythm." Am J Physiol 257(1 Pt 1): E6-14.
- Walker, J. J., J. R. Terry, et al. (2010). "Origin of ultradian pulsatility in the hypothalamic-pituitary-adrenal axis." Proc Biol Sci 277(1688): 1627-33.
- Yao, Z., D. C. DuBois, et al. (2006). "Modeling circadian rhythms of glucocorticoid receptor and glutamine synthetase expression in rat skeletal muscle." Pharm Res 23(4): 670-9.

See more of this Session: Poster Session: Engineering Fundamentals In Life Science
See more of this Group/Topical: Food, Pharmaceutical & Bioengineering Division

(625g) A Pathway-Centric View of the Brain's Homeostasis Transcriptome During Acute Sustained Hypertension

Rajanikanth Vadigepalli, Gregory Gonye and James Schwaber, Department of Pathology, Anatomy and Cell Biology, Thomas Jefferson University, Philadelphia, PA

The mechanisms of the development and origin of hypertension have been unclear. The recent evidence for "neurogenic hypertension" by resetting of brainstem blood pressure control mechanisms over long timescales is a modern scientific triumph. These physiological discoveries now provide the new opportunity to seek the underlying adaptive processes and altered mechanism in central neuronal control systems. We here describe a systems biology approach that bridges molecular mechanisms and neurophysiological behavior, with specific application to understanding the molecular basis for the setpoint control of blood pressure by the brain.

Our working hypothesis is that elevated blood pressure induces a temporal sequence of gene expression changes as an adaptive response and that this changed molecular phenotype underlies alterations in central blood pressure setpoint control. In order to identify the molecular response process we use high resolution quantitative PCR to profile large sets of functionally relevant genes across select brain nuclei in the autonomic-emotional neuraxis, in phenotypically homogeneous sets of neurons, and in single identified neurons - from *in vivo* samples. We sample four key brain regions in the autonomic-emotional neuraxis, specifically (1) the nucleus of the solitary tract - NTS, caudal and rostral ventro-lateral medulla - CVLM and RVLM, and (2) the Central Nucleus of the Amygdala (CeA). Within the NTS we are particularly focused on the noradrenergic A2 neurons. Increases in blood pressure induce broad transcriptional changes within 1h throughout the autonomic-emotional neuraxis, but with distinct expression programs in each region.

Our experimental results and dynamic response pattern analysis reveals a temporal sequence of gene regulatory changes that indicates a causative influence that starts in the NTS, subsequently followed by changes in CeA, and later evolving in the caudal and rostral VLM nuclei. We found a very early (within the first hour of the stimulus) and widespread gene regulatory response involving neuron-specific functions, the pattern of which was dynamic even in the short term (over the course of four hours). There is evidence of coordinate regulation in all the four brain regions studied, in elements of angiotensin II type 1 receptor (AT1R) mediated pathway, downstream transcriptional regulators including members of AP-1, CREB and ATF families, and ionotropic glutamate receptors. Since these functional systems are known to play prominent roles in normal homeostatic maintenance and hypertension pathobiology, we explore their adaptive regulation here in some detail.

We analyzed the data using a novel bioinformatics approach that identifies context-specific signaling network that is consistent with our gene expression data. The networks with highly significant consistency showed remarkable similarity within the treatment groups but are distinct for hypertensive vs normotensive samples. Our analysis indicated that the AT1R pathway interactions are likely to be different across the brain regions, and conditionally dependent on the hypertensive state. Results of A2 cell analysis show (1) significant systems-level gene expression response to increases in blood pressure within an hour, differential to controls, (2) that the response differs from cardiovascular NTS and other identified neurons, and (3) A2 cells are molecularly heterogeneous. The time series gene expression data also identify key receptors and transmitters affected by the blood pressure. The present broader measures provide not only simultaneous measures of many molecules-genes, but also provide context to understand the adaptive molecular process and networks. In addition, time series allows consideration of dynamics within and across distinct neuronal "nodes" forming a functional circuit. The A2 cell data radically expands our understanding of these neuron types and identification of molecular participants in the adaptive response predicts useful targets for genetic manipulation technologies.

These results reveal a dynamic transcriptional regulation that is likely to coordinately impact the homeostatic control circuits along the autonomic-emotional neuraxis.

Research Support: NIH - R01 GM083108, R33 HL088283 and R33 HL087361.

See more of this Session: Poster Session: Engineering Fundamentals In Life Science
See more of this Group/Topical: Food, Pharmaceutical & Bioengineering Division

(625h) Development of a Model-Based Noninvasive Continuous-Time Glucose Monitoring Device for Non-Insulin Dependent People

Derrick K. Rollins Sr.[1], Lucas Beverlin[2], Kaylee Kotz[1], Nisarg Vyas[3] and Dave Andre[4], (1)Chemical and Biological Engineering, Iowa State University, Ame, IA, (2)Department of Statistics, Iowa State University, Ames, IA, (3)BodyMedia Inc., Pittsburgh, PA, (4)BodyMedia, Inc.

Continuous-time glucose monitoring (CGM) effectivelyimproves glucose control by providing frequently sampled information thatallows the user to associate changes in their glucose levels with changes intheir behavior. For example, a user is able to see immediately how the size ofa meal affects how high their glucose level becomes as well as the duration ofelevated glucose levels. Currently,the most widely used and effective CGM devices rely on a sensor that isinserted invasively under the skin. Sensors cost from $35 to $60 and last 3days to a week. Due to invasiveness and cost, the primary users of current CGMdevices are insulin dependent people (type 1 and some type 2 diabetics). Thus,the primary goal of this research is the development of a non-invasive CGMdevice that would be used by health conscious non-insulin dependent people(including non-diabetics) that would help reduce obesity, the onset of type 2 diabetes, as well as the progression of type 2diabetes.

To accomplish this objective, this work is focusing on thedevelopment of a device that: 1. has a simple or no reliance for on food entry;2. has a relatively short calibration period; 3. requires few to no lancetmeasurements for calibration and; 4. has an accuracy that is comparable tolancet meters. The approach of this research is to use a novel modeling methodtechnique to infer glucose concentration using non-invasive input measurementsfrom variables representing food, activity, circadian rhythm, and stressvariations. The main component of this system will be a theBodyMedia armband that will automatically collectthe activity and stress data. The food information will be entered manually bythe user via the time stamp button on the armband. The model and modeldevelopment algorithm will reside in the armband and will use a Wiener approachbased on the work of Rollins et al. (2010). Data from a lancet glucose meter will beentered automatically or manually andwill be used to develop a subject specific model for the person wearing thedevice. After the model is completely developed, lancet measurements will nolonger be needed for calibration. An interface device will be connected to thearmband to display the glucose concentration in every five minutes.

Using about 20 test subjects with 4 weeks of data collectioneach, results have been obtained to support the modeling viability necessary tobuild an armband monitoring device. Before using these data sets the foodquantities were converted to food indices to mimic time stamping and a lancetsampling rate of only four values per day. Several improvements to the Wienermodeling method developed by Rollins et al.

were made to address the infrequentnature of the data (i.e., four glucose measurements/day), frequent start- ups(i.e., taking the armband on and off and restarting it frequently), and thelack of steady state data. The accomplishments of this work include the abilityto develop subject specific models under these restrictions. Results arepresented for models developed after three days, 2 weeks and four 4 weeks thatsupport an initial calibration period of three days with accuracy improvingover time and no need for lancet measurements after 3 to 4 weeks. Thus, sincethe model does not drift, this device would not appear to need glucosemeasurements for calibration once it is fully calibrated to the user wearingthe device.

Rollins, D. K., N. Bhandari,J. Kleinedler, K. Kotz, A. Strohbehn, L. Boland, M. Murphy, D. Andre, N. Vyas,G. Welk and W. Franke, "Free-living inferential modeling of blood glucoselevel using only noninvasive inputs," Journal of Process Control **20** 95-107 (2010).

See more of this Session: Poster Session: Engineering Fundamentals In Life Science
See more of this Group/Topical: Food, Pharmaceutical & Bioengineering Division

(625i) The Development of A General Mathematical Model for the Prediction of Tumor Growth Based On Experimental Data

Bradley P. Hambly, Chemical Engineering, Tenneessee Tech. University, Cookeville, TN and Mario Oyanader, Chemical Engineering, Tenneessee Tech University, Cookeville, TN

Mathematical modeling of growth kinetics is important in predicting the behavior of many biological systems such as tumor size, population dynamics and disease progression. In many cases it is important to accurately describe the growth of biological system, such as the planning of chemotherapeutics for cancer patients or tracking the growth of bio-film in industrial pipes. Logistic and Gompertz are two models in particular that have been studied to a great extent in biological and medical systems. A series of hyperbolastic models H1, H2, and H3 were proposed by Tabataba that more accurately describe both systems of Ehrlich tumor growth and broiler growth kinetics then previous models. Upon observation of the hyperbolastic models, it appears that each took a specific form of a generalized model. It was found that a generalized form could become the proposed models with the addition of two or three assumptions. We perform a comparative study of the previously proposed hyperbolastic models and a generalized form H_0. The results of this comparison show that the assumptions that make up the earlier proposed hyperbolastic models are not necessarily correct; as the generalized model produced the smallest error in modeling both Ehrlich tumor growth and broiler growth kinetics.

See more of this Session: Poster Session: Engineering Fundamentals In Life Science
See more of this Group/Topical: Food, Pharmaceutical & Bioengineering Division

(625j) Cytomic Analysis and Fatty Acids Profiling of the Snow Algae Chlamydomonas Nivalis by NaCl Stress Using Laser Scanning Confocal Microscope, Flow Cytometry and GC-MS

Dong Wei, Na Lv and Xiao-li Jiang, School of Light Industry and Food Sciences, South China University of Technology, Guangzhou, China

The snow algae *Chlamydomonas nivalis*, a typical microalgal species discovered on snow field at Antarctic and similar extreme environments on earth, can adapt to very low temperature, high light intensity and strong UV radiation for survival by forming aplanospore rich in astaxanthin, a brown red pigment with super anti-oxidative effect. *C. nivalis* is realized not only a model species to investigate the mechanism of cell response and adaptation to stress conditions, but also a potential alternative bioresource for astaxanthin production by microalgal biotechnology. The response and resistance of microalgae to stress environment depend on a series of essential regulation of metabolic pathway, especially on lipid biosynthesis and cell wall structure. However, current research of *C. nivalis* mostly focuses on the classification, life cycle, modes of reproduction, regional distribution and cell ultra-structure. Cytomic analysis and fatty acid profiling of *C. nivalis* in resistant physiology has paid little attention.

In the present work, the 6-day culture of *C. nivalis* UTEX LB 2824 in Bold 1NV medium under autotrophic growth was treated by NaCl with final concentrations of 0.25%, 0.50%, 0.75%, 1.00%, 1.25%, 1.50% for 1, 2, 3, 5, 7, 11, 15, 24 or 48 h, respectively. The content of neutral and polar lipid referred as fluorescent intensity in stressed cells after stained with specific dye Nile Red was quantified by cytomic analysis based on laser scanning confocal microscope (LSCM, Ex: 543 nm; scan: 560-615 nm) and flow cytometry (FCM, Ex: 488 nm, FL1 at 530 nm, FL2 at 575 nm and SSC signal). After extraction and derivation of triacylglycerol (TG) and free fatty acids (FFA) from stressed cells, the profiles of FFA and FA in TG were analyzed by GC-MS. In general, the content of neutral and polar lipid showed the similar trends with total lipid in cells. The fluorescent intensities of lipids constantly increased to the maximum during 7 h, then dropped to a stable level within 24 to 48 h. The trend was not rigorous within all NaCl concentrations, but the highest increase of 211% for total lipid at 0.75% NaCl for 7 h, 68 times increase for neutral lipid at 1.00% NaCl for 7 h and 10 times increase for polar lipid at 1.25% NaCl for 5 h were observed, respectively. The cell size in all groups of NaCl stress enlarged quickly during 3 to 5 h, then stopped to become large. Big changes of composition and content of several FFA (C12:0, C16:0, C16:3, C18:0, C18:1 and C18:3) were observed with variation for 4.61, 3.70, 4.37, 2.68, 2.64, 3.14 times, respectively. C16:0, C18:0, C18:1 and C18:3 in TG also varied much for 2.67, 2.53, 4.91, 2.42 times, respectively. The results indicated *C. nivalis* could synthesize, accumulate lipid and change fatty acids profile very quickly during several hours by stress induction. The content of C16:0, C18:0, C18:1 and C18:3 fatty acids could change the degree of lipid unsaturation

(DLU) and play an important role on regulating the membrane fluidity, which result an enhancement of cell response and resistance to NaCl stress. This is the first report to demonstrate the lipid changes corresponding to NaCl stress by "omics" technology, and figure out the regulation mechanism of lipid biosynthesis by NaCl induction in *C. nivalis*.

See more of this Session: Poster Session: Engineering Fundamentals In Life Science
See more of this Group/Topical: Food, Pharmaceutical & Bioengineering Division

(625k) Worm Array Device for Studying Gene Expression Analysis In C. Elegans

Hyewon Lee and Hang Lu, School of Chemical & Biomolecular Engineering, Georgia Institute of Technology, Atlanta, GA

Gene expression is a complex process affected by genetic and environmental factors, and it often stochastically fluctuates. *C. elegans* has been broadly used for gene-expression analysis because of its transparent tissue, completely sequenced genome, mapped neuron wiring, and the ease of applying genetics. To dissect complex regulatory mechanisms and to get statistically meaningful data, one needs to image a large number of animals at single-cell resolution. Conventional experimental methods are usually time-consuming and data may not be reliable because of human intervention. Recent development of microfluidic systems increases the throughput of experiments but it usually requires large off-chip equipment and complications in fabrication. To overcome these limits, we present here a simple microdevice for imaging *C. elegans*. We fabricated the device in polydimethylsiloxane which is easy to make. This microarray does not include any active component so it is easy to be implanted without any complex off-chip components. Worms can be loaded by syringe manually and they are passively guided to the imaging channels. Their loading efficiency was tuned by controlling fluidic resistance with electric circuit model and COMSOL model. This device is more useful when imaging small size worms like L1 developmental stages. Current L1 imaging devices are high susceptible to be clogged even after using filters because the small size and number of imaging channels. However, this array has multiple imaging channels in one device and clogging in one single channel does not fail the entire experiment. Using the developed microfluidic worm arrays, we can image large number of worms and compare differences gene expression pattern in different developmental stages of worms.

See more of this Session: Poster Session: Engineering Fundamentals In Life Science
See more of this Group/Topical: Food, Pharmaceutical & Bioengineering Division

(625l) Expression Profiling Analysis Following Cecal Ligation and Puncture (CLP) Treatment In Rat Liver

Qian Yang[1], Mehmet A. Orman[2], Francois Berthiaume[3], Marianthi G. Ierapetritou[4] and Ioannis (Yannis) P. Androulakis[1], (1)Department of Chemical and Biochemical Engineering, Rutgers - The State University of New Jersey, Piscataway, NJ, (2)Department of Chemical and Biochemical Engineering, Rutgers- The State University of New Jersey, Piscataway, NJ, (3)Department of Biomedical Engineering, Rutgers University, Piscataway, NJ, (4)Department of Chemical and Biochemical Engineering, Rutgers, The State University of New Jersey, Piscataway, NJ

Severe sepsis is defined as a potentially deadly medical condition which is characterized by a whole-body inflammatory state [1]. Cecal ligation and puncture (CLP) model is an animal model that mimics the physiological changes of human sepsis, which is considered as the gold standard for sepsis researches [2]. Following CLP, animals generally develop bacteremia, hypothermia, hypotension, and hyper-metabolic and catabolic state at whole body level. Given the fact that liver controls the metabolic activity of body and produces acute phase proteins during the systemic inflammation, it is one of the target organs to understand the underlying molecular mechanisms of the septic state and to propose therapeutic approaches. Therefore, genome-wide microarray technology has been already applied to reveal transcriptional changes in liver following the CLP treatment [3-5]. However, these studies either focus on a single time point (24 h post-CLP) which probably misses the critical early response or do not take the time scale into account, i.e. the inherent ordering and spacing provided by the time points are ignored. Thus, the aim of this study is to get a better understanding of the hepatic transcriptional response to CLP by generating a rich time series as well as utilizing bioinformatics tools to identify the dynamic patterns of CLP induced-genes showing distinct expression profiles over time compared to a time dependent control.

In this study, the infection was induced by applying CLP treatment. Rats were first anesthetized, and then the analgesic buprenorphine (0.01 to 0.05 mg/kg) and Bupivicaine (0.125% to 0.25%) were given subcutaneously. The abdominal cavity was cut open by a 2 cm midline incision. The cecum of the rat was exposed and ligated just below the ileocecal valve so that intestinal obstruction was not produced. We took care to not ligate the cecal branch of the ileocecal artery, thus preserving viability of the cecum itself, in order to increase the survival rate. The cecum was punctured 4 times with a 20 gauge needle and replaced in the peritoneum. The abdominal incision was then sutured in layers using interrupted monofilament sutures. The animal received 10 mL/kg saline intraperitoneally for resuscitation. Negative controls (sham CLP or SCLP) consist of animals treated identically without receiving cecal ligation and puncture. Rats were single caged after the treatments and given standard rat chow and water ad libitum until sacrifice. Animals are sacrificed (starting at 9am) at different time points (0, 2, 4, 8, 12, 16, 20 and 24hr post-treatment) and liver tissues were collected and frozen for microarray analysis (n=3 per time point per group). The tissues were lysed and homogenized using Trizol, and the RNAs were further purified and treated with DNase using RNeasy columns (Qiagen). Then cRNAs prepared from the RNAs of liver tissues using protocols provided by Affymetrix were utilized to hybridize Rat Genome 230 2.0 Array (GeneChip, Affymetrix) comprised of more than 31,000 probe sets.

In order to analyze SCLP and CLP expression profiles, gene expression data analysis includes normalization, filtering, combining the datasets and clustering. First, DNA chip analyzer (dChip) software was used with invariant-set normalization and perfect match (PM) model to generate expression values. Then normalized two data sets corresponding to CLP and SCLP groups were investigated to identify the temporally and differentially expressed probesets by comparing the overall-AUC (area under the gene expression–time curve in CLP group) with the baseline-AUC (area under the gene expression–time curve in SCLP group) for each gene [6] in order to explore significant net responses of gene expression profiles following the CLP. Finally concatenated data sets corresponding to differentially expressed probesets in CLP and SCLP groups were combined to form one single matrix, which was then clustered using the approach of "consensus clustering" [7]. The goal was to identify subsets of transcripts with coherent expression pattern in CLP and SCLP. In total 1987 CLP-responsive probesets in 4 clusters were identified. We characterized the biological relevance of the intrinsic responses by evaluating the enrichment of the corresponding subsets by using KEGG database through ARRAYTRACK [8] as well as analyzed the functions of each individual gene [9].

Consensus clustering identifies 4 statistically dominant transcriptional clusters composed of 474, 565, 287, 661 probe sets respectively. Functional characterization elucidated that these responses are mainly associated with inflammation (cluster 1 and cluster 2) and xenobiotics biodegradation (cluster 3 and cluster 4). The results show that the first 2 clusters exhibit up-regulation following CLP treatment when compared to the control group, which are both related to inflammatory response. More specifically, genes in cluster 1 are important in natural killer cell mediated cytotoxicity (*Gzmb, Icam1, Nfat5, Nfatc4, Shc2*), MAPK signaling pathway (*Cacna1a, Dusp1, Fgf12, Fos, Il1a, Nfatc4, Rasgrp3, Taok1, Tp53*), and B-cell receptor signaling pathway (*Fos, Nfat5, Nfatc4, Rasgrp3*). Genes involved in cluster 2 participate in Jak-STAT signaling pathway (*Cblc, Grb2, Ifnar1, Ifnb1, Il10rb, Il7r, Jak3, Lepr, Socs3*), B cell receptor signaling pathway (*Bcl10, Fos, Grb2, Nfkb1, Nfkbie, Syk*), and Toll-like receptor pathway (*Fos, Ifnar1, Ifnb1, Lbp, Nfkb1, Tlr2*). On the other hand, the last 2 clusters showing down-regulation following the CLP compared to control are associated with liver metabolism. They are both enriched in xenobiotics biodegradation pathway (*Cyp1a2, Cyp2b1, Cyp2c12, Cyp3a23/3a1, Cyp1a2, Cyp2b1, Cyp2c12, Cyp3a23/3a1, Adh1, Cyp2c22, Cyp2c23, Cyp2e1, Cyp3a2, Ephx1, Gsta3, mGsta4, Gstm1, Gstm2, Gstm4, Gstz1, Mgst3, Ugt1a6, Yc2*).

Interestingly, though the first two clusters are involved in similar functions, their dynamic profiles are significantly different. Our results indicate that the proinflammatory response in cluster 2 is rapidly up-regulated within the first 2 hours following CLP injury. Similarly, the inflammatory response in cluster 1 is also activated compared to the time-dependent control, which would otherwise not be considered as an upregulation if control group was assumed to be a constant value (0 time point). Following CLP injury, it was observed that the genes involved in xenobiotics degradation and metabolism in cluster 3 and 4 were suppressed starting around 2h post-CLP compared to dynamic gene expression profiles of control group. The fact that xenobiotics metabolism associated genes are observed to be downregulated in response to inflammation indicates an impairment in the liver function of detoxification, implying a CLP induced challenge to the drug-metabolism [10].

In conclusion, the methodology used in this study which allows to characterize the dynamic patterns of both CLP and SCLP group elucidated that CLP resulted in activation of certain signaling pathways as well as the impairment of the xenobiotics metabolism in the liver.

References

1. Bone RC, Balk RA, Cerra FB, Dellinger RP, Fein AM, Knaus WA, Schein RM, Sibbald WJ. Definitions for sepsis and organ failure and guidelines for the use of innovative therapies in sepsis. The ACCP/SCCM Consensus Conference Committee. American College of Chest Physicians/Society of Critical Care Medicine. 1992. *Chest*. 2009; 136(5 Suppl): e28.

2. Wichterman KA, Baue AE, Chaudry IH. Sepsis and septic shock--a review of laboratory models and a proposal. *J Surg Res*. 1980; 29(2): 189-201.

3. Cobb JP, Laramie JM, Stormo GD, Morrissey JJ, Shannon WD, Qiu Y, Karl IE, Buchman TG, Hotchkiss RS. Sepsis gene expression profiling: murine splenic compared with hepatic responses determined by using complementary DNA microarrays. *Crit Care Med*. 2002; 30(12): 2711-21.

4. Li ZJ, Li YP, Gai HR, Xue YL, Feng XZ. [Research of gene expression profile of liver tissue in rat sepsis model]. *Zhongguo Wei Zhong Bing Ji Jiu Yi Xue*. 2007; 19(3): 156-9.

5. Chinnaiyan AM, Huber-Lang M, Kumar-Sinha C, Barrette TR, Shankar-Sinha S, Sarma VJ, Padgaonkar VA, Ward PA. Molecular signatures of sepsis: multiorgan gene expression profiles of systemic inflammation. *Am J Pathol*. 2001; 159(4): 1199-209.

6. Scheff J, Almon R, DuBois D, Jusko W, Androulakis I. Assessment of Pharmacologic Area Under the Curve When Baselines are Variable. *Pharmaceutical Research*. 2011: 1-9.

7. Nguyen TT, Nowakowski RS, Androulakis IP. Unsupervised selection of highly coexpressed and noncoexpressed genes using a consensus clustering approach. *Omics*. 2009; 13(3): 219-37.

8. Tong W, Cao X, Harris S, Sun H, Fang H, Fuscoe J, Harris A, Hong H, Xie Q, Perkins R. ArrayTrack--supporting toxicogenomic research at the US Food and Drug Administration National Center for Toxicological Research. *Environmental health perspectives*. 2003; 111(15): 1819-1826.

9. Twigger S, Lu J, Shimoyama M, Chen D, Pasko D, Long H, Ginster J, Chen CF, Nigam R, Kwitek A, Eppig J, Maltais L, Maglott D, Schuler G, Jacob H, Tonellato PJ. Rat Genome Database (RGD):

mapping disease onto the genome. *Nucleic Acids Res*. 2002; 30(1): 125-8.

10. Morgan ET. Regulation of cytochrome p450 by inflammatory mediators: why and how? *Drug Metab Dispos*. 2001; 29(3): 207-12.

See more of this Session: Poster Session: Engineering Fundamentals In Life Science
See more of this Group/Topical: Food, Pharmaceutical & Bioengineering Division

(625m) Effects of Small-Molecule Inhibition of Central Metabolism In a B-Cell Lymphoma Model of Myc-Driven Cancer

Taylor A. Murphy and Jamey Young, Chemical and Biomolecular Engineering, Vanderbilt University, Nashville, TN

Cancer metabolism has, in recent years, become an attractive avenue for targeted therapeutics. Anabolic and catabolic processes are tightly regulated within any proliferating cell, with a cancerous cell being no different. Central metabolism supplies the cell with intermediate metabolites that satisfy energy, redox homeostasis, and biosynthetic requirements. The overall metabolic state is determined by the oncogenic mutations and microenvironmental conditions that characterize a particular cancer cell. The transcription factor c-Myc is an oncoprotein that exhibits deregulated expression in over 30% of all cancers. In addition to its role as a global regulator of cell cycle and apoptosis, c-Myc exerts significant control over cellular metabolism. Targeting deregulated metabolic pathways with small-molecule inhibitors has become an attractive therapeutic strategy because of recent findings that suggest these metabolic alterations may play a critical role in promoting cancer progression.

Our recent work using a B-cell line with regulatable expression of c-Myc reveals a strong coupling between c-Myc expression and metabolic state. We applied metabolic flux analysis (MFA) using ^{13}C-labeled glucose tracers and GC-MS profiling to compare High and Low Myc states. Growth was inhibited by 50% under the Low Myc state, while specific rates of glucose uptake and lactate secretion were decreased by 20% and uptake of several amino acids such as glutamine and serine were decreased by over 50% in comparison to the High Myc state. We have also treated High and Low Myc expressing cells with selective metabolic inhibitors to determine their effects on growth and metabolism. We hypothesized that the High Myc cells would be more sensitive to these inhibitors than the Low Myc cells. Early results indicate that High Myc cells are more sensitive to oxamate, which is a specific inhibitor of lactate dehydrogenase, with an IC50 an order of magnitude lower than that of the Low Myc cells. Ongoing work is being conducted to quantify the metabolic alterations that occur in response to this and other metabolic inhibitors using ^{13}C-MFA, and how they differ depending on the level of c-Myc expression.

See more of this Session: Poster Session: Engineering Fundamentals In Life Science
See more of this Group/Topical: Food, Pharmaceutical & Bioengineering Division

(625n) Dynamics of the Systemic Response to Experimental Burn Injury and Sepsis: Hepatic Metabolic Flux Distribution

Mehmet A. Orman[1], John Mattick[1], Marianthi G. Ierapetritou[1], Ioannis P. Androulakis[1] and Francois Berthiaume[2], (1)Department of Chemical and Biochemical Engineering, Rutgers, The State University of New Jersey, Piscataway, NJ, (2)Department of Biomedical Engineering, Rutgers, The State University of New Jersey, Piscataway, NJ

Severe burns and trauma, as "primary" injury, lead to a systemic hypercatabolic and hypermetabolic response characterized by a significant loss of lean body mass and a negative nitrogen balance. Patients without subsequent complications may fully recover; however, often times critically ill patients end up being subjected to a "second hit," usually superimposed infections that prolong the inflammatory response and the associated systemic hypercatabolic and hypermetabolic response. Such metabolic abnormalities, when lingering, are correlated with a dramatic loss of lean body mass and increased risk of death by multiple organ dysfunction syndrome. There is currently little basic information on the effect of the second hit on the metabolic response of the host. In the current studies, we are investigating the effect of a standard burn injury followed by experimentally induced infection, namely cecal ligation and puncture (CLP), on the hepatic metabolic response. Since the liver orchestrates many of the critical pathways responsible for carbohydrate and amino acid metabolism in the body, it provides a unique window into the systemic metabolic response. More specifically, rats were subjected to a full-thickness scald burn on the dorsal skin corresponding to 20% of the total body surface area (TBSA), followed by CLP two days later. Some animals received a "sham-burn," where animals were anesthetized and prepped but not burned, as well as a "sham-CLP" where animals underwent laparotomy but no cecal damage. Totally four different groups were investigated: Burn/CLP, Burn/Sham-CLP, Sham-Burn/CLP and Sham-Burn/Sham-CLP. Livers were retrieved at different time points extending up to 10 days from each animal group (n=3 per group), and perfused in an ex vivo perfusion system to measure hepatic metabolic rates in a well-defined environment, thus providing a comprehensive picture of intrinsic changes in hepatic metabolism [1-3]. An additional variable, overnight fasting prior to perfusion, was also investigated in a subset of animals because all prior published studies used fasted animals. Net metabolic fluxes for the important metabolites entering and exiting the liver were used as the primary measurements and applied to a stoichiometric based methodology incorporating flux balance and metabolic pathway analyses (elementary mode analysis) in order to determine intracellular fluxes and active pathways. Pathway-based thermodynamic constraints were further introduced into the model.

Burn-only animals (prior to CLP induction) that were in a fed state prior to perfusion exhibited a significant down-regulation in urea production, and a slight increase in beta-hydroxybutyrate production and oxygen consumption. Fasting prior to perfusion did not affect the burn-induced responses resulting in increased beta-hydroxybutyrate production and oxygen consumption, but reversed the urea response such that burn significantly increased urea synthesis. In addition, in burned animals, fasting prior to perfusion led to increased glutamine uptake. Finally, fasting increased lactate uptake in sham-treated animals, but not in burned animals. These results suggest that fasting alters specific aspects – but not all - of the metabolic response to burn injury, and care must be taken to define the nutritional state of the animal model used in such studies. Conversely, burn injury alters the response to fasting, which indicates that the normal adaptations to the periodicity of food intake are altered by burn injury. We are in the process of extending these studies to the sequential burn-CLP double hit model. We plan to identify intracellular fluxes and critical pathways in burn and CLP livers compared to their sham counterparts, in order to both quantify and qualify the effects of these traumas on liver metabolism. These studies will provide clues about specific nutrients which can be used to diminish the physiologic stress and the hypermetabolic response.

References

1. Banta, S., et al., *Contribution of gene expression to metabolic fluxes in hypermetabolic livers induced through burn injury and cecal ligation and puncture in rats.* Biotechnology and Bioengineering, 2007. **97**(1): p. 118-137.

2. Lee, K., et al., *Metabolic Flux Analysis of Postburn Hepatic Hypermetabolism.* Metabolic Engineering, 2000. **2**(4): p. 312-327.

3. Lee, K., et al., *Profiling of dynamic changes in hypermetabolic livers.* Biotechnology and Bioengineering, 2003. **83**(4): p. 400-415.

See more of this Session: Poster Session: Engineering Fundamentals In Life Science
See more of this Group/Topical: Food, Pharmaceutical & Bioengineering Division

(625o) Stochastic Origins of Macrophage Polarization

Yishan Chuang, Chemical and Biologial Engineering, Northwestern University, Evanston, IL and Joshua N. Leonard, Chemical and Biological Engineering, Northwestern University, Evanston, IL

Innate immune cells such as macrophages play a central role in regulating local immune networks, and are involved in many chronic diseases ranging from cancer to autoimmune diseases such as diabetes. However, the process by which these networks are coordinated and regulated remains poorly understood. To maintain a balance between homeostasis and protective inflammation, macrophages can adopt either immunostimulatory (M1) or immunosuppressive (M2) phenotypes, which are referred to as functional *polarization*. Tumors manipulate this balance by inducing the production of immunosuppressive stimuli (such as IL-10) in the tumor microenvironment, which shifts macrophages into a tumor-promoting M2 phenotype. Such local immune dysfunction is recognized as one of the major barriers to cancer immunotherapy. Macrophages are also functionally *plastic*, meaning that they can interconvert between functional states. In particular, it has been reported that tumor-associated macrophages (TAMs) can be converted from a M2 phenotype to a M1 phenotype following the application of stimuli such as interleukin-12 (IL-12). To date, macrophage polarization has been studied in the context of coherent stimuli (either pro-M1 or pro-M2 stimuli alone). However, immunostimulatory and immunosuppressive stimuli rarely exist independently *in vivo*, especially at tumor sites; macrophages' responses to such contradictory inputs are not known.

The primary goal of this study was to determine how macrophages "calculate" a response when presented with contradictory stimuli. Macrophages were pre-polarized by incubation with combinations of IL-10 and IL-12, and then cells were "activated" with bacterial endotoxin (or lipopolysaccharide, LPS), which is known to enhance either M1- or M2-type responses in pre-polarized cells. Cellular responses were assessed using a quantitative real-time PCR (QPCR) panel. Contrary to previous reports, the presence of IL-10 prevented IL-12-mediated promotion of an M1 phenotype. Moreover, M2-type responses increased with IL-10 dose and were largely independent of IL-12 co-treatment. From these bulk measurements, one cannot determine whether these trends indicate the presence of intermediate polarization states or, instead, a heterogeneous mixture of M1- and M2-polarized cells. Therefore, to investigate polarization at the individual cell level, we next analyzed cells for characteristic M1 and M2 markers by flow cytometry. Using this methodology, we observed M1 and M2 co-existing in single populations. The probability of polarization towards an M2 state increased with IL-10 dose, and this probability was independent of IL-12 co-treatment. Interestingly, some cells remained non-responsive to LPS-mediated activation, and this probability of activation was not modulated by cytokine pretreatment. These data represent the first evidence to date that macrophage polarization is a stochastic process.

This intriguing and exciting discovery motivates many questions, including the following: Do individual macrophages exhibit functional plasticity? Do some cells possess an intrinsic bias for polarization to a particular phenotype? How might polarization bias and plasticity be interrelated? To investigate such questions, we have developed a fluorescent reporter-based platform for dynamically tracking macrophage polarization states in individual cells over time. This system incorporates a microfluidic platform for precisely controlling and modulating the extracellular environment. These tools allow us to probe and investigate macrophage polarization and plasticity at an unprecedented level of detail, allowing us to address such fundamental questions as those described above. Here, we present the results of investigations into the stochastic origins of macrophage polarization. Such new insights promise to transform our understanding of local immune responses and should identify novel therapeutic targets and strategies, which may overcome existing barriers to treatment for cancer and other diseases involving chronic immune dysfunction.

See more of this Session: Poster Session: Engineering Fundamentals In Life Science
See more of this Group/Topical: Food, Pharmaceutical & Bioengineering Division

(625p) Antibacterial Activity of Doxycycline Loaded Chitosan/Poly(L-lactic acid)/ Polycaprolactone Blend for Wound Dressing Device

Wasinee Boonkong, Program in Petrochemistry, Chulalongkorn University, Chachoengsao, Thailand

Antibacterial activity of doxycycline loaded Chitosan/Poly(L-lactic acid)/Polycaprolactone blend for wound dressing device

Wasinee Boonkong[a], Amorn Petsom[b], and Nuttha Thongchul[b]

[a]*Program in Petrochemistry , Faculty of Science, Chulalongkorn University, Bangkok 10330, Thailand*

[b]*Institute of Biotechnology and Genetic Engineering, Chulalongkorn University, Bangkok 10330 Thailand*

Keywords : Chitosan, , Poly(L-lactic acid), doxycycline, Antibacterial, and wound dressing

Chitosan and polylactic acid (PLA) are the renewable, compostable, and biocompatible polymers that provide many useful applications in various areas while polycaprolactone (PCL) that is used to improve processing characteristics of resulting plastics when blended or copolymerized with other polymers. Recent works report the antibacterial activity of doxycycline loaded Chitosan/Poly(L-lactic acid)/ Polycaprolactone blend film against *Staphylococcus aureus* and *Escherichia coli*, as model Gram-positive and Gram-negative bacteria, respectively. In this work, we synthesized low molecular weight PLA via direct polycondensation. Low molecular weight PLA was then blended with polycaprolactone to enhance impact resistance. Later, the PLA/PCL blend was mixed with chitosan solution and then casted into the film using solvent evaporation technique. Glycerol was added in order to promote casting. The blended film was studied as a model device to deliver drug to targeted human organ which in this case was the skin with opened wound. The model drug used in this study, doxycycline, was loaded into the blend before casting. Film properties and the delivery rate of doxycycline were investigated. The mechanical properties including swelling, tensile and T-peel tests were also observed. The loaded chitosan/Poly(L-lactic acid)/Polycaprolactone film delivery study gave a high % release rate which possibly reached 100% loaded. In vitro an antibacterial test of the virgin chitosan film was demonstrated its ability to inhibit common bacterial growth (*S. aureus and E. coli*) for a prolonged duration. The blended film without doxycycline was unable to inhibit both bacteria while the presence of doxycycline in the film led to the highest degree of inhibition in both bacterial cultures.

(625q) PDMS Surface Modification with Non-Fouling Zwitterionic Materials

Andrew J. Keefe, Chemical Engineering, University of Washington, Seattle, WA

See more of this Session: Poster Session: Engineering Fundamentals In Life Science
See more of this Group/Topical: Food, Pharmaceutical & Bioengineering Division

(625r) Preparation and Confirmation of Fla-B Conjugated Capsular Polysaccharide of Streptococcus Pneumoniae

Minh N. Bui[1], Sang-Il Yoon[2], Seung-Jin Jung[1], Eun-Seong Seo[2] and Doman Kim[3], (1)Interdisciplinary Program of Graduate School for Bioenergy and Biomaterials, Chonnam National University, Gwangju 500-757, South Korea, (2)School of Biological Sciences and Technology & Research Institute for Catalysis, Chonnam National University, Gwangju 500-757, South Korea, (3)Interdisciplinary Program of Graduate School for Bioenergy and Biomaterials, School of Biological Sciences and Technology & Research Institute for Catalysis, Chonnam National University, Gwangju 500-757, South Korea

Streptococcus pneumoniae causes many types of pneumococcal infection; especially, the incidence of pneumococcal disease which is the highest in infants under 2 years and in people over 60 years of age. Capsular polysaccharides (CPS) are the primary basis for pathogenicity, antigenic and serotypes. The large numbers of serotypes (over 90 characterized serotypes are known until now) have required the development of vaccine products composed of polysaccharides. Two types of vaccines are currently in clinical use: polysaccharide vaccine, and pneumococcal conjugate vaccines. Among the more than 90 different *S. pneumoniae* serotypes, serotype 19A, 4 and 3 are globally very prevalent. In this study, we investigated a simple process for the production and purification by four simple steps of 19A CPS to make conjugated CPS. A simplified purification procedure including adjustment of cell lysate pH to 4.5, fractionation with 50%–80% ethanol and dialysis obtained capsular polysaccharide (CPS) in a yield of 31.32 ± 3.11 mg from 1 L culture (75% recovery after lyses). The product contained only 69.6 μ g protein (99.78% purity) and 0.8 mg (sum of the precipitants from 50~60, 60~70 and 70~80%) nucleic acid (97.45% purity). The size of capsular polysaccharide was reduced by 10 mM acid hydrolysis and separated to different sizes using ethanol fractionation (30~70%). Each product showed elevated purity. Furthermore we conjugated CPS serotypes of 19A, 4, or 3 with Flagellin B (FlaB), respectively, and confirmed conjugation by Western blot analysis using FlaB antibody and CPS antibody.

See more of this Session: Poster Session: Engineering Fundamentals In Life Science
See more of this Group/Topical: Food, Pharmaceutical & Bioengineering Division

(625s) Poly (lactide-co-glycolide)/Carbon Nanotube Composite Scaffolds for Bone Tissue Engineering Applications

Qingsu Cheng, Biomedical Engineering Program, University of South Carolina Columbia, Columbia, SC and **Ehsan Jabbarzadeh**, Department of Chemical Enginnering and Biomedical Engineering Program, University of South Carolina Columbia, Columbia, SC

There is a growing need for bone regeneration strategies due to trauma and musculoskeletal diseases. To meet this need, tissue engineering approaches have been developed in which a three dimensional (3-D) porous scaffold can be used with or without living cells and growth factors to replace or regenerate damaged bone. In tissue engineering strategies, scaffolds have an inevitable influence on the organization, growth and differentiation of cells in tissue engineered constructs. In this study, poly (lactide-co-glycoide) (PLGA) and carbon nanotube were used to make novel composite scaffolds for bone tissue engineering applications. We used scanning electron microscopy (SEM) to examine the morphology of and pore structure of composite scaffolds. Next, we studied the role of carbon nanotube length, density and orientation in mechanical behavior of composite scaffolds. Finally, we examined the attachment, growth and osteogenic differentiation of human mesenchymal stem cells (hMSCs) on PLGA/ carbon nanotube composite scaffolds. In this presentation, we demonstrate the ability to increase osteoconductivity of scaffolds by tuning the density and orientation of embedded carbon nanotubes.

See more of this Session: Poster Session: Engineering Fundamentals In Life Science
See more of this Group/Topical: Food, Pharmaceutical & Bioengineering Division

(625t) A Novel Vaccine for Streptococcus Pneumoniae Using Polyanhydride Nanoparticle Adjuvants

Shannon Haughney[1], Latrisha Petersen[1], Janice King[2], Amanda Ramer-Tait[3], Amy Schoofs[3], David Briles[2], Michael J. Wannemuehler[3] and Balaji Narasimhan[1], (1)Department of Chemical and Biological Engineering, Iowa State University, Ames, IA, (2)Department of Microbiology, University of Alabama Birmingham, Birmingham, AL, (3)Department of Veterinary Microbiology and Preventive Medicine, Iowa State University, Ames, IA

The World Health Organization estimates that pneumonia causes 1.6 million deaths every year, the majority in children in developing countries. *Streptococcus pneumoniae* is a leading cause of bacterial pneumonia in children. The current polysaccharide conjugate vaccine (PCV) has been shown to be effective in reducing cases of pneumonia, however it has several limitations. The vaccine is expensive and complicated to manufacture, leading to limited availability in developing countries, it does not provide cross-protection across pneumococcal serotypes, it has not been shown to prevent nasopharyngeal carriage, and the polysaccharide-based vaccine leads to poor immunogenicity in young children.

The pneumococcal surface protein PspA has demonstrated the ability to provide protection against challenge with *S. pneumoniae*, and has been found to be present on all known isolates of pneumonia, eliminating the need to include multiple serotypes in a vaccine. Additionally, the recombinant PspA protein is inexpensive to produce, which could lead to increased availability of a vaccine. It is also thought to play a role in nasopharyngeal carriage through its ability to bind to and prevent killing by apolactoferrin; therefore, vaccination with PspA could also provide protection against colonization. However PspA has not been proven to elicit an immune response without the addition of an adjuvant, and aluminum hydroxide, a commonly used adjuvant, requires a three-dose parenteral immunization regime to provide protection.

Polyanhydrides are a class of biomaterials that have shown much promise in drug and vaccine delivery because of their biocompatibility, controlled protein release kinetics, protein stabilization, and immune modulation. They can be fabricated into nanoparticles for administration via inhalation or injection and work effectively as vaccine adjuvants. Encapsulation of F1-V protein into nanoparticles made from copolymers of 1,6-bis(p-carboxyphenoxy)hexane (CPH) and 1,8-bis(p-carboxyphenoxy)-3,6-dioxaoctane (CPTEG) recently demonstrated the ability to provide protection against intranasal challenge with *Yersinia pestis*. Furthermore, CPTEG:CPH nanoparticles degrade through a combination of bulk and surface erosion providing a sustained release of protein, while maintaining the protein's structure and function. Encapsulation of the PspA antigen into the amphiphilic 50:50 CPTEG:CPH nanoparticles has demonstrated a sustained release of the protein over 60 days, and the released protein retained its primary structure as indicated by SDS-PAGE. Upon vaccination with the PspA-loaded (1μ g/dose) 50:50 CPETG:CPH nanoparticles plus 2μ g soluble PspA, in a single dose, mice elicited a humoral immune response indicated by the production of PspA specific antibodies, which was only slightly lower than mice immunized with the 3 dose regimen positive control (1μ g PspA adsorbed to alum). These results provide evidence supporting the capabilities of amphiphilic polyanhydride nanoparticle-based adjuvants for the prevention and treatment of infectious respiratory diseases.

See more of this Session: Poster Session: Engineering Fundamentals In Life Science
See more of this Group/Topical: Food, Pharmaceutical & Bioengineering Division

(625u) Central Nervous System Development In Planaria As a Model for Studying the Dynamics of Fetal Alcohol Syndrome

Jesse R. Lowe[1], William C. Platt[2], Tyler D. Mahool[1], Robert D. Cusack[2], Kevin Grazel[2] and Mary M. Staehle[1], (1)Rowan University, Glassboro, NJ, (2)Chemical Engineering, Rowan University, Glassboro, NJ

Fetal Alcohol Syndrome (FAS) affects approximately 3-5 children of every 1,000 live births in the United States. It manifests as an underdeveloped central nervous system with symptoms such as cognitive deficiencies, learning disabilities, and craniofacial abnormalities. FAS is thought to be caused by an alcohol-induced disturbance to the dynamic process of central nervous system (CNS) development. However, the molecular mechanisms underlying this disturbance have been challenging to study using traditional animal models, and therefore relatively little is known about the dynamic molecular etiology of FAS. We present a preliminary study on the effects of alcohol exposure on CNS development in a new animal model, *Schmidtea mediterranea* planaria. Planaria are small freshwater flatworms considered to be the simplest organism with a CNS. They have the remarkable ability to regenerate complete, fully-functional planaria from any fragment within approximately 7 days.

To assess CNS developmental dynamics during alcohol exposure, we conducted behavioral and functional assays throughout development in head-regenerating planaria. Worms were transferred to alcohol-containing solutions (1-4% v/v) following head removal. On each of the subsequent 9 days, the planarian movement and light avoidance tendencies were quantified. While these functions were not different in intact worms exposed to low levels of alcohol and alcohol-naïve intact worms, alcohol exposure during head regeneration induced a delay and retardation of CNS development. The severity of these effects increased in a statistically significant dose-dependent manner, indicating that there is a dynamic response to alcohol exposure during CNS development in head-regenerating planaria. Our results suggest that head-regenerating planarian fragments exposed to low levels of alcohol can be used as a model system for studying the alcohol-induced disturbance of fetal CNS development.

See more of this Session: Poster Session: Engineering Fundamentals In Life Science
See more of this Group/Topical: Food, Pharmaceutical & Bioengineering Division

(625v) Control of Cell Adhesion and Migration Using Nanoengineered Substrates

Greg M. Harris, Department of Chemical Engineering, University of South Carolina, Columbia, SC and Ehsan Jabbarzadeh, Department of Chemical Enginnering and Biomedical Engineering Program, University of South Carolina Columbia, Columbia, SC

Control of Cell Adhesion and Migration using Nanoengineered Substrates

G Harris and E.Jabbaradeh

Department ofChemical Engineering, University of South Carolina, Columbia, SC

Abstract

Mesenchymal stem cells (MSCs) are uniquely positionedas the most promising cell source for tissue engineering and cell transplantstrategies due to their unique capability of self-renewal aswell as being able to differentiate into diverse cell types. However, their use as a therapy is hampered dueto the limited understanding of the mechanisms by which cells integrateenvironmental stimuli. Asignificant step towards this goal can be achieved through the development of platforms with controllable physical and chemical propertiesthat allow for further exploration of the co-operative involvement of thesesignals. The focus of this study was the development of nanoengineered substrates with finelycontrollable ligand density, topographical features and surface stiffness forthe study of cell-extracellular matrix (ECM) interactions. To this end, dip-pen-nanolithography (DPN) was employed to create islands ofcollagen and fibronectin with different size, shape, and spacing. Time-lapsemicroscopy in combination with immunofluorescent staining was used to monitor stemcell adhesion and migratory characteristics on nanoengineered substrates. In thispresentation, we discuss how the interplay between various ECM propertiescontrols cell signaling characteristics that are important in ultimate definingof stem cell lineage commitment.

See more of this Session: Poster Session: Engineering Fundamentals In Life Science
See more of this Group/Topical: Food, Pharmaceutical & Bioengineering Division

(625w) Determining the Effect of Spacing In Protection of Staphylococcus Aureus by Pseudomonas Aeruginosa

Leonela A. Villegas, Jinzi Deng and Leslie M. Shor, Department of Chemical, Materials and Biomolecular Engineering, University of Connecticut, Storrs, CT

Biofilm-causing bacteria such as *Pseudomonas aeruginosa* and *Staphylococcus aureus* are associated with adverse clinical outcomes due to several antibiotic resistance mechanisms inherent to biofilms. Both bacterial species are opportunistic pathogens that can be found individually or in co-culture with other organisms on catheters, skin, mucosa, or the respiratory tract. Recent studies have investigated the effect of the aminoglycoside antibiotic Tobramycin and its production of a new *S. aureus* phenotype, Small Colony Variants (SCV). An exoproduct, 4-hydroxy-2-heptylquinoline-*N*-oxide (HQNO), secreted from *P. aeruginosa* suppresses the electron transport inhibiting the growth of wild-type *S. aureus* and therefore protecting it from antibiotics. However, the effect of the distance between *P. aeruginosa* and *S. aureus* and specific fluxes of HQNO needed to produce SCVs are unknown. Studies have used Petri Dish culture methods to determine the concentrations needed, but this does not address the rate of delivery, or flux. Here we employ a microfluidic device designed to deliver spatially and temporally varying concentrations and fluxes to a central field. Using the device, concentration and fluxes of HQNO versus susceptibility to Tobramycin are determined as a function of position and time. We confirm that SCVs are derived from HQNO exposure by comparing the size of the colonies grown with the antistaphylococcal substance against the SCVs grown in co-culture with *P. aeruginosa.* When introducing .5 ug/ml HQNO and .55 ug/ml Tobramycin into the QUAD the lawn shows to have an increase in SVCs near the HQNO diffusing source and are more resistant than the wildtype *S. aureus* at above MIC levels. Pathogenic co-cultures may have different antibiotic susceptibilities versus pure culture infections. Our results demonstrate the use of microfluidic devices and mass transport modeling for more quantitatively understanding the mechanism for antibiotic resistance. Future work may utilize a similar approach for mechanistic studies as well as screening combinatorial therapies.

See more of this Session: Poster Session: Engineering Fundamentals In Life Science
See more of this Group/Topical: Food, Pharmaceutical & Bioengineering Division

(625x) Chitosan Scaffold Morphology Optimized for Ultrasonically Mediated Articular Cartilage Synthesis

Tobias Louw[1], Sanjukta Guhathakurta[2], Anuradha Subramanian[2] and Hendrik Viljoen[2], (1)University of Nebraska, Lincoln, NE, (2)Chemical and Biomolecular Engineering, University of Nebraska, Lincoln, NE

Cell seeded scaffolds provide the platform for effective tissue engineering. These scaffolds must be designed to optimize the environment for increased cell proliferation and viability. Our research group uses porous chitosan scaffolds for the ultrasonically mediated synthesis of articular cartilage. Chitosan is obtained from the deacetylation of chitosan, the second most abundant natural polymer. It is non-toxic and biodegradable, allowing for the direct implantation in patients and the high concentration of primary amine groups on chitosan makes it ideal for surface functionalization to enhance cell growth.

Chitosan scaffolds are prepared through a two-step freezing, lyophilizing process. Dissolved chitosan exposed to subzero temperatures experience phase separation during the solidification of the solvent. The solvent is then selectively removed through freeze drying. Scanning electron microscopy reveals a spatial variation in pore diameter. Our central hypothesis is that the pore size is dependent on the spatial and temporal temperature profile during the freezing process. We have developed and validated a simple yet accurate finite difference model to predict the internal energy of the chitosan solution, including non-linearities due to temperature dependent material properties. We then correlate the model results with the scaffold morphology as determined by scanning electron microscopy.

Furthermore, the use of ultrasound to enhance the physiological properties of the synthetic tissue must be investigated. We investigate the propagation of the acoustic wave through the porous scaffold and propose an engineered structure to maximize the beneficial effects of ultrasound.

See more of this Session: Poster Session: Engineering Fundamentals In Life Science
See more of this Group/Topical: Food, Pharmaceutical & Bioengineering Division

NOVEL METHOD FOR QUALITY CONTROL OF MYOBLAST SHEET TOWARD CLINICAL TRANSPLANTATION

Masahiro Kino-oka, Ngo Xuan Trung and Masahito Taya, Osaka University, Osaka, Japan

Introduction

The cell sheet engineering is emerging as an advanced technique with preparation of scaffold-free three-dimensional (3-D) tissue (1), not only for transplantations, but also for *in vitro* researches. Temperature-responsive surface of poly-N-isopropylacrylamide (PIPAAm) can form the monolayer cell sheet without any enzymatic digestion to keep intact vital cell-cell junctions and extracellular matrix (ECM) which plays a role of glue for sheet rapidly assembling as well as the host after transplantation (1). Sasagawa *et al.* previously constructed the multi-layered structure of skeletal muscle myoblast cells in which prevascular formation by endothelial migration was observed (2).

Recently autologous transplantation of multilayered myoblast sheet has been attracting attention as a new technique for curing myocardial infarction, which is associated with the dysfunction of cardiomyocytes and irreversible cell loss. This method can overcome the disadvantages such as less take ratio of transplanted cells through the direct injection of myoblast suspension. Skeletal myoblasts, which are easy to be harvested from patients, have ability to become active, self-renew and differentiate, permitting muscle regeneration upon muscle injury. As shown in Fig.1, the sheet of myoblasts also has ability to source the cytokines which improve heart function due to paracrine system including the facilitation of angiogenesis and the attraction of progenitors on affected part.

From manufacturing point of view, the system development for process and quality controls is important to be concerned, leading to the active commercialization using the cell sheets. Many researches have been tackled concerning cell source exploring, cell culture, sheet assembling, and *in vivo* animal tests. However, the method for quality control of myoblast sheet, especially for transplant efficacy, has not been systematized.

Many models have been proposed for angiogenesis. The *in vivo* tests are one of the practical methods to estimate overall efficacy of transplants, *ie.* sheet products. However, the conventional technique can make insufficient estimation of product quality for the autologous transplantation due to less quantitative evaluation, non-human estimation as well as the patient dependence. Thus, manufacturers long for creating a novel method suitable to *in vitro* quantitative estimation of the transplants.

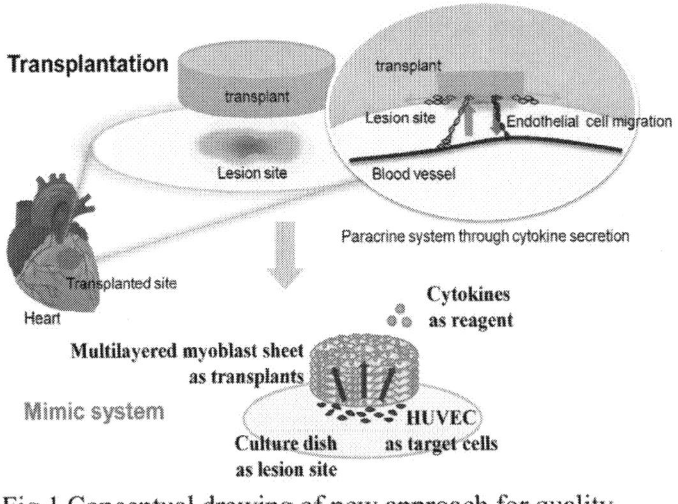

Fig.1 Conceptual drawing of new approach for quality evaluations of multilayer sheet through angiogenesis

In addition, the cell migration in 3-D construct plays an important role in physiological and pathological phenomena such as embryonic development, cell alignment, immune reaction, angiogenesis, metastasis and so on (3). Understanding the mechanisms of cell migration will be useful for designing biomimetic structures and functional engineered tissues. Although the behavior of cells on 2-D culture surface has been investigated extensively, spatial cell movement in 3-D tissue, especially in vertical migration inside the tissue, is not well considered due to the absence of methods *in vitro* enabling quantitative and reproducible measurements.

In the present study, five-layered myoblast sheet was fabricated as a 3-D model to evaluate vertical cell migration by confocal laser scanning microscopy with image processing. And we establish the mimic system of transplantation which consists of endothelial cells (Human Umbilical Vein Endothelial Cells, HUVEC) on culture dish as target cells on lesion site and five-layered myoblast sheet as transplants, and focus on understanding angiogenesis procedure as post transplantation.

Materials and Methods

Human skeletal muscle myoblasts were used through the experiments. The subcultures of them were conducted at 37 °C under an atmosphere of 5% CO_2 in Dulbecco's modified Eagle's medium (DMEM) containing 10% fetal bovine serum (FBS) and antibiotics (100 U/cm^3 penicillin G, 0.1 mg/cm^3 streptomycin and 0.25 mg/cm^3 amphotericin B), and the laminin-coated surface employed HSMM's subculture. The starter cells harvested from the subcultures were stained for the live cell imaging by CellTracker Green and CellTracker Orange, exhibiting fluorescently green and orange cells, respectively, according to commercially recommended protocol which describes the staining condition of 5 μM for 15 min.

The stained cells employed the fabrication of multilayer sheet according to newly developed procedures as follow (Fig. 2). Myoblast cells were seeded to be 2.3×10^5 cells/cm^2 in wells (diameter of 1.9 cm^2/well) with temperature-responsive surface, and 24 h-incubation at 37 °C in 5% CO_2 was

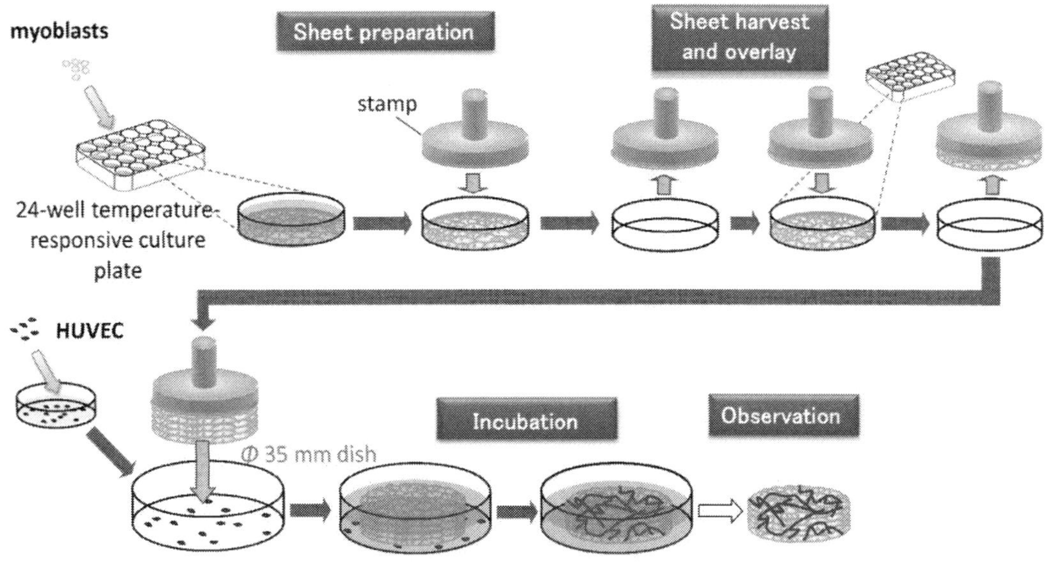

Fig.2 Procedure of preparation of mimic system with multilayer sheet

conducted to form monolayer sheet. Here the medium depth was set to be 2 mm through the experiments. To harvest the monolayer sheet, the stamp with gelatin gel was put on the monolayer sheet in a well at 37 °C and the temperature was shifted to 20 °C. After 30 min, the stamp was lifted to harvest the monolayer sheet from the bottom surface of the well. Further repetition for the sequential harvests of monolayer sheets forms the multilayer structure on the stamp. The multilayered sheet with gelatin gel was separated from the stamp to put on the 35 mm culture dish. If necessary, HUVEC were incubated on the dish in advance. After addition of medium, temperature was shifted to 37 °C to conduct the 1 h incubation for gelatin melting, and the medium was changed with fresh one. In the present study, the fabricated culture system with five-layered sheet employed the analyses of sheet behaviors.

Results and discussion

As a typical culture system, the five-layered sheets consisting of basal layers (green) and other layers (orange) stained by CellTracker Green and CellTracker Orange, respectively, were prepared to observe tepmpo-spatial cell distribution by using confocal laser scanning microscopes. To determine spatial distribution of target cells, population of green and orange cells in each layer at 0 and 48 h of incubation time was observed and quantitatively analyzed by the image processing of colored voxels. As shown in Fig.3, at initial of incubation time, the green cells existed at bottom in the sheet, and active cellular migration occurred toward horizontal as well as vertical directions at any positions in the sheet, revealing the sheet fluidity. In addition, green cells migrated toward upper layers with elapsed incubation time. To understand the extent of sheet fluidity, the vertical distribution of green cells, which is assumed to be equivalent to the distribution of green voxels, were estimated.

In the culture system with HUVEC, the HUVEC behaviors in multilayered myoblast sheets were observed on the basis of confocal laser microscopy. The tempo-spatial observation revealed the sheet fluidity as well as the active migration of HUVEC, compared with the myoblast fluid in the sheet. In addition, as shown in Fig.4, the spontaneous migration of HUVEC caused the local network at the middle sheet layer within 24h. With elapsed time, the maturity of the network was seemed. However, further incubation over 120 h led to the decay of the network formation. In the present study, the minimal culture medium without any additional cytokines was applied. Then, the network formation was suggested to occur by paracrine of VEGF and HGF secreted from the myoblast sheet.

The quantitative analysis was conducted by estimating the total length and tip number of HUVEC and their ratio. The total length increased at initial phase and was maintained to be the same level after 48 h, suggesting the morphological change from spherical and spindle shape at initial phase until 48 h as well as no cell division through the incubation due to the minimal culture medium. On the other hand, the tip number decreased

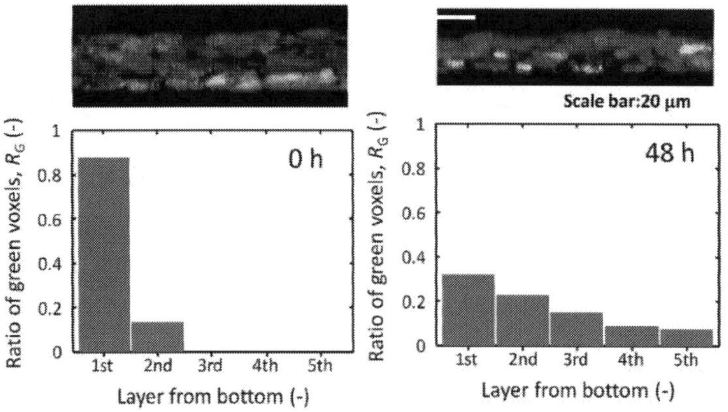

Fig.3 Sheet fluidity and green cell distribution in multilayered sheet

through the incubation, although the slight increase in the tip number occurred. The ratio increased and maximum level reached at 96 h, concluding that this ratio is available to represent the extent of network formation in the sheet.

In conclusion, our proposed mimic system possess the operational variables such as incubation time, initial density of HUVEC, myoblast density in sheet, contamination population of skeletal fibroblasts and so on. The population of skeletal myoblasts and fibroblasts, which depends on cell sources from the patients, drastically affected the HUVEC connectivity (data not shown). These results suggest our mimic system can make *in vitro* quantitative estimation of angiogenesis potential for the transplants from the patients.

References

1. Yang, J., Yamato, M., Kohno, C., Nishimoto, A., Sekine, H., Fukai, F. and Okano, T. (2005), "Cell sheet engineering: recreating tissues without biodegradable scaffolds", *Biomaterials* **26**, 6415-6422.
2. Horwitz, R. and Webb, D. (2003), "Cell migration", *Curr. Biol.*, **13**, 756-759.
3. Sasagawa, T., Shimizu, T., Sekiya, S., Haraguchi, Y., Yamato, M., Sawa, Y. and Okano, T. (2010) , "Design of prevascularized three-dimensional cell-dense tissues using a cell sheet stacking manipulation technology", *Biomaterials*, **31**, 1646-1654.

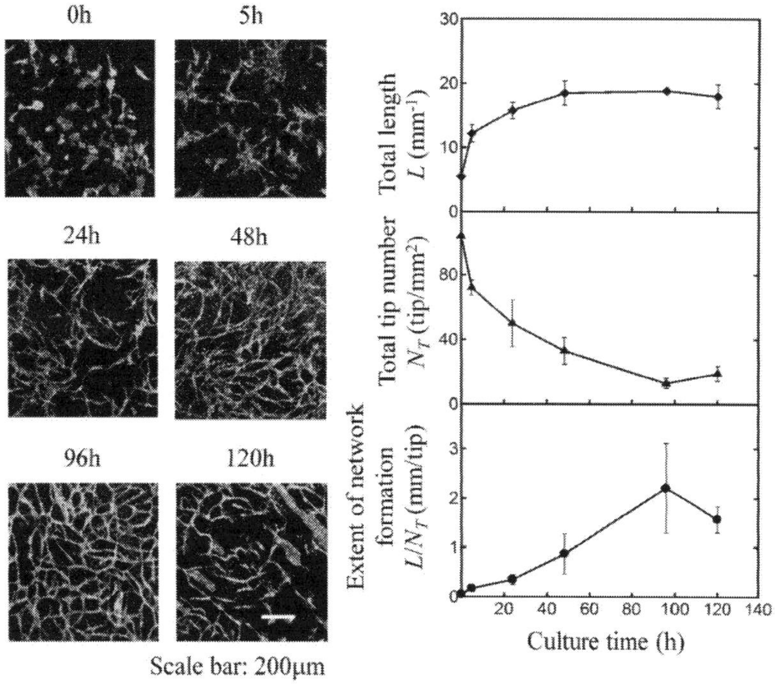

Fig.4 Profile of network formation of HUVEC in multilayered sheet

(625z) Binding and Dissociation of Biomarkers for Alzheimer's Disease On Biosensor Surfaces

Ajit Sadana, Chemical Engineering, University of Mississippi, University, MS, Lokesh Taneja, University of Mississippi, University, MS and Kennon C. Shelton, Chemical Engineering Department, University of Mississippi, University, MS

A fractal analysis is presented for the binding and dissociation (if applicable) of Alzheimer's disease biomarkers on biosensor surfaces. The Alzheimer's disease biomarkers analyzed include: (a) the binding of different concentrations of APP_{770} (amyloid precursor protein) in complete medium to the DE antibodies immobilized on a solid surface using total internal reflection elliposmetry (Mustafa et al., 2010), (b) binding of different concentrations of PrP (prion protein) variant, ARQ (in nM) in solution to asolectine liposome antibodies immobilized on a SPR biosensor chip (Steunou et al., 2010), (c) binding of different concentrations of O_3 (in nM) in solution to asolectine liposomes immobilized on the surface of a SPR biosensor chip (Steunou et al., 2010), (d) binding of 100 nM ARQ-Cu (II) to chip immobilized PC liposomes with or without lipid rafts (Steunou et al., 2010), (e) binding of different concentrations of Ab (in mM) in solution to sonicated fAb immobilized on a flow cell (Hasegawa et al., 2002), (f) binding of different concentrations of Ab (in mM) in solution to fAb immobilized on a flow cell (repetition runs) (Hasegawa et al., 2002), (g) binding of acetylcholine and AChE inhibitor to acetylcholinesterase and choline oxidase immobilized on a working electrode using gel-entrapment (Lenigk et al., 2000).

Alzheimer's disease (AD) is a very serious pathological disease. Any insights that may be gained into the further progression of AD will lead to significant management, control of, and alleviation of this very serious debilitating disease. Different approaches may be used, and are being aggressively researched and applied to better understand this disease. The attempt made in this presentation to better understand the binding kinetics of AD biomarkers to different biosensor surfaces is a step in this direction.

See more of this Session: Poster Session: Engineering Fundamentals In Life Science
See more of this Group/Topical: Food, Pharmaceutical & Bioengineering Division

(626a) Pseudo-Continuous Fermentation Using a Novel Bioreactor to Facilitate the Study of a Co-Culture System for Ethanol Production

Min Hea Kim[1], Meng Liang[1], Qinghua He[2] and Jin Wang[1], (1)Department of Chemical Engineering, Auburn University, Auburn, AL, (2)Department of Chemical Engineering, Tuskegee University, Tuskegee, AL

Keywords: bioreactor,co-culture, fermentation, ethanol production, co-fermentation, bioethanol, lignocellulose, mixed substrate, *P.stipitis*, *S. cerevisiae*

The bioethanol industry hasdeveloped rapidly in recent years to cope with the rising cost of fossil fuel. The conversion of lignocellulosic materials into ethanol has been regarded asone of the most promising renewable energy sources. This is not only because ofthe moderate cost and high availability of lignocellulosic biomass, but alsobecause lignocellulosic ethanol process can greatly reduce CO_2emission [1]. Glucose and xylose are two major components in thelignocellulosic hydrolysates. While glucose fermentation is well established,xylose fermentation remains an unsolved problem in the industrializedlignocellulosic ethanol process [2]. In existing studies of ethanolicfermentation, besides recombinant strategy, co-culture strategy has drawn muchresearch interest in the past few decades [3-7]. However, currently there is alack of systematic study on the properties of the co-culture system.

In this work, we focus on theco-culture system consisting of *Saccharomycescerevisiae* and *Pichia stipitis*. One of the major difficulties of the co-culturestrategy for simultaneous glucose/xylose fermentation is the inability toprovide the optimal fermentation conditions for two different strains [8-10].Most of studies on co-culture process reported that while the fermentation ofglucose in the mixed substrate proceeded efficiently with a glucose-fermentingyeast strain, the fermentation of xylose with a xylose-fermenting strain oftenexhibited low ethanol productivities due to the differences in the fermentativeregulation for ethanol production [11-15]. Another difficulty of co-culturestrategy is the diauxic growth of *P.stipitis* due to its preferential utilization of glucose over xylose whencultured on mixed substrates [7, 11,16].

In order to address thesedifficulties and to improve the co-culture process efficiency, we havedeveloped a novel bioreactor which separates the two different yeast strains,and enables different fermentative conditions for different strains. Inaddition, we have built a cell retention module which enables thepseudo-continuous fermentation mode, i.e. continuous fermentation with cellretention. With pseudo-continuous operation, we can completely eliminate thepossible wash-out associated with continuous fermentation and can easily testdifferent operation conditions. In this study, we show that pseudo-continuousoperation using the developed bioreactor provides an effective tool to studyco-culture systems, which allow us to investigate the dynamic of co-culturesystem for lignocellulosic ethanol production systemically.

References

1. IEA (2007), IEA energy technology essentials: Biofuel production, OECD/IEA-ETE02, 1-4. Available from: http://www.iea.org/techno/essentials2.pdf
2. Li Y. et al., (2009), Influence of metal addition on ethanol production with *Pichia stipitis* ATCC 58784, *J. Indust. Microbiol. Biotechol.*, 36, 491-497
3. Kordowska-waiter M, Targonski Z, (2002), Ethanol fermentation on glucose/xylose mixture by co-cultivation of restricted glucose catabolite repressed mutants of *Pichia stipitis* with respiratory deficient mutants of *Saccharomyces cerevisiae*, *Acta. Microbiol. Pol.*, 51, 345-352
4. Lebeau T, Jouenne T, Junter GA, (1997), Simultaneous fermentation of glucose and xylose by pure and mixed cultures of *Saccharomyces cerevisiae* and *Candida shehatae* immobilized in a two-chambered bioreactor, *Enzyme Microb. Technol.*, 21, 265-272
5. Grootjen DRj, Meijlink LHHM, Vleesenbeek R, Van der Lans RGJM., Luyben KCAM, (1991), Cofermentation of glucose and xylose with immobilized *Pichia stipitis* in combination with *Saccharomyces cerevisiae*, *Enzyme Microb. Technol.*, 13:530-536
6. Laplace JM, Delgenes JP, Moletta R, Navarro JM, (1993), Cofermentation of glucose and xylose to ethanol by a respiratory-deficient mutant of *Saccharomyces cerevisiae* co-cultivated with a xylose-fermenting yeast, *J. Ferment. Bioeng.*, 75, 207-212
7. Grootjen DRJ, Jansen ML, Van der Lans RGJM Luyben KCA, (1991), Reactors in series for the complete conversion of glucose/xylose mixtures by *Pichia stipitis* and *Saccharomyces* cerevisiae, *Enxyme Microb. Technol.*, 12, 828-833
8. Laplace JM, Delgenes JP, Moletta R, Navarro JM, (1992), Fermentation of lignocellulosic sugars to ethanol: selection of mutants of *Pichia stipitis* affected for D-glucose utilization, *Enzyme Microb. Technol.*, 14, 644-648
9. Grootjen DRJ, Meijlink LHHM, Vleesenbeek R, Van der Lans RGJM., Luyben KCAM, (1990), Cofermentation of glucose and xylose with immobilized *Pichia stipitis* and *Saccharomyces cerevisiae*, *Enzyme Microb. Technol.*, 12, 860-864
10. Laplace JM, Delgenes JP. Moletta R, Navarro JM, (1993), Effects of culture conditions on the co-fermentation of a glucose and xylose mixture to ethanol by a mutant of *Saccharomyces diastaticus* associated with *Pichia stipitis*, *Appl. Microb. Bitechnool.*, 39, 760-763
11. Hamidimotlagh R, Nahvi I, Emtiazi G, Abedinifar S, (2007), Mixed sugar fermentation by *Pichia stipitis*, *Saccharomyces cerevisiae*, and an isolated xylose fermenting *Kluyveromyces marxianus* and their cocultures, *Afr. J. Biotechnol.*, 6, 1110-1114
12. Jeffries TW et al., (2007), Genome sequence of the lignocellulose-bioconverting and xylose-fermenting yeast *Pichia stipitis*, *Nature Biotechnology*, 25, 319-326
13. Taniguchi M, Takahiro I, Toshihiro T, Michhiro F, (1997), Ethanol production from a mature of glucose and xylose by a novel co-culture system with two fermenters and two microfiltration modules, *J Ferment. Bioeng.*, 1, 59-64
14. Delgenes JP, Moletta R, Navaro JM, (1989), Fermentation of d-xylose, D-glucose, L-arabinose mixture by Pichia stipitis: effect of the oxygen transfer rate on fermentation performance, *Biotechnol. Bioeng*. 34, 398-402

15. Lebeau T, Jouenne T, Junter GA, (2007), Long term incomplete xylose fermentation, after glucose exhaustion, with *Candida shehatae* co-immobilized with *Saccharomyces cerevisiae*, *Microb. Research*, 162, 211-218

See more of this Session: Poster Session: Food Engineering and Bioprocessing
See more of this Group/Topical: Food, Pharmaceutical & Bioengineering Division

(626aa) Genetic Regulation of the Cellulosome In Clostridium Acetobutylicum

Hadi Nazem-Bokaee, Biological Systems Engineering, Virginia Tech, Blacksburg, VA and Ryan S. Senger, Biological Systems Engineering Department, Virginia Tech, Blacksburg, VA

The engineering of a microorganism to produce advanced biofuels in a consolidated bioprocess remains an important challenge. The bacterium *Clostridium acetobutylicum* ATCC 824 is an industrial organism that undergoes acetone, biobutanol, and ethanol (ABE) fermentation. While *C. acetobutylicum* is able to produce biobutanol from both hexose and pentose sugars, it is unable to utilize cellulose due to the lack of a functional cellulosome, an extracellular multi-enzyme complex containing functional cellulase enzymes. *C. acetobutylicum* contains genes that encode for a potentially effective cellulosome, but this cellulosomal operon remains transcriptionally repressed, even in the presence of cellulose. Several research groups have presented hypotheses regarding the repression of the cellulosome in *C. acetobutylicum*. Here, we use a combinatorial sampling of the cellulosome promoter region as a method to determine location of potential repressor binding. These are then compared to computational predictions of repressor binding in this promoter region. The results of this research have provided extensive insight into the regulatory mechanisms responsible for suppressing the cellulosome in this highly solventogenic industrial bacterium. The results presented here are currently being used to design effective metabolic engineering strategies.

See more of this Session: Poster Session: Food Engineering and Bioprocessing
See more of this Group/Topical: Food, Pharmaceutical & Bioengineering Division

(626ab) Rigorous Proton Balancing and Other New Tools to Reduce the Phenotypic Solution Space of a Genome-Scale Model

Ryan S. Senger, Biological Systems Engineering Department, Virginia Tech, Blacksburg, VA and Mike McAnulty, Biological Systems Engineering, Virginia Tech, Blacksburg, VA

The usefulness of genome-scale modeling using flux balance analysis is well known. The resulting phenotypic solution space consists of all possible flux solutions to the flux balancing problem. Here, three methods are discussed that significantly reduce the size of the phenotypic solution space and allow the user to zero in on unique metabolic solutions. The first is by considering the rate at which protons enter and leave the cell. This is largely dictated by the mechanisms of proton exchange inherent to the organism of study and the extracellular environment. This is currently leading to purely kinetic models to describe this process in several organisms. The second method presented here is a more rigorous method of proton balancing inside the cell. The issues of proton balancing given reaction notations common to biochemical databases (e.g., KEGG) are well-documented. However, the solutions presented in the literature are also not rigorously correct. Proton balancing should be done considering mixtures of protonated and de-protonated states for compounds, as dictated by their pKa values. Here, it is shown how this treatment impacts overall quantitative solutions to genome-scale models. A tutorial is also presented regarding how to obtain and predict necessary model constants for a large metabolic network of complex molecules. Finally, optimization of the biomass equation has shown to dramatically impact genome-scale model solutions. For example, artificially setting the ATP maintenance coefficient has been found to let a user achieve a wide number of quantitative solutions. These issues will be discussed in detail and a variety of possible solutions and guidelines will be presented. Results will focus on the production of biobutanol from *Clostrdidium acetobutylicum* ATCC 824.

See more of this Session: Poster Session: Food Engineering and Bioprocessing
See more of this Group/Topical: Food, Pharmaceutical & Bioengineering Division

Effect of thermoplastic extrusion of maize on the efficiency of bioethanol production of high concentred worts fermented with *Zymomonas mobilis*.

Peralta-Contreras, M.[1], Aguilar –Zamarripa, E. E.[1], Pérez-Carrilo, E.[1], Escamilla-Garcia, E.[2], Serna-Saldívar, S. O.[1]*.

[1]*Departamento de Biotecnología e Ingeniería de Alimentos, Tecnológico de Monterrey, Monterrey, Nuevo León, México.*

[2] *Centro de investigaciones y Desarrollo en Ciencias Odontológicas en Ciencias de la Salud, UANL*

*Corresponding author. Ave. Eugenio Garza Sada 2501 Sur, Col. Tecnológico C.P. 64849, Monterrey, N.L., México; Tel: +52 81 8158-2269; E-mail: sserna@itesm.mx

---

Abstract

The growing demand for efficient processes and fermentation technologies for fuel ethanol production have been the driving force for the implementation of thermoplastic extrusion. This technology allows the quick gelatinization of starch. The aim of this research was to compare the use of extruded maize meals with dried ground maize in terms of characteristics of worts obtained after conventional liquefaction and saccharification. The materials have shown a 17.54% and 7.652% WSI for the extruded and ground maize, respectively. The notorious difference is attributed to structural changes due the mechanical and thermal stress produced during extrusion. Worts were evaluated in terms of Brix content (12 and 20 °P), pH and FAN content. Results indicated that worts contained a similar initial pH (5.51 ±0.09) and FAN content. The rate of consumption of FAN compounds present in worts adjusted to 12°P it was not significant different among treatments ($p>0.05$).However, the thermoplastic extruded maize increased approximately 15.45% the rate of FAN consumption after 8h. The initial fermentable sugar content of worts adjusted two 20°P was 2.14% higher in the extruded maize wort compared to the regular ground maize. *Zymomonas mobilis* was capable of fermenting with good efficiency worts adjusted to 20ºP. The 20°P worts produced from extruded meals had approximately 21.2% less fermentable sugars after 8 h fermentation compared to counterparts produced from regular maize meal. This research demonstrated that

Zymomonas mobilis was capable of fermenting 20°P worts and that the bacterium was more efficient fermenting liquefied and saccharified extruded meals.

Key words: *Zymomonas mobilis*, Extrusion, Maize, Bioethanol fermentation.

Introduction:

The growing energy demand has been a driving force to develop research in renewable sources for several countries as well as for the use of affordable raw materials (Kumar et al., 2007). The fermentation for bio-ethanol production is a consolidated microbiological process which has different development stages from experimental development to industrial scale; also it has a different starting raw material (Antoni et al. 2007). Ethanol can be produced using any biological source with a considerable sugar amount or starting material that can be converted into sugars such as starch (2,5).

Starch structure during enzymatic process generates dextrins which after saccharification generates sugars that can be converted during fermentation (4) . Maize is one of the most important cereals with the highest starch contents reported (4,5). Actually maize is the most employed substrate for ethanol production in the US (5,19). Despite the tension between the production of biofuels and the importance of cereal grains as a human foodstuff, bio-ethanol is a growing industry which produces 1/3 of bioethanol, 1/3 of distilled grains and 1/3 of CO_2. The type the co-products, process and equipment used, end products is a future concern, some of the possible options for the corn ethanol industry : fuels, extrusion, pellets, plastics and human food applications (19).

Maize is a cereal based model to develop bio-ethanol production because the starch content and the endosperm type, several pre-treatment technologies for gelatinization have been reported. Kumar et al. (2006) reported that considering its advantages as a substrate, maize produces the highest ethanol yield (360 L of ethanol per ton of raw material). However, in Mexico, maize is a critical element due to its importance because it is a base commodity in the population's diet. This reality generates controversy between its use for bio-fuels production and human nutrition. Nevertheless, the corn-based industry is a growing sector to produce bioethanol which is currently considered as predominant liquid biofuel in US (21,5). The alternatives to achieve a process in less time and in an efficient way have been studied by different authors for changing different parameters (21).

Extrusion is a promising technology because it changes the physicochemical properties of starch. The raw material is processed under cooking conditions with a combination of factors such as moisture, pressure, temperature and mechanical stress (14, 17) in order to pre-gelatinized starch. Under certain conditions the rupture degrees of the starch framework might be affected by the composition of the sample; however maize flour has been reported to produce the most expanded product (14).

The temperature required to complete gelatinization is between 80 and 100 °C, however, the screw speed promotes a complete gelatinization for the rupture of starch grain. Therefore, the substrate can hydrolyze more efficiently increasing the bio-disponibility of sugars from the raw materials (3).

Extrusion has been a technology with a wide range of applications such as glucose syrup production. The influence of extrusion based on degermed corn has been related to parameters such as moderate temperature, α-amilase amount, amyloglucosidase amount and liquefaction time (11). The high crystalinity and the strong molecular and intramolecular interactions which are present in Zein proteins and polysaccharides are modified by applying heat and stress lead to thermal degradation (13).

Therefore, the material properties change as well as increase its potential application. Many investigations have been devoted to the physicochemical, molecular and nutritional changes occurring under a wide range of extrusion conditions (14).

Zymomonas mobilis is one of the microorganisms most often used for bioethanol production because its characteristic physiology and metabolism makes it a remarkable ethalogenic bacterium for future industrial applications. Some of these characteristics include the 95% conversion of glucose to ethanol with a maximum 3% cell mass incorporation, a high rate growth, acid conditions tolerance; in addition, it doesn't require oxygen conditions and therefore diminishes aeretation conditions. It also requires less nutrients for the media culture and it has a higher osmotolerance to substrate and product because of the presence of hopanoids in its cell membrane. This feature, the hopanoids, acts as stabilizers in the cell membrane (23).

Some of the main features of this ethalogenic bacteria support the potential applications for the microorganism. *Z. mobilis* is part of the Alpha-Protobacter one of the most studied groups among microorganisms which includes well-known bacteria for several applications such as medical, agronomical and industrial (12). The *Zymomonas* gender is characterized by big bacillus, gramm-negative, fermentative anaerobe which is able to achieve a vigorous fermentation from sugars to ethanol, these characteristics can be considered for its implementation in industrial process. The production of the ethanol based on technology of first generation has been revisited for several autors (4,5) but the correlations of a pretreatment technology such as extrusion have been scarcely explored in the fermentation technology.

The aim of the current research is to evaluate the effect of thermoplastic extrusion in maize and its physicochemical properties for worts and fermentation process in the performance of *Zymomonas mobilis*. The varying parameters are the pretreatment technology to produce wort (extrusion vs ground milled) and the concentration for the maize worth (12°P and 20 °P) to evaluate in fermentation conditions.

Materials and methods

Microorganims

Zymomonas mobilis (ATCC 31823) is the bacteria selected to perform fermentation in the saccharified dried ground and extruded worts.

Culture maintenance

The activation of the microorganism was achieved in a synthetic media RM, with the following composition: 20.0 g glucose,10.0 g yeast extract, 2.0 g KH_2PO_4 and a pH of 6.0, per liter. The inoculum was cultivated at 30°C until it reaches ¾ of the exponential phase. Plate count and optical density were performed to verify the microorganism growth.

Physical properties of maize worts

Dry milled and extrusion were the process to modify the starting material to produce the maize worts. The extrusion was performed under the following conditions: final temperature of 120 °C with a mechanical torque of 80 % and a screw speed of 250 RPM. The physical properties of the material were compared before fermentation. The determinations for this purpose are water solubility index (WSI) and water absorption index (WAI) for the treatments evaluated in the current research (1). A remarkable property is WSI because increase its value in the case of the extruded material.

Particle size characterization of maize worts

A 100 g sample was placed in a rotab during a 5 min period to classify the fractions of particles for every treatment (4).

Maize wort production

Maize dried ground and extruded were liquefied with α-amilase and saccharified with AMG according to literature (4). The resulting saccharified worts were filtered with a sieve 1mm of screen (U.S. standard sieves). The Brix content was adjusted to 12 and 20 P with a digital refractometer (ATAGO HRS 500, Bellevue WA, USA). After that the worts were adjusted with HCl 0.1 N to pH value of 6. Finally, the worts were pasteurized according to reported literature (5).

Inoculum preparation

The bacteria criotubes were preserved at -80 °C and they were anaerobically activated in a synthetic sterile RM media. A 10% of volume of inoculums was required in synthetic media in sterile Erlenmeyer flasks, incubated at 30 °C without shaker until the cells achieved ¾ exponential phase. For the determination of this activity, the culture was monitored measuring turbidity at 560 nm in a spectrophotometer (UV-Visible Spectro photomerter Spectronic GENESYS) during 12-14 hours. Some samples were conserved for further study.

Fermentation process

The saccharified maize worts were pasteurized and placed into a laminar flow chamber (Labconco Model 36212, Kansas City MI, USA) for the fermentor assembling. The fermentator design was integrated with the following devices: a CO_2 injection port (1) previously filter with a 0.2 μm membrane (2) (PALL Co, Port Washington NY, USA) which generates a positive pressure inside of the bioreactor which facilitates the exit of the sample by the sampling port (3) without open the system and break the anaerobiosis conditions. Every piece of the system was sterilized previously to the assembling. After the fermentors were inoculated with *Z. mobilis*, every system was placed at 30 °C inside of a incubator chamber (Lab-Line Model 3526).

pH quantification

The pH measurements were performed at the moment which the sample was taken in a potenciometer (Orion, model 1230 , Germany). The dispositive was previously calibrated with buffers at pH 4.0 and 7.0.

Brix content

The Brix content was determined placing an 80 μL of aliquout according to Davila et al (2009) during the fermentation.

FAN content

The FAN content was determined according to the AOAC (1991) 945.30L method.This determination is based on the free aminoacids reaction with a ninhydrin reagent to quantify amount of free aminocid in the worts.

Design of Experiments

A factorial with 2^2 was considered for the preliminary study .Concentration (12 and 20 P) and process (extrusion and dried grounded). The data analysis was analyzed with Minitab.

Results and discussions

Physical and chemical properties from the extruded materials.

The particle size was affected in a significant manner for the extrusion process. The change of particle size distribution can be found in Table 2. Comparing the two processes, a higher percentage of fine particles can be found in the extruded treatment. Finer raw materials tend to give products with softer textures and smaller cell structures, while the coarse particles produce larger cell texture.

Moisture is one of the properties that allow an appropriate expansion during based on several factors. Water affects the interactions of the raw material and the dissipation of mechanical energy. This parameter is important because the viscosity of melt decreases with increasing content (14,1). For the purposes of the current study the moisture content was determined with the first expanded samples of the process and it was found that the moisture was similar with the dried grounded material.

Z. mobilis culture

Zymomonas mobilis growth curve in the experimental conditions for the current work were determine by turbidity technique with absorbance (wavelength 560 nm). The results for the bacterium are reported in figure 3. In order to identify the possible growth conditions, the bacterium was evaluated with (200 rpm) and without shaker conditions at 30 °C in a RM synthetic media. The experimental data was compared with Kruskal wallis ($\alpha = 0.05$, df = 15). After the analysis it was found that there was no significant difference between the growing conditions for the inoculum. Therefore, *Z. mobilis* growth was established without mixing conditions for the rest of the experiments with maize worts.

Based on preliminary results, and according literature the medium culture for the bacteria was selected (Sahm et al. 2006). The preparation was establish to set up conditions for the fermentation inoculums to evaluated extruded and dried grounded treatments at the two different levels. According to Zhang and Fheng (2010) the optimum volume for fermentation inoculums is between the 7.5 and 10% v/v for *Z. mobilis* considering the maximum volume for every glass Erlenmeyer flask, 300ml for our experimental design. The advantages of *Z. mobilis* have been described with respect to its higher specific rate of glucose uptake and its behavior tolerance with substrates between 100 and 250 g/L (Panda et al. 1997).

The inoculums aliquot for further experiments was establish at 1.6 of absorbance value at 560 nm wavelength corresponding to ¾ of the exponential growth phase of this microorganism. In these conditions, a count of 1.2 x 10^7 cells/mL was found corresponding to the recommended inoculums in order to start the fermentation (6).

pH profile in fermentation

The pH range achieve for *Z. mobilis* was between 4.6-4.9 during the 72 hours of fermentation for all the treatments. In figure 2 depicts this behavior. This bacterium during fermentation can reach a pH 4-5.5 which is similar to reported literature (19). One of the most important factors in bioethanol production is pH. In the case of values under 3.5 during fermentation is a indicator that the metabolism of *Z. mobilis* for over production of lactic acid and other organic acids decreases pH during the fermentative process. It also affects the behavior and bacterium physiology, achieving a lower ethanol yield (15) . In all the cases, a maximum decrease was observed of the pH value between the 10 and 25 h of the fermentation which is an indicator that the process of conversion to ethanol is finished (23).

In order evaluate the process effects (dried ground and extruded) and the wort concentration (12 and 20 °P), a statistical analysis was performed ANOVA (Table 3). Interesting information and statistical evidence were found that support the factor that has influence in the final pH (p-value=0.002) in the fermentations related to the natural process for the worts; in addition the concentration was not a relevant factor.

The effect of process type over the pH can be related with the particle size present in the wort. In the extrusion process, a particle size distribution has been found oriented to the medium (149-420nm) and the small (<149 nm) size in the case of the ground maize with the milling process the particle size was oriented to the coarse fraction size (Table 2).The type and the size of the particle is related to the bio-disponibility of sugars that the bacterium is able to use for its growth and metabolism. In the case of the extrusion process, the material was over thermo-mechanic stress conditions and therefore it has an influence in the denaturalization of the starch structure (8). The starch changes can affect the soluble sugars presents in the wort and impact in a direct manner the pH and ethanol yield (15).

Brix consumption during fermentation.

For all the treatments, a significant decrease was observed between 10 and 25 h which coincide with pH decrease. Fermentations based on starch starting materials, however, have been reported with ethanol production starting at hour 24; it can be possible estimate a final stage in the fermentation process on 25h. In fermentation, Brix monitoring is a key element.

The Brix content is strongly related with the amount of soluble solids in a liquid sample because the change of refraction index. A Brix degree is the percentage of the total amount of solids in solution expressed in solute grams/100g of solution (g/g) (20). For food analysis the Brix content is related with the sugar content because it is a remarkable characteristics from the carbohydrates its ability to affect the refraction index in solution. Therefore, the Brix content, in a calibrated refractometer, is a reliable measurement for sugar content in a sample (9).

The statistical analysis ANOVA was applied for the experiments (Table 4) and the brix content was found to be affected by the type of process and initial content. Also the interaction between factors was found statistical significant. This fact can be related to the process that were placed every raw material affecting biodisponibility and sugar content (8).

A negative effect was observed with the fermentation with worts at 20 P for the *Z. mobilis* because the final concentration of Brix was higher at the end respect to 12 P (fig. 5) which implies a sugar amount without fermentation might reduce the efficiency process. It was observed a better behavior at 12 °P which has been reported a extensively study in literature comparing maize with another substrates such as sorghum and decorticated sorghum (16).

FAN consumption during fermentation

In the case of fermentation based on starch substrate, it is important to take in consideration not only the carbohydrate amount but also the nitrogen compounds in amount and diversity because they have impact on the yield process (23). The FAN optimum content is 150mg/L for a good fermentation performance (9,22). The saccharified for this study have 85mg/mL for the 20 °B and 65 mg/mL for the saccharified worts at 12 °B which can affect on the performance of the fermentation. However, in *Z. mobilis* this parameter might have less effect because the amount of nitrogen compounds is less required in comparison that other microorganisms. In future work might be estimated the optimum FAN content for a fermentation process with *Z. mobilis*.

Even the initial FAN content was lower for the evaluated worts, most of the *Zymomonas* strains are known to be autotrophic foe the vitamin pantothenic acid and although there are several reports in the literature of defined and minimal media formulations achieving high performance fermentations by *Zymomonas mobilis* relatively little is understood concerning the specific influence on growth and fermentation of individual components. In some cases *Z. mobilis* and its lower nutrients media culture requirements can assimilate nitrogen potentially could significantly reduce the level of corn steep liquor (CSL) or another type of vitamins in the media culture required to support growth and fermentation (7).The current experiments were performed without additional FAN supplementation and the bacteria was able to develop growth and tolerate higher concentrations of substrate.

In figure 4, a FAN consumption profile similar in the four treatments is depicted. A major decrease in FAN concentration was observed between 8 and 18h period of time that coincide with the bacterium growth (Fig.2) and the consumption of sugars (Fig. 3). The total consumption of FAN in the saccharified worts was in its lowest point at 38 h; it can be related with the death stage of the inoculum and the autolysis of *Z. mobilis* that contribute to the amino acids in the media and an increase of FAN content (10).

Conclusions

Extrusion as pre-treatment for the yellow maize has an effect in sugar consumption and final pH of the fermentation of *Z. mobilis*, especially when the soluble solid concentration for the saccharified worts is 20 °P. The fermentations for the saccharified extruded worts (12 and 20 °P) have shown a decrement more pronounced respect the sugar consumption for the dried grounded maize. This effect can be related to the particle size difference generated for the process of each wort, also due to the physicochemical changes of the starch during they extrusion which impact the bio-disponibility of the sugars as substrate during fermentation. However, in both pre-treatments it was observed a negative effect in sugar consumption at 20 °P because the final sugar concentration was higher and it might reduce the efficiency process for the incomplete sugar conversion.

The pH range achieve by *Z. mobilis* it was held between 4.6-4.9 during the 72 h of fermentation for all the treatments being an optimum condition. However, the statistical analysis confirmed a significant effect due to process extrusion over the pH of fermentation. In addition, pH and Brix content consumption are correlated with the fermentation stage and in both factors the extrusion process impacts the performance and these aspect is important to determine the type of pretreatment for the wort production. Considering the initial and final FAN content, this was not affected by the type of process applied to the yellow maize, the FAN amount was not the optimum reported for another microorganism but the bacterium was able to reach part of its metabolism (7). In future work, it would be necessary to determine the nitrogen compounds demand for *Z. mobilis*. Also, in order to determine the type of treatment and soluble solid concentration for the wort, it is necessary to consider more analytical methods to detect ethanol production during the bacteria metabolism in order to evaluate the efficiency of the overall process.

Tables

Table 1. Physical and chemical properties for processed maize flour used for the fermentation worts.

	Dried Grounded	Extruded 120 °C
Moisture %	12.621 ± 0.097	12.90 ±0.102
Apparent density (kg/100 L)	51.067 ± 0.23	54.48 ±0.62
WAI %	2.823 ± 0.189	4.424 ±1.26
WSI %	7.652 ± 0.268	17.54 ±6.68
Ash content %	1.181 ± 0.055	1.179 ±0.025
Protein %	8.841 ± 0.50	8.094 ±0.18
Phenolic compound [b]	376.39 ± 68.18	238.37 ±25.19
Crude Fat %	3.129 ± 0.26	0.417 ±0.084
Total starch[a]	64.643 ± 0.5997	65.125±1.32

[a] Total Starch procedure according to AOAC Method 996.11 and AACC Method 76.13
[b] Expressed in μg of ac galic equivalents/ dried gram

Table 2. Particle size of the treated maize flours analyzed with rotap before the enzymatic process.

Treatment	Particle size fraction		
	Coarse (>420 mm)	Medium (149-420 mm)	Fine (<149 mm)
Dried grounded	61.68%	35.91%	1.65%
Extruded 120°C	37.74%	47.66%	14.58%

Figures

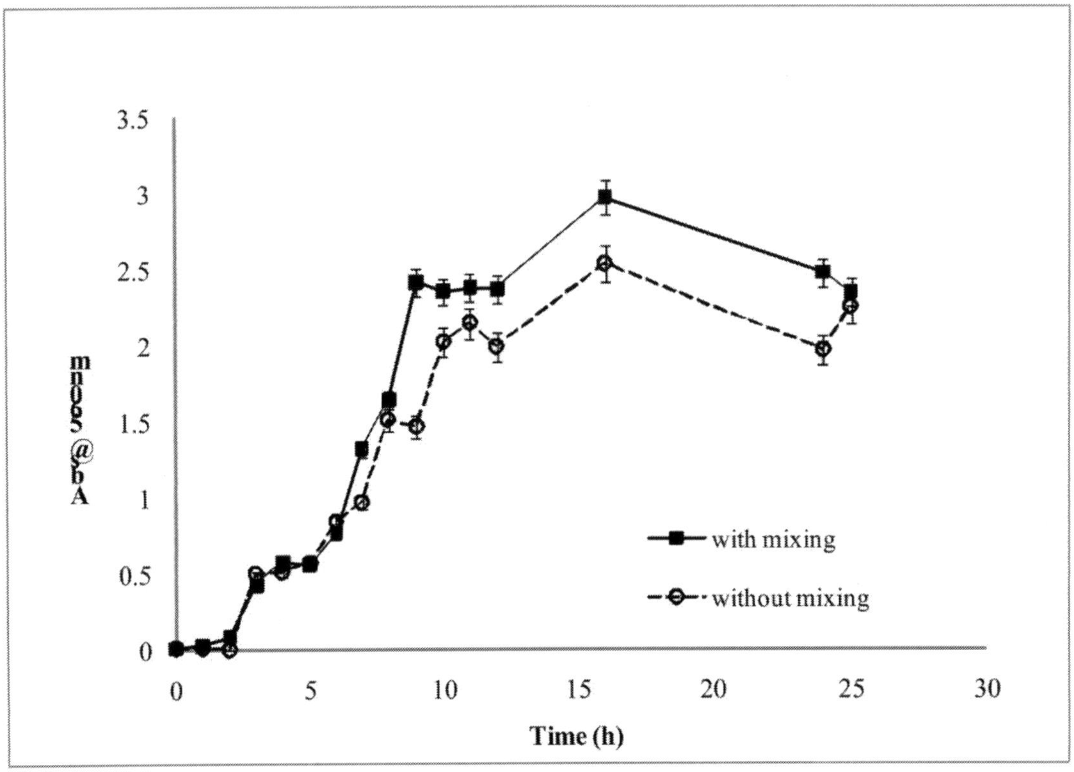

Figure 1.

Z. mobilis growth curve with (200 rpm) and without mixing correlated with optical density in RM synthetic media incubated at 30 °C.

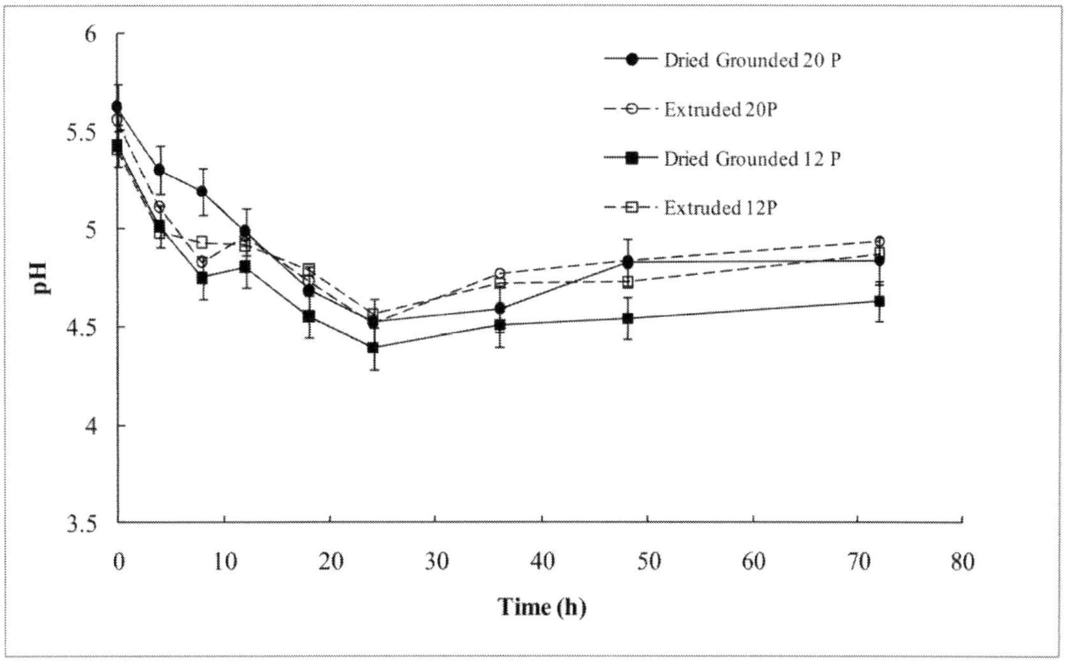

Figure 2.

pH profile of *Z. mobilis* during fermentation with maize milled worth at 12°P y 20°P and maize extruded wort at 12°P y 20°P.

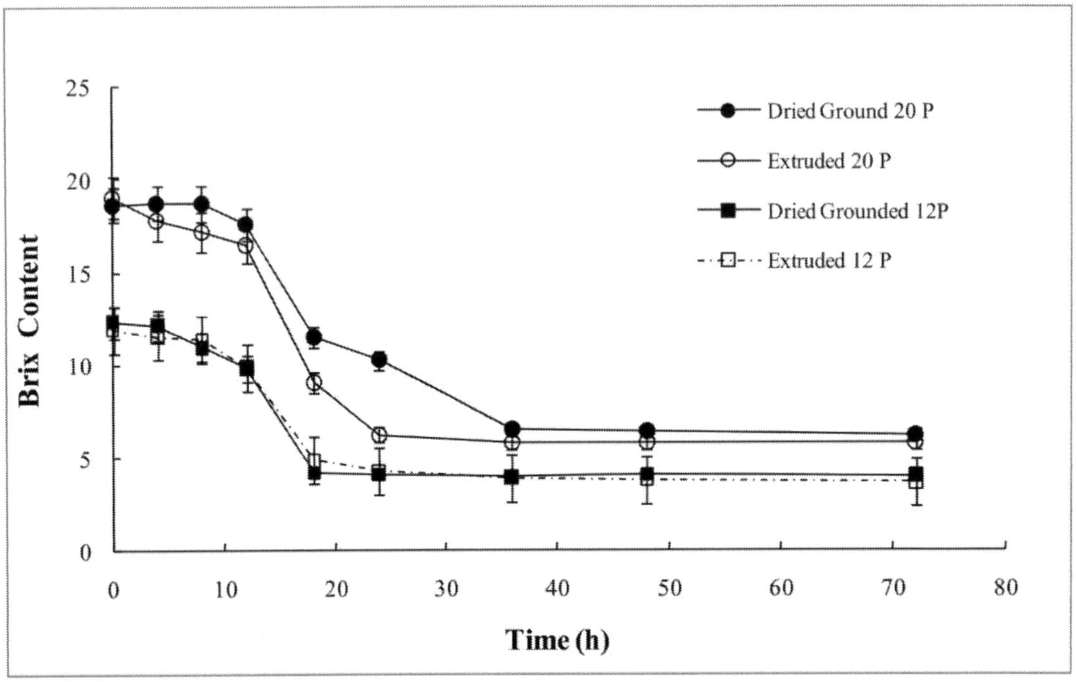

Figure 3.

Brix consumption *Z. mobilis* during fermentation. Maize grounded wort at 12 and 20 °P is compared extruded maize wort at 12 and 20 °P.

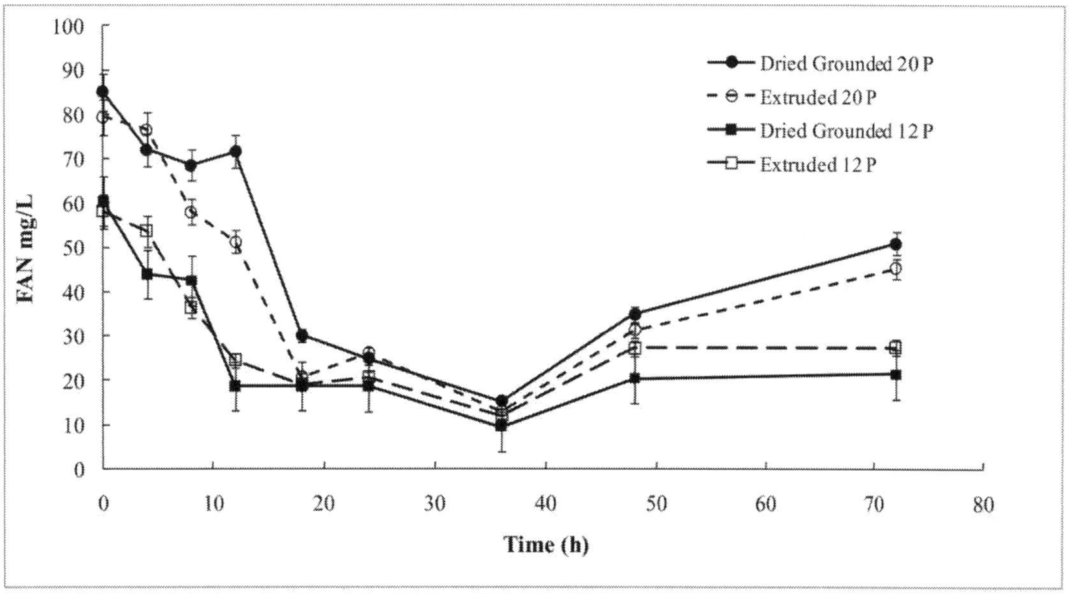

Figure 4.
FAN consumption profile of *Z. mobilis* during fermentation in the worts treatments: grounded maize at 12 and 20 °P and extruded maize at 12 and 20 °P.

References

1. Anderson, R.A. 1982. Water absorption and solubility and amylograph characteristics of Roll-Cooked small grain products. Cereal Chem 59 (4): 265-269.
2. Antoni, D., Zverlov, V. V. and Schwarz, W. H. 2007. Biofuels from microbes. Applied Microbiology and Biotechnology 77:23-35.
3. Baks, T., Kappen, F. H. J., Janssen, A. E. M. and Boom, R. M. 2008. Towards an optimal process for gelatinisation and hydrolysis of highly concentrated starch-water mixtures with alpha-amylase from B. licheniformis. Journal of Cereal Science 47:214-225.
4. Chuck-Hernandez, C., Perez-Carrillo, E. and Serna-Saldivar, S. O. 2009. Production of bioethanol from steam-flaked sorghum and maize. Journal of Cereal Science 50:131-137.
5. Davila-Gomez, F. J., Chuck-Hernandez, C., Perez-Carrillo, E., Rooney, W. L. and Serna-Saldivar, S. O. 2011. Evaluation of bioethanol production from five different varieties of sweet and forage sorghums (Sorghum bicolor (L) Moench). Industrial Crops and Products 33:611-616.
6. Erten, H., Tanguler, H. and Cakiroz, H. 2007. The effect of pitching rate on fermentation and flavour compounds in high gravity brewing. Journal of the Institute of Brewing 113:75-79.
7. Fein, J. E., Zawadzki, B. C., Lawford, H. G. and Lawford, G. R. 1983. Controlling morphological instability of Zymomonas mobilis strains in continuous culture. Applied and Environmental Microbiology 45:1899-1904.
8 .Govindasamy, S., Campanella, O. H. and Oates, C. G. 1997. Enzymatic hydrolysis of sago starch in a twin-screw extruder. Journal of Food Engineering 32:403-426.
9. Ingledew, W. M. 1995. The biochemistry of alcohol production in: The alcohol textbook, 1. K. T.P., D.R. y Murtagh, J.E, ed. Nottingham University Press: Nottingham.
10. Leiper, K. A., Schlee, C., Tebble, I. and Stewart, G. G. 2006. The fermentation of beet sugar syrup to produce bioethanol. Journal of the Institute of Brewing 112:122-133.
11. Ma, C. and Shen, D. 2010. Experiment on extrusion parameters of producing glucose syrup with extruded degermed maize added moderate temperature amylase. Nongye Jixie Xuebao/Transactions of the Chinese Society of Agricultural Machinery 41:126-130.
12. Madigan, M., Matinko, J., Dunlap, P., y Clark, P. 2009. Biologia de los microorganismos. Pearson.
13. Maio, E. D., Mali, R. and Iannace, S. 2010. Investigation of Thermoplasticity of Zein and Kafirin Proteins: Mixing Process and Mechanical Properties. Journal of Polymers and the Environment 18:626-633.
14. Majzoobi, M. and Farahnaky, A. 2010. Comparison of the effects of extrusion cooking on some cereal starches. International Journal of Food Engineering 6.
15. Panesar, P. S., Marwaha, S. S. and Kennedy, J. F. 2006. Zymomonas mobilis: An alternative ethanol producer. Journal of Chemical Technology and Biotechnology 81:623-635.
16. Perez-Carrillo, E., Serna-Saldivar, S., Alvarez, M. and and Cortes-Callejas, M. L. 2008. Effect of Sorghum Decortication and Use of Protease before Liquefaction with

Thermoresistant alfa-Amylase on Efficiency of Bioethanol Production. Cereal Chemistry 85:792-798.

17. Riaz, M. 2000. Extruders in Food Applications. Technomic: Pennsylvania.
18. Sahm, H., Bringer-Meyer, S. and Sprenger, G. 2006. The Genus <i>Zymomonas</i>. The Prokaryotes:201-221.
19. Saunders, J. A. and Rosentrater, K. A. 2009. Survey of US fuel ethanol plants. Bioresource Technology 100:3277-3284.
20. Son, H. S., Hong, Y. S., Park, W. M., Yu, M. A. and Lee, C. H. 2009. A novel approach for estimating sugar and alcohol concentrations in wines using refractometer and hydrometer. Journal of Food Science 74:C106-C111.
21. Swana, J., Yang, Y., Behnam, M. and Thompson, R. 2011. An analysis of net energy production and feedstock availability for biobutanol and bioethanol. Bioresource Technology 102:2112-2117.
22. Thomas, K. C. and Ingledew, W. M. 1990. Fuel alcohol production: Effects of free amino nitrogen on fermentation of very-high-gravity wheat mashes. Applied and Environmental Microbiology 56:2046-2050.
23. Towo, E., Matuschek, E. and Svanberg, U. 2006. Fermentation and enzyme treatment of tannin sorghum gruels: Effects on phenolic compounds, phytate and in vitro accessible iron. Food Chemistry 94:369-376.

(626b) Comparison of Bioethanol Production with Saccharomyces Cerevisiae Immobilized On Different Supports

Ravi Dhabhai, Chemical Engineering, Malaviya National Institute of Technology Jaipur, Jaipur, India, **S. P. Chaurasia**, Chemical Engineering, Malaviya National Institute of Technology, Jaipur, India and Ajay K. Dalai, Chemical Engineering, University of Saskatchewan, Saskatoon, SK, Canada

Use of immobilized yeast in bioethanol fermentation generally increases ethanol yield and productivity compared to free cell fermentation. In present work, different supports (wheat straw, pine wood chips, granular activated carbon, and cotton balls) were compared for their immobilization ability for yeast *Saccharomyces cerevisiae*. The immobilization supports were characterized for their surface properties like BET surface area, pore volume and average pore size. The immobilization was carried out by culturing 48 h grown *S. cerevisiae* culture with immobilization support for 24 h with gentle shaking. The immobilization was confirmed by optical microscope and by the efficiency of immobilized cells to ferment glucose faster compared to free culture.

Fermentation experiments with glucose as sole substrate (initial concentration 155.0±5.0 g/l) were carried out using immobilized yeast and free cells. Pine wood chips (PWC) and cotton balls (CB) proved to be better supports for immobilization and bioethanol production as evident by their higher volumetric ethanol productivity (Q_p) of 1.5 g/L.h and 1.44 g/L.h respectively while with wheat straw, higher Q_p (1.4 g/L.h) was observed but desorption of cells from support into medium was also high. Granular activated carbon (GAC) proved to be least efficient as a yeast support with lowest Q_p (0.684 g/L.h) and sugar utilization (68%). Except GAC, immobilized cells produced ethanol 1.4-1.6 times faster compared to free cells. PWC seemed to be better and efficient support for yeast immobilization.

See more of this Session: Poster Session: Food Engineering and Bioprocessing
See more of this Group/Topical: Food, Pharmaceutical & Bioengineering Division

(626c) Microbial Production of Docosahexaenoic Acid by Marine Fungi: How to Convert Glucose to Lipid Effectively

Lujing Ren, College of Biotechnology and Pharmaceutical Engineering, Nanjing University of Technology, Nanjing, China

DHA(Docosahexaenoic acid, 22:6¢$^{4.7.10.13.16.19}$)is an important kind of •-3 polyunsaturated fatty acid (PUFAs). As the necessary component of cell membranes, DHA, known as "Brain Gold", is regarded to be essential for the development of visual and neurological development of infants. In addition, it also could reduce or inhibit risk factors involved in various diseases like cardiovascular diseases. So, it is widely used in food and medical industry recently. The traditional source of DHA is fish oil, but the quality of the fish oil is variable and depends on fish species, season and location of catching sites. In addition to the high content of EPA and cholesterol, the application of fish oil in food and pharmaceutical industries was limited severely. Therefore, microbial production of DHA is the inevitable choice to ensure current DHA usage. This presentation will introduce the research results involving how to convert glucose to lipid effectively by marine fungi, *Schizochytrium* sp. in our group.

High cell density is the premise of the high yield of intracellular products, so the first research objective is about how to improve cell growth rate and cell yield coefficient. As we know, cell growth need abundant nutrition but lipid accumulation only happened in the nutrition limitation or environmental stress conditions. So the second part will concern this contradiction and investigated the metabolic mechanism of *Schizochytrium* sp. in response to nutrition limitation and environmental stress, aiming at obtaining lipids as much as possible. Thirdly, high DHA percentage in total fatty acids is the premise for high oil quality, which concerning the supply of reducing power, NADPH. A regulation strategy of adding specific activators at different fermentation stages to reinforce NADPH supply was proposed by tracking the changes of cell growth, DHA accumulation and key enzymes for NADPH synthesis during the fermentation stages. Finally, based on the above research, the results were applied to the amplification process in the industry, which will lay the foundation for the realization the goal of "high yield ", "high quality "and "stable production" of DHA bio-refining process, at the same time support the reference of methodology for other PUFAs production.

See more of this Session: Poster Session: Food Engineering and Bioprocessing
See more of this Group/Topical: Food, Pharmaceutical & Bioengineering Division

Improvement of L(+)-lactic acid production by in vivo regulation of alcohol dehydrogenase in *Rhizopus oryzae*

Sitanan Thitiprasert[1], Sarintip Sooksai[2], and Nuttha Thongchul[2]*
[1]Biotechnology Program, Faculty of Science, Chulalongkorn University, Bangkok, Thailand
[2]Institute of Biotechnology and Genetic Engineering, Chulalongkorn University, Bangkok, Thailand

Introduction

Lactic acid is a specialty chemical that has many applications in chemical, food, and pharmaceutical industries. Recently, there has been an expanding use of an optically pure lactic acid for synthesizing polylactic acid. It is expected that in the near future, the non environmental friendly petroleum-based plastic will be substituted by polylactic acid due to the promising renewable and biodegradable properties [1-3]. Currently, lactic acid is commercially produced via fermentation because it leads to the production of the optically pure lactic acid while chemical synthesis requires drastic conditions and results in the racemates [4-5]. To develop more efficient lactic acid fermentation process for polylactic acid, the major concern is to improve the productivity and to reduce the cost of raw materials [6-8]. Many microorganisms can produce lactic acid including bacteria (*Lactobacillus* and *Bacillus* species, for example) and fungi. Interestingly, filamentous fungus *Rhizopus oryzae* can produce an optically pure L-lactic acid from simple media such as starchy materials and lignocellulosic products without prior treatment; therefore, can reduce the production cost [2,9-10].

Unlike homofermentative lactic acid bacteria, fermentation by *R. oryzae* results in 4 major end products including L-lactic acid, cell biomass, fumaric acid, and ethanol. The formation of the end products depends on the cultivation conditions [11-13]. Under sufficient oxygen supply, *R. oryzae* converts glucose to pyruvate and pyruvate is then entered tricarboxylic acid (TCA) cycle where ATP and NADH are generated. The energy and cofactors produced in TCA cycle facilitate the conversion of pyruvate to L-lactic acid by lactate dehydrogenase (*LDH*). Therefore, oxygen promotes L-lactic acid production.

Several methods have been used to prevent byproduct formation; thus, increase L-lactic acid production. In free cell culture, high oxygen concentration is achieved by means of agitation and aeration [11-12]. However, rigorous agitation causes high power input and mycelial damage resulting in low L-lactic acid production. It was reported that improved oxygen transfer in fungal fermentation was obtained by better morphological control [14-22]. Several immobilization techniques have been used for controlling morphology including immobilization in polyurethane foam cubes, immobilization in calcium alginate gels, immobilization on the fibrous matrix, and immobilization on polymer supports. However, the ethanol formation as an evidence of further cell growth was still observed in long-term cultivation of immobilized cells. Therefore, none of them could absolutely limit ethanol production.

It is believed that prevention of pyruvate flux toward undesirable pathways can be achieved by inhibiting the enzymes in those pathways. Previous studies revealed that ethanol was still found as the major byproduct in lactic acid fermentation by *R. oryzae* NRRL395 in the cultures with rigorous oxygen supply. Under insufficient oxygen microenvironment, *R. oryzae* produces ethanol by the conversion of pyruvate to acetaldehyde and CO_2 by pyruvate decarboxylase (*PDC*) followed by the oxidation of acetaldehyde to ethanol by alcohol dehydrogenase (*ADH*). Many studies on in vitro inhibition of yeast *ADH* (purified/partially purified) have been reported. Use of substrate analog such as 2,2,2-trifluoroethanol or modification of *ADH* conformation by addition of sulfhydryl reagents showed significant inhibition of *ADH* activity [23-25]. Nevertheless, little is known about *R. oryzae ADH* inhibition especially in the in vivo test. In this study, suppression of *ADH* during lactic acid fermentation by living *R. oryzae* was attempted. As a result, more pyruvate flux could shift toward lactic acid production pathway. The effects of different *ADH* inhibitors under different concentrations on fermentation kinetics of immobilized *R. oryzae* in a shaken flask culture as well as *ADH* and *LDH* activities were investigated in this study.

Materials and methods

Lactic acid fermentation with the presence of *ADH* inhibitor was investigated in the shaken flask culture of immobilized *R. oryzae* on a cotton matrix (5cm×5cm) at 30°C, 200 rpm. The fermentation kinetics as well as the enzyme activities (both *ADH* and *LDH*) when the inhibitor was present during the fermentation were compared with those without the inhibitor (control).

Microorganism and inoculum preparation

Rhizopus oryzae NRRL 395, a filamentous fungus producing L(+)-lactic acid obtained from the Northern Regional Research Center, Peoria, IL, U.S.A., was used in this study. The sporangiospores were collected from the 7-day culture on potato dextrose agar plate by shaving and extracting the spores with sterile water. The spore concentration was determined by spore counting using a haemacytometer. The spore suspension was then adjusted to 10^6 spores/mL by dilution with sterile water.

Growing R. oryzae in the fermentation medium containing different ADH inhibitors

Lactic acid fermentation by *R. oryzae* consisted of 2 phases i.e. growth phase for spore germination and initial cell growth and production phase for producing lactic acid by cells obtained in the growth phase. This fermentation usually suffers from ethanol production during both growth and production phases; thus, resulting in low lactic acid yield than the theoretical value. To inhibit ethanol production during fermentation, *ADH* inhibitors were added during the fermentation.

During the growth phase, the medium consisted of 50 g/L glucose and 5 g/L yeast extract was used for spore germination and initial cell growth where the control of pH was not necessary during this phase. Fifty milliliter of the growth medium containing a preweighed cotton cloth (5cm×5cm) in a 250 mL Erlenmeyer flask was inoculated with 0.5 mL *R. oryzae* spore suspension (10^6 spores/mL). The spores were allowed to germinate and immobilize onto the cotton matrix during the growth phase. After the growth phase (48 h), the growth medium was removed from the culture flask and 50 mL of the production medium was added for enhancing lactic acid production. The production medium contained 70 g/L glucose, 0.6 g/L KH_2PO_4, 0.25 g/L $MgSO_4{\cdot}7H_2O$, 0.088 g/L $ZnSO_4{\cdot}7H_2O$, and 0.3 g/L urea. 5 M NaOH was added periodically for pH control during the production phase. Fermentation was carried out until glucose depletion. Samples were taken every 6 h for analyses of glucose, lactic acid, and ethanol as well as for enzyme assays. The effects of the inhibitor on fungal growth and metabolism during fermentation were investigated. Firstly, in order to compare the results from this work with the in vitro *ADH* inhibition reported in the previous works, 6 *ADH* inhibitors including 1,2-diazole or pyrazole (PZ), 2,2,2-trifluoroethanol (TFE), Iodoacetic acid (IAA), N-ethylmaleimide (NEM), 4,4'-dithiodipyridine (DSDP), and 4-hydroxymercury benzoic acid or *p*-chloromercuribenzoate (PCMB) at 1 mM were initially added during the growth phase and later they were added after changing the medium during the production phase [25]. The effects of these 6 inhibitors on fungal metabolism were also investigated during the production phase of the pregrown *R. oryzae* in the simple growth medium (without the inhibitor). After screening the inhibitors which did not harm to fungal growth and metabolism in the in vivo test, the effects of such inhibitor concentration were also studied.

Product yields and volumetric productivities

Amounts of cell growth and products formed during fermentation were analyzed. Product yield was determined from the ratio of product formed to glucose consumed during fermentation. Volumetric productivity was defined as the total amount of product formed per volume per time.

Cell biomass

Cell concentration in the fermentation broth was determined from the dry weight of cell biomass. Immobilized cell biomass on the cotton towel was harvested, washed, and dried at 105°C until reaching a constant weight. Cell concentration (in g/L) was calculated by the following equation:

$$\text{Cell concentration} = \frac{\text{dry weight of immobilized cell biomass on cotton towel (g)} - \text{dry weight of preweighed cotton cloth (g)}}{\text{volume of fermentation broth (L)}} \qquad \text{(eq.1)}$$

Substrate and product analyses

High performance liquid chromatography (HPLC) was used to analyze the organic compounds (glucose, lactic acid, fumaric acid, and ethanol) in the samples. Sample from fermentation broth was centrifuged and diluted with double distilled water. Twenty microliter of diluted particle-free sample was injected into an organic acid analysis column (Biorad, Aminex HPX-87H ion exclusion organic acid column; 300mm×7.8mm) maintained at 45°C in a column oven. 0.005M H_2SO_4 was used as an eluant at 0.6 mL/min flow rate. A refractive index detector was used to detect the organic compounds. A standard containing 2 g/L of each component was injected as a reference to determine the sample concentration. The peak area was used for the comparison basis. It is noted that HPLC can detect both L(+) and (D)-lactic acids.

Cell extraction, partial enzyme purification, and protein determination

The sample of fresh fungal mycelia immobilized on the cotton cloth was washed thoroughly with sterile distilled water 3 times before eliminating the free water. The free water remained at the sample was adsorbed onto the filter papers by putting the sample in between 2 filter papers. Filter-dry immobilized mycelia were carefully peeled off the cotton cloth. 2 sets of 1 g filter-dry mycelia were prepared: one for determination of the moisture content in the filter-dry mycelia and the other one subjected to enzyme extraction process. To determine the moisture content, 1 g sample was dried at 105°C until reaching the constant weight. The moisture content (in %) was calculated by the equation below:

$$\text{Moisture content (\%)} = \frac{\text{weight of filter-dry sample (1 g)} - \text{dry weight of sample (g)}}{\text{dry weight of sample (g)}} \times 100 \qquad \text{(eq.2)}$$

Another 1 g filter-dry sample for enzyme extraction was frozen at -20°C for 1 h before grinding in an ice-cold mortar for 5 min. One point five milliliter of 50 mM Tris HCl buffer containing 10 mM mercaptoethanol was added into the ground mycelia. The suspension of the ground mycelia with the glass beads was homogenized in an ultrasonic disruptor for 16.75 min (30 s interval with 45 s break). Later, the glass beads were removed by filtration and the suspension was centrifuged at 12,000 rpm, 4°C for 30 min. The supernatant, so called later the mycelial extract, was used for protein determination and enzyme assays. Protein concentration in the mycelial extract was determined by Lowry method using bovine serum albumin as a standard [26].

Enzyme assays

The mycelial extract obtained from the extraction of the fresh fungal mycelia collected during fermentation with or without the inhibitors was analyzed for the activities of the enzymes of interest including *ADH* (both forward and backward directions) and *LDH* to investigate the consequences of the presence of the inhibitor in the living *R. oryzae*.

Alcohol dehydrogenase (*ADH*(b) – conversion of ethanol into acetaldehyde) was assayed spectrophotometrically by a coupled assay system at 340 nm, depending on NAD^+ reduction. The reaction mixture contained 250 µmol ethanol, 4 µmol NAD^+, 0.2 mL mycelial extract, and 250 µmol Tris HCl buffer (pH 7.7) in a total volume of 3.15 mL [27]. The reaction was started when adding the mycelial extract into the reaction mixture. The activity was detected for 5 min at 30°C (similar temperature to the culture temperature). *ADH*(b) activity was monitored at 340 nm expressed as µmol NAD^+ reduced per min using the calibration curve (absorbency at 340 nm versus NAD^+ concentration in µmol).

The conversion of acetaldehyde into ethanol was also observed by following the oxidation of NADH at 340 nm. The reaction mixture contained 250 µmol acetaldehyde, 4 µmol NADH, 0.2 mL mycelial extract, and 250 µmol Tris HCl buffer (pH 7.7) in the total volume of 3.15 mL. The reaction was started with the addition of the mycelial extract into the reaction mixture. *ADH* forward (ADH(f)) activity was detected for 5 min at 30°C (similar temperature to the culture temperature) by following the change in absorbency at 340 nm which was later converted into µmol NADH oxidized using the calibration curve (absorbency at 340 nm versus NADH concentration in µmol).

In addition, lactate dehydrogenase (*LDH*) was assayed by following the oxidation of NADH at 340 nm in the reaction mixture containing 25 µmol sodium pyruvate, 1 µmol NADH, 0.2 mL mycelial extract, and 250 µL phosphate buffer (pH6.5) in the total volume of 3.15 mL [27-28]. The reaction was started with the same manner as that described in the *ADH*(f) activity assay.

Results and discussions

Morphology and metabolic responses of R. oryzae grown in the medium containing ADH inhibitors

Similar morphology was observed when culturing *R. oryzae* in the medium containing 1,2-diazole or 2,2,2-trifluoroethanol as compared with those grown in the simple glucose medium (control) (Fig. 1). Spores germinated and completely immobilized onto the cotton cloth provided. This resulted in cell-free fermentation broth. On the other hand, cell partially immobilized on the cotton cloth when 4,4'-dithiodipyridine was present during the growth phase (Fig. 2). No growth was observed in the fermentation when 4-hydroxymercury benzoic acid was added into the medium (data not shown) while prolonged growth was found in the fermentation with the presence of iodoacetic acid or N-ethylmaleimide in the medium during the growth phase (Fig. 2).

Product yields and productivities of fermentation by *R. oryzae* in the glucose medium with the presence of *ADH* inhibitors during cultivation are given in Table 1. Similar cell, lactate, and ethanol yields were obtained from fermentations with 1,2-diazole or 2,2,2-trifluoroethanol as compared with the fermentation without inhibitor during the growth phase while more cell biomass with less lactic acid and ethanol production was obtained with the presence of 4,4'-dithiodipyridine in the growth medium. Interestingly, although *ADH* inhibitor, including 1,2-diazole, 2,2,2-trifluoroethanol, or 4,4'-dithiodipyradine, was initially present since the growth phase, *R. oryzae* still produced ethanol as a major product during this phase. This was perhaps due to the affinity of these inhibitors to *ADH* at the concentration studied could not compete with the enzyme-substrate affinity under the operating conditions during the growth phase [29]. Moreover, yeast extract present in the growth medium for promoting spore germination and initial cell growth favored ethanol production simultaneously [30].

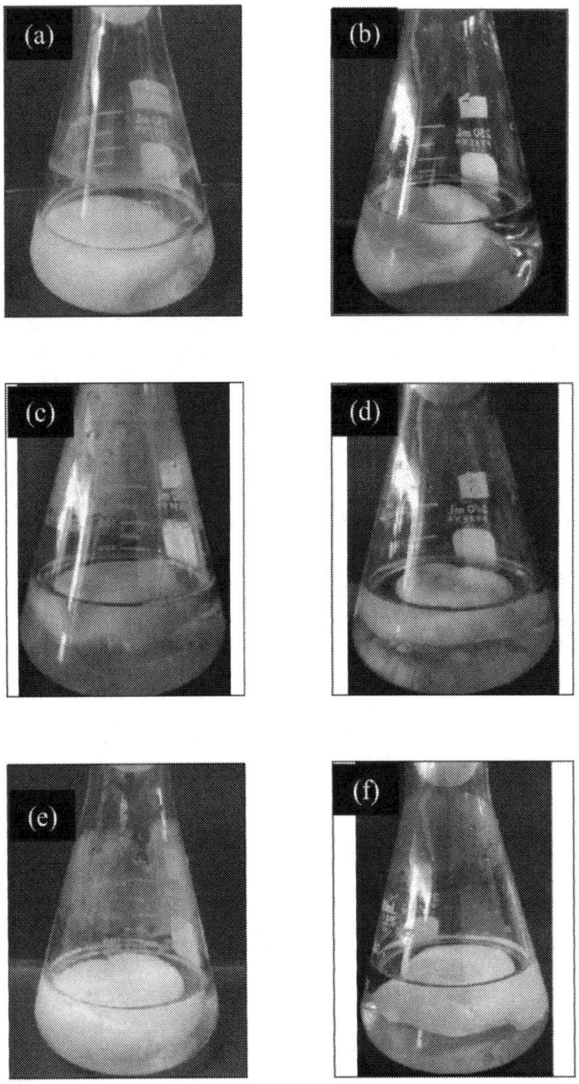

Fig. 1 Morphology of immobilized R. oryzae on cotton towel grown in the medium (a) without the inhibitor at 24-h growth phase; b without the inhibitor at 24-h production phase; c containing 1 mM 1,2-diazole at 24-h growth phase; d containing 1 mM 1,2-diazole at 24-h production phase; e containing 1 mM 2,2,2-trifluoroethanol at 24-h growth phase; and f containing 1 mM 2,2,2-trifluoroethanol at 24-h production phase

Except for the fermentations containing Iodoacetic acid, N-ethylmaleimide, 4,4'-dithiodipyridine, or 4-hydroxymercury benzoic acid, after 48 h culture in the growth medium containing *ADH* inhibitor, the broth was discarded and replaced with the freshly new production medium containing the inhibitor (1 mM). During the production phase, almost similar cell yields from the fermentation containing 1,2-diazole or 2,2,2-trifluoroethanol in the medium to that from the fermentation without the inhibitor were obtained. It was evident that 1,2-diazole and 2,2,2-trifluoroethanol did not greatly affect the fermentation productivities as observed from the similar values to those of the fermentation without the inhibitor along the cultivation period. Only slightly increasing lactate productivity was observed when 2,2,2-trifluoroethanol was present during the production phase. For the prolonged culture in the growth medium containing N-ethylmaleimide or 4,4'-dithiodipyridine, almost twice higher cell and ethanol yields as well as the productivities were observed. *R. oryzae* growth and product formation from fermentations with 6 *ADH* inhibitors compared with those without the inhibitor can be summarized as follows:

Cell: PCMB < IAA < NEM < CT = PZ = TFE < DSDP
Lactate: PCMB < IAA < DSDP < NEM < CT < PZ < TFE
Ethanol: PCMB < IAA < CT = PZ = TFE < NEM < DSDP

Remark: the strongest inhibiting effects of the tested *ADH* inhibitor towards the strongest promoting effects on cell, lactate, and ethanol production by the inhibitor were given from the minimum to the maximum (from the left to the right). Control fermentation using the simple medium was also compared along with the fermentations with different *ADH* inhibitors in the medium.

To prevent the negative effect of *ADH* inhibitors that was found particularly in the fermentations containing Iodoacetic acid, N-ethylmaleimide, or 4-hydroxymercury benzoic acid, on initial cell growth during the growth phase, *R. oryzae* was pregrown in the simple growth medium for 48 h. During the production phase, product yields and productivities correspondent to the regulation of *ADH*s and *LDH* by 6 inhibitors were observed. Slightly increasing lactate and ethanol yields to those of the fermentation without inhibitor were observed in the fermentation containing 1,2-diazole or 2,2,2-trifluoroethanol while the similar cell biomass yield was obtained (Table 2). It was found that 1,2-diazole and 2,2,2-trifluoroethanol did not greatly influence the fermentation productivity during the production phase. On the other hand, Iodoacetic acid, N-ethylmaleimide, 4,4'-dithiodipyridine, and 4-hydroxymercury benzoic acid strongly repressed *R. oryzae* metabolism as observed from low product yields and productivities. The effects of 6 *ADH* inhibitors on pregrown *R. oryzae* metabolism during the production phase can be summarized as follows:

Cell: DSDP < NEM < PCMB < CT = PZ = TFE = IAA
Lactate: PCMB < IAA = NEM < DSDP < CT < PZ < TFE
Ethanol: IAA < DSDP < NEM < PCMB < CT < TFE < PZ

Remark: the strongest inhibiting effects of the tested *ADH* inhibitor towards the strongest promoting effects on cell, lactate, and ethanol production by the inhibitor were given from the minimum to the maximum (from the left to the right). Control fermentation using the simple medium was also compared along with the fermentations with different *ADH* inhibitors in the medium.

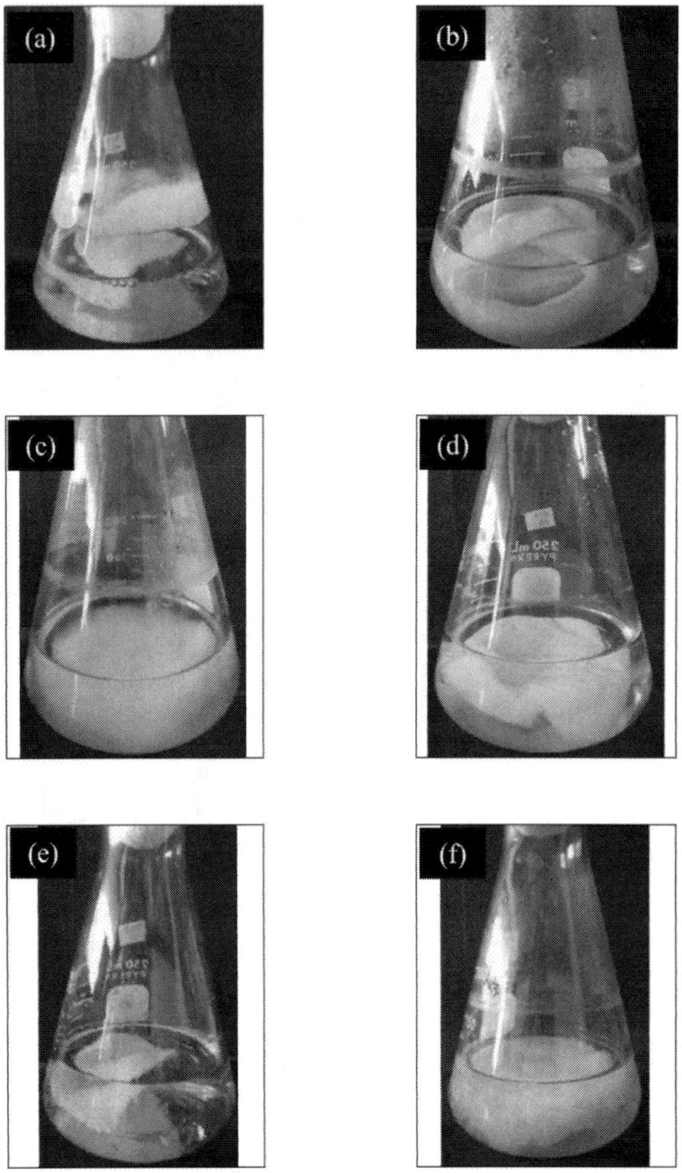

Fig. 2 Appearance of *R. oryzae* with the presence of different *ADH* inhibitors (a) free cells grown in the medium containing 1 mM 4,4'-dithiodipyridine at 24 h growth phase; (b) prolonged free cell growth in the medium containing 1 mM 4,4'-dithiodipyridine at 120 h growth phase; (c) growth in the medium containing 1 M Iodoacetic acid at 24 h growth phase; (d) growth in the medium containing 1 mM Iodoacetic acid at 24 h production phase; (e) growth in the medium containing 1 mM *N*-ethylmaleimide at 48 h growth phase; (f)

Strong inhibitory effects of Iodoacetic acid, N-ethylmaleimide, 4,4'-dithiodipyridine, and 4-hydroxymercury benzoic acid on ethanol production could be described from the irreversible binding of such inhibitors to the sulfhydryl group at the active site of *ADH* so that it could not serve as a ligand for the zinc atom at the catalytic site [31]. Iodoacetic acid and N-ethylmaleimide irreversibly alkylated the S-S bond at the active site; therefore, preventing correct conformation for substrate binding [32]. 4,4'-dithiodipyridine also irreversibly bound with the sulfhydryl group at the active site while Hg(II) complexation with the S-S bond at the active site was found with the presence of 4-hydroxymercury benzoic acid [24,31].

In contrary to Iodoacetic acid, N-ethylmaleimide, 4,4'-dithiodipyridine, and 4-hydroxymercury benzoic acid, 1,2-diazole (cofactor analog) and 2,2,2-trifluoroethanol (substrate/product analog) reversibly bound to the enzyme; therefore, the inhibition was competitive. Although 1,2-diazole and 2,2,2-trifluoroethanol did not show significant reduction of ethanol formation but they gave the little improvement in lactic acid production when compared with the fermentation without the inhibitor. It was believed that ethanol production could be completely inhibited when the appropriate amount of 1,2-diazole or 2,2,2-trifluoroethanol present in the medium; therefore, improved lactic acid production could be achieved.

1,2-diazole and 2,2,2-trifluoroethanol as the competitive ADH inhibitors

Competitive inhibition by 1,2-diazole or 2,2,2-trifluoroethanol was observed in the fermentations with different concentrations (0.01 mM – 10 mM) of these 2 inhibitors. During the growth phase, it was found that varied concentrations of 1,2-diazole did not significantly influence the fungal metabolism as observed from the similar yields of cell biomass (~0.04-0.07 g/g glucose), lactate (~0.08-0.12 g/g glucose), and ethanol (~0.36-0.42 g/g glucose) (Table 3). But increasing lactic acid yield with decreasing ethanol yield was observed during the production phase when increasing 1,2-diazole concentration from 0 mM to 0.1 mM and 2,2,2-trifluoroethanol concentration from 0 mM to 1 mM. During the production phase, the highest lactic acid yield of 0.44 g/g glucose was observed from the fermentation with 0.1 mM 1,2-diazole. Along the cultivation, almost similar cell biomass, lactate, and ethanol productivities were found in the fermentations with 1,2-diazole and 2,2,2-trifluoroethanol at different concentrations when compared with those in the fermentation without the inhibitor (Table3).

Table 1 Product yields and productivities from fermentations by *R. oryzae* in the glucose medium with the presence of *ADH* inhibitors at 1 mM during both growth and production phases.

Inhibitors	Growth phase						Production phase					
	Product yield (g/g glucose)			Productivity (g/L·h)			Product yield (g/g glucose)			Productivity (g/L·h)		
	Cell	Lactate	Ethanol	Cell	Lactate	Ethanol	Cell	Lactate	Ethanol	Cell	Lactate	Ethanol
CT	0.06±0.03	0.09±0.01	0.37±0.03	0.07±0.01	0.11±0.00	0.45±0.06	0.03±0.02	0.34±0.01	0.12±0.01	0.02±0.01	0.40±0.00	0.14±0.12
PZ(1)	0.06±0.01	0.10±0.01	0.36±0.03	0.07±0.01	0.12±0.00	0.43±0.07	0.02±0.00	0.37±0.01	0.14±0.02	0.02±0.00	0.38±0.01	0.15±0.02
TFE(1)	0.06±0.00	0.12±0.01	0.38±0.03	0.07±0.00	0.14±0.01	0.45±0.05	0.01±0.00	0.43±0.01	0.16±0.06	0.02±0.00	0.44±0.01	0.17±0.08
IAA(1)	0.01±0.00	0	0	0	0	0	0	0.01±0.00	0.06±0.00	0	0.01±0.00	0.04±0.00
PCMB(1)	0	0	0	0	0	0	0	0	0	0	0	0
NEM(1)	0	0	0	0	0	0	0.05±0.00	0.15±0.00	0.27±0.00	0.04±0.00	0.16±0.00	0.26±0.00
DSDP(1)	0.08±0.03	0.06±0.00	0.25±0.00	0.03±0.00	0	0.07±0.00	0.04±0.00	0.09±0.00	0.29±0.00	0.06±0.00	0.10±0.00	0.21±0.00

Remarks: CT = control; PZ = 1,2-diazole; TFE = 2,2,2-trifluoroethanol; PCMB = 4-hydroxymercury benzoic acid; NEM = N-ethylmaleimide; DSDP = 4,4'-dithiodipyridine; IAA = Iodoacetic acid; (1) = concentration of the inhibitor at 1 mM present in the medium

Table 2 Product yields and productivities during the production phase with the presence of *ADH* inhibitors at the concentration of 1 mM of pregrown *R. oryzae* in the simple glucose medium.

Inhibitors	Production phase					
	Product yield (g/g glucose)			Productivity (g/L·h)		
	Cell	Lactate	Ethanol	Cell	Lactate	Ethanol
CT	0.05±0.01	0.34±0.01	0.12±0.01	0.02±0.01	0.40±0.00	0.14±0.01
PZ(1)	0.05±0.03	0.36±0.08	0.16±0.01	0.01±0.02	0.36±0.05	0.16±0.00
TFE(1)	0.05±0.02	0.37±0.04	0.15±0.05	0.03±0.00	0.40±0.03	0.16±0.05
IAA(1)	0.05±0.00	0.02±0.00	0.02±0.00	0.01±0.00	0	0
PCMB(1)	0.04±0.00	0	0.08±0.00	0.01±0.00	0	0.01±0.00
NEM(1)	0.01±0.00	0.02±0.00	0.06±0.00	0.02±0.00	0.003±0.00	0.02±0.00
DSDP(1)	0	0.11±0.00	0.04±0.00	0.01±0.00	0.12±0.00	0.01±0.00

Remarks:

- In the growth phase, *R. oryzae* was pregrown in the simple growth medium containing 50 g/L glucose and 5 g/L yeast extract for 48 h at 30°C and 200 rpm
- CT = control; PZ = 1,2-diazole; TFE = 2,2,2-trifluoroethanol; PCMB = 4-hydroxymercury benzoic acid; NEM = N-ethylmaleimide; DSDP = 4,4'-dithiodipyridine; IAA = Iodoacetic acid; (1) = concentration of the inhibitor at 1 mM present in the medium

The effect of 1,2-diazole and particularly 2,2,2-trifluoroethanol on lactate yield was more evident when they were added during the production phase of pregrown *R. oryzae* (Table 4). The highest lactate yield of 0.47 g/g glucose was found at the lowest 2,2,2-trifluoroethanol concentration of 0.01 mM used in this work which was about 38% increasing yield compared with that of the control fermentation (0.34 g/g glucose). Increasing 1,2-diazole concentration in the production medium slightly influenced product yields. Nevertheless, ethanol (~0.12-0.19 g/g glucose) and cell (~0.04-0.05 g/g glucose) yields from the fermentations with 1,2-diazole or 2,2,2-trifluoroethanol were not lower than those from the simple glucose fermentation as expected. It was found that productivities during the production phase were slightly affected by varying the concentration of *ADH* inhibitor (Table 4). The highest lactate productivity of 0.43 g/L·h was achieved when 0.01 mM 2,2,2-trifluoroethanol was present in the production medium.

It was expected that lower ethanol yield while increasing lactic acid yield should have been achieved from fermentations when 1,2-diazole or 2,2,2-trifluroethanol were present; however, the kinetics data revealed that the in vivo regulation of lactic acid and ethanol production by these 2 inhibitors during cultivation of immobilized *R. oryzae* was not clearly seen. It could be explained that the presence of these 2 inhibitors in the living *R. oryzae* culture with complex enzyme system caused metabolic shift not only at the targeted alcohol fermentation pathway as previously expected. Better understanding in metabolic flux balance when these 2 inhibitors are present in the living *R. oryzae* culture is required in order to achieve limited ethanol production and enhanced lactic acid production.

Table 3 Product yields and productivities from fermentations by *R. oryzae* in the glucose medium containing 1,2-diazole or 2,2,2-trifluoroethanol at different concentrations during both growth and production phases.

Inhibitors	Growth phase						Production phase					
	Product yield (g/g glucose)			Productivity (g/L·h)			Product yield (g/g glucose)			Productivity (g/L·h)		
	Cell	Lactate	Ethanol	Cell	Lactate	Ethanol	Cell	Lactate	Ethanol	Cell	Lactate	Ethanol
CT	0.06±0.03	0.09±0.01	0.37±0.03	0.07±0.01	0.11±0.00	0.45±0.06	0.03±0.02	0.34±0.01	0.12±0.01	0.02±0.01	0.40±0.00	0.14±0.12
PZ(0.01*)	0.07±0.00	0.11±0.01	0.42±0.07	0.07±0.01	0.13±0.03	0.46±0.03	0.08±0.05	0.36±0.15	0.13±0.04	0.04±0.02	0.33±0.15	0.12±0.04
TFE(0.01)	0.04±0.04	0.09±0.03	0.39±0.03	0.07±0.01	0.10±0.04	0.41±0.00	0.06±0.07	0.32±0.14	0.13±0.02	0.01±0.00	0.36±0.11	0.15±0.00
PZ(0.1)	0.05±0.01	0.11±0.01	0.40±0.05	0.07±0.01	0.13±0.02	0.49±0.08	0.02±0.01	0.44±0.10	0.18±0.04	0.03±0.00	0.44±0.07	0.18±0.02
TFE(0.1)	0.06±0.00	0.09±0.02	0.40±0.03	0.07±0.00	0.12±0.02	0.49±0.01	0.04±0.03	0.39±0.02	0.17±0.01	0.04±0.04	0.43±0.09	0.18±0.00
PZ(1)	0.06±0.01	0.10±0.00	0.36±0.03	0.07±0.01	0.12±0.00	0.43±0.07	0.02±0.00	0.37±0.01	0.14±0.02	0.02±0.00	0.38±0.01	0.15±0.02
TFE(1)	0.06±0.00	0.12±0.01	0.38±0.03	0.07±0.00	0.14±0.01	0.45±0.05	0.01±0.00	0.43±0.01	0.16±0.06	0.02±0.00	0.44±0.01	0.17±0.08
PZ(10)	0.05±0.00	0.11±0.02	0.41±0.02	0.06±0.01	0.12±0.02	0.49±0.06	0.01±0.01	0.32±0.07	0.19±0.01	0.01±0.01	0.33±0.05	0.19±0.00
TFE(10)	0.06±0.01	0.08±0.01	0.39±0.10	0.06±0.01	0.09±0.03	0.42±0.02	0.02±0.02	0.33±0.03	0.16±0.03	0.02±0.02	0.29±0.02	0.14±0.05

Remarks: CT = control; PZ = 1,2-diazole; TFE = 2,2,2-trifluoroethanol; (*) = concentration of the inhibitor in mM present in the medium

Table 4 Product yields and productivities during the production phase with the presence of various concentrations of 1,2-diazole or 2,2,2-trifluoroethanol of pregrown *R. oryzae* in the simple glucose medium.

Inhibitors	**Production phase**					
	Product yield (g/g glucose)			**Productivity (g/L·h)**		
	Cell	**Lactate**	**Ethanol**	**Cell**	**Lactate**	**Ethanol**
CT	0.05±0.01	0.34±0.01	0.12±0.01	0.02±0.01	0.40±0.00	0.14±0.01
PZ(0.01*)	0.04±0.02	0.37±0.08	0.14±0.07	0.01±0.01	0.37±0.06	0.15±0.08
TFE(0.01)	0.04±0.02	0.47±0.03	0.15±0.03	0.02±0.01	0.43±0.00	0.14±0.02
PZ(0.1)	0.04±0.03	0.38±0.04	0.16±0.01	0.01±0.01	0.37±0.06	0.16±0.01
TFE(0.1)	0.04±0.03	0.42±0.01	0.19±0.03	0.04±0.03	0.42±0.04	0.19±0.02
PZ(1)	0.05±0.03	0.36±0.08	0.16±0.01	0.01±0.02	0.36±0.05	0.16±0.00
TFE(1)	0.05±0.02	0.37±0.04	0.15±0.05	0.03±0.00	0.40±0.03	0.16±0.05
PZ(10)	0.04±0.04	0.34±0.01	0.19±0.06	0.07±0.06	0.35±0.00	0.19±0.05
TFE(10)	0.04±0.04	0.32±0.00	0.13±0.01	0.02±0.00	0.34±0.02	0.13±0.01

Remarks:

- In the growth phase, *R. oryzae* was pregrown in the simple growth medium containing 50 g/L glucose and 5 g/L yeast extract for 48 h at 30°C and 200 rpm
- CT = control; PZ = 1,2-diazole; TFE = 2,2,2-trifluoroethanol; (*) = concentration of the inhibitor in mM present in the medium

In vivo effects of ADH inhibitors on enzyme activities during lactic acid fermentation

Fig. 3 represents changes in *LDH* and *ADH* activities throughout the cultivation period when 1,2-diazole or 2,2,2-trifluoroethanol were present in both growth and production phases. During the growth phase (0-48 h), enzyme activities increased with the increasing time. The enzyme activities continuously dropped during the production phase (48-120 h). The highest activities were found at 48 h cultivation. Similar results were found by Chotisubha-anandha and coworkers [33]. They observed the increase in the activity of lactate dehydrogenase with the increasing fermentation time (from 0 h to 48 h). After that the activity dropped due to glucose depletion. When 2,2,2-trifluoroethanol was present, the activities of *LDH, ADH(f), and ADH(b)* were dramatically increased while similar activities were found in fermentation with 1,2-diazole as compared with those in the fermentation in the simple glucose medium. The chemical structure of 2,2,2-trifluoroethanol is analogous to ethanol. This might lead to the increasing enzyme expression level due to cellular response to the substrate/product analog. While 1,2-diazole competed with the cofactors NAD^+/NADH. Low enzyme activities were found in fermentations with the presence of Iodoacetic acid, N-ethylmaleimide, 4,4'-dithiodipyridine, and 4-hydroxymercury benzoic acid along the cultivation period (data not shown).

Fig. 3(a)

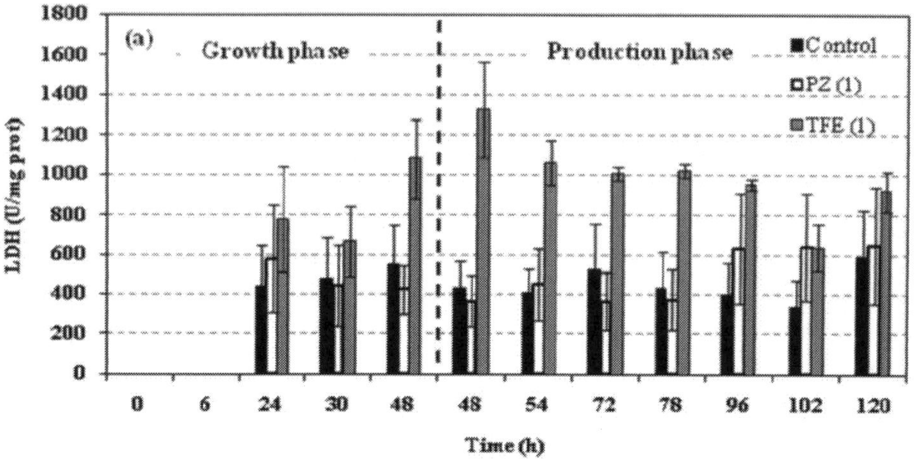

Fig. 3(b)

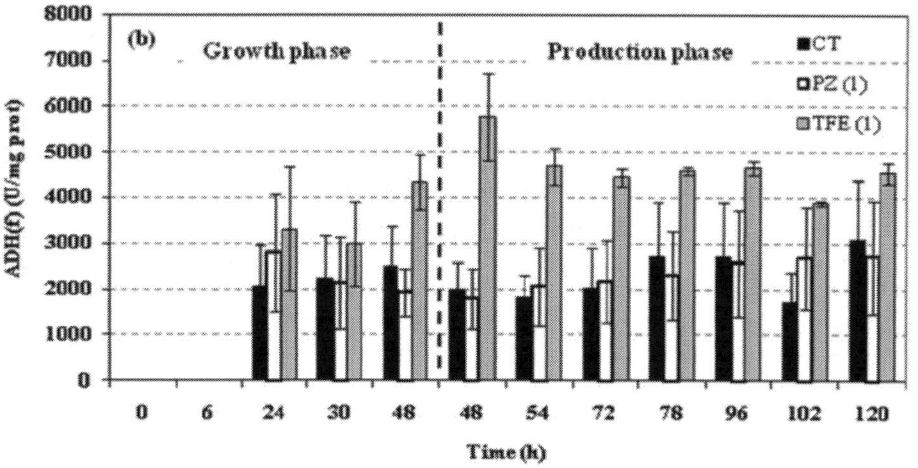

Fig. 3(c)

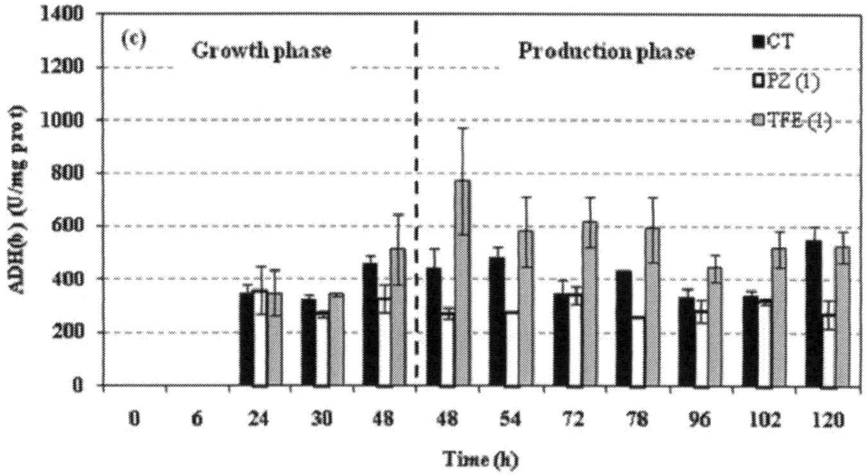

Fig. 3 Enzyme activities during the fermentation by *R. oryzae* in the glucose medium containing 1,2-diazole (PZ) or 2,2,2-trifluoroethanol (TFE) at 1 mM; (a) lactate dehydrogenase (*LDH*); (b) alcohol dehydrogenase (*ADH(f)*) - acetaldehyde to ethanol; and (c) alcohol dehydrogenase (*ADH(b)*) - ethanol to acetaldehyde.

Similar trends of changes in enzyme activities were found in the fermentations of pregrown *R. oryzae* compared with those fermentations when the inhibitors was added since the growth phase and that without the inhibitor. First, *R. oryzae* was pregrown in the simple glucose medium during the growth phase. Later in the production phase the inhibitor was added into the medium. Fig. 4 shows the enzyme activities at 48 h cultivation. No enzyme activity was detected when Iodoacetic acid, N-ethylmaleimide, 4,4'-dithiodipyridine, or 4-hydroxymercury benzoic acid were added into the production medium. While the fermentation with 2,2,2-trifluoroethanol led to the increasing *LDH* and *ADH*s activities. The presence of 1,2-diazole in the medium seemed not to affect enzyme expression (Fig. 4). As mentioned in previous studies, the binding of Iodoacetic acid, N-ethylmaleimide, 4,4'-dithiodipyridine, or 4-hydroxymercury benzoic acid to the enzyme active site are irreversible [23-25]. Therefore, the presence of these 4 inhibitors during cultivation of living *R. oryzae* prevented further enzyme-substrate binding. This eventually led to no enzyme activity found when performing the enzyme assays of the mycelial extract later.

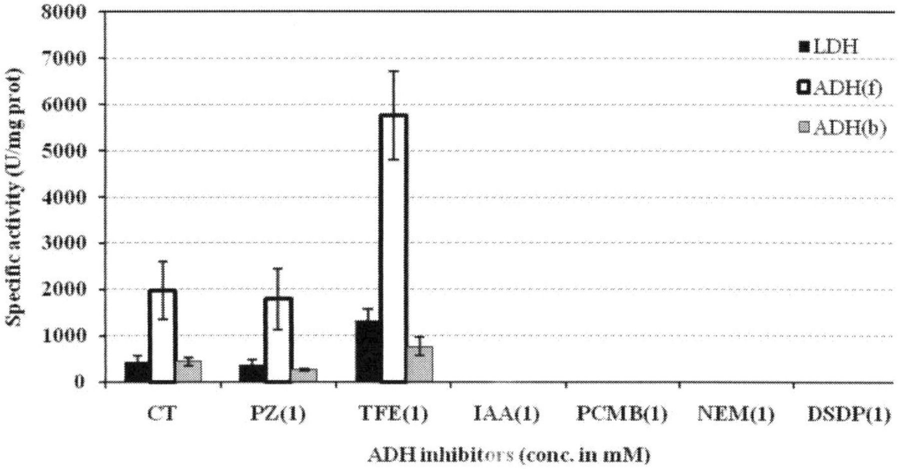

Fig. 4 Activities of lactate dehydrogenase (*LDH*) and alcohol dehydrogenase (both *ADH(f)* and *ADH(b)*) at 48 h. During growth and production phases, *R. oryzae* was cultivated in the medium containing various *ADH* inhibitors at 1 mM.

It is obvious that the addition of the inhibitor at an early stage in fermentation caused metabolic repression (Figs. 4 and 5). When the inhibitor was added only during the production phase, higher expression levels were found compared with those with the inhibitors since the growth phase (Figs. 4 and 5). Almost 50 times increased *LDH* activity was observed during the production phase with the presence of 2,2,2-trifluoroethanol of *R. oryzae* culture that was pregrown in the simple glucose medium. Although *ADH(f)* activity was lower than *LDH* activity in the fermentations containing the inhibitors, it was still higher than that found in the control fermentation. The increasing enzyme activity when the inhibitor was present only in the production phase suggested that addition of *ADH* inhibitor during the production phase is sufficient for improving lactic acid production. Also, the chance of cell damage by adding the inhibitor only in the production phase can be prevented during the earlier growth stage.

In this study, the inhibition of *ADH* by the inhibitors which corresponded with the lower *ADH* activity and eventually led to less ethanol produced during cultivation of *R. oryzae* was expected. But in the living cells, addition of the potential *ADH* inhibitors led to the highly increasing specific enzyme activities of *LDH* and *ADH*s compared to the specific activities observed in the glucose fermentation without the inhibitor (control). These results were presumably explained by the influence of the inhibitor added itself as well as the effects of other environmental factors such as local nutrient/oxygen transport. For example, for all cases (with or without the inhibitors) at some local area likely at the central core of the immobilized cell matrix, oxygen might be limited; thus, *R. oryzae* cells in such area readily entered alcohol fermentative pathway. With the presence of the inhibitor, some *ADH*s produced were bound with the inhibitor reversibly or irreversibly depending on the type of inhibition. Presumably, cells responded to this phenomenon by increasing *ADH* expression. As a consequence, largely increasing *ADH* activities were observed as compared to those in

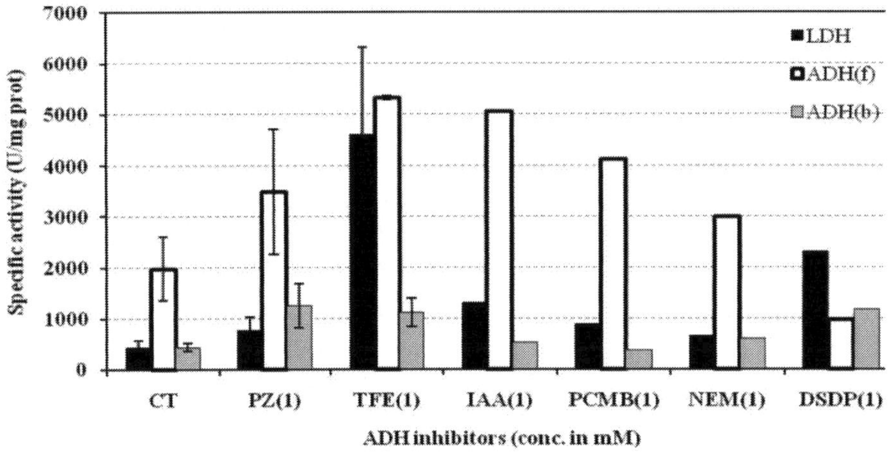

Fig. 5 Activities of lactate dehydrogenase (*LDH*) and alcohol dehydrogenase (both *ADH(f)* and *ADH(b)*) at 48 h with various *ADH* inhibitors at 1 mM. During the growth phase, *R. oryzae* was pregrown at the simple glucose medium at 30°C, 200 rpm.

the control experiment. Increasing *LDH* specific activity was also observed. This might be explained by the metabolic shift at pyruvate branch point when the *ADH* inhibitor was present because *ADH* inhibitors controlled flows through the metabolic pathways. The concentrations of these inhibitors were the regulating signals which govern the activities of the controlling enzymes at the pyruvate branch point.

Conclusion

This study reveals that ethanol production dominated during the growth phase while lactic acid production took over during the production phase. The presence of *ADH* inhibitors earlier since the growth phase gave strong inhibition of cell growth and metabolism of *R. oryzae*. Slightly increasing lactic acid production but significant increasing *LDH* activity was obtained in fermentations with 2,2,2-trifluoroethanol during the production phase. In addition, *ADH* activities in the fermentation with 2,2,2-trifluoroethanol were dramatically increased but the similar ethanol production to the fermentation without the inhibitor was obtained. It is expected that with the presence of *ADH* inhibitor, less pyruvate shifts toward ethanol production; thus, an increase in lactic acid or cell biomass production should have been observed. However, the results shown in this study implied that in vivo inhibition of *ADH* by adding the inhibitors during fermentation did not only alter the conversion of acetaldehyde to ethanol but it also shifted the pyruvate flux through other related pathway.

Acknowledgments

This work has been supported by Thailand Research Fund and the Graduate School of Chulalongkorn University via the TRF Master Research Grant (MRG-WII51E053). Partial funding from the National Research University Project of Commission on Higher Education (CHE) and the Ratchadapiseksomphot Endowment Fund (AM007I) is also acknowledged. The authors appreciated US Department of Agriculture for providing the fungal strain used in this study.

References

1. Wang, L., Zhao, B., Liu, B., Yu, B., Ma, C., Su, F., Hua, D., Li, Q., Ma, Y. and Xu, P. (2010). Efficient production of L-lactic acid from corncob molasses, a waste by-product in xylitol production, by a newly isolated xylose utilizing *Bacillus* sp. strain. *Bioresource Technology*, 101, 7908-7915.
2. Thongchul, N. and Yang, S.-T. (2003). Controlling filamentous fungal morphology by immobilization on a rotating fibrous matrix to enhance oxygen transfer and L(+)-lactic acid production by *Rhizopus oryzae*. In *Fermentation Biotechnology*, ed. B.C. Saha. Oxford University Press, MA, 2003, pp. 36-51.
3. Wang, L., Zhao, B., Li, F., Xu, K., Ma, C., Tao, F., Li, Q. and Xu, P. (2010). Highly efficient production of D-lactate by Sporolactobacillus sp. CASD with simultaneous enzymatic hydrolysis of peanut meal. *Applied Microbiology and Biotechnology* (*in press*) DOI: 10.1007/s00253-010-2904-9.
4. Hofvendahl, K. and Hahn-Hagerdal, B. (2000). Factor affecting the fermentative lactic acid production from renewable resources. *Enzyme and Microbial Technology*, 26, 87-107.
5. Wang, L., Zhao, B., Liu, B., Yang, C., Yu, B., Li, Q., Ma, C., Xu, P. and Ma, Y. (2010). Efficient production of L-lactic acid from cassava powder by *Lactobacillus rhamnosus*. *Bioresource Technology*, 101, 7895-7901.
6. Hsieh, C.M., Yang, F.-C. and Iannotti, E.L. (1999). The effect of soy protein hydrolyzates on fermentation by *Lactobacillus amylovorus*. *Process Biochemistry*, 34, 173-179.
7. Silva, E.M. and Yang, S.-T. (1995). Continuous production of lactic acid from acid whey by *Lactobacillus helveticus* in a fibrous-bed bioreactor. *Journal of Biotechnology*, 41, 59-70.
8. Tsai, S.P. and Moon, S.-H. (1998). An integrated bioconversion process for production of L-lactic acid from starchy potato feedstocks. *Applied Biochemistry and Biotechnology*, 70-72, 417-428.
9. Woiciechowski, A.L., Soccol, C.R., Ramos, L.P. and Pandey, A. (1999). Experimental design to enhance the production of L-(+)-lactic acid from steam-exploded wood hydrolysate using *Rhizopus oryzae* in a mixed-acid fermentation. *Process Biochemistry*, 34, 949-955.
10. Zhou, Y., Dominguez, J.M., Cao, N., Du, J. and Tsao, G.T. (1999). Optimization of L-lactic acid production from glucose by *Rhizopus oryzae* ATCC52311. *Applied Biochemistry and Biotechnology*, 77-79, 401-407.

11. Cui, Y.Q., van der Lans, R.G.J.M. and Luyben, K.C.A.M. (1997). Effect of agitation intensities on fungal morphology of submerged fermentation. *Biotechnology and Bioengineering*, 55, 715-726.
12. Cui, Y.Q., van der Lans, R.G.J.M. and Luyben, K.C.A.M. (1998). Effects of dissolved oxygen tension and mechanical forces on fungal morphology in submerged fermentation. *Biotechnology and Bioengineering*, 57, 409-419.
13. Skory, C.D., Freer, S.N. and Bothast, R.J. (1998). Production of L-lactic acid by *Rhizopus oryzae* under oxygen limiting conditions. *Biotechnology Letters*, 20, 191-194.
14. Bai, F., Wang, L., Huang, H., Xu, J., Caesar, J., Ridgway, D., Gu, T. and Moo-Young, M. (2001). Oxygen mass-transfer performance of low viscosity gas-liquid-solid system in a split-cylinder airlift bioreactor. *Biotechnology Letters*, 23, 1109-1113.
15. Bai, D.-M., Jia, M.-Z., Zhao, X.-M., Ban, R., Shen, F., Li, X.-G. and Xu, S.-M. (2003). L(+)-lactic acid production by pellet-form *Rhizopus oryzae* R1021 in a stirred tank fermentor. *Chemical Engineering Science*, 58, 785-791.
16. Dong, X.Y., Bai, S. and Sun, Y. (1996). Production of L(+)-lactic acid with *Rhizopus oryzae* immobilized in polyurethane foam cubes. *Biotechnology Letters*, 18, 225-228.
17. Hang, Y.D., Hamamci, H. and Woodams, E.E. (1989). Production of L(+)-lactic acid by *Rhizopus oryzae* immobilized in calcium alginate gels. *Biotechnology Letters*, 11, 119-120.
18. Park, E.Y., Kosakai, Y. and Okabe, M. (1998). Efficient production of L-(+)-lactic acid using mycelial cotton-like flocs of *Rhizopus oryzae* in an air-lift bioreactor. *Biotechnology Progress*, 14, 699-704.
19. Sun, Y., Li, Y.-L. and Bai, S. (1999). Modeling of continuous L(+)-lactic acid production with immobilized *R. oryzae* in an airlift bioreactor. *Biochemical Engineering Journal*, 3, 87-90.
20. Tamada, M., Begum, A.A. and Sadi, S. (1992). Production of L(+)-lactic acid by immobilized cells of *Rhizopus oryzae* with polymer supports prepared by gamma-ray-induced polymerization. *Journal of Fermentation Bioengineering*, 74, 379-383.
21. Yang, C.W., Lu, Z. and Tsao, G.T. (1995). Lactic acid production by pellet-form *Rhizopus oryzae* in a submerged system. *Applied Biochemistry and Biotechnology*, 51-52, 57-71.
22. Yin, P., Yahiro, K., Ishigaki, T., Park, Y. and Okabe, M. (1998). L(+)-lactic acid production by repeated batch culture of *Rhizopus oryzae* in air-lift bioreactor. *Journal of Fermentation and Bioengineering*, 85, 96-100.
23. Taber, R.L. (1998). The competitive inhibition of yeast alcohol dehydrogenase by 2,2,2-trifluoroethanol. *Biochemical Education*, 26, 239-242.
24. Zheng, S.-Y., Xu, D., Wang, H.-R., Li, J. and Zhou, H.-M. (1997). Kinetics of irreversible inhibition of yeast alcohol dehydrogenase during modification by 4,4'-dithiodipyridine. *International Journal of Biological Macromolecules*, 20, 307-313.
25. Yoneya, T. and Sato, Y. (1979). Alcohol dehydrogenase from *Rhizopus javanicus*. *Applied and Environmental Microbiology*, 37, 1073-1078.
26. Lowry, O.H., Rosebrough, N.J., Farr, A.L. and Randall, R.J. (1951). Protein measurement with the Folin phenol reagent. *Journal of Biological Chemistry*, 193, 265-275.
27. Pritchard, G.G. (1971). An NAD^+-independent L-lactate dehydrogenase from *Rhizopus oryzae*. *Biochimica et Biophysica acta*, 250, 25-34.
28. Skory, C.D. (2000). Isolation and expression of lactate dehydrogenase genes from *Rhizopus oryzae*. *Applied and Environmental Microbiology*, 66, 2343-2348.
29. Singlevich, T.E. and Barboriak, J.J. (1971). Pyrazole-induced inhibition of yeast alcohol dehydrogenase. *Biochemical Pharmacology*, 20, 2087-2089.

30. Thongchul, N., Navankasattusas, S. and Yang, S.T. (2010). Production of lactic acid and ethanol by *Rhizopus oryzae* integrated with cassava pulp hydrolysis. *Bioprocess and Biosystems Engineering*, 33, 407-416.
31. Springman, E.B., Angleton, E.L., Birkedal-Hansen, H. and Van Wart, H.E. (1990). Multiple modes of activation of latent human fibroblast collagenase: Evidence for the role of a Cys73 active-site zinc complex in latency and a "cysteine switch" mechanism for activation. *Biochemistry. Proceedings of the National Academy of Sciences. USA*, 87, 364-368.
32. Winslow, J.W. (1981). The reaction of sulfhydryl groups of sodium and potassium ion-activated adenosine triphosphatase with N-ethylmaleimide. The relationship between ligand-dependent alterations of nucleophilicity and enzymatic conformation states. *The Journal of Biological Chemistry*, 256, 9522-9531.
33. Chotisubha-anandha, N., Thitiprasert, S., Tolieng, V. and Thongchul, N. (2011). Improved oxygen transfer and increased L-lactic acid production by morphology control of *Rhizopus oryzae* in a static bed bioreactor. *Bioprocess and Biosystems Engineering (in press*

(626e) Evaluation of Lipid Accumulation From Cellulosic Sugars by Mortierella Isabellina for Biodiesel Production

Zhenhua Ruan[1], Mike Zanotti[1], Xiaoqing Wang[1], Chad Ducey[2] and Yan Liu[1], (1)Biosystems and Agricultural Engineering, Michigan State University, East Lansing, MI, (2)E-Biofuels, Middletown, IN

Mortierella isabellina ATCC42613, a filamentous fungus, was used to assess the lipid production on different carbon sources including glucose, xylose, sulfuric acid and sodium hydroxide treated corn stover enzymatic hydrolysate. The yield of lipid accumulated in fungal biomass kept constant for both glucose (0.41-0.44 g/g biomass) and xylose (0.39-0.44 g/g biomass) as carbon sources with the initial glucose and xylose concentrations varying from 28g/L to 92g/L. The data showed lipid accumulation was independent of carbon to nitrogen ratio from 66 to 309 indicating that this fungus can be cultivated under nitrogen rich medium to increase productivity since microorganisms grow slower in nutrient limited medium. In addition, the lipid conversion yield from xylose was high resulting in a lipid production of 8.8g·L-1, the highest production yet for fungal lipid grown on xylose medium to date. For mixed substrates, diauxic growth was observed as glucose and xylose were consumed subsequently. Lipid accumulation from sulfuric acid and sodium hydroxide treated corn stover enzymatic hydrolysate using a filamentous fungus *Mortierella isabellina* was investigated for the first time. Sulfuric acid pretreated corn stover enzymatic hydrolysate exhibited quite similar biomass and lipid production compared to synthetic medium. While sodium hydroxide pretreated corn stover enzymatic hydrolysates provided a bit inferior performance compared with mimic sugar medium. Lipid profile analysis showed that intracellular fatty acid composition remained relatively constant at each incremental sugar level for both xylose and glucose cultures. The major fatty acids in each culture were C16:0, C18:1, and C18:2. Predicted biodiesel properties such as iodine value, saponification, and higher heating values analyses indicated the consistence of intracellular fatty acid compositions with jatropha, palm, soybean, and sunflower oils. These results elucidate the great possibility of utilization of filamentous fungal fermentation to accumulate lipids from lignocellulosic biomass for biodiesel production.

See more of this Session: Poster Session: Food Engineering and Bioprocessing
See more of this Group/Topical: Food, Pharmaceutical & Bioengineering Division

(626f) Production of Total Lipid and Fatty Acids From Cassava Bagasse Hydrolysate by Heterotrophic Chlorella Protothecoides Under N-Stavation

Dong Wei and Junhui Chen, School of Light Industry and Food Sciences, South China University of Technology, Guangzhou, China

With the rising price of crude oil, biodiesel from microalgae has become attractive and popular as an alternative biofuels in recent years. Heterotrophic fermentation of green microalgae *Chlorella* has been successfully validated for biomass and total lipid production from various biomass hydrolysate. Cassava bagasse is the most abundant byproduct from starch industry in south China and Southeastern Asian countries, consisting of 35-45% residual starch and 16-18% cellulose plus hemicellulose. The goal of this work was to evaluate the feasibility of using cassava bagasse as a cheap feedstock for the production of total lipid and fatty acids by heterotrophic *Chlorella protothecoides* under nitrogen starvation.

In this study, 10% (w/v) superfine powder of cassava bagasse was hydrolyzed by cellulase ACCELLERASE 1500 with dosage of 0.05mL/g biomass at pH 4.8, 50°C for 36 h, and the hydrolysate contained 27.4g/L glucose. The Basal medium containing 20g/L reducing sugar and 1mM urea was prepared from the hydrolysate and used for the culture of *Chlorella* at 28±0.5°C, pH 6.5 and DO>50% in 5L bioreactor, while the control medium contained 20g/L glucose as carbon source. To compare the effect of two media in the batch culture for 8 days, the biomass, total lipid content and the ratio of total fatty acids in lipid reached 6.9g/L, 53.6%DW and 67.4% in the hydrolysate medium, while 7.7g/L, 64.7%DW and 62.3% were obtained in control medium, respectively, resulting a similar productivities of total fatty acids (0.31 and 0.39 g/L/d). For the production of total lipid and fatty acids, repeated fed-batch culture was performed for 15 days with renewable rate at 20% in 5L bioreactor. The fed medium containing 200g/L reducing sugar and 10mM urea was prepared from the hydrolysate and fed repeatedly to keep reducing sugar concentration between 8-20g/L. The final biomass and total lipid content reached 9.95g/L and 58.3%DW, while the ratio of total fatty acids in lipid reached 66.7%, resulting in a lower productivity of total fatty acid reached 0.26 g/L/d. C16 and C18 fatty acids were over 85% of total fatty acid in cells, and the content of C18 fatty acids was increasing during the culture. Several inhibitors including acetic acid, furfural and benzoic acid were found in the hydrolysate by GC-MS, they probably affected the growth of *Chlorella* and lipid accumulation in long-term operation. Our study demonstrated the possibility of using cassava bagasse hydrolysate as carbon source instead of glucose for lipid and fatty acids production by heterotrophic *Chlorella*, and it is worthy of further study for the large-scale production.

See more of this Session: Poster Session: Food Engineering and Bioprocessing

Modeling the Structure and Transition Kinetics of Small, Aggregation-Prone Proteins: Identifying Key Atomic Interactions via Efficient Simulation Methods

Camilo Velez-Vega, Fernando Escobedo & Ernesto Borrero

In this work, several schemes were developed and used to increase the simulation efficiency of two proteins associated with human diseases, and give insights on their structure and the kinetics and mechanism of conformational transitions. Our studies focus on the recognition of particular structural markers that may lead to protein misfolding and aggregation. Two systems are studied, one where the main challenge is to characterize the varied structural motifs in a relatively floppy protein (amyloid β-42), and the other where distinct conformers exist and the main challenge is to elucidate the transition mechanism (insulin chain B).

Concerning structural identification, the conformational behavior of the wild-type amyloid β-42 (Aβ-42) monomer and two of its mutants was explored by a novel technique based on the replica exchange method, to identify structural features that may promote or deter early-stage oligomerization in water. The markers used for this purpose indicate that while the three peptides are relatively flexible they have distinct preferential structures and degree of rigidity. In particular, we found that one mutant that remains in the monomeric state in experiments displays a characteristic N-terminal structure that significantly enhances its rigidity. This finding is consistent with various studies that have detected a reduction in oligomerization frequency and Aβ-related toxicity upon sequence-specific antibody or ligand binding to the N-terminal tail of wild-type monomers, likely leading to the stabilization of this region. In general, our results highlight a potential role of the N-terminal segment on Aβ oligomerization and give insights into specific interactions that may be responsible for promoting the pronounced structural changes observed upon introducing point mutations on the wild-type Aβ-42 peptide.[1]

Regarding kinetic and mechanistic characterization, an optimized Forward Flux Sampling Method is applied to the study of the unfolding transitions for insulin chain B monomer in explicit solvent at physiological conditions. Specifically, the simulations provide valuable insights on the biological relevance of key structural changes in the peptide's N-terminal, for insulin's dissociation and binding mechanisms.

1. Velez-Vega & Escobedo, *J. Phys. Chem. B* 2011, Vol. 115, No. 17, pp. 4900-4910. doi:10.1021/jp1086575

(626h) Media Optimization for Enhanced Propionic Acid Production by Propionibacterium Acidipropionici

Jianxin Sun and Shang-Tian Yang, The Ohio State University, Columbus, OH

Propionic acid is consumed as human food and animal feed preservatives due to its inhibition effect on the growth of mold and some bacteria. Besides petroleum-based chemical synthesis, an alternative way to produce propionic acid is through fermentation, which is safe, sustainable and environmentally benign. Propionibacterium acidipropionici has been the most used species for propionic acid production from sugars. In this study, we compared various carbon sources (glucose, fructose and sucrose) and found no significant differences in propionic acid production. Furthermore, industrial grade substrate and nutrients as sugarcane juice or molasses were used as basal media to lower the feedstock cost. Yields (0.5g/g glucose) are comparable for both sugarcane juice and synthetic medium. Different nitrogen sources were added to the basal media to improve the production of propionic acid. NH_3, ammonium chloride, ammonium sulfate were good candidates for inorganic nitrogen sources, while corn steep liquor (CSL), yeast, yeast extract, for organic nitrogen sources. Minerals and vitamins as Ca-pantothenate, thiamine, biotin were tested to enhance the propionic acid final titer and yield. Salt effect (KCl, NaCl) and propanol inhibition were also studied: 0-80g/L NaCl or 0-100g/L KCl) significantly inhibited cell growth and propionic acid production; and when the propanol in the media is over 25g/L, the inhibition effect became significant.

See more of this Session: Poster Session: Food Engineering and Bioprocessing
See more of this Group/Topical: Food, Pharmaceutical & Bioengineering Division

(626i) Identification of the Cellulase Component That Contributes to the Efficient Saccharification of Cellulose At Low Enzyme Loadings

Makoto Ikeo[1], Yoshiki Ueno[2] and Daisuke Taneda[2], (1)Technology Development Department, JGC Corporation, Higashiibaraki-gun, Ibaraki Prefecture, Japan, (2)Technology Development, JGC Corporation, Higashiibaraki-gun, Ibaraki Prefecture, Japan

At the 2010 AIChE annual meeting, JGC Corporation reported on a method that enables efficient saccharification of cellulosic biomass at low enzyme loadings. In the report, the solution to the following phenomenon, which had been frequently observed in past studies, was presented: "Under a constant substrate loading, the saccharification of cellulose will reach its maximum yield when sufficient enzyme loading is used. However, when the enzyme loading is reduced, the yield of the reaction plateaus before it reaches the maximum yield." This phenomenon will be termed as the "Plateau Phenomenon" in this paper. JGC Corporation demonstrated that the Plateau Phenomenon can be avoided by putting the saccharification reaction under the static condition. When the saccharification of 100ml 10% filter paper slurry was conducted under agitated condition, the resulting glucose concentration was 90g/L at the enzyme loading of 600mg Cellulase SS (Nagase Chemtex). At the enzyme loading of 60mg, the glucose concentration plateaued at 60g/L. However, by putting the reaction under the static condition, the final glucose concentration of 90g/L was achieved even at the enzyme loading of 60mg. This avoidance of the Plateau Phenomenon at lower enzyme loading and under the static condition was defined as the "Static Effect," and the inactivation of the enzyme by agitation was the suspected cause.

In this study, JGC Corporation identified a particular component of the cellulase that causes the Static Effect. This component is drastically inactivated by agitation. When the enzyme solution without any substrate was agitated at 115rpm and a temperature of 50°C, 80% of this component was lost from the solution within the first day, indicating its inactivation. The saccharification of cellulosic biomass was conducted using this "component-inactivated" enzyme solution. As a result, the enzyme solution in which this component is inactivated didn't show a significant difference between the static and agitated condition. In other words, the enzyme solution without this component lost its ability to exhibit the Static Effect. Thus, this component was deduced to have a key role in cellulose saccharification, and it was identified by LC/MS analysis. This component will be defined as the "Key Enzyme" from here on. (The identity of this enzyme will not be given in the abstract since the patent application is currently being filed)

To confirm the effect of the Key Enzyme, the enzyme solution was fractionated using a HiLoad Q Sepharose HP 26/10 column. Some of the fractions obtained consisted mainly of the Key Enzyme. Using these fractions, enzyme solutions with varied ratios of the Key Enzyme to CBH1 was prepared, and saccharification experiments under a constant protein loading and the static condition were conducted. The results showed that, as the ratio of the Key Enzyme against CBH1 increased, the rate and the yield of the reaction increased. Moreover, when only the CBH1 was used for saccharification, the Plateau Phenomenon was observed, but by addition of the Key Enzyme, the Static Effect was observed. When a sufficient amount of the Key Enzyme was added, the Plateau Phenomenon was avoided. These results confirmed that the synergism between the Key Enzyme and CBH1 can overcome the Plateau Phenomenon.

A number of past studies reported the synergism between the individual cellulase components, especially those with CBH1. However, the synergism between CBH1 and the Key Enzyme reported in this study can only be confirmed under the static condition. Thus, in the past studies which were conducted under agitated conditions, the function of the key enzyme and its degree of synergism with CBH1 may have been underrated since the enzyme was easily inactivated by the shear force exerted by agitation.

Further investigation of the role and properties of the Key Enzyme is necessary for effective saccharification of cellulose, and commercialization of the enzymatic saccharification process.

See more of this Session: Poster Session: Food Engineering and Bioprocessing
See more of this Group/Topical: Food, Pharmaceutical & Bioengineering Division

(626j) Production of Fumaric Acid In Fermentation by Rhizopus Oryzae: Process Design and Economic Analysis

Kun Zhang and Shang-Tian Yang, Chemical and Biomolecular Engineering, The Ohio State University, Columbus, OH

The production of fumaric acid from renewable resources by fungal fermentation has received much attention due to its extensive applications in polymer, medicine and food industries and the rapidly increased production cost of petroleum-based chemicals. However, current fungal fermentation process suffers from low product yield and productivity mainly due to difficulties in controlling cell morphology and mass transfer during fermentation in conventional stirred-tank bioreactors. By optimizing the seed culture and fermentation conditions, we achieved a product yield of ~80% (w/w), productivity of ~0.9 g/L·h, and product titer of ~64 g/L from glucose in batch fermentation with *Rhizopus oryzae*. Based on these results, process scale-up design and economic analysis were performed using SuperPro Designer. For a plant with annual production of 30,000 tons fumaric acid, three fermenters each with a volume of ~1,350 m^3 running in staggered mode would be required. The product cost is estimated at $1.36/kg. With the projected capital investment of ~$75 million and annual revenue of $64 million based on the current market price of $2.2/kg for the petroleum-based fumaric acid, the payback time would be 3.57 years (return on investment of 28%). This study suggested that the fermentation route for the bio-based fumaric acid would be economically competitive and could replace current petroleum-based chemical synthesis method.

See more of this Session: Poster Session: Food Engineering and Bioprocessing
See more of this Group/Topical: Food, Pharmaceutical & Bioengineering Division

(626k) Ethanol Production Using "Alkalibaculum Bacchi" Strains CP11, CP13 and CP15 From Syngas

Kan Liu[1], **Hasan K. Atiyeh**[1], Ralph S. Tanner[2], Mark R. Wilkins[1] and Raymond L. Huhnke[1], (1) Biosystems and Agricultural Engineering, Oklahoma State University, Stillwater, OK, (2)Department of Botany and Microbiology, University of Oklahoma, Norman, OK

Ethanol production from syngas components CO and H_2 using "*Alkalibaculum bacchi*" strains CP11, CP13 and CP15 was investigated in bottle fermentations containing 100 mL of yeast extract medium at 37°C and pH 8.0. Two commercial syngas mixtures (Syngas I: 20% CO, 15% CO_2, 5% H_2, 60% N_2) and (Syngas II: 40% CO, 30% CO_2, 30% H_2) were used. Strains CP11, CP13 and CP15 were fed syngas every 24 h for 360 h. Liquid samples were collected to measure pH, optical density (OD) and product concentrations. Gas samples were collected to measure the gas composition in the head space and calculate the product yields and conversion efficiencies of syngas components to ethanol. The results showed that the three strains were able to convert CO and H_2 into ethanol and acetic acid.

Maximum ethanol concentration (1.7 g/L) and yield (61%) were obtained in fermentations with strain CP15 and Syngas II. Strain CP15 produced over twofold more ethanol in Syngas I, compared to strains CP11 and CP13. In addition, CP15 produced 18% and 71% more ethanol than strains CP11 and CP13, respectively, with Syngas II. Strain CP13 produced over two fold more acetic acid than strains CP11 and CP15 in both gas mixtures after 360 h of fermentation. These results showed the potential of the new "*Alkalibaculum bacchi*" strains CP11, CP13 and CP15 to produce ethanol from CO and H_2.

See more of this Session: Poster Session: Food Engineering and Bioprocessing
See more of this Group/Topical: Food, Pharmaceutical & Bioengineering Division

(626l) Comparative Analysis On the Cytoplasmic and Membrane Proteomes of Clostridium Acetobutylicum Revealed the Molecular Basis for Butanol Tolerance

Yin Li, Shaoming Mao and Yanping Zhang, Institute of Microbiology, Chinese Academy of Sciences, Beijing, China

The solventogenic bacterium *Clostridium acetobutylicum* is the most important species of *Clostridium* used in the fermentation industry. However, the intolerance to butanol hampers the efficient production of solvents. Butanol toxicity has been attributed to the effect on cellular proteins, but the molecular mechanism underpinning butanol tolerance remains unclear. Here we report a systemic comparative analysis on the cytoplasmic and membrance proteomes of the wild type strain DSM 1731 and its butanol-tolerant mutant Rh8. We established the most comprehensive reference cytoplasmic and membrane proteome map for *C. acetobutylicum*. Furthermore, we identified 102 differentially expressed cytoplasmic proteins and 73 membrane proteins. Hierarchical clustering analysis revealed that Rh8 cells have evolved a more stabilized membrane structure, a more economic energy metabolism, and a more sensitive regulation on protein expression, to cope with the butanol challenge. This study allows us to obtain a systemic understanding on the molecular basis of butanol tolerance in *C. acetobutylicum*.

See more of this Session: Poster Session: Food Engineering and Bioprocessing
See more of this Group/Topical: Food, Pharmaceutical & Bioengineering Division

(626m) Biological Conversion of Carbon Dioxide to Chemicals

Im Gyu Kim[1], Dong Gyun Kang[1], Byung Hoon Jo[2], Yoo Seong Choi[3] and **Hyung Joon Cha**[1], (1) Department of Chemical Engineering, Pohang University of Science and Technology, Pohang, South Korea, (2)School of Interdisciplinary Bioscience and Bioengineering, Pohang University of Science and Technology, Pohang, South Korea, (3)Department of Chemical Engineering, Chungnam National University, Daejeon, South Korea

In recent years, several carbon dioxide capture and storage technologies are being developed. One of the most promising biological carbon dioxide sequestration methods is to use carbonic anhydrase (CA). CA is a zinc-containing metalloenzyme catalyzing the reversible hydration of carbon dioxide (CO_2 + $H_2O \rightarrow HCO_3^-$). CA plays an important role for sequestrating carbon dioxide since uncatalyzed interconversion between carbon dioxide and bicarbonate is slow. In the present work, we expressed recombinant CA originated from *Neisseria gonorrhoeae* in *Escherichia coli.* We purified recombinant CA with 28 kDa molecular weight including periplasmic signal sequence. The purified recombinant CA showed high biological activity. In addition, calcium carbonate precipitation was successfully observed in carbon dioxide-saturated water solution by addition of calcium chloride. We found that kinetics for forming crystal morphology was much faster than the case without enzyme. Collectively, we could confirm that the possibility of carbon dioxide sequestration using CA and successful conversion of carbon dioxide into calcium carbonate.

See more of this Session: Poster Session: Food Engineering and Bioprocessing
See more of this Group/Topical: Food, Pharmaceutical & Bioengineering Division

(626n) Engineering E. Coli for 2,3-Butanediol Production From Cellodextrins

Hyun-Dong Shin[1], Murat Johnson[1], Seon-Won Kim[2] and Rachel Ruizhen Chen[1], (1)Chemical & Biomolecular Engineering, Georgia Institute of Technology, Atlanta, GA, (2)Division of Applied Life Science, Gyeongsang National University, Jinju, South Korea

2,3-Butanediol(2,3-BDO) is a bulk chemical with broad applications including as fuel and fuel additives. A sustainable technology for synthesizing 2,3-BDO from renewable biomass is important to alleviate the dependence on oil supply for the production of bulk chemicals and fuels. While microbial production of 2,3-BDO was shown previously with engineered *E. coli* from glucose, consolidated production of 2,3-BDO directly from cellulosic material offers an opportunity to bring the technology to be competitive with petroleum-based production route.

In this presentation, we describe an effort in our laboratores toward a cellulosic 2,3-BDO. We have successfully engineered an *E. coli* to utilize cellodextrins, the hydrolysis intermediates from cellulose depolymerization by periplasmic expression of a cellodextrinase. The resulting strain grew well on a cellodextrins with varying degrees of polymerization. A synthetic operon for BDO biosynthesis was constructed and inserted in the cellodextrinase-expressing cells, resulting in a biocatalyst capable of converting cellodextrin to 2,3-BDO with 84% conversion yield without exogenous β-glucosidase.

To the best of our knowledge, this is the first successful engineering effort that led to cellodextrin-utilizing *E. coli* for 2,3-BDO production. This is an important step toward consolidated bioprocessing for production of biofuel and biorefinery products from lignocelluloses. The ability for the biocatalyst to directly use cellodextrins eliminates the need for exogenous β-glucosidase and removes from hydrolysate cellobiose and cellodextrins, potential inhibitors for the cellulases. Further, by combining these catalysts with other strains of *E. coli* or other microorganisms that produces cellulases, a true-consolidated bioprocess could result, leading to a more cost-effective technology.

See more of this Session: Poster Session: Food Engineering and Bioprocessing
See more of this Group/Topical: Food, Pharmaceutical & Bioengineering Division

(626o) Improvement of 2,3-Butanediol Production In Enterobacter Aerogenes Via Lactate Dehydrogenase Deletion

Min-Kyu Oh, Moo-Young Jung and Chiam-Yu Ng, Chemical and Biological Engineering, Korea university, Seoul, South Korea

2,3-Butanediol (2,3-BD) is a promising bulk chemical due to its extensive industrial applications. Through chemical reaction such as dehydration and esterification, 2,3-BD can be converted into various chemical compounds including 1,3-butadiene, diacetyl, methyl ethyl ketone (MEK; butan-2-one). These chemical compounds are used to make highly valued materials such as synthetic rubber, antifreeze agent, flavoring agent. Several microorganism species such as *Klebsiella pneumonia, Klebsiella oxytoca* and *Enterobacter aerogenes* are natural producers of 2,3-BD. Among these microorganisms, E. aerogenes is a facultative anaerobe, which is able to produce significant amounts of 2,3-BD in mixed fermentations, while generating a number of soluble and gaseous products. Furthermore, it has a high growth rate, high vitality in different types of environment, and is able to utilize diverse range of carbon sources. First, we performed medium optimization and pH control to obtain the optimal culture condition for high cell growth and high production of 2,3-BD product. In addition, we have successfully performed lactate dehydrogenase (ldh) gene knockout by using the Red lambda recombination method. ldh gene knock out strain showed improved yield of 2,3-BD when compare with wild type. We also overexpressed budA, budB, budC and bdh gene of E.aerogenes 2,3-BD pathway operon separately and the whole operon collectively to characterize the role of each gene and to enhance the production of 2,3-BD.

See more of this Session: Poster Session: Food Engineering and Bioprocessing
See more of this Group/Topical: Food, Pharmaceutical & Bioengineering Division

(626p) Quality Control and Flavor Maintenance of Cantonese Sausage by Multiple Natural Antioxidants

Dandan Cai[1], Pengli Xu[1], Wei-Lun Chang[2], **Kaijun Xiao**[3] and Zhaomei Wang[1], (1)South China University of Technology, Guangzhou, China, (2)Biotechnology, Bioprocessing Innovative Company, Dublin, OH, (3)College of Light Industry and Food Science, South China University of Technology, Guangzhou, China

Cantonese sausage, famous for the unique flavors in Guangzhou, easily loses its intrinsic flavor and quality due to the lipid oxidation during storage. The inhibitive effects of single and multiple natural antioxidants (NAs), and single and multiple synthetic antioxidants (SAs) on the quality changes of Cantonese sausage were investigated by detecting acid value (AV) and Peroxide value (POV). The best multiple NA to maintain Cantonese sausage quality was selected among the NAs and SAs. The characteristic flavor of Cantonese sausage with and without NAs was determined by Dynamic Headspace-Gas Chromatography-Mass Spectrometry (DHS-GC-MS). Results showed that the antioxidative effect of the selected antioxidants was in the order: VC>rosemary>VE>tea polyphenols>AOB≈TBHQ>BHA>PG>BHT>phytic acids, when the maximum permitted amount in Chinese National Standard are implemented. The antioxidative effects on Cantonese sausage with multiple NAs are all better than those with single NAs and SAs. AV of Cantonese sausage added VC-rosemary and VE-rosemary both are 3.63 (mg KOH/g fat) after stored at room temperature for 120 days, while the POV (g/100g fat) added VC-rosemary and VE-rosemary is 0.389 and 0.396 after stored at room temperature for the same period, respectively, which are below the Chinese National standard (AV: 4mg KOH/1g fat, POV: 0.50g/100g fat). DHS-GC-MS analysis showed that ethanol and ethyl esters of Cantonese sausage with the both multiple NAs and SAs are the dominant volatile components, which is similar with the control group. Therefore, multiple NAs have a good prospective in the application of Cantonese sausage preservation.

See more of this Session: Poster Session: Food Engineering and Bioprocessing
See more of this Group/Topical: Food, Pharmaceutical & Bioengineering Division

(626q) Design and Optimization of a Pilot-Scale Digester for Biogas Production

Kelsi R. Andrus, Stephen Dufreche, Mark Zappi and Rakesh Bajpai, Chemical Engineering, University of Louisiana at Lafayette, Lafayette, LA

In many industrial situations, wastes are produced that can be harmful to the environment. Many of these organic wastes can be digested anaerobically to produce a methanogenic biogas rich in energy. Treatment of such wastes through digestion can reduce BOD below allowable limits for discharge. Preliminary results utilizing lab-scale bioreactors for digestion of sewage sludge and shrimp processing waste were very promising in regard to methane content and total gas production. However, little to no data exists on larger reactors for many suitable feedstocks. A pilot-scale 250 gallon, jacketed reactor was purchased an set up on a trailer for mobile operation, with the intent to process multiple sources of local waste. These wastes include sewage sludge, shrimp and crab wastes, beer wastes, wood chips, and other food processing plants' wastes.

Considerations in the design and operation of the digester will be presented, as well as optimization and production data from various waste sources.

See more of this Session: Poster Session: Food Engineering and Bioprocessing
See more of this Group/Topical: Food, Pharmaceutical & Bioengineering Division

(626r) Analysis of Fermentation Process for Scfv Production by Proteomics

Yuya Fujiki[1], Yoichi Kumada[2] and Michimasa Kishimoto[2], (1)Chemistry and Materials Technology, Kyoto Institute of Technology, Kyoto, Japan, (2)Bio-molecular Engineering, Kyoto Institute of Technology, Kyoto, Japan

During the production phase of the culture for the foreign protein production by recombinant *Pichia pastoris*, the control of methanol feeding is important because methanol is used not only for protein production but also for the growth of the strain. The starvation of the carbon source inhibits the growth of the strain and the production of foreign protein. Furthermore, excess methanol concentration in culture broth severely affected the metabolism in the strain. Therefore, we constructed the methanol feeding control system with the aid of semiconductor gas sensor. Using this system, we increased the production amount of a model foreign protein, e.g. single-chain Fv antibody (scFv). As a result of the fed-batch culture by a 2L jar-fermenter, we obtained 73 g/L of the dry cell weight (DCW) and 1.5g/L of scFv concentration, which is 5 times higher than the upper limit of the product concentration obtained in the flask cultivation. The scFv production started soon after the start up of methanol feeding operation. The production continued around 100 hours and the production rate was kept almost constant. Then, we analyzed the cell in the growth and production phase with two-dimensional electrophoresis (2-DE). The images obtained from the cells in different phases are quite different, which suggested that the expression protein of the cells seriously changed with the phase of cultivation.

See more of this Session: Poster Session: Food Engineering and Bioprocessing
See more of this Group/Topical: Food, Pharmaceutical & Bioengineering Division

(626s) A New Drug Screening Method for Chinese Herbal Medicines As Potential Enhancers for Embryonic Stem Cell Proliferation

Ding Li, William G. Lowrie Department of Chemical and Biomolecular Engineering, The Ohio State University, Columbus, OH, Jufang Wang, School of Biosciences & Bioengineering, South China University of Technology, Guangzhou, China and Shang-Tian Yang, Department of Chemical and Biomolecular Engineering, Ohio State University, Columbus, OH

In this work, we developed a new drug screening method based on three-dimensional (3-D) cell culture for Chinese herbal medicines and for the first time discovered the potential application of a traditional herbal medicine as a new proliferation enhancer of embryonic stem cell culture through this method. This new screening method used nonwoven polyethylene terephthalate (PET) fibers as scaffolds for the 3-D culture of cells expressing enhanced green fluorescent protein (EGFP) in the medium with herbal medicine extracts. The extracts of three traditional Chinese herbal medicines, *Epimedium dacidii Franch* (EDF), *Ganoderma lucidum* spore (GLS), and *Ginkgo biloba* (GB), were selected as the screening targets and their effects on the proliferations of three cell lines (i.e., mouse embryonic stem (mES) D3 cells, colon cancer HT-29 cells, and breast cancer MCF-7 cells) were evaluated by measuring EGFP fluorescence intensities at a certain wave length. For each herbal medicine, the extracts were dissolved in DMEM cell culture medium at the concentrations of 0.01%, 0.1%, and 1% (w/v). Cell culture in the medium containing deoxycholic acid (DCA) (0.1 mM, 1 mM, and 10 mM) was used as the positive control for screening. The cell morphologies under different culture conditions were also characterized via scanning electron microscopy (SEM). Results showed that the growth of all three cell lines were significantly inhibited in the medium with 1% of herbal medicine exacts, which indicated that all three herbal medicines demonstrated strong cytotoxicity at 1% concentration regardless of the cell type. At a concentration of 0.1%, the herbal medicines showed different effects on cell proliferation depending on the cell type: GLS displayed a promotional effect on mES D3 cell proliferation but a reverse effect on the two cancer cell lines, while both EDF and GB inhibited the proliferation of mES D3 cells but showed no apparent impact on the two cancer cells. Those results were also supported by cell morphologies on SEM images. Based on the above results, we selected GLS as a new proliferation enhancer and for the first time used DMEM medium containing 0.1% GLS for stem cell culture. The new medium enhanced the specific growth rate and the maximum cell density of mES D3 cells by 28.6% and 29.1%, respectively, compared with the cell culture in regular DMEM medium. In summary, this 3-D cell culture-based screening method is feasible and efficient in evaluating the effects of herbal medicines on cell growth. GLS selected by this method shows its potential applications in mass production of embryonic stem cells and for treating cancers.

See more of this Session: Poster Session: Food Engineering and Bioprocessing
See more of this Group/Topical: Food, Pharmaceutical & Bioengineering Division

A COMPREHENSIVE EXPERIMENTAL INVESTIGATION ON THE DISTILLATION PROCESS OF CHINESE LIQUOR

Hailong Li [a], *Weixing Huang* [a], *Caihong Shen* [b], *Bin Yi* [b]

(a. School of Chemical Engineering, Sichuan University, Chengdu, China;
b. Key Laboratory of Liquor Making Bio. Tech. & Appl. of Sichuan Province, Luzhou, China)

Introduction

Chinese liquor is one of the oldest distillates in the world. Compared with other spirits, such as vodka, whisky and brandy, Chinese liquor has higher ethanol concentration (normally 46-60% by volume) [1]. Chinese liquor has a consumption of more than 4 million kilograms and sale revenue of about 500 billion Chinese Yuan(CNY) each year [2]. According to the aroma characteristics, Chinese liquor can be generally divided into five styles: the strong aroma style, light aroma style, sauce aroma style, rice aroma style and miscellaneous aroma style [3]. Based on the characteristics of strong aroma, mellow and agreeable sweetness, fragrant accord and long-lasting aftertaste, the strong fragrance type liquor takes about 70% of total liquor production [2,4]. The manufacturing of Chinese liquor consists of three consecutive processes: fermentation, distillation and blending. The fermentation is typically operated at 28-32 °C for a few months under anaerobic conditions in a solid state. Afterwards, the fermented grains are distilled with steam to extract ethanol and other flavor compounds. Finally, after being aged for a period of time, these distillates are blended to yield the final product.

The distillation plays a critical role to link the initial fermentation with the final blending. The goal of distillation is to extract the flavor compounds in the fermented grains. However, the distillation process has been less investigated and currently applied distillation operation still follows the way that it has been done in ancient time. The quality of the liquor is much determined by the favor compounds, which takes up only about 2%(w/w) [5]. Study about the variation of flavor compounds during the distillation process of Chinese liquor has few been reported. The operational conditions, such as steam flowrate, bed height and bulk density of fermented grains will surely affect the extractions of ethanol and flavor compounds during the distillation process. The cross-steaming has been stated to be an effective technique to enhance the yield of the liquor [6,7]. However, the effect of the cross-steaming on the extractions of flavor compounds and the concentrations of flavor compounds in the distilled liquor is poorly known. In addition, the liquor-vapor leaking is found to usually leak from the water sealing in the early-stage of the distillation process, but the effect of the leaking on the yield and the concentrations of flavor compounds in the distilled liquor has not been experimentally investigated.

In this work, comprehensive experiments were carried out in an industrial-scale distiller to investigate the variation of flavor compounds during the distillation process, and the effects of some operational conditions on the concentrations of flavor compounds in the distilled liquor and the liquor yield. The experiments focus on the distillation process of the strong fragrance type liquor, but the findings in this work also have important reference for other aroma liquors on account of the essential similarity for the distillation processes. This work provides experimental base for understanding of distillation process and optimization of the distillation operation of Chinese liquor.

Experimental

A schematic diagram of the experimental distiller used in this study is shown in Figure 1. It consists of a steam supply pipe, a steam boiler, a steam distributor, a barrel for the fermented grains, a lid, a liquor-vapor tube and a condenser for the liquor-vapor. The linkage between the steam boiler and barrel is packing seal, and the other points are sealed by water. The experimental distiller is in industrial-scale, where the inner diameter and height of the barrel are 1000 and 1250 mm, respectively.

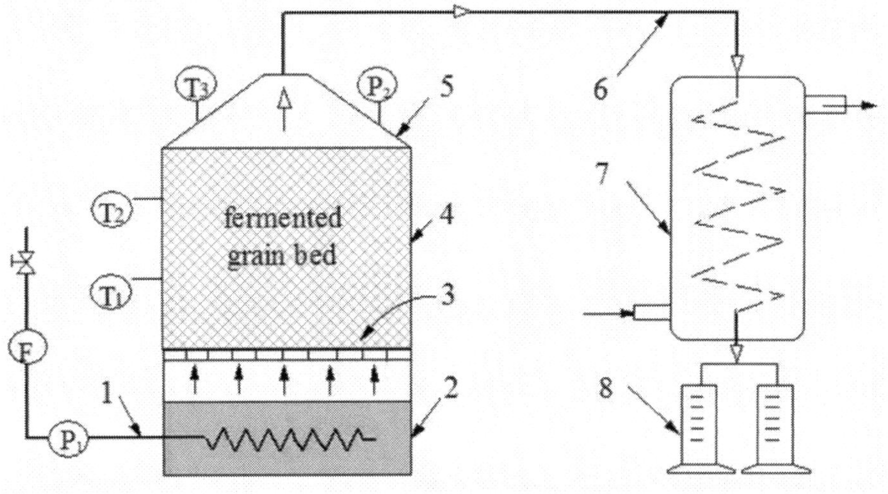

Figure 1. Schematic diagram of the experimental distiller
1. Steam pipe; 2. Steam boiler; 3. Steam distributor; 4. Barrel; 5. Lid;
6. Liquor-vapor tube; 7. Condenser; 8. Sampling bottle

The distillates were continuously collected through the whole distillation process. To investigate the variation of flavor compounds during the distillation process, the concentrations of flaovr compounds in each sample was determined by gas chromatography. To assess the distillation results, the distillate samples were divided into two parts based on their ethanol concentrations in line with the actual production: the distillates with ethanol concentration higher than 45% by volume were the effective products, while those with ethanol concentration lower than 45% by volume were the non-effective products for poor quality. The effective products are the distilled liquor and called 'liquor' for short in this study. To determine the effect of operational conditions on the change of flavor compounds during the distillation, the concentration of flavor compounds in the distilled liquor were tested. All the analyses for the concentrations of flavor compounds were performed in three replications and the average is reported. To research the effect operational conditions on the yield of the liquor, the yield in the study is calculated based on 100 kg fermented grains, in which the ethanol concentration is 60% by volume.

To set the effect of the steam flowrate on the concentrations of flavor compounds, six levels of the flowrates, i.e., 1.20, 1.60, 1.90, 2.30, 2.50 and 3.30 kg/min, were introduced, with keeping the bed height and bulk density fixed at 1250 mm and 389.7 kg/m^3, respectively. The effect of the bulk density were considered at three levels of the densities, i.e., 389.7, 443.2 and 557.6 kg/m^3, with keeping the bed height and steam flowrate fixed at 1250 mm and 1.60 kg/min, respectively. The effect of the bed height were studied at seven levels of the heights, i.e., 600, 700, 800, 900, 1000, 1100 and 1250 mm, with keeping the steam flowrate fixed at 1.60 kg/min. The effect of the cross-steaming were discussed at five volume levels of edible alcohol, i.e., 0, 2, 4, 6 and 8 L, with keeping the bed height, steam flowrate and bulk density fixed at 1250 mm, 1.60 kg/min and 389.7 kg/m^3, respectively. The added edible alcohol is an admixture of ethanol and water, and the ethanol concentration is 95% by volume.

Ethanol concentration was determined by a density meter (DMA 5000, Anton Paar, Graz, Austria), based upon the relationship between the density of water-ethanol mixtures and the ethanol concentration in the water-ethanol mixtures. An Agilent 6980N (America) gas chromatography fitted with a flame ionization detector (FID) and a HP-INNOWax column of 60m × 0.25 mm × 0.25 μm (Hewlett-Packard, Palo Alto, CA) were used to quantify the flavor compounds. Conditions were as follows: GC column temperature, at 40 °C (4 min), at 2.5 °C/min to 60 °C, and then 25 °C/min to 120 °C, and increased from 10 °C/min to 240 °C, and held at 240 °C for 4 min; carrier gas, nitrogen at the flowrate of 1.8 ml/min; split ratio, 1:50; injector and detector temperatures, 240 °C.

Results and Discussion

Variation of important flavor compounds during the distillation process

To investigate the variation of flavor compounds during the distillation process, experiments were carried out and the steam flowrate, bed height and bulk density were 1.60 kg/min, 1250 mm and 390 kg/m^3, respectively. The distillates were continuously collected through the whole distillation process and the concentrations ethanol and flavor compounds in the liquor were determined by density meter and gas chromatography. The concentrations for more than 50 kinds of flavor compounds were obtained. For simplification, the concentrations of important flavor compounds are given in Table 1.

Table 1. The concentrations of ethanol and flavor compounds in the distillates (g/L)

Time (min)	Ethanol (%v/v)	1-Propanol	1-Butanol	Isobutanol	1-Hexanol	Ethyl acetate
0.33	73.892	1.392	1.420	0.886	0.977	43.232
1.00	74.241	1.382	1.423	0.777	0.969	27.945
1.67	74.349	1.312	1.420	0.681	0.958	20.453
2.33	73.343	1.241	1.318	0.617	0.952	16.275
3.00	72.331	1.204	1.253	0.554	0.947	12.758
3.67	71.249	1.149	1.161	0.506	0.930	11.090
4.33	69.502	1.055	1.105	0.465	0.869	9.773
5.00	67.615	0.951	0.986	0.419	0.835	8.326
5.67	65.009	0.894	0.926	0.379	0.804	7.205
6.33	61.711	0.812	0.844	0.332	0.728	5.655
7.00	58.074	0.753	0.780	0.299	0.671	4.935
7.67	53.945	0.692	0.716	0.270	0.642	4.106
8.33	49.340	0.621	0.642	0.239	0.602	3.699
9.00	44.054	0.546	0.566	0.208	0.600	2.531
9.67	38.738	0.481	0.500	0.181	0.565	2.136
10.50	33.135	0.414	0.430	0.154	0.504	1.869
11.50	26.936	0.335	0.347	0.125	0.455	1.640
12.50	20.470	0.259	0.276	0.116	0.435	2.589
13.50	15.051	0.206	0.219	0.103	0.391	2.249
14.50	11.343	0.180	0.169	0.081	0.361	2.101
15.50	8.473	0.142	0.133	0.075	0.321	1.827

Values are presented as mean of three replicates.

Ethanol concentration decreases gradually during the distillation process, thought the concentration in the distillates obtained in the early-stage of distillation (the first sample). The lower ethanol concentration in the distillates obtained in the early-stage of distillation may be due to that some residual water in the condenser enters into the distillate sample. It can be concluded from Table 1 that the concentrations of alcohol-soluble flavor compounds decrease continuously during the distillation process, such as, 1-propanol, 1-butanol, isobutanol, 1-hexanol, ethyl acetate, ethyl butanoic and ethyl hexanoate; while the water-soluble flavor compounds, such as, ethyl lactate, acetic acid, butanoic acid and hexanoic acid, show decreased concentrations during the distillation process. Also, the different decreasing/increasing rates for these alcohol- compounds/water-compounds can be attributed to different compatibility with ethanol. The result implies that the variations of flavor compounds in the liquor during the distillation process are closely related to their dissolving properties in ethanol.

It is generally known that the quality of the liquor is largely determined by the concentrations of alcohol-soluble flavor compounds, and the liquor with higher concentrations of alcohol-soluble compounds possesses better quality. The analysis result for the concentrations of flavor compounds in Table 1 implies the quality of the distillates drops down through the distillation process. Therefore, to enhance the quality of the liquor, the goal of the distillation should be aimed at creating favorable condition for the extraction of alcohol-soluble compounds.

Table 1. (Continued)

Time (min)	Ethyl butanoic	Ethyl hexanoate	Ethyl lactate	Acetic acid	Butanoic acid	Hexanoic acid
0.33	15.195	101.402	13.248	3.353	3.054	3.175
1.00	9.998	65.815	13.146	3.233	2.599	3.225
1.67	7.616	53.119	14.948	3.253	2.653	3.385
2.33	6.324	46.451	17.471	3.391	2.806	4.945
3.00	5.281	41.382	20.781	3.500	3.394	6.981
3.67	4.657	37.888	25.213	3.785	3.555	8.473
4.33	4.201	35.082	30.680	4.089	4.831	7.369
5.00	3.694	32.236	37.820	4.444	4.922	7.512
5.67	3.272	29.628	47.077	4.985	5.410	8.370
6.33	2.793	26.941	59.378	5.530	6.078	9.711
7.00	2.533	25.455	72.962	6.315	7.123	11.601
7.67	2.306	24.507	87.548	7.453	8.345	13.888
8.33	2.077	22.930	102.573	8.256	9.643	16.185
9.00	1.642	20.349	113.551	9.309	10.923	18.848
9.67	1.402	18.261	131.623	10.530	12.485	21.726
10.50	1.228	16.025	149.217	11.285	14.454	27.459
11.50	1.091	13.246	158.499	12.954	16.014	28.110
12.50	1.090	12.072	163.460	13.852	17.305	34.716
13.50	0.872	9.229	178.861	15.047	18.530	39.698
14.50	0.818	8.177	182.339	16.271	20.030	41.239
15.50	0.661	6.267	189.858	22.225	21.529	46.533

Values are presented as mean of three replicates.

Changes of liquor yields and flavor compound concentrations in the liquors with steam flowrate

Figure 2 shows the effect of the steam flowrate q_m on the liquor yield m and quality. The increase of steam flowrate from 1.20 to 1.60 kg/min leads to an increased liquor yield, but with further increasing the steam flowrate, the yield turns to decrease, especially after the steam flowrate exceeds to 2.30 kg/min. Therefore, there exists an optimal steam flowrate from the viewpoints of higher yield.

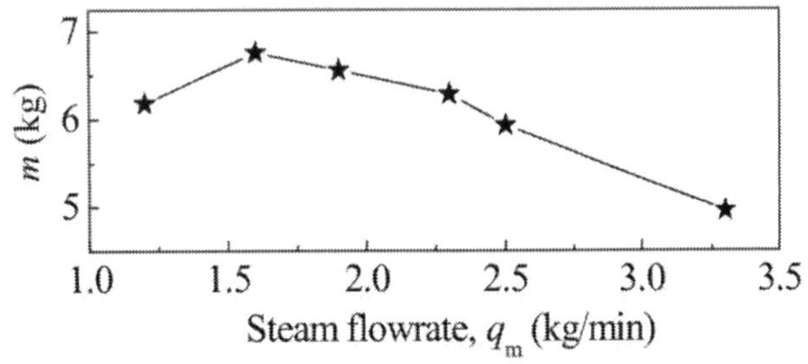

Figure 2: Chang of liquor yield with steam flowrate

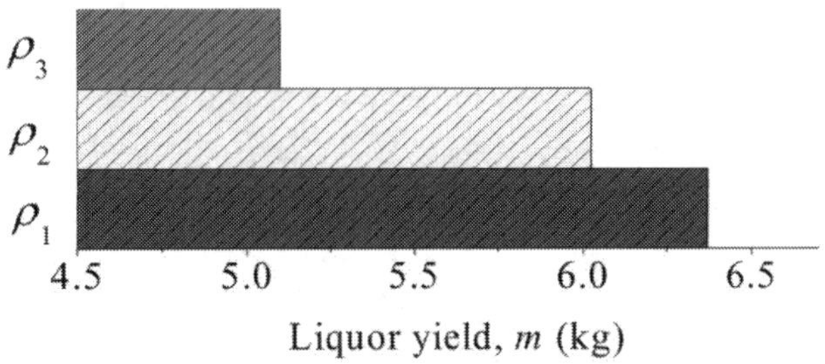

Figure 3: Effect of the bulk density on liquor yield

The concentrations of flavor compounds in the liquors obtained at different steam flowrates are listed is Table 2. It can be seen from this figure that the concentrations of ethanol and alcohol-soluble compounds present parabolic changes with the steam flowrate. The maximum concentrations are obtained at the steam flowrate of 1.60 kg/min. However, the changes of the concentrations of water-soluble compounds are opposite to that of alcohol-soluble compounds. For example, the concentrations of ethyl lactate, acetic acid and butanoic acid decrease with increasing the steam flowrate from 1.20 to 1.60 kg/min, while they turn to increase gradually for further increasing steam flowrate. The concentration of alcohol-soluble compounds reaches the maximum value at 1.60 kg/min, which suggests the liquor distilled with the steam flowrate possesses best quality. Therefore, it can be concluded that there exists an optimal steam flowrate for the distillation.

Table 2. The concentrations of flavor compounds in the liquors obtained at different steam flowrates (g/L)

Compounds	Steam flowrate (kg/min)					
	1.20	**1.60**	**1.90**	**2.30**	**2.50**	**3.30**
Ethanol (%v/v)	61.008	64.925	63.875	62.322	59.123	53.324
1-Propanol	1.144	1.645	1.922	1.498	1.287	1.132
1-Butanol	2.425	2.584	2.816	2.676	2.380	2.128
Isobutanol	0.393	0.513	0.473	0.413	0.362	0.397
1-Hexanol	2.668	3.290	3.127	2.644	2.226	1.903
Ethyl acetate	17.367	19.167	18.815	16.024	13.128	10.255
Ethyl butanoic	5.591	5.844	5.713	5.523	5.213	4.804
Ethyl hexanoate	39.290	42.042	40.126	38.930	37.747	35.834
Ethyl lactate	52.759	44.233	50.081	55.438	58.785	63.811
Acetic acid	4.302	4.250	4.350	4.422	4.5061	4.679
Butanoic acid	2.844	2.858	3.011	3.165	3.297	3.350
Hexanoic acid	6.566	6.504	6.960	7.898	8.929	9.058

Values are presented as mean of three replicates.

Change of the concentration of flavor compounds in the liquor and liquor yield with bulk density

The fermented grains are compressible to a certain extent. Different loading modes result in different bulk densities. The effect of the bulk density on the liquor yield m is presented in Figure 3, in which ρ_1, ρ_2 and ρ_3 denote the bulk density of 389.7, 443.2 and 557.6 kg/m^3, respectively. It can be observed from this figure that the increase in the bulk density leads to a steady downward trend of the liquor yield.

Table 3 shows the change of the concentrations of flavor compounds in the liquors with bulk density. With the increase of bulk density, the concentrations of alcohol-soluble compounds in the liquors decrease gradually, such as, 1-propanol, 1-butanol, isobutanol, 1-hexanol, ethyl acetate, ethyl butanoic and ethyl hexanoate; while the concentrations of alcohol-soluble compounds exhibit upward trends, such as, ethyl lactate, acetic acid, butanoic acid and hexanoic acid. The result indicates that lower bulk density leads to better quality for higher concentration of alcohol-soluble compounds. Therefore, from the view point of higher yield and better quality, lower bulk density is recommended.

Table 3. The concentrations of flavor compounds in the liquors obtained with different bulk densities of fermented grains (g/L)

Compounds	**Bulk density (kg/m^3)**		
	389.7	**443.2**	**557.6**
Ethanol (%v/v)	66.322	62.606	54.583
1-Propanol	1.645	1.535	1.280
1-Butanol	2.584	2.403	2.052
Isobutanol	0.513	0.440	0.326
1-Hexanol	3.290	2.897	2.338
Ethyl acetate	19.167	16.555	13.525
Ethyl butanoic	5.844	5.669	5.104
Ethyl hexanoate	42.042	39.324	36.016
Ethyl lactate	44.233	48.367	53.902
Acetic acid	4.250	4.432	4.585
Butanoic acid	2.858	3.046	3.281
Hexanoic acid	6.504	7.604	8.953

Values are presented as mean of three replicates.

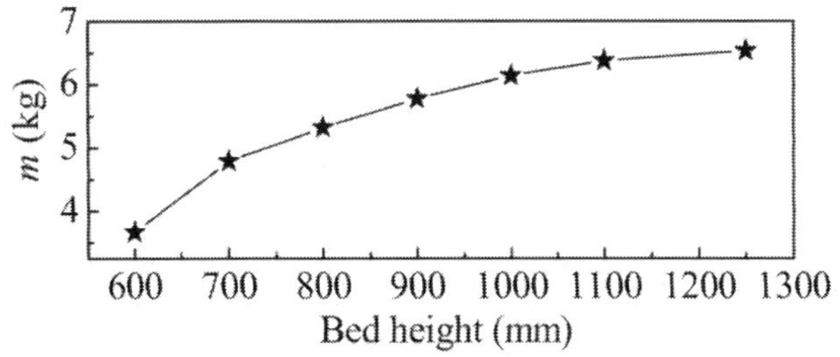

Figure 4: Change of liquor yield with the bed height

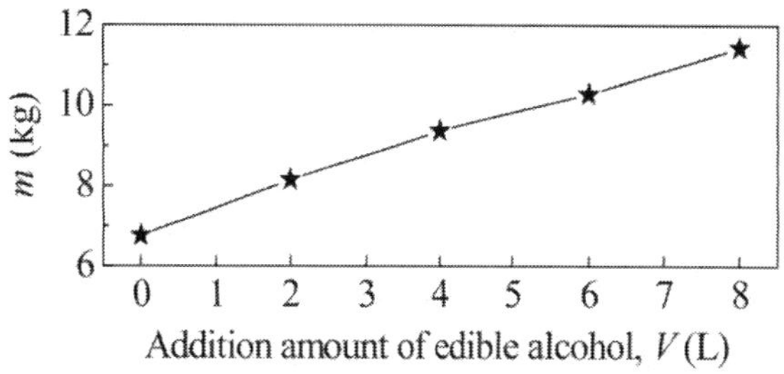

Figure 5: Effect of the cross-steaming on liquor yield

Change of the concentrations of flavor compounds in the liquor and liquor yield with bed height

The changes of liquor yield m with the bed height of fermented grains H is shown in Figure 4. The increase in bed height from 600 to 1250 mm leads to steady enhancement in liquor yield. This may be due to the fact that higher grain bed provides longer flow-path for steam, which benefits for the accumulation of ethanol to produce effective products (ethanol concentration >45% by volume). The result suggests that when the bed height increases from the present 800 mm (the bed height currently used in industrial operation) to 1250 mm, the yield is enhanced about 30%, implying great potential for the improvement of current distillation equipment.

Table 4 gives the concentrations of flavor compounds in the liquors distilled at different bed height. It is clear from this table that the concentrations of alcohol-soluble flavor compounds increases continuously with increasing the bed height, while the increase of bed height causes the concentrations of water-soluble flavor compounds. This finding implies that the quality of the liquor is enhanced by increasing the bed height. Therefore, the present bed height (700~900 mm) is not the optimal height for the distillation, and it should be increased to 1250 mm at least.

Table 4. The concentrations of flavor compounds in the liquors distilled at different bed heights (g/L)

Compounds	Bed height (mm)						
	600	**700**	**800**	**900**	**1000**	**1100**	**1250**
Ethanol (%v/v)	52.753	56.324	61.336	63.108	64.475	65.673	66.322
1-Propanol	1.042	1.187	1.254	1.376	1.492	1.569	1.645
1-Butanol	1.680	1.925	2.016	2.173	2.407	2.525	2.584
Isobutanol	0.221	0.290	0.337	0.372	0.414	0.476	0.513
1-Hexanol	1.904	2.584	2.760	2.947	3.149	3.206	3.290
Ethyl acetate	9.362	11.603	14.452	16.332	17.565	18.769	19.167
Ethyl butanoic	4.668	4.903	5.337	5.557	5.640	5.752	5.844
Ethyl hexanoate	33.324	35.644	37.685	39.020	40.373	41.830	42.042
Ethyl lactate	58.756	55.788	53.081	49.446	46.648	44.580	44.233
Acetic acid	4.804	4.674	4.550	4.473	4.382	4.275	4.250
Butanoic acid	3.487	3.246	3.116	3.068	2.945	2.887	2.858
Hexanoic acid	9.568	9.158	8.264	7.782	7.050	6.643	6.504

Values are presented as mean of three replicates.

Table 5. The concentrations of flavor compounds in the liquors obtained at different addition amounts of edible alcohol in the boiler (g/L)

Compounds	Edible alcohol volume (L)				
	0	**2**	**4**	**6**	**8**
1-Propanol	1.645	1.490	1.92	0.655	0.470
1-Butanol	2.584	2.439	2.300	1.673	1.087
Isobutanol	0.513	0.477	0.382	0.259	0.229
1-Hexanol	3.290	3.325	3.041	2.648	2.223
Ethyl acetate	19.167	18.814	16.384	12.524	9.121
Ethyl butanoic	5.844	5.415	4.889	3.523	2.514
Ethyl hexanoate	42.042	40.766	37.423	30.257	23.386
Ethyl lactate	44.233	48.836	40.087	35.436	30.788
Acetic acid	4.250	4.453	4.228	3.971	3.505
Butanoic acid	2.858	3.052	2.815	2.560	2.317
Hexanoic acid	6.504	6.837	6.466	5.975	5.620

Values are presented as mean of three replicates.

Effect of the cross-steaming technique on the concentrations of flavor compounds and liquor yield

Five levels of edible alcohol with the volume of 0, 2, 4, 6 and 8 L were added into the steam boiler to investigate the effect of the cross-steaming technique on liquor yield and concentrations of flavor compounds in the liquors. The liquor yield m is positively correlated with the increase of the amount of added edible alcohol (Figure 5), suggesting the cross-steaming enhances the liquor yield.

The concentrations of flavor compounds in the liquors obtained from different addition amounts in the boiler are presented in Table 5. It can be observed from this table that the concentrations of flavor compounds has inconspicuous change when a small amount of edible alcohol (<2 L) is added, while an excess amount results in sharp degradation of the concentration of flavor compounds, especially for the alcohol-soluble compounds. The results imply that adding small amount of edible alcohol into the boiler will increase the liquor yield and does not affect the quality of the liquor; while when excess amount of edible alcohol is added, the obtained liquor has poor quality (effective products). Therefore, the key is to determine appropriate addition amount.

Effect of the liquor-vapor leaking on the concentrations of flavor compounds in the liquor

The liquor-vapor leaking is a phenomenon commonly occurred in liquor distillation process, especially in the initial distillation stage. We have found that the products produced in the initial distillation stage are high quality products. Hence, the liquor-vapor leaking may leads to the loss of high quality products and decreases the quality of the distilled liquor in theory, but few experiments have been reported. To investigate the effect of the leaking on the distillation results, the concentrations of flavor compounds in the liquors obtained from the leaking and no-leaking cases have been tested, shown in Table 6. The leaking decreases substantially the concentrations of alcohol-soluble compounds, such as, 1-propanol, 1-butanol, ethyl acetate, ethyl butanoic, and ethyl hexanoate. While the concentrations of water-soluble compounds in the liquor obtained at the leaking case are higher. The results indicate that the leaking drops down obviously the quality of the liquor. In addition, it is also found that, when the leaking happens, the liquor yield is about 5 % less than the yield obtained at the no-leaking case. Thus, the leaking must be avoided in the distillation process of Chinese liquor.

Table 6. Effect of the liquor-vapor leaking on the concentrations of flavor compounds in the liquors obtained (g/L)

Compounds	Leaking	No leaking	Compounds	Leaking	No leaking
1-Propanol	1.032	1.645	Ethyl hexanoate	35.578	42.042
1-Butanol	1.072	2.584	Ethyl lactate	49.504	44.233
Isobutanol	0.478	0.513	Acetic acid	5.362	4.250
1-Hexanol	0.805	3.290	Butanoic acid	5.058	2.858
Ethyl acetate	10.960	19.167	Hexanoic acid	8.598	6.504
Ethyl butanoic	4.255	5.844			

Values are presented as mean of three replicates.

Conclusions

Comprehensive experiments were carried out in an industrial-scale distiller to investigate the distillation process of Chinese liquor. It is found that the variations of flavor compounds during the distillation process are closely related to their dissolving properties in ethanol. The concentrations of alcohol-soluble flavor compounds in the distillates decrease gradually during the distillation process, while the concentrations of water-soluble compounds in the distillates show upward trends. The concentrations of alcohol-soluble compounds present parabolic changes with the steam flowrate and the highest concentrations are obtained at the steam flowrate of 1.60 kg/min. From the view points of larger yield and higher concentrations of alcohol-soluble compounds, there exists an optimal steam flowrate for the distillation. Both the liquor yield and the concentrations of alcohol-soluble flavor compounds increase continuously with decreasing the bulk density and/or increasing the bed height, implying that the fermented grains should be loosely packed and the present bed height (700~900 mm) should be increased. Adding appropriate amount of edible alcohol in the steam boiler can increase the liquor yield has not obvious influence on the concentrations of flavor compounds, while excess amounts of edible alcohol is added, the concentrations of alcohol-soluble and water-soluble compounds are decreased sharply. The liquor-vapor leaking decreases the liquor yield and the concentrations of alcohol-soluble flavor compounds, suggesting the leaking is deleterious to the distillation of Chinese liquor.

References

1. Shi E., Huang W.X., Xiao Z.Y., Tang X.Y., Xu R.Q. and Zeng F.J., (2007), Study on improvement of the quality in Chinese new-type liquor by pervaporation with polydimethylsiloxane membrane. J Food Process Eng, 30: 38-50.
2. China Brewing Industry Association and China Alcoholic Drinks Industry Editorial Board, (2008). Yearbook of China alcoholic drinks industry (2008): Light Industry Publishing House of China, Beijing.
3. Fan W.L. and Qian M.C., (2006), Characterization of aroma compounds of Chinese Wuliangye" and "Jiannanchun" liquors by aroma extract dilution analysis. J Agric Food Chem., 54: 2695-2704.
4. Zhang S.Y., Shen C.H. and Xu D.F., (2009), Review on the technological development of Luzhou-flavor style Liquor. Liquor Making, 36 (1): 8-10 (in Chinese).
5. Wang Z.Y and Yi C.S., (2000), Effects of the content and ratio relationship of the chromatic spectrum structure components on liquor flavor style and quality, Liquor-making Sci. & Technol., 102: 93-96 (in Chinese).
6. Du G. and Song K., (2004), Application of fermented condensed pit-flavor cross-steaming liquor in liquor production. Liquor-making Sci. & Technol., 126: 114-115 (in Chinese).
7. Zhang B., Wu X.J. and Chen Q.R., (2003), Investigation on the technique of cross-steaming by addition alcohol at the bottom of the pots. Liquor-making Sci. & Technol., 116: 47-48 (in Chinese).

PREPERATION OF STEVIOL WITH SPECIFIC ENZYMATIC HYDROLYSIS OF STEVIOSIDE USING B-GALACTOSIDASE FROM *SULFOLOBUS SOLFATARICUS*

Hui-da Wan, Yong-mei Xia[1]
State Key Laboratory of Food Science and Technology, School of Chemical and Materials Engineering, Jiangnan University, Wuxi, Jiangsu, China

Abstract: Steviol is a rare component in extract of *Stevia rebaudiana* Bertoni leaves, which could be prepared by acidic hydrolysis or enzymatic hydrolysis of stevioside. *β*-Galactosidase from *sulfolobus solfataricus* was found to hydrolyze all the glycosidic bonds of St. The highest St conversion reached 99.0% in 6h with steviol as the only product. By contrast, the mixture of steviol and isosteviol was obtained by HCl hydrolysis. Therefore, enzymatic hydrolysis of stevioside by *β*-galactosidase from *sulfolobus solfataricus* is a efficient process for selective preparation of steviol.

Keywords: stevioside; *β*-galactosidase; steviol

Introduction

Stevia glycosides are a group of low-calorie natural sweeteners isolated from leaves of *Stevia rebaudiana* Bertoni, all of which have the same aglycon(steviol, Figure 1)[1-3]. The main component of stevia glycosides is stevioside (13-O-*β*-sophorosyl-19-O-*β*-D-glucosyl-steviol, St) and rebaudiosides A (RA), which are 200-350 times sweeter than sugar[4]. Besides of being sweeteners, they exhibit potential pharmaceutical properties, such as anti-inflammatory, anti-hyperglycemic, anti-hypertensive, anti-tumor, and immunomodulatory actions[3, 5-7]. However, residual bitterness of St affects its qualitative characteristics, meanwhile the amount of which is largest in dry leaves[8]. This drawback restricts the application in functional foods, pharmaceutical industry and many other fields. In order to overcome the problem, more recently, transglycosylation of St by glycosyltransferase is a more appropriate choice[9-12]. Moreover, steviol, isosteviol, steviolbioside, rubusoside and so on can be obtained through hydrolysis of St using the biocatalyst or chemical catalyst. Although the sweetness is decrease, yet they including their derivatives possess valuable biological properties[13-15]. *β*-Galactosidases (*β*-D-galactoside galactohydrolase, EC3.2.1.23) can cleave *β*(1-4) linkage of lactose, also referred to as lactases, and some of which with transglycosylation activity transfer galactose formed from lactose cleavage onto the donor, such as lactose, sucrose, et al[16, 17]. Consequently, *β*-galactosidases are used to produce low-lactose milk and manufacture galactooligosaccharides in dairy industry. In this article, *β*-galactosidase from *sulfolobus solfataricus* prepared in the laboratory was studied as biocatalyst to hydrolyze St. To our knowledge, there are few reports about enzymatic hydrolysis of St by *β*-galactosidase.

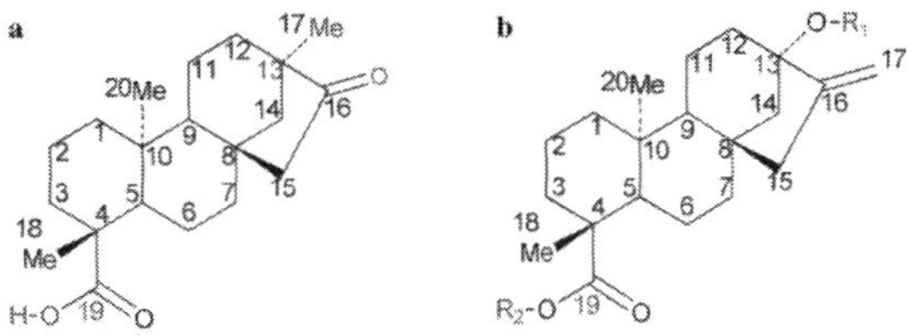

[1] Corresponding author: ymxia@jiangnan.edu.cn

Figure 1. Schematic structure of steviol, isosteviol, St and RA
a: isosteviol; b: steviol, R_1=H, R_2=H; St, R_1=-Glu-Glu; R_2= -Glu; RA, R_1=-Glu-$(Glu)_2$, R_2=-Glu; Glu is the abbreviation for glucose;

Materials and methods

Materials

β-Galactosidase from *sulfolobus solfataricus* was produced in the laboratory (50 U/mL, Jing Wu, Jiangnan University, China). Stevioside (≥97%, HPLC, $C_{38}H_{60}O_{18}$) was purchased from GLG life tech Co., Ltd. Rebaudioside A (≥99%, HPLC, $C_{44}H_{70}O_{23}$) was provided Changzhou Niutang Chemical Plant Co., Ltd. Acetonitrile(HPLC grade) was obtained from Jiangsu Hanbon Science & Technology Co., Ltd. All other reagents were analytical grade unless otherwise stated.

Hydrolytic activity of β-galactosidase[18]

100 mM o-nitrophenyl-*β*-D-galactopyranoside (oNPG) in acetate buffer(pH 4.0) was used as substrate, and reaction was initiated by addition 100 μL enzyme solution in the same buffer at 75 °C for 15min, which was stopped by adding 100 μL of 2 M Na_2CO_3. The absorbance at 420 nm was measured owing to the release of o-nitrophenol (oNP). One unit (U) of enzyme activity was defined as the amount of enzyme required to liberate the equivalent of 1 μmol of oNP per min under the above reaction conditions.

Hydrolysis of stevioside by β-galactosidase

In a typical reaction, 0.4g St, 20mL deionized water and enzyme (625 U/g St) were mixed in a 50 mL Erlenmeyer flask, shaken(150 rpm) at 75 °C for 3 h. After various time intervals, samples were withdrawn and determined by an HPLC system. The product precipitated out during the reaction and was filtered or centrifuged.

HPLC analysis for stevioside

St was determined by the method of our previous report with some modifications[19]. The composition of the stevioside were determined with HPLC (Waters 1525 and 2996, United States) using a NH_2 column(APS-2 HYPERSIL, 250mm×4.6mm, Thermo, United States), equipped with a UV detector and recorded at 210 nm. A mixture of acetonitrile and water was used as an eluent, gradient from 75: 25 *v/v* (2 min) to 50: 50 *v/v* (25 min), the flow rate was 1 mL/min. The experiments were performed in triplicate. The conversion of St was calculated as following:

$$\text{St coversion} = \frac{C_0 - C_t}{C_0} \times 100\%$$

The C_0 represents the initial St concentration (g/L), C_t represents the real time St concentration in the reaction mixture (g/L). The St concentration was determined using HPLC with a standard calibration curve.

Results and discussions

Owing to low solubility in water, the product can precipitate out automatically from reaction mixture for 0.5h approximately. This phenomenon in favor of positive hydrolysis reaction and is convenient for product separation by filter or centrifugation at 15,000 for 10 minutes. Precipitate was purified by recrystallization twice from acetone and petroleum ether[20], then characterized by LC-MSMS and ^{13}CNMR. A unique peak in LC was detected, and the mass spectrum showed the molecular ion with a m/z 317 $[M-H]^-$, compatible with the molecular formula $C_{20}H_{30}O_3$(M 318). Combination of the ^{13}CNMR spectra (table 1), product was confirmed as steviol not isosteviol.

Table 1. ^{13}CNMR chemical shift of product

C	Isosteviol[a]	Steviol[b]	product
1	40.72	41.1	41.0
2	20.13	19.8	19.8
3	38.48	38.6	38.6
4	44.64	43.9	43.8
5	58.16	57.1	56.9
6	22.94	22.6	22.6
7	42.47	41.8	41.8
8	48.16	41.8	41.8
9	55.88	54.3	54.2
10	39.25	39.8	39.8
11	21.44	20.8	20.7
12	39.08	40.7	40.7
13	41.05	79.8	79.7
14	55.18	47.4	47.4
15	49.85	48.1	48.1
16	225.08	157.6	157.7
17	20.20	102.9	102.8
18	29.54	29.3	29.3
19	181.5	180	180.0
20	13.97	15.9	15.8

a, b: data reported in literatures[21, 22];

Catalyzed by β-galactosidase (625U/g St) for 6 h, the highest conversion of stevioside of 99.0% was obtained with 20mg/mL of the St in water at 75 °C. Therefore, β-galactosidase from *sulfolobus solfataricus* is a highly efficient enzyme for hydrolysis of glycosidic bonds of St. When lactose or galactose was used as glycosyl donor and St as acceptor, there were no transglycosylation products prepared, but St was converted to steviol by β-galactosidase completely.

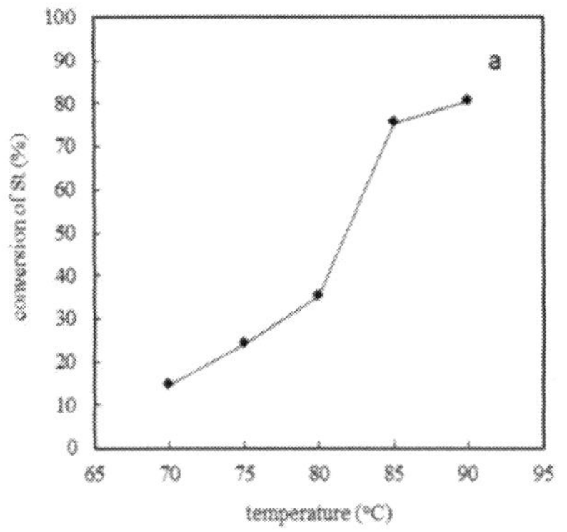

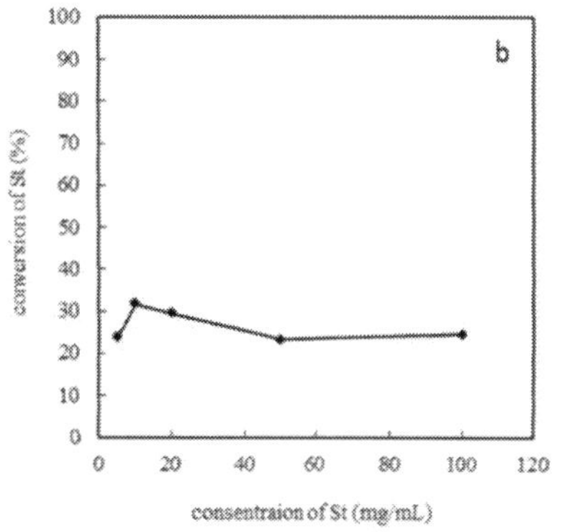

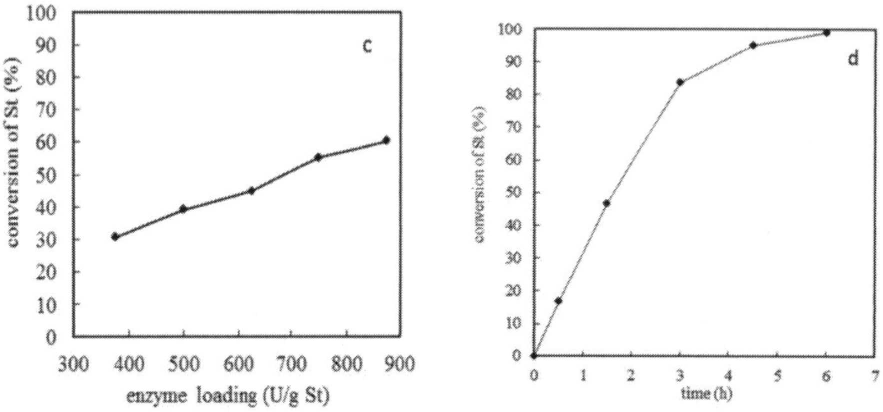

Figure 2. Optimization of enzymatic hydrolysis reaction conditions by *β*-galactosidase from *sulfolobus solfataricus*

a, St(25mg), 375U/g St in 5mL water for 1.5h at different temperature, respectively; b, St(25mg to 100mg), 375U/g St in 5mL water at 75°C for 1.5h; c, St(100mg) in 5mL water at 75°C for 1.5h under each enzyme loading (375U/g St to 875U/St); d, St(0.4g), 625U/g St in 20mL water at 75°C

Hydrolysis of St with HCl or H_2SO_4 affords Wagner-Meerwein rearrangement product isosteviol and enzymatic hydrolysis of St usually affords steviol[20]. However, St can be hydrolyzed by *Pancreatin* efficiently, isosteviol and steviol were generated under pH 4.0 and pH 7.0, respectively[13]. Therefore, effect of pH on St conversion and product composition by *β*-galactosidase was studied. Figure 3 shows enzyme is stable under pH 4-6. At the same time, from TLC plate (Figure 4, ethyl acetate/methanol 9:1, by vol, spots were visualized by spraying with sulfuric acid/ethanol 5:95, by vol), only steviol was acquired even in different pH ranges.

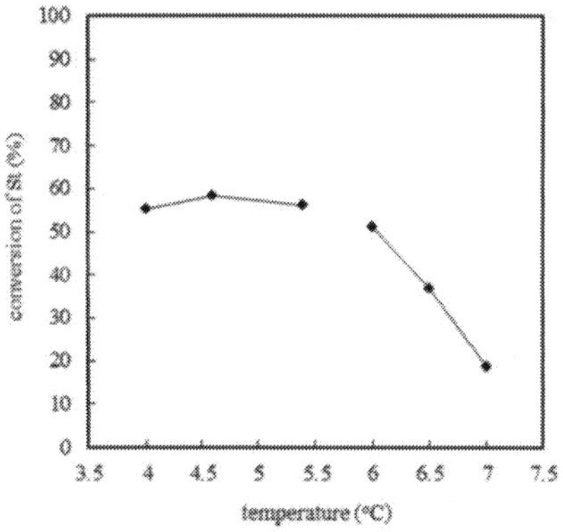

Figure 3. Effects of pH on hydrolysis of St using *β*-galactosidase from *sulfolobus solfataricus*

The buffer system used was acetate buffer (pH 4.0-5.4, 50mM) and phosphate buffer (pH 6.0-7.0, 50mM); other reaction conditions were same, 75 °C, 625U/g St, 20mg/mL,1,5h;

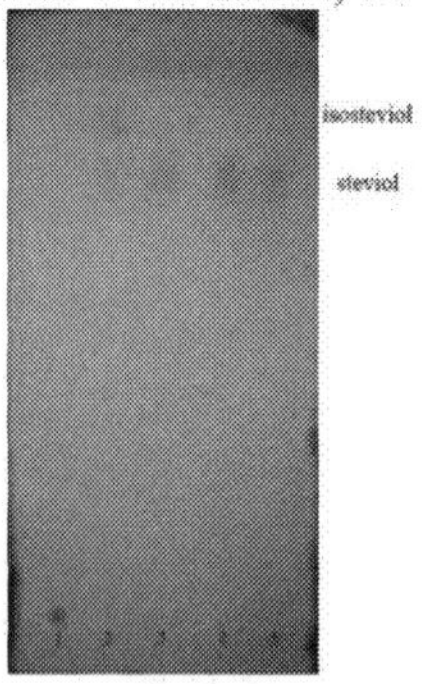

Figure 4. Comparison of the products of HCl hydrolysis with the products of enzymatic hydrolysis

1, St; 2: products of HCl hydrolysis (according to literature [18]); 3-5: products by enzymatic hydrolysis (deionized water, pH 4 acetate buffer and pH 4 potassium biphthalate buffer as solvent, respectively);

Conclusions

Preparation of steviol by enzymatic hydrolysis of stevioside using *β*-galactosidase from *sulfolobus solfataricus* was exploited in this article. Under currently optimum reaction conditions, St conversion was 99.0% in 6h. Primary product steviol was obtained finally that can precipitate out automatically from the reaction mixture, and pH of the reaction mixture didn't affect the product composition herein. *β*-Galactosidase from *sulfolobus solfataricus* was reported to hydrolyze glycosidic bonds in St firstly, which is therefore an potential extraordinary and green method for preparation of steviol.

Acknowledgments

The authors are thankful to financial support from the Program for Innovation Team of Jiangnan University (2008CXTD01), Innovation Fund for the Industry-University Cooperative Research in Jiangsu Province (BY2010115) and National Natural Science Foundation of China (31171752). We also thank Dr. Wu Jing at Jiangnan University for generously providing *β*-galactosidase.

References

1. Brahmachari, G., Mandal, L. C., Roy, R., Mondal, S. and Brahmachari, A. K. (2011), "Stevioside and Related Compounds-Molecules of Pharmaceutical Promise: A Critical Overview," *Archiv Der Pharmazie*, *344*, pp. 5-19.
2. Mishra, P. K., Singh, R., Kumar, U. and Prakash, V. (2010), "Stevia rebaudiana - a magical sweetener," *Global J Biotechnol Biochem*, *5*, pp. 62-74.
3. Chatsudthipong, V. and Muanprasat, C. (2009), "Stevioside and related compounds: Therapeutic benefits beyond sweetness," *Pharmacology & Therapeutics*, *121*, pp. 41-54.
4. Heerranz-Lopez, M., Barrajon-Catalan, E., Beltran-Debon, R., Joven, J. and Micol, V. (2010), "Stevia is a source for alternative sweeteners: potential medicinal effects," *Agro Food Ind Hi-Tech*, *21*, pp. 38-42.
5. Geeraert, B., Crombe, F., Hulsmans, M., Benhabiles, N., Geuns, J. M. and Holvoet, P. (2010), "Stevioside inhibits atherosclerosis by improving insulin signaling and antioxidant defense in obese insulin-resistant mice," *International Journal Of Obesity*, *34*, pp. 569-577.
6. Atteh, J. O., Onagbesan, O. M., Tona, K., Decuypere, E., Geuns, J. M. C. and Buyse, J. (2008), "Evaluation of supplementary stevia (Stevia rebaudiana, bertoni) leaves and stevioside in broiler diets: effects on feed intake, nutrient metabolism, blood parameters and growth performance," *Journal Of Animal Physiology And Animal Nutrition*, *92*, pp. 640-649.
7. Sharipova, R. R., Strobykina, I. Y., Mordovskoi, G. G., Chestnova, R. V., Mironov, V. F. and Kataev, V. E. (2011), "Antituberculosis activity of glycosides from Stevia rebaudiana and hybrid compounds of steviolbioside and pyridinecarboxylic acid hydrazides," *Chemistry Of Natural Compounds*, *46*, pp. 902-905.
8. Goyal, S. K., Samsher and Goyal, R. K. (2010), "Stevia (Stevia rebaudiana) a bio-sweetener: a review," *International Journal Of Food Sciences And Nutrition*, *61*, pp. 1-10.
9. Jaitak, V., Kaul, V., Bandna, Kumar, N., Singh, B., Savergave, L., Jogdand, V. and Nene, S. (2009), "Simple and efficient enzymatic transglycosylation of stevioside by β-cyclodextrin glucanotransferase from Bacillus firmus,"

Biotechnology Letters, *31*, pp. 1415-1420.

10. Kochikyan, V., Markosyan, A., Abelyan, L., Balayan, A. and Abelyan, V. (2006), "Combined enzymatic modification of stevioside and rebaudioside A," *Applied Biochemistry and Microbiology*, *42*, pp. 31-37.
11. Abelyan, V. H., Balayan, A. M., Ghochikyan, V. T. and Markosyan, A. A. (2004), "Transglycosylation of stevioside by cyclodextrin glucanotransferases of various groups of microorganisms," *Hayastani Kensabanakan Handes*, *56*, pp. 3-9.
12. Xu, Z. W., Li, Y. Q., Wang, Y. H., Yang, B. and Ning, Z. X. (2009), "Production of *β*-fructofuranosidase by Arthrobacter sp. and its application in the modification of stevioside and rebaudioside A," *Food Technol Biotechnol*, *47*, pp. 137-143.
13. Milagre, H. M. S., Martins, L. R. and Takahashi, J. A. (2009), "Novel Agents for Enzymatic and Fungal Hydrolysis of Stevioside," *Brazilian Journal of Microbiology*, *40*, pp. 367-372.
14. Li, H. P., Dong, C. M., Hou, Y. M. and Liu, H. M. (2009), "Synthesis of 17-OH-Isosteviol," *Chemical Research In Chinese Universities*, *25*, pp. 116-117.
15. Okamoto, K., Nakano, H., Yatake, T., Kiso, T. and Kitahata, S. (2000), "Purification and some properties of a beta-glucosidase from Flavobacterium johnsonae," *Biosci Biotechnol Biochem*, *64*, pp. 333-340.
16. Park, A.-R. and Oh, D.-K. (2010), "Galacto-oligosaccharide production using microbial beta-galactosidase: current state and perspectives," *Applied Microbiology and Biotechnology*, *85*, pp. 1279-1286.
17. Otieno, D. O. (2010), "Synthesis of β-Galactooligosaccharides from Lactose Using Microbial β-Galactosidases," *Comprehensive Reviews in Food Science and Food Safety*, *9*, pp. 471-482.
18. Park, H. Y., Kim, H. J., Lee, J. K., Kim, D. and Oh, D. K. (2008), "Galactooligosaccharide production by a thermostable β-galactosidase from *Sulfolobus solfataricus*," *World Journal of Microbiology and Biotechnology*, *24*, pp. 1553-1558.
19. Wan, H. D., Cai, Y. and Xia, Y. M.: *2010 AIChE Annual Meeting, 10AIChE, November 7, 2010 - November 12, 2010* 2010.
20. Avent, A. G., Hanson, J. R. and De Oliveira, B. H. (1990), "Hydrolysis of the diterpenoid glycoside, stevioside," *Phytochemistry*, *29*, pp. 2712-2715.
21. deOliveira, B. H. and Strapasson, R. A. (1996), "Biotransformation of isosteviol by Fusarium verticilloides," *Phytochemistry*, *43*, pp. 393-395.
22. Kohda, H., Kasai, R., Yamasaki, K., Murakami, K. and Tanaka, O. (1976), "New sweet diterpene glucosides from Stevia rebaudiana," *Phytochemistry*, *15*, pp. 981-983.

(626v) An Optimization Algorithm for Estimating Microbial Survival Parameters

Guibing Chen, Center for Excellence in Post-Harvest Technologies, North Carolina A&T State University, Kannapolis, NC

The microbial survival curve can be described by an explicit linear or nonlinear equation of time when microorganisms are exposed to a *lethal agent* (e.g., temperature and high pressure) at constant intensity. Each model parameter involved in the equation is a function of the agent intensity, where parameters used to describe the model parameters are called microbial survival parameters. If the microbial survival parameters are known, then survival curves under changing agent intensity can be predicted. Therefore, the objective of this study was to develop an algorithm based on the steepest decent optimization method to estimate microbial survival parameters from survival data under known agent profiles, i.e., agent intensity versus time. The developed algorithm was validated by using the published thermal inactivation data of Salmonella. The data showed that Salmonella's survival curves followed the Weibull model, a popular nonlinear model. Using the data, Salmonella's survival parameters were identified from final survival ratios of three survival curves, single and multiple whole survival curves under non-isothermal conditions, respectively. The results indicated that microbial survival parameters could be accurately estimated from multiple whole survival curves under changing lethal agent intensity. In principle, the developed algorithm can also be applied to studies on nutrients preservation during food processing where degradation of food nutritional compounds induced by the processing is governed by certain chemical reaction kinetics.

See more of this Session: Poster Session: Food Engineering and Bioprocessing
See more of this Group/Topical: Food, Pharmaceutical & Bioengineering Division

(626w) High Protein Content Distillers Dried Grains with Solubles

Song Li, Wei Liu, Kent D. Rausch, M. E. Tumbleson and Vijay Singh, Agricultural and Biological Engineering, University of Illinois at Urbana-Champaign, Urbana, IL

Distillers dried grains with solubles (DDGS) with 30% protein content and 40% neutral detergent fiber content (1), is a coproduct of conventional dry grind ethanol plant. Due to its high fiber content, DDGS is used primarily as ingredient in ruminant animal diets. Reducing fiber and increasing protein content will allow DDGS use as an ingredient in poultry and swine diets. Increase in protein and decrease in fiber content of DDGS can be achieved by modified dry grind corn process (E-Mill process) that fractionates corn kernel (to recover germ and fiber as coproducts) prior to fermentation (2). Further increase in protein content and decrease in fiber content can be achieved by extraction of oligosaccharides and other soluble carbohydrates or by acid precipitation of the protein from E-Mill DDGS.

Two corn hybrids were fractionated and dry grind processed to produce E-Mill DDGS. E-Mill DDGS was further processed to produce protein isolate and concentrate products. The yield of isolate product was too low (<2%) to be feasible economically. However, concentrate yield was about 43 to 52%. Protein content of concentrate product for one of the hybrids was 60%.

1. Liu, K. 2011. Chemical composition of distillers grains, a review. Agric. Food Chem. 59: 1508-1526.
2. Singh, V., Johnston, D. B., Naidu, K., Rausch, K. D., Belyea, R. L., and Tumbleson, M. E. 2005. Comparison of modified dry-grind corn processes for fermentation characteristics and DDGS composition. Cereal Chem. 82:187-19.

See more of this Session: Poster Session: Food Engineering and Bioprocessing
See more of this Group/Topical: Food, Pharmaceutical & Bioengineering Division

(626x) Carbon Footprint of a Dairy Feed Mill In the U.S

Felix Adom[1], Charles Workman[1], Greg Thoma[2] and David Shonnard[1], (1)Chemical Engineering, Michigan Technological University, Houghton, MI, (2)Department of Chemical Engineering, University of Arkansas, Fayetteville, AR

A carbon footprint analysis was conducted for a single dairy feed mill located in Michigan with the aim of developing a preliminary assessment of feed mill operations in the US. This choice of milling site is most representative of a northerly US mill which usually has its feed ingredients arriving on site with relatively high moisture content and needing drying. The main goal is to determine the greenhouse (GHG) emissions on a basis of 1 kg of milled dairy feed (an "average" feed formulation for dairy animal nutrition).

The major inputs identified in this cradle-to-factory gate analysis specifically includes; the feed ingredients (grains, some processed crop input as well as other inputs such as vitamins, minerals, and supplements), on-site energy usage, and transportation of feed input to both the milling site and various dairy farms after milling. Data collection has been a combination of a site-visit, peer reviewed journal articles, mill survey and direct communication with feed mill manager. With the feed mill reporting close to about 170 different types of feed ingredients, different modeling approaches were required for analyzing the GHG impact of the input data. Upstream GHG burdens for the inputs to the feed mill were estimated using a combination of EcoInvent database unit processes and economic input-output modeling using Economic Input/output Lifecycle Assessment (EIO-LCA) tool. About 84% (by weight) of feed ingredients (inputs) for the mill was analyzed using GHG emission factors from unit processes in the EcoInvent database while the remaining 16% was estimated by employing EIO-LCA database with Simapro as the computational tool. SimaPro was also used in estimating both the onsite energy and transportation impacts.

The relative contribution of the feed mill dairy feed to global warming potential was estimated to be 0.568 kg CO2 equivalents/kg milled dairy rations (economic allocation). The majority of the GHG footprint of mill feed is due to input crops and other major ingredients to the mill (approximately 85% of mill inputs by mass). The next largest category for GHG emissions is the large number of mineral ingredients, contributing approximately 10% to the total mill carbon footprint. Supplements, which include highly processed ingredients like vitamins and amino acids, contribute about 4% to the carbon footprint. Onsite energy consumption contributes only about 3% while transportation impact for both the delivery and distribution of feed to dairies accounted approximately 9% of the footprint.

Input crops and other major ingredients like distillers grain soluble (DDGS), corn gluten, soybean meal among others were the largest contributors to the overall footprint. The supplements and minerals together contributes a total of about 14% towards the overall footprint, this is not surprising given that they arrive at the site in a much more processed form. Finally, natural gas for crop drying accounts for 80% of the total onsite energy consumption.

See more of this Session: Poster Session: Food Engineering and Bioprocessing
See more of this Group/Topical: Food, Pharmaceutical & Bioengineering Division

Natural Prebiotics as Encapsulating Agents for Probiotic Microencapsulation by Spray Drying

Paola Hernández-Carranza, María Teresa Jiménez and Nelly Ramírez

Universidad de las Américas Puebla. Departamento de Ingeniería Química, Alimentos y Ambiental. ExHacienda Sta. Catarina Mártir S/N, Cholula, Puebla. Mexico. 72810. mariat.jimenez@udlap.mx

Abstract

Microencapsulation of bioactive compounds (antioxidants, vitamins, minerals, omega-3 lipids, probiotics, among others) has been increasingly studied in the last years due to the actual interest in nutraceutical components and functional food. The main objective of this technique is to protect the bioactive compounds from environmental conditions, which may diminish their functionality, such as oxygen, pH, humidity, light or temperature exposure. Among the different microencapsulation processes, spray drying produces a final powder product with good quality properties for distribution, transportation and storage. Different encapsulation agents have been studied for microencapsulation by spray drying in behalf of increasing the viability of the bioactive compounds and to promote an additional functionality to the final product as well, such is the case of prebiotics. Prebiotics are soluble non-digestible carbohydrates by humans but which selectively enhances *Bifidobacterium* and *Lactobacillus* growth (commonly microorganisms present in the human gut). Some examples of prebiotics are inulin, fructans (fructo-oligosacharides) or galacto-sacharides. Besides, several microorganisms (probiotics) have demonstrated benefic effects in humans which have been attributed to the lactic acid and short chain-fatty acids production, which reduces de pH of the colon, causing a decrease of the pathogenic bacteria survival.

In the present study, extracts of different natural sources (orange and grapefruit peel, artichoke and asparagus) are evaluated to determine their fructans content and use them as encapsulating agents for spray drying encapsulation of the probiotic *Lactobacillus casei*.

Previous to the extraction process with hot water at 85°C, the samples were cut in slices of 1cm width, dried in a hot air dryer at 35 °C for 48 h. Fructans content were determined by the AOAC methods 991.43 and 997.03 respectively. Aqueous solutions were prepared with the extracts and maltodextrin to achieve 25% p/p. The extracts and the solutions prepared with the extracts were subjected to spray drying process using an inlet air temperature (Tin) of 145 °C and evaluating two mass flux (QL) 10 and 15 g/min. Physical characterization of the encapsulated powder was obtained by determining moisture, water activity and moisture gain.

Results demonstrated that the vegetable studied have 0.5 – 11% of fructans. The lower outlet temperature for the spray drying process was of 63°C with QL of 15 g/min, obtaining a final probiotic content of 10^7 cfu/g. The final product showed good physical properties of water content (2%) and a_w (<0.3), smaller gain of moisture (3-4 times less) was achieved.

The final results show that natural prebiotics can be used as encapsulating agents in the spray drying process using maltodextrin to assist probiotic growth and be used as functional ingredients in new food product development.

(626z) Generating Synthetic Pathways for Valuable Bio-Commodities

Advait Apte, Biological Systems Engineering, Virginia Tech, Blacksburg, VA and Ryan S. Senger, Biological Systems Engineering Department, Virginia Tech, Blacksburg, VA

Several high-value chemical intermediates, commodities, and specialty chemicals are still produced from petroleum sources. Since several have no known biosynthetic routes, these pathways must be created through synthetic biology, should these chemicals be produced from renewable resources. Pathway generator algorithms are available in the literature; however, the problem of combinatorial explosion severely limits the size of synthetic pathways produced by algorithms based on enzyme targeting of functional groups. To address this issue, we have built a synthetic pathway discovery algorithm (SyPath) that is based on one-dimensional unique SMILES chemical representation and regular expression encoded reaction rules. Specific catalytic actions of key metabolic enzymes are well-documented in online databases. However, these datasets only account for experimentally observed enzymatic reactions and those predicted through homology modeling. It is hypothesized that several more possibilities exist, but many will only be feasible with enzyme engineering. While algorithms similar in nature to SyPath exist in the literature, here we show the novelty of our approach by providing possible long multi-step synthetic pathways to several bio-commodities of interest.

See more of this Session: Poster Session: Food Engineering and Bioprocessing
See more of this Group/Topical: Food, Pharmaceutical & Bioengineering Division

(631a) Modeling of Power-Blending Using Real-Time NIR Data

Niranjan Kottala, Bristol-Myers Squibb Co., New Brunswick, NJ, Dimuthu A. Jayawickrama, Analytical R&D, Bristol-Myers Squibb, New Brunswick, NJ and Howard Stamato, Dpst, Bristol-Myers Squibb Company, New Brunswick, NJ

Powder blending is one of the most widely used unit operation in the pharmaceutical manufacturing, but it is also the least understood unit operation. The main objective of this poster is to present the mathematical model and to discuss its application to predict the number of rotations necessary to achieve an acceptably low RSD using real-time component concentration data. Real-time data to develop the model was obtained using on-line NIR spectrophotometer attached to the blender. Component concentration data obtained as a function of blender rotations during the blending operation was used to determine the model parameter fraction mixed, and determine the minimum number of revolutions necessary to achieve a uniform blend. A good agreement was observed between the calculated and measured rotations to achieve the content uniformity during the blending operation. This model can be used as a predictive tool to quantify the powder blending upon scale-up

See more of this Session: Poster Session: Pharmaceutical Engineering
See more of this Group/Topical: Food, Pharmaceutical & Bioengineering Division

(631b) Hydrophobic Charge Induction Displacement Chromatography Visualized with Confocal Laser Scanning Microscopy

Guofeng Zhao, Lin Zhang, **Xiaoyan Dong** and Yan Sun, Department of Biochemical Engineering, School of Chemical Engineering and Technology, Tianjin University, Tianjin, China

Displacement chromatography is a separation technique with high capacity and high resolving power. Mass transfer kinetics is an important factor that determines the separation efficiency of displacement chromatography, but few studies have been reported on the kinetics of the displacement process at a particle level. In this work, the kinetics of protein and displacer adsorptions, protein displacement and adsorbent regeneration in displacement chromatography were studied by both stirred-batch experiments and online visualization with confocal laser scanning microscopy (CLSM). Stirred-batch experiments showed that small-molecule displacers had higher effective diffusivities and mass transfer rates than proteins. For the displacement systems used in this work, the rate of displacement was dependent on the mass transfer and adsorption rate of the displacer. Direct visualization of the displacement process was achieved by labeling the protein with fluorescein isothiocyanate and by using a fluorescent displacer, rhodamine 6G. The concentration profiles obtained by CLSM clearly revealed the higher intraparticle mass transfer rate of the displacer and subsequent high rate of displacement. The observations by CLSM were mostly consistent with the results of stirred-batch experiments. However, although the direct visualizations by CLSM has provided valuable information of the microscopic processes in displacement chromatography, efforts are still needed to develop an improved displacement model system (proteins and displacers) that can provide more accurate and quantitative information of displacement processes.

See more of this Session: Poster Session: Pharmaceutical Engineering
See more of this Group/Topical: Food, Pharmaceutical & Bioengineering Division

(631c) Application of Computational Fluid Dynamics (CFD) Method for the Development of Bioreactor Scale-Down Model

An Zhang[1], Valerie Tsang[1], Helena Yusuf-Makagiansar[1], Brian Bell[2] and Ankit Adhiya[2], (1)Biogen Idec, San Diego, CA, (2)ANSYS, Inc., Canonsburg, PA

A robust and representative bioreactor scale down model is becoming a fundamental tool to facilitate rapid bioprocess development. At the early stage of development process, the scale down model is critical for process definition and prototyping in order that the cell performance, product yield, and product quality defined in small scale optimization studies are repeatable in larger scale. At the later stage of development life cycle, the scale down model is a key for process characterization and development of design space of the process. Computational fluid dynamics (CFD) has been widely used to solve and analyze problems that involve fluid flows by simulating the interaction of liquids and gases with surfaces defined by boundary conditions. In this study, the CFD techniques were used to develop a scale down model to investigate mixing, oxygen mass transfer characteristics and turbulence, strain rate, and bubble size distribution in a 5L stirred-tank bioreactor based on the engineering environment for the cell growth and metabolism. In addition, the validated small scale CFD model was compared to the large scale model (2K/15K) to verify its scalability in terms of energy dissipation (power per volume) or mass transfer coefficient ($k_L a$).

See more of this Session: Poster Session: Pharmaceutical Engineering
See more of this Group/Topical: Food, Pharmaceutical & Bioengineering Division

(631d) Hot-Melt Extrusion: Case Study and the Fundamentals of Miscibility

Adeyinka Adegoke, Ashish Sarode and **Peng Wang**, Department of Chemical Engineering, University of Rhode Island, Kingston, RI

Traditional blending processes in the pharmaceutical industry are operated at room temperature. In recent years, melt mixing has received a great deal of interest, mainly due to the challenge associated with the ever increasing number of poorly soluble drugs. In addition to the dissolution enhancement, melt mixing also possesses several other advantages. It is a solvent free process; it can be realized by using a continuous extruder, which is commonly used in the plastic and processed food industries. Our experimental data show that API and polymeric excipients can form one phase at the elevated temperature. The drug may remain in amorphous form or recrystallize after the mixture is cooled to room temperature, depending mostly on the formulation. In addition, our results suggest that melt mixing can be applied to significantly increase or slow down drug's release rate, which cannot be achieved if the same components are blended at room temperature.

See more of this Session: Poster Session: Pharmaceutical Engineering
See more of this Group/Topical: Food, Pharmaceutical & Bioengineering Division

Synthesis and Anti-aggregatory of Baicalein derivatives functionalized on the B-ring

Jing Wang, Xiang Liu* and Baoping Dong
School of Chemical and Material Engineering, Jiangnan University, Wuxi 214122, China

Abstract: A concise and efficient synthesis of baicalein derivatives was introduced. Firstly, substituted cinnamoyl chlorides were synthesized starting from benzaldehyde derivatives. Then 3,4,5-trimethoxyphenol were directly acylated with substituted cinnamoyl chlorides to give corresponding chalcones by Fries rearrangement. Under the action of I_2/DMSO, 5,6,7-trimethoxylbaicalein derivatives were feasible obtained by intramolecular cyclization of chalcones. Finally, the demethylation was performed to give the desired baicalein derivatives functionalized on the B-ring in the ideal yields. The final products possess a wide range of pharmacological activity, notably anti-aggregatory,anticancer and antioxidation.

***Keywords**:* baicalein derivatives, 3,4,5-trimethoxyphenol, demethylate, anti-aggregatory

1. Introduction

Baicalein(5,6,7-trihydroxyflavone,**1a**) is the major flavonoid that isolate from *Scutellaria Baicalensis Georgi*, a famous traditional Chinese herb widely used to cure hypertension, inflammatory and aggregatory diseases[1] since ancient times in China. Recently investigated show that baicalein also has anti-HIV effect[2].The main source of baicalein results from plants. Until now, several synthetic routes of this compound have been reported. In 1973, Agasimundin and Siddappa[3] reported a route make use of 1-(2,4,6-trihydroxyphenyl)ethanone and 1-ethyl-3-phenyl-1,3-propanedione to get the intermediate 6-Acetyl-5,7-dihydroxyflavone,which was treated with H_2O_2 to give baicalein. In 2010, Jian Yang and co-workers also reported a reaction of 3,4,5-trimethoxyphenol, acetic anhydride and benzaldehyde to get 2′-Hydroxy-4′,5′,6′-trimethoxychalcone and then treated with HBr–HOAc to obtain baicalein[4]. Righi[5] utilized a procedure synthesis of baicalein starting from crysin by Wesley–Moser type rearrangement.

But the synthesis of baicalein derivatives is rarely reported. Herein, we report a concise and efficient procedure to synthesize baicalein derivatives functionalized on the B-ring starting from benzaldehyde derivatives and 3,4,5-trimethoxyphenol.

E-mail address: liuxiang@jiangnan.edu.cn (X. Liu)

2. *Results and discussion*

A concise and efficient synthesis of baicalein derivatives functionalized on the B-ring was carried out by the reactions outlined in scheme1. Firstly,substituted cinnamic acids **3** were obtained from corresponding benzaldehyde derivatives **2** by Perkin Reaction. Then cinnamic acids **3** were easily generated chlorides **4** with the chlorinating agent. Under the action of BF_3–Et_2O complex, substituted cinnamoyl chlorides were directly acylated with 3,4,5-trimethoxyphenol **5** followed by Fries rearrangement to give chalcones **6**. The following intramolecular oxidative cyclization of chalcones was achieved with the catalytic amounts of iodine in DMSO[6] to acquire 5,6,7-trimethoxylbaicalein derivatives 7. Finally, **7** was treated with 47% HBr/HOAc (1 : 2) for demethylation to get the desired baicalein derivatives **1** functionalized on the B-ring in the ideal yields.

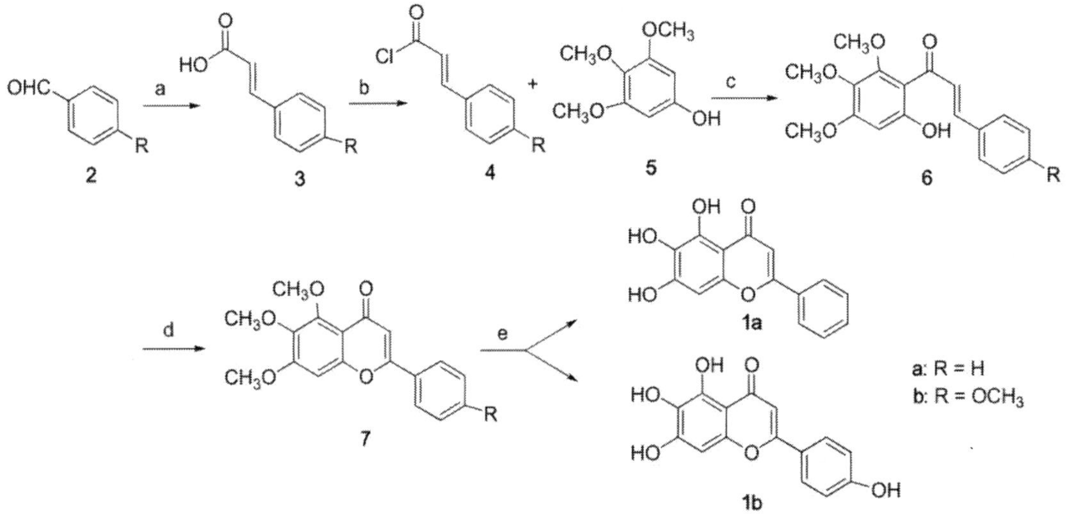

Scheme 1. Reagents and conditions: a. $(CH_3CO)_2O$, CH_3COOK,heat; b. $SOCl_2$, CH_2Cl_2; c. BF_3-Et_2O, heat; d. I_2, DMSO, heat; e. 47% HBr/HOAc, heat.

3. *Experimental*

All reagents were commercial materials and were used directly unless otherwise noted. DMF was dehydrated over 4Å molecular sieves. All melting points were measured on a SPSIC WRS-1B melting apparatus and were uncorrected. IR spectra were recorded on a Boman FTLA2000 IR spectrometer. ^{1}H NMR spectra were recorded on a 400 MHz Bruker Avance III 400 NMR spectrometer. Mass spectra were recorded on a Waters Acquity UPLCSQD Liquid Mass spectrometer.

General synthesis procedure

3.1 substituted cinnamic acids 3

In a 100 mL three-necked flask fitted with a mechanical stirrer,10 mL(0.11 mol) of Acetic Anhydride and 8.6 mL(0.085 mol) of benzaldehyde derivative **2** were added. Potassium acetate as catalyst, the reaction mixture was heated under reflux for 1 h and then added 50 mL hot water, a little sodium carbonate to adjust the solution alkaline. The mixture was steam distilled untill the distillate has no oil-free beads, vacuum filtration, collected the filtrates and added HCl to give a precipitate, which was crystallized from ethanol to give the desired product **3**.

cinnamic acid(3a) According to the synthesis procedure the title compound was obtained as a white crystals(8.2 g,65%): mp 135-136□; IR(KBr) v cm^{-1}=3074 (CH_3) , 1685 (C=O), 1626 (C=C); ^{1}H-NMR(400 MHz, DMSO-d_6) δ 12.40 (s, 1H, –COOH), 7.69 (dd, *J* = 6.6,2.9 Hz, 2H, H–2,6), 7.60 (d, *J* = 16.0 Hz, 1H, H–*α*), 7.42 (dd, *J* = 6.5,3.7 Hz, 3H, H–3,4,5), 6.54 (d, *J* = 16.0 Hz, 1H, H–*β*).

4-methoxy cinnamic acid(3b) According to the synthesis procedure the title compound was obtained as a white crystals(10.6 g,70%): mp 171-172□; IR(KBr) v cm^{-1} = 2957(CH_3), 1686 (C=O), 1598 (C=C); ^{1}H-NMR(400 MHz, DMSO-d_6) δ 12.24 (s, 1H ,–COOH), 7.64 (d, *J* = 8.7Hz, 2H, H–2,6), 7.54 (d, *J* = 16.0 Hz, 1H, H–*α*), 6.97 (d, *J* = 8.7 Hz, 2H, H–3,5), 6.38 (d, *J* = 16.0 Hz, 1H, H–*β*), 3.79 (s, 3H, 4–OCH_3).

3.2 Chalcones 6

A solution of a cinnamic acid derivative **3** in dichloromethane (20 mL) was added in a dry round-bottomed flask. Sulfone chloride was added *via* syringe followed by a trace amount of anhydrous DMF. After maintaining at room temperature for 4 h, the reaction mixture was concentrated in vacuum to give the corresponding chloride **4**, which was used directly without further purification. To a mixture of the aryl chloride **4** and 3,4,5-trimethoxyphenol **5** (20 mmol) was slowly added BF_3–Et_2O (20 ml) *via* syringe. The reaction mixture was heated for 30 min and then quenched with excess of water. Filtration and recrystallization from ethanol gave Chalcone **6**.

9-Hydroxy-5,6,7-trimethoxychalcone(6a) According to the synthesis procedure the title compound was obtained as an orange solid(4.90 g,78%): mp 86-87□; IR(KBr) v cm^{-1}=1630 (C=O), 1561 (C=C); ^{1}H-NMR(400 MHz, DMSO-d_6) δ 12.16 (s, 1H, –OH), 7.73 (dd, *J* = 6.6, 2.8 Hz, 2H, H–2',6'),7.59 (d, *J* = 4.2 Hz, 2H, H–2,3), 7.45 (m, 3H, H–3', 4',5'), 6.39 (s, 1H, H–8), 3.84 (d, *J* = 4.2 Hz, 6H, 5,7–OCH_3), 3.69 (s, 3H, 6–OCH_3).

9-Hydroxy-5,6,7,4′-tetramethoxychalcone(6b) According to the synthesis procedure the title compound was obtained as a golden yellow solid(5.72 g,83%): mp 137-138□; IR(KBr) v cm^{-1}=1622 (C=O), 1552 (C=C); ^{1}H-NMR(400 MHz, DMSO-d_6) δ 12.29 (s, 1H, –OH), 7.69 (d, *J* = 8.8 Hz, 2H, H–2',6'), 7.60 (d,*J* = 15.8 Hz, 1H, H–3), 7.46 (d, *J* = 15.7 Hz, 1H, H–2), 7.01 (d, *J* = 8.8 Hz, 2H, H–3',5'), 6.37 (s,1H, H–8), 3.83 (d, *J* = 2.9 Hz, 6H, 5,7–OCH_3), 3.81 (s, 3H, 6–OCH_3), 3.69 (s, 3H, 4'–OCH_3).

3.3 Flavonoids 7

A mixture of a chalcone derivative **6** (10 mmol) and iodine(0.2 g, 0.79 mmol) in DMSO(30 mL) was heated to refluxed for 2 h, and then carefully poured into 20% sodium sulfite solution. The mixture was filtered and the cake was recrystallized from ethanol to give compound **7**.

5,6,7-Trimethoxylbaicalein(7a) According to the synthesis procedure the title compound was obtained as a yellow solid(2.85 g, 90%): mp 156-157□; IR(KBr) v cm^{-1}=1637 (C=O); ^{1}H-NMR(400 MHz, DMSO-d_6) δ 8.08 (dd, *J* = 7.6, 1.9 Hz,2H, H–2',6'), 7.58 (m, 3H, H–3', 4',5'), 7.26 (s, 1H, H–3), 6.84 (s, 1H, H–8), 3.96 (s, 3H, 7–OCH_3), 3.81 (s, 3H, 5–OCH_3), 3.77 (s, 3H, 6–OCH_3). EI-MS: *m/z* $[M]^+$313.

4′-Methoxy-5,6,7-trimethoxylbaicalein(7b) According to the synthesis procedure the title compound was obtained as a dark yellow solid(2.3 g, 68%): mp 147-148□; IR(KBr) v cm^{-1}=1636

(C=O), 1603 (C=C); ^{1}H-NMR(400 MHz, DMSO-d_6) δ 8.03 (d, *J* = 8.7 Hz, 2H, H–2',6'), 7.23 (s, 1H, H–3), 7.11(d, *J* = 8.8 Hz, 2H, H–3',5'), 6.73 (s, 1H, H–8), 3.95 (s, 3H, 7–OCH_3), 3.85 (s, 3H, 5–OCH_3), 3.80 (s, 3H, 6–OCH_3), 3.76 (s, 3H, 4'–OCH_3). EI-MS: *m/z* $[M]^+$343.

3.4 Baicalein derivatives 1

A solution of a flavonoid derivative**7**(3.2 mmol) in 47% HBr (25 ml) and glacial acetic acid (50 ml) was heated to refluxed for 20 h and then carefully poured onto crushed ice (500 g). The resulting yellow precipitate was filtered and collected. Recrystallization from ethanol afforded the desired baicalein derivative **1**.

Baicalein(1a) According to the synthesis procedure the title compound was obtained as a yellow crystal(0.8 g ,90%): mp 257-258□; IR(KBr) ν cm^{-1} =3000-3400 (OH),1660 (C=O); ^{1}H-NMR(400 MHz, DMSO-d_6) δ 12.65 (s, 1H, 5–OH), 10.55 (s, 1H, 7–OH), 8.80(s, 1H, 6–OH), 8.13 (d, *J* = 6.5 Hz, 2H, H–2',6'), 7.62 (m, 3H, H–3',4',5'), 6.93 (s, 1H, H–3), 6.62 (s, 1H, H–8). EI-MS: *m/z* $[M]^+$271.

4′-Hydroxy Baicalein(1b) According to the synthesis procedure the title compound was obtained as a yellow crystal(0.83 g ,93%): mp 288-290□; IR(KBr) ν cm^{-1} =3000-3400 (OH),1665 (C=O); ^{1}H-NMR(400 MHz, DMSO-d_6) δ 12.79 (s, 1H, 4'–OH), 10.46 (s, 1H, 5–OH), 10.32 (d, *J* = 13.4 Hz, 1H, 7–OH), 8.74 (s, 1H , 6–OH), 7.93 (dd, *J* = 17.0,8.7 Hz, 2H, H–2',6'), 6.92 (d, *J* = 8.2 Hz, 2H, H–3',5'), 6.77 (d, *J* = 27.0 Hz, 1H, H–3), 6.57 (s, 1H, H–8). EI-MS: *m/z* $[M]^+$287.

References

1. Kimara Y. Yokoi K. Matsushita N. and Okuda H(1997),"Effects of flavonoids isolated from Scutellariae radix on the production of tissue-type plasminogen activator and plasminogen activator and plasminogen activator inhibitor induced by thrombin and thrombin receptor agonist peptide in cultured human umbilical ven endothlial cell," *J Pharm Pharmacol*, 49(8), pp.816–822.
2. Wu JA. Attle AS. and Zhang L. *etal*(2001),"Anti-HIV activity of medicinal herbs:usage and potenitial development,"*Am J Chin Med*, 29(1), pp.69–81.
3. Y.S. Agasimundin and S.Siddappa(1973),"A Convenient Method for the Synthesis of Baicalein and 4'-Methylgamatin," *J.C.S. Perkin I*, pp.503–505.
4. Duo-Zhi Chen. Jian Yang. and Bo Yang. *etal*(2010),"Total synthesis of baicalein," *Journal of Asian Natural Products Research*, 12(2), pp.124–128.
5. Giuliana Righi. Roberto Antonioletti. and Ilaria Proietti Silvestri. *etal*(2010),"Convergent synthesis of mosloflavone, negletein and baicalein from crysin," *Tetrahedron,* 66, pp.1294–1298.
6. Pinto Diana C. G. A. Silva Artur M. S.and Cavaleiro Jose A. S (1996),*J. Heterocyclic Chem*, 33, pp.1887–1893.

(631f) Recycling Homogeneous Catalysts: Tunable Solvents for the Hydroformylation of p-Methylstyrene

Wilmarie Medina-Ramos[1], Ali Z. Fadhel[2], Rani Jha[3], Pamela Pollet[3], Charles L. Liotta[3] and Charles A. Eckert[1], (1)Chemical and Biomolecular Engineering, Georgia Institute of Technology, Atlanta, GA, (2)Georgia Institute of Technology, Atlanta, GA, (3)Chemistry and Biochemistry, Georgia Institute of Technology, Atlanta, GA

We use **O**rganic-**A**queous **T**unable **S**olvents (OATS) for the rhodium-catalyzed hydroformylation of *p*-methylstyrene. OATS combine the benefits of homogeneous and heterogeneous catalysis while overcoming the limitations of both: Reactions are carried out under homogeneous conditions, followed by a CO_2-induced heterogeneous separation. Then the polar protic component (aqueous-rich phase) containing the catalyst phase separates from the relatively nonpolar aprotic component (organic-rich phase, usually THF or acetonitrile) containing the reactant, facilitating the separation and the recycle of catalyst. The homogeneous reactions provide fast, highly-selective reactions with excellent yields. Upon the addition of moderate pressures (around 3 MPa) of CO_2, the hydrophobic product partitions into the organic-rich phase with more than 99% removal efficiency and the hydrophilic catalyst is retained in the aqueous-rich phase with 99.9% retention efficiency. In addition, we were able to recycle the catalyst for five consecutive reactions without significant loss of catalytic activity. Also, we explored applications of monophasic mixtures of organics and ionic-liquids to combine homogeneous reactions with CO_2-induced heterogeneous separations in these systems.

See more of this Session: Poster Session: Pharmaceutical Engineering
See more of this Group/Topical: Food, Pharmaceutical & Bioengineering Division

(631g) Understanding Organic Scale Deposition In Microfluidic Systems for Fine Chemical Synthesis

W. Jeffrey Horne and Ryan L. Hartman, Department of Chemical and Biological Engineering, University of Alabama, Tuscaloosa, AL

Active pharmaceutical intermediates (APIs) have been discovered, studied, and produced using batch processes in the laboratory and industrial settings. However, the synthesis and production of important compounds in continuous processes can offer economic advantages and safer routes to preparation. Moving from batch to continuous flow systems is not without challenges in real-world scenarios. The preparation of useful compounds often requires multiple reaction and separation steps. Additionally, the majority of reactions important to fine chemicals involve the formation of solids, which can deposit on reactor, equipment, and catalyst support surfaces. Understanding organic scale deposition in continuous processing, a largely unaddressed issue, is the main focus of this project.

The study of reaction pathways on the micro-scale can offer advantages, including smaller amounts of reactants and solvents, significantly higher surface-to-volume ratios in the reactor, and enhanced mass- and heat-transfer characteristics. In this work, we investigate the microfluidic deposition and dissolution of biaryl compounds commonly encountered in fine chemical synthesis in attempt to understand the molecular and physical interactions of organic compounds with the surfaces of the reactors and other process equipment. These results will be presented and discussed.

See more of this Session: Poster Session: Pharmaceutical Engineering
See more of this Group/Topical: Food, Pharmaceutical & Bioengineering Division

(631h) Effect of Composition and Temperature On the Structural Evolution and Dissolution Behavior of Drug/Polymer Systems

Qing Zhu[1], Michael T. Harris[1], Lynne Taylor[2] and Hsin-Yun Hsu[1], (1)Chemical Engineering, Purdue University, West Lafayette, IN, (2)Industrial and Physical Pharmacy, Purdue University, West Lafayette, IN

In this study, it is found that both the composition and crystallization temperature played important roles in determining the size and location of the drug domains in the polymer matrix and hence affect the dissolution behavior. By optimizing the preparation parameters, the formulation with desired performance can be designed.

See more of this Session: Poster Session: Pharmaceutical Engineering
See more of this Group/Topical: Food, Pharmaceutical & Bioengineering Division

(631i) Dispersion In Plug Flow Tube Reactors Used for Pharmaceutical Intermediates

Brian D. Haeberle[1], Martin D. Johnson[2], Nikolay Zaborenko[2] and Adam D. McFarland[3], (1)Eli Lilly and Company, Indianapolis, IN, (2)Chemical Product Research and Development, Eli Lilly and Company, Indianapolis, IN, (3)Analytical Sciences Research and Development, Eli Lilly and Company, Indianapolis, IN

Flow tube reactors are gaining widespread use in process development groups at Eli Lilly. Characterizing axial dispersion is important to evaluate reactor design and is used to predict conversion and product distribution along the length of the reactor. Lilly has utilized tube reactors for continuous asymmetric hydrogenations, hydroformylations, thermal rearrangements, thermal couplings, cyclizations, azide chemistries, direct hydroxylation, cyclopropanation, and oxidation reactions. Dispersion in both coiled tube reactors and pipes in series geometries were investigated. The volume of the reactors ranged from 70 ml to 7 L. The residence time distribution was used to determine the dispersion of each reactor. To determine the RTD a step change in a solution of 40/60% THF/Toluene to 40/60% THF/m-Xylene was implemented. The concentration of m-Xylene in the reactor effluent was monitored by an in line Raman probe positioned at the outlet of the reactor. From the F curve the dispersion coefficient was calculated. The reactors investigated in the study were designed for low axial dispersion and resulted in dispersion numbers on the order of 0.001.

See more of this Session: Poster Session: Pharmaceutical Engineering
See more of this Group/Topical: Food, Pharmaceutical & Bioengineering Division

(631j) Study Effect of Hydroxypropyl Cellulose Grade and Foam Quality On Foam Granulation of a High Drug Load Formulation

Chandrakant Patel, Otilia Koo and Faranak Nikfar, Drug Product Science and Technology, Bristol-Myers Squibb Company, New Brunswick, NJ

Purpose: To investigate the influences of HPC grade and foam quality on foam granulation of a high drug load formulation.

Methods: Two grades of HPC (LF and EXF) with different viscosities were used to produce foam of varied quality (90%, 50%, and 0%). The foam quality is expressed as the ratio of the air to the liquid on volume basis. A pilot scale foam generator was employed to make foam of desired quality. The viscosities of the HPC solutions were determined using Brookfield$^{\grave{O}}$ viscometer. A high drug load (80.5% w/w) formulation was granulated in a Diosna$^{\grave{O}}$ granulator fitted with 4L bowl (batch size: 400 g, HPC solution delivered as foam: 25% w/w, foam addition rate: 75 g/min, impeller tip speed: 5.6 m/s, high chopper, 95 s granulation time, 30 s wet massing time). The power profiles were monitored during granulation. The granules were dried using a fluid bed dryer at 75°C and further milled by passing through a conical mill fitted with 0.045 inch round screen. Granule properties such as particle size, flow and bulk density were measured. The compactibility of the granules was assessed using Stylcam$^{\grave{O}}$.

Results: The viscosities of HPC LF and EXF solutions used to prepare the foam were 282 and 141 cP. At similar foam qualities, HPC LF foam resulted in higher power consumption than HPC EXF foam. There was no clear trend on power consumption with decreasing foam quality during granulation. Granulations made with different HPC grade and foam quality did not exhibit any significant difference in drying behavior. The granulation prepared with HPC EXF 0% foam quality exhibited the highest % fines among all the granulations. Granulations prepared with 50% foam quality showed poorer flow compared to 90% and 0% foam quality, regardless of HPC grade. Bulk density was similar among the granulations. Compactibility of the granulations was similar, although there were some difference when dwell time was decreased.

Conclusions: HPC grade or the foam quality did not have major effect during the foam granulation process and on the granulation properties when granulated at optimal foam level of 25% w/w. However, HPC grade and foam quality can be chosen to tailor the desired granulation properties.

(631k) Modeling of Batch Distillation for Scale-up

Andrew S. Cosbie and Seth Huggins, Chemical Process R&D, Amgen, Thousand Oaks, CA

Modeling a batch distillation before scaling up can provide useful insight to the time and materials required to reach the desired end point. Commercially available software offers models that have already been developed and simply require the user to enter parameters specific to their equipment and process streams. We present several distillation models developed in DynoChem using activity coefficients from different thermodynamic models. We discuss the accuracy of these models on the kilo-scale through multiple successful applications. We also demonstrate the regression of NRTL parameters for a binary solvent system with the presence of a dissolved, non-volatile component.

See more of this Session: Poster Session: Pharmaceutical Engineering
See more of this Group/Topical: Food, Pharmaceutical & Bioengineering Division

(631l) Fluidized Bed Drying of Pharmaceutical Powders and Granules

Xue Liu, Rutgers University, Piscataway, NJ, Fernando. J Muzzio, Chemical and Biochemical Engineering, Rutgers University, Piscataway, NJ, Johannes G. Khinast, Institute for Process and Particle Engineering, Graz University of Technology, Graz, Austria and Benjamin Glasser, Dept. of Chemical and Biochemical Engineering, Rutgers University, Piscataway, NJ

Fluidized beds are extensively used in the pharmaceutical and other chemical industries either as a batch or continuous process for drying moist powders and granular solids because of good mixing of solids and intensive heat and mass transfer between the solid and hot gas phases in the system. Generally, high energy input is required during this process to provide the latent heat of water evaporation, and thus on-line measurements and computational simulations of fluidized bed drying become important to determine the optimal operation conditions in order to minimize energy usage. In the pharmaceutical industry, the drying step in many pharmaceutical production lines is often a bottleneck because it is difficult to determine the time required to dry the materials as well as analytically verify that a predetermined end point has been reached. If the active pharmaceutical ingredient has a low melting point, controlling the drying process becomes more critical due to the sensitivity of the formulation to the temperature and the risk of compromising product quality and performance during drying.

In this work, moist pharmaceutical powders and granules, such as lactose, avicel and dibasic calcium phosphate anhydrous (DCPA), have been dried using a Glatt GPCG-1 spray fluidized bed dryer/granulator. The pressure drop across the bed, the inlet and outlet air temperatures and the product temperature are measured as a function of time. The moisture content in the product during drying is measured using two approaches: 1) Taking samples from the bed every 3 minutes, drying the samples in an oven and calculate the sample moisture based on the mass loss before and after drying. 2) Using NIR to on-line measure the sample moisture. We have compared the NIR on-line measurement and the reference samples (dried in the oven), and a good match can be obtained. The NIR method can be used to determine the end point of the drying process. We varied the initial moisture content inside the pharmaceutical materials from 5% to 20% because in pharmaceutical manufacturing fluidized bed drying is generally carried out after the wet granulation step and the moisture content in the wet granules is normally in this range. We have classified the powders/granules into fines, medium, coarse, and examined the flow hydrodynamics, the drying rate and the agglomeration behaviors for each size cut as well as mixtures in the bed. The impact of air flow and air temperature on the drying process was also investigated. This provides us useful information for the drying mechanisms given some specific powder/granule materials.

To optimize the fluidized bed drying process, we have carried out simulations using a commercial software, STAR CCM+ by CD-adapco. The simulation results are compared with experimental measurements. The goal is to characterize performance of fluidized bed dryer systems and to develop a predictive model for drying of granulations with different size distributions and different moisture contents.

See more of this Session: Poster Session: Pharmaceutical Engineering
See more of this Group/Topical: Food, Pharmaceutical & Bioengineering Division

(631m) Effect of Chistoan Molecular Weight On the Transport of Solutes Across a Model Pulmonary Epithelium

Radovan Dimovski[1], Daniel Brewer[1], Balaji Bharatwaj[2] and Sandro R. P. da Rocha[2], (1)Chemical Engineering and Materials' Science, Wayne State University, Detroit, MI, (2)Chemical Engineering and Materials Science, Wayne State University, Detroit, MI

Due to its proximity to the bloodstream, oral inhalation (OI) of therapeutics is a viable non-invasive drug delivery strategy for targeting not only the diseases of the lung, but also for systemic delivery. Characteristics such as its large surface area, thin cellular barrier, and low proteolytic activity makes the lung an attractive organ for administering therapeutics capable of successfully permeating into systemic circulation. The intercellular tight junctions populating the pulmonary epithelia are among the several biological hindrances that need to be overcome in order to successfully transport relevant therapeutic moieties. A strategy that has been widely employed for opening the tight junctions is the use of permeation enhancers. One such permeation enhancer that has been successful shown to modulate the tight junctional properties of pulmonary epithelium is high molecular weight Chitosan (CS). CS is a biocompatible, biodegradable polysaccharide that typically exhibits low cytotoxicity.

Several previous studies have focused on the permeation enhancement effects of CS at high molecular weights, and have concluded that CS was able to considerably enhance the transport of macromolecules across pulmonary epithelia. Here, we discuss the permeation effects of both low and high MW CS on the transport across a pulmonary epithelium, Calu-3. CS was depolymerized to different MWs. Calu-3 cells were cultured on Transwell® inserts under an air-interface, grown to confluence and characterized. The transport of solutes was evaluated in presence of CS with different MWs.

Keywords: Pulmonary drug delivery; Calu-3; permeation enhancement; transport of therapeutics

See more of this Session: Poster Session: Pharmaceutical Engineering
See more of this Group/Topical: Food, Pharmaceutical & Bioengineering Division

(631n) Free Interface Diffusion Based Microfluidic Platform for Solid Form Screening of Candidate Drugs

Sachit Goyal[1], Michael R. Thorson[1], Yuchuan Gong[2], Geoff G.Z. Zhang[2], Charles F. Zukoski[1] and Paul J.A. Kenis[1], (1)Department of Chemical & Biomolecular Engineering, University of Illinois at Urbana Champaign, Urbana, IL, (2)Solid State Sciences, Abbott Laboratories, North Chicago, IL

Screening for salts or cocrystals of candidate drugs (CDs) with appropriate physicochemical properties is an expensive and critical step in the drug development process. We present a microfluidic platform for solid form screening (salt and co-crystal) of candidate drugs using limited sample amounts (< 10 ug). Traditional automated well-plate based screening methods [1] require 5-10 mg quantities of compound per condition. Using these methods solid form screening is delayed, as limited amount (≈ 10 mg) of CD is available early in the development phase. Using a microfluidic approach enables reduction in the reaction volumes, increase in number of conditions that can be screened and improvement in control over mixing [2]. Controlled mixing provides control over local supersaturation levels, thereby affecting the likelihood of salt formation.

A high throughput platform employing free interface diffusion has been reported for protein crystallization [2]. The platform's limitation like solvent absorption and incompatibility with organic solvents restricts its utility to the pharmaceutical industry. Here, we present a microfluidic platform compatible with a wide range of solvents used in the pharmaceutical industry. This platform [3] enables combinatorial mixing of CDs and counterions (salt or co-crystal formers) solutions on-chip in 48 wells arrays (~ 75 nL each). Several model compounds, carbamazepine, piroxicam, and caffeine, have been used to validate the platform's ability to screen for cocrystals or salts. On-chip Raman analysis was used to confirm the identity of the crystalline forms.

References

1. http://symyx.desantisbreindel.com/downloads/Polymorph%20Workflow.pdf
2. C. L. Hansen, E. Skordalakes, J. M. Berger and S. R. Quake, PNAS, **2002**, 99, 16531-16536
3. M.R. Thorson, S. Goyal, Y. Gong, G.G.Z. Zhang, C.F. Zukoski and P.J.A. Kenis, Lab on a Chip (to be submitted in May, **2011**)

See more of this Session: Poster Session: Pharmaceutical Engineering
See more of this Group/Topical: Food, Pharmaceutical & Bioengineering Division

(631o) PEG-Grafted Chitosan As Biodegradable Gene Carriers to the Pulmonary Epithelium

Daniel Brewer[1], Denise S. Conti[1], Fernando Cassio[2], Reinaldo Bazito[2] and Sandro R. P. da Rocha[1], (1) Chemical Engineering and Materials Science, Wayne State University, Detroit, MI, (2)Institute of Chemistry, University of So Paulo

Several pulmonary disorders may be targeted with gene therapy, including lung cancer, cystic fibrosis, asthma, and chronic obstructive pulmonary disease.[1] However, there are many challenges that need to be overcome in order to successfully deliver genetic material to the respiratory tract. The mucus layer (upper airways), lung surfactant (alveoli), and alveolar macrophages are some of the most important extracellular barriers. Uptake, cellular trafficking, and release of the genetic material constitute the main intracellular barriers. In order to overcome such barriers, various gene carriers have been proposed. Polymeric gene carriers (polyplexes) have demonstrated advantages over viruses and lipids, but the polymer chemistry and the physical properties of the polyplexes are known to greatly affect the transfection process. In addition, the aerosolisation technique can also play a role in transfection efficiency, since the amount and location where the polyplexes are delivered depend on the device and on particle engineering. The formulation of such carriers is thus of great relevance as well.

In this work will discuss the transfection efficiency of polyethylene glycol grafted chitosan (CS-*g*-PEG) polyplexes in airway and alveolar cells *in vitro*. Those results will be compared with the efficiency of non-modified CS polyplexes. PEG has been selected, as it has been shown to reduce the clearance from alveolar macrophages, provide faster diffusion and less aggregation through the mucus layer, to interact less with pulmonary surfactant, to diminish serum protein coating, and also to enhance the stability of dispersed nanoparticles.[2] The transfection results will be discussed based on transport studies of the polyplexes across a model mucosal layer, and their interaction with lung surfactant.

Key Words: Oral Inhalation, Gene Delivery, Lungs, *In Vitro* Transfection, Polyplexes, Mucus Layer, Lung Surfactant.

References:

1. Roy, I. and N. Vij, *Nanodelivery in airway diseases: Challenges and therapeutic applications.* Nanomedicine, 2010. **6**(2): p. 237-244.

2. Sanders, N., et al., *Extracellular barriers in respiratory gene therapy.* Adv. Drug Deliv. Rev., 2009. **61**(2): p. 115-127.

(631p) Formation and Characterization of PLA/ PCL Particles by Supercritical Fluid Technology for Release of Active Agents

Priscila S. C. Sacchetin[1], Rafaela F. Setti[2], Paulo T. V. Rosa[2] and **ngela M. Moraes**[1], (1)School of Chemical Engineering, State University of Campinas, Campinas, Brazil, (2)Institute of Chemistry, State University of Campinas, Campinas, Brazil

Polycaprolactone or simply PCL stands out among the polymers that have been used in the constitution of biomaterials. When compared to other compounds of the same category, it has the appropriate profile as a carrier of drugs that need to be released for long periods. Its biodegradability can be controlled by associating this material with other synthetic polymers such as poly lactic acid, or PLA. The polymer blends formed by the combination of these materials may show improvement in mechanical properties and also on active agents permeability in drug delivery systems. The formation of particles through the use of supercritical fluid technology has been evaluated over the last years, however there has been a major difficulty in the formation of polymeric particulate systems when using only PCL. This fact can be attributed to PCL tendency of forming fibers and films under supercritical conditions. A promising alternative is the association of PCL with PLA, since the use of PLA can successfully result in microparticles formation in supercritical conditions. In this work, the influence of different concentration ratios of PCL and PLA (ranging from 1:9 to 1:1 in polymer solutions at 1% w/v in dichloromethane) was studied on particle formation under controlled conditions of pressure (8 MPa), polymer solution flow rate (0.5 mL/min), CO_2 flow rate (18 g/mL.h) and temperature (40 °C), through the technique of polymer precipitation by SAS, or Supercritical Anti-Solvent method. PCL/PLA particles were successfully obtained particles using this approach. Residual solvent (accessed by head space gas chromatography) was lower than 600 ppm. The mean particle size increased (from 15 to 54 um) increasing PCL concentration from 10 to 50%. It was also noticed that the low miscibility between PLA and PCL, as confirmed by differential scanning calorimetry (DSC), resulted in the formation of aggregated particles. The degree of crystallinity evaluated by X-ray diffraction showed that as the PCL concentration increased in the polymer solution, a reduction of PLA crystallinity was observed. Samples prepared in all tested conditions showed low zeta potential values, regardless of the variation in the PLA/ PCL proportion. Samples prepared at low PCL to PLA ratios seemed, then, more appropriated to be used in pharmaceutical applications.

See more of this Session: Poster Session: Pharmaceutical Engineering
See more of this Group/Topical: Food, Pharmaceutical & Bioengineering Division

(631q) Development and Characterization of Porous Chitosan/Xanthan Gum Scaffolds for Animal Cells Culture

Mrcia Z. Bellini and **ngela M. Moraes**, School of Chemical Engineering, State University of Campinas, Campinas, Brazil

Polymeric biodegradable materials are usually employed in the tissue engineering area as constituents of supports for cell adhesion, maintenance and growth. Properties such as biocompatibility, pore shapes and sizes, behavior in biological fluids, as well as the characteristics of the tissue to be regenerated itself, are important issues for the selection of the materials to be used in the production of this kind of supports. Systems based on the biocompatible biopolymers chitosan and xanthan may be a promising alternative for the preparation of scaffolds intended for use in skin healing. The purpose of this work was then to develop and evaluate the characteristics of chitosan and xanthan porous scaffolds obtained in the presence or not of the surfactants Tween 80 and Pluronic F68. The results show that the proportion of the polymers and the addition of the surfactants affect the scaffolds thickness, porosity, mechanical resistance and cytotoxicity, as well their ability to absorb distinct aqueous solutions and their stability regarding mass loss in the same solutions. Supports without surfactants showed thickness from 117 to 123 μm, high aqueous solutions absorption capacity (from 6 to 86 g^{-1} of solution g^{-1} of dry membrane), maximum mass loss of 23% when exposed to serum-supplemented RPMI culture medium for 144 h, tensile strength up to 25 MPa and volume expansion around 8 times when hydrated. The addition of surfactants resulted in thickness increase to up to 809 μm, decreasing, however, volume expansion after hydration in around 75%, tensile strenght to around 1 MPa and aqueous solutions absorption to 7 to 34 g^{-1} of solution g^{-1} of dry membrane. *In vitro* tests performed with L929 cells cultured in serum-supplemented RPMI showed that the supports prepared without surfactants and with Pluronic F68 are not cytotoxic, while samples obtained with Tween 80 induced high cell death, despite the fact that both surfactants should not affect cell behavior at the employed concentration.

See more of this Session: Poster Session: Pharmaceutical Engineering
See more of this Group/Topical: Food, Pharmaceutical & Bioengineering Division

(631r) The Effect of Tablet Porosity and Droplet Infilitration On the Release Rates of API Solutions Deposited Onto Methyl Cellulose Tablets

Marlena Brown, Chemical and Biochemical Engineering/Food Science, Rutgers University, Ocean, NJ

Spreading and infiltration of inkjet-printed droplets of ibuprofen has been studied. Small droplets of an aqueous solution (8.8 % w/w ibuprofen, ethanol, and ethylene glycol monoethyl ether) were deposited onto porous methylcellulose tablets compressed at 10kN, 15kN, and 20kN pressures. The wetting dynamics, spreading, and infiltration of the droplets were characterized to determine the effect on API dissolution profiles. Mass transfer of the droplet was also examined to determine the coupling between the processes of the spreading, infiltration, and evaporation dynamics. It was found that increased values of pressure in tablet compression resulted in diminished API dissolution rates. Additionally, it was determined that the infiltration of the drug into the tablet pores decreased directly as tablet compression increased.

See more of this Session: Poster Session: Pharmaceutical Engineering
See more of this Group/Topical: Food, Pharmaceutical & Bioengineering Division

(631s) Dissolution of Non-Disintigrating Solid Dosage Forms In a Modified Dissolution Testing Apparatus 2

Xiangming Wu and **Piero M. Armenante**, Otto H. York Department of Chemical, Biological and Pharmaceutical Engineering, New Jersey Institute of Technology, Newark, NJ

In the pharmaceutical industry, dissolution testing is routinely carried out to determine the dissolution rate of oral solid dosage forms. The United States Pharmacopoeia (USP) Dissolution Apparatus 2 is the most commonly used device used in dissolution testing. Although the equipment and procedures are listed in the USP, errors and test failures still remain.

Hydrodynamics effects have been shown to play a major role on testing performance, and tablet location in the dissolution vessel can have a significant effect on the dissolution profile. Changing the location of the impeller can change the hydrodynamics effects in the bottom area of the vessel. Thus, a Modified Dissolution Testing Apparatus 2 is being proposed here, whose impeller is located off center.

In this work, the dissolution profiles of non-disintegrating calibrator tablets containing salicylic acid were experimentally determined using two systems, i.e., a Standard USP Dissolution Testing Apparatus 2 (Standard System) and a Modified Standard USP Dissolution Testing Apparatus 2 (Modified System) in which the impeller was located 8 mm off the vessel centerline. The dissolving tablets were located at different off-center positions on the vessel bottom to test the effect of tablet location in these two systems.

Tablet dissolution in the Standard System was found to be strongly dependent on tablet location, as previously reported by this and other research groups. This apparatus appears to generate variable results that may not be associated with the tablets undergoing testing but with the hydrodynamic characteristics of the apparatus itself and the location of the tablet on the vessel bottom. However, when the same experiments were conducted in the Modified System, the dissolution profiles for the same tablets were found to be nearly completely insensitive to tablet location.

The dissolution process in the Modified System was faster than that in the Standard System because of the improved mixing performance of the Modified System resulting from the non-symmetrical placement of the impeller. However, when the Modified System was operated at 35 rpm, the dissolution profiles for centrally located tablets were found to be very similar to those for the Standard System operating at 50 rpm. Unlike the Standard System however, the dissolution profiles obtained at 35 rpm in the Modified System were found to be insensitive to tablet location.

It can be concluded that the newly proposed Modified System for dissolution testing is a simple and yet robust and valid alternative to the current dissolution testing practice using the Standard USP Dissolution Testing Apparatus.

See more of this Session: Poster Session: Pharmaceutical Engineering
See more of this Group/Topical: Food, Pharmaceutical & Bioengineering Division

(631t) Evaluation of the Biopharmaceutical Potential of the Lichen Isolate Xanthoparmelia Somloensis

Lew P. Christopher[1], Vasudeo Zambare[2] and Archana Zambare[2], (1)Center for Bioprocessing Research and Development and Department of Chemical and Biological Engineering, South Dakota School of Mines & Technology, Rapid City, SD, (2)Center for Bioprocessing Research & Development, South Dakota School of Mines & Technology, Rapid City, SD

The Black Hills of South Dakota offer a biodiversity in lichen species with a great potential for biopharmaceutical applications. The present study was carried out to evaluate the antibacterial and antioxidant activities of the lichen isolate *Xanthoparmelia somloensis,* native to the Black Hills. The antibacterial activity was assayed against four clinical strains using an agar well diffusion method. Except for *Escherichia coli*, the antibacterial and antioxidant extracts were found inhibitory to *Streptomyces aureus, Streptococcus pyogenes* and *Steptococcus agalactiae* with minimum inhibitory concentration values in the range 0.7-0.9 mg/ml. The antioxidant activity of the lichen extracts was assessed using the 1,1-diphenyl-2-picryl-hydrazyl free radical scavenging assay. The lipid peroxidation reaction of acetone and methanol extracts was inhibited 85% and 81%, respectively A free radical scavenging activity of 77% (acetone extract) and 65% (methanol extract) was determined. It was demonstrated that both the antibacterial and antioxidant activities correlated well with the protein to polysaccharide ratio rather than the polyphenol content of the lichen extracts. The results reported here warrant further investigations to establish the usefulness of *X. somloensis* in biomedical applications such as treatment of respiratory and urinary tract infections.

See more of this Session: Poster Session: Pharmaceutical Engineering
See more of this Group/Topical: Food, Pharmaceutical & Bioengineering Division

(635a) Comparison of General Metabolism of Rhizopus Oryzae Using Glycerol and Glucose As Carbon Sources Via Metabolic Flux Analysis

Xiaoqing Wang, Zhenhua Ruan, Yan (Susie) Liu and Wei Liao, Department of Biosystems and Agricultural Engineering, Michigan State University, East Lansing, MI

Crude glycerol is one of the main biodiesel waste streams. Effective utilization of the by-product glycerol can alleviate the environmental impacts of crude glycerol, and further improve substrate utilization efficiency of biofuels production. It has been reported that glycerol is a good carbon source for many microorganisms such as fungi and bacteria for value-added fuels and chemical production. Until now, two main routes for glycerol metabolism have been identified for eukaryotic micro-organisms. One route is that glycerol is phosphorylated first and subsequently dehydrogenated. The other is that glycerol is first oxidized by a NAD^+-dependent dehydrogenase to dihydroxyacetone, and then phosphorylated. Both routes lead to glycolysis pathway to form pyruvate. In this study, we explore the utilization of glycerol in Rhizopus oryzaes. *R. oryzae* is a fungus that lives worldwide. Our preliminary results show that with crude glycerol as sole carbon resource, *R. oryzae* has higher biomass yield and lower lactic acid production compared with glucose or pyruvate growth cultures. However, less is known about the metabolism of glycerol utilization by *R. oryzae*. In this study, we focus on the investigation of glycerol metabolism in *R. oryzae*. Using a combined enzyme assays and ^{13}C-isotopomer-labeling approaches, we examine the central metabolic pathways and quantify the flux distributions at pyruvate branch point. The knowledge obtained from this research provides important insights to the potential of novel fungal species for biotechnology applications.

See more of this Session: Advances In Fermentation and Biological Conversion
See more of this Group/Topical: Food, Pharmaceutical & Bioengineering Division

(635b) Understanding the Effect of Redox Potential On Microbial Metabolism Through Comparative Proteomic Profiling: Using Microbial Production of 1,3-Propandiol As An Example

Yanping Zhang[1], Chenyu Du[2], Zhu'an Cao Sr.[2] and Yin Li[1], (1)Institute of Microbiology, Chinese Academy of Sciences, Beijing, China, (2)Department of Chemical Engineering, Tsinghua University, Beijing, China

Anaerobic and micro-aerobic fermentation processes are widely applied for the microbial production of fuels and chemicals. Redox potential (oxidoreduction potential, ORP) level was shown to be a critical factor affecting the growth and metabolism of (facultative) anaerobes, but the molecular mechanism remains largely unknown. We investigated the effect of ORP control on the fermentative production of 1,3-propanediol, a bulk chemical that is produced under micro-aerobic or anaerobic conditions. We developed a novel redox potential-based screening strategy and isolated *Klebsiella oxytoca* mutants with enhanced 1,3-propanediol-producing capability. Moreover, we used comparative proteomic profiling method to understand the metabolic shift of *K. oxytoca* when the extracellular ORP was altered. Comparative analysis of the cytoplasmic proteome of *K. oxytoca* under two different ORP levels revealed ~40 differentially expressed proteins, one third of which are involved in the central glycerol metabolism, amino acid and nucleotide biosynthesis pathways, thus providing new targets for further metabolic engineering of 1,3-propandiol producers.

See more of this Session: Advances In Fermentation and Biological Conversion
See more of this Group/Topical: Food, Pharmaceutical & Bioengineering Division

(635c) Model of Microbial Fatty Acid Production Under Different Growth Regimes

J. Tyler Youngquist[1], Wesley D. Marner II[2] and Brian Pfleger[1], (1)Chemical and Biological Engineering, University of Wisconsin Madison, Madison, WI, (2)U.S. Department of Energy Great Lakes Bioenergy Research Center, Madison, WI

Model of microbial fatty acid production under different growth regimes

J. Tyler Youngquist, Chemical and Biological Engineering, University of Wisconsin, Madison, WI, Wesley D. Marner II, University of Wisconsin, Madison, WI, and Brian F. Pfleger, Chemical and Biological Engineering, University of Wisconsin, Madison, WI

Microbial fatty acids are promising precursors for producing drop-in biofuels such as alkanes, olefins, and esters. However, for this process to become economically viable, engineered microorganisms must be able to approach the maximum theoretical yield of fatty acids from a given renewable carbon feedstock. While significant effort has focused on modifying the organisms, little published work has described efforts to understand the cultivation conditions that maximize fatty acid production. Here, we present a mathematical model for microbial fatty acid production in an engineered strain of *Escherichia coli* based on changes in controlled growth rate and nutrient composition in the media. To estimate parameter values for the model, growth and production experiments were performed in a lab scale chemostat under numerous growth regimes and media compositions. To facilitate this effort, a strain containing three chromosomally located copies of a gene encoding a medium chain thioesterase under the control of the P_{trc} promoter was constructed. The resulting strain could be cultivated in the absence of antibiotics with fatty acid overproduction induced with IPTG. Three independent chemostat experiments were maintained over a period of one to two weeks during which time multiple dilution rates and media compositions were tested. Fatty acid titers, sugar concentrations, and key nutrient concentrations were measured at each steady state. Higher titers and yields were observed under low dilution rates, however productivity was diminished. Using the data obtained from the chemostat cultures, the model was able to identify an optimal growth regime that predicts the highest yield of fatty acids on a single carbon substrate for our current overproducing strain. Such a model will prove useful for scaling up production of fatty acid products from new engineered microorganisms.

See more of this Session: Advances In Fermentation and Biological Conversion
See more of this Group/Topical: Food, Pharmaceutical & Bioengineering Division

(635d) Combining Gene Inactivation and Gene Amplification to Increase Propionic Acid Production

Zhongqiang Wang, Chemical & Biomolecular Engineering, The Ohio State University, Columbus, OH and Shang-Tian Yang, The Ohio State University, Columbus, OH

Propionibacterium freudenreichii subsp. *shermanii* has been used to produce propionic acid via fermentation processes. This organism forms propionic acid, acetic acid and succinic acid as main fermentation products. In order to decrease the production of acetic acid and divert more carbon flux toward propionic acid forming steps, phosphotransacetylase gene (*pta*) and acetate kinase gene (*ack*) were inactivated using two strategies. First, *pta* was disrupted by insertion of an integrational nonreplicative plasmid containing chloramphenicol resistance gene (*cmlA*/*cmxA*) through single crossover recombination. Since *pta* is the upstream gene in the *pta-ack* operon, acetate kinase activity was also attenuated. Second, *pta-ack* was completely substituted on the chromosome by *cmlA*/*cmxA* through double crossover recombination. The resulting two mutants, named cm-PTA and cm-PTAK, produced less acetic acid compared with the wildtype. However, due to the loss of the ATP-forming acetate kinase activity, the specific growth rates of both mutants were lower. The insertion of antibiotics resistance gene was confirmed by PCR and the enzyme inactivation was tested by enzyme assay and SDS-PAGE. Propionyl-CoA:succinate-CoA transferase (CoA T) is the last and limiting enzyme in the propionic acid pathway. In order to enhance propionic acid production while reducing acetic acid level, an additional copy of CoA T gene was introduced into chromosome together with *cmlA*/*cmxA* via double crossover recombination, replacing *pta-ack* operon. The resulting mutant cm-COAT was characterized on molecular level. All the three mutants were used in batch fermentation and fed-batch fermentation in 5-L bioreactor to evaluate their acid production performance.

See more of this Session: Advances In Fermentation and Biological Conversion
See more of this Group/Topical: Food, Pharmaceutical & Bioengineering Division

(635e) Hydrogenase Inhibition AS the Mechanism of Product Selectivity by Clostridium Thermocellum IN High-Pressure Cultures

Hsin Fen Li[1], Barbara L. Knutson[1], Sue E. Nokes[2], Bert C. Lynn Jr.[3] and Michael D. Flythe[4], (1) Chemical and Materials Engineering, University of Kentucky, Lexington, KY, (2)Department of Biosystems and Agricultural Engineering, University of Kentucky, Lexington, KY, (3)Chemistry, University of Kentucky, Lexington, KY, (4)Department of Animal and Food Sciences, University of Kentucky, Lexington, KY

HYDROGENASE INHIBITION AS THE MECHANISM OF PRODUCT SELECTIVITY BY *CLOSTRIDIUM THERMOCELLUM* IN HIGH-PRESSURE CULTURES

Hsin-Fen Li [1], Barbara L. Knutson[1], Sue E. Nokes [2], Bert C. Lynn [4], Michael D. Flythe [4, 5]

[1]University of Kentucky, Department of Chemical and Materials Engineering, Lexington, Kentucky 40506

[2]University of Kentucky, Department of Biosystems and Agricultural Engineering, Lexington, Kentucky 40506

[3]University of Kentucky, Department of Chemistry, Lexington, Kentucky 40506

[4]USDA, Agricultural Research Service, Forage-Animal Production Research Unit; Lexington, Kentucky 40546

[5]University of Kentucky, Department of Animal and Food Sciences, Lexington, Kentucky 40546

Clostridium thermocellum is a cellulose-degrading thermophile that produces ethanol, but its utility in fuel production from biomass is limited by co-production of acetate and lactate. Our previous investigations of continuous fermentations of *C. thermocellum* at elevated pressure (7.0 MPa, and 13.0 MPa) reveal that when the solubility of product gas (H_2) increases under elevated hydrostatic pressure, the acetate:ethanol ratios decrease as a result of a simultaneous increase in ethanol production and a decrease in acetate production. While the decreased acetate production is broadly attributed to the inhibition of bacterial hydrogenases by the end-product H_2, the dramatic increase in ethanol production is unexplained. With a goal of probing the underlying mechanism of the effect of H_2 partial pressure on ethanol production and manipulating the fermentation towards high ethanol selectivity, we examine the growth and end product formation of *C. thermocellum* in cultures perturbed with H_2 and several known hydrogenase inhibitors (carbon monoxide and methyl viologen). The effect of these inhibitors in continuous cultures, batch cultures and high-density washed cell cultures (in the absence of growth phase-dependent substrate and energy use) is compared to fermentations in the presence of N_2 headspace. The addition of exogenous H_2 with a hydrogenase inhibitor (methyl viologen) caused approximately 80% decrease in acetate production. Likewise, batch *C. thermocellum* grew in the presence of the hydrogenase inhibitor, carbon monoxide (CO), but less than 1% acetate was produced. When the sparge gas used in continuous cultures was switched from N_2 to H_2, acetate production decreased approximately 35%, and ethanol production increased more than 350%. The conversion of NADH to NAD^+ is coupled to the production of H_2, lactic acid, or ethanol. Thus, the result provides evidence for compensatory production of lactate and/or ethanol occurred when hydrogenases were inhibited. The contribution of substrate uptake rate to enhanced ethanol production was examined.

See more of this Session: Advances In Fermentation and Biological Conversion
See more of this Group/Topical: Food, Pharmaceutical & Bioengineering Division

(635f) Improved Production of Biofuel Precursors From CO2 Fermentation Via Growth and Physiology Optimization of Moorella Thermoacetica

Hamid Rismani-Yazdi, Department of Chemical Engineering,, Massachusetts Institute of Technology, Cambridge, MA, **Peng Hu**, Department of Chemical Engineering, Massachusetts Institute of Technology, Cambridge, MA and Gregory Stephanopoulos, Chemical Engineering, Massachusetts Institute of Technology, Cambridge, MA

The first two authors equally contributed to this work.

Approximately half of the glucose feedstock used during the biofuel (e.g. ethanol) fermentation process is lost as CO2. This is the result of carbon oxidation that is necessary to generate the reducing equivalents required to lower the oxygen content of an initial glucose molecule (C6H12O6) to that of the more reduced products (e.g. ethanol). It has been suggested that if the emitted CO2 could be captured using hydrogen from a non-fossil fuel source, the amount of land required to produce a given amount of biofuel would be reduced by two thirds. While thermochemical processes were contemplated in the above CO2 fixation concept, biological methods are just as applicable. Furthermore, the latter will have a higher overall yield as they operate closer to equilibrium and are consequently more efficient.

A group of anaerobic bacteria known as acetogens, have the ability to grow on CO_2 and H_2 and produce various chemical (e.g. acetate) and biofuels (e.g. ethanol). However, a major challenge facing this process is the poor solubility of gaseous substrates and consequently the low productivity of the process. In this presentation we describe the growth and physiology optimization of an improved strain of a thermophilic acetogen, *Moorella thermoacetica*, and the design and operation of a high pressure bioreactor to increase the productivity of H_2/CO_2 conversion to acetate. The produced acetate could be then used as a precursor by oleaginous microorganisms for producing biodiesel. The results on effects of various components of the medium and bioreactor operating conditions on growth, productivity and adaptation of *M. thermoacetica* to convert H_2/CO_2 to acetate will be presented.

See more of this Session: Advances In Fermentation and Biological Conversion
See more of this Group/Topical: Food, Pharmaceutical & Bioengineering Division

(635g) Highly Improved Plasmid DNA Production by Optimizing the Plasmid Construct and Fermentation Conditions

Ram N. Trivedi, Chemical and Petroleum Engineering; Microbiology and Molecular Genetics, University of Pittsburgh, Pittsburgh, PA, Patrick R. Bartlow, Bioengineering, University of Pittsburgh, Pittsburgh, PA, Jonathan Meade, Chemical Engineering, Carnegie Mellon, Pittsburgh, PA, Saleem A. Khan, Microbiology and Molecular Genetics, University of Pittsburgh, Pittsburgh, PA, Michael M. Domach, Chemical Engineering, Carnegie Mellon University, Pittsburgh, PA and Mohammad M. Ataai, Chemical and Petroleum Engineering, University of Pittsburgh, Pittsburgh, PA

This presentation details our effort to develop a combined host-plasmid system that provides plasmid copy numbers that approach 10^4 copies/cell. The work is driven by DNA vaccines and gene delivery vectors which are becoming a growing segment of new clinical trials. Over 20% of ongoing trials rely on plasmid vectors spanning all disease categories and including nine trials currently in Phase III. Moreover, four DNA vaccines have been approved for veterinary use since 2005.

Given the large number of potential vaccines and the size of the target animal and human populations, large amounts of plasmid DNA will be required. Our ultimate goal is to develop a production platform that has high volumetric productivity when inexpensive and simple minimal media is used. To this end, we employ plasmid pNTC8485 in DH5α that possess a temperature-responsive ColE1 origin of replication, encodes reporter EGFP, and has a major advantage of reduced metabolic load by encoding a small antisense RNA that mitigates chromosomally-endowed toxicity from *sacB* when sucrose is present in the media. Replication region contains two promoters which direct the synthesis of two RNA molecules (RNA I and RNA II) involved in control of plasmid copy number. We have modified the origin of replication, yielding approximately 8000 plasmid copies/cell, a 6-fold increase in plasmid copy number relative to unaltered DH5α pNTC8485. Further increase in process productivity was achieved by improving the fermentation conditions that enables the selective agent (sucrose) to be metabolized once the catabolic resources are exhausted from the medium. After reaching stationary phase in media with 8% sucrose, treatment with invertase releases glucose that spurs added cell growth without noticeable plasmid dilution, nearly tripling DNA volumetric productivity. Overall, about a 20-fold increase in volumetric productivity was achieved by combining the productive mutations of origin of replication and improved fermentation conditions.

See more of this Session: Advances In Fermentation and Biological Conversion
See more of this Group/Topical: Food, Pharmaceutical & Bioengineering Division

(638a) Bubble Behavior and Oxygen Transport In a Taylor-Couette Bioreactor

Jian Qiao, Eldin Wee Chuan Lim and **Chi-Hwa Wang**, Department of Chemical and Biomolecular Engineering, National University of Singapore, Singapore, Singapore

Taylor vortices can be formed by subjecting a viscous liquid to a shear stress within an annular space between two rotating cylinders and a stationary bottom surface. Taylor vortex flow is found in many practical applications such as reaction, filtration and extraction. In this study, Taylor vortex flow is applied in a bioreactor to culture cells that are seeded in degradable porous scaffolds. This choice is with reference to its advantages of low shear stress, high mass transfer rate and easy scaling-up characteristics. It is important to understand the transport phenomenon inside the bioreactor system for optimization on the performance. To avoid the shear stress caused by sparging, air bubbles are injected to form bubble rings inside the system. A high-speed camera is used to measure the bubble behavior, in conjunction with an oxygen sensor for in-situ measurements of the oxygen concentration in the flow field. The effect of flow pattern of bubbles on the oxygen consumption rate in the bioreactor is investigated for designing the optimal bubble size and bubble number for cell culture and tissue engineering applications.

See more of this Session: Bioreactors In Tissue Engineering
See more of this Group/Topical: Food, Pharmaceutical & Bioengineering Division

(638b) Design of a Radial Bioreactor for Tissue Regeneration

Prasana Raja Bhaskar and Sundararajan. V. Madihally, School of Chemical Engineering, Oklahoma State University, Stillwater, OK

Design of a Radial Bioreactor for Tissue Regeneration

Tissue Engineering deals with regenerating tissues and organs to replace or support the function of defective body parts.Fabrication of tissues in the laboratory environment requires engineering of scaffolds, the three dimensional synthetic frame structures which serve as a mimic of extracellular matrix for cell adhesion, migration and proliferation. Bioreactors of different configurations and flow systems have been utilized for regenerating tissues in seeded technique. However, quality of regenerated tissues is poor, attributed to incomplete understanding of fluid flow characteristics.

In this study, fluid flow characteristics in radial flow bioreactors were investigated. First, simulations were performed using Computational Fluid Dynamics (CFD) technique to understand the effect of i) inlet/outlet diameters ii) diameter of the porous structure, thickness of the porous structure, and iv) flow rate on i) nutrient distribution, ii) pressure drop and iii) shear stress developed in the system. In addition, baffle design was also considered to improve nutrient distribution. Michaels-Mentens equation was used to describe the nutrient (oxygen and glucose)consumption for the smooth muscle cell, using literature data. As a starting point, minimum volumetric flow rate was calculated using previously established relationship [1].Fluid flow through the porous structure was described using Brinkman Equation, with the characteristics of chitosan-gelatin scaffolds formed using freeze drying. No slip at the walls and outlet pressure of 1 atm were the boundary conditions.

The study was performed using COMSOL 4.1Multiphysics software with reaction engineering module for different i) inlet diameters(3, 6 and 9 mm)and outlet diameters(10, 20 and 30 mm), ii) scaffold diameter (20, 50, 100 mm), iii) volumetric flow rates(0.001, 0.01, 0.1 and 1 mL/min) and iv) scaffold thickness(2,4 and 6mm).To validate the simulation results, one of the bioreactor configurations was built in-house and experiments were conducted using porous chitosan-gelatin scaffolds. Measured pressure drop and residence time distribution characteristics validated the simulated results. Based on these findings, important factors to consider are discussed.

References

[1] Devarapalli M, Lawrence BJ, Madihally SV. Modeling Nutrient Consumptions in Large Flow-Through Bioreactors for Tissue Engineering.Biotechnology/ Bioengineering. 103(5):1003-1015, 2009.

(638c) Perfusion Modulated Macromolecule Distribution for Enhanced Osteogenesis In Chitosan-Based Composite Scaffolds

Katelyn Sellgren and Teng Ma, Chemical and Biomedical Engineering, Florida State University, Tallahassee, FL

Biomimetic scaffolds provide a multitude of chemical and structural cues that direct the differentiation of human mesenchymal stem cells, and are essential for influencing 3D construct development and functional outcome. Perfusion flow has the capability to augment this potential by creating a dynamic and physiologically relevant growth environment that has been shown to enhance the osteogenic potential of hMSCs. A wide range of bioreactor systems have been used to obtain these results with the most common being spinner flasks, rotating wall vessels, and perfusion systems. Of these, only the direct perfusion bioreactor systems perfuse media throughout the scaffold while the others are limited to surface flow. These flow variations could significantly bias the macromolecule distribution within the 3D constructs we therefore aim to study the ability of perfusion flow pattern to modulate the microenvironment and impact the differentiation potential of hydroxyapatite-chitosan-gelatin (HCG) constructs. The HCG scaffolds were housed in an in-house modular perfusion bioreactor system that allows hMSCs to be dynamically seeded using the same inoculm and then subsequently cultured under different flow conditions (Zhao and Ma 2005, Zhao *et al.*, 2009). The media flow pattern was found to profoundly impact macromolecule distribution within the scaffold and to enhance cellular spatial distribution, hMSC proliferation and osteogenic differentiation potential as assessed by alkaline phosphatase, alizarin red, osteopontin, osteocalcin, and BMP-2 protein expression, and microCT bone volume analysis (Sellgren and Ma 2011). Our results show that transverse perfusion flow significantly influences the extracellular matrix (ECM) microenvironment, over that of parallel flow that in turn impact hMSC proliferation and osteogenic differentiation. The results highlight the potential for combining biomimetic HCG scaffolds with the perfusion bioreactor system as a streamlined fabrication process for bone construct development from hMSCs.

References:

Sellgren K, and Ma T. 2011, Perfusion Conditioning of Hydroxyapatite-Chitosan-Gelatin Scaffolds for Bone Tissue Regeneration from Human Mesenchymal Stem Cells, J Tissue Eng Regen Med, DOI: 10.1002/term.396

Zhao F, and Ma T. 2005, Perfusion Bioreactor System for Human Mesenchymal Stem Cell Tissue Engineering: Dynamic Cell Seeding and Construct Development, *Biotechnol Bioeng*, **91**: 482-493

Zhao F, Grayson W L, Ma T, Irsigler A. 2009, Perfusion Affects the Tissue Development Patterns of

Human Mesenchymal Stem Cells in 3D scaffolds, *J Cell Physiol*, **219**: 421-429

See more of this Session: Bioreactors In Tissue Engineering
See more of this Group/Topical: Food, Pharmaceutical & Bioengineering Division

(638d) Dynamics of Diffusivity and Pressure Drop In Flow-Through and Parallel Flow Bioreactors During Tissue Regeneration

Jagdeep T. Podichetty, Dhananjay V. Dhane and Sundararajan. V. Madihally, School of Chemical Engineering, Oklahoma State University, Stillwater, OK

Utilization of bioreactors to regenerate tissues outside the body has been intensely investigated in functional tissue engineering. Different configurations of bioreactors have been explored to regenerate tissues including flow-through (also referred as transverse flow), parallel-flow, rotating wall, and microgravity. For regenerating large tissues, flow-through and parallel-flow configurations are preferred due to uniform support offered to the scaffold. However, transport characteristics in high aspect ratio scaffolds (for example, 100 mm diameter and 2 mm thick) during tissue regeneration are not well understood.

In this study, two designs i) flow-through (Design 6) and ii) parallel flow (Design 8) reactors were utilized to analyze the effect of tissue development. First both designs were simulated using COMSOL 3.5 Multiphysics (COMSOL, Inc., Burlington, MA) with properties of 2%-2% chitosan-gelatin porous structures, similar to previous publication [1]. The fluid flow was defined by Brinkman equation on the porous regions using pore characteristics of Chitosan-Gelatin scaffold formed at -80°C (55 μm and 318 pores/mm^2). These results were then compared with experiments performed using similar dimension reactors that could hold 100 mm diameter and 2 mm porous structures. Pressure drop values across the reactor were calculated at flow rates 5 mL/min to 25 mL/min with 5 mL/min increment. At low flow rates experimental results were in agreement with the simulation results for both reactor configurations. However, increased flow rate in Design 6, showed channeling at circumference of the reactor leading to lower pressure drop relative to simulation.

Since effective diffusivity data for oxygen and glucose through different chitosan gelatin porous structures was unavailable in literature, effective diffusivity of glucose was experimentally determined using custom-built apparatus. The pore size of scaffolds with different composition was also obtained from experiments using scanning electron micrographs. The experimental results were then used for computer simulation, to study nutrient distribution and consumption by smooth muscle cells across the scaffold for both reactors. To understand whether nutrient transfer is limited by diffusion or convection, Peclet number inside the scaffold were determined. Velocities obtained from simulations along with diffusivity values from experiment were used to calculate Peclet number for both reactors. These results showed that Design 6 is convective flow dominant whereas Design 8 is diffusion-dependent. In Design 6, highest flow rate that can be used depend on mechanical strength of the porous scaffold. While in Design 8, reduced porosity (<20%) significantly restricted nutrient distribution. In order to enhance nutrient distribution, additional inlets and outlets were included into the reactor. These new designs improved nutrient distribution for 2 mm scaffolds with variable cell densities. However, it was found that parallel-flow configuration is not suitable for porous structures greater than 4 mm thick scaffolds, especially at very low porosities.

References

[1] Devarapalli M, Lawrence BJ, Madihally SV. Modeling Nutrient Consumptions in Large Flow-Through Bioreactors for Tissue Engineering. Biotechnology/ Bioengineering. 103(5):1003-1015, 2009.

See more of this Session: Bioreactors In Tissue Engineering
See more of this Group/Topical: Food, Pharmaceutical & Bioengineering Division

(638e) Effect of Scaffold Positioning On Tissue Thickness In a Batch Tissue Culture Reactor

Anirudh R. Patrachari, Jong Kyu Hong, Arland H. Johannes and Sundararajan. V. Madihally, School of Chemical Engineering, Oklahoma State University, Stillwater, OK

Porous scaffolds used in various tissue engineering cultures are grown by inserting the seeded construct into the bottom of a tissue culture well, except skin grafts and hepatocytes. Skin grafts are lifted to the air-liquid interface while hepatocytes are cultured on collagen gels in a sandwich configuration. This study evaluates the effects of raising the elevation of the scaffold within the static culture system, on nutrient distribution, primarily oxygen and glucose. Further, the effects of scaffold positioning on maximum cultivable tissue thickness was evaluated in static culture conditions. A parametric analysis of oxygen and glucose diffusion through the culture media is modeled using the Fick's second law. Finite element simulations were performed on COMSOL Multiphysics 3.5a to determine the concentration profiles and hence, the oxygen and glucose transport rates to the scaffold with i) changing porosity and effective diffusivity calculated using Mackie-Meares relationship and ii) consumption kinetics using Michaelis-Menten kinetics. The study demonstrates that altering the elevation of the scaffold within a static culture well influences the maximum cultivable tissue thickness.

See more of this Session: Bioreactors In Tissue Engineering
See more of this Group/Topical: Food, Pharmaceutical & Bioengineering Division

(638f) Engineered Arteries Developed In a Multi-Graft Flow-Stretch Bioreactor with In Vivo Evaluation

Zeeshan Syedain, University of Minnesota- Twin Cities, Minneapolis, MN and Robert T. Tranquillo, Biomedical Engineering and Chemical Engineering & Materials Science, University of Minnesota, Minneapolis, MN

Introduction: Tissue engineering provides a means to create arterial grafts that can maintainvascular function comparable to native vessels. To date, successful vasculargrafts have been developed using the cell-sheet method as well as synthetic polymer-basedapproaches. Mechanical conditioning such as cyclic conditioning in bioreactorhas been shown to improve collagen production and mechanical properties ofengineered tissue (Syedain et al 2008). Nutrient transport also plays animportant role in tissue development. Here we present a novel bioreactordesigned to apply cyclic stretch combined with transmuralflow (Pulse flow stretch (PFS) bioreactor) to arterial constructs. Further,arteries developed in the bioreactor were evaluated in vivo in a rat model.

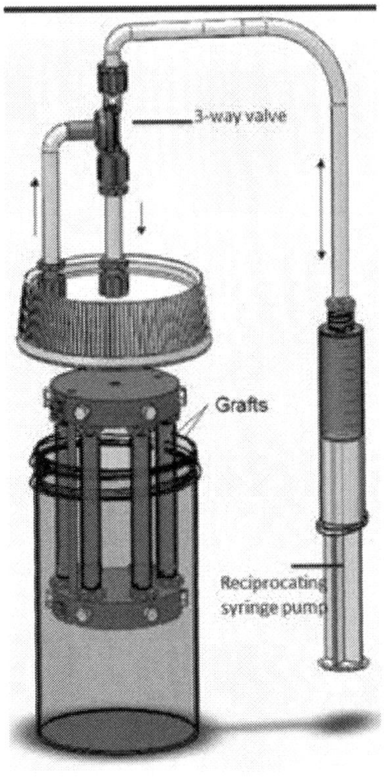

Figure 1: Schematic of Pulse Flow Stretch Bioreactor

Material and Methods: Fibrin-basedengineered arteries were fabricated using 4 mg/mL fibrin and 1 million neonatalhuman dermal fibroblasts (nHDF) per mL. After 2 weeks, they were mounted in PFS bioreactor for5 wk at 7% stretch. Grafts conditioned in bioreactorwere tested for mechanical properties and collagen production. Further, 1.5 mmID grafts were implanted interpositionally in theabdominal aorta in 10 wk old HSD:RH-Fox^{rnu} (nude) rats (Harlin)with end-to-end anastomosis using 10-0 nylon sutures. Rats were maintainedpost-operatively on an anti-coagulant regiment of Aspirin and Plavix. Thegrafts were harvested at 12 and 24 weeks post-implantation.

Result andDiscussion: A bioreactor was developed to conditionup to 6 grafts in a single manifold with applied cyclic flow leading tosimultaneous circumferential stretching and transmuralflow in order to promote tissue growth (Figure 1). The pressure and diameterwere non-invasively monitored and used to calculate stiffness in the range of80-120 mmHg. Using pressure-diameter correlations, a method was validatedto non-invasively predict burst strength during bioreactor culture. Thecell induced axial shortening of the construct was also accommodated to achievecircumferential alignment and mechanical anisotropy comparable to nativearteries. Grafts implanted in Rat were evaluated at 12 and 24 weeks. At 12weeks, grafts (n=2) were aneurysmal. Several other grafts indicatedclotting at 12 weeks. At 24 weeks (n=1), a graft from a separate fabrication showedno dilation or substantial intimal hyperplasia (Fig. 2a,c).The elastin concentration increased dramatically, becoming comparable to thenative artery (Fig. 2b). Histology showed organized elastin similar to thenative artery as well as complete graft endothelialization(Fig. 2c).

Conclusion: Fibrin-basedtissue engineering provides a means to create completely biological engineeredarteries in short duration (7 weeks of incubation). Here, we present abioreactor system to condition artery to achieve physiological compliance andburst strength. While successful implants have been accomplished usingsynthetic scaffolds and a cell sheet-based method, arterial implants have notbeen reported to date for a completely biological scaffold. To this end, wepresent preliminary findings of arterial implants showing function to 6 months *invivo* is possible using completely biological tissue-engineered arteriesfabricated from fibrin and human cells.

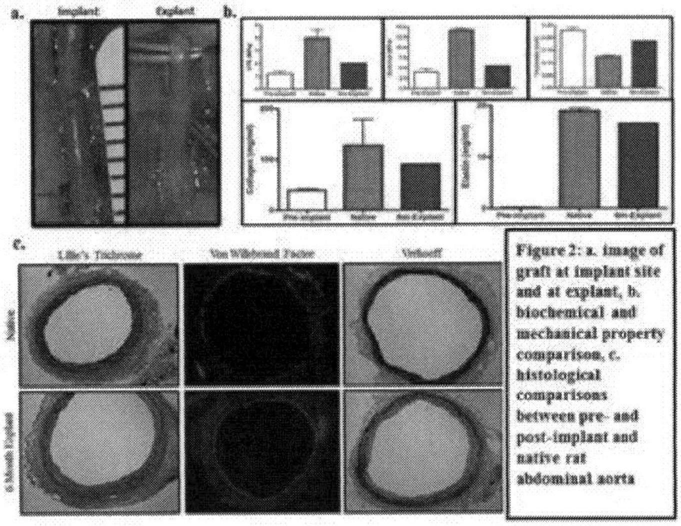

Figure 2: a. image of graft at implant site and at explant, b. biochemical and mechanical property comparison, c. histological comparisons between pre- and post-implant and native rat abdominal aorta

References:

Syedainet al. PNAS 2008: 105(18) p6537-6542.

Acknowledgment:

Fundingfrom NIH R01 HL083880 (to R.T.T.)

See more of this Session: Bioreactors In Tissue Engineering
See more of this Group/Topical: Food, Pharmaceutical & Bioengineering Division

(638g) Combining Systems Biology and Bioprocessing: Opportunity for Cancer Stem Cell Bioprocessing

Yonghyun (John) Kim, Samsung Biomedical Research Institute, Seoul, South Korea

In the emerging era of personalized medicine, the dream is to provide individualized translational research so that each patient could receive a more personalized, targeted therapy. This requires doing translational research using the patients' own tumor samples, but a major hurdle is the scarcity of the cells obtainable from freshly dissociated patient samples. Although difficult-to-attain, such cells provide a more clinically relevant and meaningful information than established cell lines. Yet expansion by serial, long-term culture of these cells typically changes their original phenotype and even genotype. Therefore, it remains critical to develop a quick, efficient expansion system of newly dissociated patient samples in order to perform adequate personalized preclinical translational research.

Unlike traditional cell lines that are often grown in adherent manner in complex media containing animal sera, the original characteristics of the patient-derived cells are better maintained in sphere-forming, stem cell culture media. As cells divide, they generate tumor cell spheres that retain the original characteristics, but when spheres grow too large in typical static cultures, it leads to mass transfer limitations that cause cell death. A well-controlled culture environment will be important, especially since it is the cells themselves that are the desired products (as opposed to, e.g., protein production). These challenges provide a unique opportunity for biochemical engineers to develop a sphere-forming submerged culture bioprocessing system, and a prototype is proposed here.

In addition, utilizing the hypothesis of cancer stem cells can aid in the proposed in vitro expansion. The hypothesis states that in the heterogenous tumor mass, there exists stem-cell like tumor cells that are uniquely endowed to form tumor, have self-renewal capacity and are inherently resistant to chemotherapy and/or radiation. For some solid tumors, particular surface markers have been identified that allow prospective isolation of cancer stem cells. For example in glioblastoma, the most malignant form of brain tumor, CD133(+) cells were originally proposed to correspond with cancer stem cells. However, recent reports are starting to challenge this original characterization, and other markers are starting to be considered, such as CD15(+) and/or HGFR(c-Met)(+). Also in breast cancer, the original CD24(-/low) CD44(+) cancer stem cell markers are now considered insufficient and the more recent studies include adding Lin(-), EpCAM(+) and/or ALDH(+). In addition, with many of these surface markers, it remains unclear as to what actual biological function they provide to the cancer stem cells. Therefore, a systems biology approach would be appropriate in better characterizing the stem cell population of the tumor mass. In particular, membrane subproteomics approach of patient-derived tumor cells is proposed.

The system proposed here of combining systems biology approach with bioprocess development for cancer stem cell engineering will function as a bridge-builder between the traditional preclinical translational researchers and bioprocess engineers. Because the conditions for growing cancer stem cells are similar to stem cell culture, our proposed work has broad applicability in the future to the field of cancer research as well as stem cell bioprocessing.

See more of this Session: Bioreactors In Tissue Engineering
See more of this Group/Topical: Food, Pharmaceutical & Bioengineering Division

(639b) The Detection of Cancer Biomarkers On Biosensor Surfaces

Lokesh Taneja, University of Mississippi, University, MS, Ajit Sadana, Chemical Engineering, University of Mississippi, University, MS and Kennon C. Shelton, Chemical Engineering Department, University of Mississippi, University, MS

A fractal analysis is presented for the binding and dissociation (if applicable) of different cancer biomarkers to appropriate receptors immobilized on different biosensor surfaces. Some of the cancer biomarkers analysed include: (a) the binding of 10 Molv NIG 3T3 cancer cells in hyposomotic solution to a microelectrode (de la Rica et al, 2009), (b) binding of different concentrations CEA (carcinoembryonic antigen) in BSA-PBS solution to the anti-CEA antibody functionalized to a microring resonator biosensor surface (Washburn et al., 2009), (c) detection of telomerase over-expression for cancer diagnostics by an optical fiber nanobiosensor (Zheng and Li, 2010), (d) binding of 150 ng/mL the cancer biomarker interleukin-8 (IL-8) in solution to the anti-IL-8 antibody immobilized on a microring resonator biosensor surface (Washburn et al., 2009), (e) binding of the cancer biomarker, prostate specific antigen (PSA) in solution to the anti-PSA immobilized on a microring resonator biosensor surface (Washburn et al., 2009), (f) binding of the cancer biomarker, tumor-necrosis factor-a (TN-a) in solution to the anti-TNF-a antibody immobilized on the microring biosensor surface (Washburn at al, 2009), (g) binding of the cancer biomarker, carbohydrate antigen 125 (CA 125) to a nanoparticle-based sandwich electrochemical immunosensor (Tang et al., 2010), and (h) detection of the cancer biomarkers, galectin-3, galectin-4, and galectin-8 in solution to a protruding galactoside surface (Yoshioka et al., 2010).

Predictive relationships are developed for the binding and dissociation of the different biomarkers to the appropriate receptors immobilized on different biosensor surfaces. There is an urgent need and focus to help develop more sensitive and selective biosensors for cancer biomarker detection.

See more of this Session: Biosensors, Biodiagnosis and Bioprocess Monitoring
See more of this Group/Topical: Food, Pharmaceutical & Bioengineering Division

(639c) Real-Time, Label-Free Identification and Differentiation of Environmental Responses In Living Biofilms

Michael S. Waters[1], Santanu Kundu[2], Irene Calizo[3], Angela R. H. Walker[4], Nancy J. Lin[5] and Sheng Lin-Gibson[5], (1)Polymers Division, Biomaterials Group, National Institute of Standards and Technology, Gaithersburg, MD, (2)Department of Polymer Science and Engineering, University of Massachusetts-Amherst, Amherst, MA, (3)Optical Technology Division, National Institute of Standards and Technology (NIST), Gaithersburg, MD, (4)Physics Lab, National Institute of Standards and Technology, Gaithersburg, MD, (5)NIST, Gaithersburg, MD

The current primary means of identifying biofilms mediated disease is by identifying the presence of specific bacteria. Blame is then assigned to these bacteria. This approach, however can lead to misdiagnosis. In fact, many bacteria associated with pathogenesis exist in a non-pathogenic form in the natural human microbial flora. Current methods in evaluating bacterial biofilm pathogenicity are focused on evaluating overall biofilm presence and/or viability. This approach implements bulk evaluation that limits the ability discern if a live biofilm is exhibiting pathogenic properties, or is surviving in a non-pathogenic capacity. In an attempt to evolve analytical techniques to distinguish living biofilms in pathogenic and non-pathogenic modes, acidogenic dental pathogen *Streptococcus mutans* was used as a model. In this study, sucrose was used to control the exhibition of *S. mutans* pathogenic properties, which were then evaluated by surface enhanced Raman spectroscopy (SERS) to identify unique chemical signatures. These techniques require minimal sample preparation and offer a rapid, low-cost, high throughput system for clinical evaluations, or evaluating of the capacity of various materials to affect bacterial-mediated biofilm pathogenesis.

See more of this Session: Biosensors, Biodiagnosis and Bioprocess Monitoring
See more of this Group/Topical: Food, Pharmaceutical & Bioengineering Division

(639d) Quantifying Tyrosine Kinase Activity In Cancer Cell Lysates Via Phosphorylation of Hydrogel-Immobilized Substrates

Alicia D. Powers, Bi Liu, Andrew G. Lee and Sean P. Palecek, Chemical and Biological Engineering, University of Wisconsin-Madison, Madison, WI

Many types of cancer are caused by overactive or overexpressed kinases. Over the past decade, the use of kinase inhibitors has offered a new approach for treating cancers associated with increased or disregulated kinase activity. Many patients, however, are initially resistant to these inhibitors or develop resistance during treatment. Additionally, diagnosis of a particular type of cancer does not provide insight into what kinase therapies may be effective. This research seeks to develop a kinase activity assay to determine which kinase inhibitors will be most effective for individual patients and to quickly identify if a patient has developed resistance to a particular inhibitor. Specifically, we are currently working to detect Met tyrosine kinase activity in lung cancer as well as to multiplex the assay to simultaneously detect the activity of multiple kinases, including Bcr-Abl kinase. To accomplish these goals, we immobilize spots of a phosphorylation substrate for Met kinase in a microchannel on a glass slide. We then expose the substrate to a kinase reaction mixture containing lysate from a cancer cell line. A kinase inhibitor may also be included in the reaction mixture. A phosphotyrosine antibody followed by a fluorescent secondary antibody is then used to quantify substrate phosphorylation and kinase activity. Our results have demonstrated increases in substrate phosphorylation with increasing lysate and immobilized substrate concentrations. Further results also show that a Met kinase inhibitor decreases substrate phosphorylation. Immunodepleting Met kinase from the cell lysate also decreases substrate phosphorylation. Together, these results show that this assay can detect specific kinase activity in cancer cell lines. This assay offers promise as a way to quickly and relatively cheaply determine which kinase inhibitors will be effective for which patients, allowing patients to more quickly receive an effective treatment.

See more of this Session: Biosensors, Biodiagnosis and Bioprocess Monitoring
See more of this Group/Topical: Food, Pharmaceutical & Bioengineering Division

(639e) High-Throughput Characterization of Single Antigen-Specific T Lymphocytes by Microengraving

Jonghoon Choi[1], Elizabeth M. Bradshaw[2], Sally C. Kent[3], Qing Han[4], David A. Hafler[5] and J. Christopher Love[1], (1)Department of Chemical Engineering, The David H. Koch Institute for Integrative Cancer Research, Massachusetts Institute of Technology, Cambridge, MA, (2)Neurology, Brigham and Women's Hospital, Boston, MA, (3)Division of Diabetes, University of Massachusetts Medical School, Worcester, MA, (4)Chemical Engineering, The David H. Koch Institute for Integrative Cancer Research, MIT, Cambridge, MA, (5)Department of Neurology and Immunobiology, Yale University School of Medicine, New Haven, CT

$CD4^+$ T lymphocytes (T helper cells) are critical for coordinating immune responses against infections and in autoimmune conditions. $CD4^+$ T cells sustain the activation and proliferation of $CD8^+$ cytotoxic T cells, as well as induce class switching of immunoglobulins and induction of memory B cells. Given the significant role assumed by $CD4^+$ T cells, evaluating the number and functional activities of antigen-specific $CD4^+$ T cells is, therefore, critical for monitoring the progression of pathological conditions and responses to interventions like vaccines. Here, we describe a new approach for identifying and assessing the functional responses of single, antigen-specific $CD4^+$ T cells directly *ex vivo* using antigen-presenting cells (APCs) to induce activation. Central to this approach, a process called microengraving enables the generation of microarrays for detection of secreted proteins captured from a large number of single cells (~10^5 cells per assay). Using this approach, it is possible to measure the rate of secretion for up to eight cytokines released, and by automated microscopy, the characteristic phenotypes of single $CD4^+$ T cells stimulated with activated DCs. Our data demonstrate the ability to link $CD4^+$ T cell phenotypes to functional responses against pools of peptides derived from pathogenic agents restricted through MHC-II presentation. This integrated approach for single-cell analysis suggests a new strategy for *ex vivo*, high-throughput screening of antigen-specific human $CD4^+$ T cells and may provide new insights in understanding the contributions of rare immune cells in defense of various immune diseases.

See more of this Session: Biosensors, Biodiagnosis and Bioprocess Monitoring
See more of this Group/Topical: Food, Pharmaceutical & Bioengineering Division

(639f) Using QCM-D and Ellipsometry to Determine the Orientation and State of Hydration of Antibodies Adsorbed On a Hydrophobic Surface

Meredith E. Wiseman and Curtis W. Frank, Chemical Engineering, Stanford University, Stanford, CA

Understanding and controlling the structure and orientation of proteins at interfaces is critical in designing biosensors where the capture of an analyte is dependent on the correct orientation and conformation of a surface-tethered protein. Quartz Crystal Microbalance with Dissipation (QCM-D) is a technique often used to monitor protein adsorption due to its excellent sensitivity, ease of experimental design, and label-free detection method. One important and often underexploited feature of QCM-D in the liquid phase is that it measures a hydrated protein mass, including the mass of mechanically coupled water in the measurement output. In practice, this means that either viscoelastic modeling or *in-situ* combination with a complementary measurement technique is needed to understand the effects of water content on the QCM-D response and to separate dry from hydrated protein mass in a quantitative manner.

In this study, we demonstrate the utility of QCM-D in quantifying adsorbed protein amount as well as characterizing orientation by examining the case of a human monoclonal antibody adsorbed onto a model hydrophobic, CH_3-terminated surface. An analysis of the protein-binding ability of the antibody combined with an examination of its mechanical coupling to the surface suggests that the antibodies initially adsorb in a 'flat-on' orientation, unable to capture their antigen, until they reach a surface coverage at which they no longer fit 'flat-on' and begin to adopt a mixture of 'end-on' orientations, with both Fab and Fc portions of the antibody accessible on the surface at monolayer coverage. By combining QCM-D with simultaneous ellipsometry measurements and by varying the density of the surrounding buffer, we demonstrate that the antibody is less hydrated at low surface coverages, while at high surface coverage the mass of hydration is quite significant, comprising more than 50% of the total mass measured by QCM-D alone. Based on this example, we detail strategies for how to use QCM-D to achieve quantitative measures of protein surface excess as well as qualitative orientational information, with and without the added benefit of a complementary measurement technique.

See more of this Session: Biosensors, Biodiagnosis and Bioprocess Monitoring
See more of this Group/Topical: Food, Pharmaceutical & Bioengineering Division

(639g) Fluorescence Lifetime Sorting with A Frequency Domain Analysing Flow Cytometer

Patrick Jenkins and Jessica P. Houston, Chemical Engineering, New Mexico State University, Las Cruces, NM

Background

The realm of chemical engineering is expanding rapidly and has incorporated into it principles of bio-engineering and biological techniques. One such method is flow cytometry and is used in research and diagnostic medicine to deduce information about cells or microparticles based upon their relative size, internal complexity and fluorescence character. This is determined by using hydrodynamic properties to focus a flowing stream of a sample of interest into a single flowline and individually interrogate across a Gaussian profile laser beam. From the scattered and fluoresced light we can determine things about the cells such as its place in the cell cycle, protein expression levels and surface marker characteristics. The "killer app" for flow cytometry came in the 1980's when CD4+T-cell level determination led to a rapid and accurate determination of when HIV infection became full blown AIDS.

These measurements of fluoresced photons by a photodetector have served the cytometry community well in the past yet suffer from a very critical flaw. Fluorophores emit light over a broad spectral range and experimenters will often encounter spectral overlap from two emitting fluorescent species. When spectral overlap between fluorophores emitting in regions that exhibit cellular autofluorescence occurs there is a compounding issue whereby the exogenous signal cannot be separated from the endogenous fluorescence. Although mathematical compensation methods exist, they are only numerical subtractions and not true measurements.

In previous word, we have shown that the phase shift of each fluorophore can be measured to overcome the issues of spectral overlap by capturing the fluorescent molecule's excited state lifetime. Our methods digitally capture amplified voltage waveforms from a cytometer's detector. A Fourier transform and digital processing is done to extract the excited state lifetimes of fluorophores on flowing cells. In this contribution we present results for implementing the digital methods with sorting of cells and fluorescence microspheres

Material & Methods

The testbed for this experiment was a modified commercial sorting flow cytometer. We replaced the stock laser lines with solid state lasers that can either be directly modulated or modulated using an electro-optic modulator to impart periodic wave characteristics into the beam. This allows for single and multiple harmonic information to be imparted into the signal for Fourier analysis. Signals are routed from the cytometer photodetectors through high frequency amplifiers and into a specialized data acquisition system. It is within this system that standard cytometric data is recorded and the Fourier transform on the signals is done to extract out the phase information at signal harmonics of interest. The recorded phase, at a particular frequency, is proportional to the excited state fluorescence lifetime of a fluorophore.

The phase shift is a measure of how much a periodic signal lags to a reference. The degree of lag of an excited state fluorescence process only equates to a measure in the tens of nanoseconds, and a high frequency signal is needed in order to determine a phase shift that can be discerned by sampling equipment.

Our investigation is of a set of two fluorescent microspheres whose lifetimes have previously been individually measured on our system. By using modulation levels in the megahertz range we are able to quantify and distinguish between the excited state lifetimes of the two microspheres and set up gating regions to send back to the sorting cytometer. This instrument is then able to utilize an electro-static deflection system to deflect droplets created by a piezo-electric resonator from the flowing stream of the cytometer, into specific bins. These samples are then tested for sort purity by analysis on a separate dedicated analysis cytometer.

Results

With these approaches we demonstrate that using frequency domain techniques can lead to the ability to distinguish different fluorophores based solely on their excited state fluorescence lifetime. Additionally we show that it is possible to not only distinguish in software, but to redirect signals back to a sorting cytometer and be able to sort microparticles based upon fluorescence lifetime characteristics. Ultimately these techniques can be implemented in commercial cytometry and expand the functionality of instruments by allowing for intensity and temporal characteristics to be used to distinguish between fluorophores in cellular systems to aid in research and diagnostics.

See more of this Session: Biosensors, Biodiagnosis and Bioprocess Monitoring
See more of this Group/Topical: Food, Pharmaceutical & Bioengineering Division

(648a) Perfusion-Decellularization of Pancreatic Matrix A Scaffold for Bio-Engineered Pancreas

Saik Kia Goh, Bioengineering, University of Pittsburgh, Pittsburgh, PA, Suzanne Bertera, Department of Pediatrics, Children's Hospital of Pittsburgh, Pittsburgh, PA and Ipsita Banerjee, Chemical and Petroleum Engineering, University of Pittsburgh, Pittsburgh, PA

Introduction: Type I Diabetes affects over 1 million people in the United States. While islet transplantation has proved to be a promising therapeutic strategy, there are still limitations both in maintaining a viable culture of primary islets and a suitable scaffold for transplantation to ensure islet viability and functionality. We hypothesize that a 3-D scaffold recapitulating the native geometry and microenvironment would retain cellular functionality *in-vitro* and *in-vivo*. To this effect, we propose that perfusion-decellularization of whole pancreas could yield an acellular scaffold suitable for pancreatic tissue engineering.

Materials and Method: Cadaveric pancreata were isolated from adult, heparinized mice (n=8) and mounted on a perfusion apparatus. Ionic detergent, 0.1% SDS was perfused via the pancreatic duct until tissues were translucent and white in color (24-28 hours). To evaluate the scalability of this technique to a human-sized organ, decellularization of an adult cow pancreas was also performed. The resulting acellular matrices were analyzed for integrity of vasculature tree via trypan blue injections. Cell nuclei post-decellularization was quantified via dapi staining. The cytoskeletal element, actin was examined using phalloidin. Immunofluorescence and histology were done to characterize the retention of the extracellular matrix (ECM) protein after decellularization of the pancreas. For *in-vitro* culture, acellular matrices were seeded with Beta-TC-6 cells or mouse ES cells ($2x10^5$) to examine the proliferation, survival and differentiation (n=3) of the seeded cells. For whole organ recellularization, Beta-TC-6 or mouse ES cells ($30x10^6$) were introduced into the acellular matrices via the pancreatic duct in 3 steps, with 20 min interval between each step. Recell constructs (n=3) were then mounted on a perfusion apparatus to allow dynamic culture for 3 days. Both *in-vitro* and whole organ recell constructs were analyzed using histology, TUNEL, immunofluorescence staining and qRT-PCR.

Results and Discussion: Perfusion via the pancreatic duct with 0.1% SDS resulted in complete decellularization of both murine and cow pancreata. The vascular architecture was retained in acellular pancreas. No dapi staining of intact nuclei or cytoskeletal element were detected in acellular pancreas. Microstructure and major ECM component such as collagen I, collagen IV, fibronectin and laminin were preserved in acellular pancreas. Insulin response assay showed preserved insulin response to glucose challenge in the Beta-TC-6 seeded construct in both *in-vitro* culture and whole organ culture. In ES cells seeded construct, qRT-PCR results showed elevated gene expression of pancreatic differentiation markers. Immunohistochemistry from whole organ recell demonstrated better viability (<5% TUNEL positive).

Conclusion: Perfusion-decellularization of cadaveric pancreata efficiently removes cells while retaining ECM protein and microstructure. This perfusable 3-D whole organ scaffold yields functional tissue construct with preserved endocrine function and favored endocrine differentiation. Further studies are underway to establish the suitability of acellular pancreatic ECM as a scaffold to transplant donor islets and achieve euglycaemia in diabetic recipients.

See more of this Session: Extracellular Matrix In Tissue Engineering
See more of this Group/Topical: Food, Pharmaceutical & Bioengineering Division

(648g) A Novel Tissue-Engineered Liver Using Heparin-Immobilized Decellularized Liver Matrix

Yeonhee Kim[1], Maria-Louisa Izamis[2], Martin L. Yarmush[3], Korkut Uygun[4] and Basak Uygun[4], (1) Surgery, Massachusetts General Hospital / Harvard Medical School, Boston, MA, (2)Center for Engineering in Medicine, Massachusetts General Hospital, Harvard Medical School, Shriners Hospital for Children, Boston, MA, (3)Massachusetts General Hospital, Harvard Medical School, Boston, MA, (4)Center for Engineering in Medicine, MGH/Harvard Medical School, Boston, MA

Orthotopic liver transplantation is the only definitive treatment for end-stage liver disease. This procedure is limited by the shortage of available donor organs. Recently, we developed methods to prepare transplantable liver grafts using decellularized whole liver that can serve as a functional alternative for donor liver transplantation. The decellularized liver matrix (DLM) is the most suitable bioscaffold for designing the tissue-engineered liver due to its fascinating characteristics including preserved extracellular microenvironment of intact liver and functional vascular structure, and ability to be repopulated with primary cells. The use of decellularized matrix accelerated engineering whole organ via recellularization with primary hepatocytes through portal vein perfusion and exhibited preservation and enhancement of liver-specific phenotypes of the engineered liver graft. The predominant obstacle for engineered liver transplantation is thrombogenicity that induces platelet activation and ultimately hinders the delivery of nutrients and oxygen within the entire engineered tissue mass. In this study, we fabricated heparin-immobilized decellularized matrices that can prevent thrombosis and improve the survival of the graft in vivo. The thromboresistant surface modification of the acellular three-dimensional scaffold was achieved via layer-by-layer self-assembly and was optimized through *ex vivo* diluted blood perfusion. Preliminary results showed that only heparin-immobilized decellularized matrices exhibited even and positive staining for toluidine blue over the entire biocaffold as compared to control, and prevented thrombus formation during ex vivo blood perfusion. The heparin-immobilized DLM will be utilized to recellularize the scaffold with primary hepatocytes to generate functional liver grafts that can be used as a favorable replacement for liver transplantation.

See more of this Session: Extracellular Matrix In Tissue Engineering
See more of this Group/Topical: Food, Pharmaceutical & Bioengineering Division

(648c) Do Sensory Neurons Synthesize Extracellular Matrix Molecules and Remodel Their 3D Environment?

Swarnalatha Balasubramanian, Naga Shilpa Vadala and Dr. Jennie B. Leach, Chemical and Biochemical Engineering, University of Maryland Baltimore County, Baltimore, MD

Nerve injuries can be catastrophic as neurons are the only cell type in adult human that does not proliferate. Current repair treatments of peripheral nerve injuries use transplanted nerve grafts and are generally successful but suffer several disadvantages such as multiple surgeries, loss of function at the graft donor site, and low success rates for large injuries. A potential "off-the-shelf" alternative is the nerve guidance channel, but it is unclear how to best design these tubular synthetic devices. We hypothesize that this lack of knowledge stems from an incomplete understanding of how neurons behave in three-dimensional (3D) structures or matrices. While two-dimensional (2D) substrates such as petri plates have been incredibly valuable in revealing intricacies of cell biology, recent works have established that the response of non-neuronal cells is dramatically altered in a 3D matrix as compared to 2D culture [1]. Thus, our group strives to advance the understanding of how neurons respond to 3D matrices and we seek a biological basis upon which to rationally design new devices to promote nerve repair.

We have made significant progress in this direction by establishing that the cell-matrix signaling pathways responsible for sensory neuron morphology and development are altered in 3D versus 2D culture [2]. We have also demonstrated that 3D gel properties are dramatically altered in the presence of encapsulated neurons, as observed from increased gel stiffness, decreased swelling and slower degradation rate vs acellular gels [3]. This presentation will overview these findings and also poses the hypothesis that like non-neuronal cells, neurons actively sense and remodel their 3D microenvironment. Thus, the aim of our current work is to determine how neurons interact with these structures via mechanisms of matrix degradation, synthesis and remodeling. We have found that extracellular matrix synthesis is altered in a 3D microenvironment as compared to 2D substrates. These findings provide a biological basis for the design and testing of new peripheral nerve repair therapies.

References:

1. Irons H, et al., J Neural Eng 2008, 5:333. 2. Ribeiro A., et al. Substrate Three-Dimensionality Induces Elemental Morphological Transformation of Sensory Neurons on a Physiologic Timescale, submitted; Ribeiro A., et al. β1-integrin Cytoskeletal Signaling Regulates Sensory Neurons Response to Matrix Dimensionality, submitted. 3. Ribeiro A., et al. Degradable scaffolds for neural progenitor cell delivery. submitted.

(648d) Gelatin Microspheres As Scaffolds for Adipose Derived Stem Cells and Liver Tissue Engineering

Yen Wah Tong, Department of Chemical and Biomolecular Engineering, National University of Singapore, Singapore, Singapore and Anjaneyulu Kodali, Chemical and Biomolecular Engineering, National University of Singapore, Singapore, Singapore

At present, organ transplantation is the only means for treating various kinds of tissue failures. Though this approach has been employed for a long time now, it is still associated with issues such as donor organ shortage and immune rejections. Stem cell based tissue engineering seems to be an exciting and revolutionary strategy which can replace organ transplantation in treating damaged tissues. Embryonic stem cells and various adult stem cells are being investigated for this purpose and each of them has their own advantages and disadvantages. Adipose derived stem cells (ADSCs) that are isolated from lipoaspirates are gaining interest in the recent times because of their ease of availability in large numbers and their multi-potential differentiation ability.

However, there are quite a few challenges which need to be overcome for the viable usage of such stem cell therapies for regenerative medicine applications. One such major challenge is to find a compatible platform for the *in vitro* growth and *in vivo* delivery of these cells in a viable and functional state. In order to closely mimic the *in vivo* environment three dimensional cell culture matrices are being preferred over the traditional two dimensional cell culture practices. In the present work we aim to investigate the suitability of three dimensional Gelatin Microsphere (GM) scaffold systems as a platform for liver tissue reconstruction and for the growth and differentiation of these newly emerging ADSCs. Since gelatin is derived from collagen, GMs can act as replacement for extracellular matrix of the body.

A successful protocol has been established to isolate ADSCs from the adipose tissues harvested from human patients with signed consent in compliance with procedures approved by the institutional review board. Gelatin microspheres fabricated using a water in oil emulsion technique were tested for their suitability for *in vitro* growth and differentiation of ADSCs by studying their proliferation using MTT assay and by differentiating them towards adipogenic lineage. Our results show that Gelatin microspheres were able to support the growth and differentiation of ADSCs.

In a simultaneous study we have also investigated the suitability of GMs for liver tissue engineering by culturing human hepatoma cells. According to our results the liver cells were able to proliferate and function better on GMs compared to the 2D tissue culture plates and were able to form tissue like cell-microsphere constructs.

Thus our results until now show that GMs are good platforms for culturing ADSCs and for liver tissue engineering. In future we aim to combine these two strategies to develop a ADSCs therapy for liver tissue reconstruction using gelatin microspheres as platforms.

See more of this Session: Extracellular Matrix In Tissue Engineering
See more of this Group/Topical: Food, Pharmaceutical & Bioengineering Division

(648e) Effect of Chitosan Fiber Stiffness On Mesenchymal Stem Cell Attachment and Proliferation

Oksana Blowytsky, Chemical Engineering, Wayne State University, Warren, MI, Mohammad Z. Albanna, Department of Biomedical Engineering, Wayne State University, Detroit, MI, Henry L. Walters, Cardiovascular Surgery, Children's Hospital of Michigan, Detroit, MI and Howard WT Matthew, Chemical Engineering & Materials Science, Wayne State University, Detroit, MI

Introduction: A key challenge in stem cell research is the ability to direct stem cell growth and differentiation towards specific tissue lineages. While complex interactions between soluble factors and extracellular matrix (ECM) have been investigated extensively, physical properties such as substrate elasticity and stiffness are only recently gaining attention as tools for influencing stem cell differentiation and proliferation. It has been demonstrated that substrate stiffness can be used to direct mesenchymal stem cell (MSC) differentiation into mature neuronal, muscle and bone tissues. In this study, we hypothesized that increasing substrate stiffness would improve MSC attachment and proliferation on chitosan fibers.

Materials and Methods: Chitosan fibers were produced by extrusion of chitosan solutions into aqueous ammonia, followed by air drying. Fiber stiffness was modified through various combinations of thermal annealing and chemical crosslinking. Fibers were either dried at room temperature, annealed at 195 °C, or crosslinked with heparin. Monotonic tensile testing was done to determine tensile strength, elastic modulus, and breaking strain of processed fibers. Ovine bone marrow-derived MSCs were isolated, characterized and cultured on fibers with stiffness values of 7, 20, 91 and 128 MPa for 9 days. MSC proliferation was assessed using Alamar Blue, viability was evaluated using Calcein AM and cytoskeletal organization was visualized using rhodamine-phalloidin.

Results and Discussion: MSCs maintained more than 95% cell viability on all fibers during the culture. Results also demonstrated that MSC attachment and proliferation rate increased in proportion to fiber stiffness. Fibers with the lowest mechanical properties, those dried at room temperature, exhibited a lower rate of cell proliferation as compared to cells seeded on fibers with the highest mechanical properties, i.e. annealed at 195°C. Interestingly, the covalent incorporation of heparin onto fibers reduced fiber stiffness but improved MSC attachment and proliferation, probably through growth factor and ECM protein binding. Actin fiber density and alignment increased with chitosan fiber stiffness.

Conclusion: Results demonstrated that increasing chitosan fiber stiffness significantly improved MSC attachment, proliferation and cytoskeletal organization. Growing MSC on chitosan fibers with tunable stiffness may provide an alternate method for influencing MSC performance in tissue engineering scaffolds.

Acknowledgment: The authors would like to thank the Perri Foundation for partial funding of this project through the Department of Cardiovascular Surgery at Children's Hospital of Michigan.

See more of this Session: Extracellular Matrix In Tissue Engineering
See more of this Group/Topical: Food, Pharmaceutical & Bioengineering Division

(648f) Developing Biocompatible Polycations for Cell Encapsulation

Minglin Ma, Wendy Liu and Daniel Anderson, Koch Institute for Integrative Cancer Research, Massachusetts Institute of Technology, Cambridge, MA

In cell encapsulation, a cationic polymer is often used to form a semi-permeable polyelectrolyte complex membrane. Many currently used polycations such as poly(L-lysine) (PLL) often cause severe immune responses. Using combinatorial polymerization, we synthesized a new library of cationic polymers. When used as surface coatings, these polymers elicited different levels of foreign body responses depending on their chemical structures. We identified the polymer coatings that inhibited macrophage activation *in vitro*, reduced the recruitment of inflammatory cells 24 hours and the fibrosis 30 days after subcutaneous implantation of the carboxylated polystyrene microparticles. These polymers offer great alternatives to the typically used PLL for cell encapsulation and many other applications.

See more of this Session: Extracellular Matrix In Tissue Engineering
See more of this Group/Topical: Food, Pharmaceutical & Bioengineering Division

Managing Uncertainty in Design Space

Pat Whitcomb
Stat-Ease, Inc.
2021 E. Hennepin Ave
Suite 480
Minneapolis, MN 55413
Ph 612.746.2036
fax 612.746.2056
pat@statease.com

Mark Anderson
Stat-Ease, Inc.
2021 E. Hennepin Ave
Suite 480
Minneapolis, MN 55413

2011 AIChE Annual Meeting

Managing Uncertainty in Design Space

Objective: Define an operating window in process factor space where we have 95% confidence that 99% of the population meet (or exceed) specifications.

- Use empirical DOE to model the responses as functions of the process factors.
- Use a tolerance interval to "back off" (provide a buffer) from the specifications.
- Size the DOE for required half-width of tolerance interval.

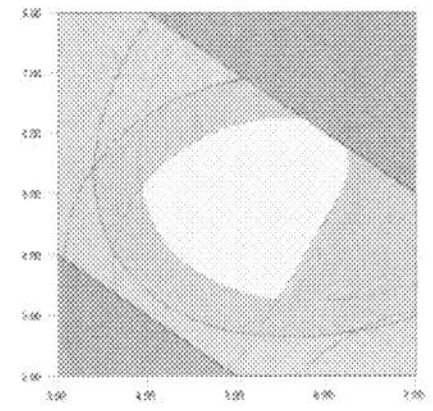

Illustrative Example
Tableting Process

This case study illustrates how response surface DOE and tolerance intervals can be used to set an operating window where specifications are consistently met.

- An optimal design is run on two process parameters (granulation time and lubrification time) in a tableting process.
- Three responses (dissolution, friability and hardness) are measured.
- The specifications are:
  - Dissolution ≥ 75%
  - Friability ≤ 0.5%
  - Hardness ≥ 10 kP

Initial Operating Window
Specifications as Bounds

y_1 = Dissolution %
(specification ≥ 75%)

y_2 = Friability %
(specification ≤ 0.5%)

y_3 = Hardness kP
(specification ≥ 10 kP)

Specifications on Graphical Overlay

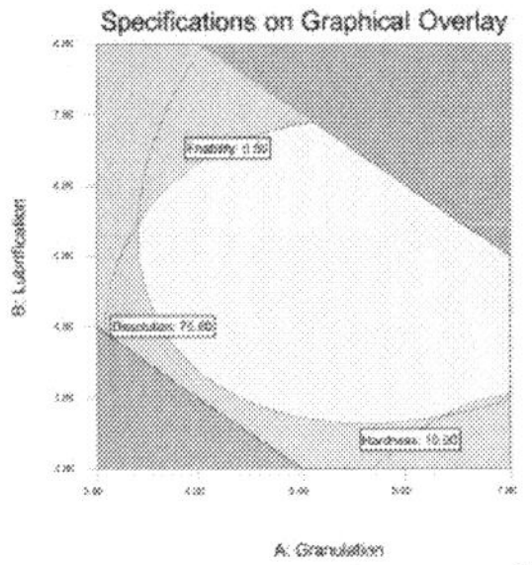

Uncertainty is a BIG Problem
Specifications as Bounds

If the tableting process is operated on a boundary, then 50% of tablets produced fall outside the specification.

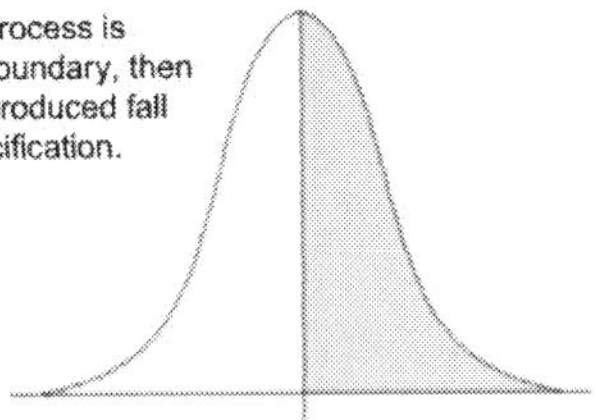

Managing Uncertainty in Design Space.

If we want more confidence that individual units are within specifications:

- Then we should back off using a tolerance interval rather than a confidence interval.
- Specify the confidence level; e.g. 95%.
- Specify the portion of the population to be within specifications; e.g. 99%.

Note:

- *A confidence interval (CI) pads our design space to give us confidence that the process mean is within boundaries.*
- *A tolerance interval (TI) give us confidence that a stated proportion of the population is within specifications.*

Tolerance Interval
Definition

Tolerance Interval: An interval that covers a fixed proportion of outcomes from the population with a pre-determined confidence level for estimating the population mean and standard deviation.

$$\bar{y} - k_2(s) \le P\% \text{ of population} \le \bar{y} + k_2(s)$$

$$TI = \bar{y} \pm k_2(s)$$

$$TI = \bar{y} \pm d$$

d is the half width of the tolerance interval

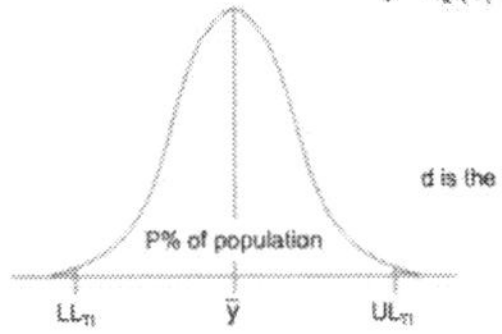

AIChE 2011 7

Interval Calculations
Single Sample & Normal Distribution

Confidence Interval: $\bar{y} \pm t\left(\frac{s}{\sqrt{n}}\right)$

Tolerance Interval: $\bar{y} \pm k_2 s$

Note:

- TI multipliers (k_2) are larger than those for a CI (t); especially for small samples.
- As the sample size increases:
  - CIs shrink towards zero.
  - TIs tend towards a fixed value: z(s).

df	$t_{\alpha=0.05}$	$k_{2\ \alpha=0.05,\ P=0.95}$
1	12.706	37.674
2	4.303	9.916
3	3.182	6.370
4	2.776	5.079
5	2.571	4.414
10	2.228	3.259
25	2.060	2.612
50	2.009	2.375
100	1.984	2.232
∞	1.960	1.960

Larger sample sizes are needed for TIs than CIs.

AIChE 2011 8

Tolerance Interval
Calculation for a DOE

TI calculation for a Normal Sample

$$TI = \bar{y} \pm k_2 s$$

TI calculation for a DOE[3]

$$TI = \hat{y}_0 \pm s\left[t_{\alpha,df}\sqrt{x_0'(X'X)^{-1}x_0} + \Phi^{-1}(P)\sqrt{\frac{n-p}{\chi^2(n-p,1-\alpha)}}\right]$$

$$TI = \hat{y}_0 \pm (s)(t_{\alpha,df})(\text{TI multiplier})$$

$$TI = \hat{y}_0 \pm d$$

d is the half width of the tolerance interval

AIChE 2011 9

TI Interval Multipliers
Single Sample versus Two-Factor DOE

Comparison*:

single sample $k_{2(\alpha=0.05,\ P=99\%)}$

versus

$(t_{\alpha=0.05,df})(\text{TI multiplier}_{\alpha=0.05,\ P=99\%})$

Take home message:

Tolerance intervals that cover a design volume are wider than those at a single point in space.

* Basis is average TI multiplier for a two-factor unconstrained IV optimal design for a quadratic model with 6+ model points, 5 lack of fit points and 5 replicates.

df	$k_{2\ \alpha=0.05,\ P=0.99}$	(t)(TIM)
10	4.277	5.013
11	4.150	4.860
12	4.044	4.739
13	3.955	4.636
14	3.878	4.532
15	3.812	4.460
20	3.577	4.155
25	3.432	3.966
30	3.344	3.826
50	3.121	3.518
100	2.933	3.224

*TIM = TI multiplier

AIChE 2011 10

Interval Multipliers for "s"
TI DOE, TI sample and CI sample

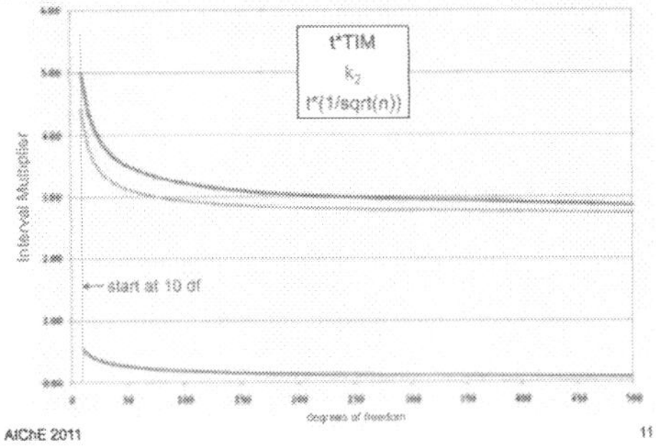

AIChE 2011 11

Back off from Specification
One-Sided 95% tolerance interval (TI)

One-sided TI calculation for DOE

$$TI = \hat{y}_0 + d \quad \text{or} \quad TI = \hat{y}_0 - d$$

d is the offset for the tolerance interval

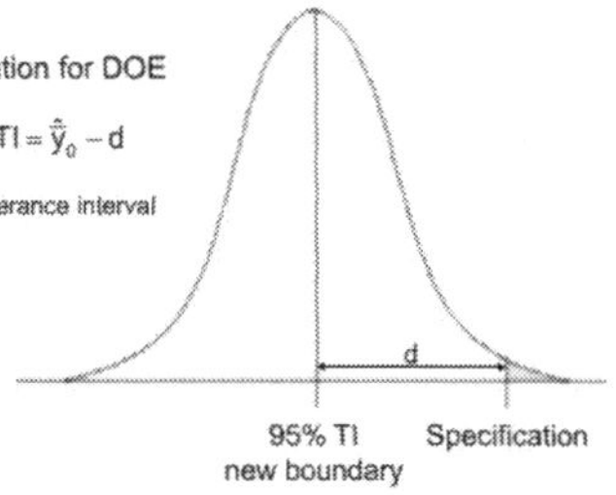

AIChE 2011 12

RSM DOE Process *(1 of 2)*
Tableting Process

1. Identify opportunity and define objective.
   Define an operating window in process factor space where we have 95% confidence that 99% of the population meets (or exceeds) specifications.
2. State objective in terms of measurable responses.
   a. Define the precision that is required for each response.
      Tolerance interval (α=0.05, P=99%) half-width (d):
      dissolution d = 9, friability d = 0.3, hardness d = 0.7
   b. Estimate experimental error (σ) for each response.
      dissolution s = 2, friability s = 0.03, hardness s = 0.14
   c. Use fraction of design space (FDS) to quantify precision.

RSM DOE Process *(2 of 2)*
Tableting Process

3. Select the input factors and ranges to study.
   (Consider both your region of interest and region of operability.)
   A = Granulation time: 3 – 7 minutes
   B = Lubrification time: 2 – 8 minutes
   7 ≤ A+B (total mixing time) ≤ 12
4. Choose a polynomial to estimate: *quadratic*
5. Select a design and:
   - Size design for precision needed *(next slide)*
   - Examine the design layout to ensure all the factor combinations are safe to run and are likely to result in meaningful information (no disasters).

DOE with Tolerance Intervals
Sizing for Precision Requirements

Define the precision required as the half-width of a Tolerance Interval that contains 99% of the population with 95% confidence:

Response	d	s	d/s
dissolution	9.0	2.00	4.5
friability	0.2	0.03	6.7
hardness	0.7	0.14	5

Size the DOE to the smallest (d/s) ratio.

Sizing for Precision Requirements
Tolerance Intervals *(page 1 of 3)*

16 runs (6 model, 5 lack of fit, 5 replicates)

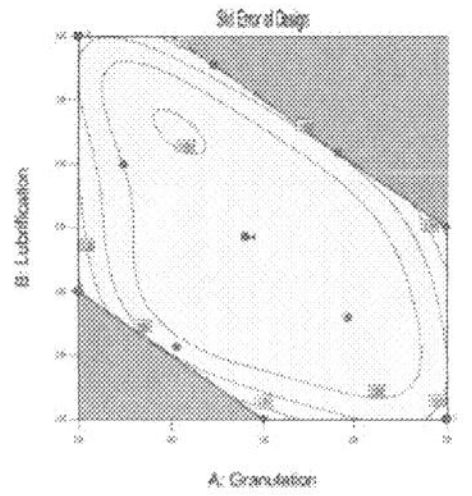

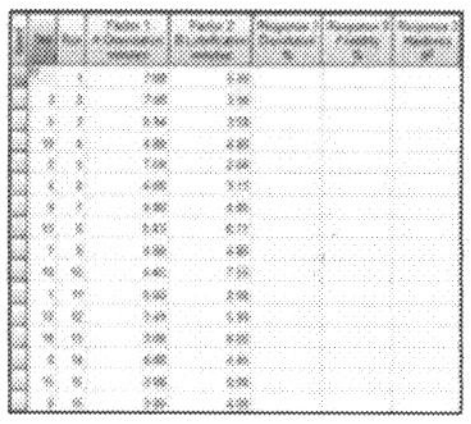

Sizing for Precision Requirements
Tolerance Intervals *(page 2 of 3)*

16 runs (6 model, 5 lack of fit, 5 replicates) FDS = 0%

d = 9
s = 2
a = 0.05
P = 0.99

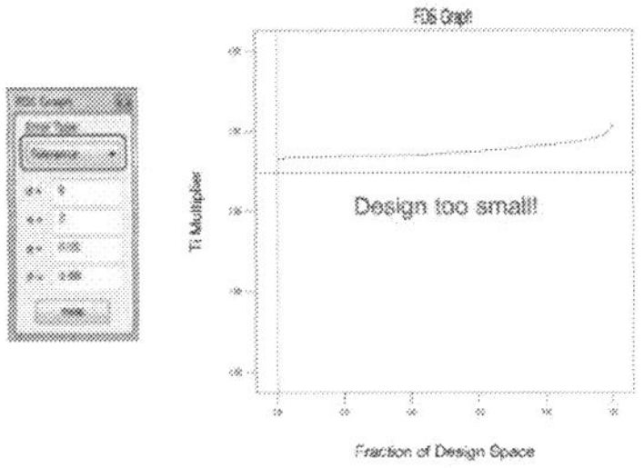

Sizing for Functional Design
Confidence Intervals

16 runs (6 model, 5 lack of fit, 5 replicates) FDS = 96%

d = 3
s = 2
a = 0.05
P = 0.99

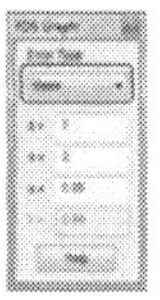
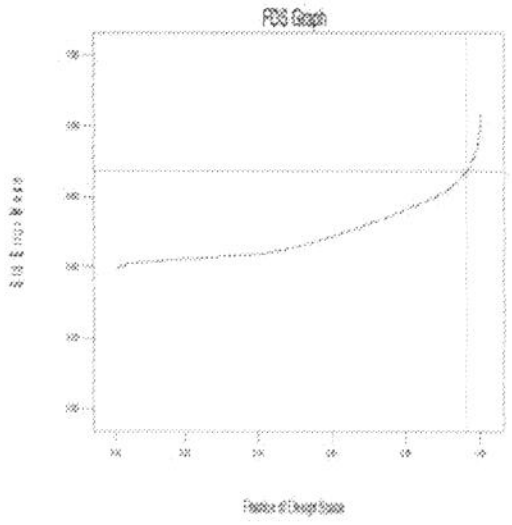

Sizing for functional design at:
http://asq.org/cpi/

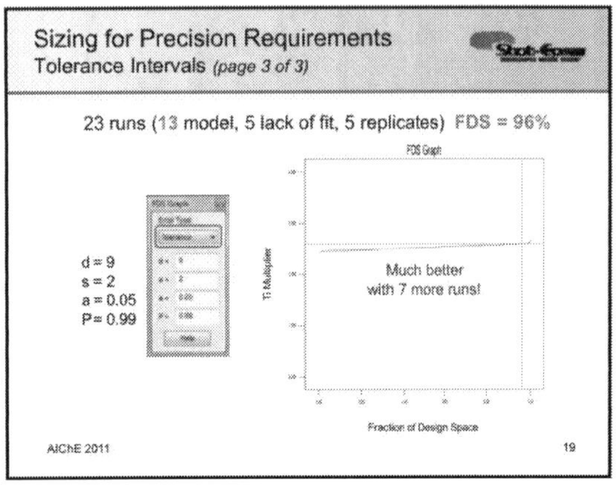
Sizing for Precision Requirements
Tolerance Intervals (page 3 of 3)
23 runs (13 model, 5 lack of fit, 5 replicates) FDS = 96%
d = 9
s = 2
a = 0.05
P = 0.99
Much better
with 7 more runs!
Fraction of Design Space
AIChE 2011
19

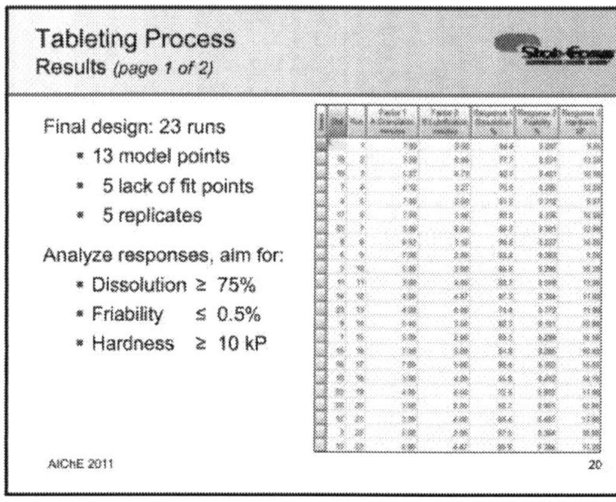
Tableting Process
Results (page 1 of 2)
Final design: 23 runs
13 model points
5 lack of fit points
5 replicates
Analyze responses, aim for:
Dissolution ≥ 75%
Friability ≤ 0.5%
Hardness ≥ 10 kP
AIChE 2011
20

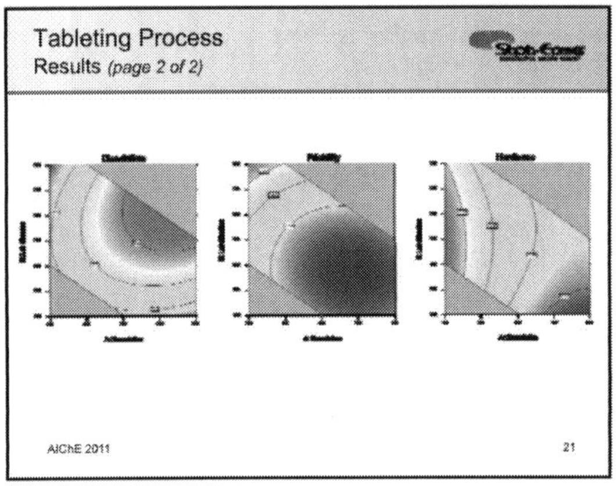
Tableting Process
Results (page 2 of 2)
AIChE 2011
21

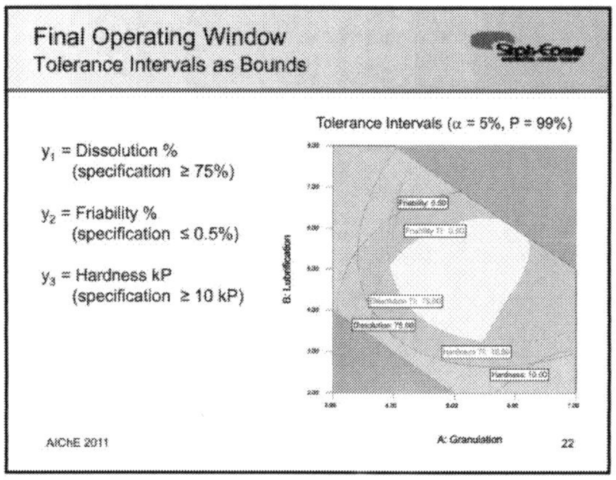
Final Operating Window
Tolerance Intervals as Bounds
y1 = Dissolution %
(specification ≥ 75%)
y2 = Friability %
(specification ≤ 0.5%)
y3 = Hardness kP
(specification ≥ 10 kP)
Tolerance Intervals (α = 5%, P = 99%)
B: Lubrication
A: Granulation
AIChE 2011
22

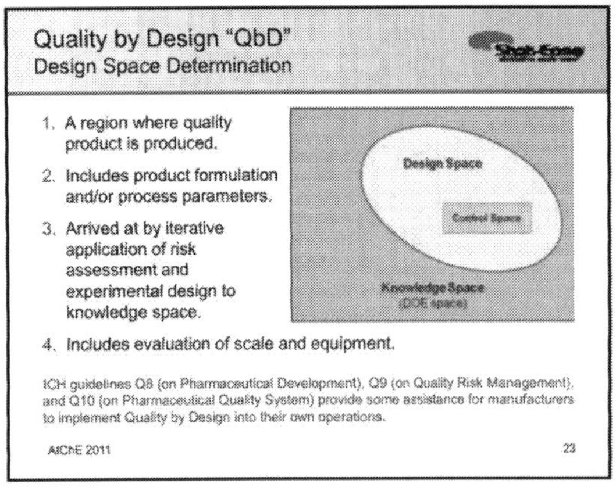
Quality by Design "QbD"
Design Space Determination
1. A region where quality product is produced.
2. Includes product formulation and/or process parameters.
3. Arrived at by iterative application of risk assessment and experimental design to knowledge space.
4. Includes evaluation of scale and equipment.
Design Space
Knowledge Space
(DOE space)
ICH guidelines Q8 (on Pharmaceutical Development), Q9 (on Quality Risk Management), and Q10 (on Pharmaceutical Quality System) provide some assistance for manufacturers to implement Quality by Design into their own operations.
AIChE 2011
23

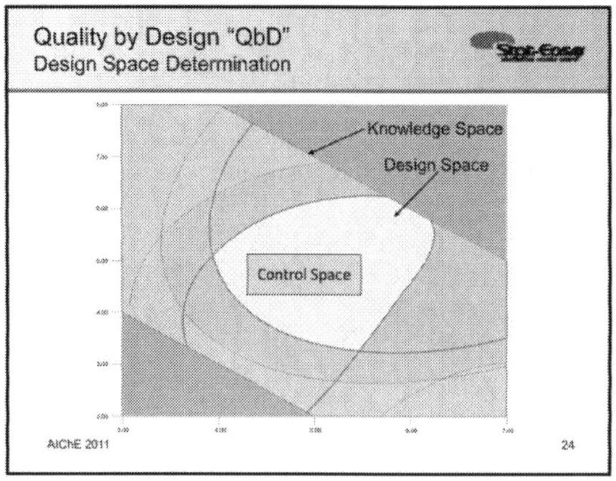
Quality by Design "QbD"
Design Space Determination
Knowledge Space
Design Space
Control Space
AIChE 2011
24

DOE with Tolerance Intervals
How many Runs

Quadratic model (TI for d = 9 and s = 2) FDS ≥ 90%

Factors	p + 5 + 5*	Extra runs	Increase
2	16	7	44%
4	25	10	40%
6	38	13	34%
8	55	15	27%
10	76	16	21%

* p coefficients + 5 lack of fit + 5 replicates

AIChE 2011 25

Managing Uncertainty in Design Space
Summary

1. Size your DOE appropriately for its propose:
   - Functional design – size via confidence intervals.
     A confidence interval pads our design space to give us confidence that the process mean is within boundaries.
   - Verify specifications – size via tolerance intervals.
     A tolerance interval give us confidence that a stated portion of the population is within specifications.
2. TI multipliers are larger than those for a CI.
   - As the sample size increases:
     - CIs shrink towards zero.
     - TIs tend towards a fixed value.
3. Tolerance intervals require a large design (sample size).

AIChE 2011 26

References

1. Alysa R. Zahran, Christine M. Anderson-Cook and Raymond H. Myers, "Fraction of Design Space to Assess Prediction", *Journal of Quality Technology*, Vol. 35, No. 4, October 2003.
2. Heidi B. Goldfarb, Christine M. Anderson-Cook, Connie M. Borror and Douglas C. Montgomery, "Fraction of Design Space plots for Assessing Mixture and Mixture-Process Designs", *Journal of Quality Technology*, Vol. 36, No. 2, October 2004.
3. Steven DE Gryze, Ivan Langhans and Martina Vandebroek, "Using the Intervals for Prediction: A Tutorial on Tolerance Intervals for Ordinary Least-Squares Regression", *Chemometrics and Intelligent Laboratory Systems*, 87 (2007) 147 – 154.
4. Gerald J. Hahn and William Q. Meeker, *Statistical Intervals; A Guide for Practitioners*, (1991) John Wiley and Sons, Inc,
5. Gary Oehlert and Patrick Whitcomb (2001), Sizing Fixed Effects for Computing Power in Experimental Designs, *Quality and Reliability Engineering International*, July 27, 2001.

AIChE 2011 27

(655b) An Integrated Design Space Involving All Unit Operations in Tablet Manufacturing

Theodora Kourti, GSK, Hamilton, ON, Canada

A design space can be determined per unit operation, per cluster of unit operations or for the entire process.

By linking unit operations, and therefore recognizing a continuum in drug production that spans from the Drug Substance Properties to the Drug Product, it is possible to create a more versatile and robust Design Space. The final product that delivers the active pharmaceutical ingredient to a patient, is indeed the result of a multidimensional combination of raw material attributes and process parameters that span several unit operations including those of the drug substance production (like reaction and crystallization), the ones from drug product production (like granulation and compression) but also packaging. Each one of them can impact one or more final quality characteristics, as well as stability.

Therefore, if the Design Space is addressed in a holistic form, then the quality of the final product, including its stability, should be expressed as the result of the values and the interactions of these parameters. Also by treating the design space in a holistic way it would provide the manufacturer with the most cost efficient operation and provide for high yield and low operating costs, because problems at later unit operations would most likely be anticipated and corrected in earlier operations. In other words the control strategy will be part of the product design, such that it can be implemented in the most cost effective way. In this model, the tablet manufacturing process is mapped from start to end, and both design space and control strategy can be more dynamic as an observation at the extreme ends of the design space can be offset elsewhere in the process to provide conforming material.

This works gives an example where all unit operations are integrated in the Design space. Variables from all unit operations are incuded for continued process verification. The work includes different strengths of tablets

See more of this Session: Integrated QbD Approach for Design Space Development Including Real Time Release
See more of this Group/Topical: Topical I: Comprehensive Quality by Design in Pharmaceutical Development and Manufacture

(655c) Design Space Definition Using a Variational Bayes' Approximation

Linas Mockus[1], **Jos Miguel Lanez**[2], Lee E. Kirsch[3] and Gintaras V. Reklaitis[1], (1)Purdue University, West Lafayette, IN, (2)School of Chemical Engineering, Purdue University, West Lafayette, IN, (3) Division of Pharmaceutics, The University of Iowa, Iowa City, IA

The design space may be interpreted as the constrained region of the key manufacturing parameters space which provides assurance of quality of a drug given that manufacturing occurs in this region. The metric used to represent this assurance should not be deterministic, rather it should be stated in probabilistic terms. In this work, we report on the use of Bayesian methods to develop a suitable risk metric based on statistical models of the manufacturing processes and product properties. The Bayesian inference is carried out to determine the posterior distribution of the probability of the product meeting quality specifications. Here, we propose an alternative optimization based procedure, the variational Bayes' approximation, to obtain the posterior distribution. A sequential methodology is proposed to deal with non-linear models and the consideration of covariance. The results are compared to the widely used but computationally intensive Markov Chain Monte Carlo method. The proposed approached is illustrated with information drawn from a QbD study on Gabapentin.

See more of this Session: Integrated QbD Approach for Design Space Development Including Real Time Release
See more of this Group/Topical: Topical I: Comprehensive Quality by Design in Pharmaceutical Development and Manufacture

(655d) A Data-Driven Modeling Approach Using the Design of Dynamic Experiments Methodology In Calculating the Design Space of Batch Pharmaceutical Processes

Christos Georgakis, Systems Research Institute and Chemical and Biological Engineering, Tufts University, Medford, MA

The development of fundamental models for fed-batch pharmaceutical processes range from very challenging to almost impossible as not all the inner workings of the process are quantitatively known. Thus, the accurate calculation of the Design Space in which the process will be operated is a substantial challenge. One approach that has been recently proposed [1] is to use the Design of Experiments (DoE) methodology to develop data-driven models of the process. A major limitation of this approach is its inability to account for time varying operations that are often more optimal than those that are constant with time. Such time-varying operation include but are not limited to the varying the reaction temperature, the crystallizer cooling, the feeding rate of a sugar source to a fermentation. This dynamic character is even more important if the process will operate under feedback control. In such a case the operating conditions are changed by the controller to ensure the product quality at the end of the batch.

We present here an approach that utilizes the recently postulated Design of Dynamic Experiments (DoDE) methodology [2] that generalizes the classical DoE approach [3, 4]. This is achieved by systematically designing a set of experiments with respect to the time-varying decision variables. Using the collected data, a response surface model (RSM) is estimated that quantifies the impact that uncontrolled and controlled inputs (disturbances and manipulated variables, respectively) have on the quality of the product at the end of the batch. Process disturbances include the variability in feedstock properties as well as the variability in the operating conditions. The manipulated variables include those that can be changed with time such as the reactor temperature or the crystallizer cooling rate, or the substrate feed rate in fermentation, as well as others. A feedback controller aims to change the manipulated variables so that the variability in the final product is minimized.

The data-driven models derived from the DoDE data are used to quantify the design and control spaces of the process. Several simulated processes and an experimental asymmetric hydrogenation process of an API are examined to illustrate the general methodological concept that can be used in many batch pharmaceutical processes that avail the possibility of time varying conditions.

References:

1. A-Mab: a Case Study in Bioprocess Development CMC Biotech Working Group, Ver. 2.1

October 30, 2009, Internet Source: http://www.casss.org/associations/9165/files/A-Mab_Case_Study_Version_2-1.pdf
2. Georgakis, C. (2009); "A Model- Free Methodology for the Optimization of Batch Processes: Design of Dynamic Experiments" *Proceedings of the 8th IFAC International Symposium on Dynamics and Control of Process Systems (DYCOPS)* Istanbul, Turkey, July 2011
3. Box, G. E., & Draper, N. R. (2007); *Response Surfaces, Mixtures, and Ridge Analyses* (2nd Edition) Wiley- Interscience.
4. Montgomery, D. C. (2005); *Design and Analysis of Experiments* (6th Edition ed.) John Wiley & Sons, Inc.

See more of this Session: Integrated QbD Approach for Design Space Development Including Real Time Release
See more of this Group/Topical: Topical I: Comprehensive Quality by Design in Pharmaceutical Development and Manufacture

(655e) Dynamic Design Space As An Integrated Component of Quality by Design

Benoit Igne, **Carl A. Anderson** and James K. Drennen III, School of Pharmacy, Duquesne University, Pittsburgh, PA

The Food and Drug Administration requires from pharmaceutical companies the development of extensive knowledge databases regarding the control of the manufacturing process of a drug product. Through development and validation, drug manufacturers discover their process boundaries and identify a range of process parameters for each unit operation, the design space. However, limited work is done to study the effect of changing raw material on the robustness of the design space. Furthermore, once approved by the FDA, a process cannot be changed without filing post-approval changes that might require adding new experiments to the knowledge space. In the present study, the development of a design space for the creation of Excedrin (acetaminophen, caffeine) tablets through direct compression was investigated. A design of experiment including different excipent ratio of microcrystaline cellulose and lactose, two cros-carmelose sodium levels, and four compression forces was created using an industrial size press to define a knowledge space. Classical critical quality attributes (CQAs) (disintegration time, dissolution, radial tensile strength, friability) were measured and a design space identified. In order to test the robustness of the design space, raw material properties of acetaminophen (particle size) and lactose (ratio anhydrous to monohydrate) were modified. Resulting tablets were analyzed for relevant CQAs (disintegration time and radial tensile strength) to investigate how common variability in the raw material impacts the design space.

See more of this Session: Integrated QbD Approach for Design Space Development Including Real Time Release
See more of this Group/Topical: Topical I: Comprehensive Quality by Design in Pharmaceutical Development and Manufacture

(655g) Dissolution Testing of Prednisone and Salicylic Acid Calibrator Tablets At Different Tablet Locations

Anandhavalavan Arulmozhi and **Piero M. Armenante**, Otto H. York Department of Chemical, Biological and Pharmaceutical Engineering, New Jersey Institute of Technology, Newark, NJ

Dissolution testing is routinely carried out in the pharmaceutical industry to determine the rate of dissolution of solid dosage forms. This test is one of the several tests that pharmaceutical companies typically conduct on oral dosage formulations (e.g., tablets) to determine compliance. The USP Dissolution Testing Apparatus 2 is the most common of the apparatuses listed in the USP. However, it has been shown previously that the dissolution profile of a tablet undergoing dissolution in the USP Dissolution Apparatus 2 can be affected by the tablet location in the apparatus.

In this work, the dissolution rates of both non-disintegrating tablets (salicylic acid) and disintegrating tablets (Prednisone) were experimentally determined for many different tablet locations, both centered on the vessel bottom and off-center. The location of the tablet was experimentally varied in very small increments in order to determine the exact location where a transition in the dissolution profile occurred. It was found that in a small region (2-4 mm in radius) centered around the vessel centerline just below the impeller the dissolution profiles were similar to those observed with a centered tablet. However, outside this region the dissolution profiles were found to be significantly different, as indicated by the values of the Similarity Factor f_1 and the Difference Factor f_2. These finding are consistent with previous hydrodynamic investigations that showed the existence of a poorly mixed zone below the USP Apparatus 2 impeller. The results of this work can guide the practitioner on when to accept dissolution testing results based on tablet location.

See more of this Session: Integrated QbD Approach for Design Space Development Including Real Time Release
See more of this Group/Topical: Topical I: Comprehensive Quality by Design in Pharmaceutical Development and Manufacture

(656a) Induction of a Lipid Phase Change In the Host Cell Membrane by a Bacterial Toxin, *Aggregatibacter Actinomycetemcomitans* leukotoxin (LtxA)

Angela C. Brown, Department of Chemical and Biological Engineering, Drexel University, Philadelphia, PA, Kathleen Boesze-Battaglia, Department of Biochemistry, University of Pennsylvania, School of Dental Medicine, Philadelphia, PA and Edward T. Lally, Department of Pathology, University of Pennsylvania, School of Dental Medicine, Philadelphia, PA

Pathogenic bacteria have developed numerous strategies to evade the host immune response, one of which is the secretion of protein toxins and other virulence factors to produce an environment within the host that is more conducive to their survival. The leukotoxin (LtxA) produced by *Aggregatibacter actinomycetemcomitans* performs this task by selectively killing human white blood cells, allowing the bacteria to flourish in the host and disease progression (localized aggressive periodontitis) to occur. This toxin is a member of the repeats-in-toxin (RTX) family of proteins, each of which is produced by a virulent Gram-negative bacterium, including *Escherichia coli* and *Bordetella pertussis* (the causative agent of whooping cough). It has long been established that the RTX toxins are cytolytic; however, recent evidence has shown that the interactions of these toxins with cell membranes may be more complex than simple pore-formation. We therefore initiated this study to determine the mechanism of this early but important step in the pathogenicity of the bacterium.

To characterize the interaction of LtxA with membranes, we used a variety of biochemical and biophysical techniques, including scanning electron microscopy (SEM), fluorescence spectroscopy, differential scanning calorimetry (DSC), ^{31}P nuclear magnetic resonance (^{31}P NMR), and surface plasmon resonance (SPR). Most of this work was performed with model membranes (liposomes) to determine the lipid-specific interactions that occur during the toxin-membrane interaction.

Using fluorescence spectroscopy, we showed that LtxA disrupts membranes only when they contain a nonlamellar lipid, such as phosphoethanolamine (PE). These types of lipids are involved in the formation of nonlamellar lipid phases, including inverted hexagonal (H_{II}) phases. This membrane disruption was inhibited by cholesterol sulfate and a single-chained phospholipid (lysoPC), both of which inhibit H_{II} phase formation, suggesting that the toxin may promote the formation of this nonlamellar lipid phase. DSC demonstrated that the toxin lowers the bilayer-to-H_{II} phase transition temperature (T_H) of nonlamellar lipids, indicating that LtxA does promote the formation of the nonbilayer (H_{II}) phase. ^{31}P NMR verified this promotion of the H_{II} phase and demonstrated that this process occurs through the formation of an isotropic intermediate lipid structure.

We compared these results to standard models of pore formation, including the barrel stave pore, the toroidal pore, and the carpet mechanism, and determined that these results represent a previously undescribed mechanism of membrane disruption. Importantly, several RTX toxins, as well as other bacterial and eukaryotic toxins, have been shown, to some extent, to interact with nonlamellar lipids, suggesting that this newly detailed mechanism may, in fact, be widely conserved among protein toxins.

In this work, we have determined a new example of a mechanism by which a pathogen hijacks certain properties of the lipids in the host cell membrane to facilitate its attack and cause disease. Specifically, we have shown that LtxA cytotoxicity occurs through a direct interaction of the toxin with the lipids of the membrane, resulting in a lipid phase change from a stable bilayer phase to a weaker nonlamellar phase. This mechanism of membrane damage may be well-conserved and therefore represents an important target of future therapeutic approaches.

See more of this Session: Intermolecular Interactions
See more of this Group/Topical: Food, Pharmaceutical & Bioengineering Division

(656b) Characterizing Interactions Between Thrombin, Antithrombin III, and An Anti-Thrombin Antibody by Composition-Gradient Multi-Angle Light Scattering

Sophia Kenrick and Daniel Some, Wyatt Technology Corporation, Santa Barbara, CA

Macromolecular interactions influence all phases of biopharmaceutical development from drug discovery and target validation to the characterization of stable drug formulations at therapeutic doses. Composition-gradient multi-angle static light scattering (CG-MALS) is a powerful, label-free technique for quantifying reversible interactions and can be applied to many stages of the drug development process. Because assays are performed in solution, complex interactions can be observed without the influence of tags or surface immobilization. This not only provides for absolute characterization of affinity and stoichiometry but also enables study of the effects of solvent composition, pH, and molecular conformation on a particular interaction. Of particular interest are interactions between therapeutic targets, such as enzymes and growth factors, their natural ligands, and therapeutic antibodies. Using CG-MALS, the interaction between thrombin-α (Thr) and an anti-thrombin antibody (Ab) was measured, yielding the expected 1:2 Ab:Thr stoichiometry as well as a binding affinity of 15 nM at each binding site. In addition, the time-dependent, irreversible association of thrombin and antithrombin III (AT) was observed and a second-order rate constant calculated, $k_{on} = 6x10^{-3}$ $M^{-1}s^{-1}$.

Finally, CG-MALS was able to identify the recognition of the Ab for the Thr-AT complex to determine whether the antibody bound the thrombin active site.

See more of this Session: Intermolecular Interactions
See more of this Group/Topical: Food, Pharmaceutical & Bioengineering Division

(656c) Distinct Characteristics of CD44s Binding to Fibrin(ogen) Investigated Using Force Spectroscopy At the Single Molecule Level

Phrabha Shalini Raman, Christina S Alves, Denis Wirtz and Konstantinos Konstantopoulos, Chemical and Biomolecular Engineering, Johns Hopkins University, Baltimore, MD

CD44 and fibrin(ogen) play crucial roles in the hematogenous dissemination of tumor cells. The standard isoform of CD44, CD44s, from HL60 human myeloid leukemia cells binds to both immobilized fibrin and fibrinogen under flow, mediating firm adhesion in the low shear regime and at high substrate concentrations. We utilized single molecule force spectroscopy to delineate the distinct kinetic and micromechanical properties of CD44s binding to fibrin and fibrinogen. To this end, fibrin or fibrinogen immobilized on the cantilever tip was brought in contact with immunopurified CD44s incorporated into lipid vesicles and layered on a polyethyleneimine (PEI)-cushioned glass slide. The CD44s-fibrin pair has a lower unstressed dissociation rate and a higher tensile strength than the CD44s-fibrinogen bond, which may explain the ability of CD44s coated microspheres to bind immobilized fibrin more efficiently than fibrinogen at a shear stress level of 0.5 dyne/cm^2. Incubating the CD44s incorporated lipid vesicles with function blocking anti-CD44 monoclonal antibody Hermes-1 did not affect the frequency of binding to fibrin and fibrinogen, indicating the presence of an alternate binding site on CD44s for fibrin(ogen). Increasing the duration of receptor-ligand contact from 2 to 200 milliseconds did not affect the micromechanical properties of CD44s-fibrin bond. Furthermore, the molecular recognition of CD44s by fibrin(ogen) requires the presence of both N-linked glycans and the glycosaminoglycans namely chondroitin sulfate and dermatan sulfate at all tested contact durations. Thus, our single molecule characterization of the biophysical properties of receptor-ligand pairs in conjunction with biochemical studies provides a strong quantitative understanding to the interactions of CD44, which is an aberrantly expressed protein on human metastatic tumor cells, with fibrin and fibrinogen that play vital roles in the facilitation of cancer metastasis.

See more of this Session: Intermolecular Interactions
See more of this Group/Topical: Food, Pharmaceutical & Bioengineering Division

(656d) High-Throughput Analysis of Concentration-Dependent Antibody Self-Association

Shantanu V. Sule, Anna Marie Marcelino-Cruz, Tyler R. Sample and **Peter M. Tessier**, Dept of Chemical & Biological Engineering, Rensselaer Polytechnic Institute, Troy, NY

Monoclonal antibodies are typically monomeric and non-viscous at low concentrations, yet they display highly variable associative and viscous behavior at elevated concentrations. Although measurements of antibody self-association are critical for understanding this complex behavior, traditional biophysical methods are not capable of characterizing concentration-dependent protein self-association in a high-throughput manner. In this presentation, we will describe a nanoparticle-based method (self-interaction nanoparticle spectroscopy) capable of rapidly measuring concentration-dependent self-interactions for three human monoclonal antibodies with unique solution behaviors. We find that gold nanoparticles conjugated with antibodies at low protein concentrations display self-association behavior (as measured by the interparticle distance-dependent Plasmon wavelength) that is well correlated with static light scattering measurements obtained at orders of magnitude higher antibody concentrations. Using this methodology, we find that the several monoclonal antibodies display complex pH-dependent self-association behavior that is strongly influenced by the solution ionic strength. Importantly, we find that a polyclonal human antibody is non-associative for all solution conditions evaluated for the monoclonal antibodies, suggesting that antibody self-association is more specific than previously realized. We expect that our findings will accelerate the analysis of sequence and structural determinants of antibody self-interactions.

See more of this Session: Intermolecular Interactions
See more of this Group/Topical: Food, Pharmaceutical & Bioengineering Division

Polyphenols Modify Soluble Amyloid-β Aggregate Distributions to Impact Amyloid-β Induced Changes in Blood-brain Barrier Function

A compromised blood-brain barrier (BBB) can cause neurological damage either directly, as in the case of hemorrhagic stroke, or indirectly by increasing the concentrations of blood plasma proteins in the brain parenchyma. These plasma proteins are associated with amyloid plaques that are believed to promote the pathogenesis of Alzheimer's disease (AD). Although AD diagnosis requires evidence of parenchymal plaques, cerebrovascular plaque deposits that result in cerebral amyloid angiopathy (CAA) present in 90% of AD patients. Both cerebrovascular deposits, characteristic of CAA, and neuritic plaques, hallmarks of AD, are composed of fibrillar amyloid-β protein (Aβ). Monomeric Aβ self-assembles to create soluble aggregates, which serve as nucleation sites for further growth into fibrillar plaques.

Our lab has previously developed a model of the BBB using monolayers of human brain microvascular endothelial cells (HBMVEC). Monolayers challenged with soluble Aβ aggregates exhibit increased permeability, reflective of the *in vivo* BBB breakdown in AD. Further experimentation has identified the involvement of NF-κB signaling pathways in the stimulation of monolayer permeability by soluble Aβ aggregates. Ideal therapeutics would be able to hinder both aggregate formation and aggregate-induced cell responses. Epidemiological studies demonstrate that diets rich in polyphenolic compounds correlate with a lower incidence of AD. As a result of their antioxidant capabilities, polyphenols have the potential to attenuate NF-κB signaling; however, their ability to modulate Aβ aggregation is not well described. We have investigated the ability of polyphenols with a flavone base structure to mitigate Aβ aggregation and the subsequent adverse effects on the microvasculature. Aβ monomer aggregations prepared in the presence and absence of polyphenol produce different size distributions of soluble Aβ aggregates. We have characterized aggregates produced as a result of interaction with polyphenols before introducing them into the cellular environment. Following this analysis, NF-κB signaling and tight junction integrity were quantified in HBMVEC monolayers challenged with the Aβ aggregates. Evaluating the size distributions provides a more rigorous framework for understanding the observed changes in BBB function.

Single molecule characterization of designed ankyrin-repeat protein/ligand interaction

Luthur Siu-Lun Cheung, Manu Kanwar, Marc Ostermeier and Konstantinos Konstantopoulos

Department of Chemical and Biomolecular Engineering, The Johns Hopkins University, Baltimore, MD 21218, USA

A fundamental understanding of biomolecular recognition is essential to the determination and optimization of important design parameters for developing the next generation of biosensors and biothreat detection systems. Antibodies are the most widely used binding molecules in biomedical research with high specificity and affinity to the targets of interest; however, they are generally expensive to produce and require disulfides for stability. To overcome these limitations, non-antibody scaffolds such as designed ankyrin repeat proteins (DARPins) have been developed as binding proteins. DARPins have been engineered to detect diverse target proteins with pM dissociation constant, thus offering an attractive alternative to antibodies. Combining protein engineering techniques with single-molecule force spectroscopy, we have developed techniques for characterizing the binding kinetics of the DARPin off7 and maltose binding protein (MBP) bond under external forces. MBP was directionally oriented onto a lipid bilayer platform by attaching the CD44 transmembrane domain to its N-terminus. Atomic force microscopy (AFM) was performed to measure the rupture force and dissociation kinetic constants of the interaction between off7 and MBP. Rupture force (f_b) of a single off7-MBP bond was determined to be a function of loading rate and was in the range of 50-80 pN. The unstressed off-rate (k^0_{off}) and reactive compliance (x_β) of off7-MBP bond were calculated to be 6.1×10^{-3} /s and 0.59 nm, respectively. By analyzing the crystal structure of the off7-MBP complex using virtual alanine scanning in the Robetta server, binding energy at the interfacial amino acids of the DARPin off7 protein was estimated. Based on this simulation result, we selectively mutated interfacial amino acids and measured their effects on the binding kinetics. Our data show that mutations at two specific amino acids on DARPin off7 essentially abrogate binding. Therefore, the DARPin off7-MBP binding interface is characterized by a hot-spot motif in which binding energy is largely concentrated only to a few amino acids. In summary, we have developed a novel experimental approach to study the kinetics (k^0_{off}) and micromechanical properties (f_b and x_β) of DARPin off7-MBP bond under forced dissociation. Our data provide informative molecular cues to understand how the engineered protein interacts with the target molecule.

(656g) Hierarchical Self-Assembly of Peptide Amphiphiles Into Fibre Bundles Mediated by Cell-Binding Motif RGDS

Yen Wah Tong, Department of Chemical and Biomolecular Engineering, National University of Singapore, Singapore, Singapore and Jingnan Luo, Chemical and Biomolecular Engineering, National University of Singapore, Singapore, Singapore

The study of self-assembling peptide fibrous nanostructures was inspired by the natural biological system and is of great interest to many material designers for its extensive applications as scaffolds in tissue engineering, vehicles for drug delivery and even as templates for the mineralization of heavy metal nanoparticles. Various design strategies for self-assembling peptides into nanofibres have been explored. The design rules are either based on protein structures, in the form of alpha helices and beta-sheets, or based on peptide derivatives such as peptide amphiphiles and π-stacking systems. Among of them, the single-tail peptide amphiphiles (PA) discovered by Stupp and his c-workers is a versatile system with high degrees of control over the nanofibre's structure and chemical functionality.

While peptide nanofibres and its application have been well investigated, fabrication of higher hierarchical order of such nanofibres has not received similar research attention. Only recently, two novel pathways of peptides self-assembly into aligned fibre bundles were reported by Stupp's group. The pathways of X-ray irradiation or temperature elevation have been found to enable the formation of bundles of aligned nanofibres. In this work, we report a new pathway to form bundles of nanofibres by inducing and controlling inter-nanofibre interactions using the well-known bioactive sequence RGD. RGD has been found to bind to various types of integrins including $\alpha5\beta1$, $\alpha V\beta3$, $\alpha8\beta1$, $\alpha V\beta1$, $\alpha V\beta6$, and $\alpha IIb\beta3$and is frequently used to improve cell adhesion, cell proliferation, cell migration and cell differentiation. In our system, RGD as a cell binding sequence was found to take part in the structural thickening of the fibre bundles. The findings that RGDS type sequences function not only as bioactive motifs but also as key structural units in the lateral assembly of fibres could aid our understanding of fibrillogenesis in nature.

See more of this Session: Intermolecular Interactions
See more of this Group/Topical: Food, Pharmaceutical & Bioengineering Division

(666a) Particle Engineering Based Framework for Cohesion Reduction of Fine Pharmaceutical Powders Via Dry Coating

Rajesh N. Dave, Chemical Engineering, New Jersey Institute of Technology, Newark, NJ

Flow and handling of fine powders is a topic of great interest to pharmaceutical industry. High cohesion observed in fine pharmaceutical powders lead to problems such as, agglomeration, poor flowability, electrostatic charging and low bulk density. These also lead to downstream problems such as poor content uniformity, marginal improvement in dissolution rate of fine API powders, and need for extra processing steps such as granulation. Some of these flow problems may be mitigated or kept under control via use of flow-aids and glidants. However, dry coating based surface modification is a more rigorous approach due to its ability to achieve precise surface coating of nano-silica and other flow-aides; making it very popular in recent academic research.

In this presentation, we provide results using several devices, both based on batch and scalable continuous operation, for pharmaceutically relevant powders where the surface coating includes not only silica but also several types of "guest" materials. Dry coating devices are categorized in two approaches; the first approach is suitable for powders in 150 down to 10 micron sizes where attrition of powder size is minimal. In the second approach, suitable for even finer particles, including those used in the inhaler application having aerodynamic diameters in the size range of 2 ~ 5 microns, simultaneous micronization and surface modification is considered. Results of the bulk properties for these powders are presented to illustrate the improvement in flow, fluidization, dispersion, lack of agglomeration, bulk density, and electrostatic tendency, all leading to potentially significant cost benefits in industrial processing, handling, storage and transportation. Finally, we propose a 2-D phase-map based on key bulk powder properties to help make manufacturing decisions regarding the formulation strategy for solid pharmaceutical dosages.

See more of this Session: Particle Engineering as Applied to Pharmaceutical Formulations I
See more of this Group/Topical: Food, Pharmaceutical & Bioengineering Division

(666b) Investigation of Tablet Coating Processes Using Discrete Element Method Simulations

Gregor Toschkoff[1], **Daniele Suzzi**[1], Sarah Just[2], Klaus Knop[2], Peter Kleinebudde[2] and Johannes G. Khinast[3], (1)Research Center Pharmaceutical Engineering (RCPE), Graz, Austria, (2)Institute of Pharmaceutics and Biopharmaceutics, Heinrich Heine University, Dsseldorf, Germany, (3)Institute for Process and Particle Engineering, Graz University of Technology, Graz, Austria

Investigationof tablet coating processes using Discrete Element Method simulations

G. Toschkoff[1], D. Suzzi[1],S. Just[2], K. Knop[2], P. Kleinebudde[2], J. G.Khinast[1,3]

[1] *ResearchCenter Pharmaceutical Engineering GmbH, Graz, Austria*

[2] *Instituteof Pharmaceutics and Biopharmaceutics, Heinrich Heine University, Dsseldorf,Germany*

[3] *Institutefor Process and Particle Engineering, Graz University of Technology, Graz,Austria*

Inthe pharmaceutical industry, drum coating is a widely used unit operation toproduce tablet films of different purposes. In this process, a rotating drumaccounts for the necessary mixing of the tablets, and a coating solution isinjected from above by means of an atomizing nozzle. The appliedcoating layer(s) fulfill different functions, e.g. taste masking and coloring,control of the release of the active pharmaceutical ingredient (API) from thecore of a tablet, application of an additional API, or protection of the tabletcore from environmental influences.

Forall aspects of the coating mentioned above, the uniformity of coating is of greatimportance. This includes both, inter-tablet uniformity and intra-tabletuniformity. In fact, as inhomogeneity in the coating thickness can lead to serioushazards (e.g., significant variations in APIs delivery rate), a single tabletthat fails testing can lead to the rejection of the whole batch.

Althoughdrum coating is a widespread technology in the pharmaceutical industry, processdesign is more often than not based on trial-and-error practices and operatorexperience. In recent years, parallel to an increased effort in experimentalwork [1], numerical simulations of particle motion using the Discrete ElementsMethod (DEM) have proven to be an important tool in the detailed investigationof the tablet coating process, as well as in other particle-basedpharmaceutical processes [2,3].

***Figure 1**: ExamplesofDEM simulations in two tablet coating apparatus that were used in this work.Left: Driam Driaconti cycled continuous coater, right: Bohle BFC5 lab-scalecoater.*

The aim of this work is toanalyze and understand the effects of parameters like tablet form, fill volume orpan rotation speed on the intra-tablet coating variability [3] in differentcoating devices. To this end, DEM simulations are performed to numericallyreproduce the tablet motion inside different coating machines. Materialparameters needed for the simulation are gathered from measurements. For eachgeometry, different tablet shapes, namely bi-convex, oval and/or round, aremodeled by the "glued spheres" approach. Further parameter variations includedifferent fill volumes or different rotational speeds.

Important process attributesthat one wants to know are, e.g., residence time of the tablets under thecoating spray, intra-tablet coating variability, or tablets velocities pattern.While many of these quantities are fastidious or even impossible to get byexperimentation, they are readily extracted from the DEM simulations results.

In the following, twoexamples of the acquired information are given. Figure 2 shows a matrix of themean velocity of the tablets, for different fill levels and different tabletshapes, allowing a direct qualitative and quantitative comparison of thedifferent setups.

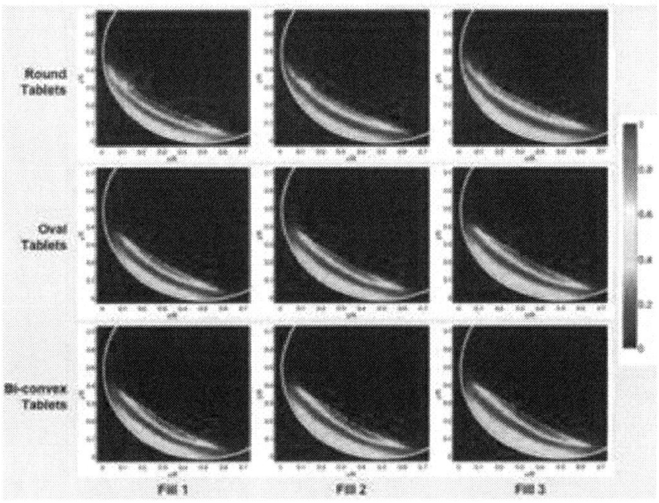

***Figure 2**: Normalizedtime-averaged tablet velocity on the grid for round, oval and bi-convex tabletsat the different coater fill ratios for a vertical slice in the middle of thecoating apparatus. From [4]*

Figure 3 shows the fractional residence time,i.e., the ratio between the time spent by a tablet in the spray zone located atthe top of the surface and the total coating time, for the Driam continuouscoater for a simulation time of 60 seconds. It can be seen that both tabletshape (quantified as length-to-height ratio) and fill level have an influenceon the time that a single tablet spends exposed to the spray. From thisinformation, an expected coating variability and in the end coating processtime can be estimated.

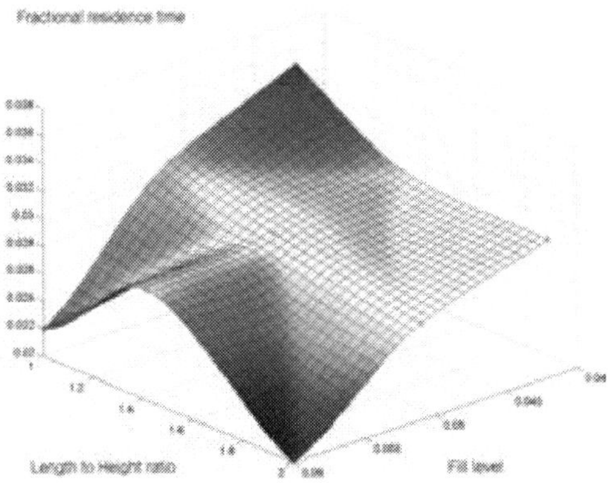

***Figure 3** Averagefractional RT in the spray zone for round, oval and bi-convex tablets at thecoater fill ratios.*

Conclusion

The DEMsimulation has proven to be a valuable tool to gain understanding the dynamicalbehavior of the tablets. The gathered information is essential to e.g. obtain asatisfactory intra-tablet coating homogeneity, which in turn is necessary to minimizethe number of tablet batches that have to be rejected. The outcomes of thiswork aims at demonstrating the utility of numerical simulation in thedevelopment and the design of pharmaceutical tablet coating processes.

References

1. Tobiska,S., Kleinebudde, P., **2003**. *Coating uniformity and coating efficiencyin a Bohle Lab-Coater using oval tablets.* European Journal of Pharmaceuticsand Biopharmaceutics 56, 3-9.

2. Adam,S., Suzzi, D., Radeke, C., Khinast, J.G., **2010**. *An integrated Qualityby Design (QbD) approach towards design space definition of a blending unitoperation by Discrete Element Method (DEM) simulation.* European Journal ofPharmaceutical Sciences, In Press.

3. Ketterhagen,W. R.; am Ende, M. T. & Hancock, B. C., **2009,** *Process modeling inthe pharmaceutical industry using the discrete element method,* Journal ofPharmaceutical Sciences, , *98*, 442-470

4. Suzzi,D.; Toschkoff, G; Radl, S.; Machold, D.; Fraser, S. D.; Glasser B. J.; Khinast,J. G, **2011,** *DEM Simulation of Continuous Tablet Coating: Effects ofTablet Shape and Fill Level on Inter-Tablet Coating Variability.* Submittedto Chemical Engineering Science.

See more of this Session: Particle Engineering as Applied to Pharmaceutical Formulations I
See more of this Group/Topical: Food, Pharmaceutical & Bioengineering Division

(666c) Triboelectric Behavior of Mannitol Powder with Respect to Important Influencing Factors

Stefan Karner, Institute for process and particle engineering, TU Graz, Graz, Austria and Nora Anne Urbanetz, TU Graz, TU Graz, Graz, Austria

Triboelectric behavior of mannitol powder with respect to important influencing factors

Stefan Karner, Nora Anne Urbanetz

Institute for Process and Particle Engineering, Graz University of Technology/Austria

Since most pharmaceutical powders and especially inhalation powders consist of small particles of insulating material electrostatic charge arises when contact of the particles among themselves or with other surfaces happens in powder handling processes. This triboelectric charge carried by the powder can cause problems like material loss, poor flowability and dosing behavior. The phenomenon of the emergence of charge and its impact on the characteristics of the powder are widely uninvestigated. Several scientists such as Murtomaa et al.[2], Zhu et al.[5], Matsusaka et al.[1], Telko et al.[4] and Rowley et al.[3] examined the emergence of charge in common powder handling processes. The above mentioned problem of uniform dosing is particularly of high interest in dry powder inhalation manufacturing. Therefore the aim of this work is to investigate the triboelectric behavior of the new alternative dry powder inhaler (DPI) carrier material mannitol.

Spray dried mannitol was donated by Roquette Frères (Lestrem, France). It shows the characteristic diameters of d_{10}=80 µm, d_{50}=149 µm and d_{90}=209 µm. Charging experiments were carried out by blending the powder in a tumble blender (T2F Turbula®, Willy A. Bachofen AG - Maschinenfabrik, Switzerland) and measuring the charge in a Faraday cup connected to a high resistance electrometer (Keithley 6517 B, Keithley Instruments, USA). First the dependence of the arising charge on the mixing time was determined. Charge saturation was found after about 20 minutes mixing time. Sign and magnitude of this saturation charge can be influenced by varying the mixing process parameters. Therefore subsequent studies focused on the influences of the particle size, the mixing container size and the fines fraction present in the blend.

In contrast to other studies this work uses statistical tools like design of experiments to investigate the significance of the effects and the interactions of more than one influencing factors. The experimental data revealed strong influences of all investigated factors and also some interesting interactions between them.

In addition to the charging investigations also some discharging tests were performed. This was done by using design of experiments as well to check several influencing factors.

The obtained data are highly useful to improve the understanding of the emergence and impact of electrostatic charge on pharmaceutical powders, especially inhalation powders, and will help to enhance the performance of DPIs.

References

(1) Matsusaka, S.; Masuda, H. Electrostatics of particles. Advanced Powder Technology. 2003;14:143-166

(2) Murtomaa, M.; Savolainen, M.; Christiansen, L.; Rantanen, J.; Laine, E. & Yliruusi, J. Static electrification of powders during spray drying. Journal of Electrostatics. 2004;62:63-72

(3) Rowley, G. Quantifying electrostatic interactions in pharmaceutical solid systems. International Journal of Pharmaceutics. 2001; 227:47-55

(4) Telko, M. J.; Kujanpää, J.; Hickey, A. J. Investigation of triboelectric charging in dry powder inhalers using electrical low pressure impactor (ELPI(TM)). International Journal of Pharmaceutics. 2007;336:352-360

(5) Zhu, K.; Tan, R.B.H.; Chen, F.; Ong, K.H; Heng, P.W. Influence of particle wall adhesion on particle electrification in mixers. International Journal of Pharmaceutics. 2007;328:22-34

See more of this Session: Particle Engineering as Applied to Pharmaceutical Formulations I
See more of this Group/Topical: Food, Pharmaceutical & Bioengineering Division

(666d) Analysis of Non-Linear Tensile Strength Variation In Binary Mixtures Using Multi-Particle Finite Element Method (MPFEM)

Tuhin Sinha, School of Mechanical Engineering, Purdue University, West Lafayette, IN, Jennifer Sinclair Curtis, University of Florida, Gainesville, FL, Bruno C. Hancock, Pfizer Inc., Groton, CT, James D. Litster, Chemical Engineering, Purdue University, West Lafayette, IN and Carl R. Wassgren, Mechanical Engineering, Purdue University, West Lafayette, IN

Pharmaceutical tablets are comprised of a mixture of powders with varying properties. To predict the compaction performance of a drug formulation, it is essential to understand the interaction between the constituents of a mixture that will affect the bulk mechanical properties of the resulting mixture. It has been observed experimentally that mechanical properties of a mixture vary non-linearly with the mass fraction of the constituents, and this non-linear interaction could be either positive or negative. Hence, it is difficult to predict the bulk mechanical properties of mixtures, e.g., tensile strength and Young's Modulus, using simple linear mixing rules. Previous researchers have tried both experiments and analytical models to comprehend the causes of this observed non-linearity but have been unable to provide a proper explanation of it.

The multi-particle finite element method (MPFEM) is used to study the micromechanics of particle interactions and their effect on the macroscopic mechanical properties of binary mixtures. In MPFEM, the particles are modeled in the FEM framework with a given set of material and contact properties. The particles can deform based on the applied boundary conditions and contact interactions. Although the current MPFEM models cannot quantitatively predict the bulk mechanical properties of a mixture,they can give a reasonable explanation behind the experimentally observed non-linear behavior. This talk will focus on analyzing the variation of tensile strength in binary mixtures. Simulations will also be compared to experimental results.

See more of this Session: Particle Engineering as Applied to Pharmaceutical Formulations I
See more of this Group/Topical: Food, Pharmaceutical & Bioengineering Division

(666e) Formulation and Manufacture of Pharmaceuticals by Fluidized Bed Impregnation of APIs Onto Porous Carriers

Plamen I. Grigorov, Benjamin J. Glasser and Fernando J. Muzzio, Chemical and Biochemical Engineering, Rutgers, the State University of New Jersey, Piscataway, NJ

Approximately65% of all prescription drugs are manufactured as solid dosage forms, whichinclude tablets and capsules. In most of these cases the final formulation is aphysical blend consisting of an excipient and an active pharmaceuticalingredient (API), which preferably is homogeneously distributed throughout theexcipient. Mixing of powders almost always poses great challenges toformulators, especially in cases where there is a pronounced difference betweenthe powders' particle size, flowability, surface properties, etc. Thesechallenges often translate into un-homogeneously distributed API. For verypotent drugs the amount of API in the solid dosage form can be as low as 0.1%by weight. This very low API loading poses one of the biggest problems inpharmaceutical product development: the control of blend uniformity. Low API contentvariability in the blend or high blend homogeneity are highly desired andstrictly enforced by the U.S. Food and Drug Administration (FDA). In thecommonly available approaches for blend uniformity control (direct blending,wet/dry granulation) as the API concentration decreases, the variability of theblend increases; this makes it very difficult to meet FDA's requirements forlow drug loadings. Therefore a process or method that is able to tightlycontrol API variability in blends, regardless of drug loadings or physicalproperties of the API, has become very desirable.

In this paper wedescribe a novel method for the formulation and manufacture of pharmaceuticalsthat circumvents all difficulties associated with powder mixing and all negativeconsequences associated with such blends. The method involves fluidized bed(FB) impregnation of APIs onto porous carriers. The main idea behind thismethod is the deposition of the drug material inside the excipient particles.This impregnation is achieved by spraying an API solution onto porous excipientin a fluidized bed, achieving what could be seen as continuous wet impregnationand drying process. Overall quality of resulting impregnated blend depends onprocess parameters such as product drying temperature and spray rate of APIsolution.

Impregnated excipients exhibit one main difference when compared toregular blends. That is, the API is located inside carrier particles whichresults in powder properties that are almost identical to those of the pure excipient, thus eliminating API/excipientsegregation. Combining all this with the facts that FBs exhibit very goodpowder mixing and that each excipient particle is impregnated "one at atime", results in highly homogeneous API distribution. We also discoveredthat consequent milling of the impregnated excipientfurther improves blend uniformity. Included table below shows blend uniformity(as % RSD) for impregnated blends (milled and um-milled) of various loadingsusing acetaminophen (APAP) as the active pharmaceutical ingredient.

Other advantagesof proposed formulation method include: simple and efficient (eliminatesseveral unit operations from drug substance and drug product development), safe(minimizes API powder handling), depends only on excipient'sproperties (independent of API nature).

% APAP Loading	**Blend Uniformity, % RSD**
8.82 % (un-milled)	1.05
8.82 % (milled)	0.54
1.02 % (un-milled)	0.70
1.02 % (milled)	0.56
0.1 % (un-milled)	0.99
0.1 % (milled)	0.42

See more of this Session: Particle Engineering as Applied to Pharmaceutical Formulations I
See more of this Group/Topical: Food, Pharmaceutical & Bioengineering Division

(666f) Attrition and Abrasion Resistance of Particles Coated with Pre-Mixed Polymer Coating Systems

Giacomo Perfetti, Chemical Engineering, NanoStructured Materials, Delft University of Technology, Delft, Netherlands, Frederic Depypere, Laboratory of Food Technology and Engineering, Ghent University, Faculty of Bioscience Engineering, Ghent, Belgium, Sheida Zafari, Nanostructured Materials Group, Faculty of Applied Sciences, Delft University of Technology, Delft, Netherlands and Gabrie Meesters, DSM, NA, Netherlands

Nowadays many particulate products are coated to avoid dust formation and breakage when subjected to forces with different magnitude and direction. In order to distinguish the influence of each force, different testers are needed. In this work Purox-S and Cellets 1000 core particles were coated in a top spray fluid bed coater by three different pre-mixed polymer coating systems (Opadry®, Kollicoat®IR and Instacoat AQUA). The repeated impact tester (RIT) was used to measure their strength towards attrition caused by low magnitude normal forces. Tangential forces may also cause rupture to the particles, so an abrasion tester was used to measure this amount of damage caused to the particles. This work provides information on the resistance of the core particles coated with different mixed polymers against attrition and abrasion.

See more of this Session: Particle Engineering as Applied to Pharmaceutical Formulations I
See more of this Group/Topical: Food, Pharmaceutical & Bioengineering Division

(667a) A Population Balance Model Based Distribution Shaping Control for Tailored Dissolution of Crystalline Products

Zoltan K. Nagy, Chemical Engineering Department, Loughborough University, Loughborough, United Kingdom and Nahid Sanzida, Chemical Engineering, Loughborough University, Loughborough, United Kingdom

A novel methodology for the dynamic optimization of the crystal size distribution (CSD) considering growth, nucleation and dissolution mechanisms for batch cooling crystallization processes is presented. The model based distribution shaping control can be used to design the crystal size distribution of solid products to achieve a desired dissolution profile. By simultaneously optimizing the seed recipe and the supersaturation profile for the crystallization process a tailored dissolution can be achieved. Simulation and experimental studies will be presented to illustrate the potential benefits of the proposed product engineering approach. The population balance model which incorporates both the dissolution and growth mechanism is developed and is solved using a combined quadrature method of moments and method of characteristics approach, which provides a computationally efficient solution of the model, suitable for optimization. The optimization problem is solved to design a fast and a slow dissolution profile, and the operating conditions are evaluated experimentally. The experimental setup includes process analytical technology probes for the monitoring of the crystallization process, including FBRM and ATR-UV/Vis. The approach can have significant potential for application in tailored drug delivery, where the release rate of an active ingredient can be designed by optimizing the dissolution rate. The same approach can have application in case of pesticides for more efficient and environmentally friendly application. Tailoring the particle size distribution of the crystallization product, so that an optimal dissolution rate (heat release rate) is achieved for the case of exothermic reactive processes for which the dissolution is followed by reactions, can improve the safe operation of these systems. Similarly the same strategy can also be applied for reactions, which produce a significant amount of gasses and induce excessive liquid rise, which can lead to the explosion of reactors (two-phase flow can occur in the connection pipes and pressure can build-up).

See more of this Session: Particle Formation and Crystallization Processes From Liquids, Slurries and Emulsions
See more of this Group/Topical: Particle Technology Forum

CONTINUOUS CRYSTALLIZATION FOR DRUG NANOSUSPENSIONS

T. Panagiotou, R. J. Fisher[a], R. Bruno[b]

Microfluidics International, Newton, Massachusetts, USA
(mpanagiotou@idexcorp.com.com)
[a] Dept. of Chemical Engineering, Massachusetts Institute of Technology, Cambridge, MA USA
[b] RP Bruno and Associates, Newton, MA, USA

Keywords: nanoparticles, crystallization, bioavailability.

Presented here is a novel technology (PureNano^{TM}) for continuous crystallization and large scale production of drug mirco- or nano- particles with tailored properties, in a highly efficient matter. The technology allows for turbulent mixing of crystallization precursors (solvent, antisolvent, seeds, etc.), the formation of turbulence eddies in the range of 25-50nm, and therefore high spatial uniformity of supersaturation. In addition, PureNano^{TM} allows for precise control of mixing times and intensity, and temperature history. Therefore it allows for precise control of the crystallization process. The energy for crystallization required by this method may be orders of magnitude lower than the energy demands of conventional methods for creating micro- and nano- suspensions, which involve breaking large crystals to smaller entities. In addition, crystallization times may be greatly reduced through minimizing of diffusion limitations.

Background. Estimates suggest that each year roughly half of all newly formulated drugs having potentially high pharmacological value will have little chance to make it beyond the laboratory and into marketplace. Drug formulations exist basically as either emulsions or suspensions, the latter consisting of non- water soluble (hydrophobic) active pharmaceutical ingredients (APIs) suspended in a liquid. Drug suspensions are of great interest because of their high bioavailability which significantly reduces the amount of drug that needs to be delivered as compared to drug emulsions. However, formulation difficulties may still exist since the particle size of the solid API plays a crucial role in 1) the stability of the suspension and 2) the safe delivery of the drug.

Certain crystallization processing technologies claiming to produce stable nano-particles are mainly experimental and may have major shortcomings. As an example, low pressure impinging jet technology[1] is characterized by low flow rates, low jet velocities, and therefore low energy. These jet velocities are over two orders of magnitude lower than velocities used in PureNano^{TM}. Low pressures and low energy levels are associated with large particles and inability of the technology to produce stable particles and handle high solid loadings without plugging or fouling. Another example is use of supercritical carbon dioxide spray processing technology. This technology has been proven to be impractical for manufacturing because it is not scalable and is highly energy intensive.
Recent breakthroughs in continuous crystallization technology now enable pharmaceutical and biotechnology companies to overcome issues related to stability and

safe drug delivery. Microfluidics International Corporation is marketing this technology as their PureNano™ Continuous Crystallizer and associated process protocols. Consequently such systems take on an important role in the formulation and the continuous manufacturing of high purity stable nano-suspensions which is not always achievable with conventional particle size reduction methods.The PureNano™ Crystallizer, which incorporates a scalable mixing chamber, is an energy-efficient method. A case study is presented later that demonstrates the success of this system in formulating a nano-suspension cancer drug that conventional particle size reduction methods were unable to achieve.

Principal of Operation/Applications. Applications best suited for the PureNano™ Crystallizer are dependent upon the length of time for crystallization to occur. These "resident" times are critical to the crystallization process and are dependent upon the API involved, which may require residence times that range from a few hundred microseconds to several hundred milliseconds. To accommodate this range of times, multiple system configurations are available.

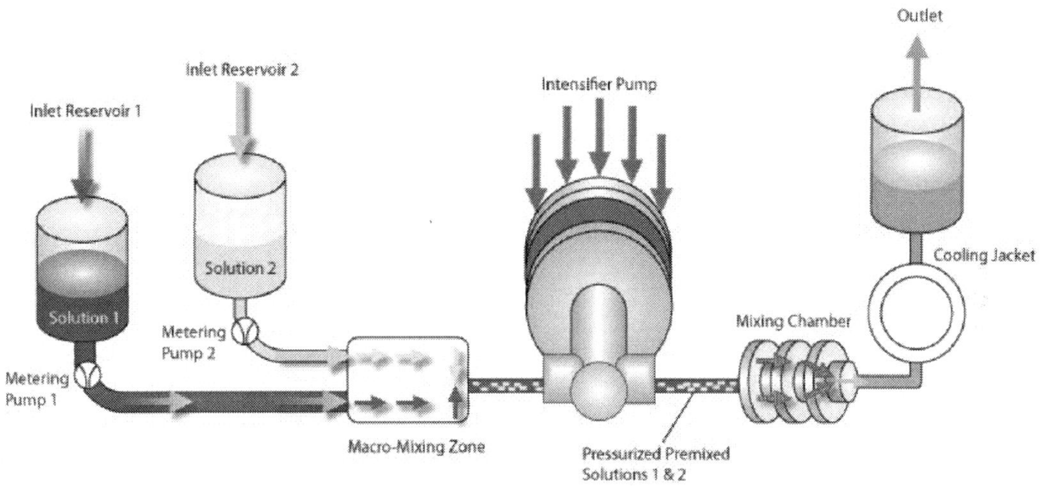

Figure 1. The PureNano Crystallizer configured with coaxial feed

To illustrate the principal of operation we selected the coaxial feed system as depicted in Fig 1. The API is first dissolved in a predetermined solvent, Solution 1, and is placed in one of the two inlet reservoirs. Solution 2, the "anti-solvent," which in many cases may be water, is placed in the second reservoir as shown. This feed configuration allows controlled pre-mixing of the two streams within a "macro-mixing" zone prior to entering the mixing chamber. This pre- mixing occurs for a predetermined period of time within the macro-mixing zone, usually in the order of several milliseconds, creating a small number of product nuclei for "seeding". The feed rates, solution ratios and mixing intensities within the coaxial feed are precisely controlled with metering pumps. This pre-mixed solution is then pressurized and subsequently enters the mixing chamber as a single stream which is internally split into two jets that collide. This forces interact at the nano-scale resulting in a continuous output flow of a stable high purity nano-suspension.

As previously reported by Microfluidics, the PureNano™ Crystallizer has been successful in producing stable nano-suspensions in the laboratory[2,3] where conventional particle size reduction methods were proven to be unsuccessful. In all PureNano™ Crystallizer configurations, post processing may be necessary to prevent crystal growth or to alter crystal habit and/or morphology.

<u>Case Study.</u> Presented here are preliminary results from an ongoing collaborative effort with a well known pharmaceutical company in an effort to successfully formulate a cancer drug identified here as Compound V. To increase bioavailability, it was determined that the drug needed to be delivered intravenously in the form of a nano-dispersion. .

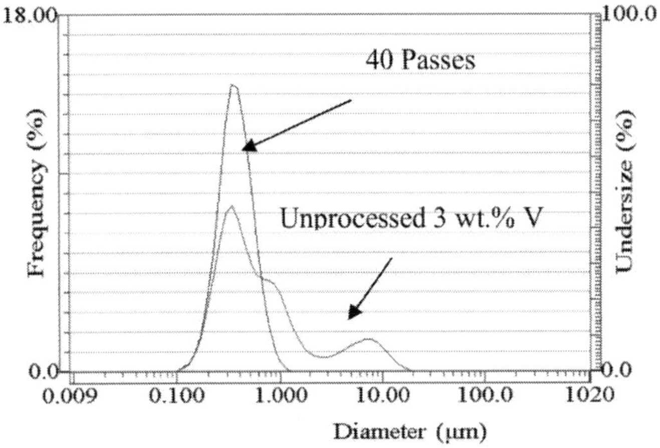

Passes	Particle Size (microns)			
	d10	d50	d95	Mean
Unprocessed	0.220	0.470	7.283	1.401
40	0.227	0.379	0.734	0.410

Figure 2. Results with conventional particle size reduction method

Microfluidics conventional "top down" particle size reduction technology and PureNano were used for the production of the nano-suspension. The required particle size was a median size less than 200 nanometers. Using the top down method, a suspension of the drug in water was prepared and processed using a Model 110EH-30 Microfluidizer processor. After 40 passes through the processor, the mean particle size of the API was reduced from 1.401 microns to 401 nanometers with no further particle size reduction resulting with additional passes (Fig. 2). The resulting suspension was unacceptable; it did not meet the stability requirements.

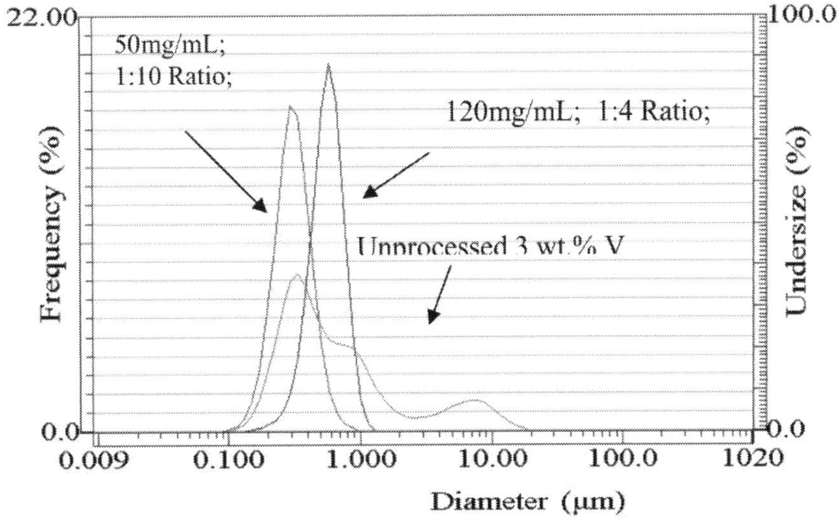

Conc. (mg/mL)	Ratio (Sol:Anti)	Particle Size (microns)			
		d10	d50	d95	Mean
20mg/mL	1:10	0.109	0.177	0.329	0.257
50mg/mL	1:10	0.058	0.102	0.277	0.126
120mg/mL	1:10	0.350	0.529	0.812	0.539
120mg/mL	1:4	0.188	0.287	0.489	0.302

Figure 3. Results with the PureNano Crystallizer

Using a PureNano™ Crystallizer in conjunction with our nano-suspensions development protocol, a continuous crystallization process was developed. Compound V was dissolved in a solvent, polyethylene glycol (PEG), at various concentrations (Fig. 3), in the range of 20-120mg/ml. This stream was mixed with the antisolvent stream (water) at the appropriate proportions using a PureNano™ Crystallizer processor. The ratio of the antisolvent to solvent streams varied from 1:4 to 1:10 producing various degrees of supersaturation. A stable nano-suspension with a median particle size of 102 nanometers was achieved in a single pass. The particle size distribution was well within specifications.

References:

[1] Johnson, B.; Prud'homme, R. Chemical Processing and Micromixing in Confined Impinging Jets. *AICHE J.* **2003,** 49, 2264-2282.

[2] Panagiotou, T.; Fisher, R. Form Nanoparticles via Controlled Crystallization. *CEP*, 2008, 33-39C.

[3] Panagiotou, T.; Mesite, S.; Fisher, R. Production of Norfloxacin Nanosuspensions Using Microfluidics Reaction Technology through Solvent/Antisolvent Crystallization *Ind. Eng. Chem. Res.*, **2009**, 48 (4), 1761-1771.

(667c) High Pressure Homogenization As Particle Size Reduction Technology In the Pharmaceutical Industry: Optimization, Modeling and Isolation Techniques

Plamen Grigorov[1], Christopher P. Long[2] and Azzeddine Lekhal[1], (1)Chemical Process Development & Commercialization, Merck and Co. Inc., Rahway, NJ, (2)College of Engineering, Cornell University, Ithaca, NY

High pressurehomogenization (HPH) is a technology widely used in various industries. In thepharmaceutical industry HPH is used in broad range of applications such as cellrupture, dispersions, emulsions and particle size reduction. Many of today'sactive pharmaceutical ingredients (APIs) exhibit low solubility andpermeability. In order to improve their bioavailability, these APIs are oftenmilled to small sizes. Milling techniques traditionally used in thepharmaceutical industry can be divided into two main categories: wet and dry.Each category has its benefits and drawbacks. Dry milling (e.g. jet milling) can achieve micron size particles with no agglomeration but it is not preferredfor very potent compounds and can introduce amorphous content into the APIs. Wetmilling on the other hand is safe for handling potent compounds but usually cannot achieve very small particles (e.g. rotor-stator mills) or has potential forextraneous matter to be introduced (e.g. media mills). Producing smallparticles by a wet milling technique is also associated with particleaggregation and/or difficult isolation.

High pressurehomogenization can be viewed as a wet micronization technique capable ofachieving micron and sub-micron particle sizes. It is a top-down approach forcontrolling particles size during which slurry of API in a non-solvent media isforced through a very small nozzle under very high pressures. This enormouspressure drop causes cavitation, high shear forces and particle-particlecollisions, which are the main factors for particle breakage. Proper selectionof operation parameters (pressure, nozzle size, flow configuration, number ofpasses) and process parameters (non-solvent media, solid content) can yieldvery narrow particle size distributions (PSD).

In this paper wedescribe our investigational work evaluating HPH as a particle size reductiontechnique and its potential for use in API manufacturing. The work presented consistsof three main parts. In the first part we investigate the effect of process andoperation parameters on particle breakage performance. Very important processdesign considerations for API processing are also presented. In the second partwe propose a mathematical model capable of describing particle breakage and HPHperformance as a function of operation parameters for the specific equipment athand. The proposed model helps with the answer of one very important question:how to scale-up HPH? In the third part we evaluate various particle isolationand drying techniques. These include the following: conventional filtration anddrying, spray drying and freeze-drying. Investigated are the effects theseisolation protocols have on the API's physical attributes such as final PSD,specific surface area (SSA) and bulk density. Finally, a simple decision treeis presented for the selection of these isolation techniques depending on finalAPI formulation method.

See more of this Session: Particle Formation and Crystallization Processes From Liquids, Slurries and Emulsions

See more of this Group/Topical: Particle Technology Forum

(667d) Dynamic Change In Polymer Particle Shape and Size by Simple Emulsion-Diffusion Process

Shaowen Ji, Devesh Srivastava and Ilsoon Lee, Chemical Engineering and Materials Science, Michigan State University, East Lansing, MI

Polymer hollow microspheres have attracted growing attentions in drug delivery, catalyst carriers, and microencapsulation. Here we report a simple preparation of hollow poly(lactic acid) (PLA) microspheres by single emulsion-diffusion method in a turbulent shear flow system. Modified Taylor-Couette mixer was introduced to generate uniform high shear fluid flow, facilitating the formation of hollow structured particles. The critical formation transition of polymer nanospheres to hollow microspheres was studied and controlled. An increasing addition of glycerol into the aqueous phase changed the flow from turbulent inertial to turbulent viscous regime. When the operation temperature was higher than the glass transition temperature of PLA, multiple emulsion droplets were obtained by the inclusion of water droplets during oil drops coalescence. Fast polymer solidification and water entrapping during diffusion process caused the formation of hollow PLA particles. Under harsh conditions, these hollow microspheres collapsed to open structure due to water escaping. Higher mixing speed rate and longer mixing time resulted in more uniform particle size and size distribution. The effects of water to oil ratio, pH of aqueous phase, and electrolyte were investigated for further engineered control of particle shape, size, and size distribution.

See more of this Session: Particle Formation and Crystallization Processes From Liquids, Slurries and Emulsions
See more of this Group/Topical: Particle Technology Forum

(667e) An Investigation of the Applicability of Micro-Calorimetry for the Measurement of Supersaturation During Batch Crystallization

Lotfi Derdour, Drug Product Science and Technology, Bristol-Myers Squibb co., New Brunswick, NJ, Frederic Buono, P R&D, Bristol-Myers Squibb Co, New Brunswick, NJ and Chiajen Lai, Drug Product Science and Technology, Research and Development, Bristol-Myers Squibb Company, New Brunswick, NJ

A new application of reaction micro-calorimetry for early stage crystallization development is described. Using a multi- channel Insight RT-10 reaction calorimeter, critical crystallization information such as detection of primary nucleation can be obtained with minute material quantities (<2 g) . The micro-calorimetry approach was successfully applied to determine the variation of the transient concentration and supersaturation for both cooling and anti-solvent crystallizations. These data can provide information about crystal growth rate at an early stage of the development which permits tailoring robust crystallization processes. Additionally, the ability to analyze up to 8 crystallization processes in parallel, the ease of instrument set-up and sample preparation and the no calibration requirement make micro-calorimetry an ideal tool for crystallization development for early stage projects.

See more of this Session: Particle Formation and Crystallization Processes From Liquids, Slurries and Emulsions
See more of this Group/Topical: Particle Technology Forum

(667f) Ag/Cu Particle Generation by Cosolvent Spray Pyrolysis: Effect of Cosolvent

Kai Zhong[1], George Peabody[1], Amanda Lafleur[1], Howard Glicksman[2] and Sheryl H. Ehrman[1], (1) Department of Chemical and Biomolecular Engineering, University of Maryland, College Park, MD, (2) DuPont Electronic Technologies, Research Triangle Park, NC

A cosolvent spray pyrolysis process was developed for the generation of various single component and multi-component metal particles that require a reducing atmosphere to promote the formation of oxide free metal particles. During the process, a cosolvent was added to a precursor solution, and it decomposed in the furnace. The original work on the cosolvent process utilized ethanol as the cosolvent. To make the process amenable to commercialization, other cosolvents were investigated. Ethylene glycol with a much higher flash point was found to produce oxide free metal particles. We found some differences when ethylene glycol(EG) was used as the cosolvent versus the use of ethanol (ET). In order to understand the effect of the cosolvent on the particle properties as well as to elucidate reaction pathways during the pyrolysis process, single component Cu and two component Cu/Ag particles were generated using either ethylene glycol or ethanol as the cosolvent. Residence times were adjusted from 0.2 s to 3 s by changing the carrier gas flow rate from 2.5 L/min to 10 L/min and furnace temperature from 400 °C to 1000 °C. The powder particles were characterized by SEM, XRD, FTIR and He pycnometry. Experimental results showed that $Cu(NO_3)_2$ first decomposes to $Cu_2(OH)_3(NO_3)$ and CuO and then is reduced to form pure Cu particles. Higher density particles were generated with the cosolvent of ET rather than with EG. The use of ET also led to an alternate morphology of product particles. For two component Cu/Ag particle generation, it was observed that pure Ag precipitated out in less than 0.2 s, and promoted the transition from $Cu(NO_3)_2$ to Cu after the evaporation of solvent.

Different phase separation behavior was observed for ET and EG, which can be attributed to the different solubility and reactivity of the nitrates in these solvents.

See more of this Session: Particle Formation and Crystallization Processes From Liquids, Slurries and Emulsions
See more of this Group/Topical: Particle Technology Forum

(667g) Experimental Investigation and Population Balance Equation Modeling of Solid Lipid Nanoparticle Aggregation Dynamics

Michael A. Henson[1], Yihui Yang[1] and Alessandro Corona III[2], (1)Department of Chemical Engineering, University of Massachusetts Amherst, Amherst, MA, (2)Fabric and Home Care Innovation Center, Procter & Gamble - Household Care, Cincinnati, OH

Solid lipid nanoparticles (SLNs) have applications in drug delivery and the encapsulation of bioactive, lipophilic compounds such as those required for functional foods. SLNs are commonly obtained by first preparing a lipid emulsion via high pressure homogenization or microfluidization followed by cooling in an ice bath or other low temperature environment such that the lipid emulsion drops crystallize and solid particles are formed. A major obstruction to the industrial use of SLNs is their tendency to aggregate and form large particles and eventually gels when stored at room temperature. The prevailing theory is that aggregation is driven by the lipid crystals undergoing a polymorphic transformation from the thermodynamically unstable a form to stable b form. A large increase in surface area occurs as the spherical a particles are transformed into platelet-like b particles, causing a substantial decrease in surfactant coverage on the hydrophobic surfaces and inducing particle aggregation.

We tested this theory by performing a series of aggregation experiments and by developing a population balance equation (PBE) model for prediction of the aggregation dynamics. Oil-in-water emulsions were prepared by high pressure homogenization with 5 wt% lipid (tristearin) and 1 wt% surfactant (mixture of Tween 60 and Span 60) at 85°C. The lipid emulsions were cooled in an ice bath for 8 hours to obtain SLNs and the resulting solutions were stored at 20°C. Aggregation dynamics were monitored daily by measuring the particle size distributions with light scattering and by examining polymorphic transformation behavior with differential scanning calorimetry. Our experimental results suggest that the polymorphic transformation rather than particle aggregation is the rate determining step of the aggregation process in this system.

The PBE model was formulated to predict the particle volume percent distribution under the assumption that particles collide due to Brownian motion and that only b particles could aggregate. The polymorphic transformation was modeled to be first-order in the concentration of a particles and first-order in the particle surface coverage such that transformation rate was zero if the particle had maximum coverage. An aggregation efficiency was introduced by assuming b particle aggregation was dependent on the surface coverage of each particle, with no aggregation possible when both particles had maximum coverage. The model was able to qualitatively predict our experimental data as well as other SLN aggregation data available in the literature. Our preliminary modeling results further suggest that the polymorphic transformation is the rate determining step of the aggregation process and support the hypothesis that aggregation is driven by the creation of new surface area due to the transformation.

(671a) Supported Manganese Triazacyclononane Cis-Dihydroxylation Catalysts

Nicholas Schoenfeldt, Andrew Korinda and **Justin M. Notestein**, Chemical and Biological Engineering, Northwestern University, Evanston, IL

Manganese complexes of1,4,7-trimethyl-1,4,7-triazacyclononane are highly active and selective for thelow temperature oxidation of alkenes and other substrates with environmentallybenign oxidants such as H_2O_2. Previous investigations onthe homogeneous catalyst revealed that carboxylate co-catalysts could beemployed to raise catalyst productivity by reducing the inherent catalaseactivity of the materials, as well as tune the reaction selectivity towardeither epoxidation or *cis*-dihydroxylation. *Cis*-dihydroxylation isa relatively unusual but powerful reaction, primarily catalyzed by expensive Rhcomplexes or extremely toxic OsO_4. There is thus an incentive forthe development of novel catalysts, especially solid catalysts. These Mncomplexes use mild reaction conditions (room temperature and below), benignoxidants (H_2O_2) and less toxic metal centers (Mn),demonstrating potential utility for the organic transformations necessary in pharmaceuticalsynthesis.

We have found that functionalization of common catalystsupports with carboxylate moieties allows the *in-situ* assembly andactivation of the Mn complex onto the support under oxidation reactionconditions. The immobilized catalyst requires orders of magnitude lesscarboxylate co-catalyst than the soluble analog and requires up to 8x lowerresidence times for a given conversion. The immobilization technique also removescumbersome synthetic requirements typically required to create amine ligandsthat can be covalently attached to a support. For example, we have found thatthe cis-diol selectivity can be increased from <50% to ~75% by changing onlythe nature of the supported carboxylate. Supporting characterization by X-rayabsorption fine structure and diffuse reflectance UV-visible spectroscopy helpsus further understand the nature of this supported catalyst.

See more of this Session: Reaction Engineering In Pharmaceuticals and Fine Chemicals
See more of this Group/Topical: Catalysis and Reaction Engineering Division

(671b) Reaction Development for Bench Scale, Continuous Synthesis of a Drug Substance

Patrick L. Heider[1], Soubir Basak[1], Louis Buchbinder[2], Rachael C. Hogan[1], Ketan Pimparkar[1], Aaron D. Wolfe[2] and Klavs F. Jensen[1], (1)Chemical Engineering, Massachusetts Institute of Technology, Cambridge, MA, (2)Department of Chemical Engineering, Northeastern University, Boston, MA

Recently the pharmaceutical industry has shown an increased interest in implementing continuous manufacturing processes. While this movement has been supported by the Food and Drug Administration, there have been limited reports of continuous production systems. We have worked within the Novartis-MIT Center for Continuous Manufacturing to develop a fully integrated, bench scale process for taking chemical intermediates to a coated tablet. This talk will highlight the reaction development and operation for the two chemical transformations performed. Translating unit operations from batch to continuous required the development of different reaction conditions. Reaction times were significantly decreased and solvent usage was reduced or even eliminated. The increased heat transfer available to continuous processes was utilized to perform a rapid, exothermic quench not possible in batch operation.

Operating a continuous synthesis process at the bench scale required addressing several issues not present in smaller, isolated systems. A thorough understanding of mixing was important for operation at the bench scale where neither the short diffusion times of microreactors nor the turbulent flow of pilot plants were present. This was addressed in multiple ways including modeling the residence time distribution and making use of *in situ* gas formation to promote mixing. Monitoring of chemical composition was required to maintain reaction stoichiometry and was achieved using a variety of online measurements including UV and IR spectroscopy and density. Lastly, the bench scale system provided a challenging scale up for membrane based liquid-liquid separators originally developed for microscale continuous multistep synthesis (1).

References:

1. H. R. Sahoo, J. G. Kralj, K. F. Jensen, Angew. Chem. Int. Ed. 2007, 46, 5704.

See more of this Session: Reaction Engineering In Pharmaceuticals and Fine Chemicals
See more of this Group/Topical: Catalysis and Reaction Engineering Division

(671c) Practical and Selective Oxidation and Amination of Alcohols In Flow

Klaus Hellgardt[1], K.K. Hii[2], N. Zotova[3] and F. Roberts[3], (1)Chemical Engineering, Imperial College London, London, United Kingdom, (2)Chemistry, Imperial College London, London, United Kingdom, (3)Imperial College London, London, United Kingdom

The development of clean, selective and atom-efficient reactions has become the new paradigm in organic synthesis, driven by a desire to reduce economic, as well as environmental costs of chemical production. To this effect, the oxidation of alcohols to aldehydes and ketones may be considered one of the most challenging reactions, ubiquitous in organic chemistry, yet least efficient. The lack of chemoselective and atom-economical oxidants often make these reactions unattractive, particularly for the large scale synthesis of complex molecules. In this work, we demonstrate a number of new ways of oxidising alcohols using either H_2O_2 or O_2 as an oxidant by performing catalysis in a continuous flow high temperature/pressure reactor, using supported Au and Ru cataysts systems, thus fulfilling several criteria for practical and 'green' synthesis. We further demonstrate the utilisation of the intermediate aldehyde in the alkylation of amines without preactivation of alcohols.

See more of this Session: Reaction Engineering In Pharmaceuticals and Fine Chemicals
See more of this Group/Topical: Catalysis and Reaction Engineering Division

(671d) Rotating Foam Reactors for Fine Chemical Synthesis

T. Alexander Nijhuis, M.A. Leon Mattheus, R. Tschentscher, J. van der Schaaf and J.C. Schouten, Chemical Reactor Engineering, Eindhoven University of Technology, Eindhoven, Netherlands

The production of fine chemicals and pharmaceuticals is commonly performed in batch reactors, using slurry catalysts, due to their high flexibility. However, the catalyst particles need to be filtered from the slurry when the reaction is finished, which is a costly and time consuming operation and makes re-using the catalyst difficult. In this presentation we will be showing the development of new three-phase reactor configurations based on the application of solid foam packings. These materials have a good mechanical stability, a high porosity, and a high specific surface area, which offers great potential for their use as catalyst packings. The good mechanical strength, allows us to use the material as both catalyst support and stirrer at the same time. The foam surface area can be further increased by deposition of a catalyst wash coat or using sol-gel methods.

We will be showing that the solid foam stirrer reactor has an efficient mixing behavior and high mass transfer rates. Additionally, the reaction mixture can be drained off easily, without the need to filter the catalyst particles. Using different cell-density foams (different geometrical surface areas) and different washcoat thicknesses it is possible to optimize the performance of the reactor in cases where the catalyst activity and mass transfer are effecting the selectivity of a reaction, for example in the selective hydrogenation of an alkyne to an alkene. We will be showing the performance of this novel reactor type for different gas-liquid reactions catalyzed by a heterogeneous catalyst coated on the stirrer. We will be comparing its performance to conventional slurry catalysts in a reactor equipped with a conventional Rushton stirrer.

See more of this Session: Reaction Engineering In Pharmaceuticals and Fine Chemicals
See more of this Group/Topical: Catalysis and Reaction Engineering Division

(671e) Economic and Process Engineering Evaluations of Polymer-Grade Terephthalic Acid Production In A Spray Reactor

Meng Li, Department of Chemical and Petroleum Engineering, University of Kansas, Lawrence, KS, Fenghui Niu, Center for Environmentally Beneficial Catalysis, University of Kansas, Lawrence, KS, Xiaobin Zuo, Department of Chemistry, University of Kansas, Lawrence, KS, Daryle H. Busch, CEBC/Chemistry, University of Kansas, Lawrence, KS and Bala Subramaniam, Chemical and Petroleum Engineering, Center for Environmentally Beneficial Catalysis, University of Kansas, Lawrence, KS

There is emerging interest in developing renewable plastics derived from feedstocks (such as renewable *p*-xylene) from plant-based biomass. In such cases, it is imperative that green conversion technologies are also employed. CEBC researchers have been investigating a spray process concept as a greener alternative to the conventional Mid-Century (MC) process for terephthalic acid (TPA) production from *p*-xylene. Compared to the MC process in which the air is dispersed into the stirred liquid phase containing *p*-xylene and the Co/Mn/Br based catalyst dissolved in acetic acid, the spray process uses a nozzle to disperse the reaction mixture as fine droplets into a gas phase dominated by saturated acetic acid vapor and also containing O_2 (in slight excess over the stoichiometric amount needed for *p*-xylene oxidation to TPA) and CO_2 (as carrier for O_2). The increased gas/liquid interfacial area was aimed at eliminating O_2 starvation in the liquid phase, thereby avoiding incomplete oxidation to achieve improved selectivity and purity of the TPA product. In the MC process, the crude TPA from the stirred reactor contains significant amount of impurities such as approximately 1400 ppmw 4-carboxybenzaldehyde (4-CBA) and 600 ppmw *p*-toluic acid due to the incomplete oxidation. To obtain polymer-grade TPA, a purification process by hydrogenation associated with a significant number of additional processing steps is required. Employing a semi-continuous spray process (*p*-xylene sprayed continuously but no continuous product withdrawal), it was confirmed by reliable analytical methods that the solid TPA product meets the polymer-grade purity specification of 25 ppmw 4-CBA or less, implying the hydrogenation step is not required. Comparative economic analyses show that the total capital investment and utility costs associated with the spray process are roughly half of that in the MC process, thus significantly reducing the cost of the polymer grade TPA product. Preliminary experimental results from continuous operation of the spray reactor show that *p*-xylene was being converted mostly to TPA and that the purest solid TPA product obtained in the reactor contains 3 ppmw 4-CBA. Based on reactor modeling, factors affecting the control of reactor temperature and open-loop stability of the reactor are also elucidated.

See more of this Session: Reaction Engineering In Pharmaceuticals and Fine Chemicals
See more of this Group/Topical: Catalysis and Reaction Engineering Division

(680a) The Visualization Evolution In Real-Time (VERT) Method As a Tool In the Study of Complex Phenotypes In Microbial Systems

Luis H. Reyes, Department of Chemical Engineering, Texas A&M University, College Station, TX and Katy Kao, Chemical Engineering, Texas A&M University, College Station, TX

Inhibitor tolerance in biocatalysts is a complex phenotype. The rational design of more robust microbial systems generally requires knowledge of the genetic determinants involved in the desired trait. Towards this goal, we have developed a novel experimental adaptive evolution method, Visualizing Evolution in Real-Time (VERT), for the evolutionary engineering of microbial systems for complex phenotypes. VERT allows the identification of adaptive events and helps to facilitate the isolation of adaptive mutants from the evolving population for further analysis and characterization.

We applied VERT to evolve *Escherichia coli* for enhanced tolerance to the biofuel, butanol. Butanol is highly toxic to microorganisms, with most strains being able to tolerate up to 2% (v/v) of this organic solvent. Wild-type *E. coli* can tolerate up to approximately 1% (v/v) of n-butanol. We successfully evolved E. coli for enhanced butanol tolerance using VERT. The butanol tolerant mutants were isolated from the population and analyzed to elucidate the molecular mechanisms involved in tolerance.

See more of this Session: Advances In Metabolic Engineering and Bioinformatics III - Experimental
See more of this Group/Topical: Food, Pharmaceutical & Bioengineering Division

(680b) Implementation of New Routes and Redesign of Native Pathways for Synthesis of Bioplastics In *E. Coli*

Daniel E. Agnew, Chemical and Biological Engineering, University of Wisconsin - Madison, Madison, WI and Brian Pfleger, Chemical and Biological Engineering, University of Wisconsin Madison, Madison, WI

Polyhydroxyalkanoates (PHA) are a class of microbially derived polyesters that are biodegradable, renewable alternatives to traditional, petroleum derived polymers (plastics). While PHA have already shown promise at an industrial scale, their use has been limited in part by cost but also by the physical properties of the polymer (*e.g.*,elasticity). Like all polymers, these properties stem from the make-up of the individual monomer units, such as the length and composition of the side chain as well as the composition of the polymer itself. We have explored the possibility of creating a general approach to producing monomers *in vivo* from unrelated carbon sources in *Escherichia coli* via two distinct metabolic pathways. One approach was to exploit the modularity and assembly line characteristics of polyketide synthases (PKS) to generate PHA monomers from basic cellular building blocks. To this end, we have assembled a proof-of-concept system and tested the viability of the use of a PKS monomer synthase in *E. coli*. A second approach involved the engineering of native metabolic pathways to produce PHA based on intermediates scavenged from fatty acid metabolism. Our goal here is to be able to produce a PHA homopolymer by generating a single chain-length fatty acid pool and blocking the iterative nature of fatty acid β-oxidation.

See more of this Session: Advances In Metabolic Engineering and Bioinformatics III - Experimental
See more of this Group/Topical: Food, Pharmaceutical & Bioengineering Division

(680c) Genome-Wide Approach to Engineering Inhibitor Tolerance In E. Coli for Biofuel Applications

Tirzah Y. Glebes, Nicholas R. Sandoval, Sophie J. Weiss, Philippa J. Reeder and Ryan T. Gill, Chemical and Biological Engineering, University of Colorado at Boulder, Boulder, CO

The microbial conversion of lignocellulosic biomass to biofuels is a promising field for renewable energy. Large-scale implementation of this technology is partially thwarted by biocatalyst sensitivity to hydrolysate components like furfural and acetate. *Escherichia coli* is a widely used model organism for the conceptual understanding of complex phenotypes and is used here to explore inhibitor toxicity of hydrolysate components. We report the use of genome-wide tools developed by our lab, namely SCalar Analysis of Library Enrichments (SCALEs) and TRackable Multiplex Recombineering (TRMR), to determine genetic targets for the engineering of tolerant clones. One such target that has been identified is the overexpression of D-sedoheptulose 7-phosphate isomerase (*lpcA*), which improves growth by 430% in 1 g/l furfural. This overexpression also increases growth in acetate (124% at 15 g/l) and ethanol (180% at 25 g/l), potentially lending itself to robust applications. We use the insight gained from multiple selections to elucidate synergistic tolerant functionalities while also revealing novel mechanisms of toxicity. The advantage of this approach grants trackable knowledge that might otherwise be missed by directed engineering or untraced mutational studies.

See more of this Session: Advances In Metabolic Engineering and Bioinformatics III - Experimental
See more of this Group/Topical: Food, Pharmaceutical & Bioengineering Division

(680d) High-Throughput Screen of Extracellular Metabolites Using Microfluidics

Adel Ghaderi, Benjamin L. Wang, Hang Zhou and Gregory Stephanopoulos, Chemical Engineering, Massachusetts Institute of Technology, Cambridge, MA

High-throughput screen of extracellular metabolites using microfluidics

Cellular metabolism is defined by the biochemical reactions taking place within the cell as well as the uptake and secretion of metabolites in the extracellular medium. Combinatorial approaches for strain improvement aim to randomly modulate the metabolic profile within a population followed by selection of a clone with enhanced phenotype. However, conventional methods for the high-throughput screening of cellular metabolism (e.g. FACS) are unable to retain the association between a particular clone and its extracellular metabolites. In an industrial setting, the metabolites of interest are frequently extracellular (e.g. substrate or secreted product). As such, the ability to screen populations based on their extracellular metabolism is highly desirable.

We have developed a high-throughput screening method based on the measurement of the extracellular metabolites of single cells. Using microfluidics, we encapsulate single cells in the droplets of a water-in-oil emulsion. The electromagnetic properties of these emulsions allow us to coalesce droplets containing cells with assay droplets and later sort the combined droplets based on the assay result. We have demonstrated this technology for the measurement of glucose, xylose, and lactate, though other analytes are possible. Sorting populations for xylose uptake has provided us with crucial insight regarding the metabolism of recombinant xylose-consuming *Saccharomyces cerevisiae*. This platform can be extended to various cell types, as we have demonstrated with mammalian cells (Jurkat, clone E6-1). In this latter context, lactate measurement may provide insight regarding cancer cell metabolism that would otherwise be difficult to access. We have also shown the ability to distinguish between enantiomers, an important capability with regards to pharmaceutical production. The broad range of detectable analytes and cells that can be cultured using this powerful approach promises to significantly enhance metabolic engineering efforts.

See more of this Session: Advances In Metabolic Engineering and Bioinformatics III - Experimental
See more of this Group/Topical: Food, Pharmaceutical & Bioengineering Division

(680e) Developing Metabolic Engineering Tools for a Cellulolytic Marine Bacterium Saccharophagus Degradans 2-40T

Hyun-Dong Shin, Justin Harper and Rachel Ruizhen Chen, Chemical & Biomolecular Engineering, Georgia Institute of Technology, Atlanta, GA

Saccharophagus degradans 2-40^T is a gram-negative marine bacterium with a rare capability of degrading many types of complex polysaccharides, including cellulose. Analyses of its recently sequenced genome and subsequent experimental studies suggest that its cellulolytic system is unique in several aspects, notably the lack of exo-acting glucanases and significant surface features present only when metabolizing complex polysaccharides, signify a departure from known cellulolytic systems that rely on secreted soluble cellulases or complex cellulosome. Thus the *Saccharophagus* system provides a unique model to study how gram-negative bacteria overcome transport limitations in accessing a polymeric substrate, a key aspect of cellulose metabolism. Additionally, the cellulolytic *Saccharophagus degradans* could be exploited for consolidated production of biofuel and other useful products. The fundamental and applied research requires reliable genetic tools, which are lacking for this marine bacterium.

We describe in this presentation our research toward establishing metabolic engineering tools in this microorganism. Starting from a broad range host plasmid, we succeeded in constructing a plasmid that was replicated and stably maintained by the bacterium. Electroporation protocols were developed and optimized, which led to transformation efficacy of about 1000 colonies per micrograms of DNA. Both kanamycin and chloramphenicol were found to be suitable as selection markers. We also tested several promoters for inducible expression of heterologous proteins, a key step in metabolic engineering. Using red fluorescent protein as the first model protein, we are testing several promoters including those native to the organism, which will result in constitutive and inducible expression of heterologous proteins at varied levels.

To the best of our knowledge, this is the first expression plasmid system ever established for this microorganism. The genetic tools from the research could be used to uncover novel *Saccharophagus* cellulolytic mechanism diverged significantly from other known systems. Our understanding on cellulose metabolism from this bacterium offers the first gram-negative template for synthetic biologists and metabolic engineers to develop more efficient biocatalysts for biofuel and other valuable products.

See more of this Session: Advances In Metabolic Engineering and Bioinformatics III - Experimental
See more of this Group/Topical: Food, Pharmaceutical & Bioengineering Division

(680f) Improving the Growth Rate of Cellulolytic Clostridia Through Genomic Library Enrichment

Benjamin G. Freedman, Biological Systems Engineering, Virginia Tech, Blacksburg, VA and Ryan S. Senger, Biological Systems Engineering Department, Virginia Tech, Blacksburg, VA

Clostridium cellulolyticum (ATCC 35319) contains a fully functional cellulosome enabling the degradation of cellulose and hemicellulose for fermentation. We ultimately plan to use this organism for biofuels production in a consolidated bioprocess. However, substantial improvements in cell growth rate are still needed before the organism can prove useful for an industrial fermentation. Creating a *C. cellulolyticum* genomic DNA (gDNA) library and enriching collections of mutants through a growth competition assay allows for rapid self-selection of genes, regulatory sequences, or DNA fragments that yield increased cell growth. Here, we also report the findings of transgenic library enrichment. A genomic library was prepared from the industrially-viable *C. acetobutylicum* (ATCC 824) gDNA and enriched in *C. cellulolyticum* cultures. We have used the results from dual library enrichments and natural selection through continuous culture bioreactor allowed us to identify a similar genomic fragment within both organisms which, when overexpressed in *C. cellulolyticum*, gives a significant growth advantage over organisms containing control plasmids. These represent significant steps in developing an economically viable cellulolytic organism for the production of biofuels in a consolidated bioprocess. Here, we provide significant details of the enrichment results and assess the capabilities of *C. cellulolyticum* in a consolidated bioprocess.

See more of this Session: Advances In Metabolic Engineering and Bioinformatics III - Experimental
See more of this Group/Topical: Food, Pharmaceutical & Bioengineering Division

(683a) Noninvasive, Optical, Continuous, Real-Time Molecular Sensing and Kinetic Modeling Using a Novel, near-Infrared, Implantable, Microfabricated VCSEL Based-Biosensor

Natesh Parashurama[1], Tom O'Sullivan[2], Adam De la Zerda[1], Simon Liu[1], James Harris[3], Pascale Kallassi[3], Ofer Levi[4] and Sanjiv Sam Gambhir[5], (1)Department of Radiology- Molecular Imaging Program at Stanford, Stanford University, Stanford, CA, (2)Beckman Laser Institute and Medical Clinic, UC Irvine, Irvine, CA, (3)Electrical Engineering, Stanford University, (4)Institute for Biomaterials and Biomedical Engineering, University of Toronto, Toronto, CA, (5)Radiology and Molecular Imaging Program, Stanford University Schools of Engineering and Medicine, Stanford, CA

Molecular imaging employing optical imaging instrumentation is a widespread, robust technique for in vivo imaging in preclinical models. Nevertheless, current limitations of optical imaging instrumentation include bulky, expensive equipment, the requirements for anesthetized, immobilized subjects, limitations to deep tissue imaging and temporal resolution, and the inability to perform long term (days) continuous imaging in one field of view. To complement whole body imaging approaches, we have recently developed a novel microfabricated, implantable biosensor, consisting of a Vertical Cavity Surface Emitting Laser (VCSEL), photodetector and optical filter optimized for Cy5.5 (675 excitation, 5 nm bandwidth) excitation and emission (O'Sullivan, Parashurama et al. Optics Express 2010). Here we utilize the same miniature biosensor, housing two detectors with one as a reference, to perform continuous molecular sensing of an established molecular probe pair, cRGD-Cy5.5 (n=6) and cRAD-Cy5.5 (n=3) in U87 tumors in nude mice. After injection, continuous sensing was achieved for every 5 seconds per detector as long as six hours, and major sources of noise were replicated with a reference detector. Peak activity was measured between 700-1000s after injection for both RAD and RGD. With background subtraction, the signal to background ratio for cRGD at 30 min, 1 h, 1.5h, and 2h was 1.69 ± 0.37, 1.73 ± 0.28, 1.75 ± 0.37, and 1.83 ± 0.42, respectively (N=6). On the other hand, the signal to background ratio of cRAD at 30 min, 1 h, 1.5h, and 2h was 0.68 ± 0.20, 0.77 ± 0.20, 0.84 ± 0.27, 0.91 ± 0.27, respectively (N=3), and the differences between were statistically significant (30 min ($P<0.001$), 1h ($P<0.001$), 1.5h ($P<0.002$), 2h ($P<0.010$) and correlated well with CCD camera data. We then analyzed continuous data by performing 4 parameter curve fitting of normalized, continuous data for individual mice, comparing control sites to tumor sites. Our results quantitatively demonstrate that in certain subgroups of mice, loss of signal (probe) from tumor occurs at a different rate then loss of signal (probe) from control site. We conclude that our biosensor works in a comparable fashion to CCD camera, can be used to continuously sense dynamics of molecular probe in a disease model, can reveal individual difference between mice, and can be adapted for both as a new enabling tool in noninvasive sensing and minimally invasive sensing, in both preclinical models and potentially clinical settings.

(683b) Dye-Conjugated Dendrimers As Bright and Photostable Nanoprobes for Fluorescence Microscopy

Younghoon Kim[1], Sunghoon Kim[2], Melikhan Tanyeri[3], John A. Katzenellenbogen[2] and **Charles M. Schroeder**[3], (1)Department of Chemical and Biomolecular Engineering, University of Illinois at Urbana-Champaign, Urbana, IL, (2)Department of Chemistry, University of Illinois at Urbana-Champaign, Urbana, IL, (3)Chemical and Biomolecular Engineering, University of Illinois at Urbana-Champaign, Urbana, IL

We demonstrate dye-conjugated dendrimers as nanoscale imaging probes for high-resolution fluorescence microscopy. Fluorescence-based imaging techniques currently rely on a limited set of fluorescent probes, including fluorescent proteins, organic dyes and quantum dots. However, there is a strong need for development of bright and photostable nanoprobes to enable biological imaging with enhanced spatiotemporal resolution. Recent advances in single molecule fluorescence microscopy have broken the diffraction barrier for optical imaging, thereby enabling observation of biological processes at nanometer-scale resolution. Super-resolution imaging techniques such as photoactivated localization microscopy (PALM) and stochastic optical reconstruction microscopy (STORM) rely on either photoactivable fluorescent dye or fluorescent proteins and photoswitchable organic dyes. However, the performance of these super-resolution imaging techniques is limited by the photophysical properties of the current set of fluorescent probes.

In this work, we use dye-conjugated dendrimers as a new class of nanoprobes for fluorescence microscopy, which enables high-resolution fluorescence imaging. Generation-5 polyamidoamine (PAMAM) dendrimers were engineered to contain sets of multiple cyanine dyes, and the photophysical properties were characterized by single molecule fluorescence microscopy using total internal reflection microscopy (TIRF-M). We observed that collective effects in the multichromophoric nanoprobe system result in highly extended photobleaching lifetimes and enhanced localization precision compared to single organic dyes such as Cy5. In addition, we engineered photoswitchable dye-conjugated dendrimers as enhanced nanoprobes for super-resolution imaging. Dye-conjugated dendrimers are bright, photostable and nanometer-sized macromolecular probes that can be synthesized to contain a variety of surface chemical functionalities for biological labeling and imaging. Based on these advantages, we anticipate that dye-conjugated dendrimers will pervasive application as a new class of nanoprobes for fluorescence microscopy.

See more of this Session: Bioimaging and Diagnostics I
See more of this Group/Topical: Food, Pharmaceutical & Bioengineering Division

(683c) Imaging the Lipid and Cholesterol Distribution within the Plasma Membranes of Intact Cells with High-Resolution Imaging Mass Spectrometry

Mary L. Kraft, Chemical & Biomolecular Engineering, University of Illinois at Urbana-Champaign, Urbana, IL

Domains enriched with cholesterol and sphingolipids, often referred to as lipid rafts, are believed to be present in the plasma membrane and correlate important cellular functions. Characterizing the sizes, shapes, and precise compositions of these membrane domains is especially challenging due to the lack of methods to acquire chemically specific information with high lateral resolution. Here we use a combination of stable isotope labeling and high resolution imaging mass spectrometry to chemically image the sphingolipid and cholesterol distribution within the plasma membrane with at least 100-nm-lateral resolution. Statistical methods are used to quantitatively assess the presence of cholesterol and sphingolipid-enriched domains in the plasma membrane.

See more of this Session: Bioimaging and Diagnostics I
See more of this Group/Topical: Food, Pharmaceutical & Bioengineering Division

(683d) Forster Resonance Energy Transfer Probes for Measuring Biochemical Activities: A Flexible Cloning Library and Suitability for Quantitative Experiments

Marc R. Birtwistle, Katarzyna Kida, Alexander von Kreigsheim, Drieke Vandamme and Walter Kolch, Systems Biology, University College Dublin, Belfield, Ireland

Numerous genetically-encoded, Forster Resonance Energy Transfer (FRET) probes for monitoring biochemical activities in live cells have been developed over the past decade. In general, these probes are a single protein where an acceptor/donor fluorescent protein (FP) pair is connected by a "sensing unit" that is conformationally reactive to the biochemical activity of interest. Changes in this biochemical activity alter the distance and/or orientation between the acceptor and donor, and thus are observable as changes in FRET. As these probes allow for collection of high frequency, spatially resolved data on signaling events in live cells, they are an attractive technology for developing quantitative models of spatiotemporal signaling dynamics. We are exploiting the general, modular FRET probe structure in combination with multi-site gateway cloning technology to create a library of sensing units and FPs with various localization tags. This library allows us to construct probes "on-demand" (within ~1 week) for measuring biochemical activities of interest, in cellular compartments of interest, all while controlling which parts of the light spectrum are taken for FRET analysis (to the extent possible by existing FRET-compatible FPs). However, to be useful for quantitative modeling purposes, the observed FRET from such probes should also have a large linear range of responsiveness. Here we report that measuring FRET by intensity-based, ratiometric methods, as is currently standard practice, yields data that scales non-linearly with respect to the biological quantity of interest. A log transformation of such ratiometric data improves the effective linear range, and paradoxically, some overlap between donor and acceptor emission spectra, which is usually avoided, should be beneficial for increasing the effective linear range of log-transformed ratiometric data. An alternative way to measure FRET is by analyzing fluorescence lifetimes, and we find that measuring FRET via fluorescence lifetimes is inherently linear with respect to the biological quantity of interest. Overall, our results provide a rapid and flexible cloning system for creating custom FRET probes and also suggest that fluorescence lifetime imaging is the preferred technique for obtaining quantitative data from FRET probes.

See more of this Session: Bioimaging and Diagnostics I
See more of this Group/Topical: Food, Pharmaceutical & Bioengineering Division

COMPARISION OF HYPER-SPECTRAL WIDE-FIELD AND CONFOCAL FLUORESCENCE MICROSCOPIC TECHNIQUES

Naga Srilakshmi Annamdevula[1], Ashley Stringfellow[2], Diego Alvarez[3,2], Thomas C.Rich[2] , Silas J. Leavesley[1,2]

[1]Chemical and Biomolecular Engineering, University of South Alabama, Mobile, AL, [2]Pharmacology, College of Medicine, University of South Alabama, Mobile, AL, [3]Internal Medicine, College of Medicine, University of South Alabama, Mobile, AL

Acute lung injury (ALI) is characterized by alterations in epithelial and endothelial cell function, pulmonary edema, and atelectasis. Pulmonary microvascular endothelial cells (PMVECs) are believed to play an important role in reconstructing the damaged endothelium during lung injury. Hence, there is a need to track PMVECs to understand their role in reforming damaged endothelium. Fluorescent protein expression is a potentially powerful tool for labeling and tracking these cells. However, attempts to detect fluorescently-labeled cells in lung tissue using standard single-wavelength fluorescence microscopy approaches have been unsuccessful, due to the high autofluorescence of the pulmonary vasculature and surrounding lung tissues.

We have previously used hyperspectral imaging and linear spectral unmixing analysis to detect, track, and quantify cells labeled with green fluorescent protein (GFP). We have found that spectral imaging and analysis can be used to detect GFP expressing cells in highly autofluorescent lung tissues. In contrast, standard single-wavelength fluorescence microscopy techniques could not accurately discriminate between GFP and autofluorescence. Spectral imaging and analysis studies were performed on an inverted, wide-field, spectral fluorescence microscope. We have conducted similar spectral studies using the Nikon A1R laser scanning confocal fluorescence microscope. The wide-field and confocal systems differ in their sample illuminations, detector configurations and spectral filtering methods. The wide-field inverted spectral fluorescence microscope is equipped with an acoustic-optic tunable filter (AOTF) and electron-multiplying charge coupled device (EMCCD) camera. The Nikon A1 laser scanning confocal microscope is equipped with a diffraction grating and a 32-channel photomultiplier tube (PMT) array. The goal of our present work is to study the effectiveness of both spectral fluorescence microscope systems in identifying GFP-expressing PMVECs in lung tissues.

To evaluate the effectiveness of each system, we have performed a sensitivity study analysis. We have considered several independent and dependent parameters. Independent parameters evaluated for confocal spectral microscopy were confocal pinhole size, laser power, PMT gain and rate of photobleaching. Independent parameters evaluated for wide-field spectral fluorescence microscopy were CCD camera gain and arc-lamp intensity. To determine the effect of each independent parameter on spectral imaging and analysis, we used signal-to-noise ratio (SNR), root-mean-square error (RMS), and intensity of the signal as dependent parameters. SNR is one of the important characteristic properties of an image and while RMS error gives the error in an image after spectral unmixing analysis. Hence, we calculated SNR and RMS error values to study the effect of each independent parameter on spectral imaging and analysis. We used MATLAB, ENVI, and Nikon Elements software packages to carry out this analysis.

The sensitivity studies performed allowed us to study the performance of various dependent and independent parameters on spectral imaging and unmixing analysis. We have also discussed methods to optimize the confocal pinhole, laser power, and PMT gain for the present study. This study could help in selecting and optimizing the independent parameters for a range of spectral assays. For instance, a low PMT value with high laser power reduces the noise and enhances SNR and could be used for the sample with weak GFP signal and strong autofluorescence. This could help in discriminating GFP from autofluorescence spectrum. SNR and RMS error varied significantly as a function of the independent parameters. For example, 30 % increase in PMT gain (increase from 148 volts to 198 volts) increases the RMS error by approximately 66%. Our results indicate that GFP photobleaches more rapidly than auotofluorescence. These results should be applicable to many types of *in-vitro* and *ex-vivo* spectral assays.

(683f) Tailoring Fluorescence Emission Level of Optical Contrast Agents by Gold Nanoparticles

Jianting Wang[1], Michael H. Nantz[2], Samuel Achilefu[3], Jin Z. Zhang[4] and **Kyung A. Kang**[1], (1) Chemical Engineering, University of Louisville, Louisville, KY, (2)Chemistry, University of Louisville, Louisville, KY, (3)Department of Radiology, Washington University in St. Louis, St. Louis, MO, (4) Chemistry and Biochemistry, University of California, Santa Cruz, Santa Cruz, CA

The fluorescence of a fluorophore can be artificially altered by influencing excitation/emission states of its electrons, such as placing it in an electromagnetic (EM) field. This phenomenon can be beneficially used in developing an efficacious fluorescent contrast agent for biosensing and molecular imaging. Good candidates for providing EM field in a local level are nano metal structures generating strong EM (plasmon) field upon receipt of light. Gold nanoparticle (GNP) is ideal for the purpose because of its inert and non-toxic nature, and its surface properties accommodating easy conjugation of biomolecules.

The three main groups of the parameters affecting fluorescence emission level by GNP are: (1) GNP properties: the size of the GNP (core-to-shell ratio if the particle is hollow) and the wavelength of the incident light [in a practical sense, it is the excitation (Ex) light wavelength of the fluorophore]; (2) Fluorophore properties: the wavelength of the emission peak (Em) and the quantum yield; and (3) The distance between a fluorophore and a GNP. We have theoretically analyzed the fluorescence change by GNP, with respect to these parameters, for the fluorophores that are frequently used in biomedical sensing/imaging. For the fluorophores with Ex/Em wavelengths near the plasmon resonance peak of the GNP (~520 nm) [e.g., FITC, AF532, Cy3, etc.], the fluorescence is extensively quenched when the fluorophore is placed within 10 nm from GNP. This means that, for these fluorophores, GNP can be used as an effective quencher. For those with the Ex/Em wavelengths away from the resonance peak [e. g., AF647, AF700, ICG, etc.] the fluorescence is quenched when the fluorophore is very close (less than a few nanometers) to a GNP. At a particular distance between the fluorophore and a GNP, the fluorescence can be significantly enhanced, and the level of the enhancement can be strongly dependent on the GNP size. For bigger GNPs, the distances for both quenching and the maximum enhancement are shorter, and the enhancement level is greater.

With the knowledge obtained from the theoretical analysis, we have designed optical contrast agents of Cypate, an NIR fluorophore, with spherical solid GNPs and hollow gold nanospheres (HGNs). Cypate is excellent for *in vivo* application because it is a derivative from FDA approved flurophore, ICG. Most natural fluorescence generated by tissues is in UV or visible wavelength and, using a NIR fluorophore can avoid the tissue generated fluorescence. Also NIR can penetrate into the tissue deeper than visible light. Cypate was conjugated to GNPs/ HGNs of various sizes via spacers at various lengths, and its fluorescence level was observed. Utilizing the GNP's fluorescence quenching effect when fluorophore is very close to GNPs, an enzyme-triggered contrast agent was developed by placing Cypate to GNP/ HGNs via a very short and substrate containing spacer. The enzyme used for our system was urokinase type tissue plasminogen activator (uPA), an enzyme secreted by highly invasive breast cancer. This agent emitted little fluorescence. In the presence of uPA, the spacer was cleaved and the fluorescence was restored. In addition, to utilize the GNP's fluorescence-enhancement property, Cypate was conjugated to GNPs/HGNs *via* spacers at particular lengths and we were able to enhance Cypate emission level extensively.

Fluorescence tailoring using GNPs can be realized by appropriately selecting fluorophores and manipulating the parameters associated. This mechanism can be used in developing optical contrast agents that are conditionally quenching or emitting, and that are with enhanced sensitivity.

The authors acknowledge the financial support from U.S. Army Breast Cancer Program (BC074387).

See more of this Session: Bioimaging and Diagnostics I
See more of this Group/Topical: Food, Pharmaceutical & Bioengineering Division

(683g) Quantitative, Multiparameter Analysis of Negatively Enriched CTC

Brandon L. Miller[1], Alejandra Garcia Villa[1], Jeffrey J. Chalmers[1] and Maryam Lustberg[2], (1)Chemical and Biomolecular Engineering, The Ohio State University, Columbus, OH, (2)Division of Medical Oncology, The Ohio State University, Columbus, OH

CTCs have been isolated in the peripheral blood samples of patients with various kinds of solid tumor malignancies, and elevated CTC numbers correlate with adverse clinical outcomes. While there is an accumulating body of evidence supporting CTCs as prognostic biomarkers, the most important clinically relevant questions including how to best utilize CTCs in clinical decision making remain unanswered, and are actively being investigated. Even more fundamentally, the question of whether CTC numbers alone are sufficient to begin to address these clinical situations or whether additional markers for CTC characterization need to be taken into account ***remains unanswered***.

To help address these questions, recent research investigations have shifted from purely detection to phenotypic characterization of CTCs since there may be subpopulations of CTCs that have more aggressive and metastatic potential. We have assembled a collaborative team, comprised of engineers, chemists, microscopists, medical oncologists and statisticians, to develop comprehensive multiparameter characterizations of CTCs that are isolated in the peripheral blood of patients with solid tumor malignancies. Based on significant research, discussions, and experimental results, our team reached the conclusion that quantitative multiparameter expression testing of CTCs was essential for furthering our understanding of how to best utilize CTC information clinically. By multiparameter analysis, we are referring to more than four markers per individual cell. Once multiparameter technology is developed, then clinical validation studies will be performed to determine the clinical significance of CTC subtypes in addition to CTC numbers.

However, typical current epifluorescent microscopy of more than three markers has significant challenges including: 1) a judgment call on the part of the investigator as to whether the cell is "positive" or "negative" for the marker, 2) potential for bleed through or cross-talk between different "channels" presenting the possibility that a "positive" signal is actually from another probe/marker, 3) bleach out of the dye, such as in the case of FITC, 4) current limit to approximately four different colors due to dye and technological limitations, and 5) most confocal microscopes do not have the capability to quantitatively measure the intensity of the fluorescence from specific cells, and 6) in the case of the confocal system, challenges of getting enough "confocal time".

To address these challenges, we are developing multispectral imaging technology which will allow, potentially, up to 8 different markers in a quantitative manner. Our approach utilizes computer controlled microscope stage, modified commercial Nunance Multispectral Imaging technology, and highly optimized and customized antibodies and fluorescent probes. This presentation will present our progress on actual breast and head and neck cancer patients circulating tumor cells.

See more of this Session: Bioimaging and Diagnostics I
See more of this Group/Topical: Food, Pharmaceutical & Bioengineering Division

(696a) Use of Single-Use Technologies In Downstream Purification of Mabs - A Case Study

Venkatesh Natarajan, EMD Millipore, Billerica, MA

The employment of single-use technologies in the downstream processing of biologics in general and monoclonal antibodies in particular is gaining significant traction in the industry. Their use eliminates cleaning validation requirements and can potentially offer significant savings in capital costs and startup time, especially when designing new facilities. Though the advantages of single use technology have been widely talked about, there is relatively little application data, especially with respect to the entire downstream purification train. In this presentation, data will be presented on an entire mAb downstream purification train at 50 – 100 g scale, executed largely using single-use technologies including bags, tubing and filtration assemblies, devices and systems. Data will also be presented on the process development work performed to size the various unit operations.

See more of this Session: Innovations In Biopharmaceutical Processing: Factory of the Future
See more of this Group/Topical: Topical I: Comprehensive Quality by Design in Pharmaceutical Development and Manufacture

(696b) Alternative Approaches to Cell Culture Harvests

Gregory Frank, Amgen, Thousand Oaks, CA

Alteranatives to traditional centrifuge harvests for cell culture harvests are discussed.

See more of this Session: Innovations In Biopharmaceutical Processing: Factory of the Future
See more of this Group/Topical: Topical I: Comprehensive Quality by Design in Pharmaceutical Development and Manufacture

(696c) Scale and Bioreactor Design Translation for Rocking Bag Bioreactors

John Bowers, BioProcess Development, Merck, Union, NJ

Scale and Bioreactor Design Translation for RockingBag Bioreactors.

The translations of scale and bioreactor designs for disposable rockingbag bioreactors are currently performed intuitively or with trial an errorapproaches. Conditions are typically basedon manufacturer recommendations or historical usage. A momentum based fluiddynamic analysis of rocking bag bioreactors provides the foundation forestimating the mixing conditions, the gas transfer rates and the shearenvironment in the bioreactor. Few assumptions are needed, and fundamental,well established understandings of fluid flow provide quantitative descriptionsof various phenomena.

This work focused on WAVE

The results obtained from the approach was compared with literaturedata and observations of rocking bag bioreactors from 2-L to 500-L, includethose used in manufacturing. The analysispredicted well most important bioreactor aspects of bioreactor performance,including oxygen transfer limitations with scale, mixing issues with fillvolume, similarity of conditions to stirred tank bioreactors and micro-carrierattached cell damage. With thisunderstanding of the fluid mechanics, performance variations during scale-upand site changes can be anticipated and addressed, and proper operatingparameters can be set.

Development scientist andengineers who are using, or are evaluating rocking bag bioreactors. Any one who needs to understand the operatingprinciples of thistype of disposable bioreactor.

Education:

Ph. D. ChemicalEngineering, Princeton University; B. S. Chemical Engineering, Carnegie-Mellon University

Experience:

18 years ofbio-pharmaceutical equipment and process design, process scale-up and processmodeling experience at Merck / Schering-Plough in cell culture and downstreampurification.

See more of this Session: Innovations In Biopharmaceutical Processing: Factory of the Future
See more of this Group/Topical: Topical I: Comprehensive Quality by Design in Pharmaceutical

(696d) Control of Virus Contamination In Cell Culture

Gregory S. Stimpfl, Research and Development, Amgen, Inc., Thousand Oaks, CA

Abstract to be submitted following publication approval.

See more of this Session: Innovations In Biopharmaceutical Processing: Factory of the Future
See more of this Group/Topical: Topical I: Comprehensive Quality by Design in Pharmaceutical Development and Manufacture

(696e) PAT Applications In Recombinant Protein Production Cell Culture Processes

Veronica Carvalhal, Lscc-PTD, Genentech, South San Francisco, CA

The regulatory agency's intent to make Process Analytical Technology (PAT) an industry-regulatory agreement makes this a very challenging and exciting time for both the pharmaceutical and biotechnology industries. In order to achieve the desired process understanding and control, the necessary effort should be used in developing tools and applying them either in the development or industrial stages. PAT is not a new strategy for the biotechnology industry or cell culture processes. Research and process development teams have long invested resources and expertise in process understanding. Nevertheless, PAT is a unique strategy where the final process adapts in real-time has not been applied frequently. This may be due to concerns with the challenges of implementing post-market process changes or costly and complex new technologies. In recombinant protein production processes, product quality is typically determined primarily by the selected cell clone and fermentation process conditions. Thus, upstream PAT process monitoring and control strategies may significantly benefit recombinant protein production processes. Fermentation pH or pO_2 controls are examples of PAT applications that have been applied and improved in the last couple of decades. More difficult to apply PAT strategies involve new sensor technologies or complex feedback approaches. Examples of nutrient controlled feeding and innovative air sparging strategies will be described as PAT applications that permit better process understanding and product quality control.

See more of this Session: Innovations In Biopharmaceutical Processing: Factory of the Future
See more of this Group/Topical: Topical I: Comprehensive Quality by Design in Pharmaceutical Development and Manufacture

(696f) At-Line NIR Spectroscopy As a Simple and Effective PAT Monitoring Technique In Mab Cultivations During Process Development and Manufacturing

Christian Hakemeyer[1], Silke Werz[1], Jose Cardoso Menezes[2], Francisca Folque[3], Gledson Jose[3] and **Ulrike Strauss**[1], (1)Pharmaceutical Biotech Production & Development, Roche Diagnostics GmbH, Penzberg, Germany, (2)Chemical & Biological Engineering Department, I.S.T., Lisbon, Portugal, (3) Technology Services & Products, 4Tune Engineering Ltd, Lisbon, Portugal

The experience of using at-line NIR transmittance spectroscopy on supernatant samples from a Mab cultivation process spanning several scales, clones and development years, from 2.5 l to 10,000 l is described. The collected spectra after preprocessing were used to (1) do process state estimation and (2) to obtain several culture parameters. Multivariate process trajectories were computed from NIR spectra acquired at-line and were used to: enhance process understanding across different scales up to industrial scale, assess batch-to-batch variability and scale-up issues, and examine the relative importance (criticality) of different sources of variability. The proposed approach demonstrates that NIR monitoring carried out at-line on samples already taken for reference methods analysis, (1) is very effective in replacing in-process monitoring of some critical process parameters by reference methods, (2) enables batch trajectories as defined by multiple culture parameters to be computed, (3) can be used to do guided sampling (i.e., near harvest or to investigate a deviation from the nominal trajectory by established bioanalytical methods) and (4) has potential use in implementing process corrections or even control strategies. The proposed approach is applicable throughout process development as well as to existing large-scale validated CHO bioprocesses, doesn't require more complex and expensive set-ups (viz., in-situ, on-line probes and process equipment) and is capable of supporting continuous improvement efforts on legacy and new but non-QbD Mab manufacturing processes.

See more of this Session: Innovations In Biopharmaceutical Processing: Factory of the Future
See more of this Group/Topical: Topical I: Comprehensive Quality by Design in Pharmaceutical Development and Manufacture

Challenges, Opportunities, and Scientific Aspects of Implementing PAT in Biopharmaceuticals

Huiquan Wu and Mansoor Khan, Food and Drug Administration, Silver Spring, MD
([*]Corresponding author: Huiquan.wu@fda.hhs.gov; Tel: 301-796-0022)

Abstract: As an essential milestone for both the cGMP for the 21st Century and the recent pharmaceutical quality regulatory initiatives (Q8/Q9/Q10), the 2004 FDA Guidance for Industry "PAT—A Framework for Innovative Pharmaceutical Development, Manufacturing, and Quality Assurance" has proposed a science-based and risk-based regulatory pathway for pharmaceutical industry. Over the past 7 years the pace of implementation of PAT in small molecule drugs and biopharmaceuticals has been quite different. In this presentation, challenges and opportunities, technologies available as well as some scientific aspects of PAT real time process monitoring and control in biotech and biopharmaceutical area will be discussed based on several available case studies. In authors' opinion, when the pharmaceutical community is ready to embrace the challenges with enthusiasm, the benefits from implementing PAT in the biotech and biopharmaceutical area can be realized in the coming years.

1. Introduction

The pharmaceutical community has been facing significant challenges, such as increasing complexity not only in the aspect of product design such as dosage form and drug delivery system designs, but also in the aspect of pharmaceutical process design, drug manufacturing science and technologies. This requires more sophisticated and better controlled manufacturing process than simple, immediate release dosage form manufacturing process. The traditional pharmaceutical manufacturing mode has been batch processing, which relies on some tests of in-process samples and full tests of final products. This traditional mode, while can provide quality products to the consumers, it has some inherent disadvantages. For example, in the event a product issue is identified via final product testing, typically it will be hard to find out the root cause. Even if the root cause can be identified, it will be too late as the batch is completed by the time the analysis results become available. As late Nobel Laureate Dr. Werner Heisenberg said: "a single point can only tell so much; it doesn't give direction if only a single value is available". In this regard, real-time process monitoring and control offers significant advantages, as it can provide a scientific basis for timely critical process decision making and ultimately ensure that product quality is built-in.

As an essential milestone for both the cGMP for the 21st Century and the recent pharmaceutical quality regulatory initiatives (Q8/Q8(R2)/Q9/Q10)[1-4], the 2004 FDA Process Analytical Technology (PAT) Guidance[5] for Industry "PAT—A Framework for Innovative Pharmaceutical Development, Manufacturing, and Quality Assurance" has proposed a science-based and risk-based regulatory pathway for pharmaceutical sector. Over the past 7 years, it is noticeable that the pace of implementation of PAT in small molecule drugs and biopharmaceuticals has been quite different. In this paper, challenges and opportunities, technologies available as well as some scientific aspects of PAT real time process monitoring and control in biotech and biopharmaceutical area will be discussed based on several available case studies.

2. Rationale Scientific Considerations during Developing a PAT-Based Monitoring and Controlling System for a Pharmaceutical Process

The FDA PAT Guidance defines PAT as "a system for designing, analyzing, and controlling manufacturing through timely measurements (i.e., during processing) of critical quality and performance attributes of raw and in-process materials and processes, with the goal of ensuring final product quality. It is important to note that the term analytical in PAT is viewed broadly to include chemical, physical, microbiological, mathematical, and risk analysis conducted in an integrated manner. The goal of PAT is to enhance understanding and control the manufacturing process, which is consistent with our current drug quality system: quality cannot be tested into products; it should be built-in or should **be** by design." When developing an integrated PAT approach for any pharmaceutical processes including biotech and biopharmaceutical processes, as a process engineer, it is important to keep the following questions in mind:

(1) What to measure?
(2) Where to measure?
(3) How to measure?
(4) How to analyze the data?
(5) How to develop and validate a model?
(6) How to control the process in a timely manner?
(7) Etc.

Although this list can be easily getting long, based on our limited knowledge, the first three questions are fundamental—they are more measurement process design-oriented. It is imperative for these types of questions answered during the initial planning stage of a PAT-oriented project. The remaining questions are more chemometrics and control-oriented, which requires significant expertise and resources to dig in. However, they could be addressed during the 2nd stage of the PAT-oriented project.

First, we need to decide what property/quality attribute or process parameter to be measured or analyzed. For a typical PAT project, often times the final decision to the first question is either a critical quality attribute (CQA), or a critical process parameter (CPP), or a combination of CQA and CPP. The answer to this first question will be also linked to what kind of analytical technology to be used. Then we need to think about where to measure the property/quality attribute/process parameter in the process. This will be linked again to the choice of analytical technologies and process analyzers, and what location the process analyzer will be put as well. The choice of measurement location should be both representative and informative. Thirdly we need to figure out how to measure the property of the process material and/or process parameter. Depending on the way the process and/or process equipment the process analyzer is interfaced to, the answer to this 3rd question might be at-line, in-line, on-line, or even near real time as defined in the FDA's PAT Guidance. With the first three questions answered, the answers to the questions remaining on the above list will be more technical-oriented. Therefore, answering these questions would require significant resources and expertise.

3. Challenges and Opportunities for implementing PAT in biotech and biopharmaceutical area

Over the last 7 years, PAT development and implementation in the small molecule pharmaceutical area has been making steady progress. The general philosophy of PAT has been incorporated into a broader pharmaceutical quality initiative "Pharmaceutical Quality for the 21st Century Initiative", or Quality-by-Design (QbD). The essential PAT elements, including PAT process monitoring and control, process understanding, have been become key QbD elements. According to ICH Q8 and Q8(R2), process control strategy is required for a successful QbD implementation. In a broad sense, the process understanding and process knowledge gained through PAT efforts, as well as control strategy established via PAT implementation, will collectively eliminate or mitigate the risk of delivering pharmaceutical products with poor or substandard quality to the patients and the public. Therefore, PAT is a scientific mechanism for pharmaceutical risk management. The general risk management framework is outlined in the ICH Q9 guideline.

Compared to their small molecule counterparts, biotechnology products and biopharmaceutical processes are recognized by their complexity due to facts including but not limited to, (1) variability existing in both product and process, such as compositional and structural heterogeneity, lot-to-lot variability; (2) process outcomes sensitive to foreign perturbations and process conditions, etc. Therefore, implementation of PAT in the biotech and biopharmaceuticals area would normally require more efforts and dedication. In addition, collaboration among various disciplines is expected to play a significant role.

Depending on the types of critical quality attributes (CQAs) and critical process parameters (CPPs) to be monitored and controlled, different technologies or process sensors could be used for implementation of PAT for a biotech or biopharmaceutical process. Table 1 below lists several technologies used for small molecule PAT/QbD applications based on reference[6] and authors' experience. Some of them are applicable to biotech and biopharmaceutical PAT/QbD applications, too. Taking the cell culture unit operation as one example, Table 2 blow is a list of technologies available for process monitoring and control, not meant to be an exhaustive list (adopted from reference[7]).

Table 1 Several Technologies used for small molecule PAT/QbD applications (some of them applicable to biotech and biopharmaceutical PAT/QbD applications, too)

Techniques	**Quantitative Data**	**Qualitative Date**	**Raw material ID**	**Physical Parameters**	**On-Line Capability**	**Near-Line Capability**
Mid-Infrared	Yes,with ATR and probes; usually only liquid samples	Yes, with ATR and probes; powder and liquid samples	Yes, not easily but similar to qualitative data	Yes, but not as clean as NIR	Yes, but more difficult than NIR/Raman; firbers expesnive and brittle	Yes, samples may be taken in near real time.

Near-Infrared	Yes, any physical form; transmission or refelection	Yes, large numbers of chemometric algorithms	Yes, all forms; ID and quality	Yes, hydrogen bonding makes it sensitive to all changes	Yes, either wireless units or by fiber probes from hardened units	Yes, ruggedized units are standard
Chemical Imaging	Yes, possible with chemometric algorithms	Yes.	Yes.	Yes, For example, hardness, domain size statistics, etc.	Not yet.. But possible. One system was in the pilot plant in 2003.	Not yet. Excellent capability for characterizing samples off-line (intact) quickly.
Far-Infrared or Terahertz	Possible with chemometric algorithms	Yes, sensitive to crystalline materials	Yes, but mostly for crystalline materials	Yes, intermolecular THz absorption makes it easy to detect interfaces or layer boundaries	Not yet. Recent development?	Not yet, But has excellent off-line characterization capability (both intact or destructive mode)
Raman	Yes, but semi-quantitative; without reference, not truly quantitative	Yes, insensitive to water	Yes, portable units are available	Yes, hydrogen bonding seen	Yes, mainly liquids and slurry systems.	Yes, both portable and traditional units
FBRM	Yes, but needs validation.	Yes.	Not applicable	Yes, chord length.	Yes, mainly liquids and slurry systems.	Yes, samples may be taken in near real time.
PVM	Yes.	Yes.	Not applicable	Yes, particle size and morphology.	Yes, mainly liquids and slurry systems.	Yes, samples may be taken in near real time.
In-situ AFM	Yes.	Yes.	Not applicable	Yes.	Yes, mainly solids or solids exposed to liquids	Yes.

Table 2 Possible process parameters to be monitored and/or controlled by various PAT sensors for a mammalian cell culture unit operation (adopted from reference[7])

Process parameter	PAT tool	PAT application example
Glucose concentration	Continuous FIA in conjunction with chemiluminescence assay	Control volume mixing pattern and residence time monitored to indicate metabolic condition to prevent cell starvation
Biomass concentration	Conductance and capacitance-based biomass probe	Specific growth rate and product formation monitored to determine onset of stationary phase and cell lysis
Biomass concentration	In-situ microscopes coupled with image processors and computer control	Specific growth rate used to control product formation in agitated cell broth cultivation
Oxygen uptake rate and specific oxygen uptake rate	Mass-balance technique for measuring kLa, dissolved O2 concentration, gas compositions	Control of physiological state of cells and cellular activity
Biomass, glucose, and acetate concentrations	NIR spectroscopy and electronic nose device	Control of fed batch cultivation and monitoring of the quality of the culture
Cell size, biomass concentration	Dielectric spectroscopy	Closed-loop control of medium feed rate in fluidized bed fermenters
Glucose and biomass concentrations	Multiwavelength fluorescence (MVF), software sensors and Multivariate Data Analysis	Control of specific growth rate and product quality
Gas and hydrocarbons	Gas sensor array as an electronic nose	Detection of early infections in mammalian cell cultures
Heat fluxes	Flow micro calorimeter, dielectric spectroscopy	Monitoring of specific growth rate and control of utilization of media components

It is appreciable that scientific researches in the area of biotech and biopharmaceutical PAT/QbD applications have been pretty active over the last few years. Some innovative pharmaceutical companies have also actively presented their biotech and biopharmaceutical PAT works at premier meetings such as IFPAC, AAPS Annual Meetings, AIChE Annual Meetings, etc. Most noticeably, the FDA CDER Office of Biotechnology Products (OBP) QbD pilot program[8] was launched in July 15, 2008. All of these efforts and programs together created a positive momentum for biotech and biopharmaceutical PAT/QbD to move forward. While the PAT/QbD initiative opens a new quality paradigm for biotech and biopharmaceutical development and manufacturing, numerous opportunities and challenges are co-existing. For example,

(1) How to use the PAT tools listed above and other new technologies to gain in-depth and better process understanding of the biotech and biopharmaceutical processes?
(2) How to use other modern analytical tools to gain better understanding of the product characteristics of the mAbs and other biopharmaceuticals?
(3) How to use advanced control tools to better control the biotech and biopharmaceutical process?
(4) How to use Design of Experiments (DOE) and other multivariate statistical tools to better understand and manage the variability inherent with biotech and biopharmaceutical products and processes?
(5) How to integrate the PAT/QbD methodologies into the existing subject knowledge system?
(6) How to establish the linkage among the process, product, and clinical safety and efficacy of drug products?

It calls for collaborative effort among the industry, academia, and regulatory parties. At the moment, our understanding of the link between each process parameter, each product quality attribute, and the clinical safety and efficacy of the drug is generally limited. It would be ideal if the linkage among the process, product, and patient can be established via an integrated manner in the PAT/QbD paradigm, although most likely it will be case-by-case.

4. Case Studies for Biotech and Biopharmaceutical PAT applications

For the two case studies illustrated below, some of the scientific aspects of implementing PAT/QbD in the biotech and biopharmaceutical area will be explored to certain detail.

4.1. NIR Monitoring of a Bioreactor Process (Adopted from Reference[9])

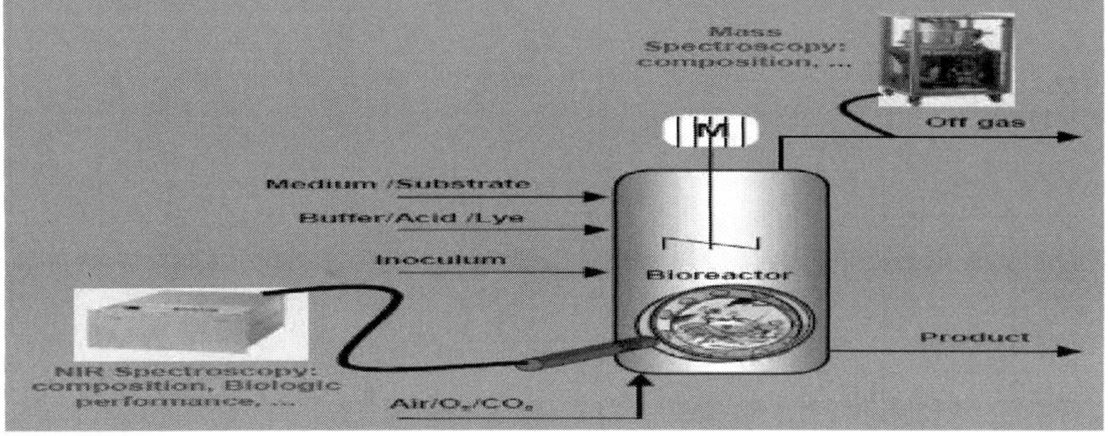

Figure 1. PAT real time monitoring of a bioreactor process

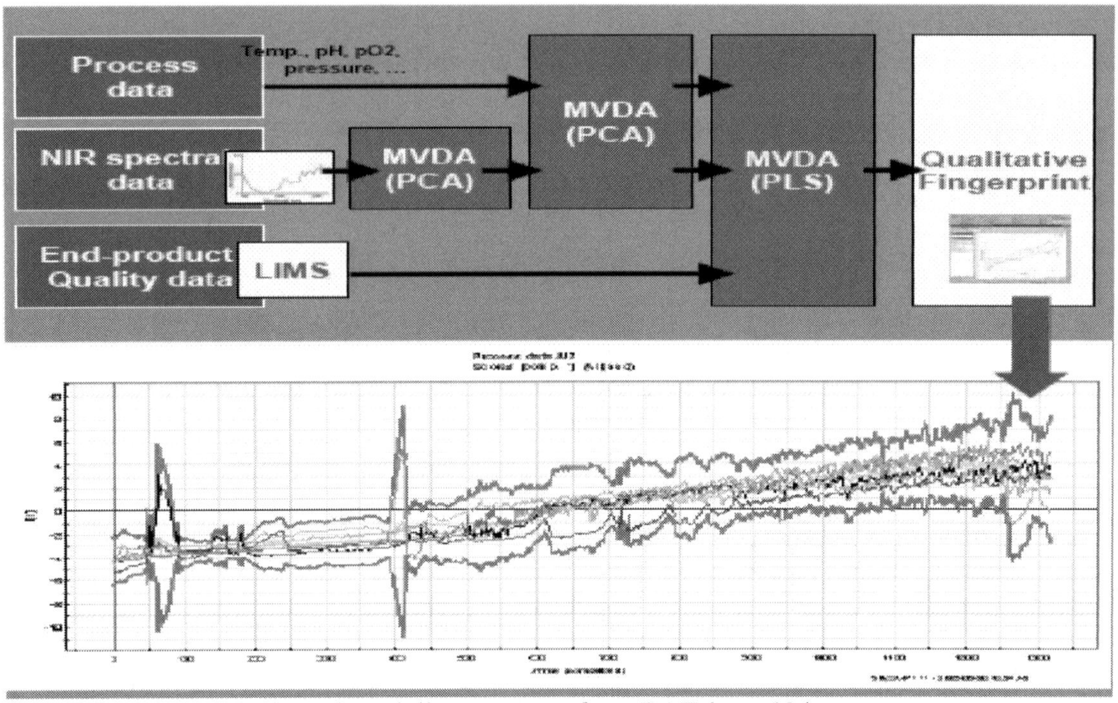

Figure 2. Data analysis and modeling strategy for a PAT-based bioreactor process

Although detailed data were not shown here, this publicly presented case study illustrated a PAT concept of using an in situ transmission probe and other process probes (pH, O_2, etc.) to online monitor a bioreactor process, and linking it to the end-product quality data via chemometrics techniques (such as Principal Component Analysis (PCA), Partial Least Square (PLS) regression). The chemometrics tools were used to identify qualitative fingerprint of the process, and to create a process trajectory for on-line process control. Similar case studies have been reported for some small molecule drug PAT applications by many researchers including authors[10-13].

4.2. Rapid Monitoring of Recombinant Antibody Production by Mammalian Cell Cultures Using Fourier Transform Infrared Spectroscopy and Chemometrics (Adopted from Reference[14])

In a recently published work[14], Fourier transform infrared (FT-IR) spectroscopy combined with multivariate statistical analyses was investigated as a physicochemical tool for monitoring secreted recombinant antibody production in cultures of Chinese hamster ovary (CHO) and murine myeloma nonsecreting 0 (NS0) cell lines. Medium samples were taken during culture of CHO and NS0 cell lines, and analyzed by FT-IR spectroscopy. It was found that FT-IR generates information on multiple aspects of the medium over the length of a culture, varying from utilization of the medium components to excretion/secretion of materials from cells into the medium.

The PCA shows that a clear trend in PC1 from left to right is observed for all cell lines. The loading plot PC1 shows that wave numbers corresponding to amide bonds (1700-1550 cm^{-1}) have a positive effect, where those corresponding to polysaccharides (1200-1000 cm^{-1}) have a negative influence. Therefore, PCA is separating the samples on the basis of both cell growth as indicated by the consumption of polysaccharides, and accumulation of antibody, as reflected by increased protein in the medium.

PLSR analysis was used to predict antibody titers, and these regression models were able to predict antibody titers accurately with low error when compared to ELISA data. PLSR was also able to predict glucose and lactate amounts in the medium samples accurately. The work demonstrates that FT-IR spectroscopy has great potential as a tool for monitoring cell cultures for recombinant protein production and offers a starting point for the on-line measurement of antibody production in industrial scale bioreactors.

In summary, FT-IR Spectroscopy as a means to monitor cells in culture has certain advantages and disadvantages. Advantages include: (1) Simultaneous assessment of mAb titer, glucose consumption, and lactose accumulation; (2) Minimal sample volume required (<20 μL); (3) Technically simple; it avoids need for chromatography. Disadvantages include: it can only scan dehydrated samples. For practical purposes, FTIR spectroscopy can be used for monitoring cell growth and determining optimal media conditions, and for selection of high-producing clones.

An internal study of using real time spectroscopy to monitor a bioreactor process has been undertaken within the FDA. Preliminary yet promising results have been obtained. However, the details of real time spectroscopy as a PAT tool for the process monitoring are to be worked out via further experiments and techniques such as chemometrics and other modeling methodologies. The study results will be released when ready.

5. Concluding Remarks

PAT is vital for implementing QbD in Pharmaceuticals, including biotech and biopharmaceutical processes and products. PAT focuses on process understanding, process control, process & product design. Chemometrics and process modeling are enabling tools to extract information and knowledge from process and product data, and to formulate and present the design space for regulatory fillings. In authors' opinion, when the pharmaceutical community is ready to embrace the challenges with enthusiasm, the benefits from implementing PAT in the biotech and biopharmaceutical area can be realized in the coming years.

Disclaimer: ***The views and opinions expressed in this paper are only those of the authors, and do not necessarily reflect the views or policies of the FDA.***

Acknowledgements:

The supports from colleagues and management at authors' division (Division of Product Quality Research) and offices(Office of Testing and Research, Office of Biotechnology Products, Office of Pharmaceutical Sciences) in CDER, FDA, as well as from various technology providers and authors of literatures cited are acknowledged.

References

1. FDA/ICH, 2006a. Guidance for Industry. Q8 Pharmaceutical Development. Available at: http://www.fda.gov/cder/guidance/6746fnl.pdf

2. FDA/ICH, 2009. Guidance for Industry, Q8(R2) Pharmaceutical Development. Draft available at:http://www.fda.gov/downloads/Drugs/GuidanceComplianceRegulatoryInformation/Guidances/ucm073507.pdf

3. FDA/ICH, 2006b. "Guidance for Industry. Q9 Quality Risk Management". Available at: http://www.fda.gov/cder/guidance/7153fnl.pdf

4. FDA/ICH, 2007. Q10. Pharmaceutical Quality System. Draft Consensus Guideline. Available at: http://www.fda.gov/cder/guidance/7891dft.pdf

5. FDA, 2004. "Guidance for Industry. PAT—A Framework for Innovative Pharmaceutical Development, Manufacturing, and Quality Assurance". Available at: http://www.fda.gov/cder/guidance/6419fnl.pdf

6. Ciurczak, E.W.. "A Spectroscopy Field Guide for PAT", Pharmaceutical Manufacturing, July/Aug. 2008, p.18-26

7. Rathore, A.S., Bhambure, R., Ghare, V. 2010. Process analytical technology (PAT) for biopharmaceutical products. Anal Bioanal Chem. DOI 10.1007/s00216-010-3781-x

8. Morris, W. and Hough, E., Parenteral Drug Association (PDA) Newsletter, "QbD Pilot Program for Biotech Seeks Similar Answers as One for Small Molecules", Nov-Dec, 2008.

9. Ingrid Maes. Application of PAT in bioreactor process development and manufacturing. the FIP 2008 PAT Workshop, Aug. 30-31, 2008. Basel, Switzerland.

10. Wu, H. and Khan, M.A., 2009. Quality-by-Design (QbD): An Integrated Approach for Evaluation of Powder Blending Process Kinetics and Determination of Powder Blending End-Point. J Pharm Sci. 98(8), 2784-2798.

11. Wu, H., Tawakkul, M., White, M., and Khan, M.A., 2009. Quality-by-Design (QbD): An Integrated Multivariate Approach for the Component Quantification in Powder Blends. Int J Pharm. 372(1-2), 39-48.

12. Wu, H. and Khan, M.A., 2010. Quality-by-Design (QbD): An Integrated Process Analytical Technology (PAT) Approach for Real-time Monitoring and Mapping the State of a Pharmaceutical Co-precipitation Process. Journal of Pharmaceutical Sciences 99(3): 1516-1534.

13. Wu, H. and Khan, M.A., 2011. Quality-by-Design (QbD): An Integrated Process Analytical Technology (PAT) Approach to Determine the Nucleation and Growth Mechanisms during a Dynamic Pharmaceutical Co-precipitation Process. Journal of Pharmaceutical Sciences 100(5): 1969-1986.

14. Sellick, C.A., et al., 2010. Rapid Monitoring of Recombinant Antibody Production by Mammalian Cell Cultures Using Fourier Transform Infrared Spectroscopy and Chemometrics. Biotech & Bioeng. 106(3):432-442.

(701a) Modeling a Reactive Crystallization with a Reduced Population Balance Equation for Dissolution and Growth to Control Critical Quality Attributes

Daniel Patience[1], Linli He[2], Dale Mowrey[2], Partha Mudipalli[3], Michael Christie[2] and Roger Bakale[2], (1)Chemical Process Research & Development, Cephalon, Malvern, SD, (2)Chemical Process R&D, Cephalon, Malvern, PA, (3)API Technologies, Cephalon, West Chester, PA

Modeling has always been a powerful weapon in a chemical engineer's arsenal. Use of modeling should be considered a key piece in the application of Quality by Design (QbD) in drug substance and drug product development. In this study, reaction, dissolution and growth kinetics are determined for a reactive crystallization of an intermediate-grade drug substance (DS) currently in development at Cephalon, Inc. Initial data showed that decreasing particle size of the starting material (in the DS formation step) increased the levels of entrapped starting material in the DS product. Smaller particles of starting material dissolve faster enhancing the rate of product formation. This leads to product coating the remaining starting material, thus increasing the level of entrapped starting material. The level of starting material in the DS is a critical quality attribute (CQA). Also, decreasing the particle size of the product led to lower levels of starting material entrapment. Breakage exposes the interior starting material to additional reagent. DS particle size is a potential CQA. The goal of this work is to model and hence predict entrapment levels and particle size as a function of dissolution, reaction, and crystallization kinetics. In the solution phase, a Mannich reaction occurs which is modeled as two second order reversible reactions in series. The model equations to describe this reactive crystallization consist of a solution phase mass balance for the reaction kinetics coupled with two population balance equations (PBEs). The PBEs describe the particle size densities for the nearly insoluble starting material and the nearly insoluble DS product during dissolution and precipitation. Due to the low solubility of the starting material and the DS product, we assume the solution phase concentration is always saturated with respect to these two materials. Therefore, a reduced fast dissolution and high growth model was developed to describe this process. The model in this study is a new model for reactive crystallizations because of the reduced nature and the ability to describe multiple crystallization phenomena and two second order reversible reactions in series with minimal parameters. Using the modeling software package Octave, the model parameters (reaction, nucleation and nuclei-disappearance rate constants) were estimated by fitting model predictions to experimental measurements including particle number (via Mettler's Lasentec FBRM probe), reactor temperature, and percent entrapped starting material in DS. This type of modeling approach is a macroscopic modeling approach for QbD and can be superior to a factorial or central-composite design approach because it can offer physical insights. Such insights include knowledge of the reaction and crystallization rate constants and the ability to extrapolate beyond previously studied experimental conditions.

(701b) Model-Based Characterisation of Organic and Aqueous Tablet Film Coating Processes: Parameter Estimation and Risk Management

Salvador Garca Muoz[1], Mary T. Am Ende[2], **Mark A. Pinto**[3] and Sean K. Bermingham[3], (1)Process Modeling & Engineering Technology, Pfizer Global Research and Development, Groton, CT, (2)Pharmaceutical Development, Pfizer Global Research and Development, Groton, CT, (3)Process Systems Enterprise, London W6 7HA, United Kingdom

Tablet coating is an important processing step in thepharmaceutical industry in which tablets are coated with an aqueous or organicfilm coating for both aesthetic and functional reasons.

Mathematical models of tablet film coating are important inpharmaceutical development as they aid in design of experiments, scale-up anddetermining optimal process conditions as formulations change. A universalsteady-state film coating model has previously been developed (am Ende,Berchielli, 2005) with the aim of providing process engineers with a means ofpredicting target operating conditions for optimization, scale-up androbustness studies. In that work, the model was validated with experimentalconditions and the predictions obtained were found to be in good agreement withdata not used for model validation.

In this study, the steady-state model was first validatedagainst steady state plant data using gSOLIDS (Process Systems Enterprise, UK).It was found that starting with one experiment, as experiments were added, the95% confidence interval decreased rapidly reducing to less than 10% with fouror more experiments.

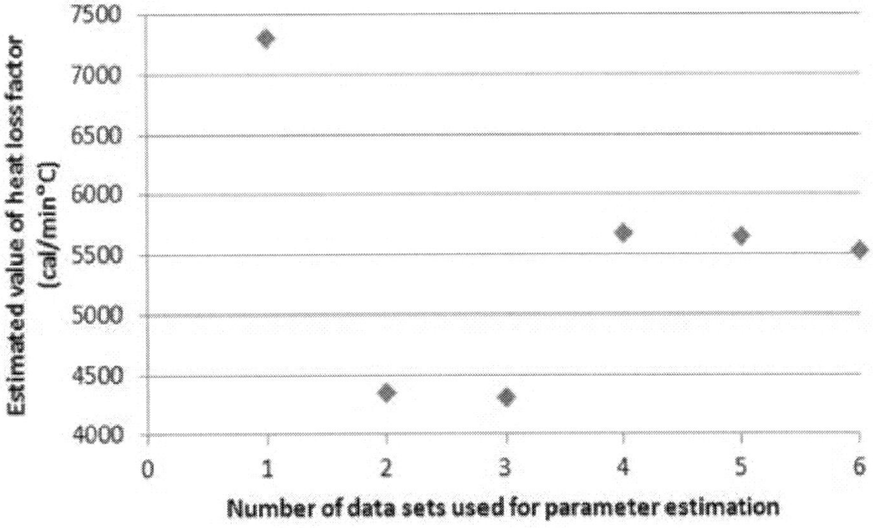

Figure 1: Change in estimated value of heat loss factor as experimental data is added

In order to quantify whether sufficient data was used formodel validation, the 95% T-value was compared to the reference T-value. (Ifthe 95% T-value is greater than the reference T-value, this indicates thatsufficient data was used to for model validation). The results obtainedindicated that at least three experiments are needed to accurately estimate theunknown model parameter. This was consistent with the results obtained for the95% confidence interval which was more than 50% when only two experiments wereused.

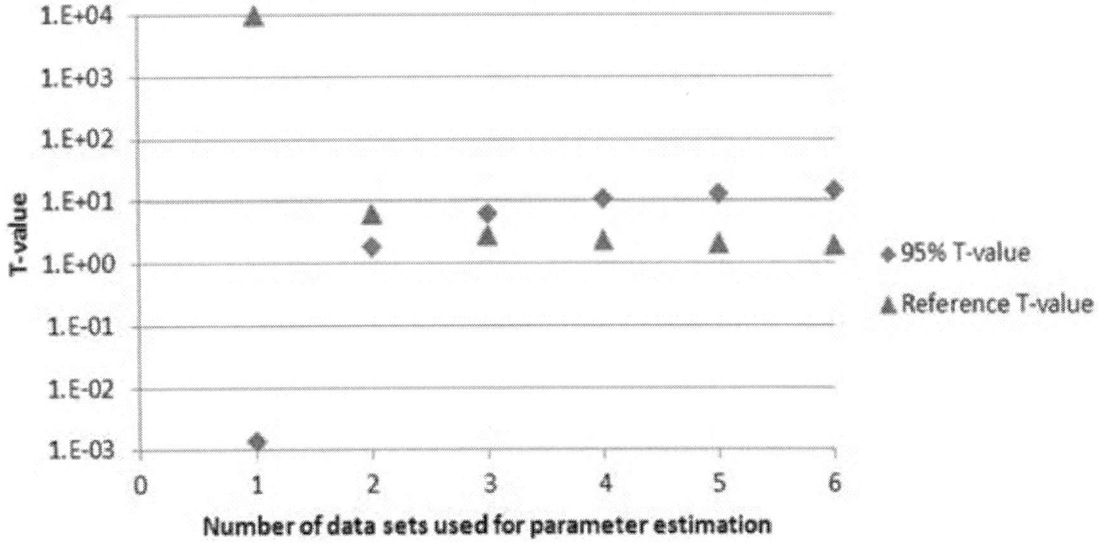

Figure 2: Changes in reference and 95% T-values as experimental data is added

As indicated by the 95% confidence intervals, the estimatedvalue of the model parameter is not a perfect estimate. Therefore in order todetermine a reasonable estimate of the model prediction a Monte Carlosimulation was carried out. In this simulation several instances of the modelwere run with each instance having the unknown model parameter sampled from anormal distribution centred at the parameter estimate and with a standarddeviation obtained from the 95% confidence interval. The results from thesenumerous simulations were then averaged to obtain an estimate of the modelprediction. In addition, the results from these simulations were used to obtaina 95% confidence interval on the model predictions. The results obtainedindicated that the model predictions were in good agreement with theexperimental data used for parameter estimation. Further, given the confidenceinterval of the estimated value of the model parameter, the standard deviationof the model prediction was much smaller than that of the data indicating thatthe predictions of the model vary very little with respect to the uncertaintyin the parameter estimate.

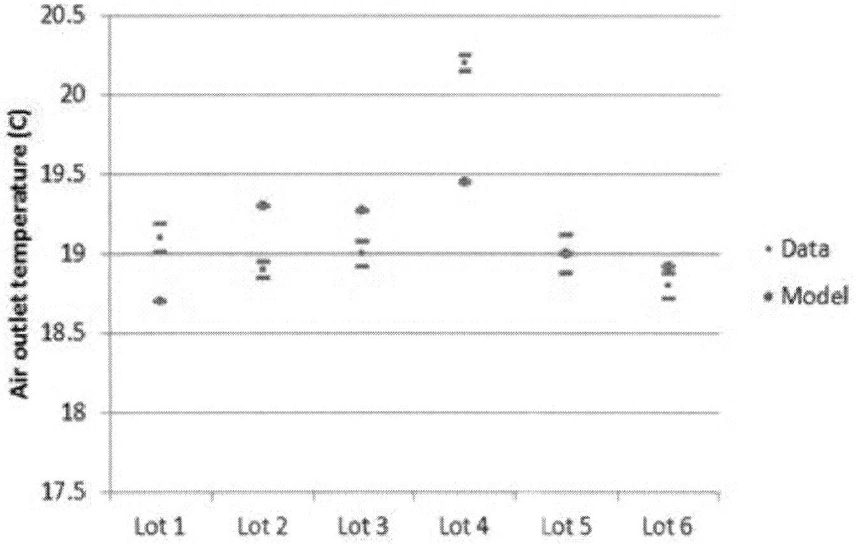

Figure 3: Comparison of model predictions with experimental data

A similar analysis was conducted using dynamic data. In thiscase however, although the data used for parameter estimation was deemedsufficient, the quality of the fit was quantified as being unsatisfactory evenif several experiments were used. This indicates insufficiencies in the modelwhich is expected as a steady state model was used to study dynamic plant data.

In summary, the results presented above indicate that withrelatively few steady state experiments, an accurate estimate can be obtainedof the unknown model parameter in the steady state film coating model. Themodel predictions were found to be in good agreement with the experimentaldata. The uncertainty in the model parameter was quantified and it was foundthat the model predictions were relatively unaffected given the uncertainty inthe parameter estimate.

The analysis presented above is generic and can be used toquantity uncertainty in any mathematical model. This information can beespecially useful as it can be used to quantify the risk associated withdecisions made using these mathematical models. Further, the rigour of theanalysis helps determine whether more experimentation is needed or whetherfurther model development is needed (or possibly a combination of both).

References

M. am Ende, A Berchielli (2005) A thermodynamic model fororganic and aqueous tablet film coating. *Pharmaceutical Development andTechnology*, 1:47-58

See more of this Session: Model Based QbD In Drug Substance or Drug Product Development
See more of this Group/Topical: Topical I: Comprehensive Quality by Design in Pharmaceutical Development and Manufacture

(701c) Design Space Verification for Tablet Dosage Form Using Modeling

Mary T. am Ende[1], Thomas P. Garcia[2], Kim Vukovinsky[3], Penelope Butterell[4], Alex Chueh[5], Bernd Schuemmelfeder[6], Vincent E. McCurdy[2] and Daniel Gierer[1], (1)Pharmaceutical Development, Pfizer Worldwide Research and Development, Groton, CT, (2)Global CMC, Pfizer Worldwide Research and Development, Groton, CT, (3)PharmaTherapeutics, Pfizer Worldwide Research and Development, Groton, CT, (4)Gqts, Pfizer Global Supply, Zaventem, Belgium, (5)Global Manufacturing Services, Pfizer Global Supply, Pearl River, NY, (6)Global Manufacturing Services, Pfizer Global Supply, Freiburg, Germany

Quality by Design principles are well established in the pharmaceutical industry. It is becoming common place to develop the design space for a drug product and include it in the regulatory filing. There are a broad range of methods to establish design space, some of which are focused on individual unit operations and others that take a holistic view of the entire product processing. A new challenge is developing approaches for verifying the design space for commercial production, especially for those defined using data obtained from modeling or small scale experiments. In this talk, a proposal for verifying the design space for a drug product with multiple CPPs, KPPs and CQAs will be discussed using principles of scale-dependencies, use of precedented models, and (when necessary) large scale confirmation batches.

See more of this Session: Model Based QbD In Drug Substance or Drug Product Development
See more of this Group/Topical: Topical I: Comprehensive Quality by Design in Pharmaceutical Development and Manufacture

(701d) New Model-Based Platform for Enhanced Content Uniformity Prediction of Oral Dosage Forms

Salvador Garca-Muoz, Process Modeling & Engineering Technology, Pfizer Worldwide Research & Development, Groton, CT and Weili Yu, Pharmaceutical Development, Pfizer Worldwide Research & Development, Groton, CT

New Model-based Platform for Enhanced Content Uniformity Prediction of Oral Dosage Forms

Being a critical quality attribute for capsules and tablets, content uniformity (CU) has received continued attention from practitioners and academics with multiples proposals to relate it with upstream properties such as the particle size distribution and true density of the active pharmaceutical ingredient (API).

In this work we propose a new model to predict the CU of a lot of tablets or capsules, based on a corrected algorithm that considers: the particle size distribution of the API and its variability, the true density of the API, the size dependant potency of any drug product intermediate and the weight variability of the lot of finished product. The model is based on a particle level mass balance and a Monte-Carlo simulation of the sampling process. The accuracy of this new model is demonstrated by showing predicted versus observed values of content uniformity (RSD) for multiples cases.

This model enables a much improved risk assessment exercise between API characteristics, production process and CU; and provides a platform to establish acceptable variability ranges and targets that are fundamental components of a Quality by Design development.

See more of this Session: Model Based QbD In Drug Substance or Drug Product Development
See more of this Group/Topical: Topical I: Comprehensive Quality by Design in Pharmaceutical Development and Manufacture

(701e) Computation of Reliabilities for Extent of Reactant Conversion Incorporating Batch Effects with Applications to ICH Q8 Design Space

Brian R. Crump[1], Rick Lewis[2], Zifang Guo[2] and John Peterson[3], (1)Process Engineering, GSK, Research Triangle Park, NC, (2)Statistical Sciences, GSK, Research Triangle Park, NC, (3)Quantitative Sciences, GSK, Upper Providence, PA

A fundamental reaction model was developed to explain rate observations under periods of both kinetic and mass transfer control. The reaction starting material was a potential genotoxin such that the residual starting material concentration in the product was strictly regulated. The primary means of controlling the starting material concentration in the product was via the extent of the reaction. Hence, it was important to be able to reliably forecast the reaction conversion as a function of process and equipment parameters. This presentation will demonstrate a means of gauging the reliability of the reaction model.

It has been shown (Peterson, 2008) that a predictive distribution for quality responses was important for process capability and ICH Q8 design space quantification. Experiments involving mechanistic models for API production often have relatively small measurement error variability. However, predicting a future quality response involved sources of variation other than just measurement error. For example, one needed to model the batch-to-batch variation, which can be substantially larger than the measurement error variation.

Within the Bayesian statistical framework, development of a predictive distribution for future quality responses is, in theory, straightforward. However, it can be computationally challenging to compute such distributions, and numerically intensive methods are often necessary. In this presentation, we use the BUGS software application to develop Bayesian predictive models with and without random variation due to batch effects. We show that failure to model the batch-to-batch variation can result in noticeable bias for process capability and ICH Q8 design space development.

See more of this Session: Model Based QbD In Drug Substance or Drug Product Development
See more of this Group/Topical: Topical I: Comprehensive Quality by Design in Pharmaceutical Development and Manufacture

(701f) Estimating Reaction Model Uncertainty with Bayesian Markov Chain Monte Carlo

Jacob Albrecht, Bristol-Myers Squibb Company, New Brunswick, NJ

The use of kinetic models to predict reaction performance is becoming an integral element of pharmaceutical process development. These models, built on fundamental mechanistic understanding and laboratory data, must be characterized by how well they can describe experimental observations. This task of quantifying uncertainty for reaction models is complicated by model non-linearity and experimental error. Because experimental error may not follow a conventional distribution, the validity of common statistical assumptions behind least-squares regression and normal error distributions should be reexamined when fitting models to data.

This paper reports a comparison of different techniques to compare parameter regression and confidence intervals for pharmaceutically relevant chemical reaction models. Simulated experiments were created by adding two types of noise (normally distributed and experimentally measured) collected using two analytical techniques (HPLC versus inline FTIR) to four types of kinetic models (linearized, exponential, coupled ODE and DAE). The best fit parameter values and confidence intervals were then calculated according to a multivariate *t*-distribution or two different Bayesian Markov-Chain Monte Carlo algorithms: one estimating error assuming a normal error distribution and a new, more adaptable non-parametric likelihood function. The uncertainty estimates were evaluated based on their ability to correctly regress the model parameter's probability distribution in scenarios common to pharmaceutical reaction engineering. The findings indicate that while multivariate *t*-distributions are sufficient for simpler non-linear models with normal error distributions, only BMCMC accurately estimates parameter ranges when errors are non-normal or when the model system is complex.

See more of this Session: Model Based QbD In Drug Substance or Drug Product Development
See more of this Group/Topical: Topical I: Comprehensive Quality by Design in Pharmaceutical Development and Manufacture

(701g) Controllability Analysis of Protein Glycosylation In CHO Cells

Melissa M. St. Amand, Kevin Tran, Devesh Radhakrishnan, Anne S. Robinson and Babatunde A. Ogunnaike, Chemical Engineering, University of Delaware, Newark, DE

Biopharmaceuticals represent the largest growing class of therapeutics with a US market of approximately $50 billion in 2009 and an expected steady increase in future sales (1). Many biopharmaceuticals, such as the therapeutic monoclonal antibodies Herceptin and Avastin, are produced as recombinant proteins from Chinese hamster ovary (CHO) cells that are cultivated in bioreactors. As with other manufactured products, these therapeutic proteins are effective only when their product quality attributes (bioactivity, potency, purity, etc.) lie within a specific range of values. Of the many factors that affect the quality and bioactivity of these proteins, arguably the most important is glycosylation, a post-translational modification in which a carbohydrate chain, termed a glycan, is added to a protein. To function as intended *in vivo,* most therapeutic protein treatments validated for human use must have a precise distribution of glycans (i.e., specific percentages of glycans with specific sugar monomers such as galactose, sialic acid, or fucose). Unlike other cellular processes such as DNA replication and protein production, however, glycosylation has no master template, and, as a result, glycan formation and attachment are subject to variability and are often non-uniform. Consequently, regulatory agencies, such as the Food and Drug Administration (FDA) and European Medicines Agency (EMEA), are now encouraging biopharmaceutical manufacturers to develop strategies to control glycosylation online during production. However, online glycosylation control, which will involve active monitoring and effective control of key process variables that are critical to product quality, is yet to be implemented in the biopharmaceutical industry for a variety of reasons, mostly attributable to the complexity of these bioprocesses, the non-availability of on-line measurements, and the lack of comprehensive control paradigms tailor-made for such processes.

Our goal is to develop—and validate experimentally—such a comprehensive strategy for effective online, real-time control of glycosylation. To achieve this goal, however, it is imperative first to assess the controllability of the process of glycosylation. Specifically, with x defined as the vector of the percentages of each glycan form present in the overall glycan pool from a batch of therapeutic protein, we must determine if protein glycosylation can be directed from any initial state $x(0)=x_0$ to any arbitrarily specified desired final state x_f, in finite time, via admissible manipulations of available process variables and operating conditions. Unfortunately, even though there is sufficient mechanistic knowledge to enable the development of a first-principles process model of glycosylation, by the very nature of the glycosylation process, with its convoluted reaction schemes and complex network architecture of approximately 23,000 reactions and almost 8,000 glycoforms, it is entirely *impossible* to carry out standard theoretical closed-form, state-space controllability analysis with the resulting mathematical model. While numerical analysis is always possible *in principle*, the intrinsic nonlinearity and sheer complexity of the mathematical model demands that even model simulations must be carried out systematically and judiciously if the required system controllability information is to be extracted efficiently from the simulation results.

In this presentation, we discuss a simulation-based procedure for assessing the controllability of glycosylation predicated upon employing statistical design of experiments to vary glycosylation enzyme and sugar nucleotide concentrations (model inputs) strategically. The resulting glycosylation responses (model outputs) are analyzed via ANOVA to identify the enzymes and sugar nucleotide donors that exert the most significant effects on the predominant glycans found in recombinant proteins produced with CHO cells. The statistical analysis results are subsequently used to assess the degree to which glycosylation is controllable. In addition to providing useful insight into the extent to which online glycosylation control is achievable in practice in an actual experimental process, we also discuss how the procedure provides a framework for developing directed approaches for the genetic engineering of cell lines to produce protein with a specific desired glycan structure.

[1] Aggarwal, S., What's fueling the biotech engine-2009-2010. Nat Biotechnol, 2010. 28(11): p. 1165-71.

See more of this Session: Model Based QbD In Drug Substance or Drug Product Development
See more of this Group/Topical: Topical I: Comprehensive Quality by Design in Pharmaceutical Development and Manufacture

(719a) Engineered Cells That Detect and Kill a Pathogen: a Novel Synthetic Biology-Based Antimicrobial Strategy

Nazanin Saeidi, Choon Kit Wong, Tat-Ming Lo, Hung Xuan Nguyen, Chueh Loo Poh and **Matthew Wook Chang**, School of Chemical and Biomedical Engineering, Nanyang Technological University, Singaproe, Singapore

Synthetic biology has allowed us to design and construct new biological systems that have the potential to resolve important issues related to healthcare. In particular, considering the stalled development of advanced antibiotics and the emergence of antibiotic-resistant pathogens, we now must strive to exploit synthetic biology approaches for designing a new treatment regimen for infectious diseases. Therefore, in this study, we aimed to engineer microbes to detect and kill a pathogen using synthetic biology principles. Briefly, we designed and constructed a genetic system, which comprises detecting and killing devices, that enables *Escherichia coli* to sense and kill a pathogenic *Pseudomonas aeruginosa* strain. We further characterized the detecting and killing devices to understand their functionalities, which subsequently helped us to construct the final system and verify its designed behavior. Finally, we showed that our engineered *E. coli* detects and kills *P. aeruginosa*, which offers a novel synthetic biology-based antimicrobial strategy that could readily be expanded to treating other infectious pathogens.

See more of this Session: Systems and Synthetic Biology of Interacting Microorganisms
See more of this Group/Topical: Food, Pharmaceutical & Bioengineering Division

(719b) Quantification of Interactions Between Viruses and Their Defective Interfering Particles

Fulya Akpinar, Chemical and Biological Engineering, University of Wisconsin-Madison, Madison, WI and John Yin, Chemical and Biological Engineering, University of Wisconsin - Madison, Madison, WI

When a virus infects a susceptible host cell, it hijacks the cell's biosynthetic machinery and resources to produce progeny virus. Infected cells may also produce virus-like particles that carry genetic material but are unable to further infect cells. One class of such virus-like particles are defective interfering particles (DIPs), which carry a defective viral genome that renders them unable to reproduce in living cells. However, when DIPs co-infect a cell with viable virus, they can interfere with virus production. Their ability to interfere with or attenuate normal viral infections makes DIPs a useful tool to probe virus-cell interactions and potentially treat viral infections. These features provide motivation to develop quantitative measures to better understand how DIP and virus particles interact. As a model system, we study the behavior of vesicular stomatitis virus (VSV) and its DIPs when cultured on baby hamster kidney (BHK21) cells. VSV is a negative-sense single-stranded RNA virus related to rabies, ebola and respiratory syncytial virus, but it is non-pathogenic to humans and it has served as a prototype for the study of virus-DIP interactions. Here we show how measures of virus and virus-like particle counts and activity level can be integrated to gain insight into the mechanism of infection interference. More broadly, our results show that non-infectious byproducts of infection can nevertheless have important consequences for the growth and spread of viruses.

See more of this Session: Systems and Synthetic Biology of Interacting Microorganisms
See more of this Group/Topical: Food, Pharmaceutical & Bioengineering Division

(719c) Droplet-Enabled Co-Cultivation of Drug-Producing Natural Microbiota Isolated From a Marine Invertebrate

Jihyang Park[1], Christopher M. Rath[2], Mark A. Burns[3], David Sherman[4] and Xiaoxia Lin[1], (1) Department of Chemical Engineering, University of Michigan, Ann Arbor, MI, (2)Life Science Institute, University of Michigan, Ann Arbor, MI, (3)Chemical Engineering, University of Michigan, Ann Arbor, MI, (4)Life Sciences Institute, University of Michigan, Ann Arbor, MI

Certain marine invertebrates have been known to harbor microbial communities that produce drug and antibiotic molecules. For instance, ET-743, an anticancer drug, is a natural product originally purified from the tunicate *Ecteinascidia turbinata* and is believed to be a secondary metabolite secreted by a symbiotic bacterium, *Candidatus Endoecteinascidia frumentensis*. Biocatalytic cluster for the synthesis of ET-743 in the microbe of interest was identified previously, but the bacterium has never been cultivated in laboratory so far. Due to extremely low yield from the natural source and an inability to culture *C. frumentensis*, mass production of ET-743 currently depends on a costly semi-synthetic pathway. In this work, we extracted drug-producing microbiota from tunicate and co-cultivated them in micro-droplets in an attempt to isolate the drug-producing species and to elucidate the microbial interactions in the tunicate microbial community. Previously, we developed a microfluidic device for highly parallel co-cultivation of microbial communities and demonstrated its effectiveness in discovering synergistic microbial interactions using a synthetic *E. coli* model system. Employing aqueous microdroplets, the device could readily encapsulate and co-cultivate various subsets of a microbial community. This platform was extended for the investigation of the tunicate microbiota. Isolated tunicate microbiota were co-cultivated in conventional rich media and seawater collected from the original habitat of tunicates. Grown co-cultures were retrieved from the device and characterized by Terminal Restriction Fragment Length Polymorphism (TRFLP) and sequencing of 16S rRNA genes. Cultures developed under conditions close to its native environment showed higher diversity compared to those derived from the TSB medium. These results showed that our droplet-enabled co-cultivation technology can effectively decompose complicated natural microbiota and thus facilitate the elucidation of underlying interaction networks. Our ongoing efforts include further optimization of the co-cultivation environment by exploring microaerobic conditions that better mimic the original habitat of the tunicate microbiota.

See more of this Session: Systems and Synthetic Biology of Interacting Microorganisms
See more of this Group/Topical: Food, Pharmaceutical & Bioengineering Division

() Microgravity and Swimming Motility Increase Biofilm Formation by Pseudomonas - During Spaceflight

Woo-Seong Kim, Farah K. Tengra, Zachary Young, Jasmine Shong, Ravindra Pangule, Jonathan S. Dordick, Joel L. Plawsky and Cynthia H. Collins, Chemical and Biological Engineering, Rensselaer Polytechnic Institute, Troy, NY

The microgravity environment encountered during spaceflight has been reported to alter a wide range of bacterial phenotypes, including cell growth, gene expression, antibiotic resistance, and virulence. However, the exact mechanism of how spaceflight alters these behaviors remains largely unknown. The goal of this work was to build upon the current understanding of microbial responses to spaceflight, focusing on the effect of microgravity on community behaviors, such as biofilm formation and quorum sensing. Bacterial biofilms were abundant on the MIR space station, where they were responsible for increasing corrosion and blocking a water purification system. Biofilms continue to be a challenge on the International Space Station (ISS). Health and safety hazards linked to the development of biofilms are also of great concern because decreased immune function has been observed in space travelers. We characterized biofilm formed by *Pseudomonas aeruginosa* PA14 during the STS-132 space shuttle mission in May 2010. We chose to study *P aeruginosa* biofilm formation during spaceflight because it forms biofilms both inside and outside of the human body, and can switch between benign and pathogenic interactions with humans. Further, *P. aeruginosa* caused astronaut Fred Haise to become sick during the Apollo 13 mission. We examined biofilm formation in artificial urine media (AUM) containing either 5 or 50 mM phosphate because water purification and recycling is essential for the ISS and other space travel endeavors. Phosphate has been previously reported to modulate microbial virulence. Direct measurements of biomass showed that *P.aeruginosa* PA14 formed 3 to 5-fold more biofilm during spaceflight compared to normal gravity, regardless of phosphate concentration. This result was confirmed using confocal laser scanning microscopy (CLSM) and image analysis with the COMSTAT software package. In contrast with the biofilm results, we found that AUM with 5 mM phosphate showed increased planktonic biomass during spaceflight, while there was no difference between spaceflight and ground controls for AUM with 50 mM phosphate. We also examined the role of swimming motility in biofilm formation during spaceflight. Motility is an important community behavior and is important for *P. aeruginosa* biofilm formation. Further, a link between motility and microbial response to spaceflight has been postulated in the literature. Unlike wild-type *P. aeruginosa*, two swimming motility mutants, *ΔflgK* and *ΔmotABCD,* did not show an increase in biofilm biomass during spaceflight. These results suggest that swimming motility may play an important role in increasing biofilm biomass during spaceflight. In conclusion, the community-level behavior of *P. aeruginosa* PA14 was significantly modulated during spaceflight. This work has set the stage for our ongoing experiment designed to elucidate the role of microgravity on cell-cell communication and polymicrobial biofilms formation.

(719d) Unstructured Modeling of a Synthetic Microbial Consortium for Consolidated Production of Ethanol

Timothy Hanly, Chemical Engineering, University of Massachusetts, Amherst, MA and Michael H. Henson, Department of Chemical Engineering, University of Massachusetts Amherst, Amherst, MA

The conversion of lignocellulosic biomass to liquid fuels such as ethanol is required for the commercialization of second generation biofuels. Reducing operating costs by combining the saccharification and fermentation steps of this process into one reactor has long been a goal of biofuels research. The typical approach to consolidated bioprocessing, engineering a single, omnipotent microbe to perform both these tasks, can result in undue metabolic burden and bottlenecks. Using a defined mixed culture of specialized microbes is a promising alternative that exploits the native capabilities of each member species. One possible consortium is composed of the filamentous fungus *Trichoderma reesei*, and two fermentative yeasts, *Saccharomyces cerevisiae* and *Pichia stipitis*. *T. reesei* is able produce high concentrations of cellulolytic enzymes. *S. cerevisiae* and *P. stipitis* efficiently produce ethanol from glucose and xylose, the predominant monomeric sugars that result from biomass hydrolysis, and have similar growth requirements as *T. reesei*.

We have developed a simple dynamic model of the proposed consortium with unstructured descriptions of enzyme synthesis, lignocellulose degradation, sugar uptake, cell growth and ethanol production. The batch growth model contained 10 ordinary differential equations with parameters obtained from the literature to the extent possible. The dynamic model equations were solved over a range of initial feedstock concentrations with 2:1 cellulose-hemicellulose ratios to identify the inoculum concentration that maximized the ethanol productivity. The model was further used to examine how the ratio of cellulose to hemicellulose in the feedstock affected ethanol productivity and the optimal inoculum. A sensitivity analysis of model parameters identified several promising targets for the improvement of ethanol production through genetic engineering. The five highest ranked targets were the hydrolysis rate constant, the basal and induced production rates of the *T. reesei* hemicellulase, the saturation constant for *S. cerevisiae* growth on glucose, and the ethanol inhibition constant for *P. stipitis*. Simultaneous implementation of 25% improvements in these parameters was predicted to yield a 15% increase in ethanol productivity compared to the original consortium. A series of batch mixed-culture experiments was performed to validate the predictive power of the model and to identify unmodeled interactions between the three microbes.

See more of this Session: Systems and Synthetic Biology of Interacting Microorganisms
See more of this Group/Topical: Food, Pharmaceutical & Bioengineering Division

(719e) A Multi-Level Programming Framework for Metabolic Modeling of Microbial Communities

Ali R. Zomorrodi and Costas D. Maranas, Chemical Engineering, The Pennsylvania State Univeristy, University Park, PA

Microorganisms in their natural environments rarely live alone but rather function in concert in integrative and interactive units, (i.e., communities). Despite the growing availability of high-throughput sequencing and metagenomic data, we still know very little about the metabolic contributions of individual microbial players within an ecological niche and the extent and directionality of interactions among them. This calls for development of efficient modeling frameworks to shed light on less understood aspects of metabolism in microbial communities. Spurred by recent advances in reconstruction and analysis of metabolic networks of individual microorganisms, researchers have started to reconstruct and analyze *in silico*, metabolic models of simple microbial consortia. Despite these efforts, a comprehensive flux balance analysis modeling procedure to analyze and characterize microbial communities of increasing size with any combination of positive or negative interactions is still lacking. Here, we introduce a comprehensive flux balance analysis framework for microbial communities, which relies on a multi-level optimization formulation to properly describe trade-offs between individual vs. community level fitness criteria. In contrast to earlier approaches that rely on a single objective function, here, we consider a separate biomass maximization problem for each species as inner problems and subsequently integrate them in the outer problem by selecting the alternate optima of the inner problems so as a community-level objective function is optimized. This modeling framework is general enough to capture any type of interactions (positive, negative or combination of both) for any number of species (or guilds) involved. We first analyze a simple and well-determined microbial community involving the syntrophic association between *D. vulgaris* and *M. maripaludis* to demonstrate its ability in recapitulating known interactions. Next, we considered modeling two more complex ecological systems to determine the relative abundance of each microorganism and uncover the direction and extent of inter-species metabolite transfers. This study demonstrates the importance of trade-offs between species- and community-level fitness criteria and lays the foundation for analysis of various types of interactions in multi-species microbial systems using genome-scale metabolic models.

See more of this Session: Systems and Synthetic Biology of Interacting Microorganisms
See more of this Group/Topical: Food, Pharmaceutical & Bioengineering Division

(719f) Computer-Aided Design of a Synthetic Yeast-Bacteria Community

Konstantinos Biliouris[1], David M. Babson[2], Claudia Schmidt-Dannert[3] and Yiannis N. Kaznessis[1], (1) Chemical Engineering and Materials Science, University of Minnesota, Minneapolis, MN, (2) Biotechnology Institute, University of Minnesota, Saint Paul, MN, (3)Dept. Biochemistry, Molecular Biology & Biophysics, University of Minnesota, Minneapolis, MN

The field of synthetic biology has greatly evolved and numerous artificially engineered functions can now be implemented by single cells carrying the appropriate genetic information. However, in order for the cells to perform powerful tasks, co-operation between them may be necessary. To this end, various synthetic systems whose function requires cell-cell communication are being designed. These systems exhibit a wide range of behaviors ranging from heterogeneous yeast communities whose growth depends on one another to bacteria that kill each other, form various patterns and behave as predator-prey ecosystems [1]. Nevertheless, to our knowledge, there is no synthetic system composed of a yeast-bacteria community where bacteria co-operate with yeast to accomplish a task.

Here, we present the *in silico* design of a synthetic yeast-bacteria community. We are designing, modeling and simulating a synthetic community composed of *S. cerevisiae* and *E.coli* cells. To implement this, we are developing a stochastic model that captures all the interactions taking place within and between the eukaryotic and prokaryotic cells. Integration of well-characterized molecular regulatory elements into the genome of those two microbes allows for communication through quorum sensing. The survival of yeast depends on the existence of bacteria and vice versa. Interesting dynamics that stem from the substantial differences underlying eukaryotes and prokaryotes are observed. The conditions under which bacteria can successfully communicate with yeast are also investigated.

This study explores the various behaviors exhibited by the cohabitation of synthetic yeast and bacteria cells. Future directions involve the experimental design of synthetic yeast-bacteria communities based on key information acquired by our model.

References

1) Brenner K, Lingchong Y, Arnold FH, Engineering microbial consortia: a new frontier in synthetic biology, Trends Biotechnol. 2008;66(9):483–489

See more of this Session: Systems and Synthetic Biology of Interacting Microorganisms
See more of this Group/Topical: Food, Pharmaceutical & Bioengineering Division

(719g) A Synthetic Microbial Community Design Based On Syntrophic Metabolite Exchange

Hans C. Bernstein and Ross P. Carlson, Chemical and Biological Engineering, Montana State University, Bozeman, MT

A novel synthetic microbial community was designed and built based on biomimicry of key ecological roles found in stable, naturally occurring microbial consortia. The engineered community paradigm exploits the syntrophic ecological relationship found in fermenter-oxidizer community motifs. The synthetic consortia were composed of metabolically engineered *Escherichia coli* cultures enabling enhanced culture performance as compared to traditional mono-culture techniques. Individual community members were designed for primary-productivity defined here as the ability to assimilate glucose or secondary-heterotrophy which specialized in oxidizing system byproducts. The study demonstrated that synthetic biology enabled division of labor permits simultaneous optimization of multiple tasks which gives rise to new emergent system properties. The benefits of this engineered community system were demonstrated using three relevant culturing systems: batch, chemostat and biofilm where glucose-based biomass productivities increased by ~10, 30 and 50% respectively compared to appropriate monocultures. The synthetic community template was also expanded to consider recombinant bioplastic synthesis in *E. coli*. The increased culture productivity was an emergent property of the consortial interactions and could not be obtained from monocultures. Additionally, biofilm community cultures showed an example of another emergent property by self assembling into strain specific spatial laminations. The study identified a metabolic engineering archetype which can serve as a biocatalyst platform and be easily adapted for many *E. coli* based bioprocesses like bioplastic production.

See more of this Session: Systems and Synthetic Biology of Interacting Microorganisms
See more of this Group/Topical: Food, Pharmaceutical & Bioengineering Division

(722a) Modulation of Wnt/β-Catenin Signaling In Human Embryonic Stem Cells Using 3-D Microwell Array System

Samira M. Azarin, Xiaojun Lian, Elise A. Larson, Juan J. de Pablo and Sean P. Palecek, Department of Chemical and Biological Engineering, University of Wisconsin-Madison, Madison, WI

Though many advances have been made in controlling human embryonic stem cell (hESC) behavior, large-scale stem cell tissue engineering applications still face two main obstacles: spontaneous differentiation during expansion in the undifferentiated state and low yields for many differentiation protocols. We have developed a 3-D microwell array that presents a tunable, homogeneous system for studying the effects of various microenvironmental cues on stem cell processes. Previous work has shown that this microwell system promotes long-term self-renewal of hESCs and can generate uniform embryoid bodies (EBs) with enhanced cardiogenic potential. Elucidating the mechanisms underlying these effects can provide effective targets for controlling self-renewal and differentiation. Here we used the microwell system to evaluate the effects of modulating cell-cell contact on Wnt/β-catenin signaling in hESCs.

Wnt signaling has been implicated in both self-renewal and cardiogenesis of hESCs, and competition for β-catenin between the Wnt pathway and cadherin-mediated cell-cell interactions can impact cell behavior. Flow cytometry results showed that hESCs cultured in 3-D microwells had higher expression of E-cadherin per cell than hESCs grown on 2-D controls. Since β-catenin must be present to stabilize cadherin-mediated contacts, the increase in E-cadherin expression in microwells limited the amount of β-catenin available to translocate to the nucleus and participate in Wnt signaling, as demonstrated by immunofluorescence data showing a lack of nuclear β-catenin localization in hESCs cultured in microwells. These results were confirmed via Western blot analysis of nuclear and cytoplasmic protein extracts from microwells and 2-D controls. Additionally, the lack of nuclear β-catenin in microwells corresponded to a downregulation of Wnt target genes *FST*, *AXIN2*, *CCND1*, *FZD7*, and *SLUG*. Although there was reduced Wnt signaling in microwells, data from a Wnt reporter line showed that EBs formed from microwells contained higher levels of Wnt signaling than EBs from 2-D controls, and gene expression analysis of the Wnt-positive cells within EBs showed an upregulation of genes associated with cardiogenesis as compared to the Wnt-negative cells. These differences in Wnt/β-catenin signaling suggest a potential mechanism by which microwell confinement affects hESC self-renewal and cardiac differentiation that could be translated into development of scalable, efficient processes for tissue engineering applications.

See more of this Session: Tissue Engineering Microenvironment I
See more of this Group/Topical: Food, Pharmaceutical & Bioengineering Division

(722b) Mechanical Properties of the Microenvironment Regulate Epithelial-Mesenchymal Transition

Qike Kyle Chen[1], KangAe Lee[1], Cecillia Lui[1], Derek C. Radisky[2] and Celeste M. Nelson[3], (1) Department of Chemical & Biological Engineering, Princeton University, Princeton, NJ, (2)Mayo Clinic Cancer Center, Jacksonville, FL, (3)Department of Chemical & Biological Engineering and Department of Molecular Biology, Princeton University, Princeton, NJ

Epithelial-mesenchymal transition (EMT) is a phenotypic alteration in which epithelial cells acquire mesenchymal traits. It is increasingly appreciated that EMT, a critical event in embryonic development, contributes to pathological conditions including fibrosis and cancer metastasis. Matrix metalloproteinase-3 (MMP3), a matrix-degrading enzyme, induces EMT in culture and fibroris and tumorigenesis in vivo. The cellular microenvironment plays a critical role in tumor progression. Tissue stiffening is a known risk factor for developing breast cancer. Here we investigated the role of tissue stiffness in MMP3-induced EMT and showed that matrix compliance is a critical determinant of mammary epithelial cell fate in response to MMP3. Soft substrata with compliances comparable to that of a normal mammary gland protect cells from MMP3-induced EMT, whereas stiffer substrata with compliances comparable to that of a breast tumor promote EMT. Matrix stiffness governs EMT response by controlling the subcellular localization of key signaling intermediates. Our results demonstrate how biomechanical properties of the cellular microenvironment regulate EMT and will help identify potential therapeutic targets for MMP-mediated fibrosis and malignancy in vivo.

See more of this Session: Tissue Engineering Microenvironment I
See more of this Group/Topical: Food, Pharmaceutical & Bioengineering Division

(722c) Micrometer Scale Guidance of Mesenchymal Stem Cells to Form Structurally Oriented Cartilage Extracellular Matrix

Chih-Ling Chou, Chemical Engineering, Case Western Reserve University, Cleveland, OH, Alexander L. Rivera, Biomedical Engineering, Case Western Reserve University, Cleveland, OH, Jean F. Welter, Department of Biology, Case Western Reserve University, Cleveland, OH and Harihara Baskaran, Chemical Engineering and Biomedical Engineering, Case Western Reserve University, Cleveland, OH

Tissue engineering has tremendous potential for long-term repair of cartilage lesions, but current tissue engineered cartilage constructs, while similar in biochemical features, have inferior mechanical properties compared to native cartilage. This problem may be due to a lack of an oriented structure in the constructs at the microscale that is present in the native tissue. The goal of this study is to test the hypothesis that microscale features on scaffolds will cause the differentiating mesenchymal stem cells (MSCs) to preferentially arrange themselves and to create a microscale-oriented extracellular matrix similar to the native tissue structure. Another goal of the study is to investigate a method to convert the findings to form larger 3D cartilage constructs. Channels of varying microscale dimensions (25-1000 micrometers) were formed in collagen-based scaffolds via microfabrication. The channel quality was confirmed through image analysis. Human MSCs were seeded in these channels, and Live/Dead staining was used to confirm the viability of cells within the channels. Selective attachment and spreading of MSCs within the channels was ensured by modifying the plateau regions with triblock copolymer F108. The chondrogenic potential of MSCs seeded in these channels was investigated by culturing them in chondrogenic medium for three weeks, then evaluating them by type II collagen immunohistochemistry, and for mechanical behavior. We show selective adhesion of MSCs in microchannels. MSCs aligned along the length of the channel at optimal channel dimensions of 25-100 micrometers. We further show mature oriented type II collagen formation in smaller channels. In addition, we show the mechanical properties (modulus of elasticity and ultimate stress) were significantly improved in scaffolds with MSCs in smaller channels. More importantly, a new method was developed to obtain large 3D constructs that contain these microscale guidance features. Microscale channels of collagen can guide MSCs and lead to improved mechanical biochemical function of the construct. In addition, we demonstrate that large cartilage constructs with such microscale guidance features can be fabricated for in vivo testing. This work paves the way for developing cartilage constructs with ultrastructure similar to native tissue.

See more of this Session: Tissue Engineering Microenvironment I
See more of this Group/Topical: Food, Pharmaceutical & Bioengineering Division

(722d) Biomimetic Polymer Scaffolds for Neural Stem Cell Adhesion & Differentiation

Jocie F. Cherry[1], Aaron L. Carlson[1], Joachim Kohn[2], Melitta Schachner[3] and Prabhas V. Moghe[1], (1) Biomedical Engineering, Rutgers University, Piscataway, NJ, (2)Chemistry and Chemical Biology, Rutgers University and The New Jersey Center for Biomaterials, Piscataway, NJ, (3)Cell Biology and Neuroscience, Rutgers University, Piscataway, NJ

Neurodegenerative diseases and neurotraumatic injuries resultin an irreplaceable cell loss and concomitant deficit in motor and sensoryfunctions. An ideal strategy for repair willallow for cell survival, migration, and integration of exogenous cells withhost tissue. Biomaterials provide a potential strategy to elicit theseresponses by presentation of specific cues to control cellular behavior. While much of current biomaterials research has focused on optimizing themechanical and soluble cues required to direct neural cell behavior, adhesivecues and, more importantly, cell-cell interactions may also play a role. In this study, we investigated therole of differential presentation of fragments of L1, a transmembrane,homophilic binding, neural cell adhesion molecule, as a biointerfacial strategyto direct neuronal cell retention and differentiation of neural stem cells onimplantable scaffolds, while inhibiting the adhesion of non-neuronalphenotypes, such as astrocytes and fibroblasts. This approach ishypothesized to not only promote the survival and outgrowth phenomena ofneuronally developed cells, but also guide the development of neural stem cellsas L1 has been implicated in the normal development of the central nervoussystem.

Two-dimensional films of poly(desaminotyrosyl tyrosine ethylester carbonate) (poly(DTE carbonate) polymers were used as thebiomaterial model for L1 functionalization as these polymers are highlybiocompatible and can be tuned to alter protein adsorption,hydration/biomechanical properties, and degradation behaviors. Variousconfigurations of L1 presentation were compared in terms of relativeconcentrations, orientations, and multivalency of L1 display, including passiveadsorption on cationic, poly-D-lysine-treated substrates or L1-Fc fusedfragment presentation from protein A-coated polymer substrates. ProteinA-based presentation of L1-Fc was hypothesized to result in the outward displayof L1 while resulting in a clustered ligand presentation of L1 and increaseligand efficacy, as well as enhance cell binding affinity and L1-mediatedintracellular signaling.

We report that the configuration of L1 presentation markedaffected neural cell behavior on the biofunctionalized polymersubstrates. When presented through an *in vivo*-likeconfiguration on protein A, L1 significantly increased neurite outgrowth ofspinal cord neurons compared to the traditional presentation of passiveadsorption on poly-D-lysine as early as 24 hours. After 72 hours, whileboth conditions supported the formation of neural networks, protein A-presentedL1 resulted in a denser population of extended neurites. Next, the roleof L1 on active differentiation of human neural stem cells was examined. Neuralstem cells derived from H9 human embryonic stem cells were used for thesestudies. Protein A presented-L1 yielded quantitative enhancement in levels ofneuronal differentiation and outgrowth of human neural stem cells in comparisonto PDL-presented L1 after 7 days in culture. This approach is beingintegrated within three-dimensional scaffolds for the design of neural stemcell-transplantable constructs for the management of spinal cord injury.

See more of this Session: Tissue Engineering Microenvironment I
See more of this Group/Topical: Food, Pharmaceutical & Bioengineering Division

(722e) Endothelial Differentiation of Mesenchymal Stem Cells Using Surface-Bound Cues

Richard J. Galas Jr., Chemical Engineering, Purdue University, West Lafayette, IN and Julie C. Liu, School of Chemical Engineering, Purdue University, West Lafayette, IN

In the emerging field of tissue engineering, there is a growing need for vascularizing tissues *in vitro*. Implants will spontaneously vascularize, but this process is limited to fractions of a micrometer per day. Without a vasculature to overcome diffusion limitations, *in vivo* cell viability is limited to 100-200 microns within the implant. Tissues can be prevascularized *in vitro*, a process that accelerates anastomosis with host vasculature *in vivo*. Endothelial cells can be used as a source for *in vitro* vascularization, but their drawbacks include donor site morbidity and low proliferative capacity. Recently, it has been shown that mesenchymal stem cells (MSCs) can be differentiated into endothelial cells via soluble vascular endothelial growth factor (VEGF). Since MSCs can be differentiated into a variety of cell types, spatial control of differentiation could allow the development of an entire tissue from a single initial cell source.

The objective of this study is to covalently attach VEGF to a surface and evaluate the use of this surface in differentiating MSCs to the endothelial cell fate. The bifunctional, heterogeneous cross-linker, sulfo-SANPAH, was used to attach VEGF to tissue-culture polystyrene. Successful reaction was confirmed by spectrophotometry and ELISA. VEGF surface densities ranging from 10 – 50 ng/ cm^2 were obtained. MSC differentiation was evaluated by immunofluorescence, qPCR, and capillary formation on Matrigel™. This study demonstrated that the VEGF-modified surface resulted in an increase in endothelial markers.

See more of this Session: Tissue Engineering Microenvironment I
See more of this Group/Topical: Food, Pharmaceutical & Bioengineering Division

(722f) Covalent Immobilization of Transforming Growth Factor-β1 (TGF-β1) for Enhanced Vascular Functionality In Vitro Perhaps Through Prolonged Activation of TGF-β1 Pathway

Mao-Shih Liang, Chemical and Biological Engineering, State University of New York at Buffalo, Amherst, NY and Stelios T. Andreadis, Department of Chemical and Biological Engineering, University at Buffalo, the State University of New York, Amherst, NY

In general, growth factor delivery to 3D tissue constructs is limited by diffusion which causes delivery in vivo limited by low efficiency, lack of targeting and side effects. Here we report a novel method to covalently conjugate transforming growth factor-β1 (TGF-β1), which involved in many physiological processes, in fibrin hydrogels. Specifically, TGF-β1 was modified to generate a fusion protein containing the factor XIII recognition sequence (NQEQVSP) and a plasmin cutting site (GYISR). During fibrin polymerization factor XIII links the glutamine (Q) in this sequence to leucine (L303) on α-chain of fibrinogen, thereby conjugating TGF-b1 covalently to the hydrogel. Two peptides NQEQVSP with and without plasmin cleavage sites GYIRSR amino acids sequences were fused to TGF-β1 between the latency associated peptide and mature-TGF-β1, and they were termed LNG-TGF-β1 and LN-TGF-β1 respectively. These fusion proteins were expressed in Chinese Hamster Ovary (CHO) cells and purified by chromatography and they had similar bioactivity to commercially derived TGF-β1 as shown by TGF-b1 dependent promoter (PAI) activity and by inhibition of cell proliferation. In addition, fusion TGF-b1 proteins bound to fibrinogen in a factor XIII dose-dependent manner and this binding prevented TGF-b1 release from fibrin hydrogels. Next we investigated the effects of TGF-b1 immobilization on the function of the vascular wall. To this end, we fabricated vascular constructs with immobilized TGF-b1 using fibrin hydrogels containing smooth muscles cells (SMC) derived from mesenchymal stem cells (MSCs). Interestingly, vascular reactivity in response to receptor and non-receptor mediated pathways was significantly higher in the constructs containing immobilized TGF-b1 as compared to those treated with the soluble growth factor. Also, the elastic modulus of the same constructs significantly increased over time during the 8-week culture period. To further investigate whether these functional effects of immobilized TGF-b1 were due to differences in the signaling response, we measured Smad2 phosphorylation as a function of time in 3D vascular constructs. Interestingly, cells in fibrin gels with conjugated TGF-b1 showed prolonged Smad2 phosphorylation at least up to 96hr, while treatment with soluble TGF-b1 induced only transient phosphorylation for one day. This result suggests that the enhanced contractility and improved mechanical properties of the vascular constructs might be attributed to prolonged activation of the TGF-b1 pathway.

See more of this Session: Tissue Engineering Microenvironment I
See more of this Group/Topical: Food, Pharmaceutical & Bioengineering Division

(722g) The In Vitro Cellular Response to Three-Dimensional Hyaluronic Acid Hydrogels Templated by Self-Assembled Poragen Architecture

Richelle C. Thomas, Chemical Engineering, University of Texas at Austin, Austin, TX and Christine E Schmidt, Biomedical Engineering, University of Texas at Austin, Austin, TX

Natural tissues boast a variety of patterns, porosities, and water contents to yield dynamic systems that regulate and support life enabling processes. When the function of these tissues goes awry, however, it is critical to implement biomedical technologies that are able to mimic the native environment and leverage their own intrinsic diversities in order to achieve sustained maintenance and replenish the native function. The wound healing process coordinates a complicated milieu of proteins, cells and other chemicals for the optimal recovery of the injured or diseased tissue. Thus, it is imperative that wound healing scaffolds combine both the chemical and physical attributes of the extracellular matrix to achieve the most desirable outcome.

Particularly in the area of nerve regeneration, there is a need for long range order to impart contact guidance for bridging large injury gaps. Current methods are not able to bridge large nerve injury gaps with efficacy to restore both motor and sensory function comparable to uninjured nerve. Three-dimensional hydrogel polymer matrices boasting internal alignment are an area of active research within tissue engineering. Hyaluronic acid is a natural polymer intimately involved in the body's wound healing process commonly employed for neural regenerative therapies. These typically amorphous hydrogels, however, do not provide any significant physical attributes beyond the inherent gel porosity and chemical regenerative support. To this end, the goal of this work is to fabricate three-dimensional biopolymer hydrogel matrices that leverage uniquely patterned crystalline networks.

A novel method of imposing an internal architecture through three-dimensional hydrogels uniquely yields both chemical and physical guidance in the injury site without the need for therapeutic additives. A colloidal crystal dissolved in the hydrogel solution either in excess of its thermal or miscibility limit allows for the formation of pores that are the negative imprint of the crystalline pattern. Restricting the hydrogel to the interstitial crystalline space is a novel method of imposing an internal architecture within three-dimensional gels to uniquely yield chemical and physical properties advantageous for a wide variety of applications. The bulk mechanical properties of these hydrogels have been characterized to demonstrate the extent to which the hydrogels are able to be modulated.

We hypothesize that by exploiting the ability to capture the negative imprint of poragen self-assembly within a hydrogel matrix, the resulting bulk material properties will promote preferential neurite extension along the path mapped by this structural template. Long range internal templating of these uniquely patterned liquid crystals is expected to improve the efficacy of regenerated neural processes. The ability to stabilize native, relevant architectures within hyaluronic acid gels yield a material can be optimized for peripheral nerve regeneration, vascular therapies and drug delivery.

See more of this Session: Tissue Engineering Microenvironment I
See more of this Group/Topical: Food, Pharmaceutical & Bioengineering Division

(724a) Modulating Protein Homeostasis to Rescue the Folding of Unstable, Degradation-Prone Secretory Proteins

Fan Wang and Laura Segatori, Chemical and Biomolecular Engineering, Rice University, Houston, TX

Lysosomal Storage Disorders (LSD) are a group ofmore than 50 metabolic diseases caused by deficiency in hydrolytic enzymes andaberrant buildup of metabolites in lysosomes. Gaucher's disease, the mostcommon LSD, is characterized by mutations in the geneencoding for lysosomal glucocerebrosidase (GC). Most mutations are single aminoacid substitutions that do not directly impair the enzyme activity, but rather destabilizeits native folding, leading to ER-associated degradation (ERAD). If mutatedenzyme variants are forced to fold into their native structure, they regaincatalytic activity. We previously reported evidence of this approach by culturingpatient-derived fibroblasts with small molecules that enhance thecellular folding capacity (*Wang, F. ACS Chem Biol. 2011 Feb 18;6 (2):158-68*).We demonstrated that the L-type Ca^{2+} channel blocker lacidipine remodelsprotein homeostasis by influencing the expression of ER chaperones and inducingmodest activation of the unfolded protein response. Specifically, we reportedthat upregulation of the main ER chaperone BiP/ GRP78 and enhanced expression ofGC encoding gene induced by lacidipine treatment play a key role in rescuingthe folding and trafficking of mutated GC (*Wang, F. Chem Biol. 2011 In press*).This suggests that prolonging ER retention facilitate folding ofdegradation-prone enzyme variants in need for higher chaperoning capacity thantheir wild type counterpart. Based on this finding, we attempted an alternativestrategy to engineer the cellular folding capacity based on ERAD inhibition. Weused small molecules that inhibit different steps of the ERAD pathway andconducted a series of mechanistic studies to further elucidate GC misfoldingpathways. The results reported contribute to our fundamental understanding ofcellular folding and provide insights for the development of molecularengineering strategies to rescue the folding of unstable, degradation-pronesubstrates.

See more of this Session: Advances In Protein Expression and Post-Translational Modification
See more of this Group/Topical: Food, Pharmaceutical & Bioengineering Division

(724b) The Role of Protein Motifs In Membrane Protein Trafficking In *S. Cerevisiae*

Zachary T. Britton[1], Chester E. Markwalter[1], Carissa L. Young[1] and **Anne Skaja Robinson**[2], (1) University of Delaware, Newark, DE, (2)Department of Chemical Engineering, University of Delaware, Newark, DE

Membrane proteins make up ~30% of the human genome, and G-protein coupled receptors (GPCRs) are the largest family of human membrane proteins, with ~800 members. In recent years, a number of crystal structures of several members of the rhodopsin-like (Class A) GPCRs have been reported, with one agonist-bound active form reported for the adenosine A_2a receptor. However, the path to obtaining crystal structures remains a major challenge, with roadblocks that include heterologous expression, stabilization in membrane-mimetics, and identification of crystallization conditions. Here we address the first hurdle – heterologous expression – in studying human and rat NK_2 receptor expression in *Saccharomyces cerevisiae.* Although these two receptors are highly similar in sequence, they display different trafficking profiles and expression yields. Using chimeric receptors, we find that cellular localization appears to be driven by a single domain. Specific trafficking motifs were tested by site-directed mutagenesis, and the effectiveness of these changes were analyzed by confocal microscopy, Western analysis, ligand-binding studies, and examination of cellular stress responses.

See more of this Session: Advances In Protein Expression and Post-Translational Modification
See more of this Group/Topical: Food, Pharmaceutical & Bioengineering Division

(724c) Expression Enhancing Introns for Increased Protein Expression In Yeast

Mitchell Tai and Gregory Stephanopoulos, Chemical Engineering, Massachusetts Institute of Technology, Cambridge, MA

In the field of metabolic engineering, strong overexpression of pathway genes is recognized as one of the principle strategies towards increasing flux of desired products. In model organisms such as *E. coli* and *S. cerevisiae*, there are a variety of well-characterized, strong-expression systems available. However, as other organisms become more relevant for industrial applications, there is a growing need for broadly applicable systems which can provide strong expression. Here we examine the use of expression-enhancing introns for strong gene overexpression. The spliceosomal intron sequence from the translation elongation factor (TEF) open reading frame produces a 15-fold increase in expression over the intron-less promoter sequence in the oleaginous yeast organism, *Yarrowia lipolytica*. We use this strong expression platform to increase the expression of endogenous diacylglycerol acyltransferase (DGA) to increase lipid accumulation in *Y. lipolytica*. Increased lipid accumulation can be used for the microbial production of triglycerides (TAG) and poly-unsaturated fatty acids (PUFAs) for biodiesel and nutraceutical applications. The strong expression-enhancement from introns demonstrates an effective platform for gene overexpression which utilizes endogenous machinery and is potentially applicable to a variety of yeast species. Furthermore it helps establish *Y. lipolytica* as an emergent organism for industrial applications in protein production and lipid accumulation.

See more of this Session: Advances In Protein Expression and Post-Translational Modification
See more of this Group/Topical: Food, Pharmaceutical & Bioengineering Division

(724d) Toward Development of a Scalable Cell-Free Protein Synthesis Platform From Yeast Extract

C. Eric Hodgman and Michael C. Jewett, Chemical and Biological Engineering, Northwestern University, Evanston, IL

As a compliment to *in vivo* protein expression platforms, cell-free protein synthesis (CFPS) systems have shown remarkable utility for making protein pharmaceuticals in recent years. The most common CFPS platforms are derived from *Escherichia coli* lysate and wheat germ extract (WGE). While both systems are useful, they do have disadvantages: *E. coli* due to its limited ability to carry out post-translational modifications and WGE in terms of scalability and consistency. To overcome these disadvantages, we are developing a novel eukaryotic CFPS platform from yeast extract. Yeast is a particularly attractive because as a model organism it has been intensely studied and therefore has a wealth of detailed knowledge of its genetics, biochemistry, and physiology. Moreover, yeast has simple, inexpensive, and scalable methods for cultivation under well controlled conditions and is widely used for recombinant DNA protein expression in the pharmaceutical industry. Here we report on efforts to streamline and scalable extract preparation for obtaining highly active CFPS from yeast extracts. We examine the effect of replacing glass bead lysis with high-pressure homogenization, replacing column chromatography with dialysis for desalting, removing nuclease treatment steps, and altering the growth conditions. Our results demonstrate improved CFPS activity and, broadly, set the stage for developing a yeast CFPS platform that provides for inexpensive, productive, and scalable expression of a variety of protein pharmaceuticals.

See more of this Session: Advances In Protein Expression and Post-Translational Modification
See more of this Group/Topical: Food, Pharmaceutical & Bioengineering Division

(724e) Engineered Biosynthesis of Glycoproteins with Eukaryotic N-Glycans In Escherichia Coli

Juan D. Valderrama-Rincon[1], Adam C. Fisher[2], Judith Merritt[2], Yao-Yun Fan[3], Craig Reading[2], Krishan Chhiba[4], Markus Aebi[3] and Matthew P. DeLisa[1], (1)Chemical and Biomolecular Engineering, Cornell University, Ithaca, NY, (2)Glycobia Inc., Ithaca, NY, (3)Department of Biology, Institute of Microbiology, ETH Zrich, Zrich, Switzerland, (4)Department of Neurobiology and Behavior, Cornell University, Ithaca, NY

Asparagine-linked (*N*-linked) glycosylation pathways have recently been discovered in bacteria. However, bacterial *N*-glycans are structurally distinct from their eukaryotic counterparts, rendering bacteria marginally useful for therapeutic applications. Here we report the engineering of a eukaryotic protein glycosylation pathway in the Gram-negative bacterium *Escherichia coli*. This involved the heterologous expression of four eukaryotic proteins, including yeast uridine diphosphate-*N*-acetylglucosamine transferases (Alg13, Alg14) and mannosyltransferases (Alg1, Alg2), and a bacterial oligosaccharyltransferase (PglB). The resulting synthetic pathway enabled a laboratory strain of *E. coli* to produce secretory glycoproteins bearing asparagine-linked eukaryotic trimannosyl core *N*-glycans (mannose$_3$-*N*-acetylglucosamine$_2$). This development opens the door for the production of therapeutic glycoproteins in *E. coli* and the biosynthesis of novel *N*-glycan structures.

See more of this Session: Advances In Protein Expression and Post-Translational Modification
See more of this Group/Topical: Food, Pharmaceutical & Bioengineering Division

(724f) Encapsulating Proteins Into Bacterial Microcompartments

Edward Y. Kim and Danielle Tullman-Ercek, Chemical Engineering, University of California, Berkeley, CA

Bacteria have been favored for the production of biologically derived chemicals due to their rapid growth, robustness to environmental conditions, simple genetics, and capacity to overexpress recombinant proteins. However, problems such as cellular toxity and undesired side reactions prohibit or greatly reduce yields for many products. Targeting the synthesis of these products to bacterial microcompartments can solve these problems by sequestering toxic intermediates and excluding non-relevant enzymes from the compartment. Furthermore, the enhanced kinetics achieved by concentrating substrates and enzymes can improve yields. Here, we show the encapsulation of GFP to the propanediol (PDU) microcompartment by fusion of an N-terminal signal peptide and evaluate the efficiency of encapsulation by flow cytometry. We then investigate the role of charged amino acids in the N-terminal signal peptides for efficient encapsulation. Understanding the encapsulation of heterologous proteins is the first step in targeting multienzyme pathways to bacterial microcompartments to serve as nanobioreactors for the synthesis of bioproducts in bacteria.

See more of this Session: Advances In Protein Expression and Post-Translational Modification
See more of this Group/Topical: Food, Pharmaceutical & Bioengineering Division

(727a) Metabolic Engineering of Bacteria for Sustainable Production of Fatty Acid Derived Products

Brian F. Pfleger, Chemical and Biological Engineering, University of Wisconsin Madison, Madison, WI

Finding a sustainable alternative for today's petrochemical industry is a major challenge facing chemical engineers and society at large. To be sustainable, routes for converting solar energy into organic compounds for use as both fuels and chemical building blocks must be identified, understood, and engineered. Before petroleum, most chemicals were isolated from natural sources or produced using biological conversions (e.g. ethanol from sugar). Advances in synthetic biology and other biological engineering disciplines have expanded the scope of what can be produced in a living organism. In my opinion, synthetic biology is the study of biology from the constructionist perspective. As in other engineering disciplines, synthetic biologists want to apply a general understanding of biology to construct complex systems from well-characterized parts. Once novel synthetic biological systems (e.g. enzymes for biofuel synthesis) are constructed, they must be engineered to function inside living cells without negatively impacting the host's physiology. In most cases first generation systems fail to meet this goal. My group uses systems biology tools to identify metabolic, regulatory, and/or physiological barriers which often can be overcome with metabolic engineering strategies. Here, I will present work to develop strains of bacteria for producing hydrocarbon compounds from sustainable feedstocks. As a biofuel, biologically synthesized hydrocarbons offer many advantages over alcohols including higher energy density, lower water solubility, and compatibility with current engines. Hydrocarbons are also a precursor to most organic chemicals. In this talk, I will describe recent work which identified and characterized enzymes capable of producing alpha-olefins and published work describing engineering efforts to produce fatty acids and fatty acid derived hydrocarbons in *Escherichia coli*. Our work has combined functional genomics analysis, synthetic biology construction techniques, bioinformatics, and metabolic modeling to metabolically engineer superior microorganisms. We have tested our strains in media formulated with biomass-derived sugars and are developing photosynthetic microorganisms to by-pass the biomass middle man.

See more of this Session: Biobased Fuels and Chemicals
See more of this Group/Topical: Food, Pharmaceutical & Bioengineering Division

(727b) Metabolic Engineering for 1-Butanol Production From Glucose or Carbon Dioxide

Ethan I. Lan[1], Claire R. Shen[2], Yasumasa Dekishima[3] and James C. Liao[2], (1)Biomedical Engineering IDP, University of California Los Angeles, Los Angeles, CA, (2)Chemical and Biomolecular Engineering, University of California Los Angeles, Los Angeles, CA, (3)Biotechnology Lab, Mitsubishi Chemical Group Science and Technology Research Center, Inc, Yokohama, Japan

Production of chemicals or fuels from renewable sources is an attractive approach to solving the energy and environmental problems. 1-Butanol, a chemical feedstock and potential fuel, has been produced by fermentation of carbohydrates, both in native *Clostridium* species and various engineered hosts. *Clostridial* or CoA-dependent 1-butanol production pathway utilizes the same chemistry as the fatty acid biosynthesis. Therefore this pathway may be extended further for the production of other even-number longer chain alcohols. We demonstrate that high titer and high yield 1-butanol production from glucose in *E. coli* is enabled by engineering driving forces. Production of 1-butanol directly from CO_2 using genetically modified cyanobacteria will also be discussed. Together, these results demonstrate the feasibility of producing even-number longer chain alcohols from renewable sources.

See more of this Session: Biobased Fuels and Chemicals
See more of this Group/Topical: Food, Pharmaceutical & Bioengineering Division

(727c) Pathway and Host Engineering for Biosynthesis of 3-Hydroxy-γ-Butyrolactone and 3,4-Dihydroxybutyric Acid In Escherichia Coli

Himanshu H. Dhamankar, Collin H. Martin and Kristala L.J. Prather, Chemical Engineering, Massachusetts Institute of Technology, Cambridge, MA

3-hydroxy-γ-butyrolactone (3-HBL), one of the top ten value added chemicals from biomass in a 2004 Department of Energy report, is a versatile chiral building block, frequently used in the pharmaceutical industry for synthesis of blockbuster drugs like Zetia®, cholesterol reducing synthetic statins such as atorvastatin and rosuvastatin, antibiotics such as carbapenems and linzolid (Zyvox®) and the nutritional supplement L-Carnitine. Our group established the first complete biosynthetic pathway towards 3-HBL and its hydrolyzed form, 3,4-dihydroxybutyric acid (DHBA) in *Escherichia coli* using glucose and glycolate as feedstocks. Glycolate itself may be synthesized from glucose in *E. coli,* using the endogenous glyoxylate shunt, eliminating the need to supply it as a separate feedstock. Integration of this endogenous glycolate synthesis pathway with the already established 3-HBL/DHBA pathway has allowed us to successfully demonstrate for the very first time the direct synthesis of 3-HBL and DHBA from glucose as a sole feedstock, opening avenues for synthesis of these valuable chemicals from other biomass derived simple sugars. Enhancing pathway productivity and titers further requires specific host and pathway engineering efforts to reduce the metabolic burden, enhance cofactor availability and to achieve optimal distribution of the supplied glucose between the endogenous glyoxylate shunt for glycolate synthesis and the *de novo* constructed 3-HBL/DHBA pathway. Work on the design and engineering of this integrated pathway and the ongoing efforts to improve pathway productivity will be presented.

See more of this Session: Biobased Fuels and Chemicals
See more of this Group/Topical: Food, Pharmaceutical & Bioengineering Division

(727d) Engineering a Synthetic Fungal-Bacterial Consortium for Cellulosic Isobutanol Production

Jeremy J. Minty[1], Marc E. Singer[1], Alissa Kerner[1], Jungho Ahn[1], James C. Liao[2] and Xiaoxia Nina Lin[1], (1)Department of Chemical Engineering, University of Michigan, Ann Arbor, MI, (2)Deparment of Chemical and Biomedical Engineering, University of California Los Angeles, Los Angeles, CA

Present approaches to microbial cellulosic biofuel production are largely based on the "superbug" paradigm of aiming to incorporate all required functionalities (optimized biofuel production pathways, cellulase production, etc.) into a single organism. Inspired by the widespread occurrence of synergistic microbial communities in nature, we explore a novel alternative approach: the design and construction of a microbial consortium consisting of multiple species which cooperate to directly convert cellulose to biofuel. Three microbial specialists are utilized for the consortium: the cellulolytic fungus *Trichoderma reesei* RUT-C30, which secretes cellulase enzymes to hydrolyze lignocellulose into component hexose and pentose mono/oligosaccharides, an engineered *E. coli* hexose specialist that ferments hexose hydrolysis products, and an engineered *E. coli* pentose specialist that ferments pentose products. The modularity of our system will allow it to be readily adapted to the large portfolio of existing *E. coli* strains metabolically engineered to produce biofuels or commodity chemicals. As proof-of-concept, we demonstrate production of isobutanol from cellulosic feedstocks by implementing the consortium with isobutanol-producing *E. coli* strains. Modeling and experimental studies are used to gain insights into the interplay between ecological interactions of the consortium members and isobutanol production. Our results suggest strategies for optimizing robustness, stability, and isobutanol production by genetically tuning interactions between the consortium members.

See more of this Session: Biobased Fuels and Chemicals
See more of this Group/Topical: Food, Pharmaceutical & Bioengineering Division

(727e) Identification of Gene Over-Expression Targets for the Development of Robust Yeast Strains for Cellulosic Ethanol Production

Pedro V. Pena, Steven Glasker and Friedrich Srienc, Chemical Engineering and Materials Science, and BioTechnology Institute, University of Minnesota, Minneapolis/St. Paul, MN

Bioethanol production from cellulosic biomass is projected to significantly reduce the environmental impact over traditional methods of producing this biofuel. However, this technology is at its infancy, and several challenges, including biomass feedstock processing, need to be addressed to improve efficiency. For instance, acetate is present at growth inhibitory concentrations in biomass hydrolysates. Moreover, inhibition of growth is intensified by the presence of increasing concentrations of the produced bioethanol. Thus, engineering of resistance to growth inhibitors, including product inhibition, is essential to improve bioethanol yields while decreasing the need to detoxify the cellulosic feedstock. Here, we apply the Cytostat cell culture technique, developed in our lab, to screen for acetate and ethanol resistance phenotypes in *Saccharomyces cerevisiae* cells transformed with a genomic library, designed to investigate the effects of gene over-expression from native promoters. The main advantages of this approach are that, first, the Cytostat selects for inhibitor specific resistance because the continuous culture is maintained at very low cell densities, and thus the composition of the growth medium remains constant throughout the experiment. This circumvents the possibility of resistance arising from the metabolic consumption or modification of the inhibitor, such that the resistance phenotype must be established by a sustainable mechanism in response to the constant challenge. Second, identifying the genetic modification only requires the sequencing of a small genomic fragment from the library, not whole genome sequencing. Using this approach, acetic acid resistant transformants, with double the growth rate of the WT strain in minimal medium supplemented with 40 mM acetic acid, were rapidly enriched in a Cytostat from a pool of cells transformed with the library. Similarly, ethanol resistant transformants were isolated in minimal medium supplemented with 2.5% v/v ethanol, with nearly 40% improved growth rates over the WT. The roles of the over-expressed genes in the resistance phenotypes, including the kinetics of glucose consumption and ethanol production, cell size homeostasis, and viability will be presented.

See more of this Session: Biobased Fuels and Chemicals
See more of this Group/Topical: Food, Pharmaceutical & Bioengineering Division

(727f) Combinatorial Optimization of a Highly Efficient Xylose Utilization Pathway In Saccharomyces Cerevisiae

Jing Du, Department of Chemical and Biomolecular Engineering, Institute for Genomic Biology, University of Illinois Urbana Champaign, Urbana, IL, Byoungjin Kim, Department of Chemical and Biomolecular Engineering, Institute for Genomic Biology, University of Illinois at Urbana-Champaign, Urbana, IL and Huimin Zhao, Department of Chemical and Biomolecular Engineering, Department of Chemistry and Biochemistry, Institute for Genomic Biology, University of Illinois at Urbana-Champaign, Urbana, IL

Bioconversion of plant derived lignocellulosic materials into biofuels has drawn increasing attention as an attractive strategy for replacement of fossil fuels. *Saccharomyces cerevisiae*, also known as baker's yeast, is considered one of the most promising organisms for ethanol production from lignocellulosic feedstock. Unfortunately, pentose sugars, which make up to 30% of biomass hydrolysate, cannot be utilized by *S. cerevisiae*. Heterologous pentose utilization pathways from bacterial and fungal resources have been introduced into *S. cerevisiae* to enable the assimilation of pentose sugars. However, pentose utilization of recombinant *S. cerevisiae* strains are inefficient due to the low expression level and activity of heterologous genes, redox imbalance resulted from different preference of cofactor for oxidation and reduction reactions, and suboptimal metabolic flux through different catalytic steps.

A lot of research has been carried out to improve the pentose utilization by *S. cerevisiae* targeting different aspects of these issues, but it is very challenging to come up with a single strategy which can solve all three problems at the same time. Here we report the combinatorial optimization of a highly efficient pentose utilization pathway in *S. cerevisiae*.

Specifically, we chose the fungal three-step xylose utilization pathway as our target pathway, and performed optimization of the pathway through shuffling of pathway enzymes or promoters. For the enzyme-based optimization, 10 to 20 different enzyme homologues from various fungal species with different catalytic efficiency and cofactor preference have been cloned for each pathway enzyme, and were assembled into a library of xylose utilizing pathways using our recently developed DNA assembler method. For the promoter-based optimization, promoter mutants with varied strength were created by error-prone PCR and randomly assembled into the pathway to generate a library of xylose utilizing pathways.

The resultant libraries exhibited a very high efficiency for correct assembly of the complete multi-gene pathway and good diversity of different homologues within the same catalytic step. Using the libraries generated, clones with better combination of enzyme homologues can be selected by faster cell growth. This method can also be applied to directly search for the pathway with a better fit in industrial yeast strains, which may have different metabolic flux patterns compared to the laboratory *S. cerevisiae* strains.

See more of this Session: Biobased Fuels and Chemicals
See more of this Group/Topical: Food, Pharmaceutical & Bioengineering Division

(727g) Role of Lipid Saturation In Modulating the Effects of n-Butanol On Membrane Phase Behavior

Yogi Kurniawan[1], Geoffrey D. Bothun[1], Keerthi P. Venkataramanan[2] and Carmen Scholz[3], (1) Department of Chemical Engineering, University of Rhode Island, Kingston, RI, (2)Biotechnology Science and Engineering Program, University of Alabama in Huntsville, Huntsville, AL, (3)Department of Chemistry, University of Alabama, Huntsville, Huntsville, AL

Butanol has been identified by the US Department of Energy as an alternative fuel and a platform chemical for biorefining. When produced by fermentation, butanol concentrations are often low because of the low butanol tolerance of microorganisms. Low tolerance is attributed in part to butanol fluidization of cell membranes. One mechanism by which microbes adapt to butanol is by changing the ratio of saturated to unsaturated membrane lipids. To date, little is known regarding how the ratio of saturated to unsaturated lipids modulates the effects of butanol fluidization.

In this research we examined the effects of n-butanol on the fluidity and phase behavior of synthetic lipid membranes composed of 1,2-dipalmitoyl-sn-glycero-3-phosphocholine (DPPC) and 1,2-dioleoyl-sn-glycero-3-phosphocholine (DOPC) using fluorescence anisotropy and differential scanning calorimetry (DSC). For this system, temperatures below the melting temperature of DPPC (T_m = 42 °C) yielded coexisting gel DPPC and fluid DOPC phases, while temperatures above T_m yielded a fluid phase. DPPC alone exhibited membrane fluidization at butanol concentrations < 10 g/l and interdigitation at concentrations > 10 g/l butanol. These results, coupled with the large melting hysteresis that was observed between heating and cooling scans due to interdigitation, are consistent with previous studies. With increasing DOPC concentration (up to a 1:3 molar ratio of DPPC/DOPC) the fluidizing effects of n-butanol were enhanced in the coexisting gel DPPC phases. This effect may be attributed to greater partitioning of n-butanol via favorable packing within DOPC phases where the acyl tails are kinked due to the double bonds. The effects of DOPC concentration on DPPC interdigitation are less clear. T_m hysteresis was reduced while the melting enthalpy hysteresis was increased with increasing DOPC concentration. This may reflect differences by which DOPC inhibits or enhances DPPC interdigitation with heating and cooling, or the partitioning behavior of n-butanol between DPPC and DOPC phases.

See more of this Session: Biobased Fuels and Chemicals
See more of this Group/Topical: Food, Pharmaceutical & Bioengineering Division

(727h) Magnetic Separation of Algal for Biofuel Production

Jie Xu[1], **Jeffrey J. Chalmers**[1] and Brad Postier[2], (1)Chemical and Biomolecular Engineering, The Ohio State University, Columbus, OH, (2)Phycal, St. Louis, MO

Phycal is an algal biofuels company focusing on a hybrid photoautotrophic/heterotrophic open pond system using freshwater algal species. To **potentially reduce costs associated with algal biofuels production and/or improve the quality of intermediate steps through improved efficiencies downstream, Phycal is d**eveloping strains that can transport and store significantly more iron than wild-type strains. The potential reward is that they can out-compete contaminating species (*e.g.*, algae, bacteria, protozoa, and *etc.*) for the available iron. Once stored, the iron is in the form of magnetite inside ferritin complexes that are paramagnetic. Thus, in addition to the advantage of iron scavenging, these strains can be manipulated magnetically with the potential of magnetic separation.

In this presentation we will present progress on our work to not only over-express genes involved in iron reduction (*Fre*1), iron transport (*Fea*1), and iron storage (*Fer*1), but also our characterization of both the amount of iron within the cells through the use of mass spectrometry and cell tracking velocimetry, CTV. CTV, measures the magnetophoretic mobility, on a cell by cell basis, by measuring the induced velocity of a cell in a well defined magnetic energy gradient. With this measured magnetophoretic mobility, the magnetic susceptibility of the contained iron can be determined, and the potential of high throughput magnetic cell separation technology to separate the magnetic algae predicted

See more of this Session: Biobased Fuels and Chemicals
See more of this Group/Topical: Food, Pharmaceutical & Bioengineering Division

(728a) Construction of Stabilized Proteolipobead Microenvironments for Intramembrane Protease Analysis and Drug Discovery

M. Lane Gilchrist[1], Kwangwook Ahn[2] and Yueming Li[2], (1)Chemical Engineering, City College and the Graduate Center of the City University of New York, New York, NY, (2)Molecular Pharmacology and Chemistry Program, Memorial Sloan-Kettering Cancer Center, New York, NY

The intramembrane protease gamma-secretase is a current target of therapeutic intervention, with pivotal pathological functions within Alzheimer's disease and cancer. Our primary objective is to expand methods for *in vitro* studies of the intramembrane proteases (IMPs) with the development of a microsphere-supported biomembrane (proteolipobead) platform. IMPs are found throughout all branches of life and their functions are extremely broad. Despite extensive studies, understanding of IMP regulation and catalysis has been hampered by membrane-associated enzymology. Rhomboids and Secretases are polytopic membrane proteases that are widely conserved in all organisms. The precise reaction mechanisms of the intermembrane proteases remain to be elucidated and furthermore, rigorous analysis of the kinetics of interfacial catalysis in these systems has not yet been undertaken.

High throughput of screening of drug candidates has been carried out in bulk assay systems using cell membrane fragments, solubilized enzymes and is underway in proteoliposomes. A critical barrier to further progress in the study and HTPS of gamma-secretase is that such bulk systems do not allow for the direct in situ quantification of enzyme, substrates, or inhibitors or their relative distributions within the structures under assay. Thus, for instance, if a candidate peptide substrate with an altered sequence does not show cleavage activity in bulk assay, this question arises: did the sequence change affect delivery to the cell membrane fragments that contain gamma-secretase or was the peptide truly catalytically inactive? This situation is amplified if the substrate candidate is a larger protein of greater hydrophobicity and more prone to aggregation. Furthermore, due to these factors and the instability of these systems there are limitations in the scope of assays possible using solubilized systems and proteoliposomes in bulk solution. We have expanded in vitro models of gamma-secretase to include proteolipobeads (microsphere-supported biomembranes), enabling: 1) characterization and verification of biomembrane loading of substrates, inhibitors and protein effectors, 2) studies of co-localization, interactions and lateral mobility of membrane-bound assay constituents, 3) the development of flow cytometry-based assays of cleavage, and 4) stability–enhanced microenvironments. Using confocal imaging in tandem with flow cytometry, we will present findings from i) direct measurements of substrate and gamma-secretase loading within supported biomembranes, ii) the localization and quantification of enzyme:substrate and enzyme:inhibitor complexes within supported biomembranes, and iii) comparisons of stability of supported assays to existing systems.

As new drug targets in cancer and infectious disease involved in regulated intramembrane proteolysis

(RIP) based cell signaling are uncovered, it is expected that other IMP enzymes can be subsequently studied and probed with HTPS with this platform.

See more of this Session: Bioimaging and Diagnostics II
See more of this Group/Topical: Food, Pharmaceutical & Bioengineering Division

(728b) Technologies to Determine the Effectiveness of Experimental Drugs: Are Circulating Tumor Cells One Possible Solution?

Jeffrey J. Chalmers[1], Alejandra Garcia Villa[1], Brandon L. Miller[1], Priya Balasubramanian[1], Bhuvaneswari Ramaswamy[2] and Maryam Lustberg[2], (1)Chemical and Biomolecular Engineering, The Ohio State University, Columbus, OH, (2)Division of Medical Oncology, The Ohio State University, Columbus, OH

Triple negative (TN) breast cancer, typically defined as estrogen- and progesterone- receptor negative and Her2 normal or not amplified, represents 10-15% of all breast cancers. This would mean that ~20,000-30,000 women will be diagnosed with TN breast cancer in the United States annually. Although TN cancers represent only a small proportion of all breast cancers it is responsible for a disproportionate number of breast cancer-related deaths. This subtype of breast cancer is unique in several ways: it typically affects younger women and is more prevalent in African-American women, particularly premenopausal AA women and in women who are carriers of BRCA-1 gene mutation. A common theme to these triple negative tumors is their impaired ability to repair DNA damage and their sensitivity to DNA damaging agents such as platinum chemotherapy. Elegant preclinical work has shown the effectiveness of further inhibiting DNA repair using PARP inhibitors in BRCA deficient tumors inducing a state of "synthetic lethality".

Several clinical trials of PARP inhibitors are currently ongoing in breast cancer, and we are using novel imaging and laboratory correlatives to answer a critical question of whether there is a dose- and schedule-dependent response in the tumor, using one of these experimental PARP inhibitors. Specifically, we are using novel imaging study, 3'-[F-18]Fluoro-3'-deoxythymidine Positron Emission Tomography (FLT-PET) scan to directly, yet non-invasively asses DNA damage in the tumor and obtaining circulating tumor cells, CTC, to determine the induction of the histone variant gH2AX as marker for increasing DNA damage. We are also using biomarkers in the tumor such as FANCD2 formation, BRCA1/2 and miR155, which could predict response to PARP inhibitors.

In this presentation, we wish to discuss the use of CTC and whether they can be used to predict the effectiveness of both the commonly used DNA damaging drugs, carboplatin alone or paclitaxel and carboplatin with an experimental, PARP inhibiting drug, ABT888.

See more of this Session: Bioimaging and Diagnostics II
See more of this Group/Topical: Food, Pharmaceutical & Bioengineering Division

(728c) Translating Stem Cells-Molecular Imaging of Stem Cell Transplantation In Porcine Myocardium Using Clinical MRI and PET-CT

Natesh Parashurama[1], Byeong-Cheol Ahn[2], Jaehoon Chung[3], Fumiaki Ikeno[3], Julia Swanson[4], Denis R. Merk[4], Keren Ziv[5], Jennifer Lyons[3], Tomohiko Teramoto[3], Srabani Bhaumik[6], Shahriar Yagoubhi[5], Michael McConnell[3], Raj Dash[3], Phil Yang[3], Robert Robbins[4], Todd Brinton[7], Paul Yock[8] and Sanjiv Sam Gambhir[9], (1)Department of Radiology- Molecular Imaging Program at Stanford, Stanford University, Stanford, CA, (2)Nuclear Medicine, Nuclear Medicine, Kyungpook National University, Daegu, Korea, South, Daegu, South Korea, (3)Cardiovascular Medicine, Stanford University, Stanford, (4)Cardiovascular Surgery, Stanford University, Stanford, (5)Department of Radiology- Molecular Imaging Program at Stanford, Stanford University, Stanford, (6)Life Science-Imaging, GE Global Research Center, Niskayuna, (7)Department of Radiology, Stanford University, Stanford, (8) Cardiovascular Medicine and Bioengineering, Stanford University, (9)Radiology and Molecular Imaging Program, Stanford University Schools of Engineering and Medicine, Stanford, CA

Imaging cells after cardiac cellular therapy could be a promising approach to optimizing this new treatment strategy. Marrying the strengths of cardiovascular MRI (high spatial resolution) and PET (high sensitivity) enhances the ability to assess the cell location, viability, survival, and myocardial response to treatment. To build on our previous swine studies with adenovirally transduced cells, rat mesenchymal stem cells (rMSC) were stably transduced with a second generation, self-inactivating lentivirus carrying a ubiquitin-driven triple fusion (TF) reporter gene (RG), comprised of green fluorescent protein (GFP), firefly luciferase (Fluc) and a mutant truncated herpes simplex virus type 1 thymidine kinase (HSV1-tsr39tk). In vitro studies indicated that sorting lentivirally transduced cells enhanced levels of TFRG levels, by sorting transduced cells that were 10^2 to 10^3 above background fluorescence. Cells were injected in vivo and demonstrated a linear relationship between when implanted subcutaneously between average radiance (photons/s/cm^2 /sr) and cell number (R^2= 0.9682). The highest expressing cells demonstrated a 3.1 fold increase in average radiance (photons/s/cm^2 /sr) from the initial cells after sorting three times and demonstrated a semi-logarithmic relationship, as expected (R^2= 0.9864). The PET reporter gene activity of MSC-TF was assessed of ^{3}H-penciclovir. At 3 hours activity was significantly higher than control C6 Rat glioma cells bearing the HSV-sr39ttk gene. For in vivo cell imaging studies, either all the MSC-TF cells, or 1/5 fraction of the total cells, were labeled with 30 µg/ml Iron bearing SPIO particles (35nm) using a protamine transfection technique, and injected into the hearts of swine after thoracotomy (n=4 swine). MRI was performed using the Fiesta or FGRE sequences allowed spatiotemporal distinction between cells and control particles, and allowed detection of cells after injection of 150-350 x10^6 cells. Next, clinical PET-CT was performed dynamically up to five hours after injection of F18-FHBG (14 mCi), and nine hours after injection of the rMSC-TF.. A linear relationship was determined between the number of cells injected and the SUV (standard uptake value) and signal to background ratio compared to cell number injected (n=3 swine). After 5 hours post injection, Mean SUV (Standard uptake value) and Signal/Background ratio was 0.063 and 2.11 for 150 x 10^6, 0.112 and 2.02 for 200 x 10^6, 0.189 and 2.34 for 250 x 10^6, and 0.227 and 2.82 for 300 x 10^6. Our results indicate that MRI and PET can be married together to localize injected cells for translational cell imaging. Future work will aim to use swine TFRG-MSC in survival models of swine infarction. Studies of PET reporter genes in large animals should help translate stem cells to the clinic and establish the role for molecular imaging in monitoring stem cell therapies.

See more of this Session: Bioimaging and Diagnostics II
See more of this Group/Topical: Food, Pharmaceutical & Bioengineering Division

(728d) Serum Antibody Profiling of Celiac Disease Using Bacterial Display for *In Vitro* Diagnostics

Bradley N. Spatola and Patrick S. Daugherty, Chemical Engineering, University of California, Santa Barbara, Santa Barbara, CA

Celiac disease(CD) is an autoimmune disease of the small intestine affecting geneticallysusceptible individuals following the ingestion of gluten, a protein found inwheat. Although the combination of a serological test with a small intestinalbiopsy is a highly effective method for CD diagnosis, there are still criticalquestions that need to be addressed with regards to early disease detection, identifyingadditional trigger antigens, and discovering ways to distinguish patients thatwill not improve with a gluten-free diet. A new serum antibody profilingstrategy has been optimized to selectively detect a pool of patients'antibodies that test positive for two established serological tests for CD andnot detect antibodies from a control patient set. A group of 18 uniquemimotopes, or peptides that mimic antigenic determinants, were isolated after 5rounds of enrichment by fluorescent activated cell sorting (FACS) of abacterial cell display library of random 15-mer peptides. In order to determinewhich mimotopes had the highest predictive value for CD, the cross-reactivityof these mimotopes was tested on an individual patient basis with the originaldiscovery patient cohort as well as a preliminary training set. A panel of 8clones correctly classified 17 out of 20 patients as Celiac, while onlymisclassifying 3 out of 20 controls. Additional CD patient cohorts will beacquired from the clinics of leading CD investigators and our refined bacterialdisplay antibody profiling strategy will be applied to each set of serumsamples. After a sufficiently large group of candidate mimotopes are obtainedfrom repeated screenings, the bacterial cells will be printed on glass slidesin a novel microarray assay that has been developed in parallel. Bacterial cellmicroarrays will provide a high-throughput method to characterize thesensitivity and specificity of the CD detector during the training andvalidation phases of the study and establish the detector is of clinicalrelevance.

See more of this Session: Bioimaging and Diagnostics II
See more of this Group/Topical: Food, Pharmaceutical & Bioengineering Division

(728e) Rational and Combinatorial Design of Peptide Affinity Ligands for Troponin I

Divya Chandra[1], Donald M. Cropek[2], Scott Banta[3] and Pankaj Karande[1], (1)Chemical and Biological Engineering, Rensselaer Polytechnic Institute, Troy, NY, (2)Construction Engineering Research Laboratory, US Army Engineer Research and Development Center, Champaign, IL, (3)Chemical Engineering, Columbia University, New York, NY

Peptides are promising affinity ligands for detection of proteins in biological samples. We describe a synergistic approach to peptide ligand design based on rational and combinatorial peptide design strategies. Troponin I, a biologically relevant marker of cardiac injury was used as a model protein analyte in this system. Under physiological conditions, endogenous Troponin C and Troponin T bind to Troponin I with very high affinities. Based on the previously published crystal structure of the complexes of Troponin I, T and C, we designed candidate libraries of short linear peptides (15 amino acids in length) for screening against Troponin I. Peptide libraries were synthesized using a high throughput peptide synthesizer, and screened for binding to Troponin I using high density peptide microarrays. We have discovered a wide range of peptides derived from Troponin C and T that bind to Troponin I with varying binding affinities (1 nM – 1 uM). These peptides are excellent candidates for use in biological assays for the quantitative determination of Troponin I from human serum samples. This study elucidates a systematic approach for the design and discovery of peptide affinity ligands for the detection of protein biomarkers of interest.

See more of this Session: Bioimaging and Diagnostics II
See more of this Group/Topical: Food, Pharmaceutical & Bioengineering Division

(728f) Site-Specific Immobilization and Activation of ScFv Antibodies by Poylstyrene-Binding Peptide for Biomarker Analysis

Yoichi Kumada and Michimasa Kishimoto, Department of Biomolecular Engineering, Kyoto institute of technology, Kyoto, Japan

Recently, for biomarker analysis, more attention has beenpaid to the construction of sensitive and economical immunoassay and proteinmicroarray systems that are based on ELISA methods. One of the cruciallyimportant aspects in the development of protein-based bio-assay systems is to immobilizeligand antibody with high density and highantigen-binding activity to the greatest possible extent. Furthermore, the use of recombinantantibody fragments such as single-chain Fv (scFv) antibodies in such ELISA systemis a next subject to increase in density of antibody molecule immobilized anddecrease in production cost of diagnostic agent. In this study, we developedsite-specific immobilization and solid-phase refolding methods for single-chainFv antibodies mediated by the novel polystyrene binding peptides (PS-tags:RIIIRRIRR) that had been originally isolated and optimized in the previousstudies. PS-tag-fused scFvs over-expressed in theinsoluble fraction of *E. coli* cellswere denatured and site-specifically immobilized onto a hydrophilic PS plate.Their antigen-binding activities were efficiently recovered by solid-phaserefolding. This refolding method mediated by PS-tag was applicable to a varietyof scFvs for biomarker analyses such as mouse anti-CRP antibody, anti-CEAantibody, human anti-ED-B antibody and so on. Incomparison with Maxisorp^{TM} immobilizedwith whole monoclonal antibody (Mab), more than 25times higher density of PS-tag-fused scFvs wereattained and consequently, 10 times higher sensitivity for CRP was detected insandwich ELISA. The solid-phase refolding of PS-tag-fused scFvsand antigen detection were successfully demonstrated on the surfaces of SPR andQCM sensor chips that were spin-coated with polystyrene thin layer.

See more of this Session: Bioimaging and Diagnostics II
See more of this Group/Topical: Food, Pharmaceutical & Bioengineering Division

(728g) Time-Resolved Non-Destructive 3D Interferometric Measurements of Tooth, Material and Biofilm Interactions

Michael S. Waters[1], Bin Yang[2], Anthony A. Giuseppetti[2], Gary Schumacher[2], Nancy J. Lin[3] and Sheng Lin-Gibson[3], (1)Polymers Division, Biomaterials Group, National Institute of Standards and Technology, Gaithersburg, MD, (2)ADA, Gaithersburg, MD, (3)NIST, Gaithersburg, MD

Evaluating the efficacy of dental materials to protect human teeth requires the capacity to measure tooth accurately and precisely decay. Current practices for determining tooth decay are destructive, qualitative to lowly quantitative, and/or measure bulk changes that have low to no spatial resolution. The combination of the highly variable nature of tooth enamel and the inability to perform serial analyses on the same spatial location limits the capacity to access reproducible information from any experimental set. This study explores the potential of interferometric optical profilometry to make rapid precision spatial measurements in 3-dimensions (3-D) over time (4-D) of human tooth enamel decay as a complementary measurement to existing techniques. In this study, using bioinert 3-D alignment translation stages, human tooth decay was measured with respect to pathogenic dental bacterial biofilms. These investigations demonstrate the ability to quantitatively determine the rate of tooth decay in previously unseen spatial and temporal scales. These new, rapid, low-cost techniques minimize effort for sample preparation and use very few consumables, opening the feasibility for high-throughput investigations of clinical dental materials to a wider international community.

See more of this Session: Bioimaging and Diagnostics II
See more of this Group/Topical: Food, Pharmaceutical & Bioengineering Division

(729a) Proteome-Based Development of Affinity Motifs and Reduced Contaminant Strains

Ellen M. Brune[1], Tammy M. Rechtin[1], Alicia Kight[2], Mohammad M. Ataai[3], Ralph Henry[2] and Robert R. Beitle[1], (1)Ralph E Martin Department of Chemical Engineering, University of Arkansas, Fayetteville, AR, (2)Department of Biological Sciences, University of Arkansas, Fayetteville, AR, (3) Chemical and Petroleum Engineering, University of Pittsburgh, Pittsburgh, PA

This effort comprises both the development of novel affinity purification tags concurrent with the construction of host cell(s) with an optimized proteome. Protein affinity tags are designed regions (both short peptides and large independent domains) that enhance a chromatographic process by utilizing the region's natural propensity to bind to a specific target immobilized on a solid matrix. Affinity tags designed by our group have the ability to promote binding to multiple stationary phases, using the adsorption and elution characteristics of naturally occurring genomic proteins as guideposts. One such affinity motif has been based on *Escherichia coli* peptidyl prolyl *cis/trans*-isomerase (*slyD*). The *slyD* gene product is a known Immobilized Metal Affinity Chromatography (IMAC) and Ion-Exchange Chromatography (IEC) contaminant that significantly reduces column capacity and complicates gradient design. This genomic protein provided a candidate amino acid sequence, for promoting adsorption during IMAC and IEC, consisting of a 55 amino acid sequence rich in histidine, glutamic acid and aspartic acid. Two affinity tags were developed and were fused to Green Fluorescent Protein (GFP) along with other proteins to analyze expression, solubility of the fusion product and isolation. This presentation will discuss the process leading to the design of the affinity sequences, present the cloning strategy for the GFP construct and the optimized cell line, and compare binding characteristics and solubility in the presence and absence of the affinity tags. Data will be presented and compared other affinity tags (e.g. His6, Arg6 and GST).

See more of this Session: Bioseparations and Downstream Processing
See more of this Group/Topical: Food, Pharmaceutical & Bioengineering Division

(729b) Intein-Mediated One-Step Purification of Escherichia Coli Secreted Human Antibody Fragments

Wan-Yi Wu[1], Keith Miller[2], **Michael J. Coolbaugh**[3] and David W. Wood[3], (1)Chemical and Biological Engineering, Princeton University, Princeton, NJ, (2)Pacific Northwest National Laboratory, Richland, WA, (3)Chemical and Biomolecular Engineering, The Ohio State University, Columbus, OH

The purification of proteins using self-cleaving affinity purification tags based on engineered inteins is an attractive alternative to conventional purification methods, including purification with His-6 tags followed by protease addition to remove the tag. Currently, self-cleaving purification tag technology is generally limited to expression in prokaryotic hosts such as *Escherichia coli.* One of the major limitations of using *E. coli* as an expression host is that it is generally not capable of producing disulfide-bond containing proteins, such as antibodies and antibody fragments, in a soluble form. The current study overcomes this limitation by secreting fusion proteins into the periplasm, where correct disulfide bond formation can occur. The target proteins used in this study were *E. coli* maltose-binding protein (MBP) and β-lactamase, as well as two disulfide-bond containing human antibody fragments. Each target protein of interest was fused downstream from a chitin-binding-domain/intein tag. Secretion of these tag-intein-fusion proteins into the periplasm was accomplished by engineering a PelB secretion leader sequence upstream of the intein-tagged fusion gene. After expression, the fusion proteins were recovered in active form from the periplasm through simple osmotic shock of cell culture. A chitin-agarose affinity resin column purified the full-length tagged fusion protein from the rest of the periplasmic proteome. Induction of intein cleavage with a simple pH shift (from pH 8.5 to 6.5 at room temperature) released the purified target protein from the column. In all cases, the purified proteins were highly active and obtained at high yield in soluble form, without refolding. This work demonstrates the feasibility of coupling the ΔI-CM intein based self-cleaving tags with periplasmic secretion in *E. coli* to greatly increase the diversity and complexity of proteins that can be purified using this system, including proteins with structurally important disulfide linkages.

See more of this Session: Bioseparations and Downstream Processing
See more of this Group/Topical: Food, Pharmaceutical & Bioengineering Division

(729c) Influence of Different Spacer Arms On Mimetic Ligand™ A2P and B14 Membranes for Human IgG Purification

Cristiana Boi[1], Simone Dimartino[1], Stefan Hofer[2], Jeannie Horak[2], Sharon Williams[3], Giulio C. Sarti[1] and Wolfgang Lindner[2], (1)University of Bologna, Bologna, Italy, (2)Department of Analytical Chemistry, University of Vienna, Vienna, Austria, (3)ProMetic Biosciences Ltd, Cambridge, United Kingdom

Microporous membranes are an attractive alternative to circumvent the typical drawbacks associated to bead-based chromatography.

In particular, the present work intends to evaluate different affinity membranes for antibody capture, to be used as an alternative to Protein A resins. To this aim, two Mimetic Ligands™ A2P and B14, were coupled onto different epoxide and azide group activated membrane supports using different spacer arms and immobilization chemistries. These new mimetic membrane materials were investigated by static and by dynamic binding capacity studies, using pure polyclonal human immunoglobulin G (IgG) solutions. Complete chromatographic cycles were also performed using a real cell culture supernatant containing monoclonal IgG_1. The best results were obtained by combining the new B14 ligand with a TRZ-spacer and an improved Epoxy 2 membrane support material. The new B14-TRZ-Epoxy2 membrane adsorbent provided binding capacities of approximately 3.1 mg/mL, besides (i) a good selectivity towards IgG, (ii) high IgG recoveries of above 90%, (iii) a high Pluronic-F68 tolerance and (iv) noB14-ligand leakage under harsh cleaning-in-place conditions (0.6M sodium hydroxide). Furthermore, foreseeable improvements in binding capacity will promote the implementation of membrane adsorbers in antibody manufacturing.

See more of this Session: Bioseparations and Downstream Processing
See more of this Group/Topical: Food, Pharmaceutical & Bioengineering Division

(729d) Preparation and Characterization of Nonwoven Membranes for Bioseparations

Haiyan Liu[1], Yong Zheng[2], Patrick V. Gurgel[2] and Ruben G. Carbonell[1], (1)Department of Chemical and Biomolecular Engineering, North Carolina State University, Raleigh, NC, (2)Department of Chemical and Biomolecular Engineering, ProMetic Life Sciences and NC State University, Mont-Royal, QC, Canada

Nonwoven fabrics are of great interest as potential materials for bioseparations due to their controllable porosity, pore size, relatively high surface area and low cost. These materials might have applications in the removal of low concentration contaminants, such as toxins and pathogens, offering the possibility of producing disposable devices, by-passing the validation and cleaning costs associated with these steps. On the other hand, grafted nonwoven materials might also be applicable for the purification of proteins present at high concentration in the feed stream, by producing grafted materials containing a conformal coating of polymer over the base fibers.

Polybutylene terephthalate (PBT) nonwoven membrane was subjected to photo-initiated graft polymerization to achieve a conformal layer of glycidyl methacrylate (GMA). The grafted material was then modified in two ways: (1) by the conversion of the epoxy groups in the polyGMA layer into amino groups, forming an ion-exchange membrane, or (2) by attaching a diethylene glycol (DEG) spacer arm, followed by the attachment of an affinity chromatography ligand, forming an affinity membrane.

The resulting nonwoven membranes were characterized using ATR-FTIR, SEM or XPS, showing that each reaction step was successful, characterizing the chemical functionality and the fiber morphology before and after modification. In order to demonstrate its ability to resist non-specific protein binding, the hydrophylized grafted material with no reactive groups was tested against several proteins, including bovine serum albumin (BSA) and lysozyme for its ability to resist non-specific binding, and showed a reduction of about 1 log in non-specific binding, when compared to the base material.

The ion exchange membrane was tested using BSA, and showed equilibrium binding capacities and dynamic binding capacities of around 170 mg protein/g membrane and 90 mg protein/g membrane, respectively. These values correspond to 80 and 42 mg protein/ml of membrane bed (in column format). A separation of a mixture of human immunoglobulin G (HIgG) and human serum albumin (HSA) yielded fractions with 93.4% purity in hIgG (as a flow-through fraction) and 99% purity in salt-eluted HSA. The results obtained for the ion-exchange membrane suggest that the nonwoven produced has a high capacity for binding of proteins, and is a promising material for use in bioseparations.

The affinity membrane was tested against a contaminant protein typically present at very low concentrations in feed-streams common to the bioprocessing industry. The modified nonwoven membrane was capable of capturing the low-concentration target protein from a complex animal tissue homogenate processed minimally, while binding significantly less of other proteins non-specifically, when compared to a resin containing a similar ligand. These results show that nonwoven fabrics have the potential to become a viable alternative to resins as a medium for protein purification.

See more of this Session: Bioseparations and Downstream Processing
See more of this Group/Topical: Food, Pharmaceutical & Bioengineering Division

(729e) Affinity Ligands Derived From Hyperthermophilic Protein Scaffolds for Chromatographic Separations

Nimish Gera[1], Andrew B. Hill[1], Dalon P. White[1], Ruben G. Carbonell[2] and **Balaji Rao**[2], (1)Chemical and Biomolecular Engineering, North Carolina State University, Raleigh, NC, (2)Department of Chemical and Biomolecular Engineering, North Carolina State University, Raleigh, NC

Affinity chromatography has great potential to reduce the cost and complexity of protein purification steps in the production of biopharmaceuticals. However, availability of robust affinity ligands for use in industrial chromatography remains a challenge. Ideally, an affinity ligand must have all of the following properties: high binding affinity and specificity for the target protein to be purified, mild elution conditions, ability to withstand repeated sterilization cycles using procedures such as "clean in place" (CIP) that involves a wash step with sodium hydroxide (NaOH) solution) and low cost. Conventional protein ligands for chromatography such as Protein A or antibodies have one or more of the following limitations: high cost, low stability, harsh elution conditions. Small ligands such as peptides can offer milder elution conditions and higher stability relative to protein ligands, in addition to lower cost. However small ligands typically have low affinities for their targets and lack the selectivity required for many applications. Here we show that the Sso7d protein from the hyperthermophilic archaeon *Sulfolobus solfataricus* can be engineered to generate highly stable affinity ligands for a wide spectrum for targets.

To validate the versatility of the Sso7d scaffold, we generated a library of 10^8 Sso7d mutants by randomizing 10 surface exposed amino acid residues on Sso7d, using yeast surface display and subsequently isolated binding proteins for a diverse set of model targets; our chosen targets included a small organic molecule (fluorescein), a 12 amino acid peptide fragment from the C-terminus of beta-catenin, the model proteins hen egg lysozyme and streptavidin, and immunoglobulins from chicken and mouse (1). Mutant proteins based on Sso7d can be expressed at high yields in the *E. coli* cytoplasm. Further, these mutants have high thermal stability, resistant to chemical denaturation by Guanidine Hydrochloride and retain their secondary structure after extended incubation in extreme pH conditions. Taken together, our results show that Sso7d-derived ligands have properties that make them likely to be well-suited for chromatographic separations.

To further explore the use of Sso7d-based ligands in the context of protein purification, we isolated Sso7d-based proteins that bind the Fc portion of human IgG (hFc). These hFc binders could be used to isolate human IgG from complex media. We systematically characterized the binding of the Sso7d-based ligands to various IgG isotypes as well the effect of glycosylation on binding. Further, using a combination of histidine scanning mutagenesis and directed evolution, we isolated Sso7d variants that can elute human IgG under milder conditions. We expect that Sso7d-based hFc binders identified in this study can be used for purification of humanized antibodies as well as human IgG from plasma for use in intravenous immunoglobulin therapy.

(1) Gera, N. et al., Highly Stable Binding Proteins Derived from the Hyperthermophilic Sso7d Scaffold, J. Mol. Biol. (2011) http://dx.doi.org/10.1016/j.jmb.2011.04.020

See more of this Session: Bioseparations and Downstream Processing
See more of this Group/Topical: Food, Pharmaceutical & Bioengineering Division

(729f) Mobile Phase Modifier Effects In Multimodal Chromatography: Fundamental Understanding and Creation of Selectivity Windows

Melissa A. Holstein[1], Siddharth Parimal[1], Ying Hou[1], Alex Freed[1], Scott A. McCallum[2] and Steven M. Cramer[1], (1)Department of Chemical and Biological Engineering, Rensselaer Polytechnic Institute, Troy, NY, (2)Rensselaer Polytechnic Institute, Troy, NY

A multifaceted approach consisting of chromatography, nuclear magnetic resonance, and isothermal titration calorimetry was used to provide fundamental insight into protein adsorption in multimodal chromatographic systems. A wide experimental space was examined using column and high-throughput batch chromatography experiments with a variety of mobile phase modifiers. These experimental techniques were used in conjunction with molecular simulations to understand protein-ligand interactions at the molecular level. The results demonstrate how the appropriate use of mobile phase modifiers and multimodal resins can enable unique separations to be performed that cannot be achieved with traditional methods.

See more of this Session: Bioseparations and Downstream Processing
See more of this Group/Topical: Food, Pharmaceutical & Bioengineering Division

(729g) High-Throughput Continuous Magnetic Sorting of Unlabeled Red Blood Cells (RBC) and Analysis by Cell Tracking Velocimetry (CTV)

Jeffrey J. Chalmers[1], **Jie Xu**[1], Xiaoxia Jin[2], Lee Moore[2] and Maciej Zborowski[3], (1)Chemical and Biomolecular Engineering, The Ohio State University, Columbus, OH, (2)Biomedical Engineering, Cleveland Clinic, Cleveland, OH, (3)Department of Biomedical Engineering, Cleveland Clinic, Cleveland, OH

We have designed and are developing a continuous magnetic sorter that can process up to 10^{11} RBCs in 24 hours from a bioreactor, producing these RBC. This research is part of a much larger project, sponsored by DARPA, to produce a portable, self-renewing supply of blood transfusions for battlefield casualties, i.e. "blood Pharming". These RBCs are produced from hematopoetic stem cells isolated from human placenta; allowed to grow in an undifferentiated state for many generations, and finally induced to produce mature RBC. Since cells grow at different rates, and do not necessarily differentiate all at the same time, mature RBCs must be continually removed from culture.

Since hemoglobin contains iron, mature RBCs are inherently magnetic and no additional labeling is required. Two primary technical challenges confront magnetic sorting of transfusable RBCs: gas exchange and scale-up.

The net electron spin of hemoglobin's heme complex is positive, thereby inducing a magnetic susceptibility, only in the absence of oxygen. Therefore, in order to sort RBCs, degassing membranes are used to quickly and efficiently remove oxygen from the cell suspension. Immature, nonmagnetic progenitor cells are quickly returned to the oxygenated bioreactor.

A continuous separation channel with one inlet and two outlets has been designed to meet the capacity of continuous processing 10^{11} RBCs in 24 hours. Using 2 by 2 four-channel parallel system design, we were able running the separation at concentration at 10^8 cells/ml, and flow rate at 0.1 ml/min. This would give the capacity of up to 10^{11} RBCs in 24 hours with two systems.

Over the last decade we have been developing and using an instrument, referred to as CTV, which allows the velocity of hundreds of cells and particles to be simultaneously determined, using microscopic imaging and computer analysis, in a well defined and nearly constant magnetic energy gradient. By plotting the settling velocity and magnetic velocity at the same time, we are able to further analyze not only the cell population, but further characterize the various fractions that are separated.

See more of this Session: Bioseparations and Downstream Processing

(748a) Low Viscosity Highly Concentrated Dispersions of Stable Protein Nanoclusters for Subcutaneous Injection

Ameya U. Borwankar[1], Maria Andrea Miller[1], Aileen K. Dinin[1], Brian Wilson[1], Mei Yan Lee[1], Tarik A. Khan[1], Thomas M. Truskett[2], Jennifer Maynard[1] and Keith P. Johnston[1], (1)Dept. Chemical Engineering, University of Texas at Austin, Austin, TX, (2)Department of Chemical Engineering and Institute for Theoretical Chemistry, The University of Texas at Austin, Austin, TX

Stabilizing proteins in solution at high concentration is of broad interest in science, technology, and human health for the treatment of a wide range of diseases. Proteins have a tendency to undergo irreversible aggregation, gelation or precipitation at high concentrations due to unfolding caused by specific short ranged forces. Herein we create a translucent dispersion of dense equilibrium nanoclusters of conformationally stable proteins which retain their stability in vitro and in vivo. The nanoclusters are formed by adding trehalose as a crowder which leads to an entropic depletion attraction between the protein monomers. This attraction results in increased inter-protein attraction and hence the trehalose concentration can be used to tune the cluster size. The high protein volume fraction within the cluster stabilizes the protein conformation through a self-crowding mechanism and the primarily repulsive electrostatic interactions between the clusters keep them colloidally stable, preventing gelation. The clusters dissociate into native conformationally stable protein molecules upon dilution *in vitro* and *in vivo* (mice). This tunablility of multi-scale interactions , specifically inter-monomer short range attraction and inter-cluster long range repulsion, to produce nanoclusters is not protein-specific. Therefore this nnaocluster concept can be applied to a wide range of therapeutic proteins without the need to engineer modified proteins through changing the amino acid sequence.

See more of this Session: Particle Engineering as Applied to Pharmaceutical Formulations II
See more of this Group/Topical: Food, Pharmaceutical & Bioengineering Division

(748b) Formation of O/W Emulsions by Static Mixers and Production of Microparticles for Pharmaceutical Applications

Nikolett Kiss[1], Gnter Brenn[2], Hannes Pucher[1], Juliana Wieser[1], Stefan Scheler[3], Herwig Jennewein[3], Daniele Suzzi[1] and Johannes G. Khinast[4], (1)Research Center Pharmaceutical Engineering (RCPE), Graz, Austria, (2)Institute of Fluid Mechanics and Heat Transfer, Graz University of Technology (TUG), Graz, Austria, (3)Sandoz GmbH, Sandoz Development Center Austria, Kundl, Austria, (4)Institute for Process and Particle Engineering, Graz University of Technology, Graz, Austria

Formation of O/W emulsions by static mixers and production of microparticles for pharmaceutical applications

N. Kiss,[a,b] G. Brenn,[a] H. Pucher,[b] J. Wieser,[b] S. Scheler,[c] H. Jennewein,[c] D. Suzzi,[b] J. Khinast[b,d]

[a] Institute of Fluid Mechanics and Heat Transfer, Graz University ofTechnology (TUG), Inffeldgasse 25/F, 8010 Graz,Austria

[b] Research Center Pharmaceutical Engineering, Inffeldgasse 21a, 8010 Graz, Austria

[c] Sandoz GmbH, Sandoz Development CenterAustria, Biochemiestrasse 10, 6250 Kundl, Austria

[d] Institute for Process andParticle Engineering, TUG, Inffeldgasse 21a, 8010 Graz, Austria

Email for correspondence: nikolett.kiss@tugraz.at

The emulsion extraction method is widely used in thepharmaceutical industry for producing controlled release polymer-based microparticles [1]. The encapsulation of the activepharmaceutical ingredient (API) in a Poly(lactic-co-glycolicacid) (PLGA) polymer carrier matrix determines several important pharmaceuticalproperties, including release profile and stability of the API [2]. Oneobjective of the present work is to study the oil-in-water emulsificationprocess, which is the first process step for the production of microparticles for pharmaceutical applications by the emulsionextraction method. Another objective of this study is to characterize theresulting solid microspheres.

The emulsification (e.g., droplet formation) stepdetermines the size and size distribution of the resulting microspheres [3]. Emulsionswere produced using SMX static mixers from Sulzer ChemTech with two different diameters (6 mm and 10 mm). Inthis work [4] we focused on multiphase blending in the laminar regime,reflected by the low Reynolds numbers in our experiments. The materials usedfor preparing the emulsions of the present study were selected based on theirpotential application as precursors for pharmaceutical particle formation.

An organic phase containing the dissolved PLGA and theAPI, and an aqueous surfactant solution were mixed by SMX static mixers. The organicphase drop size spectra in the emulsions were measured by laser diffraction,using the instrument Sympatec HELOS, H2395. Theinstrument provides the volume-based size spectrum and the Sautermean diameter of the oil droplets. The influencing parameters flow velocity, diameterand number of the mixer elements, PLGA polymer concentration in the dispersed phase, oil phase hold-up, and surfactant concentrationwere considered as important for the emulsion formation. They are, therefore,also relevant for a scale-up of the process. In the experiments they were variedin the ranges of industrial interest. The mean oil droplet size in theemulsions was found to decrease with increasing flow velocity, with increasingnumber of mixer elements, with increasing concentration of surfactant and withdecreased mixer element diameter. For the moderately concentrated systemspresented in this work, the dispersed phase hold-up was found to not impact theemulsification process. The mean oil drop size was influenced by the PLGApolymer concentration in the dispersed phase to a moderate extent. Fordeveloping an empirical correlation of the oil droplet size on the basis ofmeasurements, dimensional analysis was applied. This analytical method groupsprocess parameters into a smaller number of non-dimensional groups. Anon-dimensional correlation for the dispersed-phase drop size in the emulsion,in the form of a droplet Ohnesorge number, as afunction of the capillary number and the viscosity ratio of the two liquids isdeduced from the experimental results. The correlation allows the prediction ofthe Sauter mean oil droplet size as a function of thestatic mixer operation parameters and of the liquid properties with very highaccuracy [4], which is an essential piece of information about the emulsionproperties relevant for the industrial application.

The proposed correlation allows us to extract thescaling laws governing the oil drop formation in laminar emulsification by theSMX static mixers used.

Microsphere preparation by solvent extraction basicallyconsists of four major steps: (i) dissolution ordispersion of the bioactive compound, often in an organic solvent containingthe matrix forming material; (ii) emulsification of this organic phase in a secondcontinuous (frequently aqueous) phase immiscible with the first one; (iii) extractionof the solvent from the dispersed phase by the continuous phase transforming thedroplets into solid microspheres; (iv) harvesting and drying of themicrospheres [3]. The second objective of the present work was to study theeffect of the extraction of the solvent from the dispersed phase on the microparticle properties. The particles are formed by solventextraction from the organic droplets in a stirred vessel. For the particleformation step, two process parameters were considered important: the emulsiondroplet size produced by the SMX static mixer and the stirring rate of theanchor impeller in the 5 L tank reactor during the solvent extraction. In modelexperiments these two process parameters were varied to optimize microparticle properties.

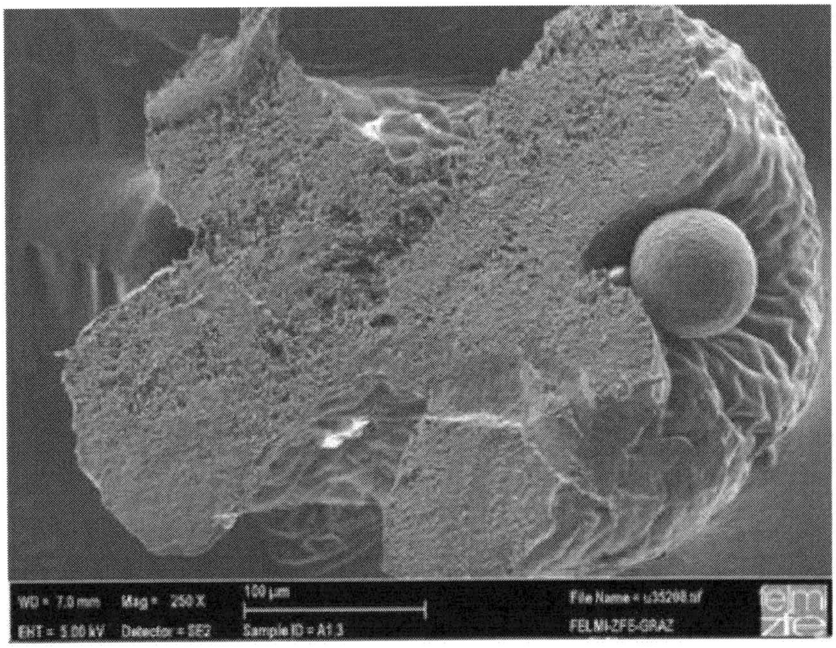

Figure 1. SEM secondaryelectron image of a microparticle cross section (micrographby FELMI Graz)

The size of the final solidmicrospheres was measured by laser diffraction. The morphology and composition (e.g.,API distribution in the microparticles) were analyzedby electron microscopy (see Figure 1). Determiningthe porosity, skeletal volume and density of the microparticlesby gas pycnometry should quantify the properties of the pore structure of the microspheresand help to understand the role of all the process parameters in the formationof the particles.

References

1. Mansour, H. M., Sohn, M., Al-Ghananeem, A., DeLuca, P. P., 2010. Materials for Pharmaceutical DosageForms: Molecular Pharmaceutics and Controlled Release Drug Delivery Aspects.International Journal of Molecular Sciences 11, 3298-3322.

2. Mora-Huertas, C.E., Fessi, H., Elaissari, A., 2010. Polymer-based nanocapsulesfor drug delivery. International Journal of Pharmaceutics 385, 113-142.

3. Freitas, S., Merkle, H., Gander, B., 2005. Microencapsulation by solvent extraction/evaporation: reviewing thestate of the art of microsphere preparation process technology. *Journal of controlled Release 102, 313-332.*

4. Kiss, N., Brenn, G., Pucher, H.,Wieser, J., Scheler, S., Jennewein, H., Suzzi, D., Khinast, J., 2011. Formation of O/W emulsions by static mixers for pharmaceuticalapplications. Submitted to Chemical Engineering Science

See more of this Session: Particle Engineering as Applied to Pharmaceutical Formulations II
See more of this Group/Topical: Food, Pharmaceutical & Bioengineering Division

(748c) Design and Analysis of a Modified Electrospray Process for Fabricating Polymeric Nano- and Microparticles for Drug Encapsulation

Alireza Rezvanpour, Chemical and Biomolecular Engineering, National University of Singapore, Singapore, Singapore, **William B. Krantz**, Chemical & Biological Engineering, University of Colorado, Boulder, CO and Chi-Hwa Wang, Department of Chemical and Biomolecular Engineering, National University of Singapore, Singapore, Singapore

Electrohydrodynamic Atomization (EHDA), also referred to as the electrospray process, has been recently developed to fabricate polymeric micro- or nano-particles as well as fibers for drug encapsulation. EHDA in the cone-jet mode can produce highly monodisperse aerosols. As a result of strong electric forces, liquid droplets can be created by the electrospray process and converted into solid particles. In this study EHDA was carried out in a glass encapsulation chamber in which solid pharmaceutical particles were fabricated after evaporating the solvent from sprayed droplets. A high voltage nozzle with a concentric ring and a grounded collection plate were enclosed in an EHDA encapsulation chamber. A modified design for the encapsulation chamber was used in this study in which the particles are collected inside the chamber. Three major improvements were made in this modified design relative to prior art. First, the modified chamber is completely cylindrical whereas prior designs had two conical ends. Second, the particles fabricated in this modified chamber are collected inside the vessel far away from the spray zone. Finally, the nitrogen inlet and outlet in the modified chamber have larger diameters than earlier designs. Less reverse flow, higher particle collection efficiency and a lower amount of residual organic solvent in the collected particles are the main advantages of the aforementioned modifications.

The effects of different parameters on the particle collection efficiency were experimentally investigated. Important factors affecting the particle collection efficiency were the flow rate of the liquid solution used to create the particles, the nitrogen flow rate and the nozzle voltage. Systematic scaling analysis of the describing equations for the electrospray process permitted collapsing all the collection efficiency data onto a universal plot as a function of only one dimensionless group that incorporates all the process parameters. This plot can be used to determine the effect on the collection efficiency of changing the liquid solution flow rate, nitrogen flow rate and nozzle voltage as well as geometrical parameters such as length and diameter of the chamber, and location of the collection plate.

See more of this Session: Particle Engineering as Applied to Pharmaceutical Formulations II
See more of this Group/Topical: Food, Pharmaceutical & Bioengineering Division

(748d) Scale-up of Polyanhydride Particle Synthesis Methods and Their Effect On Drug Release Kinetics

Timothy Brenza, Chemical and Biological Engineering, Iowa State University, Ames, IA and Balaji Narasimhan, Department of Chemical and Biological Engineering, Iowa State University, Ames, IA

Polyanhydrides are biodegradable materials which have been shown to provide sustaineddelivery of their encapsulated payloads. Copolymers based upon sebacic acid (SA), 1,6-bis-(*p*-carboxyphenoxy) hexane(CPH), and 1,8-bis-(*p*-carboxyphenoxy)-3,6-dioxaocatane (CPTEG) have previouslybeen synthesized into therapeutic loaded particles through solventprecipitation in our lab. Techniques have been established for the productionof polyanhydride particles that are polydisperse on both the micron and nanoscale. However, the solvent precipitation method is a batch operation which isnot easily scalable. Therefore, in thiswork we evaluate the synthesis of polyanhydrideparticles through solvent evaporation via spray drying, a more easily scalableprocess. The operating parameters of the spray dryer were optimized withrespect to particle yield. Model drugs were encapsulated into the polyanhydride particles to determine encapsulationefficiency and for studies on release kinetics. The resulting polymer particleswere also compared to the particles synthesized through a solvent precipitationmethod for physical particle characteristics (size, surface characteristics,shape and porosity). From these results we show that the method of particlesynthesis plays an important role in the performance of drug loaded polyanhydride particles.

See more of this Session: Particle Engineering as Applied to Pharmaceutical Formulations II
See more of this Group/Topical: Food, Pharmaceutical & Bioengineering Division

(748e) Oil-In-Water Emulsion Based Fabrication of Biodegradable Prolate Spheroids (rods) for Drug Delivery

Michael J. Heslinga and Omolola Eniola-Adefeso, Chemical Engineering, University of Michigan, Ann Arbor, MI

p.MsoNormal, li.MsoNormal, div.MsoNormal { margin: 0in 0in 0.0001pt; font-size: 10pt; ""; }div. Section1 { page: Section1; }

Biodegradable polymeric nanospheres have been widely suggested as injectable drug delivery vehicles as they are more likely than larger particles to clear the microcirculation. Nevertheless, recent publications suggest that nano-sized particles exhibit minimal localization and adhesion to the blood vessel wall from bulk blood flow typical in large vessels. Non-spherical particles may offer a higher likelihood of finding and binding to the blood vessel wall than spherical particles of the same volume due to the hydrodynamics of blood flow, leading to improved efficiency in the delivery of therapeutics to the vasculature.

We previously showed that biodegradable poly(lactic-co-glycolic acid) (PLGA) prolate spheroids can be fabricated via oil-in-water (O/W) and water-in-oil-in-water (W/O/W) emulsion solvent evaporation. We now demonstrate the flexibility of these fabrication methods to multiple biodegradable polymers and therapeutics. In addition to particles made from PLGA, we fabricated prolate spheroid rods from poly(L-lactide-co-epsilon-caprolactone) (PLC) polymer. Additionally, we now demonstrate that these techniques can encapsulate a wide range of therapeutics and imaging agents. Encapsulated hydrophobic drugs include paclitaxel, lovastatin, and TNP-470, an analogue of fumagillin, while bovine serum albumin (BSA) was loaded as a hydrophilic drug. For imaging, we demonstrate the capacity to fabricate spheroids loaded with fluorescein dye and iron oxide nanoparticles. We find that the release profile of loaded particles is geometry dependent by comparison of paclitaxel release from loaded spheroids and spheres of equivalent volumes at 37°C.

See more of this Session: Particle Engineering as Applied to Pharmaceutical Formulations II
See more of this Group/Topical: Food, Pharmaceutical & Bioengineering Division

(748f) Increasing Oral Bioavailability of SR 13668 by Using Polymeric Nanoparticles

Hao Shen[1], Aryamitra A. Banerjee[2], Alexander V. Lyubimov[2], Izet M. Kapetanovic[3], Yoon Yeo[4] and **Ying Liu**[5], (1)Department of Chemical Engineering, University of Illinois at Chicago, Chicago, IL, (2) Department of Pharmacology, University of Illinois at Chicago, Chicago, IL, (3)National Cancer Institute, Bethesda, MD, (4)Department of Industrial and Physical Pharmacy and Weldon School of Biomedical Engineering, Purdue University, West Lafayette, IN, (5)Department of Chemical Engineering and Biopharmaceutical Sciences, University of Illinois at Chicago, Chicago, IL

Increasing Oral Bioavailability of SR 13668 by Using Polymeric Nanoparticles

Hao Shen[1], Aryamitra A. Banerjee[2], Alexander V Lyubimov[2,*], Izet M. Kapetanovic[3], Yoon Yeo[4,5] and Ying Liu[1,6,*]

[1] Department of Chemical Engineering, University of Illinois at Chicago, Chicago, IL 60607

[2] Department of Pharmacology, University of Illinois at Chicago, Chicago, IL 60612

[3]National Cancer Institute, Bethesda, MD 20892

[4] Department of Industrial and Physical Pharmacy, Purdue University, West Lafayette, IN 47907

[5] Weldon School of Biomedical Engineering, Purdue University, West Lafayette, IN 47907

[6] Department of Biopharmaceutical Sciences, University of Illinois at Chicago, Chicago, IL 60607

* To whom correspondence should be addressed.

SR13668 (2,10-dicarbethoxy-6-methoxy-5,7-dihydro-indolo-(2,3-b)carbazole), is a new candidate cancer chemopreventive agent based on a naturally occurring anti-cancer agent indole-3-carbinol (I3C), which was found in cruciferous vegetables and had promising anti-cancer activity both in vitro and in vivo models. SR13668 has a safe preclinical toxicology profile (genotoxicity battery, acute and subchronic rat and dog studies). However, SR13668 showed a very poor oral bioavailability due to its limited aqueous solubility and permeability.

In this work, we have encapsulated the cancer chemopreventive drug, SR13668, into polymeric nanoparticles to increase its bioavailability, by using Flash NanoPrecipitation (FNP), a scalable process to prepare nanoparticles with controlled size distribution and a high drug loading rate. The effects of size of each block of the copolymer, ratio of the polymers to drug, supersaturation values of the polymers and the drug, presence of co-precipitator, and type of polymer on the size distribution of the nanoparticles and therapeutic loading is demonstrated. Drug loading rate up to 33% can be achieved by using FNP, which is more than twenty times higher compared to the traditional nanoprecipitation. Poly (ethylene oxide)-b-poly(epsilon-caprolactone) (PEG-b-PCL) and Poly(D,L-lactide-co-glycolide) (PLGA) were used to encapsulate SR13668. To increase the long-term stability of nanoparticles and to avoid Ostwald ripening and particle growth, nanoparticle suspensions were either freeze dried or spray dried to dry powders. Methods to redisperse the nanoparticles have been developed.

Pharmacokinetic (PK) analysis was performed in mice and dog models. High performance liquid chromatography (HPLC) - mass spectrometry (MS) was used to measure the SR13668 concentration in both whole blood and blood plasma. By comparing to SR13668 in methylcellulose formulation, polymeric nanoparticles encapsulating SR13668 dramatically enhanced the bioavailability of the drug.

See more of this Session: Particle Engineering as Applied to Pharmaceutical Formulations II
See more of this Group/Topical: Food, Pharmaceutical & Bioengineering Division

(752a) Quality by Design for DPIs -A Regulatory Perspective

Prasad Peri, DHHS/FDA/Cder/OPS/ONDQA/Dndqa I, FDA, Silver Spring, MD

Inhalation Powders have become an important dosage form in developing products for delivery to the lungs. Since the approval of the first DPI in the US in the late 1990's there has been a rapid surge in the number of drug development programs involving Dry Powder Inhalers. The draft MDI/DPI guidance, developed more than 10 years ago, discuss approaches for CMC information to be submitted in an application. An update to this guidance is being developed and the proposed regulatory recommendations for DPIs will suggest a Quality by Design (QbD) approach whereby a science and risk based understanding of the device design, its effect on the performance, and the patient interaction is sought. Specifically, the impact of interaction of the critical quality attributes of the actives, excipients, device components, on the Quality Target Product Profile (QTPP) of the finished product (e.g. delivered dose, aerodynamic performance of the drug product) is an example of enhanced development information that can be included in applications. These recommendations are in line with the Agency's continued efforts to promote the implementation of QbD principles in drug development.

See more of this Session: QbD In Dry Powder Inhaler Development : Multifaceted Powder Characterization for Product Development
See more of this Group/Topical: Topical I: Comprehensive Quality by Design in Pharmaceutical Development and Manufacture

(752b) QbD In Dry Powder Inhaler Development

David Morton, Drug Delivery, Disposition and Dynamics, Monash Institute of Pharmaceutical Sciences, Monash University, Melbourne, Melbourne, Australia

Summary

Developing products comprising un-bound fine cohesive powders throws up significant challenges. Dry powder inhalers present a prime example. Why? Because the behaviour of fine cohesive particle behaviour is notoriously complex, and notably more so than any single phase solid, liquid or gaseous material (1): where powders can exhibit the behaviour of solids, liquids or gases- and sometimes all three. Furthermore, the complexity is compounded by the need to match the nature of such a complex powder, with a small dispersion device with aerodynamically complex air flow channels intended to create some form of dispersion energy, as well as the reliance on bringing the function together by the complex and variable inspiration pattern of the patient as they inconsistently inhale through these air channels to actuate the system (2).

A few have ventured to computationally model and even develop measurement or theoretically based strategies to try to predict performance of a powder within a given device. It is proposed that none can claim success, in a way that captures effective predictive capability. Product development in this area consequently remains largely based on empirical controls of the formulation, and consequently QbD implementation in its true sense remains out of reach. Successful current development is therefore largely based on parametric control of both input materials and processes, but where the influence of parameters are not fully known, they are selected without a fundamental scientific basis, and do not yield a true design space, nor any predictive capacity. Furthermore, it is also proposed here that outcomes of the measurement of basic surface properties of real (as opposed to model and ideal) materials is either too simplistic, based on unrealistic "isolated" measurements or even beyond current scope of practical technology- for example, single particle adhesion force-contacts, electrostatic charge measurement, surface rugosity, and probing molecular surface structure cannot be either usefully or effectively measured for real pharmaceutical powders. Hence the subtitle of this paper- is a provocative challenge that past research has suffered by too great a focus much on the measurement, rather than as a problem-centred approach.

In this paper, a number of case studies will be used to illustrate apparent contradictions and failures to develop even a simple predictive capability from such measurements.

However, we have on the market highly successful and effective DPI products, and progress in understanding is ongoing. Some semi-empirical approaches have also had qualified success, such as the Cohesive-Adhesive Balance approach (CAB), despite its flaws (3). This paper will then focus on some more recent problem-driven and bulk scale approaches to make more pragmatic and hopefully effective powder property measurements and models.

References

1. Igwe GJI, 1991. “Powder technology and multiphase systems: Gas permeametry and surface area measurement”. New York: Ellis Horwood Ltd, Preface.

2. Morton, DAV, Staniforth, JN, 2005. “The Challenge of the New: Device-Formulation Matching in Dry Powder Inhaler Systems.” Pharm. Manuf. Packing Sourcer, Spring, pp. 80–83.

3. Begat P, Morton DAV, Staniforth JN, Price R, 2004. “The cohesive-adhesive balances in dry powder inhaler formulations. I: Direct quantification by atomic force microscopy.” Pharm Res 21:1591–1597.

See more of this Session: QbD In Dry Powder Inhaler Development : Multifaceted Powder Characterization for Product Development
See more of this Group/Topical: Topical I: Comprehensive Quality by Design in Pharmaceutical Development and Manufacture

(752c) The Influence of Crystals, Particles and Powders On the Performance of Dry Powder Inhalers

Jag Shur, School of Pharmacy & Pharmacology,, University of Bath, Bath, Avon, BA2 7AY, United Kingdom and Robert Price, University of Bath, Bath, Avon, BA2 7AY, United Kingdom

Summary

Dry powder inhalers (DPIs) are an attractive means of delivering drugs to the lung. The fluidization, entrainment and de-aggregation of a DPI formulation is a complex relationship between the patient's inspiratory effort, the air flow field generated within a device and the cohesive properties of the powder bed.

The inter-relationship between factors including particle size, particle shape, distribution of interparticulate and mechanical forces govern the performance and efficacy of DPI drug delivery systems. However, quantitative measurements of these properties and their under-lying relationships remain difficult to discern.

This presentation will focus on some key developments in understanding the relationships between the governing properties of DPIs and how these properties can be harnessed to develop designer formulation systems for DPIs.

INTRODUCTION

The ability to modify the flowability, deaggregation and dispersion of particles in a well-defined particle size range is a key requirement for the delivery of therapies to the lungs using dry powder inhalers (DPIs). The de-aggregation and dispersion of particles from DPIs is controlled by interfacial forces, and therefore governs the performance and efficacy of medicaments delivered to the lung by DPIs.

Several physicochemical factors can directly influence these properties, amongst which the mesoscopic surface structure of particulates plays a dominant role. However, quantitative measurements of these effects are difficult to discern, particularly as the size of particles approaches the micrometer dimensions. However, quantitative colloid-probe atomic force microscopy (AFM) measurements of the spatial distribution of the interactive forces, over and upon contact with the substrate surface have allowed changes in solid-state properties to be directly related to particle adhesion. Furthermore, this approach has enabled the ability to relate particle adhesion to the performance of DPI drug delivery systems.

Another critical component of delivery of dry powders to the lungs is related to the powder fluidisation process within the delivery device. The process of fluidization is related to many factors such as particle size, particle shape, bulk density and powder flowability. The influence of each of the components on the fluidization process is complex, however the development of a fluidisation energy measurement has provided a significant insight into specific relationship between powder flow, cohesive properties of the powder and DPI performance.

The presentation will specifically focus on the current limitations in the processing of sub-10mm particulates for DPIs, highlighting the direct influence of particle shape, surface morphology and interfacial properties on particle interactions. In addition, key developments in particle engineering techniques in modifying the particle architecture and surface geometry will be explored. The presentation will also focus on how rheometric properties of granular materials may be a critical tool in the development of powder-based formulations for inhaled drug delivery.

See more of this Session: QbD In Dry Powder Inhaler Development : Multifaceted Powder Characterization for Product Development
See more of this Group/Topical: Topical I: Comprehensive Quality by Design in Pharmaceutical Development and Manufacture

(752d) A Dry Powder Inhaler, a User Interface or a Formulation Interface? Designing for Quality Amidst Competing Requirements

James Tibbatts, Concept Flow Ltd, Cambridge, United Kingdom

A dry powder inhaler is the physical interface that sits between the patient and the powder formulation. Faced with the device, the patient must go through a number of steps to prime and prepare the dose, before inhaling on the device. During the inhalation, the patient's inspiratory energy is used to break up the powder and to deliver the powder to the lungs. Clearly, both the inner workings of the inhaler and the manner in which that inhaler is used have a significant and direct influence on the clinical efficacy of the product. However, this raises two competing design requirements which are often dealt with as separate design activities.

Laboratory testing shows that, when correctly used, the performance variation between different inhalers, and at different flow rates through the same inhaler, is significant. When user error and compliance issues are considered on top of that inherent variability, the need for quality design is abundantly clear. However, true QbD to mitigate this variability is perhaps an unrealistic proposition. Despite this, there remain a wide range of poorly understood factors which are central to the DPI performance where improved understanding and characterisation can bring us a step closer to QbD.

The physics that determines performance is multifaceted and is influenced by turbulence, shear, impaction, entrainment, deaggregation and deposition. Compounding this complexity is a lack of available methods for direct characterisation of these factors. As a result, the DPI airpath is often developed as a consequence of, and not in concert with, usability and mechanical design requirements. This approach effectively prioritises the aspects of the design that can be more readily quantified over and above the aspects of the design that are poorly understood. Instead of improving quality, this design paradigm simply hides the unknowns.

This paper will focus the characterisation of the airflow within the device and on the characterisation of the powder behaviour as a result of that airflow. The analysis will focus on new approaches for both in-silico and in-vitro data analysis that can help to unpick the causal relationships that determine DPI performance. The data also highlights the detrimental influence that competing design requirements of usability and performance can have on the performance and consistency of the product. The presented methods provide a set of tools that can be used to improve the quality of DPI design, and that can be used provide a central focus for disparate design teams.

(754a) Application of a Joint Risk Assessment Platform to Identify Suitable Drug Substance and Drug Product Process Options

Joshua D. Engstrom[1], Jonathan Brown[1], Olav Lyngberg[2], Lindsay Hobson[3] and San Kiang[1], (1)Drug Product Science and Technology, Bristol-Myers Squibb Company, New Brunswick, NJ, (2)Early Phase Chemical Development, Bristol-Myers Squibb Company, New Brunswick, NJ, (3)Late Phase Chemical Development, Bristol-Myers Squibb Company, New Brunswick, NJ

The final selection of a robust drug substance process in combination with a drug product process can be extremely challenging, especially in cases in which the crystal form and powder properties of the active pharmaceutical ingredient (API) directly impact the development of a robust and economic drug product process. For such cases, it is necessary to assess and manage the overall risk of a given drug substance (API) + drug product combination instead of analyzing drug substance and drug product risks independently. In this presentation a case study is described in which a joint risk assessment platform was organized to rank combinations of various API crystal form inputs with drug product processes. The joint risk assessment was executed in the following phases: planning, data collection for each API + drug product option, data review and scoring, and final API + drug product ranking and review. Data collected from lab feasibility studies on each option were utilized to help more accurately score the perceived processing risks in each unit operation. The risks were scored using a failure modes and effects analysis (FMEA) quality risk assessment tool. Through the joint risk assessment approach, appropriate API + drug product options were identified to help the drug substance and drug product teams effectively direct resources toward processes that have the highest chance of successfully delivering a commercially manufactured drug product.

See more of this Session: Scale-up of Pharmaceutical Manufacturing Processes: Toward a QbD Approach
See more of this Group/Topical: Topical I: Comprehensive Quality by Design in Pharmaceutical Development and Manufacture

(754b) Improved Process Understanding Through Implementation of QbD Methodologies

Shane T. Grosser[1], Aaron Moment[2], Angela Spartalis[2], George Zhou[3] and Paul Fernandez[2], (1) Chemical Process Development and Commercialization, Merck & Co., Inc., Rahway, NJ, (2)Chemical Process Development and Commerciliazation, Merck & Co., Inc., Rahway, NJ, (3)Analytical Development and Commercialization and Development - Specialty, Merck & Co., Inc., Rahway, NJ

Purpose: The work presented here was conducted to develop a process for the pure step of an ongoing developmental program that simultaneously provides operational flexibility and sufficient process robustness inline with QbD filing strategies. Particular challenges of this process are the strong dependence of final API PSD on seed temperature and the significant impact that solvent composition has on that temperature. Additional steps have been investigated to limit the variability of residual solvents going into the pure step and therefore simplify the seed point determination. Unfortunately these steps have both yield and time-cycle implications and could unnecessarily impact productivity for batches when residual solvent levels may already be low. As a result, the project team has decided to develop a mechanism by which the optimal seed temperature is determined in real-time based upon batch composition assays, therefore enabling increased flexibility in solvent composition without comprising batch quality.

Methods: This work will demonstrate the application of risk assessment strategies for targeted experimental planning and implementation of statistically designed experiments (DoE) for investigation of interacting process parameters. We will also discuss the use of pilot-scale confirmatory runs of DoE results and how large scale experimentation was used to drive further DoE investigation.

We will present the use of PAT technologies for the rapid collection of solubility data to aid in mapping API solubility as a function of solvent composition for a 4-solvent system.

Lastly, we will discuss the use of thermodynamic and empirical modeling tools for the translation of the acquired solubility map into operational parameter ranges and assay specifications to be implemented in the full-scale demonstration and subsequent filing.

Results: Guided by a formal risk assessment, significant DoE experimentation was performed aimed at understanding the effect that 9 operational parameters have upon the API PSD, purity and form. None of the investigated parameters showed evidence of a purity or form impact while PSD was primarily impacted by seed temperature. Although the PSD was not directly influenced by solvent composition considerations, it was readily observable that solvent composition had a profound impact on the optimal seed temperature so by extension, needed to be considered to ensure API PSD within specification.

To facilitate understanding of the relationship between solvent composition and optimal seed temperature, the crystallization platform was employed to construct an API solubility map across the 4-dimensional solvent composition space. With the use of the crystallization platform, the collection of solubility as a function of temperature was automated for each solvent composition. The use of PAT technologies helped to attain the desired data while avoiding the complication of offline sampling of a solution susceptible of crystallization upon cooling, thus greatly reduced the time and effort required to construct this solubility map.

With the relationship between solvent composition and API solubility measured at discrete solvent compositions, the next step was to convert that map into a model that allowed investigation of the continuum of solvent compositions studied. Two strategies were investigated. In the first strategy, a thermodynamic model was used to predict the API solubility. The solubility model employed was based upon the Unifac activity coefficient model and the data obtained through the PAT investigation was used to regress the binary interaction parameters between the API and the solvent subgroups. Secondly, an empirical (polynomial with three variables) model was also developed and found to offer only slightly less accurate predictions of API solubility. Due to the expected ease of implementation on the factory floor and within the filing, the decision was made to utilize the polynomial model long term. With a firm understanding of the relationship between composition and API solubility and an eye to expected process variabilities, the allowable seed temperature range could be determined and ranges on the allowable solvent levels could be specified.

Conclusions: Through efficient application of QbD methodologies, the extended development team has established a process that enables significant operational flexibility and process robustness for an otherwise challenging seed point determination. Although the seed temperature is critical in ensuring the final API PSD, a procedure has been put in place to ensure an appropriate seed temperature while simultaneously leveraging the flexibility in solvent composition afforded through QbD.

See more of this Session: Scale-up of Pharmaceutical Manufacturing Processes: Toward a QbD Approach
See more of this Group/Topical: Topical I: Comprehensive Quality by Design in Pharmaceutical Development and Manufacture

(754c) Harnessing Computational Fluid Dynamics (CFD) to Guide Reliable Scale-up of Pharmaceutical Processes

Ann M. Czyzewski, Process Engineering, Abbott Laboratories, North Chicago, IL, Hsien-Hsin Tung, Global Pharmaceutical Research & Development, Abbott Laboratories, North Chicago, IL, Shailendra V. Bordawekar, Process R & D, Global Pharmaceutical Research and Development, Abbott Laboratories, North Chicago, IL and **Nandkishor K. Nere**, Process R&D, Global Pharmaceutical Reseach and Development, Abbott Laboratories, North Chicago, IL

Successful scale-up of pharmaceutical batch and fed-batch processes in stirred vessels is largely a function of mixing that is governed by underlying hydrodynamics. This in turn is a strong function of various geometric and operational parameters, namely reactor diameter, fill level, impeller (type, number, size, location, and speed), the degree of baffling, feed location and addition rate along with physicochemical properties of the fluid and processing conditions (temperature and pressure). Reactor and impeller geometric configurations as well as vessel internals in laboratory reactors are significantly different than those at the pilot and commercial scales. Hence, it is crucial to understand the differences in the hydrodynamics across scales for accurate process scale-up. To this end, computational fluid dynamics (CFD) provides an efficient and reliable means for characterizing fluid flow.

We present general cases where the information obtained from CFD simulations carried out for a variety of configurations across scales can be effectively used to develop process understanding and guide reliable scale-up of batch and fed-batch processes.

See more of this Session: Scale-up of Pharmaceutical Manufacturing Processes: Toward a QbD Approach
See more of this Group/Topical: Topical I: Comprehensive Quality by Design in Pharmaceutical Development and Manufacture

(754d) Model-Guided Quality by Design: A Case Study On a Pharmaceutical Final Intermediate

Nathaniel D. Kopp[1], Zhongmin Xu[1], James S. Bergum[2], Thomas L. Laporte[1], Jose E. Tabora[1] and Lori A. Spangler[1], (1)Chemical Development, Bristol-Myers Squibb Company, New Brunswick, NJ, (2) Global Biometric Sciences, Bristol-Myers Squibb Company, New Brunswick, NJ

This case study outlines an application of Quality by Design (QbD) principles to the development of the final intermediate manufacturing process. The final intermediate process consists of three steps: (1) deprotection, (2) alkylation and (3) recrystallization. This study will focus on the alkylation step which affords two key impurities and requires control of three process parameters: temperature, solvent ratio and reagent equivalents to produce high quality material.

Through a multi-staged experimental plan, a statistical model was developed and refined to guide the selection and verification of the manufacturing design space. Early Design of Experiments (DoE) laboratory work established the dependence of the quality attributes on specific process parameters which were investigated further to develop an initial statistical model This model was further refined through incorporation of additional laboratory and production-scale data to provide guidance for the determination of the optimal manufacturing design space. The 95% prediction intervals of the model were utilized as an initial guide for the design space. The design space was verified in the laboratory and at pilot scale. This presentation will discuss the implementation of QbD principles to maximize the manufacturing process robustness for the alkylation step through experimental design, statistical model development and design space verification.

See more of this Session: Scale-up of Pharmaceutical Manufacturing Processes: Toward a QbD Approach
See more of this Group/Topical: Topical I: Comprehensive Quality by Design in Pharmaceutical Development and Manufacture

(754e) Incorporating Computational Fluid Dynamics Methods for Quality by Design Framework for Exploring Design Space of a Fermentor

Sravan Kumar Nallamothu, ANSYS Inc, Pune, India and **R. Muralikrishnan**, Ansys, ANSYS Inc, Pune, India

In the Quality by Design (QbD) framework, there is a significant emphasis on the robust characterization of manufacturing processes and prospectively identifying the engineering design space that ensures product quality. Although the QbD approach is expected to bring about significant long-term benefits, there are concerns regarding the increased amount of characterization work that will be required upfront, which can be very resource intensive. In cases where the number of input variables is large, performing a design of experiments (DOE) with statistical significance may be impractical.

Computational fluid dynamics (CFD) provides a first principles approach to gain insight into the hydrodynamics and mass transfer. The scope of traditional CFD methods can be extended to explore the engineering design space of the mixing process by coupling to statistical tools. In this talk, the authors summarize the results of a multivariate study of a prototypical fermentation process. And while a direct link to product quality cannot be provided by CFD alone, the authors hope to illustrate how CFD coupled with statistical methods can lower the risk of this unit operation having an adverse impact on product quality.

See more of this Session: Scale-up of Pharmaceutical Manufacturing Processes: Toward a QbD Approach
See more of this Group/Topical: Topical I: Comprehensive Quality by Design in Pharmaceutical Development and Manufacture

(754f) Scaling up Strategy for Continuous Powder Mixing Process

Yijie Gao, Chem and BioChem Engineering, Rutgers, the State University of New Jersey, Piscataway, NJ, Fernando. J Muzzio, Chemical and Biochemical Engineering, Rutgers University, Piscataway, NJ and Marianthi G. Ierapetritou, Department of Chemical and Biochemical Engineering, Rutgers, The State University of New Jersey, Piscataway, NJ

Scaling upstrategy for continuous powder mixing process

Yijie Gao, Fernando Muzzio andMarianthi G. Ierapetritou

Dept. of Chemical and Biochemical Engineering, RutgersUniversity, Piscataway, NJ 08854

Continuouspowder mixing has attracted a lot of interest among pharmaceutical industry. Basedon the principle of similarity proposed by Johnstoneand Thring [1],many works have been done recently that target the scale-up of continuouspowder mixing. In this work, a quantitative scale-up strategy is described thatallows the transition from lab to industrial scale. By using the periodicsection modeling developed in our previous study [2], scale-up of the cross-sectional mixing and scale-up of theaxial mixing are separately considered. To capture the cross-sectional mixingcomponent, the conception of variance spectrum is used. The Lomb-Scargle Periodogram techniquewidely applied in astronomical studies has been introduced to estimate thevariance spectrum of unevenly distributed samples in powder mixing process. Theeffect of constant sample size on mixing performance of different scales is thuselucidated. To capture the axial mixing component, the residence timedistribution is used to characterize mixing of different scales.

Two casestudies are presented to illustrate the applicability of the developedstrategy. For non-segregating mixing case, similar scale-up characteristics areobserved for both cross-sectional and axial mixing. While satisfying the principleof similarity leads to similar decay contours of variance spectrum, a samplingsize much smaller than scales of both mixers should be used to achieve similar measurementsof mixing performance. A 2.5 power increase of flux rate is generallyobtainable in the scale-up of non-segregating materials. However, in processeswhere cohesive materials are used, segregation occurs due to a tendency ofagglomeration, and a larger increase of flow rate is observed. This resultsfrom the effect of cohesion on axial motion especially when the scale of mixingprocess is small. Our method can be used to provide specific guidance of scale-upof different powder mixing cases.

Reference

1. JohnstoneRW, Thring MW, *Pilot plants, models andscale-up methods in chemical engineering.*

1957, MacGraw-Hill: New York.

2. GaoY, Ierapetritou M, Muzzio F. Periodic section modeling of convective continuouspowder mixing processes. *AIChE Journal*.2011; . In Press: DOI: 10.1002/aic.12563.

See more of this Session: Scale-up of Pharmaceutical Manufacturing Processes: Toward a QbD Approach
See more of this Group/Topical: Topical I: Comprehensive Quality by Design in Pharmaceutical Development and Manufacture

(754g) Modified Application of Johanson's Model towards Roller Compaction

Jasmine M. Rowe and John R. Crison, Drug Product Science & Technology, Bristol-Myers-Squibb, New Brunswick, NJ

ModifiedApplication of Johanson's Model Towards Roller Compaction

Jasmine Rowe, JohnCrison

Rollercompaction (RC) is a continuous dry granulation process commonly used in thepharmaceutical industry to improve powder flow and blend uniformity via powderdensification. RC has gained substantial attention over the last couple ofdecades because neither moisture nor heat is required for granulation (as isthe case for wet granulation and hot melt extrusion, respectively). Theprocess by which densification is achieved is through powder being fed betweentwo rolls that are rotating in opposite directions of each other. Thiscounter-rotating motion of the rolls and friction between the powder and theroll surface draw the powder through the rolls and maximum compaction(densification) occurs when the separation between the rolls is at itsminimum. The resultant compact, or briquette, may then be milled to achievethe desired granule size.

While the RC mechanical process appears tobe simple, efforts to quantitatively model this process have proven challengingbecause of complex material behavior in the feeding and compaction zones. Consequently, a trial-and-error approach has typically been adopted towards RCprocess design, which is both time- and resource-intensive. In response to theneed for a more efficient approach towards RC processing, Johanson developedone of the most widely-used models for RC simulation, a powder mechanics modelbased on Jenike's steady state flow theory that predicts the physicalproperties of the final compact based on material and process parameters [Johanson, J. R. (1965).A rolling theory for granular solids. ASME, Journal of Applied Mechanics SeriesE, 32(4), 842 848]. However, practical application of Johanson's model to experimental work hasbeen limited because one of the model input parameters, the pressure beingapplied to the feed powder as it enters the rolls (P_0), is difficultto obtain experimentally without sophisticated pressure sensor-instrumentedequipment.

In this work, an alternative approach hasbeen established, which expands upon Johanson's model to enable itsincorporation into a daily work flow. This new method requires only standard,routinely-measured, process parameters as inputs and is capable of solving for P_0a priori to extensive experimentation. An excellent correlation between simulatedand experimental results has been achieved for both placebo and active blends. Additionally, a dimensionless relationship between key process input parametersand final compact properties has been elucidated, which provides valuableinformation for efficient scale-up procedures.

(761a) Dynamic Microfluidic Platform to Analyze the Effects of Spatiotemporal Gradient Changes On Neutrophil Chemotaxis

Matthew Byrne[1], Ashish Kapoor[1], Yuki Kimura[1], Fei Wang[2], Christopher V. Rao[1] and Paul J. A. Kenis[1], (1)Chemical and Biomolecular Engineering, University of Illinois, Urbana Champaign, Urbana, IL, (2)Cell and Developmental Biology, University of Illinois, Urbana Champaign, Urbana, IL

Neutrophils constitute the largest class of white blood cells and play a vital role in the innate immune response. These cells sense and migrate up concentration gradients toward sites of infection and inflammation in a process known as chemotaxis. *In vivo*, many cell types respond to infection and inflammation, creating a complex and constantly changing multiple chemoattractant landscape. Much is known of the signaling pathways and molecular interactions that regulate neutrophil response to a single static chemical cue. And while work has begun to elucidate cellular response to multiple static cues, little is known about the effect of dynamic changes in gradients on cellular response.

Previous dynamic microfluidic platforms have studied the temporal responses of neutrophil chemotaxis to changes in the concentration gradients of a single gradient [1]. We propose to examine neutrophil response to spatiotemporal changes in multiple chemoattractant gradients. Using a combination of chemoattractants, such as those produced near the source of infection (end-target chemoattractants) and those produced by inflamed host cells (endogenous intermediary chemoattractants), a dynamically changing microenvironment similar to that *in vivo* can be studied.

The goal of this work is to understand how neutrophils integrate and prioritize dynamically changing chemoattractant gradients. By analyzing neutrophil migration in response to changes in the multiple chemoattractant landscape (fMLP, IL8, and LTB4), we can begin to unravel how neutrophils respond to highly complex chemoattractant landscapes.

Reference

1. Irimia, D., Liu, S., et al. Microfluidic system for measuring neutrophils migratory responses to fast switches of chemical gradients. Lab on a Chip **6**(2006).

See more of this Session: Tissue Engineering Microenvironment II
See more of this Group/Topical: Food, Pharmaceutical & Bioengineering Division

(761b) Interstitial Flow Increases Glioma Cell Migration Via CXCR4/CXCL12 Dependent Autologous Chemotaxis

Jennifer M. Munson, School of Chemical & Biomolecular Engineering, Georgia Institute of Technoloy, Atlanta, GA, Ravi V. Bellamkonda, Biomedical Engineering, Georgia Institute of Technoloy, Atlanta, GA and Melody A. Swartz, Institute for Bioengineering, cole Polytechnique Fdrale de Lausanne (EPFL), Lausanne, Switzerland

Introduction: Malignant brain tumors have a highly invasive phenotype that makes them difficult to treat with normal therapies. The invasion of gliomas is well documented but still poorly understood and thus difficult to overcome in a clinical setting. Patterns of migration in the brain correlate with cerebrospinal fluid flow patterns. Further, gradients of CXCL12 coincide with the same flow patterns in the brain. Additionally, it has been well studied that the chemokine CXCL12 has a chemoattractant effect on many gliomas and the degree to which this chemokine attracts glioma cells correlates with degree of malignancy. Here we propose a mechanism for glioma invasion in the brain that is dependent on flow and chemokine gradients. The Swartz lab has previously described a mechanism of autologous chemotaxis by breast cancer cells under interstitial flow conditions via CCR7/CCL21 signaling (Shields, J., *et al.* 2007 *Cancer Cell*). This mechanism says that a cell secretes a chemokine that will bind to the extracellular matrix. Under non-flow conditions, the chemokine binds radially out in all directions. However, when interstitial flow occurs around the cell, this chemokine is pushed downstream and creates a gradient around the cell causing chemotaxis. We propose and show evidence that a similar mechanism of flow-induced migration is occurring via the chemokine/receptor pair CXCL12/CXCR4 which is so prevalent in brain tumors.

Methods: Several glioblastoma cell lines were used for experiments (RT2, C6, 9L and U87MG). 100,000 cells were embedded in 3-D matrices consisting of 0.08% hyaluronic acid (Glycosan) and 0.12% collagen I (BD Biosciences) at a total volume of 100ul in a tissue culture insert (Millipore). Flow was applied by adding media to the top of the transwell flowed overnight by pressure differential. Gels were removed, inserts imaged and quantified. For chemokine detection, gels, cells and media were collected, protein quantified, and CXCL12 measured using ELISA (R&D). For live imaging, cells were plated in gels in a specially designed microfluidic device with pump-controlled flow applied and imaged over 16 hours in an incubated chamber. Cell migration was analyzed using ImageJ tracking programs and MATLAB postprocessing for analysis of migration velocity and distance. Further studies were conducted in radial flow chambers to examine the mechanism by which chemotaxis occurs through immunohistochemical staining. Last, to model chemotaxis, a COMSOL model was developed from a previously used model (Fleury, M, *et al.* 2006 *Biophys. J.*) to calculate the gradients formed around cells in our *in vitro* system.

Results: RT2 migration was enhanced the most at 2.5 fold when exposed to flow in a 3-D matrix. Other cell types also showed increased invasion under flow conditions (1.5-2 fold). Transmigrated cells are expressed as a percent of total cells and this increased from an average of 0.1% of total to 0.3% of total cells ($p<0.001$). Further, addition of the small molecule CXCR4 inhibitor AMD3100 (Sigma) at a concentration of 10μM inhibited cell migration enhancement under flow back to 0.1% migrated cells ($p<0.001$). This data indicates that there is an effect of interstitial flow on the migration of glioma cells and that this enhancement is dependent on the chemokine receptor pair CXCL12-CXCR4.

Results of live imaging revealed two interesting effects: general cell migration was increased under flow conditions ($p<0.05$), and there was a general change in the trend of cell direction with velocity moving in the direction of flow in a subset of cells. Further, in follow up experiments in radial flow chambers with the chemokine CXCL12 in excess, there was an inhibition of the flow-enhanced invasion effect. These data together indicate that there is directional migration of cells and that blocking the ability of gradients to form in the gel yields a decreased effect. Results of the model indicated that based on the experimental factors involved in the studies, a 9% CXCL12 gradient would form around an individual cell at disease-state flow rates (2μm/s linear velocity).

Conclusions: We show here that flow influences the migration of glioma cells in a 3-D microenvironment and that this enhancement is dependent on CXCL12 and CXCR4. This is significant in that this chemokine and receptor are highly implicated in the malignancy of brain cancer and development of an invasive phenotype. Building upon previous research concerning autologous chemotaxis using CCL21 in tumors of the periphery, we show here that there is a similar mechanism in central nervous system tumor migration with the chemokine CXCL12. Thus, we show the first instance of flow directly affecting brain tumor cells and evidence the possibility of autologous chemotaxis in glioma.

See more of this Session: Tissue Engineering Microenvironment II
See more of this Group/Topical: Food, Pharmaceutical & Bioengineering Division

(761c) Effects of Macroscopic Flow On hMSC Microenvironment and Osteogenic Differentiation In Perfusion Bioreactor

Junho Kim and Teng Ma, Chemical and Biomedical Engineering, Florida State University, Tallahassee, FL

Human mesenchymal stem cells (hMSCs) have multi-lineage differentiation potentials and combining hMSCs with 3D scaffolds is an important strategy to enhance functional outcomes in bone tissue regeneration. In this approach, perfusion bioreactor plays a critical role in regulating 3D construct microenvironment and in providing controlled physiochemical and biomechanical environments. To date, media flow in the bioreactors is either parallel (*e.g.*, spinner flask and rotating wall vessel) or transverse to the 3D scaffolds (*e.g.*, direct perfusion). Although bioreactors of both flow conditions have been extensively used in bone tissue engineering, the impact of macroscopic flow on the ECM microenvironment and hMSC fate is unknown. Using the in-house modular perfusion bioreactor system, we investigated the effects of the macroscopic flows (*i.e.*, parallel or transverse) on the formation of construct microenvironment and its subsequent influence on hMSC proliferation and osteogenic differentiation in the porous poly(ethylene terephthalate) scaffolds. The parallel flow effectively retained ECM proteins and growth factors (*e.g.*, FGF-2) within the scaffolds, preserving hMSC proliferation potential with higher CFU-F numbers and upregulation of OCT-4 expressions. In contrast, the transverse flow enhanced hMSC osteogenic differentiation with higher ALP activity and calcium deposition. The analysis of the transcriptional profiles of the osteogenic gene (*e.g.*, ALP, Runx-2, and OSX) and the up-regulation of osteogenic proteins (*e.g.*, BMP-2, ALP, and OC) suggest the enhanced hMSC osteogenic differentiation as a combined result of shear stress stimulation and the depletion of pro-mitotic growth factors. Utilizing the media flow to influence the formation of ECM microenvironment before and after chemical induction, we are currently investigating the optimal bioreactor strategy for enhancing bone construct development in 3D scaffolds. The results will provide important insight into the role of macroscopic flow on construct microenvironment formation and bone construct function.

See more of this Session: Tissue Engineering Microenvironment II
See more of this Group/Topical: Food, Pharmaceutical & Bioengineering Division

(761d) A Novel Method to Distribute Cells within the Void Volume of a 3D Chitosan-Collagen Scaffold

Revathy Subramanian and Heather Gappa Fahlenkamp, Chemical Engineering, Oklahoma State University, Stillwater, OK

An important step in the development of 3D tissue models is the seedingof cells within a scaffold. This requires spatially uniform distribution of attached cells, thusproviding a basis for uniform tissue regeneration. A novel method to obtain a uniformcell distribution of cells within a 3D chitosan-collagen scaffold is achievedby filling the void volume of the scaffold with cells. The freeze dried chitosan-collagensolution is crosslinked with 10 wt% tripolyphosphate to produce a scaffold withinterconnected pores. The mean pore size of the dry scaffolds is between 100 to200 m and the average scaffold thicknessis approximately 500 mm. The fibroblast cells are mixed in acollagen solution to obtain a homogenous collagen-cell solution. The scaffoldsare seeded by adding the collagen-cell solution to the top of the microporousscaffold and applying a centrifugal force to distribute the cells throughoutthe void spaces of the scaffold. The scaffolds are placed into a CO_2incubator at 37 C to allow the collagen-cell solution to gel, thus trapping thecells throughout the scaffold. The collagen solution acts as an extracellularmatrix (ECM) and promotes the growth of cells in the scaffolds. The combinedproperties of the collagen-cell solution viscosity and the centrifugal forcecan be used to control the distribution of the cells throughout the scaffold. Theviscosity of the collagen solution can be varied by controlling theconcentration of the solution. The collagen concentration also has an effect onthe rate of gelation. The cells trapped in collagen gels are observed for viability,proliferation, and migration. Light microscopy images of sectioned scaffoldsshow the distribution of the collagen-cell gels in the scaffold. Cell viabilityand proliferation within the scaffolds are compared to traditional 2D cellculture on a flat surface and to cells distributed within collagen gels. Preliminaryobservations show that the cells attached to the scaffolds and are viable. Seedingefficiency on the scaffold is lower compared to 2D cell culture. However, cellproliferation in the scaffold is greater than for 2D and collagen gels cellculture. The next phase of this research is to test this novel cell seedingmethod with varying scaffold properties, such as pore size, porosity, andthickness.

See more of this Session: Tissue Engineering Microenvironment II
See more of this Group/Topical: Food, Pharmaceutical & Bioengineering Division

(761e) Influence of Hydrogel and Chondrogeneic Soluble Factors On Fibroblast Differentiation

Kenneth J. Walker and Sundararajan V. Madihally, School of Chemical Engineering, Oklahoma State University, Stillwater, OK

Recent advances in stem cell technologies demonstrate the reprogramming capability of mature fibroblasts into induced stem cells by genetic engineering. However, the effect of matrix elements and soluble factors on differentiation is not well understood. The objective of the research was to test the differentiation of Human Foreskin Fibroblasts (HFF-1) into chondrocytes using 3D structures and soluble factors. Mesenchymal stem cell (MSC) differentiation to chondrocytes was used as a positive control and same conditions were used for fibroblasts. Chitosan-gelatin hydrogels and porous structures were formed using techniques established in the literature.

HFF-1 cells are initially cultured in serum containing media for more rapid growth. Once confluent the cells were seeded in pellet form into the hydrogels and differentiated using Complete Chondrogenic Medium. Hydrogels were formed by incubating the chitosan-gelatin solution at 37°C (98.6^{o}F). The seeded hydrogels were cultured in for one week as well as twenty eight days changing the media every three to four days, similar to MSC cultures. At the time of inoculation the cells were stained with CFDA for a visual confirmation of suspension of cells in the hydrogel and spreading characteristics via fluorescent microscopy. The cells harvested after one week were used for flow cytometry analysis and supernatants of all cultures were preserved for collagen type II analysis. Cells were stained with anti-CD44-PE antibody and anti-CD151-FITC antibody and analyzed by flow cytometry. The retrieved media from the MSC and fibroblast cultures were tested for Collagen Type II using an ELISA. Histology and immunohistochemical analyses were performed on the regenerated structures after twenty eight days for visual confirmation of the production of Collagen Type II. Masson's Trichrome and Alcian blue were used for the presence of cartilage and proteoglycans.

These results show that the fibroblasts show rounded morphology in the hydrogels compared to spindle shape on porous structures and tissue culture plastic. Both conditions showed reduced cell growth relative to tissue culture plastic. ELISA analysis suggested Collagen Type II production in the first two in fibroblast cultures, similar to MSC cultures. Flow cytometry profiles demonstrate alternations in CD44 and CD151 expression. Analysis of the histology showed an increase in matrix accumulation in the structures.

See more of this Session: Tissue Engineering Microenvironment II
See more of this Group/Topical: Food, Pharmaceutical & Bioengineering Division

(761f) Microenvironment Formation and Enhancement of hMSC Osteogenesis In Hydroxyapatite-Chitosan-Gelatin Hydrogel

Kimberly T. Hunter and Teng Ma, Chemical and Biomedical Engineering, Florida State University, Tallahassee, FL

Tissue engineered bone grafts require both an osteogenic cell source and scaffolds that can support the regeneration of healthy bone. Hydroxyapatite (HA), chitosan (C), and gelatin (G) have chemical similarity to the structural components of natural bone and have been frequently used as bone scaffolds. Human mesenchymal stem cells (hMSCs) are inducible osteoprogenitors and a promising cell source for bone regeneration. In this study, the formation of extracellular microenvironment (ECM) in the composite hydroxyapatite-chitosan-gelatin (HCG) hydrogel and its impact on hMSC fate are investigated. The association of hydroxyapatite, chitosan, and gelatin forms a biodegradable hydrogel with 60 wt% water and an initial stiffness of ~ 20 kPa. Preconditioning in serum containing media results in the formation of micro/nanopores in the HCG gels, and increased ECM protein adsorption. Expression of integrin $\alpha_2\beta_1$ and $\alpha_5\beta_1$ coincides with ECM enrichment, suggesting the enhanced cell-ECM interactions that contribute to hMSC migration and proliferation. The elevated expression of bone marker proteins and genes in the absence of chemical induction further demonstrates HCG's intrinsic osteogenic inductive properties. The results show that ECM proteins can effectively functionalize HCG hydrogel and form an instructive microenvironment that promotes bone construct development from hMSCs.

See more of this Session: Tissue Engineering Microenvironment II
See more of this Group/Topical: Food, Pharmaceutical & Bioengineering Division

(761g) Injectable Bone Regeneration Therapy: Encapsulation of Mesenchymal Stem Cells and Osteoconductive Ceramic

Kevin B. Miles and Howard WT Matthew, Chemical Engineering & Materials Science, Wayne State University, Detroit, MI

Current repair and regeneration therapies for major bone loss or trauma are inadequate. Improved therapies would fill defects left by bone fractures, as well as regenerate the bone by stimulating deposition of a mineralized matrix. Ideally, this new therapy could incorporate mesenchymal stem cells (MSCs) with a resorbable, osteoconductive ceramic in an injectable formulation that would facilitate filling and repair of irregularly shaped bone defects. In this study, the encapsulation of MSCs with osteoconductive hydroxyappatite (HAP) microcarriers was investigated as a potential bone regeneration therapy. The encapsulation method, termed "complex coacervation," involved formation of an insoluble polyelectrolyte complex between the anionic glycosaminoglycan chondroitin 4-sulfate (CSA), and the cationic polysaccharide chitosan. In short, a CSA solution with suspended MSCs and HAP microcarriers was extruded as 400 micron diameter droplets and collected into stirred chitosan, thus forming an insoluble complex capsule around the CSA-MSC-HAP droplet. After washes with PBS, capsules were transferred to culture. Osteogenic differentiation was induced in the encapsulated MSC using a standard cocktail. Results were compared to dish-cultured and encapsulated MSC without HAP microcarriers as controls. The MSC formed aggregates within the microcapsules and also adhered to the capsule walls and the HAP microcarriers. After 18 days in osteogenic culture, MSCs encapsulated with HAP microcarriers exhibited extensive deposition of mineralized matrix, compared to control capsules without induced osteogenesis. Mineralization was visualized via fluorescence of tetracycline incorporated into the newly synthesized matrix. Mineral deposition was evident within cell layers, around HAP microcarriers, and within the CSA-chitosan capsule wall. Calcein AM/ethidium homodimer staining indicated that MSC encapsulated with HAP microcarriers retained viability through at least 30 days of culture. Results to date indicate that CSA-chitosan microcapsules containing HAP microcarriers support osteogenic differentiation and rapid mineralization of cultured MSCs, and may be suitable as an injectable, bone-regenerating therapeutic. These microcapsule cultures are being further analyzed to quantify MSC metabolism, matrix deposition, and cell-matrix organization in both osteogenic and expansion culture media. Additional analyses for alkaline phosphatase activity and bone-specific proteins will be used to confirm differentiation to an osteogenic lineage. Mechanical properties of the cultured capsules will also be evaluated by uniaxial compression.

See more of this Session: Tissue Engineering Microenvironment II
See more of this Group/Topical: Food, Pharmaceutical & Bioengineering Division

AUTHOR INDEX

AUTHOR INDEX

AUTHOR INDEX

AUTHOR INDEX

AUTHOR INDEX

AUTHOR INDEX

AUTHOR INDEX

AUTHOR INDEX

AUTHOR INDEX

AUTHOR INDEX

AUTHOR INDEX

AUTHOR INDEX

AUTHOR INDEX

AUTHOR INDEX

AUTHOR INDEX

AUTHOR INDEX

AUTHOR INDEX

AUTHOR INDEX

AUTHOR INDEX